MERCRUISER

Stern[drive]
2001-08 REPAIR MANUAL
ALL GASOLINE ENGINES AND DRIVES

SELOC®

Managing Partners	Dean F. Morgantini, S.A.E.
	Barry L. Beck
Executive Editor	Kevin M. G. Maher, A.S.E.
Production Managers	Melinda Possinger
	Ronald Webb

Manufactured in USA
© 2009 Seloc Publishing
104 Willowbrook Lane
West Chester, PA 19382
ISBN 10: 0-89330-068-3
ISBN 13: 978-0-89330-068-5
4567890123 8765432109

www.selocmarine.com
1-866-SELOC55

CONTENTS

1 GENERAL INFORMATION, SAFETY AND TOOLS

HOW TO USE THIS MANUAL	1-2
BOATING SAFETY	1-4
BOATING EQUIPMENT (NOT REQUIRED BUT RECOMMENDED)	1-10
SAFETY IN SERVICE	1-12
TROUBLESHOOTING	1-13
SHOP EQUIPMENT	1-16
TOOLS	1-18
FASTENERS, MEASUREMENT, AND CONVERSIONS	1-26

2 ENGINE AND DRIVE MAINTENANCE

ENGINE AND DRIVE MAINTENANCE	2-2
FLUIDS AND LUBRICANTS	2-33
LUBRICATION POINTS	2-55
BOAT MAINTENANCE	2-60
WINTER STORAGE	2-66
FIRING ORDERS	2-69
SPRING COMMISSIONING	2-69
SPECIFICATIONS	2-71

3 ENGINE MECHANICAL - GM INLINE ENGINES

ENGINE MECHANICAL	3-2
EXPLODED VIEWS	3-16
SPECIFICATIONS	3-18

4 ENGINE MECHANICAL - GM V6 AND V8 ENGINES

ENGINE MECHANICAL	4-2
EXPLODED VIEWS	4-48
SPECIFICATIONS	4-56

5 ENGINE OVERHAUL

ENGINE RECONDITIONING	5-2

6 FUEL SYSTEM - CARBURETORS

FUEL AND COMBUSTION	6-2
CARBURETED FUEL SYSTEM	6-3
MERCARB 2BBL CARB	6-8
WEBER WFB 4BBL CARB	6-17
WIRING DIAGRAMS	6-27
SPECIFICATIONS	6-30

7 FUEL SYSTEM - FUEL INJECTION (EFI)

FUEL AND COMBUSTION	7-2
ELECTRONIC FUEL INJECTION	7-4
SYSTEM DIAGNOSIS - TBI ENGINES AND EARLY 2001 MPI ENGINES	7-26
SYSTEM DIAGNOSIS - 2001-08 MPI ENGINES	7-47
SPECIFICATIONS	7-70

8 COOLING SYSTEM

COOLING SYSTEMS	8-2
MAINTENANCE AND TESTING	8-2
FLOW DIAGRAMS	8-8

CONTENTS

9 IGNITION AND ELECTRICAL SYSTEMS

UNDERSTANDING AND TROUBLESHOOTING ELECTRICAL SYSTEMS	9-2
BATTERY	9-8
CHARGING SYSTEM	9-10
STARTER CIRCUIT	9-18
ELECTRONIC SPARK SYSTEM (EST)	9-24
IGNITION SYSTEMS	9-24
THUNDERBOLT V IGNITION SYSTEM	9-29
DISTRIBUTOR ELECTRONIC IGNITION (DEI) SYSTEM - ECM 555	9-34
DISTRIBUTORLESS ELECTRONIC IGNITION (DIS) SYSTEM - PCM 555	9-36
INSTRUMENTS AND GAUGES	9-37
SENDING UNITS AND SWITCHES	9-40
WIRING DIAGRAMS	9-45
SPECIFICATIONS	9-74

10 DRIVE SYSTEMS - ALPHA

STERN DRIVE UNIT - ALPHA	10-2
GEAR HOUSING (LOWER UNIT)	10-5
DRIVESHAFT HOUSING (UPPER UNIT)	10-27
TRANSOM ASSEMBLY	10-36
SPECIFICATIONS	10-52

11 DRIVE SYSTEMS - BRAVO

STERN DRIVE UNIT - BRAVO	11-2
GEAR HOUSING (LOWER UNIT)	11-4
DRIVESHAFT HOUSING (UPPER UNIT)	11-18
TRANSOM ASSEMBLY	11-32
SPECIFICATIONS	11-46

12 DRIVE SYSTEMS - TRANSMISSIONS

VELVET TRANSMISSIONS	12-2
ZF/HURTH TRANSMISSIONS	12-6
SPECIFICATIONS	12-10

13 TRIM AND TILT

STANDARD POWER TRIM AND TILT	13-2
AUTO TRIM II SYSTEM	13-24
WIRING SCHEMATICS	13-25

14 STEERING

MANUAL STEERING SYSTEM	14-2
POWER STEERING SYSTEM	14-4
COMPACT HYDRAULIC SYSTEM	14-19

15 REMOTE CONTROLS

CONVENTIONAL REMOTE CONTROLS	15-2
SMARTCRAFT DIGITAL THROTTLE AND SHIFT (DTS)	15-7
PRECISION PILOT	15-33

MASTER INDEX

MASTER INDEX	15-38

SAFETY NOTICE

Proper service and repair procedures are vital to the safe, reliable operation of all marine engines, as well as the personal safety of those performing repairs. This manual outlines procedures for servicing and repairing engines and drive systems using safe, effective methods. The procedures contain many NOTES, CAUTIONS and WARNINGS which should be followed, along with standard procedures, to minimize the possibility of personal injury or improper service which could damage the vehicle or compromise its safety.

It is important to note that repair procedures and techniques, tools and parts for servicing these engines, as well as the skill and experience of the individual performing the work, vary widely. It is not possible to anticipate all of the conceivable ways or conditions under which the engine may be serviced, or to provide cautions as to all possible hazards that may result. Standard and accepted safety precautions and equipment should be used during cutting, grinding, chiseling, prying, or any other process that can cause material removal or projectiles.

Some procedures require the use of tools specially designed for a specific task. Before substituting another tool or procedure, you must be completely satisfied that neither your personal safety, nor the performance of the vessel, will be endangered. All procedures covered in this manual requiring the use of special tools will be noted at the beginning of the procedure by means of an **OEM symbol**

Additionally, any procedure requiring the use of an electronic tester or scan tool will be noted at the beginning of the procedure by means of a **DVOM symbol**

Although information in this manual is based on industry sources and is complete as possible at the time of publication, the possibility exists that some manufacturers made later changes which could not be included here. While striving for total accuracy, Seloc Publishing cannot assume responsibility for any errors, changes or omissions that may occur in the compilation of this data. We must therefore warn you to follow instructions carefully, using common sense. If you are uncertain of a procedure, seek help by inquiring with someone in your area who is familiar with these motors before proceeding.

PART NUMBERS

Part numbers listed in this reference are not recommendations by Seloc Publishing for any particular product brand name, simply iterations of the manufacturer's suggestions. They are also references that can be used with interchange manuals and aftermarket supplier catalogs to locate each brand supplier's discrete part number.

SPECIAL TOOLS

Special tools are recommended by the manufacturers to perform a specific job. Use has been kept to a minimum, but, where absolutely necessary, they are referred to in the text by the part number of the manufacturer if at all possible; and also noted at the beginning of each procedure with one of the following symbols: **OEM** or **DVOM.**

The **OEM** symbol usually denotes the need for a unique tool purposely designed to accomplish a specific task, it will also be used, less frequently, to notify the reader of the need for a tool that is not commonly found in the average tool box.

The **DVOM** symbol is used to denote the need for an electronic test tool like an ohmmeter, multi-meter or, on cetain later engines, a scan tool.

These tools can be purchased, under the appropriate part number, from your local dealer or regional distributor, or an equivalent tool can be purchased locally from a tool supplier or parts outlet. Before substituting any tool for the one recommended, read the SAFETY NOTICE at the top of this page.

Providing the correct mix of service and repair procedures is an endless battle for any publisher of "How-To" information. Users range from first time do-it yourselfers to professionally trained marine technicians, and information important to one is frequently irrelevant to the other. The editors at Seloc Publishing strive to provide accurate and articulate information on all facets of marine engine repair, from the simplest procedure to the most complex. In doing this, we understand that certain procedures may be outside the capabilities of the average DIYer. Conversely we are aware that many procedures are unnecessary for a trained technician.

SKILL LEVELS

In order to provide all of our users, particularly the DIYers, with a feeling for the scope of a given procedure or task before tackling it we have included a rating system denoting the suggested skill level needed when performing a particular procedure. One of the following icons will be included at the beginning of most procedures:

EASY. These procedures are aimed primarily at the DIYer and can be classified, for the most part, as basic maintenance procedures; battery, fluids, filters, plugs, etc. Although certainly valuable to any experience level, they will generally be of little importance to a technician.

MODERATE. These procedures are suited for a DIYer with experience and a working knowledge of mechanical procedures. Even an advanced DIYer or professional technician will occasionally refer to these procedures. They will generally consist of component repair and service procedures, adjustments and minor rebuilds.

DIFFICULT. These procedures are aimed at the advanced DIYer and professional technician. They will deal with diagnostics, rebuilds and internal engine/drive components and will frequently require special tools.

SKILLED. These procedures are aimed at highly skilled technicians and should not be attempted without previous experience. They will usually consist of machine work, internal engine work and gear case rebuilds.

Please remember one thing when considering the above ratings—they are a guide for judging the complexity of a given procedure and are subjective in nature. Only you will know what your experience level is, and only you will know when a procedure may be outside the realm of your capability. First time DIYer, or life-long marine technician, we all approach repair and service differently so an easy procedure for one person may be a difficult procedure for another, regardless of experience level. All skill level ratings are meant to be used as a guide only! Use them to help make a judgement before undertaking a particular procedure, but by all means read through the procedure first and make your own decision—after all, our mission at Seloc is to make boat maintenance and repair easier for everyone whether you are changing the oil or rebuilding an engine. Enjoy boating!

ALL RIGHTS RESERVED

No part of this publication may be reproduced, transmitted or stored in any form or by any means, electronic or mechanical, including photocopy, recording, or by information storage or retrieval system, without prior written permission from the publisher.

The materials contained in this manual are the intellectual property of Seloc Publishing, Inc., a Pennsylvania corporation, and are protected under the laws of the United States of America at Title 17 of the United States Code. Any efforts to reproduce any of the content of this manual, in any form, without the express written permission of Seloc Publishing, Inc. is punishable by a fine of up to $250,000 and 5 years in jail, plus the recovery of all proceeds including attorneys fees.

ACKNOWLEDGMENTS

Seloc Publishing expresses appreciation to the following companies who supported the production of this book:
- Marine Mechanics Institute—Orlando, FL
- Belks Marine—Holmes, PA

Thanks to John Hartung and Judy Belk of Belk's Marine for for there assistance, guidance, patience and access to some of the motors photographed for this manual.

Seloc Publishing would like to express thanks to the fine companies who participate in the production of all our books:
- Hand tools supplied by Craftsman are used during all phases of our vehicle teardown and photography.
- Many of the fine specialty tools used in our procedures were provided courtesy of Lisle Corporation.
- Much of our shop's electronic testing equipment was supplied by Universal Enterprises Inc. (UEI).

1

GENERAL INFORMATION, SAFETY & TOOLS

HOW TO USE THIS MANUAL	1-2
BOATING SAFETY	1-4
EQUIPMENT (NOT REQUIRED BUT RECOMMENDED)	1-10
SAFETY IN SERVICE	1-12
TROUBLESHOOTING	1-13
SHOP EQUIPMENT	1-16
SPECIFICATIONS	1-17
TOOLS	1-18
FASTENERS, MEASUREMENTS AND CONVERSIONS	1-26

BASIC OPERATING PRINCIPLES 1-13
  2-STROKE MOTORS 1-13
  4-STROKE MOTORS 1-15
  COMBUSTION 1-16
BOATING SAFETY **1-4**
  COURTESY MARINE EXAMINATIONS 1-10
  REGULATIONS FOR YOUR BOAT 1-4
  REQUIRED SAFETY EQUIPMENT 1-5
CHEMICALS 1-17
  CLEANERS 1-18
  LUBRICANTS & PENETRANTS 1-17
  SEALANTS 1-17
COMPASS 1-10
  COMPASS PRECAUTIONS 1-11
  INSTALLATION 1-11
  SELECTION 1-10
ELECTRONIC TOOLS 1-23
  BATTERY TESTERS 1-23
  BATTERY CHARGERS 1-23
  GAUGES 1-24
  MULTI-METERS (DVOMS) 1-23
EQUIPMENT (NOT REQUIRED BUT RECOMMENDED) **1-10**
  ANCHORS 1-10
  BAILING DEVICES 1-10
  COMPASS 1-10
  FIRST AID KIT 1-10
  OAR/PADDLE 1-10
  TOOLS AND SPARE PARTS 1-12
  VHF-FM RADIO 1-10
FASTENERS, MEASUREMENTS AND CONVERSIONS **1-26**
  BOLTS, NUTS AND OTHER THREADED RETAINERS 1-26
  TORQUE 1-27
  STANDARD AND METRIC MEASUREMENTS 1-27
HAND TOOLS 1-19
  BREAKER BARS 1-20
  HAMMERS 1-22
  PLIERS 1-21
  SCREWDRIVERS 1-21
  SOCKET SETS 1-19
  TORQUE WRENCHES 1-20
  WRENCHES 1-21
HOW TO USE THIS MANUAL **1-2**
  AVOIDING COMMON MISTAKES 1-3
  AVOIDING TROUBLE 1-2
  CAN YOU DO IT? 1-2
  DIRECTIONS AND LOCATIONS 1-2
  MAINTENANCE OR REPAIR? 1-2

PROFESSIONAL HELP 1-3
PURCHASING PARTS 1-3
WHERE TO BEGIN 1-2
MEASURING TOOLS 1-24
  DEPTH GAUGES 1-26
  DIAL INDICATORS 1-25
  MICROMETERS & CALIPERS 1-25
  TELESCOPING GAUGES 1-26
REGULATIONS FOR YOUR BOAT 1-4
  CAPACITY INFORMATION 1-4
  CERTIFICATE OF COMPLIANCE 1-4
  DOCUMENTING OF VESSELS 1-4
  HULL IDENTIFICATION NUMBER 1-4
  LENGTH OF BOATS 1-4
  NUMBERING OF VESSELS 1-4
  REGISTRATION OF BOATS 1-4
  SALES AND TRANSFERS 1-4
  VENTILATION 1-5
  VENTILATION SYSTEMS 1-5
REQUIRED SAFETY EQUIPMENT 1-5
  FIRE EXTINGUISHERS 1-6
  PERSONAL FLOTATION DEVICES 1-7
  SOUND PRODUCING DEVICES 1-8
  TYPES OF FIRES 1-5
  VISUAL DISTRESS SIGNALS 1-8
  WARNING SYSTEM 1-7
SAFETY IN SERVICE **1-12**
  DO'S 1-12
  DON'TS 1-12
SAFETY TOOLS 1-16
  EYE AND EAR PROTECTION 1-16
  WORK CLOTHES 1-17
  WORK GLOVES 1-16
SHOP EQUIPMENT **1-16**
  CHEMICALS 1-17
  SAFETY TOOLS 1-16
SPECIFICATIONS **1-17**
  CONVERSION FACTORS 1-27
  TYPICAL TORQUE VALUES - METRIC BOLTS 1-28
  TYPICAL TORQUE VALUES - U.S. STANDARD BOLTS 1-28
TOOLS **1-18**
  ELECTRONIC TOOLS 1-23
  HAND TOOLS 1-19
  MEASURING TOOLS 1-24
  OTHER COMMON TOOLS 1-22
  SPECIAL TOOLS 1-23
TROUBLESHOOTING **1-13**
  BASIC OPERATING PRINCIPLES 1-13

1-2 GENERAL INFORMATION, SAFETY & TOOLS

HOW TO USE THIS MANUAL

This manual is designed to be a handy reference guide to maintaining and repairing your Mercruiser. We strongly believe that regardless of how many or how few year's experience you may have, there is something new waiting here for you.

This manual covers the topics that a factory service manual (designed for factory trained mechanics) and a manufacturer owner's manual (designed more by lawyers than boat owners these days) covers. It will take you through the basics of maintaining and repairing your outboard, step-by-step, to help you understand what the factory trained mechanics already know by heart. By using the information in this manual, any boat owner should be able to make better informed decisions about what they need to do to maintain and enjoy their outboard.

Even if you never plan on touching a wrench (and if so, we hope that we can change your mind), this manual will still help you understand what a mechanic needs to do in order to maintain your engine.

Can You Do It?

If you are not the type who is prone to taking a wrench to something, NEVER FEAR. The procedures provided here cover topics at a level virtually anyone will be able to handle. And just the fact that you purchased this manual shows your interest in better understanding your outboard.

You may even find that maintaining your outboard yourself is preferable in most cases. From a monetary standpoint, it could also be beneficial. The money spent on hauling your boat to a marina and paying a tech to service the engine could buy you fuel for a whole weekend of boating. And, if you are really that unsure of your own mechanical abilities, at the very least you should fully understand what a marine mechanic does to your boat. You may decide that anything other than maintenance and adjustments should be performed by a mechanic (and that's your call), but if so you should know that every time you board your boat, you are placing faith in the mechanic's work and trusting him or her with your well-being, and maybe your life.

It should also be noted that in most areas a factory-trained mechanic will command a hefty hourly rate for off site service. If the tech comes to you this hourly rate is often charged from the time they leave their shop to the time that they return home. When service is performed at a boat yard, the clock usually starts when they go out to get the boat and bring it into the shop and doesn't end until it is tested and put back in the yard. The cost savings in doing the job yourself might be readily apparent at this point.

Of course, if even you're already a seasoned Do-It-Yourselfer or a Professional Technician, you'll find the procedures, specifications, special tips as well as the schematics and illustrations helpful when tackling a new job on a motor.

■ To help you decide if a task is within your skill level, procedures will often be rated using a wrench symbol in the text. When present, the number of wrenches designates how difficult we feel the procedure to be on a 1-4 scale. For more details on the wrench icon rating system, please refer to the information under Skill Levels at the beginning of this manual.

Where to Begin

Before spending any money on parts, and before removing any nuts or bolts, read through the entire procedure or topic. This will give you the overall view of what tools and supplies will be required to perform the procedure or what questions need to be answered before purchasing parts. So read ahead and plan ahead. Each operation should be approached logically and all procedures thoroughly understood before attempting any work.

Avoiding Trouble

Some procedures in this manual may require you to "label and disconnect . . ." a group of lines, hoses or wires. Don't be lulled into thinking you can remember where everything goes - you won't. If you reconnect or install a part incorrectly, the motor may operate poorly, if at all. If you hook up electrical wiring incorrectly, you may instantly learn a very expensive lesson.

A piece of masking tape, for example, placed on a hose and another on its fitting will allow you to assign your own label such as the letter "A", or a short name. As long as you remember your own code, you can reconnect the lines by matching letters or names. Do remember that tape will dissolve when saturated in some fluids (especially cleaning solvents). If a component is to be washed or cleaned, use another method of identification. A permanent felt-tipped marker can be very handy for marking metal parts; but remember that some solvents will remove permanent marker. A scribe can be used to carefully etch a small mark in some metal parts, but be sure NOT to do that on a gasket-making surface.

SAFETY is the most important thing to remember when performing maintenance or repairs. Be sure to read the information on safety in this manual.

Maintenance or Repair?

Proper maintenance is the key to long and trouble-free engine life, and the work can yield its own rewards. A properly maintained engine performs better than one that is neglected. As a conscientious boat owner, set aside a Saturday morning, at least once a month, to perform a thorough check of items that could cause problems. Keep your own personal log to jot down which services you performed, how much the parts cost you, the date, and the amount of hours on the engine at the time. Keep all receipts for parts purchased, so that they may be referred to in case of related problems or to determine operating expenses. As a do-it-yourselfer, these receipts are the only proof you have that the required maintenance was performed. In the event of a warranty problem (on new motors), these receipts can be invaluable.

It's necessary to mention the difference between maintenance and repair. Maintenance includes routine inspections, adjustments, and replacement of parts that show signs of normal wear. Maintenance compensates for wear or deterioration. Repair implies that something has broken or is not working. A need for repair is often caused by lack of maintenance.

For example: draining and refilling the gearcase oil is maintenance recommended by all manufacturers at specific intervals. Failure to do this can allow internal corrosion or damage and impair the operation of the motor, requiring expensive repairs. While no maintenance program can prevent items from breaking or wearing out, a general rule can be stated: MAINTENANCE IS CHEAPER THAN REPAIR.

Directions and Locations

◆ See Figure 1

Two basic rules should be mentioned here. First, whenever the Port side of the engine (or boat) is referred to, it is meant to specify the left side of the engine when you are sitting at the helm. Conversely, the Starboard means your right side. The Bow is the front of the boat and the Stern or Aft is the rear.

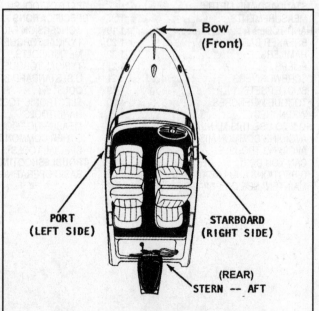

Fig. 1 Common terminology used for reference designation on boats of all size. These terms are used through out the text

Most screws and bolts are removed by turning counterclockwise, and tightened by turning clockwise. An easy way to remember this is: righty-tighty; lefty-loosey. Corny, but effective. And if you are really dense (and we have all been so at one time or another), buy a ratchet that is marked ON and OFF (like Snap-on® ratchets), or mark your own. This can be especially helpful when you are bent over backwards, upside down or otherwise turned around when working on a boat-mounted component.

Professional Help

Occasionally, there are some things when working on an outboard that are beyond the capabilities or tools of the average Do-It-Yourselfer (DIYer). This shouldn't include most of the topics of this manual, but you will have to be the judge. Some engines require special tools or a selection of special parts, even for some basic maintenance tasks.

Talk to other boaters who use the same model of engine and speak with a trusted marina to find if there is a particular system or component on your engine that is difficult to maintain.

You will have to decide for yourself where basic maintenance ends and where professional service should begin. Take your time and do your research first (starting with the information contained within) and then make your own decision. If you really don't feel comfortable with attempting a procedure, DON'T DO IT. If you've gotten into something that may be over your head, don't panic. Tuck your tail between your legs and call a marine mechanic. Marinas and independent shops will be able to finish a job for you. Your ego may be damaged, but your boat will be properly restored to its full running order. So, as long as you approach jobs slowly and carefully, you really have nothing to lose and everything to gain by doing it yourself.

On the other hand, even the most complicated repair is within the ability of a person who takes their time and follows the steps of a procedure. A rock climber doesn't run up the side of a cliff, he/she takes it one step at a time and in the end, what looked difficult or impossible was conquerable. Worry about one step at a time.

Purchasing Parts

◆ See Figures 2 and 3

When purchasing parts there are two things to consider. The first is quality and the second is to be sure to get the correct part for your engine. To get quality parts, always deal directly with a reputable retailer. To get the proper parts always refer to the model number from the information tag on your engine prior to calling the parts counter. An incorrect part can adversely affect your engine performance and fuel economy, and will cost you more money and aggravation in the end.

Just remember a tow back to shore will cost plenty. That charge is per hour from the time the towboat leaves their home port, to the time they return to their home port. Get the picture. . .$$$?

So whom should you call for parts? Well, there are many sources for the parts you will need. Where you shop for parts will be determined by what kind of parts you need, how much you want to pay, and the types of stores in your neighborhood.

Your marina can supply you with many of the common parts you require. Using a marina as your parts supplier may be handy because of location (just walk right down the dock) or because the marina specializes in your particular brand of engine. In addition, it is always a good idea to get to know the marina staff (especially the marine mechanic).

The marine parts jobber, who is usually listed in the yellow pages or whose name can be obtained from the marina, is another excellent source for parts. In addition to supplying local marinas, they also do a sizeable business in over-the-counter parts sales for the do-it-yourselfer.

Almost every boating community has one or more convenient marine chain stores. These stores often offer the best retail prices and the convenience of one-stop shopping for all your needs. Since they cater to the do-it-yourselfer, these stores are almost always open weeknights, Saturdays, and Sundays, when the jobbers are usually closed.

The lowest prices for parts are most often found in discount stores or the auto department of mass merchandisers. Parts sold here are name and private brand parts bought in huge quantities, so they can offer a competitive price. Private brand parts are made by major manufacturers and sold to large chains under a store label. And, of course, more and more large automotive parts retailers are stocking basic marine supplies.

Avoiding the Most Common Mistakes

There are 3 common mistakes in mechanical work:

1. Following the incorrect order of assembly, disassembly or adjustment. When taking something apart or putting it together, performing steps in the wrong order usually just costs you extra time; however, it CAN break something. Read the entire procedure before beginning disassembly. Perform everything in the order in which the instructions say you should, even if you can't immediately see a reason for it. When you're taking apart something that is very intricate, you might want to draw a picture of how it looks when assembled at one point in order to make sure you get everything back in its proper position. When making adjustments, perform them in the proper order; often, one adjustment affects another, and you cannot expect satisfactory results unless each adjustment is made only when it cannot be changed by subsequent adjustments.

■ Digital cameras are handy. If you've got access to one, take pictures of intricate assemblies during the disassembly process and refer to them during assembly for tips on part orientation.

2. Over-torquing (or under-torquing). While it is more common for over-torquing to cause damage, under-torquing may allow a fastener to vibrate loose causing serious damage. Especially when dealing with plastic and aluminum parts, pay attention to torque specifications and utilize a torque wrench in assembly. If a torque figure is not available, remember that if you are using the right tool to perform the job, you will probably not have to strain yourself to get a fastener tight enough. The pitch of most threads is so slight

Fig. 2 By far the most important asset in purchasing parts is a knowledgeable and enthusiastic parts person

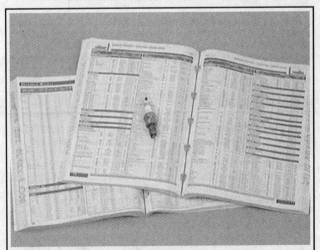

Fig. 3 Parts catalogs, giving application and part number information, are provided by manufacturers for most replacement parts

1-4 GENERAL INFORMATION, SAFETY & TOOLS

that the tension you put on the wrench will be multiplied many times in actual force on what you are tightening.

3. Cross-threading. This occurs when a part such as a bolt is screwed into a nut or casting at the wrong angle and forced. Cross-threading is more likely to occur if access is difficult. It helps to clean and lubricate fasteners, then to start threading with the part to be installed positioned straight inward. Always start a fastener, etc. with your fingers. If you encounter resistance, unscrew the part and start over again at a different angle until it can be inserted and turned several times without much effort. Keep in mind that some parts may have tapered threads, so that gentle turning will automatically bring the part you're threading to the proper angle, but only if you don't force it or resist a change in angle. Don't put a wrench on the part until it has been tightened a couple of turns by hand. If you suddenly encounter resistance, and the part has not seated fully, don't force it. Pull it back out to make sure it's clean and threading properly.

BOATING SAFETY

In 1971 Congress ordered the U.S. Coast Guard to improve recreational boating safety. In response, the Coast Guard drew up a set of regulations.

Aside from these federal regulations, there are state and local laws you must follow. These sometimes exceed the Coast Guard requirements. This section discusses only the federal laws. State and local laws are available from your local Coast Guard. As with other laws, "Ignorance of the boating laws is no excuse." The rules fall into two groups: regulations for your boat and required safety equipment on your boat.

Regulations For Your Boat

Most boats on waters within Federal jurisdiction must be registered or documented. These waters are those that provide a means of transportation between two or more states or to the sea. They also include the territorial waters of the United States.

DOCUMENTING OF VESSELS

A vessel of five or more net tons may be documented as a yacht. In this process, papers are issued by the U.S. Coast Guard as they are for large ships. Documentation is a form of national registration. The boat must be used solely for pleasure. Its owner must be a citizen of the U.S., a partnership of U.S. citizens, or a corporation controlled by U.S. citizens. The captain and other officers must also be U.S. citizens. The crew need not be.

If you document your yacht, you have the legal authority to fly the yacht ensign. You also may record bills of sale, mortgages, and other papers of title with federal authorities. Doing so gives legal notice that such instruments exist. Documentation also permits preferred status for mortgages. This gives you additional security, and it aids in financing and transfer of title. You must carry the original documentation papers aboard your vessel. Copies will not suffice.

REGISTRATION OF BOATS

If your boat is not documented, registration in the state of its principal use is probably required. If you use it mainly on an ocean, a gulf, or other similar water, register it in the state where you moor it.

If you use your boat solely for racing, it may be exempt from the requirement in your state. Some states may also exclude dinghies, while others require registration of documented vessels and non-power driven boats.

All states, except Alaska, register boats. In Alaska, the U.S. Coast Guard issues the registration numbers. If you move your vessel to a new state of principal use, a valid registration certificate is good for 60 days. You must have the registration certificate (certificate of number) aboard your vessel when it is in use. A copy will not suffice. You may be cited if you do not have the original on board.

NUMBERING OF VESSELS

A registration number is on your registration certificate. You must paint or permanently attach this number to both sides of the forward half of your boat. Do not display any other number there.

The registration number must be clearly visible. It must not be placed on the obscured underside of a flared bow. If you can't place the number on the bow, place it on the forward half of the hull. If that doesn't work, put it on the superstructure. Put the number for an inflatable boat on a bracket or fixture. Then, firmly attach it to the forward half of the boat. The letters and numbers must be plain block characters and must read from left to right. Use a space or a hyphen to separate the prefix and suffix letters from the numerals. The color of the characters must contrast with that of the background, and they must be at least three inches high.

In some states your registration is good for only one year. In others, it is good for as long as three years. Renew your registration before it expires. At that time you will receive a new decal or decals. Place them as required by state law. You should remove old decals before putting on the new ones. Some states require that you show only the current decal or decals. If your vessel is moored, it must have a current decal even if it is not in use.

If your vessel is lost, destroyed, abandoned, stolen, or transferred, you must inform the issuing authority. If you lose your certificate of number or your address changes, notify the issuing authority as soon as possible.

SALES AND TRANSFERS

Your registration number is not transferable to another boat. The number stays with the boat unless its state of principal use is changed.

HULL IDENTIFICATION NUMBER

A Hull Identification Number (HIN) is like the Vehicle Identification Number (VIN) on your car. Boats built between November 1, 1972 and July 31, 1984 have old format HINs. Since August 1, 1984 a new format has been used.

Your boat's HIN must appear in two places. If it has a transom, the primary number is on its starboard side within two inches of its top. If it does not have a transom or if it was not practical to use the transom, the number is on the starboard side. In this case, it must be within one foot of the stern and within two inches of the top of the hull side. On pontoon boats, it is on the aft crossbeam within one foot of the starboard hull attachment. Your boat also has a duplicate number in an unexposed location. This is on the boat's interior or under a fitting or item of hardware.

LENGTH OF BOATS

For some purposes, boats are classed by length. Required equipment, for example, differs with boat size. Manufacturers may measure a boat's length in several ways. Officially, though, your boat is measured along a straight line from its bow to its stern. This line is parallel to its keel.

The length does not include bowsprits, boomkins, or pulpits. Nor does it include rudders, brackets, outboard motors, outdrives, diving platforms, or other attachments.

CAPACITY INFORMATION

◆ See Figure 4

Manufacturers must put capacity plates on most recreational boats less than 20 feet long. Sailboats, canoes, kayaks, and inflatable boats are usually exempt. Outboard boats must display the maximum permitted horsepower of their engines. The plates must also show the allowable maximum weights of the people on board. And they must show the allowable maximum combined weights of people, engine(s), and gear. Inboards and stern drives need not show the weight of their engines on their capacity plates. The capacity plate must appear where it is clearly visible to the operator when underway. This information serves to remind you of the capacity of your boat under normal circumstances. You should ask yourself, "Is my boat loaded above its recommended capacity" and, "Is my boat overloaded for the present sea and wind conditions?" If you are stopped by a legal authority, you may be cited if you are overloaded.

CERTIFICATE OF COMPLIANCE

◆ See Figure 4

Manufacturers are required to put compliance plates on motorboats greater than 20 feet in length. The plates must say, "This boat," or "This equipment complies with the U. S. Coast Guard Safety Standards in effect on the date of certification." Letters and numbers can be no less than one-eighth of an inch high. At the manufacturer's option, the capacity and compliance plates may be combined.

GENERAL INFORMATION SAFETY & TOOLS 1-5

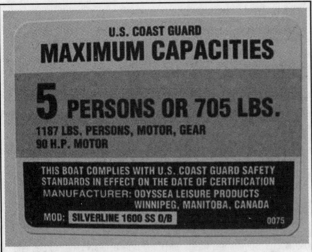

Fig. 4 A U.S. Coast Guard certification plate indicates the amount of occupants and gear appropriate for safe operation of the vessel (and allowable engine size for outboard boats)

VENTILATION

A cup of gasoline spilled in the bilge has the potential explosive power of 15 sticks of dynamite. This statement, commonly quoted over 20 years ago, may be an exaggeration; however, it illustrates a fact. Gasoline fumes in the bilge of a boat are highly explosive and a serious danger. They are heavier than air and will stay in the bilge until they are vented out.

Because of this danger, Coast Guard regulations require ventilation on many powerboats. There are several ways to supply fresh air to engine and gasoline tank compartments and to remove dangerous vapors. Whatever the choice, it must meet Coast Guard standards.

■ The following is not intended to be a complete discussion of the regulations. It is limited to the majority of recreational vessels. Contact your local Coast Guard office for further information.

General Precautions

Ventilation systems will not remove raw gasoline that leaks from tanks or fuel lines. If you smell gasoline fumes, you need immediate repairs. The best device for sensing gasoline fumes is your nose. Use it! If you smell gasoline in a bilge, engine compartment, or elsewhere, don't start your engine. The smaller the compartment, the less gasoline it takes to make an explosive mixture.

Ventilation for Open Boats

In open boats, gasoline vapors are dispersed by the air that moves through them. So they are exempt from ventilation requirements.

To be "open," a boat must meet certain conditions. Engine and fuel tank compartments and long narrow compartments that join them must be open to the atmosphere." This means they must have at least 15 square inches of open area for each cubic foot of net compartment volume. The open area must be in direct contact with the atmosphere. There must also be no long, unventilated spaces open to engine and fuel tank compartments into which flames could extend.

Ventilation for All Other Boats

Powered and natural ventilation are required in an enclosed compartment with a permanently installed gasoline engine that has a cranking motor. A compartment is exempt if its engine is open to the atmosphere. Diesel powered boats are also exempt.

VENTILATION SYSTEMS

There are two types of ventilation systems. One is "natural ventilation." In it, air circulates through closed spaces due to the boat's motion. The other type is "powered ventilation." In it, air is circulated by a motor-driven fan or fans.

Natural Ventilation System Requirements

A natural ventilation system has an air supply from outside the boat. The air supply may also be from a ventilated compartment or a compartment open to the atmosphere. Intake openings are required. In addition, intake ducts may be required to direct the air to appropriate compartments.

The system must also have an exhaust duct that starts in the lower third of the compartment. The exhaust opening must be into another ventilated compartment or into the atmosphere. Each supply opening and supply duct, if there is one, must be above the usual level of water in the bilge. Exhaust openings and ducts must also be above the bilge water. Openings and ducts must be at least three square inches in area or two inches in diameter. Openings should be placed so exhaust gasses do not enter the fresh air intake. Exhaust fumes must not enter cabins or other enclosed, non-ventilated spaces. The carbon monoxide gas in them is deadly.

Intake and exhaust openings must be covered by cowls or similar devices. These registers keep out rain water and water from breaking seas. Most often, intake registers face forward and exhaust openings aft. This aids the flow of air when the boat is moving or at anchor since most boats face into the wind when properly anchored.

Power Ventilation System Requirements

◆ See Figure 5

Powered ventilation systems must meet the standards of a natural system, but in addition, they must also have one or more exhaust blowers. The blower duct can serve as the exhaust duct for natural ventilation if fan blades do not obstruct the air flow when not powered. Openings in engine compartment, for carburetion are in addition to ventilation system requirements.

Required Safety Equipment

Coast Guard regulations require that your boat have certain equipment aboard. These requirements are minimums. Exceed them whenever you can.

TYPES OF FIRES

There are four common classes of fires:
- Class A - fires are of ordinary combustible materials such as paper or wood.
- Class B - fires involve gasoline, oil and grease.
- Class C - fires are electrical.
- Class D - fires involve ferrous metals

One of the greatest risks to boaters is fire. This is why it is so important to carry the correct number and type of extinguishers onboard.

The best fire extinguisher for most boats is a Class B extinguisher. Never use water on Class B or Class C fires, as water spreads these types of fires. Additionally, you should never use water on a Class C fire as it may cause you to be electrocuted.

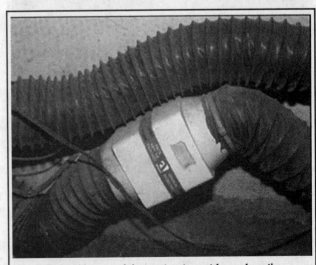

Fig. 5 Typical blower and duct system to vent fumes from the engine compartment

1-6 GENERAL INFORMATION, SAFETY & TOOLS

FIRE EXTINGUISHERS

◆ See Figure 6

If your boat meets one or more of the following conditions, you must have at least one fire extinguisher aboard. The conditions are:
- Inboard or stern drive engines
- Closed compartments under seats where portable fuel tanks can be stored
- Double bottoms not sealed together or not completely filled with flotation materials
- Closed living spaces
- Closed stowage compartments in which combustible or flammable materials are stored
- Permanently installed fuel tanks
- Boat is 26 feet or more in length.

Contents of Extinguishers

Fire extinguishers use a variety of materials. Those used on boats usually contain dry chemicals, Halon, or Carbon Dioxide (CO_2). Dry chemical extinguishers contain chemical powders such as Sodium Bicarbonate - baking soda.

Carbon dioxide is a colorless and odorless gas when released from an extinguisher. It is not poisonous but caution must be used in entering compartments filled with it. It will not support life and keeps oxygen from reaching your lungs. A fire-killing concentration of Carbon Dioxide can be lethal. If you are in a compartment with a high concentration of CO_2, you will have no difficulty breathing. But the air does not contain enough oxygen to support life. Unconsciousness or death can result.

Halon Extinguishers

Some fire extinguishers and "built-in" or "fixed" automatic fire extinguishing systems contain a gas called Halon. Like carbon dioxide it is colorless and odorless and will not support life. Some Halons may be toxic if inhaled.

To be accepted by the Coast Guard, a fixed Halon system must have an indicator light at the vessel's helm. A green light shows the system is ready. Red means it is being discharged or has been discharged. Warning horns are available to let you know the system has been activated. If your fixed Halon system discharges, ventilate the space thoroughly before you enter it. There are no residues from Halon but it will not support life.

Although Halon has excellent fire fighting properties; it is thought to deplete the earth's ozone layer and has not been manufactured since January 1, 1994. Halon extinguishers can be refilled from existing stocks of the gas until they are used up, but high federal excise taxes are being charged for the service. If you discontinue using your Halon extinguisher, take it to a recovery station rather than releasing the gas into the atmosphere. Compounds such as FE 241, designed to replace Halon, are now available.

Fire Extinguisher Approval

Fire extinguishers must be Coast Guard approved. Look for the approval number on the nameplate. Approved extinguishers have the following on their labels: "Marine Type USCG Approved, Size. . ., Type. . ., 162.208/," etc. In addition, to be acceptable by the Coast Guard, an extinguisher must be in serviceable condition and mounted in its bracket. An extinguisher not properly mounted in its bracket will not be considered serviceable during a Coast Guard inspection.

Care and Treatment

Make certain your extinguishers are in their stowage brackets and are not damaged. Replace cracked or broken hoses. Nozzles should be free of obstructions. Sometimes, wasps and other insects nest inside nozzles and make them inoperable. Check your extinguishers frequently. If they have pressure gauges, is the pressure within acceptable limits? Do the locking pins and sealing wires show they have not been used since recharging?

Don't try an extinguisher to test it. Its valves will not reseat properly and the remaining gas will leak out. When this happens, the extinguisher is useless.

Weigh and tag carbon dioxide and Halon extinguishers twice a year. If their weight loss exceeds 10 percent of the weight of the charge, recharge them. Check to see that they have not been used. They should have been inspected by a qualified person within the past six months, and they should have tags showing all inspection and service dates. The problem is that they can be partially discharged while appearing to be fully charged.

Some Halon extinguishers have pressure gauges the same as dry chemical extinguishers. Don't rely too heavily on the gauge. The extinguisher can be partially discharged and still show a good gauge reading. Weighing a Halon extinguisher is the only accurate way to assess its contents.

If your dry chemical extinguisher has a pressure indicator, check it frequently. Check the nozzle to see if there is powder in it. If there is, recharge it. Occasionally invert your dry chemical extinguisher and hit the base with the palm of your hand. The chemical in these extinguishers packs and cakes due to the boat's vibration and pounding. There is a difference of opinion about whether hitting the base helps, but it can't hurt. It is known that caking of the chemical powder is a major cause of failure of dry chemical extinguishers. Carry spares in excess of the minimum requirement. If you have guests aboard, make certain they know where the extinguishers are and how to use them.

Using a Fire Extinguisher

A fire extinguisher usually has a device to keep it from being discharged accidentally. This is a metal or plastic pin or loop. If you need to use your extinguisher, take it from its bracket. Remove the pin or the loop and point the nozzle at the base of the flames. Now, squeeze the handle, and discharge the extinguisher's contents while sweeping from side to side. Recharge a used extinguisher as soon as possible.

If you are using a Halon or carbon dioxide extinguisher, keep your hands away from the discharge. The rapidly expanding gas will freeze them. If your fire extinguisher has a horn, hold it by its handle.

Legal Requirements for Extinguishers

You must carry fire extinguishers as defined by Coast Guard regulations. They must be firmly mounted in their brackets and immediately accessible.

A motorboat less than 26 feet long must have at least one approved hand-portable, Type B-1 extinguisher. If the boat has an approved fixed fire extinguishing system, you are not required to have the Type B-1 extinguisher. Also, if your boat is less than 26 feet long, is propelled by an outboard motor, or motors, and does not have any of the first six conditions described at the beginning of this section, it is not required to have an extinguisher. Even so, it's a good idea to have one, especially if a nearby boat catches fire, or if a fire occurs at a fuel dock.

A motorboat 26 feet to less than 40 feet long, must have at least two Type B-1 approved hand-portable extinguishers. It can, instead, have at least one Coast Guard approved Type B-2. If you have an approved fire extinguishing system, only one Type B-1 is required.

A motorboat 40 to 65 feet long must have at least three Type B-1 approved portable extinguishers. It may have, instead, at least one Type B-1

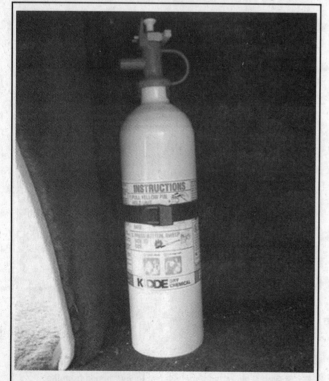

Fig. 6 An approved fire extinguisher should be mounted close to the operator for emergency use

GENERAL INFORMATION SAFETY & TOOLS 1-7

plus a Type B-2. If there is an approved fixed fire extinguishing system, two Type B-1 or one Type B-2 is required.

WARNING SYSTEM

Various devices are available to alert you to danger. These include fire, smoke, gasoline fumes, and carbon monoxide detectors. If your boat has a galley, it should have a smoke detector. Where possible, use wired detectors. Household batteries often corrode rapidly on a boat.

There are many ways in which carbon monoxide (a by-product of the combustion that occurs in an engine) can enter your boat. You can't see, smell, or taste carbon monoxide gas, but it is lethal. As little as one part in 10,000 parts of air can bring on a headache. The symptoms of carbon monoxide poisoning - headaches, dizziness, and nausea - are like seasickness. By the time you realize what is happening to you, it may be too late to take action. If you have enclosed living spaces on your boat, protect yourself with a detector.

PERSONAL FLOTATION DEVICES

Personal Flotation Devices (PFDs) are commonly called life preservers or life jackets. You can get them in a variety of types and sizes. They vary with their intended uses. To be acceptable, PFDs must be Coast Guard approved.

Type I PFDs

A Type I life jacket is also called an offshore life jacket. Type I life jackets will turn most unconscious people from facedown to a vertical or slightly backward position. The adult size gives a minimum of 22 pounds of buoyancy. The child size has at least 11 pounds. Type I jackets provide more protection to their wearers than any other type of life jacket. Type I life jackets are bulkier and less comfortable than other types. Furthermore, there are only two sizes, one for children and one for adults.

Type I life jackets will keep their wearers afloat for extended periods in rough water. They are recommended for offshore cruising where a delayed rescue is probable.

Type II PFDs

A Type II life jacket is also called a near-shore buoyant vest. It is an approved, wearable device. Type II life jackets will turn some unconscious people from facedown to vertical or slightly backward positions. The adult size gives at least 15.5 pounds of buoyancy. The medium child size has a minimum of 11 pounds. And the small child and infant sizes give seven pounds. A Type II life jacket is more comfortable than a Type I but it does not have as much buoyancy. It is not recommended for long hours in rough water. Because of this, Type IIs are recommended for inshore and inland cruising on calm water. Use them only where there is a good chance of fast rescue.

Type III PFDs

◆ See Figure 7

Type III life jackets or marine buoyant devices are also known as flotation aids. Like Type IIs, they are designed for calm inland or close offshore water where there is a good chance of fast rescue. Their minimum buoyancy is 15.5 pounds. They will **not** turn their wearers face up.

Type III devices are usually worn where freedom of movement is necessary. Thus, they are used for water skiing, small boat sailing, and fishing among other activities. They are available as vests and flotation coats. Flotation coats are useful in cold weather. Type IIIs come in many sizes from small child through large adult.

Life jackets come in a variety of colors and patterns - red, blue, green, camouflage, and cartoon characters. From purely a safety standpoint, the best color is bright orange. It is easier to see in the water, especially if the water is rough.

Type IV PFDs

◆ See Figure 8 and 9

Type IV ring life buoys, buoyant cushions and horseshoe buoys are Coast Guard approved devices called throwables. They are made to be thrown to people in the water, and should not be worn. Type IV cushions are often used as seat cushions. But, keep in mind that cushions are hard to hold onto in the water, thus, they do not afford as much protection as wearable life jackets.

The straps on buoyant cushions are for you to hold onto either in the water or when throwing them, they are **NOT** for your arms. A cushion should never be worn on your back, as it will turn you face down in the water.

Type IV throwables are not designed as personal flotation devices for unconscious people, non-swimmers, or children. Use them only in emergencies. They should not be used for, long periods in rough water.

Ring life buoys come in 18, 20, 24, and 30 in. diameter sizes. They usually have grab lines, but you will need to attach about 60 feet of polypropylene line to the grab rope to aid in retrieving someone in the water. If you throw a ring, be careful not to hit the person. Ring buoys can knock people unconscious

Type V PFDs

Type V PFDs are of two kinds, special use devices and hybrids. Special use devices include boardsailing vests, deck suits, work vests, and others. They are approved only for the special uses or conditions indicated on their labels. Each is designed and intended for the particular application shown on its label. They do not meet legal requirements for general use aboard recreational boats.

Hybrid life jackets are inflatable devices with some built-in buoyancy provided by plastic foam or kapok. They can be inflated orally or by cylinders of compressed gas to give additional buoyancy. In some hybrids the gas is released manually. In others it is released automatically when the life jacket is immersed in water.

The inherent buoyancy of a hybrid may be insufficient to float a person unless it is inflated. The only way to find this out is for the user to try it in the water. Because of its limited buoyancy when deflated, a hybrid is recommended for use by a non-swimmer only if it is worn with enough inflation to float the wearer.

If they are to count against the legal requirement for the number of life jackets you must carry, hybrids manufactured before February 8, 1995 must be worn whenever a boat is underway and the wearer must not go below decks or in an enclosed space. To find out if your Type V hybrid must be worn to satisfy the legal requirement, read its label. If its use is restricted it will say, "REQUIRED TO BE WORN" in capital letters.

Hybrids cost more than other life jackets, but this factor must be weighed against the fact that they are more comfortable than Types I, II or III life jackets. Because of their greater comfort, their owners are more likely to wear them than are the owners of Type I, II or III life jackets.

The Coast Guard has determined that improved, less costly hybrids can save lives since they will be bought and used more frequently. For these

Fig. 7 Type III PFDs are recommended for inshore/inland use on calm water (where there is a good chance of fast rescue)

Fig. 8 Type IV buoyant cushions are thrown to people in the water. If you can squeeze air out of the cushion, it should be replaced

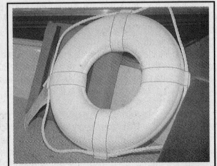

Fig. 9 Type IV throwables, such as this ring life buoy, are not designed for unconscious people, non-swimmers, or children

1-8 GENERAL INFORMATION, SAFETY & TOOLS

reasons, a new federal regulation was adopted effective February 8, 1995. The regulation increases both the deflated and inflated buoyancy's of hybrids, makes them available in a greater variety of sizes and types, and reduces their costs by reducing production costs.

Even though it may not be required, the wearing of a hybrid or a life jacket is encouraged whenever a vessel is underway. Like life jackets, hybrids are now available in three types. To meet legal requirements, a Type I hybrid can be substituted for a Type I life jacket. Similarly Type II and III hybrids can be substituted for Type II and Type III life jackets. A Type I hybrid, when inflated, will turn most unconscious people from facedown to vertical or slightly backward positions just like a Type I life jacket. Type II and III hybrids function like Type II and III life jackets. If you purchase a new hybrid, it should have an owner's manual attached that describes its life jacket type and its deflated and inflated buoyancys. It warns you that it may have to be inflated to float you. The manual also tells you how to don the life jacket and how to inflate it. It also tells you how to change its inflation mechanism, recommended testing exercises, and inspection or maintenance procedures. The manual also tells you why you need a life jacket and why you should wear it. A new hybrid must be packaged with at least three gas cartridges. One of these may already be loaded into the inflation mechanism. Likewise, if it has an automatic inflation mechanism, it must be packaged with at least three of these water sensitive elements. One of these elements may be installed.

Legal Requirements

A Coast Guard approved life jacket must show the manufacturer's name and approval number. Most are marked as Type I, II, III, IV or V. All of the newer hybrids are marked for type.

You are required to carry at least one wearable life jacket or hybrid for each person on board your recreational vessel. If your vessel is 16 feet or more in length and is not a canoe or a kayak, you must also have at least one Type IV on board. These requirements apply to all recreational vessels that are propelled or controlled by machinery, sails, oars, paddles, poles, or another vessel. Sailboards are not required to carry life jackets.

You can substitute an older Type V hybrid for any required Type I, II or III life jacket provided:

1. Its approval label shows it is approved for the activity the vessel is engaged in
2. It's approved as a substitute for a life jacket of the type required on the vessel
3. It's used as required on the labels
and
4. It's used in accordance with any requirements in its owner's manual (if the approval label makes reference to such a manual.)

A water skier being towed is considered to be on board the vessel when judging compliance with legal requirements.

You are required to keep your Type I, II or III life jackets or equivalent hybrids readily accessible, which means you must be able to reach out and get them when needed. All life jackets must be in good, serviceable condition.

General Considerations

The proper use of a life jacket requires the wearer to know how it will perform. You can gain this knowledge only through experience. Each person on your boat should be assigned a life jacket. Next, it should be fitted to the person who will wear it. Only then can you be sure that it will be ready for use in an emergency. This advice is good even if the water is calm, and you intend to boat near shore.

Boats can sink fast. There may be no time to look around for a life jacket. Fitting one on you in the water is almost impossible. Most drownings occur in inland waters within a few feet of safety. Most victims had life jackets, but they weren't wearing them.

Keeping life jackets in the plastic covers they came wrapped in, and in a cabin, assure that they will stay clean and unfaded. But this is no way to keep them when you are on the water. When you need a life jacket it must be readily accessible and adjusted to fit you. You can't spend time hunting for it or learning how to fit it.

There is no substitute for the experience of entering the water while wearing a life jacket. Children, especially, need practice. If possible, give your guests this experience. Tell them they should keep their arms to their sides when jumping in to keep the life jacket from riding up. Let them jump in and see how the life jacket responds. Is it adjusted so it does not ride up? Is it the proper size? Are all straps snug? Are children's life jackets the right sizes for them? Are they adjusted properly? If a child's life jacket fits correctly, you can lift the child by the jacket's shoulder straps and the child's chin and ears will not slip through. Non-swimmers, children, handicapped persons, elderly persons and even pets should always wear life jackets when they are aboard. Many states require that everyone aboard wear them in hazardous waters.

Inspect your lifesaving equipment from time to time. Leave any questionable or unsatisfactory equipment on shore. An emergency is no time for you to conduct an inspection.

Indelibly mark your life jackets with your vessel's name, number, and calling port. This can be important in a search and rescue effort. It could help concentrate effort where it will do the most good.

Care of Life Jackets

Given reasonable care, life jackets last many years. Thoroughly dry them before putting them away. Stow them in dry, well-ventilated places. Avoid the bottoms of lockers and deck storage boxes where moisture may collect. Air and dry them frequently.

Life jackets should not be tossed about or used as fenders or cushions. Many contain kapok or fibrous glass material enclosed in plastic bags. The bags can rupture and are then unserviceable. Squeeze your life jacket gently. Does air leak out? If so, water can leak in and it will no longer be safe to use. Cut it up so no one will use it, and throw it away. The covers of some life jackets are made of nylon or polyester. These materials are plastics. Like many plastics, they break down after extended exposure to the ultraviolet light in sunlight. This process may be more rapid when the materials are dyed with bright dyes such as "neon" shades.

Ripped and badly faded fabrics are clues that the covering of your life jacket is deteriorating. A simple test is to pinch the fabric between your thumbs and forefingers. Now try to tear the fabric. If it can be torn, it should definitely be destroyed and discarded. Compare the colors in protected places to those exposed to the sun. If the colors have faded, the materials have been weakened. A life jacket covered in fabric should ordinarily last several boating seasons with normal use. A life jacket used every day in direct sunlight should probably be replaced more often.

SOUND PRODUCING DEVICES

All boats are required to carry some means of making an efficient sound signal. Devices for making the whistle or horn noises required by the Navigation Rules must be capable of a four-second blast. The blast should be audible for at least one-half mile. Athletic whistles are not acceptable on boats 12 meters or longer. Use caution with athletic whistles. When wet, some of them come apart and loose their "pea." When this happens, they are useless.

If your vessel is 12 meters long and less than 20 meters, you must have a power whistle (or power horn) and a bell on board. The bell must be in operating condition and have a minimum diameter of at least 200mm (7.9 in.) at its mouth.

VISUAL DISTRESS SIGNALS

◆ See Figure 10

Visual Distress Signals (VDS) attract attention to your vessel if you need help. They also help to guide searchers in search and rescue situations. Be sure you have the right types, and learn how to use them properly.

It is illegal to fire flares improperly. In addition, they cost the Coast Guard and its Auxiliary many wasted hours in fruitless searches. If you signal a distress with flares and then someone helps you, please let the Coast Guard or the appropriate Search And Rescue (SAR) Agency know so the distress report will be canceled.

Recreational boats less than 16 feet long must carry visual distress signals on coastal waters at night. Coastal waters are:
- The ocean (territorial sea)
- The Great Lakes
- Bays or sounds that empty into oceans
- Rivers over two miles across at their mouths upstream to where they narrow to two miles.

Recreational boats 16 feet or longer must carry VDS at all times on coastal waters. The same requirement applies to boats carrying six or fewer passengers for hire. Open sailboats less than 26 feet long without engines are exempt in the daytime as are manually propelled boats. Also exempt are boats in organized races, regattas, parades, etc. Boats owned in the United States and operating on the high seas must be equipped with VDS.

A wide variety of signaling devices meet Coast Guard regulations. For pyrotechnic devices, a minimum of three must be carried. Any combination can be carried as long as it adds up to at least three signals for day use and

GENERAL INFORMATION SAFETY & TOOLS 1-9

at least three signals for night use. Three day/night signals meet both requirements. If possible, carry more than the legal requirement.

■ The American flag flying upside down is a commonly recognized distress signal. It is not recognized in the Coast Guard regulations, though. In an emergency, your efforts would probably be better used in more effective signaling methods.

Types of VDS

VDS are divided into two groups; daytime and nighttime use. Each of these groups is subdivided into pyrotechnic and non-pyrotechnic devices.

Daytime Non-Pyrotechnic Signals

A bright orange flag with a black square over a black circle is the simplest VDS. It is usable, of course, only in daylight. It has the advantage of being a continuous signal. A mirror can be used to good advantage on sunny days. It can attract the attention of other boaters and of aircraft from great distances. Mirrors are available with holes in their centers to aid in "aiming." In the absence of a mirror, any shiny object can be used. When another boat is in sight, an effective VDS is to extend your arms from your sides and move them up and down. Do it slowly. If you do it too fast the other people may think you are just being friendly. This simple gesture is seldom misunderstood, and requires no equipment.

Daytime Pyrotechnic Devices

Orange smoke is a useful daytime signal. Hand-held or floating smoke flares are very effective in attracting attention from aircraft. Smoke flares don't last long, and are not very effective in high wind or poor visibility. As with other pyrotechnic devices, use them only when you know there is a possibility that someone will see the display.

To be usable, smoke flares must be kept dry. Keep them in airtight containers and store them in dry places. If the "striker" is damp, dry it out before trying to ignite the device. Some pyrotechnic devices require a forceful "strike" to ignite them.

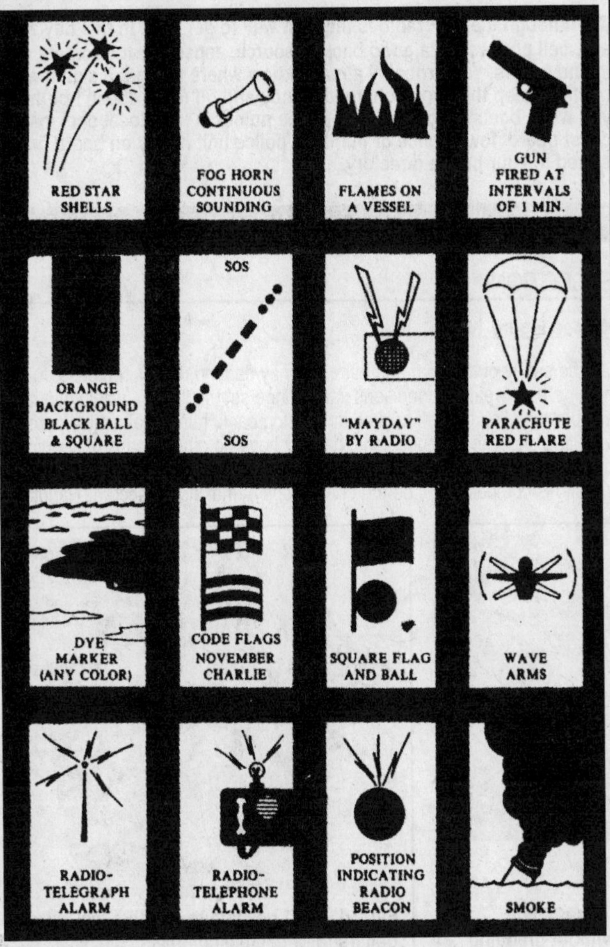

All hand-held pyrotechnic devices may produce hot ashes or slag when burning. Hold them over the side of your boat in such a way that they do not burn your hand or drip into your boat.

Nighttime Non-Pyrotechnic Signals

An electric distress light is available. This light automatically flashes the international Morse code SOS distress signal (••• — •••). Flashed four to six times a minute, it is an unmistakable distress signal. It must show that it is approved by the Coast Guard. Be sure the batteries are fresh. Dated batteries give assurance that they are current.

Under the Inland Navigation Rules, a high intensity white light flashing 50-70 times per minute is a distress signal. Therefore, use strobe lights on inland waters only for distress signals.

Nighttime Pyrotechnic Devices

◆ See Figure 11

Aerial and hand-held flares can be used at night or in the daytime. Obviously, they are more effective at night.

Currently, the serviceable life of a pyrotechnic device is rated at 42 months from its date of manufacture. Pyrotechnic devices are expensive. Look at their dates before you buy them. Buy them with as much time remaining as possible.

Like smoke flares, aerial and hand-held flares may fail to work if they have been damaged or abused. They will not function if they are or have been wet. Store them in dry, airtight containers in dry places. But store them where they are readily accessible.

Aerial VDSs, depending on their type and the conditions they are used in, may not go very high. Again, use them only when there is a good chance they will be seen.

A serious disadvantage of aerial flares is that they burn for only a short time; most burn for less than 10 seconds. Most parachute flares burn for less than 45 seconds. If you use a VDS in an emergency, do so carefully. Hold hand-held flares over the side of the boat when in use. Never use a road hazard flare on a boat; it can easily start a fire. Marine type flares are specifically designed to lessen risk, but they still must be used carefully.

Aerial flares should be given the same respect as firearms since they are firearms! Never point them at another person. Don't allow children to play with them or around them. When you fire one, face away from the wind. Aim it downwind and upward at an angle of about 60 degrees to the horizon. If there is a strong wind, aim it somewhat more vertically. Never fire it straight up. Before you discharge a flare pistol, check for overhead obstructions that might be damaged by the flare. An obstruction might deflect the flare to where it will cause injury or damage.

Fig. 10 Internationally accepted distress signals

Fig. 11 Moisture-protected flares should be carried onboard any vessel for use as a distress signal

1-10 GENERAL INFORMATION, SAFETY & TOOLS

Disposal of VDS

Keep outdated flares when you get new ones. They do not meet legal requirements, but you might need them sometime, and they may work. It is illegal to fire a VDS on federal navigable waters unless an emergency exists. Many states have similar laws.

Emergency Position Indicating Radio Beacon (EPIRB)

There is no requirement for recreational boats to have EPIRBs. Some commercial and fishing vessels, though, must have them if they operate beyond the three-mile limit. Vessels carrying six or fewer passengers for hire must have EPIRBs under some circumstances when operating beyond the three-mile limit. If you boat in a remote area or offshore, you should have an EPIRB. An EPIRB is a small (about 6 to 20 in. high), battery-powered, radio transmitting buoy-like device. It is a radio transmitter and requires a license or an endorsement on your radio station license by the Federal Communications Commission (FCC). EPIRBs are either automatically activated by being immersed in water or manually by a switch.

Courtesy Marine Examinations

One of the roles of the Coast Guard Auxiliary is to promote recreational boating safety. This is why they conduct thousands of Courtesy Marine Examinations each year. The auxiliarists who do these examinations are well-trained and knowledgeable in the field.

These examinations are free and done only at the consent of boat owners. To pass the examination, a vessel must satisfy federal equipment requirements and certain additional requirements of the coast guard auxiliary. If your vessel does not pass the Courtesy Marine Examination, no report of the failure is made. Instead, you will be told what you need to correct the deficiencies. The examiner will return at your convenience to redo the examination.

If your vessel qualifies, you will be awarded a safety decal. The decal does not carry any special privileges, it simply attests to your interest in safe boating.

BOATING EQUIPMENT (NOT REQUIRED BUT RECOMMENDED)

Although not required by law, there are other pieces of equipment that are good to have onboard.

Oar/Paddle (Second Means of Propulsion)

All boats less than 16 feet long should carry a second means of propulsion. A paddle or oar can come in handy at times. For most small boats, a spare trolling or outboard motor is an excellent idea. If you carry a spare motor, it should have its own fuel tank and starting power. If you use an electric trolling motor, it should have its own battery.

Bailing Devices

All boats should carry at least one effective manual bailing device in addition to any installed electric bilge pump. This can be a bucket, can, scoop, hand-operated pump, etc. If your battery "goes dead" it will not operate your electric pump.

First Aid Kit

◆ See Figure 12

All boats should carry a first aid kit. It should contain adhesive bandages, gauze, adhesive tape, antiseptic, aspirin, etc. Check your first aid kit from time to time. Replace anything that is outdated. It is to your advantage to know how to use your first aid kit. Another good idea would be to take a Red Cross first aid course.

Anchors

◆ See Figure 13

All boats should have anchors. Choose one of suitable size for your boat. Better still, have two anchors of different sizes. Use the smaller one in calm water or when anchoring for a short time to fish or eat. Use the larger one when the water is rougher or for overnight anchoring.

Carry enough anchor line, of suitable size, for your boat and the waters in which you will operate. If your engine fails you, the first thing you usually should do is lower your anchor. This is good advice in shallow water where you may be driven aground by the wind or water. It is also good advice in windy weather or rough water, as the anchor, when properly affixed, will usually hold your bow into the waves.

VHF-FM Radio

Your best means of summoning help in an emergency or in case of a breakdown is a VHF-FM radio. You can use it to get advice or assistance from the Coast Guard. In the event of a serious illness or injury aboard your boat, the Coast Guard can have emergency medical equipment meet you ashore.

■ Although the VHF radio is the best way to get help, in this day and age, cell phones are a good backup source, especially for boaters on inland waters. You probably already know where you get a signal when boating, keep the phone charged, handy and off (so it doesn't bother you when boating right?). Keep phone numbers for a local dockmaster, coast guard, tow service or maritime police unit handy on board or stored in your phone directory.

Compass

SELECTION

◆ See Figure 14

The safety of the boat and her crew may depend on her compass. In many areas, weather conditions can change so rapidly that, within minutes, a skipper may find himself socked in by a fog bank, rain squall or just poor visibility. Under these conditions, he may have no other means of keeping to his desired course except with the compass. When crossing an open body of water, his compass may be the only means of making an accurate landfall.

Fig. 12 Always carry an adequately stocked first aid kit on board for the safety of the crew and guests

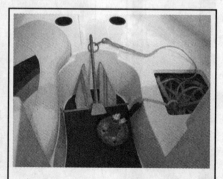

Fig. 13 Choose an anchor of sufficient weight to secure the boat without dragging

Fig. 14 Don't hesitate to spend a few extra dollars for a reliable compass

GENERAL INFORMATION SAFETY & TOOLS

Fig. 15 The compass is a delicate instrument which should be mounted securely in a position where it can be easily observed by the helmsman

During thick weather when you can neither see nor hear the expected aids to navigation, attempting to run out the time on a given course can disrupt the pleasure of the cruise. The skipper gains little comfort in a chain of soundings that does not match those given on the chart for the expected area. Any stranding, even for a short time, can be an unnerving experience.

A pilot will not knowingly accept a cheap parachute. By the same token, a good boater should not accept a bargain in lifejackets, fire extinguishers, or compass. Take the time and spend the few extra dollars to purchase a compass to fit your expected needs. Regardless of what the salesman may tell you, postpone buying until you have had the chance to check more than one make and model.

Lift each compass, tilt and turn it, simulating expected motions of the boat. The compass card should have a smooth and stable reaction.

The card of a good quality compass will come to rest without oscillations about the lubber's line. Reasonable movement in your hand, comparable to the rolling and pitching of the boat, should not materially affect the reading.

INSTALLATION

◆ See Figure 15

Proper installation of the compass does not happen by accident. Make a critical check of the proposed location to be sure compass placement will permit the helmsman to use it with comfort and accuracy. First, the compass should be placed directly in front of the helmsman, and in such a position that it can be viewed without body stress as he sits or stands in a posture of relaxed alertness. The compass should be in the helmsman's zone of comfort. If the compass is too far away, he may have to bend forward to watch it; too close and he must rear backward for relief.

Second, give some thought to comfort in heavy weather and poor visibility conditions during the day and night. In some cases, the compass position may be partially determined by the location of the wheel, shift lever and throttle handle.

Third, inspect the compass site to be sure the instrument will be at least two feet from any engine indicators, bilge vapor detectors, magnetic instruments, or any steel or iron objects. If the compass cannot be placed at least two feet (six feet would be better but on a small craft, let's get real two feet is usually pushing it) from one of these influences, then either the compass or the other object must be moved, if first order accuracy is to be expected.

Once the compass location appears to be satisfactory, give the compass a test before installation. Hidden influences may be concealed under the cabin top, forward of the cabin aft bulkhead, within the cockpit ceiling, or in a wood-covered stanchion.

Move the compass around in the area of the proposed location. Keep an eye on the card. A magnetic influence is the only thing that will make the card turn. You can quickly find any such influence with the compass. If the influence cannot be moved away or replaced by one of non-magnetic material, test to determine whether it is merely magnetic, a small piece of iron or steel, or some magnetized steel. Bring the north pole of the compass near the object, then shift and bring the south pole near it. Both the north and south poles will be attracted if the compass is demagnetized. If the object attracts one pole and repels the other, then the compass is magnetized. If your compass needs to be demagnetized, take it to a shop equipped to do the job PROPERLY.

After you have moved the compass around in the proposed mounting area, hold it down or tape it in position. Test everything you feel might affect the compass and cause a deviation from a true reading. Rotate the wheel from hard over-to-hard over. Switch on and off all the lights, radios, radio direction finder, radio telephone, depth finder and, if installed, the shipboard intercom. Sound the electric whistle, turn on the windshield wipers, start the engine (with water circulating through the engine), work the throttle, and move the gear shift lever. If the boat has an auxiliary generator, start it.

If the card moves during any one of these tests, the compass should be relocated. Naturally, if something like the windshield wipers causes a slight deviation, it may be necessary for you to make a different deviation table to use only when certain pieces of equipment are operating. Bear in mind, following a course that is off only a degree or two for several hours can make considerable difference at the end, putting you on a reef, rock or shoal.

Check to be sure the intended compass site is solid. Vibration will increase pivot wear.

Now, you are ready to mount the compass. To prevent an error on all courses, the line through the lubber line and the compass card pivot must be exactly parallel to the keel of the boat. You can establish the fore-and-aft line of the boat with a stout cord or string. Use care to transfer this line to the compass site. If necessary, shim the base of the compass until the stile-type lubber line (the one affixed to the case and not gimbaled) is vertical when the boat is on an even keel. Drill the holes and mount the compass.

COMPASS PRECAUTIONS

◆ See Figures 16, 17 and 18

Many times an owner will install an expensive stereo system in the cabin of his boat. It is not uncommon for the speakers to be mounted on the aft bulkhead up against the overhead (ceiling). In almost every case, this position places one of the speakers in very close proximity to the compass, mounted above the ceiling.

Fig. 16 This compass is giving an accurate reading, right?

Fig. 17 ...well think again, as seemingly innocent objects may cause serious problems...

Fig. 18 ...a compass reading off by just a few degrees could lead to disaster

1-12 GENERAL INFORMATION, SAFETY & TOOLS

You probably already know that a magnet is used in the operation of the speaker. Therefore, it is very likely that the speaker, mounted almost under the compass in the cabin will have a very pronounced effect on the compass accuracy.

Consider the following test and the accompanying photographs as proof:
First, the compass was read as 190 degrees while the boat was secure in her slip.

Next, a full can of soda in an aluminum can was placed on one side and the compass read as 204 degrees, a good 14 degrees off.

Next, the full can was moved to the opposite side of the compass and again a reading was observed, this time as 189 degrees, 11 degrees off from the original reading.

Finally, the contents of the can were consumed, the can placed on both sides of the compass with NO effect on the compass reading.

Two very important conclusions can be drawn from these tests.
- Something must have been in the contents of the can to affect the compass so drastically.
- Keep even innocent things clear of the compass to avoid any possible error in the boat's heading.

■ **Remember, a boat moving through the water at 10 knots on a compass error of just 5 degrees will be almost 1.5 miles off course in only ONE hour. At night, or in thick weather, this could very possibly put the boat on a reef, rock or shoal with disastrous results.**

Fig. 19 A flashlight with a fresh set of batteries is handy when repairs are needed at night. It can also double as a signaling device

Tools and Spare Parts

◆ See Figures 19 and 20

Carry a few tools and some spare parts, and learn how to make minor repairs. Many search and rescue cases are caused by minor breakdowns that boat operators could have repaired. Carry spare parts such as propellers, fuses or basic ignition components (like spark plugs, wires or even ignition coils) and the tools necessary to install them.

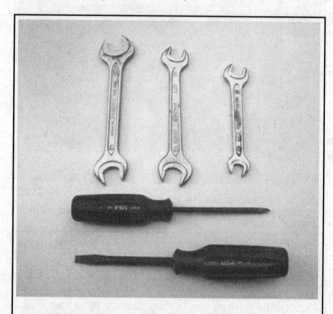

Fig. 20 A few wrenches, a screwdriver and maybe a pair of pliers can be very helpful to make emergency repairs

SAFETY IN SERVICE

It is virtually impossible to anticipate all of the hazards involved with maintenance and service, but care and common sense will prevent most accidents.

The rules of safety for mechanics range from "don't smoke around gasoline," to "use the proper tool(s) for the job." The trick to avoiding injuries is to develop safe work habits and to take every possible precaution. Whenever you are working on your boat, pay attention to what you are doing. The more you pay attention to details and what is going on around you, the less likely you will be to hurt yourself or damage your boat.

Do's

- Do keep a fire extinguisher and first aid kit handy.
- Do wear safety glasses or goggles when cutting, drilling, grinding or prying, even if you have 20-20 vision. If you wear glasses for the sake of vision, wear safety goggles over your regular glasses.
- Do shield your eyes whenever you work around the battery. Batteries contain sulfuric acid. In case of contact with the eyes or skin, flush the area with water or a mixture of water and baking soda; then seek immediate medical attention.
- Do use adequate ventilation when working with any chemicals or hazardous materials.
- Do disconnect the negative battery cable when working on the electrical system. The secondary ignition system contains EXTREMELY HIGH VOLTAGE. In some cases it can even exceed 50,000 volts. Furthermore, an accidental attempt to start the engine could cause the propeller or other components to rotate suddenly causing a potentially dangerous situation.
- Do follow manufacturer's directions whenever working with potentially hazardous materials. Most chemicals and fluids are poisonous if taken internally.
- Do properly maintain your tools. Loose hammerheads, mushroomed punches and chisels, frayed or poorly grounded electrical cords, excessively worn screwdrivers, spread wrenches (open end), cracked sockets, or slipping ratchets can cause accidents.
- Likewise, keep your tools clean; a greasy wrench can slip off a bolt head, ruining the bolt and often harming your knuckles in the process.
- Do use the proper size and type of tool for the job at hand. Do select a wrench or socket that fits the nut or bolt. The wrench or socket should sit straight, not cocked.
- Do, when possible, pull on a wrench handle rather than push on it, and adjust your stance to prevent a fall.
- Do be sure that adjustable wrenches are tightly closed on the nut or bolt and pulled so that the force is on the side of the fixed jaw. Better yet, avoid the use of an adjustable if you have a fixed wrench that will fit.
- Do strike squarely with a hammer; avoid glancing blows.
- Do use common sense whenever you work on your boat or motor. If a situation arises that doesn't seem right, sit back and have a second look. It may save an embarrassing moment or potential damage to your beloved boat.

Don'ts

- Don't run the engine in an enclosed area or anywhere else without proper ventilation - EVER! Carbon monoxide is poisonous; it takes a long time to leave the human body and you can build up a deadly supply of it in your system by simply breathing in a little every day. You may not realize you are slowly poisoning yourself.
- Don't work around moving parts while wearing loose clothing. Short sleeves are much safer than long, loose sleeves. Hard-toed shoes with neoprene soles protect your toes and give a better grip on slippery surfaces. Jewelry, watches, large belt buckles, or body adornment of any kind is not

GENERAL INFORMATION SAFETY & TOOLS

safe working around any craft or vehicle. Long hair should be tied back under a hat.

• Don't use pockets for toolboxes. A fall or bump can drive a screwdriver deep into your body. Even a rag hanging from your back pocket can wrap around a spinning shaft.

• Don't smoke when working around gasoline, cleaning solvent or other flammable material.

• Don't smoke when working around the battery. When the battery is being charged, it gives off explosive hydrogen gas. Actually, you shouldn't smoke anyway, it's bad for you. Instead, save the cigarette money and put it into your boat!

• Don't use gasoline to wash your hands; there are excellent soaps available. Gasoline contains dangerous additives that can enter the body through a cut or through your pores. Gasoline also removes all the natural oils from the skin so that bone dry hands will suck up oil and grease.

• Don't use screwdrivers for anything other than driving screws! A screwdriver used as a prying tool can snap when you least expect it, causing injuries. At the very least, you'll ruin a good screwdriver.

TROUBLESHOOTING

Troubleshooting can be defined as a methodical process during which one discovers what is causing a problem with engine operation. Although it is often a feared process to the uninitiated, there is no reason to believe that you cannot figure out what is wrong with a motor, as long as you follow a few basic rules.

To begin with, troubleshooting must be systematic. Haphazardly testing one component, then another, **might** uncover the problem, but it will more likely waste a lot of time. True troubleshooting starts by defining the problem and performing systematic tests to eliminate the largest and most likely causes first.

Start all troubleshooting by eliminating the most basic possible causes. Begin with a visual inspection of the boat and motor. If the engine won't crank, make sure that the kill switch or safety lanyard is in the proper position. Make sure there is fuel in the tank and the fuel system is primed before condemning the carburetor or fuel injection system. On electric start motors, make sure there are no blown fuses, the battery is fully charged, and the cable connections (at both ends) are clean and tight before suspecting a bad starter, solenoid or switch.

The majority of problems that occur suddenly can be fixed by simply identifying the one small item that brought them on. A loose wire, a clogged passage or a broken component can cause a lot of trouble and are often the cause of a sudden performance problem.

The next most basic step in troubleshooting is to test systems before components. For example, if the engine doesn't crank on an electric start motor, determine if the battery is in good condition (fully charged and properly connected) before testing the starting system. If the engine cranks, but doesn't start, you know already know the starting system and battery (if it cranks fast enough) are in good condition, now it is time to look at the ignition or fuel systems. Once you've isolated the problem to a particular system, follow the troubleshooting/testing procedures in the section for that system to test either subsystems (if applicable, for example: the starter circuit) or components (starter solenoid).

Basic Operating Principles

◆ See Figures 21 and 22

Before attempting to troubleshoot a problem with your motor, it is important that you understand how it operates. Once normal engine or system operation is understood, it will be easier to determine what might be causing the trouble or irregular operation in the first place. System descriptions are found throughout this manual, but the basic mechanical operating principles for both 2-stroke engines (like most of the outboards covered here) and 4-stroke engines (like some outboards and like your car) are given here. A basic understanding of both types of engines is useful not only in understanding and troubleshooting your outboard, but also for dealing with other motors in your life.

All motors covered by this manual (and probably MOST of the motors you own) operate according to the Otto cycle principle of engine operation. This means that all motors follow the stages of intake, compression, power and exhaust. But, the difference between a 2- and 4-stroke motor is in how many times the piston moves up and down within the cylinder to accomplish this. On 2-stroke motors (as the name suggests) the four cycles take place in 2 movements (one up and one down) of the piston. Again, as the name suggests, the cycles take place in 4 movements of the piston for 4-stroke motors.

2-STROKE MOTORS

The 2-stroke engine differs in several ways from a conventional four-stroke (automobile or marine) engine.

1. The intake/exhaust method by which the fuel-air mixture is delivered to the combustion chamber.
2. The complete lubrication system.
3. The frequency of the power stroke.

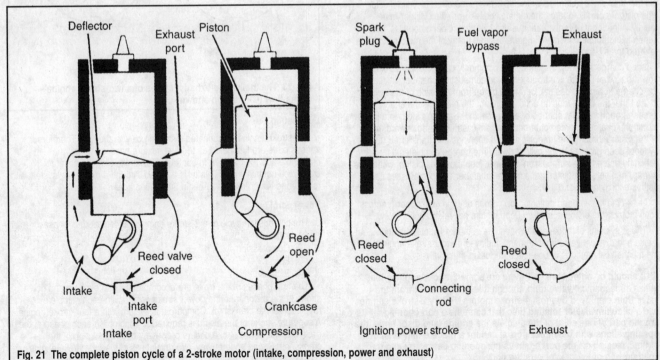

Fig. 21 The complete piston cycle of a 2-stroke motor (intake, compression, power and exhaust)

1-14 GENERAL INFORMATION, SAFETY & TOOLS

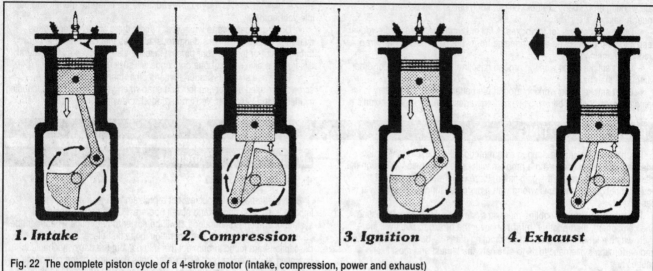

Fig. 22 The complete piston cycle of a 4-stroke motor (intake, compression, power and exhaust)

Let's discuss these differences briefly (and compare 2-stroke engine operation with 4-stroke engine operation.)

Intake/Exhaust

◆ See Figures 23 thru 26

Two-stroke engines utilize an arrangement of port openings to admit fuel to the combustion chamber and to purge the exhaust gases after burning has been completed. The ports are located in a precise pattern in order for them to be open and closed off at an exact moment by the piston as it moves up and down in the cylinder. The exhaust port is located slightly higher than the fuel intake port. This arrangement opens the exhaust port first as the piston starts downward and therefore, the exhaust phase begins a fraction of a second before the intake phase.

Actually, the intake and exhaust ports are spaced so closely together that both open almost simultaneously. For this reason, 2-stroke engines from some manufacturers utilize deflector-type pistons. This design of the piston top serves two purposes very effectively.

First, it creates turbulence when the incoming charge of fuel enters the combustion chamber. This turbulence results in a more complete burning of the fuel than if the piston top were flat. The second effect of the deflector-type piston crown is to force the exhaust gases from the cylinder more rapidly. Although this configuration is used in many older outboards, it is generally not found only many motors. Instead many of the motors could be referred to as Loop charged.

Loop charged motors, or as they are commonly called "loopers", differ in how the air/fuel charge is introduced to the combustion chamber. Instead of the charge flowing across the top of the piston from one side of the cylinder to the other the use a looping action on top of the piston as the charge is forced through irregular shaped openings cut in the piston's skirt. In a looper motor, the charge is forced out from the crankcase by the downward motion of the piston, through the irregular shaped openings and transferred upward by long, deep grooves in the cylinder wall. The charge completes its looping action by entering the combustion chamber, just above the piston, where the upward motion of the piston traps it in the chamber and compresses it for optimum ignition power.

Unlike the knife-edged deflector top pistons used in non-looper (or cross flow) motors, the piston domes on Loop motors are relatively flat.

These systems of intake and exhaust are in marked contrast to individual intake and exhaust valve arrangement employed on four-stroke engines (and the mechanical methods of opening and closing these valves).

■ It should be noted here that there are some 2-stroke engines that utilize a mechanical valve train, though it is very different from the valve train employed by most 4-stroke motors. Rotary 2-stroke engines use a circular valve or rotating disc that contains a port opening around part of one edge of the disc. As the engine (and disc) turns, the opening aligns with the intake port at and for a predetermined amount of time, closing off the port again as the opening passes by and the solid portion of the disc covers the port.

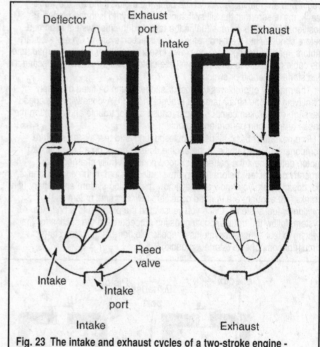

Fig. 23 The intake and exhaust cycles of a two-stroke engine - Cross flow (CV) design shown

Lubrication

A 2-stroke engine is lubricated by mixing oil with the fuel. Therefore, various parts are lubricated as the fuel mixture passes through the crankcase and the cylinder. In contrast, four-stroke engines have a crankcase containing oil. This oil is pumped through a circulating system and returned to the crankcase to begin the routing again.

Power Stroke

The combustion cycle of a 2-stroke engine has four distinct phases.
1. Intake
2. Compression
3. Power
4. Exhaust

The four phases of the cycle are accomplished with each up and down stroke of the piston, and the power stroke occurs with each complete revolution of the crankshaft. Compare this system with a four-stroke engine. A separate stroke of the piston is required to accomplish each phase of the cycle and the power stroke occurs only every other revolution of the crankshaft. Stated another way, two revolutions of the four-stroke engine crankshaft are required to complete one full cycle, the four phases.

GENERAL INFORMATION SAFETY & TOOLS 1-15

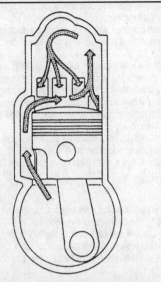

Fig. 24 Cross-sectional view of a typical loop-charged cylinder, showing charge flow while piston is moving downward

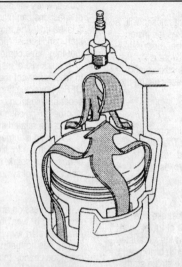

Fig. 25 Cutaway view of a typical loop-charged cylinder, depicting exhaust leaving the cylinder as the charge enters through 3 ports in the piston

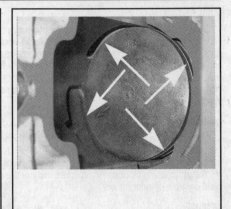

Fig. 26 The combustion chamber of a typical looper, notice the piston is far enough down the cylinder bore to reveal intake and exhaust ports

Physical Laws

◆ See Figure 27

The 2-stroke engine is able to function because of two very simple physical laws.

One: Gases will flow from an area of high pressure to an area of lower pressure. A tire blowout is an example of this principle. The high-pressure air escapes rapidly if the tube is punctured.

Two: If a gas is compressed into a smaller area, the pressure increases, and if a gas expands into a larger area, the pressure is decreased.

If these two laws are kept in mind, the operation of the 2-stroke engine will be easier understood.

Actual Operation

◆ See Figure 21

■ The engine described here is of a carbureted type. EFI and DFI/Optimax motors operate similarly for intake of the air charge and for exhaust of the unburned gasses. Obviously though, the very nature of fuel injection changes the actual delivery of the fuel/oil charge.

Beginning with the piston approaching top dead center on the compression stroke: the intake and exhaust ports are physically closed (blocked) by the piston. During this stroke, the reed valve is open (because as the piston moves upward, the crankcase volume increases, which reduces the crankcase pressure to less than the outside atmosphere (creates a vacuum under the piston). The spark plug fires; the compressed fuel-air mixture is ignited; and the power stroke begins.

As the piston moves downward on the power stroke, the combustion chamber is filled with burning gases. As the exhaust port is uncovered, the gases, which are under great pressure, escape rapidly through the exhaust ports. The piston continues its downward movement. Pressure within the crankcase (again, under the piston) increases, closing the reed valves against their seats. The crankcase then becomes a sealed chamber so the air-fuel mixture becomes compressed (pressurized) and ready for delivery to the combustion chamber. As the piston continues to move downward, the intake port is uncovered. The fresh fuel mixture rushes through the intake port into the combustion chamber striking the top of the piston where it is deflected along the cylinder wall. The reed valve remains closed until the piston moves upward again.

When the piston begins to move upward on the compression stroke, the reed valve opens because the crankcase volume has been increased, reducing crankcase pressure to less than the outside atmosphere. The intake and exhaust ports are closed and the fresh fuel charge is compressed inside the combustion chamber.

Pressure in the crankcase (beneath the piston) decreases as the piston moves upward and a fresh charge of air flows through the carburetor picking up fuel. As the piston approaches top dead center, the spark plug ignites the

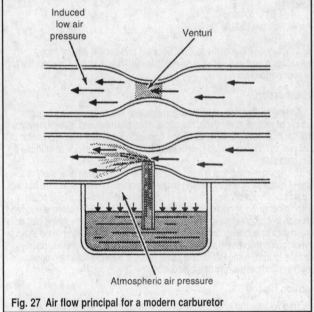

Fig. 27 Air flow principal for a modern carburetor

air-fuel mixture, the power stroke begins and one complete Otto cycle has been completed.

4-STROKE MOTORS

◆ See Figure 22

The 4-stroke motor may be easier to understand for some people either because of its prevalence in automobile and street motorcycle motors today or perhaps because each of the four strokes corresponds to one distinct phase of the Otto cycle. Essentially, a 4-stroke engine completes one Otto cycle of intake, compression, ignition/power and exhaust using two full revolutions of the crankshaft and four distinct movements of the piston (down, up, down and up).

Intake

The intake stroke begins with the piston near the top of its travel. As crankshaft rotation begins to pull the piston downward, the exhaust valve closes and the intake opens. As volume of the combustion chamber increases, a vacuum is created that draws in the air/fuel mixture from the intake manifold.

GENERAL INFORMATION, SAFETY & TOOLS

Compression

Once the piston reaches the bottom of its travel, crankshaft rotation will begin to force it upward. At this point the intake valve closes. As the piston rises in the bore, the volume of the sealed combustion chamber (both intake and exhaust valves are closed) decreases and the air/fuel mixture is compressed. This raises the temperature and pressure of the mixture and increases the amount of force generated by the expanding gases during the Ignition/Power stroke.

Ignition/Power

As the piston approaches top dead center (the highest point of travel in the bore), the spark plug will fire, igniting the air/fuel mixture. The resulting combustion of the air/fuel mixture forces the piston downward, rotating the crankshaft (causing other pistons to move in other phases/strokes of the Otto cycle on multi-cylinder motors).

Exhaust

As the piston approaches the bottom of the Ignition/Power stroke, the exhaust valve opens. When the piston begins its upward path of travel once again, any remaining unburned gasses are forced out through the exhaust valve. This completes one Otto cycle, which begins again as the piston passes top dead center, the intake valve opens and the Intake stroke starts.

COMBUSTION

Whether we are talking about a 2- or 4-stroke engine, all Otto cycle, internal combustion engines require three basic conditions to operate properly,

1. Compression
2. Ignition (Spark)
3. Fuel

A lack of any one of these conditions will prevent the engine from operating. A problem with any one of these will manifest itself in hard-starting or poor performance.

Compression

An engine that has insufficient compression will not draw an adequate supply of air/fuel mixture into the combustion chamber and, subsequently, will not make sufficient power on the power stroke. A lack of compression in just one cylinder of a multi-cylinder motor will cause the motor to stumble or run irregularly.

But, keep in mind that a sudden change in compression is unlikely in 2-stroke motors (unless something major breaks inside the crankcase, but that would usually be accompanied by other symptoms such as a loud noise when it occurred or noises during operation). On 4-stroke motors, a sudden change in compression is also unlikely, but could occur if the timing belt or chain was to suddenly break. Remember that the timing belt/chain is used to synchronize the valve train with the crankshaft. If the valve train suddenly ceases to turn, some intake and some exhaust valves will remain open, relieving compression in that cylinder.

Ignition (Spark)

Traditionally, the ignition system is the weakest link in the chain of conditions necessary for engine operation. Spark plugs may become worn or fouled, wires will deteriorate allowing arcing or misfiring, and poor connections can place an undue load on coils leading to weak spark or even a failed coil. The most common question asked by a technician under a no-start condition is: "do I have spark and fuel" (as they've already determined that they have compression).

A quick visual inspection of the spark plug(s) will answer the question as to whether or not the plug(s) is/are worn or fouled. While the engine is shut **OFF** a physical check of the connections could show a loose primary or secondary ignition circuit wire. An obviously physically damaged wire may also be an indication of system problems and certainly encourages one to inspect the related system more closely.

If nothing is turned up by the visual inspection, perform the Spark Test provided in the Ignition System section to determine if the problem is a lack of or a weak spark. If the problem is not compression or spark, it's time to look at the fuel system.

Fuel

If compression and spark is present (and within spec), but the engine won't start or won't run properly, the only remaining condition to fulfill is fuel. As usual, start with the basics. Is the fuel tank full? Is the fuel stale? If the engine has not been run in some time (a matter of months, not weeks) there is a good chance that the fuel is stale and should be properly disposed of and replaced.

■ **Depending on how stale or contaminated (with moisture) the fuel is, it may be burned in an automobile or in yard equipment, though it would be wise to mix it well with a much larger supply of fresh gasoline to prevent moving your driveability problems to that motor. But it is better to get the lawn tractor stuck on stale gasoline than it would be to have your boat motor quit in the middle of the bay or lake.**

For hard starting motors, is the choke or primer system operating properly. Remember that the choke/prime should only be used for **cold** starts. A true cold start is really only the first start of the day, but it may be applicable to subsequent starts on cooler days, if the engine sat for more than a few hours and completely cooled off since the last use. Applying the primer to the motor for a hot start may flood the engine, preventing it from starting properly. One method to clear a flood is to crank the motor while the engine is at wide-open throttle (allowing the maximum amount of air into the motor to compensate for the excess fuel). But, keep in mind that the throttle should be returned to idle immediately upon engine start-up to prevent damage from over-revving.

Fuel delivery and pressure should be checked before delving into the carburetor(s) or fuel injection system. Make sure there are no clogs in the fuel line or vacuum leaks that would starve the motor of fuel.

Make sure that all other possible problems have been eliminated before touching the carburetor. It is rare that a carburetor will suddenly require an adjustment in order for the motor to run properly. It is much more likely that an improperly stored motor (one stored with untreated fuel in the carburetor) would suffer from one or more clogged carburetor passages sometime after shortly returning to service. Fuel will evaporate over time, leaving behind gummy deposits. If untreated fuel is left in the carburetor for some time (again typically months more than weeks), the varnish left behind by evaporating fuel will likely clog the small passages of the carburetor and cause problems with engine performance. If you suspect this, remove and disassemble the carburetor following procedures under Fuel System.

SHOP EQUIPMENT

Safety Tools

WORK GLOVES

◆ See Figure 28

Unless you think scars on your hands are cool, enjoy pain and like wearing bandages, get a good pair of work gloves. Canvas or leather gloves are the best. And yes, we realize that there are some jobs involving small parts that can't be done while wearing work gloves. These jobs are not the ones usually associated with hand injuries.

A good pair of rubber gloves (such as those usually associated with dish washing) or vinyl gloves is also a great idea. There are some liquids such as solvents and penetrants that don't belong on your skin. Avoid burns and rashes. Wear these gloves.

And lastly, an option. If you're tired of being greasy and dirty all the time, go to the drug store and buy a box of disposable latex gloves like medical professionals wear. You can handle greasy parts, perform small tasks, wash parts, etc. all without getting dirty! These gloves take a surprising amount of abuse without tearing and aren't expensive. Note however, that some people are allergic to the latex or the powder used inside some gloves, so pay attention to what you buy.

EYE AND EAR PROTECTION

◆ See Figures 29 and 30

Don't begin any job without a good pair of work goggles or impact resistant glasses! When doing any kind of work, it's all too easy to avoid eye injury through this simple precaution. And don't just buy eye protection and leave it on the shelf. Wear it all the time! Things have a habit of breaking,

GENERAL INFORMATION SAFETY & TOOLS   1-17

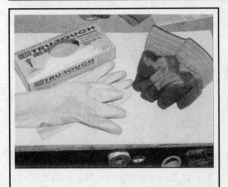

Fig. 28 Three different types of work gloves. The box contains latex gloves

Fig. 29 Don't begin major repairs without a pair of goggles for your eyes and earmuffs to protect your hearing

Fig. 30 Things have a habit of, splashing, spraying, splintering and flying around during repairs

chipping, splashing, spraying, splintering and flying around. And, for some reason, your eye is always in the way!

If you wear vision-correcting glasses as a matter of routine, get a pair made with polycarbonate lenses. These lenses are impact resistant and are available at any optometrist.

Often overlooked is hearing protection. Engines and power tools are noisy! Loud noises damage your ears. It's as simple as that! The simplest and cheapest form of ear protection is a pair of noise-reducing ear plugs. Cheap insurance for your ears! And, they may even come with their own, cute little carrying case.

More substantial, more protection and more money is a good pair of noise reducing earmuffs. They protect from all but the loudest sounds. Hopefully those are sounds that you'll never encounter since they're usually associated with disasters.

WORK CLOTHES

Everyone has "work clothes." Usually these consist of old jeans and a shirt that has seen better days. That's fine. In addition, a denim work apron is a nice accessory. It's rugged, can hold some spare bolts, and you don't feel bad wiping your hands or tools on it. That's what it's for.

When working in cold weather, a one-piece, thermal work outfit is invaluable. Most are rated to below freezing temperatures and are ruggedly constructed. Just look at what local marine mechanics are wearing and that should give you a clue as to what type of clothing is good.

Chemicals

There is a whole range of chemicals that you'll find handy for maintenance and repair work. The most common types are: lubricants, penetrants and sealers. Keep these handy. There are also many chemicals that are used for detailing or cleaning.

When a particular chemical is not being used, keep it capped, upright and in a safe place. These substances may be flammable, may be irritants or might even be caustic and should always be stored properly, used properly and handled with care. Always read and follow all label directions and be sure to wear hand and eye protection!

LUBRICANTS & PENETRANTS

◆ See Figure 31

Anti-seize is used to coat certain fasteners prior to installation. This can be especially helpful when two dissimilar metals are in contact (to help prevent corrosion that might lock the fastener in place). This is a good practice on a lot of different fasteners, BUT, NOT on any fastener that might vibrate loose causing a problem. If anti-seize is used on a fastener, it should be checked periodically for proper tightness.

Lithium grease, chassis lube, silicone grease or a synthetic brake caliper grease can all be used pretty much interchangeably. All can be used for coating rust-prone fasteners and for facilitating the assembly of parts that are a tight fit. Silicone and synthetic greases are the most versatile.

■ Silicone dielectric grease is a non-conductor that is often used to coat the terminals of wiring connectors before fastening them. It may sound odd to coat metal portions of a terminal with something that won't conduct electricity, but here is it how it works. When the connector is fastened the metal-to-metal contact between the terminals will displace the grease (allowing the circuit to be completed). The grease that is displaced will then coat the non-contacted surface and the cavity around the terminals, SEALING them from atmospheric moisture that could cause corrosion.

Silicone spray is a good lubricant for hard-to-reach places and parts that shouldn't be gooped up with grease.

Penetrating oil may turn out to be one of your best friends when taking something apart that has corroded fasteners. Not only can they make a job easier, they can really help to avoid broken and stripped fasteners. The most familiar penetrating oils are Liquid Wrench® and WD-40®. A newer penetrant, PB Blaster® works very well (and has become a mainstay in our shops). These products have hundreds of uses. For your purposes, they are vital!

Before disassembling any part, check the fasteners. If any appear rusted, soak them thoroughly with the penetrant and let them stand while you do something else (for particularly rusted or frozen parts you may need to soak them a few days in advance). This simple act can save you hours of tedious work trying to extract a broken bolt or stud.

SEALANTS

◆ See Figures 32 and 33

Sealants are an indispensable part for certain tasks, especially if you are trying to avoid leaks. The purpose of sealants is to establish a leak-proof bond between or around assembled parts. Most sealers are used in conjunction with gaskets, but some are used instead of conventional gasket material.

The most common sealers are the non-hardening types such as Permatex® No.2 or its equivalents. These sealers are applied to the mating surfaces of each part to be joined, then a gasket is put in place and the parts are assembled.

■ A sometimes overlooked use for sealants like RTV is on the threads of vibration prone fasteners.

One very helpful type of non-hardening sealer is the "high tack" type. This type is a very sticky material that holds the gasket in place while the parts are being assembled. This stuff is really a good idea when you don't have enough hands or fingers to keep everything where it should be.

The stand-alone sealers are the Room Temperature Vulcanizing (RTV) silicone gasket makers. On some engines, this material is used instead of a gasket. In those instances, a gasket may not be available or, because of the shape of the mating surfaces, a gasket shouldn't be used. This stuff, when used in conjunction with a conventional gasket, produces the surest bonds.

RTV does have its limitations though. When using this material, you will have a time limit. It starts to set-up within 15 minutes or so, so you have to assemble the parts without delay. In addition, when squeezing the material out of the tube, don't drop any glops into the engine. The stuff will form and

1-18 GENERAL INFORMATION, SAFETY & TOOLS

Fig. 31 Keep a supply of anti-seize, penetrating oil, lithium grease, electronic cleaner and silicone spray

Fig. 32 Sealants are essential for preventing leaks

Fig. 33 On some engines, RTV is used instead of gasket material to seal components

set and travel around a cooling passage, possibly blocking it. Also, most types are not fuel-proof. Check the tube for all cautions.

CLEANERS

◆ See Figures 34 and 35

There are two basic types of cleaners on the market today: parts cleaners and hand cleaners. The parts cleaners are for the parts; the hand cleaners are for you. They are **not** interchangeable.

There are many good, non-flammable, biodegradable parts cleaners on the market. These cleaning agents are safe for you, the parts and the environment. Therefore, there is no reason to use flammable, caustic or toxic substances to clean your parts or tools.

As far as hand cleaners go; the waterless types are the best. They have always been efficient at cleaning, but they used to all leave a pretty smelly odor. Recently though, most of them have eliminated the odor and added stuff that actually smells good. Make sure that you pick one that contains lanolin or some other moisture-replenishing additive. Cleaners not only remove grease and oil but also skin oil.

■ Most women already know to use a hand lotion when you're all cleaned up. It's okay. Real men DO use hand lotion too! Believe it or not, using hand lotion before your hands are dirty will actually make them easier to clean when you're finished with a dirty job. Lotion seals your hands, and keeps dirt and grease from sticking to your skin.

Fig. 34 Citrus hand cleaners not only work well, but they smell pretty good too. Choose one with pumice for added cleaning power

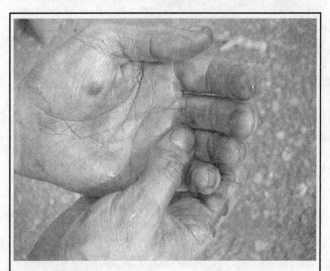

Fig. 35 The use of hand lotion seals your hands and keeps dirt and grease from sticking to your skin

TOOLS

◆ See Figure 36

Tools; this subject could fill a completely separate manual. The first thing you will need to ask yourself, is just how involved do you plan to get. If you are serious about maintenance and repair you will want to gather a quality set of tools to make the job easier, and more enjoyable. BESIDES, TOOLS ARE FUN!!!

Almost every do-it-yourselfer loves to accumulate tools. Though most find a way to perform jobs with only a few common tools, they tend to buy more over time, as money allows. So gathering the tools necessary for maintenance or repair does not have to be an expensive, overnight proposition.

When buying tools, the saying "You get what you pay for ..." is absolutely true! Don't go cheap! Any hand tool that you buy should be drop forged and/or chrome vanadium. These two qualities tell you that the tool is strong enough for the job. With any tool, go with a name that you've heard of before, or, that is recommended buy your local professional retailer. Let's go over a list of tools that you'll need.

Most of the world uses the metric system. However, some American-built engines and aftermarket accessories use standard fasteners. So, accumulate your tools accordingly. Any good DIYer should have a decent set of both U.S. and metric measure tools.

■ Don't be confused by terminology. Most advertising refers to "SAE and metric", or "standard and metric." Both are misnomers. The Society of Automotive Engineers (SAE) did not invent the English system of measurement; the English did. The SAE likes metrics just fine. Both English (U.S.) and metric measurements are SAE approved. Also, the current "standard" measurement IS metric. So, if it's not metric, it's U.S. measurement.

GENERAL INFORMATION SAFETY & TOOLS 1-19

Hand Tools

SOCKET SETS

◆ See Figures 37 thru 43

Socket sets are the most basic hand tools necessary for repair and maintenance work. For our purposes, socket sets come in three drive sizes: 1/4 inch, 3/8 inch and 1/2 inch. Drive size refers to the size of the drive lug on the ratchet, breaker bar or speed handle.

A 3/8 inch set is probably the most versatile set in any mechanic's toolbox. It allows you to get into tight places that the larger drive ratchets can't and gives you a range of larger sockets that are still strong enough for heavy-duty work. The socket set that you'll need should range in sizes from 1/4 inch through 1 inch for standard fasteners, and a 6mm through 19mm for metric fasteners.

You'll need a good 1/2 inch set since this size drive lug assures that you won't break a ratchet or socket on large or heavy fasteners. Also, torque wrenches with a torque scale high enough for larger fasteners are usually 1/2 inch drive.

Plus, 1/4 inch drive sets can be very handy in tight places. Though they usually duplicate functions of the 3/8 in. set, 1/4 in. drive sets are easier to use for smaller bolts and nuts.

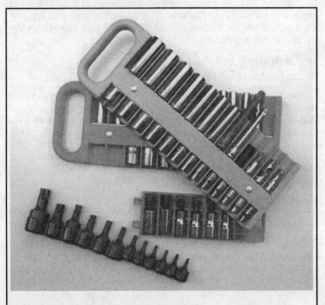

Fig. 36 Socket holders, especially the magnetic type, are handy items to keep tools in order

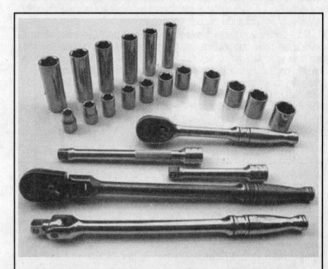

Fig. 37 A 3/8 in. socket set is probably the most versatile tool in any mechanic's tool box

Fig. 38 A swivel (U-joint) adapter (left), a wobble-head adapter (center) and a 1/2 in.-to-3/8 in. adapter (right)

Fig. 39 Ratchets come in all sizes and configurations from rigid to swivel-headed

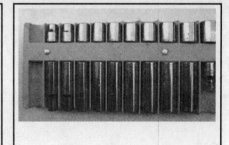

Fig. 40 Shallow sockets (top) are good for most jobs. But, some bolts require deep sockets (bottom)

Fig. 41 Hex-head fasteners require a socket with a hex shaped driver

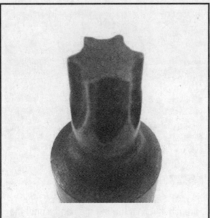

Fig. 42 Torx® drivers . . .

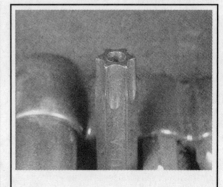

Fig. 43 . . . and tamper resistant drivers are required to remove special fasteners

1-20 GENERAL INFORMATION, SAFETY & TOOLS

As for the sockets themselves, they come in shallow (standard) and deep lengths as well as 6 or 12 point. The 6 and 12 points designation refers to how many sides are in the socket itself. Each has advantages. The 6 point socket is stronger and less prone to slipping which would strip a bolt head or nut. 12 point sockets are more common, usually less expensive and can operate better in tight places where the ratchet handle can't swing far.

Standard length sockets are good for just about all jobs, however, some stud-head bolts, hard-to-reach bolts, nuts on long studs, etc., require the deep sockets.

Most marine manufacturers use recessed hex-head fasteners to retain many of the engine parts. These fasteners require a socket with a hex shaped driver or a large sturdy hex key. To help prevent torn knuckles, we would recommend that you stick to the sockets on any tight fastener and leave the hex keys for lighter applications. Hex driver sockets are available individually or in sets just like conventional sockets.

More and more, manufacturers are using Torx® head fasteners, which were once known as tamper resistant fasteners (because many people did not have tools with the necessary odd driver shape). Since Torx® fasteners have become commonplace in many DIYer tool boxes, manufacturers designed newer tamper resistant fasteners that are essentially Torx® head bolts that contain a small protrusion in the center (requiring the driver to contain a small hole to slide over the protrusion. Tamper resistant fasteners are often used where the manufacturer would prefer only knowledgeable mechanics or advanced Do-It-Yourselfers (DIYers) work.

TORQUE WRENCHES

◆ See Figure 44

In most applications, a torque wrench can be used to ensure proper installation of a fastener. Torque wrenches come in various designs and most stores will carry a variety to suit your needs. A torque wrench should be used any time you have a specific torque value for a fastener. Keep in mind that because there is no worldwide standardization of fasteners, so charts or figure found in each repair section refer to the manufacturer's fasteners. Any general guideline charts that you might come across based on fastener size (they are sometimes included in a repair manual or with torque wrench packaging) should be used with caution. Just keep in mind that if you are using the right tool for the job, you should not have to strain to tighten a fastener.

Beam Type

◆ See Figures 45 and 46

The beam type torque wrench is one of the most popular styles in use. If used properly, it can be the most accurate also. It consists of a pointer attached to the head that runs the length of the flexible beam (shaft) to a scale located near the handle. As the wrench is pulled, the beam bends and the pointer indicates the torque using the scale.

Click (Breakaway) Type

◆ See Figures 47 and 48

Another popular torque wrench design is the click type. The clicking mechanism makes achieving the proper torque easy and most use a ratcheting head for ease of bolt installation. To use the click type wrench you pre-adjust it to a torque setting. Once the torque is reached, the wrench has a reflex signaling feature that causes a momentary breakaway of the torque wrench body, sending an impulse to the operator's hand. But be careful, as continuing the turn the wrench after the momentary release will increase torque on the fastener beyond the specified setting.

BREAKER BARS

◆ See Figure 49

Breaker bars are long handles with a drive lug. Their main purpose is to provide extra turning force when breaking loose tight bolts or nuts. They come in all drive sizes and lengths. Always take extra precautions and use the proper technique when using a breaker bar (pull on the bar, don't push, to prevent skinned knuckles).

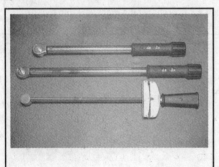

Fig. 44 Three types of torque wrenches. Top to bottom: a 3/8 in. drive beam type that reads in inch lbs., a 1/2 in. drive clicker type and a 1/2 in. drive beam type

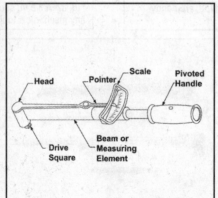

Fig. 45 Parts of a beam type torque wrench

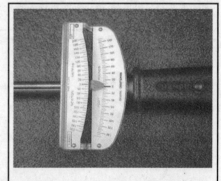

Fig. 46 A beam type torque wrench consists of a pointer attached to the head that runs the length of the flexible beam (shaft) to a scale located near the handle

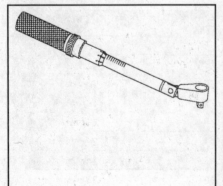

Fig. 47 A click type or breakaway torque wrench - note this one has a pivoting head

Fig. 48 Setting the torque on a click type wrench involves turning the handle until the specification appears on the dial

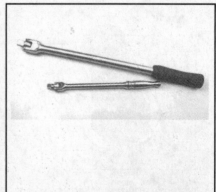

Fig. 49 Breaker bars are great for loosening large or stuck fasteners

GENERAL INFORMATION SAFETY & TOOLS   1-21

WRENCHES

◆ See Figures 50 thru 54

Basically, there are 3 kinds of fixed wrenches: open end, box end, and combination.

Open-end wrenches have 2-jawed openings at each end of the wrench. These wrenches are able to fit onto just about any nut or bolt. They are extremely versatile but have one major drawback. They can slip on a worn or rounded bolt head or nut, causing bleeding knuckles and a useless fastener.

■ Line wrenches are a special type of open-end wrench designed to fit onto more of the fastener than standard open-end wrenches, thus reducing the chance of rounding the corners of the fastener.

Box-end wrenches have a 360° circular jaw at each end of the wrench. They come in both 6 and 12 point versions just like sockets and each type has some of the same advantages and disadvantages as sockets.

Combination wrenches have the best of both. They have a 2-jawed open end and a box end. These wrenches are probably the most versatile.

As for sizes, you'll probably need a range similar to that of the sockets, about 1/4 in. through 1 in. for standard fasteners, or 6mm through 19mm for metric fasteners. As for numbers, you'll need 2 of each size, since, in many instances, one wrench holds the nut while the other turns the bolt. On most fasteners, the nut and bolt are the same size so having two wrenches of the same size comes in handy.

■ Although you will typically just need the sizes we specified, there are some exceptions. Occasionally you will find a nut that is larger. For these, you will need to buy ONE expensive wrench or a very large adjustable. Or you can always just convince the spouse that we are talking about safety here and buy a whole (read expensive) large wrench set.

One extremely valuable type of wrench is the adjustable wrench. An adjustable wrench has a fixed upper jaw and a moveable lower jaw. The lower jaw is moved by turning a threaded drum. The advantage of an adjustable wrench is its ability to be adjusted to just about any size fastener.

The main drawback of an adjustable wrench is the lower jaw's tendency to move slightly under heavy pressure. This can cause the wrench to slip if it is not facing the right way. Pulling on an adjustable wrench in the proper direction will cause the jaws to lock in place. Adjustable wrenches come in a large range of sizes, measured by the wrench length.

PLIERS

◆ See Figure 55

Pliers are simply mechanical fingers. They are, more than anything, an extension of your hand. At least 3 pairs of pliers are an absolute necessity - standard, needle nose and slip joint.

In addition to standard pliers there are the slip-joint, multi-position pliers such as ChannelLock® pliers and locking pliers, such as Vise Grips®.

Slip joint pliers are extremely valuable in grasping oddly sized parts and fasteners. Just make sure that you don't use them instead of a wrench too often since they can easily round off a bolt head or nut.

Locking pliers are usually used for gripping bolts or studs that can't be removed conventionally. You can get locking pliers in square jawed, needle-nosed and pipe-jawed. Locking pliers can rank right up behind duct tape as the handy-man's best friend.

SCREWDRIVERS

You can't have too many screwdrivers. They come in 2 basic flavors, either standard or Phillips. Standard blades come in various sizes and

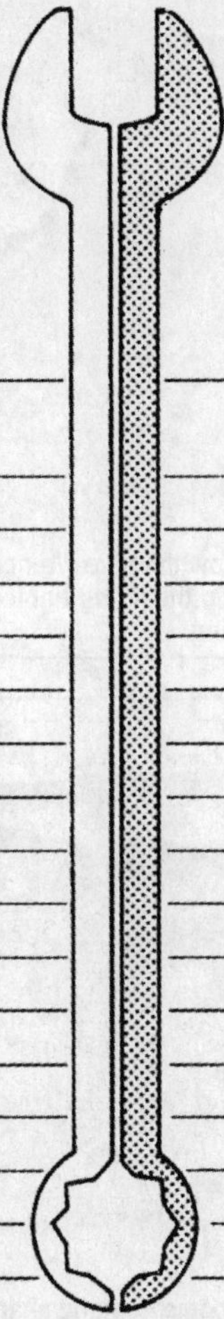

INCHES	DECIMAL	DECIMAL	MILLIMETERS
1/8"	.125	.118	3mm
3/16"	.187	.157	4mm
1/4"	.250	.236	6mm
5/16"	.312	.354	9mm
3/8"	.375	.394	10mm
7/16"	.437	.472	12mm
1/2"	.500	.512	13mm
9/16"	.562	.590	15mm
5/8"	.625	.630	16mm
11/16"	.687	.709	18mm
3/4"	.750	.748	19mm
13/16"	.812	.787	20mm
7/8"	.875	.866	22mm
15/16"	.937	.945	24mm
1"	1.00	.984	25mm

Fig. 50 Comparison of U.S. measure and metric wrench sizes

thickness for all types of slotted fasteners. Phillips screwdrivers come in sizes with number designations from 1 on up, with the lower number designating the smaller size. Screwdrivers can be purchased separately or in sets.

HAMMERS

◆ See Figure 56

You need a hammer for just about any kind of work. You need a ball-peen hammer for most metal work when using drivers and other like tools. A plastic hammer comes in handy for hitting things safely. A soft-faced deadblow hammer is used for hitting things safely and hard. Hammers are also VERY useful with non air-powered impact drivers.

Other Common Tools

There are a lot of other tools that every DIYer will eventually need (though not all for basic maintenance). They include:
- Funnels
- Chisels
- Punches
- Files
- Hacksaw
- Portable Bench Vise
- Tap and Die Set
- Flashlight
- Magnetic Bolt Retriever
- Gasket scraper
- Putty Knife
- Screw/Bolt Extractors
- Prybars

Hacksaws have just one use - cutting things off. You may wonder why you'd need one for something as simple as maintenance or repair, but you never know. Among other things, guide studs to ease parts installation can be made from old bolts with their heads cut off.

A tap and die set might be something you've never needed, but you will eventually. It's a good rule, when everything is apart, to clean-up all threads, on bolts, screws or threaded holes. Also, you'll likely run across a situation in which you will encounter stripped threads. The tap and die set will handle that for you.

Gasket scrapers are just what you'd think, tools made for scraping old gasket material off of parts. You don't absolutely need one. Old gasket material can be removed with a putty knife or single edge razor blade. However, putty knives may not be sharp enough for some really stubborn gaskets and razor blades have a knack of breaking just when you don't want them to, inevitably slicing the nearest body part! As the old saying goes, "always use the proper tool for the job". If you're going to use a razor to scrape a gasket, be sure to always use a blade holder.

Putty knives really do have a use in a repair shop. Just because you remove all the bolts from a component sealed with a gasket doesn't mean it's going to come off. Most of the time, the gasket and sealer will hold it tightly. Lightly inserting a putty knife at various points between the two parts will break the seal without damage to the parts.

A small - 8-10 in. (20-25cm) long - prybar is extremely useful for removing stuck parts.

■ Never use a screwdriver as a prybar! Screwdrivers are not meant for prying. Screwdrivers, used for prying, can break, sending the broken shaft flying!

Screw/bolt extractors are used for removing broken bolts or studs that have broken off flush with the surface of the part.

Fig. 51 Always use a backup wrench to prevent rounding flare nut fittings

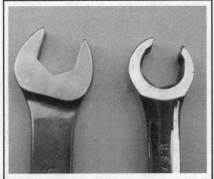

Fig. 52 Note how the flare wrench jaws are extended to grip the fitting tighter and prevent rounding

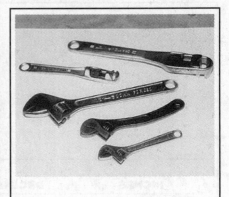

Fig. 53 Several types and sizes of adjustable wrenches

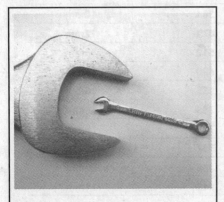

Fig. 54 You may find a nut that requires a particularly large or small wrench (that is usually available at your local tool store)

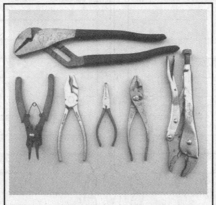

Fig. 55 Pliers come in many shapes and sizes. You should have an assortment on hand

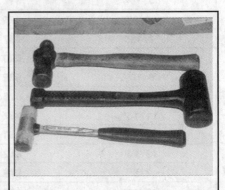

Fig. 56 Three types of hammers. Top to bottom: ball peen, rubber dead-blow, and plastic

GENERAL INFORMATION SAFETY & TOOLS 1-23

Special Tools

◆ See Figure 57

Almost every marine engine around today requires at least one special tool to perform a certain task. In most cases, these tools are specially designed to overcome some unique problem or to fit on some oddly sized component.

When manufacturers go through the trouble of making a special tool, it is usually necessary to use it to ensure that the job will be done right. A special tool might be designed to make a job easier, or it might be used to keep you from damaging or breaking a part.

Don't worry, MOST maintenance procedures can either be performed without any special tools OR, because the tools must be used for such basic things, they are commonly available for a reasonable price. It is usually just the low production, highly specialized tools (like a super thin 7-point star-shaped socket capable of 150 ft. lbs. (203 Nm) of torque that is used only on the crankshaft nut of the limited production what-dya-callit engine) that tend to be outrageously expensive and hard to find. Hopefully, you will probably never need such a tool.

Special tools can be as inexpensive and simple as an adjustable strap wrench or as complicated as an ignition tester. A few common specialty tools are listed here, but check with your dealer or with other boaters for help in determining if there are any special tools for YOUR particular engine. There is an added advantage in seeking advice from others, chances are they may have already found the special tool you will need, and know how to get it cheaper (or even let you borrow it).

Electronic Tools

BATTERY TESTERS

The best way to test a non-sealed battery is using a hydrometer to check the specific gravity of the acid. Luckily, these are usually inexpensive and are available at most parts stores. Just be careful because the larger testers are usually designed for larger batteries and may require more acid than you will be able to draw from the battery cell. Smaller testers (usually a short, squeeze bulb type) will require less acid and should work on most batteries.

Electronic testers are available and are often necessary to tell if a sealed battery is usable. Luckily, many parts stores have them on hand and are willing to test your battery for you.

BATTERY CHARGERS

◆ See Figure 58

If you are a weekend boater and take your boat out every week, then you will most likely want to buy a battery charger to keep your battery fresh. There are many types available, from low amperage trickle chargers to electronically controlled battery maintenance tools that monitor the battery voltage to prevent over or undercharging. This last type is especially useful if you store your boat for any length of time (such as during the severe winter months found in many Northern climates).

Even if you use your boat on a regular basis, you will eventually need a battery charger. The charger should be used anytime the boat is going to be in storage for more than a few weeks or so. Never leave the dock or loading ramp without a battery that is fully charged.

Also, some smaller batteries are shipped dry and in a partial charged state. Before placing a new battery of this type into service it must be filled and properly charged. Failure to properly charge a battery (which was shipped dry) before it is put into service will prevent it from ever reaching a fully charged state.

MULTI-METERS (DVOMS)

◆ See Figure 59

Multi-meters or Digital Volt Ohmmeter (DVOMs) are an extremely useful tool for troubleshooting electrical problems. They can be purchased in either analog or digital form and have a price range to suit any budget. A multi-meter is a voltmeter, ammeter and ohmmeter (along with other features) combined into one instrument. It is often used when testing solid state circuits because of its high input impedance (usually 10 mega ohms or more). A brief description of the multi-meter main test functions follows:

- Voltmeter - the voltmeter is used to measure voltage at any point in a circuit or to measure the voltage drop across any part of a circuit. Voltmeters usually have various scales and a selector switch to allow the reading of different voltage ranges. The voltmeter has a positive and a negative lead. To avoid the possibility of damage to the meter, whenever possible, connect the negative lead to the negative (-) side of the circuit (to ground or nearest the ground side of the circuit) and connect the positive lead to the positive (+) side of the circuit (to the power source or the nearest power source). Luckily, most quality DVOMs can adjust their own polarity internally and will indicate (without damage) if the leads are reversed. Note that the negative voltmeter lead will always be black and that the positive voltmeter will always be some color other than black (usually red).

- Ohmmeter - the ohmmeter is designed to read resistance (measured in ohms) in a circuit or component. Most ohmmeters will have a selector switch which permits the measurement of different ranges of resistance (usually the selector switch allows the multiplication of the meter reading by 10, 100, 1,000 and 10,000). Some ohmmeters are "auto-ranging" which means the meter itself will determine which scale to use. Since the meters are powered by an internal battery, the ohmmeter can be used like a self-powered test light. When the ohmmeter is connected, current from the ohmmeter flows through the circuit or component being tested. Since the ohmmeter's internal resistance and voltage are known values, the amount of current flow through the meter depends on the resistance of the circuit or component being tested. The ohmmeter can also be used to perform a continuity test for suspected open circuits. In using the meter for making continuity checks, do not be concerned with the actual resistance readings. Zero resistance, or any ohm reading, indicates continuity in the circuit. Infinite resistance indicates an opening in the circuit. A high resistance reading where there

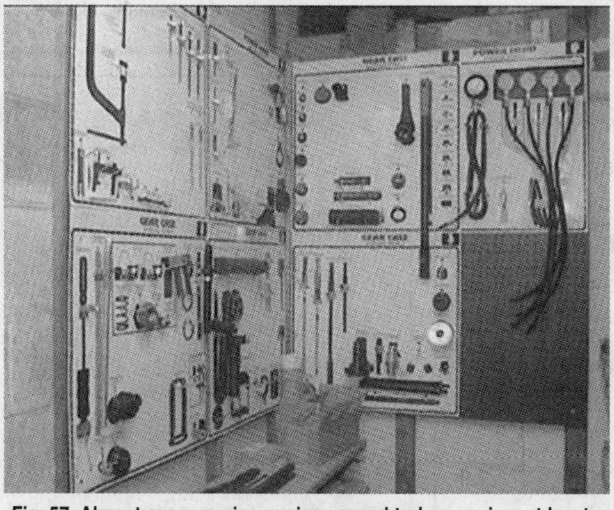

Fig. 57 Almost every marine engine around today requires at least one special tool to perform a certain task

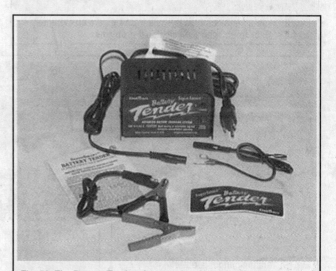

Fig. 58 The Battery Tender® is more than just a battery charger, when left connected, it keeps your battery fully charged

1-24 GENERAL INFORMATION, SAFETY & TOOLS

should be little or none indicates a problem in the circuit. Checks for short circuits are made in the same manner as checks for open circuits, except that the circuit must be isolated from both power and normal ground. Infinite resistance indicates no continuity, while zero resistance indicates a dead short.

✱✱WARNING

Never use an ohmmeter to check the resistance of a component or wire while there is voltage applied to the circuit.

- Ammeter - an ammeter measures the amount of current flowing through a circuit in units called amperes or amps. At normal operating voltage, most circuits have a characteristic amount of amperes, called "current draw" which can be measured using an ammeter. By referring to a specified current draw rating, then measuring the amperes and comparing the two values; one can determine what is happening within the circuit to aid in diagnosis. An open circuit, for example, will not allow any current to flow, so the ammeter reading will be zero. A damaged component or circuit will have an increased current draw, so the reading will be high. The ammeter is always connected in series with the circuit being tested. All of the current that normally flows through the circuit must also flow through the ammeter; if there is any other path for the current to follow, the ammeter reading will not be accurate. The ammeter itself has very little resistance to current flow and, therefore, will not affect the circuit, but, it will measure current draw only when the circuit is closed and electricity is flowing. Excessive current draw can blow fuses and drain the battery, while a reduced current draw can cause motors to run slowly, lights to dim and other components to not operate properly.

GAUGES

Compression Gauge

◆ See Figure 60

An important element in checking the overall condition of your engine is to check compression. This becomes increasingly more important on outboards with high hours. Compression gauges are available as screw-in types and hold-in types. The screw-in type is slower to use, but eliminates the possibility of a faulty reading due to pressure escaping by the seal. A compression reading will uncover many problems that can cause rough running. Normally, these are not the sort of problems that can be cured by a tune-up.

Vacuum/Pressure Gauge/Pump

◆ See Figures 61, 62 and 63

Vacuum gauges are handy for discovering air leaks, late ignition or valve timing, and a number of other problems. A hand-held vacuum/pressure pump can be purchased at many automotive or marine parts stores and can be used for multiple purposes. The gauge can be used to measure vacuum or pressure in a line, while the pump can be used to manually apply vacuum or pressure to a solenoid or fitting. The hand-held pump can also be used to power-bleed trailer and tow vehicle brakes.

Measuring Tools

Eventually, you are going to have to measure something. To do this, you will need at least a few precision tools.

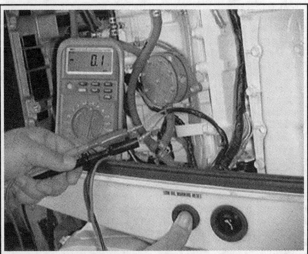

Fig. 59 Multi-meters, such as this one from UEI, are an extremely useful tool for troubleshooting electrical problems

Fig. 60 Cylinder compression test results are extremely valuable indicators of internal engine condition

Fig. 61 Vacuum gauges are useful for troubleshooting including testing some fuel pumps

Fig. 62 You can also use the gauge on a hand-operated vacuum pump for tests

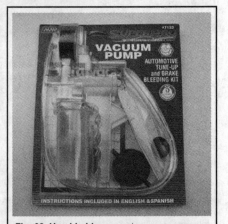

Fig. 63 Hand-held vacuum/pressure pumps are available at most parts stores

MICROMETERS & CALIPERS

Micrometers and calipers are devices used to make extremely precise measurements. The simple truth is that you really won't have the need for many of these items just for routine maintenance. But, measuring tools, such as an outside caliper can be handy during repairs. And, if you decide to tackle a major overhaul, a micrometer will absolutely be necessary.

Should you decide on becoming more involved in boat engine mechanics, such as repair or rebuilding, then these tools will become very important. The success of any rebuild is dependent, to a great extent on the ability to check the size and fit of components as specified by the manufacturer. These measurements are often made in thousandths and ten-thousandths of an inch.

Micrometers

◆ See Figures 64 and 65

A micrometer is an instrument made up of a precisely machined spindle that is rotated in a fixed nut, opening and closing the distance between the end of the spindle and a fixed anvil. When measuring using a micrometer, don't over-tighten the tool on the part as either the component or tool may be damaged, and either way, an incorrect reading will result. Most micrometers are equipped with some form of thumbwheel on the spindle that is designed to freewheel over a certain light touch (automatically adjusting the spindle and preventing it from over-tightening).

Outside micrometers can be used to check the thickness of parts such shims or the outside diameter of components like the crankshaft journals. They are also used during many rebuild and repair procedures to measure the diameter of components such as the pistons. The most common type of micrometer reads in 1/1000 of an inch. Micrometers that use a vernier scale can estimate to 1/10 of an inch.

Inside micrometers are used to measure the distance between two parallel surfaces. For example, in powerhead rebuilding work, the "inside mike" measures cylinder bore wear and taper. Inside mikes are graduated the same way as outside mikes and are read the same way as well.

Remember that an inside mike must be absolutely perpendicular to the work being measured. When you measure with an inside mike, rock the mike gently from side to side and tip it back and forth slightly so that you span the widest part of the bore. Just to be on the safe side, take several readings. It takes a certain amount of experience to work any mike with confidence.

Metric micrometers are read in the same way as inch micrometers, except that the measurements are in millimeters. Each line on the main scale equals 1mm. Each fifth line is stamped 5, 10, 15 and so on. Each line on the thimble scale equals 0.01 mm. It will take a little practice, but if you can read an inch mike, you can read a metric mike.

Calipers

◆ See Figures 66, 67 and 68

Inside and outside calipers are useful devices to have if you need to measure something quickly and absolute precise measurement is not necessary. Simply take the reading and then hold the calipers on an accurate steel rule. Calipers, like micrometers, will often contain a thumbwheel to help ensure accurate measurement.

DIAL INDICATORS

◆ See Figure 69

A dial indicator is a gauge that utilizes a dial face and a needle to register measurements. There is a movable contact arm on the dial indicator. When the arm moves, the needle rotates on the dial. Dial indicators are calibrated to show readings in thousandths of an inch and typically, are used to measure end-play and runout on various shafts and other components.

Dial indicators are quite easy to use, although they are relatively expensive. A variety of mounting devices are available so that the indicator can be used in a number of situations. Make certain that the contact arm is always parallel to the movement of the work being measured.

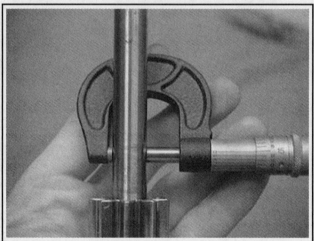

Fig. 64 Outside micrometers measure thickness, like shims or a shaft diameter

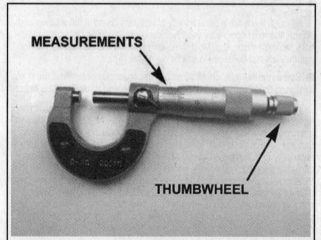

Fig. 65 Be careful not to over-tighten the micrometers always use the thumbwheel

Fig. 66 Calipers are the fast and easy way to make precise measurements

Fig. 67 Calipers can also be used to measure depth . . .

Fig. 68 . . . and inside diameter measurements, to 0.001 in. accuracy

1-26 GENERAL INFORMATION, SAFETY & TOOLS

TELESCOPING GAUGES

◆ See Figure 70

A telescope gauge is really only used during rebuilding procedures (NOT during basic maintenance or routine repairs) to measure the inside of bores. It can take the place of an inside mike for some of these jobs. Simply insert the gauge in the hole to be measured and lock the plungers after they have contacted the walls. Remove the tool and measure across the plungers with an outside micrometer.

DEPTH GAUGES

◆ See Figure 71

A depth gauge can be inserted into a bore or other small hole to determine exactly how deep it is. One common use for a depth gauge is measuring the distance the piston sits below the deck of the block at top dead center. Some outside calipers contain a built-in depth gauge so you can save money and buy just one tool.

Fig. 69 This dial indicator is measuring the end-play of a crankshaft during a powerhead rebuild

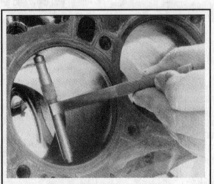

Fig. 70 Telescoping gauges are used during powerhead rebuilding procedures to measure the inside diameter of bores

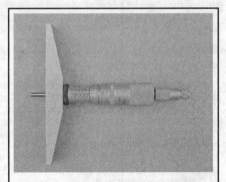

Fig. 71 Depth gauges are used to measure the depth of bore or other small holes

FASTENERS, MEASUREMENTS AND CONVERSIONS

Bolts, Nuts and Other Threaded Retainers

◆ See Figures 72 and 73

Although there are a great variety of fasteners found in the modern boat engine, the most commonly used retainer is the threaded fastener (nuts, bolts, screws, studs, etc.). Most threaded retainers may be reused, provided that they are not damaged in use or during the repair.

■ Some retainers (such as stretch bolts or torque prevailing nuts) are designed to deform when tightened or in use and should not be reused.

Whenever possible, we will note any special retainers which should be replaced during a procedure. But you should always inspect the condition of a retainer when it is removed and you should replace any that show signs of damage. Check all threads for rust or corrosion that can increase the torque necessary to achieve the desired clamp load for which that fastener was originally selected. Additionally, be sure that the driver surface itself (on the fastener) is not compromised from rounding or other damage. In some cases a driver surface may become only partially rounded, allowing the driver to catch in only one direction. In many of these occurrences, a fastener may be installed and tightened, but the driver would not be able to grip and loosen the fastener again. (This could lead to frustration down the line should that component ever need to be disassembled again).

If you must replace a fastener, whether due to design or damage, you must always be sure to use the proper replacement. In all cases, a retainer of the same design, material and strength should be used. Markings on the heads of most bolts will help determine the proper strength of the fastener. The same material, thread and pitch must be selected to assure proper installation and safe operation of the motor afterwards.

Thread gauges are available to help measure a bolt or stud's thread. Most part or hardware stores keep gauges available to help you select the proper size. In a pinch, you can use another nut or bolt for a thread gauge. If the bolt you are replacing is not too badly damaged, you can select a match by finding another bolt that will thread in its place. If you find a nut that will thread properly onto the damaged bolt, then use that nut as a gauge to help select the replacement bolt. If however, the bolt you are replacing is so badly damaged (broken or drilled out) that its threads cannot be used as a gauge, you might start by looking for another bolt (from the same assembly or a similar location) which will thread into the damaged bolt's mounting. If so, the other bolt can be used to select a nut; the nut can then be used to select the replacement bolt.

In all cases, be absolutely sure you have selected the proper replacement. Don't be shy, you can always ask the store clerk for help.

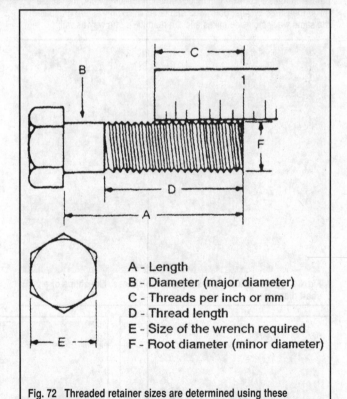

A - Length
B - Diameter (major diameter)
C - Threads per inch or mm
D - Thread length
E - Size of the wrench required
F - Root diameter (minor diameter)

Fig. 72 Threaded retainer sizes are determined using these measurements

✵✵WARNING

Be aware that when you find a bolt with damaged threads, you may also find the nut or tapped bore into which it was threaded has also been damaged. If this is the case, you may have to drill and tap the hole, replace the nut or otherwise repair the threads. Never try to force a replacement bolt to fit into the damaged threads.

Torque

Torque is defined as the measurement of resistance to turning or rotating. It tends to twist a body about an axis of rotation. A common example of this would be tightening a threaded retainer such as a nut, bolt or screw. Measuring torque is one of the most common ways to help assure that a threaded retainer has been properly fastened.

When tightening a threaded fastener, torque is applied in three distinct areas, the head, the bearing surface and the clamp load. About 50 percent of the measured torque is used in overcoming bearing friction. This is the friction between the bearing surface of the bolt head, screw head or nut face and the base material or washer (the surface on which the fastener is rotating). Approximately 40 percent of the applied torque is used in overcoming thread friction. This leaves only about 10 percent of the applied torque to develop a useful clamp load (the force that holds a joint together). This means that friction can account for as much as 90 percent of the applied torque on a fastener.

Standard and Metric Measurements

Specifications are often used to help you determine the condition of various components, or to assist you in their installation. Some of the most common measurements include length (in. or cm/mm), torque (ft. lbs., inch lbs. or Nm) and pressure (psi, in. Hg, kPa or mm Hg).

In some cases, that value may not be conveniently measured with what is available in your toolbox. Luckily, many of the measuring devices that are available today will have two scales so U.S. or Metric measurements may easily be taken. If any of the various measuring tools that are available to you do not contain the same scale as listed in your specifications, use the conversion factors that are provided in the Specifications section to determine the proper value.

The conversion factor chart is used by taking the given specification and multiplying it by the necessary conversion factor. For instance, looking at the first line, if you have a measurement in inches such as "free-play should be 2 in." but your ruler reads only in millimeters, multiply 2 in. by the conversion factor of 25.4 to get the metric equivalent of 50.8mm. Likewise, if a specification was given only in a Metric measurement, for example in Newton Meters (Nm), then look at the center column first. If the measurement is 100 Nm, multiply it by the conversion factor of 0.738 to get 73.8 ft. lbs.

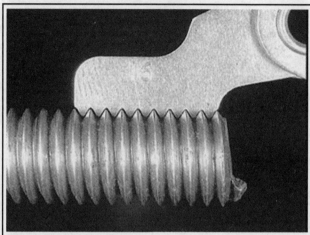

Fig. 73 Thread gauges measure the threads-per-inch and the pitch of a bolt or stud's threads

SPECIFICATIONS

CONVERSION FACTORS

LENGTH-DISTANCE
- Inches (in.) x 25.4 = Millimeters (mm) x .0394 = Inches
- Feet (ft.) x .305 = Meters (m) x 3.281 = Feet
- Miles x 1.609 = Kilometers (km) x .0621 = Miles

VOLUME
- Cubic Inches (in3) x 16.387 = Cubic Centimeters x .061 = in3
- IMP Pints (IMP pt.) x .568 = Liters (L) x 1.76 = IMP pt.
- IMP Quarts (IMP qt.) x 1.137 = Liters (L) x .88 = IMP qt.
- IMP Gallons (IMP gal.) x 4.546 = Liters (L) x .22 = IMP gal.
- IMP Quarts (IMP qt.) x 1.201 = US Quarts (US qt.) x .833 = IMP qt.
- IMP Gallons (IMP gal.) x 1.201 = US Gallons (US gal.) x .833 = IMP gal.
- Fl. Ounces x 29.573 = Milliliters x .034 = Ounces
- US Pints (US pt.) x .473 = Liters (L) x 2.113 = Pints
- US Quarts (US qt.) x .946 = Liters (L) x 1.057 = Quarts
- US Gallons (US gal.) x 3.785 = Liters (L) x .264 = Gallons

MASS-WEIGHT
- Ounces (oz.) x 28.35 = Grams (g) x .035 = Ounces
- Pounds (lb.) x .454 = Kilograms (kg) x 2.205 = Pounds

PRESSURE
- Pounds Per Sq. In. (psi) x 6.895 = Kilopascals (kPa) x .145 = psi
- Inches of Mercury (Hg) x .4912 = psi x 2.036 = Hg
- Inches of Mercury (Hg) x 3.377 = Kilopascals (kPa) x .2961 = Hg
- Inches of Water (H₂O) x .07355 = Inches of Mercury x 13.783 = H₂O
- Inches of Water (H₂O) x .03613 = psi x 27.684 = H₂O
- Inches of Water (H₂O) x .248 = Kilopascals (kPa) x 4.026 = H₂O

TORQUE
- Pounds-Force Inches (in-lb) x .113 = Newton Meters (N·m) x 8.85 = in-lb
- Pounds-Force Feet (ft-lb) x 1.356 = Newton Meters (N·m) x .738 = ft-lb

VELOCITY
- Miles Per Hour (MPH) x 1.609 = Kilometers Per Hour (KPH) x .621 = MPH

POWER
- Horsepower (Hp) x .745 = Kilowatts x 1.34 = Horsepower

FUEL CONSUMPTION*
- Miles Per Gallon IMP (MPG) x .354 = Kilometers Per Liter (Km/L)
- Kilometers Per Liter (Km/L) x 2.352 = IMP MPG
- Miles Per Gallon US (MPG) x .425 = Kilometers Per Liter (Km/L)
- Kilometers Per Liter (Km/L) x 2.352 = US MPG

*It is common to covert from miles per gallon (mpg) to liters/100 kilometers (1/100 km), where mpg (IMP) x 1/100 km = 282 and mpg (US) x 1/100 km = 235.

TEMPERATURE
- Degree Fahrenheit (°F) = (°C x 1.8) + 32
- Degree Celsius (°C) = (°F − 32) x .56

1-28 GENERAL INFORMATION, SAFETY & TOOLS

SAE Bolts

SAE Grade Number	1 or 2			5			6 or 7		

Bolt Markings

Manufacturers' marks may vary—number of lines always two less than the grade number.

Usage	Frequent			Frequent			Infrequent		
Bolt Size (inches)—(Thread)	Maximum Torque			Maximum Torque			Maximum Torque		
	Ft-Lb	kgm	Nm	Ft-Lb	kgm	Nm	Ft-Lb	kgm	Nm
1/4—20	5	0.7	6.8	8	1.1	10.8	10	1.4	13.5
—28	6	0.8	8.1	10	1.4	13.6			
5/16—18	11	1.5	14.9	17	2.3	23.0	19	2.6	25.8
—24	13	1.8	17.6	19	2.6	25.7			
3/8—16	18	2.5	24.4	31	4.3	42.0	34	4.7	46.0
—24	20	2.75	27.1	35	4.8	47.5			
7/16—14	28	3.8	37.0	49	6.8	66.4	55	7.6	74.5
—20	30	4.2	40.7	55	7.6	74.5			
1/2—13	39	5.4	52.8	75	10.4	101.7	85	11.75	115.2
—20	41	5.7	55.6	85	11.7	115.2			
9/16—12	51	7.0	69.2	110	15.2	149.1	120	16.6	162.7
—18	55	7.6	74.5	120	16.6	162.7			
5/8—11	83	11.5	112.5	150	20.7	203.3	167	23.0	226.5
—18	95	13.1	128.8	170	23.5	230.5			
3/4—10	105	14.5	142.3	270	37.3	366.0	280	38.7	379.6
—16	115	15.9	155.9	295	40.8	400.0			
7/8—9	160	22.1	216.9	395	54.6	535.5	440	60.9	596.5
—14	175	24.2	237.2	435	60.1	589.7			
1—8	236	32.5	318.6	590	81.6	799.9	660	91.3	894.8
—14	250	34.6	338.9	660	91.3	849.8			

Metric Bolts

Relative Strength Marking	4.6, 4.8			8.8		

Bolt Markings

Usage	Frequent			Infrequent		
Bolt Size	Maximum Torque			Maximum Torque		
Thread Size x Pitch (mm)	Ft-Lb	Kgm	Nm	Ft-Lb	Kgm	Nm
6 x 1.0	2–3	.2–.4	3–4	3–6	.4–.8	5–8
8 x 1.25	6–8	.8–1	8–12	9–14	1.2–1.9	13–19
10 x 1.25	12–17	1.5–2.3	16–23	20–29	2.7–4.0	27–39
12 x 1.25	21–32	2.9–4.4	29–43	35–53	4.8–7.3	47–72
14 x 1.5	35–52	4.8–7.1	48–70	57–85	7.8–11.7	77–110
16 x 1.5	51–77	7.0–10.6	67–100	90–120	12.4–16.5	130–160
18 x 1.5	74–110	10.2–15.1	100–150	130–170	17.9–23.4	180–230
20 x 1.5	110–140	15.1–19.3	150–190	190–240	26.2–46.9	160–320
22 x 1.5	150–190	22.0–26.2	200–260	250–320	34.5–44.1	340–430
24 x 1.5	190–240	26.2–46.9	260–320	310–410	42.7–56.5	420–550

2

ENGINE & DRIVE MAINTENANCE

ENGINE AND DRIVE MAINTENANCE .. 2-2
FLUIDS AND LUBRICANTS 2-33
LUBRICATION POINTS 2-55
BOAT MAINTENANCE 2-60
WINTER STORAGE 2-66
FIRING ORDERS 2-69
SPRING COMMISSIONING 2-69
SPECIFICATIONS 2-71

ANODES (ZINCS). 2-61
  INSPECTION 2-62
  LOCATIONS 2-62
  SERVICING 2-61
BATTERIES 2-63
  CHARGERS 2-66
  CHECKING SPECIFIC GRAVITY 2-65
  CLEANING 2-63
  REPLACING CABLES 2-66
  SAFETY PRECAUTIONS 2-66
  STORAGE 2-66
  TERMINALS 2-65
BELTS. 2-8
  ADJUSTMENT 2-9
  INSPECTION 2-8
  REMOVAL & INSTALLATION 2-10
  SERPENTINE BELT ROUTING 2-10
BOAT MAINTENANCE **2-60**
  ANODES (ZINCS) 2-61
  BATTERIES 2-63
  CONTINUITY CIRCUIT 2-63
  INSIDE THE BOAT 2-60
  THE BOAT'S EXTERIOR 2-60
COOLING SYSTEM 2-38
  DRAINING & FILLING 2-42
  FLUSHING THE CLOSED SYSTEM 2-42
  FLUSHING THE SEAWATER SYSTEM -
  INBOARDS 2-40
  FLUSHING THE SEAWATER SYSTEM -
  STERN DRIVES 2-39
  LEVEL CHECK 2-38
  PRESSURE CAP TESTING 2-38
  PRESSURE TEST 2-39
CONTINUITY CIRCUIT 2-63
CYLINDER COMPRESSION 2-20
  CHECKING 2-20
ENGINE & DRIVE MAINTENANCE **2-2**
  BELTS. 2-8
  CYLINDER COMPRESSION 2-20
  FLAME ARRESTOR 2-3
  FUEL FILTER 2-6
  IDLE SPEED & MIXTURE 2-27
  IGNITION TIMING 2-25
  IMPELLER/WATER PUMP 2-16
  PCV VALVE 2-29
  PROPELLER 2-30
  SEAWATER STRAINER 2-16
  SERIAL NUMBER IDENTIFICATION ... 2-2
  SPARK PLUG WIRES 2-25
  SPARK PLUGS 2-21
  THERMOSTAT 2-13
  VALVE ADJUSTMENT 2-29
  WATER PUMP/IMPELLER 2-16
ENGINE COUPLER/U-JOINT SPLINES .. 2-59
FIRING ORDERS **2-69**
FLAME ARRESTOR 2-3
  DESCRIPTION & OPERATION 2-3
  REMOVAL & INSTALLATION 2-4
FLUID DISPOSAL 2-33
FLUIDS & LUBRICANTS **2-33**
  COOLING SYSTEM 2-38
  FLUID DISPOSAL 2-33
  FUEL & OIL RECOMMENDATIONS 2-33
  POWER STEERING 2-37
  POWER TRIM PUMP 2-54
  STERN DRIVE UNIT 2-52
  TRANSMISSION FLUID 2-50
FUEL & OIL RECOMMENDATIONS 2-33
  ENGINE OIL 2-33
  FUEL 2-33
  OIL & FILTER CHANGE 2-35
  OIL LEVEL CHECK 2-34
  PRIMING 2-33
FUEL FILTER 2-6
  REMOVAL & INSTALLATION 2-6
IDLE SPEED & MIXTURE 2-27
  ADJUSTMENT 2-27

IGNITION TIMING 2-25
  ADJUSTMENT 2-25
IMPELLER/WATER PUMP 2-16
  REMOVAL & INSTALLATION 2-16
INSIDE THE BOAT 2-60
LUBRICATION POINTS **2-55**
  PCV VALVE 2-29
  INSPECTION 2-29
  REMOVAL & INSTALLATION 2-29
POWER STEERING 2-37
  BLEEDING 2-37
  FLUID LEVEL 2-37
POWER TRIM PUMP 2-54
  FLUID LEVEL 2-54
PROPELLER 2-30
  REMOVAL & INSTALLATION 2-30
PROPELLER SHAFT 2-60
SEAWATER STRAINER 2-16
  CLEANING & INSPECTION 2-16
SERIAL NUMBER IDENTIFICATION 2-2
  ENGINE 2-2
  STERN DRIVE 2-2
  TRANSMISSION 2-3
  TRANSOM ASSEMBLY 2-3
SHIFT CABLE & TRANSMISSION
LINKAGE PIVOT POINTS 2-55
  SPARK PLUG WIRES 2-25
  REMOVAL & INSTALLATION 2-25
  TESTING 2-25
SPARK PLUGS 2-21
  HEAT RANGE 2-21
  INSPECTION & GAPPING 2-25
  READING PLUGS 2-23
  REMOVAL & INSTALLATION 2-22
  SERVICE 2-22
SPECIFICATIONS **2-71**
  CAPACITIES 2-77
  ENGINE MODEL APPLICATIONS 2-71
  MAINTENANCE INTERVALS 2-79
  TUNE-UP 2-74
SPRING COMMISSIONING **2-69**
  SPRING COMMISSIONING CHECKLIST . 2-69
SPRING COMMISSIONING CHECKLIST .. 2-69
STEERING SYSTEM 2-57
  ALPHA 2-57
  BRAVO 2-57
  TIE BAR PIVOT POINTS 2-57
STERN DRIVE UNIT 2-52
  CHECKING FOR WATER 2-53
  DRAIN AND REFILL 2-53
  FLUID LEVEL 2-52
THERMOSTAT 2-13
  REMOVAL & INSTALLATION 2-13
  TESTING 2-16
THROTTLE CABLE 2-55
  3.0L 2-55
  4.3L V6 & 5.0L/5.7L/6.2L V8 2-55
  8.1L V8 2-55
TRANSMISSION FLUID 2-50
  DRAINING FLUID 2-50
  FLUID LEVEL 2-50
TRANSOM, GIMBAL ASSEMBLY,
HINGE PINS & GIMBAL BEARING 2-58
VALVE ADJUSTMENT 2-29
WATER PUMP/IMPELLER 2-16
  REMOVAL & INSTALLATION 2-16
WINTER STORAGE **2-66**
  GENERAL INFORMATION 2-66
  WINTER STORAGE CHECKLIST 2-66
WINTER STORAGE CHECKLIST 2-66
  CARBURETED & TBI ENGINES 2-66
  4.3L V6, 5.0L/5.7L/6.2L & 2006-08 8.1L MPI
  ENGINES 2-67
  2001-05 8.1L MPI ENGINES 2-68
ZINCS (ANODES) 2-61
  INSPECTION 2-62
  LOCATIONS 2-62
  SERVICING 2-61

2-2 ENGINE AND DRIVE MAINTENANCE

ENGINE AND DRIVE MAINTENANCE

At Seloc, we estimate that 75% of engine repair work can be directly or indirectly attributed to lack of proper care for the engine. This is especially true of care during the off-season period. There is no way on this green earth for a mechanical engine to be left sitting idle for an extended period of time, say for six months, and then be ready for instant satisfactory service.

Imagine, if you will, leaving your car or truck for six months, and then expecting to turn the key, having it roar to life, and being able to drive off in the same manner as a daily occurrence.

Therefore it is critical for an engine to either be run (at least once a month), preferably, in the water and properly maintained between uses or for it to be specifically prepared for storage and serviced again immediately before the start of the season.

Only through a regular maintenance program can the owner expect to receive long life and satisfactory performance at minimum cost.

Many times, if an engine is not performing properly, the owner will "nurse" it through the season with good intentions of working on the unit once it is no longer being used. As with many New Year's resolutions, the good intentions are not completed and the engine may lie for many months before the work is begun or the unit is taken to the marine shop for repair.

Imagine, if you will, the cause of the problem being a blown head gasket. And let us assume water has found its way into a cylinder. This water, allowed to remain over a long period of time, will do considerably more damage than it would have if the unit had been disassembled and the repair work performed immediately. Therefore, if an engine is not functioning properly, do not stow it away with promises to get at it when you get time, because the work and expense will only get worse, the longer corrective action is postponed. In the example of the blown head gasket, a relatively simple and inexpensive repair job could very well develop into major overhaul and rebuild work.

OK, perhaps no one thing that we do as boaters will protect us from risks involved with enjoying the wind and the water on a powerboat. But, each time we perform maintenance on our boat or motor, we increase the likelihood that we will find a potential hazard before it becomes a problem. Each time we inspect our boat and motor, we decrease the possibility that it could leave us stranded on the water.

In this way, performing boat and engine service is one of the most important ways that we, as boaters, can help protect ourselves, our boats, and the friends and family that we bring aboard.

Serial Number Identification

◆ See Figures 1, 2 and 3

An engine specifications decal can generally be found on top of the flame arrestor or the side of the rocker arm cover, on most models - all pertinent serial number information can be found here, unfortunately this decal is not always legible on older boats so please refer to the following procedures for each individuals unit's serial number location.

ENGINE

◆ See Figures 1 thru 7

■ Serial numbers tags are frequently difficult to see when the engine is installed in the boat; a mirror can be a handy way to read all the numbers.

In addition to the flame arrestor cover or coolant reservoir, the engine serial number is stamped on the starboard rear side of the engine; above and behind the starter motor for all stern drive applications.

On inboards, it can be found in essentially the same place, stamped on a flange at the rear of the engine, but the starter is actually slightly above and behind it.

STERN DRIVE

◆ See Figures 8 and 9

The stern drive serial number decal can be found on the upper port side of the unit. The serial number will be on the right side of the decal, while the gear ratio will be on the left side. Make sure you don't confuse the two!

You may also come across some models that omit the ratio figure.

Fig. 1 The engine serial number sticker can look like this...

Fig. 2 ...or this...

Fig. 3 ...or this

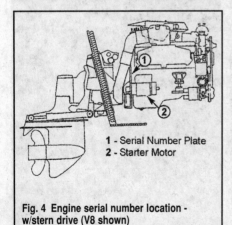

1 - Serial Number Plate
2 - Starter Motor

Fig. 4 Engine serial number location - w/stern drive (V8 shown)

Fig. 5 A good look at the id tag on stern drive models

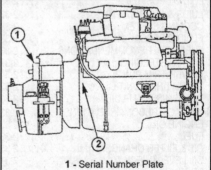

1 - Serial Number Plate
2 - Starter Motor

Fig. 6 Engine serial number location - w/inboard (V8 shown)

ENGINE AND DRIVE MAINTENANCE

Fig. 7 A good look at the id tag on inboard models

Fig. 8 The stern drive serial number and ratio can usually be found here...

Fig. 9 ...although some may only show the serial number

TRANSMISSION

◆ See Figures 10 thru 13

The transmission identification plate can be found on a boss on the port side of all Velvet Drive transmissions, facing upward. ZF/Hurth units will have their id plates attached to an indent on the top center of the unit.

TRANSOM ASSEMBLY

◆ See Figure 14

The transom assembly serial number decal can be found on the upper end of the unit, facing away from the stern.

Flame Arrestor

DESCRIPTION & OPERATION

In a marine engine compartment, the minimal amount of dust and dirt in the air mean that a marine air filter requires less maintenance than its counterpart in the automotive world. However, the maintenance of a marine air filter is equally important.

The marine filter prevents dirt from entering the engine and scoring the cylinder walls. This lessens oil consumption and extends the engine's life. The air filter on some engines is also used as an intake silencer to quiet the intake air sound as it rushes into the cylinder head from the intake ports.

Over time, the air filter element will become clogged with dirt and oil, decreasing the amount of air entering the engine and lowering engine output. If an excessive amount of oil is clogging the filter, this could be an indication of worn cylinders or piston ring failure causing high pressure in the crankcase.

The maintenance interval for the flame arrestor cleaning is at the end of the first boating season, and then every 100 hours of engine operation or once a year, whichever comes first. On Horizon models, the interval is increased to every 300 hours or three years, whichever comes first.

Transmission Identification

Make	Model	Model Number	Ratio	ID Plate Color Code
Velvet	5000A Down Angle	20-01-002	1.25:1	Black
		20-01-003	1.5:1	Black
		20-01-004	2.1:1	Black
		20-01-005	2.5:1	Black
		20-01-006	2.8:1	Black
	5000V V-Drive	20-02-003	1.5:1	Blue
		20-02-004	2.0:1	Blue
		20-02-005	2.5:1	Blue
	71C Direct	10-17-004	1.0:1	Red
	71C In-Line	10-17-006	1.5:1	Red
		10-17-012	1.5:1	Red
	72C V-Drive	10-05-011	1.5:1	Green
		10-05-002	2.0:1	Green
		10-05-005 (Gear)	2.5:1	Green
		10-05-004 (Chain)	2.5:1	Green
	72C Direct	10-18-002	1.0:1	Green
	72C In-Line	10-18-004	1.5:1	Green
		10-18-006	2.0:1	Green
		10-18-010	2.5:1	Green
		10-18-012	3.0:1	Green
ZF/Hurth	63A/630A	HSW630A	1.5:1	-
			2.1:1	-
			2.5:1	-
			2.7:1	-
	63IV/630V	HSW630V	1.55:1	-
			2.0:1	-
			2.5:1	-
	80A/800A	HSW800A	2.85:1	-
	800A2	HSW800A2	2.85:1	-

Fig. 10 Transmission Identification Chart

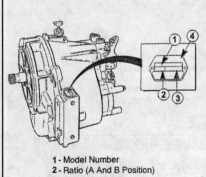

1 - Model Number
2 - Ratio (A And B Position)
3 - Serial Number
4 - Identification Plate Model Color Code

Fig. 11 The transmission identification plate - Velvet 5000 Down Angle

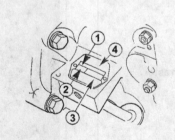

1 - Model Number
2 - Ratio (A And B Position)
3 - Serial Number
4 - Identification Plate Model Color Code

Fig. 12 The transmission identification plate - Velvet V-drive and In-Line

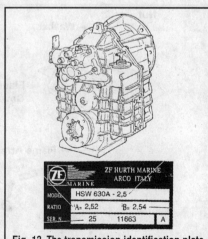

Fig. 13 The transmission identification plate - ZF/Hurth

2-4 ENGINE AND DRIVE MAINTENANCE

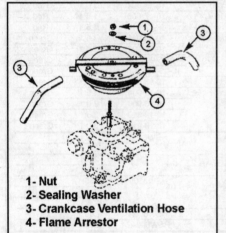

Fig. 14 Transom assembly serial number location. Note also the engine designation sticker.

REMOVAL & INSTALLATION

All Models W/Carburetor Or TBI

◆ See Figures 15 thru 19

1. Remove or open the engine compartment cover.
2. Tag and disconnect the crankcase ventilation hoses (if equipped) from the arrestor and the rocker arm covers.
3. Remove the nut and washer securing the flame arrestor to the carburetor or TBI unit. Many models have a decorative cover over the arrestor.
4. Lift off the flame arrestor.
5. Clean the arrestor in solvent and dry with compressed air if possible; otherwise make sure that it dries completely by air. Clean the hoses and then inspect them for cracks or deterioration. Replace if necessary.
6. Position the arrestor over the stud and reconnect the ventilation hoses.
7. Install the washer and nut. Tighten securely. Close the engine compartment.

4.3L, 5.0L, 5.7L & 6.2L MPI Engines

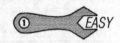

◆ See Figures 20 thru 25

1. Remove or open the engine compartment hatch.
2. Loosen the mounting nut/knob on the flame arrestor cover and remove the cover. Some models may not have a cover.

Fig. 15 Removing the flame arrestor - 3.0L engines

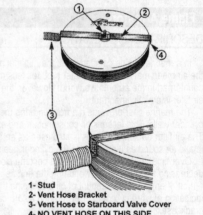

Fig. 16 Removing the flame arrestor - V6 and V8 engines w/carburetor

Fig. 17 Models with the Bodensee/SAV1 emission system have a vent hose bracket on the arrestor

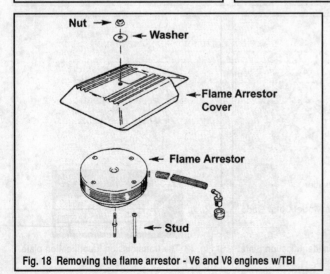

Fig. 18 Removing the flame arrestor - V6 and V8 engines w/TBI

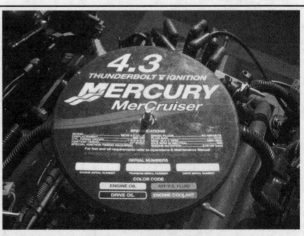

Fig. 19 Remember that most models will have a cover over the actual flame arrestor

ENGINE AND DRIVE MAINTENANCE 2-5

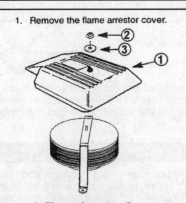

Fig. 20 Remove the flame arrestor cover on early 350 Mag and 6.2L engines...

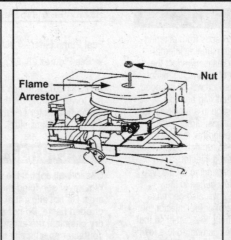

Fig. 21 ...and then remove the stud nut to lift off the arrestor

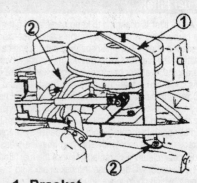

Fig. 22 Some early models will have a bracket over the flame arrestor...

Fig. 23 ...which means that the arrestor is probably secured by a clamp (but sometimes uses a nut instead)

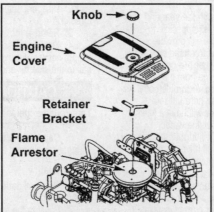

Fig. 24 All later models use a knob to secure the cover...

Fig. 25 ...and a 3-sided retainer for the arrestor itself

3. If equipped with an arrestor bracket (early models only), remove the mounting nuts on the fuel rail and lift off the bracket. Make sure you don't pull out the fuel rail stud.

4. The flame arrestor is mounted in one of 3 ways - a clamp, a nut or a retainer. Most models will use the 3-sided retainer; either way, loosen the nut or bolt(s) and lift off the arrestor.

5. On early models, clean the arrestor in solvent and dry with compressed air if possible; otherwise make sure that it dries completely by air. On all later models, use either steam or clean water; drying with compressed air if at all possible.

6. Position the arrestor on the throttle body and tighten the clamp bolt or mounting nut securely. If your model uses the 3-sided retainer, tighten the 3 nuts to 12 ft. lbs. (108 inch lbs.).

7. Install the bracket, if equipped, over the arrestor and onto the fuel rail. Install the thin washer, then the thick one and finally the nut.

8. Install the cover, tightening the nut or knob securely. Remember that the cover is usually plastic, so don't tighten too tight!

8.1L MPI Engines

◆ See Figures 26 and 27

1. Remove or open the engine compartment cover.
2. Remove the 3 fasteners retaining the engine cover and lift it off.
3. Loosen the retaining clamp at the throttle body and slide off the arrestor.
4. Clean the arrestor in solvent and dry with compressed air if possible; otherwise make sure that it dries completely by air.
5. Position the arrestor and tighten the retaining clamp securely (approx. 42 inch lbs. {4.7 Nm}).
6. Install the engine cover, tightening the fasteners securely (remember, its plastic, so not too tight) and close the engine compartment.

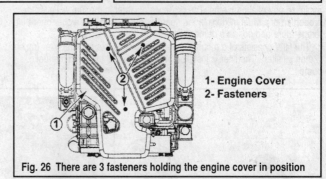

Fig. 26 There are 3 fasteners holding the engine cover in position

Fig. 27 A good shot of the flame arrestor and clamp

2-6 ENGINE AND DRIVE MAINTENANCE

Fuel Filter

A fuel filter is designed to keep particles of dirt and debris from entering the carburetors or the fuel injection system and clogging the tiny internal passages of either. A small speck of dirt or sand can drastically affect the ability of the fuel system to deliver the proper amount of air and fuel/oil to the engine. If a filter becomes clogged, the flow of gasoline will be impeded. This could cause lean fuel mixtures, hesitation and stumbling and idle problems in carburetors. Although a clogged fuel passage in a fuel injected engine could also cause lean symptoms and idle problems, dirt can also prevent a fuel injector from closing properly. A fuel injector that is stuck partially open by debris would likely cause the engine to run rich due to the unregulated fuel constantly spraying from the pressurized injector.

Regular cleaning or replacement of the fuel filter (depending on the type or types used) will decrease the risk of blocking the flow of fuel to the engine, which could leave you stranded on the water. It will also decrease the risk of damage to the small passages of a carburetor or fuel injector that could require more extensive and expensive replacement. Keep in mind that fuel filters are usually pretty inexpensive (at lease when compared to a tow) and replacement is a simple task. Service your fuel filter on a regular basis to avoid fuel delivery problems. All filters should be replaced no less than once a season or every 100 hours of operation, whichever comes first; although halving this interval is cheap insurance!

The type of fuel filter used on your engine will vary with the year and model. Because of the number of possible variations it is impossible to accurately give instructions based on model. Instead, we will provide instructions for the different types of filters the manufacturer used on various families of motors or systems with which they are equipped. To determine what filter(s) are utilized by your engine, trace the fuel line from the tank to the fuel pump and then from the pump to the carburetors or throttle body. As a general rule of thumb, the majority of engines covered here utilize a canister-type in-line water separating filter. 3.0L engines will have a filter incorporated into the fuel pump; but may also have an in-line water separating filter. Additionally, most carbureted engines will further utilize a small filter/screen in the carburetor; and, EFI engines will have a small inline filter downstream of the larger one if they use a boost pump.

As mentioned previously, most new engines have a factory-installed water separating fuel filter. This type filter is also available as an accessory for all other engines and should be installed at the earliest possible convenience. Such a kit is not expensive, and contains instructions for correct installation.

A water separating filter, as its name suggests, removes water and other fuel system contaminants before they reach the carburetor and helps minimize potential problems. The presence of water in the fuel will alter the proportion of air/fuel mixture to the "lean" side, resulting in a higher operating temperature and possible damage to pistons, if not corrected.

The filter consists of a mounting plate and disposable canister filter (much like an oil filter). The filter is installed between the fuel tank and the fuel pump.

REMOVAL & INSTALLATION

Fuel Pump Filter - 3.0L Engines Only

◆ See Figures 28, 29 and 30

■ MOST 3.0L engines are equipped with this style filter, but not all - if the lower casting on the pump body is solid then you know to look for a standard water separating fuel filter (usually at the front of the engine on the starboard side).

✱✱ CAUTION

Observe all applicable safety precautions when working around fuel. Whenever servicing the fuel system, always work in a well-ventilated area. Do not allow fuel spray or vapors to come in contact with a spark or open flame. Do not smoke while working around gasoline. Keep a dry chemical fire extinguisher near the work area. Always keep fuel in a container specifically designed for fuel storage; also, always properly seal fuel containers to avoid the possibility of fire or explosion.

1. Remove or open the engine compartment hatch.
2. Disconnect the negative battery cable and then remove the flame arrestor.
3. Remove the safety wire from the screw at the bottom of the pump.
4. Loosen the screw and release the filter bowl bail from the housing. It's not necessary to remove the bail completely, you can usually swing the bail to the side far enough to allow removal of the bowl/canister.

✱✱ CAUTION

There will be fuel in the bowl, so have plenty of rags available.

5. Carefully pry the bowl from the pump housing and remove the spring, filter element and gasket.

To install:

6. Clean all parts carefully and check for any cracks, deterioration or other damage.
7. Position the spring, new filter and gasket into the bowl. Make sure that the open end faces the pump.
8. Hold the bowl in position on the pump and snap the retaining bail into place.
9. Tighten the screw securely, but not too tight, and then the safety wire (if equipped).
10. Reconnect the battery cable, install the flame arrestor and start the engine. Check that there are no fuel leaks and then install or close the engine compartment hatch.

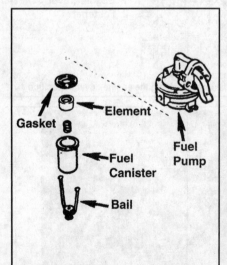

Fig. 28 Exploded view of the fuel pump filter - 3.0L engines

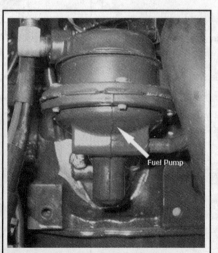

Fig. 29 Not all 3.0L engines utilize an in-pump filter. If your fuel pump looks like this, than your engine doesn't have one and will use a normal water separating filter in addition to the inlet filter in the carb

Fig. 30 Here's a good shot of what the pump will look like if it has an internal filter

ENGINE AND DRIVE MAINTENANCE

Carburetor Fuel Inlet Filter - 2 bbl

◆ See Figure 31

 MODERATE

✱✱ CAUTION

Observe all applicable safety precautions when working around fuel. Whenever servicing the fuel system, always work in a well-ventilated area. Do not allow fuel spray or vapors to come in contact with a spark or open flame. Do not smoke while working around gasoline. Keep a dry chemical fire extinguisher near the work area. Always keep fuel in a container specifically designed for fuel storage; also, always properly seal fuel containers to avoid the possibility of fire or explosion.

1. Remove or open the engine compartment cover.
2. Disconnect the negative battery cable.
3. Remove the flame arrestor as detailed in this section.
4. Position an open end wrench on the fuel inlet filter nut at the carburetor. Position another wrench on the fuel line nut. Loosen the fuel line nut while holding the inlet nut with the other wrench and pull out the fuel line. Make sure you plug the line to prevent any fuel from spilling and carefully position it out of the way.
5. Loosen and then carefully remove the inlet nut from the carburetor body. Pull out the gasket(s) (one or two), the filter element and the spring.

To install:

6. Although most elements can be cleaned and reused, we recommend replacement with a new one whenever possible.
7. Insert the spring into the carburetor and then slide in the filter element. Be sure that the open end of the filter faces out (toward the inlet nut).
8. Position the large gasket over the inlet nut threads and the small one inside the nut. Screw the nut into the carburetor and tighten it to 18 ft. lbs. (24 Nm).
9. Clean the threads and position the fuel line nut into the inlet, seat it a few turns with your fingers and then tighten it to 18 ft. lbs. (24 Nm).
10. Install the flame arrestor and connect the battery cable. Start the engine and check for fuel leaks.
11. Install the engine cover and close the engine compartment.

Carburetor Fuel Inlet Filter - 4 bbl

◆ See Figure 32

 MODERATE

Models that are equipped with a Weber 4 bbl carburetor have a fuel inlet filter incorporated into the carburetor body. Disassembly of a substantial portion of the carburetor is required - please refer to the Fuel System section found later in this manual for detailed procedures.

Inline Fuel Filter - EFI Engines

 EASY

✱✱ CAUTION

Observe all applicable safety precautions when working around fuel. Whenever servicing the fuel system, always work in a well-ventilated area. Do not allow fuel spray or vapors to come in contact with a spark or open flame. Do not smoke while working around gasoline. Keep a dry chemical fire extinguisher near the work area. Always keep fuel in a container specifically designed for fuel storage; also, always properly seal fuel containers to avoid the possibility of fire or explosion.

■ As of 2002, all EFI engines utilizing a fuel boost pump are equipped with an additional inline filter between the fuel tank and the boost pump. 2001 models with a boost pump can be retrofitted with this filter (#35-864572A1) if the owner so desires.

1. Remove or open the engine compartment hatch. Disconnect the negative battery cables.
2. Trace the fuel line until you find the small automotive-like inline filter.
3. Secure the filter fitting nut with a wrench and use a 2nd wrench to loosen the line fitting. Repeat this for the other end and remove the filter.
4. Remove the line fittings from the fuel lines if necessary.
5. Install the fittings if removed and then thread them into the new filter and tighten securely.

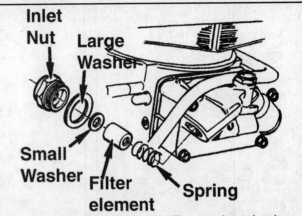

Fig. 31 Exploded view of the fuel inlet filter on carbureted engines - 2 bbl

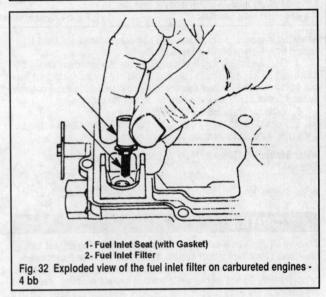

1 - Fuel Inlet Seat (with Gasket)
2 - Fuel Inlet Filter

Fig. 32 Exploded view of the fuel inlet filter on carbureted engines - 4 bb

Water Separating Filter - All Engines Exc. 2004-08 V8 w/Gen III Cool Fuel Module

◆ See Figures 33, 34 and 35

EASY

✱✱ CAUTION

Observe all applicable safety precautions when working around fuel. Whenever servicing the fuel system, always work in a well-ventilated area. Do not allow fuel spray or vapors to come in contact with a spark or open flame. Do not smoke while working around gasoline. Keep a dry chemical fire extinguisher near the work area. Always keep fuel in a container specifically designed for fuel storage; also, always properly seal fuel containers to avoid the possibility of fire or explosion.

■ 2004 and later engines may not always utilize a standard water separating fuel filter. If you cannot find this filter on the front Starboard side of your engine (looks like an oil filter) than your engine is most likely equipped with a Gen III Cool Fuel module - please refer to the appropriate procedures for this application.

1. Remove or open the engine compartment hatch. Disconnect the negative battery cables.
2. On models equipped with a filter cover, unsnap the latch and remove the upper and lower covers from the filter.
3. Remove the canister filter from the mounting plate by rotating the canister counterclockwise. An oil filter wrench may be necessary to break the filter free. Keep the filter upright to avoid spilling fuel. Properly dispose of the fuel and fuel saturated canister. Make sure you have plenty of rags

2-8 ENGINE AND DRIVE MAINTENANCE

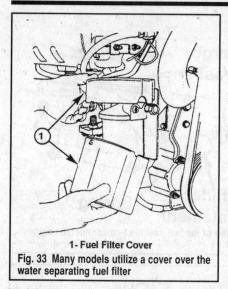

Fig. 33 Many models utilize a cover over the water separating fuel filter

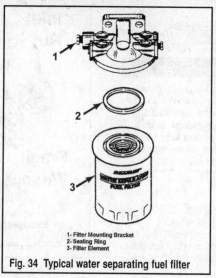

1- Filter Mounting Bracket
2- Sealing Ring
3- Filter Element

Fig. 34 Typical water separating fuel filter

Fig. 35 A good look at a typical installation

handy just in case! The old canister filter cannot be cleaned and used a second time. Never attempt to reuse the filter!

4. Coat the sealing ring(s) of a new canister filter with clean engine oil (there may be 2 rings so make sure that the old one comes out and the new one goes in!). Install the filter onto the mounting plate and tighten it securely by hand. Never use an oil filter wrench to tighten the canister.

5. Install the filter covers if so equipped.

6. Reconnect the battery cables and start the engine. Check that there are no fuel leaks and then close the engine hatch.

Water Separating Filter - 2004-08 V8 Engines w/Gen III Cool Fuel Module

◆ See Figure 36

✱✱ CAUTION

Observe all applicable safety precautions when working around fuel. Whenever servicing the fuel system, always work in a well-ventilated area. Do not allow fuel spray or vapors to come in contact with a spark or open flame. Do not smoke while working around gasoline. Keep a dry chemical fire extinguisher near the work area. Always keep fuel in a container specifically designed for fuel storage; also, always properly seal fuel containers to avoid the possibility of fire or explosion.

■ 2004 and later engines may not always utilize a standard water separating fuel filter. If you were unable to find this filter on the front Starboard side of your engine (looks like an oil filter) than your engine is most likely equipped with a Gen III Cool Fuel module and this is the appropriate procedure for your engine.

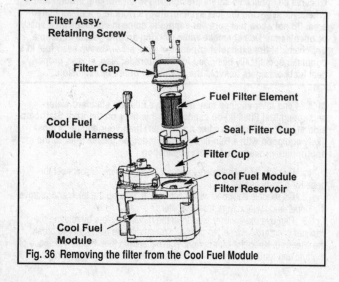

Fig. 36 Removing the filter from the Cool Fuel Module

1. Mercury recommends that the engine has been shut down for at least 12 hours prior to performing the following steps, so make sure the engine is cold!

2. Open the engine compartment and disconnect the battery cables. Close the fuel supply if equipped with a shut-off valve.

3. Locate the fuel module and tag and disconnect the module harness.

4. Turn the key to the **START** position and allow the engine to crank for 5 seconds until fuel system pressure has been relieved. Turn off the engine.

5. Grab the filter handle and gently pull upward until the filter unseats itself. Do not remove the assembly yet.

6. Loosen the 3 filter assembly retaining screws until they disengage from the module. Do not remove the screws from the cap.

7. Allow a few minutes for any residual fuel in the filter assembly to drain into the reservoir. Hold the filter cup securely while turning the handle clockwise until you can remove the handle. Lift out the filter element.

To install:

8. Clean any water or debris from the filter cup, dry it out and position a new filter element into the cup. Make sure that you press the filter in until it is fully seated.

9. Install a new O-ring on the cup and then turn on (counterclockwise) the cap until it locks into place.

10. Slowly lower the assembly into the module so that the screws in the cap line up with their respective holes. Remember that there's probably fuel in the reservoir, so be careful not to spill any. Tighten the retaining screws until they are just snug.

11. Make sure that filter cap is firmly, and correctly, seated against the module and then tighten the screws to 53 inch lbs. (6 Nm).

12. Open the fuel shut-off valve if equipped and then connect the module harness connector.

13. Connect the battery cables and run the engine while checking for any leaks. Don't forget a water supply if the boat is out of the water!!

Belts

INSPECTION

◆ See Figures 37 thru 42

V-belts and serpentine belts should be inspected on a regular basis for signs of glazing or cracking. A glazed belt will be perfectly smooth from slippage, while a good belt will have a slight texture of fabric visible. Cracks will usually start at the inner edge of the belt and run outward. All worn or damaged drive belts should be replaced immediately. It is best to replace all drive belts at one time, as a preventive maintenance measure, during this service operation.

Inspect the alternator, power steering and water pump V-belts every 100 hours or 12 months (whichever comes first) for evidence of wear such as cracking, fraying, and incorrect tension.

ENGINE AND DRIVE MAINTENANCE

Inspect the serpentine belt on 2001 Standard models every 100 hours or 12 months (whichever comes first) for evidence of wear such as cracking, fraying, and incorrect tension.

Inspect the serpentine belt on 2001 Horizon models and all 2002-08 models every 300 hours or 36 months (whichever comes first) for evidence of wear such as cracking, fraying, and incorrect tension.

Determine the V-belt tension at a point halfway between the pulleys by pressing on the belt with moderate thumb pressure. The belt should deflect 1/4 in. (6mm). If the defection is found to be too much or too little, make adjustments as necessary.

Determine serpentine belt tension at the halfway point of the longest span between pulleys - usually between the idler pulley and the alternator or sea water pump. Pressing on the belt with moderate thumb pressure it should deflect 1/4 in. (6mm) on all except the newer (2006ish) 8.1L engines where it should be 1/2 in. (13mm). If the defection is found to be too much or too little, loosen the mounting bolts and make adjustments as necessary.

■ While using thumb pressure to check belt deflection is a good way to get an idea of correct belt tension, we highly recommend using a belt tension gauge.

■ When replacing belts, we recommend cleaning the inside of the belt pulleys to extend the service life of the belts. Never use automotive belts, marine belts used on your engine are heavy duty and not interchangeable.

ADJUSTMENT

V-Belts

◆ See Figures 43 and 44

Before you attempt to adjust any of your engine's belts, apply penetrating oil to the bracket fasteners to make them easier to loosen.
1. Remove or open the engine compartment hatch.
2. Disconnect the negative battery cable.
3. Loosen the component pivot bolt.
4. Loosen the component pump pulley adjustment bracket bolt.
5. Using a wooden lever, pry the component toward or away from the engine until the proper tension is achieved.

** WARNING

Do not over-tighten the drive belts; the pump and/or alternator bearings can be damaged.

6. Tighten the component pulley adjustment bracket bolt to 16 ft. lbs. (28 Nm).
7. Tighten the component pivot bolt to 35 ft. lbs. (48 Nm).
8. Reconnect the negative battery cable.

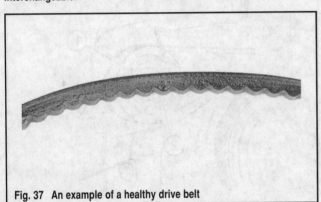

Fig. 37 An example of a healthy drive belt

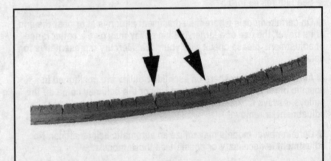

Fig. 38 Deep cracks in this belt will cause flex, building up heat that will eventually lead to belt failure

Fig. 39 The cover of this belt is worn, exposing the critical reinforcing cords to excessive wear

Fig. 40 Installing too wide a belt can result in serious belt wear and/or breakage

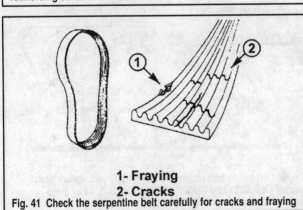

1- Fraying
2- Cracks
Fig. 41 Check the serpentine belt carefully for cracks and fraying

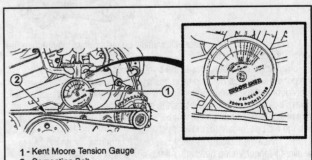

1 - Kent Moore Tension Gauge
2 - Serpentine Belt

Fig. 42 Using a belt tension gauge

2-10 ENGINE AND DRIVE MAINTENANCE

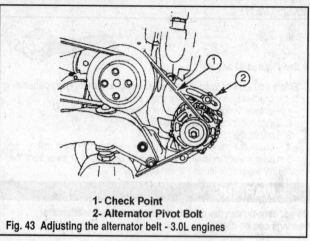

Fig. 43 Adjusting the alternator belt - 3.0L engines
1- Check Point
2- Alternator Pivot Bolt

Serpentine Belts

◆ See Figures 45 thru 57

■ On certain models where the adjustment pinion is at or near the end of its travel, the use of a larger 108mm pulley may give a better range of adjustment, please check with your local Mercury representative for details.

■ All brackets and washers on each idler pulley are positioned in a specific order and must remain that way or the belt may come off the pulleys. Always make sure that you note their orientation during adjustment or removal.

■ Certain newer models may utilize an automatic belt tensioner. No adjustment is necessary or possible on these models.

1. Remove or open the engine hatch and disconnect the battery cables.
2. Loosen the tensioner pulley adjustment stud locknut with a 5/8 in. wrench. Leave the wrench on the nut.
3. Put on 5/16 in. socket over the adjustment stud and turn it until belt deflection equals 1/4 in. at the longest span between two pulleys (usually between the idler pulley and the alternator or seawater pump). Hold the socket on the stud to maintain tension and tighten the locknut securely.

■ If available, we highly recommend using a belt tension gauge as a final check.

4. Connect the battery cables and run the engine for a few minutes. Recheck the belt tension.

SERPENTINE BELT ROUTING

◆ See Figures 46 thru 57

REMOVAL & INSTALLATION

V-Belts

The replacement of the inner belt on multi-belted engines may require the removal of the outer belts.

To replace a drive belt, loosen the pivot and mounting bolts of the component which the belt is driving, then, using a wooden lever or equivalent, pry the component inward to relieve the tension on the drive belt; always be careful where you locate the prybar, or damage to components may result. Slip the belt off the component pulley, and match the new belt with the old belt for length and width.

1. These measurements must be equal. It is normal for an old belt to be slightly longer than a new one. After a new belt is installed correctly, properly adjust the tension.

■ When removing more than one belt, be sure to mark them for identification. This will help avoid confusion when replacing the belts.

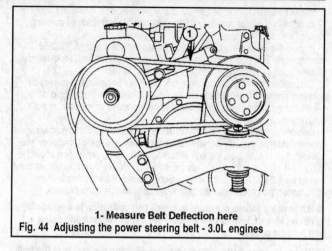

Fig. 44 Adjusting the power steering belt - 3.0L engines
1- Measure Belt Deflection here

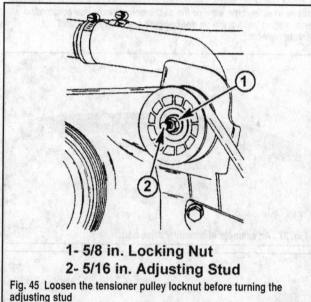

Fig. 45 Loosen the tensioner pulley locknut before turning the adjusting stud
1- 5/8 in. Locking Nut
2- 5/16 in. Adjusting Stud

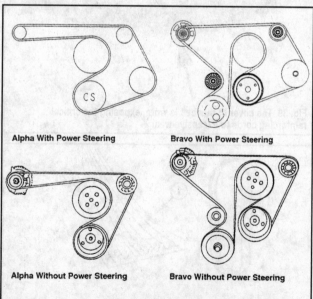

Fig. 46 Serpentine belt routing - 4.3L V6 engines between serial number OL619084 and OM322780; 5.0L/5.7L/6.2L V8 engines between OL619000 and OM299999

ENGINE AND DRIVE MAINTENANCE

Serpentine Belts

All Engines Exc. 8.1L

◆ See Figures 46 thru 48 and 53 thru 56

■ All brackets and washers on each idler pulley are positioned in a specific order and must remain that way or the belt may come off the pulleys. Always make sure that you note their orientation during adjustment or removal.

1. Remove or open the engine hatch and disconnect the battery cables.
2. Loosen the tensioner pulley adjustment stud locknut with a 5/8 in. wrench.
3. Position a 5/16 in. socket over the adjustment stud and turn it until the belt is loose. Remove the belt.
4. Install the belt over the pulleys as detailed in the routing diagrams.
5. Position a 5/16 in. socket over the adjustment stud and turn it until belt deflection equals 1/4 in. at the longest span between two pulleys. Hold the socket on the stud to maintain tension and tighten the locknut securely.

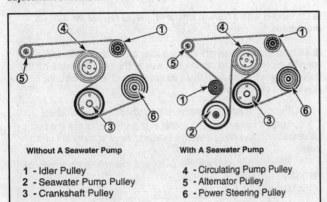

Fig. 47 Serpentine belt routing (stern drive) - 4.3L V6 engines above serial number OM322781; 5.0L/5.7L/6.2L V8 engines above OM300000

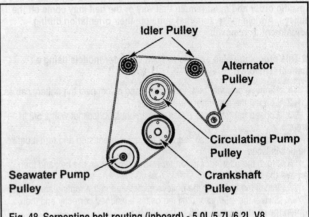

Fig. 48 Serpentine belt routing (inboard) - 5.0L/5.7L/6.2L V8 engines above OM300000

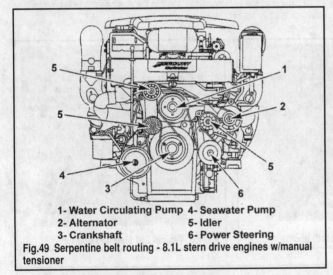

1- Water Circulating Pump   4- Seawater Pump
2- Alternator   5- Idler
3- Crankshaft   6- Power Steering

Fig. 49 Serpentine belt routing - 8.1L stern drive engines w/manual tensioner

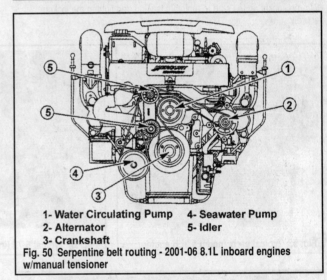

1- Water Circulating Pump   4- Seawater Pump
2- Alternator   5- Idler
3- Crankshaft

Fig. 50 Serpentine belt routing - 2001-06 8.1L inboard engines w/manual tensioner

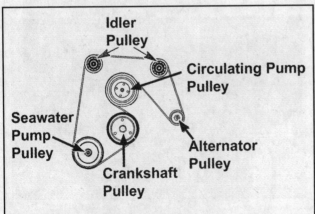

Fig. 51 Serpentine belt routing - 2006-08 8.1L inboard engines w/manual tensioner

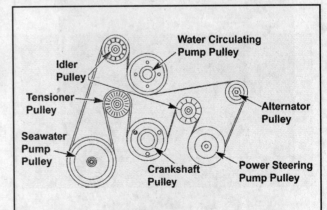

Fig. 52 Serpentine belt routing - 2006-08 8.1L engines w/automatic tensioner

2-12 ENGINE AND DRIVE MAINTENANCE

■ If available, we highly recommend using a belt tension gauge as a final check.

6. Connect the battery cables and run the engine for a few minutes. Recheck the belt tension.

2001-06 8.1L Engines

◆ See Figures 49 thru 52, 57 and 58

■ All brackets and washers on each idler pulley are positioned in a specific order and must remain that way or the belt may come off the pulleys. Always make sure that you note their orientation during adjustment or removal.

■ This procedure also applies to 2006 and later models using a manual tensioner.

1. Remove or open the engine hatch and disconnect the battery cables.
2. Loosen the alternator mounting bolt.
3. Loosen the tensioner pulley adjustment stud locknut with a 5/8 in. wrench.
4. Position a 5/16 in. socket over the adjustment stud and turn it until the belt is loose.
5. Move the alternator to further relieve tension on the belt and then remove the belt.
6. Install the belt over the pulleys as detailed in the routing diagrams.
7. Swivel the alternator until the belt is tensioned correctly and then tighten the mounting bolt to 35 ft. lbs. (48 Nm).
8. Position a 5/16 in. socket over the adjustment stud and turn it until belt deflection equals 1/4 in. (6mm) at the longest span between two pulleys. Hold the socket on the stud to maintain tension and tighten the locknut securely.

■ If available, we highly recommend using a belt tension gauge as a final check.

9. Connect the battery cables and run the engine for a few minutes. Recheck the belt tension.

2006-08 8.1L Engines W/Automatic Tensioner

◆ See Figures 49 thru 52 and 59

■ On models with a manual tensioner, please refer to the 2001-06 procedures.

■ All brackets and washers on each idler pulley are positioned in a specific order and must remain that way or the belt may come off the pulleys. Always make sure that you note their orientation during adjustment or removal.

1. Remove or open the engine hatch and disconnect the battery cables.
2. Install a breaker bar and socket over the tensioner bolt and then rotate the bar clockwise to move the tensioner away from the belt until it stops.
3. Slide the belt off of the idler pulley and then relieve the tension on the breaker bar.
4. Remove the belt.
5. Install the belt and use the tool to move the tensioner into position so you can get the belt over the last idler.
6. Slowly relieve the pressure on the tensioner until it is released and tensioning the belt properly.
7. Connect the battery cables and run the engine for a few minutes. Recheck the belt tension (1/2 in.).

Fig. 53 Typical early model 4.3LH engine

Fig. 54 Typical 5.7L (inboard) engine

Fig. 55 Typical 350 Mag MPI engine

Fig. 56 Typical 6.2L engine

Fig. 57 Typical 8.1L engine

ENGINE AND DRIVE MAINTENANCE 2-13

Fig. 58 A good shot of the alternator bolt

Thermostat

The thermostat is a simple temperature sensitive valve that opens and closes to control cooling water flow through the engine. In operation, the thermostat hovers somewhere between open and closed. As engine load and temperature increase, the thermostat opens to allow more cooling water into the engine. As temperature and load decrease, the thermostat closes.

A sticking thermostat will either allow the temperature to rise well above the normal operating temperature before it opens, or, if stuck in the open position, will never allow the engine to reach operating temperature.

All thermostats are rated based on the temperature at which they open and this rating should always be stamped somewhere on the thermostat, usually on the flange area. All engines covered here utilize a 160°F thermostat except for a few early 3.0L engines that use a brass thermostat, which is rated at 143°F. All 3.0L engines using a standard stainless thermostat (and this is probably all models covered here) are rated at 160°F.

■ On all engines covered here, the thermostat housing can be found on the front, top of the engine (in the crossover on the 8.1L) - easily identifiable by the large hoses attached to it.

✷✷ CAUTION

Serious damage may result from operating your engine without a thermostat! Don't even consider this!

✷✷ CAUTION

NEVER use an automotive thermostat in a marine engine. No matter how tempting this may seem, forget it!!

REMOVAL & INSTALLATION

3.0L Engines

◆ See Figures 60 thru 63

1. Open or remove the engine hatch cover and disconnect the negative battery cables.
2. Drain all water from the cylinder block and exhaust manifold as detailed in the Cooling System section.
3. Loosen the hose clamps and then wiggle the coolant hoses off of the thermostat housing.
4. Remove the mounting bolts (2) with their lock washers and then remove the thermostat housing cover. Some models may have a lifting eye incorporated in the housing - take note of its positioning.
5. Lift out the thermostat and discard it. If you are not sure that it is inoperable, perform the testing procedures outlined in this section.

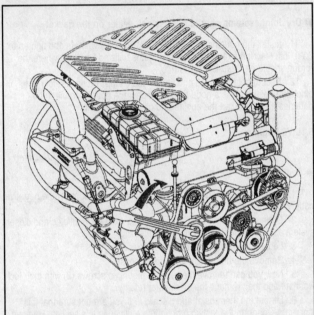

Fig. 59 Use a breaker bar to relieve the tension on later models with an automatic tensioner

Fig. 60 Installing the thermostat

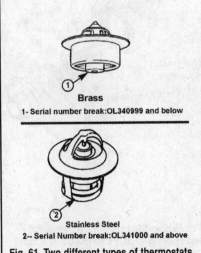

Fig. 61 Two different types of thermostats

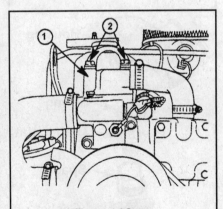

1- Thermostat Cover
2- Attaching Bolts

Fig. 62 Removing the thermostat housing cover on models with the brass thermostat - 3.0L engines

2-14 ENGINE AND DRIVE MAINTENANCE

Fig. 63 A good look at the thermostat housing

3. Loosen the hose clamps and then wiggle the coolant hoses off of the thermostat housing.

4. Remove the mounting bolts (usually 2) with their lock washers and then remove the thermostat housing or cover. Some models may have a lifting eye incorporated in the housing - take note of its positioning.

5. Lift out the thermostat and discard it. If you are not sure that it is inoperable, perform the testing procedures outlined in this section.

6. Make sure that you carefully scrape off all remaining gasket material from the thermostat housing and cover.

7. On seawater systems, position the O-ring into the housing and then insert a new thermostat into the housing with the element end toward the housing bottom and the flange on the unit seated into the recess in the housing.

8. Slide in the sleeve so it aligns with the groove in the housing bore.

9. On closed systems, insert a new thermostat into the housing with the element end toward the housing bottom and the flange on the unit seated into the recess in the housing.

10. Coat both sides of a new housing gasket with Quicksilver Perfect Seal (or similar) and position it onto the housing so that the holes line up. If your engine is equipped with an audio warning temperature switch, the thermostat will have continuity rivets, do not use a sealer or the alarm will not work properly.

6. Make sure that you carefully scrape off all remaining gasket material from the thermostat housing and cover.

7. Insert a new thermostat into the housing with the element end toward the housing bottom and the flange on the unit seated into the recess in the housing.

■ Remember that some early 3.0L models came equipped with one of two thermostats. Models up to and including serial number 0L34099 use a brass thermostat rated at 143°F, models with serial number 0L341000 and above come equipped with a stainless steel thermostat rated at 160°F.

8. Coat both sides of a new housing gasket with Quicksilver Perfect Seal (or similar) and position it onto the housing so that the holes line up. If your engine is equipped with an audio warning temperature switch, the thermostat will have continuity rivets, do not use a sealer or the alarm will not work properly.

9. Install the housing cover and mounting bolts/washers and tighten to 30 ft. lbs. (41 Nm).

10. Reconnect the hoses and tighten the clamps being careful not to pinch the hose. This is a good time to inspect the hoses! Connect the batteries and then start the engine and check for leaks.

4.3L V6 And 5.0L/5.7L/6.2L V8 Engines

◆ See Figures 64 thru 73

1. Open or remove the engine hatch cover and disconnect the negative battery cables.

2. Drain all water from the cylinder block and exhaust manifolds as detailed in the Cooling System section.

■ Dry Joint systems should not use any sealer on the gasket.

11. Install the housing/cover and mounting bolts/washers and tighten to 30 ft. lbs. (41 Nm) except on closed systems utilizing Dry Joint where the torque should be 22 ft. lbs. (30 Nm).

12. Reconnect the hoses and tighten the clamps being careful not to pinch the hose. This is a good time to inspect the hoses! Connect the batteries and then start the engine and check for leaks.

8.1L V8 Engines

◆ See Figure 74

1. Open or remove the engine hatch cover and disconnect the negative battery cables.

2. Drain all water from the cylinder block, exhaust manifolds and closed system as detailed in the Cooling System section.

3. Loosen the thermostat retainer screws.

4. Remove the heat exchanger (for access).

5. Now you can remove the retainer mounting screws (2) with their lock washers and then remove the thermostat retainer.

6. Lift out the thermostat and discard it. If you are not sure that it is inoperable, perform the testing procedures outlined in this section. Remove and discard the O-ring.

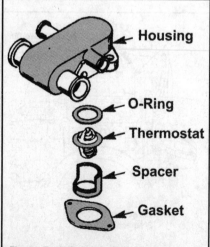

Fig. 64 Exploded view of the thermostat and housing - 4.3L V6 and 5.0L/5.7L/6.2L V8 carb or TBI engines w/seawater system

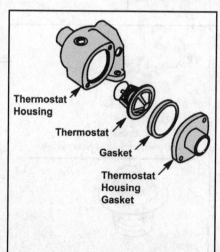

Fig. 65 Exploded view of the thermostat and housing - 4.3L V6 and 5.0L/5.7L/6.2L V8 carb or TBI engines w/closed system

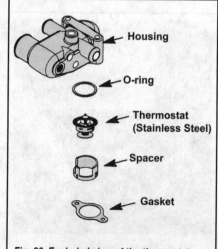

Fig. 66 Exploded view of the thermostat and housing - 4.3L V6 and 5.0L/5.7L/6.2L V8 MPI engines w/seawater system

ENGINE AND DRIVE MAINTENANCE 2-15

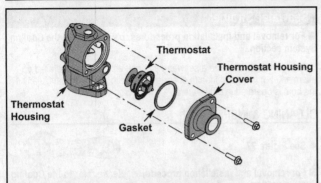

Fig. 67 Exploded view of the thermostat and housing - 4.3L V6 and 5.0L/5.7L/6.2L V8 MPI engines w/closed system

7. Make sure that you carefully scrape off all remaining gasket material from the thermostat retainer and crossover, and then position a new O-ring into the crossover.

8. Insert a new thermostat into the crossover. The element must be pointing into the housing (this means the pointed side is facing up) and the flange on the unit should be seated into the recess in the housing.

9. Install the retainer and mounting bolts/washers and tighten them finger-tight.

10. Install the heat exchanger and then tighten the retainer screws to 12 ft. lbs. (16 Nm).

11. Fill the closed system with a 50/50 mixture of antifreeze and water. Connect the batteries and then start the engine and check for leaks.

✱✱ CAUTION

If the boat is out of the water, make sure that a flushing kit is installed before starting the engine.

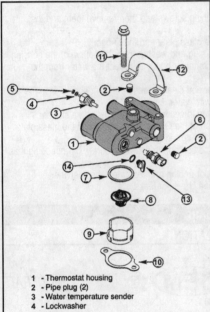

1 - Thermostat housing
2 - Pipe plug (2)
3 - Water temperature sender
4 - Lockwasher
5 - Nut
6 - Temperature sensor
7 - Thermostat gasket
8 - Thermostat
9 - Thermostat sleeve
10 - Thermostat housing gasket to intake
11 - Screw (2)
12 - Lifting eye
13 - Air vent plug (for use when draining seawater)
14 - O-ring

Fig. 68 Exploded view of the thermostat and housing - 5.0L/5.7L/6.2L V8 MPI engines w/seawater system. Dry Joint w/single point and manual 3 point drain systems

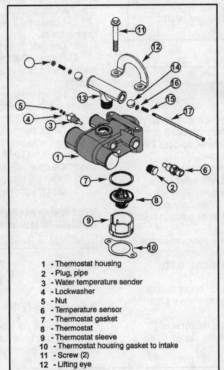

1 - Thermostat housing
2 - Plug, pipe
3 - Water temperature sender
4 - Lockwasher
5 - Nut
6 - Temperature sensor
7 - Thermostat gasket
8 - Thermostat
9 - Thermostat sleeve
10 - Thermostat housing gasket to intake
11 - Screw (2)
12 - Lifting eye
13 - Tee-fitting
14 - Check ball (2)
15 - Spring (2)
16 - Washer (2)
17 - Screw
18 - Locknut

Fig. 69 Exploded view of the thermostat and housing - 5.0L/5.7L/6.2L V8 MPI engines w/seawater system. Dry Joint w/multi-point drain system

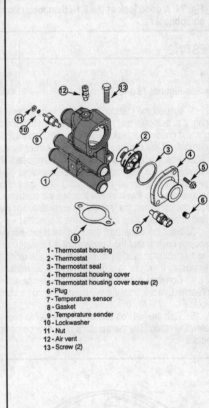

1 - Thermostat housing
2 - Thermostat
3 - Thermostat seal
4 - Thermostat housing cover
5 - Thermostat housing cover screw (2)
6 - Plug
7 - Temperature sensor
8 - Gasket
9 - Temperature sender
10 - Lockwasher
11 - Nut
12 - Air vent
13 - Screw (2)

Fig. 70 Exploded view of the thermostat and housing - 5.0L/5.7L/6.2L V8 MPI engines w/closed systems using Dry joint

Fig. 71 A good look at the V6 thermostat housing (carb model)

Fig. 72 A good look at the 5.7L thermostat housing (closed system)

Fig. 73 A good look at the 6.2L thermostat housing (seawater system)

2-16 ENGINE AND DRIVE MAINTENANCE

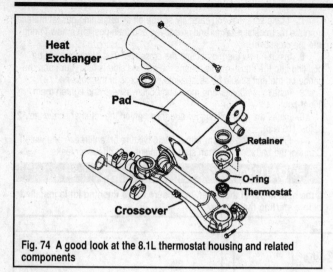

Fig. 74 A good look at the 8.1L thermostat housing and related components

TESTING

◆ See Figures 75 and 76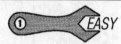

1. Inspect the thermostat at room temperature. If the thermostat is fully open, it is defective and must be replaced. Hold the thermostat up to the light and check it for leaks. A light leak around the perimeter indicates the thermostat is not closing, and therefore, it must be replaced.
2. Attach a length of thread to the thermostat. Now, suspend the thermostat and a thermometer inside a container filled with water (do not use distilled water or ethylene glycol!). Take care to be sure neither the thermostat or the thermometer touches the container. If either one does touch the container, the test will be unreliable. Stir the water occasionally to avoid direct heat being applied to the thermostat
3. Heat the water until the thermostat just begins to open - when this happens confirm that the temperature is the same as the thermostat rating. The thermometer reading must agree with the rating stamped on the thermostat. If the unit fails the test, it must be replaced.
4. Continue to heat the water until a temperature 25° above the rating is reached. At this time the thermostat should be completely open; if not, replace it.
5. Turn the heat off and allow the water to cool to a temperature 10° below the rating. The thermostat should now be completely closed; if not, replace it.

Seawater Strainer

■ For removal and installation procedures, please refer to the Cooling System section.

Many models will utilize a seawater strainer, either factory-installed or aftermarket. It is a good idea to check the unit with regularity, particularly if the boat is operated in dirty waters.

CLEANING & INSPECTION

◆ See Figure 77

■ For removal and installation procedures, please refer to the Cooling System section.

1. If your vessel is equipped with a seacock, close it. If not equipped, disconnect the seawater inlet line at the strainer and plug it securely to prevent water overflow.
2. Loosen the 2 retaining screws and then remove them and the strainer cover.
3. Carefully lift off the glass plate and the O-ring.
4. Remove the drain plug and washer from the bottom of the housing. You may want to have a container handy to collect the water from the housing.
5. Lift out the strainer and clear the mesh of any debris. Rinse the strainer and the housing with water.
6. Screw the drain plug in securely and then insert the strainer.
7. Check the condition of the O-ring and install it and the glass plate. We suggest replacing the O-ring regardless of its condition.
8. Install the cover and tighten the screws securely, but not so tight as to warp the cover.
9. Open the seacock or unplug and reconnect the inlet line.
10. Start the engine and check for leaks.

Impeller/Water Pump

REMOVAL & INSTALLATION

Alpha Drives

◆ See Figures 78 thru 86

1. Remove the lower unit.
2. Pull the water tube and coupling seal out of the water pump body. Separate the tube and coupling, and discard the O-rings

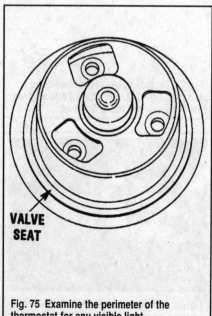

Fig. 75 Examine the perimeter of the thermostat for any visible light

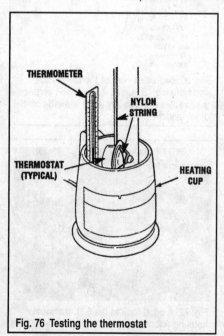

Fig. 76 Testing the thermostat

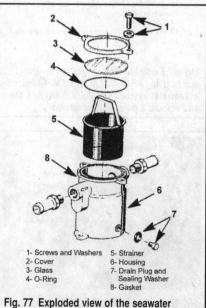

Fig. 77 Exploded view of the seawater strainer

ENGINE AND DRIVE MAINTENANCE

3. Slide the water seal up and free of the driveshaft. Discard the seal. Remove the bolts securing the water pump body, and then slide the body up and free of the driveshaft. It may be necessary to use a few small prybars - one on each mounting flange - to gently persuade the pump body to "break" loose from the plate.

4. Slide the impeller up and free of the driveshaft. Remove the Woodruff key from the cutout in the driveshaft.

5. Remove the gasket from the under side of the body. It is possible this gasket will remain on the plate when the body is removed.

■ Although not strictly necessary, we recommend performing the following step as well. You've got the thing apart already, why takes chances?

6. Slide the face plate and gaskets up and free of the driveshaft.

To install:

※※ **WARNING**

The water pump impeller must be in very good condition for satisfactory service. The pump performs an extremely important function by supplying sufficient water to properly cool the stern drive and the engine. Therefore, good shop practice dictates - replace the water pump impeller whenever the unit is disassembled, but then that's why you're here isn't it?

7. Inspect the water tube coupling for wear or damage. It's always a good idea to replace the 2 O-rings.

8. Inspect the impeller for any wear or damage. Replace as necessary.

9. Slide the small hole gasket down the driveshaft, followed by the face plate and the large hole gasket. Holes in the gaskets and plate will only align with each other and the holes in the lower unit one way. If the holes do not align, one or more of the items is upside down. Correct the situation by turning one or more of the items over.

10. Apply just a "dab" of grease to the Woodruff key, and then place it in the driveshaft keyway. Slide the impeller down the driveshaft and onto the face plate with the cutout in the impeller indexed over the Woodruff key. If an old impeller with a "set" to the blades is being installed, face the curl of the blades in a counterclockwise direction. If the direction is reversed, premature impeller failure will surely occur.

11. Slide the pump body down the driveshaft and just to the top of the impeller. Keeping the gasket, impeller, and body mounting holes aligned is not an easy task. However, on early models if a couple of pins are inserted down through just two opposite holes, as shown, the holes will stay aligned while the body is worked down over the impeller. The pins can be old drill bits, small diameter bolts, rod, whatever is handy. Exert some downward pressure on the pump body and at the same time rotate the driveshaft clockwise and the impeller blades will "set" in the proper direction.

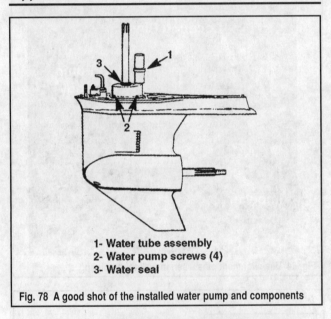

1- Water tube assembly
2- Water pump screws (4)
3- Water seal

Fig. 78 A good shot of the installed water pump and components

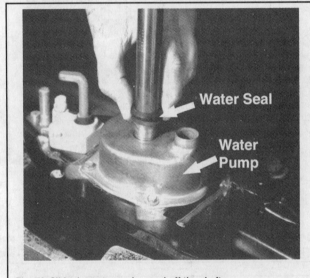

Fig. 79 Slide the water seal up and off the shaft. . .

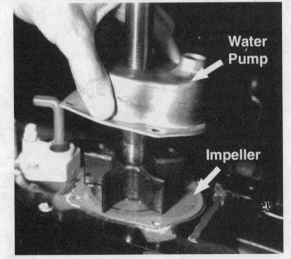

Fig. 80 . . .lift off the pump housing. . .

Fig. 81 . . .and then remove the old gasket

ENGINE AND DRIVE MAINTENANCE

12. Start a couple of the body mounting bolts (coat the threads with Perfect seal), and then remove the 2 pins. Install the remaining mounting bolts, and tighten them all to 60 inch lbs. (7.9 Nm).

■ A special water pump seal seating tool is required to properly seat the seal on top of the water pump case. This tool is only available from MerCruiser in a kit (#26-81657A2).

13. Apply a light coating of lubricant to the driveshaft. Slide the water pump face seal down the driveshaft until it is about 1/2 way down the shaft. Obtain special seal seating tool from the kit identified above. Slide the tool down the driveshaft and onto the seal. Push the seal down with the tool until the tool makes contact with the pump body. If the tool is not available, a large washer with an inside diameter slightly larger than the driveshaft can be used. Push the seal down evenly onto the water pump body until the seal face makes light contact with the pump body - while pulling up on the driveshaft. Remove the tool.

14. Coat new O-rings lightly with Quicksilver 2-4-C Marine Lubricant and install them in the coupling.

15. Slide the tube sleeve into the coupling and then insert the assembly into the pump housing. Make sure you do not damage the O-rings.

16. Install the lower unit.

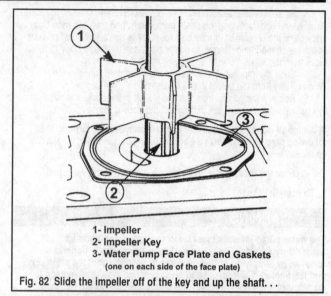

1- Impeller
2- Impeller Key
3- Water Pump Face Plate and Gaskets
(one on each side of the face plate)

Fig. 82 Slide the impeller off of the key and up the shaft...

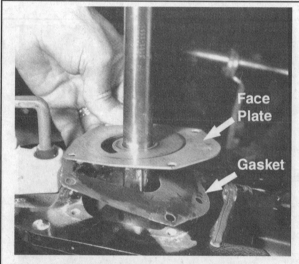

Fig. 83 ...and then lift off the face plate and gaskets

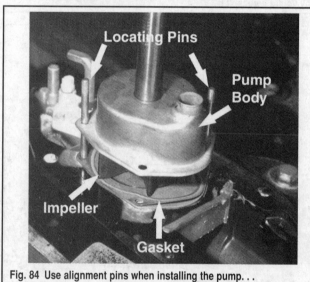

Fig. 84 Use alignment pins when installing the pump...

Fig. 85 ...and then tighten the bolts

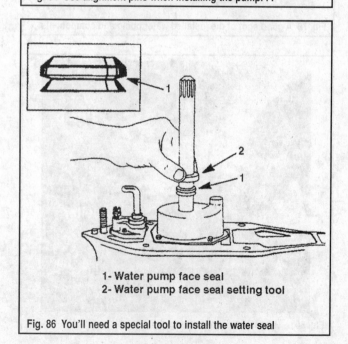

1- Water pump face seal
2- Water pump face seal setting tool

Fig. 86 You'll need a special tool to install the water seal

ENGINE AND DRIVE MAINTENANCE

Inboards & Bravo Drives - 2001 350 Mag MPI/6.2L (Up to Serial #299999) and All Carbureted/TBI Engines

◆ See Figures 87, 88 and 89

If performing the following while boat is in the water, close the seacock. If your boat is not equipped with a seacock, disconnect and plug the seawater inlet line to prevent water from entering the system.

■ The engine should be OFF and cool.

1. Drain the seawater system completely.
2. Disconnect the inlet and outlet hoses at the back or the pump.
3. Relieve tension and then remove the serpentine belt. All engines covered here should be equipped with a serpentine belt, but if you have one of the stragglers: loosen the power steering pump bracket and move the pump until tension is relieved on the drive belt, now remove the power steering belt and then remove the seawater pump belt.
4. Loosen the pump bracket or brace mounting bolts (usually 2) at the cylinder block. Pull the bolts out and remove the pump assembly. Most models will come off with the brackets and brace attached although there are certain applications where you simply unbolt the pump from the bracket and brace. Either way, your goal is to eventually get the pump free of the bracket and brace, so you be the judge based on what you see.
5. Remove the 5 pump housing mounting bolts and their washers from the rear of the pump and lift the housing off the pump body. Remove the gasket and/or quad ring.
6. If it didn't come out when you separated the pump halves, remove the impeller. On a few early models you will need to pull out the rubber plug and then lift out the impeller.

To install:

7. Inspect the pump assembly for cracks, damage or other signs of wear.
8. Coat the impeller with a little soapy water, position it over the bore in the pump housing and then turn it slowly in its normal direction of rotation while pressing down on it until it seats into the bore.
9. Position the wear plate over the bearing shaft if it was removed.
10. Position a new quad or O-ring into the lip of the housing, align the flats on the impeller with the bearing shaft and slide the assembly onto the shaft.
11. If the pump bracket was removed, install it now - make sure that the 87outlet fitting is on top! Install 2 bolts into the alignment holes to ensure that the two halves line up and then install the remaining bolts. Tighten them all to 30 ft. lbs. (41 Nm).
12. Install the pump assembly to the cylinder block and tighten the bracket/brace bolts to 30 ft. lbs. (41 Nm).
13. Reconnect the water hoses, outlet on top, and tighten the hose clamps securely.
14. Install the serpentine belt (or drive belts) and adjust tension accordingly.
15. Open the seacock or reconnect the inlet line. Start the engine and check for leaks or overheating.

Inboards & Bravo Drives - MPI Engines (Exc. 2001 350 Mag/6.2L)

◆ See Figures 90, 91 and 92

If performing the following while boat is in the water, close the seacock. If your boat is not equipped with a seacock, disconnect and plug the seawater inlet line to prevent water from entering the system.

■ The engine should be OFF and cool.

1. Drain the seawater system completely.
2. Disconnect the inlet and outlet hoses at the back or the pump.
3. Relieve tension and then remove the serpentine belt. On certain later 8.1L engines you may also have to remove the idler pulley bracket or the automatic tensioner bracket.
4. On models utilizing an air actuated drain valve:
   a. Disconnect the vent line at the top of the drain valve (attached to back of pump).
   b. Push in the plastic ring around each air hose and, while holding the ring in, pull out the air hose. Repeat on the other hose. Take note of which went where as they are each a different size.
5. Ease tension on the serpentine belt as detailed previously and then remove the belt.
6. Loosen the pump bracket/brace mounting bolts (2 or 3) at the cylinder block. Pull the bolts out and remove the pump assembly. Most models will come off with the brackets and brace attached although there are certain applications where you can simply unbolt the pump from the bracket and brace. Either way, your goal is to eventually get the pump free of the bracket and brace, so you be the judge based on what you see.
7. Remove the 6 bolts from the rear of the back plate and separate the plate from the housing. Remove the O-ring and throw it away. Mercruiser

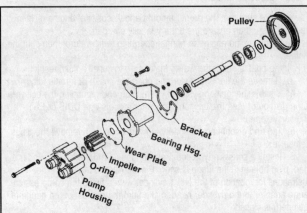

Fig. 87 Exploded view of the water pump - 2001 350 Mag MPI/6.2L (Up to Serial #299999) and All Carbureted/TBI Engines

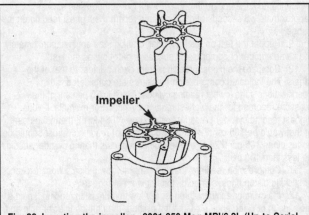

Fig. 88 Inserting the impeller - 2001 350 Mag MPI/6.2L (Up to Serial #299999) and All Carbureted/TBI Engines

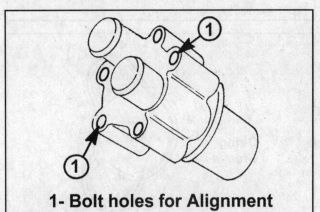

1- Bolt holes for Alignment

Fig. 89 Insert the alignment bolts into these holes - 2001 350 Mag MPI/6.2L (Up to Serial #299999) and All Carbureted/TBI Engines

2-20 ENGINE AND DRIVE MAINTENANCE

suggests pressing of the pump pulley, but we think it's not necessary if only replacing the impeller.

8. If it didn't come out when you separated the pump halves, remove the impeller.

■ **Further disassembly of the pump is detailed in the Cooling System section.**

To install:

9. Inspect the pump assembly for cracks, damage or other signs of wear.

10. Coat the impeller with a little soapy water, position it over the bore in the pump housing and then turn it slowly in its normal direction of rotation while pressing down on it until it seats into the bore.

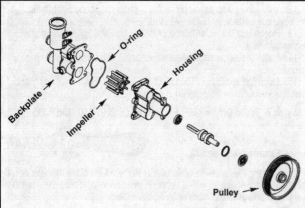

Fig. 90 Exploded view of the water pump - MPI Engines (Exc. 2001 350 Mag/6.2L)

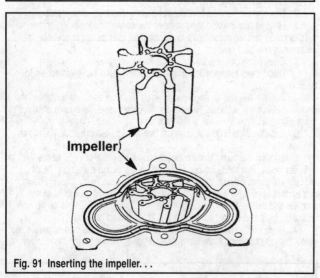

Fig. 91 Inserting the impeller...

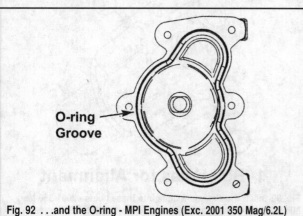

Fig. 92 ...and the O-ring - MPI Engines (Exc. 2001 350 Mag/6.2L)

11. Install a new O-ring into the groove and align the housing assembly with the back plate assembly. Install the mounting bolts and tighten to 88 inch lbs. (9.9 Nm).

12. If you removed the bracket from the pump, re-position it and tighten all bolts to 88 inch lbs. (9.9 Nm).

13. Position the pump assembly to the cylinder block (or bracket if you got lucky!) and tighten the mounting bolts to 30 ft. lbs. (41 Nm) on all models but the 8.1L, where you need to make them 37 ft. lbs. (50 Nm).

14. Press the pulley back onto the pump shaft if you chose to remove it.

15. Reconnect the water hoses, outlet on top, and tighten the hose clamps securely.

16. Reconnect the vent line and air hoses on air actuated drain models. Remember that the air hoses are different sizes and then just insert them into the fitting until they stop. Pull back on the hose to ensure that they are connected correctly.

17. Install the idler or tensioner bracket if you had to remove one of them. Install the serpentine belt and adjust tension accordingly.

18. Open the seacock or reconnect the inlet line. Start the engine and check for leaks or overheating.

Cylinder Compression

Cylinder compression test results are extremely valuable indicators of internal engine condition. The best marine mechanics automatically check an engine's compression as the first step in a comprehensive tune-up. A compression test will uncover many mechanical problems that can cause rough running or poor performance.

CHECKING COMPRESSION

◆ See Figures 93 and 94

1. Make sure that the proper amount and viscosity of engine oil is in the crankcase, then ensure the battery is fully charged.

2. Warm-up the engine to normal operating temperature, then shut the engine **OFF**. If the boat is out of the water, make sure to install a flush test kit.

3. Remove the flame arrestor and open the choke or throttle fully.

4. Disable the ignition system by removing and grounding the coil wire at the distributor (see the Ignition section for models with DDIS or EST ignition systems).

5. Tag and disconnect all spark plug wires and then remove the plugs themselves.

6. Install a screw-in type compression gauge into the No. 1 cylinder spark plug hole until the fitting is snug. Please refer to the firing order illustrations for location of the No. 1 cylinder. When fitting the compression gauge adapter to the cylinder head, make sure the bleeder of the gauge (if equipped) is closed.

7. According to the tool manufacturer's instructions, connect a remote starting switch to the starting circuit.

8. With the ignition switch in the **OFF** position, use the remote starting switch to crank the engine through at least five compression strokes (approximately 5 seconds of cranking) and record the highest reading on the gauge.

9. Repeat the test on each cylinder, cranking the engine approximately the same number of compression strokes and/or time as the first.

10. Compare the highest readings from each cylinder to that of the others. The indicated compression pressures are considered within specifications if the lowest reading cylinder is within 70 percent of the pressure recorded for the highest reading cylinder. For example, if your highest reading cylinder pressure was 150 psi (1034 kPa), then 70 percent of that would be 105 psi (724 kPa). So the lowest reading cylinder should be no less than 105 psi (724 kPa). Mercruiser suggests that no cylinder should be less than 100 psi.

11. Compression readings that are generally low indicate worn, broken, or sticking piston rings, scored pistons or worn cylinders.

12. If a cylinder exhibits an unusually low compression reading, squirt a tablespoon of clean engine oil into the cylinder through the injector hole and repeat the compression test. If the compression rises after adding oil, it means that the cylinder's piston rings and/or cylinder bore are damaged or

ENGINE AND DRIVE MAINTENANCE 2-21

worn. If the pressure remains low, the valves may not be seating properly (a valve job is needed), or the head gasket may be blown near that cylinder.

13. If compression in any two adjacent cylinders is low (with normal compression in the other cylinders), and if the addition of oil doesn't help raise compression, there is leakage past the head gasket. Oil and coolant in the combustion chamber, combined with blue or constant white smoke from the tailpipe, are symptoms of this problem. However, don't be alarmed by the normal white smoke emitted from the tailpipe during engine warm-up during cold weather. There may be evidence of water droplets on the engine oil dipstick and/or oil droplets in the cooling system if a head gasket is blown.

Fig. 93 When checking the compression, always use a quality gauge

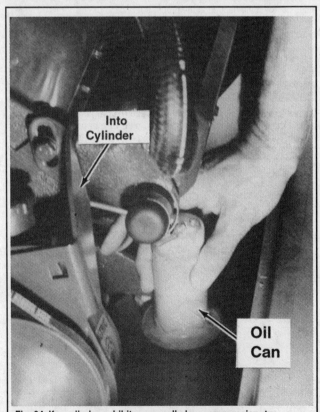

Fig. 94 If a cylinder exhibits unusually low compression, try squirting about a tablespoon of oil into it

Spark Plugs

The spark plug performs four main functions:
- It fills a hole in the cylinder head.
- It acts as a dielectric insulator for the ignition system.
- It provides spark for the combustion process to occur.
- It removes heat from the combustion chamber.

It is important to remember that spark plugs do not create heat, they help remove it. Anything that prevents a spark plug from removing the proper amount of heat can lead to pre-ignition, detonation, premature spark plug failure and even internal engine damage.

In the simplest of terms, the spark plug acts as the thermometer of the engine. Much like a doctor examining a patient, this "thermometer" can be used to effectively diagnose the amount of heat present in each combustion chamber.

Spark plugs are valuable tuning tools, when interpreted correctly. They will show symptoms of other problems and can reveal a great deal about the engine's overall condition. By evaluating the appearance of the spark plug's firing tip, visual cues can be seen to accurately determine the engine's overall operating condition, get a feel for air/fuel ratios and even diagnose driveability problems.

As spark plugs grow older, they lose their sharp edges and material from the center and ground electrodes is slowly eroded away. As the gap between these two points grows, the voltage required to bridge this gap increases proportionately. The ignition system must work harder to compensate for this higher voltage requirement and hence there is a greater rate of misfires or incomplete combustion cycles. Each misfire means lost horsepower, reduced fuel economy and higher emissions. Replacing worn out spark plugs with new ones (with sharp new edges) effectively restores the ignition system's efficiency and reduces the percentage of misfires, restoring power, economy and reducing emissions.

How long spark plugs last will depend on a variety of factors, including engine compression, fuel used, gap, center/ground electrode material and the conditions in which the engine is operated.

SPARK PLUG HEAT RANGE

◆ See Figure 95

Spark plug heat range is the ability of the plug to dissipate heat from the combustion chamber. The longer the insulator (or the farther it extends into the engine), the hotter the plug will operate; the shorter the insulator (the closer the electrode is to the block's cooling passages) the cooler it will operate.

Selecting a spark plug with the proper heat range will ensure that the tip will maintain a temperature high enough to prevent fouling, yet be cool enough to prevent pre-ignition. A plug that absorbs little heat and remains too cool will quickly accumulate deposits of oil and carbon since it is not hot enough to burn them off. This leads to plug fouling and consequently to misfiring. A plug that absorbs too much heat will have no deposits but, due to the excessive heat, the electrodes will burn away quickly and might possibly lead to pre-ignition or other ignition problems.

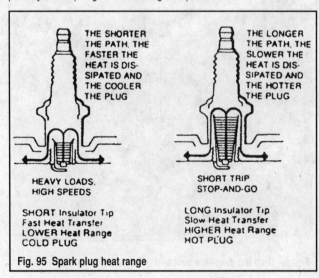

Fig. 95 Spark plug heat range

2-22 ENGINE AND DRIVE MAINTENANCE

Pre-ignition takes place when plug tips get so hot that they glow sufficiently to ignite the air/fuel mixture before the actual spark occurs. This early ignition will usually cause a pinging during heavy loads and if not corrected will result in severe engine damage. While there are many other things that can cause pre-ignition, selecting the proper heat range spark plug will ensure that the spark plug itself is not a hot-spot source.

SPARK PLUG SERVICE

■ New technologies in spark plug and ignition system design have pushed the recommended replacement interval higher and higher. However, this depends on usage and conditions.

Spark plugs should only require replacement once a season. The electrode on a new spark plug has a sharp edge but with use, this edge becomes rounded by wear, causing the plug gap to increase. As the gap increases, the plug's voltage requirement also increases. It requires a greater voltage to jump the wider gap and about two to three times as much voltage to fire a plug at high speeds than at idle.

Tools needed for spark plug replacement include: a ratchet, short extension, spark plug socket (there are two types; either 13/16 in. or 5/8 in., depending upon the type of plug), a combination spark plug gauge and gapping tool and a can of anti-seize type compound.

REMOVAL & INSTALLATION

◆ See Figures 96 and 97

1. When removing spark plugs, work on one at a time. Don't start by removing the plug wires all at once, because unless you number them, they may become mixed up. Take a minute before you begin and number the wires with tape.
2. Disconnect the negative battery cable or turn the battery switch **OFF**.
3. If the engine has been run recently, allow the engine to thoroughly cool. Attempting to remove plugs from a hot cylinder head could cause the plugs to seize and damage the threads in the cylinder head, especially on aluminum heads!
4. Carefully twist the spark plug wire boot to loosen it, then pull the boot using a twisting motion to remove it from the plug. Be sure to pull on the boot and not on the wire, otherwise the connector located inside the boot may become separated from the high-tension wire.

■ A spark plug wire removal tool is recommended as it will make removal easier and help prevent damage to the boot and wire assembly.

5. Using compressed air (and safety glasses), blow debris from the spark plug area to assure that no harmful contaminants are allowed to enter the combustion chamber when the spark plug is removed. If compressed air is not available, use a rag or a brush to clean the area. Compressed air is available from both an air compressor or from compressed air in cans available at photography stores.

■ Remove the spark plugs when the engine is cold, if possible, to prevent damage to the threads. If plug removal is difficult, apply a few drops of penetrating oil to the area around the base of the plug and allow it a few minutes to work.

6. Using a spark plug socket that is equipped with a rubber insert to properly hold the plug, turn the spark plug counterclockwise to loosen and remove the spark plug from the bore.

※※ WARNING

Avoid the use of a flexible extension on the socket. Use of a flexible extension may allow a shear force to be applied to the plug. A shear force could break the plug off in the cylinder head, leading to costly and frustrating repairs. In addition, be sure to support the ratchet with your other hand - this will also help prevent the socket from damaging the plug.

7. Evaluate each cylinder's performance by comparing the spark plug condition. Check each spark plug to be sure they are all of the same manufacturer and have the same heat range rating. Inspect the threads in the spark plug opening of the block and clean the threads before installing the plug.
8. When purchasing new spark plugs, always ask the dealer if there has been a spark plug change for the engine being serviced. Many times manufacturers will update the type of spark plug used in an engine to offer better efficiency or performance.
9. Crank the engine through several revolutions to blow out any material that might have become dislodged during cleaning. Always use a new gasket (if applicable), but never use gaskets on taper seat plugs. The gasket must be fully compressed on clean seats to complete the heat transfer process and to provide a gas tight seal in the cylinder.
10. Inspect the spark plug boot for tears or damage. If a damaged boot is found, the spark plug boot and possible the entire wire will need replacement.
11. Check the spark plug gap prior to installing the plug. Most spark plugs do not come gapped to the proper specification.
12. Apply a thin coating of anti-seize on the thread of the plug. This is extremely important on aluminum head engines.
13. Carefully thread the plug into the bore by hand. If resistance is felt before the plug completely bottomed, back the plug out and begin threading again.

※※ WARNING

Do not use the spark plug socket to thread the plugs. Always carefully thread the plug by hand or using an old plug wire to prevent the possibility of cross-threading and damaging the cylinder head bore.

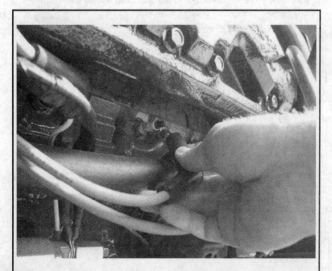

Fig. 96 Grab the plug wire boot and twist while removing it

Fig. 97 Removing the spark plug

ENGINE AND DRIVE MAINTENANCE 2-23

14. Carefully tighten the spark plug. If the plug you are installing is equipped with a crush washer, tighten the plug until the washer seats, then turn it 1/4 turn to crush the washer. Whenever possible, spark plugs should be tightened to the factory torque specification:
- 3.0L engines - 22 ft. lbs. (30 Nm)
- 4.3L V6 engines (carb or TBI) - 15 ft. lbs. (20 Nm)
- 4.3L V6 engines (MPI) - 11 ft. lbs. (15 Nm); exc. on new cylinder heads which should have an initial torque of 22 ft. lbs. (30 Nm)
- 5.0L/5.7L/6.2L V8 engines (carb or TBI) - 15 ft. lbs. (20 Nm); exc. on new cylinder heads which should have an initial torque of 22 ft. lbs. (30 Nm)
- 5.0L/5.7L/6.2L V8 engines (MPI) - 11 ft. lbs. (15 Nm); exc. on new cylinder heads which should have an initial torque of 22 ft. lbs. (30 Nm)
- 8.1L V8 engines - New plugs: 22 ft. lbs. (30 Nm); Used plugs: 15 ft. lbs. (20 Nm)

15. Apply a small amount of silicone dielectric grease to the end of the spark plug lead or inside the spark plug boot to prevent sticking, then install the boot to the spark plug and push until it clicks into place. The click may be felt or heard. Gently pull back on the boot to assure proper contact.
16. Connect the negative battery cable or turn the battery switch **ON**.
17. Start the engine and insure proper operation.

READING SPARK PLUGS

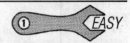

◆ See Figures 98 thru 104

Reading spark plugs can be a valuable tuning aid. By examining the insulator firing nose color, you can determine much about the engine's overall operating condition.

In general, a light tan/gray color tells you that the spark plug is at the optimum temperature and that the engine is in good operating condition.

Dark coloring, such as heavy black wet or dry deposits usually indicate a fouling problem. Heavy, dry deposits can indicate an overly rich condition, too cold a heat range spark plug, possible vacuum leak, low compression, overly retarded timing or too large a plug gap.

If the deposits are wet, it can be an indication of a breached head gasket, oil control from ring problems or an extremely rich condition, depending on what liquid is present at the firing tip.

Look for signs of detonation, such as silver specs, black specs or melting or breakage at the firing tip.

Fig. 98 A normally worn spark plug should have light tan or gray deposits on the firing tip (electrode)

Fig. 99 A carbon-fouled plug, identified by soft, sooty black deposits, may indicate an improperly tuned engine

Fig. 100 A physically damaged spark plug may be evidence of severe detonation in that cylinder. Watch the cylinder carefully between services, as a continued detonation will not only damage the plug but will most likely damage the engine

Fig. 101 An oil-fouled spark plug indicates an engine with worn piston rings

Fig. 102 This spark plug has been left in the engine too long, as evidenced by the extreme gap. Plugs with such an extreme gap can cause misfiring and stumbling accompanied by a noticeable lack of power

Fig. 103 A bridged or almost bridged spark plug, identified by the build-up between the electrodes caused by excessive carbon or oil build-up on the plug

2-24 ENGINE AND DRIVE MAINTENANCE

Compare your plugs to the illustrations shown to identify the most common plug conditions.

Fouled Spark Plugs

A spark plug is fouled when the insulator nose at the firing tip becomes coated with a foreign substance, such as fuel, oil or carbon. This coating makes it easier for the voltage to follow along the insulator nose and leach back down into the metal shell, grounding out, rather than bridging the gap normally.

Fuel, oil and carbon fouling can all be caused by different things but in any case, once a spark plug is fouled, it will not provide voltage to the firing tip and that cylinder will not fire properly. In many cases, the spark plug cannot be cleaned sufficiently to restore normal operation. It is therefore recommended that fouled plugs be replaced.

Signs of fouling or excessive heat must be traced quickly to prevent further deterioration of performance and to prevent possible engine damage.

Overheated Spark Plugs

When a spark plug tip shows signs of melting or is broken, it usually means that excessive heat and/or detonation was present in that particular combustion chamber or that the spark plug was suffering from thermal shock.

Since spark plugs do not create heat by themselves, one must use this visual clue to track down the root cause of the problem. In any case, damaged firing tips most often indicate that cylinder pressures or temperatures were too high. Left unresolved, this condition usually results in more serious engine damage.

Detonation refers to a type of abnormal combustion that is usually preceded by pre-ignition. It is most often caused by a hot spot formed in the combustion chamber.

As air and fuel is drawn into the combustion chamber during the intake stroke, this hot spot will "pre-ignite" the air fuel mixture without any spark from the spark plugs.

Detonation

Detonation exerts a great deal of downward force on the pistons as they are being forced upward by the mechanical action of the connecting rods. When this occurs, the resulting concussion, shock waves and heat can be severe. Spark plug tips can be broken or melted and other internal engine components such as the pistons or connecting rods themselves can be damaged.

Left unresolved, engine damage is almost certain to occur, with the spark plug usually suffering the first signs of damage.

■ **When signs of detonation or pre-ignition are observed, they are symptom of another problem. You must determine and correct the situation that caused the hot spot to form in the first place.**

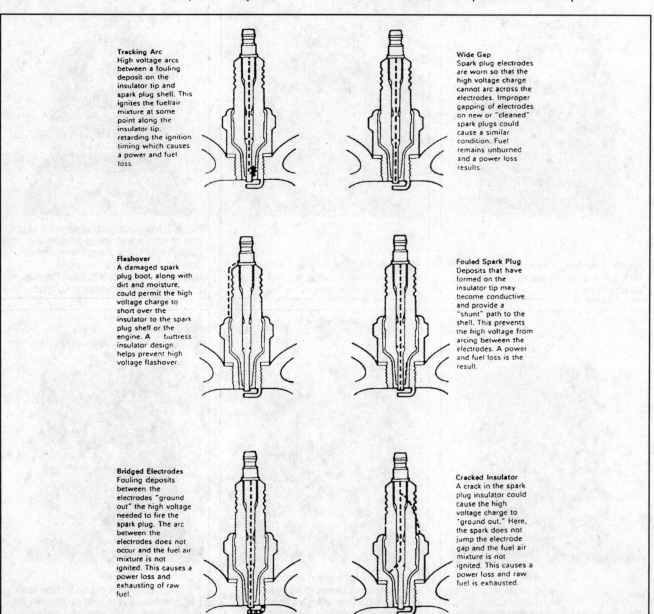

Fig. 104 Typical spark plug problems showing damage that may indicate engine problems

ENGINE AND DRIVE MAINTENANCE 2-25

INSPECTION & GAPPING

◆ See Figures 105 and 106

A particular spark plug might fit hundreds of engines and although the factory will typically set the gap to a pre-selected setting, this gap may not be the right one for your particular engine.

Insufficient spark plug gap can cause pre-ignition, detonation, even engine damage. Too much gap can result in a higher rate of misfires, noticeable loss of power, plug fouling and poor economy.

Check the spark plug gap before installation. The ground electrode (the L-shaped one connected to the body of the plug) must be parallel to the center electrode and the specified size wire gauge must pass between the electrodes with a slight drag.

Do not use a flat feeler gauge when measuring the gap on a used plug, because the reading may be inaccurate. A round wire-type gapping tool is the best way to check the gap. The correct gauge should pass through the electrode gap with a slight drag. If you're in doubt, try a wire that is one size smaller or larger. The smaller gauge should go through easily, while the larger one shouldn't go through at all.

Wire gapping tools usually have a bending tool attached. Use this tool to adjust the side electrode until the proper distance is obtained. Never attempt to bend the center electrode. Also, be careful not to bend the side electrode too far or too often as it may weaken and break off within the engine, requiring removal of the cylinder head to retrieve it.

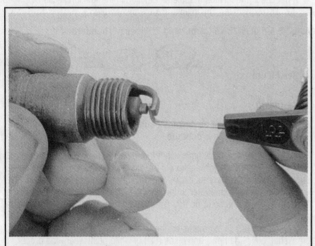

Fig. 105 Using a wire-type spark plug gapping tool to check the distance between center and ground electrodes

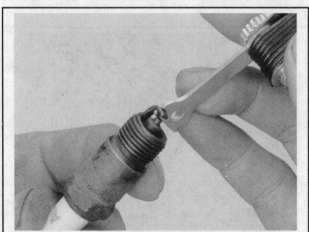

Fig. 106 Most spark plug gapping tools have an adjusting tool used to bend the ground electrode. USE IT! This tool greatly reduces the chance of breaking off the electrode and is much more accurate

Spark Plug Wires

TESTING

Each time you remove the engine cover, visually inspect the spark plug wires for burns, cuts or breaks in the insulation. Check the boots on the coil and at the spark plug end. Replace any wire that is damaged.

Once a year, usually when you change your spark plugs, check the resistance of the spark plug wires with an ohmmeter. Wires with excessive resistance will cause misfiring and may make the engine difficult to start. In addition worn wires will allow arcing and misfiring in humid conditions.

Remove the spark plug wire from the engine. Test the wires by connecting one lead of an ohmmeter to the coil end of the wire and the other lead to the spark plug end of the wire. Resistance should measure approximately 7000 ohms per foot of wire. If a spark plug wire is found to have excessive (high) resistance, the entire set should be replaced.

REMOVAL & INSTALLATION

◆ See Figure 96

When installing a new set of spark plug wires, replace the wires one at a time so there will be no confusion. Coat the inside of the boots with dielectric grease to prevent sticking. Install the boot firmly over the spark plug until it clicks into place. The click may be felt or heard. Gently pull back on the boot to assure proper contact. Repeat the process for each wire.

■ It is important to route the new spark plug wire the same as the original and install it in a similar manner on the engine. Improper routing of spark plug wires may cause engine performance problems.

Ignition Timing

ADJUSTMENT

3.0L Engines w/EST Ignition Systems

◆ See Figures 107 thru 110

■ Failure to follow the timing procedure instructions exactly will result in improper timing and cause performance problems at the least and possibly severe engine damage.

1. Connect a suitable timing light to the No. 1 spark plug lead (see firing order illustrations for location of the No. 1 cylinder). Connect the power supply lead to the battery as detailed in the light manufacturer's instructions.

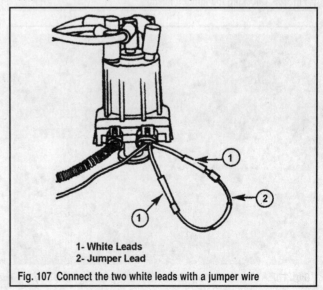

1- White Leads
2- Jumper Lead

Fig. 107 Connect the two white leads with a jumper wire

2-26 ENGINE AND DRIVE MAINTENANCE

2. Connect a tachometer to the engine as detailed by the manufacturer. Do not use the tachometer on the instrument panel as it will not provide the necessary accuracy.

3. Locate the timing marks on the engine timing cover (just above the crankshaft pulley) and place a bit of white paint where the proper mark should be (serial #0L096999 and below - 1° BTDC, #0L097000-0L0340999 - 1° ATDC, #0L0341000 and above - 2° ATDC). Timing marks are generally shown in 2° increments from TDC.

4. Start the engine and allow it to reach normal operating temperature at idle.

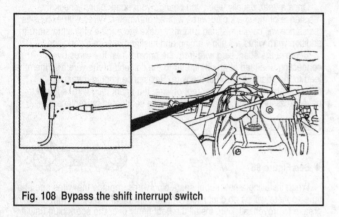

Fig. 108 Bypass the shift interrupt switch

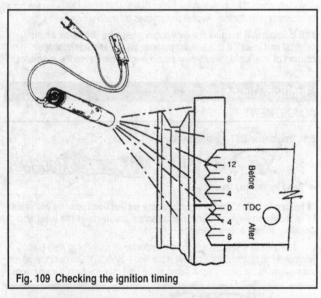

Fig. 109 Checking the ignition timing

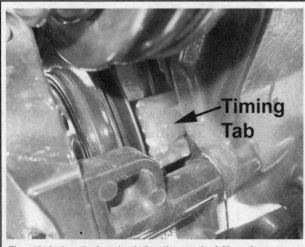

Fig. 110 A close look at the timing plate on the 3.0L engine

5. Disable the timing advance system by installing a jumper wire across the two white leads on the distributor 3-pin connector. This jumper wire is available from your local dealer (#91-818812A1), or you can easily fabricate one with a 6 in. length of 16 gauge wire and 2 bullet terminal ends.

6. Bypass the shift interrupt switch by disconnecting the two wires at the switch and connecting them together. DO NOT FORGET TO RECCONNECT THESE TWO WIRES!

7. While still idling, point the light at the timing marks. The strobe will make it appear that the mark on the tab and the mark on the pulley stand still in alignment.

8. If the timing requires adjustment, loosen the clamp bolt at the base of the distributor (DDIS models have a motion sensor in place of the distributor) and then carefully rotate the distributor or sensor until the correct marks line up.

9. Tighten the clamp bolt to 20 ft. lbs. (27 Nm) and check the timing one last time.

10. Reconnect the 2 shift interrupt switch leads to the switch and then remove the jumper wire at the base of the distributor.

11. With the timing light still connected and the engine at idle, check that the timing advanced to 10-14° BTDC after the jumper was removed on models with initial timing of 1° BTDC; 12-16° BTDC on models with initial timing of 1° ATDC; or, 13-17° on models with initial timing of 2° ATDC.

12. Run the engine up to 2400-2800 rpm and check the timing again. Total advance should be:
- 21-25° on models w/initial timing of 1° BTDC
- 23-27° on models w/initial timing of 1° ATDC
- 24-28° on models w/initial timing of 2° ATDC

If not check the ignition module as detailed in the Ignition section.

13. Disconnect the timing light and the tachometer.

4.3L V6 & 5.0L/5.7L V8 Carbureted Engines w/Thunderbolt V

◆ See Figure 111

Base Timing

■ The idle speed must be correctly adjusted and within specifications before performing this procedure.

1. Connect a suitable timing light to the No. 1 spark plug lead (see firing order illustrations for location of the No. 1 cylinder). Connect the power supply lead to the battery as detailed in the light manufacturer's instructions.

2. Connect a tachometer to the engine as detailed by the manufacturer. Do not use the tachometer on the instrument panel, as it will not provide the necessary accuracy.

3. Locate the timing marks on the engine timing cover (just above the crankshaft pulley/damper) and place a bit of white paint where the proper mark should be (10° BTDC). Timing marks are generally shown in 2° increments from TDC.

4. Locate the ignition system timing lead and connect it to a good ground (-) to lock the ignition module into base timing mode. The lead

Fig. 111 A close look at the timing plate on the 4.3L engine, other models similar

ENGINE AND DRIVE MAINTENANCE

should be a purple and white wire near the distributor or in front of the engine near the fuel line.

5. Start the engine and allow it to reach normal operating temperature at idle.

6. While still idling, point the light at the timing marks. The strobe will make it appear that the mark on the tab and the mark on the damper/pulley stand still in alignment.

7. If the timing requires adjustment, loosen the clamp bolt at the base of the distributor and then carefully rotate the distributor until the correct marks line up.

8. Tighten the clamp bolt to 18 ft. lbs. (25 Nm).

9. Disconnect the jumper wire from the timing lead and ground.

10. Disconnect the light and tachometer.

Ignition Timing

■ The idle speed and dwell must be correctly adjusted and within specifications before performing this procedure.

1. Connect a suitable timing light to the No. 1 spark plug lead (see firing order illustrations for location of the No. 1 cylinder). Connect the power supply lead to the battery as detailed in the light manufacturer's instructions.

2. Connect a tachometer to the engine as detailed by the manufacturer. Do not use the tachometer on the instrument panel, as it will not provide the necessary accuracy.

3. Locate the timing marks on the engine timing cover (just above the crankshaft pulley/damper) and place a bit of white paint where the proper mark should be (10° BTDC). Timing marks are generally shown in 2° increments from TDC.

4. Locate the ignition system timing lead and connect it to a good ground (-) to lock the ignition module into base timing mode. The lead should be a purple and white wire near the distributor or in front of the engine near the fuel line.

5. Start the engine and run it at 1300 rpm until it reaches normal operating temperature.

6. Disconnect the throttle cable at the carburetor.

7. With the engine idling, adjust the idle speed screw to ensure proper idle speed.

8. With the engine now idling at the correct speed, point the light at the timing marks. The strobe will make it appear that the mark on the tab and the mark on the damper/pulley stand still in alignment.

9. If the timing requires adjustment, loosen the clamp bolt at the base of the distributor and then carefully rotate the distributor until the correct marks line up.

10. Tighten the clamp bolt to 18 ft. lbs. (25 Nm).

11. Adjust the idle mixture screw - inward is Lean and outward is Rich.

12. Recheck the timing.

13. Shut down the engine and disconnect the light and the jumper wire from the timing lead and ground.

14. Reinstall and adjust the throttle cable. Open and close the remote control lever, ensuring that the carburetor throttle lever in coming in contact with the idle speed screw each time.

15. Restart the engine, increase the engine speed to 1300 rpm and then allow it to roll back to idle speed. Shut off the engine. Make sure the throttle lever is touching the idle screw

16. Disconnect the tachometer.

4.3L V6 & 5.0L/5.7L/6.2L V8 Engines w/TBI and MPI

■ Mercury is a little confusing here. Although it appears the following procedure is valid for TBI models, it's questionable whether it is appropriate for early MPI models; and certainly IS NOT valid for 2002 and later MPI models which use the ECM to control all ignition timing adjustments.

Although ignition timing is adjustable on these models, it is generally controlled by the EFI electronic control module. In order to adjust the timing, the ECM must be forced to enter into its service mode by using a scan tool. This done, the ECM will stabilize the base timing to allow for adjustment by conventional means of rotating the distributor.

■ The idle speed and dwell must be correctly adjusted and within specifications before performing this procedure.

1. Open or remove the engine compartment hatch.

2. Connect a suitable timing light to the No. 1 spark plug lead (see firing order illustrations for location of the No. 1 cylinder). Connect the power supply lead to the battery as detailed in the light manufacturer's instructions.

3. Connect a tachometer to the engine as detailed by the manufacturer. Do not use the tachometer on the instrument panel, as it will not provide the necessary accuracy.

4. Locate the timing marks on the engine timing cover (just above the crankshaft pulley/damper) and place a bit of white paint where the proper mark should be as detailed in the Tune-Up Specifications chart. Timing marks are generally shown in 2° increments from TDC.

5. Start the engine and allow it to reach normal operating temperature at idle.

6. Turn off the engine, locate the data link connector (DLC) on the EFI/MPI main harness and plug in a scan or timing tool - please refer to the Fuel Injection section for further details on this.

7. Restart the engine and allow the idle to stabilize. On TBI or MPI/MEFI-1 engines, adjust the throttle lever so that the idle moves to about 1200 rpm. On MPI/MEFI-2 and -3 engines the computer will automatically adjust the idle to 1200 rpm when the scan tool is set to the service mode.

8. While still idling, point the light at the timing marks. The strobe will make it appear that the mark on the tab and the mark on the damper/pulley stand still in alignment.

9. If the timing requires adjustment, loosen the clamp bolt at the base of the distributor and then carefully rotate the distributor until the correct marks line up.

10. Tighten the clamp bolt to 18 ft. lbs. (25 Nm) and then recheck the timing.

11. Disconnect the scan/timing tool. On engines where the throttle lever was adjusted up to 1200 rpm, adjust it back to the normal idle position.

12. Run the engine up to 1300 rpm and then slowly allow it to settle back down to idle. Recheck the idle and adjust the throttle cable if necessary.

13. Check the timing one last time. If still correct, disconnect the light and tachometer.

8.1L V8 Engines

Ignition timing on these models is controlled by the ECM and not adjustable.

Idle Speed & Mixture

ADJUSTMENT

Preliminary Idle - 2 bbl Carburetors

◆ See Figures 112, 113 and 114

■ The following adjustments will provide sufficient idle speed and mixture settings to get your engine started. Final adjustments must always be made with the engine running.

1. Locate the idle speed screw on your carburetor and turn it in or out until it is just resting on the idle cam, but is not moving it. The idle screw is threaded into the throttle linkage on the side of the carburetor and has a spring between the screw head and the linkage.

2. Turn the screw inward (clockwise) two additional turns.

3. Locate the idle mixture screw on the base of the carburetor and turn it in (clockwise) until it just lightly seats itself and then back it out the appropriate number of turns (see Tune-Up Specifications chart).

■ ON 2004 and later models with a TKS carburetor, turning the idle mixture screw will require a special adjusting tool (#91-866201)

✱✱ CAUTION

Be careful not to turn the idle mixture screw in past the seated position or you risk damaging the seat or needle.

■ On models equipped with emissions carburetors, the idle mixture screw is sealed. Fuel mixture is not adjustable on these models.

2-28 ENGINE AND DRIVE MAINTENANCE

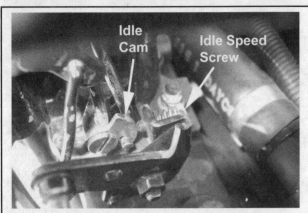

Fig. 112 The idle speed adjusting screw should rest on the idle cam - 3.0L engine

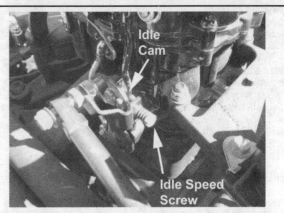

Fig. 113 The idle speed adjusting screw should rest on the idle cam - 5.0L/5.7L engines

Fig. 114 Never tighten the idle mixture screw against its seat - 3.0L shown

Preliminary Idle - 4 bbl Carburetors

◆ See Figures 115 and 116

■ The following adjustments will provide sufficient idle speed and mixture settings to get your engine started. Final adjustments must always be made with the engine running.

1. Locate the idle speed screw on your carburetor and turn it in until it is just resting on the throttle lever cam, but is not moving it. The idle screw is threaded into a block on the side of the carburetor and has a spring between the screw head and the linkage.

2. Locate the idle mixture screws on the base of the carburetor and turn them in (clockwise) until they just lightly seat themselves and then back each one out the appropriate number of turns (see the Tune-Up Specifications chart).

✳✳ CAUTION

Be careful not to turn the idle mixture screw in past the seated position or you risk damaging the seat or needle.

■ On models equipped with emissions carburetors, the idle mixture screw is sealed. Fuel mixture is not adjustable on these models.

Final Idle - 2 bbl Carburetors

◆ See Figures 112, 113 and 114

1. With the boat in the water and the engine running, connect a tachometer as per the manufacturer's instructions.

2. With the drive in Neutral, run the engine at approximately 1500 rpm until it reaches normal operating temperature.

3. Shift the drive into Forward gear, at idle. Make sure there's someone at the helm and you're in open water when doing this!

4. Being careful not to lose the spacer or anchor stud, disconnect the throttle cable barrel at the anchor stud.

5. Check the idle speed on the tachometer. If not within specifications (see Tune-Up Specifications chart) adjust the idle speed screw until the proper rpm is achieved.

6. When the engine is idling at the proper speed, turn the idle mixture screw, on non-emissions models, clockwise until the engine speed begins to drop due to a lean mixture. Remember the setting.

■ ON 2004 and later models with a TKS carburetor, turning the idle mixture screw will require a special adjusting tool (#91-866201)

7. Now turn the idle mixture screw counterclockwise until the idle speed begins to drop due to a rich mixture. Remember the setting.

8. Turn the mixture screw to a point midway between the settings from the two previous steps until the engine runs smoothly. Check the idle speed once again and readjust if not within specifications.

9. Turn off the engine, remove the tachometer and reconnect the throttle cable at the anchor stud.

Final Idle - 4 bbl Carburetors

◆ See Figures 115 and 116

1. Connect a tachometer as per the manufacturer's instructions.

2. Locate the ignition timing lead near the distributor and connect a jumper wire between it and a good ground (-). This lead should be a purple and white wire; grounding it will force the ignition module into base timing mode.

3. Start the engine. With the drive in Neutral, run the engine at approximately 1500 rpm until it reaches normal operating temperature.

4. Disconnect the throttle cable from the carburetor.

5. Shift the drive into Forward gear, at idle. Make sure you're in open water when doing this!

6. Check the idle speed on the tachometer. If not within specifications (see Tune-Up Specifications chart) adjust the idle speed screw until the proper rpm is achieved.

7. When the engine is idling at the proper speed, check that the idle mixture screws are backed out the appropriate number of turns on non-emissions models.

8. Install and adjust the throttle cable.

9. Turn off the engine, remove the tachometer and disconnect the jumper wire from the timing lead and ground.

ENGINE AND DRIVE MAINTENANCE   2-29

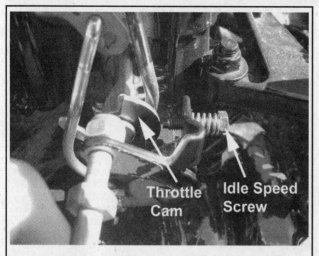

Fig. 115 The idle speed adjusting screw should rest on the throttle lever - 4.3L engine

Fig. 116 Never tighten the idle mixture screws against their seats - 4.3L engine

4.3L Engines - Sav1 Emission Carburetor

Idle speed and mixture are set at the factory on these models. Mixture screws are sealed. Idle is adjustable only via the propane-enrichment method - we recommend taking the boat to an authorized service facility with the proper tools in order to complete this procedure.

V6 & V8 Engines With TBI/MPI

Idle speed is constantly monitored by the electronic control module (ECM) and controlled by the idle air control valve (IAC). Idle speed and mixture are not adjustable. Please refer to the Fuel System section for further information on the fuel injection system.

PCV Valve

◆ See Figure 117

With the exception of the 8.1L, most V6 and V8 engines are equipped with a positive crankcase ventilation (PCV) circuit that utilizes a PCV valve in the Port rocker cover in order to ventilate unburned crankcase gases back into the engine via the intake manifold in order that they can be re-burned. The PCV valve should be replaced every boating season or 100 hours of operation (whichever comes first) on carbureted and TBI engine; 300 hours or every 3rd year on MPI engines.

A PCV system that is malfunctioning can cause rough running or idle, and also increased fuel consumption. Do not attempt to disconnect or bypass the system.

REMOVAL & INSTALLATION

◆ See Figure 118

1. Locate the PCV valve in the Port cylinder head cover.
2. Carefully wiggle it back and forth while pulling upward on the valve itself until it pops out of the cover.
3. Loosen the clamp (if equipped) and disconnect the breather hose from the valve.
4. Reconnect the hose and press the valve back into the cylinder head.

INSPECTION

◆ See Figure 119

Start the engine and allow it to reach normal operating temperature. Pop out the PCV valve as detailed above and cover the opening with your thumb. You should be able to feel significant vacuum; if not replace the valve. With the valve still in your fingers, shake it back and forth a few times; you should be able to hear the inside components moving around.

Valve Adjustment

All engines covered in this manual are equipped with hydraulic valve lifters and do not require periodic valve adjustment. Adjustment to zero lash is maintained automatically by the hydraulic pressure in the lifters. For initial adjustment procedures after cylinder head work, please refer to the appropriate Engine Mechanical section.

Fig. 117 The PCV valve is located in the port cylinder head cover

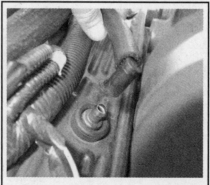

Fig. 118 Pop the PCV valve out of the cylinder head cover

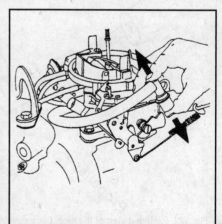

Fig. 119 Testing the PCV valve

2-30 ENGINE AND DRIVE MAINTENANCE

Propeller

REMOVAL & INSTALLATION

Alpha & Bravo I And II Drives

◆ See Figures 120 thru 130

CAUTION

If the drive unit or the lower unit is not disconnected or removed from the boat it is imperative that you disconnect the battery cables and remove the ignition key prior to performing this procedure.

1. Bend the locking tabs on the tab washer out and away from the grooves in the splined washer. Never pry on the end of the propeller hub because any small distortion will affect propeller performance or possibly break the hub off the propeller.
2. Place a block of wood between one blade of the propeller and the anti-cavitation plate to prevent the shaft from rotating. Place the correct size socket and breaker bar over the propeller nut and remove the nut from the propeller shaft.
3. On Alpha (w/o Flo-Torque) and Bravo II drives, slide the tab washer, splined washer, continuity washer (Alpha only), propeller, and thrust hub (washer) off of the propeller shaft.
4. On Alpha (w/Flo-torque) and Bravo I drives, slide the tab washer, drive sleeve adapter, propeller, drive hub and thrust hub (washer) off of the propeller shaft.
5. If the propeller is frozen to the shaft, heat must be applied to the shaft so as to melt out the rubber inside the hub. Using heat will destroy the hub, but there is no other way. As heat is applied, the rubber will expand and the propeller will actually be blown from the shaft. Therefore, stand clear to avoid personal injury.
6. Use a knife and cut the hub off the inner sleeve.
7. The sleeve can be removed by cutting it off with a hacksaw or cold chisel, or it can be removed with a puller. Again, if the sleeve is frozen, it may be necessary to apply heat.
8. Remove the thrust hub from the propeller or shaft.

To install:

9. Check to be sure the splines on the propeller shaft are in good condition. If there is any visible sign of corrosion on the shaft, the corrosion must be removed before assembly is completed. Verify the threaded end of the propeller shaft is clean and the threads are in good condition.
10. Verify the splined steel hub inside the propeller is free of corrosion. Check to be sure the rubber hub is tight and secured to the propeller inner hub. Verify there is no sign of the rubber hub spinning on the inside of the propeller.
11. Slide the propeller thrust washer (hub) onto the shaft with the smaller side of the washer facing out - aft. Coat the propeller shaft and splines with one of the following lubricants in the order of preference; Quicksilver Special Lubricant 101, 2-4-C Lubricant or Perfect Seal.
12. If it came off separate from the propeller, slide the drive hub into the propeller on Alpha (w/Flo-torque) and Bravo I drives.
13. Slide the propeller onto the shaft followed by the continuity washer (Alpha w/o Flo-Torque), splined washer or drive sleeve adapter, tab washer and then the propeller nut, as indicated in the accompanying illustrations. These components must be assembled in the correct order.

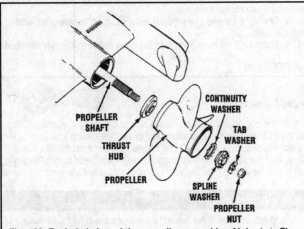

Fig. 120 Exploded view of the propeller assembly - Alpha (w/o Flo-Torque drive hub)

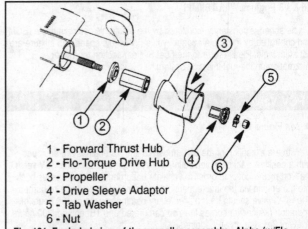

1 - Forward Thrust Hub
2 - Flo-Torque Drive Hub
3 - Propeller
4 - Drive Sleeve Adaptor
5 - Tab Washer
6 - Nut

Fig. 121 Exploded view of the propeller assembly - Alpha (w/Flo-Torque drive hub) and Bravo I and X

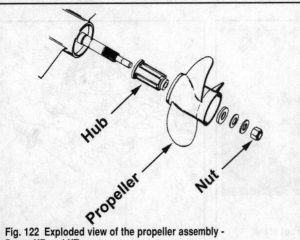

Fig. 122 Exploded view of the propeller assembly - Bravo XR and XZ

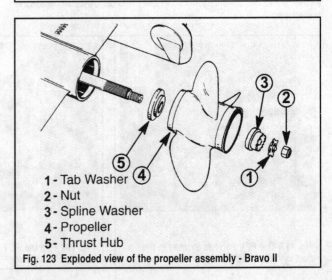

1 - Tab Washer
2 - Nut
3 - Spline Washer
4 - Propeller
5 - Thrust Hub

Fig. 123 Exploded view of the propeller assembly - Bravo II

ENGINE AND DRIVE MAINTENANCE

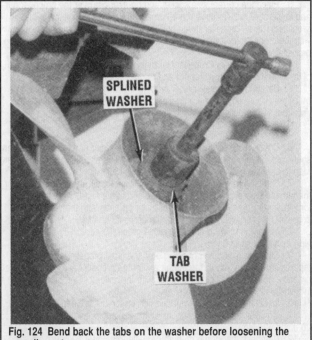

Fig. 124 Bend back the tabs on the washer before loosening the propeller nut

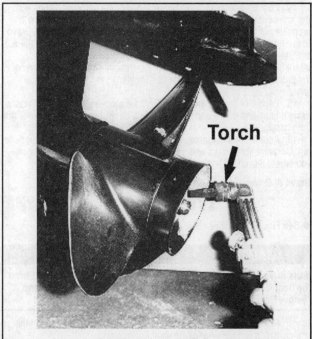

Fig. 125 Heat the propeller shaft to loosen the frozen hub

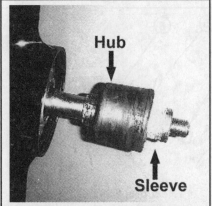

Fig. 126 The rubber hub rides on the sleeve, inside the propeller flange

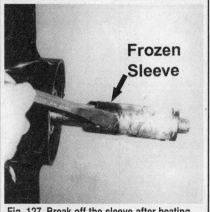

Fig. 127 Break off the sleeve after heating the shaft again...

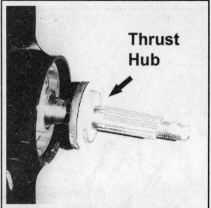

Fig. 128 ...and then remove the thrust washer

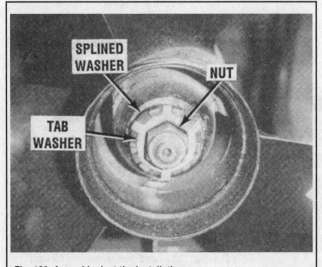

Fig. 129 A good look at the installation

Fig. 130 Don't forget to bend those tabs over

2-32 ENGINE AND DRIVE MAINTENANCE

14. Position a block of wood between one of the propeller blades and the anti-cavitation plate to keep the propeller from rotating. Tighten the propeller nut to 55 ft. lbs. (75 Nm) on the Alpha, Bravo I and 2004-08 Bravo II; or 60 ft. lbs. (82 Nm) on the 2001-03 Bravo II. Continue tightening the nut until three tabs on the tab washer align with the grooves in the splined washer. There should be a minimum of two threads showing on the end of the propeller shaft after the nut has been tightened to the proper torque value.

15. Bend 3 of the tab washer tabs into the grooves of the splined washer using a blunt-end punch and hammer. The tabs will prevent the propeller nut from backing off.

16. After the first operation, bend out the tabs and re-torque the propeller nut. Continue tightening it until you are able to bend in 3 tabs. Make sure you repeat this check again after 20 hrs. of operation.

Bravo III Drives

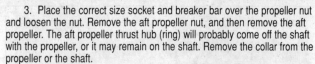

◆ See Figures 131 thru 135

⁂ CAUTION

If the drive unit or the lower unit is not disconnected or removed from the boat it is imperative that you disconnect the battery cables and remove the ignition key prior to performing this procedure.

1. Place a block of wood between one blade of the propeller and the anti-cavitation plate to prevent the shaft from rotating.
2. On 2004 and later drives, remove the bolt and washers securing the propeller nut anode. Pull off the anode.
3. Place the correct size socket and breaker bar over the propeller nut and loosen the nut. Remove the aft propeller nut, and then remove the aft propeller. The aft propeller thrust hub (ring) will probably come off the shaft with the propeller, or it may remain on the shaft. Remove the collar from the propeller or the shaft.
4. Obtain an Inner Propeller Nut tool (#91-805457). Move the tool into place over the inner propeller nut and remove the nut.
5. Carefully remove the inner propeller. The inner propeller thrust hub (ring) may come off with the propeller or remain on the shaft. Remove the ring from the propeller or from the shaft.

To install:

6. Check to be sure the splines and the threaded end of both propeller shafts are clean and in good condition. Apply a liberal coating of Quicksilver Marine Lubricant 2-4-C to both the inner and outer propeller shafts.
7. Slide the forward thrust hub/ring onto the outer propeller shaft with the tapered side of the hub facing the aft end of the shaft. Slide the forward/inner propeller onto the outer propeller shaft with the splines in the propeller indexed with the splines of the propeller shaft.
8. Install the forward propeller nut. Tighten the nut to 100 ft. lbs. (136 Nm) using the special tool.
9. Slide the rear thrust hub/ring onto the inner propeller shaft with the tapered side of the hub facing the aft end of the propeller shaft. Install the aft propeller onto the inner propeller shaft with the splines in the propeller indexed with the splines on the shaft.
10. Install the aft propeller nut and tighten the nut to 60 ft. lbs. (81 Nm).
11. On 2004 and later drives, install the nut anode and tighten the bolt to 20 ft. lbs. (27 Nm).
12. Re-check the torque on both propeller nuts after 20 hours of operation.

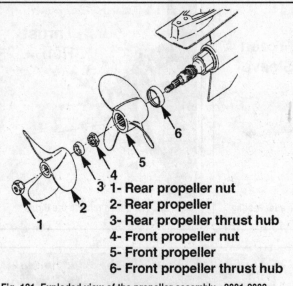

1- Rear propeller nut
2- Rear propeller
3- Rear propeller thrust hub
4- Front propeller nut
5- Front propeller
6- Front propeller thrust hub

Fig. 131 Exploded view of the propeller assembly - 2001-2003 models

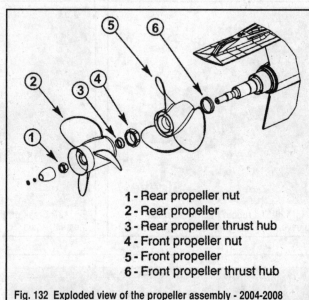

1- Rear propeller nut
2- Rear propeller
3- Rear propeller thrust hub
4- Front propeller nut
5- Front propeller
6- Front propeller thrust hub

Fig. 132 Exploded view of the propeller assembly - 2004-2008 models

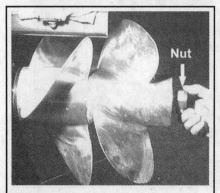

Fig. 133 Removing the rear propeller

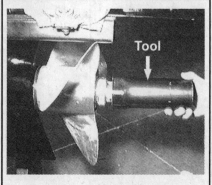

Fig. 134 Install the special tool. . .

Fig. 135 . . .and then remove the forward propeller

ENGINE AND DRIVE MAINTENANCE

FLUIDS & LUBRICANTS

Fluid Disposal

Used fluids such as engine oil, gear oil, antifreeze and power steering fluid are hazardous waste and must be disposed of properly and responsibly. Before draining any fluids it is always a good idea to check with your local authorities; in many areas there are recycling programs available for easy disposal. Service stations, Parts stores and Marinas also often will accept waste fluids for recycling.

Be sure of your local recyclers policies before draining any fluids, as many will not accept fluids that have been mixed together.

Fuel & Oil Recommendations

FUEL

All engine covered in this manual are designed to run on unleaded fuel. Never use leaded fuel in your boat's engine. The minimum octane rating of fuel being used for your engine must be at least 87, which means regular unleaded, but some engines may require higher octane ratings. Fuel should be selected for the brand and octane that performs best with your engine.

The use of a fuel too low in octane (a measure of anti-knock quality) will result in spark knock. Newer systems have the capability to adjust the engine's ignition timing to compensate to some extent, but if persistent knocking occurs, it may be necessary to switch to a higher grade of fuel. Continuous or heavy knocking may result in engine damage.

ENGINE OIL

◆ See Figure 136

Nothing affects the performance and durability of an engine more than the engine oil. If inferior oil is used, or if your engine oil is not changed regularly, the risk of piston seizure, piston ring sticking, accelerated wear of the cylinder walls or liners, bearings and other moving components increases significantly.

Maintaining the correct engine oil level is one of the most basic (and essential) forms of engine maintenance. Get into the habit of checking your oil on a regular basis; all engines naturally consume small amounts of oil, and if left neglected, can consume enough oil to damage the internal components of the engine. Assuming the oil level is correct because you "checked it the last time" can be a costly mistake.

If your engine has not been operated for more than 6 months it should be primed prior to starting.

When shutting the engine down, always let the engine idle a few minutes to bring engine temperature down to a normal level. Since the engine is, at least in part, cooled by engine oil, it is necessary to allow the engine oil temperature to stabilize prior to shutdown. Not allowing the temperature to stabilize can damage vital engine components.

Every container of engine oil for sale in the U.S. should have a label describing what standards it meets. Engine oil service classifications are designated by the American Petroleum Institute (API), based on the chemical composition of a given type of oil and testing of samples. The ratings include "S" (normal gasoline engine use) and "C" (commercial, fleet and diesel) applications. Over the years, the S rating has been supplemented with various letters, each one representing the latest and greatest rating available at the time of its introduction. During recent years these ratings have changed and most recently (at the time of publication), the rating is SH or CH-4. Each successive rating usually meets all of the standards of the previous alpha designation, but also meets some new criteria, meets higher standards and/or contains newer or different additives. Since oil is so important to the life of your engine, you should obviously NEVER use an oil of questionable quality. Oils that are labeled with modern API ratings, including the "energy conserving" donut symbol, have been proven to meet the API quality standards. Always use the highest grade of oil available. The better quality of the oil, the better it will lubricate the internals of your engine.

In addition to meeting the classification of the API, your oil should be of a viscosity suitable for the outside temperature in which your engine will be operating. Oil must be thin enough to get between the close-tolerance moving parts it must lubricate. Once there, it must be thick enough to separate them with a slippery oil film. If the oil is too thin, it won't separate the parts; if it's too thick, it can't squeeze between them in the first place - either way, excess friction and wear takes place. To complicate matters, cold-morning starts require a thin oil to reduce engine resistance, while high speed boating requires a thick oil which can lubricate vital engine parts at temperatures.

According to the Society of Automotive Engineers' (SAE) viscosity classification system, an oil with a high viscosity number (such as SAE 40 or SAE 50) will be thicker than one with a lower number (SAE 10W). The "W" in 10W indicates that the oil is desirable for use in winter operation, and does not stand for "weight". Through the use of special additives, multiple-viscosity oils are available to combine easy starting at cold temperatures with engine protection at high speeds. For example, a 10W40 oil is said to have the viscosity of a 10W oil when the engine is cold and that of a 40 oil when the engine is warm. The use of such an oil will decrease engine resistance and improve efficiency.

Mercury Marine has run through various recommendations with regard to multi-viscosity oils over the years so be certain to check you owner's manual before purchasing your oil.

All models here come with the recommendation to use Quicksilver Marine oil in a special 25W-40 viscosity, but continue to discourage anything but temporary use of other multi-weight oils (20W-40 or 20W-50 only). Mercruiser is a little confusing when it comes to using synthetic oils - they're pretty specific on engine into the early 2000's that they DO NOT recommend a synthetic oil, but then by the mid-2000's they are suggesting that you SHOULD use a synthetic. Our best recommendation is to check the Owner's Manual and/or the tag on the engine, if it's still there, and go with that.

Priming

■ Anytime your boat's engine has not been run for more than 6 months it is a good idea to prime the engine prior to starting it.

1. Check that the proper amount of oil is in the crankcase (see the Capacities chart).

2. Remove the spark plugs.

3. With the ignition key in the OFF position, connect a remote starter according to the manufacturer's instructions. If a remote starter is unavailable, disconnect the Purple wire from the ignition coil and then use the ignition key to turn the engine over. Make sure that you tape the terminal on the Purple wire to prevent it from touching ground.

4. Crank the engine for 15 seconds and then allow the starter to cool for 1 minute. Repeat this sequence 2 more times until a total cranking time of 45 seconds has been reached.

5. Remove the remote starter or reconnect the Purple wire to the coil.

6. Install the spark plugs and start the engine.

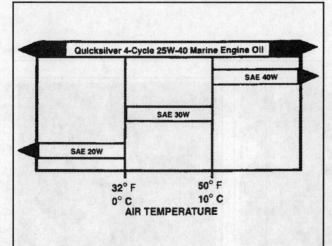

Fig. 136 The ambient temperature dictates the required viscosity of engine oil

2-34 ENGINE AND DRIVE MAINTENANCE

OIL LEVEL CHECK

◆ See Figures 137 thru 144

✳✳ CAUTION

The EPA warns that prolonged contact with used engine oil may cause a number of skin disorders, including cancer! You should make every effort to minimize your exposure to used engine oil. Protective gloves should be worn when changing the oil. Wash your hands and any other exposed skin areas as soon as possible after exposure to used engine oil. Soap and water, or waterless hand cleaner should be used.

When checking the oil level, it is best that the boat be level and the oil be at operating temperature. Checking the level immediately after stopping the engine will give a false reading; always wait about at least 5 minutes before checking.

It is normal for an engine to naturally consume oil during the course of operation, particularly during break-in on a new engine. You should not be alarmed if the oil level in your engine drops slightly between inspections. In fact certain of MerCruiser's high performance engines may use up to a quart of oil every 5 hours when operated at full throttle.

Also the color of the oil is usually a pitch black color. Smelling the oil is a better indicator of oil condition than the color. If the oil smells burned, it should be replaced immediately.

Over-filled crankcases can cause a fluctuation or drop in oil pressure, and, particularly on MerCruiser engines, clattering from the rocker arms. Take great care in checking and filling your engine with oil. Always maintain it between the **ADD** and **FULL**, **OK RANGE** or **OP RANGE** markings on the dipstick. Beware of false readings by checking the level too soon after adding oil.

■ It takes a little while for fresh oil poured into the engine to reach the crankcase. Wait for about 3 minutes and then check the oil level again.

Oil level should be checked each day the engine is operated and it is best to check it while in the water.

Fig. 137 The dipstick on the 3.0L engine is on the Starboard side

Fig. 138 The dipstick on V6 and 5.0L/5.7L/6.2L V8 engines is on the Starboard side, at the front. . .

Fig. 139 . . .or on the Port side, at the front

Fig. 140 There may be 3 dipsticks on the 8.1L V8 engine, 2 on the rear of the engine (each side) and this one at the front

Fig. 141 A good shot of the dipstick

Fig. 142 The oil filler cap can look like this. . .

Fig. 143 . . .or this. . .

Fig. 144 . . .or even this

ENGINE AND DRIVE MAINTENANCE   2-35

1. Run the engine until it reaches normal operating temperature. Shut it off and allow the oil to settle for at least 5 minutes.

2. Locate the engine oil dipstick. The 8.1L engine may have three, one on each side of the rear of the engine and one at the front, but on most later engines, the front one is for stern drive engines and the back two are for inboards.

3. Clean the area around the dipstick to prevent dirt from entering the engine.

4. Remove the dipstick and note the color of the oil. Wipe the dipstick clean with a rag.

5. Insert the dipstick fully into the tube and remove it again. Hold the dipstick horizontal and read the level on the dipstick. The level should always be at the upper limit. If the oil level is below the upper limit, sufficient oil should be added to restore the proper level of oil in the crankcase. Most dipsticks are marked with **ADD** and **FULL**, **OK RANGE** or **OP RANGE** gradations.

6. See the recommendations under Engine Oil for the proper viscosity and type of oil, but for the most part, its 25W-40.

7. Oil is added through the filler port cap in the top of the cylinder head cover (except on the 8.1L, where the fill tube is at the front of the engine). Add oil slowly and check the level frequently to prevent overfilling the engine.

✱✱ WARNING

Do not overfill the engine. If the engine is overfilled, the crankshaft will whip the engine oil into a foam causing loss of lubrication and severe engine damage.

OIL & FILTER CHANGE

◆ See Figures 145 thru 154

A few precautions can make the messy job of oil and filter maintenance much easier. By placing oil absorbent pads, available at industrial supply stores, into the area below the engine, you can prevent oil spillage from reaching the bilge.

It is a good idea to warm the engine oil first so it will flow better; this will also ensure that many of the contaminates in the bottom of the pan will be suspended in the oil. This is accomplished by starting the engine and allowing it to reach normal operating temperature.

Changing engine oil is sometimes complicated by the location of the drain plug. Although most of the engines covered here should be equipped with MerCruiser's Quick Drain system, some may not be. Most boats equipped with inboard engines use an evacuation pump to remove the used engine oil through the dipstick tube. If you don't have a permanently mounted oil suction pump in your engine compartment, you may want to consider installing one. This pump sucks waste oil out through either the dipstick tube or a connection on the oil drain plug and is available from your Mercruiser dealer or from any number of aftermarket sources. They come in a variety of configurations: motorized, hand-pumped or attached to an ordinary household drill.

The maintenance interval for oil and filter change is every 100 hours of engine operation or annually, whichever comes first. This interval should be strictly kept; in fact we recommend cutting the interval in half!

Fig. 145 The 3.0L filter is on the Starboard side, toward the front

Fig. 146 Older V6 and V8 engines will have the filter on the Port side, toward the rear

Fig. 147 But most newer V6 and V8 engines will have a remote filter on the Port side

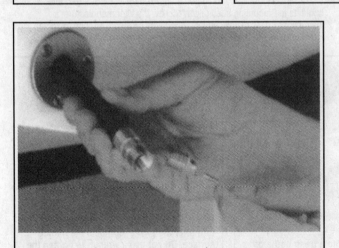

Fig. 148 Most models should be equipped with a Quick Drain

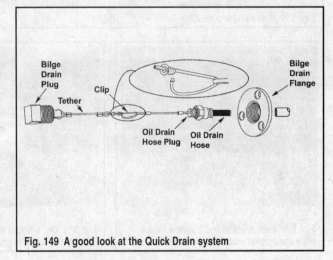

Fig. 149 A good look at the Quick Drain system

2-36 ENGINE AND DRIVE MAINTENANCE

■ Since you will be hanging into the engine compartment, gather all the tools and spare parts necessary for the job. Don't forget plenty of rags to clean up any spills, and most importantly, remember a container or plastic milk jug to drain the oil into.

1. Start the engine and run it until it reaches normal operating temperature. Turn the engine off and remove the dipstick.

✱✱ CAUTION

Use caution around the hot oil; when at operating temperature, it is hot enough to cause a severe burn.

Models with Quick Drain

2. If equipped with the Quick Drain oil change system, remove the bilge plug and pull the tether through the hole until the drain hose comes through the hull. Position a container under the hose and then remove the drain hose plug.

3. Drain the oil completely and then replace the drain line plug. Feed the hose and tether back through the hull and install the plug. Skip to Step 9.

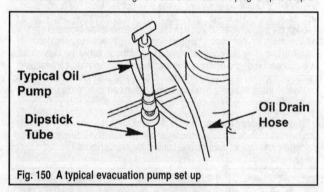

Fig. 150 A typical evacuation pump set up

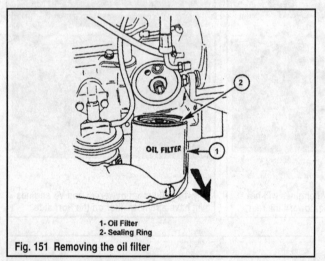

Fig. 151 Removing the oil filter
1- Oil Filter
2- Sealing Ring

Models without Quick Drain

4. If your vessel is not equipped with the drain system, connect the evacuation pump hose to the dipstick tube or insert it into the tube.

5. Position the other hose, or the outlet at the bottom of the pump in your container. Keep in mind that the fast flowing oil, which will spill out of the pump hose will flow with enough force that it could miss the container and end up all over the deck or in the bilge. Position the container accordingly and be ready to move it if necessary.

6. Allow the oil to drain until nothing but a few drops come out of the pump. It should be noted that depending on the angle of the engine, some oil may be left in the crankcase. This is normal and should not cause the engine harm.

7. Remove evacuation pump and reinsert the dipstick.

All models

8. Position a drain pan, or a cut-down milk jug under the oil filter. Some filters are mounted horizontally and some vertically. In either case, there is usually oil left in the filter. When the filter is removed, oil will flow out of the engine and the filter. If you are not prepared, you will have a mess on your hands. It may also be necessary to use a funnel or fashion some type of drain shield to guide the oil into the drain pan. This can be a simple as a recycled oil bottle with the bottom cut off or an elaborate creation made of tin. In any case, its purpose is to prevent oil from spilling all over the engine and bilge.

■ Another trick when removing the oil filter is to secure a large Ziplock bad around the filter and mounting base. Unscrew the filter through the bag and allow the bag to catch most of the oil.

9. To remove the filter, you will probably need an oil filter strap wrench (available at almost any place that sells marine or automotive parts and fluids). Heat from the engine tends to tighten even a properly installed filter and makes it difficult to remove. Place the wrench on the filter as close to the engine as possible, while still leaving room to work. This will put the wrench at the strongest part of the filter (near the threaded end) and prevent crushing the filter. Loosen the filter with the wrench using a counterclockwise turning motion.

■ Most models will utilize a remote oil filter, but it still comes off in the same manner.

10. Once loosened, wrap a rag around the filter and unscrew it from the boss on the engine. Make sure that the drain pan and shield are positioned properly before you start unscrewing the filter. Should some of the hot oil happen to get on your hands and burn you, dump the filter into the drain pan.

11. Wipe the base of the mounting boss with a clean, dry lint-free rag. If the filter is installed vertically, you may want to fill the filter about half way with oil prior to installation. This will prevent oil starvation when you fire the engine up again. Pre-filling the oil filter is usually not possible with horizontally mounted filters, nor with remote filters that are mounted upside down. Smear a little bit of fresh oil on the filter gasket to help it seat properly on the engine.

12. Install the filter and tighten it approximately 1/4 of a turn after it contacts the mounting boss (always follow the filter manufacturer's instructions). This usually equals "hand tight." Using a wrench to tighten the filter is not required and should not be considered.

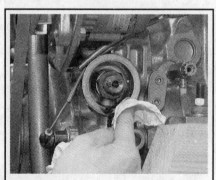

Fig. 152 Before installing the new filter, wipe the gasket mating surface clean

Fig. 153 Dip a finger into a fresh bottle of oil, and lubricate the gasket of the new filter

Fig. 154 Tighten the new filter 1/2 to 3/4 of a turn after the gasket contacts the mating surface. Do not use a wrench to tighten the filter!

ENGINE AND DRIVE MAINTENANCE

⚠ WARNING

Never operate the engine without engine oil. Severe and costly engine damage will result in a matter of seconds without proper lubrication.

13. If any oil has gotten into the bilge, remove it using an oil absorbent pad. These pads are specially formulated to only absorb oil and will not soak up any water in the bilge. Perform a visual inspection and make sure all connections are tight.

14. Carefully remove the drain pan from under the oil filter and transfer the oil into a suitable container for recycling.

15. Refill the engine with the proper quantity and quality of oil **immediately** (see the Capacities chart). You may laugh at the severity of this warning, but if you wait to refill the engine and someone unknowingly tries to start it, severe and costly engine damage will result.

16. Refill the engine crankcase slowly through the filler cap on the valve cover. Use a funnel as necessary to prevent spilling oil. Check the level often. You may notice that it usually takes less than the amount of oil listed in the Capacities chart to refill the crankcase. This is only until the engine is started and the oil filter is filled with oil.

17. To make sure the proper level is obtained, run the engine to normal operating temperature, turn the engine OFF, allow the oil to drain back into the oil pan and recheck the level after about 5 minutes. Top off the oil to the correct mark on the dipstick.

Power Steering

FLUID LEVEL

◆ See Figures 155 thru 159

■ Power steering fluid level should always be checked with the engine hot if at all possible.

■ Models using the Compact Hydraulic Steering system are detailed in the Steering section.

1. Start the engine and run it until it reaches normal operating temperature.
2. Turn the engine OFF and position the drive unit so that it is straight back (centered).
3. Locate the dipstick cap on top of the power steering fluid reservoir. Unscrew the cap and remove the dipstick. Wipe down the dipstick with a clean rag and reinsert it into the reservoir making sure that the cap is screwed all the way down.
4. Remove the dipstick again and check that the fluid level is up to the **FULL HOT** mark (**FULL COLD** if the level is being checked cold). If the level is between the **FULL** mark and the indent on the stick, it's OK. Below the indent or hash mark and you must add fluid. Use either Quicksilver Power Steering Fluid or Dexron(r) II ATF fluid. Do Not Overfill!
5. It is always a good idea to bleed the system after adding fluid.

BLEEDING THE SYSTEM

All Models Except 8.1L V8

■ Models using the Compact Hydraulic Steering system are detailed in the Steering section.

■ It is important that all air be removed from the system after filling. If air is left in the system, the fluid in the pump may foam during operation causing discharge or spongy steering.

1. With the engine stopped and the drive unit positioned straight back (centered), check that the fluid level is at the **FULL COLD** mark on the dipstick.
2. Turn the steering wheel from lock to lock several times in succession (at least 5 cycles) and then check the fluid level again. Add fluid if necessary.
3. Install the cap/dipstick onto the reservoir, start the engine (install a flushing device if not in the water) and run it at idle on Alphas or fast idle on Bravos (1300 rpm) until it reaches normal operating temperature. At the same time you should be turning the wheel from lock to lock again.

Fig. 155 Locate the power steering pump and reservoir...

Fig. 156 ...and then remove the dipstick

Fig. 157 Certain 8.1L V8 engines utilize a remote reservoir that looks like this...

Fig. 158 ...or this

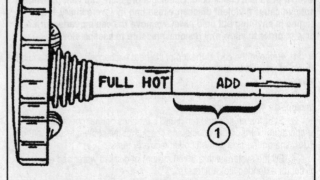

Fig. 159 Typical power steering fluid dipstick - models with the round reservoir will differ slightly

2-38 ENGINE AND DRIVE MAINTENANCE

4. Move the drive unit so that it is straight back and turn off the engine. Remove the reservoir cap allowing any foam in the pump to escape.

5. Reinsert the dipstick and check the fluid again, making sure that it is at the **FULL HOT** mark this time. Add fluid as necessary.

6. If the fluid is still foamy, repeat the procedure several times until the foam is gone and the level remains constant.

8.1L V8 Engines

■ It is important that all air be removed from the system after filling. If air is left in the system, the fluid in the pump may foam during operation causing discharge or spongy steering.

1. With the engine stopped and the drive unit positioned straight back (centered), check that the fluid level is at the **FULL HOT** mark on the dipstick.

2. Slowly turn the steering wheel from lock to lock several times in succession and then check the fluid level again. Add fluid if necessary.

3. Install the cap/dipstick onto the reservoir, start the engine and run it at a idle (650 rpm) until it reaches normal operating temperature. At the same time you should be turning the wheel from lock to lock again.

4. Turn off the engine and make sure the drive is centered. Remove the reservoir cap allowing any foam in the pump to escape.

5. Reinsert the dipstick and check the fluid again, making sure that it is at the **FULL HOT** mark this time. Add fluid as necessary.

6. If the fluid is still foamy, repeat the procedure several times until the foam is gone and the level remains constant.

Cooling System

All engines covered in this service utilize one of two cooling systems, or variations thereof. Simply put, there is a Seawater Cooling system and a Closed Cooling system.

Seawater systems are just that. . .they utilize the water that the boat is operating in to cool and lubricate the drive and engine. Water is drawn in through the stern drive unit and circulated to, and through, the engine and its components.

Closed cooling systems are actually 2 systems working together; a seawater system and a closed system consisting of antifreeze. The two systems work together in a variety of ways.

Complete information on individual system operation and component repair procedures is detailed in Cooling Systems section. In this section we will only deal with checking fluid levels and flushing/draining the systems.

LEVEL CHECK

Closed Cooling System Only

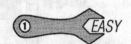

◆ See Figures 160 and 161

■ Always allow the engine to cool down before checking the coolant level. If possible, check cold. Opening a pressure cap with the engine hot can cause a violent discharge resulting in severe injury. If the engine is anything but cold always remove the cap a quarter turn at a time in order to allow any residual pressure to escape slowly.

1. Remove the cap at the top of the heat exchanger. Coolant should be visible at the bottom of the filler neck, or within 1 in. (25mm) of the bottom of the neck.

2. Reinstall the cap on the heat exchanger making sure that seats on its stops on the filler neck.

3. Start the engine and run it until it reaches normal operating temperature. Turn off the engine and check that the coolant level is to the **FULL** line on the side of the coolant recovery tank.

4. Fill the system with a 50/50 mixture of distilled water and ethylene glycol (or extended life) antifreeze.

■ Mercruiser does not recommend the use of propylene glycol antifreeze.

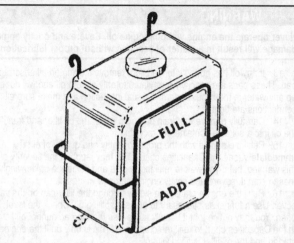

Fig. 160 Check that the coolant level is to the FULL line on the coolant recovery tank when the engine is hot. Most tanks look like this. . .

Fig. 161 . . .but the tank on the 8.1L V8 engines looks like this

PRESSURE CAP TESTING

Closed System Only

◆ See Figures 162, 163 and 164

The pressure cap on the coolant reservoir is designed to maintain closed system pressure at approximately 14 psi when the engine is at normal operating temperature. Clean, inspect and pressure test the cap at the end of your first season and then every 100 hours (or once a year) thereafter.

※※ **WARNING**

Make sure the engine is cool before attempting to remove the cap. To be safe, always turn the cap 1/4 turn and allow any residual pressure to escape before removing it completely. Use a heavy rag or wear gloves.

1. Remove the cap and wash it thoroughly to remove any debris from the sealing surfaces.

2. Inspect the gasket and rubber seal for tears, cuts or cracks. If you notice anything, we recommend replacing the entire cap although you can replace the gasket on some older engines if you like.

3. Make sure that the locking tabs on the cap are not bent or damaged in any way. If so, replace the cap.

4. Use a cooling system pressure tester and, following the manufacturer's instructions, install the cap. Check that the cap relieves

ENGINE AND DRIVE MAINTENANCE   2-39

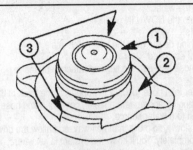

1- Rubber Seal
2- Gasket
3- Locking Tabs

Fig. 162 Check the pressure cap carefully...

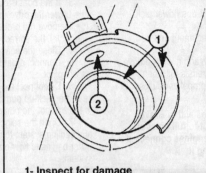

1- Inspect for damage
2- Clean coolant recovery pasages

Fig. 163 ...and then check the filler neck

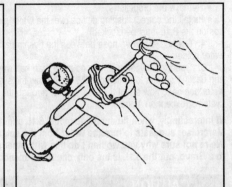

Fig. 164 Using a pressure tester to check the cap

pressure at 16 psi (15 psi on the 8.1L) and holds pressure for 30 seconds without falling below 11 psi. Replace the cap if it fails.

5. Check the inside of the filler neck for debris; it should be completely smooth. Check that the lock flanges on the filler neck are not bent or damaged and then install the cap.

PRESSURE TEST

Closed System Only

⁂ WARNING

Make sure the engine is cool before attempting to remove the cap. To be safe, always turn the cap 1/4 turn and allow any residual pressure to escape before removing it completely. Use a heavy rag or wear gloves.

1. Remove the pressure cap from the reservoir or heat exchanger.
2. Perform the pressure cap Testing procedure. If the cap is bad, this may be your problem. Replace the cap and ensure that the problem does not go away; if it persists, move to the next step.
3. Add coolant so that the level is within one inch of the bottom of the filler neck. Attach a cooling system pressure tester to the filler neck and pressurize the system to 20 psi (19 psi on the 8.1L).
4. Watch the gauge for about 2 minutes. If the pressure remains steady, you're OK. If the pressure drops, move to the next step.
5. Maintain the specified pressure and check the entire closed system for any leaks - hoses, plugs, petcocks, pump seals, etc.). Also, listen very closely for any hissing or bubbling.
6. If you've still not found any leaks, test the heat exchanger as detailed in the Cooling section.
7. If the exchanger is OK then you most likely have an problem with loose head bolts, a bad head gasket or a warped cylinder head, not the cooling system.

FLUSHING THE SEAWATER SYSTEM - STERN DRIVES

■ If your boat is operated in saltwater, heavy mineral fresh water or severely polluted water, flush the cooling system regularly - after each usage if at all possible! Always flush the system before draining and/or winter storage.

Your boat can be equipped with a number of different types of water pick-ups...through the hull, through the transom or through the stern drive unit. Flushing procedures for these can be different.

Stern Drive Water Pick-Ups

◆ See Figures 165 and 166

■ Mercruiser stern drive units can be equipped with any of 3 types of water pick-ups - dual water pick-ups, low water pick-up or side water pick-up. Each will require a slightly different type of flushing device as shown in the illustration.

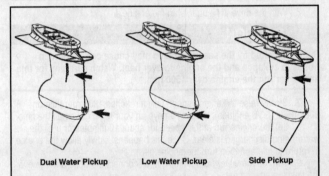

Dual Water Pickup   Low Water Pickup   Side Pickup

Fig. 165 Mercruiser drive units require flushing devices matched to the type of pick-ups they use

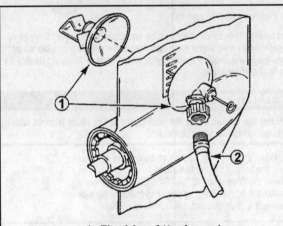

1- Flushing Attachment
2- Water Hose

Fig. 166 Attach a flushing device to the drive unit when flushing the cooling system

⁂ CAUTION

NEVER run the engine when the boat is out of the water without water being supplied to the stern drive unit.

1. Drain the seawater section of the cooling system.
2. If the boat is still in the water:
• Raise the drive unit into the TRAILER position
• Install the correct flushing device over the water pick-ups (lower pick-ups on the 8.1L must be blocked)
• Lower the drive unit into the DOWN/IN position
3. If the boat is out of the water:
• Lower the drive unit into the DOWN/IN position

2-40 ENGINE AND DRIVE MAINTENANCE

- Remove the propeller(s)
- Install the correct flushing device over the water pick-ups (lower pick-ups on the 8.1L must be blocked)

4. Connect a garden hose between the flushing attachment and a water spigot.

5. Open the spigot(s) slowly, no more than half way, and allow the drive and cooling system to fill completely. You'll know the system is full when water begins to flow out of the drive unit, or through the propeller on closed systems (except on Alphas).

■ Interestingly, when flushing late model 8.1L engines with a Bravo, Mercruiser suggests to turn the water from the spigot to full pressure. We're not sure why you wouldn't do this with other engines connected to a Bravo, but the 8.1L is the only one recommending it.

✱✱ CAUTION

Never allow the water to reach the flushing device at anything more than low pressure.

6. With the drive in Neutral, start the engine.

✱✱ CAUTION

Suction created by the seawater pump may cause certain garden hoses to collapse and the engine to over heat. Watch carefully for this and never run the engine over 1500 rpm.

7. Slowly raise the engine speed until it reaches 1300 rpm (idle on 2001-05 8.1L V8 engines), keeping an eye on your temp gauge all the time.

8. Let the engine run at this speed for about 10 minutes or until the water being discharged is clear. Once this happens, slowly allow the engine to return to idle and then turn the engine off.

9. Turn off the water from the spigot, disconnect the hose and remove the flushing attachment.

Alternate Water Pick-Ups

◆ See Figures 165, 166 and 167

■ Mercruiser stern drive units can be equipped with any of 3 types of water pick-ups - dual water pick-ups, low water pick-up or side water pick-up. Each will require a slightly different type of flushing device as shown in the illustration.

✱✱ CAUTION

NEVER run the engine when the boat is out of the water without water being supplied to the stern drive unit.

1. Drain the seawater section of the cooling system.
2. If the boat is still in the water:
- Raise the drive unit into the TRAILER position
- Install the correct flushing device over the water pick-ups
- Lower the drive unit into the DOWN/IN position

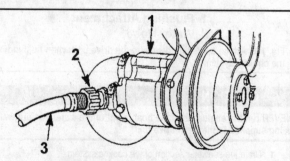

1- Seawater Pickup Pump
2- Adaptor
3- Tap Water Hose

Fig. 167 Disconnect the inlet hose, install an adapter and attach a hose to the inlet neck on models with a seawater pick-up pump

3. If the boat is out of the water:
- Lower the drive unit into the DOWN/IN position
- Remove the propeller(s)
- Install the correct flushing device over the water pick-ups

4. Close the seacock if equipped.

5. Connect a garden hose between the flushing attachment and a water spigot.

6. Disconnect the water inlet hose (usually the upper) at the back of the seawater pick-up pump. Using an appropriate adapter, connect a 2nd hose to the back of the pump and to a water source.

7. Open the spigot(s) slowly, no more than half way, and allow the drive and cooling system to fill completely. You'll know the system is full when water begins to flow out of the drive unit, or through the propeller on closed systems.

■ Interestingly, when flushing late model 8.1L engines with a Bravo, Mercruiser suggests to turn the water from the spigot to full pressure. We're not sure why you wouldn't do this with other engines connected to a Bravo, but the 8.1L is the only one recommending it.

✱✱ CAUTION

Never allow the water to reach the flushing device at anything more than low pressure.

8. With the drive in Neutral, start the engine.

✱✱ CAUTION

Suction created by the seawater pump may cause certain garden hoses to collapse and the engine to over heat. Watch carefully for this and never run the engine over 1500 rpm.

9. Slowly raise the engine speed until it reaches 1300 rpm (idle on 2001-05 8.1L V8 engines), keeping an eye on your temp gauge all the time.

10. Let the engine run at this speed for about 10 minutes or until the water being discharged is clear. Once this happens, slowly allow the engine to return to idle and then turn the engine off.

11. Turn off the water from the spigots, disconnect the hoses and remove the flushing attachments.

FLUSHING THE SEAWATER SYSTEM - INBOARDS

Carbureted & TBI Models

◆ See Figure 166

■ If your boat is operated in saltwater, heavy mineral fresh water or severely polluted water, flush the cooling system regularly - after each usage if at all possible! Always flush the system before draining and/or winter storage.

✱✱ CAUTION

If your boat has a seacock (a water inlet valve), it must remain closed during this procedure in order to prevent water from flowing back into the boat. If your boat does not have a seacock, locate the water inlet hose at the seawater pick-up pump, disconnect it and plug it. We highly recommend that you leave a note to yourself in the vicinity of the ignition key reminding of the fact that this procedure has been done so that you, or someone else, doesn't start the engine after flushing without reopening the seacock or reconnecting the inlet hose. Seem silly? Do it anyway!

1. If the boat is in the water, close the seacock (if equipped) or disconnect and plug the seawater inlet line to prevent seawater from entering the boat.

2. Drain the seawater system.

3. Disconnect the inlet line at the seawater pump and, using the appropriate adapter, connect a garden hose to the pump.

4. Connect the other end of the hose to a water source and open the tap - no more than 1/3 force.

✱✱ CAUTION

Never use full water pressure!

ENGINE AND DRIVE MAINTENANCE 2-41

5. With the transmission in Neutral, start the engine and allow it to idle for 10 minutes or so. As soon as you notice that the discharge water is clear, turn the engine Off.

6. Turn off the water supply and disconnect the hose from the spigot and the back of the pump. Reconnect the inlet line to the pump and tighten the clamp securely.

■ If the boat is in the water, do not open the seacock until the next time you intend to start the engine or else you risk contaminated water flowing back into the engine. If you disconnected the inlet line instead, leave it plugged until you start the boat for the same reason.

5.7L/6.2L V8 MPI Inboard Models

◆ See Figure 168

■ If your boat is operated in saltwater, heavy mineral fresh water or severely polluted water, flush the cooling system regularly - after each usage if at all possible! Always flush the system before draining and/or winter storage.

✱✱ CAUTION

If your boat has a seacock (a water inlet valve), it must remain closed during this procedure in order to prevent water from flowing back into the boat. If your boat does not have a seacock, locate the water inlet hose at the seawater pick-up pump, disconnect it and plug it. We highly recommend that you leave a note to yourself in the vicinity of the ignition key reminding of the fact that this procedure has been done so that you, or someone else, doesn't start the engine after flushing without reopening the seacock or reconnecting the inlet hose. Seem silly? Do it anyway!

1. If the boat is in the water, close the seacock (if equipped) or disconnect and plug the seawater inlet line to prevent seawater from entering the boat.
2. Drain the seawater system.
3. Attach a quick connect fitting to the end of a water hose and open the spigot fully.
4. Locate the flush valve connected to a black hose on the exhaust elbow. There should be a flush socket attachment with a blue cap connected to it.
5. Snap the hose and fitting into the socket and then start the engine. Make sure the transmission is in Neutral.

✱✱ CAUTION

Do not let the hose run for more than 15 seconds without starting the engine.

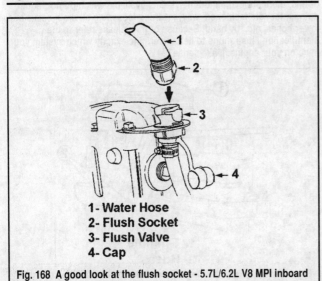

Fig. 168 A good look at the flush socket - 5.7L/6.2L V8 MPI inboard models

1- Water Hose
2- Flush Socket
3- Flush Valve
4- Cap

6. Allow the engine to run between 600 and 800 rpm until it reaches normal operating temperature. Be sure to keep an eye on your temperature gauge in case overheating occurs.

7. Allow the engine to flush for about 10 minutes or until the discharge water is coming our clean and clear.

8. Shut off the engine and IMMEDIATELY turn off the water supply. Press the release button and disconnect the hose from the flush socket.

9. Snap the blue cap into position and open the seacock or unplug the inlet line.

5.7L/6.2L V8 MPI Tow Sports Models

◆ See Figures 169 and 170

■ If your boat is operated in saltwater, heavy mineral fresh water or severely polluted water, flush the cooling system regularly - after each usage if at all possible! Always flush the system before draining and/or winter storage.

✱✱ CAUTION

If your boat has a seacock (a water inlet valve), it must remain closed during this procedure in order to prevent water from flowing back into the boat. If your boat does not have a seacock, locate the water inlet hose at the seawater pick-up pump, disconnect it and plug it. We highly recommend that you leave a note to yourself in the vicinity of the ignition key reminding of the fact that this procedure has been done so that you, or someone else, doesn't start the engine after flushing without reopening the seacock or reconnecting the inlet hose. Seem silly? Do it anyway!

1. If the boat is in the water, close the seacock (if equipped) or disconnect and plug the seawater inlet line to prevent seawater from entering the boat.

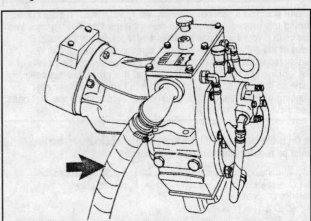

Fig. 169 Disconnect the line at the transmission on Walter V-drives - 5.7L/6.2L V8 MPI Tow Sports models

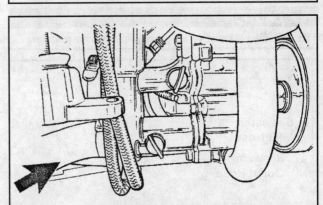

Fig. 170 On other transmissions, disconnect the inlet line at the seawater pump - 5.7L/6.2L V8 MPI inboard/Tow Sports models

2-42 ENGINE AND DRIVE MAINTENANCE

2. Drain the seawater system.
3. On models using a Walter V-Drive transmission, disconnect the water inlet line at the transmission.
4. Attach an appropriate connector to the end of water hose and connect it to the transmission. Skip to Step 7.
5. On all other transmissions, disconnect the inlet line (upper) at the aft end of the seawater pump.
6. Attach an appropriate connector to the end of a water hose and attach it to the inlet fitting on the pump.

※※ CAUTION

If the boat is out of the water, make sure you remove the propeller.

7. Turn on the water source.
8. Move the remote control to Neutral and start the engine. Allow the engine to idle for about 10 minutes or until the discharge water is coming out clean and clear.
9. Shut off the engine and turn off the water supply.
10. Disconnect the hose from the transmission or seawater pump and reconnect the original lines, tightening the clamps securely.
11. Open the seacock or unplug and reconnect the inlet line.
12. Replace the propeller if removed.

8.1L V8 MPI Inboard Models

◆ See Figure 171

■ If your boat is operated in saltwater, heavy mineral fresh water or severely polluted water, flush the cooling system regularly - after each usage if at all possible! Always flush the system before draining and/or winter storage.

※※ CAUTION

If your boat has a seacock (a water inlet valve), it must remain closed during this procedure in order to prevent water from flowing back into the boat. If your boat does not have a seacock, locate the water inlet hose at the seawater pick-up pump, disconnect it and plug it. We highly recommend that you leave a note to yourself in the vicinity of the ignition key reminding of the fact that this procedure has been done so that you, or someone else, doesn't start the engine after flushing without reopening the seacock or reconnecting the inlet hose. Seem silly? Do it anyway!

■ On boats with water lift exhaust collectors or mufflers, the engine must be running while flushing the system.

1. If the boat is in the water, close the seacock (if equipped) or disconnect and plug the seawater inlet line to prevent seawater from entering the boat.
2. Drain the seawater system.
3. Disconnect the inlet line (upper) at the aft end of the seawater pump.
4. Attach an appropriate adapter to the end of a water hose and attach it to the inlet fitting on the pump.

※※ CAUTION

If the boat is out of the water, make sure you remove the propeller.

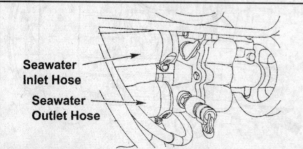

Fig. 171 Disconnect the inlet line at the seawater pump - 8.1L V8 engines

5. Turn on the water source to not more than half capacity on 2001-05 engines, or full pressure on 2006 and later engines.
6. Move the remote control to Neutral and start the engine. Allow the engine to idle for about 10 minutes or until the discharge water is coming out clean and clear. On 2006 and later engines, run the engine up to about 1300 rpm.
7. Shut off the engine and turn off the water supply.
8. Disconnect the hose from the seawater pump and reconnect the original line, tightening the clamp securely.
9. Open the seacock or unplug and reconnect the inlet line.
10. Replace the propeller if removed.

■ If the boat is in the water, do not open the seacock until the next time you intend to start the engine or else you risk contaminated water flowing back into the engine. If you disconnected the inlet line instead, leave it plugged until you start the boat for the same reason.

FLUSHING THE CLOSED SYSTEM

The very name of this systems implies that it is not necessary to flush it. If for some reason you absolutely must flush the closed portion of this system, simply follow the drain and fill procedures. Remember though that the closed system uses a conventional seawater cooling system in conjunction with its closed portion - this system MUST be flushed on a regular basis and all procedures are detailed previously under Seawater Systems.

DRAINING & FILLING

◆ See Figures 172 thru 176

The cooling system should be drained, cleaned, and refilled each season although Mercury's recommendations are for every 2 years on normal anti-freeze systems and 5 years or 1000 hours on systems with Extended Life coolant. We think its cheap insurance to do it every season, but you certainly can't go wrong by following the factory's suggestion. The bow of the boat must be higher than the stern to properly drain the cooling system. If the bow is not higher than the stern, water will remain in the cylinder block and in the exhaust manifold. Insert a piece of wire into the drain holes, but not in the petcock to ensure sand, silt, or other foreign material is not blocking the drain opening.

If the engine is not completely drained for winter storage, trapped water can freeze and cause severe damage. The water in the oil cooler - if so equipped with power steering - must also be drained.

■ Although all models covered here should be equipped with drain systems - certainly all MPI engines, and we think all others as well - there may be a few engines not equipped with one of these systems. Additionally, these systems sometimes fail, so we will endeavor to provide manual draining procedures as well whenever possible. On top of this, there are manual drain systems available which do not require doing it the 'old fashioned' way; meaning loosening all drain cocks and lower hoses, etc., by hand. Because of this, please refer to the identification illustrations to help determine exactly which system you have on your engine. Make sense??

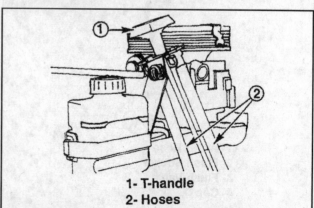

1- T-handle
2- Hoses

Fig. 172 Example of the single point drain system used on all the 3.0L engine

ENGINE AND DRIVE MAINTENANCE

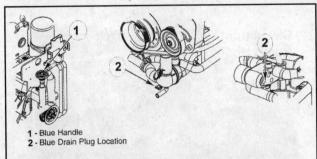

Fig. 173 Example of the single point drain system used on all V6 and V8 engines but the 8.1L V8

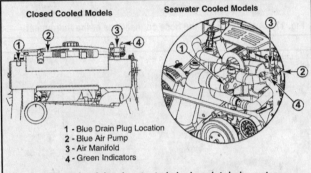

Fig. 174 Example of the air actuated single point drain system used on all V6 and V8 engines but the 8.1L V8

Fig. 175 Example of the 3 point drain system used on all V6 and V8 engines but the 8.1L V8

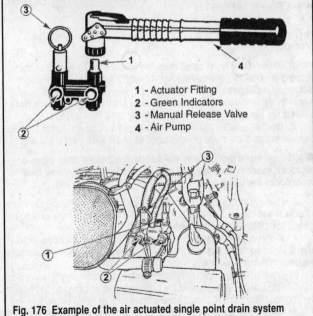

Fig. 176 Example of the air actuated single point drain system used 8.1L V8 engines

3.0L Engines - Manual Draining

◆ See Figures 177, 178 and 179

■ The engine must be as level as possible while performing this procedure.

1. If the boat is in the water, close the seacock to prevent water from entering the cooling system. If the boat is not equipped with a seacock, disconnect the seawater inlet line and plug it. Make sure you leave yourself a note by the ignition switch reminding yourself to turn the valve back on or connect the line, especially if the boat is going to sit for a period of time.
2. Position suitable containers under all drain plugs and hoses (if space permits) to catch any water being drained or else it will collect in the bilge.
3. Remove the drain plug on the port side of the cylinder block.
4. Remove the drain plug on the exhaust manifold or elbow. If your engine is not equipped with a drain plug for some reason, simply disconnect the hose from the lower fitting.
5. On engines equipped with risers, remove the drain plug if equipped.
6. On models equipped with power steering, disconnect the lower seawater hose at the cooler - some models may actually have a drain plug on the bottom side of the cooler.
7. On engines with closed cooling systems, remove the drain plug from the bottom of the heat exchanger. If there is no plug, disconnect the seawater inlet hose at the bottom of the exchanger.
8. Disconnect and remove the lower end of all hoses and hold them as low as possible while draining. Reconnect when all water has stopped draining. On models with closed cooling, refer to the flow diagrams at the rear of this section so that you do not disconnect any hoses used for the closed portion of the system.
9. Using a stiff piece of wire, clean out all drain holes. Don't forget any drain holes in the drive unit!
10. Turn the ignition switch **ON** and crank the engine over a few times to force out any residual water in the block or seawater pump. DO NOT START THE ENGINE!
11. Coat the threads of all drain plugs with Perfect Seal and then reinstall them. Tighten the plugs and petcocks securely.
12. If you are winterizing the boat and its going to sit for a few months, it's a good idea to remove the thermostat housing and fill the cylinder block and head with a mixture of water and anti-freeze (please use propylene rather than ethylene). Remove the inlet line at the exhaust manifold and do the same. Make sure you drain all coolant into a suitable container prior to refilling the system with seawater next season.
13. If you closed the seacock or disconnected the water inlet line, make sure you open or reconnect it PRIOR to restarting the engine.

3.0L Engines - Single Point Drain System

◆ See Figures 179, 180 and 181

■ The engine must be as level as possible while performing this procedure.

1. If the boat is in the water, close the seacock to prevent water from entering the cooling system. If the boat is not equipped with a seacock, disconnect the seawater inlet line and plug it. Make sure you leave yourself a note by the ignition switch reminding yourself to turn the valve back on or connect the line, especially if the boat is going to sit for a period of time.
2. Position suitable containers under all hoses (if space permits) to catch any water being drained or else it will collect in the bilge.
3. Locate the blue drain hoses and disconnect the 2 lines from their top fittings.
4. Bend them down so they are lower than their bottom connections. If they are difficult to lower, use the T-handle to persuade them. If the hoses do not drain sufficiently, use a thin gauge wire to unblock them. Reconnect the lines at the handle bracket.
5. On models equipped with power steering, disconnect the lower seawater hose at the cooler - some models may actually have a drain plug on the bottom side of the cooler.
6. On engines with closed cooling systems, remove the drain plug from the bottom of the heat exchanger. If there is no plug, disconnect the seawater inlet hose at the bottom of the exchanger.
7. Although Mercruiser does no mention this, we think it a good idea to disconnect and remove the lower end of all hoses and hold them as low as possible while draining. Reconnect when all water has stopped draining. On

2-44 ENGINE AND DRIVE MAINTENANCE

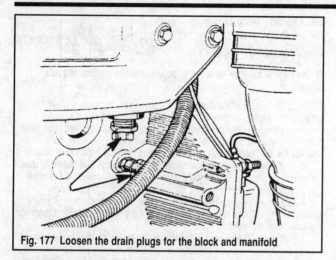

Fig. 177 Loosen the drain plugs for the block and manifold

Fig. 178 A good shot of the hose connections at the thermostat

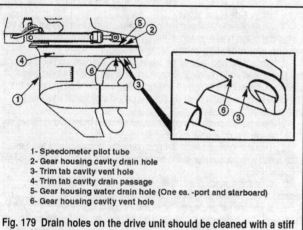

1- Speedometer pilot tube
2- Gear housing cavity drain hole
3- Trim tab cavity vent hole
4- Trim tab cavity drain passage
5- Gear housing water drain hole (One ea. -port and starboard)
6- Gear housing cavity vent hole

Fig. 179 Drain holes on the drive unit should be cleaned with a stiff wire - Alpha

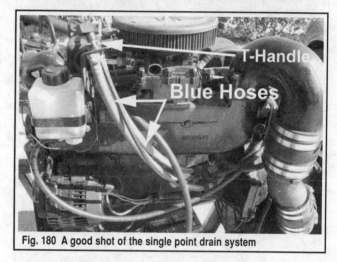

Fig. 180 A good shot of the single point drain system

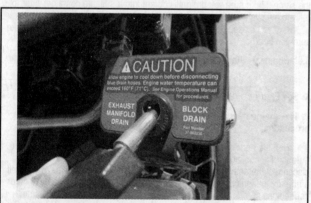

Fig. 181 The T-handle can be used to help push the hoses below their drain points

models with closed cooling, refer to the flow diagrams at the rear of this section so that you do not disconnect any hoses used for the closed portion of the system.

8. Using a stiff piece of wire, clean out all drain holes. Don't forget any drain holes in the drive unit!

9. Coat the threads of all drain plugs with Perfect Seal and then reinstall them. Tighten the plugs and petcocks securely.

10. If you are winterizing the boat and its going to sit for a few months, it's a good idea to remove the thermostat housing and fill the cylinder block and head with a mixture of water and anti-freeze (please use propylene rather than ethylene). Remove the inlet line at the exhaust manifold and do the same. Make sure you drain all coolant into a suitable container prior to refilling the system with seawater next season.

11. If you closed the seacock or disconnected the water inlet line, make sure you open or reconnect it PRIOR to restarting the engine.

4.3L V6 and 5.0L/5.7L/6.2L V8 Engines - Manual Draining

◆ See Figures 182 thru 191

 MODERATE

■ The engine must be as level as possible while performing this procedure.

1. If the boat is in the water, close the seacock to prevent water from entering the cooling system. If the boat is not equipped with a seacock, disconnect the seawater inlet line and plug it. Make sure you leave yourself a note by the ignition switch reminding yourself to turn the valve back on or connect the line, especially if the boat is going to sit for a period of time.

2. Position suitable containers under all drain plugs and hoses (if space permits) to catch any water being drained or else it will collect in the bilge.

■ Although some models may still be using standard drain plugs or petcocks, most models covered here should be using the new blue drain cocks instead.

3. Remove the drain plugs on each side of the cylinder block. On many models, the starboard side plug will share a Y-fitting with a knock sensor; be careful not to disturb the sensor while draining.

4. Remove the drain plugs on the exhaust manifolds or elbows. If your engine is not equipped with a drain plug for some reason, simply disconnect the hose from the lower fitting.

5. On engines equipped with risers, remove the drain plug if equipped.

6. Remove the large hose at the engine circulating pump on the front of the engine. Many models should have a drain cock here, so simply use that.

7. Check for a drain cock at the water tube on carbureted models or the fuel cooler on EFI models.

ENGINE AND DRIVE MAINTENANCE

8. On models with a seawater pump (Bravo), remove the lower hose at the pump. Many newer models will have drain cocks on both pump hoses, so simply use these.

9. On models equipped with power steering, disconnect the lower seawater hose at the cooler - some models may actually have a drain plug on the bottom side of the cooler.

10. On engines with closed cooling systems, remove the drain plug from the bottom of the heat exchanger. If there is no plug, disconnect the seawater inlet hose at the bottom of the exchanger. It's also a good idea to remove the end caps and allow all of the tubes to drain.

11. In addition to the above, we think it's a good idea to disconnect and remove the lower end of all hoses and hold them as low as possible while draining. Reconnect when all water has stopped draining. On models with closed cooling, refer to the flow diagrams at the rear of this section so that you do not disconnect any hoses used for the closed portion of the system.

12. Using a stiff piece of wire, clean out all drain holes. Don't forget any drain holes in the drive unit!

13. Turn the ignition switch **ON** and crank the engine over a few times to force out any residual water in the block or seawater pump. DO NOT START THE ENGINE!

14. Coat the threads of all drain plugs with Perfect Seal and then reinstall them. Tighten the plugs and petcocks securely.

15. If you are winterizing the boat and its going to sit for a few months, it's a good idea to remove the thermostat housing and fill the cylinder block

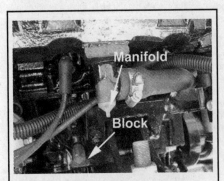

Fig. 182 Loosen the drain plugs for the block and manifold, starboard side

Fig. 183 Loosen the drain plug for the manifold, port side

Fig. 184 Loosen the drain plug for the block, port side

Fig. 185 The engine circulating pump hose on many models will also have a drain plug

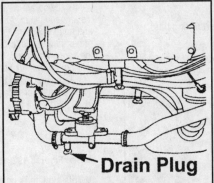

Fig. 186 Check the water tube for a drain plug on carbureted models

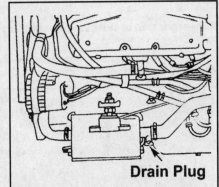

Fig. 187 Check the fuel cooler for a drain plug on EFI models

Fig. 188 If equipped with a seawater pump, remove the lower hose, or open the drain cock

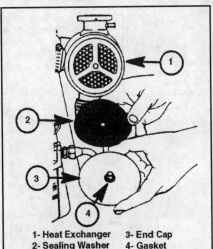

Fig. 189 It's a good idea to remove the end caps on the heat exchanger (closed system only) and let the tubes drain

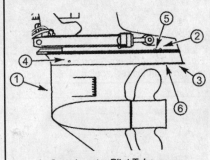

Fig. 190 Be sure to clean out all drain and vent holes on the Alpha

2-46 ENGINE AND DRIVE MAINTENANCE

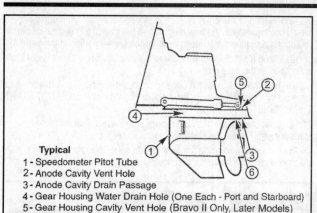

Typical
1 - Speedometer Pitot Tube
2 - Anode Cavity Vent Hole
3 - Anode Cavity Drain Passage
4 - Gear Housing Water Drain Hole (One Each - Port and Starboard)
5 - Gear Housing Cavity Vent Hole (Bravo II Only, Later Models)
6 - Gear Housing Cavity Drain Hole (Not All Models)

Fig. 191 Be sure to clean out all drain and vent holes on the Bravo

and head with a mixture of water and anti-freeze (please use propylene rather than ethylene). Remove the inlet line at the exhaust manifold and do the same. Make sure you drain all coolant into a suitable container prior to refilling the system with seawater next season.

16. If you closed the seacock or disconnected the water inlet line, make sure you open or reconnect it PRIOR to restarting the engine.

8.1L V8 Engines - Manual Draining

◆ See Figures 171, 189, and 191 thru 195

■ The engine must be as level as possible while performing this procedure.

1. If the boat is in the water, close the seacock to prevent water from entering the cooling system. If the boat is not equipped with a seacock, disconnect the seawater inlet line and plug it. Make sure you leave yourself a note by the ignition switch reminding yourself to turn the valve back on or connect the line, especially if the boat is going to sit for a period of time.

2. Position suitable containers under all drain cocks and hoses (if space permits) to catch any water being drained or else it will collect in the bilge.

3. Remove the large hose at the engine circulating pump on the front of the engine. Many models should have a drain cock here, so simply use that.

4. Remove the drain cocks on each side of the fuel cooler if equipped.

5. Remove blue drain cock from the air actuated water drain assembly on the front port side.

6. Remove the blue drain cocks from each hose at the seawater pump.

7. On models equipped with power steering, disconnect the lower seawater hose at the cooler - some models may actually have a drain plug on the bottom side of the cooler.

8. On engines with closed cooling systems, remove the drain plug from the bottom of the heat exchanger. If there is no plug, disconnect the seawater inlet hose at the bottom of the exchanger. It's also a good idea to remove the end caps and allow all of the tubes to drain.

9. In addition to the above, we think it's a good idea to disconnect and remove the lower end of all hoses and hold them as low as possible while draining. Reconnect when all water has stopped draining. On models with closed cooling, refer to the flow diagrams at the rear of this section so that you do not disconnect any hoses used for the closed portion of the system.

10. Using a stiff piece of wire, clean out all drain holes. Don't forget any drain holes in the drive unit!

11. Turn the ignition switch **ON** and crank the engine over a few times to force out any residual water in the block or seawater pump. DO NOT START THE ENGINE!

12. Coat the threads of all drain cocks with Perfect Seal and then reinstall them. Tighten the plugs and petcocks securely.

13. If you are winterizing the boat and its going to sit for a few months, it's a good idea to remove the thermostat housing and fill the cylinder block and head with a mixture of water and anti-freeze (please use propylene rather than ethylene). Remove the inlet line at the exhaust manifold and do the same. Make sure you drain all coolant into a suitable container prior to refilling the system with seawater next season.

14. If you closed the seacock or disconnected the water inlet line, make sure you open or reconnect it PRIOR to restarting the engine.

V6 & V8 Engines - Single Point Drain System (Manual)

◆ See Figures 192, 193, 194 and 196 thru 198

■ This system is used predominantly on models with an Alpha stern drive unit.

■ The engine must be as level as possible while performing this procedure.

1. **If the boat is in the water**, close the seacock to prevent water from entering the cooling system. If the boat is not equipped with a seacock, disconnect the seawater inlet line and plug it. Make sure you leave yourself a note by the ignition switch reminding yourself to turn the valve back on or connect the line, especially if the boat is going to sit for a period of time.

2. Position a suitable container under the distribution housing drain fitting to catch any water being drained or else it will collect in the bilge.

Fig. 192 Although not recommended by Mercury, you may also wish to drain the fuel cooler...

Fig. 193 ...and seawater pump...

Fig. 194 ...or engine circulating pump separately as a safe guard

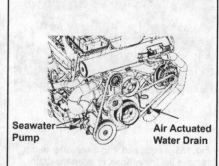

Fig. 195 Blue drain plugs - 8.1L engines

ENGINE AND DRIVE MAINTENANCE   2-47

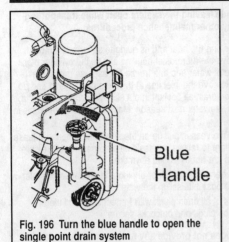

Fig. 196 Turn the blue handle to open the single point drain system

Fig. 197 Remove the drain cock in the housing to vent the system (if the boat is still in the water)

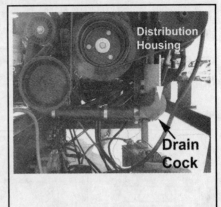

Fig. 198 If the water is not draining, you will need to remove this drain cock as well

3. Locate the round blue handle at the front of the engine and turn it counterclockwise approximately 2 turns until it stops - the red marking on the shaft will indicate that the system is open.

■ Do not force the drain handle, as the threads will strip easily!

4. **If the boat is still in the water**, IMMEDIATELY remove the blue drain cock from the side of the thermostat housing to ensure that the system is properly vented. You must remove this plug within 30 seconds of opening the drain system (turning the blue handle).

5. You should be able to observe water draining from the fitting (either orange or red) on the bottom of the distribution housing. If not, remove the blue drain cock on the side of the housing.

6. Allow the system to drain for at least 5 minutes and then reinstall the drain cock in the thermostat housing and/or the distribution housing if either one was removed.

■ Mercury recommends leaving the system open while transporting the boat, or during any other maintenance procedures.

7. Slowly turn the blue handle clockwise until it stops; once again, don't force it. When the handle is fully seated, you should not be able to see any red.

8. Coat the threads of all drain plugs with Perfect Seal and then reinstall them. Tighten the plugs and petcocks securely.

9. If you are winterizing the boat and its going to sit for a few months, it's a good idea to remove the thermostat housing and fill the cylinder block and head with a mixture of water and anti-freeze (please use propylene rather than ethylene). Remove the inlet line at the exhaust manifold and do the same. Make sure you drain all coolant into a suitable container prior to refilling the system with seawater next season. Make sure the drive unit is in the DOWN/IN position.

10. Open the seacock or unplug and reinstall the water inlet line if the boat is in the water.

■ Although Mercury does not provide any further procedures for draining the system, we feel it might not hurt to drain any other components and hoses as well (similar to the manual drain procedures) if they have their own drain cocks. Fuel coolers, power steering coolers, seawater pump, engine circulating pump hose, etc all might be things you choose to drain additionally, particularly if you are winterizing.

V6 & V8 Engines (Exc. 8.1L) - Single Point Drain System (Air Actuated)

◆ See Figures 192, 193, 194 and 199 thru 203

■ The engine must be as level as possible while performing this procedure.

1. **If the boat is in the water**, close the seacock to prevent water from entering the cooling system. If the boat is not equipped with a seacock, disconnect the seawater inlet line and plug it. Make sure you leave yourself a note by the ignition switch reminding yourself to turn the valve back on or connect the line, especially if the boat is going to sit for a period of time.

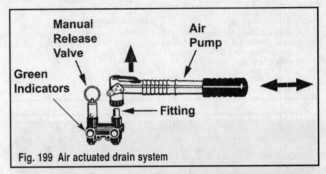

Fig. 199 Air actuated drain system

Fig. 200 A good look at the air pump

2. Position a suitable container under the distribution housing drain fitting and the seawater pump fitting to catch any water being drained or else it will collect in the bilge.

3. Remove the blue air pump from its snap holder. Make sure the pump lever is flush with the handle and then install the pump onto the threaded air fitting. Once on the fitting, flip the small lever up to the vertical position.

4. Slowly pump air into the system until both of the green indicators are extended and water is draining from the drain fittings on the distribution housing and seawater pump. The distribution housing will most likely begin draining first.

5. **If the boat is in the water**, IMMEDIATELY remove the blue drain cock from the side of the thermostat housing (seawater) or heat exchanger (closed) to vent the system. This must be done within 30 seconds of pressurizing the system.

6. Confirm that you actually have water draining from the system at each of the drains - if not, you must now follow all steps for the 3 Point Drain System procedure. If the system is draining, move on to the next step.

2-48 ENGINE AND DRIVE MAINTENANCE

7. Allow the system to drain for at least 5 minutes or until the water stops draining. Feel free to continue pumping air into the system so that the green indicators remain extended.

8. Carefully crank the engine over a few times to force out any trapped water in the seawater pump. DO NOT allow the engine to actually start!

9. Reinstall the drain cock in the thermostat housing or heat exchanger if they were removed.

10. Flip the lever down and remove the air pump. Reinstall it on its retainer.

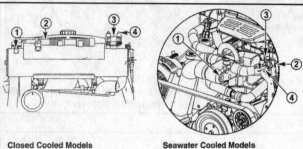

Fig. 201 Air actuated drain system components

Closed Cooled Models
1 - Blue Drain Plug Location
2 - Blue Air Pump
3 - Air Manifold
4 - Green Indicators

Seawater Cooled Models

Fig. 202 Remove the air valve cover...

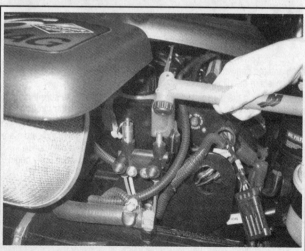

Fig. 203 ...and then install the pump

■ Mercury recommends leaving the system open while transporting the boat, or during any other maintenance procedures.

11. If you are winterizing the boat and its going to sit for a few months, it's a good idea to remove the thermostat housing and fill the cylinder block and head with a mixture of water and anti-freeze (please use propylene rather than ethylene). Remove the inlet line at the exhaust manifold and do the same. Make sure you drain all coolant into a suitable container prior to refilling the system with seawater next season. Make sure the drive unit is in the DOWN/IN position.

12. Before re-using the vessel, pull up on the small ring atop the release valve (next to the air fitting) to relieve any residual air pressure. Make sure that the green indicators are not extended anymore.

13. Don't forget to open the seacock or unplug and reconnect the inlet line before restarting the boat if its still in the water.

14. Coat the threads of all drain plugs with Perfect Seal and then reinstall them. Tighten the plugs and petcocks securely.

■ Although Mercury does not provide any further procedures for draining the system, we feel it might not hurt to drain any other components and hoses as well (similar to the manual drain procedures) if they have their own drain cocks. Fuel coolers, power steering coolers, seawater pump, engine circulating pump hose, etc all might be things you choose to drain additionally, particularly if you are winterizing.

8.1L V8 Engines - Single Point Drain System (Air Actuated)

◆ See Figures 199 thru 205

■ The engine must be as level as possible while performing this procedure.

1. **If the boat is in the water**, close the seacock to prevent water from entering the cooling system. If the boat is not equipped with a seacock, disconnect the seawater inlet line and plug it. Make sure you leave yourself a note by the ignition switch reminding yourself to turn the valve back on or connect the line, especially if the boat is going to sit for a period of time.

2. Position a suitable container under the drains to catch any water being drained or else it will collect in the bilge.

3. Remove the blue air pump from its snap holder. Make sure the pump lever is flush with the handle and then install the pump onto the threaded air fitting. Once on the fitting, flip the small lever up to the vertical position.

4. Slowly pump air into the system until both of the green indicators are extended and water is draining from the drains on both sides of the engine.

5. Confirm that you actually have water draining from the system at each of the drains - if not, you must now follow all steps the manual procedure. If the system is draining, move on to the next step.

6. Allow the system to drain for at least 5 minutes or until the water stops draining. Feel free to continue pumping air into the system so that the green indicators remain extended.

7. Most water will drain from the system, however there may be some residual water in the heat exchanger. Remove the end caps and allow all tubes to drain completely.

■ Mercury recommends leaving the system open while transporting the boat, or during any other maintenance procedures.

8. If you are winterizing the boat and its going to sit for a few months, it's a good idea to remove the thermostat housing and fill the cylinder block and head with a mixture of water and anti-freeze (please use propylene rather than ethylene). Remove the inlet line at the exhaust manifold and do the same. Make sure you drain all coolant into a suitable container prior to refilling the system with seawater next season. Make sure the drive unit is in the DOWN/IN position.

9. Before re-using the vessel, pull up on the small ring atop the release valve (next to the air fitting) to relieve any residual air pressure. Make sure that the green indicators are not extended anymore and the release valve seats completely.

10. Don't forget to open the seacock or unplug and reconnect the inlet line before restarting the boat if its still in the water.

ENGINE AND DRIVE MAINTENANCE

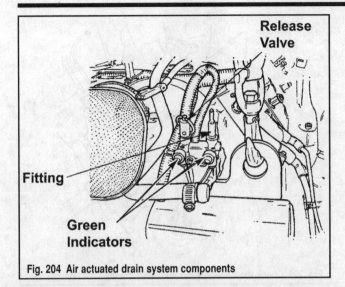

Fig. 204 Air actuated drain system components

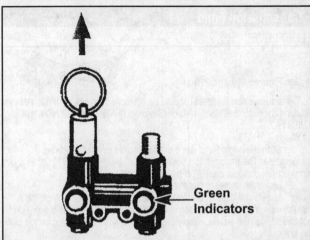

Fig. 205 Make sure that you pull the release valve and check to see that the green indicator is no longer extended

■ Although Mercury does not provide any further procedures for draining the system, we feel it might not hurt to drain any other components and hoses as well (similar to the manual drain procedures) if they have their own drain cocks. Fuel coolers, power steering coolers, seawater pump, engine circulating pump hose, etc all might be things you choose to drain additionally, particularly if you are winterizing.

V6 and V8 Engines - 3 Point Drain System (Manual)

◆ See Figures 192, 194 and 206 thru 208

■ Always use this procedure if the single point drain systems (manual or air actuated) fail.

■ The engine must be as level as possible while performing this procedure.

1. **If the boat is in the water**, close the seacock to prevent water from entering the cooling system. If the boat is not equipped with a seacock, disconnect the seawater inlet line and plug it. Make sure you leave yourself a note by the ignition switch reminding yourself to turn the valve back on or connect the line, especially if the boat is going to sit for a period of time.

2. Position a suitable container under the distribution housing drain and the seawater pump drains to catch any water being drained or else it will collect in the bilge.

3. Remove the blue drain cocks from the distribution housing and the seawater pump (2 drains).

4. **If the boat is still in the water**, IMMEDIATELY remove the blue drain cock from the side of the thermostat housing to ensure that the system is properly vented. You must remove this plug within 30 seconds of opening the drain cocks. Not all engines will utilize this plug.

5. You should be able to observe water draining from the fitting (either orange or red) on the bottom of the distribution housing and the seawater pump.

6. Allow the system to drain for at least 5 minutes or until the water stops draining.

7. Carefully crank the engine over a few times to force out any trapped water in the seawater pump. DO NOT allow the engine to actually start! It's probably a good idea to pull the stop lanyard or disable the ignition to ensure a no-start.

■ Mercury recommends leaving the system open while transporting the boat, or during any other maintenance procedures.

8. Coat the threads of all drain cocks with Perfect Seal and then reinstall them. Tighten the plugs and petcocks securely.

9. If you are winterizing the boat and its going to sit for a few months, it's a good idea to remove the thermostat housing and fill the cylinder block and head with a mixture of water and anti-freeze (please use propylene rather than ethylene). Remove the inlet line at the exhaust manifold and do the same. Make sure you drain all coolant into a suitable container prior to refilling the system with seawater next season. Make sure the drive unit is in the DOWN/IN position.

10. Open the seacock or unplug and reinstall the water inlet line if the boat is in the water.

■ Although Mercury does not provide any further procedures for draining the system, we feel it might not hurt to drain any other components and hoses as well (similar to the manual drain procedures) if they have their own drain cocks. Fuel coolers, power steering coolers, engine circulating pump hose, etc all might be things you choose to drain additionally, particularly if you are winterizing.

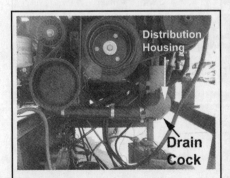

Fig. 206 Remove the drain cock in the distribution housing...

Fig. 207 ...and the seawater pump (2)...

Fig. 208 ...before removing this one in the thermostat housing (but only if the boat is in the water)

2-50 ENGINE AND DRIVE MAINTENANCE

Transmission Fluid

FLUID LEVEL

◆ See Figures 209 and 210

The transmission fluid level should be checked on a weekly basis. Always fill using hydraulic oil (Mobil 424) or Dexron(r) III ATF fluid. DO NOT MIX THE TWO.

Warm

1. Start the engine and run it until it reaches normal operating temperature. The transmission fluid must be at least 190° F (87° C), as the incorrect temperature can affect oil level greatly.
2. Move the shifter to Neutral and turn off the engine.
3. Locate the dipstick on the transmission and remove it. Wipe the dipstick with a clean, lint-free rag and re-insert it into the transmission - do not thread it all the way in, simply press it in firmly so it is resting on top of the hole.

■ Unlike engine oil, you need to check the transmission fluid immediately after turning off the engine. Do not wait for the fluid to settle.

4. Remove the dipstick and check that the level is up to the **FULL** mark. If not, carefully add the appropriate fluid through the dipstick hole until the correct level is reached.

Cold

We figure that if you can check the fluid level without having to start the engine every time, you'll probably be more likely to check it regularly so here's a way you can do just that.

5. Follow the procedure for checking the oil with the engine hot and make sure that the level is correct.
6. Allow the boat to sit overnight and then remove the dipstick with the engine/transmission cold.
7. Take note of the oil level, wipe off the dipstick with a clean, lint-free rag and scribe a mark on the stick where the oil level had been.
8. Although this does not obviate the need to check the fluid level when hot on a regular basis, it allows you a good, quick idea that your level is OK in between normal checks.

DRAINING FLUID

Velvet - Except 5000 Series

◆ See Figures 211 thru 214

1. Clean the area around the cooler hose connection on the lower side of the housing.

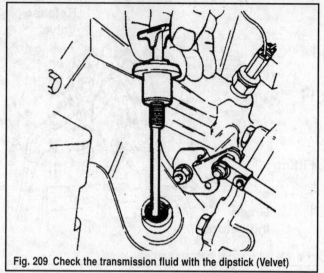

Fig. 209 Check the transmission fluid with the dipstick (Velvet)

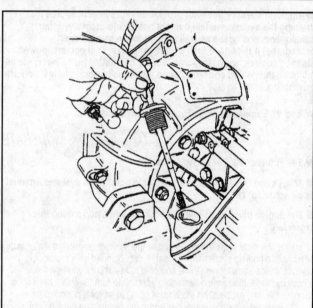

Fig. 210 Check the transmission fluid with the dipstick (ZF/Hurth)

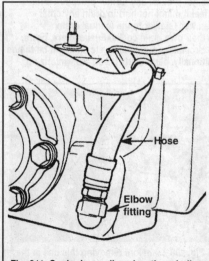

Fig. 211 Cooler hose elbow location - in-line units

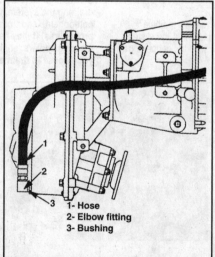

Fig. 212 Cooler hose elbow location - V-drive units

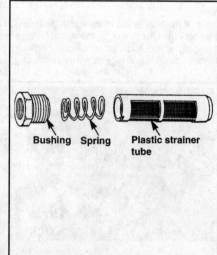

Fig. 213 All units utilize a small strainer (note the notch used for positioning)

ENGINE AND DRIVE MAINTENANCE

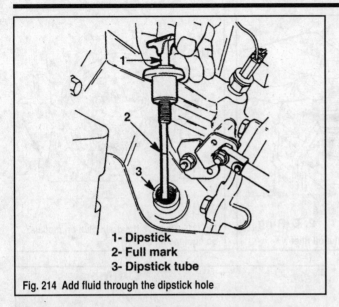

Fig. 214 Add fluid through the dipstick hole

1- Dipstick
2- Full mark
3- Dipstick tube

2. Loosen the fitting nut and disconnect the hose at the elbow. Have some rags handy and be prepared to plug the hose end immediately.

3. Position a suitable container under the elbow and then remove the elbow from the bushing and allow the oil to drain.

4. Remove the bushing and pull out the spring and plastic strainer tube. Clean the strainer in solvent.

5. Unplug the cooler hose and drain the cooler into the container also.

6. Check the oil for large metal or rubber chips indicating a failing unit or damaged hose. Small metal particles are normal and should not be cause for alarm.

7. Insert the plastic strainer into the bushing hole so the notch in the strainer is down and facing out toward the side of the housing. Position the spring and then screw in the bushing. On all transmissions, coat the threads with Perfect Seal and tighten the bushing to 25 ft. lbs. (34 Nm).

8. Coat the threads of the elbow with Perfect Seal and screw it into the bushing so it is secure.

9. Reconnect the cooler hose and tighten the fitting securely.

10. Remove the dipstick and fill the transmission with fluid until it registers **FULL** on the stick. Use Dexron III(r) ATF in the quantity indicated in the Capacities chart

11. Install the dipstick, start the engine and allow it to run at 1500 rpm for about 2 min. Stop the engine and recheck the fluid level immediately. Add fluid as necessary until it meets the **FULL** mark.

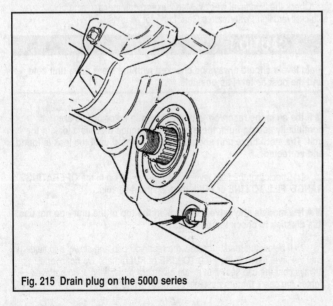

Fig. 215 Drain plug on the 5000 series

Velvet 5000 Series

◆ See Figures 215 and 216

1. Clean the area around the drain plug on the rear of the unit, below and starboard of the propeller shaft.

2. Position a suitable container under the plug and then remove the plug from the housing. Removing the dipstick will aid in draining.

3. Check the oil for large metal or rubber chips indicating a failing unit or damaged hose. Small metal particles are normal and should not be cause for alarm.

4. Coat the threads of the drain plug with Perfect Seal, thread it into the housing and tighten to 25 ft. lbs. (34 Nm).

5. Remove the dipstick completely now and fill the transmission with fluid (Dexron III ATF) until it registers **FULL** on the stick.

6. Install the dipstick, start the engine and allow it to run at 1500 rpm for about 2 min. Stop the engine and recheck the fluid level immediately. Add fluid as necessary until it meets the **FULL** mark. Please refer to the Capacities chart for exact quantities

7. Reinstall the dipstick, making sure to tighten the T-handle securely.

ZF/Hurth

◆ See Figures 217 thru 221

1. Clean off the exterior of the transmission around the filter and the drain plug.

2. Loosen the oil filter set screw with a 6mm Allen wrench. Turn the filter cover while lifting up and remove the filter. There may be a few early models that do not use the set screw.

3. Separate the filter from the cap and throw away the 2 O-rings.

4. Position a suitable drain pan under the drain plug on the 630V and 63IV. Remove the plug and drain the fluid. Reinstall the drain plug and tighten it securely.

5. On the other models, insert a suction tube into the filter hole and remove the fluid.

6. Check the oil for large metal or rubber chips indicating a failing unit or damaged hose. Small metal particles are normal and should not be cause for alarm.

7. Fill the transmission through the oil filter hole, while checking the level regularly with the dipstick.

8. Coat new O-rings with clean transmission fluid and position them in their respective grooves. Install a new filter to the cap. Reinstall the oil filter and tighten the set screw securely.

9. Install the dipstick, start the engine and allow it to run at 1500 rpm for about 2 min. Stop the engine and recheck the fluid level immediately. Add fluid as necessary until it meets the **FULL** mark. Please refer to the Capacities chart for exact quantities

10. Reinstall the dipstick, making sure to tighten the T-handle securely.

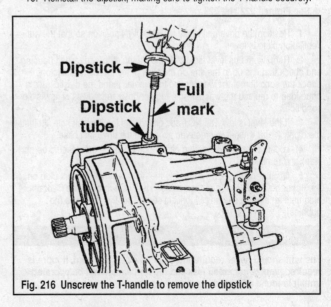

Fig. 216 Unscrew the T-handle to remove the dipstick

2-52 ENGINE AND DRIVE MAINTENANCE

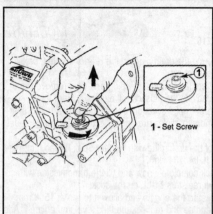

Fig. 217 On the ZF/Hurth, remove the small set screw and turn the cover...

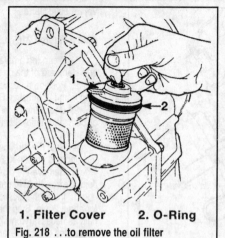

1. Filter Cover 2. O-Ring
Fig. 218 ...to remove the oil filter

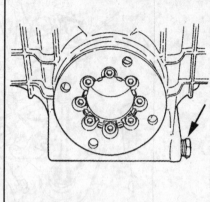

Fig. 219 Remove the drain plug on models so equipped...

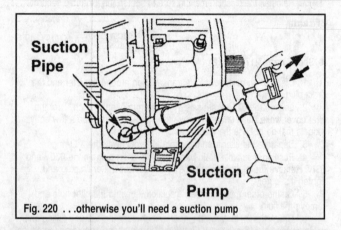

Fig. 220 ...otherwise you'll need a suction pump

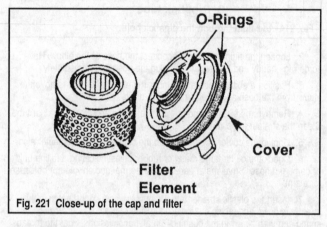

Fig. 221 Close-up of the cap and filter

Stern Drive Unit

FLUID LEVEL

All units covered here should be equipped with a drive lube monitor on the forward port side of the engine, but we have also included procedures for models without a monitor just incase.

Alpha Drives Without An Oil Reservoir

◆ See Figures 222 and 223

1. Position the drive unit in the full IN/DOWN position so that the anti-ventilation plate is level.
2. Remove the oil level dipstick found on top of the driveshaft housing and check that it is up to the line on the stick. If the level is satisfactory, check the condition of the washer and then install it and the dipstick. It's a good idea to replace the washer on a regular basis regardless of its visible condition.
3. If the level is low, DO NOT add oil through the dipstick hole. Reinstall the dipstick and washer to create an air lock in the drive unit case.
4. Locate the oil fill/drain plug on the bottom of the unit (should be port side) and remove it.
5. Quickly install a lube pump into the hole, remove the vent plug on the upper housing (Port) and then pump in lubricant through the fill/drain plug until an air-free stream of lubricant comes out the vent hole (no bubbles).

※※ **CAUTION**

The unit should never require more than 2 oz. of lubricant. If more is required, you've got an oil leak and the unit should not be operated until it is found and fixed.

6. With the lube pump still attached, reinstall the oil vent plug and then quickly remove the pump and install the fill/drain plug. Both plugs should be tightened to 40 inch lbs. (4 Nm).
7. Clean any excess oil from the housing. Recheck the oil level on the dipstick again and then recheck it a final time after the next use.

All Models With A Drive Lube Monitor

◆ See Figures 224 and 225

Check the fluid level weekly. MerCruiser recommends using only Quicksilver High Performance Gear Lube for all units.

※※ **CAUTION**

Fluid levels should always be checked with the stern drive unit cold and the boat as level as possible.

■ If the oil in the reservoir is milky brown in color or you are continually adding fluid, there is a good chance there is a leak in the unit. The stern drive unit should not be operated until the leak is found and corrected.

1. Check that the fluid level in the reservoir is up to the **OPERATING RANGE**, **FILL TO LINE** or **FULL** mark on the reservoir.

■ A few models may have a dipstick in the top of the unit - do not use this dipstick to check the oil level.

2. If below the mark, unscrew the reservoir cap and slowly add fluid until the level reaches the **FILL TO LINE** or **FULL** mark. Do not overfill. Always coat the cap seal with a little oil before reinstalling it; do not over-tighten the cap - 1/4 turn past initial seating is fine.

ENGINE AND DRIVE MAINTENANCE 2-53

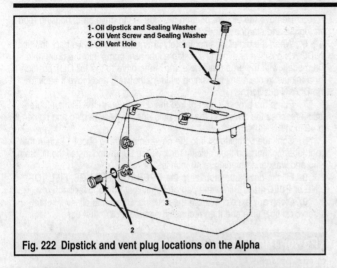

Fig. 222 Dipstick and vent plug locations on the Alpha

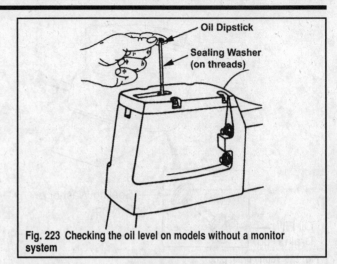

Fig. 223 Checking the oil level on models without a monitor system

Fig. 224 A good look at the gear lube monitor...

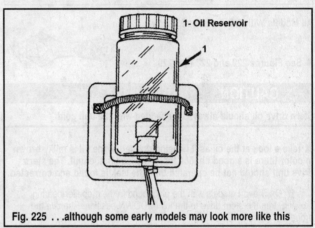

Fig. 225 ...although some early models may look more like this

✷✷ CAUTION

The unit should never require more than 2 oz. (59 ml) of lubricant. If more is required, you've got an oil leak and the unit should not be operated until it is found and fixed.

CHECKING FOR WATER

✷✷ CAUTION

Fluid levels should always be checked with the stern drive unit cold.

Take a look at the oil reservoir, if the oil in the reservoir is milky brown in color there is a good chance there is a leak in the unit. The stern drive unit should not be operated until the leak is found and corrected.

If the drive unit has sat overnight, or if your are working on a model without a reservoir, trim the unit to the full UP/OUT position and briefly remove the drain/fill plug at the bottom of the unit to take a sample of lubricant. If the oil sample is milky brown in color there is a good chance there is a leak in the unit. The stern drive unit should not be operated until the leak is found and corrected.

DRAIN AND REFILL

The oil in the stern drive should be changed at least every 100 hours of operation or once a season, whichever comes first. It's never a bad idea to change the fluid in your unit more frequently, particularly if you use your boat in severe service conditions. MerCruiser recommends using only Quicksilver High Performance Gear Lube for all units.

All units covered here should be equipped with a drive lube monitor on the forward port side of the engine, but we have also included procedures for models without a monitor just incase.

Alpha Drives Without An Oil Reservoir

◆ See Figures 222 and 226

✷✷ CAUTION

Stern drive oil should always be changed with the unit cold.

■ Take a look at the oil as it is being drained, if the oil is milky brown in color there is a good chance there is a leak in the unit. The stern drive unit should not be operated until the leak is found and corrected.

1. Trim the stern drive to the full OUT position for models with the fill/drain plug in the bottom of the lower housing.
2. Remove the upper oil vent plug from the side of the drive shaft housing (Port).
3. Position an oil drain plan or an old plastic milk jug under the drain hole on the bottom of the drive unit and then remove the fill/drain plug, keeping a slight inward pressure on it until it is completely unthreaded.
4. When the old oil has completely drained, trim the drive back to the IN/DOWN position and install a suitable lubricant pump into the drain hole.
5. Pump the proper lubricant into the drive through the fill/drain hole until it comes out of the vent hole with no bubbles.
6. With the lube pump still attached, reinstall the oil vent plug, tighten it to 40 inch lbs. (4 Nm) and then quickly remove the pump and install the fill/drain plug and washer; tighten it to the same specs as the vent plug.
7. Clean any excess oil on the housing. Check the oil level a final time and then recheck it again after the first use.

2-54 ENGINE AND DRIVE MAINTENANCE

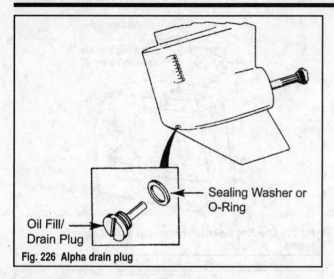

Fig. 226 Alpha drain plug

All Models With A Drive Lube Monitor

◆ See Figures 222 and 227 thru 229

✳✳ CAUTION

Stern drive oil should always be changed with the unit cold.

■ Take a look at the oil as it is being drained, if the oil is milky brown in color there is a good chance there is a leak in the unit. The stern drive unit should not be operated until the leak is found and corrected.

1. On Bravo I models with the plug found in the propeller bearing housing, trim the stern drive to the full IN/DOWN position. Remove the propeller.
2. On all other models, trim the stern drive to the full OUT position.
3. Unbolt the drive lube monitor from its mounting bracket, unscrew the cap and pour out all oil into a suitable container. Clean the reservoir and install it back into the mounting bracket. Do not refill it yet!
4. Position an oil drain pan or an old plastic milk jug under the drain hole on the bottom of the drive unit (or under the prop housing on Bravo I) and then remove the fill/drain plug, keeping a slight inward pressure on it until it is completely unthreaded.
5. Remove the oil vent plug on the side of the drive shaft housing (port on Alphas and starboard on Bravos).
6. When the old oil has completely drained, trim the drive back to the IN/DOWN position and install a suitable lubricant pump into the drain hole. On models with the prop drain hole, move the drive to the full OUT position for a minute to drain any remaining oil in the housing and move it back to the DOWN position.
7. Pump the proper lubricant into the drive through the fill/drain hole until it reaches the bottom of the vent hole. Install the vent plug and tighten it to 40 inch lbs. (4 Nm).
8. Continue pumping oil into the drive until there is about 1 inch in the gear lube monitor and then quickly remove the pump and install the fill/drain plug and washer, tightening to 40 inch lbs. (4 Nm).
9. Fill the drive lube monitor to the **OPERATING RANGE**, **FILL TO LINE** or **FULL** mark. Screw the cap on tightly, but do not over-tighten.
10. Clean any excess oil from the housing. Check the oil level in the reservoir a final time and then recheck it again after the first use.

Power Trim Pump

FLUID LEVEL

Early 2001 Models

◆ See Figure 230

Fluid levels should be checked weekly. Use Quicksilver Power Trim and Steering fluid. SAE 10W-30 or 10W-40 motor oil may be used also.

1. Move the stern drive to the full IN/DOWN position.
2. Check that the oil level is between the **MAX** and **MIN** marks on the side of the reservoir. Add oil through the filler hole, no higher than the bottom of the filler neck.
3. Move the drive unit through its full range 6-10 times in order to purge any air that may be in the system. Lower the drive a final time and recheck the oil level; repeat the process if necessary.
4. Install the fill cap. Some models utilize a vent screw located at the bottom of the pump body, just above the reservoir; if equipped, tighten the screw fully and then back it out 2 complete turns so that the system is properly vented. Models without a vent screw use a vented fill cap, make sure that the filler neck seal is removed so that the cap vents properly.

All Other Models

◆ See Figure 231

Fluid levels should be checked weekly. Use Quicksilver Power Trim and Steering fluid.

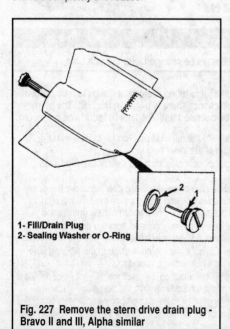

Fig. 227 Remove the stern drive drain plug - Bravo II and III, Alpha similar

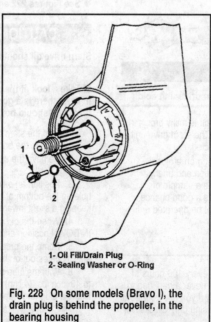

Fig. 228 On some models (Bravo I), the drain plug is behind the propeller, in the bearing housing

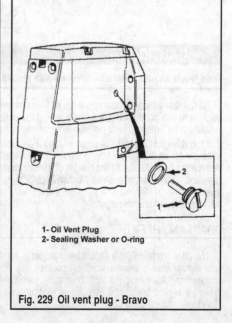

Fig. 229 Oil vent plug - Bravo

ENGINE AND DRIVE MAINTENANCE   2-55

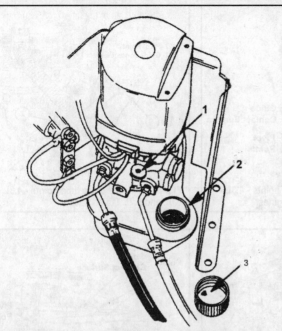

1- If there is a vent screw, it must be backed out 2 full turns. If there is no vent screw, remove the Fill Neck Seal to vent the pump.
2- Maintain the oil level between the "Max" and "Min" marks on the side of the reservoir. When adding oil, fill to the bottom lip on the Fill Neck, as shown.
3- Vent Hole

Fig. 230 Power trim pump reservoir - early 2001 models

1. Move the stern drive to the full IN/DOWN position.
2. Check that the oil level is between the **MAX** and **MIN** marks on the side of the reservoir.
3. Remove the fill cap and make sure that the cap plug (if equipped) has previously been removed - if not, remove it and throw it away (you needn't replace it). Check the oil level is up to the bottom of the filler neck, filling through the hole as necessary.
4. Move the drive unit through its full range 6-10 times (Bravo) or 3-5 times (Alpha) in order to purge any air that may be in the system. Lower the drive a final time and recheck the oil level; repeat the process if necessary.
5. Install the fill cap. Some models utilize a vent screw located at the bottom of the pump body (usually Alphas), just above the reservoir; if equipped, tighten the screw fully and then back it out 2 complete turns so that the system is properly vented. Models without a vent screw use a vented fill cap, make sure that the filler neck seal is removed so that the cap vents properly.

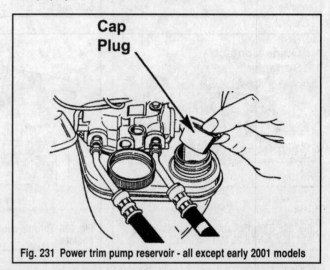

Fig. 231 Power trim pump reservoir - all except early 2001 models

LUBRICATION POINTS

Throttle Cable

3.0L ENGINES

◆ See Figure 232

Lubricate the cable pivot points every 100 hours or 12 months, whichever comes first. If the vessel is being operated in salt water, it can't hurt to lubricate every 50 hours or 6 months whichever comes first. Always use SAE 30W engine oil on the pivot points, or Quicksilver   2-4-C Marine Lubricant with Teflon on the cable guide contact surfaces.

4.3L V6 & 5.0L/5.7L/6.2L V8 ENGINES

◆ See Figures 233, 234 and 235

Lubricate the cable pivot points every 100 hours or 12 months, whichever comes first. If the vessel is being operated in salt water, it can't hurt to lubricate every 50 hours or 6 months whichever comes first. Always use SAE 30W engine oil on the pivot points and cable guide contact surfaces.

8.1L V8 ENGINES

◆ See Figure 236

Lubricate the cable pivot points every 100 hours or 12 months, whichever comes first. If the vessel is being operated in salt water, it can't hurt to lubricate every 50 hours or 6 months whichever comes first. Always use SAE 30W engine oil on the pivot points and cable guide contact surfaces.

Shift Cable & Transmission Linkage Pivot Points

3.0L ENGINES

◆ See Figure 237

Lubricate the cable pivot points every 100 hours or 12 months, whichever comes first. If the vessel is being operated in salt water, lubricate every 50 hours or 6 months whichever comes first. Always use SAE 30W engine oil, or Quicksilver   2-4-C Marine Lubricant.

4.3L V6 & 5.0L/5.7L/6.2L V8 ENGINES

◆ See Figures 238 thru 242

Lubricate the cable pivot points every 100 hours or 12 months, whichever comes first. If the vessel is being operated in salt water, lubricate every 50 hours or 6 months whichever comes first. Always use SAE 30W engine oil.

8.1L V8 ENGINES

◆ See Figures 214, 242 and 243

Lubricate the cable pivot points every 100 hours or 12 months, whichever comes first. If the vessel is being operated in salt water, lubricate every 50 hours or 6 months whichever comes first. Always use SAE 30W engine oil.

2-56 ENGINE AND DRIVE MAINTENANCE

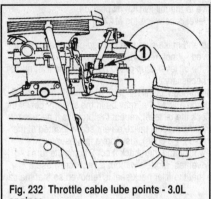

Fig. 232 Throttle cable lube points - 3.0L engines

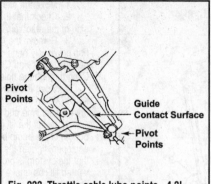

Fig. 233 Throttle cable lube points - 4.3L, 5.0L and 5.7L carbureted engines

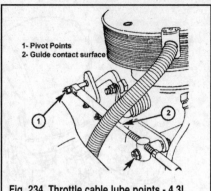

Fig. 234 Throttle cable lube points - 4.3L, 5.0L and 5.7L TBI engines

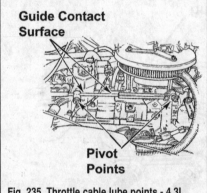

Fig. 235 Throttle cable lube points - 4.3L, 5.0L, 5.7L and 6.2L MPI engines

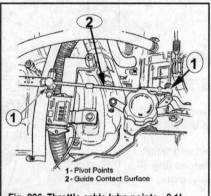

Fig. 236 Throttle cable lube points - 8.1L engines

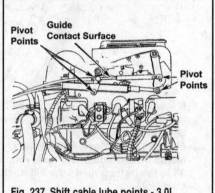

Fig. 237 Shift cable lube points - 3.0L engines

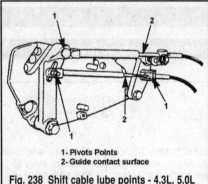

Fig. 238 Shift cable lube points - 4.3L, 5.0L and 5.7L engines (carb and TBI)

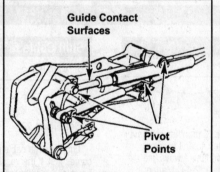

Fig. 239 Shift cable lube points - 4.3L, 5.0L, 5.7L and 6.2L engines (MPI)

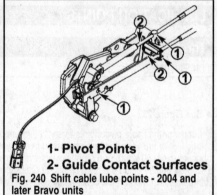

Fig. 240 Shift cable lube points - 2004 and later Bravo units

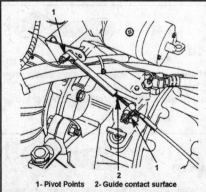

Fig. 241 Shift cable lube points - inboard engines

Fig. 242 Don't forget the detent balls in the shift lever on in-line transmissions

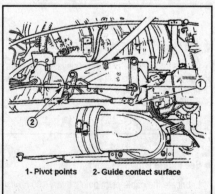

Fig. 243 Shift cable lube points - 8.1L engines

ENGINE AND DRIVE MAINTENANCE

Steering System

ALPHA UNITS

◆ See Figures 244 and 245

■ Always check the steering system and its components for loose, missing or damaged components and fittings before performing your lubrication procedures.

Lubricate the various steering system components every 50 hours or 2 months, whichever comes first when the vessel is being operated in fresh water. In salt water, lubricate every 25 hours or 30 days whichever comes first.

The transom end of the steering cable must be fully retracted into the cable housing before applying Quicksilver 2-4-C Marine Lube to the fitting.

The steering cable end should be lubricated with Quicksilver Special Lube 101, as should the pivot bolt.

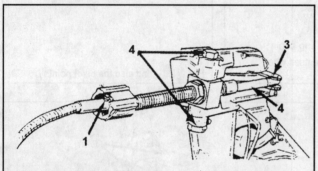

1- Steering cable grease fitting - 2-4-C Marine Lubricant
2- Steering - Cable end and exposed portion - Special Lubricant 101
3- Pivot Points - SAE 20 or 30 engine oil
4- Pivot Bolts - Special Lubricant 101

Fig. 244 Lubricate the steering system at these locations - Alpha w/manual steering

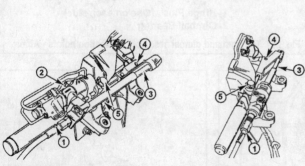

Earlier Style Control Valve   **Later Style Control Valve**
1 - Steering Cable Grease Fitting - 2-4-C Marine Lubricant With Teflon
2 - Control Valve Grease Fitting - 2-4-C Marine Lubricant With Teflon
3 - Steering Cable End - Special Lubricant 101
4 - Pivot Point - Sae 20 Or 30 Engine Oil
5 - Pivot Bolts - Special Lubricant 101

Fig. 245 Lubricate the steering system at these locations - Alpha w/power steering

Lubricate all pivot points on the system with SAE 30W motor oil.

The control valve grease fitting takes Quicksilver 2-4-C Marine Lube.

BRAVO UNITS

◆ See Figures 246 thru 249

■ Always check the steering system and its components for loose, missing or damaged components and fittings before performing your lubrication procedures.

Lubricate the system components every 100 hours or once a season, whichever comes first.

2001-03 Models

The transom end of the steering cable must be fully retracted into the cable housing before applying Quicksilver 2-4-C Marine Lube to the fitting.

The steering cable end should be lubricated with Quicksilver Special Lube 101, as should the pivot bolt.

Lubricate all pivot points on the system with SAE 30W motor oil.

The control valve grease fitting takes Quicksilver 2-4-C Marine Lube.

2004 & Later Models

On models with a grease fitting, the transom end of the steering cable must be fully retracted into the cable housing before applying Quicksilver Special Lube 101 to the fitting. Approximately 3 pumps from a normal hand operated grease gun should do the trick.

The steering cable end should be lubricated with Quicksilver Special Lube 101 once the cable has been fully extended, as should the pivot bolt.

Lubricate all pivot points on the system with Quicksilver Special Lube 101.

The control valve grease fitting (if equipped) takes Quicksilver 2-4-C Marine Lube.

TIE BAR PIVOT POINTS

◆ See Figure 250

Lubricate all pivot points at least every 100 hours or once a season, whichever comes first. Use only SAE 30W engine oil.

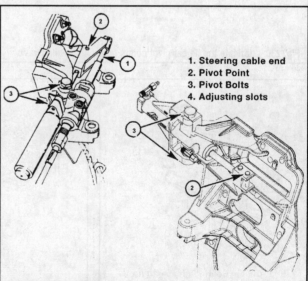

1. Steering cable end
2. Pivot Point
3. Pivot Bolts
4. Adjusting slots

Fig. 246 Lubricate the steering system at these locations - Bravo through about 2003

2-58 ENGINE AND DRIVE MAINTENANCE

Transom and Gimbal Assembly, Hinge Pins and Gimbal Bearing

◆ See Figures 251 thru 254

On Alpha units, lubricate every 100 hours or 6 months, whichever comes first when the vessel is being operated in fresh water. In salt water, lubricate every 50 hours or 2 months whichever comes first. Always use Quicksilver 2-4-C Marine Lubricant on the hinge pins and U-Joint/Gimbal Bearing Grease on the bearing.

On Bravo Standard units, lubricate every 100 hours or once a season, whichever comes first. On Horizon models, lubricate every 200 hours or every third season, whichever comes first. Always use U-Joint/Gimbal Bearing Grease on the bearing.

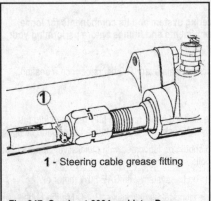

Fig. 247 On about 2004 and later Bravo models, check for a steering cable grease fitting on certain models...

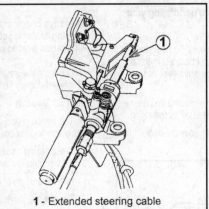

Fig. 248 ...lube the cable itself when extended...

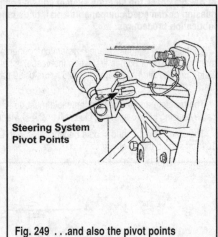

Fig. 249 ...and also the pivot points

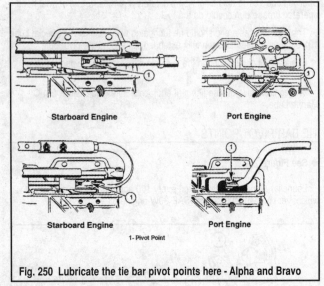

Fig. 250 Lubricate the tie bar pivot points here - Alpha and Bravo

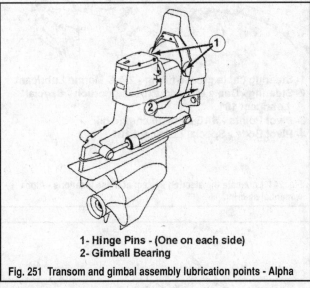

Fig. 251 Transom and gimbal assembly lubrication points - Alpha

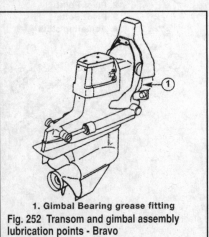

Fig. 252 Transom and gimbal assembly lubrication points - Bravo

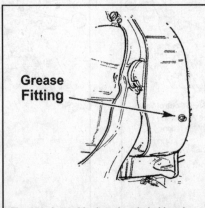

Fig. 253 A good look at the gimbal bearing grease fitting (all models)...

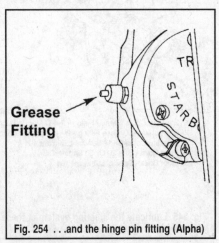

Fig. 254 ...and the hinge pin fitting (Alpha)

ENGINE AND DRIVE MAINTENANCE   2-59

Engine Coupler/U-Joint Splines

◆ See Figures 255 thru 258

Lubricate every 100 hours or once a season whichever comes first. On Horizon models, lubricate every 200 hours or every third season, whichever comes first. Always use Quicksilver Engine Coupler Spline grease except on extension models where you'll need U-Joint & Gimbal Bearing grease.

■ If the vessel is operated at idle for prolonged periods of time, it's a good idea to lubricate the coupler every 50 hours or so.

If your U-Joint cross bearing are equipped with grease fittings, use 3-6 pumps of grease from a standard hand-operated grease gun.

On driveshaft extension models, use 3-4 pumps of grease from a standard hand-operated grease gun on the shaft grease fittings and 10-12 pumps on the transom end and engine end fittings.

When lubricating the engine coupler splines with the grease fitting, use 8-10 pump of grease from a standard hand-operated grease gun or until grease comes out of the fitting. On Bravo models, it is not necessary to remove the drive.

Starter Motor - Inboard Engines Only

◆ See Figure 259

Disconnect the ignition coil/coil pack lead(s) and then lubricate the front bushing on the starter motor while cranking the engine. Use SAE 30W motor oil only.

Pull the plastic plug out of the hole in the flywheel housing a drip a few drops of SAE 30W motor oil through the hole and onto the starter motor shaft.

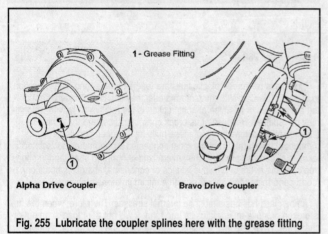

Fig. 255 Lubricate the coupler splines here with the grease fitting

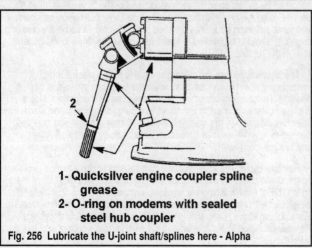

Fig. 256 Lubricate the U-joint shaft/splines here - Alpha

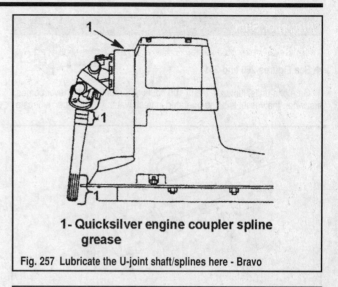

Fig. 257 Lubricate the U-joint shaft/splines here - Bravo

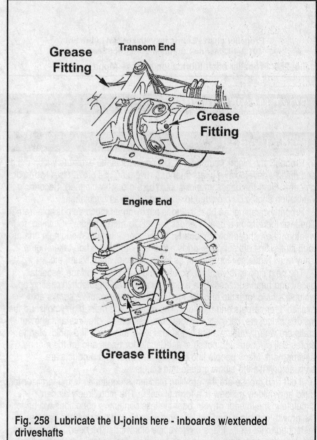

Fig. 258 Lubricate the U-joints here - inboards w/extended driveshafts

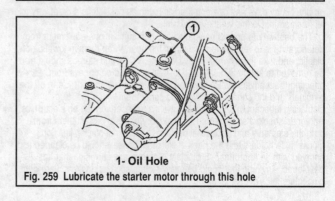

Fig. 259 Lubricate the starter motor through this hole

2-60 ENGINE AND DRIVE MAINTENANCE

Propeller Shaft

◆ See Figures 260 and 261

On Alpha units, lubricate every 100 hours or 6 months, whichever comes first when the vessel is being operated in fresh water. In salt water, lubricate every 50 hours or 2 months whichever comes first. Always use Quicksilver Special Lube 101, Quicksilver 2-4-C Marine Lubricant or Quicksilver Perfect Seal. All three are listed in order of their relative effectiveness.

On Bravo units, lubricate every 4 months when the vessel is being operated in fresh water. In salt water, lubricate every 2 months. Always use Quicksilver Special Lube 101 or Quicksilver 2-4-C Marine Lubricant. The two are listed in order of their relative effectiveness.

Remove the propeller and apply lubricant to the shaft splines.

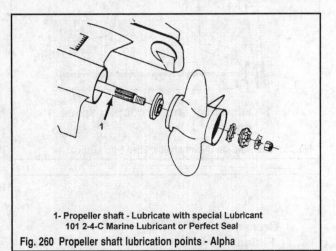

1- Propeller shaft - Lubricate with special Lubricant 101 2-4-C Marine Lubricant or Perfect Seal

Fig. 260 Propeller shaft lubrication points - Alpha

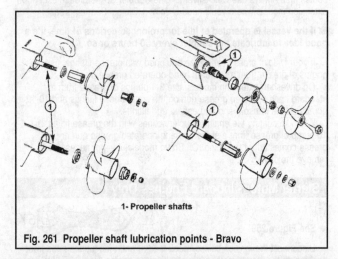

1- Propeller shafts

Fig. 261 Propeller shaft lubrication points - Bravo

BOAT MAINTENANCE

Inside The Boat

Probably the biggest surprise for boat owners is the extent to which mold and mildew develop in a boats interior. Preventing this growth is a two-fold process. First, the boat's interior should be thoroughly cleaned. Second, ventilation should be provided to allow adequate air circulation.

Properly cleaning the boat includes removing as much as possible from it. The less there is on a boat, the less there is to attract mildew. Clothing, foul weather gear, shoes, books, charts, paper goods, leather, bedding, curtains, food stuffs, first aid supplies, odds and ends, etc., should be taken home. They're all fertile soil for mildew, as is dirt, grease, soap scum, etc.

The next step is to vacuum, scrub and polish every surface, especially galley and head surfaces. Use a polish that leaves a protective coating on which it's hard for mold to get a foothold. Mildew-preventive sprays are available for carpets and furniture. The cleaner and more highly polished the insides of lockers, cabinets, refrigerators, icemakers, drawers and shower stalls are when you leave them, the less likely it is you'll find mold. Marine stores sell cleaners and polishes specifically for boats and for the marine environment. Many people find that regular household products are satisfactory. It's the elbow grease that counts.

If dirt and grease are the growing medium, stagnant air is the fertilizer for mold and mildew (mildew is a form of mold). The more freely air can circulate in the interior of your boat the less conducive conditions will be for the growth of mold. Openings or vents at each end near the top can be fashioned to let air in. Flaps or stovepipe elbows can be used to keep rain out. Hatches and windows can then be left partially open. Visiting the boat on dry, sunny days and opening it up as much as possible is a great way to let fresh air in. Do this whenever possible.

The best way to protect navigation and communications equipment from deteriorating is to remove the units from the boat. Wrap them in towels, not plastic, and take them home. Television sets, VCRs and stereos should also be removed to keep them safe and to keep cold and dampness from affecting them adversely. Before storing electronics equipment, clean off the terminals and the connecting plugs and spray them with a moisture-displacing lubricant that specifically says it's intended for use on electronics. Antennas should also be stored in a safe, dry place to protect them from both the elements and accidental damage. All terminal blocks, junction blocks, fuse holders and the back of electrical panels should be cleaned and sprayed with an appropriate protectant. Remove any corrosion that has developed. Anything that could be easily stolen, such as anchors, flare guns, binoculars, etc., is best taken home.

The Boat's Exterior

HULL

◆ See Figure 262

Fiberglass reinforced plastic hulls are tough, durable, and highly resistant to impact. However, like any other material they can be damaged. One of the advantages of this type of construction is the relative ease with which it may be repaired. Because of its break characteristics, and the simple techniques used in restoration, these hulls have gained popularity throughout the world. From the most congested urban marina, to isolated lakes in wilderness areas, to the severe cold of northern seas, and in sunny tropic remote rivers of primitive islands or continents, fiberglass boats can be found performing their daily task with a minimum of maintenance.

A fiberglass hull has almost no internal stresses. Therefore, when the hull is broken or stove-in, it retains its true form. It will not dent to take an out-of-shape set. When the hull sustains a severe blow, the impact will either be absorbed by deflection of the laminated panel or the blow will result in a definite, localized break. In addition to hull damage, bulkheads, stringers, and other stiffening structures attached to the hull may also be affected and therefore, should be checked. Repairs are usually confined to the general area of the rupture.

The best way to care for a fiberglass hull is to first wash it thoroughly. Immediately after hauling the boat, while the bottom is still wet, is best, if possible. Remove any growth that has developed on the bottom. Use a pressure cleaner or a stiff brush to remove barnacles, grass, and slime. Pay particular attention to the waterline area. A scraper of some sort may be needed to attack tenacious barnacles. Pot scrubbers work well. Attend to any blisters; don't wait!

Remove cushions and all weather curtains and enclosures and take them home. Make sure they are clean before storing them, and don't store them tightly rolled. After washing the topsides, remove any stains that have developed with one of the fiberglass stain removers sold in marine stores. For stubborn stains, wet-sanding with 600-grit paper may be necessary. Remove oxidation and stains from metal parts. Apply a coat of wax to everything.

ENGINE AND DRIVE MAINTENANCE

BELOW WATERLINE

◆ See Figures 263 thru 266

A foul bottom can seriously affect boat performance. This is one reason why racers, large and small, both powerboat and sail, are constantly giving attention to the condition of the hull below the waterline.

In areas where marine growth is prevalent, a coating of vinyl, anti-fouling bottom paint should be applied. If growth has developed on the bottom, it can be removed with a solution of Muriatic acid applied with a brush or swab and then rinsed with clear water. Always use rubber gloves when working with Muriatic acid and take extra care to keep it away from your face and hands. The fumes are toxic. Therefore, work in a well-ventilated area, or if outside, keep your face on the windward side of the work.

Barnacles have a nasty habit of making their home on the bottom of boats that have not been treated with anti-fouling paint. Actually they will not harm the fiberglass hull, but can develop into a major nuisance.

If barnacles or other crustaceans have attached themselves to the hull, extra work will be required to bring the bottom back to a satisfactory condition. First, if practical, put the boat into a body of fresh water and allow it to remain for a few days. A large percentage of the growth can be removed in this manner. If this remedy is not possible, wash the bottom thoroughly with a high-pressure fresh water source and use a scraper. Small particles of hard shell may still hold fast. These can be removed with sandpaper.

Anodes (Zincs)

The idea behind anodes is simple: When dissimilar metals are dunked in water and a small current is leaked between or amongst them, the less-noble metal (galvanically speaking) is sacrificed.

The zinc alloy that the anodes are made of is designed to be less noble than the aluminum alloy your drive unit is made from. If there's any electrolysis and there almost always is, the inexpensive zinc anodes are consumed in lieu of the expensive drive.

These zincs need a little attention in order to do their job. Make sure they're there, solidly attached to a clean mounting site and not covered with any kind of paint or wax.

Periodically inspect them to make sure they haven't eroded too much. At a certain point in the erosion process, the mounting holes start to enlarge, which is when the zinc might fall off. Obviously, once that happens your drive no longer has any protection.

SERVICING

◆ See Figures 267 thru 271

Depending on what kind of drive your boat has, you might have any number of zincs. Regardless of the number, there are some fundamental rules to follow that will give your boat's sacrificial anodes the ability to do the best job protecting your boat's underwater hardware that they can.

The first thing to remember is that zincs are electrical components and like all electrical components, they require good clean connections. So after you've undone the mounting hardware and removed last year's zincs, you want to get the zinc mounting sites clean and shiny.

Get a piece of coarse emery cloth or some 80-grit sandpaper. Thoroughly rough up the areas where the zincs attach (there's often a bit of corrosion residue in these spots). Make sure to remove every trace of corrosion.

Zincs are attached with stainless steel machine screws that thread into the mounting for the zincs. Over the course of a season, this mounting hardware is inclined to loosen. Mount the zincs and tighten the mounting hardware securely. Tap the zincs with a hammer hitting the mounting screws squarely. This process tightens the zincs and allows the mounting hardware to become a bit loose in the process. Now, do the final tightening. This will insure your zincs stay put for the entire season.

Fig. 262 The best way to care for a fiberglass hull is to wash it thoroughly to remove any growth that has developed on the bottom

Fig. 263 In areas where marine growth is prevalent, a coating of vinyl, anti-fouling bottom paint should be applied

Fig. 264 This anti-foul paint has seen better days and must be replaced

Fig. 265 This hull is in even worse condition and should be sand blasted and repainted

Fig. 266 This beautiful new fiberglass hull will not stay this good looking for long if it is not protected with anti-foul paint

2-62 ENGINE AND DRIVE MAINTENANCE

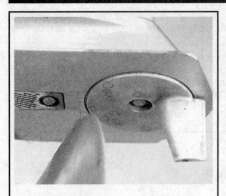

Fig. 267 What a trim tab should look like when it's in good condition

Fig. 268 Although many stern drives use the trim tab as an anode...

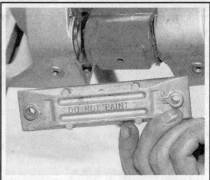

Fig. 269 ...other types of anodes are also used throughout the drive, like this one mounted on a stern bracket...

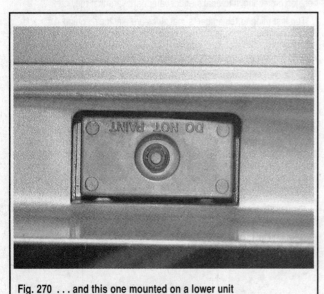

Fig. 270 ...and this one mounted on a lower unit

Fig. 271 Most anodes are easily removed by loosening and removing their attaching fasteners

Although all engines, drives and boats differ, here is a list of where you might expect to find anodes on your application:
- Bottom of the anti-cavitation plate
- Underside of the gimbal housing (also where the Mercathode System electrode assembly is found)
- Transom of the boat
- Ends of the trim cylinders
- Bearing carrier (Alpha and Bravo I).

LOCATIONS

◆ See Figures 272 thru 279

INSPECTION

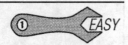

◆ See Figure 280

If you use your boat in salt water, and your zincs never wear, inspect them carefully. Paint or wax on zincs prevents them from working properly. They must be left bare. If the zincs are installed properly and not painted or waxed, inspect around them for signs of corrosion. If corrosion is found, strip it off immediately and repaint with a rust inhibiting paint. If in doubt, replace the zincs.

On the other hand, if your zinc seems to erode in no time at all, this may be a symptom of the zinc itself. Each manufacturer uses a specific blend of metals in their zincs. If you are using zincs with the wrong blend of metals, they may erode more quickly or leave you with diminished protection.

1 - Gearcase anodic plate
2 - Ventilation plate anode
3 - Drive mounted anodic block
4 - MerCathode system
5 - Anode kit
6 - Trim cylinder anodes
7 - Bearing carrier anode
8 - Propshaft anode

Fig. 272 Here's a good look at where you can normally find an anode, although not all drives will have anodes in all locations

ENGINE AND DRIVE MAINTENANCE 2-63

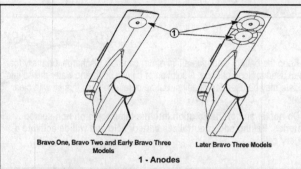

Fig. 273 Gearcase anodic plate - all drives will have one or two anodes on the underside of the back of the lower unit...

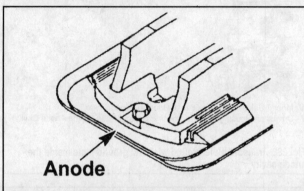

Fig. 274 ...while most will also have one located on the top side of the leading edge of the lower unit - ventilation plate anode

Continuity Circuit
◆ See Figures 281 thru 287

In order to ensure proper electrical continuity between the engine, transom assembly and drive unit, the transom assembly and drive unit are equipped with special ground circuit wires. Without this system, the anodes and the MerCathode system would be useless. All grounds should be checked for loose connections and damaged wires every 3 years or 200 hours, whichever comes first.

Batteries

Difficulty in starting accounts for almost half of the service required on boats each year. A survey by a major engine parts company indicated that roughly one third of all boat owners experienced a "won't start" condition in a given year. When an engine won't start, most people blame the battery when, in fact, it may be that the battery has run down in a futile attempt to start an engine with other problems.

Maintaining your battery in peak condition may be thought of as either tune-up or maintenance material. Most wise boaters will consider it to be both. A complete check up of the electrical system in your boat at the beginning of the boating season is a wise move. Continued regular maintenance of the battery will ensure trouble free starting on the water.

Complete battery service procedures are included in this section. The following are a list of basic electrical system service checks that should be performed.
- Check the battery for solid cable connections
- Check the battery and cables for signs of corrosion damage
- Check the battery case for damage or electrolyte leakage
- Check the electrolyte level in each cell
- Check to be sure the battery is fastened securely in position
- Check the battery's state of charge and charge as necessary
- Check battery voltage while cranking the starter. Voltage should remain above 9.5 volts

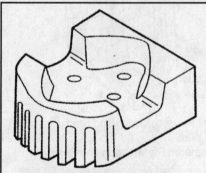

Fig. 275 Drive mounted anodic block - this will usually be found on the underside of the gimbal housing; and connected to the MerCathode system if equipped

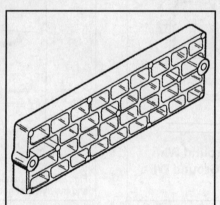

Fig. 276 Anode kit - will be mounted to transom if equipped

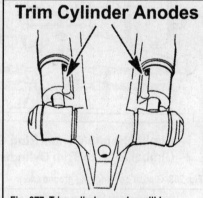

Fig. 277 Trim cylinder anodes will be mounted right where you'd expect

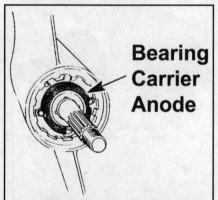

Fig. 278 Bearing carrier anodes are used on Alpha and Bravo I drives and found between the propeller and the housing

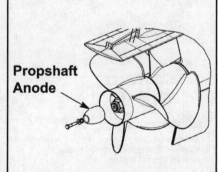

Fig. 279 Propeller shaft anodes are used on most Bravo III drives

Fig. 280 Such extensive erosion of a trim tab compared with a new tab suggests an electrolysis problem or complete disregard for periodic maintenance

2-64 ENGINE AND DRIVE MAINTENANCE

- Clean the battery, terminals and cables
- Coat the battery terminals with dielectric grease or terminal protector
- Check the tension on the alternator belt.

Batteries that are not maintained on a regular basis can fall victim to parasitic loads (small current drains which are constantly drawing current from the battery, like clocks, small lights, etc.). Normal parasitic loads may drain a battery on boat that is in storage and not used frequently. Boats that have additional accessories with increased parasitic load may discharge a battery sooner. Storing a boat with the negative battery cable disconnected or battery switch turned **OFF** will minimize discharge due to parasitic loads.

CLEANING

Keep the battery clean; as a film of dirt can help discharge a battery that is not used for long periods. A solution of baking soda and water mixed into a paste may be used for cleaning, but be careful to flush this off with clear water.

■ Do not let any of the solution into the filler holes on non-sealed batteries. Baking soda neutralizes battery acid and will de-activate a battery cell.

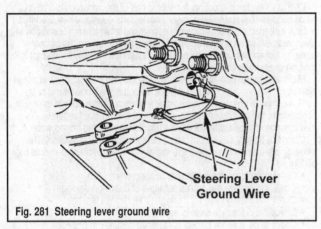

Fig. 281 Steering lever ground wire

1 - Inner Transom Plate To Gimbal Housing Ground Wire
2 - Driveshaft Housing To Gear Housing Ground Plate (Inside Anode Cavity)

Fig. 282 Transom plate ground wire and ground plate inside the anode cavity

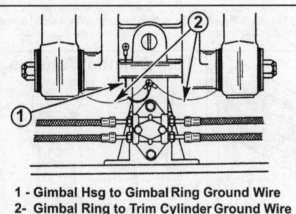

1 - Gimbal Hsg to Gimbal Ring Ground Wire
2 - Gimbal Ring to Trim Cylinder Ground Wire

Fig. 283 Gimbal housing/ring ground wires

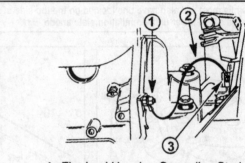

1 - Flywheel Housing Grounding Stud
2 - Ground Wire
3 - Inner Transom Plate Grounding Screw

Fig. 284 Inner transom grounding mechanisms between the plate and the flywheel housing

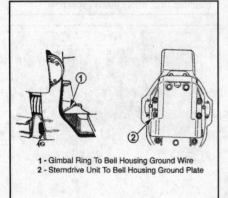

1 - Gimbal Ring To Bell Housing Ground Wire
2 - Sterndrive Unit To Bell Housing Ground Plate

Fig. 285 Bell housing ground wire and plate

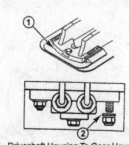

1 - Driveshaft Housing To Gear Housing Anodic Plate
2 - Hydraulic Connector Block To Gimbal Housing Ground Washer

Fig. 286 Drive connections

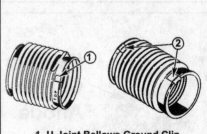

1 - U-Joint Bellows Ground Clip
2 - Exhaust Bellows Ground Clips

Fig. 287 U-Joint and exhaust bellows ground clips, exhaust tube will be similar

ENGINE AND DRIVE MAINTENANCE 2-65

CHECKING SPECIFIC GRAVITY

The electrolyte fluid (sulfuric acid solution) contained in the battery cells will tell you many things about the condition of the battery. Because the cell plates must be kept submerged below the fluid level in order to operate, maintaining the fluid level is extremely important. In addition, because the specific gravity of the acid is an indication of electrical charge, testing the fluid can be an aid in determining if the battery must be replaced. A battery in a boat with a properly operating charging system should require little maintenance, but careful, periodic inspection should reveal problems before they leave you stranded.

✱✱ CAUTION

Battery electrolyte contains sulfuric acid. If you should splash any on your skin or in your eyes, flush the affected area with plenty of clear water. If it lands in your eyes, get medical help immediately.

As stated earlier, the specific gravity of a battery's electrolyte level can be used as an indication of battery charge. At least once a year, check the specific gravity of the battery. It should be between 1.20 and 1.26 on the gravity scale. Most parts stores carry a variety of inexpensive battery testing hydrometers. These can be used on any non-sealed battery to test the specific gravity in each cell.

Conventional Battery

◆ See Figure 288

A hydrometer is required to check the specific gravity on all batteries that are not maintenance-free. The hydrometer has a squeeze bulb at one end and a nozzle at the other. Battery electrolyte is sucked into the hydrometer until the float or pointer is lifted from its seat. The specific gravity is then read by noting the position of the float/pointer. If gravity is low in one or more cells, the battery should be slowly charged and checked again to see if the gravity has come up. Generally, if after charging, the specific gravity of any two cells varies more than 50 points (0.50), the battery should be replaced, as it can no longer produce sufficient voltage to guarantee proper operation.

Check the battery electrolyte level at least once a month, or more often in hot weather or during periods of extended operation. Electrolyte level can be checked either through the case on translucent batteries or by removing the cell caps on opaque-case types. The electrolyte level in each cell should be kept filled to the split ring inside each cell, or the line marked on the outside of the case.

■ Never use mineral water or water obtained from a well. The iron content in these types of water is too high and will shorten the life of or damage the battery.

If the level is low, add only distilled water through the opening until the level is correct. Each cell is separate from the others, so each must be checked and filled individually. Distilled water should be used, because the chemicals and minerals found in most drinking water are harmful to the battery and could significantly shorten its life.

If water is added in freezing weather, the battery should be warmed to allow the water to mix with the electrolyte. Otherwise, the battery could freeze.

Maintenance-Free Batteries

Although some maintenance-free batteries have removable cell caps for access to the electrolyte, the electrolyte condition and level is usually checked using the built-in hydrometer "eye." The exact type of eye varies between battery manufacturers, but most apply a sticker to the battery itself explaining the possible readings. When in doubt, refer to the battery manufacturer's instructions to interpret battery condition using the built-in hydrometer.

The readings from built-in hydrometers may vary, however a green eye usually indicates a properly charged battery with sufficient fluid level. A dark eye is normally an indicator of a battery with sufficient fluid, but one that may be low in charge. In addition, a light or yellow eye is usually an indication that electrolyte supply has dropped below the necessary level for battery (and hydrometer) operation. In this last case, sealed batteries with an insufficient electrolyte level must usually be discarded.

BATTERY TERMINALS

◆ See Figures 289 and 290

At least once a season, the battery terminals and cable clamps should be cleaned. Loosen the clamps and remove the cables, negative cable first. On batteries with top mounted posts, the use of a puller specially made for this purpose is recommended. These are inexpensive and available in most parts stores.

Clean the cable clamps and the battery terminal with a wire brush, until all corrosion, grease, etc., is removed and the metal is shiny. It is especially important to clean the inside of the clamp thoroughly (a wire brush is useful here), since a small deposit of foreign material or oxidation there will prevent a sound electrical connection and inhibit either starting or charging. It is also a good idea to apply some dielectric grease to the terminal, as this will aid in the prevention of corrosion.

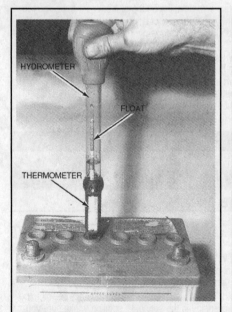

Fig. 288 The best way to determine the condition of a battery is to test the electrolyte with a battery hydrometer

Fig. 289 A battery post cleaner is used to clean the battery posts. . .

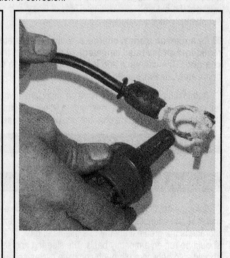

Fig. 290 . . .and the battery terminals

2-66 ENGINE AND DRIVE MAINTENANCE

After the clamps and terminals are clean, reinstall the cables, negative cable last; do not hammer the clamps onto battery posts. Tighten the clamps securely, but do not distort them. To retard corrosion, give the clamps and terminals a thin external coating of clear polyurethane paint after installation.

Check the cables at the same time that the terminals are cleaned. If the insulation is cracked or broken, or if its end is frayed, the cable should be replaced with a new one of the same length and gauge.

BATTERY & CHARGING SAFETY PRECAUTIONS

Always follow these safety precautions when charging or handling a battery.

1. Wear eye protection when working around batteries. Batteries contain corrosive acid and produce explosive gas a byproduct of their operation. Acid on the skin should be neutralized with a solution of baking soda and water made into a paste. In case acid contacts the eyes, flush with clear water and seek medical attention immediately.
2. Avoid flame or sparks that could ignite the hydrogen gas produced by the battery and cause an explosion. Connection and disconnection of cables to battery terminals is one of the most common causes of sparks.
3. Always turn a battery charger **OFF**, before connecting or disconnecting the leads. When connecting the leads, connect the positive lead first, then the negative lead, to avoid sparks.
4. When lifting a battery, use a battery carrier or lift at opposite corners of the base.
5. Ensure there is good ventilation in a room where the battery is being charged.
6. Do not attempt to charge or load-test a maintenance-free battery when the charge indicator dot is indicating insufficient electrolyte.
7. Disconnect the negative battery cable if the battery is to remain in the boat during the charging process.
8. Be sure the ignition switch is **OFF** before connecting or turning the charger **ON**. Sudden power surges can destroy electronic components.
9. Use proper adapters to connect the charger leads to batteries with non-conventional terminals.

BATTERY CHARGERS

Before using any battery charger, consult the manufacturer's instructions for its use. Battery chargers are electrical devices that change Alternating Current (AC) to a lower voltage of Direct Current (DC) that can be used to charge a marine battery. There are two types of battery chargers - manual and automatic.

A manual battery charger must be physically disconnected when the battery has come to a full charge. If not, the battery can be overcharged, and possibly fail. Excess charging current at the end of the charging cycle will heat the electrolyte, resulting in loss of water and active material, substantially reducing battery life.

■ As a rule, on manual chargers, when the ammeter on the charger registers half the rated amperage of the charger, the battery is fully charged. This can vary, and it is recommended to use a hydrometer to accurately measure state of charge.

Automatic battery chargers have an important advantage - they can be left connected (for instance, overnight) without the possibility of overcharging the battery. Automatic chargers are equipped with a sensing device to allow the battery charge to taper off to near zero as the battery becomes fully charged. When charging a low or completely discharged battery, the meter will read close to full rated output. If only partially discharged, the initial reading may be less than full rated output, as the charger responds to the condition of the battery. As the battery continues to charge, the sensing device monitors the state of charge and reduces the charging rate. As the rate of charge tapers to zero amps, the charger will continue to supply a few milliamps of current - just enough to maintain a charged condition.

REPLACING BATTERY CABLES

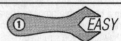

Battery cables don't go bad very often, but like anything else, they can wear out. If the cables on your boat are cracked, frayed or broken, they should be replaced.

When working on any electrical component, it is always a good idea to disconnect the negative (-) battery cable. This will prevent potential damage to many sensitive electrical components

Always replace the battery cables with one of the same length, or you will increase resistance and possibly cause hard starting. Smear the battery posts with a light film of dielectric grease, or a battery terminal protectant spray once you've installed the new cables. If you replace the cables one at a time, you won't mix them up.

■ Any time you disconnect the battery cables, it is recommended that you disconnect the negative (-) battery cable first. This will prevent you from accidentally grounding the positive (+) terminal when disconnecting it, thereby preventing damage to the electrical system.

Before you disconnect the cable(s), first turn the ignition to the **OFF** position. This will prevent a draw on the battery which could cause arcing. When the battery cable(s) are reconnected (negative cable last), be sure to check all electrical accessories are all working correctly.

STORAGE

If the boat is to be laid up for the winter or for more than a few weeks, special attention must be given to the battery to prevent complete discharge or possible damage to the terminals and wiring. Before putting the boat in storage, disconnect and remove the batteries. Clean them thoroughly of any dirt or corrosion and then charge them to full specific gravity reading. After they are fully charged, store them in a clean cool dry place where they will not be damaged or knocked over, preferably on a couple blocks of wood. Storing the battery up off the deck, will permit air to circulate freely around and under the battery and will help to prevent condensation.

Never store the battery with anything on top of it or cover the battery in such a manner as to prevent air from circulating around the filler caps. All batteries, both new and old, will discharge during periods of storage, more so if they are hot than if they remain cool. Therefore, the electrolyte level and the specific gravity should be checked at regular intervals. A drop in the specific gravity reading is cause to charge them back to a full reading.

In cold climates, care should be exercised in selecting the battery storage area. A fully-charged battery will freeze at about 60° below zero. A discharged battery, almost dead, will have ice forming at about 19° above zero.

WINTER STORAGE

General Information

Taking extra time to store the boat properly at the end of each season will increase the chances of satisfactory service at the next season. Remember, storage is the greatest enemy of a marine engine. In a perfect world the unit should be run on a monthly basis. The steering and shifting mechanism should also be worked through complete cycles several times each month. But who lives in a perfect world!

For most of us, if a small amount of time is spent in winterizing our beloved boats, the reward will be satisfactory performance, increased longevity and greatly reduced maintenance expenses. Remember that this is also the perfect time to review the Maintenance Interval charts and perform many of the procedures detailed in this section - do it now, because when it gets warm again, you want to go boating, not work on your engine!

Winter Storage Checklist

CARBURETED & TBI ENGINES

Proper winterizing involves adequate protection of the unit from physical damage, rust, corrosion, and dirt. The following steps provide guide to winterizing your marine engine at the end of a season.

Always keep a note or a checklist of just what maintenance needs to be done prior to starting the engine in the spring. In addition, make sure you mark down everything that was done before or during winter lay-up.

1. Keeping the fuel tank full over the winter will help cut down on condensation. Using fuel additives, like Quicksilver Gasoline stabilizer, that

control bacterial growth or prevent gelling in cold climates are also popular with many boaters. Always refer to the instructions on the container for the proper amount of stabilizer to add when filling your tanks.

2. Start the engine and allow it to idle until it reaches normal operating temperature - this will allow the stabilizer to mix with the fuel thoroughly. Turn off the engine. Make sure you are connected to a water source if the boat is out of the water!

3. Replace the oil and oil filter. Normal combustion produces corrosive acids that are absorbed by the oil. Leaving dirty oil in the engine for an extended time allows these acids to attack and damage bearing surfaces.

4. Change the drive/transmission oil.

5. Flush the sea water system thoroughly as detailed else where in this section.

6. Close the fuel shut-off valve, if equipped, to stop all flow of fuel from the tanks to the engine. If not equipped with a shut-off valve, carefully disconnect the inlet line and plug it (and the fitting) securely.

7. On EFI engines, remove the water separating fuel filter and pour in about 2 oz. of 2 cycle outboard oil. Reinstall the filter, start the engine and run it until it dies. Remove the filter again and install a new one.

8. On carbureted engines, remove the flame arrestor and start the engine. Run the engine at fast idle (1000-1500 rpm) and squirt about 8 oz. of Quicksilver Storage Seal into the carburetor. Just as the engine is starting to sputter, squirt the remaining 2 oz. of Seal into the carb and allow the engine to die. You can also use SAE 20W motor oil if no Storage Seal is available.

9. Disconnect the inlet line at the seawater pump.

10. Clean the flame arrestor and crankcase ventilation lines thoroughly and then re-install them.

11. Perform all applicable lubrication procedures detailed in this section.

12. Drain the seawater system, taking special care to empty all the low spots. We strongly recommend filling it with propylene glycol antifreeze!

13. On closed cooling systems, replace the coolant with a clean, fresh, 50/50 mixture of water and antifreeze. Always mix the antifreeze and water mixture prior to pouring it into the engine.

14. Fully charge the batteries. Disconnect all leads. Unattended, a battery naturally discharges over a period of several weeks. The electrolyte on a discharged battery can freeze at 20°F (-7°C), so keep the batteries fully charged or, better still, remove them to a warmer storage area. Small automatic trickle chargers work well.

15. Treat battery and cable terminals with petroleum jelly, silicone grease, or a heavy-duty corrosion inhibitor.

16. Protect external surfaces with a heavy-duty corrosion inhibitor.

17. Lubricate the steering system.

18. Lubricate the gimbal bearing, propeller shaft and U-joints.

19. Check the engine alignment and inspect the drive unit for obvious signs of damage. Clean the exterior surfaces and repaint any bare metal surfaces.

Although not specifically recommended by the factory, you may also want to perform the following steps:

20. Inspect all hoses for signs of softening, cracking or bulging, especially those routinely exposed to high heat. Check hose clamps for tightness and corrosion.

21. Remove the spark plugs and pour about an ounce of Storage Seal into each cylinder. You can also use SAE 20W motor oil. Crank the engine with the starter for a few seconds to make sure that the oil coats the cylinder bores completely.

22. Loosen the adjusting bolts or tensioners on all drive or serpentine belts so the tension is relieved.

23. Grease all greaseable points on the drivetrain.

24. Carefully coat the alternator and starter with a light lubricant to disperse the water.

25. Ensure that all drain holes on the stern drive are open by inserting a piece of wire into them.

26. Make sure the stern drive is in the full IN/DOWN position.

27. Cover the engine with a waterproof sheet in case there are any leaks from above. Some boaters will take the time to seal all openings to the engine (air inlet, breathers, exhaust), however this traps moisture and may do more harm than good.

28. Place a checklist in a handy spot to remind you of just what maintenance needs to be done prior to starting the engine in the spring - and also what you just completed!

29. If you visit the boat during the winter, additional protection can be achieved using the starter to turn the engine over and circulate oil to the bearings and cylinder walls.

30. If the boat is to be stored out of the water, always remove the drain plug(s) if equipped. We suggest you attach the plug to the ignition key as added insurance that you remember to reinstall it in the spring - sound silly? Why take the chance! If the boat is in the water, make sure that the seacock is fully closed.

■ **Special inhibiting oils are available that provide greater protection. Use these to replace the standard engine oil; run the engine briefly to coat all surfaces, and then drain the oil. Protection remains good, provided the engine is not turned.**

4.3L V6, 5.0L/5.7L/6.2L & 2006-08 8.1L MPI ENGINES

Proper winterizing involves adequate protection of the unit from physical damage, rust, corrosion, and dirt. The following steps provide guide to winterizing your marine engine at the end of a season.

Always keep a note or a checklist of just what maintenance needs to be done prior to starting the engine in the spring. In addition, make sure you mark down everything that was done before or during winter lay-up.

1. Fill the fuel tanks with fresh fuel and a sufficient quantity of Fuel System Treatment & Stabilizer:

   a. Obtain a 6 gal. fuel can or remote fuel tank.

   b. Add 5 gal. of regular unleaded fuel.

   c. Add 2 qts. of Premium Plus 2-cycle outboard oil; yes that's right, 2-stroke outboard oil.

   d. Add 5 oz. of Fuel System Treatment & Stabilizer, or 1 oz. of the concentrate.

   e. Make sure the engine is cool and release the fuel pressure as detailed in the Fuel System section.

   f. Close the fuel shut-off valve, if equipped, to stop all flow of fuel from the tanks to the engine. If not equipped with a shut-off valve, carefully disconnect the inlet line and plug it (and the fitting) securely.

   g. Now connect the remote tank with the special mixture to the fuel inlet fitting. Make sure there is a water supply and start the engine. Run the engine at 1300 rpm for 5 min., slowly return it to normal idle and turn it off.

■ **Never run the remote fuel tank dry during this procedure. If this happens, add more and start the process over again.**

   h. Remove the water separating fuel filter and install a new one. Remember this is a pressurized system, so don't forget to relieve the pressure as detailed in the appropriate Fuel Systems section if you haven't already done so.

2. Replace the oil and oil filter as detailed elsewhere in this section. Normal combustion produces corrosive acids that are absorbed by the oil. Leaving dirty oil in the engine for an extended time allows these acids to attack and damage bearing surfaces.

3. Change the drive/transmission oil as detailed elsewhere in this section.

4. Close the fuel shut-off valve, if equipped, to stop all flow of fuel from the tanks to the engine. If not equipped with a shut-off valve, carefully disconnect the inlet line and plug it (and the fitting) securely.

5. Flush the seawater system thoroughly as detailed else where in this section or in the Cooling System section.

6. Perform all applicable lubrication procedures detailed in this section.

7. Drain the seawater system, as detailed in this section and also in the Cooling System section; taking special care to empty all the low spots. We strongly recommend filling it with propylene glycol antifreeze!

■ **For additional insurance, some owners fill the seawater system with a mixture of propylene glycol anti-freeze (not ethylene like used in the closed system) and water to protect the engine against the lowest anticipated temperatures (as detailed by the antifreeze manufacturer). If you do this, just make sure to drain it completely (and safely) when you re-commission in the spring.**

8. On closed cooling systems, replace the coolant with a clean, fresh, 50/50 mixture of water and antifreeze (ethelene glycol). Always mix the antifreeze and water mixture prior to pouring it into the engine.

9. Fully charge the batteries. Disconnect all leads. Unattended, a battery naturally discharges over a period of several weeks. The electrolyte on a discharged battery can freeze at 20°F (-7°C), so keep the batteries fully charged or, better still, remove them to a warmer storage area. Small automatic trickle chargers work well.

ENGINE AND DRIVE MAINTENANCE

10. Treat battery and cable terminals with petroleum jelly, silicone grease, or a heavy-duty corrosion inhibitor.
11. Protect external surfaces with a heavy-duty corrosion inhibitor.
12. Lubricate the steering system.
13. Lubricate the gimbal bearing, propeller shaft and U-joints.
14. Check and inspect the drive unit for obvious signs of damage. Clean the exterior surfaces and repaint any bare metal surfaces.

Although not specifically recommended by the factory, you may also want to perform the following steps:

15. Inspect all hoses for signs of softening, cracking or bulging, especially those routinely exposed to high heat. Check hose clamps for tightness and corrosion.
16. Remove the spark plugs and pour about an ounce of Storage Seal into each cylinder. You can also use SAE 20W motor oil. Crank the engine with the starter for a few seconds to make sure that the oil coats the cylinder bores completely.
17. Loosen the adjusting bolts or tensioners on all drive or serpentine belts so the tension is relieved.
18. Grease all greaseable points on the drivetrain.
19. Carefully coat the alternator and starter with a light lubricant to disperse the water.
20. Ensure that all drain holes on the stern drive are open by inserting a piece of wire into them.
21. Make sure the stern drive is in the full IN/DOWN position.
22. Cover the engine with a waterproof sheet in case there are any leaks from above. Some boaters will take the time to seal all openings to the engine (air inlet, breathers, exhaust), however this traps moisture and may do more harm than good.
23. Place a checklist in a handy spot to remind you of just what maintenance needs to be done prior to starting the engine in the spring - and also what you just completed!
24. If you visit the boat during the winter, additional protection can be achieved using the starter to turn the engine over and circulate oil to the bearings and cylinder walls.
25. If the boat is to be stored out of the water, always remove the drain plug(s) if equipped. We suggest you attach the plug to the ignition key as added insurance that you remember to reinstall it in the spring - sound silly? Why take the chance! If the boat is in the water, make sure that the seacock is fully closed.

■ **Special inhibiting oils are available that provide greater protection. Use these to replace the standard engine oil; run the engine briefly to coat all surfaces, and then drain the oil. Protection remains good, provided the engine is not turned.**

2001-05 8.1L MPI ENGINES

■ **Things are a little fuzzy here - although the following are the procedures detailed by the factory for these models, we would suggest following the 2006-08 procedures instead. Both will work fine, but we just think the newer way is better. . .you be the judge.**

Proper winterizing involves adequate protection of the unit from physical damage, rust, corrosion, and dirt. The following steps provide guide to winterizing your marine engine at the end of a season.

Always keep a note or a checklist of just what maintenance needs to be done prior to starting the engine in the spring. In addition, make sure you mark down everything that was done before or during winter lay-up.

1. Fill the fuel tanks with fresh fuel and a sufficient quantity of Fuel System Treatment & Stabilizer. Be sure to follow the instructions on the container regarding the quantity necessary.
2. Start the engine and allow it to idle until it reaches normal operating temperature. This will ensure the stabilizer mixing properly with the fuel. Turn off the engine.
3. Replace the oil and oil filter as detailed elsewhere in this section. Normal combustion produces corrosive acids that are absorbed by the oil. Leaving dirty oil in the engine for an extended time allows these acids to attack and damage bearing surfaces.
4. Change the drive/transmission oil as detailed elsewhere in this section.
5. Flush the seawater system thoroughly as detailed elsewhere in this section or in the Cooling System section.
6. Close the fuel shut-off valve, if equipped, to stop all flow of fuel from the tanks to the engine. If not equipped with a shut-off valve, carefully disconnect the inlet line and plug it (and the fitting) securely.
7. Allow the engine to cool completely and then remove the water separating fuel filter. Pour in about 2 oz. of 2 cycle outboard oil; yes, that right, 2-stroke engine oil! Reinstall the filter.
8. Tag and disconnect the harness connections at the fuel pump and boost pump.
9. Start the engine and run it at idle until it dies. Remove the filter again and install a new one.
10. Reconnect the pump harnesses.
11. Perform all applicable lubrication procedures detailed in this section.
12. Drain the seawater system, as detailed in this section and also in the Cooling System section; taking special care to empty all the low spots. We strongly recommend filling it with propylene glycol antifreeze.

■ **For additional insurance, some owners fill the seawater system with a mixture of propylene glycol anti-freeze (not ethylene like used in the closed system) and water to protect the engine against the lowest anticipated temperatures (as detailed by the antifreeze manufacturer). If you do this, just make sure to drain it completely (and safely) when you re-commission in the spring.**

13. On closed cooling systems, replace the coolant with a clean, fresh, 50/50 mixture of water and antifreeze (ethylene glycol). Always mix the antifreeze and water mixture prior to pouring it into the engine.
14. Fully charge the batteries. Disconnect all leads. Unattended, a battery naturally discharges over a period of several weeks. The electrolyte on a discharged battery can freeze at 20°F (-7°C), so keep the batteries fully charged or, better still, remove them to a warmer storage area. Small automatic trickle chargers work well.
15. Treat battery and cable terminals with petroleum jelly, silicone grease, or a heavy-duty corrosion inhibitor.
16. Protect external surfaces with a heavy-duty corrosion inhibitor.
17. Lubricate the steering system as detailed in the Lubrication procedures found in this section.
18. Lubricate the gimbal bearing, propeller shaft and U-joints as detailed in the Lubrication procedures found in this section.
19. Check and inspect the drive unit for obvious signs of damage. Clean the exterior surfaces and repaint any bare metal surfaces.

Although not specifically recommended by the factory, you may also want to perform the following steps:

20. Inspect all hoses for signs of softening, cracking or bulging, especially those routinely exposed to high heat. Check hose clamps for tightness and corrosion.
21. Remove the spark plugs and pour about an ounce of Storage Seal into each cylinder. You can also use SAE 20W motor oil. Crank the engine with the starter for a few seconds to make sure that the oil coats the cylinder bores completely.
22. Loosen the adjusting bolts or tensioners on all drive or serpentine belts so the tension is relieved.
23. Grease all greaseable points on the drivetrain.
24. Carefully coat the alternator and starter with a light lubricant to disperse the water.
25. Ensure that all drain holes on the stern drive are open by inserting a piece of wire into them.
26. Make sure the stern drive is in the full IN/DOWN position.
27. Cover the engine with a waterproof sheet in case there are any leaks from above. Some boaters will take the time to seal all openings to the engine (air inlet, breathers, exhaust), however this traps moisture and may do more harm than good.
28. Place a checklist in a handy spot to remind you of just what maintenance needs to be done prior to starting the engine in the spring - and also what you just completed!
29. If you visit the boat during the winter, additional protection can be achieved using the starter to turn the engine over and circulate oil to the bearings and cylinder walls.
30. If the boat is to be stored out of the water, always remove the drain plug(s) if equipped. We suggest you attach the plug to the ignition key as added insurance that you remember to reinstall it in the spring - sound silly? Why take the chance! If the boat is in the water, make sure that the seacock is fully closed.

■ **Special inhibiting oils are available that provide greater protection. Use these to replace the standard engine oil; run the engine briefly to coat all surfaces, and then drain the oil. Protection remains good, provided the engine is not turned.**

ENGINE AND DRIVE MAINTENANCE 2-69

SPRING COMMISSIONING

Satisfactory performance and maximum enjoyment can be realized if a little time is spent in commissioning your boat in the spring. Assuming you have followed the steps we recommended to winterize your vessel (in addition to any the manufacturer specifies) and the unit has been properly stored, a minimum amount of work should be required to prepare it for use.

After performing the spring commissioning and testing the boat on the water, it is a good idea to perform a full tune-up. Remember, you are relying on your engine to get you where you want to go. Treat it good now and it will treat you good later.

Spring Commissioning Checklist

The following steps outline a logical sequence of tasks to be performed before starting your engine for the first time in a new season.

1. Pick up the checklist you made to remind yourself of just what maintenance needs to be done prior to starting the engine in the spring. You did remember to write yourself a checklist. . .right?
2. INSTALL THE DRAIN PLUG IF REMOVED!
3. Remove the cover placed over the engine last winter. Unseal any engine openings (air inlet, breathers, exhaust) previously sealed.
4. Replace all zincs.
5. If you took our advice, you removed the battery for the winter. While it was in storage, you should have kept it fully charged and it should be ready to go. So install the battery and connect the battery cables. Treat battery and cable terminals with petroleum jelly, silicone grease, or a heavy-duty corrosion inhibitor. Capacity test the batteries.
6. Tighten/adjust any drive or serpentine belts.
7. As you did in the winter, inspect all hoses for signs of softening, cracking or bulging, especially those routinely exposed to high heat. Check hose clamps for tightness and corrosion.
8. Ensure the exhaust manifolds are tight.
9. On closed system engines, check the condition of the coolant mixture with a coolant tester and adjust the mixture by adding antifreeze or water.
10. Bleed the fuel system of air.
11. Open the seacock or unplug the inlet line.
12. If you added coolant mixture to the seawater system, please make sure that you flush the system.
13. Start the engine and allow it to reach operating temperature.
14. Once running, check the oil pressure, the raw water discharge and the engine for oil and water leaks.
15. Drain water from the filter bowls. If an excessive amount of water is noted in the fuel system, you may want to consider a fuel conditioner or replcing the old fuel with fresh fuel.
16. After testing the boat on the water, it is a good idea to change the engine oil and filter. The drive/transmission oil should also be changed. Yes, we know you probably changed it last winter, but do it again anyway! This eliminates the adverse effects of moisture in the oil.
17. Run through your winterizing check list and make sure that anything disconnected, closed or removed is put back in place.

FIRING ORDERS

◆ See Figures 291 thru 298

■ To avoid confusion, ALWAYS label the spark plug wires with a piece of tape before disconnecting them from the plug or distributor cap. Always use the firing order you observe on your engine over the one depicted here!

■ It has come to our attention that there are a certain number of 3.0L engines out there that may detail the firing order as 1-4-3-2 on the spark arrestor. Although there is nothing in Mercruiser literature to confirm this, we have provided pictures of these engines as well. If your engine has this odd firing order marking, please be sure to confirm its correctness by observing the distributor cap and spark plug wire routing.

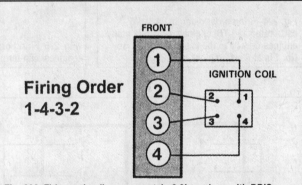

Fig. 292 Firing order diagram - certain 3.0L engines with DDIS ignition systems, please confirm with the firing order stamping on your engine and the cap-to-plug wire routing

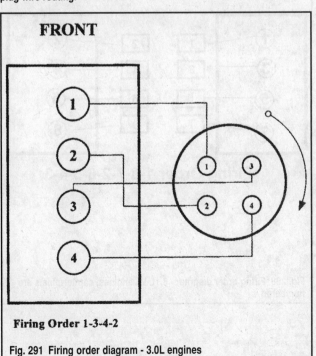

Fig. 291 Firing order diagram - 3.0L engines

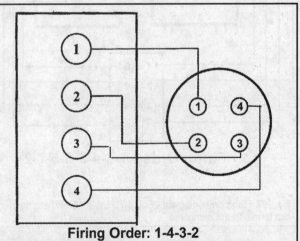

Fig. 293 Firing order diagram - certain 3.0L engines with EST ignition systems, please confirm with the firing order stamping on your engine and the cap-to-plug wire routing

2-70 ENGINE AND DRIVE MAINTENANCE

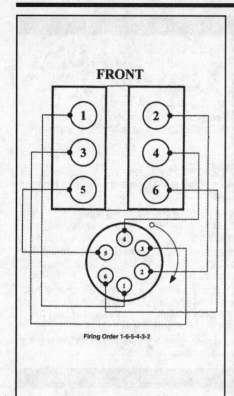

Fig. 294 Firing order diagram - 4.3L V6 carbureted and TBI engines. Note that many engines may have the cap rotated 180° so No. 1 is at the 12 o'clock position

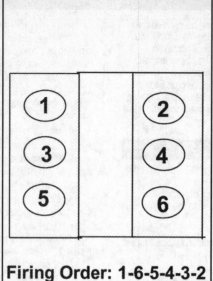

Firing Order: 1-6-5-4-3-2

Fig. 295 Firing order diagram - 4.3L V6 MPI engines, cap terminals are numbered

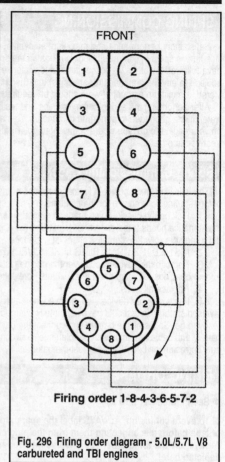

Firing order 1-8-4-3-6-5-7-2

Fig. 296 Firing order diagram - 5.0L/5.7L V8 carbureted and TBI engines

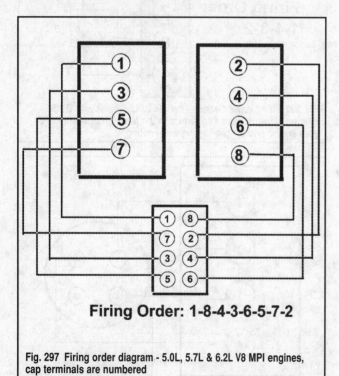

Firing Order: 1-8-4-3-6-5-7-2

Fig. 297 Firing order diagram - 5.0L, 5.7L & 6.2L V8 MPI engines, cap terminals are numbered

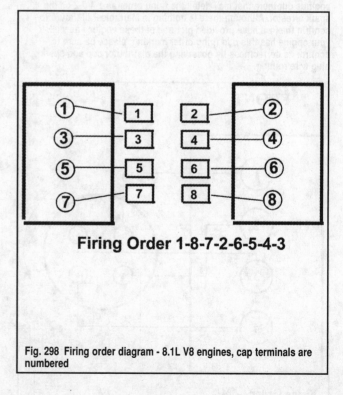

Firing Order 1-8-7-2-6-5-4-3

Fig. 298 Firing order diagram - 8.1L V8 engines, cap terminals are numbered

ENGINE AND DRIVE MAINTENANCE

SPECIFICATIONS

ENGINE MODELS

Year	Engine	Model	Displacement L/Cu. In.	Engine Type	HorsePower (At Prop.)	Fuel Delivery	Drive System
2001	3.0L	3.0L	3.0/181	L4	135	2 bbl	I/O: Alpha
	4.3L	4.3L	4.3/262	V6	190	2 bbl	I/O: Alpha, Bravo II, III
		4.3LH	4.3/262	V6	205	4 bbl	I/O: Alpha, Bravo II, III
		4.3L EFI	4.3/262	V6	210	TBI or MPI	I/O: Alpha, Bravo II, III
	5.0L	5.0L	5.0/305	V8	220	2 bbl	I/O: Alpha, Bravo I, II, III
		5.0L EFI	5.0/305	V8	240	TBI or MPI	I/O: Alpha, Bravo I, II, III
	5.7L	5.7L	5.7/350	V8	250	2 bbl	I/O: Alpha, Bravo I, II, III
			5.7/350	V8	260	2 bbl	I/B: Velvet, Hurth
			5.7/350	V8	270	2 bbl	I/B: Velvet
		5.7L EFI	5.7/350	V8	260	TBI or MPI	I/O: Alpha, Bravo I, II, III
		350 Mag MPI	5.7/350	V8	300	MPI	I/O: Alpha, Bravo I, II, III
			5.7/350	V8	300	MPI	I/B: Velvet, Hurth
			5.7/350	V8	315	MPI	I/B: Velvet
		350 Mag MPI/Horizon	5.7/350	V8	300	MPI	I/O: Alpha, Bravo I, II, III
			5.7/350	V8	300	MPI	I/B: Velvet, Hurth
	6.2L	Black Scorpion	6.2/377	V8	330	MPI	I/B: Velvet
		MX6.2 MPI	6.2/377	V8	320	MPI	I/O: Bravo I, II, III
			6.2/377	V8	320	MPI	I/B: Velvet, Hurth
	8.1	8.1S HO	8.1/496	V8	420	MPI	I/B: Velvet, Hurth
		8.1S Horizon	8.1/496	V8	370	MPI	I/B: Velvet, Hurth
		496 Mag	8.1/496	V8	375	MPI	I/O: Bravo I, II, III
		496 Mag HO	8.1/496	V8	425	MPI	I/O: Bravo I, II, III, X, XR, XZ
2002	3.0L	3.0L	3.0/181	L4	135	2 bbl	I/O: Alpha
	4.3L	4.3L	4.3/262	V6	190	2 bbl	I/O: Alpha, Bravo II, III
		4.3L EFI	4.3/262	V6	210	TBI	I/O: Alpha, Bravo II, III
		4.3L MPI	4.3/262	V6	220	MPI	I/O: Alpha, Bravo II, III
	5.0L	5.0L	5.0/305	V8	220	2 bbl	I/O: Alpha, Bravo I, II, III
		5.0L EFI	5.0/305	V8	240	TBI	I/O: Alpha, Bravo I, II, III
		5.0L MPI	5.0/305	V8	260	MPI	I/O: Alpha, Bravo I, II, III
	5.7L	5.7L	5.7/350	V8	250	2 bbl	I/O: Alpha, Bravo I, II, III
			5.7/350	V8	260	2 bbl	I/B: Velvet, Hurth
			5.7/350	V8	270	2 bbl	I/B: Velvet
		5.7L EFI	5.7/350	V8	260	TBI or MPI	I/O: Alpha, Bravo I, II, III
		350 Mag MPI	5.7/350	V8	300	MPI	I/O: Alpha, Bravo I, II, III
			5.7/350	V8	300	MPI	I/B: Velvet, Hurth
			5.7/350	V8	315	MPI	I/B: Velvet
		350 Mag MPI/Horizon	5.7/350	V8	300	MPI	I/O: Bravo I, II, III
			5.7/350	V8	300	MPI	I/B: Velvet, Hurth
	6.2L	Black Scorpion	6.2/377	V8	330	MPI	I/B: Velvet
		MX6.2 Black Scorpion	6.2/377	V8	340	MPI	I/B: Velvet
		MX6.2 MPI	6.2/377	V8	320	MPI	I/O: Bravo I, II, III
			6.2/377	V8	320	MPI	I/B: Velvet, Hurth
		MX6.2 MPI/Horizon	6.2/377	V8	320	MPI	I/O: Bravo I, II, III
			6.2/377	V8	320	MPI	I/B: Velvet, Hurth
	8.1L	8.1S HO	8.1/496	V8	420	MPI	I/B: Velvet, Hurth
		8.1S Horizon	8.1/496	V8	370	MPI	I/B: Velvet, Hurth
		496 Mag	8.1/496	V8	375	MPI	I/O: Bravo I, II, III
		496 Mag HO	8.1/496	V8	425	MPI	I/O: Bravo I, II, III, X, XR, XZ
2003	3.0L	3.0L	3.0/181	L4	135	2 bbl	I/O: Alpha
	4.3L	4.3L	4.3/262	V6	190	2 bbl	I/O: Alpha, Bravo II, III
		4.3L MPI	4.3/262	V6	220	MPI	I/O: Alpha, Bravo II, III
	5.0L	5.0L	5.0/305	V8	220	2 bbl	I/O: Alpha, Bravo I, II, III
		5.0L MPI	5.0/305	V8	260	MPI	I/O: Alpha, Bravo I, II, III
	5.7L	5.7L	5.7/350	V8	250	2 bbl	I/O: Alpha, Bravo I, II, III
			5.7/350	V8	260	2 bbl	I/B: Velvet, ZF
			5.7/350	V8	270	2 bbl	I/B: Velvet

2-72 ENGINE AND DRIVE MAINTENANCE

ENGINE MODELS

Year	Engine	Model	Displacement L/Cu. In.	Engine Type	HorsePower (At Prop.)	Fuel Delivery	Drive System
2003 (cont'd)	5.7L	350 Mag MPI	5.7/350	V8	300	MPI	I/O: Alpha, Bravo I, II, III
			5.7/350	V8	300	MPI	I/B: Velvet, ZF
			5.7/350	V8	315	MPI	I/B: Velvet
		350 Mag MPI/Horizon	5.7/350	V8	300	MPI	I/O: Bravo I, II, III
			5.7/350	V8	300	MPI	I/B: Velvet, ZF
	6.2L	Black Scorpion	6.2/377	V8	330	MPI	I/B: Velvet
		MX6.2 Black Scorpion	6.2/377	V8	340	MPI	I/B: Velvet
		MX6.2 MPI	6.2/377	V8	320	MPI	I/O: Bravo I, II, III
			6.2/377	V8	320	MPI	I/B: Velvet, ZF
		MX6.2 MPI/Horizon	6.2/377	V8	320	MPI	I/O: Bravo I, II, III
			6.2/377	V8	320	MPI	I/B: Velvet, ZF
	8.1L	8.1S HO	8.1/496	V8	420	MPI	I/B: Velvet, ZF
		8.1S Horizon	8.1/496	V8	370	MPI	I/B: Velvet, ZF
		496 Mag	8.1/496	V8	375	MPI	I/O: Bravo I, II, III
		496 Mag HO	8.1/496	V8	425	MPI	I/O: Bravo I, II, III, X, XR, XZ
2004	3.0L	3.0L	3.0/181	L4	135	2 bbl	I/O: Alpha
	4.3L	4.3L	4.3/262	V6	190	2 bbl	I/O: Alpha, Bravo II, III
		4.3L MPI	4.3/262	V6	220	MPI	I/O: Alpha, Bravo II, III
	5.0L	5.0L	5.0/305	V8	220	2 bbl	I/O: Alpha, Bravo I, II, III
		5.0L MPI	5.0/305	V8	260	MPI	I/O: Alpha, Bravo I, II, III
	5.7L	5.7L	5.7/350	V8	250	2 bbl	I/O: Alpha, Bravo I, II, III
			5.7/350	V8	260	2 bbl	I/B: Velvet, ZF
			5.7/350	V8	270	2 bbl	I/B: Velvet
		350 Mag MPI	5.7/350	V8	300	MPI	I/O: Alpha, Bravo I, II, III
			5.7/350	V8	300	MPI	I/B: Velvet, ZF
			5.7/350	V8	315	MPI	I/B: Velvet
		350 Mag MPI/Horizon	5.7/350	V8	300	MPI	I/O: Bravo I, II, III
			5.7/350	V8	300	MPI	I/B: Velvet, ZF
	6.2L	Black Scorpion	6.2/377	V8	330	MPI	I/B: Velvet
		MX6.2 Black Scorpion	6.2/377	V8	340	MPI	I/B: Velvet
		MX6.2 MPI	6.2/377	V8	320	MPI	I/O: Bravo I, II, III
			6.2/377	V8	320	MPI	I/B: Velvet, ZF
		MX6.2 MPI/Horizon	6.2/377	V8	320	MPI	I/O: Bravo I, II, III
			6.2/377	V8	320	MPI	I/B: Velvet, ZF
	8.1L	8.1S HO	8.1/496	V8	420	MPI	I/B: Velvet, ZF
		8.1S Horizon	8.1/496	V8	370	MPI	I/B: Velvet, ZF
		496 Mag	8.1/496	V8	375	MPI	I/O: Bravo I, II, III
		496 Mag HO	8.1/496	V8	425	MPI	I/O: Bravo I, II, III, X, XR, XZ
2005	3.0L	3.0L	3.0/181	L4	135	2 bbl	I/O: Alpha
	4.3L	4.3L	4.3/262	V6	190	2 bbl	I/O: Alpha, Bravo II, III
		4.3L MPI	4.3/262	V6	220	MPI	I/O: Alpha, Bravo II, III
	5.0L	5.0L	5.0/305	V8	220	2 bbl	I/O: Alpha, Bravo I, II, III
		5.0L MPI	5.0/305	V8	260	MPI	I/O: Alpha, Bravo I, II, III
	5.7L	5.7L	5.7/350	V8	250	2 bbl	I/O: Alpha, Bravo I, II, III
			5.7/350	V8	260	2 bbl	I/B: Velvet, ZF
			5.7/350	V8	270	2 bbl	I/B: Velvet, ZF
		350 Mag MPI	5.7/350	V8	300	MPI	I/O: Alpha, Bravo I, II, III
			5.7/350	V8	300	MPI	I/B: Velvet, ZF
			5.7/350	V8	315	MPI	I/B: Velvet, ZF
		350 Mag MPI/Horizon	5.7/350	V8	300	MPI	I/O: Bravo I, II, III
			5.7/350	V8	300	MPI	I/B: Velvet, ZF
	6.2L	Black Scorpion	6.2/377	V8	330	MPI	I/B: Velvet, ZF
		MX6.2 Black Scorpion	6.2/377	V8	340	MPI	I/B: Velvet, ZF
		MX6.2 MPI	6.2/377	V8	320	MPI	I/O: Bravo I, II, III
			6.2/377	V8	320	MPI	I/B: Velvet, ZF
		MX6.2 MPI/Horizon	6.2/377	V8	320	MPI	I/O: Bravo I, II, III

ENGINE AND DRIVE MAINTENANCE 2-73

ENGINE MODELS

Year	Engine	Model	Displacement L/Cu. In.	Engine Type	HorsePower (At Prop.)	Fuel Delivery	Drive System
2005 (cont'd)	6.2	MX6.2 MPI/Horizon	6.2/377	V8	320	MPI	I/B: Velvet, ZF
	8.1	8.1S HO	8.1/496	V8	420	MPI	I/B: Velvet, ZF
		8.1S Horizon	8.1/496	V8	370	MPI	I/B: Velvet, ZF
		496 Mag	8.1/496	V8	375	MPI	I/O: Bravo I, II, III, X, XR, XZ
		496 Mag HO	8.1/496	V8	425	MPI	I/O: Bravo I, II, III, X, XR, XZ
2006	3.0L	3.0L	3.0/181	L4	135	2 bbl	I/O: Alpha
	4.3L	4.3L	4.3/262	V6	190	2 bbl	I/O: Alpha, Bravo II, III
		4.3L MPI	4.3/262	V6	220	MPI	I/O: Alpha, Bravo II, III
	5.0L	5.0L	5.0/305	V8	220	2 bbl	I/O: Alpha, Bravo I, II, III
		5.0L MPI	5.0/305	V8	260	MPI	I/O: Alpha, Bravo I, II, III
	5.7L	5.7L	5.7/350	V8	250	2 bbl	I/O: Alpha, Bravo I, II, III
		350 Mag MPI	5.7/350	V8	270	MPI	I/B: Velvet, ZF
			5.7/350	V8	300	MPI	I/O: Alpha, Bravo I, II, III
			5.7/350	V8	300	MPI	I/B: Velvet, ZF
		350 Mag MPI/Horizon	5.7/350	V8	315	MPI	I/B: Velvet, ZF
			5.7/350	V8	300	2 bbl	I/B: Velvet, ZF
	6.2L	Black Scorpion	6.2/377	V8	330	MPI	I/B: Velvet, ZF
		MX6.2 Black Scorpion	6.2/377	V8	340	MPI	I/B: Velvet, ZF
		MX6.2 MPI	6.2/377	V8	320	MPI	I/O: Bravo I, II, III
		MX6.2 MPI/Horizon	6.2/377	V8	320	MPI	I/B: Velvet, ZF
	8.1L	8.1S HO	8.1/496	V8	420	MPI	I/B: Velvet, ZF
		8.1S Horizon	8.1/496	V8	370	MPI	I/B: Velvet, ZF
		496 Mag	8.1/496	V8	375	MPI	I/O: Bravo I, II, III, X, XR
		496 Mag HO	8.1/496	V8	425	MPI	I/O: Bravo I, II, III, X, XR
2007	3.0L	3.0L TKS	3.0/181	L4	135	2 bbl	I/O: Alpha
	4.3L	4.3L TKS	4.3/262	V6	190	2 bbl	I/O: Alpha, Bravo II, III
		4.3L MPI	4.3/262	V6	220	MPI	I/O: Alpha, Bravo II, III
	5.0L	5.0L TKS	5.0/305	V8	220	2 bbl	I/O: Alpha, Bravo I, II, III
		5.0L MPI	5.0/305	V8	260	MPI	I/O: Alpha, Bravo I, II, III
	5.7L	5.7L TKS	5.7/350	V8	250	2 bbl	I/O: Alpha, Bravo I, II, III
			5.7/350	V8	260	MPI	I/B: Velvet, ZF
		350 Mag MPI	5.7/350	V8	300	MPI	I/O: Alpha, Bravo I, II, III
			5.7/350	V8	300	MPI	I/B: Velvet, ZF
		350 Mag MPI/Horizon	5.7/350	V8	300	MPI	I/B: Velvet, ZF
		Scorpion 350	5.7/350	V8	330	MPI	I/B: Velvet, Walter, ZF
		Tow Sport 5.7 MPI	5.7/350	V8	315	MPI	I/B: Velvet, Walter, ZF
		Tow Sport 5.7 TKS	5.7/350	V8	270	2 bbl	I/B: Velvet, ZF
	6.2L	MX6.2 MPI	6.2/377	V8	320	MPI	I/O: Bravo I, II, III
			6.2/377	V8	320	MPI	I/B: Velvet, ZF
		MX6.2 MPI/Horizon	6.2/377	V8	320	MPI	I/B: Velvet, ZF
		Scorpion 377	6.2/377	V8	340	MPI	I/B: Velvet, Walter, ZF
	8.1L	8.1S HO	8.1/496	V8	420	MPI	I/B: Velvet, ZF
		8.1S Horizon	8.1/496	V8	370	MPI	I/B: Velvet, ZF
		496 Mag MPI	8.1/496	V8	375	MPI	I/O: Bravo I, II, III, X, XR
		496 Mag HO MPI	8.1/496	V8	425	MPI	I/O: Bravo I, II, III, X, XR

ENGINE MODELS

Year	Engine	Model	Displacement L/Cu. In.	Engine Type	HorsePower (At Prop.)	Fuel Delivery	Drive System
2008	3.0L	3.0L TKS	3.0/181	L4	135	2 bbl	I/O: Alpha
		3.0L MPI ①	3.0/181	L4	135	MPI	I/O: Alpha
	4.3L	4.3L TKS	4.3/262	V6	190	2 bbl	I/O: Alpha, Bravo II, III
		4.3L MPI	4.3/262	V6	220	MPI	I/O: Alpha, Bravo II, III
	5.0L	5.0L TKS	5.0/305	V8	220	2 bbl	I/O: Alpha, Bravo I, II, III
		5.0L MPI	5.0/305	V8	260	MPI	I/O: Alpha, Bravo I, II, III
	5.7L	5.7L TKS	5.7/350	V8	250	2 bbl	I/O: Alpha, Bravo I, II, III
			5.7/350	V8	260	MPI	I/B: Velvet, ZF
		5.7L MPI	5.7/350	V8	300	MPI	I/B: Velvet, ZF
		5.7L MPI Horizon	5.7/350	V8	300	MPI	I/B: Velvet, ZP
			5.7/350	V8	300	MPI	I/O: Alpha, Bravo I, II, III
		350 Mag MPI	5.7/350	V8	300	MPI	I/B: Velvet, ZF
		350 Mag MPI/Horizon	5.7/350	V8	300	MPI	I/B: Velvet, ZF
		Scorpion 350	5.7/350	V8	330	MPI	I/B: Velvet, Walter, ZF
		Tow Sport 5.7 MPI	5.7/350	V8	315	MPI	I/B: Velvet, Walter, ZF
		Tow Sport 5.7 TKS	5.7/350	V8	270	2 bbl	I/B: Velvet, ZF
	6.2L	MX6.2 MPI/377 Mag	6.2/377	V8	320	MPI	I/O: Bravo I, II, III
			6.2/377	V8	320	MPI	I/B: Velvet, ZF
		MX6.2 MPI/Horizon	6.2/377	V8	320	MPI	I/B: Velvet, Walter, ZF
	8.1L	8.1S HO/HO Horizon	8.1/496	V8	420	MPI	I/B: Velvet, ZF
		8.1S Horizon	8.1/496	V8	370	MPI	I/B: Velvet, ZF
		496 Mag MPI	8.1/496	V8	375	MPI	I/O: Bravo I, II, III, X, XR
		496 Mag HO MPI	8.1/496	V8	425	MPI	I/O: Bravo I, II, III, X, XR

I/B — Inboard
I/O — Sterndrive
MPI — Multi-Port Injection
TBI — Throttle Body Injection
TKS — Turn Key Start

① California Emissions models only; no information available at time of publication

2-74 ENGINE AND DRIVE MAINTENANCE

TUNE-UP SPECIFICATIONS

Year	Engine	Model	Spark Plug Type	Gap In. (mm)	Ignition Timing @ Idle (deg)	Idle Speed Rpm (Neutral)	Idle Mixture (Turns Out)	Oil Pressure @ 2000 rpm (psi)	Fuel Pressure @ 1800 rpm (psi)	Compression Pressure (psi)	Full Throttle Operation (rpm)
2001	3.0L	3.0L	①	0.035 (0.9)	②	700	1 1/4	30	6-8	100 min.	4400-4800
	4.3L	4.3L, 4.3LH	①	0.045 (1.1)	10 BTDC	650	1 1/4	30 min.	3-7	100 min.	4400-4800
		4.3L MPI	AC 41-932	0.060 (1.52)	8 BTDC	600 ③	NA	18 min.	43	100 min.	4400-4800
		4.3L TBI	①	0.045 (1.1)	8 BTDC	600 ③	NA	30 min.	30	100 min.	4400-4800
	5.0L	5.0L	①	0.045 (1.1)	10 BTDC	650	1 1/4	30 min.	3-7	100 min.	4400-4800
		5.0L MPI	AC 41-932	0.060 (1.52)	NA	600 ③	NA	18 min.	43	100 min.	4600-5000
		5.0L TBI	①	0.045 (1.1)	8 BTDC	600 ③	NA	30 min.	30	100 min.	4400-4800
	5.7L	5.7L	①	0.045 (1.1)	10 BTDC	650	1 1/4	30 min.	3-7	100 min.	4400-4800
		5.7L EFI	①	0.045 (1.1)	8 BTDC	600 ③	NA	30 min.	30	100 min.	4400-4800
		350 Mag MPI I/O	AC 41-932	0.060 (1.52)	NA	600 ③	NA	18 min.	43	100 min.	4600-5000
		350 Mag MPI I/B	AC 41-932	0.060 (1.52)	NA	600 ③	NA	18 min.	43	100 min.	4400-4800
		350 Mag MPI I/B Tow	AC 41-932	0.060 (1.52)	NA	600 ③	NA	18 min.	43	100 min.	4600-5000
	6.2L	Black Scorpion	④	0.035 (0.9)	8 BTDC	600 ③	NA	30 min.	30	100 min.	4600-5000
		MX6.2 MPI I/O	①	0.045 (1.1)	8 BTDC	600 ③	NA	30 min.	30	100 min.	4800-5200
		MX6.2 MPI I/B	①	0.045 (1.1)	8 BTDC	600 ③	NA	30 min.	30	100 min.	4800-5200
	8.1L	8.1S Horizon	⑤	0.060 (1.52)	NA	650	NA	60	6 ⑧	100 min.	4200-4600
		8.1S HO	⑤	0.060 (1.52)	NA	650	NA	60	6 ⑧	100 min.	4400-4800
		496 Mag	⑤	0.060 (1.52)	NA	650	NA	60	6 ⑧	100 min.	4400-4800
		496 Mag HO	⑤	0.060 (1.52)	NA	650	NA	60	6 ⑧	100 min.	4600-5000
2002	3.0L	3.0L	①	0.035 (0.9)	②	700	1 1/4	30	6-8	100 min.	4400-4800
	4.3L	4.3L	①	0.045 (1.1)	10 BTDC	650	1 1/4	30 min.	3-7	100 min.	4400-4800
		4.3L MPI	AC 41-932	0.060 (1.52)	8 BTDC	600 ③	NA	18 min.	43	100 min.	4400-4800
		4.3L TBI	①	0.045 (1.1)	8 BTDC	600 ③	NA	30 min.	30	100 min.	4400-4800
	5.0L	5.0L	①	0.045 (1.1)	10 BTDC	650	1 1/4	30 min.	3-7	100 min.	4400-4800
		5.0L MPI	AC 41-932	0.060 (1.52)	NA	600 ③	NA	18 min.	43	100 min.	4600-5000
		5.0L TBI	①	0.045 (1.1)	8 BTDC	600 ③	NA	30 min.	30	100 min.	4400-4800
	5.7L	5.7L	①	0.045 (1.1)	10 BTDC	650	1 1/4	30 min.	3-7	100 min.	4400-4800
		5.7L EFI	①	0.045 (1.1)	8 BTDC	600 ③	NA	30 min.	30	100 min.	4400-4800
		350 Mag MPI I/O	AC 41-932	0.060 (1.52)	NA	600 ③	NA	18 min.	43	100 min.	4600-5000
		350 Mag MPI I/B	AC 41-932	0.060 (1.52)	NA	600 ③	NA	18 min.	43	100 min.	4400-4800
		350 Mag MPI I/B Tow	AC 41-932	0.060 (1.52)	NA	600 ③	NA	18 min.	43	100 min.	4600-5000
	6.2L	Black Scorpion	AC 41-932	0.060 (1.52)	NA	600 ③	NA	30 min.	43	100 min.	4600-5000
		MX6.2 MPI I/O	AC 41-932	0.060 (1.52)	NA	600 ③	NA	30 min.	43	100 min.	4800-5200
		MX6.2 MPI I/B	AC 41-932	0.060 (1.52)	NA	600 ③	NA	30 min.	43	100 min.	4600-5000
	8.1L	8.1S Horizon	⑤	0.060 (1.52)	NA	650	NA	60	6 ⑧	100 min.	4200-4600
		8.1S HO	⑤	0.060 (1.52)	NA	650	NA	60	6 ⑧	100 min.	4400-4800
		496 Mag	⑤	0.060 (1.52)	NA	650	NA	60	6 ⑧	100 min.	4400-4800
		496 Mag HO	⑤	0.060 (1.52)	NA	650	NA	60	6 ⑧	100 min.	4600-5000
2003	3.0L	3.0L	①	0.035 (0.9)	②	700	1 1/4	30	6-8	100 min.	4400-4800
	4.3L	4.3L	①	0.045 (1.1)	10 BTDC	650	1 1/4	30 min.	3-7	100 min.	4400-4800
		4.3L MPI	AC 41-932	0.060 (1.52)	8 BTDC	600 ③	NA	18 min.	43	100 min.	4400-4800
	5.0L	5.0L	①	0.045 (1.1)	10 BTDC	650	1 1/4	30 min.	3-7	100 min.	4400-4800
		5.0L MPI	AC 41-932	0.060 (1.52)	NA	600 ③	NA	18 min.	43	100 min.	4600-5000
	5.7L	5.7L	①	0.045 (1.1)	10 BTDC	650	1 1/4	30 min.	3-7	100 min.	4400-4800
		350 Mag MPI I/O	AC 41-932	0.060 (1.52)	NA	600 ③	NA	18 min.	43	100 min.	4600-5000
		350 Mag MPI I/B	AC 41-932	0.060 (1.52)	NA	600 ③	NA	18 min.	43	100 min.	4400-4800
		350 Mag MPI I/B Tow	AC 41-932	0.060 (1.52)	NA	600 ③	NA	18 min.	43	100 min.	4600-5000
	6.2L	Black Scorpion	AC 41-932	0.060 (1.52)	NA	600 ③	NA	30 min.	43	100 min.	4800-5200
		MX6.2 Black Scorpion	AC 41-932	0.060 (1.52)	NA	600 ③	NA	30 min.	43	100 min.	4800-5200
		MX6.2 MPI I/O	AC 41-932	0.060 (1.52)	NA	600 ③	NA	30 min.	43	100 min.	4800-5200
		MX6.2 MPI I/B	AC 41-932	0.060 (1.52)	NA	600 ③	NA	30 min.	43	100 min.	4800-5200
		MX6.2 MPI I/B Tow	AC 41-932	0.060 (1.52)	NA	600 ③	NA	30 min.	43	100 min.	4800-5200
	8.1L	8.1S Horizon	⑤	0.060 (1.52)	NA	650	NA	60	6 ⑧	100 min.	4200-4600
		8.1S HO	⑤	0.060 (1.52)	NA	650	NA	60	6 ⑧	100 min.	4400-4800
		496 Mag	⑤	0.060 (1.52)	NA	650	NA	60	6 ⑧	100 min.	4400-4800
		496 Mag HO	⑤	0.060 (1.52)	NA	650	NA	60	6 ⑧	100 min.	4600-5000

ENGINE AND DRIVE MAINTENANCE 2-75

TUNE-UP SPECIFICATIONS

Year	Engine	Model	Spark Plug Type	Spark Plug Gap In. (mm)	Ignition Timing @ Idle (deg)	Idle Speed Rpm (Neutral)	Idle Mixture (Turns Out)	Oil Pressure @ 2000 rpm (psi)	Fuel Pressure @ 1800 rpm (psi)	Compression Pressure (psi)	Full Throttle Operation (rpm)
2004	3.0L	3.0L (w/TKS)	①	0.035 (0.9)	②	800	3 1/8	30	6-8	100 min.	4400-4800
		3.0L (w/o TKS)	①	0.035 (0.9)	②	700	1 1/4	30	6-8	100 min.	4400-4800
	4.3L	4.3L (w/TKS Gen 1)	①	0.045 (1.1)	10 BTDC	650	2 5/8	30 min.	3-7	100 min.	4400-4800
		4.3L (w/TKS Gen 2)	①	0.045 (1.1)	10 BTDC	650	3	30 min.	3-7	100 min.	4400-4800
		4.3L (w/o TKS)	①	0.045 (1.1)	10 BTDC	650	1 1/4	30 min.	3-7	100 min.	4400-4800
		4.3L MPI	AC 41-932	0.060 (1.52)	8 BTDC	600 ③	NA	18 min.	43	100 min.	4400-4800
	5.0L	5.0L (w/TKS Gen 1)	①	0.045 (1.1)	10 BTDC	650	2 1/2	30 min.	3-7	100 min.	4400-4800
		5.0L (w/TKS Gen 2)	①	0.045 (1.1)	10 BTDC	650	2 1/4	30 min.	3-7	100 min.	4400-4800
		5.0L (w/o TKS)	①	0.045 (1.1)	10 BTDC	650	1 1/4	30 min.	3-7	100 min.	4400-4800
		5.0L MPI	AC 41-932	0.060 (1.52)	NA	600 ③	NA	18 min.	43	100 min.	4600-5000
	5.7L	5.7L (w/TKS Gen 1)	①	0.045 (1.1)	10 BTDC	650	2 7/8	30 min.	3-7	100 min.	4400-4800
		5.7L (w/TKS Gen 2)	①	0.045 (1.1)	10 BTDC	650	3	30 min.	3-7	100 min.	4400-4800
		5.7L (w/o TKS)	①	0.045 (1.1)	10 BTDC	650	1 1/4	30 min.	3-7	100 min.	4400-4800
		350 Mag MPI I/O	AC 41-932	0.060 (1.52)	NA	600 ③	NA	18 min.	43	100 min.	4800-5200
		350 Mag MPI I/B	AC 41-932	0.060 (1.52)	NA	600 ③	NA	18 min.	43	100 min.	4600-5000
		350 Mag MPI I/B Tow	AC 41-932	0.060 (1.52)	NA	600 ③	NA	18 min.	43	100 min.	4600-5000
	6.2L	Black Scorpion	AC 41-932	0.060 (1.52)	NA	600 ③	NA	30 min.	43	100 min.	4800-5200
		MX6.2 Black Scorpion	AC 41-932	0.060 (1.52)	NA	600 ③	NA	30 min.	43	100 min.	4800-5200
		MX6.2 MPI I/O	AC 41-932	0.060 (1.52)	NA	600 ③	NA	30 min.	43	100 min.	4800-5200
		MX6.2 MPI I/B	AC 41-932	0.060 (1.52)	NA	600 ③	NA	30 min.	43	100 min.	4600-5000
		MX6.2 MPI I/B Tow	AC 41-932	0.060 (1.52)	NA	600 ③	NA	30 min.	43	100 min.	4800-5200
	8.1L	8.1S Horizon	⑤	0.060 (1.52)	NA	650	NA	60	6 ⑥	100 min.	4200-4600
		8.1S HO	⑤	0.060 (1.52)	NA	650	NA	60	6 ⑥	100 min.	4400-4800
		496 Mag	⑤	0.060 (1.52)	NA	650	NA	60	6 ⑥	100 min.	4400-4800
		496 Mag HO	⑤	0.060 (1.52)	NA	650	NA	60	6 ⑥	100 min.	4600-5000
2005	3.0L	3.0L	①	0.035 (0.9)	②	800	3 1/8	30	6-8	100 min.	4400-4800
	4.3L	4.3L (w/TKS Gen 1)	①	0.045 (1.1)	10 BTDC	650	2 5/8	30 min.	3-7	100 min.	4400-4800
		4.3L (w/TKS Gen 2)	①	0.045 (1.1)	10 BTDC	650	3	30 min.	3-7	100 min.	4400-4800
		4.3L MPI	AC 41-932	0.060 (1.52)	8 BTDC	600 ③	NA	18 min.	43	100 min.	4400-4800
	5.0L	5.0L (w/TKS Gen 1)	①	0.045 (1.1)	10 BTDC	650	2 1/2	30 min.	3-7	100 min.	4400-4800
		5.0L (w/TKS Gen 2)	①	0.045 (1.1)	10 BTDC	650	2 1/4	30 min.	3-7	100 min.	4400-4800
		5.0L MPI	AC 41-932	0.060 (1.52)	NA	600 ③	NA	18 min.	43	100 min.	4600-5000
	5.7L	5.7L (w/TKS Gen 1)	①	0.045 (1.1)	10 BTDC	650	2 7/8	30 min.	3-7	100 min.	4400-4800
		5.7L (w/TKS Gen 2)	①	0.045 (1.1)	10 BTDC	650	3	30 min.	3-7	100 min.	4400-4800
		350 Mag MPI I/O	AC 41-932	0.060 (1.52)	NA	600 ③	NA	18 min.	43	100 min.	4800-5200
		350 Mag MPI I/B	AC 41-932	0.060 (1.52)	NA	600 ③	NA	18 min.	43	100 min.	4600-5000
		350 Mag MPI I/B Tow	AC 41-932	0.060 (1.52)	NA	600 ③	NA	18 min.	43	100 min.	4600-5000
	6.2L	Black Scorpion	AC 41-932	0.060 (1.52)	NA	600 ③	NA	30 min.	43	100 min.	4800-5200
		MX6.2 Black Scorpion	AC 41-932	0.060 (1.52)	NA	600 ③	NA	30 min.	43	100 min.	4800-5200
		MX6.2 MPI I/O	AC 41-932	0.060 (1.52)	NA	600 ③	NA	30 min.	43	100 min.	4800-5200
		MX6.2 MPI I/B	AC 41-932	0.060 (1.52)	NA	600 ③	NA	30 min.	43	100 min.	4600-5000
		MX6.2 MPI I/B Tow	AC 41-932	0.060 (1.52)	NA	600 ③	NA	30 min.	43	100 min.	4800-5200
	8.1L	8.1S Horizon	⑤	0.060 (1.52)	NA	650	NA	60	6 ⑥	100 min.	4200-4600
		8.1S HO	⑤	0.060 (1.52)	NA	650	NA	60	6 ⑥	100 min.	4400-4800
		496 Mag	⑤	0.060 (1.52)	NA	650	NA	60	6 ⑥	100 min.	4400-4800
		496 Mag HO	⑤	0.060 (1.52)	NA	650	NA	60	6 ⑥	100 min.	4600-5000
2006	3.0L	3.0L	①	0.035 (0.9)	②	800	3 1/8	30	6-8	100 min.	4400-4800
	4.3L	4.3L (w/TKS Gen 1)	①	0.045 (1.1)	10 BTDC	650	2 5/8	30 min.	3-7	100 min.	4400-4800
		4.3L (w/TKS Gen 2)	①	0.045 (1.1)	10 BTDC	650	3	30 min.	3-7	100 min.	4400-4800
		4.3L MPI	AC 41-932	0.060 (1.52)	8 BTDC	600 ③	NA	18 min.	43	100 min.	4400-4800
	5.0L	5.0L (w/TKS Gen 1)	①	0.045 (1.1)	10 BTDC	650	2 1/2	30 min.	3-7	100 min.	4400-4800
		5.0L (w/TKS Gen 2)	①	0.045 (1.1)	10 BTDC	650	2 1/4	30 min.	3-7	100 min.	4400-4800
		5.0L MPI	AC 41-932	0.060 (1.52)	NA	600 ③	NA	18 min.	43	100 min.	4600-5000
	5.7L	5.7L (w/TKS Gen 1)	①	0.045 (1.1)	10 BTDC	650	2 7/8	30 min.	3-7	100 min.	4400-4800
		5.7L (w/TKS Gen 2)	①	0.045 (1.1)	10 BTDC	650	3	30 min.	3-7	100 min.	4400-4800
		350 Mag MPI I/O	AC 41-932	0.060 (1.52)	NA	600 ③	NA	18 min.	43	100 min.	4800-5200
		350 Mag MPI I/B	AC 41-932	0.060 (1.52)	NA	600 ③	NA	18 min.	43	100 min.	4600-5000
		350 Mag MPI I/B Tow	AC 41-932	0.060 (1.52)	NA	600 ③	NA	18 min.	43	100 min.	4600-5000
	6.2L	Black Scorpion	AC 41-932	0.060 (1.52)	NA	600 ③	NA	30 min.	43	100 min.	4800-5200
		MX6.2 Black Scorpion	AC 41-932	0.060 (1.52)	NA	600 ③	NA	30 min.	43	100 min.	4800-5200
		MX6.2 MPI I/O	AC 41-932	0.060 (1.52)	NA	600 ③	NA	30 min.	43	100 min.	4800-5200
		MX6.2 MPI I/B	AC 41-932	0.060 (1.52)	NA	600 ③	NA	30 min.	43	100 min.	4600-5000
		MX6.2 MPI I/B Tow	AC 41-932	0.060 (1.52)	NA	600 ③	NA	30 min.	43	100 min.	4800-5200

ENGINE AND DRIVE MAINTENANCE

TUNE-UP SPECIFICATIONS

Year	Engine	Model	Spark Plug Type	Spark Plug Gap In. (mm)	Ignition Timing @ Idle (deg)	Idle Speed Rpm (Neutral)	Idle Mixture (Turns Out)	Oil Pressure @ 2000 rpm (psi)	Fuel Pressure @ 1800 rpm (psi)	Compression Pressure (psi)	Full Throttle Operation (rpm)
2006 (cont'd)	8.1L	8.1S Horizon	⑤	0.060 (1.52)	NA	650	NA	60	6 ⑥	100 min.	4200-4600
		8.1S HO	⑤	0.060 (1.52)	NA	650	NA	60	6 ⑥	100 min.	4400-4800
		496 Mag	⑤	0.060 (1.52)	NA	650	NA	60	6 ⑥	100 min.	4400-4800
		496 Mag HO	⑤	0.060 (1.52)	NA	650	NA	60	6 ⑥	100 min.	4600-5000
2007	3.0L	3.0L TKS	①	0.035 (0.9)	②	800	3 1/8	30	6-8	100 min.	4400-4800
	4.3L	4.3L (w/TKS Gen 1)	①	0.045 (1.1)	10 BTDC	650	2 5/8	30 min.	3-7	100 min.	4400-4800
		4.3L (w/TKS Gen 2)	①	0.045 (1.1)	10 BTDC	650	3	30 min.	3-7	100 min.	4400-4800
		4.3L MPI	AC 41-932	0.060 (1.52)	8 BTDC	600 ③	NA	18 min.	43	100 min.	4400-4800
	5.0L	5.0L (w/TKS Gen 1)	①	0.045 (1.1)	10 BTDC	650	2 1/2	30 min.	3-7	100 min.	4400-4800
		5.0L (w/TKS Gen 2)	①	0.045 (1.1)	10 BTDC	650	2 1/4	30 min.	3-7	100 min.	4400-4800
		5.0L MPI	AC 41-932	0.060 (1.52)	NA	600 ③	NA	18 min.	43	100 min.	4600-5000
	5.7L	5.7L I/O (w/TKS Gen 1)	①	0.045 (1.1)	10 BTDC	650	2 7/8	30 min.	3-7	100 min.	4400-4800
		5.7L I/O (w/TKS Gen 2)	①	0.045 (1.1)	10 BTDC	650	3	30 min.	3-7	100 min.	4400-4800
		5.7L I/B (w/TKS)	①	0.045 (1.1)	10 BTDC	650	3	30 min.	3-7	100 min.	4200-4600
		350 Mag MPI I/O	AC 41-932	0.060 (1.52)	NA	600 ③	NA	18 min.	43	100 min.	4800-5200
		350 Mag MPI I/B	AC 41-932	0.060 (1.52)	NA	600 ③	NA	18 min.	43	100 min.	4600-5000
		Scorpion 350 I/B Tow	AC 41-932	0.060 (1.52)	NA	600 ③	NA	18 min.	43	100 min.	4800-5200
		Tow Sport 5.7 MPI I/B	AC 41-932	0.060 (1.52)	NA	600 ③	NA	18 min.	43	100 min.	4600-5000
		Tow Sport 5.7L TKS I/B	①	0.045 (1.1)	10 BTDC	650	3	30 min.	3-7	100 min.	4400-4800
	6.2L	MX6.2 MPI I/O	AC 41-932	0.060 (1.52)	NA	600 ③	NA	30 min.	43	100 min.	4800-5200
		MX6.2 MPI I/B	AC 41-932	0.060 (1.52)	NA	600 ③	NA	30 min.	43	100 min.	4600-5000
		Scorpion 377 I/B Tow	AC 41-932	0.060 (1.52)	NA	600 ③	NA	30 min.	43	100 min.	4800-5200
	8.1L	8.1S Horizon	⑨	0.060 (1.52)	NA	650 ③	NA	30 min.	6 ⑥	100 min.	4200-4600
		8.1S HO	⑨	0.060 (1.52)	NA	650 ③	NA	30 min.	6 ⑥	100 min.	4400-4800
		496 Mag MPI	⑨	0.060 (1.52)	NA	650 ③	NA	30 min.	6 ⑥	100 min.	4400-4800
		496 Mag HO MPI	⑨	0.060 (1.52)	NA	650 ③	NA	30 min.	6 ⑥	100 min.	4600-5000
2008	3.0L	3.0L TKS	①	0.035 (0.9)	②	800	3 1/8	30	6-8	100 min.	4400-4800
	4.3L	4.3L (w/TKS Gen 1)	①	0.045 (1.1)	10 BTDC	650	2 5/8	30 min.	3-7	100 min.	4400-4800
		4.3L (w/TKS Gen 2)	①	0.045 (1.1)	10 BTDC	650	3	30 min.	3-7	100 min.	4400-4800
		4.3L MPI	AC 41-932	0.060 (1.52)	8 BTDC	600 ③	NA	18 min.	43	100 min.	4400-4800
	5.0L	5.0L (w/TKS Gen 1)	①	0.045 (1.1)	10 BTDC	650	2 1/2	30 min.	3-7	100 min.	4400-4800
		5.0L (w/TKS Gen 2)	①	0.045 (1.1)	10 BTDC	650	2 1/4	30 min.	3-7	100 min.	4400-4800
		5.0L MPI	AC 41-932	0.060 (1.52)	NA	600 ③	NA	18 min.	43	100 min.	4600-5000
	5.7L	5.7L I/O (w/TKS Gen 1)	①	0.045 (1.1)	10 BTDC	650	2 7/8	30 min.	3-7	100 min.	4400-4800
		5.7L I/O (w/TKS Gen 2)	①	0.045 (1.1)	10 BTDC	650	3	30 min.	3-7	100 min.	4400-4800
		5.7L I/B (w/TKS)	①	0.045 (1.1)	10 BTDC	650	3	30 min.	3-7	100 min.	4200-4600
		350 Mag MPI I/O	AC 41-932	0.060 (1.52)	NA	600 ③	NA	18 min.	43	100 min.	4800-5200
		5.7/350 Mag MPI I/B	AC 41-932	0.060 (1.52)	NA	600 ③	NA	18 min.	43	100 min.	4600-5000
		Scorpion 350 I/B Tow	AC 41-932	0.060 (1.52)	NA	600 ③	NA	18 min.	43	100 min.	4800-5200
		Tow Sport 5.7 MPI I/B	AC 41-932	0.060 (1.52)	NA	600 ③	NA	18 min.	43	100 min.	4600-5000
		Tow Sport 5.7L TKS I/B	①	0.045 (1.1)	10 BTDC	650	3	30 min.	3-7	100 min.	4400-4800
	6.2L	MX6.2/377 MPI I/O	AC 41-932	0.060 (1.52)	NA	600 ③	NA	30 min.	43	100 min.	4800-5200
		MX6.2 MPI I/B	AC 41-932	0.060 (1.52)	NA	600 ③	NA	30 min.	43	100 min.	4600-5000
		Scorpion 377 I/B Tow	AC 41-932	0.060 (1.52)	NA	600 ③	NA	30 min.	43	100 min.	4800-5200
	8.1L	8.1S Horizon	⑨	0.060 (1.52)	NA	650 ③	NA	30 min.	6 ⑥	100 min.	4200-4600
		8.1S HO/Horizon	⑨	0.060 (1.52)	NA	650 ③	NA	30 min.	6 ⑥	100 min.	4400-4800
		496 Mag MPI	⑨	0.060 (1.52)	NA	650 ③	NA	30 min.	6 ⑥	100 min.	4400-4800
		496 Mag HO MPI	⑨	0.060 (1.52)	NA	650 ③	NA	30 min.	6 ⑥	100 min.	4600-5000

NA Not adjustable
I/O Sterndrive
I/B Inboard
ATDC After Top Dead Center
BTDC Before Top Dead Center
TKS Turn Key Start system

① Champ: RS12YC, AC: MR43LTS, NGK: BPR6EFS
② Up to s/n 0L096999: 1 BTDC
  s/n 0L0097000 thru 0L034099: 1 ATDC
  s/n 0L0341000 and above: 2 ATDC
  s/n 0L0341000 and above: 2 ATDC
③ Checking only, not adjustable
④ NGK BPR5EFVX or BPR6EFS
⑤ Denso TJ14R-P15
⑥ Gen II: 6 psi; Gen III: 45 psi
⑦ Champ: RN2C, NGK: BR9ES
⑧ All cyl. w/in 14.5 psi of each other
⑨ AC: 41-983

ENGINE AND DRIVE MAINTENANCE

CAPACITIES

Year	Engine	Model	Engine Crankcase (With Filter) Qts. (L) ①	Cooling System Closed Qts. (L)	Cooling System Seawater Qts. (L) ②	Stern Drive Qt. (L) ③ Alpha	Bravo I	Bravo II	Bravo III	Inboard Trans. Qts. (L)
2001	3.0L	All	4 (3.8)	9 (8.5)	9 (8.5)	2.0 (1.9)	-	-	-	-
	4.3L	Carb.	4.5 (4.3)	20 (19)	15 (14.1)	2.0 (1.9)	-	3.26 (3.0)	3.0 (2.8)	-
		TBI	4.5 (4.3)	20 (19)	15 (14.1)	2.0 (1.9)	-	3.26 (3.0)	3.0 (2.8)	-
		MPI	5.0 (5.2)	20 (19)	21 (20)	2.0 (1.9)	-	3.26 (3.0)	3.0 (2.8)	-
	5.0L	Carb.	5.5 (5.25)	20 (19)	15 (14.1)	2.0 (1.9)	2.75 (2.6)	3.26 (3.0)	3.0 (2.8)	-
		TBI	5.5 (5.25)	20 (19)	15 (14.1)	2.0 (1.9)	2.75 (2.6)	3.26 (3.0)	3.0 (2.8)	-
		MPI	5.5 (5.25)	20 (19)	21 (20)	2.0 (1.9)	2.75 (2.6)	3.26 (3.0)	3.0 (2.8)	-
	5.7L	Carb.	5.5 (5.25)	20 (19)	15 (14.1)	2.0 (1.9)	2.75 (2.6)	3.26 (3.0)	3.0 (2.8)	④
		TBI	5.5 (5.25)	20 (19)	15 (14.1)	2.0 (1.9)	2.75 (2.6)	3.26 (3.0)	3.0 (2.8)	④
		MPI	5.5 (5.25)	20 (19)	21 (20)	2.0 (1.9)	2.75 (2.6)	3.26 (3.0)	3.0 (2.8)	⑤
	6.2L	All	5.5 (5.25)	20 (19)	21 (20)	2.0 (1.9)	2.75 (2.6)	3.26 (3.0)	3.0 (2.8)	⑤
	8.1L	All	9.0 (8.5)	20 (19)	19 (18)	-	2.75 (2.6)	3.26 (3.0)	3.0 (2.8)	⑥
2002	3.0L	All	4 (3.8)	9 (8.5)	9 (8.5)	2.0 (1.9)	-	-	-	-
	4.3L	Carb.	4.5 (4.3)	20 (19)	15 (14.1)	2.0 (1.9)	-	3.26 (3.0)	3.0 (2.8)	-
		TBI	4.5 (4.3)	20 (19)	15 (14.1)	2.0 (1.9)	-	3.26 (3.0)	3.0 (2.8)	-
		MPI	5.0 (5.2)	20 (19)	21 (20)	2.0 (1.9)	-	3.26 (3.0)	3.0 (2.8)	-
	5.0L	Carb.	5.5 (5.25)	20 (19)	15 (14.1)	2.0 (1.9)	2.75 (2.6)	3.26 (3.0)	3.0 (2.8)	-
		TBI	5.5 (5.25)	20 (19)	15 (14.1)	2.0 (1.9)	2.75 (2.6)	3.26 (3.0)	3.0 (2.8)	-
		MPI	5.5 (5.25)	20 (19)	21 (20)	2.0 (1.9)	2.75 (2.6)	3.26 (3.0)	3.0 (2.8)	-
	5.7L	Carb.	5.5 (5.25)	20 (19)	15 (14.1)	2.0 (1.9)	2.75 (2.6)	3.26 (3.0)	3.0 (2.8)	④
		TBI	5.5 (5.25)	20 (19)	15 (14.1)	2.0 (1.9)	2.75 (2.6)	3.26 (3.0)	3.0 (2.8)	④
		MPI	5.5 (5.25)	20 (19)	21 (20)	2.0 (1.9)	2.75 (2.6)	3.26 (3.0)	3.0 (2.8)	⑤
	6.2L	All	5.5 (5.25)	20 (19)	21 (20)	2.0 (1.9)	2.75 (2.6)	3.26 (3.0)	3.0 (2.8)	⑤
	8.1L	All	9.0 (8.5)	20 (19)	19 (18)	-	2.75 (2.6)	3.26 (3.0)	3.0 (2.8)	⑥
2003	3.0L	All	4 (3.8)	9 (8.5)	9 (8.5)	2.0 (1.9)	-	-	-	-
	4.3L	Carb.	4.5 (4.3)	20 (19)	15 (14.1)	2.0 (1.9)	-	3.26 (3.0)	3.0 (2.8)	-
		MPI	5.0 (5.2)	20 (19)	21 (20)	2.0 (1.9)	-	3.26 (3.0)	3.0 (2.8)	-
	5.0L	Carb.	5.5 (5.25)	20 (19)	15 (14.1)	2.0 (1.9)	2.75 (2.6)	3.26 (3.0)	3.0 (2.8)	-
		MPI	5.5 (5.25)	20 (19)	21 (20)	2.0 (1.9)	2.75 (2.6)	3.26 (3.0)	3.0 (2.8)	-
	5.7L	Carb.	5.5 (5.25)	20 (19)	15 (14.1)	2.0 (1.9)	2.75 (2.6)	3.26 (3.0)	3.0 (2.8)	④
		MPI	5.5 (5.25)	20 (19)	21 (20)	2.0 (1.9)	2.75 (2.6)	3.26 (3.0)	3.0 (2.8)	⑤
	6.2L	All	5.5 (5.25)	20 (19)	21 (20)	2.0 (1.9)	2.75 (2.6)	3.26 (3.0)	3.0 (2.8)	⑤
	8.1L	All	9.0 (8.5)	20 (19)	19 (18)	-	2.75 (2.6)	3.26 (3.0)	3.0 (2.8)	⑥
2004	3.0L	All	4 (3.8)	9 (8.5)	9 (8.5)	2.0 (1.9)	-	-	-	-
	4.3L	Carb.	4.5 (4.3)	20 (19)	15 (14.1)	2.0 (1.9)	-	3.26 (3.0)	3.0 (2.8)	-
		MPI	5.0 (5.2)	20 (19)	21 (20)	2.0 (1.9)	-	3.26 (3.0)	3.0 (2.8)	-
	5.0L	Carb.	5.5 (5.25)	20 (19)	15 (14.1)	2.0 (1.9)	2.75 (2.6)	3.26 (3.0)	3.0 (2.8)	-
		MPI	5.5 (5.25)	20 (19)	21 (20)	2.0 (1.9)	2.75 (2.6)	3.26 (3.0)	3.0 (2.8)	-
	5.7L	Carb.	5.5 (5.25)	20 (19)	15 (14.1)	2.0 (1.9)	2.75 (2.6)	3.26 (3.0)	3.0 (2.8)	④
		MPI	5.5 (5.25)	20 (19)	21 (20)	2.0 (1.9)	2.75 (2.6)	3.26 (3.0)	3.0 (2.8)	⑤
	6.2L	All	5.5 (5.25)	20 (19)	21 (20)	2.0 (1.9)	2.75 (2.6)	3.26 (3.0)	3.0 (2.8)	⑤
	8.1L	All	9.0 (8.5)	20 (19)	19 (18)	-	2.75 (2.6)	3.26 (3.0)	3.0 (2.8)	⑥
2005	3.0L	All	4 (3.8)	9 (8.5)	9 (8.5)	2.0 (1.9)	-	-	-	-
	4.3L	Carb.	4.5 (4.3)	20 (19)	15 (14.1)	2.0 (1.9)	-	3.26 (3.0)	3.0 (2.8)	-
		MPI	5.0 (5.2)	20 (19)	21 (20)	2.0 (1.9)	-	3.26 (3.0)	3.0 (2.8)	-
	5.0L	Carb.	5.5 (5.25)	20 (19)	15 (14.1)	2.0 (1.9)	2.75 (2.6)	3.26 (3.0)	3.0 (2.8)	-
		MPI	5.5 (5.25)	20 (19)	21 (20)	2.0 (1.9)	2.75 (2.6)	3.26 (3.0)	3.0 (2.8)	-
	5.7L	Carb.	5.5 (5.25)	20 (19)	15 (14.1)	2.0 (1.9)	2.75 (2.6)	3.26 (3.0)	3.0 (2.8)	④
		MPI	5.5 (5.25)	20 (19)	21 (20)	2.0 (1.9)	2.75 (2.6)	3.26 (3.0)	3.0 (2.8)	⑤
	6.2L	All	5.5 (5.25)	20 (19)	21 (20)	2.0 (1.9)	2.75 (2.6)	3.26 (3.0)	3.0 (2.8)	⑤
	8.1L	All	9.0 (8.5)	20 (19)	19 (18)	-	2.75 (2.6)	3.26 (3.0)	3.0 (2.8)	⑥

ENGINE AND DRIVE MAINTENANCE

CAPACITIES

Year	Engine	Model	Engine Crankcase (With Filter) Qts. (L) ①	Cooling System Closed Qts. (L)	Cooling System Seawater Qts. (L) ②	Alpha	Bravo I	Stern Drive Qt. (L) ③ Bravo II	Bravo III	Inboard Trans. Qts. (L)
2006	3.0L	All	4 (3.8)	9 (8.5)	9 (8.5)	2.0 (1.9)	-	-	-	-
	4.3L	Carb.	4.5 (4.3)	20 (19)	15 (14.1)	2.0 (1.9)	-	3.26 (3.0)	3.0 (2.8)	-
		MPI	5.0 (5.2)	20 (19)	21 (20)	2.0 (1.9)	-	3.26 (3.0)	3.0 (2.8)	
	5.0L	Carb.	5.5 (5.25)	20 (19)	15 (14.1)	2.0 (1.9)	2.75 (2.6)	3.26 (3.0)	3.0 (2.8)	-
		MPI	5.5 (5.25)	20 (19)	21 (20)	2.0 (1.9)	2.75 (2.6)	3.26 (3.0)	3.0 (2.8)	-
	5.7L	Carb.	5.5 (5.25)	20 (19)	15 (14.1)	2.0 (1.9)	2.75 (2.6)	3.26 (3.0)	3.0 (2.8)	④
		MPI	5.5 (5.25)	20 (19)	21 (20)	2.0 (1.9)	2.75 (2.6)	3.26 (3.0)	3.0 (2.8)	⑤
	6.2L	All	5.5 (5.25)	20 (19)	21 (20)	2.0 (1.9)	2.75 (2.6)	3.26 (3.0)	3.0 (2.8)	⑤
	8.1L	All	9.0 (8.5)	20 (19)	19 (18)	-	2.75 (2.6)	3.26 (3.0)	3.0 (2.8)	⑥
2007	3.0L	All	4 (3.8)	9 (8.5)	9 (8.5)	2.0 (1.9)	-	-	-	-
	4.3L	Carb.	4.5 (4.3)	20 (19)	15 (14.1)	2.0 (1.9)	-	3.26 (3.0)	3.0 (2.8)	-
		MPI	5.0 (5.2)	20 (19)	21 (20)	2.0 (1.9)	-	3.26 (3.0)	3.0 (2.8)	
	5.0L	Carb.	5.5 (5.25)	20 (19)	15 (14.1)	2.0 (1.9)	2.75 (2.6)	3.26 (3.0)	3.0 (2.8)	-
		MPI	5.5 (5.25)	20 (19)	21 (20)	2.0 (1.9)	2.75 (2.6)	3.26 (3.0)	3.0 (2.8)	-
	5.7L	Carb.	5.5 (5.25)	20 (19)	15 (14.1)	2.0 (1.9)	2.75 (2.6)	3.26 (3.0)	3.0 (2.8)	④
		MPI	5.5 (5.25)	20 (19)	21 (20)	2.0 (1.9)	2.75 (2.6)	3.26 (3.0)	3.0 (2.8)	⑤
	6.2L	All	5.5 (5.25)	20 (19)	21 (20)	2.0 (1.9)	2.75 (2.6)	3.26 (3.0)	3.0 (2.8)	⑤
	8.1L	All	9.0 (8.5)	21 (20)	19 (18)	-	⑦	⑦	⑦	⑧
2008	3.0L	All	4 (3.8)	9 (8.5)	9 (8.5)	2.0 (1.9)	-	-	-	-
	4.3L	Carb.	4.5 (4.3)	20 (19)	15 (14.1)	2.0 (1.9)	-	3.26 (3.0)	3.0 (2.8)	-
		MPI	5.0 (5.2)	20 (19)	21 (20)	2.0 (1.9)	-	3.26 (3.0)	3.0 (2.8)	
	5.0L	Carb.	5.5 (5.25)	20 (19)	15 (14.1)	2.0 (1.9)	2.75 (2.6)	3.26 (3.0)	3.0 (2.8)	-
		MPI	5.5 (5.25)	20 (19)	21 (20)	2.0 (1.9)	2.75 (2.6)	3.26 (3.0)	3.0 (2.8)	-
	5.7L	Carb.	5.5 (5.25)	20 (19)	15 (14.1)	2.0 (1.9)	2.75 (2.6)	3.26 (3.0)	3.0 (2.8)	④
		MPI	5.5 (5.25)	20 (19)	21 (20)	2.0 (1.9)	2.75 (2.6)	3.26 (3.0)	3.0 (2.8)	⑤
	6.2L	All	5.5 (5.25)	20 (19)	21 (20)	2.0 (1.9)	2.75 (2.6)	3.26 (3.0)	3.0 (2.8)	⑤
	8.1L	All	9.0 (8.5)	21 (20)	19 (18)	-	⑦	⑦	⑦	⑧

① All capacities are approximate. Always use the dipstick to determine exact quantity of oil required
② Seawater cooling system capacities are for winterization only
③ All capacities are approximate and for models with a Drive Lube Monitor. On models without a reservoir, fill to the bottom of the vent hole. On models with a reservoir, fill to the 'FULL line.
④ All capacities are approximate. Always use the dipstick to determine exact quantity of oil required
Velvet 71C: 1.5 (1.33); 72, V: 3.0 (2.75); 5000A: 3.0 (2.84); 5000V: 3.5 (3.3)
Walter: 0.75 (0.5)
ZF/Hurth 630V: 4.25 (4.0); 630A: 3.25 (3.0); 800A: 5.75 (5.5)
⑤ All capacities are approximate. Always use the dipstick to determine exact quantity of oil required
Velvet 71C inline or V: 1.75 (1.66); 71C gear reduction ; 3.0 (2.84); 72C inline, gear reduction or remote V: 1.75 (1.66); 71C V; 4.0 (3.79);5000A: 2.5 (2.4); 5000V: 3.0 (2.75)
ZF/Hurth 630V: 4.5 (4.0); 630A: 4.5 (4.0)
⑥ All capacities are approximate. Always use the dipstick to determine exact quantity of oil required
Velvet 5000A: 2.5 (2.4); 5000V: 3.0 (2.75)
ZF/Hurth 630V: 4.25 (4.0); 630A: 3.25 (3.0); 800A: 5.75 (5.5)
⑦ Bravo I: 92.5 oz. (2736 ml), Bravo II: 108.5 oz. (3209 ml), Bravo III: 100.5 oz. (2972 ml)
⑧ All capacities are approximate. Always use the dipstick to determine exact quantity of oil required
Velvet 5000A: 3.0 (2.84); 5000V: 3.5 (3.3)
ZF/Hurth 63A: 4.5 (4.0); 63V: 4.5 (4.0); 80A: 6.0 (5.5)
⑨ Vazer: 64 oz. (1892 ml)

ENGINE AND DRIVE MAINTENANCE 2-79

Maintenance Intervals Chart
3.0L Engines

Component	Procedure		Daily/ Before Use	Weekly	50 Hours/ 60 Days	100 Hours/ 120 Days	100 Hours/ Yearly
Anodes	Inspect			X			
Battery	Check Fluid Level			X			
	Inspect			X			
Cooling System	Check Level (Closed System)			X			
	Flush Seawater		①				
	Inspect Hoses And Clamps						X
	Replace Coolant		②				
Drive Belts	Inspect					X	
Electrical System	Check Wiring						X
Engine Mounts	Check Torque Rear						X
Flame Arrestor	Clean/Inspect						X
Fuel System	Check Pump Sight Tube				X		
Gimbal Ring	Lubricate						X
	Retorque						X
Ignition System	Clean/Inspect						X
	Inspect Timing						X
Ignition Timing	Inspect/Adjust						X
Impeller	Inspect/Replace						X
Oil	Check Level	Crankcase		X			
		Stern Drive		X			
		Trim Pump		X			
	Change	Crankcase					X
		Stern Drive					X
Power Package	Spray w/Rust Preventative	Saltwater			X		
		Freshwater				X	
	Clean/Paint		③				
Propeller	Lubricate Shaft	Saltwater			X		
		Freshwater				X	
Seawater Pump	Inspect/Replace						X
Seawater Strainer	Check/Clean		X				
Shift/Throttle Cable	Lubricate/Inspect						X
Steering System	Check Fluid Level			X			
	Lubricate/Inspect						X
Water Pick-Ups	Check/Clean		X				

① After each use
② Every 5 years or 1000 hours
③ Once a year

ENGINE AND DRIVE MAINTENANCE

Maintenance Intervals Chart - V6 And V8 Engines (Carb/TBI)

Component	Procedure		Daily/ Before or After Use	Weekly	50 Hours/ 60 Days	100 Hours/ Yearly	200 Hours/ 3 Years	300 Hours/ 3 Years
Air Filter	Clean				X			
Anodes	Inspect			X				
Battery	Check Fluid Level			X				
	Inspect			X				
Cooling System	Check Level (Closed System)			X				
	Flush Seawater		①					
	Inspect Hoses/Clamps	Standard				X		
		Horizon						X
	Replace Coolant		②					
Drive Belts	Inspect	Standard				X		
		Horizon						X
Electrical System	Check Wiring	Standard				X		
		Horizon						X
Engine Coupler	Lubricate	Standard				X ④		
		Horizon					X	
Engine Mounts	Check Torque	Standard				X		
		Horizon						X
Flame Arrestor	Clean/Inspect	Standard				X		
		Horizon						X
	Replace PCV Valve	Standard				X		
		Horizon						X
Fuel System	Check Sight Tube			X				
	Replace Filter					X		
Gimbal Bearing	Lubricate	Standard				X		
		Horizon					X	
Idle Speed	Check/Adjust	Standard				X		
		Horizon						X
Ignition System	Check/Inspect: Plugs, wires, cap, timing & idle	Standard				X		
		Horizon						X
Ignition Timing	Check/Adjust	Standard				X		
		Horizon						X
Impeller	Inspect/Replace	Standard				X		
		Horizon						X
Mercathode System	Test Output	Standard				X		
		Horizon					X	
Oil	Check Level	Crankcase	X					
		Stern Drive		X				
		Transmission		X				
		Trim Pump		X				
	Change Oil/Filter	Crankcase				X		
		Stern Drive				X		
		Transmission				X		
PCV Valve	Replace	Standard				X		
		Horizon						X
Power Package	Spray w/Rust Preventative	Saltwater		X				
		Freshwater				X		
	Clean/Paint		③					
Precision Pilot	Inspect	Impellers		X				
		Motor Tunnels		X				
		Safety Guard	X					
	Remove/Inspect Impellers					X		
	Rinse/Flush Motor Tunnels		X					
Propeller	Lubricate Shaft And Retorque				X			
Seawater Pump	Inspect/Replace	Standard				X		
		Horizon						X
Seawater Strainer	Check/Clean			X				
Shift/Throttle Cable	Lubricate/Inspect					X		
Spark Plugs/Wires	Check/Inspect:	Standard				X		
		Horizon						X
Steering System	Check Fluid Level			X				
	Lubricate/Inspect					X		
U-Joints	Lubricate/Inspect	Standard				X		
		Horizon					X	
Water Pick-Ups	Check/Clean			X				

① After each use
② Every 2 years for Standard, every 5 years for Horizon
   Interval will be reduced if not using extended life coolant
③ Once a year
④ Every 50 hrs if operated at idle for prolonged periods

ENGINE AND DRIVE MAINTENANCE

Maintenance Intervals Chart - 4.3L V6, 5.0L, 5.7L & 6.2L V8 MPI Engines (Sterndrive)

Component	Procedure		Daily/ Before or After Use	Weekly	50 Hours/ 60 Days	100 Hours/ Yearly	200 Hours/ 3 Years	300 Hours/ 3 Years
Anodes	Inspect			X				
Battery	Check Fluid Level				X			
	Inspect				X			
Bellows	Inspect						X	
Cooling System	Check Level (Closed System)			X				
	Flush Seawater		①					
	Inspect Hoses/Clamps							X
	Replace Coolant		②					
Crankcase Vent Hoses	Replace							X
Distributor Cap	Check/Inspect:							X
Drive Belts	Inspect							X
Electrical System	Check Wiring							X
Engine Alignment	Inspect						X	
Engine Coupler	Lubricate						X ④	
Engine Mounts	Check Torque							X
Exhaust System	Inspect Hoses/Clamps							X
Flame Arrestor	Clean/Inspect							X
	Replace PCV Valve							X
Fuel System	Replace Filter					X		
Gauges	Clean/Inspect				X ⑥			
Gimbal Ring-to-Steering Shaft	Retorque					X		
Gimbal Bearing	Lubricate						X	
Ignition System	Check/Inspect: Plugs, wires, cap, timing & idle							X
Impeller	Inspect/Replace							X
Mercathode System	Test Output						X	
Oil	Check Level							
		Crankcase	X					
		P/S Pump		X				
		Stern Drive		X				
		Trim Pump		X				
	Change Oil/Filter							
		Crankcase				X		
		Stern Drive				X		
PCV Valve	Replace							X
Power Package	Spray w/Rust Preventative							
		Saltwater			X			
		Freshwater				X		
	Clean/Paint		③					
Propeller	Lubricate Shaft And Retorque				X ⑤			
Precision Pilot	Inspect	Impellers		X				
		Motor Tunnels		X				
		Safety Guar	X					
	Remove/Inspect Impellers					X		
	Rinse/Flush Motor Tunnels		X					
Seawater Pump	Inspect/Replace							X
Seawater Strainer	Check/Clean				X			
Shift/Throttle Cable	Lubricate/Inspect					X		
Spark Plugs/Wires	Check/Inspect:							X
Steering System	Check Fluid Level			X				
	Lubricate/Inspect					X		
U-Joints	Lubricate/Inspect						X	
Water Pick-Ups	Check/Clean			X				

① After each use
② Every 2 years for Standard, every 5 years for Horizon Interval will be reduced if not using extended life coolant
③ Once a year
④ Every 50 hrs if operated at idle for prolonged periods
⑤ If in freshwater, every 4 months
⑥ If in saltwater, every 30 days or 25 hrs

2-82 ENGINE AND DRIVE MAINTENANCE

Maintenance Intervals Chart - 5.0L, 5.7L & 6.2L V8 MPI Engines
(Inboard)

Component	Procedure		Daily/ Before or After Use	Weekly	50 Hours/ 60 Days	100 Hours/ Yearly	200 Hours/ 3 Years	300 Hours/ 3 Years
Battery	Check Fluid Level				X			
	Inspect				X			
Cooling System	Check Level (Closed System)			X				
	Flush Seawater		①					
	Inspect Hoses/Clamps							X
	Replace Coolant		②					
Crankcase Vent Hoses	Replace							X
Distributor Cap	Check/Inspect:							X
Drive Belts	Inspect							X
Electrical System	Check Wiring							X
Engine-to-propeller Shaft Alignment	Inspect		③					
Engine Mounts	Check Torque							X
Exhaust System	Inspect Hoses/Clamps							X
Flame Arrestor	Clean/Inspect							X
	Replace PCV Valve							X
Fuel System	Replace Filter						X	
Gauges	Clean/Inspect				X ④			
Ignition System	Check/Inspect: Plugs, wires, cap, timing & idle							X
Impeller	Inspect/Replace							X
Oil	Check Level							
		Crankcase	X					
		Transmission		X				
	Change Oil/Filter							
		Crankcase				X		
		Transmission				X		
PCV Valve	Replace							X
Power Package	Spray w/Rust Preventative					X		
Precision Pilot	Inspect	Impellers		X				
		Motor Tunnels		X				
		Safety Guard	X					
	Remove/Inspect Impellers					X		
	Rinse/Flush Motor Tunnels		X					
Seawater Pump	Inspect/Replace							X
Seawater Strainer	Check/Clean			X				
Shift/Throttle Cable	Lubricate/Inspect					X		
Spark Plugs/Wires	Check/Inspect:							X
Steering System	Check Fluid Level			X				
	Lubricate/Inspect					X		
Water Pick-Ups	Check/Clean			X				

① After each use
② Every 2 years for Standard, every 5 years for Horizon Interval will be reduced if not using extended life coolant
③ As per Manufacturer
④ If in saltwater, every 30 days or 25 hrs

ENGINE AND DRIVE MAINTENANCE 2-83

Maintenance Intervals Chart - 8.1L V8 Engines (Sterndrive)

Component	Procedure		Daily/ Before or After Use	Weekly	50 Hours/ 60 Days	100 Hours/ Yearly	200 Hours/ 3 Years	300 Hours/ 3 Years
Anodes	Inspect			X				
Battery	Check Fluid Level				X			
	Inspect				X			
Bellows/Clamps	Inspect						X	
Cooling System	Check Level (Closed System)			X				
	Flush Seawater		①					
	Inspect Hoses/Clamps							X
	Replace Coolant		②					
Drive Belts	Inspect							X
Electrical System	Check Wiring							X
Engine Alignment	Inspect						X	
Engine Coupler	Lubricate						X ④	
Engine Mounts	Check Torque							X
Exhaust System	Inspect Hoses/Clamps							X
Flame Arrestor	Clean/Inspect							X
Fuel System	Replace Filter					X		
Gimbal Ring-to-Steering Shaft	Retorque					X		
Gimbal Bearing	Lubricate						X	
Ignition System	Check/Inspect: Plugs & wires							X
Impeller	Inspect/Replace							X
Mercathode System	Test Output						X	
Oil	Check Level	Crankcase	X					
		P/S Pump		X				
		Stern Drive		X				
		Trim Pump		X				
	Change Oil/Filter	Crankcase				X		
		Stern Drive				X		
Power Package	Spray w/Rust Preventative					X		
	Clean/Paint		③					
Precision Pilot	Inspect	Impellers		X				
		Motor Tunnels		X				
		Safety Guards	X					
	Remove/Inspect Impellers					X		
	Rinse/Flush Motor Tunnels		X					
Propeller	Lubricate Shaft And Retorque				X ⑤			
Seawater Pump	Inspect/Replace							X
Seawater Strainer	Check/Clean				X			
Shift/Throttle Cable	Lubricate/Inspect					X		
Spark Plugs/Wires	Check/Inspect:							X
Steering System	Check Fluid Level				X			
	Lubricate/Inspect					X		
U-Joints/Splines	Lubricate/Inspect						X	
Water Pick-Ups	Check/Clean				X			

① After each use
② Every 2 years for Standard, every 5 years for Horizon
   Interval will be reduced if not using extended life coolant
③ Once a year
④ Every 50 hrs if operated at idle for prolonged periods
⑤ Every 2 months

ENGINE AND DRIVE MAINTENANCE

Maintenance Intervals Chart - 8.1L V8 Engines (Inboard)

Component	Procedure		Daily/ Before or After Use	Weekly	50 Hours/ 60 Days	100 Hours/ Yearly	200 Hours/ 3 Years	300 Hours/ 3 Years
Battery	Check Fluid Level				X			
	Inspect				X			
Cooling System	Check Level (Closed System)			X				
	Flush Seawater		①					
	Inspect Hoses/Clamps							X
	Replace Coolant		②					
Drive Belts	Inspect							X
Electrical System	Check Wiring							X
Engine Mounts	Check Torque							X
Engine-to-Propeller Shaft Alignment	Inspect					③		
Exhaust System	Inspect Hoses/Clamps							X
Flame Arrestor	Clean/Inspect							X
Fuel System	Replace Filter					X		
Ignition System	Check/Inspect: Plugs & wires							X
Impeller	Inspect/Replace							X
Oil	Check Level	Crankcase	X					
		Transmission		X				
	Change Oil/Filter	Crankcase				X		
		Transmission				X		
Power Package	Spray w/Rust Preventative					X		
Precision Pilot	Inspect	Impellers		X				
		Motor Tunnels		X				
		Safety Guard	X					
	Remove/Inspect Impellers					X		
	Rinse/Flush Motor Tunnels		X					
Seawater Pump	Inspect/Replace							X
Seawater Strainer	Check/Clean			X				
Shift/Throttle Cable	Lubricate/Inspect					X		
Spark Plugs/Wires	Check/Inspect:							X
Steering System	Check Fluid Level			X				
	Lubricate/Inspect					X		
Water Pick-Ups	Check/Clean			X				

① After each use
② Every 2 years for Standard, every 5 years for Horizon
   Interval will be reduced if not using extended life coolant
③ Once a year

3

ENGINE MECHANICAL - GM IN-LINE ENGINES

ENGINE MECHANICAL 3-2
EXPLODED VIEWS 3-16
SPECIFICATIONS 3-18

CAMSHAFT, BEARINGS AND GEAR 3-13
  CHECKING LIFT 3-13
  REMOVAL & INSTALLATION............. 3-14
COMBINATION MANIFOLD................ 3-7
  REMOVAL & INSTALLATION............. 3-7
CYLINDER HEAD....................... 3-15
  REMOVAL & INSTALLATION............. 3-15
CYLINDER HEAD COVER................. 3-5
  REMOVAL & INSTALLATION............. 3-5
ENGINE 3-2
  PRIMING THE ENGINE................. 3-4
  REMOVAL & INSTALLATION............. 3-2
ENGINE COUPLER AND FLYWHEEL........ 3-11
  REMOVAL & INSTALLATION............. 3-11
ENGINE IDENTIFICATION................ 3-2
ENGINE MECHANICAL................. 3-2
  CAMSHAFT, BEARINGS AND GEAR 3-13
  COMBINATION MANIFOLD................ 3-7
  CYLINDER HEAD 3-15
  CYLINDER HEAD COVER 3-5
  ENGINE 3-2
  ENGINE COUPLER AND FLYWHEEL....... 3-11
  ENGINE IDENTIFICATION.............. 3-2
  EXHAUST ELBOW..................... 3-9
  EXHAUST HOSE (BELLOWS) 3-9
  EXHAUST PIPE(S) 3-9
  EXHAUST VALVE (FLAPPER/SHUTTER).... 3-9
  FRONT COVER AND OIL SEAL 3-12
  FRONT ENGINE MOUNTS 3-4
  GENERAL INFORMATION 3-2
  HYDRAULIC VALVE LIFTER............. 3-7
  OIL FILTER BYPASS VALVE 3-11
  OIL PAN 3-10
  OIL PUMP 3-10
  PUSH ROD COVER.................... 3-5
  REAR ENGINE MOUNTS................ 3-5
  REAR MAIN SEAL 3-11
  ROCKER ARMS AND PUSH RODS 3-6
TORSIONAL DAMPER 3-12
EXHAUST ELBOW...................... 3-9
  REMOVAL & INSTALLATION............. 3-9
EXHAUST HOSE (BELLOWS)............. 3-9
  REMOVAL & INSTALLATION............. 3-9
EXHAUST PIPE(S).................... 3-9
  REMOVAL & INSTALLATION............. 3-9
EXHAUST VALVE (FLAPPER/SHUTTER).... 3-9
  REPLACEMENT 3-9
EXPLODED VIEWS 3-16
  CYLINDER BLOCK 3-17
  CYLINDER HEAD 3-16
FRONT COVER AND OIL SEAL 3-12
  REMOVAL & INSTALLATION............ 3-12
FRONT ENGINE MOUNTS 3-4
  REMOVAL & INSTALLATION............. 3-4
HYDRAULIC VALVE LIFTER 3-7
  REMOVAL & INSTALLATION............. 3-7
OIL FILTER BYPASS VALVE 3-11
  REMOVAL & INSTALLATION............ 3-11
OIL PAN 3-10
  REMOVAL & INSTALLATION............ 3-10
OIL PUMP........................... 3-10
  REMOVAL & INSTALLATION............ 3-10
PUSH ROD COVER..................... 3-5
  REMOVAL & INSTALLATION............. 3-5
REAR ENGINE MOUNTS 3-5
  REMOVAL & INSTALLATION............. 3-5
REAR MAIN SEAL 3-11
  REMOVAL & INSTALLATION............ 3-11
ROCKER ARMS AND PUSH RODS 3-6
  REMOVAL & INSTALLATION............. 3-6
  VALVE ADJUSTMENT 3-6
SPECIFICATIONS 3-18
  TORQUE - 3.0L ENGINES.............. 3-18
  ENGINE - 3.0L ENGINES 3-19
TORSIONAL DAMPER................... 3-12
  REMOVAL & INSTALLATION............ 3-12

3-2 ENGINE MECHANICAL - GM IN-LINE ENGINES

ENGINE MECHANICAL

General Information

The MerCruiser 3.0L, 181 cubic inch displacement engine is manufactured by GM and has been a favorite combination when mated to the MerCruiser sterndrive.

This in-line four-cylinder powerplant uses a full pressure lubrication system with a disposable flow thru oil filter cartridge. Oil pressure is furnished by a gear-type oil pump, driven by the distributor, which is driven by a helical gear on the camshaft. A regulator on the oil pump controls the amount of oil pressure output. The oil pump scavenges oil from the bottom of the oil pan and feeds it through the oil filter and then to the main oil gallery in the block. Drilled passages in the block and crankshaft distribute oil to the camshaft and crankshaft to lubricate the rod, main and camshaft bearings. The main oil gallery also feeds oil to the valve lifters, which pump oil up through the hollow pushrods to the rocker arms to lubricate the valve train in the cylinder head. Cylinder numbering and firing order is identified in the illustrations at the end of the Maintenance section.

All 3.0L engines are left hand (counterclockwise) rotation when viewed from the stern of the boat. This does not necessarily indicate that your prop rotation is the same - always check them both!

NEVER, NEVER attempt to use standard automotive parts when replacing anything on your engine. Due to the uniqueness of the environment in which they are operated in, and the levels at which they are operated at, marine engines require different versions of the same part; even if they look the same. Stock and most aftermarket automotive parts will not hold up for prolonged periods of time under such conditions. Automotive parts may appear identical to marine parts, but be assured, Mercury marine parts are specially manufactured to meet Mercury marine specifications. Most marine items are super heavy-duty units or are made from special metal alloy to combat against a corrosive salt water atmosphere.

Mercury marine electrical and ignition parts are extremely critical. In the United States, all electrical and ignition parts manufactured for marine application must conform to stringent U.S. Coast Guard requirements for spark or flame suppression. A spark from a non-marine cranking motor solenoid could ignite an explosive atmosphere of gasoline vapors in an enclosed engine compartment.

Engine Identification

◆ See Figure 1

All engines can be identified by a code number stamped into the lower starboard side of the engine block, aft of the distributor mounting hole - the last two letters of the code designate the engine. 2001 and later models carry **RX** on engines below serial number 0L096999 (w/12 3/4 in. flywheel), and **RP** on engines with serial number 0L097000 and above (w/14 in. flywheel). This code is stamped on all original equipment engines and all partial replacement engines (but NOT on replacement cylinder blocks).

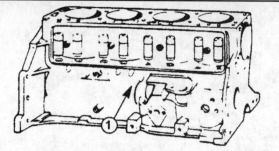

1- 4 cylinder-in-line engine code (next to distributor)

Fig. 1 The engine identification code can be found here. The last two letters are the engine designation

In the event that the engine serial number plates or stickers are missing from the engine, this is a good way to ensure that you know the exact engine in your vessel when ordering parts, etc.

Engine

REMOVAL & INSTALLATION

◆ See Figures 2, 3 and 4

1. Remove the stern drive unit as detailed in the appropriate section.
2. Open or remove the engine hatch cover.
3. Disconnect the battery cables (negative first) at the battery and then disconnect them from the engine block and starter.
4. Loosen the hose clamp on the engine wiring harness connector and unplug the instrumentation wiring harness. Label it and move it aside.
5. Disconnect the fuel inlet line at the fuel pump and quickly plug it (the line and the pump). Make sure you have rags handy as there will be some spillage.
6. Pull the cotter pin from the clevis, remove the lock nut and disconnect the shift assist assembly. Disconnect the shift cables and position them out of the way.
7. Disconnect the throttle cable at the carburetor and position it out of the way.
8. Carefully loosen and remove the two lines at the power steering control valve. Plug the hose ends and the control valve fittings. Tag the lines and fittings and move them aside.

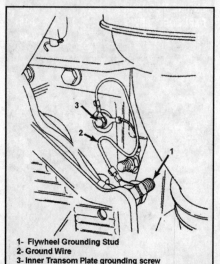

1- Flywheel Grounding Stud
2- Ground Wire
3- Inner Transom Plate grounding screw

Fig. 2 Tag and disconnect the engine ground wire.

1- Engine Wiring Harness Receptacle
2- Instrumentation Wiring Harness
3- Hose Clamp

Fig. 3 Disconnect the instrument wiring harness

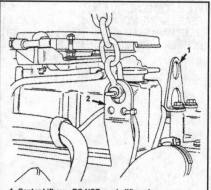

1- Center Lift eye- DO NOT use to lift engine
2- Use this lifting eye

Fig. 4 Attach an engine hoist at the lifting eyes.

ENGINE MECHANICAL - GM IN-LINE ENGINES

9. Loosen the hose clamps and disconnect the exhaust elbow bellows from the upper pipe. Remove it.
10. Tag and disconnect the trim sender wires at the engine harness. Follow the wires back to the transom and disconnect or remove them from any clamps so they may be moved out of the way.
11. Tag and disconnect the shift cut-out switch wires (near the arrestor).
12. On models equipped with the MerCathode system, tag and disconnect the wires at the controller.
13. Loosen the hose clamp and remove the water inlet line at the water tube where it exits from the transom. Plug the hose and the fitting.
14. Tag and disconnect the engine ground wire at the stud.
15. Tag and disconnect any remaining lines, wires or hoses at the engine. Carefully move them out of the way.
16. Attach a suitable engine hoist to the lifting eyes and take up any slack until it is just taught.

■ DO NOT use the front lifting eye attached to the thermostat housing as this used only for alignment purposes.

17. Locate the front engine mount and remove the two 3/8 inch lag bolts.
18. Locate the rear engine mount and remove the two mounting bolts.
19. Slowly and carefully, lift out the engine. Try not to hit the power steering control valve; or anything else for that matter.

To install:

◆ See Figures 5 thru 10

■ An engine alignment tool (#91-805475A1) is necessary to reinstall the engine. Even if the mounts have not been removed from the engine it is still a good idea to re-align the unit.

20. Make sure that the fiber washers are in position on the inner transom plate. Take a look at their condition and replace them if they look damaged or worn.
21. Install a double wound lockwasher into the fiber washer and make sure that the engine mount locknuts are in position as shown in the illustration.

■ Starting in 2003 Mercruiser has replaced the double wound lockwasher with a single stainless steel washer.

22. Lubricate the exhaust bellows with soapy water for later installation. Apply Quicksilver Engine Coupler grease to the splines of the engine coupler.
23. Attach the engine to a hoist if you disconnected it and move it into position in the vessel, over the mounting spots. Align the rear mounts over the holes in the transom plate, making sure exhaust tubes and pipe hoses are lined up.
24. Slide a large steel washer and a metal spacer onto the mounting bolt and then run them through the mount and transom plate. Tighten the bolts to 38 ft. lbs (52 Nm).

25. Carefully lower the engine fully until the front mount is resting on the stringers. Relieve the hoist tension, disconnect it and then reconnect it to the lifting eye in the thermostat housing
26. Slide the solid end of the alignment tool into the center of the gimbal housing bearing and then into the engine coupler splines - you may have to pivot the bearing slightly. DO NOT FORCE THE TOOL! Use the hoist to raise (or lower) the front of the engine until the tool slides completely into the coupler (with no binding).
27. Loosen the jam and lock nuts on the front mount(s) and then turn the adjusting nut (top of the mount) so that the mount base sits true on the stringer. Check that the alignment tool still slides freely between the engine and gimbal housing. Install the mount base lag bolts and tighten them securely. Tighten the adjusting nut and the jam nut.
28. Recheck that the alignment tool still slides freely and then remove it. Remove the engine hoist.
29. Reconnect the exhaust bellows and tighten the clamps securely (2 on each side).
30. Reconnect the water inlet hose and tighten the hose clamp securely.
31. Reconnect the instrument harness and tighten the clamp securely.
32. Carefully, and quickly, connect the power steering lines and tighten to 23 ft. lbs. (31 Nm). Don't forget to bleed the system when you are finished with the installation.
33. Reconnect the trim position sender leads, the engine ground wire, the battery cables and all other wires, lines of hoses that were disconnected during removal.
34. Unplug the fuel line and pump fitting and reconnect them.
35. Connect the MerCathode lines to the controller and coat them with liquid neoprene. If you forgot to tag the wires as we suggested, refer to the illustration for proper hook-up.
36. Move the remote control handle to the Neutral position. Install the cable end guide on the throttle lever and then push the barrel toward the end of the throttle lever to create a slight preload, reducing cable slack. Position the barrel so that it aligns with the anchor stud.
37. Secure the cable. Tighten the locknut on the end guide until it just makes contact and then back it off one full turn. Tighten the anchor screw securely, but not too tight.
38. Move the remote control lever to the full throttle position and confirm that the throttle valves in the carburetor are fully open and the throttle shaft lever is just coming in contact with the carburetor body.
39. Now move the control lever to the Neutral position and check that the throttle lever is touching the end of the idle speed screw.
40. Install and adjust the shift cables.
41. Install the drive unit. Connect the battery cables
42. Check and refill all fluids and go have fun!

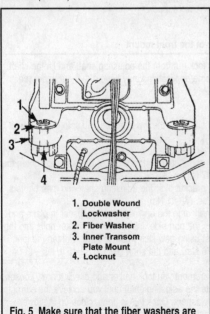

Fig. 5 Make sure that the fiber washers are in position

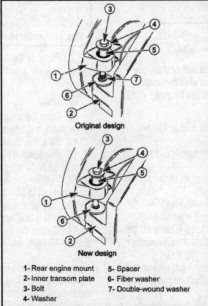

Fig. 6 Details of the rear engine mount

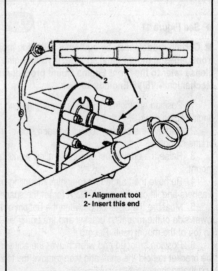

Fig. 7 Slide and alignment tool through the gimbal bearing...

3-4 ENGINE MECHANICAL - GM IN-LINE ENGINES

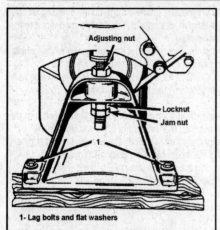

Fig. 8 ...and then adjust the front mount so that the tool slides freely between the gimbal housing and the coupler

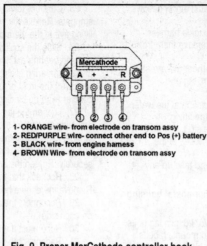

Fig. 9 Proper MerCathode controller hook-up

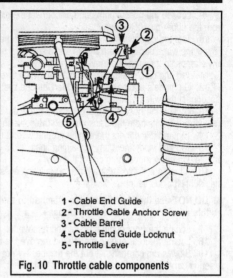

Fig. 10 Throttle cable components

PRIMING THE ENGINE

■ All engines that have not been run for more than 6 months, and any engines that have been replaced, or rebuilt should be primed with oil before initial start-up.

1. Ensure that the engine is filled with the proper amount of oil.
2. Remove the spark plugs and make sure the ignition is in the **OFF** position.
3. Connect a remote starter switch (#91-52024A1, or equivalent) to the positive (red) battery cable and the red/yellow wire at the starter motor.

■ If a remote starter switch is unavailable, disconnect the Purple wire at the ignition coil and tape it so as to prevent it from touching ground. Now you can use the ignition key.

4. Crank the engine for 15 seconds and then allow the starter to cool down for about a minute. Repeat this process 2 more times.
5. Remove the starter switch or reconnect the wire at the ignition coil.
6. Install the spark plugs.

Front Engine Mounts

REMOVAL & INSTALLATION

◆ See Figure 11

■ Certain models may be equipped with side mounts rather than the front pedestal mount detailed here. If your engine has side mounts, please refer to the Front Engine Mount procedures in the Engine Mechanical - V6/V8 Engines section.

1. Position an engine hoist over the engine and hook it up to the two engine lifting eyes.
2. Remove the two lag screws on each side of the mount where it rests on the stringer.
3. Raise the engine just enough to allow working room for removing the mount.
4. Remove the four mount-to-engine mounting bolts with their lock washers and lift out the mount. Don't forget the spacer on the port side.
5. Measure the distance between the bottom of the large washer on the lower side of the mounting bracket and the upper side of the smaller washer on top of the mount itself. Record it.
6. Position an open end wrench over the adjusting shaft just underneath the bracket to hold the shaft and then remove the top nut. Lift off the bracket and the spacer.
7. Remove the two bolts attaching the rubber mount to the upper bracket and remove it.

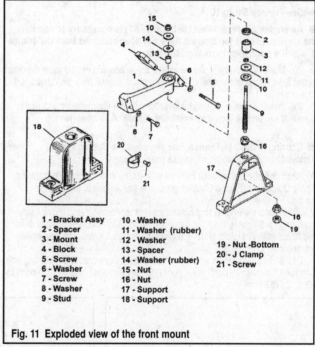

Fig. 11 Exploded view of the front mount

8. Remove the lower lock nut from the adjusting shaft and lift the shaft out of the lower mount. Some applications may have a jam nut under the locknut.

To install:

9. Drop the adjusting shaft into the lower mount and spin on the lower locknut until it is finger-tight.
10. Slide the spacer over the top end of the shaft.
11. Position the rubber mount back into the upper bracket and tighten the bolts to 30-35 ft. lbs. (41-47 Nm).
12. Position the upper bracket over the adjusting shaft and then tighten the upper nut to 50-60 ft. lbs. (68-81 Nm).
13. Position the assembly onto the engine, hold the spacer in place and install the two long bolts on the port side. Install the two shorter bolts into the starboard side. We suggest using new lock washers here. Tighten the long bolts to 48-56 ft. lbs. (65-76 Nm) and the short bolts to 32-40 ft. lbs. (43-54 Nm).
14. Retrieve the measurement you took earlier and check it now. Loosen the lower locknut and rotate the adjusting shaft until you achieve the correct distance between the two washers. Once this is done, cinch up the lower locknut until it is tight against the lower surface of the lower mount and then tighten the adjusting nut to 115-140 ft. lbs. (156-190 Nm).

ENGINE MECHANICAL - GM IN-LINE ENGINES   3-5

15. Position the mount over the lag screw holes and then slowly lower the engine until all weight is off the hoist. Install and tighten the lag screws securely.
16. If you're confident that your measurements and subsequent adjustment place the engine exactly where it was prior to removal, then you are through. If you're like us though, you may want to check the engine alignment before you fire up the engine.

Rear Engine Mounts

REMOVAL & INSTALLATION

 DIFFICULT

1. Remove the engine as previously detailed.
2. Loosen the two bolts and remove the mount from the transom plate.
3. Hold the square nut with a wrench and remove the shaft bolt. Be sure to take note of the style and positioning of the two mount washers as you are removing the bolt. Mark them, lay them out, or write it down, but don't forget their orientation!!

To install:

4. Slide the lower of the two washers onto the mount bolt, exactly as it came off.
5. Slide the bolt into the flat side of the rubber mount, install the remaining washer (as it came off!) and then spin on the square nut. Do not tighten it yet.

■ Incorrect washer installation will cause excessive vibration during engine operation.

6. Turn the assembly upside down and clamp the nut in a vise. Spin the assembly until the holes in the mounting plate are directly opposite any two of the flat sides on the nut. This is important, otherwise the slot on the engine pad will not engage the mount correctly. Secure the mount in this position and tighten the bolt to 44-52 ft. lbs. (60-71 Nm).
7. Remove the mount from the vise and position it on the transom plate. Install the bolts and washers and tighten each to 20-25 ft. lbs. (27-34 Nm).
8. Install the engine, making sure that the slot in the engine pad engages the square nut correctly. Install the two washers and locknut and tighten it to 28-30 ft. lbs. (38-41 Nm).

Cylinder Head Cover

REMOVAL & INSTALLATION

◆ See Figure 12

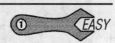 EASY

1. Open or remove the engine compartment hatch. Disconnect the negative battery cable.
2. Loosen the clamp and remove the crankcase ventilation hose at the cover (if equipped). Carefully move it out of the way.
3. Tag and disconnect the shift cut-out switch leads at the terminal block.
4. If your engine has a spark plug wire retainer attached to the cover, unclip the wires or remove the retainer.
5. Remove the fuel line at the pump (and carburetor) and plug it, and the fitting, to avoid spills and system contamination. Remove the overflow hose also, plugging it and moving it out of the way, Have some rags nearby, as there will be some seepage either way. Move the line out of the way of the cover.
6. Remove the circuit breaker bracket and carefully position it out of the way.
7. You may also want to remove the flame arrestor and disconnect the throttle cable at the carburetor as well.
8. Loosen the cover mounting bolts and lift off the cylinder head cover. Take note of any harness or hose retainers and clips that might be attached to certain of the mounting bolts; you need to make sure they go back in the same place.

To install:

9. Clean the cylinder head and cover mounting surfaces of any residual gasket material with a scraper or putty knife.

10. Position a new gasket on the cylinder head and then position the cover (don't forget the J-clip retainers!). Tighten the mounting bolts to 40 inch lbs. (4.5 Nm). Make sure any retainers or clips that were removed are back in their original positions.
11. Reconnect the fuel line. Check for leaks now, and after you restart the engine.
12. Connect the crankcase ventilation hose and the cut-out switch leads. Check that there were no other wires or hoses you may have repositioned in order to gain access to the cover.
13. Connect the battery cables.

Push Rod Cover

REMOVAL & INSTALLATION

 MODERATE

◆ See Figure 13

1. Remove the distributor with the spark plugs wires attached.
2. Remove the oil dipstick.
3. Remove, disconnect or re-route all wires, lines, or leads in the way of the cover.
4. Remove the 4 cover bolts and lift off the cover and gasket.
5. Clean the mating surfaces thoroughly and install a new gasket.
6. Position the cover and tighten the bolts to 40 inch lbs. (4.5 Nm).
7. Install the distributor and any other lines that were disconnected.

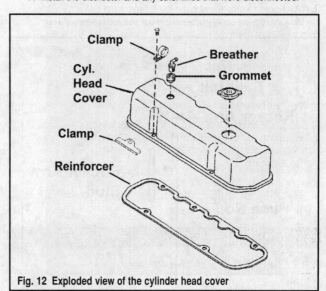

Fig. 12 Exploded view of the cylinder head cover

Fig. 13 There are 4 mounting bolts for the pushrod cover

3-6 ENGINE MECHANICAL - GM IN-LINE ENGINES

Rocker Arms and Push Rods

REMOVAL & INSTALLATION

♦ See Figures 14 and 15

1. Open or remove the engine hatch cover and disconnect the negative battery cable. Remove the cylinder head cover as detailed previously.
2. Bring the piston in the No. 1 cylinder to TDC. If servicing only one arm, bring the piston in that cylinder to TDC. The No. 1 cylinder on 3.0L engines is the first cylinder at the front of the engine.
3. Loosen and remove the rocker arm nuts and lift out the balls. Lift the arm itself off of the mounting stud and pull out the pushrod. It is very important to keep each cylinder's component parts together as an assembly. We suggest drilling a set of holes in a 2X4 and positioning the pieces in the holes.

To install:

4. Clean and inspect the rocker assemblies.
5. Coat all bearing surfaces of the rocker assembly with engine oil.
6. Slide the push rods into their holes. Make sure that each rod seats in its socket on the lifter.
7. Position the rocker arm over the stud so that the cupped side rides on the push rod. Slide the ball over the stud, install the nut and tighten it securely or until all play in the pushrod is taken up.
8. Adjust the valves and install the cylinder head cover.
9. Connect the battery cable and check the idle speed.

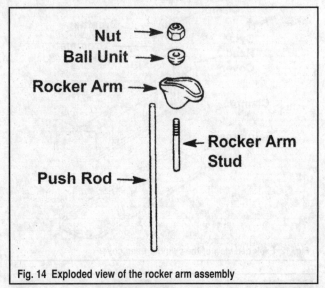

Fig. 14 Exploded view of the rocker arm assembly

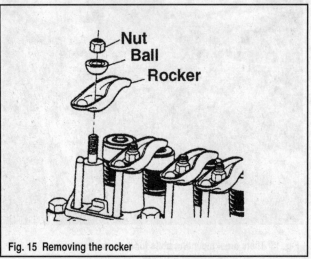

Fig. 15 Removing the rocker

VALVE ADJUSTMENT

Engine Not Running

♦ See Figure 16

These engines utilize hydraulic valve lifters, although there is no need for periodic valve adjustment, it is necessary to perform a preliminary adjustment after any work on the valve train or rocker assembly. All adjustment should be undertaken while the lifter is on the base circle of the camshaft lobe for that particular cylinder. This means the opposite side of the pointy part of each lobe.

1. Rotate the crankshaft, or bump the engine with the starter until the No. 1 cylinder is at TDC. Note that the notch or mark on the damper pulley will be lined up with the **0** mark on the timing scale. Be careful here though, this could mean that either the No. 1 or the No. 4 piston is at TDC. Place your hand on the No. 1 cylinder's valve and check that it does not move as the mark on the pulley is approaching the **0** mark on the tab. If it does not move, you're ready to proceed; if it does move, you are on the No. 4 cylinder and need to rotate the engine an additional full turn. This is important so make sure you've gotten it right!

2. Now that the No. 1 cylinder is at TDC, you can adjust the following valves:

- No. 1 cylinder: intake and exhaust
- No. 2 cylinder: intake
- No. 3 cylinder: exhaust
- No. 4 cylinder: intake

3. Loosen the adjusting nut on the rocker until you can feel lash (play in the push rod) and then tighten the nut until the lash has been removed. Carefully jiggle the push rod while tightening the nut until it won't move anymore - this is zero lash. Tighten the nut an additional 3/4 turn to set the lifter and then you're done. Perform this procedure on each of the valves listed above.

4. Slowly rotate the engine an additional full turn and this will bring the No. 4 piston to TDC. The pulley notch/mark should once again be in line with the **0** on the timing tab. You can now adjust the remaining valves as just detailed:

- No. 2 cylinder: exhaust
- No. 3 cylinder: intake
- No. 4 cylinder: exhaust

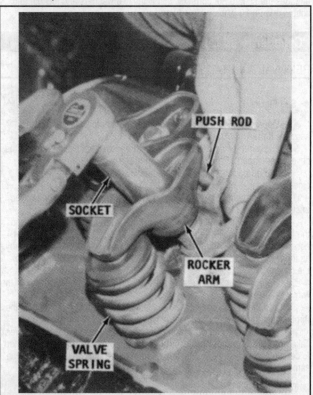

Fig. 16 Wiggle the push rod slowly while tightening the rocker adjusting nut

ENGINE MECHANICAL - GM IN-LINE ENGINES 3-7

Engine Running

1. Run the engine until it reaches normal operating temperature and then remove the cylinder head cover (while the engine is still running).
2. Loosen the rocker arm nut until the rocker starts to 'clatter' and then tighten the nut until the noise stops - this is zero lash.
3. Now tighten the nut and additional 1/4 turn and then wait about 10 seconds while the engine smoothes out. Repeat this 2 more times until the rocker nut has been tightened a total of 3/4 turns past zero lash.

※※ **CAUTION**

It is imperative that this step is not rushed and the engine has time to adjust in between each 1/4 turn!

4. Repeat the procedure on each of the remaining valves.
5. Turn off the engine and install the cover.

Hydraulic Valve Lifter

REMOVAL & INSTALLATION

1. Remove the cylinder head cover.
2. Using compressed air, thoroughly clean all dirt and grit from the cylinder head and related components.

※※ **WARNING**

If compressed air is not available, we highly recommend that you DO NOT proceed with this procedure. It is EXTREMELY important that no dirt gets into the lifter recesses before completing the installation.

3. Loosen the rocker arms and pivot the rockers off of the pushrods.
4. Disconnect the spark plug wires at the plugs.
5. Remove the high tension lead at the coil. Tag and disconnect both electrical connectors and then remove the coil and its bracket from the cylinder head.
6. Remove the distributor cap and mark the position of the rotor on the body of the distributor housing. Matchmark the distributor housing to the cylinder block and then remove the distributor, covering the hole carefully.
7. Remove the pushrod cover and gasket from the side of the cylinder block. On most later models, this will require removing the shift bracket assembly and moving it aside.
8. Remove the pushrods and lifters from the block, being very careful to keep track of where each one came from. Once again, we suggest using a 2x4 with holes drilled in it to store the rods and lifters; you'd be amazed at how quickly this job will fall apart if someone walks in and kicks the components that you have laid out on the floor!

To install:

9. Inspect the camshaft contact surface on the bottom of each lifter for excessive wear, galling or other damage. Discard the lifter if any of the above conditions are found. You'll probably also want to check out the camshaft lobe for damage also.
10. Coat the bottom of each lifter with engine oil and then carefully install each one into its respective recess.
11. Install the pushrods and then install the pushrod cover with a new gasket. Tighten the retaining bolts securely. If you removed the shift bracket previously, install it now.
12. Install the distributor so the marks on the housing and block are in alignment. Confirm that the rotor orientation mark made on the housing still aligns with the direction that the rotor is pointing. If not, or if the engine has been rotated for some reason during this procedure, please refer to the Ignition System section for complete details.
13. Install the distributor cap.
14. Install the coil and its bracket. Reconnect all plug wires and electrical leads.
15. Move the rockers into position and then adjust the valve lash as detailed previously.
16. Install the cylinder head cover.

※※ **CAUTION**

If any, or all, of the lifters has been replaced with a new one it is very important that you add GM Engine Oil Supplement to the crankcase BEFORE starting the engine.

Combination Manifold

REMOVAL & INSTALLATION

◆ See Figures 17, 18 and 19

■ These engines incorporate the intake and exhaust manifolds into one unit called a combination manifold. It is serviced as a unit.

1. Open or remove the engine compartment hatch. Disconnect the negative battery cable.
2. Drain all water from the engine, manifold and exhaust elbow as detailed in the Maintenance or Cooling sections.
3. Remove the flame arrestor and set it aside.
4. Disconnect the throttle cable from the carburetor, remove the anchor bolt and position the cable out of the way. Mark which anchor stud the cable was attached to for installation.
5. Disconnect the fuel line at the carburetor and fuel pump. Plug the pump inlet fitting.
6. Remove the crankcase ventilation hose. Tag and disconnect any vacuum or electrical lines and then remove the carburetor.
7. Disconnect the shift cables.
8. Disconnect the water inlet line.
9. Tag and disconnect any wires or hoses that may interfere with manifold removal.
10. Loosen the clamps and remove the exhaust pipe at the elbow bellows.
11. Loosen and remove the manifold mounting bolts from the center outward and remove the manifold from the cylinder head - you may have to provide a little friendly persuasion.

■ You may find that removing the alternator will allow for easier access to some of the manifold mounting nuts.

12. Remove the exhaust elbow.

To install:

13. Carefully clean all residual gasket material from the head, manifold and elbow mating surfaces with a scraper or putty knife. Inspect all gasket surfaces for scratches, cuts or other imperfections.
14. Position a new gasket on the cylinder head and install the manifold; making sure that everything is aligned properly. Tighten all nuts until they are just tight and then torque them to 20-25 ft. lbs. (27-34 Nm), starting in the center and working your way out to the ends of the manifold. The outer bolts should only be tightened to 15-20 ft. lbs. (20-27 Nm).

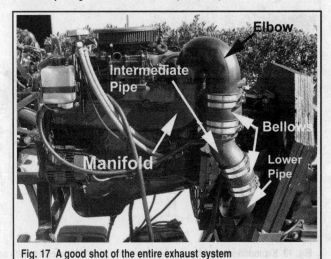

Fig. 17 A good shot of the entire exhaust system

3-8 ENGINE MECHANICAL - GM IN-LINE ENGINES

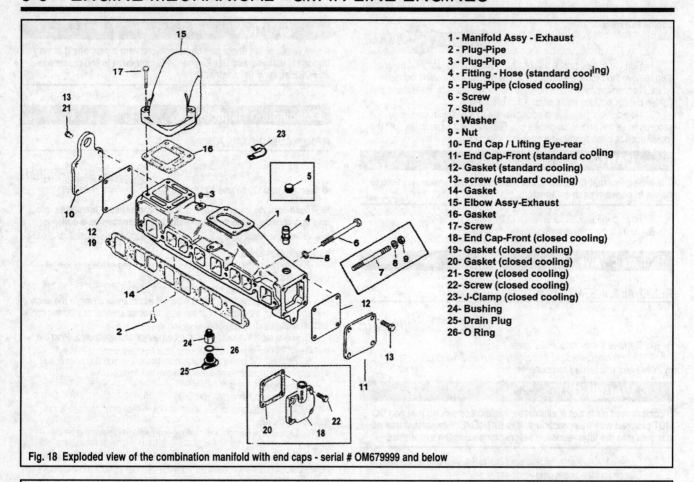

1 - Manifold Assy - Exhaust
2 - Plug-Pipe
3 - Plug-Pipe
4 - Fitting - Hose (standard cooling)
5 - Plug-Pipe (closed cooling)
6 - Screw
7 - Stud
8 - Washer
9 - Nut
10- End Cap / Lifting Eye-rear
11- End Cap-Front (standard cooling)
12- Gasket (standard cooling)
13- screw (standard cooling)
14- Gasket
15- Elbow Assy-Exhaust
16- Gasket
17- Screw
18- End Cap-Front (closed cooling)
19- Gasket (closed cooling)
20- Gasket (closed cooling)
21- Screw (closed cooling)
22- Screw (closed cooling)
23- J-Clamp (closed cooling)
24- Bushing
25- Drain Plug
26- O Ring

Fig. 18 Exploded view of the combination manifold with end caps - serial # OM679999 and below

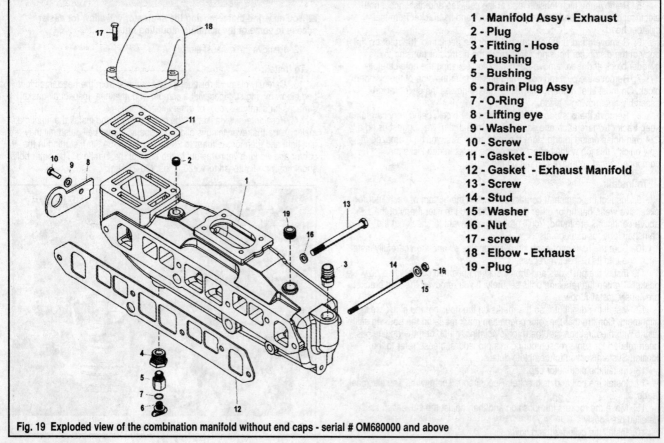

1 - Manifold Assy - Exhaust
2 - Plug
3 - Fitting - Hose
4 - Bushing
5 - Bushing
6 - Drain Plug Assy
7 - O-Ring
8 - Lifting eye
9 - Washer
10 - Screw
11 - Gasket - Elbow
12 - Gasket - Exhaust Manifold
13 - Screw
14 - Stud
15 - Washer
16 - Nut
17 - screw
18 - Elbow - Exhaust
19 - Plug

Fig. 19 Exploded view of the combination manifold without end caps - serial # OM680000 and above

ENGINE MECHANICAL - GM IN-LINE ENGINES 3-9

15. Position a new gasket on the manifold, making sure that the indents line up and then install the elbow. Tighten the mounting bolts to 25 ft. lbs. (34 Nm). Connect the exhaust pipe/bellows and tighten the clamps securely.
16. Connect the water inlet line and the shift cables.
17. Install the carburetor and reconnect the throttle cable, choke wire and any vacuum lines. Remember which anchor stud the throttle was attached to.
18. Connect the fuel line to the carb and the fuel pump (remember to unplug the pump fitting) and then install the flame arrestor.
19. Make sure that any miscellaneous lines or hoses that you may have moved or disconnected during removal are reconnected and routed properly.
20. Fill the system with water, connect the battery cable and start the engine. When the engine reaches normal operating temperature, turn it off and re-torque the manifold bolts.

Exhaust Elbow

REMOVAL & INSTALLATION

◆ See Figures 17, 18 and 19

1. Drain the cooling system.
2. Loosen the hose clamps and slide off the upper exhaust hose (bellows).
3. Remove the two bolts and washers from the elbow flange where it mates with the manifold and lift it off the manifold. A little friendly persuasion with a soft rubber mallet may be necessary! Be careful though, no need to take out all your aggressions on the poor thing.
4. Remove the gasket and/or restrictor plate. You can throw away the gaskets, but keep the plate if equipped.

To install:

5. Clean the mating surfaces of the manifold and elbow thoroughly, coat both sides of a new gasket with Gasket Sealing compound and position it onto the manifold flange.
6. Position the restrictor plate (if equipped) onto the manifold.
7. Position the elbow on the manifold, install the 2 bolts and then tighten to 25 ft. lbs. (34 Nm).
8. Slide the exhaust hose, while wiggling it, all the way onto the elbow and tighten the clamp screws securely.

Exhaust Hose (Bellows)

REMOVAL & INSTALLATION

◆ See Figures 17 thru 20

1. Loosen all four hose clamps, two on top of the hose and two on the bottom.
2. Drizzle a soapy water solution over the top of hose where it mates with the exhaust elbow and let it sit for a minute.
3. Grasp the hose with both hands and wiggle it side-to-side while pulling down on it until it separates from the elbow.
4. Now wiggle it while pulling upwards until it pops off the intermediate or exhaust pipe.
5. Check the hose for wear, cracks and deterioration. Coat the inside of the lower end of the hose with soapy water and wiggle it into position on the pipe. Remember to install the two clamps before sliding it over the end of the pipe.
6. Slide two clamps over the upper end, lubricate the inside with soapy water and wiggle the hose over the elbow (or intermediate pipe) until it is in position.
7. Tighten all four clamps securely.

Exhaust Pipe(s)

REMOVAL & INSTALLATION

◆ See Figures 17 thru 20

■ Although not necessary, some people find this easier if they remove the engine.

1. Loosen the hose clamps and wiggle the upper hose off of the exhaust elbow.
2. Loosen the next 2 clamps and remove the upper hose/bellows from the intermediate pipe.
3. Loosen the upper clamps on the lower hose/bellows and pull the intermediate pipe out of it.
4. Loosen the bottom 2 clamps and pull the lower hose off of the exhaust tube.
5. Loosen the four retaining bolts at the gimbal housing and then remove the exhaust pipe. Carefully scrape any remnants of the seal from the pipe and transom mounting surfaces.
6. Coat a new seal with 3M Adhesive and position it into the groove on the transom shield mating surface. Coat the mounting bolts with Gasket Sealing Adhesive. Position the exhaust pipe, insert the bolts and tighten them to 20-25 ft. lbs. (27-34 Nm).
7. Install the engine if removed.

Exhaust Valve (Flapper/Shutter)

REPLACEMENT

◆ See Figure 20

■ Although Mercury calls this an exhaust valve, many people also call it a flapper valve or a shutter valve. Whatever you call it, this is the small valve in the top of the exhaust pipe that prevents sea water backflow.

1. Remove the upper exhaust hose from the elbow and intermediate pipe. Now remove the intermediate pipe and the lower hose. The shutter is located in the upper end of the lower exhaust tube.
2. The valve is held in place by means of a pin running through two grommets in the sides of the pipe. Position a small punch over one end of the pin and carefully press the pin out of the pipe. Some people have had luck simply pulling upward on the assembly very carefully and popping the shutter and grommets right out of the pipe.

■ Make sure you secure the valve while removing it so nothing falls down into the exhaust pipe.

3. Press out the two grommets and discard them. Coat two new grommets with Scotch Grip Rubber Adhesive and press them back into the sides of the pipe.

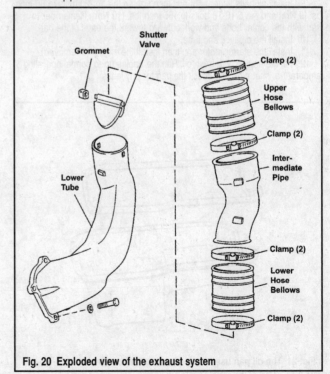

Fig. 20 Exploded view of the exhaust system

3-10 ENGINE MECHANICAL - GM IN-LINE ENGINES

4. Position the new valve into the pipe with the long side DOWN. When the valve is in place, coat the pin lightly with engine oil and slowly slide it through one of the grommets, through the two retaining holes on the valve and then through the opposite grommet. Make sure the pin ends are flush with the sides of the pipe on both sides.

5. Install the lower hose, intermediate pipe and upper hose.

Oil Pan

REMOVAL & INSTALLATION

◆ See Figure 21

■ More times than not this procedure will require the removal of the engine. Your boat and its unique engine installation will determine this, but the procedure is almost always easier with the engine removed from the boat.

1. Remove the engine as previously detailed in this section.
2. If you haven't already drained the engine oil, do it now. Make sure you have a container and lots of rags available.
3. Remove the starter motor as detailed in the Electrical section.
4. Remove the dipstick and tube.
5. Loosen and remove the oil pan retaining bolts, starting with the center bolts and working out toward the pan ends. Lightly tap the pan with a rubber mallet to break the seal and then lift it off the cylinder block. If your engine stand will allow for rotating the engine, you'll find that this will be easier with the pan facing up.

To install:

6. Clean the pan mating surfaces of any residual gasket material with a scraper or putty knife. Make sure that no old gasket material has been pressed into the retaining bolt holes in the pan, block or front cover. Clean the pan itself thoroughly with solvent.
7. Position a new pan gasket onto the pan being very careful to line up all the holes - do not use RTV sealant with this gasket. To be safe, you can apply a little RTV sealant to the joints at the rear oil seal retainer and the front cover, but remember that RTV sealer sets up in about 15 minutes so don't dally here.
8. Move the pan and gasket onto the block. It is very important that you ensure all the holes line up correctly; sometimes a few bolts inserted through the pan and gasket will help the gasket stay in place.
9. Install all bolts finger tight and then tighten the 1/4-20 bolts to 80 inch lbs. (9 Nm) and the 5/16-18 bolts to 165 inch lbs. (19 Nm). Remember to start with the center bolts and work outward toward the ends of the pan.
10. Install the dipstick and tube.
11. Install the starter motor and then install the engine (if removed).
12. Fill the engine with fresh oil. Run the engine up to normal operating temperature, shut it off and check the pan for any leaks.

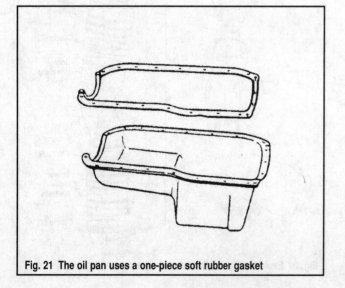

Fig. 21 The oil pan uses a one-piece soft rubber gasket

Oil Pump

REMOVAL & INSTALLATION

◆ See Figures 22 and 23

■ The two-piece oil pump utilizes two pump gears and a pressure regulator valve. A baffled pick-up tube is press-fit into the body of the pump.

1. Remove the oil pan as previously detailed. Remember that you probably need to remove the engine for this procedure.
2. Loosen and remove the pump pick-up tube support bracket bolt. The tube is pressed into the pump housing and should not be removed unless replacement is necessary.
3. Loosen and remove the two pump mounting bolts and lift off the pump assembly.
4. Check that the pump and block mating surfaces are clean and then position the pump over the block so that the pump drive shafts are aligned with the distributor tang. Make sure that the flange covers the alignment bushing. Do not use a gasket or RTV sealant.
5. Tighten the pump mounting bolts to 120 inch lbs. (14 Nm). Position the pick-up tube bracket and tighten the bolt to 60 inch lbs. (7 Nm).
6. Install the oil pan and engine.

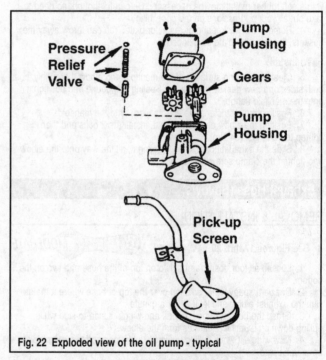

Fig. 22 Exploded view of the oil pump - typical

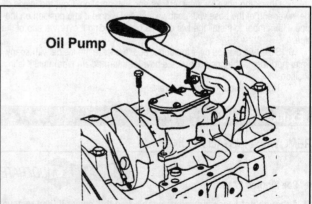

Fig. 23 Removing the oil pump

ENGINE MECHANICAL - GM IN-LINE ENGINES 3-11

Oil Filter Bypass Valve

REMOVAL & INSTALLATION

♦ See Figure 24

After removing the oil filter, check the spring and small fiber valve for proper operation. Any signs of incorrect operation, or wear and deterioration will necessitate replacement.

1. Drain the oil as detailed in the Maintenance section.
2. Remove the oil filter.
3. Using a small prybar, remove the valve.
4. Install a new valve and press it in by placing a 9/16 in. deep socket over it and tapping the socket lightly with a hammer.
5. Install the oil filter and refill the engine with the appropriate oil.

Engine Coupler And Flywheel

REMOVAL & INSTALLATION

♦ See Figures 25, 26 and 27

1. Remove the engine from the boat as detailed previously in this section.
2. Although not strictly necessary, we recommend removing the starter.
3. Loosen the retaining bolts and lift off the flywheel housing (bell housing in automotive parlance). There are 6 bolts, 4 are 1 1/4 in, 1 is 1 3/4 in. and the last is 2 in.
4. Remove the flywheel housing cover bolts and lift off the cover.
5. Loosen the three engine coupler mounting bolts and then remove the coupler. Don't lose the rubber bumpers pressed into the inner side of the coupler; there are 3 of these as well
6. Loosen the six flywheel mounting bolts gradually and as you would the lug nuts on your car or truck - that is, in a diagonal star pattern.
7. Thoroughly clean the flywheel mating surface and check it for any nicks, cracks or gouges. Check for any broken teeth.
8. Install the flywheel over the dowel on the crankshaft and tighten the mounting bolts to 75 ft. lbs. (100 Nm). Once again use the star pattern while tightening the bolts.
9. Attach a dial indicator to the engine block and take reading around the outer edge of the flywheel; pushing in on the flywheel to remove any crankshaft end play. Maximum run-out should not exceed 0.008 in. (0.203mm).
10. Insert the rubber bumpers into the coupler. Position the engine coupler and tighten the mounting bolts to 35 ft. lbs. (48 Nm). If you are re-using the old bolts, make sure you coat them with Loctite® 271. Lubricate the coupler splines with Quicksilver 2-4-C Marine Lubricant.
11. Install the flywheel housing and tighten the bolts to 30 ft. lbs. (41 Nm).
12. Install the engine.

Rear Main Seal

REMOVAL & INSTALLATION

♦ See Figures 28 and 29

■ It is not necessary to remove the engine or rear main bearing cap when removing the one-piece oil seal on these engines, although you may find it easier to do just that.

1. Remove the flywheel housing and cover as detailed in this section.
2. Remove the engine coupler and flywheel from the engine as detailed in this section.
3. Insert a small prybar into one of the three slots in the edge of the seal retainer and slowly pry the seal out of the retainer.
4. Thoroughly clean the retainer surface.
5. Apply Perfect Seal to the inner mating surface of the seal retainer. Spread a small amount of grease around the outside edge of a new seal and position it over its slot in the retainer.
6. Position a seal driver (#J-26817-A) over the seal and strike it with a mallet until the seal is fully seated in its bore.
7. Install the flywheel and engine coupler. Install the cover and flywheel housing.

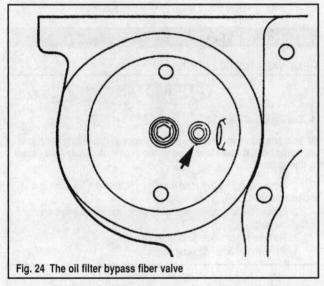

Fig. 24 The oil filter bypass fiber valve

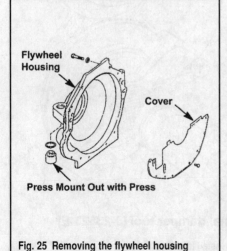

Fig. 25 Removing the flywheel housing

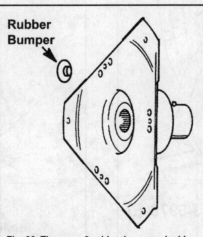

Fig. 26 There are 3 rubber bumpers inside the engine coupler

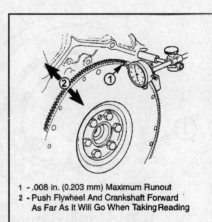

1 - .008 in. (0.203 mm) Maximum Runout
2 - Push Flywheel And Crankshaft Forward As Far As It Will Go When Taking Reading

Fig. 27 Checking the flywheel run-out

3-12 ENGINE MECHANICAL - GM IN-LINE ENGINES

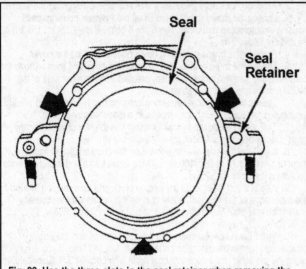

Fig. 28 Use the three slots in the seal retainer when removing the rear main seal

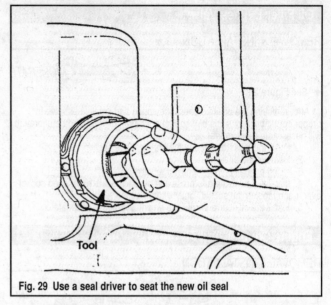

Fig. 29 Use a seal driver to seat the new oil seal

Torsional Damper

REMOVAL & INSTALLATION

◆ See Figures 30 and 31

1. If the engine is in the boat, install an engine hoist and tighten the chain so that the engine's weight is removed from the front engine mount. Do not use the lifting eye attached to the thermostat housing. Remove the front engine mount.
2. Remove the drive or serpentine belt as detailed in the Maintenance section.
3. Remove the damper retaining bolt and install special tool (#J-6978-E) onto the damper. Tighten the tool press bolt and remove the damper. MerCruiser suggests that you do not use a conventional gear puller for this procedure.

To install:

4. Coat the front cover oil seal lip with clean engine oil and then install the damper with a proper installation tool. Be sure that you thread the tool into the crankshaft at least 1/2 inch to protect the threads. In a pinch, you can use a block of wood and a plastic mallet, but be careful that the pulley does not shift on its mountings while you're hammering.

5. Install the retaining bolt and tighten it to 70 ft. lbs. (95 Nm). Tighten the drive pulley to 35 ft. lbs (48 Nm).
6. Install the drive/serpentine belt and make sure that it is adjusted properly.
7. Install the front mount and unhook the engine hoist.

Front Cover And Oil Seal

REMOVAL & INSTALLATION

◆ See Figures 32 and 33

■ This procedure may require engine removal, depending upon your particular boat. If necessary, remove the engine as detailed previously in this section.

1. Open the drain valves and drain the coolant from the block and exhaust manifold.
2. Loosen the idler pulley to provide slack, and then remove the serpentine belt.
3. Remove the torsional damper assembly.
4. Remove the water circulation pump.

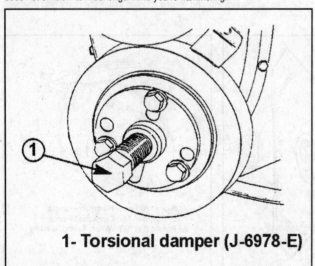

1- Torsional damper (J-6978-E)

Fig. 30 Use a special puller to remove the torsional damper...

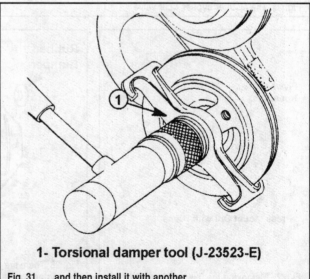

1- Torsional damper tool (J-23523-E)

Fig. 31 ...and then install it with another

ENGINE MECHANICAL - GM IN-LINE ENGINES 3-13

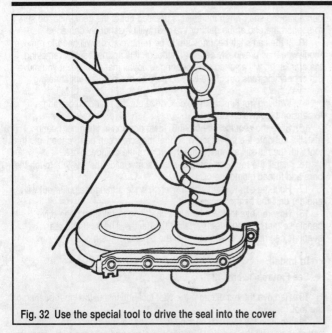

Fig. 32 Use the special tool to drive the seal into the cover

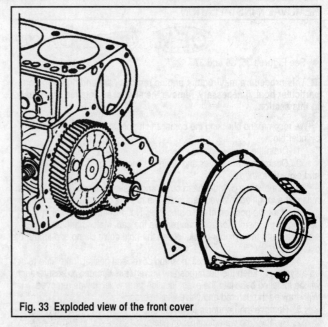

Fig. 33 Exploded view of the front cover

5. Secure the engine with a hoist and remove the front engine mount if you haven't already removed the engine.
6. Drain the oil and remove the oil pan.
7. Loosen all of the front cover retaining screws (6) and then pull off the cover and gasket. Clean all gasket material from the cover and block mating surfaces.
8. Pry the oil seal from the timing gear cover with a large drift or screwdriver - from the rear.

To install:

9. Position a new oil seal into the cover so that the lip (open side) faces into the cover. With the cover on a clean flat surface, position a seal installer (#J-35468) over the seal and drive it into place with a hammer.
10. Clean all old gasket material from the cover and block mating surfaces. Coat a new gasket lightly with Perfect Seal and stick it into position on the cylinder block.
11. Insert an alignment tool (#J-23042) through the oil seal hole. Coat a new gasket with Perfect Seal and install the front cover. Position the front cover so that the holes in the cover align correctly with the small dowel pins sticking out of the block. Install the mounting screws and tighten them to 27-35 inch lbs. (3-4 Nm). Tighten the screws evenly and gradually.

■ There is some confusion here on Mercruiser's part; although they state through out their specifications that the front cover bolt torque is 27-35 inch lbs. (3-4 Nm), they also mention that it should be 100 inch lbs. (11 Nm). We believe it to be the former rather than the latter, but were unable to find a consistant answer from our sources in the field.

12. Install the water circulation pump and the harmonic balancer/hub.
13. Install the oil pan.
14. Install the serpentine belt and check the tension.
15. Reinstall the engine mount; or if you removed the engine, install it now.
16. Refill the engine with oil and coolant. Hook the engine to a water supply, start it and check for leaks.

Camshaft, Bearings And Gear

CHECKING LIFT

◆ See Figure 34

We recommend checking the camshaft lift prior to removing it from the cylinder block.
1. Tag and disconnect the electrical connectors at the ignition coil.
2. Remove the cylinder head cover and rocker arms as detailed previously.

3. Using a special adaptor (#J-8520), connect a dial indicator so that its tip is positioned on the end of the pushrod - the adaptor should screw onto the end of the rocker stud.
4. Slowly rotate the crankshaft in the direction of engine rotation until the valve lifter is riding on the heel (back side of lobe) of the camshaft lobe. The pushrod should be at its lowest point when the lifter is on the heel.

■ A remote starter works well for turning the engine over in this situation.

5. Set the indicator to **0** and then rotate the engine until the pushrod is at the highest point of its travel. Camshaft lift should be 0.2529 in. (6.425mm).
6. Continue rotating the engine until the pushrod is back at its lowest position - make sure that the indicator still reads **0**.
7. Repeat this procedure for the remaining pushrods.
8. Install the rocker arms and adjust the valve clearance.
9. Install the cylinder head cover and reconnect the coil leads.

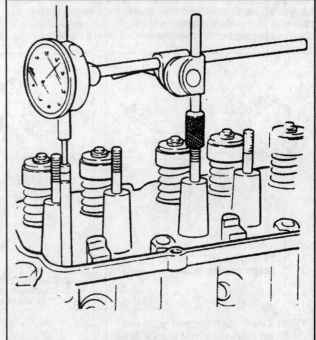

Fig. 34 Use a dial indicator when checking the camshaft lift

3-14 ENGINE MECHANICAL - GM IN-LINE ENGINES

REMOVAL & INSTALLATION

◆ See Figures 35, 36 and 37

■ This procedure may require engine removal, depending upon your particular boat. If necessary, remove the engine as detailed previously in this section.

We recommend checking the camshaft lift prior to removing it from the cylinder block.

1. Drain the engine oil.
2. Open the drain valves and drain the coolant from the cylinder block and exhaust manifold.
3. Remove the cylinder head cover and gasket. Loosen the rocker arm nuts just enough so that you can rotate the rockers off of the pushrod ends.
4. Mark the position of the distributor's No. 1 cylinder terminal on the housing of the distributor and then remove the cap. Matchmark the distributor and the cylinder block, loosen the hold down clamp and lift out the distributor.
5. Remove the ignition coil, pushrod cover and gasket. Take time to set up a system to keep the push rods and valve lifters in order, to ensure each will be installed back into the exact location from which it was removed. Withdraw each push rod and valve lifter in order.
6. Remove the harmonic balancer.
7. Remove the water circulation pump.
8. Secure the engine with a hoist and remove the front engine mount if you haven't already removed the engine.
9. Remove the oil pan.
10. Remove the front cover and gasket. Clean all gasket material from the cover and block mating surfaces.
11. Rotate the camshaft gear until the holes in the gear are aligned with the thrust plate screws. Remove the two screws. Carefully withdraw the camshaft gear and camshaft by pulling the gear straight forward and the shaft out of the block.

✱✱ CAUTION

Be very careful not to damage the camshaft bearings while removing the camshaft.

12. Check the gear and thrust plate end-play. This clearance should be 0.001-0.005 in. (0.03-0.1mm) max.. If the decision is made to replace the camshaft gear, or the thrust plate, the gear must be pressed from the shaft. Gear removal from the camshaft requires the use of camshaft gear removal tool (#J-791) and an arbor press. Place the end of the removal tool onto the table of an arbor press, and then press the shaft free of the gear. The thrust plate must be positioned to prevent the woodruff key in the shaft from damaging the shaft when the shaft is pressed out of the gear. Also, be sure to support the end of the gear or the gear will be seriously damaged.

13. If the camshaft bearings are to be removed, you will need to remove the flywheel as previously detailed. Although it is not necessary, removing the crankshaft will also facilitate bearing removal, make sure that you move the connecting rods out of the way so they do not interfere with bearing removal.
14. Working from inside the block, drive out the rear cam bearing expansion plug (welch plug).
15. Slide the pilot tool (#J-6098-01) into position in the inner bearing. Install a nut onto the puller screw so that the screw can be threaded into the tool with the nut still extending out the front of the cylinder block.
16. Install the remover onto the puller screw, slide the screw through the bore and thread it onto the pilot.
17. Hold the screw shaft with a wrench while turning the puller nut with another until the bearing comes out.
18. Now remove the pilot from the shaft and install it onto the drive handle so that the shoulder is against the handle. The front and rear bearings can now be driven out from the outside of the block.

To install:

◆ See Figures 38 and 39

19. Remove the handle from the pilot tool and install the inner bearing on the tool.
20. Position the tool and bearing to the rear of the inner bore. Install the screw shaft, with the remover, through the block and onto the pilot - from the front of the cylinder block.
21. Align the oil hole in the bearing with the oil gallery hole and then snug the puller nut up against the adaptor. Using two wrenches again, hold the screw shaft with one while turning the puller with the other until the bearing is in position.

■ The oil hole is on the top of the bearing and will not be visible during installation. To make installation easier, align the two holes and then mark the opposite side of the bearing/block.

22. Attach the drive handle to the tool and position the new front bearing onto the tool. Align the oil holes and drive the bearing in from the front (outside) of the cylinder block. Be sure that the bearing is driven in past the edge of the block at least 1/8 in. in order to expose the oil hole for the timing gear nozzle.
23. Repeat the last step for the rear bearing, but remember there is no oil nozzle hole in this bore so the bearing must be flush with the block.
24. Install a new expansion plug at the rear bearing. The plug must be flush, or no more than 1/32 in. (0.8mm) deep.
25. To assemble the camshaft parts, first firmly support the camshaft at the back of the front journal in an arbor press. Next, place the gear spacer ring and the thrust plate over the end of the shaft, and install the Woodruff

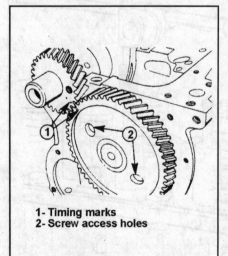

1- Timing marks
2- Screw access holes

Fig. 35 Line up the timing gear marks and the access holes should rest over the thrust plate screws (engine upside down)

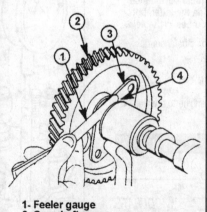

1- Feeler gauge
2- Camshaft gear
3- Camshaft thrust plate
4- Take measurement here

Fig. 36 Checking the camshaft endplay

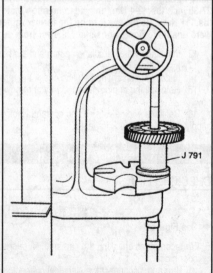

Fig. 37 The camshaft gear must be pressed off of the camshaft

ENGINE MECHANICAL – GM IN-LINE ENGINES

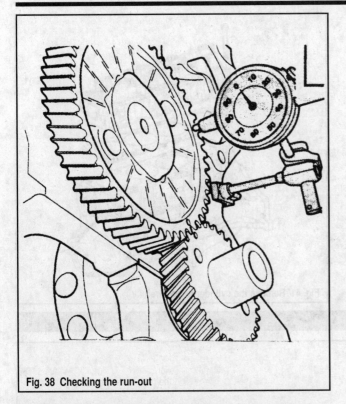

Fig. 38 Checking the run-out

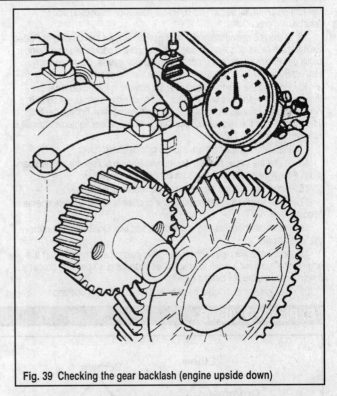

Fig. 39 Checking the gear backlash (engine upside down)

key in the shaft keyway. Install the camshaft gear and press it onto the shaft until it bottoms against the gear spacer ring. Check the end-play again.

26. If you removed the crankshaft, install it now.

27. Coat the camshaft lobes with G.M. Super Engine Oil Supplement and then pour the remainder of the can into the crankcase when you refill the engine with oil later in this procedure.

28. Slowly slide the camshaft and gear into the block, being careful not to damage the lobes or bearings. When the shaft is almost all the way in, rotate the camshaft and crankshafts until the timing marks on the teeth of each gear line up and then slide the camshaft in until the thrust plate meets the cylinder block and the holes in the plate line up with those in the block. Slowly rotate the crankshaft until the access holes in the cam gear are over the thrust plate holes. Install the two retaining screws and tighten them to 80 inch lbs. (9 Nm).

29. Connect a dial indicator to the front of the cylinder block with an adaptor so that the needle on the indicator is in contact with the face of the camshaft gear. Rotate the engine 360° checking the run-out as you go. Repeat this procedure for the crankshaft gear. Camshaft gear run-out should not exceed 0.004 in. (0.102mm). Crankshaft gear run-out should not exceed 0.003 in. (0.076mm). If run-out exceeds specification, remove the gear and check for burrs, otherwise replace the gear.

30. Move the dial indicator so that the needle is now riding on a gear tooth. Wiggle the shaft back and forth while checking the backlash reading on the indicator. Backlash should be 0.004-0.006 in. (0.102-0.152mm).

31. Install the oil pan. Install the front cover.

32. Install the harmonic balancer and water pump.

33. Installation of the remaining components is in the reverse order of removal. Don't forget to add the remainder of the Oil Supplement when refilling the engine with oil.

Cylinder Head

REMOVAL & INSTALLATION

◆ See Figures 40 and 41

1. Drain the water from the cylinder block and manifold.
2. Remove the fuel line support brackets. Disconnect the fuel line at the carburetor and fuel pump, plug the fitting holes and remove the line.
3. Remove the combination manifold as previously detailed in this section; you can leave the carburetor/throttle body attached if you like.
4. Disconnect the coolant hoses at the thermostat housing and move them out of the way. Have some rags available, as there will still be some coolant/water in the hoses.
5. Tag and disconnect the temperature sending unit lead at the thermostat housing, loosen the mounting bolts and then remove the housing and thermostat.
6. Tag and disconnect all wires at the ignition coil and then remove the mounting bracket bolt and lift off the coil.
7. Tag and disconnect the spark plug wires at the plugs; move them out of the way. Although not necessary, it's a good idea to remove the plugs themselves also.
8. Remove the circuit breaker bracket and then unbolt and remove the engine lifting bracket.
9. Remove the cylinder head cover and rocker assemblies as detailed previously in this section.
10. Loosen the cylinder head bolts, from the center bolts and working out to the ends of the head and then carefully lift the head off the block. You may need to persuade it with a rubber mallet - be careful! Set the head down carefully; do not sit it on cement.

To install:

11. Carefully, and thoroughly, remove all residual head gasket material from the cylinder head and block mating surfaces with a scraper or putty knife. Check that the mating surfaces are free of any nicks or cracks. Make sure there is no dirt or old gasket material in any of the bolt holes. Refer to

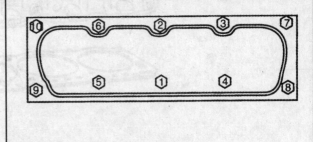

Fig. 40 Cylinder head tightening sequence

3-16 ENGINE MECHANICAL - GM IN-LINE ENGINES

the Engine Overhaul section for complete details on inspection and refurbishing procedures.

12. If using a ribbed stainless steel gasket, apply a thin coating of Perfect Seal to both sides of a new gasket and position the gasket over the cylinder block dowel pins. If using a graphite composition gasket, do not use any Perfect Seal. DO NOT use automotive-type steel gaskets.

13. Position the cylinder head over the dowels in the block. Coat the threads of the head bolts with Perfect Seal and install them finger tight. It never hurts to use new bolts, although it's not necessary. Tighten the bolts, a little at a time, in the sequence illustrated, until the proper tightening torque is achieved - 90 ft. lbs. (122 Nm).

14. Install rocker assemblies and the cylinder head cover.

15. Install the circuit breaker and engine lifting brackets. Install the spark plugs if they were removed and then connect the plug wires.

16. Install the coil and reconnect all the electrical leads.

17. Install the thermostat housing, the coolant hoses and the temperature sending lead.

18. Install the manifold and connect the fuel line. Don't forget to remove the fitting plugs.

19. Add coolant/water, connect the battery and check the oil. Start the engine and run it for a while to ensure that everything is operating properly. Keep an eye on the temperature gauge.

20. Re-tighten the cylinder head bolts after 20 hours of operation.

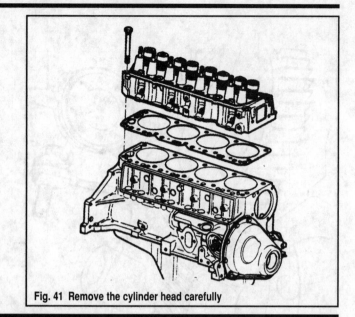

Fig. 41 Remove the cylinder head carefully

EXPLODED VIEWS

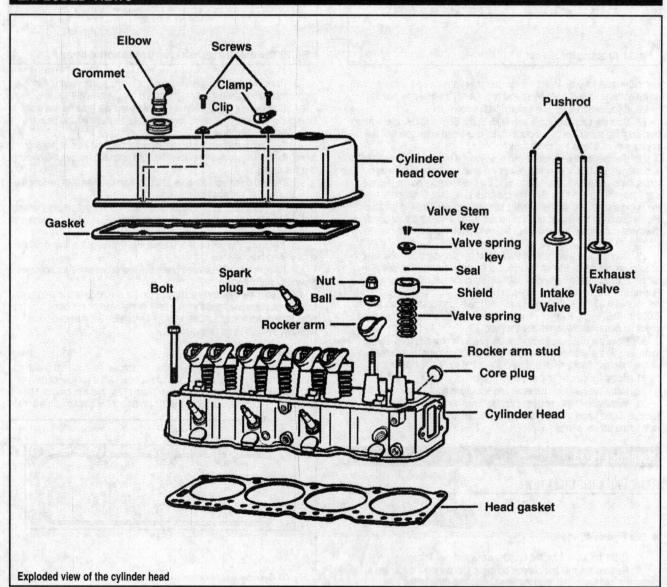

Exploded view of the cylinder head

ENGINE MECHANICAL - GM IN-LINE ENGINES

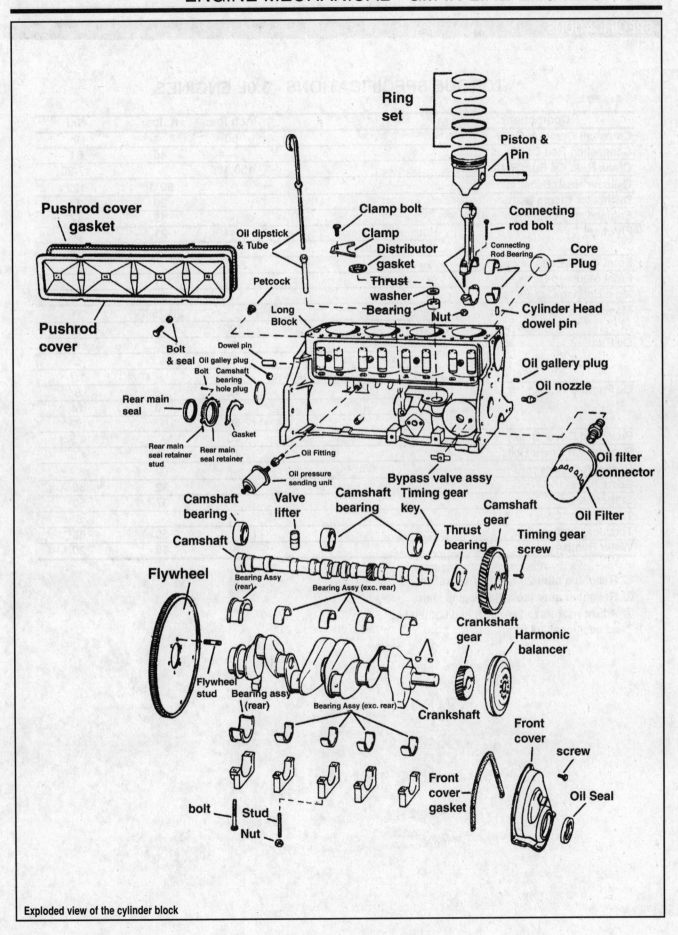

Exploded view of the cylinder block

ENGINE MECHANICAL - GM IN-LINE ENGINES

SPECIFICATIONS

TORQUE SPECIFICATIONS - 3.0L ENGINES

Component		Inch lbs.	ft. lbs.	Nm
Camshaft Sprocket Bolts		80	-	9
Connecting Rod Cap Nuts		-	45	61
Crank Rear Oil Seal Retainer Nuts		150-180	-	17-20
Cylinder Head Bolts		-	90 ①	122
Distributor Clamp Bolt		-	20	27
Engine Coupler-to-Flywheel		-	35	47
Flywheel	Housing-to-Block	-	21	28
	To Crankshaft Bolts	-	65	88
Front Cover Bolts		27-35	-	3.4
Front Mount-to-Block		-	21	28
Main Bearing Cap Bolts		-	65	88
Manifold-to-Cylinder Head	Center	-	20-25 ②	31
	Outer	120	-	25
Oil Pan	Crankcase Bolts	80	-	9
	Front Cover Bolts	45	-	5
	Studs to Oil Seal or Crankcase	15	-	1.7
Oil Pump	Cover	72	-	8
	Block	-	10	14
	Pick-up	60	-	7
Push Rod Cover Bolts		40	-	4.5
Rocker Arm Cover Bolts		40	-	4.5
Rocker Arm Nuts		-	③	
Spark Plugs		-	22	30
Starter Motor		-	37	50
Timing Gear Cover		72	-	8
Torsional Damper Bolt		-	50	68
Water Pump		-	15	20

① Retighten after 20 hours of operation
② Retighten after the initial engine start
③ Adjust rockers to zero lash and tighten bolt an additional 3/4 turn

ENGINE MECHANICAL - GM IN-LINE ENGINES 3-19

ENGINE SPECIFICATIONS - 3.0L

Component				Standard (in.) ①	Metric (mm) ①
Camshaft	Lobe Lift	Intake		0.2529 Max	6.425 Max
		Exhaust		0.2529 Max	6.425 Max
	Journal Diameter			1.8677-1.8697	47.440-47.490
	End Play			0.003-0.008	0.08-0.2032
Crankshaft	Main Journal	Diameter		2.2979-2.2994	58.367-58.404
		Taper	Production	0.0002 Max	0.005 Max
			Service Limit	0.001 Max	0.02 Max
		Out-of-Round	Production	0.0002 Max	0.005 Max
			Service Limit	0.001 Max	0.02 Max
	Main Bearing Clearance	Production	#1 thru 4	0.001-0.0024	0.025-0.060
			#5	0.0016-0.0035	0.041-0.088
		Service Limit	#1 thru 4	0.001-0.0025	0.03-0.06
			#5	0.002-0.0035	0.05-0.08
	Crankshaft End Play			0.002-0.006	0.05-0.15
	Crankpin	Diameter		2.0980-2.0995	53.289-53.327
		Taper	Production	0.0003 Max	0.007 Max
			Service Limit	0.001 Max	0.02 Max
		Out-of-Round	Production	0.0002 Max	0.051 Max
			Service Limit	0.001 Max	0.02 Max
	Connecting Rod	Bearing Clearance	Production	0.0017-0.0027	0.044-0.686
			Service Limit	0.003 Max	0.07 Max
		Side Clearance		0.006-0.017	0.153-0.4318
Cylinder Bore	Diameter			3.9995-4.0025	101.588-101.790
	Out-of-round	Production		0.0005 Max	0.012 Max
		Service Limit		0.002 Max	0.05 Max
	Taper	Production	Thrust Side	0.0005 Max	0.012 Max
			Relief Side	0.0005 Max	0.012 Max
		Service Limit		0.001 Max	0.02 Max
Cylinder Head	Gasket Surface Flatness			②	②
Piston	Clearance	Production		0.0025-0.0035 ③	0.064-0.088 ③
		Service Limit		0.0035 Max	0.08 Max
	Compression Rings Groove Clearance	Production	Top	0.0012-0.0029	0.030-0.073
			2nd	0.0012-0.0029	0.030-0.073
		Service Limit		0.003 Max	0.09 Max
	Compression Rings Gap	Production	Top	0.010-0.020	0.254-0.508
			2nd	0.010-0.020	0.254-0.508
		Service Limit		0.035 Max	0.88 Max
	Oil Ring	Groove Clearance	Production	0.001-0.006	0.026-0.152
			Service Limit	0.007 Max	0.17 Max
		Gap	Production	0.010-0.030	0.254-0.762
			Service Limit	0.040 Max	1.01 Max
	Piston pin	Diameter		0.9270-0.9271	23.546-23.550
		Clearance	Production	0.0003-0.0006	0.008-0.016
			Service Limit	0.001 Max	0.02 Max
		Interference Fit		0.0008-0.0019	0.020-0.050
Valve System	Lifter			Hydraulic	Hydraulic
	Rocker Arm Ratio			1.75:1	
	Tappet gap colapse			④	④
	Face Angle	Intake		45 deg.	
		Exhaust		45 deg.	
	Seat Angle			46 deg.	

ENGINE MECHANICAL - GM IN-LINE ENGINES

ENGINE SPECIFICATIONS - 3.0L

Component				Standard (in.) ①	Metric (mm) ①
Valve System (Cont'd)	Seat Runout			0.002 Max	0.05 Max
	Seat Width	Intake		0.0625	1.6
		Exhaust		0.07	1.8
	Stem Clearance	Production	Intake	0.0010-0.0027	0.025-0.069
			Exhaust	0.0007-0.0027	0.018-0.0686
		Service Limit	Intake	0.003 Max	0.09 Max
			Exhaust	0.004 Max	0.11 Max
	Spring w/Internal Damper Removed	Free Length		0.063	52
		Pressure	Valve open	208-222 ft. lbs. @ 1.22 in.	282-300 Nm @ 32mm
			Valve close	100-110 ft. lbs. @ 1.610 in.	136-149 Nm @ 41mm
		Installed Height ⑤	Intake	1.66	42
			Exhaust	1.66	42
	Internal Damper			None	None

① Unless otherwise noted
② 0.003 in. (0.07mm) across any six inches; or 0.007 in. (0.15mm) overall
③ Max
④ One half to one full turn tighter from zero lash
⑤ +/- 0.031 in. (0.8mm)

4

ENGINE MECHANICAL - GM V6 AND V8 ENGINES

ENGINE MECHANICAL 4-2
EXPLODED VIEWS 4-48
SPECIFICATIONS 4-56

BALANCE SHAFT 4-41
  REMOVAL & INSTALLATION 4-41
CAMSHAFT & BEARINGS 4-41
  CHECKING LIFT 4-41
  REMOVAL & INSTALLATION 4-42
CRANKSHAFT PULLEY 4-39
  REMOVAL & INSTALLATION 4-39
CROSSOVER 4-47
  REMOVAL & INSTALLATION 4-47
CYLINDER HEAD 4-43
  REMOVAL & INSTALLATION 4-43
    ALL EXC. 8.1L V8 4-43
    8.1L V8 4-44
CYLINDER HEAD COVER 4-19
  REMOVAL & INSTALLATION 4-19
ENGINE 4-3
  REMOVAL & INSTALLATION
    CARBURETED/TBI ENGINES 4-3
    MPI ENGINES 4-9
ENGINE IDENTIFICATION 4-2
  ALL EXC. 8.1L V8 4-2
  8.1L V8 4-3
ENGINE MECHANICAL 4-2
  BALANCE SHAFT 4-41
  CAMSHAFT & BEARINGS 4-41
  CRANKSHAFT PULLEY 4-39
  CROSSOVER 4-47
  CYLINDER HEAD 4-43
  CYLINDER HEAD COVER 4-19
  ENGINE 4-3
  ENGINE COUPLER/DRIVE PLATE 4-38
  ENGINE IDENTIFICATION 4-2
  EXHAUST ELBOW/RISER 4-32
  EXHAUST HOSES (BELLOWS) &
  INTERMEDIATE EXHAUST PIPE 4-34
  EXHAUST MANIFOLD 4-28
  EXHAUST VALVE (FLAPPER/SHUTTER) . 4-34
  FLYWHEEL 4-38
  FRONT COVER & OIL SEAL 4-39
  FRONT ENGINE MOUNTS 4-18
  GENERAL INFORMATION 4-2
  HARMONIC BALANCER 4-39
  HEAT EXCHANGER 4-46
  HYDRAULIC VALVE LIFTER 4-22
  INTAKE MANIFOLD 4-24
  LOWER EXHAUST PIPE (Y-PIPE) 4-34
  OIL FILTER BYPASS VALVE 4-37
  OIL PAN 4-35
  OIL PUMP 4-36
  REAR MAIN SEAL 4-44
  REAR MAIN SEAL RETAINER 4-46
  ROCKER ARM COVER 4-19
  ROCKER ARMS & PUSH RODS 4-19
  TIMING CHAIN & SPROCKETS/GEARS .. 4-40
  TORSIONAL DAMPER 4-39
  VALVE COVER 4-19
  WATER (ENGINE) CIRCULATING PUMP . 4-46
ENGINE COUPLER/DRIVE PLATE 4-38
  REMOVAL & INSTALLATION 4-38
EXHAUST ELBOW/RISER 4-32
  REMOVAL & INSTALLATION 4-32
    ALL EXC. DRY JOINT & 8.1L 4-32
    DRY JOINT SYSTEMS 4-33
    8.1L 4-33
EXHAUST HOSES (BELLOWS) 4-34
  REMOVAL & INSTALLATION 4-34
EXHAUST MANIFOLD 4-28
  REMOVAL & INSTALLATION 4-28
    ALL EXC. 8.1L V8 4-28
    8.1L V8 ENGINES 4-31
EXHAUST VALVE (FLAPPER/SHUTTER) ... 4-34
  REPLACEMENT 4-34
EXPLODED VIEWS 4-48
  4.3L V6 4-48
  5.0L, 5.7L & 6.2L V8 4-50
  8.1L V8 4-53
FLYWHEEL 4-38
  REMOVAL & INSTALLATION 4-38
FRONT COVER & OIL SEAL 4-39
  REMOVAL & INSTALLATION 4-39
FRONT ENGINE MOUNTS 4-18
  REMOVAL & INSTALLATION 4-18
GENERAL INFORMATION 4-2
  4.3L V6 ENGINES 4-2
  V8 ENGINES 4-2
HEAT EXCHANGER 4-46
  REMOVAL, DISASSEMBLY & INSTALLATION .. 4-46
HYDRAULIC VALVE LIFTER 4-22
  NOISY LIFTERS 4-22
  REMOVAL & INSTALLATION 4-22
INTAKE MANIFOLD 4-24
  REMOVAL & INSTALLATION 4-24
    ALL CARB & TBI 4-24
    V6/V8 MPI (EXC. 8.1L) 4-25
    8.1L V8 4-28
INTERMEDIATE EXHAUST PIPE 4-34
  REMOVAL & INSTALLATION 4-34
LOWER EXHAUST PIPE (Y-PIPE) 4-34
  REMOVAL & INSTALLATION 4-34
OIL FILTER BYPASS VALVE 4-37
  REMOVAL & INSTALLATION 4-37
OIL PAN 4-35
  REMOVAL & INSTALLATION 4-35
OIL PUMP 4-36
  REMOVAL & INSTALLATION 4-36
REAR MAIN SEAL 4-44
  REMOVAL & INSTALLATION 4-44
REAR MAIN SEAL RETAINER 4-46
  REMOVAL & INSTALLATION 4-46
ROCKER ARM COVER 4-19
  REMOVAL & INSTALLATION 4-19
ROCKER ARMS & PUSH RODS 4-19
  REMOVAL & INSTALLATION 4-19
    V6 4-19
    5.0L, 5.7L & 6.2L V8 4-20
    8.1L V8 4-20
VALVE ADJUSTMENT
    5.0L, 5.7L & 6.2L V8 4-21
    8.1L V8 4-22
SPECIFICATIONS 4-56
  ENGINE
    4.3L V6 4-59
    5.0L & 5.7L V8 4-61
    6.2L V8 4-63
    8.1L V8 4-65
  TORQUE
    4.3L V6 4-56
    5.0L, 5.7L & 6.2L V8 4-57
    8.1L V8 4-58
TIMING CHAIN & SPROCKETS/GEARS 4-40
  REMOVAL & INSTALLATION 4-40
TORSIONAL DAMPER 4-39
  REMOVAL & INSTALLATION 4-39
VALVE COVER 4-19
  REMOVAL & INSTALLATION 4-19
WATER (ENGINE) CIRCULATING PUMP 4-46
  REMOVAL & INSTALLATION 4-46

ENGINE MECHANICAL - GM V6 AND V8 ENGINES

ENGINE MECHANICAL

General Information

NEVER, NEVER attempt to use standard automotive parts when replacing anything on your engine. Due to the uniqueness of the environment in which they are operated in, and the levels at which they are operated at, marine engines require different versions of the same part; even if they look the same. Stock and most aftermarket automotive parts will not hold up for prolonged periods of time under such conditions. Automotive parts may appear identical to marine parts, but be assured, Mercury marine parts are specially manufactured to meet Mercury marine specifications. Most marine items are super heavy-duty units or are made from special metal alloy to combat against a corrosive saltwater atmosphere.

Mercury marine electrical and ignition parts are extremely critical. In the United States, all electrical and ignition parts manufactured for marine application must conform to stringent U.S. Coast Guard requirements for spark or flame suppression. A spark from a non-marine cranking motor solenoid could ignite an explosive atmosphere of gasoline vapors in an enclosed engine compartment.

4.3L V6 ENGINES

The MerCruiser 4.3L, 262 cubic inch displacement V6 engine is manufactured by GM. This engine is used in numerous models known as the 4.3L, 4.3LH, 4.3L EFI/MPI and 4.3 TKS.

The 4.3L model is equipped with a 2-barrel carburetor, while the 4.3LH models are equipped with a 4-barrel carburetor. Throttle body fuel injection was available on the 4.3L EFI models 2001/02, while multi point injection was available on certain 2001-02 engines and all later ones.

The lubrication system and component locations on this engine is virtually identical to the larger V8 engines, except for having only three cylinders in each bank.

A balance shaft is mounted above the camshaft on all models and extends the entire length of the block and is supported on each end by a bearing. The balance shaft is driven by gears on the end of the camshaft and equalizes the dynamic forces known as harmonic vibrations minimizing engine vibration during routine operation of a V6 engine. Cylinder numbering and firing order is identified in the illustrations at the end of the Maintenance section.

All 4.3L engines are left hand (counterclockwise) rotation when viewed from the stern of the boat. This does not necessarily indicate that your prop rotation is the same - always check them both!

V8 ENGINES

The MerCruiser 5.0L, 305 cubic inch, 5.7L, 350 cubic inch and 6.2L, 377 cubic inch displacement (small block) V8 engines are manufactured by GM. These engines are used in the following configurations:

- 5.0L
- 5.0L EFI
- 5.0L MPI
- 5.0L TKS
- 5.7L
- 5.7L EFI
- 5.7L MPI
- 5.7L TKS
- 350 MAG MPI
- 6.2L MPI
- 6.2L Black Scorpion
- Scorpion 350
- Scorpion 377
- Tow Sport 5.7 MPI/TKS

The MerCruiser 8.1L 496 cubic inch displacement (big block) V8 engines are also manufactured by GM. These engines are used in the following configurations:

- 8.1S
- 8.1S HO
- 496 Mag
- 496 Mag HO

The lubrication system is a force fed type where oil is supplied under full pressure to the crankshaft, main and connecting rod bearings, camshaft bearings and the valve lifters. Oil flow from the valve lifters is metered and pumped by the lifter through the hollow core pushrods to lubricate the rocker arms and valve train. All other components are lubricated by gravity and splash methods.

The oil pump is mounted on the rear main bearing cap and is driven by an extension shaft from the distributor - driven by the camshaft. Oil is drawn into the pump through the oil pick-up tube and screen. Should the screen become clogged, a relief valve in the screen will open and allow oil to be drawn into the pump.

Once oil reaches the pump, the pump forces oil through the lubrication system. A spring-loaded relief valve in the oil pump limits the maximum pump output pressure.

The pressurized oil flows out the pump through a full-flow disposable oil filter cartridge. On engines equipped with an oil cooler, the oil flows through the filter, out to the oil cooler via hoses and then returns to the block. Should the oil filter and/or cooler become clogged, a by-pass valve will open allowing the pressurized oil to by-pass the filter and cooler.

Some of the oil is then routed to the No. 5 crankshaft main bearing, the remainder of the oil pressure is routed to the main oil gallery. The main oil gallery is located above the camshaft and runs the full length of the block. Oil from the main gallery is routed through individual passages to the camshaft bearings, No's 1, 2, 3, and 4, crankshaft main bearings and the lifter galley's on each side of the block.

Holes in the camshaft bearings and crankshaft main bearings align with the holes in the block for oil flow. Grooves in the bearings allow oil to flow between the bearing and the component.

Oil in the lifter galleries is forced into each hydraulic lifter through a hole in the side of the lifter. Oil flowing through the lifter must pass through a metering valve in each of the lifters. The metered volume of oil then flows up through the hollow push-rods to the valve rockers. A small hole in the rocker arm allows oil to lubricate the valve train bearing surfaces. All excess oil drains back to the oil pan through oil return holes in the cylinder head.

A baffle plate or "splash pan" mounted below the main bearing caps prevents excess oil being thrown off the crankshaft from aerating the oil in the oil pan.

The distributor shaft and gear is lubricated by oil in the starboard lifter gallery. The timing chain and gears are lubricated with oil flowing out the front of the No. 1 main bearing journal. The mechanical fuel pump and pushrod is lubricated with oil thrown off the camshaft eccentric.

All V8 engines are left hand (counterclockwise) rotation when viewed from the stern of the boat except NR and MR inboards which are right hand (clockwise). This does not necessarily indicate that your prop rotation is the same - always check them both!

Cylinder numbering and firing order is identified in the illustrations at the end of the Maintenance section.

Engine Identification

ALL ENGINES EXC. 8.1L V8

◆ See Figure 1

All engines can be identified by a code number stamped into the starboard side of the front of the engine block, just below the cylinder head - the last two letters of the code designate the engine:

- LJ: 4.3L
- LK: 4.3LH and 4.3L EFI
- MH: 5.0L EFI, 5.7L/5.7L EFI (stern drive & ski), 350 Mag MPI (stern drive & ski, Black Scorpion
- MHA: 6.2L MPI (stern drive) and 6.2 Black Scorpion
- MK: 5.7L/350 Mag MPI (inboard)
- MKA: 2001 6.2L MPI (inboard)
- ZA: 5.0L
- 2LB: 4.3L MPI
- 2ML: 350 Mag MPI (inboard)
- 2ML: 6.2L MPI (inboard)
- 2MN: 350 Mag MPI (Sterndrive & tow sports)
- 2MNB: 6.2L MPI (sterndrive)
- 2ZB: 5.0L MPI

This code is stamped on all original equipment engines and all partial replacement engines (but NOT on replacement cylinder blocks).

In the event that the engine serial number plates or stickers are missing from the engine, this is a good way to ensure that you know the exact engine in your vessel when ordering parts, etc.

ENGINE MECHANICAL - GM V6 AND V8 ENGINES 4-3

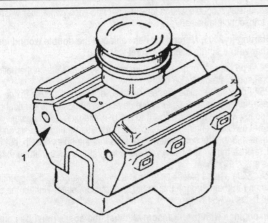

Fig. 1 The engine identification code can be found here on all engines except the 8.1L V8. The last two letters are the engine designation

1- Location of GM engine code(front starboard side,

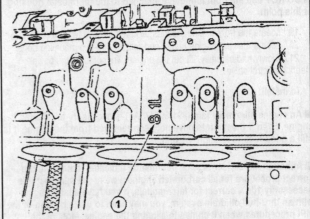

1- 8.1L identification on port side of engine block
Fig. 2 The engine identification code can be found here on the 8.1L V8

8.1L V8 ENGINES

◆ See Figure 2

The engine code **8.1L** can be found stamped on the port side of the cylinder block

Engine

REMOVAL & INSTALLATION - CARBURETED/TBI ENGINES

Sterndrives w/o Driveshaft Extension

◆ See Figures 3 thru 14

1. Remove the stern drive unit as detailed in the appropriate section.
2. Open or remove the engine hatch cover.
3. Disconnect the battery cables (negative first) at the battery and then disconnect them from the engine block and starter.
4. If quipped with a fuel shut-off valve, close it. Disconnect the fuel inlet line at the fuel pump and quickly plug it. Make sure you have rags handy as there will be some spillage.
5. Disconnect the throttle cable at the throttle lever on the carburetor/throttle body unit and position it out of the way.
6. Loosen the hose clamp on the engine wiring harness connector and unplug the instrument wiring harness. Label it and move it aside.
7. Tag and disconnect the two trim position wires at the sender unit on carbureted/TBI engines.
8. On Alpha models, tag and disconnect the engine harness wires at the shift cut-out switch harness. When you pull back the sleeves, they should be Black and White/Green, although sometimes the W/G will be Gray.
9. On models equipped with the MerCathode system, tag and disconnect the wires at the controller.
10. Loosen the hose clamp and remove the water inlet line at the gimbal housing.
11. Loosen the hose clamps and disconnect the exhaust elbow hoses/bellows from the upper pipe. Remove them.
12. Remove the shift cables at the shift plate and move them out of the way. Don't lose the hardware.
13. Carefully loosen and remove the two hydraulic lines at the power steering control valve on the transom. Plug the hose ends and the control valve fittings. Tag the lines and fittings and move them aside.
14. If equipped with a drive lube monitor, lift it out of the bracket and position it out of the way with the hose still attached.
15. Loose the clamps at the bullhorn on EFI engines
16. Tag and disconnect the engine ground wire at the stud.
17. Tag and disconnect any remaining lines, wires or hoses at the engine.
18. Attach a suitable engine hoist to the lifting eyes and take up any line slack until it is just taught.

Fig. 3 A good look at a typical engine harness connector

Fig. 4 A good look at the front engine mount

4-4 ENGINE MECHANICAL - GM V6 AND V8 ENGINES

■ DO NOT use the front lifting eye attached to the thermostat housing at this point.

19. Locate the front engine mount(s) and remove the two lag bolts.
20. Locate the rear engine mount and remove the two mounting bolts.
21. Slowly and carefully, lift out the engine. Try not to hit the power steering control valve.

To install:

■ An engine alignment tool (#91-805475A1) is necessary to reinstall the engine. Even if the mounts have not been removed from the engine it is still a good idea to re-align the unit.

■ Although Mercruiser service information indicates that this is the correct procedure for all carbureted engines, we're not sure that it is necessarily 100% correct for later models. If you have a later engine with an thru-hull oil drain system, you may want to use a portion of the MPI procedures when it comes to aligning the engine. Most of the remainder of either procedure is the same except for different lines/leads you may need to reconnect. Please read through the procedures for the MPI engines and then choose that which best fits your engine.

22. Check that the fiber washers on top of the transom plate mounting holes are still in position. If they look worn or damaged in any way, replace them. Remember, they are glued in place.

23. Install a double wound lock washer into each of the fiber washers and transom plate holes.

■ Starting in 2003, Mercruiser has replaced the double wound lock washer with a single stainless steel washer.

24. Pop the rear engine mount locknuts into their slots on the bottom of the transom plate mounting holes.
25. Apply some spline grease (Quicksilver Engine Coupler Spline Grease) to the engine coupler splines and then lubricate the exhaust bellows with soapy water - this will allow it to slip over the pipe much easier.
26. Lower the engine into the compartment so that it just rests on the washers and the transom plate. Make sure that the holes line up, but DO NOT release the hoist tension yet. Make sure at this point that the exhaust pipes are in alignment.
27. Slide a large steel washer and a metal spacer onto the mounting bolt and then run them through the mount and transom plate. Tighten the bolts to 35-40 ft. lbs (47-54 Nm).

■ On engines with thru-transom exhaust, the hoses must be connected to the elbow so that the hose is centered around the fitting in an even manner. If the fitting clearance inside the hose is not consistent, discharge water flow will be restricted.

28. Lower the front of the engine until the front mount just rests on the stringers. Unhook the hoist and connect the chain to the center lifting eye on the thermostat housing.
29. Slide the alignment tool into the center of the gimbal housing assembly bearing and then into the engine coupler splines - you may have to

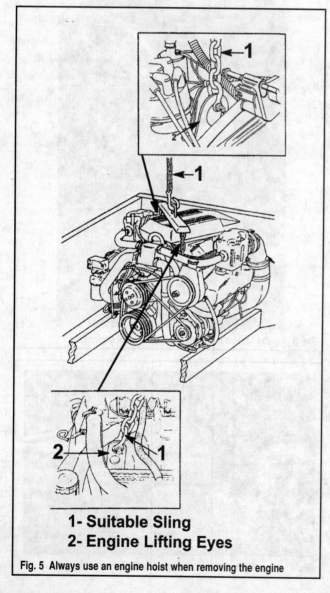

1- Suitable Sling
2- Engine Lifting Eyes

Fig. 5 Always use an engine hoist when removing the engine

Fig. 6 Remove the engine. . .

Fig. 7 . . .and install the engine on a stand after removal

ENGINE MECHANICAL - GM V6 AND V8 ENGINES 4-5

pivot the bearing slightly. Use the hoist to raise (or lower) the front of the engine until the tool slides completely into the coupler (with no binding). DO NOT force the tool through the bearing under any circumstances. Slowly raise (or lower) the front of the engine until the tool slides freely

30. Loosen the jam and lock nuts on the front mount(s) and then turn the jam nut (top nut) so that the mount base sits true on the stringer. Check that the alignment tool still slides freely between the engine and gimbal housing. Adjust by turning the adjusting nut (bottom nut, under mount bracket); counterclockwise will raise the front of the engine. Install the mount base lag bolts and tighten them securely. Tighten the adjusting nut and the jam nut.

31. Recheck that the alignment tool still slides freely and then remove it. Remove the engine hoist.

32. Reconnect the exhaust bellows and tighten the clamps securely. Refer to the illustration for positioning.

33. Reconnect the water inlet hose and tighten the hose clamp.

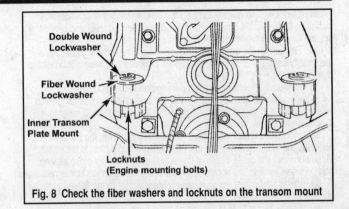

Fig. 8 Check the fiber washers and locknuts on the transom mount

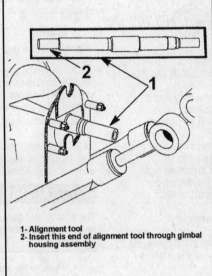

Fig. 10 Slide an alignment tool through the gimbal bearing...

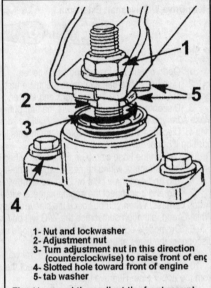

Fig. 11 ...and then adjust the front mount so that the tool slides freely between the gimbal housing and the coupler

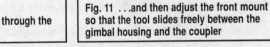

Fig. 9 Details of the rear engine mount

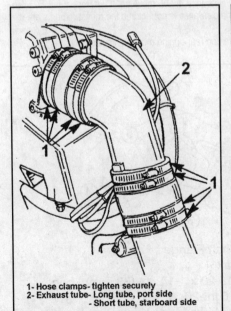

Fig. 12 Proper hose clamp location when installing the exhaust bellows (thru-prop exhaust)

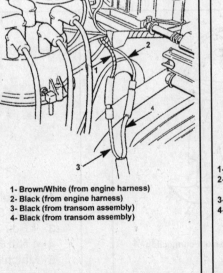

Fig. 13 Make sure the trim position sender leads are connected properly

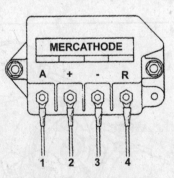

Fig. 14 Proper MerCathode controller hook-up

4-6 ENGINE MECHANICAL - GM V6 AND V8 ENGINES

34. Carefully, and quickly, connect the power steering lines. Tighten all fittings to 23 ft. lbs. (31 Nm). Don't forget check the fluid level and bleed the system when you are finished with the installation.
35. Reconnect the trim position sender leads, the engine ground wire, the battery cables and all other wires, lines of hoses that were disconnected during removal.
36. Unplug the fuel line and pump fitting and reconnect them.
37. Connect the MerCathode lines to the controller and coat them with liquid neoprene. If you forgot to tag the wires as we suggested, refer to the illustration for proper hook-up.
38. Install and adjust the throttle cable. Make sure that the barrel positions the cable so that there are no kinks or sharp bends in the cable. Make sure that the cable does not come in contact with any other moving parts. Please refer to the appropriate Fuel System section for further details.
39. Install and adjust the shift cables. Please refer to the Stern Drive section for further details.
40. Check and refill all fluids and go have fun!

Stern Drive W/Driveshaft Extension

◆ See Figures 3 thru 6 and 15

1. Open or remove the engine hatch cover.
2. Disconnect the battery cables (negative first) at the battery and then disconnect them from the engine block and starter.
3. Disconnect the fuel inlet line at the fuel pump and quickly plug it. Make sure you have rags handy as there will be some spillage.
4. Disconnect the throttle cable at the throttle lever on the unit and position it out of the way.
5. Loosen the hose clamp on the engine wiring harness connector and unplug the instrument wiring harness. Label it and move it aside.
6. Tag and disconnect the two trim position wires at the sender unit.
7. Tag and disconnect the engine harness wires at the shift cut-out switch harness. When you pull back the sleeves, they should be Black and White/Green, although sometimes the W/G will be Gray.
8. On models equipped with the MerCathode system, tag and disconnect the wires at the controller.
9. Loosen the hose clamp and remove the water inlet line at the engine.
10. Loosen the hose clamps and disconnect the exhaust elbow bellows from the upper pipe. Remove them.
11. Remove the shift cables at the shift plate and move them out of the way. Don't lose the hardware.
12. Carefully loosen and remove the 2 hydraulic lines at the power steering control valve on the transom. Plug the hose ends and the control valve fittings. Tag the lines and fittings and move them aside.
13. Tag and disconnect the engine ground wire at the stud.
14. Tag and disconnect any remaining lines, wires or hoses at the engine.
15. Loosen the 7 driveshaft shield mounting bolts/nuts and lift out the upper and lower shields at the engine-end of the shaft.
16. Paint matchmarks across the driveshaft U-joint yoke and engine output shaft flanges where they mate and then disconnect the driveshaft from the output shaft.
17. Attach a suitable engine hoist to the lifting eyes and take up any line slack until it is just taught.

■ DO NOT use the front lifting eye attached to the thermostat housing at this point.

18. Locate the front engine mount(s) and remove the 2 lag bolts.
19. Locate the rear engine mount and remove the 2 mounting bolts.

■ DO NOT TAMPER WITH THE ENGINE MOUNT ADJUSTMENT BOLTS or you will need to check the engine/shaft alignment on installation - and we do not provide this information here.

20. Slowly and carefully, lift out the engine. Try not to hit the power steering control valve.

To install:

◆ See Figures 8 thru 14 and 16 thru 18

21. Lower the engine into the compartment so that it just rests on the washers and the transom plate or stringers. Make sure that the holes line up and DO NOT release the hoist tension yet. Make sure at this point that the exhaust pipes are in alignment.
22. Lubricate the U-joints with Quicksilver U-Joint & Gimbal Bearing Grease.
23. Reconnect the driveshaft to the engine output shaft and tighten the bolts to 50 ft. lbs. (68 Nm). Be certain that the alignment marks applied earlier are in complete alignment.

✷✷ CAUTION

The pilot on the driveshaft flanges must be engaged in the input shaft and output shaft flanges. The flanges must be flush before tightening the bolts.

24. Release the tension on the engine hoist and then slide the engine slowly until you can obtain 1/4 in. (6mm) clearance between the flange shoulder and the extension shaft housing bearing. Measure the diagonal distance between the two points shown in the illustration; move the front or rear sides of the engine slightly until the two measurements are equal. Recheck the driveshaft clearance. If OK, move on. If not, repeat this procedure again until the clearance is correct and the measurements are equal.
25. Tighten the front and rear engine mount bolts securely.

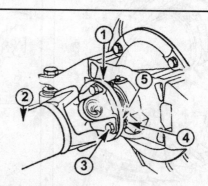

1 - Output Shaft Flange
2 - Drive Shaft
3 - 4 Bolt 7/16-20 x 1-1/2 In. (38 mm) Long
4 - 4 Nut 7/16 In.-20
5 - Matching Marks Made Upon Disassembly

Fig. 16 The matchmarks must be lined up and the flanges must be flush before tightening anything!

1- Matching marks on flange and drive shaft connection
2- Extension drive shaft U-Joint yoke
3- flange, output

Fig. 15 Matchmark the driveshaft and output shaft flanges before removal

ENGINE MECHANICAL - GM V6 AND V8 ENGINES 4-7

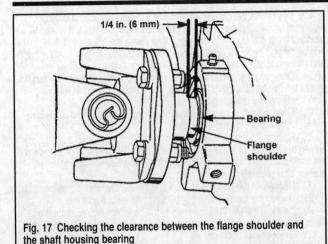

Fig. 17 Checking the clearance between the flange shoulder and the shaft housing bearing

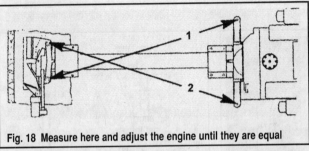

Fig. 18 Measure here and adjust the engine until they are equal

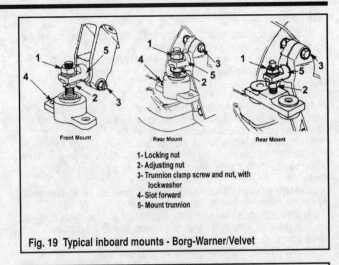

1- Locking nut
2- Adjusting nut
3- Trunnion clamp screw and nut, with lockwasher
4- Slot forward
5- Mount trunnion

Fig. 19 Typical inboard mounts - Borg-Warner/Velvet

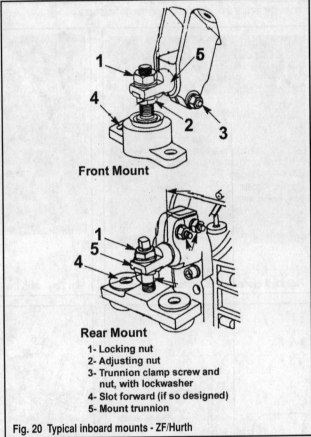

Front Mount

Rear Mount

1- Locking nut
2- Adjusting nut
3- Trunnion clamp screw and nut, with lockwasher
4- Slot forward (if so designed)
5- Mount trunnion

Fig. 20 Typical inboard mounts - ZF/Hurth

26. Coat the driveshaft shield bolts with Loctite 271 and install the lower shield. Tighten the three bolts to 30 ft. lbs. (41 Nm). Install the top shield and tighten the four bolts/nuts to 30 ft. lbs. (41 Nm).

27. Reconnect the exhaust bellows and tighten the clamps securely. Refer to the illustration for positioning.

28. Reconnect the water inlet hose and tighten the hose clamp securely.

29. Carefully, and quickly, connect the power steering lines. Tighten to 23 ft. lbs. (31 Nm). Don't forget to fill and bleed the system when you are finished with the installation.

30. Reconnect the trim position sender leads, the engine ground wire, the battery cables and all other wires, lines of hoses that were disconnected during removal.

31. Unplug the fuel line and pump fitting and reconnect them.

32. Connect the MerCathode lines to the controller and coat them with liquid neoprene. If you forgot to tag the wires as we suggested, refer to the illustration for proper hook-up.

33. Install and adjust the throttle cable. Make sure that the barrel positions the cable so that there are no kinks or sharp bends in the cable. Make sure that the cable does not come in contact with any other moving parts. Please refer to the Fuel Systems section for further details.

34. Install and adjust the shift cables. Please refer to the Drive Systems section later for further details.

35. Check and refill all fluids; and go have fun!

Inboards

◆ See Figures 5 and 19 thru 29

1. Open or remove the engine hatch cover.
2. Disconnect the battery cables (negative first) at the battery and then disconnect them from the engine block and starter.
3. Disconnect the fuel inlet line at the fuel pump and quickly plug it. Make sure you have rags handy as there will be some spillage.
4. Disconnect the throttle cable at the throttle lever on the unit and position it out of the way.
5. Loosen the hose clamp on the engine wiring harness connector and unplug the instrument wiring harness. Label it and move it aside.
6. Loosen the hose clamps and disconnect the exhaust elbow bellows from the upper pipe. Remove them
7. Remove the shift cables at the shift plate and move them out of the way. Don't lose the hardware.

8. Disconnect and remove the seawater inlet hose at the engine.
9. Tag and disconnect the engine ground wire at the stud.
10. Tag and disconnect any remaining lines, wires or hoses at the engine.
11. Disconnect the propeller shaft coupler at the transmission output flange.
12. Attach a suitable engine hoist to the lifting eyes and take up any line slack until it is just taught.

■ DO NOT use the front lifting eye attached to the thermostat housing at this point.

13. Locate the front engine mount(s) and remove the 2 lag bolts.
14. Locate the rear engine mount and remove the mounting bolts.
15. Slowly and carefully, lift out the engine. Try not to hit the power steering control valve.

ENGINE MECHANICAL - GM V6 AND V8 ENGINES

To install:

16. Loosen the stop nut and move the adjustment nuts on all mounts so that they are in the center of the adjustment range.
17. All mounts have one slotted bolt hole and this should be facing forward. The large diameter of the mounting trunnion should be fully extended.
18. Lower the engine into the compartment so that it just rests on the washers and the transom plate or stringers. Make sure that the holes line up and DO NOT release the hoist tension yet. Make sure that there is at least 1/4 in. (6mm) of up or down adjustment at all mounts.
19. Position the engine so that the transmission output flange and the propeller shaft coupler are in alignment and firmly mated together. Adjust the engine bed height to true the alignment - do not use the mount adjustment screws yet.
20. Make sure that the mounts and their bolt holes are still lined up and then install the mounting bolts. Tighten them securely.
21. Center the propeller shaft in the log. Press it down as far as it will go and then pull it up as far as it will go; position the shaft at the apparent center point of its vertical movement. Now do the same thing horizontally.
22. Make certain that the shaft coupler and the engine output flange align exactly. Most engines have a shoulder on the shaft coupler that should engage the recess on the output flange with no resistance.
23. Check that the gap between the two mating surfaces is 0.003 in. (0.07mm) when checked at 90° positions on the couplers (12, 3, 6 and 9 o'clock). If the gap is other than suggested, move the adjustment nuts on the engine mounts until you achieve the proper clearance. Always turn the front and rear mount adjusters equally to maintain level on your engine.
24. Once you are comfortable that everything is in alignment, tighten the mount bracket bolts to 50 ft. lbs. (68 Nm). Bend the tab on the washer over onto the flat of the mounting bolt head.
25. Make sure that the mount trunnion shaft is not exposed more than 3/4 in. (20mm) from the mount bracket.
26. Reconnect the propeller shaft to the transmission output shaft and tighten the bolts to 50 ft. lbs. (68 Nm). Some shafts may have setscrews and the shaft should be dimpled at their locations. Use safety wire on the set screws after securely tightening them.
27. Reconnect the water inlet hose and tighten the hose clamp securely.
28. Unplug the fuel line and pump fitting and reconnect them.
29. Connect the MerCathode lines to the controller and coat them with liquid neoprene. If you forgot to tag the wires as we suggested, refer to the illustration for proper hook-up.
30. Install and adjust the throttle cable. Make sure that the barrel positions the cable so that there are no kinks or sharp bends in the cable. Make sure that the cable does not come in contact with any other moving

Fig. 21 A close look at the rear mount...

Fig. 22 ...and the front mount

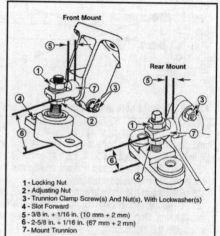

Fig. 23 Setting up the mount prior to installation

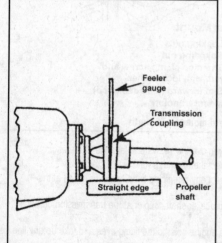

Fig. 24 Transmission and mating surfaces must have a uniform gap when measured in four positions - Borg-Warner/Velvet

Fig. 25 Transmission and mating surfaces must have a uniform gap when measured in four positions - ZF/Hurth

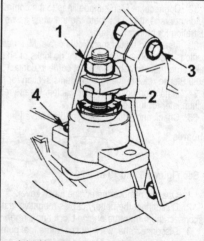

Fig. 26 Typical inboard mount with adjustment nut and slot - Borg-Warner/Velvet

ENGINE MECHANICAL - GM V6 AND V8 ENGINES

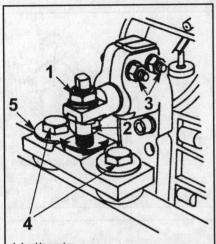

1- Locking nut
2- Adjusting nut
3- Clamping screws and nuts, with lockwashers
4- Lag screws (or bolts)
5- Slot forward (if so designed - NOT slotted on this style rear mount)

Fig. 27 Typical inboard mount with adjustment nut and slot - ZF/Hurth

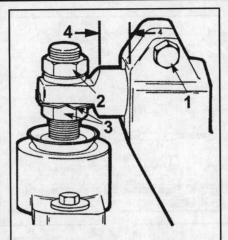

1- Tighten clamping screw and nut on all four mount brackets to 50 lb ft (68 Nm)
2- Tighten locking nut on all four mounts securely
3- Bend one of the tabs on tab washer down onto flat of adjusting nut
4- Maximum extension of large diameter of trunnion - 3/4 in. (20 mm)

Fig. 28 The adjusting trunnion should not extend from the mounting bracket more than 3/4 in. - Borg-Warner/Velvet

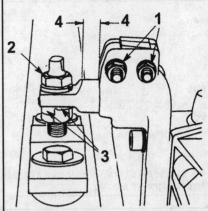

1- Tighten clamping screws and nuts on all four mount brackets to 50 lb ft (68Nm)
2- Tighten locking nut on all four mounts securely
3- Bend one of the tabs on tab washer down onto flat of adjusting nut
4- Maximum extension of large diameter of trunnion 3/4 in. (20 mm)

Fig. 29 The adjusting trunnion should not extend from the mounting bracket more than 3/4 in. - ZF/Hurth

parts. Please refer to the Fuel Systems or Remote Controls section for further details.

31. Install and adjust the shift cables. Please refer to the Transmission or Remote Controls section for further details.

32. Check and refill all fluids and go have fun!

REMOVAL & INSTALLATION - MPI ENGINES

Sterndrives - 5.0L, 5.7L & 6.2L Engines

◆ See Figures 3 thru 7, 9 thru 12, 14 and 30 thru 36

1. Remove the stern drive unit as detailed in the appropriate section.

■ Hint, before removing the drive, take a measurement from the edge of the oil drain hole in the transom to the edge of the propeller. Subtract 6 in. and record the measurement.

2. Open or remove the engine hatch cover.
3. Disconnect the battery cables (negative first) at the battery and then disconnect them from the engine block and starter.
4. If equipped with a fuel shut-off valve, close it. Disconnect the fuel inlet line at the fuel pump and quickly plug it. Make sure you have rags handy as there will be some spillage. Don't forget to relieve the fuel system pressure as detailed in the appropriate Fuel Systems section!
5. Disconnect the throttle cable at the throttle lever on the throttle body unit and position it out of the way.
6. Loosen the hose clamp on the engine wiring harness connector and unplug the instrument wiring harness. Label it and move it aside.
7. Tag and disconnect the 2 trim position wires at the sender unit.
8. On models equipped with the MerCathode system, tag and disconnect the wires at the controller - you may need to remove the thermostat housing hose to get at all the bolts.
9. Loosen the hose clamp and remove the water inlet line at the gimbal housing.
10. Loosen the hose clamps and disconnect the exhaust elbow hoses/bellows from the upper pipe. Remove them.
11. Remove the shift cables at the shift plate and move them out of the way. Don't lose the hardware.
12. Carefully loosen and remove the 2 hydraulic lines at the power steering control valve on the transom. Plug the hose ends and the control valve fittings. Tag the lines and fittings and move them aside.

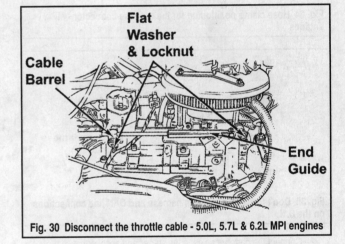

Fig. 30 Disconnect the throttle cable - 5.0L, 5.7L & 6.2L MPI engines

13. If equipped with a drive lube monitor, lift it out of the bracket and position it out of the way with the hose still attached.
14. Loose the clamps at the bullhorn.
15. Tag and disconnect the engine ground wire at the stud.
16. Tag and disconnect any remaining lines, wires or hoses at the engine.
17. Attach a suitable engine hoist to the lifting eyes and take up any line slack until it is just taught.

■ DO NOT use the front lifting eye attached to the thermostat housing at this point.

18. Locate the front engine mount(s) and remove the 2 lag bolts.
19. Locate the rear engine mount and remove the 2 mounting bolts.
20. Slowly and carefully, lift out the engine. Try not to hit the power steering control valve.

To install:

■ An engine alignment tool (#91-805475A1) is necessary to reinstall the engine. Even if the mounts have not been removed from the engine it is still a good idea to re-align the unit.

21. Re-attach the engine hoist and lower the engine into the vessel so it is in the general position.

4-10 ENGINE MECHANICAL - GM V6 AND V8 ENGINES

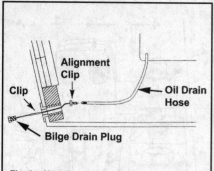

Fig. 31 Use the oil drain hose to start off the engine alignment process

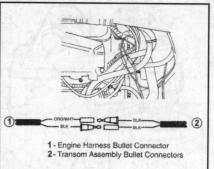

Fig. 32 Reconnecting the trim sender leads - MPI engines

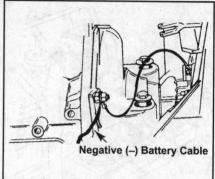

Fig. 33 Don't forget the ground wire

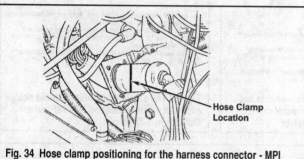

Fig. 34 Hose clamp positioning for the harness connector - MPI engines

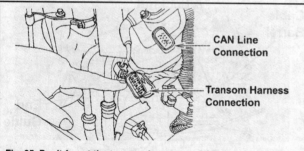

Fig. 35 Don't forget the transom harness and CAN line connections on the 8.1L

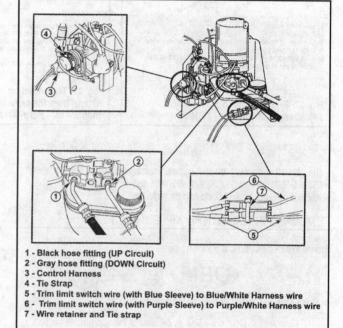

1 - Black hose fitting (UP Circuit)
2 - Gray hose fitting (DOWN Circuit)
3 - Control Harness
4 - Tie Strap
5 - Trim limit switch wire (with Blue Sleeve) to Blue/White Harness wire
6 - Trim limit switch wire (with Purple Sleeve) to Purple/White Harness wire
7 - Wire retainer and Tie strap

Fig. 36 The power trim pump has a number of connections on installation - MPI engines

22. Connect the battery cables to the starter and apply a thin coat of Liquid Neoprene. Slide the rubber boot over the positive terminal and then drape the cables over the engine so they are out of the way.
23. If equipped with a Quick Drain oil system, push the end of the oil drain hose out and through the flange. Continue pulling on the hose until it is 6 in. (152mm) from the propeller - yes, we know the drive has been removed, but these are the instructions, hopefully you paid attention in the removal process and took the measurement we suggested. If not equipped with the system, skip to Step 27.
24. Move the alignment clip on the drain hose until it is just inside the hull and against the flange. Squeeze it into position.
25. Now connect the bilge drain plug to the drain hose plug using the clip.

■ If the quick drain oil fitting is within 1/2 in. (13mm) of the bottom of the boat, remove the fitting and install the drain plug from the parts bag directly into the oil pan.

26. Push the drain hose through the flange and into the hull. Install the bilge plug in the hull.
27. Apply some spline grease (Quicksilver Engine Coupler Spline Grease) to the engine coupler splines and then lubricate the exhaust bellows with soapy water - this will allow it to slip over the pipe much easier.
28. Lower the engine into the compartment fully so that it just rests on the washers and the transom plate. Make sure that the holes line up, but DO NOT release the hoist tension yet. Make sure at this point that the exhaust pipes are in alignment.

29. Slide a large steel washer and a metal spacer onto the mounting bolt and then run them through the mount and transom plate. Tighten the bolts to 38 ft. lbs (51 Nm).

■ Make sure that the quick drain oil fitting is more than 1/2 in. (13mm) from the boat bottom.

30. Lower the front of the engine until the front mount just rests on the stringers. Unhook the hoist and connect the chain to the center lifting eye on the thermostat housing.
31. Slide the alignment tool into the center of the gimbal housing assembly bearing and then into the engine coupler splines - you may have to pivot the bearing slightly. Use the hoist to raise (or lower) the front of the engine until the tool slides completely into the coupler (with no binding). DO NOT force the tool through the bearing under any circumstances. Slowly raise (or lower) the front of the engine until the tool slides freely
32. Loosen the jam and lock nuts on the front mount(s) and then turn the jam nut (top nut) so that the mount base sits true on the stringer. Check that the alignment tool still slides freely between the engine and gimbal housing. Adjust by turning the adjusting nut (bottom nut, under mount bracket); counterclockwise will raise the front of the engine. Install the mount base lag bolts and tighten them securely. Tighten the adjusting nut and the jam nut.
33. Recheck that the alignment tool still slides freely and then remove it. Remove the engine hoist.
34. Reconnect the exhaust bellows and tighten the clamps securely. Refer to the illustration for positioning.
35. Reconnect the water inlet hose and tighten the hose clamp.

ENGINE MECHANICAL - GM V6 AND V8 ENGINES

36. Carefully, and quickly, connect the power steering lines. Tighten all fittings to 23 ft. lbs. (31 Nm). Don't forget check the fluid level and bleed the system when you are finished with the installation.

37. Reconnect the trim position sender leads, the engine ground wire, the battery cables and all other wires, lines of hoses that were disconnected during removal.

38. Unplug the fuel line and pump fitting and reconnect them.

39. Connect the MerCathode lines to the controller and coat them with liquid neoprene. If you forgot to tag the wires as we suggested, refer to the illustration for proper hook-up.

40. Tighten the power trim pump hydraulic line connections to 125 inch lbs. (14 Nm).

41. When installing the gear lube monitor, make sure the hose is routed correctly and does no contact any steering components, the engine coupler or any drive line components.

42. Install and adjust the throttle cable. Make sure that the barrel positions the cable so that there are no kinks or sharp bends in the cable. Make sure that the cable does not come in contact with any other moving parts. Please refer to the appropriate Fuel System or Remote Controls section for further details.

43. Install and adjust the shift cables. Please refer to the Stern Drive or Remote Controls section for further details.

44. Check and refill all fluids and go have fun!

Sterndrives - 8.1L Engines

◆ See Figures 3, 4, 14 and 37 thru 48

1. Remove the stern drive unit as detailed in the appropriate section.

■ **Hint, before removing the drive, take a measurement from the edge of the oil drain hole in the transom to the edge of the propeller. Subtract 6 in. and record the measurement.**

2. Open or remove the engine hatch cover.

3. Disconnect the battery cables (negative first) at the battery and then disconnect them from the engine block and starter.

4. Loosen the hose clamp on the engine wiring harness connector and unplug the 10 or 14 pin instrument wiring harness. Label it and move it aside.

5. If equipped with a fuel shut-off valve, close it. Disconnect the fuel inlet line at the fuel pump boost pump on Gen II systems, or at the Cool Fuel Module fitting on Gen III models, and quickly plug it. Make sure you have rags handy as there will be some spillage. Make sure that you relieve the fuel system pressure as detailed in the appropriate Fuel Systems section first!

6. On non-DTS models, disconnect the throttle cable at the throttle lever on the throttle body unit and position it out of the way. Don't lose the fasteners. You may find more information on the throttle cable in the appropriate Fuel Systems or Remote Controls section.

7. On non-DTS models, disconnect the shift cables at the shift plate and move them out of the way. Don't lose the hardware. You may find more information on the shift cables in the appropriate Drive Systems or Remote Controls section.

8. Carefully loosen and remove the 2 hydraulic lines at the power steering control valve on the transom. Plug the hose ends and the control valve fittings. Tag the lines and fittings and move them aside, but make sure that you take note of their routing - write it down or draw a picture please!.

■ **On later models equipped with quick-connect fittings, simply connect the two lines together instead of plugging them.**

9. If equipped with a gear lube monitor, lift it out of the bracket and position it out of the way with the hose still attached. Disconnect the monitor hose. Tag and disconnect the 2 trim position wires at the engine harness.

10. Disconnect the transom wiring harness.

11. Tag and disconnect the CAN link harness.

12. Tag and disconnect the continuity circuit wires and position them out of the way.

13. Tag and disconnect the clean power harness and/or battery isolator wires if equipped.

14. On models equipped with the MerCathode system, tag and disconnect the wires at the controller - you may need to remove the thermostat housing hose to get at all the bolts. If equipped, disconnect the quick-connect fitting as well.

15. Loosen the hose clamp and remove the seawater inlet line at the gimbal housing. On later models with a quick-connect fitting, move the retainer clip to the open position.

16. Disconnect the water heater lines if equipped.

17. Loosen the hose clamps and disconnect the exhaust elbow hoses/bellows from the upper pipe. Remove them.

18. Remove the boat drain plug - obviously, make sure that the boat is out of the water. Remove the oil drain hose lanyard from the plug.

19. Tag and disconnect any remaining lines, wires or hoses at the engine.

20. Attach a suitable engine hoist with a lifting arm to the lifting eyes and take up any line slack until it is just taught. Make sure that the engine will be level.

21. Remove the front and rear engine mounting bolts. Try to avoid disturbing the engine mounts other than the mounting bolts.

22. Slowly and carefully, lift out the engine. Try not to hit the power steering control valve.

To install:

■ **An engine alignment tool (#91-805475A1) is necessary to reinstall the engine. Even if the mounts have not been removed from the engine it is still a good idea to re-align the unit.**

23. If the engine mounts have been disturbed, make sure that the adjusting nuts on the front mounts are positioned midway on the studs so that you will have adequate up/down adjustment.

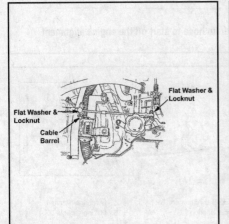

Fig. 37 Disconnect the throttle cable (non-DTS) - 8.1L engines

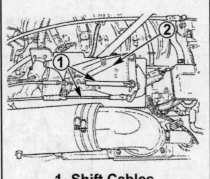

1- Shift Cables
2 - Shift Plate

Fig. 38 Disconnect the shift cables (non-DTS) - 8.1L engines

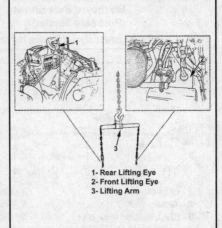

1- Rear Lifting Eye
2- Front Lifting Eye
3- Lifting Arm

Fig. 39 Use a lifting arm with the hoist and sling when removing the engine

4-12 ENGINE MECHANICAL - GM V6 AND V8 ENGINES

24. Check to see which type of rear engine mount your unit uses. Early style mounts had a smooth bottom mounting surface, while later models utilize a knurled surface. Both designs uses a fiber washer on the inner transom plate, but the earlier design also uses a double-wound lock washer.

   a. On early designs, position a fiber washer on top of each transom plate mount and then insert the small double-wound lock washer.

   b. On later designs, simply position the fiber washer on the top of each transom plate mount.

25. If removed, re-attach the engine hoist and lower the engine into the vessel so it is in the general position.

26. If equipped with a Quick Drain oil system, the quick drain oil fitting must have clearance of 1/2 in. (13mm) of the engine compartment and the bottom of the boat.

27. If the quick drain oil fitting is within 1/2 in. (13mm) or less of the bottom of the boat, remove the fitting and install the drain plug from the parts bag directly into the oil pan.

28. If the quick drain oil fitting clearance is greater than 1/2 in. (13mm):

   a. Push the end of the oil drain hose out and through the flange. Continue pulling on the hose until it is 12 in. (30.48 cm) from the flange.

   b. Move the alignment clip on the drain hose until it is just inside the hull and against the flange. Squeeze it into position.

   c. Now connect the bilge drain plug to the drain hose plug using the clip.

   d. Push the drain hose through the flange and into the hull. Install the bilge plug in the hull.

29. Apply some spline grease (Quicksilver Engine Coupler Spline Grease) to the engine coupler splines and then lubricate the exhaust bellows with soapy water - this will allow it to slip over the pipe much easier.

30. Lower the engine into the compartment fully so that it just rests on the washers and the transom plate. Make sure that the holes line up, but DO NOT release the hoist tension yet. Make sure at this point that the exhaust pipes are in alignment.

31. Slide a large steel washer and a metal spacer onto the mounting bolt and then run them through the rear mount and transom plate. Tighten the bolts to 38 ft. lbs. (51 Nm).

32. Lower the front of the engine until the front mount just rests on the stringers. Unhook the hoist.

■ The finished boat stringer must position the front mount so that a minimum of 1/4 in. (6mm) adjustment (up or down) exists after the mount contacts the stringer. If this is not the case, you will have to make alterations to the stringer.

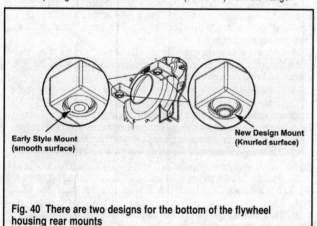

Fig. 40 There are two designs for the bottom of the flywheel housing rear mounts

Starboard Side Shown (Port Side Similar)

Fig. 42 ...while the newer style only uses the fiber washer

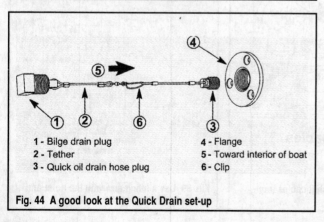

1 - Bilge drain plug
2 - Tether
3 - Quick oil drain hose plug
4 - Flange
5 - Toward interior of boat
6 - Clip

Fig. 44 A good look at the Quick Drain set-up

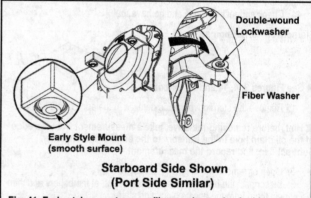

Starboard Side Shown (Port Side Similar)

Fig. 41 Early style mounts use a fiber washer and a double-wound lock washer. . .

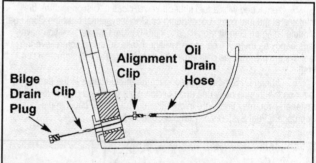

Fig. 43 Use the oil drain hose to start off the engine alignment process

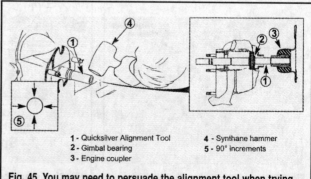

1 - Quicksilver Alignment Tool
2 - Gimbal bearing
3 - Engine coupler
4 - Synthane hammer
5 - 90° increments

Fig. 45 You may need to persuade the alignment tool when trying to get it through the bearing

ENGINE MECHANICAL - GM V6 AND V8 ENGINES 4-13

33. Slide the solid end of the alignment tool into the center of the gimbal housing assembly bearing and then into the engine coupler splines - you may have to pivot the bearing slightly. If necessary, strike the sides of the tool with a rubber mallet at 90° increments to help align the bearing to the coupler. If the tool does not fit into the coupler, remove it and adjust the front mounts.

■ **Always turn both adjustment nuts an equal amount at a time when adjusting the mounts.**

34. Loosen the lock nuts on each front mount. Adjust by turning the adjusting nut (bottom nut, under mount bracket); counterclockwise will raise the front of the engine, clockwise will lower it. Do this a little at a time until you are able to fully insert the alignment tool through the bearing and into the coupler.
35. Install the mount base lag bolts and tighten them securely. Tighten the lock nuts to 59 ft. lbs (80 Nm).
36. Recheck that the alignment tool still slides freely and then remove it.
37. If the front mount cannot be lowered enough to allow for proper alignment there is a Spacer Kit available (#12-892619A01). Reattach the engine hoist and sling, remove the rear mount bolts and then lift the engine slightly until it is level. Install the stainless steel washer from the kit (which will raise the rear of the engine) inside the inner diameter of each fiber washer and then lower the engine again and install the rear mount bolts. Remove the engine hoist. Repeat the alignment steps again until you are able to install the alignment tool correctly, and freely.
38. Coat the exposed threads and nuts on all mounts with a little bit of Perfect Seal.
39. Reconnect the exhaust bellows/hose and tighten the clamps securely. Refer to the illustration for positioning.
40. Reconnect the water inlet hose at the water tube and tighten the hose clamp. If you have a newer model with a quick-connect fitting:
    a. Ensure that the retainer clip is in the closed position and install it onto the water tube so the center of the clip and the decal are facing toward the engine.
    b. Align the slots on the fitting with the tabs on the water tube, make sure that the center line of the tube and the center of the retainer clip are facing the engine, and then press the fitting onto the tube until it snaps into place.
    c. Grab the seawater hoses carefully near the fitting and give it a good stiff tug (approx. 25 ft. lbs.) to ensure the fitting is operating correctly. If it comes off, repeat the installation steps.

■ **If your engine utilizes an extension hose between the seawater hose and the water tube, simply follow the previous steps for connecting the extension hose to the water tube.**

41. Connect the seawater line to the water pump.
42. Connect the battery cables to the starter and apply a thin coat of Liquid Neoprene. Slide the rubber boot over the positive terminal and then drape the cables over the engine so they are out of the way.
43. Carefully, and quickly, connect the power steering lines. Tighten all fittings to 23 ft. lbs. (31 Nm) making sure that they are not cross-threaded or over-tightened. If you have quick-connect fittings, disconnect the lines from each other and then pop them into position on the control valve so they snap into place. It is imperative that the lines are routed as they were before disconnecting and that they do not come into contact with anything so don't hesitate to use a couple plastic ties if need be.

■ **Don't forget the O-rings at the fittings. Don't forget to check the fluid level and bleed the system when you are finished with the installation.**

44. When installing the gear lube monitor, make sure the hose is routed correctly and does no contact any steering components, the engine coupler or any drive line components. If you removed the hose from the transom assembly, press the quick-connect fitting into position until it snaps in place.
45. Unplug the fuel line and pump fitting and reconnect them.
46. Install and adjust the throttle cable. Make sure that the barrel positions the cable so that there are no kinks or sharp bends in the cable. Make sure that the cable does not come in contact with any other moving parts. Please refer to the appropriate Fuel System or Remote Controls section for further details.
47. Install and adjust the shift cables. Please refer to the appropriate Drive Systems or Remote Controls section for further details.
48. Route the instrumentation harness back to the engine in a manner that ensures it will not rub or get pinched. It's a good idea that the harness is secured to the boat at least every 18 in.
    a. On 10-pin and SmartCraft harnesses; position the hose clamp, connect to the engine harness plug and tighten the clamp securely. Reconnect the transom harness and the CAN line.
    b. On 14-pin harnesses; position the hose clamp, connect to the engine harness plug and tighten the clamp securely. Reconnect the transom harness, depth transducer, paddle wheel and tank level connectors and then the Clean Power line. All 4 connectors should be arrayed right next to the main harness connector on top of the engine.
49. Reconnect the trim position sender leads, the engine ground wire, the battery cables, continuity circuit wire and all other wires, lines of hoses that were disconnected during removal.

✱✱ CAUTION

Never connect any accessory ground wires to the transom plate ground point - they should always be connected to the engine ground stud.

■ **Models with a 14-pin harness utilize a separate transom harness which includes a continuity ground lead. Oddly, they may also have a separate continuity circuit wire as well. The ground lead in the harness should be connected to the transom plate and the circuit wire (if equipped) should run between the transom plate and the ground stud on the flywheel housing.**

50. Connect the MerCathode lines to the controller and coat them with liquid neoprene. If you forgot to tag the wires as we suggested, refer to the illustration for proper hook-up.
51. Tighten the power trim pump hydraulic line connections to 125 inch lbs. (14 Nm).
52. Check and refill all fluids and go have fun!

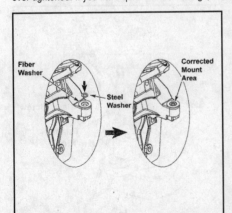

Fig. 46 The Spacer Kit (#12-892619A01) uses a stainless washer to raise the rear of the engine slightly

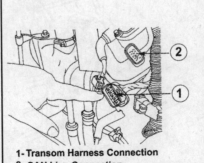

1- Transom Harness Connection
2- CAN Line Connection

Fig. 47 Connecting the transom harness and CAN line

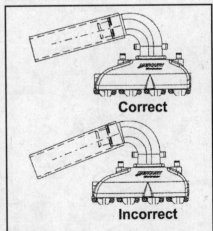

Fig. 48 Correct exhaust tube positioning is important

4-14 ENGINE MECHANICAL - GM V6 AND V8 ENGINES

Inboards - 5.7L & 6.2L Engines

◆ See Figures 21, 22, 23, 31, 34 and 49 thru 57

1. Open or remove the engine hatch cover.
2. Disconnect the battery cables (negative first) at the battery and then disconnect them from the engine block and starter.
3. If equipped with a fuel shut-off valve, close it. Disconnect the fuel inlet line at the fuel boost pump and quickly plug it. Make sure you have rags handy as there will be some spillage. Don't forget to relieve the fuel system pressure before disconnecting the line; as detailed in the appropriate Fuel Systems section.
4. Disconnect the throttle cable at the throttle lever on the throttle body unit and position it out of the way. Keep the hardware.
5. Loosen the hose clamp on the engine wiring harness connector and unplug the instrument wiring harness. Label it and move it aside.
6. Loosen the hose clamp and remove the water inlet line (upper) at the seawater pump.
7. Loosen the hose clamps and disconnect the exhaust elbow hoses/bellows from the upper pipe. Remove them. Disconnect the shaft log seal hose at the fitting on the manifold.
8. Disconnect the shift cable at the transmission on non-DTS models. Don't lose the hardware.
9. Loosen the clamps at the bullhorn on models so equipped.
10. Tag and disconnect the engine ground wire at the stud.
11. Tag and disconnect any remaining lines, wires or hoses at the engine.
12. Disconnect the propeller shaft coupler at the transmission output flange.
13. Attach a suitable engine hoist to the lifting eyes and take up any line slack until it is just taught.

■ DO NOT use the front lifting eye attached to the thermostat housing at this point.

14. Locate the front engine mount(s) and remove the 2 lag bolts.
15. Locate the rear engine mount and remove the mounting bolts.
16. Slowly and carefully, lift out the engine assembly.

To install:

17. Loosen the stop nut and move the adjustment nuts on all mounts so that they are in the center of the adjustment range.
18. All mounts have one slotted bolt hole and this should be facing forward. The large diameter of the mounting trunnion should be fully extended.
19. Connect the battery cables to the starter and apply a thin coat of Liquid Neoprene. Slide the rubber boot over the positive terminal and then drape the cables over the engine so they are out of the way.
20. Lubricate the coupler splines with Engine Coupler Spline Grease.

Models With Velvet or ZF/Hurth 8 Degree Down Angle Transmissions

21. Lower the engine assembly into the vessel so it is just about in position. Install the quick drain oil hose plug in the oil drain hose.
22. Lower the engine fully so that it is positioned in the bed with the transmission output flange aligned and butted up against the propeller shaft coupler - no visible gap between the faces.

■ If this step requires raising or lowering the assembly, adjust the engine bed; do not use the mount adjusters at this time.

23. If you must use the engine bed to alter assembly height, make sure that there is at least 1/4 in. (6mm) of up or down adjustment at all mounts after everything is lined up.
24. Once the pre-alignment tasks are complete, ensure that the mounts are all still positioned correctly, install the lag bolts (10mm) and tighten them securely.
25. Disconnect the hoist and remove it.

Models With V-Drive Transmissions

26. Lower the engine assembly into the vessel so it is just about in position. Install the quick drain oil hose plug in the oil drain hose.
27. Position the assembly so that just enough propeller shaft is protruding through the transmission and output flange so that the propeller shaft coupler will mate. Now install the coupler and position the engine so the faces are butted against each other and there is no gap.

■ If this step requires raising or lowering the assembly, adjust the engine bed; do not use the mount adjusters at this time.

28. If you must use the engine bed to alter assembly height, make sure that there is at least 1/4 in. (6mm) of up or down adjustment at all mounts after everything is lined up.
29. Once the pre-alignment tasks are complete, ensure that the mounts are all still positioned correctly, install the lag bolts (10mm) and tighten them securely.
30. Disconnect the hoist and remove it.

On All Models

31. Push the end of the oil drain hose out and through the flange. Continue pulling on the hose until it is 6 in. (152mm) from the propeller.
32. Move the alignment clip on the drain hose until it is just inside the hull and against the flange. Squeeze it into position.
33. Now connect the bilge drain plug to the drain hose plug using the clip.

■ If the quick drain oil fitting is within 1/2 in. (13mm) of the bottom of the boat, remove the fitting and install the drain plug from the parts bag directly into the oil pan.

34. Push the drain hose through the flange and into the hull. Install the bilge plug in the hull.
35. Lubricate the exhaust bellows with soapy water - this will allow it to slip over the pipe much easier. Make sure at this point that the exhaust pipes are in alignment and tighten the clamps securely. If your engine is equipped with S-pipes, it is important that they are routed underneath the transmission mounts.
36. Install the harness connector and position the hose clamp as shown in the illustration.

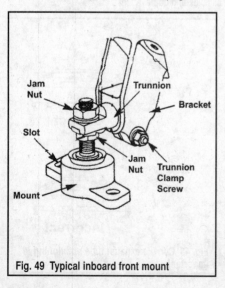

Fig. 49 Typical inboard front mount

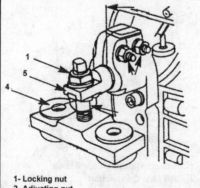

1- Locking nut
2- Adjusting nut
3- Trunnion clamp screw and nut, with lockwasher
4- Slot forward (if so designed)
5- Mount trunnion

Fig. 50 Typical inboard rear mounts - most MPI engines except. . .

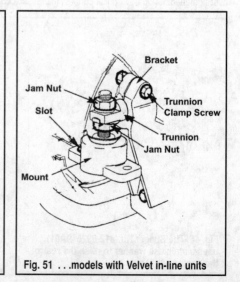

Fig. 51 . . .models with Velvet in-line units

ENGINE MECHANICAL - GM V6 AND V8 ENGINES

37. Reconnect the water inlet hose and tighten the hose clamp securely.
38. Unplug the fuel line and pump fitting and reconnect them.
39. Install and adjust the throttle cable. Make sure that the barrel positions the cable so that there are no kinks or sharp bends in the cable. Make sure that the cable does not come in contact with any other moving parts. Please refer to the Fuel Systems section for further details.
40. Install and adjust the shift cables. Please refer to the Transmission or Remote Controls section for further details.
41. Check and refill all fluids. Once done, it is necessary to perform a final engine alignment with the vessel in the water, full fuel tanks and a normal load.

Final Engine Alignment

42. The engine must now be aligned so that the transmission and the propeller shaft coupler centerlines are fully aligned and that there is no more than a 0.003 in. (0.07mm) gap between the coupling faces; regardless of whether you have a solid or flexible coupling.
43. Center the propeller shaft in the log. Press it down as far as it will go and then pull it up as far as it will go; position the shaft at the apparent center point of its vertical movement. Now do the same thing horizontally.
44. Make certain that the shaft coupler and the engine output flange align exactly.
45. Check that the gap between the two mating surfaces is 0.003 in. (0.07mm) when checked at 90° positions on the couplers (12, 3, 6 and 9 o'clock). If the gap is other than suggested, move the adjustment nuts on the engine mounts until you achieve the proper clearance. Always turn the front and rear mount adjusters equally to maintain level on your engine.
46. To raise the unit up of down, loosen the upper locknut on the mount and then turn the adjusting nut in the appropriate direction to raise or lower the engine.
47. To move the unit side to side, loosen the clamping bolts and nuts on all four mount brackets and move the engine left or right until correct alignment is achieved. There is also a small amount of adjustment leeway available via the slotted front hole on some mounts - loosen the lag bolts and move the engine. Tighten the lag bolts securely when finished.

■ The large diameter of the trunnion should not be extended more than 1 3/4 in. (45mm) from the mount bracket.

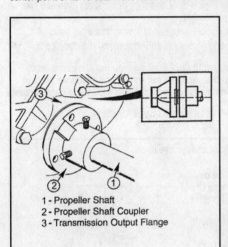

Fig. 52 The coupler and flange must align correctly on models with Velvet or ZF/Hurth 8 degree down angle transmissions

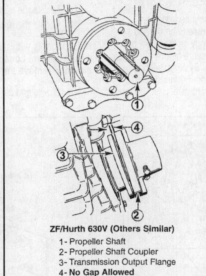

Fig. 53 The coupler and flange must align correctly on models with V-drive transmissions

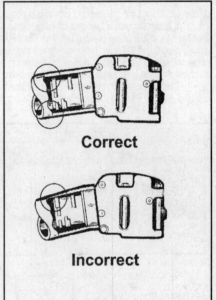

Fig. 54 Make sure the hoses are positioned correctly on the fittings

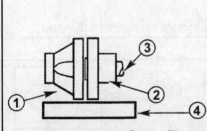

Fig. 55 Coupler-to-flange alignment is very important

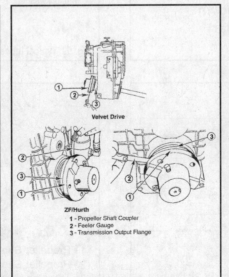

Fig. 56 Checking for parallel between the coupler and flange

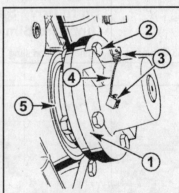

Fig. 57 Always safety wire the set screws if equipped

4-16 ENGINE MECHANICAL - GM V6 AND V8 ENGINES

48. Once the engine is correctly aligned, tighten the clamp bolts an all mounts to 50 ft. lbs. (68 Nm).
49. Tighten the upper locknut on each mount securely and then bend a tab down and over a flat on the adjusting nut.
50. Now you can secure the coupler finally; tighten the bolts and nuts to 50 ft. lbs. (68 Nm).
51. Your propeller shaft coupler may have set screws, if so: remove the screws and mark the dimple locations with a small punch. Remove the coupler and drill shallow dimples where you marked with the punch. Install the coupler again and re-tighten the bolts. Install the set screws and tighten them securely, adding safety wire as a precaution.
52. Connect the shaft seal log.

Inboards - 8.1L Engines

◆ See Figures 37 and 58 thru 68

1. Open or remove the engine hatch cover.
2. Disconnect the battery cables (negative first) at the battery and then disconnect them from the engine block and starter.
3. If equipped with a fuel shut-off valve, close it. Disconnect the fuel inlet line at the fuel boost pump (Gen II) or Cool Fuel module (Gen III) and quickly plug it and the inlet. Make sure you have rags handy as there will be some spillage. Don't forget to relieve the fuel system pressure before disconnecting the line; as detailed in the appropriate Fuel Systems section.
4. Disconnect the throttle cable at the throttle lever on the throttle body unit and position it out of the way (non-DTS). Keep the hardware.

5. Loosen the hose clamp on the engine wiring harness connector and unplug the instrument wiring harness. Label it and move it aside.
6. Loosen the hose clamp and remove the water inlet line (upper) at the seawater pump.
7. Loosen the hose clamps and disconnect the exhaust elbow hoses/bellows from the upper pipe. Remove them. Disconnect the shaft log seal hose at the fitting on the manifold.
8. Disconnect the shift cable at the transmission on non-DTS models. Don't lose the hardware.
9. Tag and disconnect the engine ground wire at the stud.
10. Tag and disconnect any remaining lines, wires or hoses at the engine.
11. Disconnect the propeller shaft coupler at the transmission output flange.
12. On models with a V-Drive, remove the propeller shaft flange from the shaft and then slide the propeller away from the transmission until you have suitable clearance.
13. Attach a suitable engine hoist and sling to the lifting eyes and take up any line slack until it is just taught.
14. Locate the front engine mounts and remove the 2 lag bolts.
15. Locate the rear engine mounts and remove the mounting bolts.

■ On models with the Gen III Cool Fuel system, the starboard mount will be hidden. Loosen the captured nut on the hose retainer bracket and then remove the water line at the Cool Fuel module to gain access.

16. Slowly and carefully, lift out the engine assembly.

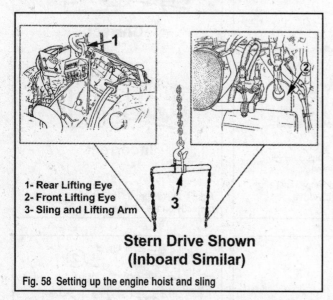

1- Rear Lifting Eye
2- Front Lifting Eye
3- Sling and Lifting Arm

Stern Drive Shown (Inboard Similar)

Fig. 58 Setting up the engine hoist and sling

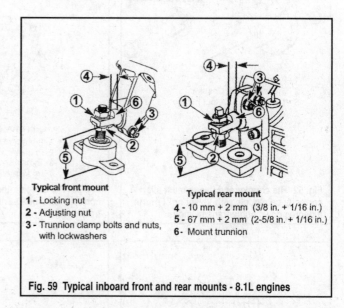

Typical front mount
1 - Locking nut
2 - Adjusting nut
3 - Trunnion clamp bolts and nuts, with lockwashers

Typical rear mount
4 - 10 mm + 2 mm (3/8 in. + 1/16 in.)
5 - 67 mm + 2 mm (2-5/8 in. + 1/16 in.)
6 - Mount trunnion

Fig. 59 Typical inboard front and rear mounts - 8.1L engines

1- Captured Nut
2- Retainer Bracket
3- Water Hose

Fig. 60 Disconnect the water hose on Gen III systems to gain access to the starboard mount - 8.1L engines

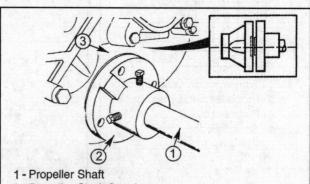

1 - Propeller Shaft
2 - Propeller Shaft Coupler
3 - Transmission Output Flange

Fig. 61 The coupler and flange must align correctly on models with Velvet or ZF/Hurth 8 degree down angle transmissions

ENGINE MECHANICAL - GM V6 AND V8 ENGINES 4-17

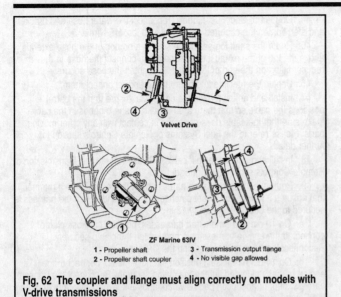

Fig. 62 The coupler and flange must align correctly on models with V-drive transmissions

1 - Propeller shaft
2 - Propeller shaft coupler
3 - Transmission output flange
4 - No visible gap allowed

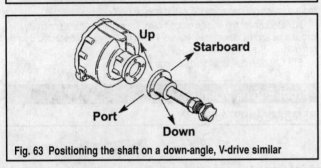

Fig. 63 Positioning the shaft on a down-angle, V-drive similar

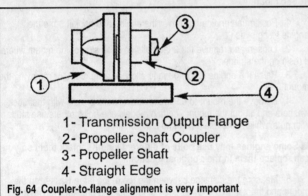

1 - Transmission Output Flange
2 - Propeller Shaft Coupler
3 - Propeller Shaft
4 - Straight Edge

Fig. 64 Coupler-to-flange alignment is very important

To install:

■ Prior to installation, it is important to have the mounts adjusted as follows so the adjustment is centered and all mounts are starting at a uniform height.

17. Loosen the lock nut and turn the adjustment nut on all mounts so that they are in the center of the adjustment range - this means that the measurement from the bottom of the mount surface to the bottom of the trunnion should be 2 5/8 in (67mm) +/- 1/16 in. (2mm).

18. Now loosen the clamp bolt and nut (1 on front mounts and 2 on the rear mounts) and ensure that the trunnion extension is 3/8 in. (10mm) +/- 1/16 in. (2mm). Once within specification, tighten the bolts and nuts to hold everything in position.

19. Most mounts have one slotted bolt hole and this should be facing forward. The large diameter of the mounting trunnion should be fully extended to the specification previously mentioned.

20. Connect the battery cables to the starter and apply a thin coat of Liquid Neoprene. Slide the rubber boot over the positive terminal and then drape the cables over the engine so they are out of the way.

21. Lubricate the splines with Engine Coupler Spline Grease.

Initial Alignment - Models With Velvet or ZF/Hurth 8 Degree Down Angle Transmissions

22. Lower the engine assembly into the vessel so it is just about in position.

23. Lower the engine fully so that it is positioned in the bed with the transmission output flange aligned and butted up against the propeller shaft coupler - no visible gap between the faces.

■ If this step requires raising or lowering the assembly, adjust the engine bed; do not use the mount adjusters at this time.

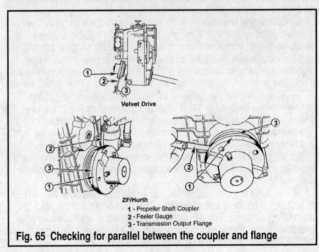

1 - Propeller Shaft Coupler
2 - Feeler Gauge
3 - Transmission Output Flange

Fig. 65 Checking for parallel between the coupler and flange

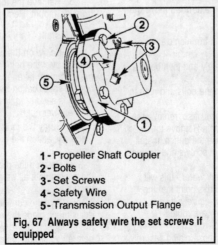

1 - Measurement - 45 mm (1-3/4 inches)
2 - Clamping bolts and nuts
3 - Locknuts on all 4 mounts
4 - Tab on tab washer

Fig. 66 Checking the trunnion extension - inboards

1 - Propeller Shaft Coupler
2 - Bolts
3 - Set Screws
4 - Safety Wire
5 - Transmission Output Flange

Fig. 67 Always safety wire the set screws if equipped

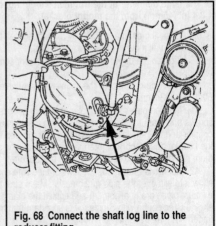

Fig. 68 Connect the shaft log line to the reducer fitting

24. If you must use the engine bed to alter assembly height, make sure that there is at least 1/4 in. (6mm) of up or down adjustment at all mounts after everything is lined up.

25. Once the pre-alignment tasks are complete, ensure that the mounts are all still positioned correctly, install the lag bolts (10mm) and tighten them securely.

26. Disconnect the hoist and remove it.

Initial Alignment - Models With V-Drive Transmissions

27. Lower the engine assembly into the vessel so it is just about in position.

28. Position the assembly so that just enough propeller shaft is protruding through the transmission and output flange so that the propeller shaft coupler will mate. Now install the coupler and position the engine so the faces are butted against each other and there is no gap.

■ If this step requires raising or lowering the assembly, adjust the engine bed; do not use the mount adjusters at this time.

29. If you must use the engine bed to alter assembly height, make sure that there is at least 1/4 in. (6mm) of up or down adjustment at all mounts after everything is lined up.

30. Once the pre-alignment tasks are complete, ensure that the mounts are all still positioned correctly, install the lag bolts (10mm) and tighten them securely.

31. Disconnect the hoist and remove it.

Final Alignment - All Models

■ Engine alignment must still be rechecked when the boat is in the water with a normal load on board and the tanks full.

32. The engine must now be aligned so that the transmission and the propeller shaft coupler centerlines are fully aligned and that there is no more than a 0.003 in. (0.07mm) gap between the coupling faces; regardless of whether you have a solid or flexible coupling.

33. Center the propeller shaft in the log. Press it down as far as it will go and then pull it up as far as it will go; position the shaft at the apparent center point of its vertical movement. Now do the same thing horizontally.

34. Make certain that the shaft coupler and the engine output flange align exactly. On most models this is easily determined because there will be a shoulder on the shaft coupler that will fit into a recess on the output flange.

35. Check that the gap between the two mating surfaces is 0.003 in. (0.07mm) when checked at 90° positions on the couplers (12, 3, 6 and 9 o'clock). If the gap is other than suggested, move the adjustment nuts on the engine mounts until you achieve the proper clearance. Always turn the front and rear mount adjusters equally to maintain level on your engine.

36. To raise the unit up of down, loosen the upper locknut on the mount and then turn the adjusting nut in the appropriate direction to raise or lower the engine.

37. To move the unit side to side, loosen the clamping bolts and nuts on all four mount brackets and move the engine left or right until correct alignment is achieved. There is also a small amount of adjustment leeway available via the slotted front hole on some mounts - loosen the lag bolts and move the engine. Tighten the lag bolts securely when finished.

■ The large diameter of the trunnion should not be extended more than 1 3/4 in. (45mm) from the mount bracket.

38. Once the engine is correctly aligned, tighten the trunnion clamp bolts/nuts on all mounts to 50 ft. lbs. (68 Nm).

39. Tighten the upper locknut on each mount securely and then bend a tab down and over a flat on the adjusting nut.

40. Now you can secure the coupler finally; tighten the bolts and nuts to 50 ft. lbs. (68 Nm).

41. Your propeller shaft coupler may have set screws, if so: remove the screws and mark the dimple locations with a small punch. Remove the coupler and drill shallow dimples where you marked with the punch. Install the coupler again and re-tighten the bolts. Install the set screws and tighten them securely, adding safety wire as a precaution.

42. Lubricate the exhaust bellows or tube with soapy water - this will allow it to slip over the pipe much easier. Make sure at this point that the exhaust pipes are in alignment and tighten the clamps securely. If your engine is equipped with S-pipes, it is important that they are routed underneath the transmission mounts.

43. Reconnect the water inlet hose and tighten the hose clamp securely.

44. On Gen III Cool Fuel models, reconnect the water hose and bracket and then tighten the captured nut to 168 inch lbs. (19 Nm).

45. Route the shaft log seal hose so that a portion of the hose extends above the top of the exhaust elbows and then connect the hose to the reducer fitting on the end of the manifold. Tighten the hose securely.

46. Unplug the fuel line and pump fitting and reconnect them.

47. Install and adjust the throttle cable. Make sure that the barrel positions the cable so that there are no kinks or sharp bends in the cable. Make sure that the cable does not come in contact with any other moving parts. Please refer to the Fuel Systems or Remote Controls section for further details.

48. Install and adjust the shift cables. Please refer to the Transmission or Remote Controls section for further details.

49. Route the instrumentation harness back to the engine in a manner that ensures it will not rub or get pinched. It's a good idea that the harness is secured to the boat at least every 18 in.

   a. On 10-pin and SmartCraft harnesses; position the hose clamp, connect to the engine harness plug and tighten the clamp securely. Reconnect the transom harness and the CAN line.

   b. On 14-pin harnesses; position the hose clamp, connect to the engine harness plug and tighten the clamp securely. Reconnect the transom harness, depth transducer, paddle wheel and tank level connectors and then the Clean Power line. All 4 connectors should be arrayed right next to the main harness connector on top of the engine.

50. Reconnect the engine ground wire, the battery cables, continuity circuit wire and all other wires, lines of hoses that were disconnected during removal.

51. Check and refill all fluids and go have fun!

Front Engine Mounts

REMOVAL & INSTALLATION

◆ See Figure 4

1. Position an engine hoist over the engine and hook it up to the 2 engine lifting eyes.

2. Loosen and remove the 2 lag bolts on each side of the mount where it rests on the stringer.

3. Raise the engine just enough to allow working room for removing the mount.

4. Remove the mount-to-engine mounting bolts or nuts with their lock washers and lift out the mount - certain models use bolts, others use studs with nuts, while some use a bolt and a stud.

■ Some engines may use bolts of different lengths; mark them so you can replace them in their original locations.

5. Measure the distance between the top of the large washer on the mount and the flat on the lower side of the mounting bracket. Record it.

6. Remove the top jam nut and washer, and then separate the bracket from the mount. Remove the lower nut and the tab washer.

To install:

7. Screw on the washer and lower nut and then position the mount bracket over the bolt. Install the washer and upper nut, and check the measurement taken in Step 5. Move the lower nut up or down until the correct specification is achieved and then tighten the upper nut down so the mount is secure against the bracket. Bend the tabs on the washer over a flat on the lower nut.

8. Spray the mounting bolts/nuts with Loctite and allow them to air dry. Once dry, attach the mount to the engine. Tighten the bolts to 30 ft. lbs. (41 Nm).

9. Position the mount over the lag screw holes and then slowly lower the engine until all weight is off the hoist. Install and tighten the lag screws securely.

10. If you're confident that your measurements and subsequent adjustment, place the engine exactly where it was prior to removal, then you are through. If you're like us though, you may want to check the alignment before you fire up the engine.

ENGINE MECHANICAL - GM V6 AND V8 ENGINES   4-19

Fig. 69 Remove the bolts (5.7L shown)...

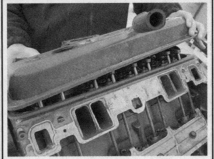
Fig. 70 ...and then lift off the cover

Fig. 71 Position a new gasket on the head

Cylinder Head Cover

REMOVAL & INSTALLATION

EASY

◆ See Figures 69 thru 72, 181, 182, 186 and 192

■ In order to perform this procedure efficiently, we recommend removing the exhaust manifold in order to have sufficient working room to remove the cylinder head cover. Although not completely necessary, it's worth the extra effort to avoid the aggravation of working around the manifolds. Please refer to the manifold procedure later in this section.

1. Open or remove the engine compartment hatch. Disconnect the negative battery cable.
2. Loosen the clamp and remove the crankcase ventilation hose at the cover. Carefully move it out of the way.
3. Tag and disconnect any lines, leads or hoses that might be in the way of removal. In some instances you may be able to simply secure them out of the way without disconnecting them - you be the judge.
4. If your engine has a spark plug wire retainer attached to the cover, unclip the wires or remove the retainer and then tag and disconnect the plug wires at the spark plugs.
5. On the 8.1L V8, remove the coolant reservoir, plug wires at the coils and the remote oil lines.

■ The 8.1L engine utilizes individual coils for each cylinder and they are attached to the cylinder head cover. It is not normally necessary to remove the coils for cylinder head cover removal, but if you need to remove the actual coils, you can find the procedures in the Ignition System section.

6. Loosen the cover mounting bolts (3 on the V6, 4 on the 5.0/5.7/6.3L V8 and 7 on the 8.1L) and lift off the cylinder head cover. Take note of any harness or hose retainers and clips that might be attached to certain of the mounting bolts; you need to make sure they go back in the same place.

To install:

7. Clean the cylinder head and cover mounting surfaces of any residual gasket material with a scraper or putty knife.
8. Position a new gasket on the cylinder head and then position the cover (don't forget the J-clips if they were used!). Tighten the mounting bolts to 106 inch lbs. (12 Nm).
9. Make sure any retainers or clips that were removed are back in their original positions.
10. Connect the crankcase ventilation hose and any other lines or hoses that may have been disconnected. Check that there were no other wires or hoses you may have repositioned in order to gain access to the cover.
11. Install the coolant tank and remote oil lines on the 8.1L.
12. Install the exhaust manifold.
13. Connect the battery cables.

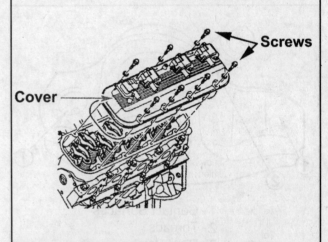
Fig. 72 On the 8.1L, the mounting bolts are on the side of the cover, rather than through the top

Rocker Arms & Push Rods

REMOVAL & INSTALLATION

V6 Engines

MODERATE

◆ See Figures 73 thru 77

1. Open or remove the engine hatch cover and disconnect the negative battery cable.
2. Remove the cylinder head cover as detailed previously.
3. Bring the piston in the No. 1 cylinder to TDC. If servicing only one arm, bring the piston in that cylinder to TDC. The No. 1 cylinder is the first cylinder at the port side of the engine.
4. Loosen the rocker bolt and unthread the rocker/bolt assembly from the support.
5. Pull out the pushrod. It is very important to keep each cylinder's component parts together as an assembly. We suggest drilling a set of holes in a 2X4 and positioning the pieces in the holes. What ever you do, don't mix them up!
6. Repeat this procedure for each rocker.
7. Remove the rocker arm support.

To install:

8. Clean and inspect the rocker assemblies for excessive wear or scoring. Inspect the bolt threads and pivot for damage.
9. Coat all bearing surfaces of the rocker assembly with engine oil.
10. Check the pushrod bores for any restriction, coat them with oil and then slide the push rods into their holes. Make sure that each rod seats in its socket on the lifter.
11. Install the rocker support with the word UP on the top, position the rocker assembly and tighten it to 22 ft. lbs. (30 Nm) in the order shown.

4-20 ENGINE MECHANICAL - GM V6 AND V8 ENGINES

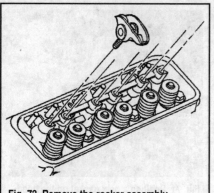
Fig. 73 Remove the rocker assembly...

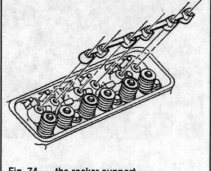
Fig. 74 ...the rocker support...

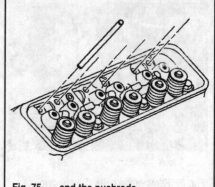
Fig. 75 ...and the pushrods

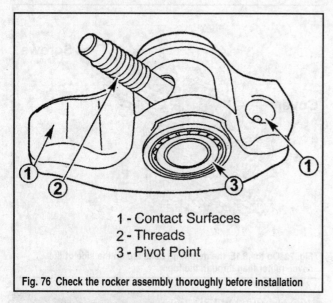

1 - Contact Surfaces
2 - Threads
3 - Pivot Point

Fig. 76 Check the rocker assembly thoroughly before installation

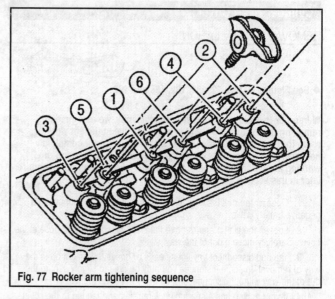
Fig. 77 Rocker arm tightening sequence

12. These are net lash engines, no additional adjustment of the valves is necessary as the lash is set automatically when the rocker is tightened to specifications.
13. Install the cylinder head cover, connect the battery cable and check the idle speed.

5.0L, 5.7L & 6.2L V8 Engines

◆ See Figure 78

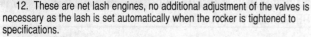 MODERATE

1. Open or remove the engine hatch cover and disconnect the negative battery cable. Remove the cylinder head cover as detailed previously.
2. Bring the piston in the No. 1 cylinder to TDC. If servicing only one arm, bring the piston in that cylinder to TDC. The No. 1 cylinder is the first cylinder at the port side of the engine.
3. Loosen and remove the rocker arm nuts and lift out the balls. Lift the arm itself off of the mounting stud and pull out the pushrod. It is very important to keep each cylinder's component parts together as an assembly. We suggest drilling a set of holes in a 2 x 4 and positioning the pieces in the holes. What ever you do, don't mix them up!

To install:

4. Clean and inspect the rocker assemblies.
5. Coat all bearing surfaces of the rocker assembly with engine oil.
6. Slide the push rods into their holes. Make sure that each rod seats in its socket on the lifter.
7. Position the rocker arm over the stud so that the cupped side rides on the push rod. Slide the ball over the stud, install the nut and tighten it until zero lash is present - meaning the rod will not wiggle.
8. Adjust the valves as detailed else where in this section.
9. Install the cylinder head cover, connect the battery cable and check the idle speed.

8.1L V8 Engines

◆ See Figure 79

 MODERATE

1. Open or remove the engine hatch cover and disconnect the negative battery cable. Remove the cylinder head cover as detailed previously.
2. Bring the piston in the No. 1 cylinder to TDC. If servicing only one arm, bring the piston in that cylinder to TDC. The No. 1 cylinder on is the first cylinder at the port side of the engine.
3. Loosen and remove the rocker arm nuts and lift out the balls. Lift the arm itself off of the mounting stud and pull out the pushrod. It is very important to keep each cylinder's component parts together as an assembly. We suggest drilling a set of holes in a 2 x 4 and positioning the pieces in the holes. What ever you do, don't mix them up!

To install:

4. Clean and inspect the rocker assemblies.
5. Coat all bearing surfaces of the rocker assembly with engine oil.
6. Slide the push rods into their holes. Make sure that each rod seats in its socket on the lifter and is fed through the restrictor plate attached to the stud.
7. Position the rocker arm over the stud so that the cupped side rides on the push rod. Slide the ball over the stud, install the nut and tighten it to 19 ft. lbs. (25 Nm), making sure that each cylinder is at TDC and the valves are closed.
8. No additional adjustment of the valves is necessary as the lash is set automatically when the rocker is tightened to specifications.
9. Install the cylinder head cover, connect the battery cable and check the idle speed.

ENGINE MECHANICAL - GM V6 AND V8 ENGINES 4-21

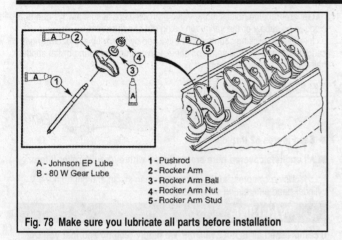

A - Johnson EP Lube
B - 80 W Gear Lube

1 - Pushrod
2 - Rocker Arm
3 - Rocker Arm Ball
4 - Rocker Arm Nut
5 - Rocker Arm Stud

Fig. 78 Make sure you lubricate all parts before installation

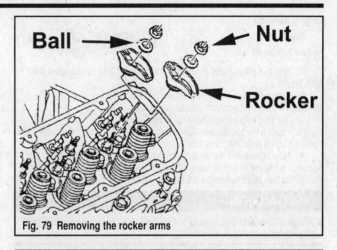

Fig. 79 Removing the rocker arms

VALVE ADJUSTMENT - 5.0L, 5.7L & 6.2L V8 ENGINES

Engine Not Running

 MODERATE

◆ See Figure 80

These engines utilize hydraulic valve lifters, although there is no need for periodic valve adjustment, it is necessary to perform a preliminary adjustment after any work on the valve train/rocker assembly. All adjustment should be undertaken while the lifter is on the base circle of the camshaft lobe for that particular cylinder. This means the opposite side of the pointy part of each lobe.

1. Remove the cylinder head covers.
2. Rotate the crankshaft, or bump the engine with the starter until the No. 1 cylinder is at TDC. Note that the notch or mark on the damper pulley will be lined up with the **0** mark on the timing scale. Be careful here though, this could mean that either the No. 1 or the No. 6 piston is at TDC. Place your hand on the No. 1 cylinder's valve and check that it does not move as the mark on the pulley is approaching the **0** mark on the tab. If it does not move, you're ready to proceed; if it does move, you are on the No. 6 cylinder and need to rotate the engine an additional full turn. This is important so make sure you've gotten it right!
3. Now that the No. 1 cylinder is at TDC, you can adjust the following valves:
   - No. 1 cylinder: intake and exhaust
   - No. 2 cylinder: intake
   - No. 3 cylinder: exhaust
   - No. 4 cylinder: exhaust
   - No. 5 cylinder: intake
   - No. 7 cylinder: intake
   - No. 8 cylinder: exhaust
   Or these valves on RH rotation engines:
   - No. 1 cylinder: intake and exhaust
   - No. 2 cylinder: exhaust
   - No. 3 cylinder: intake
   - No. 4 cylinder: intake
   - No. 5 cylinder: exhaust
   - No. 7 cylinder: exhaust
   - No. 8 cylinder: intake
4. Loosen the adjusting nut on the rocker until you can feel lash (play in the push rod) and then tighten the nut until the lash has been removed. Carefully jiggle the push rod while tightening the nut until it won't move anymore - this is zero lash. Tighten the nut an additional full turn (360°) to set the lifter and then you're done. Perform this procedure on each of the valves listed above.
5. Slowly rotate the engine an additional full turn and this will bring the No. 6 piston to TDC. The pulley notch/mark should once again be in line with the **0** on the timing tab. You can now adjust the remaining valves:
   - No. 2 cylinder: exhaust
   - No. 3 cylinder: intake
   - No. 4 cylinder: intake
   - No. 5 cylinder: exhaust
   - No. 6 cylinder: intake and exhaust
   - No. 7 cylinder: exhaust

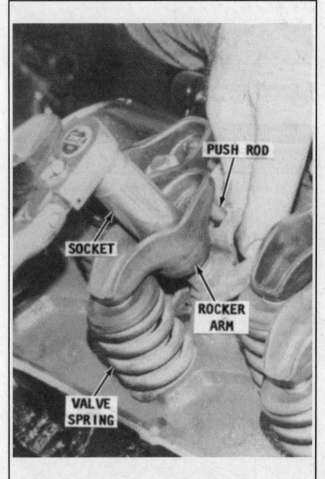

Fig. 80 Wiggle the push rod slowly while tightening the rocker arm adjusting nut

   - No. 8 cylinder: intake
   Or these valves on RH rotation engines:
   - No. 2 cylinder: intake
   - No. 3 cylinder: exhaust
   - No. 4 cylinder: exhaust
   - No. 5 cylinder: intake
   - No. 6 cylinder: intake and exhaust
   - No. 7 cylinder: intake
   - No. 8 cylinder: exhaust
6. Install the cylinder head covers.
7. Check the idle speed and mixture.

ENGINE MECHANICAL - GM V6 AND V8 ENGINES

Engine Running

◆ See Figure 81

1. Run the engine until it reaches normal operating temperature and then shut it off.
2. Remove the cylinder head cover and install rocker stoppers (#9166273) over the pushrod side of the rocker.

Start the engine again and allow it to idle.

3. Loosen the rocker arm nut until the rocker starts to 'clatter' and then tighten the nut until the noise stops - this is zero lash.
4. Now tighten the nut an additional 1/4 turn and then wait about 10 seconds while the engine smoothes out. Repeat this 3 more times until the rocker nut has been tightened a total of 3/4 turns past zero lash.

※※ CAUTION

It is imperative that this step is not rushed and the lifter has time to adjust in between each 1/4 turn!

5. Repeat the procedure on each of the remaining valves.
6. Turn off the engine, remove the stoppers and install the cover.
7. Check the idle speed and mixture.

VALVE ADJUSTMENT - 8.1L V8 ENGINES

■ Valve lash is automatically set when the rocker arm nuts are tightened down - no adjustment is necessary.

Hydraulic Valve Lifter

NOISY LIFTERS

A quick and easy way to locate a noisy lifter is to cut a piece of garden hose to about 4 ft. Have someone hold one end of the hose near the end of each intake/exhaust valve while you cup the other end of the hose, and your ear, in your hand. This should help cut out some of the other noise and pinpoint the culprit.

Another means of finding a noisy lifter is to, carefully, place your finger on the face of the valve spring retainer. If the lifter is mal-functioning, you will feel a very distinct shock (force, not electrical!) through your finger when the valve seats itself.

A hard rapping noise from the lifter is frequently caused by the plunger sticking in the lifter body bore, not allowing the spring to push it back into position. Check the lifter, particularly the plunger, for varnish or carbon deposits. Galling, usually caused by foreign debris getting in between the plunger and lifter body, can also contribute to this.

A moderate rapping noise can be caused by an abnormally high leak-down rate, a leaky check valve seat or simply, an out of adjustment valve.

General rapping noises in the valve train itself are almost always due to a clogged oil supply or the valvetrain being out of adjustment.

Intermittent clicking sounds can be caused by a small piece of dirt or grit caught between the ball seat and check valve, an out-of-round or flat spotted ball (rare) or bad adjustment.

REMOVAL & INSTALLATION

◆ See Figures 82 thru 89

■ All engines covered here are equipped with hydraulic roller lifters.

1. Using compressed air, thoroughly clean all dirt and grit from the cylinder head and related components.

※※ WARNING

If compressed air is not available, we highly recommend that you DO NOT proceed with this procedure. It is EXTREMELY important that no dirt gets into the lifter recesses before completing the installation.

2. Remove the cylinder head cover.
3. Remove the intake and exhaust manifolds.
4. Loosen the rocker arms and remove the rockers.
5. Remove the pushrods from the block, being very careful to keep track of where each one came from. Make a holding block as mentioned in the Removal procedure.

■ All engines utilize a lifter guide (some call it a restrictor) which situates each pair of lifters in their respective bore in the cylinder block. Some engines use a lifter retainer which sits over the lifters and guides - one on each side of the block and still utilizing a lifter guide around every pair of lifters (most likely V6 MPI engines). On other engines, the lifters sit in their bore in the block and the lifter guides are held in position by a lifter restrictor - a plate that looks as if it has 6 or 8 fingers extending from it (most likely V8 engines or older carb/TBI V6 engines). And, on yet other configurations, a restrictor and a retainer may be used to position and hold the lifters and guides. To compound matters, the words restrictor and retainer are frequently used interchangeably. We will attempt to provide illustrations of each type, but you will have to make the identification on your own and then follow the appropriate steps below.

6. OK, are you still with us?
7. Remove the splash shield on 8.1L engines.

If your engine uses a restricter:

8. Look into the engine between the cylinder banks. If you see a metal plate with 6 or 8 fingers coming off of it, loosen and remove the mounting bolts (2, 3 or 4, depending on the engine) and then pull off the lifter guide restrictor.

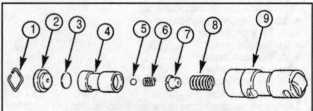

1 - Push Rod Seat Retainer
2 - Push Rod Seat
3 - Metering Valve
4 - Plunger
5 - Check Ball
6 - Check Ball Spring
7 - Check Ball Retainer
8 - Plunger Spring
9 - Lifter Body

Fig. 81 Make sure that you use rocker stoppers if adjusting the valves with the engine running

Fig. 82 Exploded view of the valve lifter

ENGINE MECHANICAL - GM V6 AND V8 ENGINES 4-23

9. Now look at the lifters, if they are in a recess in the block with a lifter guide over each pair, **matchmark each pair to the guide and remove them as pairs,** as necessary. Make sure that you have a storage system that will allow you to return each lifter to its companion lifter guide and then return each pair to their correct positions in the block.

10. Move to the installation steps.

■ If the lifters/guides are difficult to remove, a special valve lifter remover tool (#J3049-A) is available.

If your engine uses a retainer:

11. If the lifters are covered by a retainer (there will be one on each bank, holding 3 or 4 pairs of lifters each), matchmark each lifter to its respective slot in the retainer and mark the front of the retainer itself. Unbolt the retainer and carefully lift it out and off of the lifters. Matchmark each lifter to the lifter guide as well. Now you can remove the lifter(s). Make sure that you have a storage system that will allow you to return each lifter to its proper bore, retainer seat and companion lifter guide. The retainer should work fine for this.

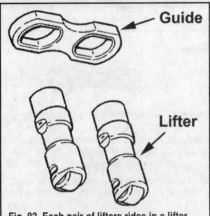

Fig. 83 Each pair of lifters rides in a lifter guide

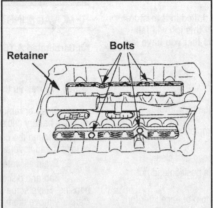

Fig. 84 Some engines (usually V6) utilize a retainer to secure the lifters...

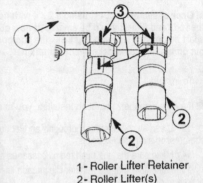

1 - Roller Lifter Retainer
2 - Roller Lifter(s)
3 - Matching Marks

Fig. 85 ...and it should be match-marked to each lifter prior to removal

Fig. 86 Other engines (usually V8) utilize a restrictor to secure the lifters/guides...

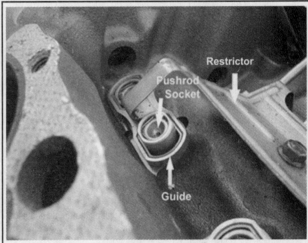

Fig. 87 ...notice the guide and the pushrod socket in the lifter

Fig. 88 The lifter should come out easily (restrictor shown in position)...

Fig. 89 ...and look like this (roller lifter shown)

4-24 ENGINE MECHANICAL - GM V6 AND V8 ENGINES

12. Ok, this step is going to sound very much like the last one, but it is slightly different because some engines use a different retainer - only you can tell for sure. If you did not see the restrictor plate described in Step 8, your lifters will be held in position by a retainer, each pair still positioned by a lifter guide. Matchmark each lifter to its respective retainer seat and the front of the retainer itself. Unbolt the retainer and carefully remove it. Matchmark each lifter pair to their lifter guide as well. Now you can remove the lifter guide(s) and the lifter(s). Make sure that you have a storage system that will allow you to return each lifter to its correct bore, retainer seat and companion lifter guide. The retainer should work fine for this.

■ Once again, we suggest using a 2 x 4 with holes drilled in it to store the rods and lifters; you'd be amazed at how quickly this job will fall apart if someone walks in and kicks the components that you have laid out on the floor or knocks over your retainer!

To install:

■ If a new camshaft has been installed, you must use new lifters.

13. Clean all components thoroughly and let dry completely, use compressed air if at all possible.
14. Make sure that the push rod oil passages are clean and clear.
15. Make sure that the retainer clip in each lifter is in position and not broken or damaged.
16. Inspect the camshaft contact surface on the bottom of each lifter for excessive wear, galling or other damage. Discard the lifter if any of the above conditions are found. You'll probably want to check out the camshaft lobe for damage also.
17. Coat the roller of each lifter that has been removed with engine oil and then carefully install each one into its respective recess.
18. Install the lifter guide over each pair of lifters that were removed, making sure that your matchmarks line up.
19. Install the retainer (if equipped), making sure that all matchmarks are aligned. There are flats on the lifter which also correspond to the retainer slot. Tighten the retainer bolts to 12 ft. lbs. (16 Nm) on early carbureted/TBI engines, or 19 ft. lbs. (25 Nm) on all other engines.
20. Install the restrictor plate (if equipped) and tighten the bolts to 12 ft. lbs. (16 Nm) on early carbureted/TBI V6 engines, or 19 ft. lbs. (25 Nm) on all other engines.

■ On engines with restrictors, the retainer fingers MUST contact ALL lifter guides. If bent, and it does not make contact, replace it with a new one. DO NOT attempt to bend it back into position.

21. Install the pushrods into the sockets on the lifters.
22. Move the rockers into position and then tighten the nut as detailed in your engine's respective Rocker Arm procedure as detailed previously.
23. Install the splash shield on the 8.1L.
24. Install the intake and exhaust manifolds.
25. Install the cylinder head cover.

※※ CAUTION

If any, or all, of the lifters has been replaced with a new one it is very important that you add an EP lube such as GM Cam and Lifter Prelube to the crankcase BEFORE starting the engine. It is also a good idea to change the oil.

Intake Manifold

REMOVAL & INSTALLATION

All Carbureted & TBI Engines

◆ See Figures 90 thru 93

1. Open or remove the engine compartment hatch. Disconnect the negative battery cable.
2. Loosen the clamp and remove the crankcase ventilation hose at the cylinder head covers. Carefully move them out of the way.
3. Drain all water from the cylinder block and manifolds.
4. Tag and disconnect all water hoses at the manifold and thermostat housing. Have some rags handy, since there will still be some water in them. Carefully move them out of the way.
5. Remove the flame arrestor and then disconnect the throttle cable at the carburetor/throttle body. Disconnect the fuel line also and plug the line end and the carb/throttle body/fuel rail fitting. If you have a non-flexible line, disconnect it at the fuel pump also. Move both the cable and fuel line out of the way.
6. Tag and disconnect any lines, leads or hoses that might be in the way of removal. In some instances you may be able to simply secure them out of the way without disconnecting them - you be the judge.
7. If your engine has a spark plug wire retainer attached to the cylinder head cover, unclip the wires or remove the retainer and then tag and disconnect the plug wires at the spark plugs.
8. Remove the distributor cap with the leads still connected. Mark the position of the distributor rotor to the distributor body, loosen the clamp bolt and remove the distributor. Please refer to the Electrical section for further details on distributor removal. DO NOT turn the engine over once the distributor has been removed.
9. Tag and disconnect any leads at the ignition module and move them out of the way.
10. Tag and disconnect the wire at the oil sending unit and then remove the unit itself.
11. Loosen and remove the manifold mounting bolts in the reverse order of the tightening sequence and then remove the manifold (with carb or throttle body attached). There is a good likelihood you will need to pry the manifold off the block; be very careful that you don't scratch or mar the mating surfaces on the block, manifold or heads.

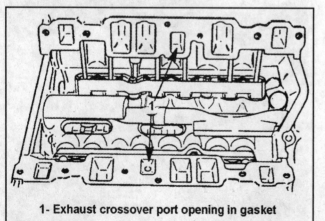

1- Exhaust crossover port opening in gasket

Fig. 90 Certain engines utilize an extra opening in the gasket for the exhaust cross-over port - V6

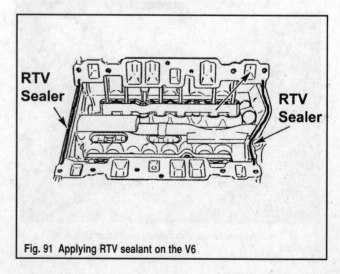

Fig. 91 Applying RTV sealant on the V6

ENGINE MECHANICAL - GM V6 AND V8 ENGINES 4-25

To install:

12. Carefully remove all remaining gasket material from the manifold mating surfaces with a scraper or putty knife. Be careful that you don't accidentally drop any old gasket material into the crankcase or intake ports on the cylinder head.
13. Inspect the manifold and all mating surfaces for any cracks or nicks.
14. Apply Perfect Seal to both sides of each new gasket and position them on the cylinder heads. Note the following precautions:

   a. Manifold gaskets are identical for both sides; always make sure that they are installed with the marked side facing UP.

   b. On V6 engines, if you have an engine with a 2 bbl carburetor make sure that you remove the metal insert from the gasket going on the starboard side cylinder head before you coat with sealer and lay into position. This will allow for heat pipe clearance on the intake manifold.

   c. When ordering your gaskets, make sure you get the right ones for your engine. Engines that utilize an automatic choke require an additional opening in the gasket for the exhaust cross-over port in the intake manifold. Without this hole the choke will not operate correctly and could cause rough operation.

15. Apply a 3/16 in. (5mm) bead of RTV sealant to the forward and aft edges of the cylinder block mating surface. Make sure that you run the bead at least a 1/2 in. up onto the gaskets.

■ There is an oil sending unit hole on the edge of many rear cylinder block mating surfaces. DO NOT get RTV sealer into this hole!

16. Install the manifold into place so that all the bolt holes line up, insert the bolts and tighten them to:
    - 11 ft. lbs. (15 Nm) on V6 engines
    - 18 ft. lbs. (24 Nm) on V8 engines

Check the Torque Specifications chart at the end of this section for any further details on tightening needs. Tighten all bolts in the order shown in the illustration. On carbureted or TBI V8 engines, make sure you coat the threads of the four inner bolts with Loctite Pipe Thread Sealant.

17. Install the oil sending unit and connect the wire.
18. Connect the ignition module leads and install the distributor as detailed in the Electrical section. Put the cap back on and reconnect the plug wires to the spark plugs.
19. Install all water hoses and tighten their clamps securely.
20. Install the fuel line at the carburetor/throttle body and fuel pump. Make sure you remove any plugs you may have inserted on removal.
21. Install the throttle cable and adjust it as detailed in the Fuel System section. Install the flame arrestor.
22. Connect all other lines, leads and hoses that may have been removed to facilitate manifold removal.
23. Connect the battery cable and start the engine. Check the ignition timing and idle speed. Check all hoses and seals for leaks.

V6/V8 MPI Engines (Exc. 8.1L)

◆ See Figures 94 thru 103, 185, 189 and 190

■ The upper and lower intake manifolds may be removed as a unit; it is not necessary to separate the two unless there is a suspected problem.

1. Open the engine compartment and disconnect the battery cables.
2. Remove the flame arrestor and cover. Disconnect the throttle cable and move it out of the way.
3. Drain the cooling systems completely.
4. Disconnect all hoses at the thermostat housing - make sure you have a container handy as there may still be some residual water in the system.
5. Tag, disconnect or move all electrical lines or leads that may interfere with manifold removal.
6. Remove the crankcase breather hoses from the cylinder head covers.
7. Disconnect and plug the fuel line at the rail and move it out of the way; if using a non-flexible line, remove it completely. Disconnect the harness at the fuel injectors.

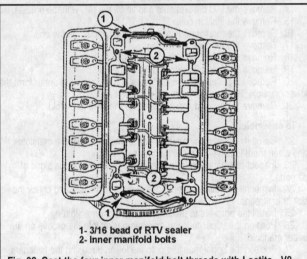

1- 3/16 bead of RTV sealer
2- Inner manifold bolts

Fig. 92 Coat the four inner manifold bolt threads with Loctite - V8 engines

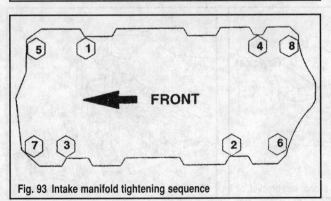

Fig. 93 Intake manifold tightening sequence

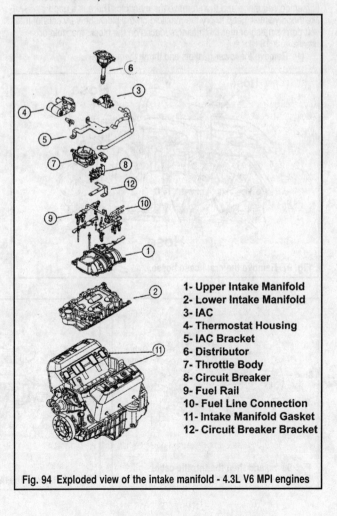

1- Upper Intake Manifold
2- Lower Intake Manifold
3- IAC
4- Thermostat Housing
5- IAC Bracket
6- Distributor
7- Throttle Body
8- Circuit Breaker
9- Fuel Rail
10- Fuel Line Connection
11- Intake Manifold Gasket
12- Circuit Breaker Bracket

Fig. 94 Exploded view of the intake manifold - 4.3L V6 MPI engines

ENGINE MECHANICAL - GM V6 AND V8 ENGINES

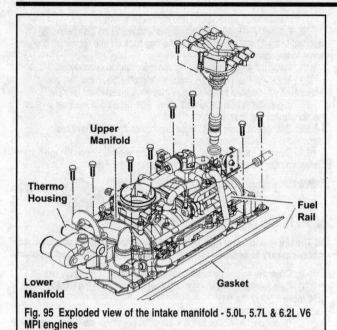

Fig. 95 Exploded view of the intake manifold - 5.0L, 5.7L & 6.2L V6 MPI engines

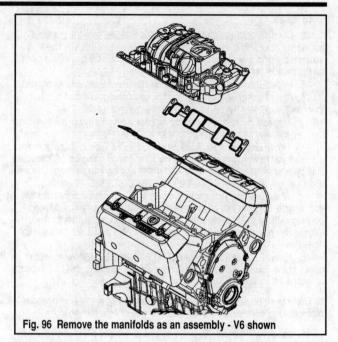

Fig. 96 Remove the manifolds as an assembly - V6 shown

8. Disconnect the distributor wires at the plugs and coil. Mark the distributor to the block, remove the clamp bolt and lift out the distributor. DO NOT engage the starter or turn the engine over once the distributor has been removed.
9. Take a look at the engine and remove any other lines, hoses or components that will interfere with manifold removal.
10. Remove the intake manifold bolts in the opposite order of the tightening sequence and then remove the manifold. There is a good likelihood you will need to pry the manifold off the block; be very careful that you don't scratch or mar the mating surfaces on the block, manifold or heads.
11. Remove the lower gaskets and throw them away.

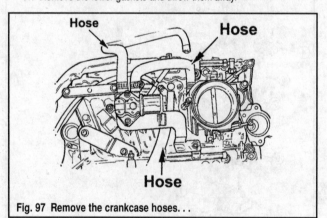

Fig. 97 Remove the crankcase hoses...

■ If simply removing the manifold assembly, please skip to the Installation procedures, otherwise proceed with disassembly.

To disassemble:

12. Remove the 2 crankcase ventilation hoses from the fuel system hardware.
13. Remove the throttle cable bracket nuts - the stud at the rear of the manifold may loosen before the nut does.
14. Remove the IAC bracket nuts and lift off the IAC with the bracket.
15. Remove the ignition coil.
16. Remove the throttle body attaching studs and lift off the unit.
17. Remove the thermostat housing.
18. Remove the MAPT and discard the seal.
19. Remove the fuel rail studs and lift off the rail.
20. Remove the 6 upper manifold studs and separate the upper form the lower manifold. Remove the gasket and throw it away.
21. Remove the brackets on each side of the lower manifold.

To assemble:

22. Carefully remove any residual gasket material from the manifolds and clean them in solvent. Dry with compressed air.
23. Inspect the manifolds for cracks, wear or other obvious signs of damage.
24. Inspect the cooling system passages for restriction and check the bolt hole threads.
25. Install the brackets to the sides of the lower manifold.
26. Position the upper-to-lower manifold gasket into the groove on the upper manifold.
27. Position the upper manifold onto the lower. If reusing the fasteners, coat the stud threads with Loctite 242. Install them and tighten to 44 inch lbs. (5 Nm). Then go back and tighten them in a 2nd pass to 89 inch lbs. (10 Nm).

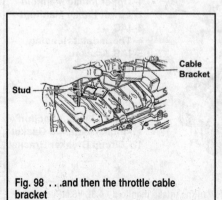

Fig. 98 ...and then the throttle cable bracket

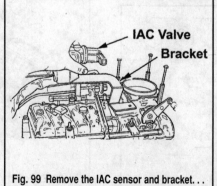

Fig. 99 Remove the IAC sensor and bracket...

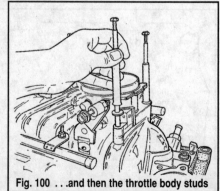

Fig. 100 ...and then the throttle body studs

ENGINE MECHANICAL - GM V6 AND V8 ENGINES

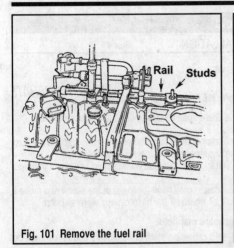

Fig. 101 Remove the fuel rail

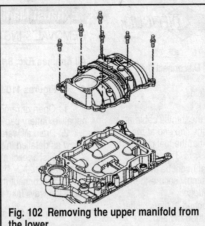

Fig. 102 Removing the upper manifold from the lower

Fig. 103 Always use a new seal on the MAPT

28. Install the fuel rail and injectors onto the manifold. Unless using new fasteners, coat the threads of the fuel rail bolt with Loctite 242. Install the studs and tighten the nuts to 27 inch lbs. (3 Nm).

29. Apply one drop of clean engine oil to a new MAPT sensor seal and install on the sensor. Install the sensor and tighten to 53 inch lbs. (6 Nm).

30. Install the coil and studs, tightening to 106 inch lbs. (12 Nm).

31. Install the thermostat housing with a new gasket.

32. Install a new sealing ring into the groove in the throttle body. Line up the dowels and position the throttle body on the manifold. Unless they are new, coat the studs with Loctite 242, install them and tighten to 80 inch lbs. (9 Nm).

33. Install the IAC bracket (with the valve) and tighten the nuts to 132 inch lbs. (15 Nm).

34. Install the throttle cable bracket and tighten the nuts to 168 inch lbs. (19 Nm). Make sure that the stud and nut are threaded correctly!

To install:

◆ See Figures 104, 105 and 106

■ If you separated the manifolds in the disassembly procedure, you can skip the next few cleaning and inspection steps since you've probably already preformed them.

35. Carefully remove all remaining gasket material from the manifold mating surfaces with a scraper or putty knife. Be careful that you don't accidentally drop any old gasket material into the crankcase or intake ports on the cylinder head.

36. Check for and clean any varnish buildup or contamination evident in any of the intake passages.

37. Inspect the manifold and all mating surfaces for any cracks or nicks.

38. Apply a 4mm (5/32 in.) bead of Ultra-Black Loctite 5900 adhesive at each end of the lower gasket, on the cylinder head side. Align the gaskets with the locator pins and position them on the cylinder heads. Make sure you do this quickly, while the sealer is still wet.

✳✳ CAUTION

Never apply too much adhesive. Do not get sealer into the oil sending unit hole.

39. Now you can apply a 5mm (13/64 in.) wide bead of the same adhesive to the front and rear of the engine block, extending the bead 13mm (1/2 in.) up onto the lower manifold gaskets on the heads.

40. Position the manifold assembly on the blocks and heads. Coat the threads of each bolt with Loctite 242, install them and tighten to 27 inch lbs. (3 Nm) in the order shown. Now tighten them again, in order, to 106 inch lbs. (12 Nm). Finally, make a 3rd pass, in order, to 132 inch lbs. (15 Nm).

41. Connect all ignition and electrical leads.

42. Connect the hoses to the thermostat housing and then connect any other hoses or lines that were removed.

43. Reconnect the fuel line.

44. Install the distributor in the position from which it was removed and tighten the clamp bolt to 18 ft. lbs. (25 Nm).

45. Install or connect any other components involved in the removal process.

46. Refill the cooling system and connect the battery cables.

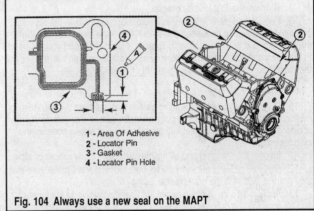

1 - Area Of Adhesive
2 - Locator Pin
3 - Gasket
4 - Locator Pin Hole

Fig. 104 Always use a new seal on the MAPT

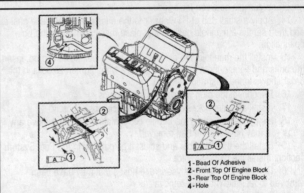

1 - Bead Of Adhesive
2 - Front Top Of Engine Block
3 - Rear Top Of Engine Block
4 - Hole

Fig. 105 Apply sealer to the fore and aft block surfaces - V6 shown, V8 identical

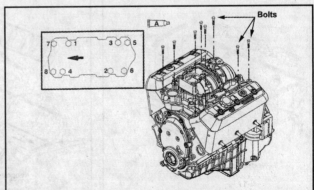

Fig. 106 Intake manifold tightening sequence - V6/V8 MPI engines (exc. 8.1L)

ENGINE MECHANICAL - GM V6 AND V8 ENGINES

8.1L V8 Engines

◆ See Figures 107, 108, 109 and 195

1. Open or remove the engine compartment hatch. Disconnect the negative battery cable.
2. Drain all water from the cylinder block and manifolds.
3. Remove the gear lube monitor.
4. Remove the flame arrestor and then disconnect the throttle cable at the throttle body. Disconnect the fuel line also and plug the line end at the throttle body. If you have a non-flexible line, disconnect it at the fuel pump also. Move both the cable and fuel line out of the way.
5. Tag and disconnect any lines, leads or hoses that might be in the way of removal. In some instances you may be able to simply secure them out of the way without disconnecting them - you be the judge.
6. If your engine has a spark plug wire retainer attached to the cylinder head cover, unclip the wires or remove the retainer and then tag and disconnect the plug wires at the spark plugs.
7. Remove the shift plate assembly.
8. Remove the PCM/Relay bracket.
9. Loosen and remove the 10 manifold mounting bolts in the reverse order of the tightening sequence and then remove the manifold (with the throttle body and fuel rail attached). There is a good likelihood you will need to pry the manifold off the block; be very careful that you don't scratch or mar the mating surfaces on the block, manifold or heads.

To install:

10. Carefully remove all remaining gasket material from the manifold mating surfaces with a scraper or putty knife. Be careful that you don't accidentally drop any old gasket into the crankcase or intake ports on the cylinder head.
11. Inspect the manifold and all mating surfaces for any cracks or nicks.
12. Manifold gaskets are identical for both sides; always make sure that they are installed with the marked side facing UP.
13. Apply a 3/16 in. (5mm) bead of 3M Brand adhesive to the forward and aft edges of the cylinder block mating surface and then position the 2 neoprene gaskets between the heads.
14. Apply a small dab of RTV sealer to the ends of the neoprene gaskets and then set the 2 manifold gaskets into place making sure that the bolt holes align.
15. Install the manifold into place so that all the bolt holes line up, insert the bolts and tighten them to 106 inch lbs. (12 Nm). Tighten all bolts in the order shown in the illustration.
16. Reconnect the plug wires to the spark plugs.
17. Install all water hoses and tighten their clamps securely.
18. Install the fuel line at the throttle body. Make sure you remove any plugs you may have inserted on removal.
19. Install the throttle cable and adjust it as detailed in the Fuel System section. Install the flame arrestor.
20. Connect all other lines, leads and hoses that may have been removed to facilitate manifold removal.
21. Connect the battery cable and start the engine. Check the ignition timing and idle speed. Check all hoses and seals for leaks.

Exhaust Manifold

REMOVAL & INSTALLATION

All Engines Exc. 8.1L V8

◆ See Figures 110 thru 116

1. Open or remove the engine compartment hatch. Disconnect the negative battery cable.
2. Drain all water and/or coolant from the engine, manifold and exhaust elbow as detailed in the Maintenance or Cooling System section.
3. Disconnect all water/coolant lines at the manifold and move them out of the way.
4. Disconnect the exhaust pipe hose (bellows) at the elbow and move it out of the way. Take note of where all the hose clamps were situated.

If removing the starboard manifold:

5. Disconnect the shift cables on stern drive models and move them out of the way.
6. Tag and disconnect the electrical lead at the shift cut-out switch on V6 and V8 Alpha models and then remove the shift plate assembly on the elbow.
7. On certain models, remove the water separating fuel filter bracket bolt (if equipped) and position it out of the way.

If removing the port manifold:

8. Disconnect the electrical leads from the ECU and then remove the ECU and bracket.
9. On certain models, remove the remote oil filter and bracket on engines so equipped.

On all manifolds:

10. Tag, disconnect or remove any other leads, lines hoses or components that may interfere with manifold removal.
11. Remove the exhaust elbow and exhaust riser (if equipped), if necessary. If working on a Dry Joint system, make sure that you keep the washers with their specific bolts.
12. Loosen the manifold retaining bolts from the center outward and then pry off the manifold/elbow assembly. There should be 6 bolts on the V6 engine and 4 bolts on the V8.

To install:

13. Carefully clean all residual gasket material from the head, manifold and elbow mating surfaces with a scraper or putty knife. Inspect all gasket surfaces for scratches, cuts or other imperfections.
14. Position a new gasket on the cylinder head and install the manifold; making sure that everything is aligned properly. Tighten all bolts/nuts until they are just tight and then torque them to 25 ft. lbs. (34 Nm). Starting in the center, and working your way out to the ends of the manifold.

■ **If working on a Dry Joint system, tighten the mounting bolts to 32 ft. lbs. (43 Nm).**

15. If you removed the manifold or elbow plugs for some reason, make sure that the threads are coated with Perfect Seal before screwing them back in.

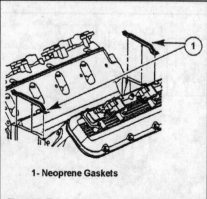

Fig. 107 The 8.1L uses 2 small neoprene gaskets. . .

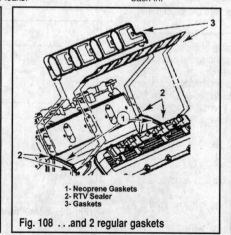

Fig. 108 . . .and 2 regular gaskets

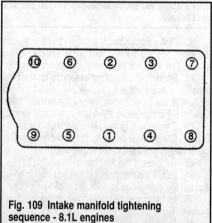

Fig. 109 Intake manifold tightening sequence - 8.1L engines

ENGINE MECHANICAL - GM V6 AND V8 ENGINES 4-29

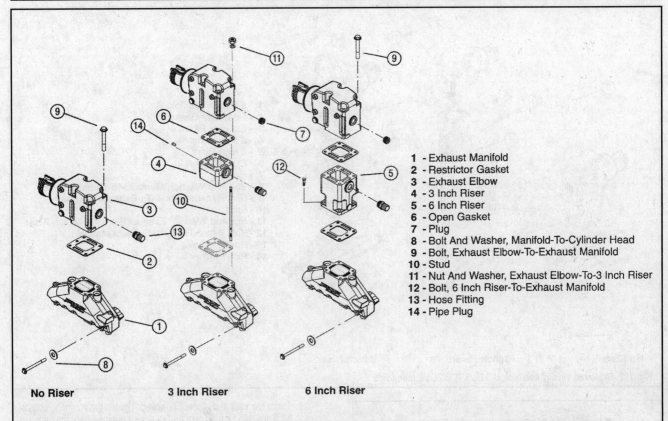

Fig. 110 Exploded view of the typical 4.3L V6 manifolds - seawater system. Models with mufflers will use an additional spacer (not shown) between the elbow and riser

1 - Exhaust Manifold
2 - Restrictor Gasket
3 - Exhaust Elbow
4 - 3 Inch Riser
5 - 6 Inch Riser
6 - Open Gasket
7 - Plug
8 - Bolt And Washer, Manifold-To-Cylinder Head
9 - Bolt, Exhaust Elbow-To-Exhaust Manifold
10 - Stud
11 - Nut And Washer, Exhaust Elbow-To-3 Inch Riser
12 - Bolt, 6 Inch Riser-To-Exhaust Manifold
13 - Hose Fitting
14 - Pipe Plug

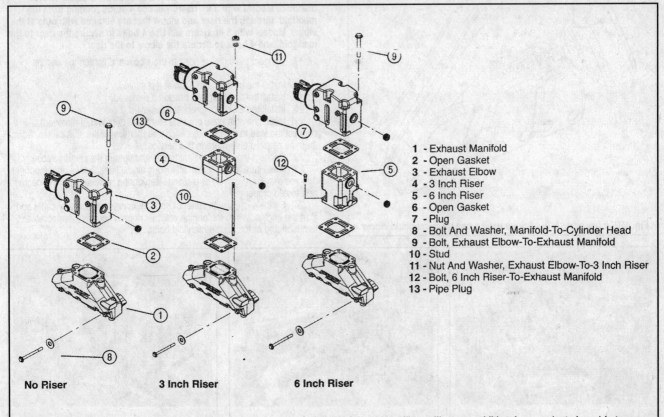

Fig. 111 Exploded view of the typical 4.3L V6 manifolds - closed system. Models with mufflers will use an additional spacer (not shown) between the elbow and riser

1 - Exhaust Manifold
2 - Open Gasket
3 - Exhaust Elbow
4 - 3 Inch Riser
5 - 6 Inch Riser
6 - Open Gasket
7 - Plug
8 - Bolt And Washer, Manifold-To-Cylinder Head
9 - Bolt, Exhaust Elbow-To-Exhaust Manifold
10 - Stud
11 - Nut And Washer, Exhaust Elbow-To-3 Inch Riser
12 - Bolt, 6 Inch Riser-To-Exhaust Manifold
13 - Pipe Plug

4-30 ENGINE MECHANICAL - GM V6 AND V8 ENGINES

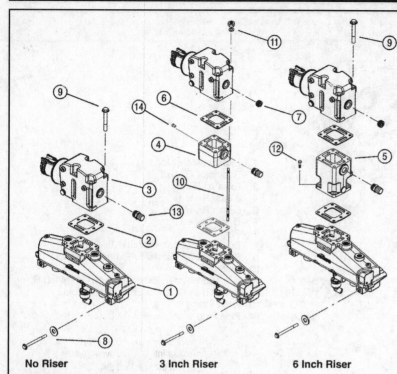

1 - Exhaust Manifold
2 - Restrictor Gasket
3 - Exhaust Elbow
4 - 3 Inch Riser
5 - 6 Inch Riser
6 - Open Gasket
7 - Plug
8 - Bolt And Washer, Manifold-To-Cylinder Head
9 - Bolt, Exhaust Elbow-To-Exhaust Manifold
10 - Stud
11 - Nut And Washer, Exhaust Elbow-To-3 Inch Riser
12 - Bolt, 6 Inch Riser-To-Exhaust Manifold
13 - Hose Fitting
14 - Pipe Plug

Fig. 112 Exploded view of the typical 5.0L/5.7L/6.2L V8 manifolds

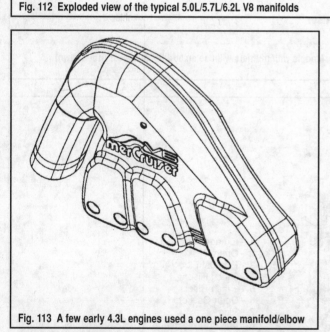

Fig. 113 A few early 4.3L engines used a one piece manifold/elbow

16. Install the riser and elbow if removed. Tighten the mounting bolts or nuts to 33 ft. lbs. (45 Nm), except on Dry Joint systems where they are tightened in two passes, first to 84 inch lbs. (9 Nm) and then to 45 ft. lbs. (61 Nm). Remember that on the Dry Joints, the washers and bolts are matched.

■ Models with no riser will use 4 bolts to secure the elbow to the manifold. Models with 3 in. risers will use 4 studs running from the manifold, through the riser and elbow that are secured with nuts at the elbow. Models with 6 in. risers will use 4 bolts to secure the riser to the manifold, and 4 bolts to secure the elbow to the riser.

17. Connect the exhaust hose to the elbow and tighten the clamps securely.
18. Install the oil filter and bracket if removed.
19. Install the fuel filter and bracket if removed.
20. Install the ECU and bracket if removed.
21. Install the shift plate onto the starboard manifold (if removed), connect the lead to the cut-out switch and then install the shift cables. Adjust them as detailed elsewhere in this service.
22. Connect the water/coolant hoses and tighten the clamps securely.
23. Make sure that any miscellaneous leads, lines, hoses or components that you may have moved or disconnected during removal are reconnected and routed properly.
24. Fill the system with water or coolant, connect the battery cable and start the engine. When the engine reaches normal operating temperature, turn it off and re-torque the manifold bolts.

Fig. 114 A good shot of a typical manifold

Fig. 115 Most models will have the shift control plate attached to the elbow...

Fig. 116 ...as well as the ECU and bracket

ENGINE MECHANICAL - GM V6 AND V8 ENGINES 4-31

8.1L V8 Engines

◆ See Figures 117 thru 120

■ Your engine may be equipped with one of two manifolds- aluminum or cast iron. Aluminum manifolds will have two 3/4 in. drain plugs, one on each end of the manifold. The cast iron manifold will have a single 3/8 in. drain plug and will also have an expansion plug located in a casting boss on the upper edge near the riser.

1. Open or remove the engine compartment hatch. Disconnect the negative battery cable.
2. Drain all water and/or coolant from the engine, manifold and exhaust elbow as detailed in the Maintenance or Cooling System sections.
3. Disconnect all water/coolant lines at the manifold and move them out of the way.
4. Disconnect the exhaust hose (bellows) from the elbow. Take note of where all the hose clamps were situated.
5. Remove the elbow and/or riser if necessary.
6. Remove the water rail if equipped (and necessary) - you may wish to do this anyway for easier access to the manifold bolts. There are 2 sizes of bolts, so keep track of where they came from. Pull out the O-rings and throw them away.

■ Some models will use studs and nuts on the water rail.

7. Loosen the manifold retaining bolts from the center outward and then carefully pry off the manifold. There are two sizes of bolts, so keep track of where they came from.
8. Loosen the stud nuts or cap screws and remove the elbow or elbow/riser if necessary.

To install:

9. Carefully clean all residual gasket material from the head, manifold and elbow mating surfaces with a scraper or putty knife. Inspect all gasket surfaces for scratches, cuts or other imperfections.
10. Position a new gasket on the cylinder head and install the manifold; making sure that everything is aligned properly. Coat the threads of the nuts with Loctite 271 and then tighten them until they are just tight. Now tighten them to 26 ft. lbs. (35 Nm) starting in the center and working your way out to the ends of the manifold.
11. Position new O-rings into the water rail and then install it. If yours is secured with bolts, coat the threads with Loctite 271 and tighten them to 7 ft. lbs. (10 Nm). If your engine used studs, coat the manifold-side stud threads with Loctite 271 and the nut-side threads with Loctite 242; tighten the studs to 7 ft. lbs. (10 Nm) and the nuts to 6 ft. lbs. (8 Nm)
12. To install an elbow without a riser:
    a. On manifolds with water rails, and aluminum manifolds without water rails, coat the threads of the elbow stud (manifold side) with Loctite 271 and then install them in the manifold, tightening to 15 ft. lbs. (20 Nm).
    b. On manifolds with water rails, aluminum manifolds without water rails and cast iron manifolds; install a new elbow gasket, turbulator, 2nd gasket and then install the elbow. Coat the stud threads with Loctite 242 and then tighten the nuts to 12 ft. lbs. (16 Nm) on aluminum manifolds; or tighten the cap screws on cast iron manifolds to 12 ft. lbs. (16 Nm).
13. To install an elbow with a riser:
    a. Coat the coarse threads of the elbow and riser studs with Loctite 271 and install them into the manifold, tightening to 15 ft. lbs. (20 Nm).
    b. Position the 2 elbow gaskets and the turbulator over the studs and onto the manifold. Now slide the riser into position over the studs.
    c. Position the gasket on top of the riser and then the elbow. Tighten the stud nuts to 12 ft. lbs. (16 Nm).
14. Connect the water/coolant hoses and tighten the clamps securely.

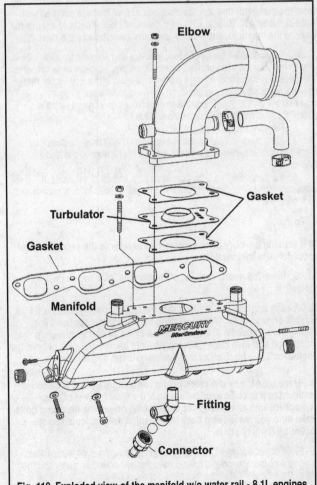

Fig. 117 Exploded view of the manifold w/water rail - 8.1L engines

Fig. 118 Exploded view of the manifold w/o water rail - 8.1L engines

4-32 ENGINE MECHANICAL - GM V6 AND V8 ENGINES

15. Make sure that any miscellaneous lines or hoses that you may have moved or disconnected during removal are reconnected and routed properly.
16. Fill the system with water or coolant, connect the battery cable and start the engine. When the engine reaches normal operating temperature, turn it off and re-torque the manifold bolts.

Exhaust Elbow/Riser

REMOVAL & INSTALLATION

All Engines Exc. Dry Joint Systems & 8.1L

◆ See Figures 110, 111, 112, 114, 115, 116 and 121

1. Drain the cooling system.
2. Loosen the 2 hose clamps and slide off the exhaust hose (bellows). Grasp it with both hands and pull it off while wiggling it from side to side. If it sticks, drip a little bit of soapy water around the lip.
3. If removing the starboard elbow, disconnect the shift cables on stern drive models and move them out of the way. Tag and disconnect the electrical lead at the shift cut-out switch on V6 and V8 Alpha models and then remove the entire shift plate assembly on the elbow.
4. If removing the port elbow, disconnect the electrical leads from the ECU and then remove the ECU and bracket. On certain models, the remote oil filter and bracket may be attached to one or two of the elbow bolts - be sure that you carefully support the filter assembly when removing the elbow bolts (you may want to tie it off). Additionally, certain engines may also have the gear lube monitor bracket attached to the elbow as well - if so, pay attention to how you support it as well.

■ **Models with no riser will use 4 bolts (cap screws) to secure the elbow to the manifold. Models with 3 in. risers will use 4 studs running from the manifold, through the riser and elbow, that are secured with nuts at the elbow. Models with 6 in. risers will use 4 bolts to secure the riser to the manifold, and 4 bolts to secure the elbow to the riser.**

5. Remove the 4 bolts/nuts, lock washers and washers from the elbow and lift it off the manifold or riser. Use an X pattern, from corner to corner as the loosening sequence. A little friendly persuasion with a soft rubber mallet may be necessary! Be careful though, no need to take out all your aggressions on the poor thing. In some instances on engines with 3 in. risers, the stud may come loose before the nut.
6. Remove the gasket(s) and discard it.
7. Early models V6 engines that use a riser and have mufflers may have a spacer in between the elbow and the riser - remove this and the gasket.
8. If equipped with a 3 in. riser, slide it up and off of the studs. If equipped with a 6 in. riser, remove the 4 mounting bolts in an X pattern and lift it off of the manifold.

To install:

■ **If working on a Dry Joint system, please refer to the separate set of procedures in this section.**

9. Clean the mating surfaces of the manifold, riser and elbow thoroughly.

■ **On early engines it is unlikely that the replacement gaskets will look like the original ones. When installing these gaskets, apply a 1/8 in. (3mm) bead of Loctite 510 to both sides of the gasket as shown in the illustration and then install the components immediately. Do not run the engine for a few hours after installation until the sealant sets up.**

■ **Please read the next 4 steps before continuing, as they are each pertinent to a specific application. Also, if a filter bracket of other component was attached to the elbow using one of the mounting bolts, make sure you move them back into position prior to installing the mounting bolts or studs.**

10. If not equipped with a riser, position a new gasket on the manifold, making sure that the indents line up (on seawater cooled engines) and then install the elbow. Tighten the mounting bolts to 33 ft. lbs. (45 Nm) in an X pattern, corner to corner.

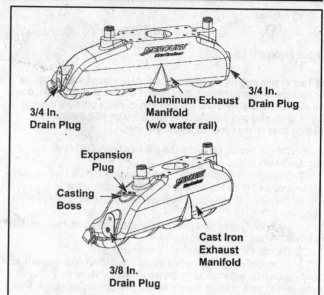

Fig. 119 Your engine may use an aluminum or a cast iron manifold - 8.1L engines

Fig. 120 A good shot of a typical 8.1L manifold (inboard)

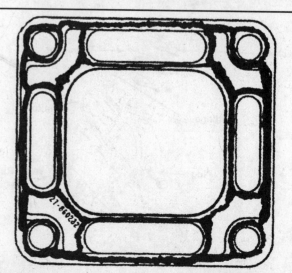

Fig. 121 Use Loctite on early engines with different replacement gaskets

ENGINE MECHANICAL - GM V6 AND V8 ENGINES 4-33

11. If using a 3 in. riser, slide it over the studs, install another new gasket and then install the elbow. Tighten the nuts to 33 ft. lbs. (45 Nm) in an X pattern, corner to corner.

12. If using a 6 in. riser, position a new gasket on the manifold, install the riser and tighten the bolts to 33 ft. lbs. (45 Nm) in an X pattern, corner to corner. Now position another new gasket and then install the elbow. Tighten the bolts to 33 ft. lbs. (45 Nm) in an X pattern, corner to corner.

13. Coat the inner lip of the exhaust hose with a little soapy water and then slide it onto the elbow while wiggling it back and forth. Tighten the clamps securely and make sure they are positioned correctly in their slots.

14. Install or connect all leads, hoses or components that were involved during the removal procedures.

15. Fill the system with water and check the shift cable adjustment. Start the engine and check for leaks or overheating.

8.1L Engines

MODERATE

◆ See Figures 117, 118, 120 and 122

1. Drain the cooling system.
2. Loosen the 2 hose clamps and slide off the exhaust hose (bellows). Grasp it with both hands and pull it off while wiggling it from side to side. If it sticks, drip a little bit of soapy water around the lip.
3. If necessary, disconnect the shift cables on stern drive models and move them out of the way. Remove the entire shift plate assembly on the elbow if necessary.
4. If necessary, disconnect the electrical leads from the ECU and then remove the ECU and bracket. On certain models, the remote oil filter and bracket may be attached to one or two of the elbow bolts - be sure that you carefully support the filter assembly when removing the elbow bolts (you may want to tie it off). Additionally, certain engines may also have the gear lube monitor bracket attached to the elbow as well - if so, pay attention to how you support it as well.

■ 8.1L engines use studs and nuts, regardless of the configuration.

5. Remove the 4 bolts (cast iron manifold) or nuts (all others), lock washers and washers from the elbow and lift it off the manifold or riser. Use an X pattern, from corner to corner as the loosening sequence. A little friendly persuasion with a soft rubber mallet may be necessary! Be careful though, no need to take out all your aggressions on the poor thing. In some instances on engines with studs, risers, the stud may come loose before the nut.
6. Remove the gasket(s) and discard it.
7. If equipped with a riser, slide it up and off of the studs. Remove the studs if needed.

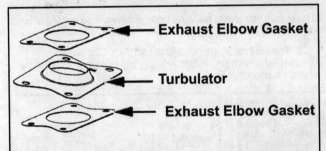
Fig. 122 Correct gasket and turbulator positioning - 8.1L engines

To install:

■ If working on a Dry Joint system, please refer to the separate set of procedures in this section.

8. Clean the mating surfaces of the manifold, riser and elbow thoroughly with a scraper or putty knife. Inspect all gasket surfaces for scratches, cuts or other imperfections.
9. To install an elbow without a riser:
   a. On manifolds with water rails, and aluminum manifolds without water rails, coat the threads of the elbow stud (manifold side) with Loctite 271 and then install them in the manifold, tightening to 15 ft. lbs. (20 Nm).
   b. On manifolds with water rails, aluminum manifolds without water rails and cast iron manifolds; install a new elbow gasket, turbulator, 2nd gasket and then install the elbow. Coat the stud threads with Loctite 242 and then tighten the nuts to 12 ft. lbs. (16 Nm) on aluminum manifolds; or tighten the cap screws on cast iron manifolds to 12 ft. lbs. (16 Nm).
10. To install an elbow with a riser:
    a. Coat the coarse threads of the elbow and riser studs with Loctite 271 and install them into the manifold, tightening to 15 ft. lbs. (20 Nm).
    b. Position the 2 elbow gaskets and the turbulator over the studs and onto the manifold. Now slide the riser into position over the studs.
    c. Position the gasket on top of the riser and then the elbow. Tighten the stud nuts to 12 ft. lbs. (16 Nm).
11. Coat the inner lip of the exhaust hose with a little soapy water and then slide it onto the elbow while wiggling it back and forth. Tighten the clamps securely and make sure they are positioned correctly in their slots.
12. Make sure that any miscellaneous lines or hoses that you may have moved or disconnected during removal are reconnected and routed properly.
13. Fill the system with water or coolant, connect the battery cable and start the engine. When the engine reaches normal operating temperature, turn it off and re-torque the bolts/nuts.
14. Check the shift cable adjustment. Check for leaks or overheating.

Dry Joint Systems

MODERATE

◆ See Figures 123, 124 and 125

■ Although removal procedures for these elbows is pretty much the same as for other engines, the installation procedures differ enough that we thought it best to separate them from the other engines.

1. If removed, coat the threads of the pipe fitting and plug on the elbow with Loctite 567, install them and tighten to 37 ft. lbs. (50 Nm).
2. Apply a little Perfect Seal to the edge of each water port on the manifold and the elbow - there are 2 on each component.
3. Position the restrictor gasket on the manifold so the side marked **UP** is, you guessed it. . .up. On seawater cooled models, make sure that the restricted water port on the gasket (the side with the little hole) is on the same side as the hose fitting on the elbow.
4. Position the elbow on the manifold. Coat the threads of the mounting bolts with Perfect Seal and install them finger-tight. Remember that the washers are matched to their bolts, so don't mix them up. Carefully line up the elbow with the exhaust hose and tighten the bolts, in an X pattern, to 84 inch lbs. (9 Nm). Now do a 2nd pass at 45 ft. lbs. (61 Nm).

■ The 7 1/2 in, 9 in. and 10 1/2 in. bolts all require that a specific washer be used in conjunction with the bolt. These are only available through Mercury's Parts/Accessories guide.

5. From here on out, the procedures are the same as for other engines; install and connect everything involved with the removal and check for leaks.

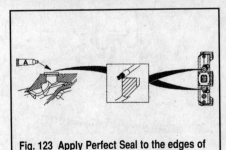
Fig. 123 Apply Perfect Seal to the edges of all 4 water ports on Dry Joint systems

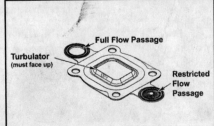
Fig. 124 With Dry Joint systems, seawater cooled engines use a restrictor gasket

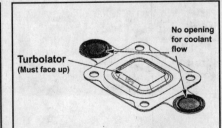
Fig. 125 With Dry Joint systems, closed cooling engines use a block off gasket

4-34 ENGINE MECHANICAL - GM V6 AND V8 ENGINES

Exhaust Hoses (Bellows) & Intermediate Exhaust Pipe

REMOVAL & INSTALLATION

 MODERATE

◆ See Figures 126 and 127

1. Starting with the upper hose, loosen all 4 hose clamps, two on top of the hose and two on the bottom.
2. Drizzle a soapy water solution over the top of the hose where it mates with the exhaust elbow and let it sit for a minute.
3. Grasp the hose with both hands and wiggle it side-to-side while pulling down on it until it separates from the elbow.
4. Now wiggle it while pulling upwards until it pops off the intermediate exhaust pipe.
5. The lower hose should be removed in the same manner as the upper.
6. Check the hose for wear, cracks and deterioration.
7. Coat the inside of the lower end of the lower hose with soapy water and wiggle it into position on the Y-pipe (lower). Remember to install the 2 clamps before sliding it over the end of the pipe.
8. Slide the 2 clamps over the upper end of the lower hose and then coat the inside with the soapy water solution and insert the bottom of the intermediate pipe into it fully until it seats on the step. Tighten the clamp screws securely.
9. Coat the inside of the lower end of the upper hose with soapy water and wiggle it into position on the intermediate pipe until it meets the step. Remember to install the 2 clamps before sliding it over the end of the pipe.
10. Slide 2 clamps over the upper end, lubricate the inside with soapy water and wiggle the hose over the elbow.
11. Tighten all 4 clamp screws securely.

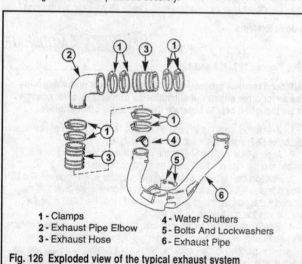

1 - Clamps
2 - Exhaust Pipe Elbow
3 - Exhaust Hose
4 - Water Shutters
5 - Bolts And Lockwashers
6 - Exhaust Pipe

Fig. 126 Exploded view of the typical exhaust system

Lower Exhaust Pipe (Y-Pipe)

REMOVAL & INSTALLATION

 DIFFICULT

◆ See Figures 126, 128 and 129

1. It is unlikely you will be able to get the pipe off without removing the engine, so remove the engine as previously detailed.
2. Remove the intermediate pipe and 2 exhaust hoses.
3. Loosen the retaining bolts at the transom shield and then remove the exhaust pipe. Carefully scrape any remnants of the seal from the pipe and transom mounting surfaces.
4. Coat a new seal with 3M Rubber Adhesive and position it into the groove on the transom shield mating surface.
5. Coat the mounting bolts with Gasket Sealing Adhesive. Position the exhaust pipe, insert the bolts and tighten them to 20-25 ft. lbs. (27-34 Nm), except on the 8.1L which is 25 ft. lbs. (34 Nm).
6. Install the intermediate pipe and hoses.
7. Install the engine.

Exhaust Valve (Flapper/Shutter)

REPLACEMENT

Thru The Propeller Exhaust Systems

 MODERATE

◆ See Figures 126 and 130

■ Although Mercury calls this an exhaust valve, many people also call it a water shutter, flapper valve or a shutter valve. Whatever you call it, this is the small valve in the top of the exhaust pipe.

1. Remove the upper exhaust hose from the elbow and intermediate pipe. Now remove the intermediate pipe and the lower hose. The shutter is located in the upper end of the lower exhaust tube (Y pipe).
2. The valve is held in place by means of a pin running through 2 grommets in the sides of the pipe. Position a small punch over one end of the pin and carefully press the pin out of the pipe. Some people have had luck simply pulling upward on the assembly very carefully and popping the shutter and grommets right out of the pipe.

■ Make sure you secure the valve while removing it so nothing falls down into the exhaust pipe.

3. Press out the 2 grommets and discard them. Coat two new grommets with Scotch Grip Rubber Adhesive and press them back into the sides of the pipe.

Fig. 127 A good look at the exhaust hoses and intermediate pipe

Fig. 128 A good look at the lower exhaust pipe

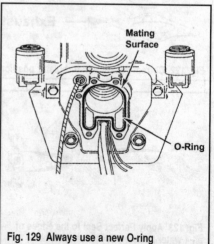

Fig. 129 Always use a new O-ring

ENGINE MECHANICAL - GM V6 AND V8 ENGINES

4. Position the new valve into the pipe with the long side DOWN. When the valve is in place, coat the pin lightly with engine oil and slowly slide it through one of the grommets, through the 2 retaining holes on the valve and then through the opposite grommet. Make sure the pin ends are flush with the sides of the pipe on both sides.

5. Install the lower hose, intermediate pipe and upper hose.

Thru The Transom Exhaust Systems

◆ See Figure 131

1. Remove the exhaust hose from the elbow and the exhaust flange.
2. Loosen the clamp and pry out the shutter
3. Install new shutter and tighten the clamp securely.
4. Connect the exhaust hose and tighten the clamps securely.

Oil Pan

REMOVAL & INSTALLATION

All Engines Exc. 8.1L V8

◆ See Figures 132, 133 and 134

■ More times than not, this procedure will require the removal of the engine. Your boat and its unique engine installation will determine this, but the procedure is almost always easier with the engine removed from the boat.

1. Remove the engine as previously detailed in this section.
2. If you haven't already drained the engine oil, do it now. Make sure you have a container and lots of rags available. Disconnect the oil drain fitting and hose if equipped.

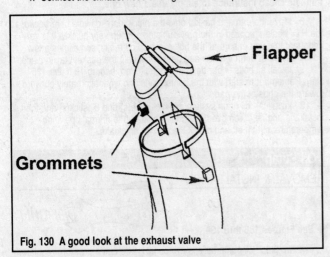

Fig. 130 A good look at the exhaust valve

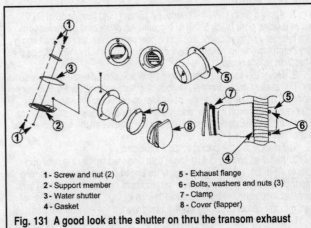

Fig. 131 A good look at the shutter on thru the transom exhaust systems

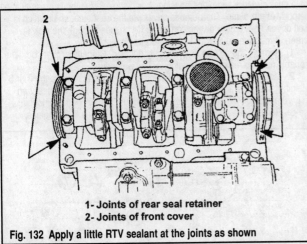

Fig. 132 Apply a little RTV sealant at the joints as shown

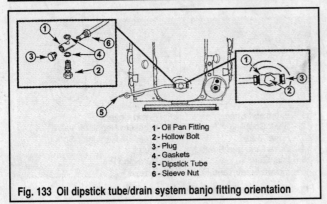

Fig. 133 Oil dipstick tube/drain system banjo fitting orientation

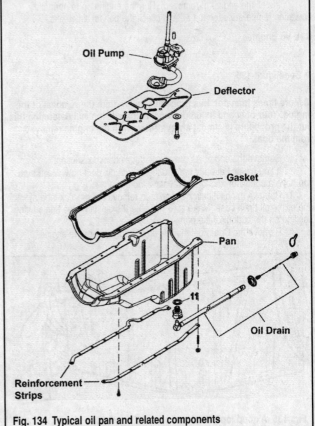

Fig. 134 Typical oil pan and related components

4-36 ENGINE MECHANICAL - GM V6 AND V8 ENGINES

3. Certain models may require removing the outlet hose at the seawater pump to provide easier access (particularly if you are attempting this without removing the engine).

4. Remove the oil dipstick and then remove the dipstick tube(s). On inboard engines or engines equipped with and oil drain system, make sure you note the dipstick tube banjo fitting position and which tube goes where.

5. Loosen and remove the oil pan retaining bolts and nuts, starting with the center bolts and working out toward the pan ends. Pry off the reinforcing strips from each side if equipped. Lightly tap the pan with a rubber mallet to break the seal and then lift it off the cylinder block. If your engine stand will allow for rotating the engine, you'll find that this will be easier with the pan facing up.

To install:

6. Clean the pan mating surfaces of any residual gasket material with a scraper or putty knife. Make sure that no old gasket material has been pressed into the retaining bolt holes in the pan, block or front cover. Clean the pan itself thoroughly with solvent.

7. Apply a 5mm (13/64 in.) wide, by 25mm (1 in.) long bead of RTV sealer (Loctite Ultra Black 5900) to the joints on either side of the rear oil seal retainer and front cover, position a new pan gasket onto the pan being very careful to line up all the holes - do not use RTV sealant with this gasket other than where noted.

8. Move the pan and gasket onto the block; don't dawdle here because the RTV sealant applied in the previous step sets up very quickly. It is very important that you ensure all the holes line up correctly; sometimes a few bolts inserted through the pan and gasket will help the gasket stay in place. Make sure that the gasket is pressed into the grooves of the front cover and rear oil seal housing.

9. Position the reinforcing strips if equipped.

10. Install all bolts and nuts finger tight and then tighten the nuts and bolts to 18 ft. lbs. (25 Nm) on the early V6 engines without pan reinforcing strips. On all other engines, tighten the stud nuts to 18 ft. lbs. (25 Nm) and the other reinforcement bolts/nuts to 106 inch lbs. (12 Nm). Remember to start with the center bolts and work outward toward the ends of the pan.

11. Install the dipstick tube(s) and dipstick. If equipped with an oil drain, install the plug bolt into the pan fitting and tighten it securely. Do the same with the sleeve nut on the line. Position the banjo correctly and tighten the hollow bolt to 180 inch lbs. (20 Nm).

12. Install the engine (if removed). Run the engine up to normal operating temperature, shut it off and check the pan for any leaks.

8.1L V8 Engines

◆ See Figure 135

■ More times than not, this procedure will require the removal of the engine. Your boat and its unique engine installation will determine this, but the procedure is almost always easier with the engine removed from the boat.

1. Remove the engine as previously detailed in this section.
2. If you haven't already drained the engine oil, do it now. Make sure you have a container and lots of rags available.
3. Remove the oil dipstick(s) and then remove the dipstick tube(s) and/or the oil drain hose. Make sure you note the dipstick tube banjo fitting position on the bottom of the pan.
4. Remove the Cool Fuel module on the port side of the engine.

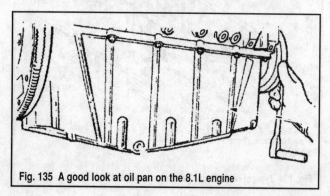

Fig. 135 A good look at oil pan on the 8.1L engine

5. Loosen and remove the oil pan retaining bolts and nuts, starting with the center bolts (alternating side to side) and working out toward the pan ends. Lightly tap the pan with a rubber mallet to break the seal and then lift it off the cylinder block. If your engine stand will allow for rotating the engine, you'll find that this will be easier with the pan facing up.

To install:

6. Clean the pan mating surfaces of any residual gasket material with a scraper or putty knife. Make sure that no old gasket material has been pressed into the retaining bolt holes in the pan, block or front cover. Clean the pan itself thoroughly with solvent.

7. Apply a small dab of RTV sealer to the joints on either side of the rear oil seal retainer and front cover, position a new pan gasket onto the pan being very careful to line up all the holes - do not use RTV sealant with this gasket other than where noted.

■ One-piece gaskets can be re-used if they are in good shape, but we think its cheap insurance to simply use a new one.

8. Move the pan and gasket onto the block; don't dawdle here because the RTV sealant applied in the previous step sets up very quickly. It is very important that you ensure all the holes line up correctly; sometimes a few bolts inserted through the pan and gasket will help the gasket stay in place.

9. Install all bolts finger tight and then tighten them to 19 ft. lbs. (25 Nm). Remember to start with the center bolts and work alternately outward toward the ends of the pan.

10. Install the oil drain system, dipstick tube(s) and dipstick(s) and then install the engine (if removed). Run the engine up to normal operating temperature, shut it off and check the pan for any leaks.

Oil Pump

REMOVAL & INSTALLATION

◆ See Figures 136 thru 139

The two-piece oil pump utilizes two pump gears and a pressure regulator valve enclosed in a two-piece housing. A baffled pick-up tube is press-fit into the body of the pump. The pump is driven via the distributor shaft which is itself driven from a gear on the camshaft. Oil passes through the pick-up screen, through the pump and then through the oil filter.

1. Remove the oil pan as previously detailed. Remember that you probably need to remove the engine for this procedure.
2. Most engines utilize an oil deflector (baffle) plate; remove the 3 or 5 retaining nuts/bolts and lift out the baffle (this should not be necessary on 8.1L V8 and 2001 V6 engines; it's there, you just don't need to remove it).
3. Remove the spacers under the baffle on the early 6.2L engines.

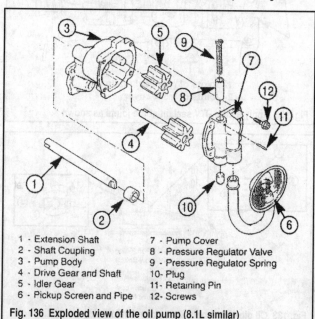

1 - Extension Shaft
2 - Shaft Coupling
3 - Pump Body
4 - Drive Gear and Shaft
5 - Idler Gear
6 - Pickup Screen and Pipe
7 - Pump Cover
8 - Pressure Regulator Valve
9 - Pressure Regulator Spring
10 - Plug
11 - Retaining Pin
12 - Screws

Fig. 136 Exploded view of the oil pump (8.1L similar)

ENGINE MECHANICAL - GM V6 AND V8 ENGINES 4-37

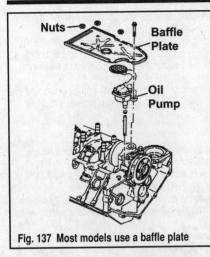

Fig. 137 Most models use a baffle plate

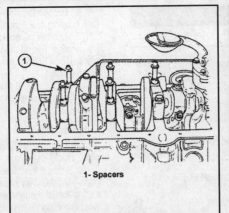

Fig. 138 There are spacers under the baffle plate on some 6.2L engines

Fig. 139 Oil pump and pick-up screen properly installed on a typical V8

4. Loosen and remove the pump mounting bolt(s) and lift off the pump assembly.

5. Check that the pump and block mating surfaces are clean. Check that the pump locator dowel pins are 0.25 in. (6.4mm) on all engines but 2001 carbureted/TBI engines.

6. Position the pump over the block so that the pump extension shaft is aligned with the distributor drive shaft. Do not use a gasket or RTV sealant.

7. Tighten the pump mounting bolts to:

- 56 ft. lbs. (75 Nm) - 8.1L engines
- 65 ft. lbs. (88 Nm) - 2001 V6 and V8 engines
- 15 ft. lbs. (20 Nm) and then an additional 65° - 2002 and later V6 and V8 engines (exc. 8.1L)

8. Install the oil baffle if used, and tighten the nuts to 30 ft. lbs. (40 Nm) on V6 and 2002 and later V8 engines; or 25 ft. lbs. (34 Nm) on early V8 engines. Use Loctite 271 for the spacers on the 6.2L engine.

9. Install the oil pan and engine. Fill the crankcase with oil, start the engine and check for leaks.

Oil Filter Bypass Valve

REMOVAL & INSTALLATION

◆ See Figures 140 and 141

All engines covered here utilize a bypass valve. After removing the oil filter or block adapter, check the spring and small fiber valve for proper operation. Any signs of incorrect operation, or wear and deterioration will necessitate replacement.

1. Drain the oil as detailed in the Maintenance section.

2. Remove the oil filter on engines without a remote filter. On engines with a remote filter, trace the lines from the filter back to the oil filter adapter on the cylinder block, remove them and then the adaptor. Remove the block adapter

3. Using a small prybar, remove the valve.

4. Install a new valve and press it in by placing a 9/16 in. deep socket over it and tapping the socket lightly with a hammer.

5. Install the oil filter on non-remote engines.

6. On remote filter engines, install the block adapter into the cylinder block and tighten it to 20 ft. lbs. (27 Nm). Install a new seal in the oil filter adapter and lubricate the sealing surfaces before installing it. Install the oil line fitting with a new quad seal and tighten it to 37 ft. lbs. (50 Nm) on all engines but the 8.1L where it should be 28 ft. lbs. (38 Nm). Install the oil lines and tighten the fittings to 20 ft. lbs. (27 Nm).

7. Refill the engine with the appropriate oil. Start the engine and check for leaks.

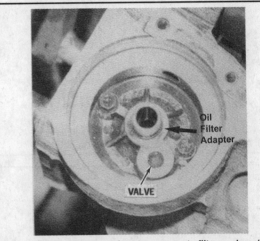

Fig. 140 Oil filter bypass valve on non-remote filter engines (remote filter similar)

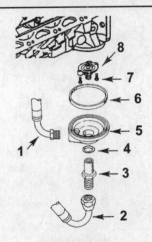

1 - Filter Adapter Oil Line
2 - Fitting/Bushing Oil Line
3 - Fitting/Bushing
4 - Quad Ring Seal
5 - Oil Filter Adapter
6 - Seal
7 - Bolt
8 - Block Adapter

Fig. 141 Exploded view of the adapters on engines with remote oil filters

4-38 ENGINE MECHANICAL - GM V6 AND V8 ENGINES

Engine Coupler/Drive Plate & Flywheel

REMOVAL & INSTALLATION

◆ See Figures 142 thru 145 and 191

1. Remove the engine from the boat as detailed previously in this section. Remove the transmission on inboards.
2. Although not strictly necessary, we recommend removing the starter.
3. On many models, you will need to disconnect the hoses and remove the power steering cooler.
4. Remove the flywheel housing cover bolts (usually 2) and lift off the cover.
5. Tag and disconnect the ground lead at the housing.

6. Loosen the retaining bolts/nuts/studs and lift off the flywheel housing (bell housing in automotive parlance).
7. Loosen the engine coupler (drive plate on inboards) mounting bolts (usually 3 or 6) and then remove the coupler. Don't lose the rubber bumpers pressed into the inner side of the coupler; there are 3 of these as well on Alpha models.
8. Loosen the 6 flywheel mounting bolts gradually and as you would the lug nuts on your car or truck - that is, in a diagonal star pattern. Lift off the flywheel.

To install:

9. Thoroughly clean the flywheel mating surface and check it for any nicks, cracks or gouges. Check for any broken teeth.
10. Inspect the guide dowels on the V6 and all V8s except the 8.1L; and make sure that they protrude 0.5 in. (13 mm).
11. Install the flywheel over the dowel on the crankshaft and tighten the mounting bolts to 75 ft. lbs. (100 Nm). Once again use the star pattern while tightening the bolts.
12. Attach a dial indicator to the engine block and take readings around the outer edge of the flywheel; pushing in on the flywheel to remove any crankshaft end play. Maximum run-out should not exceed 0.008 in. (0.203mm) or the flywheel will require replacement.
13. Insert the rubber bumpers into the coupler (Alpha only). Position the engine coupler/drive plate and tighten the mounting bolts to 35 ft. lbs. (48 Nm); or in the case of the 8.1L, 75 ft. lbs. (100 Nm). If you are re-using the old bolts, make sure you coat them with Loctite® 271. Lubricate the coupler splines with Quicksilver 2-4-C Marine Lubricant.
14. Install the flywheel housing and tighten the bolts to 30 ft. lbs. (41 Nm). Tighten the cover bolts to 80 inch lbs. (9 Nm).
15. Install the transmission if removed.
16. Install the engine.

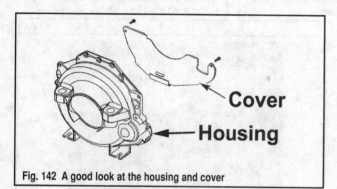

Fig. 142 A good look at the housing and cover

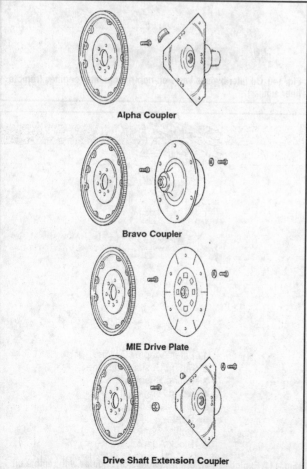

Fig. 143 Couplers are slightly different on each engine family

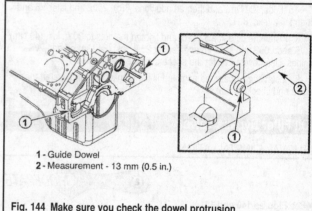

1 - Guide Dowel
2 - Measurement - 13 mm (0.5 in.)

Fig. 144 Make sure you check the dowel protrusion

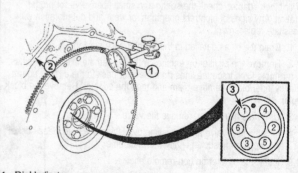

1 - Dial Indicator
2 - Push Flywheel and Crankshaft Forward as Far as It Will Go When Taking Reading
3 - Torque Sequence

Fig. 145 Tighten the flywheel bolts in sequence and then check the run-out

ENGINE MECHANICAL - GM V6 AND V8 ENGINES

Crankshaft Pulley/Torsional Damper

REMOVAL & INSTALLATION

2001 V6/V8 Engines & All 8.1L V8 Engines

◆ See Figures 146, 184 and 188

■ The torsional damper is also commonly called the crankshaft balancer.

1. Remove the serpentine belt as detailed in the Engine & Drive Maintenance section.
2. Remove the bolts and pull off the drive pulley.

■ You may have to remove the idler pulley on certain engines before removing the crankshaft pulley.

3. Remove the damper retaining bolt and install special tool (#J-23523-E) onto the damper. Tighten the tool press bolt and remove the damper; don't lose the crankshaft key.

■ MerCruiser suggests that you do not use a conventional gear puller for this procedure.

To install:

4. Inspect the crank key and then install it into the shaft using a little Loctite Ultra Black 5900.
5. Coat the seal surface with clean engine oil and then position the damper so the keyway lines up with the key.
6. Thread the rod from the installer tool into the end of the crankshaft - make sure it is in at least 1/2 in. (13mm) to prevent damage to the threads.
7. Install the plate, thrust bearing, washer and nut onto the rod and then tighten the nut slowly until the damper is fully seated. Remove the tool and its components.
8. Apply a little RTV sealant to the keyway.
9. Install the retaining bolt and tighten it to:
- 74 ft. lbs. (100 Nm) on V6 engines
- 60 ft. lbs. (81 Nm) on 5.0L/5.7L/6.2L V8 engines
- 188 ft. lbs. (255 Nm) on the 8.1L V8
10. Install the crankshaft drive pulley and tighten the bolts 35 ft. lbs. (48 Nm).
11. Install the serpentine belt and make sure that it is adjusted properly.

All 2002-08 Engines Exc. 8.1L V8

◆ See Figures 146, 147 and 188

■ The torsional damper is also commonly called the crankshaft balancer.

1. Remove the serpentine belt as detailed in the Maintenance section.
2. Remove the bolts and pull off the drive pulley.

■ You may have to remove the idler pulley on certain engines before removing the crankshaft pulley.

3. Remove the damper retaining bolt and install special tool (#J-23523-F) onto the damper. Tighten the tool press bolt and remove the damper; don't lose the crankshaft key.

■ MerCruiser suggests that you do not use a conventional gear puller for this procedure.

To install:

4. Inspect the crank key and then install it into the shaft using a little Loctite Ultra Black 5900.
5. Coat the seal surface with clean engine oil and then position the damper so the keyway lines up with the key.
6. Install the plate from the installer tool and tighten the bolts to 18 ft. lbs. (25 Nm). Thread the bolt into the end of the crankshaft and then install the bearing, washer and nut. Tighten the nut until it bottoms out and then remove the tool with all of its components.

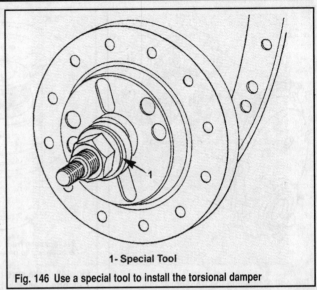

Fig. 146 Use a special tool to install the torsional damper

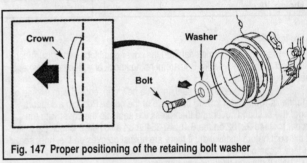

Fig. 147 Proper positioning of the retaining bolt washer

7. Install the crankshaft drive pulley and tighten the bolts 43 ft. lbs. (58 Nm).
8. Position the balancer washer onto the retaining bolt so that the crown is facing forward (toward the bolt head, not the engine). Install the retaining bolt and tighten it to 70 ft. lbs. (95 Nm).
9. Install the serpentine belt and make sure that it is adjusted properly.

Front Cover & Oil Seal

REMOVAL & INSTALLATION

◆ See Figures 148, 149, 150, 184 and 188

■ This procedure may require engine removal, depending upon your particular boat. If necessary, remove the engine as detailed previously in this section.

■ The front cover oil seal may also be removed without removing the front cover - simply pry it out (carefully!) with a small prybar and then install is with the special tool listed in the text.

1. Open the drain valves and drain the coolant from the block and exhaust manifolds.
2. Remove the serpentine belt.
3. Remove the water circulation pump.
4. Remove the heat exchanger and crossover on the 8.1L V8.
5. Remove the torsional damper.
6. Tag and disconnect the crankshaft position sensor lead on MPI engines and then remove the sensor from the cover. On 8.1L engines, this is the camshaft position sensor.
7. Although not completely necessary, we suggest that you remove the oil pan.
8. Loosen the mounting bolts (should be 6) and remove the front cover. If the oil seal needs replacement, press it out from the back (inner) side of the cover with a punch. Remove the front cover gasket.

4-40 ENGINE MECHANICAL - GM V6 AND V8 ENGINES

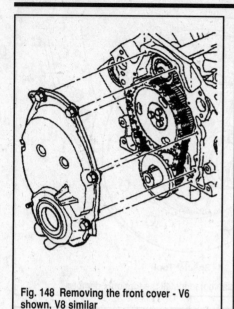

Fig. 148 Removing the front cover - V6 shown, V8 similar

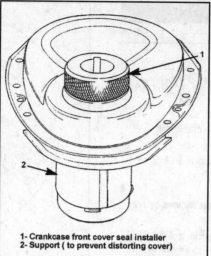

1- Crankcase front cover seal installer
2- Support (to prevent distorting cover)

Fig. 149 Make sure the front cover is supported securely before pressing in the oil seal

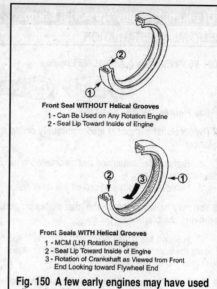

Front Seal WITHOUT Helical Grooves
1 - Can Be Used on Any Rotation Engine
2 - Seal Lip Toward Inside of Engine

Front Seals WITH Helical Grooves
1 - MCM (LH) Rotation Engines
2 - Seal Lip Toward Inside of Engine
3 - Rotation of Crankshaft as Viewed from Front End Looking toward Flywheel End

Fig. 150 A few early engines may have used oil seals without helical grooves

To install:

9. Clean all gasket material from the cover and block mating surfaces with a scraper of putty knife. Be careful not to knock any pieces of gasket in the timing assembly.

10. If you removed the oil seal, install a new one, coated with engine oil, with the lip (open end) toward the inside of the cover. Position a support under the seal and cover and then press the seal into the cover with the proper tool (#J-22102 on the 8.1L, #J-35468 on all others). Check the inside of the seal before installation; if there are helical grooves on the inner seal surface it can only be used on left hand rotation engines, if the inner surface is smooth it may be used on any engine.

11. Coat both sides of a new gasket with Perfect Seal and then position the gasket onto the engine. Install the cover so that all the bolt holes line up; there are dowel pins on the cylinder block that will help alignment. Tighten the bolts to 100 inch lbs. (11 Nm) on 2001 V6 engines, or 106 inch lbs. (12 Nm) on all other engines.

12. Install the oil pan and torsional damper.

13. Install the crankshaft or camshaft position sensor (MPI) with a new O-ring and reconnect the electrical lead. Tighten the sensor to 80 inch lbs. (9 Nm)

14. Install the heat exchanger and crossover (8.1L)

15. Install the crankshaft pulley and pull the belt back on. Check the tension adjustment.

16. Install the water circulation pump and connect the hose.

17. Install the engine if removed. Add oil and water/coolant, start the engine and check for any leaks.

Timing Chain & Sprockets/Gears

REMOVAL & INSTALLATION

◆ See Figures 151, 152 and 188

1. Remove the crankshaft pulley and torsional damper.
2. Remove the oil pan.
3. Remove the front cover. On MPI engines, make sure you remove the crankshaft (or camshaft) position sensor reluctor ring.
4. Look carefully at the camshaft and crankshaft sprockets/gears - you should notice a small indent on the front edge of one of the teeth on each sprocket/gear. Bump the engine over until these 2 marks are in alignment as shown in the illustration, a remote starter will work or you can screw the damper bolt back into the crankshaft and turn it.

■ If for some reason or another you cannot find the timing marks, ensure that the engine is at TDC on the No.1 cylinder and then mark the sprockets yourself.

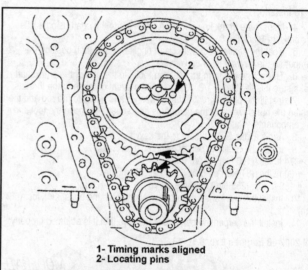

1- Timing marks aligned
2- Locating pins

Fig. 151 The two marks on each of the timing sprockets must be in alignment before removing or installing the timing chain

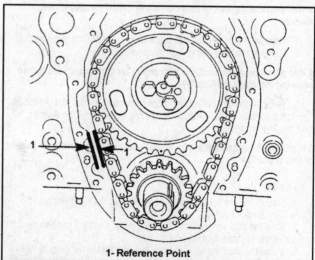

1- Reference Point

Fig. 152 When checking timing chain deflection, find a reference point on the cylinder block and use it for each measurement

ENGINE MECHANICAL - GM V6 AND V8 ENGINES

5. Dab a little paint across one of the chain links and the camshaft sprocket. Loosen the camshaft sprocket retaining bolts (3), grasp the sprocket on each side with the chain still attached and wiggle it off the shaft. It should come off readily, but if not, tap the bottom edge lightly with a rubber mallet.

6. Mount a gear puller (#J-5825-A) over the crankshaft pulley and pull it off the shaft. If you have a rh rotation engine, you can get away with a universal puller.

To install:

7. Clean the chain and sprockets/gears in solvent and let them air dry. Check the chain for wear and damage, making sure there are no loose or cracked links. Check the sprockets or gears for cracked or worn teeth.

8. Install the crankshaft sprocket/gear onto the shaft with an installation tool (#J5590).

9. Position the timing chain onto the camshaft sprocket so that the paint marks made during removal match up. If they do, and you haven't moved the engine, the timing marks on the two sprockets should also. Hold the sprocket/chain in both hands so the chain is hanging down, engage the chain around the crankshaft sprocket and then slide the cam sprocket/chain onto the camshaft. Do not force it! Tighten the 3 mounting bolts to 22 ft. lbs. (30 Nm) on 8.1L V8 engines, or 18 ft. lbs. (24 Nm) on all other engines.

10. Rotate the camshaft slightly so that it creates tension on one side of the timing chain (either side is OK). Find a reference point on the same side of the cylinder block as the side that the timing chain is tight on and then measure from this point to the outer edge of the chain.

11. Rotate the camshaft in the opposite direction until the other side of the chain is tight. Press the inner side of the chain outward until it stops and then measure from your reference point on the cylinder block (obviously, do this from the same side of the chain as you did in the previous step) to the outer edge of the chain. This is timing chain deflection and it should be no more than 3/4 in. (19mm) on 2001 engines and all 8.1L engines, or 0.4331 in. (11mm) on all other engines. If it is, replace the chain.

12. Install the front cover, the oil pan and the torsional damper.

Balance Shaft

REMOVAL & INSTALLATION

4.3L V6 Engines Only

◆ See Figures 153, 154 and 155

1. Remove the intake manifold.
2. Remove the front cover and timing chain.
3. Rotate the balance shaft until the marks on its driven gear and the camshaft drive gear are aligned.
4. Fashion a small wedge out of an old piece of wood and jam it in between the teeth of the balance shaft driven gear and the camshaft drive gear to hold the shafts from turning. Using a Torx socket (you are probably going to have to buy this one), remove the driven gear bolt.

5. Unscrew retaining stud from the drive gear on the camshaft and remove the gear itself.

6. Remove the 2 thrust plate mounting bolts from the balance shaft and lift off the plate. These are Torx fasteners also.

7. Insert a suitable prybar between the rear of the shaft and the cylinder block and carefully shimmy the shaft forward and out of the bearing housing in the rear of the block. Slide the entire shaft out through the front of the block. Be careful not to knock any debris into the inside of the cylinder block.

To install:

8. Clean the balance shaft in solvent and let it dry completely. Inspect the rear bearing for wear or damage. Inspect the front bearing for excessive side play on the shaft, wear or damage. Check the bearing bore in the front of the cylinder block for scoring.

9. Lubricate the front and rear bearing with motor oil and then carefully slide the shaft through the front bore until it feeds into the rear bearing. Tap the front edge of the shaft with a plastic mallet until the retaining ring on the front bearing seats against the cylinder block.

10. Install the thrust plate and tighten the 2 Torx fasteners to 120 inch lbs. (14 Nm).

11. Position the driven gear onto the balance shaft with the bolt finger tight. Make sure that the timing mark is at the bottom of the gear.

12. Install the camshaft drive gear so that the mark aligns with the one on the balance shaft driven gear and tighten the stud to 120 inch lbs. (16 Nm).

13. Remove the balance shaft bolt again and coat the threads with Loctite. Screw it in and tighten it to 15 ft. lbs. (20 Nm); and then turn the bolt an additional 35°.

14. Install the timing chain and front cover.
15. Install the intake manifold.

Camshaft & Bearings

CHECKING LIFT

◆ See Figures 156, 157 and 158

■ If the shaft is out of the engine, you can use a micrometer to take the heel-to-lobe measurement and the side-to-side measurement. Subtract the second measurement from the first to get lobe lift. Most times though, the shaft will still be in the cylinder block so perform the following procedure.

1. Tag and disconnect the electrical connectors at the ignition coil.
2. Remove the cylinder head cover and rocker arms as detailed previously.
3. Using a special adaptor (#J-8520-1), connect a dial indicator so that its tip is positioned on the end of the pushrod - the adaptor should screw onto the end of the rocker stud.

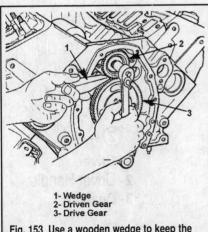

1- Wedge
2- Driven Gear
3- Drive Gear

Fig. 153 Use a wooden wedge to keep the balance and camshafts from rotating while loosening the retaining bolt.

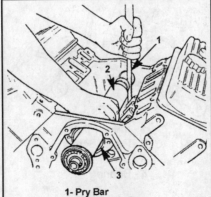

1- Pry Bar
2- Balance shaft
3- Bearing housing

Fig. 154 Carefully lever the balance shaft forward and out of the cylinder block

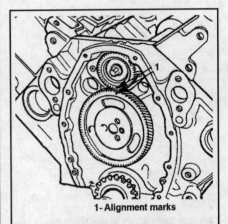

1- Alignment marks

Fig. 155 The timing marks on the balance shaft driven gear and the camshaft drive gear must be aligned on installation

4-42 ENGINE MECHANICAL - GM V6 AND V8 ENGINES

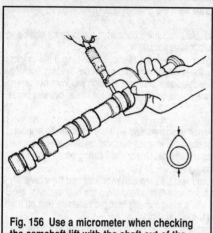

Fig. 156 Use a micrometer when checking the camshaft lift with the shaft out of the engine

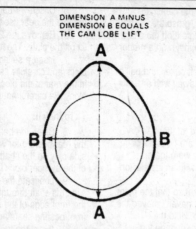

Fig. 157 Measure the camshaft lobe at these two points with the micrometer

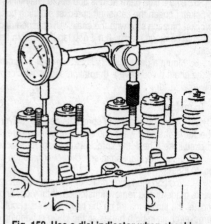

Fig. 158 Use a dial indicator when checking the camshaft lift with the shaft in the engine

4. Slowly rotate the crankshaft in the direction of engine rotation until the valve lifter is riding on the heel (back side of lobe) of the camshaft lobe. The pushrod should be at its lowest point when the lifter is on the heel.

■ A remote starter works well for turning the engine over in this situation.

5. Set the indicator to **0** and then rotate the engine until the pushrod is at the highest point of its travel. Camshaft lift should be as detailed in the Engine Specifications chart.
6. Continue rotating the engine until the pushrod is back at its lowest position - make sure that the indicator still reads 0.
7. Repeat this procedure for the remaining pushrods.
8. Install the rocker arms and adjust the valve clearance.
9. Install the cylinder head cover and reconnect the coil leads.

REMOVAL & INSTALLATION

◆ See Figures 159, 160 and 161

1. Remove the cylinder head covers and rocker assemblies as previously detailed in this section.
2. Remove the intake manifold.
3. Remove the pushrods and lifters.
4. Remove the front cover and timing chain/sprockets.
5. Remove the inner camshaft drive gear on V6 engines as detailed in the Balance Shaft section.

6. Remove the thrust plate/retainer. A few early 5.0L engines may not be equipped with the plate.
7. Thread 3 (2 on early models, just go with the number of holes in the end of the shaft) 5/16-18 x 5 in. bolts into the camshaft bolt holes and carefully pull the camshaft out of the cylinder block. You may have to wiggle it back and forth a bit, so be sure you don't lean it up or down or else you could damage the bearings.
8. If the camshaft bearings are to be removed, you will need to remove the flywheel as previously detailed. Although it is not necessary, removing the crankshaft will also facilitate bearing removal, make sure that you move the connecting rods out of the way so they do not interfere with bearing removal.
9. Working from inside the block, drive out the rear cam bearing expansion plug (welch plug).
10. Install the nut and thrust washer on the end of the camshaft bearing removal tool (#J-33049 - V6 or #J-6098-01 - 5.0L/5.7L/6.2L V8, #J-33049 - 8.1L V8).
11. Position the pilot in the first bearing (front of block) and install the puller screw through the pilot to the center bearing. Install the tool so the shoulder is toward the bearing, making sure there are a sufficient number of threads that have been engaged.
12. Hold the puller screw with one wrench and then turn the nut until the bearing comes free. Remove the tool and bearing from the screw.
13. Remove the remaining inner bearings in the same manner.
14. Position the pilot into the rear bearing and remove the rear intermediate bearing.
15. Now assemble the driver onto the driver handle and remove the front and rear bearings by driving them inward toward the center of the block.

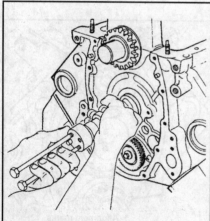

Fig. 159 Make sure you don't cant the camshaft during removal or installation. Note the 5 in. bolts screwed into the end of the shaft

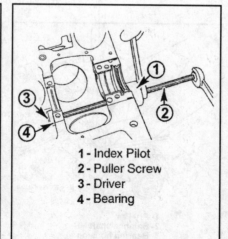

Fig. 160 Removing the center and intermediate bearings

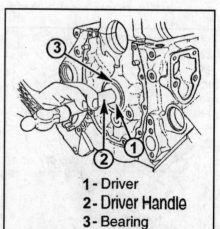

Fig. 161 Use the driver to press out the front and rear bearings

ENGINE MECHANICAL - GM V6 AND V8 ENGINES

To install:

16. If you removed the bearings, installation on these engines is essentially the reverse of removal. Take note of the following though:
- Install the front and rear bearings first, from the outside toward the inside.
- Bearing bore sizes may vary, be sure you have the correct bearings when replacing them.
- Make sure that the oil holes in the bearing align with the holes in the block, during and after installation. The front bearing must be positioned so the oil holes are an equal distance from the 6 o'clock position in the block. The intermediate and center bearings must have the oil holes at the 5 o'clock position, toward the left side of the block and at a position even with the bottom of the cylinder bore. The rear bearing must have the oil hole at the 12 o'clock position.
- Coat a new rear welch plug with sealant (Loctite 565 PST) and install it so that it is flush with the surface of the cylinder block, or no more than 1/32 in. (0.792mm) deep when measured from the outer edge of its recess on 2001 models and all 8.1L engines, or 0.347 in. (8.8mm) on all others.

17. Inspect the camshaft as detailed in the Engine Overhaul section.
18. Reinstall the long bolts and coat the journals with motor oil; coat the camshaft lobes with Johnson EP Lube or the equivalent. Carefully insert the shaft into the cylinder block and slide it all the way in.
19. Remove the installation bolts and install the thrust plate/retainer (if equipped) and tighten the bolts to 106 inch lbs. (12 Nm).
20. Install the drive gear, timing chain and front cover.
21. Drop the lifters back into their original bores so that they are aligned with the matchmarks and then install the restrictors.
22. Install the balance shaft on the V6.
23. Install the intake manifold.
24. Install the rocker assemblies and the cylinder heads covers.

Cylinder Head

REMOVAL & INSTALLATION

All Engines Exc. 8.1L V8

 DIFFICULT

◆ See Figures 162, 163, 164, 181, 182 and 186

1. Drain the water from the cylinder block and manifold.
2. Remove the fuel line support brackets. Disconnect the fuel line at the carburetor/throttle body and fuel pump, plug the fitting holes and remove the line.
3. Remove the intake and exhaust manifolds; you can leave the carburetor/throttle body attached to the intake manifold if you like.
4. Tag and disconnect the spark plug wires at the plugs; move them out of the way. Although not necessary, it's a good idea to remove the plugs themselves also.
5. Remove the cylinder head cover, rocker assemblies and pushrods.
6. Remove or relocate any components or connections that may interfere with the removal of an individual cylinder head.
7. Loosen the cylinder head bolts, in the opposite order of the tightening sequence and then carefully lift the head off the block. You may need to persuade it with a rubber mallet - be careful! Set the head down carefully; do not sit it on cement.

To install:

8. Carefully, and thoroughly, remove all residual head gasket material from the cylinder head and block mating surfaces with a scraper or putty knife. Check that the mating surfaces are free of any nicks or cracks. Make sure there is no dirt or old gasket material in any of the bolt holes. Refer to the Engine Overhaul section for complete details on inspection and refurbishing procedures.
9. Apply a THIN coating of Perfect Seal to both sides of a new ribbed stainless steel gasket and position the gasket over the cylinder block dowel pins. If your engine uses a graphite composition gasket, do not use any sealer. DO NOT use automotive-type steel gaskets.
10. Position the cylinder head over the dowels in the block. Coat the threads of the head bolts with Perfect Seal or Loctite 565 PST and install them finger tight. It never hurts to use new bolts, although it's not necessary. Tighten the bolts, a little at a time, in the sequence illustrated, until the proper tightening torque is achieved.
- Tighten all bolts, in sequence, to 22 ft. lbs. (30 Nm). Next, tighten the short bolts and additional 55°, the medium bolts an additional 65°, and the

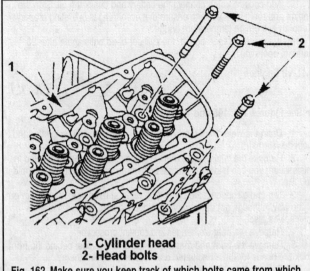

1- Cylinder head
2- Head bolts

Fig. 162 Make sure you keep track of which bolts came from which holes!

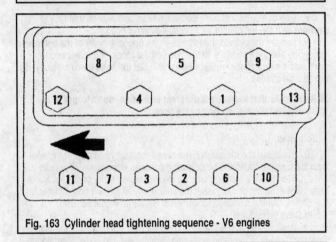

Fig. 163 Cylinder head tightening sequence - V6 engines

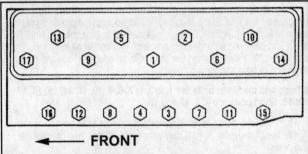

Fig. 164 Cylinder head tightening sequence - 5.0L/5.7L/6.2L V8 engines

long bolts and additional 75°. If an angle torque wrench is not available, try 26 ft. lbs. (35 Nm) on the 1st pass, 44 ft. lbs. (60 Nm) on a 2nd pass and then 66 ft. lbs. (90 Nm) on a 3rd pass.
- On the V6, short bolts are 2, 3, 6, 7, 10 and 11; medium bolts are 1 and 4; and long bolts are 5, 8, 9, 12 and 13
- On the V8, short bolts are 3, 4, 7, 8, 11, 12, 15 and 16; medium bolts are 14 and 17; and long bolts are 1, 2, 5, 6, 9, 10 and 13

11. Install rocker assemblies and the cylinder head cover.
12. Install the spark plugs if they were removed and then connect the plug wires.
13. Install the manifolds and connect the fuel line. Don't forget to remove the fitting plugs.
14. Install or connect any other components removed to facilitate getting the head off.

4-44 ENGINE MECHANICAL - GM V6 AND V8 ENGINES

15. Add coolant/water, connect the battery and check the oil. Start the engine and run it for a while to ensure that everything is operating properly. Keep an eye on the temperature gauge.

16. It never hurts to re-tighten the cylinder head bolts again after 20 hours of operation.

8.1L V8 Engines

◆ See Figures 162, 165 and 192

1. Drain the water from the cylinder block and manifold (seawater and closed systems).

2. Remove the intake and exhaust manifolds as previously detailed in this section; you can leave the throttle body attached to the intake manifold if you like.

3. Tag and disconnect the spark plug wires at the plugs; move them out of the way. Although not necessary, it's a good idea to remove the plugs themselves also.

4. Remove the heat exchanger and coolant crossover.

5. Remove the front and rear engine lifting hooks. Just behind the front hook are hoses for the air actuated drain system, remove them.

6. Remove the alternator bracket.

7. Remove the cylinder head cover and rocker assemblies as detailed previously in this section.

8. Remove or relocate any components or connections that may interfere with the removal of an individual cylinder head.

9. Loosen the cylinder head bolts, in the opposite order of the tightening sequence and then carefully lift the head off the block. You may need to persuade it with a rubber mallet - be careful! Set the head down carefully; do not sit it on cement.

■ Remember that there are 3 different size bolts, don't forget to keep track of which hole each bolt came from.

To install:

10. Carefully, and thoroughly, remove all residual head gasket material from the cylinder head and block mating surfaces with a scraper or putty knife. Check that the mating surfaces are free of any nicks or cracks. Make sure there is no dirt or old gasket material in any of the bolt holes. Refer to the Engine Overhaul section for complete details on inspection and refurbishing procedures.

11. Position the cylinder head and gasket over the dowels in the block. Coat the threads of the head bolts with Perfect Seal and install them finger tight. It never hurts to use new bolts, although it's not necessary. Tighten the bolts, a little at a time, in the sequence illustrated, until the proper tightening torque is achieved. Tighten all bolts, in sequence, to 22 ft. lbs. (30 Nm). Now take a 2nd pass with an angle torque wrench and tighten all bolts and additional 120°. Finally, tighten the short bolts an additional 30° and the long/medium bolts and additional 60°.

■ Long and mediums bolts are 1, 2, 3, 6, 7, 8, 9, 10, 11, 14, 15, 16, 17 and 18. Short bolts are 4, 5, 12 and 13.

12. Install the rocker assemblies and the cylinder head cover.

13. Install the spark plugs if they were removed and then connect the plug wires.

14. Install the manifolds and connect the fuel line. Don't forget to remove the fitting plugs.

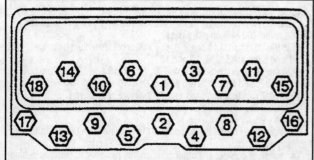

Fig. 165 Cylinder head tightening sequence - 8.1L V8 engines

15. Install or connect any other components removed to facilitate getting the head off.

16. Add coolant/water, connect the battery and check the oil. Start the engine and run it for a while to ensure that everything is operating properly. Keep an eye on the temperature gauge.

17. It never hurts to re-tighten the cylinder head bolts again after 20 hours of operation.

Rear Main Seal

REMOVAL & INSTALLATION

All Engines Exc. 8.1L

◆ See Figures 166, 167 and 168

■ This procedure had been applicable to 8.1L engines also, but sometime in 2006 they made a change and it is unclear whether that change was effective ongoing, or if it was across the board for earlier engines. Because of this, we have left references to this engine within the following procedures.

It is not necessary to remove the engine, oil pan or rear main bearing cap when removing the one-piece oil seal on these engines although you may find it easier to do just that.

1. Remove the flywheel housing and cover.
2. Remove the engine coupler and flywheel.
3. Insert a small prybar into one of the three slots in the edge of the seal retainer and slowly pry the seal out of the retainer.
4. Thoroughly clean the retainer surface.
5. Apply Perfect Seal to the inner mating surface of the seal retainer on 2001 engines and all 8.1L engines; use engine oil an all later engines. Also using engine oil, lightly lube the outer diameter of the flywheel pilot flange and locator pin.

■ The seal lip should always be facing the engine.

6. On 2001 engines, spread a small amount of grease around the outside edge of a new seal and position it over its slot in the retainer. Position a seal driver (#J-26817-A, or #J-38841 on the 8.1L) over the seal and strike it with a mallet until the seal is fully seated in its bore. Skip the next step.

■ The seal installer tool used on 2002 and later engines, should also work on the 2001 engines and on the 8.1L engines.

7. On 2002 and later engines (except 8.1L), coat the outer edge of a new seal with clean engine oil and then position it into the installer tool (#J-35621-B). Position the seal and installer over the rear of the crankshaft and tighten the tool bolts until they make contact. Turn the handle clockwise until the seal evenly seated and just about flush with the retainer. Turn the handle counterclockwise to release the seal and then remove the tool.

8. Wipe off any excess oil
9. Install the flywheel and engine coupler. Install the cover and flywheel housing.

8.1L Engines

◆ See Figures 169 thru 172

■ Sometime in 2006 Mercruiser made a change to the seal removal/installation procedures and it is unclear whether that change was effective just for ongoing engines, or if it was across the board for earlier engines as well. Because of this, we have left references to the 8.1L engine within the previous procedures. If you have an earlier model, you may choose to use the previous procedures, but the following should work as well.

It is not necessary to remove the engine, oil pan or rear main bearing cap when removing the one-piece oil seal on these engines although you may find it easier to do just that.

1. Remove the flywheel housing and cover.
2. Remove the engine coupler and flywheel.
3. Install the puller tool guide pins into the crankshaft as shown in the illustration and then install the tool over the pins.

ENGINE MECHANICAL - GM V6 AND V8 ENGINES

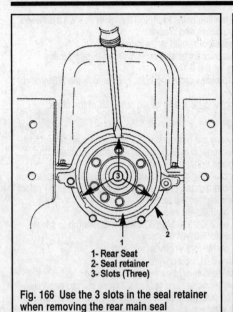

Fig. 166 Use the 3 slots in the seal retainer when removing the rear main seal

1- Rear Seat
2- Seal retainer
3- Slots (Three)

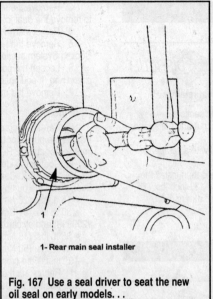

Fig. 167 Use a seal driver to seat the new oil seal on early models...

1- Rear main seal installer

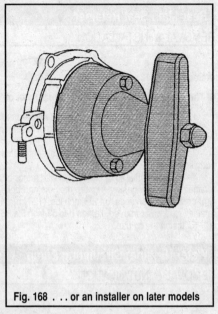

Fig. 168 ...or an installer on later models

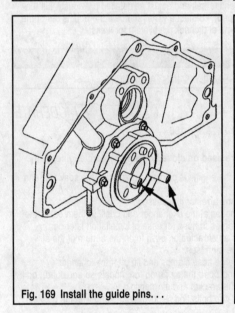

Fig. 169 Install the guide pins...

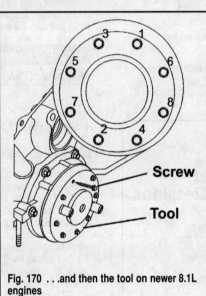

Fig. 170 ...and then the tool on newer 8.1L engines

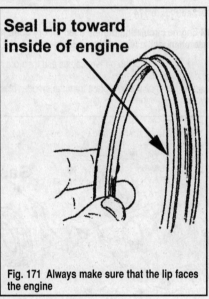

Fig. 171 Always make sure that the lip faces the engine

4. Screw in the 8 self-drilling sheet metal screws through the tool and into the seal itself using a criss-cross pattern.
5. Thread the center bolt through the tool and into the crankshaft end. Slowly tighten the bolt until the seal has pulled all of the way out of the retainer.
6. Remove the tool and guide pins.
7. Thoroughly clean the retainer surface.
8. Coat the crankshaft sealing surface with clean engine oil. Do not get any on the engine block sealing surface.
9. Position the new seal into the seal installer tool and then position them against the crankshaft so you can thread in the attaching screws.

■ The seal lip should always be facing the engine.

10. Tighten the screws securely so that the seal and tool are properly aligned on the crankshaft. Now tighten the center nut on the tool until the oil seal is fully seated.
11. Remove the tool.
12. Wipe off any excess oil
13. Install the flywheel and engine coupler. Install the cover and flywheel housing.

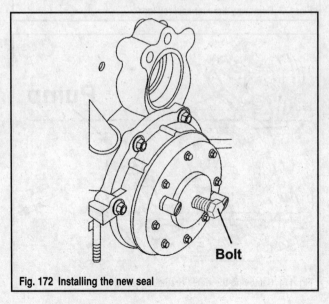

Fig. 172 Installing the new seal

4-46 ENGINE MECHANICAL - GM V6 AND V8 ENGINES

Rear Main Seal Retainer
REMOVAL & INSTALLATION

◆ See Figure 173

■ The 8.1L engine does not utilize a rear main oil seal retainer.

1. Remove the oil pan as previously detailed in this section.
2. Loosen and then remove the retainer fasteners (nut and bolts) and lift out the seal retainer.
3. Clean all mating surfaces of old gasket material with a scraper or putty knife.
4. Coat the inner lip of the oil seal with motor oil and then install the retainer with a new gasket. Tighten the nuts and bolts to 133 inch lbs. (15 Nm) on 2001 V6 engines, or 106 inch lbs. (12 Nm) all others. If the retainer stud was removed, tighten it to 53 inch lbs. (6 Nm).
5. Install the oil pan.

Water (Engine) Circulating Pump
REMOVAL & INSTALLATION

◆ See Figure 174

■ Engine circulating pumps used on Dry Joint models are not interchangeable with other models.

1. Disconnect the battery cables and then drain all water from the block and manifolds.
2. Drain all water/coolant from the cylinder block including the closed system (if equipped).

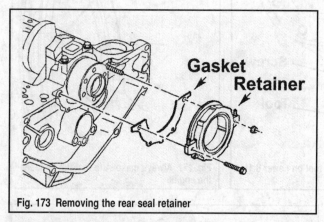

Fig. 173 Removing the rear seal retainer

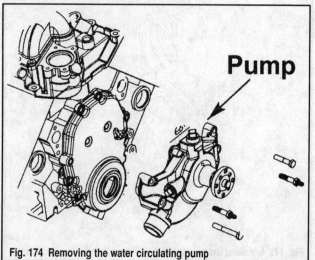

Fig. 174 Removing the water circulating pump

3. Remove the serpentine belt. On the 8.1L engine, you will also need to remove the dual idler pulleys and bracket.
4. Remove any hoses connected to the pump.
5. Remove the heat exchanger and crossover (as detailed in the Cooling System section) on the 8.1L.
6. Loosen the pump pulley mounting bolts and remove the pulley and clamp ring (if equipped).
7. Remove the mounting bolts and lift the pump off of the block.

To install:

8. Carefully scrape any old gasket material off both mounting surfaces. Inspect the pump for blockage, cracks or any other damage. Inspect the impeller for cracks. Replace either if necessary.
9. Coat both sides of a new gasket(s) with Perfect Seal and position on the cylinder block. Coat the threads of the pump mounting bolts with Perfect Seal, install the pump and tighten the bolts to:
   • 15 ft. lbs. (20 Nm) on 3.0L engines
   • 30 ft. lbs. (41 Nm) on 2001 350 Mag/6.2L MPI engines (up to serial #299999) and all carbureted/TBI V6V8 engines
   • 35 ft. lbs. (48 Nm) on MPI V8 engines (exc. 8.1L)
   • 37 ft. lbs. (50 Nm) on 8.1L V8 engines
10. Install the crossover and heat exchanger if removed.
11. Reconnect the water hoses and tighten the hose clamps securely.
12. Position the pump pulley and clamping ring on the boss. Screw the mounting bolts and lock washers in and tighten them securely.
13. Install the serpentine belt and adjust as detailed previously. Fill the closed system (if equipped), connect a water source if the boat is out of the water and start the engine and check the system for leaks.

Heat Exchanger
REMOVAL, DISASSEMBLY & INSTALLATION

◆ See Figures 175 thru 178 and 196

■ Heat exchangers are used on closed cooling system engines only.

1. Make sure that the engine is cold, and then drain the seawater and closed cooling systems.
2. Remove the flame arrestor if necessary.
3. Disconnect all hoses at the exchanger and position them out of the way. You may want to tag each hose for ease of installation later on.
4. On models with an air actuator, push in on the bottom of the air manifold and pull out the air hoses.
5. Remove the 2 large hose clamps and lift off the exchanger.
6. If you haven't unclipped the air pump (on models so equipped), do it now and then remove the bracket and air manifold.
7. Loosen the end cap bolts and remove them along with the gaskets and sealing washers.

To assemble and install:

8. Coat new end cap gaskets with Perfect Seal and then install the end caps with new sealing washers. Tighten the bolts to 36-72 inch lbs. (4-8 Nm) on carbureted and TBI engines; on MPI engines they should be tightened to 54 inch lbs. (6 Nm).

■ Make sure the O-ring is installed between the end cover and the gasket.

9. If equipped, install the air pump bracket and the air manifold, tightening the mounting bolts securely. Snap in the air pump.
10. Lower the exchanger into position on the bracket and connect all water/coolant hoses so that they are aligned and fully seated on the fittings. Tighten each of their hose clamps securely (without pinching the hose) and then install the 2 large clamps holding the exchanger to the bracket.

■ Certain newer models may utilize a rubber hose over the clamp - make sure this hose is between the clamp and the mounting bracket.

11. If your engine had the air actuator, connect the air lines to the manifold. Install the air pump over the fitting and pull the locking lever up. Fill the system with air until the green indicators extend fully - if they fail to extend, the lines are not connected properly.
12. Install the flame arrestor if removed.
13. Fill the closed system.

ENGINE MECHANICAL - GM V6 AND V8 ENGINES 4-47

Fig. 175 A good look at an exchanger on an early engine

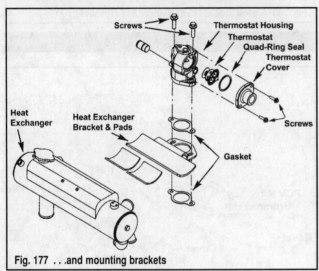

Fig. 177 ...and mounting brackets

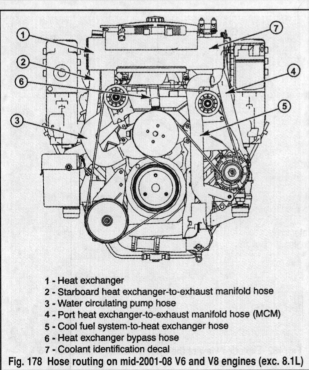

1 - Heat exchanger
2 - Starboard heat exchanger-to-exhaust manifold hose
3 - Water circulating pump hose
4 - Port heat exchanger-to-exhaust manifold hose (MCM)
5 - Cool fuel system-to-heat exchanger hose
6 - Heat exchanger bypass hose
7 - Coolant identification decal

Fig. 178 Hose routing on mid-2001-08 V6 and V8 engines (exc. 8.1L)

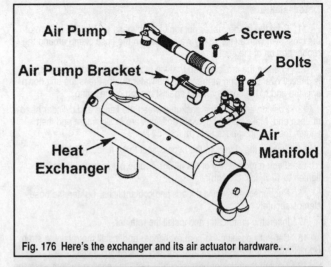

Fig. 176 Here's the exchanger and its air actuator hardware...

Crossover

REMOVAL & INSTALLATION

8.1L Engines Only

◆ See Figures 179, 180 and 196

1. Drain the cooling system.
2. Disconnect the coolant lines at the heat exchanger if you haven't already. Disconnect the hose leading to the overflow tank.
3. Remove the coolant reservoir.
4. Loosen the clamps on the water inlet seal and hose.
5. Loosen the heat exchanger mounting bolts and then loosen the hex bolt on the retaining strap.
6. Remove the retaining strap, take the mounting bolts out and then lift off the exchanger.
7. Remove the thermostat retainer and pull out the thermostat.
8. Tag and disconnect any electrical connections at the crossover.
9. Disconnect any remaining coolant lines at the crossover.
10. Remove the 2 mounting bolts on each side of the crossover at the cylinder heads and remove the crossover.

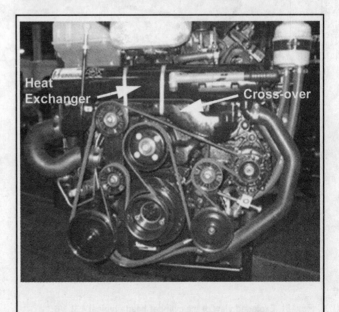

Fig. 179 The crossover is mounted underneath the heat exchanger

4-48 ENGINE MECHANICAL - GM V6 AND V8 ENGINES

To install:

11. Coat the inside rubber portion of the bypass seal on the bottom of the crossover with a little soapy water. Position the hose clamp around the rubber portion of the seal.

12. Position the crossover so that the rubber portion of the bypass seal is directly over the fitting at the top of the water pump. Press the unit down until the end of the seal is resting against the top of the pump.

13. Position a new flange gasket between the flange and cylinder head at each end. Make sure all gaskets and lines are in alignment and then tighten the mounting bolts to 37 ft. lbs. (50 Nm).

14. Tighten the clamp on the bypass seal to 16 ft. lbs. (21 Nm).

15. If you removed any of the fittings or pipe plugs, install them and tighten to 17 ft. lbs. (23 Nm).

16. Reconnect all electrical leads and coolant lines. Tighten the hose clamps securely.

17. Insert the thermostat and install the retainer.

18. Carefully position the heat exchanger and install the retaining strap. Do not tighten to tight or you could damage the holding slots.

19. Tighten the retainer bolts securely. Refill the cooling system, start the engine and check the system for leaks.

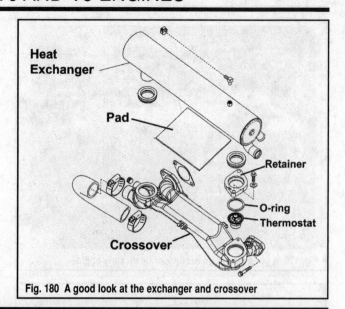

Fig. 180 A good look at the exchanger and crossover

EXPLODED VIEWS

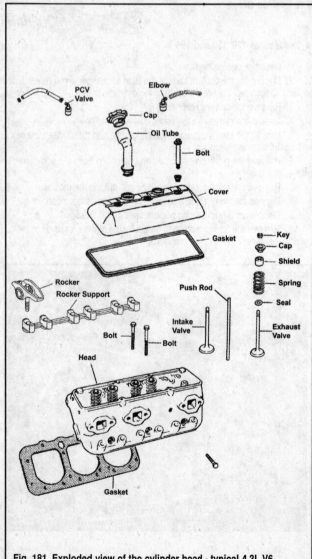

Fig. 181 Exploded view of the cylinder head - typical 4.3L V6 engine thru mid-2005

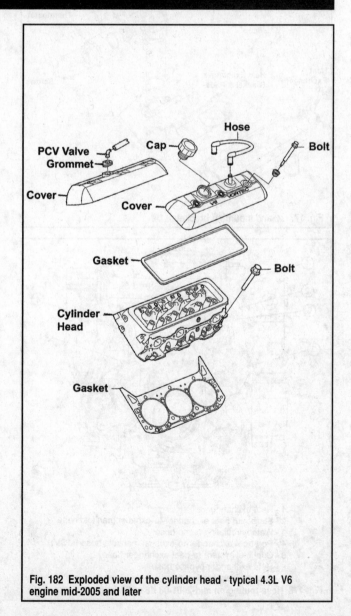

Fig. 182 Exploded view of the cylinder head - typical 4.3L V6 engine mid-2005 and later

ENGINE MECHANICAL - GM V6 AND V8 ENGINES 4-49

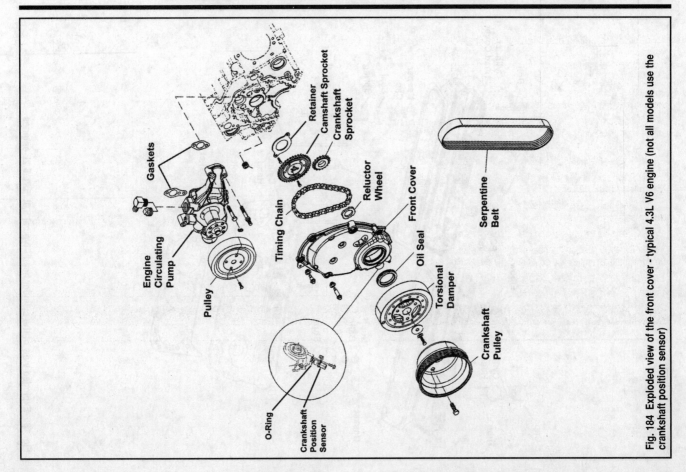

Fig. 184 Exploded view of the front cover - typical 4.3L V6 engine (not all models use the crankshaft position sensor)

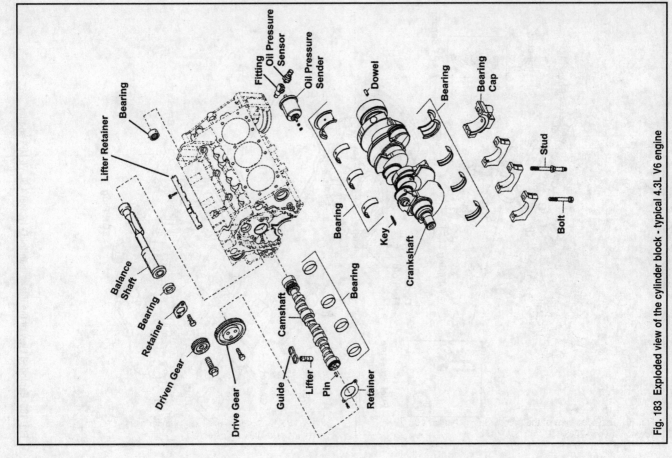

Fig. 183 Exploded view of the cylinder block - typical 4.3L V6 engine

4-50 ENGINE MECHANICAL - GM V6 AND V8 ENGINES

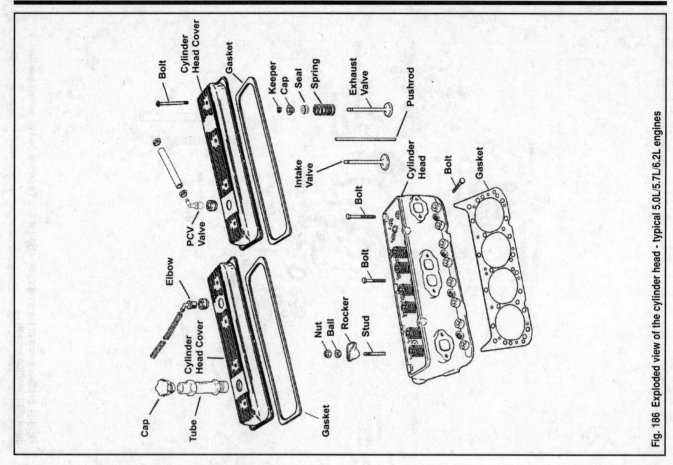

Fig. 186 Exploded view of the cylinder head - typical 5.0L/5.7L/6.2L engines

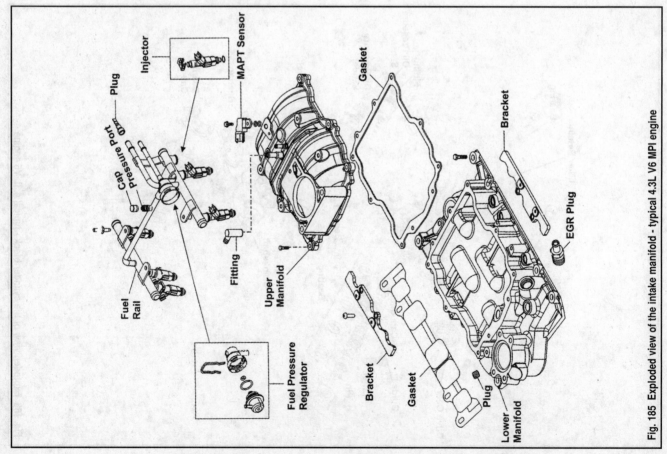

Fig. 185 Exploded view of the intake manifold - typical 4.3L V6 MPI engine

ENGINE MECHANICAL - GM V6 AND V8 ENGINES 4-51

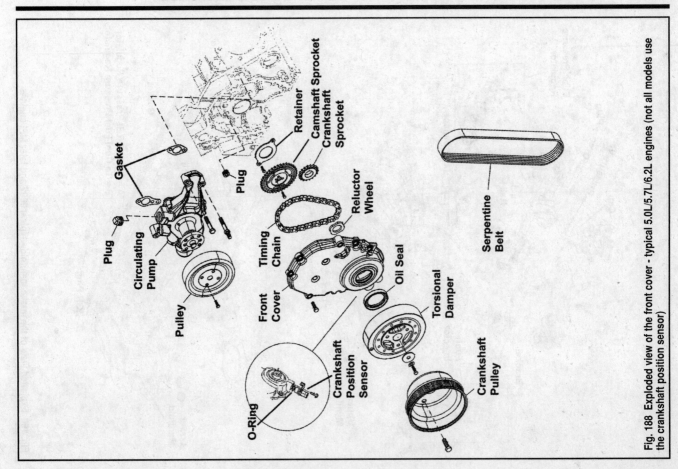

Fig. 188 Exploded view of the front cover - typical 5.0L/5.7L/6.2L engines (not all models use the crankshaft position sensor)

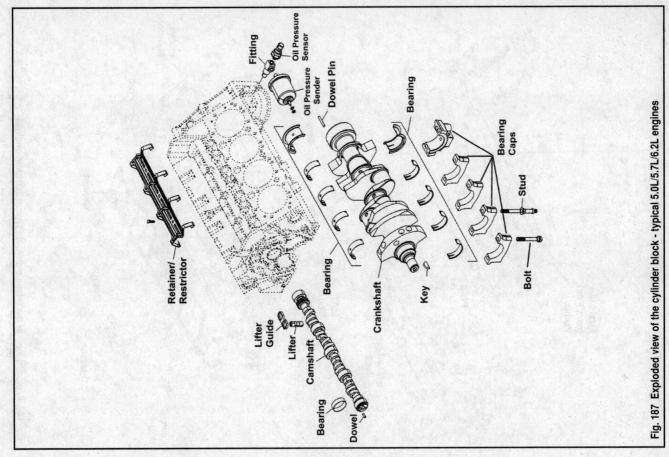

Fig. 187 Exploded view of the cylinder block - typical 5.0L/5.7L/6.2L engines

4-52 ENGINE MECHANICAL - GM V6 AND V8 ENGINES

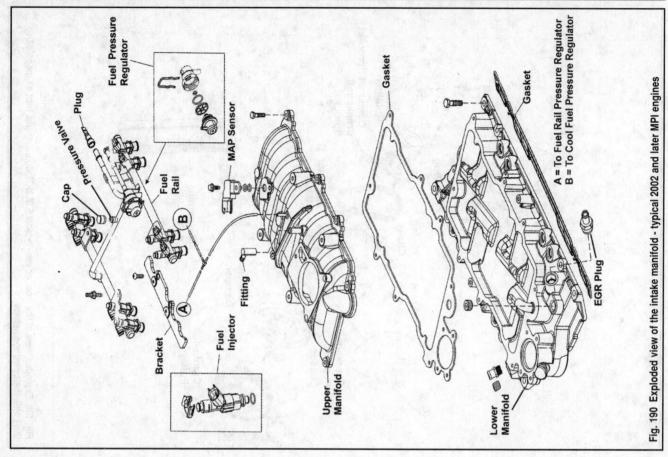

Fig. 190 Exploded view of the intake manifold - typical 2002 and later MPI engines

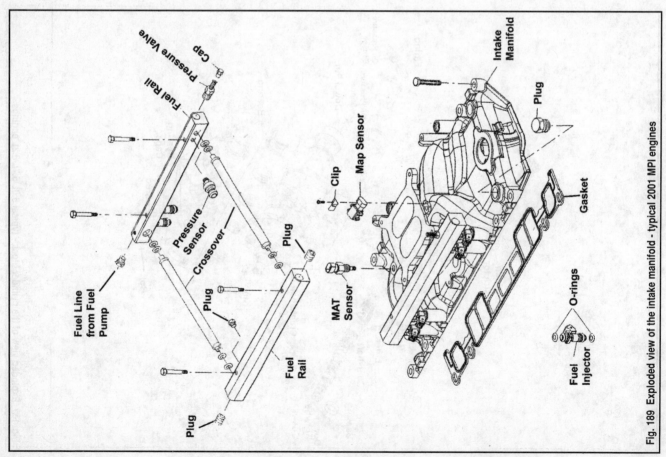

Fig. 189 Exploded view of the intake manifold - typical 2001 MPI engines

ENGINE MECHANICAL - GM V6 AND V8 ENGINES   4-53

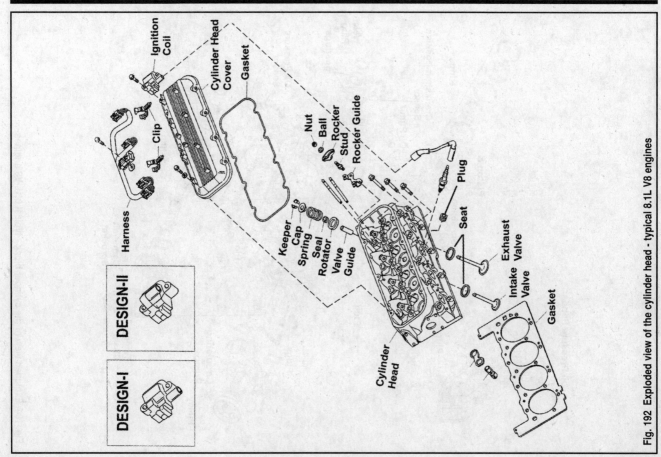

Fig. 192 Exploded view of the cylinder head - typical 8.1L V8 engines

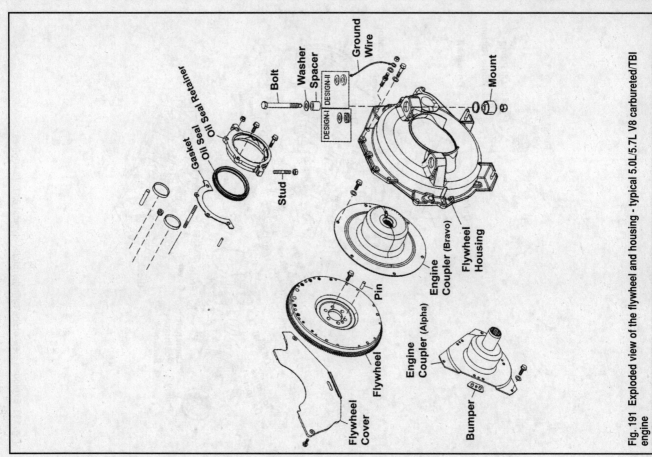

Fig. 191 Exploded view of the flywheel and housing - typical 5.0L/5.7L V8 carbureted/TBI engine

4-54 ENGINE MECHANICAL - GM V6 AND V8 ENGINES

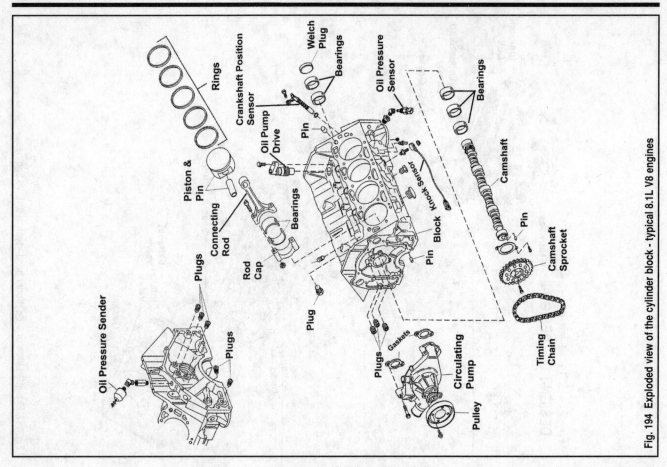

Fig. 194 Exploded view of the cylinder block - typical 8.1L V8 engines

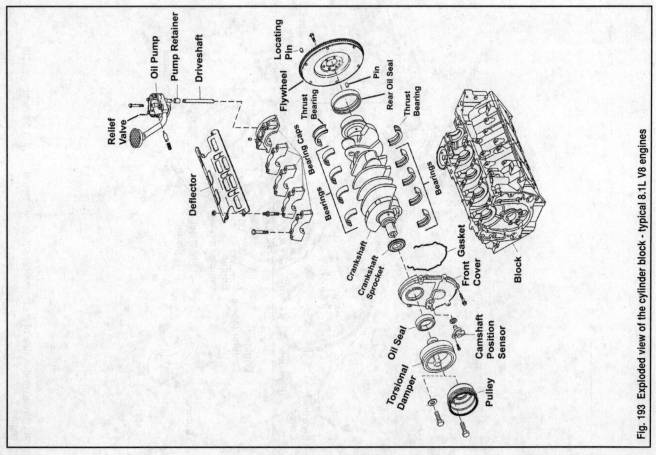

Fig. 193 Exploded view of the cylinder block - typical 8.1L V8 engines

ENGINE MECHANICAL - GM V6 AND V8 ENGINES

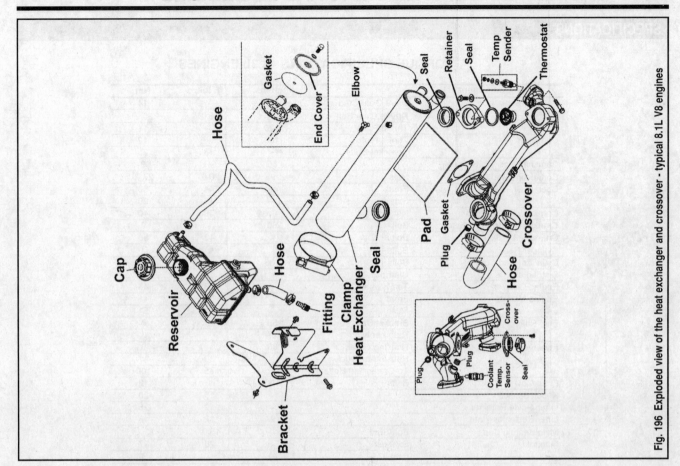

Fig. 196 Exploded view of the heat exchanger and crossover - typical 8.1L V8 engines

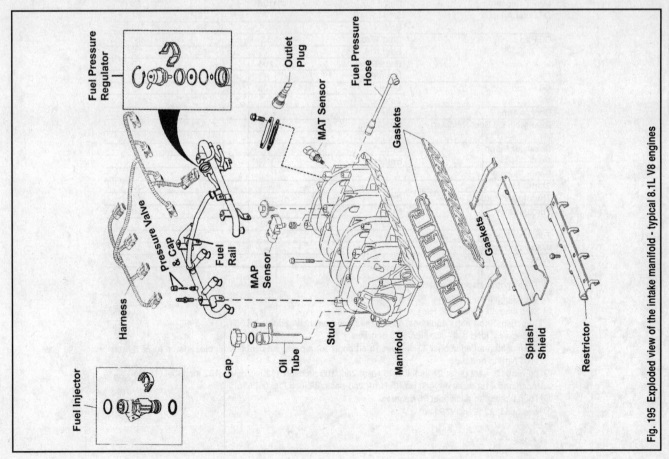

Fig. 195 Exploded view of the intake manifold - typical 8.1L V8 engines

4-56 ENGINE MECHANICAL - GM V6 AND V8 ENGINES

Specifications

TORQUE SPECIFICATIONS - 4.3L ENGINES

Component		inch lbs.	ft. lbs.	Nm
Balance Shaft	Drive Gear Stud	120	-	14
	Driven Gear Bolt	-	15 ①	20 ①
	Thrust Plate	120	-	14
Camshaft	Thrust Plate	106	-	12
	Sprocket Bolts	-	18	25
Camshaft Position Sensor		80	-	9
Connecting Rod Cap	Nuts	-	20 ②	27 ②
Coolant Drain Plug	Front	-	44	60
	Sides	-	15	20
Crankshaft Oil Deflector	Nuts	-	30	40
Crankshaft Position Sensor	Bolt	80	-	9
Crankshaft Pulley	Bolt	-	43	58
Cylinder Head	Bolts	-	③	③
Cylinder Head Cover	Bolts	106	-	12
Distributor	Clamp Bolt	-	18	25
Engine Coupler	To Flywheel	-	35	48
Engine Mounts	Front	-	30	41
	Rear	-	38	51
Fuel Rail	Bracket Bolt	53	-	6
	Retainer Nut	27	-	3
Flywheel	Housing cover	80	-	9
	Housing-to-Block	-	30	41
	To Crankshaft Bolts	-	75	100
Front Cover	Bolts	106	-	12
Knock Sensor		-	15	20
Lifter Guide Retainer		-	19	25
Lifter Restrictor Plate		-	12	16
Main Bearing Cap	Bolt/Stud	-	15 ④	20 ④
Manifold	Exhaust	-	20	27
	Intake	⑤	-	⑤
MAPT Sensor	Bolt	106	-	12
Oil Filter Adapter	5/16 in. -18	-	20	27
	Bolts	-	18	25
Oil Pan	Stud (Front)	53	-	6
	Stud Nut	-	18	25
	Bolts/Nuts	106	-	12
	Drain Plug	-	18	25
Oil Pump	Cover	106	-	12
	To Cap	-	15 ⑥	20 ⑥
Rocker Arm	Bolts	-	22	30
Rear Main Oil Seal Retainer	Nuts/Bolts	106	-	12
	Stud	53	-	6
Seawater Pump	Brace	-	30	41
	Bracket	-	30	41
Spark Plugs		-	11 ⑦	15 ⑦
Starter Motor		-	37	50
Thermostat Housing	Bolts	-	30	41
Throttle Body	Nut	88	-	10
	Stud	80	-	9
Torsional Damper	Bolt	-	70	95
Water Circulating Pump		-	35	48
Water Temperature Sender		-	20	27

① Then tighten an additional 35 degrees
② Then tighten an additional 70 degrees
③ Step One: 22 ft. lbs. (30 Nm)
   Step Two: Short bolt - additional 55 degrees; medium bolt - additional 65 degrees; long bolt - additional 75 degrees
④ Then tighten an additional 73 degrees. If an angle torque wrench is not available, tighten to 77 ft. lbs. (105 Nm) in one pass
⑤ Lower bolts - 1st pass: 27 inch lbs. (3 Nm); 2nd: 106 inch lbs. (12 Nm); 3rd: 132 inch lbs. (15 Nm)
   Upper stud - 1st pass: 44 inch lbs. (5 Nm); 2nd pass: 89 inch lbs. (10 Nm)
⑥ Then tighten an additional 65 degrees
⑦ New head: 22 ft. lbs. (30 Nm)

ENGINE MECHANICAL - GM V6 AND V8 ENGINES

TORQUE SPECIFICATIONS - 5.0L, 5.7L & 6.2L ENGINES

Component			inch lbs.	ft. lbs.	Nm
Camshaft	Thrust Plate		106	-	12
	Sprocket Bolts		-	18	25
Camshaft Position Sensor			80	-	9
Connecting Rod Cap	Nuts		-	20 ①	27 ①
Coolant Drain Plug	Front		-	44	60
	Sides		-	15	20
Crankshaft Oil Deflector	Nuts		-	30	40
Crankshaft Position Sensor	Bolt		80	-	9
Crankshaft Pulley	Bolt		-	43	58
Cylinder Head	Bolts		-	②	②
Cylinder Head Cover	Bolts		106	-	12
Distributor	Clamp Bolt		-	18	25
Drive Plate (I/B)	To Flywheel		-	35	48
Engine Coupler (I/O)	To Flywheel		-	35	48
Engine Mounts	Front		-	30	41
	Rear		-	38	51
Fuel Rail	Bracket Bolt		53	-	6
	Retainer Nut		27	-	3
Flywheel	Housing cover		80	-	9
	Housing-to-Block		-	30	41
	To Crankshaft Bolt		-	75	100
Front Cover	Bolts		106	-	12
Knock Sensor			-	15	20
Lifter Guide Retainer			-	19	25
Lifter Restrictor Plate			-	12	16
Main Bearing Cap	Bolt/Stud	Two Bolt	-	15 ③	20 ③
		Four Bolt	-	15 ④	20 ④
Manifold	Exhaust		-	20	27
	Intake		⑤	-	⑤
MAPT Sensor	Bolt		106	-	12
Oil Filter Adapter	5/16 in. -18		-	20	27
	Bolts		-	18	25
Oil Pan	Stud (Front)		53	-	6
	Stud Nut		-	18	25
	Bolts/Nuts		106	-	12
	Drain Plug		-	18	25
Oil Pump	Cover		106	-	12
	To Cap		-	15 ⑥	20 ⑥
Rocker Arm	Bolts		-	22	30
Rear Main Oil Seal Retainer	Nuts/Bolts		106	-	12
	Stud		53	-	6
Seawater Pump	Brace		-	30	41
	Bracket		-	30	41
Spark Plugs			-	11 ⑦	15 ⑦
Starter Motor			-	37	50
Thermostat Housing	Bolts		-	30	41
Transmission	To Flywheel Housing		-	50	68
Throttle Body	Nut		88	-	10
	Stud		80	-	9
Torsional Damper	Bolt		-	70	95
Water Circulating Pump	Mounting Bolt		-	35	48
	Pulley Bolt		-	19	25
Water Temperature Sender			-	20	27

① Then tighten an additional 70 degrees on the 5.0L and 5.7L, or 45 degrees on the 6.2L

② Step One: 22 ft. lbs. (30 Nm)
Step Two: Short bolt - additional 55 degrees; medium bolt - additional 65
degrees; long bolt - additional 75 degrees. If an angle torque wrench is not available, 1st pass: 26 ft. lbs.
(35 Nm), 2nd pass: 44 ft. lbs. (60 Nm), and final: 66 ft. lbs. (90 Nm)

③ Then tighten an additional 73 degrees. If an angle torque wrench is not available, tighten to 77 ft. lbs.
(105 Nm) in one pass

④ Then tighten the outboard bolt an addittional 43 degrees and the inboard bolt and stud an addittional 73
degrees. If an angle torque wrench is not available, tighten all inboard bolts to 77 ft. lbs. (105 Nm) and
the outboard bolts to 66 ft. lbs. (90 Nm)

⑤ Lower bolts - 1st pass: 27 inch lbs. (3 Nm); 2nd: 106 inch lbs. (12 Nm); 3rd: 132 inch lbs. (15 Nm)
Upper stud - 1st pass: 44 inch lbs. (5 Nm); 2nd pass: 89 inch lbs. (10 Nm)

⑥ Then tighten an additional 65 degrees

⑦ New head: 22 ft. lbs. (30 Nm)

ENGINE MECHANICAL - GM V6 AND V8 ENGINES

TORQUE SPECIFICATIONS - 8.1L ENGINES

Component		inch lbs.	ft. lbs.	Nm
Camshaft	Thrust Plate	106	-	12
	Sprocket Bolts	-	22	30
Camshaft Position Sensor		106	-	12
Connecting Rod Cap	Nuts	-	22 ①	30 ①
Coolant Drain Plug	Front	-	44	60
	Sides	-	15	20
Crankshaft Bearing Cap	Bolt/Stud	-	22 ①	30 ①
Crankshaft Oil Deflector	Nut	-	37	50
Crankshaft Position Sensor	Bolt	106	-	12
Cylinder Head	Bolts	-	②	②
Cylinder Head Cover	Bolts	106	-	12
Drive Plate (I/B)	To Flywheel	-	74	100
Engine Coupler (I/O)	To Flywheel	-	74	100
Engine Mounts	Front	-	30	41
	Rear	-	38	51
Fuel Rail	Nut	71	-	8
	Stud/Cap Screw	89	-	10
Flywheel	Housing-to-Block	-	30	41
	To Crankshaft Bolt	-	74	100
Front Cover	Bolts	106	-	12
Knock Sensor		-	15	20
Lifter Restrictor Retainer		-	19	25
Manifold	Exhaust	-	26	35
	Intake	106	-	12
Oil Pan	Bolt	-	19	25
	Drain Plug	-	21	28
Oil Pump	Cover	106	-	12
	Bolt	-	56	75
	Drive Bolt	-	19	25
Rocker Arm	Nuts	-	19	25
	Stud	-	37	50
Seawater Pump	Brace	-	30	41
	Bracket	-	30	41
Spark Plugs		-	22	30
Starter Motor		-	37	50
Thermostat Housing	Bolts	-	12	16
Transmission	To Flywheel Housing	-	50	68
Throttle Body	Nut	88	-	10
	Stud	106	-	12
Torsional Damper	Bolt	-	188	255
Water Circulating Pump	Mounting Bolt	-	37	50
	Pulley Bolt	-	19	25

① Then tighten an additional 90°, except the crankshaft bearing cap stud which is 80°

② Step One: 22 ft. lbs. (30 Nm)
   Step Two: additional 120°
   Step Three: Short bolt - additional 30°; medium and long bolt - additional 60°

ENGINE MECHANICAL - GM V6 AND V8 ENGINES 4-59

ENGINE SPECIFICATIONS - 4.3L V6

Component				Standard (in.) ①	Metric (mm) ①
Balance Shaft	Front Bearing Journal			2.1648-2.1654	55.985-55.001
	Rear Bearing Journal			1.4994-1.5000	38.084-38.100
	Rear Bearing I.D.			1.5014-1.503	37.525-37.575
	Rear Bearing O.D.			1.875-1.876	46.876-46.900
Camshaft	End Play			0.0010-0.0090	0.0254-0.2286
	Journal Diameter			1.8677-1.8697	47.440-47.490
	Journal Out-of-Round			0.0010 Max	0.025 Max
	Lobe Lift	Intake		0.283-0.287	7.20-7.30
		Exhaust		0.274-0.278	6.97-7.07
	Runout			0.0026 Max	0.065 Max
	Timing Chain Deflection			0.4331 Max	11 Max
Connecting Rod	Bearing clearance	Production		0.0015-0.0031	0.038-0.078
		Service Limit		0.0010-0.0025	0.025-0.063
	Journal	Diameter		2.2487-2.2497	57.116-57.142
		Taper	Production	0.0003 Max	0.008 Max
			Service Limit	0.001 Max	0.025 Max
		Out-of-round	Production	0.0003 Max	0.008 Max
			Service Limit	0.001 Max	0.025 Max
	Side clearance			0.0059-0.017	0.15-0.44
Crankshaft	Crankshaft	End Play		0.0020-0.079	0.05-0.20
		Run-Out		0.001	0.0254
	Main Bearing Clearance	Production	#1	0.0007-0.0021	0.018-0.053
			#2, 3, 4	0.0011-0.0023	0.028-0.058
		Service Limit	#1	0.001-0.002	0.0254-0.0508
			#2, 3, 4	0.0010-0.0025	0.0254-0.0635
	Main Bearing Journal	Diameter	#1	2.4488-2.4495	62.199-62.217
			#2, 3	2.4485-2.4494	62.191-62.215
			#4	2.4480-2.4489	62.179-62.203
		Taper	Production	0.0003 Max	0.007 Max
	Out-of-Round	Production		0.0002 Max	0.005 Max
		Service Limit		0.0010 Max	0.025 Max
Cylinder Bore	Diameter			4.0007-4.0017	101.618-101.643
	Out-of-round	Production		0.0005 Max	0.0127 Max
		Service Limit		0.002 Max	0.05 Max
	Taper	Production	Thrust side	0.0005 Max	0.0127 Max
			Relief side	0.0010 Max	0.0254 Max
		Service Limit		0.0010 Over Production	0.0254 Over Production
Cylinder Head	Surface Flatness	At Ex. Man. Deck		0.002	0.05
		At Eng. Block Deck ②		0.0039	0.1
		At In. Man. Deck		0.0039	0.1
	Intake Manifold Flatness			0.0039	0.1
Flywheel	Run-out			0.008 Max	0.203 Max
Pistons/Rings	Piston Clearance	Production		0.0007-0.0024	0.018-0.061
		Service Limit		0.0029 Max	0.075 Max
	Top Compression Ring	Groove Clearance	Production	0.0012-0.0028	0.030-0.070
			Service	0.0012-0.0033	0.030-0.085
		Gap	Production	0.010-0.016	0.25-0.40
			Service	0.010-0.020	0.25-51
	2nd Compression Ring	Groove Clearance	Production	0.0015-0.0031	0.040-0.080
			Service	0.0012-0.0033	0.030-0.085
		Gap	Production	0.015-0.023	0.38-0.58
			Service	0.015-0.031	0.38-0.80
	Oil Control Ring	Groove Clearance	Production	0.0018-0.0077	0.046-0.196
			Service	0.0018-0.0079	0.046-0.200
		Gap	Production	0.0098-0.0299	0.25-0.76
			Service	0.0002-0.0035	0.005-0.090
	Piston Pin	Diameter		0.9270-0.9271	23.545-23.548
		Clearance	Production	0.0005-0.0009	0.013-0.023
			Service	0.001 Max	0.025 Max
		Fit in Connecting Rod		0.0005-0.0019 ③	0.012-0.048 ③
Valve System	Head & Stem	Face Angle	Intake	45 deg.	
			Exhaust	45 deg.	
		Valve Diameter	Intake	1.84	46.74
			Exhaust	1.5	38.1

ENGINE MECHANICAL - GM V6 AND V8 ENGINES

ENGINE SPECIFICATIONS - 4.3L V6

Component				Standard (in.) ①	Metric (mm) ①
Valve System (cont'd)	Lash			Fixed	
	Lifter	Lift	Intake	0.414	10.527
			Exhaust	0.428	10.879
		Lifter		Roller Hydraulic	
		Rocker Arm Ratio		1.50:1	
	Seat	Angle		46 deg.	
		Correction Cut Angle	Top	30 deg.	
			Bottom	60 deg.	
		Run-Out	Intake	0.0020 Max	0.05 Max
			Exhaust	0.0020 Max	0.05 Max
		Width	Intake	0.040-0.065	1.016-1.651
			Exhaust	0.0650-0.0980	1.65-2.489
	Spring	Free Length		2.02	51.3
		Installed Ht.	Intake	1.6898-1.7098	42.92-43.43
			Exhaust	1.6898-1.7098	42.92-43.43
		Number of Coils	Approx.	4	
		Pressure	Closed	76-84 lbs. @ 1.7008 in.	338-374 N @ 43.2mm
			Open	187-203 lbs. @ 1.2717 in.	832-903 N @ 32.3mm
	Stem Clearance	Production	Intake	0.0010-0.0027	0.0254-0.0686
			Exhaust	0.0010-0.0027	0.0254-0.0686
		Service Limit	Intake	0.0010-0.0037	0.0254-0.094
			Exhaust	0.0010-0.0037	0.0254-0.094
	Stem Diameter	Production	Intake	0.3410-0.3417	8.661-8.679
			Exhaust	0.3410-0.3417	8.661-8.679
		Service Oversize	Exhaust Only	+0.0305	+0.774
	Stem Oil Seal	Installed Ht. ④		0.0394-0.0787	1-2

① Unless otherwise noted
② Within an 152mm (6 in.) area
③ Interference
④ From top of valve guide bevel to bottom of oil stem seal

ENGINE MECHANICAL - GM V6 AND V8 ENGINES

ENGINE SPECIFICATIONS - 5.0L & 5.7L V8

Component				Standard (in.) ①	Metric (mm) ①
Camshaft	End Play			0.0010-0.0090	0.0254-0.2286
	Journal Diameter			1.8677-1.8697	47.440-47.490
	Journal Out-of-Round			0.0010 Max	0.025 Max
	Runout			0.0026 Max	0.065 Max
	Timing Chain Deflection			0.4331 Max	11 Max
Connecting Rod	Bearing clearance	Production		0.0013-0.0031	0.033-0.078
		Service Limit		0.0010-0.0025	0.025-0.063
	Journal	Diameter		2.2246-2.2257	56.505-56.533
		Taper	Production	0.0003 Max	0.008 Max
			Service Limit	0.001 Max	0.025 Max
		Out-of-round	Production	0.0003 Max	0.008 Max
			Service Limit	0.001 Max	0.025 Max
	Side clearance			0.0059-0.0240	0.15-0.61
Crankshaft	Crankshaft	End Play		0.0020-0.079	0.05-0.20
		Run-Out		0.0015	0.038
	Main Bearing Clearance	Production	#1	0.0007-0.0021	0.018-0.053
			#2, 3, 4	0.0012-0.0027	0.030-0.068
			#5	0.0008-0.0024	0.020-0.060
		Service Limit	#1	0.001-0.002	0.0254-0.0508
			#2, 3, 4	0.0010-0.0025	0.0254-0.0635
			#5	0.0015-0.0025	0.038-0.063
	Main Bearing Journal	Diameter	#1	2.4484-2.4493	62.189-62.212
			#2, 3, 4	2.4481-2.4491	62.181-62.207
			#5	2.4482-2.4491	62.185-62.207
		Taper	Production	0.0002 Max	0.005 Max
			Service	0.0010 Max	0.025
	Out-of-Round		Production	0.0002 Max	0.005 Max
			Service Limit	0.0010 Max	0.025 Max
Cylinder Bore	Diameter	5.0L		3.7360-3.7381	94.894-94.947
		5.7L		4.007-4.0017	101.618-101.643
	Out-of-round	Production		0.0010 Max	0.025 Max
		Service Limit		0.002 Max	0.05 Max
Cylinder Head	Surface Flatness	At Ex. Man. Deck		0.002	0.05
		At Eng. Block Deck ②		0.0039	0.1
		At In. Man. Deck		0.0039	0.1
Flywheel	Run-out			0.008 Max	0.203 Max
Pistons/Rings	Piston Clearance	Production	5.0L	0.0007-0.0024	0.018-0.061
			5.7L	0.0007-0.0020	0.018-0.053
		Service Limit	5.0L	0.0007-0.0026	0.018-0.068
			5.7L	0.0007-0.0024	0.018-0.061
	Top Compression Ring	Groove Clearance	Production	0.0012-0.0028	0.030-0.070
			Service	0.0012-0.0035	0.030-0.090
		Gap	Production		
			5.0L	0.0098-0.0201	0.25-0.51
			5.7L	0.0098-0.0157	0.25-0.40
			Service		
			5.0L	0.0098-0.0259	0.25-0.65
			5.7L	0.0098-0.0197	0.25-0.50
	2nd Compression Ring	Groove Clearance	Production		
			5.0L	0.0012-0.0029	0.030-0.074
			5.7L	0.0015-0.0031	0.038-0.080
			Service		
			5.0L	0.0012-0.0035	0.030-0.090
			5.7L	0.0016-0.0039	0.040-0.100
		Gap	Production		
			5.0L	0.0181-0.0298	0.46-0.66
			5.7L	0.0015-0.0023	0.038-0.058
			Service		
			5.0L	0.0181-0.0354	0.46-0.90
			5.7L	0.0181-0.0315	0.46-0.80
	Oil Control Ring	Groove Clearance	Production		

ENGINE MECHANICAL - GM V6 AND V8 ENGINES

ENGINE SPECIFICATIONS - 5.0L & 5.7L V8

Component				Standard (in.) ①	Metric (mm) ①
Pistons/Rings (cont'd)			5.0L	0.0020-0.0080	0.051-0.203
			5.7L	0.0018-0.0038	0.046-0.096
		Service			
			5.0L	0.0020-0.0009	0.051-0.22
			5.7L	0.0018-0.0039	0.046-0.100
	Gap	Production		0.25-0.76	0.0098-0.0299
		Service			
			5.0L	0.0098-0.0350	0.25-0.89
			5.7L	0.0098-0.0354	0.25-0.90
	Piston Pin	Diameter		0.9270-0.9271	23.545-23.548
		Clearance	Production		
			5.0L	0.0004-0.0008	0.010-0.020
			5.7L	0.0005-0.0009	0.013-0.023
			Service	0.0005-0.0010	0.013-0.025
		Fit in Connecting Rod		0.0005-0.0020 ③	0.012-0.050 ③
Valve System	Face Angle	Intake		45 deg.	
		Exhaust		45 deg.	
	Lash			④	④
	Lifter	Lift	Intake	0.2744-0.2783	6.97-7.07
			Exhaust	0.2834-0.2874	7.20-7.30
		Lifter		Roller Hydraulic	
		Rocker Arm Ratio		1.50:1	
	Seat	Angle		46 deg.	
		Correction Cut Angle	Top	30 deg.	
			Bottom	60 deg.	
		Run-Out	Intake	0.0020 Max	0.05 Max
			Exhaust	0.0020 Max	0.05 Max
		Width	Intake		
			5.0L	0.0449-0.0701	1.14-1.78
			5.7L	0.0402-0.0650	1.02-1.65
			Exhaust		
			5.0L	0.0650-0.0980	1.65-2.49
			5.7L	0.0591-0.1008	1.50-2.56
Valve System (cont'd)	Spring	Free Length		2.02	51.3
		Installed Ht.	Intake	1.6898-1.7098	42.92-43.43
			Exhaust	1.6898-1.7098	42.92-43.43
		Number of Coils	Approx.	4	
		Pressure	Closed	76-84 lbs. @ 1.7008 in.	338-374 N @ 43.2mm
			Open	187-203 lbs. @ 1.2717 in.	832-903 N @ 32.3mm
	Stem Clearance	Production	Intake	0.0010-0.0027	0.025-0.069
			Exhaust	0.0010-0.0027	0.025-0.069
		Service Limit	Intake	0.0010-0.0037	0.025-0.094
			Exhaust	0.0010-0.0037	0.025-0.094
	Stem Diameter	Production	Intake	0.3410-0.3417	8.661-8.679
			Exhaust	0.3410-0.3417	8.661-8.679
		Service Oversize	Exhaust Only	+0.0305	+0.774
	Stem Oil Seal	Installed Ht. ⑤		0.0394-0.0787	1-2
	Valve Diameter	Intake	5.0L	1.84	46.74
			5.7L	1.94	49.28
		Exhaust		1.5	38.1

① Unless otherwise noted
② Within an 152mm (6 in.) area
③ Interference
④ 1 Full turn clockwise from zero lash
⑤ From top of valve guide bevel to bottom of oil stem seal

ENGINE MECHANICAL - GM V6 AND V8 ENGINES

ENGINE SPECIFICATIONS - 6.2L V8

Component				Standard (in.) ①	Metric (mm) ①
Camshaft	End Play			0.0010-0.0090	0.0254-0.2286
	Journal Diameter			1.8677-1.8697	47.440-47.490
	Journal Out-of-Round			0.0010 Max	0.025 Max
	Runout			0.0026 Max	0.065 Max
	Timing Chain Deflection			0.4331 Max	11 Max
Connecting Rod	Bearing clearance	Production		0.0013-0.0035	0.033-0.088
		Service Limit		0.0010-0.0030	0.025-0.076
	Big End Bore	Diameter		2.0977-2.0997	53.284-53.334
		Taper	Production	0.001	0.025
			Service Limit	0.001	0.025
		Out-of-round	Production	0.001	0.025
			Service Limit	0.001	0.025
	Journal	Diameter		2.0978-2.0998	53.284-53.334
		Taper	Production	0.0003 Max	0.007 Max
			Service Limit	0.001 Max	0.025 Max
		Out-of-round	Production	0.0003 Max	0.007 Max
			Service Limit	0.001 Max	0.025 Max
	Side clearance			0.0059-0.0240	0.15-0.61
Crankshaft	Crankshaft	End Play		0.0020-0.079	0.05-0.20
		Run-Out		0.0020-0.079	0.05-0.20
	Main Bearing Clearance	Production	#1	0.0007-0.0021	0.018-0.053
			#2, 3, 4	0.0009-0.0024	0.022-0.061
			#5	0.0010-0.0027	0.025-0.069
		Service Limit	#1	0.001-0.002	0.025-0.051
			#2, 3, 4	0.0010-0.0025	0.025-0.064
			#5	0.0015-0.0030	0.038-0.076
	Main Bearing Journal	Diameter	#1	2.4484-2.4493	62.189-62.212
			#2, 3, 4	2.4481-2.4491	62.181-62.207
			#5	2.4482-2.4491	62.185-62.207
		Taper	Production	0.0002 Max	0.005 Max
			Service	0.0010 Max	0.025
	Out-of-Round	Production		0.0002 Max	0.005 Max
Cylinder Bore	Diameter			4.0007-4.0017	101.618-101.643
	Out-of-round	Production		0.0010 Max	0.025 Max
		Service Limit		0.002 Max	0.05 Max
	Taper	Production	Thrust side	0.0005 Max	0.012 Max
			Relief side	0.0010 Max	0.025 Max
		Service Limit		0.0010 Over Production	0.025 Over Production
Cylinder Head	Surface Flatness	At Ex. Man. Deck		0.002	0.05
		At Eng. Block Deck ②		0.0039	0.1
		At In. Man. Deck		0.0039	0.1
Flywheel	Run-out			0.008 Max	0.203 Max
Pistons/Rings	Piston Clearance	Production		0.0007-0.0021	0.018-0.053
		Service Limit		0.0007-0.0027	0.018-0.068
	Top Compression Ring	Groove Clearance	Production	0.0012-0.0027	0.030-0.070
			Service	0.0012-0.0035	0.030-0.090
		Gap	Production	0.0098-0.0201	0.25-0.51
			Service	0.0012-0.0268	0.030-0.068
	2nd Compression Ring	Groove Clearance	Production	0.0015-0.0030	0.040-0.080
			Service	0.0015-0.0040	0.040-0.100
		Gap	Production	0.0181-0.0260	0.46-0.66
			Service	0.0201-0.0370	0.51-0.94
	Oil Control Ring	Groove Clearance	Production	0.0020-0.0067	0.051-0.17
			Service	0.0020-0.0076	0.051-0.195
		Gap	Production	0.0098-0.0299	0.25-0.76
			Service	0.0110-0.0319	0.28-0.81
	Piston Pin	Diameter		0.9269-0.9270	23.545-23.548
		Clearance	Production	0.0005-0.0009	0.013-0.023
			Service	0.0005-0.0010 Max	0.013-0.025 Max
		Fit in Connecting Rod		0.0008-0.0016 ③	0.021-0.040 ③
Valve System	Face Angle	Intake		45 deg.	
		Exhaust		45 deg.	
	Lash			④	④

ENGINE MECHANICAL - GM V6 AND V8 ENGINES

ENGINE SPECIFICATIONS - 6.2L V8

Component				Standard (in.) ①	Metric (mm) ①
Valve System (cont'd)	Lifter	Lift	Intake	0.3114-0.3153	7.90-8.00
			Exhaust	0.3160-0.3200	8.00-8.10
		Lifter		Roller Hydraulic	
		Rocker Arm Ratio		1.50:1	
	Seat	Angle		46 deg.	
		Correction Cut Angle	Top	30 deg.	
			Bottom	60 deg.	
		Run-Out	Intake	0.0020 Max	0.05 Max
			Exhaust	0.0020 Max	0.05 Max
		Width	Intake	0.0434-0.0650	1.02-1.65
			Exhaust	0.0591-0.1008	1.50-2.56
	Spring	Free Length		2.0197	51.3
		Installed Ht.	Intake	1.6898-1.7098	42.92-43.43
			Exhaust	1.6898-1.7098	42.92-43.43
		Number of Coils	Approx.	4	
		Pressure	Closed	76-84 lbs. @ 1.7008 in.	338-374 N @ 43.2mm
			Open	187-203 lbs. @ 1.2717 in.	832-903 N @ 32.3mm
	Stem Clearance	Production	Intake	0.0010-0.0027	0.025-0.069
			Exhaust	0.0010-0.0027	0.025-0.069
		Service Limit	Intake	0.0010-0.0037	0.025-0.094
			Exhaust	0.0010-0.0076	0.025-0.194
	Stem Diameter	Production	Intake	0.341	8.66
			Exhaust	0.341	8.66
		Service Oversize	Exhaust Only	+0.0305	+0.774
	Stem Oil Seal	Installed Ht. ⑤		0.0394-0.0787	1-2
	Valve Diameter	Intake		1.94	49.28
		Exhaust		1.5	38.1

① Unless otherwise noted
② Within an 152mm (6 in.) area
③ Interference
④ 1 Full turn clockwise from zero lash
⑤ From top of valve guide bevel to bottom of oil stem seal

ENGINE MECHANICAL - GM V6 AND V8 ENGINES

ENGINE SPECIFICATIONS - 8.1L V8

Component				Standard (in.) ①	Metric (mm) ①
Camshaft	Journal Diameter			1.9477-1.9479	49.472-49.522
	Out-of-Round			0.001 Max	0.025 Max
	Lobe Lift	Intake	8.1S/496 Mag	0.282	7.16
			8.1S HO/496 HO	0.3	7.62
		Exhaust	8.1S/496 Mag	0.284	7.22
			8.1S HO/496 HO	0.3	7.62
	Runout			0.002 Max	0.051 Max
	Timing Chain Deflection			0.375 ②	9.5 ②
Connecting Rod	Bearing Clearance	Production		0.0011-0.0029	0.0279-0.0735
	Taper	Production		0.0004	0.0102
		Service Limit		0.001 Max	0.02 Max
	Out-of-Round	Production		0.0004	0.0102
		Service Limit		0.001 Max	0.02 Max
	Side Clearance			0.015-0.027	0.384-0.686
Crankshaft	End Play			0.005-0.011	0.127-0.279
	Main Bearing Clearance	Production	#1, 2, 3, 4	0.0011-0.0024	0.028-0.061
			#5	0.0025-0.0038	0.0635-0.0965
	Main Bearing Journal	Diameter		2.7482-2.7489	69.804-69.822
		Taper		0.0004 Max	0.0102 Max
		Out-of-Round		0.0004 Max	0.0102 Max
Cylinder Bore	Diameter			4.2500-4.2507	107.95-107.97
	Out-of-Round			0.001 Max	0.025 Max
	Taper	Thrust side		0.0005 Max	0.013 Max
		Relief side		0.001 Max	0.025 Max
Cylinder Head	Surface flatness			0.004 Max ③	0.1016 Overall Max ③
Flywheel	Runout			0.008 Max	0.203 Max
Piston/Rings	Top Compression Ring	Side Clearance		0.0012-0.0029	0.031-0.074
		Gap		0.098-0.0161	0.249-0.409
	2nd Compression Ring	Side Clearance		0.0012-0.0029	0.031-0.074
		Gap		0.0177-0.0256	0.450-0.650
	Oil Control Ring	Gap		0.0098-0.0099	0.249-0.759
	Piston pin	Diameter		1.04-1.0401	26.416-26.419

ENGINE SPECIFICATIONS - 8.1L V8

Component			Standard (in.) ①	Metric (mm) ①
Piston/Rings		Clearance	0.00039-0.00066	0.010-0.017
		Fit in Rod	0.00019-0.0007 ④	0.049-0.020 ④
Valve System	Face Angle	Intake	45 deg.	
		Exhaust	45 deg.	
	Lifter	Lifter	Roller Hydraulic	
		Rocker Arm Ratio	1.70:1	
	Seat	Angle	46 deg.	
		Run-Out Intake	0.0020 Max	0.05 Max
		Run-Out Exhaust	0.0020 Max	0.05 Max
		Width Intake	0.03-0.06	0.800-1.200
		Width Exhaust	0.06-0.095	1.651-2.159
	Spring	Free Length	2.1929	55.7
		Installed Ht. Intake	1.838-1.869	46.68-47.479
		Installed Ht. Exhaust	1.838-1.869	46.68-47.479
		Pressure Closed	58-64 lbs. @ 1.838 in.	267-293 N @ 46.68mm
		Pressure Open	189-207 lbs. @ 1.34 in.	840-920 N @ 33.98mm
	Stem Clearance	Production Intake	0.001-0.0029	0.025-0.074
		Production Exhaust	0.0012-0.0031	0.030-0.079
	Stem Diameter	Production Intake	0.3715-0.3722	9.436-9.454
		Production Exhaust	0.3713-0.3720	9.431-9.449

① Unless otherwise noted
② From taut position, no more than 0.75 in. (19mm) total
③ 0.003 in. (0.76mm) with 6 in. span
④ Interference

BUY OR REBUILD?	5-3
CYLINDER HEAD	5-5
ASSEMBLY	5-9
DISASSEMBLY	5-5
GENERAL INFORMATION	5-5
INSPECTION	5-7
REFINISHING & REPAIRING	5-8
DETERMINING ENGINE CONDITION	5-2
COMPRESSION TEST	5-2
OIL PRESSURE TEST	5-2
ENGINE BLOCK	5-10
ASSEMBLY	5-13
DISASSEMBLY	5-10
GENERAL INFORMATION	5-10
INSPECTION	5-11
REFINISHING	5-13
ENGINE OVERHAUL TIPS	5-3
CLEANING	5-3
OVERHAUL TIPS	5-3
REPAIRING DAMAGED THREADS	5-4
TOOLS	5-3
ENGINE PREPARATION	5-4
ENGINE RECONDITIONING	**5-2**
BUY OR REBUILD?	5-3
CYLINDER HEAD	5-5
DETERMINING ENGINE CONDITION	5-2
ENGINE BLOCK	5-10
ENGINE OVERHAUL TIPS	5-3
ENGINE PREPARATION	5-4
ENGINE START-UP AND BREAK-IN	5-15
ENGINE START-UP AND BREAK-IN	5-15
BREAKING IT IN	5-16
KEEP IT MAINTAINED	5-16
STARTING THE ENGINE	5-15

5

ENGINE OVERHAUL

ENGINE RECONDITIONING 5-2

5-2 ENGINE OVERHAUL

ENGINE RECONDITIONING

Determining Engine Condition

Anything that generates heat and/or friction will eventually burn or wear out (for example, a light bulb generates heat, therefore its life span is limited). With this in mind, a running engine generates tremendous amounts of both; friction is encountered by the moving and rotating parts inside the engine and heat is created by friction and combustion of the fuel. However, the engine has systems designed to help reduce the effects of heat and friction and provide added longevity. The oiling system reduces the amount of friction encountered by the moving parts inside the engine, while the cooling system reduces heat created by friction and combustion. If either system is not maintained, a break-down will be inevitable. Therefore, you can see how regular maintenance can affect the service life of your engine. If you do not drain, flush and refill your cooling system at the proper intervals, deposits will begin to accumulate, thereby reducing the amount of heat it can extract from the coolant. The same applies to your oil and filter; if it is not changed often enough it becomes laden with contaminates and is unable to properly lubricate the engine. This increases friction and wear.

There are a number of methods for evaluating the condition of your engine. A compression test can reveal the condition of your pistons, piston rings, cylinder bores, head gasket(s), valves and valve seats. An oil pressure test can warn you of possible engine bearing, or oil pump failures. Excessive oil consumption, evidence of oil in the engine air intake area and/or bluish smoke from the exhaust may indicate worn piston rings, worn valve guides and/or valve seals.

COMPRESSION TEST

◆ See Figure 1

A noticeable lack of engine power, excessive oil consumption and/or poor fuel mileage measured over an extended period are all indicators of internal engine wear. Worn piston rings, scored or worn cylinder bores, blown head gaskets, sticking or burnt valves, and worn valve seats are all possible culprits. A check of each cylinder's compression will help locate the problem.

■ **A screw-in type compression gauge is more accurate than the type you simply hold against the spark plug hole.**

Although it takes slightly longer to use, it's worth the effort to obtain a more accurate reading.

1. Make sure that the proper amount and viscosity of engine oil is in the crankcase, then ensure the battery is fully charged.
2. Warm-up the engine to normal operating temperature, then shut the engine **OFF**.
3. Disable the ignition system.
4. Label and disconnect all of the spark plug wires from the plugs.
5. Thoroughly clean the cylinder head area around the spark plug ports, then remove the spark plugs.
6. Set the throttle plate to the fully open (wide-open throttle) position. You can block the throttle linkage open for this, or you can have an assistant operate the throttle lever in the boat.
7. Install a screw-in type compression gauge into the No. 1 spark plug hole until the fitting is snug.

✼✼ WARNING

Be careful not to cross-thread the spark plug hole.

8. According to the tool manufacturer's instructions, connect a remote starting switch to the starting circuit.
9. With the ignition switch in the **OFF** position, use the remote starting switch to crank the engine through at least five compression strokes (approximately 5 seconds of cranking) and record the highest reading on the gauge.
10. Repeat the test on each cylinder, cranking the engine approximately the same number of compression strokes and/or time as the first.
11. Compare the highest readings from each cylinder to that of the others. The indicated compression pressures are considered within specifications if the lowest reading cylinder is within 75 percent of the pressure recorded for the highest reading cylinder. For example, if your highest reading cylinder pressure was 150 psi (1034 kPa), then 75 percent of that would be 113 psi (779 kPa). So the lowest reading cylinder should be no less than 113 psi (779 kPa).
12. If a cylinder exhibits an unusually low compression reading, pour a tablespoon of clean engine oil into the cylinder through the spark plug hole and repeat the compression test. If the compression rises after adding oil, it means that the cylinder's piston rings and/or cylinder bore are damaged or worn. If the pressure remains low, the valves may not be seating properly (a valve job is needed), or the head gasket may be blown near that cylinder. If compression in any two adjacent cylinders is low, and if the addition of oil doesn't help raise compression, there is leakage past the head gasket. Oil and coolant in the combustion chamber, combined with blue or constant white smoke from the exhaust, are symptoms of this problem. However, don't be alarmed by the normal white smoke emitted from the exhaust during engine warm-up or from cold weather operation. There may be evidence of water droplets on the engine dipstick and/or oil droplets in the cooling system if a head gasket is blown.

OIL PRESSURE TEST

Check for proper oil pressure at the sending unit passage with an externally mounted mechanical oil pressure gauge (as opposed to relying on a factory installed dash-mounted gauge). A tachometer may also be needed, as some specifications may require running the engine at a specific rpm.

1. With the engine cold, locate and remove the oil pressure sending unit.
2. Following the manufacturer's instructions, connect a mechanical oil pressure gauge and, if necessary, a tachometer to the engine.
3. Start the engine and allow it to idle.
4. Check the oil pressure reading when cold and record the number. You may need to run the engine at a specified rpm, so check the specifications.
5. Run the engine until normal operating temperature is reached.
6. Check the oil pressure reading again with the engine hot and record the number. Turn the engine **OFF**.
7. Compare your hot oil pressure reading to that given in the chart. If the reading is low, check the cold pressure reading against the chart. If the cold pressure is well above the specification, and the hot reading was lower than the specification, you may have the wrong viscosity oil in the engine. Change the oil, making sure to use the proper grade and quantity, then repeat the test.

Low oil pressure readings could be attributed to internal component wear, pump related problems, a low oil level, or oil viscosity that is too low. High oil pressure readings could be caused by an overfilled crankcase, too high of an oil viscosity or a faulty pressure relief valve.

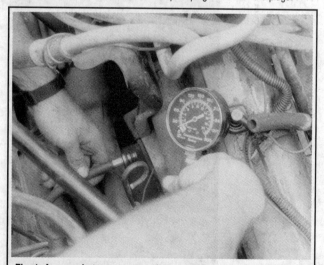

Fig. 1 A screw-in type compression gauge is more accurate and easier to use without an assistant

ENGINE OVERHAUL 5-3

Buy or Rebuild?

Now that you have determined that your engine is worn out, you must make some decisions. The question of whether or not an engine is worth rebuilding is largely a subjective matter and one of personal worth. Is the engine a popular one, or is it an obsolete model? Are parts available? Will it get acceptable gas mileage once it is rebuilt? Is the vessel it's being put into worth keeping? Would it be less expensive to buy a new engine, have your engine rebuilt by a pro, rebuild it yourself or buy a used engine? Or would it be simpler and less expensive to buy another boat? If you have considered all these matters and more, and have still decided to rebuild the engine, then it is time to decide how you will rebuild it.

■ The editors at Seloc feel that most engine machining should be performed by a professional machine shop. Don't think of it as wasting money, rather, as an assurance that the job has been done right the first time. There are many expensive and specialized tools required to perform such tasks as boring and honing an engine block or having a valve job done on a cylinder head. Even inspecting the parts requires expensive micrometers and gauges to properly measure wear and clearances. Also, a machine shop can deliver to you clean, and ready to assemble parts, saving you time and aggravation. Your maximum savings will come from performing the removal, disassembly, assembly and installation of the engine and purchasing or renting only the tools required to perform the above tasks. Depending on the particular circumstances, you may save 40-60 percent of the cost doing these yourself.

A complete rebuild or overhaul of an engine involves replacing all of the moving parts (pistons, rods, crankshaft, camshaft, etc.) with new ones and machining the non-moving wearing surfaces of the block and heads. Unfortunately, this may not be cost effective. For instance, your crankshaft may have been damaged or worn, but it can be machined undersize for a minimal fee.

So, as you can see, you can replace everything inside the engine, but, it is wiser to replace only those parts which are really needed, and, if possible, repair the more expensive ones. Later we will break the engine down into its two main components: the cylinder head and the engine block. We will discuss each component, and the recommended parts to replace during a rebuild on each.

Engine Overhaul Tips

Most engine overhaul procedures are fairly standard. In addition to specific parts replacement procedures and specifications for your individual engine, this is also a guide to acceptable rebuilding procedures. Examples of standard rebuilding practice are given and should be used along with specific details concerning your particular engine.

Competent and accurate machine shop services will ensure maximum performance, reliability and engine life. In most instances it is more profitable for the do-it-yourself mechanic to remove, clean and inspect the component, buy the necessary parts and deliver these to a shop for actual machine work.

Much of the assembly work (crankshaft, bearings, piston rods, and other components) is well within the scope of the do-it-yourself mechanic's tools and abilities. You will have to decide for yourself the depth of involvement you desire in an engine repair or rebuild.

TOOLS

The tools required for an engine overhaul or parts replacement will depend on the depth of your involvement. With a few exceptions, they will be the tools found in a mechanic's tool kit. More in-depth work will require some or all of the following:
- A dial indicator (reading in thousandths) mounted on a universal base
- Micrometers and telescope gauges
- Jaw and screw-type pullers
- Scraper
- Valve spring compressor
- Ring groove cleaner
- Piston ring expander and compressor
- Ridge reamer
- Cylinder hone or glaze breaker
- Plastigage®
- Engine stand

The use of most of these tools is illustrated in the procedures. Many can be rented for a one-time use from a local parts jobber or tool supply house specializing in marine or automotive work.

Occasionally, the use of special tools is called for. See the information on Special Tools and the Safety Notice in the front of this manual before substituting another tool.

OVERHAUL TIPS

Aluminum has become extremely popular for use in engines, due to its low weight. Observe the following precautions when handling aluminum parts:
- Never hot tank aluminum parts (the caustic hot tank solution will eat the aluminum.
- Remove all aluminum parts (identification tag, etc.) from engine parts prior to the tanking.
- Always coat threads lightly with engine oil or anti-seize compounds before installation, to prevent seizure.
- Never over-tighten bolts or spark plugs especially in aluminum threads.

When assembling the engine, any parts that will be exposed to frictional contact must be pre-lubed to provide lubrication at initial start-up. Any product specifically formulated for this purpose can be used, but engine oil is not recommended as a pre-lube in most cases.

When semi-permanent (locked, but removable) installation of bolts or nuts is desired, threads should be cleaned and coated with Loctite® or another similar, commercial non-hardening marine sealant.

CLEANING

◆ See Figures 2 thru 5

Before the engine and its components are inspected, they must be thoroughly cleaned. You will need to remove any engine varnish, oil sludge and/or carbon deposits from all of the components to insure an accurate inspection. A crack in the engine block or cylinder head can easily become overlooked if hidden by a layer of sludge or carbon.

Fig. 2 Use a gasket scraper to remove the old gasket material from the mating surfaces

Fig. 3 Use a ring expander tool to remove the piston rings

Fig. 4 Clean the piston ring grooves using a ring groove cleaner tool, or . . .

5-4 ENGINE OVERHAUL

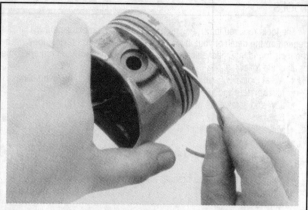

Fig. 5 ... use a piece of an old ring to clean the grooves. Be careful, the ring can be quite sharp

Most of the cleaning process can be carried out with common hand tools and readily available solvents or solutions. Carbon deposits can be chipped away using a hammer and a hard wooden chisel. Old gasket material and varnish or sludge can usually be removed using a scraper and/or cleaning solvent. Extremely stubborn deposits may require the use of a power drill with a wire brush. If using a wire brush, use extreme care around any critical machined surfaces (such as the gasket surfaces, bearing saddles, cylinder bores, etc.). Use of a wire brush is NOT RECOMMENDED on any aluminum components. Always follow any safety recommendations given by the manufacturer of the tool and/or solvent. You should always wear eye protection during any cleaning process involving scraping, chipping or spraying of solvents.

An alternative to the mess and hassle of cleaning the parts yourself is to drop them off at a local marina or machine shop (or even an automotive garage). They will, more than likely, have the necessary equipment to properly clean all of the parts for a nominal fee.

✱✱ CAUTION

Always wear eye protection during any cleaning process involving scraping, chipping or spraying of solvents.

Remove any oil galley plugs, freeze plugs and/or pressed-in bearings and carefully wash and degrease all of the engine components including the fasteners and bolts. Small parts such as the valves, springs, etc., should be placed in a metal basket and allowed to soak. Use pipe cleaner type brushes, and clean all passageways in the components. Use a ring expander and remove the rings from the pistons. Clean the piston ring grooves with a special tool or a piece of broken ring. Scrape the carbon off of the top of the piston. You should never use a wire brush on the pistons. After preparing all of the piston assemblies in this manner, wash and degrease them again.

✱✱ WARNING

Use extreme care when cleaning around the cylinder head valve seats. A mistake or slip may cost you a new seat.

When cleaning the cylinder head, remove carbon from the combustion chamber with the valves installed. This will avoid damaging the valve seats.

REPAIRING DAMAGED THREADS

◆ See Figures 6 thru 10

Several methods of repairing damaged threads are available. Heli-Coil® (shown here), Keenserts® and Microdot® are among the most widely used. All involve basically the same principle - drilling out stripped threads, tapping the hole and installing a pre-wound insert - making welding, plugging and oversize fasteners unnecessary.

Two types of thread repair inserts are usually supplied: a standard type for most inch coarse, inch fine, metric course and metric fine thread sizes and a spark lug type to fit most spark plug port sizes. Consult the individual tool manufacturer's catalog to determine exact applications. Typical thread repair kits will contain a selection of pre-wound threaded inserts, a tap (corresponding to the outside diameter threads of the insert) and an installation tool. Spark plug inserts usually differ because they require a tap equipped with pilot threads and a combined reamer/tap section. Most manufacturers also supply blister-packed thread repair inserts separately in addition to a master kit containing a variety of taps and inserts plus installation tools.

Before attempting to repair a threaded hole, remove any snapped, broken or damaged bolts or studs. Penetrating oil can be used to free frozen threads. The offending item can usually be removed with locking pliers or using a screw/stud extractor. After the hole is clear, the thread can be repaired, as shown in the series of accompanying illustrations and in the kit manufacturer's instructions.

Engine Preparation

To properly rebuild an engine, you must first remove it from the vessel, then disassemble and diagnose it. Ideally you should place your engine on an engine stand. This affords you the best access to the engine components. Follow the manufacturer's directions for using the stand with your particular engine. Remove the flywheel or coupler before installing the engine to the stand.

Now that you have the engine on a stand, and assuming that you have drained the oil and coolant from the engine, it's time to strip it of all but the necessary components. Before you start disassembling the engine, you may want to take a moment to draw some pictures, or fabricate some labels or containers to mark the locations of various components and the bolts and/or studs which fasten them. Modern day engines use a lot of little brackets and clips which hold wiring harnesses and such, and these holders are often mounted on studs and/or bolts that can be easily mixed up. The manufacturer spent a lot of time and money designing your engine/boat, and they wouldn't have wasted any of it by haphazardly placing brackets, clips or fasteners on the boat. If it's present when you disassemble it, put it back

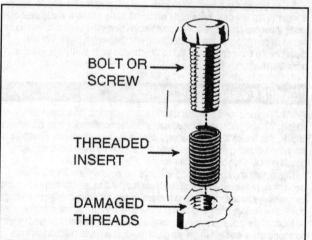

Fig. 6 Damaged bolt hole threads can be replaced with thread repair inserts

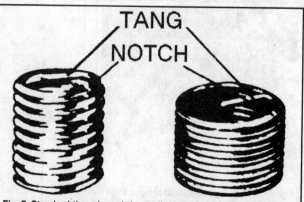

Fig. 7 Standard thread repair insert (left), and spark plug thread insert

ENGINE OVERHAUL

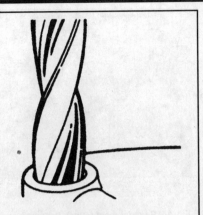

Fig. 8 Drill out the damaged threads with the specified size bit. Be sure to drill completely through the hole or to the bottom of a blind hole

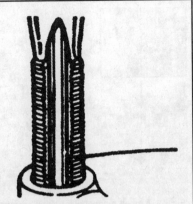

Fig. 9 Using the kit, tap the hole in order to receive the thread insert. Keep the tap well oiled and back it out frequently to clean out the chips

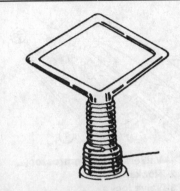

Fig. 10 Screw the insert onto the installer tool until the tang engages the slot. Thread the insert into the hole until it is 1/4-1/2 turn below the top surface, then remove the tool and break off the tang using a punch

when you assemble, you will regret not remembering that little bracket which holds a wire harness out of the path of a rotating part.

You should begin by unbolting any accessories still attached to the engine, such as the water pump, power steering pump, alternator, etc. Then, unfasten any manifolds (intake or exhaust) which were not removed during the engine removal procedure. Finally, remove any covers remaining on the engine such as the rocker arm, front or timing cover and oil pan. Some front covers may require the balancer and/or crank pulley to be removed beforehand. The idea is to reduce the engine to the bare necessities (cylinder head(s), valve train, engine block, crankshaft, pistons and connecting rods), plus any other 'in block' components such as oil pumps, balance shafts and auxiliary shafts.

Finally, remove the cylinder head(s) from the engine block and carefully place on a bench. Disassembly instructions for each component follow later.

Cylinder Head

GENERAL INFORMATION

There are two basic types of cylinder heads used on today's engines: the Overhead Valve (OHV) and the Overhead Camshaft (OHC). The latter can also be broken down into two subgroups: the Single Overhead Camshaft (SOHC) and the Dual Overhead Camshaft (DOHC). Generally, if there is only a single camshaft on a head, it is just referred to as an OHC head. Also, an engine with an OHV cylinder head is also known as a pushrod engine - all engines covered here are OHV.

Most cylinder heads these days are made of an aluminum alloy due to its light weight, durability and heat transfer qualities. However, cast iron was the material of choice in the past, and is still used on many engines today. Whether made from aluminum or iron, all cylinder heads have valves and seats. Most use two valves per cylinder, while the more hi-tech engines will utilize a multi-valve configuration using 3, 4 and even 5 valves per cylinder. When the valve contacts the seat, it does so on precision machined surfaces, which seals the combustion chamber. All cylinder heads have a valve guide for each valve. The guide centers the valve to the seat and allows it to move up and down within it. The clearance between the valve and guide can be critical. Too much clearance and the engine may consume oil, lose vacuum and/or damage the seat. Too little, and the valve can stick in the guide causing the engine to run poorly if at all, and possibly causing severe damage. The last component all cylinder heads have are valve springs. The spring holds the valve against its seat. It also returns the valve to this position when the valve has been opened by the valve train or camshaft. The spring is fastened to the valve by a retainer and valve locks (sometimes called keepers). Aluminum heads will also have a valve spring shim to keep the spring from wearing away the aluminum.

An ideal method of rebuilding the cylinder head would involve replacing all of the valves, guides, seats, springs, etc. with new ones. However, depending on how the engine was maintained, often this is not necessary. A major cause of valve, guide and seat wear is an improperly tuned engine. An engine that is running too rich, will often wash the lubricating oil out of the guide with gasoline, causing it to wear rapidly. Conversely, an engine which is running too lean will place higher combustion temperatures on the valves and seats allowing them to wear or even burn. Springs fall victim to the operating habits of the individual. A driver who often runs the engine rpm to the redline will wear out or break the springs faster then one that stays well below it. Unfortunately, 'hours of operation' takes it toll on all of the parts. Generally, the valves, guides, springs and seats in a cylinder head can be machined and re-used, saving you money. However, if a valve is burnt, it may be wise to replace all of the valves, since they were all operating in the same environment. The same goes for any other component on the cylinder head. Think of it as an insurance policy against future problems related to that component.

Unfortunately, the only way to find out which components need replacing, is to disassemble and carefully check each piece. After the cylinder head(s) are disassembled, thoroughly clean all of the components.

DISASSEMBLY

◆ See Figures 11 thru 21

Before disassembling the cylinder head, you may want to fabricate some containers to hold the various parts, as some of them can be quite small (such as keepers) and easily lost. Also keeping yourself and the components organized will aid in assembly and reduce confusion. Where possible, try to maintain a components original location; this is especially important if there is not going to be any machine work performed on the components.

1. If you haven't already removed the rocker arms and/or shafts, do so now.
2. Position the head so that the springs are easily accessed.
3. Use a valve spring compressor tool, and relieve spring tension from the retainer.

■ Due to engine varnish, the retainer may stick to the valve locks. A gentle tap with a hammer may help to break it loose.

Fig. 11 When removing a valve spring, use a compressor tool to relieve the tension from the retainer. . .

5-6 ENGINE OVERHAUL

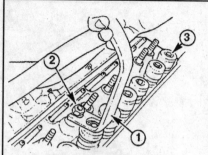

1- Valve spring compressor
2- Rocker arm nut
3- Valve locks

Fig. 12 ...you may also find a compressor that looks like this

Fig. 13 A small magnet will help in removal of the valve locks

Fig. 14 Be careful not to lose the small valve locks (keepers or keys)

Fig. 15 Remove the valve seal from the valve stem - O-ring type seal shown

Fig. 16 Removing an umbrella/positive type seal

Fig. 17 Invert the cylinder head and withdraw the valve from the valve guide bore

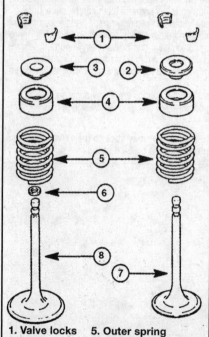

1. Valve locks
2. Retainer
3. Rotator
4. Cap
5. Outer spring
6. Valve stern oil seal
7. Intake valve
8. Exhaust valve

Fig. 18 Exploded view of the valve train - a few early 3.0L and 4.3L engines

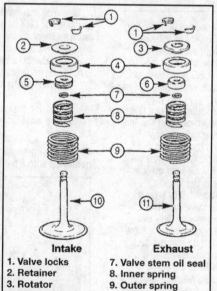

Intake
1. Valve locks
2. Retainer
3. Rotator
4. Cap
5. Valve guide oil seal
6. Oil shield

Exhaust
7. Valve stem oil seal
8. Inner spring
9. Outer spring
10. Intake valve
11. Exhaust valve

Fig. 19 Exploded view of the valve train - a few early V8 engines (carb/TBI)

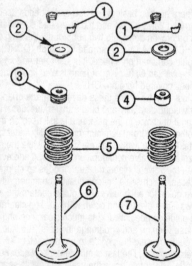

1 - Valve Locks
2 - Cap
3 - Intake Valve Stem Oil Seal
4 - Exhaust Valve Stem Oil Seal
5 - Spring
6 - Intake Valve
7 - Exhaust Valve

Fig. 20 Exploded view of the valve train - most 4 cyl., V6 and V8 engines (exc. 8.1L)

ENGINE OVERHAUL

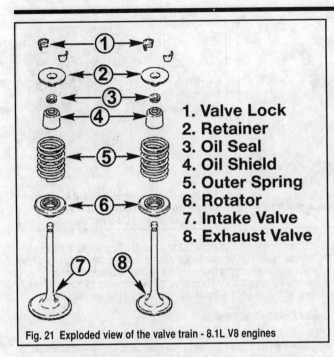

1. Valve Lock
2. Retainer
3. Oil Seal
4. Oil Shield
5. Outer Spring
6. Rotator
7. Intake Valve
8. Exhaust Valve

Fig. 21 Exploded view of the valve train - 8.1L V8 engines

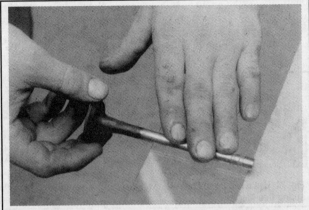

Fig. 22 Valve stems may be rolled on a flat surface to check for bends

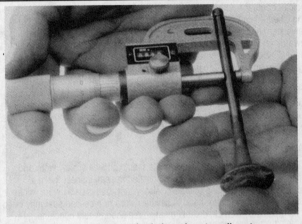

Fig. 23 Use a micrometer to check the valve stem diameter

4. Remove the valve locks from the valve tip and/or retainer. A small magnet may help in removing the locks.
5. Lift the valve spring(s), tool and all, off of the valve stem.
6. Remove the valve seal from the stem and guide. If the seal is difficult to remove with the valve in place, try removing the valve first, then the seal. Follow the steps below for valve removal.
7. Position the head to allow access for withdrawing the valve.

■ Cylinder heads that have seen a lot of miles and/or abuse may have mushroomed the valve lock grove and/or tip, causing difficulty in removal of the valve. If this has happened, use a metal file to carefully remove the high spots around the lock grooves and/or tip. Only file it enough to allow removal.

8. Remove the valve from the cylinder head.
9. If equipped, remove the valve spring shim. A small magnetic tool or screwdriver will aid in removal.
10. Repeat Steps 3 though 9 until all of the valves have been removed.

INSPECTION

Now that all of the cylinder head components are clean, it's time to inspect them for wear and/or damage. To accurately inspect them, you will need some specialized tools:

- A 0-1 in. micrometer for the valves
- A dial indicator or inside diameter gauge for the valve guides
- A spring pressure test gauge

If you do not have access to the proper tools, you may want to bring the components to a shop that does.

Valves

◆ See Figures 22 and 23

The first thing to inspect are the valve heads. Look closely at the head, margin and face for any cracks, excessive wear or burning. The margin is the best place to look for burning. It should have a squared edge with an even width all around the diameter. When a valve burns, the margin will look melted and the edges rounded. Also inspect the valve head for any signs of tulipping. This will show as a lifting of the edges or dishing in the center of the head and will usually not occur to all of the valves. All of the heads should look the same; any that seem dished more than others are probably bad. Next, inspect the valve lock grooves and valve tips. Check for any burrs around the lock grooves, especially if you had to file them to remove the valve. Valve tips should appear flat, although slight rounding with high mileage engines is normal. Slightly worn valve tips will need to be machined flat. Last, measure the valve stem diameter with the micrometer. Measure the area that rides within the guide, especially towards the tip where most of the wear occurs. Take several measurements along its length and compare them to each other. Wear should be even along the length with little to no taper. If no minimum diameter is given in the specifications, then the stem should not read more than 0.001 in. (0.025mm) below the unworn area of the valve stem. Any valves that fail these inspections should be replaced.

Springs, Retainers and Valve Locks

◆ See Figures 24 and 25

The first thing to check is the most obvious, broken springs. Next check the free length and squareness of each spring. If applicable, insure to distinguish between intake and exhaust springs. Use a ruler and/or carpenter's square to measure the length. A carpenter's square should be used to check the springs for squareness. If a spring pressure test gauge is available, check each springs rating and compare to the specifications chart. Check the readings against the specifications given. Any springs that fail these inspections should be replaced.

The spring retainers rarely need replacing, however they should still be checked as a precaution. Inspect the spring mating surface and the valve lock retention area for any signs of excessive wear. Also check for any signs of cracking. Replace any retainers that are questionable.

Valve locks should be inspected for excessive wear on the outside contact area as well as on the inner notched surface. Any locks which appear worn or broken and its respective valve should be replaced.

Valve Guides

◆ See Figure 26

Now that you know the valves are good, you can use them to check the guides, although a new valve, if available, is preferred. Before you measure anything, look at the guides carefully and inspect them for any cracks, chips or breakage. Also if the guide is a removable style (as in most aluminum

5-8 ENGINE OVERHAUL

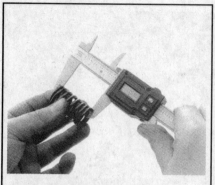
Fig. 24 Use a caliper to check the valve spring free-length

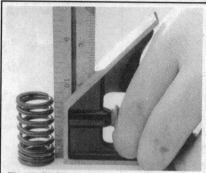
Fig. 25 Check the valve spring for squareness on a flat surface; a carpenter's square can be used

Fig. 26 A dial gauge may be used to check valve stem-to-guide clearance; read the gauge while moving the valve stem

heads), check them for any looseness or evidence of movement. All of the guides should appear to be at the same height from the spring seat. If any seem lower (or higher) from another, the guide has moved. Mount a dial indicator onto the spring side of the cylinder head. Lightly oil the valve stem and insert it into the cylinder head. Position the dial indicator against the valve stem near the tip and zero the gauge. Grasp the valve stem and wiggle towards and away from the dial indicator and observe the readings. Mount the dial indicator 90 degrees from the initial point and zero the gauge and again take a reading. Compare the two readings for a out of round condition. Check the readings against the specifications given. An Inside Diameter (I.D.) gauge designed for valve guides will give you an accurate valve guide bore measurement. If the I.D. gauge is used, compare the readings with the specifications given. Any guides that fail these inspections should be replaced or machined.

Valve Seats

A visual inspection of the valve seats should show a slightly worn and pitted surface where the valve face contacts the seat. Inspect the seat carefully for severe pitting or cracks. Also, a seat that is badly worn will be recessed into the cylinder head. A severely worn or recessed seat may need to be replaced. All cracked seats must be replaced. A seat concentricity gauge, if available, should be used to check the seat run-out. If run-out exceeds specifications the seat must be machined (if no specification is given use 0.002 in. or 0.051mm).

Cylinder Head Surface Flatness

◆ See Figures 27 and 28

After you have cleaned the gasket surface of the cylinder head of any old gasket material, check the head for flatness.

Place a straightedge across the gasket surface. Using feeler gauges, determine the clearance at the center of the straightedge and across the cylinder head at several points. Check along the centerline and diagonally on the head surface. If the warpage exceeds 0.003 in. (0.076mm) within a 6.0 in. (152mm) span, or 0.006 in. (0.152mm) over the total length of the head, the cylinder head must be resurfaced. After resurfacing the heads of a V-type engine, the intake manifold flange surface should be checked, and if necessary, milled proportionally to allow for the change in its mounting position. Always check the specifications chart for exact specifications.

Cracks And Physical Damage

Generally, cracks are limited to the combustion chamber, however, it is not uncommon for the head to crack in a spark plug hole, port, outside of the head or in the valve spring/rocker arm area. The first area to inspect is always the hottest: the exhaust seat/port area.

A visual inspection should be performed, but just because you don't see a crack does not mean it is not there. Some more reliable methods for inspecting for cracks include Magnaflux®, a magnetic process or Zyglo®, a dye penetrant. Magnaflux® is used only on ferrous metal (cast iron) heads. Zyglo® uses a spray on fluorescent mixture along with a black light to reveal the cracks. It is strongly recommended to have your cylinder head checked professionally for cracks, especially if the engine was known to have overheated and/or leaked or consumed coolant. Contact a local shop for availability and pricing of these services.

Physical damage is usually very evident. For example, a broken mounting ear from dropping the head or a bent or broken stud and/or bolt. All of these defects should be fixed or, if unrepairable, the head should be replaced.

REFINISHING & REPAIRING

Many of the procedures given for refinishing and repairing the cylinder head components must be performed by a machine shop. Certain steps, if the inspected part is not worn, can be performed yourself inexpensively. However, you spent a lot of time and effort so far; why risk trying to save a couple bucks if you might have to do it all over again?

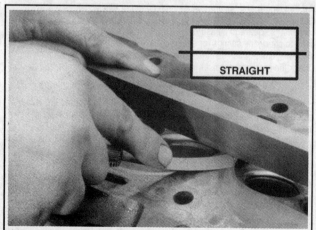
Fig. 27 Check the head for flatness across the center of the head surface using a straightedge and feeler gauge

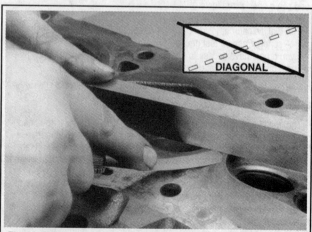
Fig. 28 Checks should also be made along both diagonals of the head surface

ENGINE OVERHAUL 5-9

Valves

Any valves that were not replaced should be refaced and the tips ground flat. Unless you have access to a valve grinding machine, this should be done by a machine shop. If the valves are in extremely good condition, as well as the valve seats and guides, they may be lapped in without performing machine work.

It is a recommended practice to lap the valves even after machine work has been performed and/or new valves have been purchased. This insures a positive seal between the valve and seat.

Lapping The Valves

■ Before lapping the valves to the seats, read the rest of the cylinder head procedure to insure that any related parts are in acceptable enough condition to continue.

■ Before any valve seat machining and/or lapping can be performed, the guides must be within factory recommended specifications.

1. Invert the cylinder head.
2. Lightly lubricate the valve stems and insert them into the cylinder head in their numbered order.
3. Raise the valve from the seat and apply a small amount of fine lapping compound to the seat.
4. Moisten the suction head of a hand-lapping tool and attach it to the head of the valve.
5. Rotate the tool between the palms of both hands, changing the position of the valve on the valve seat and lifting the tool often to prevent grooving.
6. Lap the valve until a smooth, polished circle is evident on the valve and seat.
7. Remove the tool and the valve. Wipe away all traces of the grinding compound and store the valve to maintain its lapped location.

❋❋ WARNING

Do not get the valves out of order after they have been lapped. They must be put back with the same valve seat with which they were lapped.

Springs, Retainers and Valve Locks

There is no repair or refinishing possible with the springs, retainers and valve locks. If they are found to be worn or defective, they must be replaced with new (or known good) parts.

Cylinder Head

Most refinishing procedures dealing with the cylinder head must be performed by a machine shop. Read the procedures below and review your inspection data to determine whether or not machining is necessary.

Valve Guide

■ If any machining or replacements are made to the valve guides, the seats must be machined.

Unless the valve guides need machining or replacing, the only service to perform is to thoroughly clean them of any dirt or oil residue.

There are only two types of valve guides used on automobile engines: the replaceable-type (all aluminum heads) and the cast-in integral-type (most cast iron heads). There are four recommended methods for repairing worn guides.
- Knurling
- Inserts
- Reaming oversize
- Replacing

Knurling is a process in which metal is displaced and raised, thereby reducing clearance, giving a true center, and providing oil control. It is the least expensive way of repairing the valve guides. However, it is not necessarily the best, and in some cases, a knurled valve guide will not stand up for more than a short time. It requires a special knurlizer and precision reaming tools to obtain proper clearances. It would not be cost effective to purchase these tools, unless you plan on rebuilding several of the same cylinder head.

Installing a guide insert involves machining the guide to accept a bronze insert. One style is the coil-type which is installed into a threaded guide. Another is the thin-walled insert where the guide is reamed oversize to accept a split-sleeve insert. After the insert is installed, a special tool is then run through the guide to expand the insert, locking it to the guide. The insert is then reamed to the standard size for proper valve clearance.

Reaming for oversize valves restores normal clearances and provides a true valve seat. Most cast-in type guides can be reamed to accept an valve with an oversize stem. The cost factor for this can become quite high as you will need to purchase the reamer and new, oversize stem valves for all guides which were reamed. Oversizes are generally 0.003 to 0.030 in. (0.076 to 0.762mm), with 0.015 in. (0.381mm) being the most common.

To replace cast-in type valve guides, they must be drilled out, then reamed to accept replacement guides. This must be done on a fixture which will allow centering and leveling off of the original valve seat or guide, otherwise a serious guide-to-seat misalignment may occur making it impossible to properly machine the seat.

Replaceable-type guides are pressed into the cylinder head. A hammer and a stepped drift or punch may be used to install and remove the guides. Before removing the guides, measure the protrusion on the spring side of the head and record it for installation. Use the stepped drift to hammer out the old guide from the combustion chamber side of the head. When installing, determine whether or not the guide also seals a water jacket in the head, and if it does, use the recommended sealing agent. If there is no water jacket, grease the valve guide and its bore. Use the stepped drift, and hammer the new guide into the cylinder head from the spring side of the cylinder head. A stack of washers the same thickness as the measured protrusion may help the installation process.

Valve Seats

■ Before any valve seat machining can be performed, the guides must be within factory recommended specifications.

■ If any machining or replacements were made to the valve guides, the seats must be machined.

If the seats are in good condition, the valves can be lapped to the seats, and the cylinder head assembled. See the valves procedures for instructions on lapping.

If the valve seats are worn, cracked or damaged, they must be serviced by a machine shop. The valve seat must be perfectly centered to the valve guide, which requires very accurate machining.

Cylinder Head Surface

If the cylinder head is warped, it must be machined flat. If the warpage is extremely severe, the head may need to be replaced. In some instances, it may be possible to straighten a warped head enough to allow machining. In either case, contact a professional machine shop for service.

Cracks And Physical Damage

Certain cracks can be repaired in both cast iron and aluminum heads. For cast iron, a tapered threaded insert is installed along the length of the crack. Aluminum can also use the tapered inserts, however welding is the preferred method. Some physical damage can be repaired through brazing or welding. Contact a machine shop to get expert advice for your particular dilemma.

ASSEMBLY

The first step for any assembly job is to have a clean area in which to work. Next, thoroughly clean all of the parts and components that are to be assembled. Finally, place all of the components onto a suitable work space and, if necessary, arrange the parts to their respective positions.

1. Lightly lubricate the valve stems and insert all of the valves into the cylinder head. If possible, maintain their original locations.
2. If equipped, install any valve spring shims which were removed.
3. If equipped, install the new valve seals, keeping the following in mind:
   • If the valve seal presses over the guide, lightly lubricate the outer guide surfaces.
   • If the seal is an O-ring type, it is installed just after compressing the spring but before the valve locks.
4. Place the valve spring and retainer over the stem.
5. Position the spring compressor tool and compress the spring.
6. Assemble the valve locks to the stem.

5-10 ENGINE OVERHAUL

7. Relieve the spring pressure slowly and insure that neither valve lock becomes dislodged by the retainer.
8. Remove the spring compressor tool.
9. Repeat Steps 2 through 8 until all of the springs have been installed.

Engine Block

GENERAL INFORMATION

A thorough overhaul or rebuild of an engine block would include replacing the pistons, rings, bearings, timing belt/chain assembly and oil pump. For OHV engines also include a new camshaft and lifters. The block would then have the cylinders bored and honed oversize (or if using removable cylinder sleeves, new sleeves installed) and the crankshaft would be cut undersize to provide new wearing surfaces and perfect clearances. However, your particular engine may not have everything worn out. What if only the piston rings have worn out and the clearances on everything else are still within factory specifications? Well, you could just replace the rings and put it back together, but this would be a very rare example. Chances are, if one component in your engine is worn, other components are sure to follow, and soon. At the very least, you should always replace the rings, bearings and oil pump. This is what is commonly called a "freshen up".

Cylinder Ridge Removal

Because the top piston ring does not travel to the very top of the cylinder, a ridge is built up between the end of the travel and the top of the cylinder bore.

Pushing the piston and connecting rod assembly past the ridge can be difficult, and damage to the piston ring lands could occur. If the ridge is not removed before installing a new piston or not removed at all, piston ring breakage and piston damage may occur.

■ **It is always recommended that you remove any cylinder ridges before removing the piston and connecting rod assemblies. If you know that new pistons are going to be installed and the engine block will be bored oversize, you may be able to forego this step. However, some ridges may actually prevent the assemblies from being removed, necessitating its removal.**

There are several different types of ridge reamers on the market, none of which are inexpensive. Unless a great deal of engine rebuilding is anticipated, borrow or rent a reamer.
1. Turn the crankshaft until the piston is at the bottom of its travel.
2. Cover the head of the piston with a rag.
3. Follow the tool manufacturers instructions and cut away the ridge, exercising extreme care to avoid cutting too deeply.
4. Remove the ridge reamer, the rag and as many of the cuttings as possible. Continue until all of the cylinder ridges have been removed.

DISASSEMBLY

◆ See Figures 29 and 30

The engine disassembly instructions following assume that you have the engine mounted on an engine stand. If not, it is easiest to disassemble the engine on a bench or the floor with it resting on the bell housing or transmission mounting surface. You must be able to access the connecting rod fasteners and turn the crankshaft during disassembly. Also, all engine covers (timing, front, side, oil pan, whatever) should have already been removed. Engines which are seized or locked up may not be able to be completely disassembled, and a core (salvage yard) engine should be purchased.

If not done during the cylinder head removal, remove the pushrods and lifters, keeping them in order for assembly. Remove the timing gears and/or timing chain assembly, then remove the oil pump drive assembly and withdraw the camshaft from the engine block. Remove the oil pick-up and pump assembly. If equipped, remove any balance or auxiliary shafts. If necessary, remove the cylinder ridge from the top of the bore. See the cylinder ridge removal procedure.

Rotate the engine over so that the crankshaft is exposed. Use a number punch or scribe and mark each connecting rod with its respective cylinder number. The cylinder closest to the front of the engine is always number 1. However, depending on the engine placement, the front of the engine could either be the flywheel or damper/pulley end. Generally, the front of the engine faces the bow. Use a number punch or scribe and also mark the main bearing caps from front to rear with the front most cap being number 1 (if there are five caps, mark them 1 through 5, front to rear).

✲✲ WARNING

Take special care when pushing the connecting rod up from the crankshaft because the sharp threads of the rod bolts/studs will score the crankshaft journal. Insure that special plastic caps are installed over them, or cut two pieces of rubber hose to do the same.

Again, rotate the engine, this time to position the number one cylinder bore (head surface) up. Turn the crankshaft until the number one piston is at the bottom of its travel, this should allow the maximum access to its connecting rod. Remove the number one connecting rods fasteners and cap and place two lengths of rubber hose over the rod bolts/studs to protect the crankshaft from damage. Using a sturdy wooden dowel and a hammer, push the connecting rod up about 1 in. (25mm) from the crankshaft and remove the upper bearing insert. Continue pushing or tapping the connecting rod up until the piston rings are out of the cylinder bore. Remove the piston and rod by hand, put the upper half of the bearing insert back into the rod, install the 9cap with its bearing insert installed, and hand-tighten the cap fasteners. If

Fig. 29 Place rubber hose over the connecting rod studs to protect the crankshaft and cylinder bores from damage

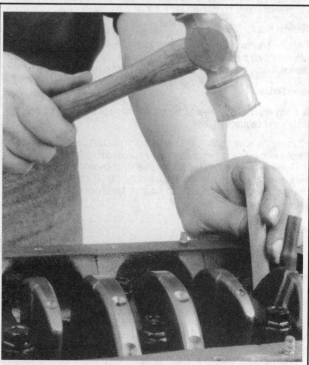

Fig. 30 Carefully tap the piston out of the bore using a wooden dowel

ENGINE OVERHAUL 5-11

the parts are kept in order in this manner, they will not get lost and you will be able to tell which bearings came form what cylinder if any problems are discovered and diagnosis is necessary. Remove all the other piston assemblies in the same manner. On V-style engines, remove all of the pistons from one bank, then reposition the engine with the other cylinder bank head surface up, and remove that banks piston assemblies.

The only remaining component in the engine block should now be the crankshaft. Loosen the main bearing caps evenly until the fasteners can be turned by hand, then remove them and the caps. Remove the crankshaft from the engine block. Thoroughly clean all of the components.

INSPECTION

Now that the engine block and all of its components are clean, it's time to inspect them for wear and/or damage. To accurately inspect them, you will need some specialized tools:
• Two or three separate micrometers to measure the pistons and crankshaft journals
• A dial indicator
• Telescoping gauges for the cylinder bores
• A rod alignment fixture to check for bent connecting rods

If you do not have access to the proper tools, you may want to bring the components to a shop that does.

Generally, you shouldn't expect cracks in the engine block or its components unless it was known to leak, consume or mix engine fluids, it was severely overheated, or there was evidence of bad bearings and/or crankshaft damage. A visual inspection should be performed on all of the components, but just because you don't see a crack does not mean it is not there. Some more reliable methods for inspecting for cracks include Magnaflux®, a magnetic process or Zyglo®, a dye penetrant. Magnaflux® is used only on ferrous metal (cast iron). Zyglo® uses a spray on fluorescent mixture along with a black light to reveal the cracks. It is strongly recommended to have your engine block checked professionally for cracks, especially if the engine was known to have overheated and/or leaked or consumed coolant. Contact a local shop for availability and pricing of these services.

Engine Block
Engine Block Bearing Alignment

Remove the main bearing caps and, if still installed, the main bearing inserts. Inspect all of the main bearing saddles and caps for damage, burrs or high spots. If damage is found, and it is caused from a spun main bearing, the block will need to be align-bored or, if severe enough, replacement. Any burrs or high spots should be carefully removed with a metal file.

Place a straightedge on the bearing saddles, in the engine block, along the centerline of the crankshaft. If any clearance exists between the straightedge and the saddles, the block must be align-bored.

Align-boring consists of machining the main bearing saddles and caps by means of a flycutter that runs through the bearing saddles.

Deck Flatness
◆ See Figure 31

The top of the engine block where the cylinder head mounts is called the deck. Insure that the deck surface is clean of dirt, carbon deposits and old gasket material. Place a straightedge across the surface of the deck along its centerline and, using feeler gauges, check the clearance along several points. Repeat the checking procedure with the straightedge placed along both diagonals of the deck surface. If the reading exceeds 0.003 in. (0.076mm) within a 6.0 in. (152mm) span, or 0.006 in. (0.152mm) over the total length of the deck, it must be machined. Always check the Specification chart for your specific engine.

Cylinder Bores
◆ See Figure 32

The cylinder bores house the pistons and are slightly larger than the pistons themselves. A common piston-to-bore clearance is 0.0015-0.0025 in. (0.0381mm-0.0635mm) - but always refer to the Specification chart for your specific engine. Inspect and measure the cylinder bores. The bore should be checked for out-of-roundness, taper and size. The results of this inspection will determine whether the cylinder can be used in its existing size and condition, or a rebore to the next oversize is required (or in the case of removable sleeves, have replacements installed).

The amount of cylinder wall wear is always greater at the top of the cylinder than at the bottom. This wear is known as taper. Any cylinder that has a taper of 0.0012 in. (0.305mm) or more, must be re-bored. Measurements are taken at a number of positions in each cylinder: at the top, middle and bottom and at two points at each position; that is, at a point 90 degrees from the crankshaft centerline, as well as a point parallel to the crankshaft centerline. The measurements are made with either a special dial indicator or a telescopic gauge and micrometer. If the necessary precision tools to check the bore are not available, take the block to a machine shop and have them mike it. Also if you don't have the tools to check the cylinder bores, chances are you will not have the necessary devices to check the pistons, connecting rods and crankshaft. Take these components with you and save yourself an extra trip.

For our procedures, we will use a telescopic gauge and a micrometer. You will need one of each, with a measuring range which covers your cylinder bore size.

1. Position the telescopic gauge in the cylinder bore, loosen the gauges lock and allow it to expand.

■ **Your first two readings will be at the top of the cylinder bore, then proceed to the middle and finally the bottom, making a total of six measurements.**

2. Hold the gauge square in the bore, 90 degrees from the crankshaft centerline, and gently tighten the lock. Tilt the gauge back to remove it from the bore.
3. Measure the gauge with the micrometer and record the reading.
4. Again, hold the gauge square in the bore, this time parallel to the crankshaft centerline, and gently tighten the lock. Again, you will tilt the gauge back to remove it from the bore.
5. Measure the gauge with the micrometer and record this reading. The difference between these two readings is the out-of-round measurement of the cylinder.
6. Repeat Steps 1 through 5, each time going to the next lower position, until you reach the bottom of the cylinder. Then go to the next cylinder, and continue until all of the cylinders have been measured.

The difference between these measurements will tell you all about the wear in your cylinders. The measurements which were taken 90 degrees from the crankshaft centerline will always reflect the most wear. That is because at this position is where the engine power presses the piston against the cylinder bore the hardest. This is known as thrust wear. Take your top, 90 degree measurement and compare it to your bottom, 90 degree measurement. The difference between them is the taper. When you measure your pistons, you will compare these readings to your piston sizes and determine piston-to-wall clearance.

Crankshaft
◆ See Figure 33

Inspect the crankshaft for visible signs of wear or damage. All of the journals should be perfectly round and smooth. Slight scores are normal for a used crankshaft, but you should hardly feel them with your fingernail. When measuring the crankshaft with a micrometer, you will take readings at the front and rear of each journal, then turn the micrometer 90 degrees and take two more readings, front and rear. The difference between the front-to-rear readings is the journal taper and the first-to-90 degree reading is the out-of-round measurement. Generally, there should be no taper or out-of-roundness found, however, up to 0.0005 in. (0.0127mm) for either can be overlooked. Also, the readings should fall within the factory specifications for journal diameters - check the chart.

If the crankshaft journals fall within specifications, it is recommended that it be polished before being returned to service. Polishing the crankshaft insures that any minor burrs or high spots are smoothed, thereby reducing the chance of scoring the new bearings.

Pistons and Connecting Rods
Pistons
◆ See Figure 34

The piston should be visually inspected for any signs of cracking or burning (caused by hot spots or detonation), and scuffing or excessive wear on the skirts. The wrist pin attaches the piston to the connecting rod. The piston should move freely on the wrist pin, both sliding and pivoting. Grasp the connecting rod securely, or mount it in a vise, and try to rock the piston back and forth along the centerline of the wrist pin. There should not be any excessive play evident between the piston and the pin. If there are C-clips

5-12 ENGINE OVERHAUL

retaining the pin in the piston then you have wrist pin bushings in the rods. There should not be any excessive play between the wrist pin and the rod bushing. Normal clearance for the wrist pin is approx. 0.001-0.002 in. (0.025-0.051mm). Please refer to the Specification chart for your specific engine.

Use a micrometer and measure the diameter of the piston, perpendicular to the wrist pin, on the skirt. Compare the reading to its original cylinder measurement obtained earlier. The difference between the two readings is the piston-to-wall clearance. If the clearance is within specifications, the piston may be used as is. If the piston is out of specification, but the bore is not, you will need a new piston. If both are out of specification, you will need the cylinder re-bored and oversize pistons installed. Generally if two or more pistons/bores are out of specification, it is best to re-bore the entire block and purchase a complete set of oversize pistons.

Connecting Rod
◆ See Figures 35 and 36

You should have the connecting rod checked for straightness at a machine shop. If the connecting rod is bent, it will unevenly wear the bearing and piston, as well as place greater stress on these components. Any bent or twisted connecting rods must be replaced. If the rods are straight and the wrist pin clearance is within specifications, then only the bearing end of the rod need be checked. Place the connecting rod into a vice, with the bearing inserts in place, install the cap to the rod and torque the fasteners to specifications. Use a telescoping gauge and carefully measure the inside diameter of the bearings. Compare this reading to the rods original crankshaft journal diameter measurement. The difference is the oil clearance. If the oil clearance is not within specifications, install new bearings in the rod and take another measurement. If the clearance is still out of specifications, and the crankshaft is not, the rod will need to be reconditioned by a machine shop.

■ You can also use Plastigage® to check the bearing clearances. The assembling procedure has complete instructions on its use.

Camshaft

Inspect the camshaft and lifters/followers as described in the Engine Mechanical section.

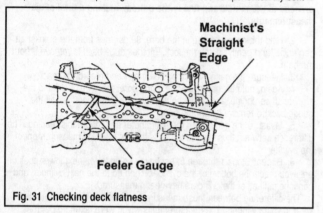
Fig. 31 Checking deck flatness

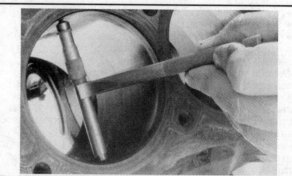
Fig. 32 Use a telescoping gauge to measure the cylinder bore diameter - take several readings within the same bore

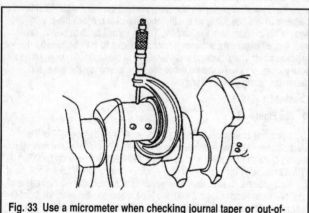
Fig. 33 Use a micrometer when checking journal taper or out-of-round

Fig. 34 Measure the piston's outer diameter, perpendicular to the wrist pin, with a micrometer

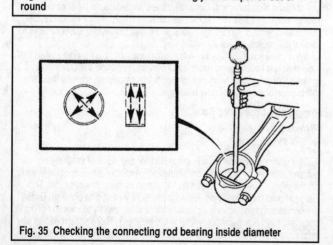
Fig. 35 Checking the connecting rod bearing inside diameter

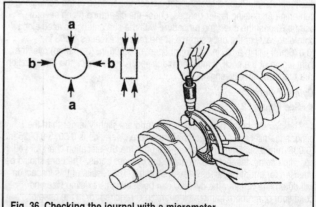
Fig. 36 Checking the journal with a micrometer

ENGINE OVERHAUL 5-13

Bearings

All of the engine bearings should be visually inspected for wear and/or damage. The bearing should look evenly worn all around with no deep scores or pits. If the bearing is severely worn, scored, pitted or heat blued, then the bearing, and the components that use it, should be brought to a machine shop for inspection. Full-circle bearings (used on most camshafts, auxiliary shafts, balance shafts, etc.) require specialized tools for removal and installation, and should be brought to a machine shop for service.

Oil Pump

■ The oil pump is responsible for providing constant lubrication to the whole engine and so it is recommended that a new oil pump be installed when rebuilding the engine.

Completely disassemble the oil pump and thoroughly clean all of the components. Inspect the oil pump gears and housing for wear and/or damage. Insure that the pressure relief valve operates properly and there is no binding or sticking due to varnish or debris. If all of the parts are in proper working condition, lubricate the gears and relief valve, and assemble the pump.

REFINISHING

◆ See Figure 37

Almost all engine block refinishing must be performed by a machine shop. If the cylinders are not to be re-bored, then the cylinder glaze can be removed with a ball hone. When removing cylinder glaze with a ball hone, use a light or penetrating type oil to lubricate the hone. Do not allow the hone to run dry as this may cause excessive scoring of the cylinder bores and wear on the hone. If new pistons are required, they will need to be installed to the connecting rods. This should be performed by a machine shop as the pistons must be installed in the correct relationship to the rod or engine damage can occur.

Pistons and Connecting Rods

◆ See Figure 38

Only pistons with the wrist pin retained by C-clips are serviceable by the home-mechanic. Press fit pistons require special presses and/or heaters to remove/install the connecting rod and should only be performed by a machine shop.

All pistons will have a mark indicating the direction to the front of the engine and the must be installed into the engine in that manner. Usually it is a notch or arrow on the top of the piston, or it may be the letter **F** cast or stamped into the piston.

C-Clip Type Pistons

1. Note the location of the forward mark on the piston and mark the connecting rod in relation.
2. Remove the C-clips from the piston and withdraw the wrist pin.

■ Varnish build-up or C-clip groove burrs may increase the difficulty of removing the wrist pin. If necessary, use a punch or drift to carefully tap the wrist pin out.

3. Insure that the wrist pin bushing in the connecting rod is usable, and lubricate it with assembly lube.
4. Remove the wrist pin from the new piston and lubricate the pin bores on the piston.
5. Align the forward marks on the piston and the connecting rod and install the wrist pin.
6. The new C-clips will have a flat and a rounded side to them. Install both C-clips with the flat side facing out.
7. Repeat all of the steps for each piston being replaced.

ASSEMBLY

Before you begin assembling the engine, first give yourself a clean, dirt free work area. Next, clean every engine component again. The key to a good assembly is cleanliness.

Mount the engine block into the engine stand and wash it one last time using water and detergent (dishwashing detergent works well). While washing it, scrub the cylinder bores with a soft bristle brush and thoroughly clean all of the oil passages. Completely dry the engine and spray the entire assembly down with an anti-rust solution such as WD-40® or similar product. Take a clean lint-free rag and wipe up any excess anti-rust solution from the bores, bearing saddles, etc. Repeat the final cleaning process on the crankshaft. Replace any freeze or oil galley plugs which were removed during disassembly.

Crankshaft

◆ See Figures 39 thru 43

1. Remove the main bearing inserts from the block and bearing caps.
2. If the crankshaft main bearing journals have been refinished to a definite undersize, install the correct undersize bearing. Be sure that the bearing inserts and bearing bores are clean. Foreign material under inserts will distort bearing and cause failure.
3. Place the upper main bearing inserts in bores with tang in slot.

■ The oil holes in the bearing inserts must be aligned with the oil holes in the cylinder block.

4. Install the lower main bearing inserts in bearing caps.
5. Clean the mating surfaces of block and rear main bearing cap.
6. Carefully lower the crankshaft into place. Be careful not to damage bearing surfaces.
7. Check the clearance of each main bearing by using the following procedure:
   a. Place a piece of Plastigage® or its equivalent, on bearing surface across full width of bearing cap and about 1/4 in. off center.

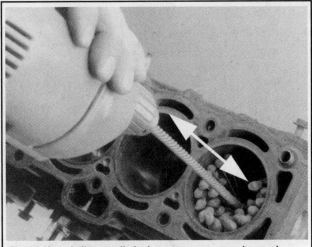

Fig. 37 Use a ball type cylinder hone to remove any glaze and provide a new surface for seating the piston rings

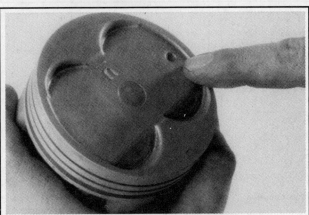

Fig. 38 Most pistons are marked to indicate positioning in the engine (usually a mark means the side facing the front)

5-14 ENGINE OVERHAUL

b. Install cap and tighten bolts to specifications. Do not turn crankshaft while Plastigage® is in place.

c. Remove the cap. Using the supplied Plastigage® scale, check width of Plastigage® at widest point to get maximum clearance. Difference between readings is taper of journal.

d. If clearance exceeds specified limits, try a 0.001 in. or 0.002 in. undersize bearing in combination with the standard bearing. Bearing clearance must be within specified limits. If standard and 0.002 in. undersize bearing does not bring clearance within desired limits, refinish crankshaft journal, then install undersize bearings.

8. Install the rear main seal.

9. After the bearings have been fitted, apply a light coat of engine oil to the journals and bearings. Install the rear main bearing cap. Install all bearing caps except the thrust bearing cap. Be sure that main bearing caps are installed in original locations. Tighten the bearing cap bolts to specifications.

10. Install the thrust bearing cap with bolts finger-tight.

11. Pry the crankshaft forward against the thrust surface of upper half of bearing.

12. Hold the crankshaft forward and pry the thrust bearing cap to the rear. This aligns the thrust surfaces of both halves of the bearing.

13. Retain the forward pressure on the crankshaft. Tighten the cap bolts to specifications.

14. Measure the crankshaft end-play as follows:

a. Mount a dial gauge to the engine block and position the tip of the gauge to read from the crankshaft end.

b. Carefully pry the crankshaft toward the rear of the engine and hold it there while you zero the gauge.

c. Carefully pry the crankshaft toward the front of the engine and read the gauge.

d. Confirm that the reading is within specifications. If not, install a new thrust bearing and repeat the procedure. If the reading is still out of specifications with a new bearing, have a machine shop inspect the thrust surfaces of the crankshaft, and if possible, repair it.

15. Rotate the crankshaft so as to position the first rod journal to the bottom of its stroke.

Pistons and Connecting Rods

◆ See Figures 44 thru 49

1. Before installing the piston/connecting rod assembly, oil the pistons, piston rings and the cylinder walls with light engine oil. Install connecting rod bolt protectors or rubber hose onto the connecting rod bolts/studs. Also perform the following:

a. Select the proper ring set for the size cylinder bore.

b. Position the ring in the bore in which it is going to be used.

c. Push the ring down into the bore area where normal ring wear is not encountered.

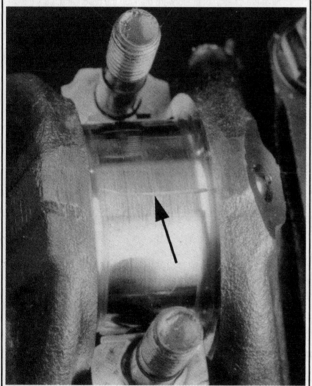

Fig. 39 Apply a strip of gauging material to the bearing journal, then install and torque the cap

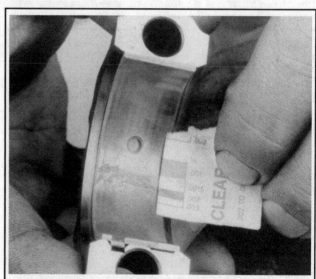

Fig. 40 After the cap is removed again, use the scale supplied with the gauging material to check the clearance

Fig. 41 A dial gauge may be used to check crankshaft end-play...

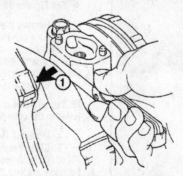

Fig. 42 ...or you can use a feeler gauge

Fig. 43 Carefully pry the crankshaft back and forth while reading the dial gauge for end-play

d. Use the head of the piston to position the ring in the bore so that the ring is square with the cylinder wall. Use caution to avoid damage to the ring or cylinder bore.

e. Measure the gap between the ends of the ring with a feeler gauge. Ring gap in a worn cylinder is normally greater than specification. If the ring gap is greater than the specified limits, try an oversize ring set.

f. Check the ring side clearance of the compression rings with a feeler gauge inserted between the ring and its lower land according to specification. The gauge should slide freely around the entire ring circumference without binding. Any wear that occurs will form a step at the inner portion of the lower land. If the lower lands have high steps, the piston should be replaced.

2. Unless new pistons are installed, be sure to install the pistons in the cylinders from which they were removed. The numbers on the connecting rod and bearing cap must be on the same side when installed in the cylinder bore. If a connecting rod is ever transposed from one engine or cylinder to another, new bearings should be fitted and the connecting rod should be numbered to correspond with the new cylinder number. The notch on the piston head goes toward the front of the engine.

3. Install all of the rod bearing inserts into the rods and caps.

4. Install the rings to the pistons. Install the oil control ring first, then the second compression ring and finally the top compression ring. Use a piston ring expander tool to aid in installation and to help reduce the chance of breakage.

5. Make sure the ring gaps are properly spaced around the circumference of the piston. Fit a piston ring compressor around the piston and slide the piston and connecting rod assembly down into the cylinder bore, pushing it in with the wooden hammer handle. Push the piston down until it is only slightly below the top of the cylinder bore. Guide the connecting rod onto the crankshaft bearing journal carefully, to avoid damaging the crankshaft.

6. Check the bearing clearance of all the rod bearings, fitting them to the crankshaft bearing journals. Follow the procedure in the crankshaft installation above.

7. After the bearings have been fitted, apply a light coating of assembly oil to the journals and bearings.

8. Turn the crankshaft until the appropriate bearing journal is at the bottom of its stroke, then push the piston assembly all the way down until the connecting rod bearing seats on the crankshaft journal. Be careful not to allow the bearing cap screws to strike the crankshaft bearing journals and damage them.

9. After the piston and connecting rod assemblies have been installed, check the connecting rod side clearance on each crankshaft journal.

10. Prime and install the oil pump and the oil pump intake tube.
11. Install the auxiliary/balance shaft/assembly if equipped.

Camshaft, Lifters And Timing Assembly

1. Install the camshaft.
2. Install the lifters/followers into their bores.
3. Install the timing gears/chain assembly.

Cylinder Head(s)

1. Install the cylinder head(s) using new gaskets.
2. Assemble the rest of the valve train (pushrods and rocker arms and/or shafts).

Engine Start-up and Break-in

STARTING THE ENGINE

Now that the engine is installed and every wire and hose is properly connected, go back and double check that all coolant and vacuum hoses are connected. Check that your oil drain plug is installed and properly tightened. If not already done, install a new oil filter onto the engine. Fill the crankcase with the proper amount and grade of engine oil. Fill the cooling system with a 50/50 mixture of coolant/water on models with a closed system.

1. Connect the battery.
2. Start the engine. Keep your eye on your oil pressure indicator; if it does not indicate oil pressure within 10 seconds of starting, turn the engine off.

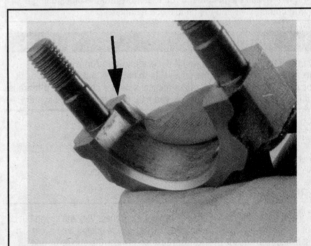

Fig. 45 The notch on the side of the bearing cap matches the tang on the bearing insert

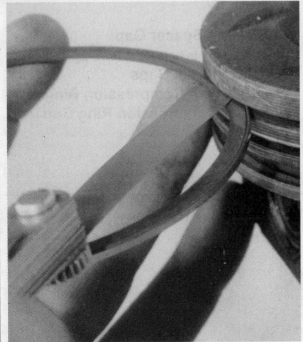

Fig. 44 Checking the piston ring-to-ring groove side clearance using the ring and a feeler gauge

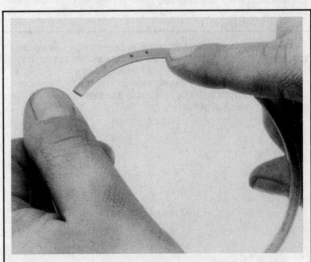

Fig. 46 Most rings are marked to show which side of the ring should face up when installed to the piston

5-16 ENGINE OVERHAUL

Fig. 47 Install the piston and rod assembly into the block using a ring compressor and the handle of a hammer

 **WARNING**

Damage to the engine can result if it is allowed to run with no oil pressure. Check the engine oil level to make sure that it is full. Check for any leaks and if found, repair the leaks before continuing. If there is still no indication of oil pressure, you may need to prime the system.

3. Confirm that there are no fluid leaks (oil or other).
4. Allow the engine to reach normal operating temperature.
5. At this point you can perform any necessary checks or adjustments, such as checking the ignition timing.
6. Install any remaining components that were removed.

BREAKING IT IN

Make the first hours on the new engine, easy ones. Vary the speed but do not accelerate hard. Most importantly, do not lug the engine, and avoid sustained high speeds until at least 20 hours. Check the engine oil and coolant levels frequently. Expect the engine to use a little oil until the rings seat. Change the oil and filter at 20 hours and then follow the normal maintenance intervals from there out.

KEEP IT MAINTAINED

Now that you have just gone through all of that hard work, keep yourself from doing it all over again by thoroughly maintaining it. Not that you may not have maintained it before, heck you could have had a couple of thousand hours on it before doing this. However, you may have bought the vehicle used, and the previous owner did not keep up on maintenance. Which is why you just went through all of that hard work. See?

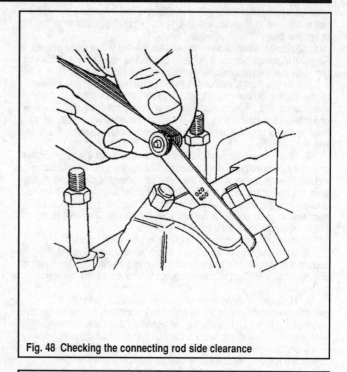

Fig. 48 Checking the connecting rod side clearance

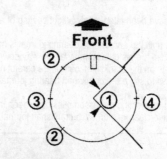

1 - **Oil Ring Spacer Gap**
   (8.1 L: Tang in hole, or slot within arc)
2 - **Oil Ring Rail Gaps**
3 - **Lower (2nd) Compression Ring Gap**
4 - **Upper Compression Ring Gap**

Fig. 49 Piston ring gap location

6
FUEL SYSTEM - CARBURETORS

FUEL AND COMBUSTION	6-2
CARBURETED FUEL SYSTEM	6-3
MERCARB 2BBLCARB	6-8
WEBER WFB 4BBL CARB	6-17
WIRING DIAGRAMS	6-27
SPECIFICATIONS	6-30

CARBURETED FUEL SYSTEM 6-3
  CARBURETOR APPLICATIONS 6-3
  ELECTRIC FUEL PUMP 6-26
  MECHANICAL FUEL PUMP 6-24
  MERCARB 2BBL CARBURETOR 6-8
  OPERATION 6-5
  TROUBLESHOOTING 6-3
  WEBER WFB 4BBL CARBURETOR ... 6-17
COMBUSTION 6-2
ELECTRIC FUEL PUMP 6-26
  DESCRIPTION.................... 6-26
  PRESSURE TEST 6-27
  REMOVAL & INSTALLATION 6-26
FUEL 6-2
  ALCOHOL-BLENDED FUELS 6-2
  HIGH ALTITUDE OPERATION 6-2
  OCTANE RATING 6-2
  RECOMMENDATIONS 6-2
  VAPOR PRESSURE 6-2
FUEL AND COMBUSTION 6-2
  COMBUSTION..................... 6-2
  FUEL 6-2
MECHANICAL FUEL PUMP.......... 6-24
  DESCRIPTION.................... 6-24
  FUEL LINE TEST................. 6-25
  PRESSURE TEST 6-25
  REMOVAL & INSTALLATION 6-25
MERCARB 2BBL CARB 6-8
  ADJUSTMENT 6-9
    ACCELERATOR PUMP LEVER 6-9
    CHOKE SETTING................ 6-11
    CHOKE UNLOADER 6-11
    FLOAT LEVEL AND DROP......... 6-11
    IDLE SPEED & MIXTURE 6-10
    PUMP ROD..................... 6-10
    THROTTLE CABLE 6-10
  ASSEMBLY 6-16

    MODELS W/O TKS................ 6-16
    MODELS W/TKS 6-17
  CLEANING & INSPECTION 6-16
  DESCRIPTION.................... 6-8
  DISASSEMBLY 6-12
    MODELS W/O TKS............... 6-12
    MODELS W/TKS 6-15
  REMOVAL & INSTALLATION 6-9
OPERATION....................... 6-5
  BASIC FUNCTIONS 6-5
  CARBURETOR CIRCUITS 6-7
  FUEL & AIR METERING 6-5
SPECIFICATIONS 6-30
  CARB APPLICATIONS 6-3
  CARB - 2 BBL MERCARB 6-30
  CARB - 4 BBL WEBER 6-30
TROUBLESHOOTING 6-3
  COMMON PROBLEMS - W/O TKS..... 6-4
  COMBUSTION RELATED PISTON
    FAILURES 6-4
  TKS DIAGNOSTICS 6-4
WEBER WFB 4BBL CARB........... 6-17
  ADJUSTMENT 6-18
    ACCELERATOR PUMP LEVER 6-18
    ACCELERATOR PUMP 6-18
    ELECTRIC CHOKE 6-18
    FLOAT LEVEL AND DROP......... 6-18
    IDLE SPEED & MIXTURE 6-19
    PORTED VACUUM SWITCH (PVS).. 6-19
    THROTTLE CABLE 6-20
  DESCRIPTION.................... 6-17
  DISASSEMBLY & ASSEMBLY........ 6-20
  REMOVAL & INSTALLATION 6-17
WIRING DIAGRAMS 6-27
  TKS SYSTEM, 3.0L 6-27
  TKS SYSTEM, 4.3L/5.0L W/ALPHA ... 6-28
  TKS SYSTEM, 5.0L/5.7L W/BRAVO ... 6-29

6-2 FUEL SYSTEM - CARBURETORS

FUEL AND COMBUSTION

Fuel

Fuel recommendations have become more complex as the chemistry of modern gasoline changes. The major driving force behind the changes in gasoline chemistry is the search for additives to replace lead as an octane booster and lubricant. These new additives are governed by the types of emissions they produce in the combustion process. Also, the replacement additives do not always provide the same level of combustion stability, making a fuel's octane rating less meaningful.

In the 1960's and 1970's, leaded fuel was common. The lead served two functions. First, it served as an octane booster (combustion stabilizer) and second, in 4-stroke engines, it served as a valve seat lubricant. For 2-stroke engines, the primary benefit of lead was to serve as a combustion stabilizer. Lead served very well for this purpose, even in high heat applications.

Today, all lead has been removed from the refining process. This means that the benefit of lead as an octane booster has been eliminated. Several substitute octane boosters have been introduced in the place of lead. While many are adequate in an engine, most do not perform nearly as well as lead did, even though the octane rating of the fuel is the same.

OCTANE RATING

A fuel's octane rating is a measurement of how stable the fuel is when heat is introduced. Octane rating is a major consideration when deciding whether a fuel is suitable for a particular application. For example, in an engine, we want the fuel to ignite when the spark plug fires and not before, even under high pressure and temperatures. Once the fuel is ignited, it must burn slowly and smoothly, even though heat and pressure are building up while the burn occurs. The unburned fuel should be ignited by the traveling flame front, not by some other source of ignition, such as carbon deposits or the heat from the expanding gasses. A fuel's octane rating is known as a measurement of the fuel's anti-knock properties (ability to burn without exploding).

Usually a fuel with a higher octane rating can be subjected to a more severe combustion environment before spontaneous or abnormal combustion occurs. To understand how two gasoline samples can be different, even though they have the same octane rating, we need to know how octane rating is determined.

The American Society of Testing and Materials (ASTM) has developed a universal method of determining the octane rating of a fuel sample. The octane rating you see on the pump at a fuel dock is known as the pump octane number. Look at the small print on the pump. The rating has a formula. The rating is determined by the R+M/2 method. This number is the average of the research octane reading and the motor octane rating.

- The Research Octane Rating is a measure of a fuel's anti-knock properties under a light load or part throttle conditions. During this test, combustion heat is easily dissipated.
- The Motor Octane Rating is a measure of a fuel's anti-knock properties under a heavy load or full throttle conditions, when heat buildup is at maximum.

VAPOR PRESSURE

Fuel vapor pressure is a measure of how easily a fuel sample evaporates. Many additives used in gasoline contain aromatics. Aromatics are light hydrocarbons distilled off the top of a crude oil sample. They are effective at increasing the research octane of a fuel sample but can cause vapor lock (bubbles in the fuel line) on a very hot day. If you have an inconsistent running engine and you suspect vapor lock, use a piece of clear fuel line to look for bubbles, indicating that the fuel is vaporizing.

One negative side effect of aromatics is that they create additional combustion products such as carbon and varnish. If your engine requires high octane fuel to prevent detonation, de-carbon the engine more frequently with an internal engine cleaner to prevent ring sticking due to excessive varnish buildup.

ALCOHOL-BLENDED FUELS

When the Environmental Protection Agency mandated a phase-out of the leaded fuels in January of 1986, fuel suppliers needed an additive to improve the octane rating of their fuels. Although there are multiple methods currently employed, the addition of alcohol to gasoline seems to be favored because of its favorable results and low cost. Two types of alcohol are used in fuel today as octane boosters, methanol (wood alcohol) or ethanol (grain alcohol).

When used as a fuel additive, alcohol tends to raise the research octane of the fuel. There are, however, some special considerations due to the effects of alcohol in fuel.

- Since alcohol contains oxygen, it replaces gasoline without oxygen content and tends to cause the air/fuel mixture to become leaner.
- On older engines, the leaching affect of alcohol may, in time, cause fuel lines and plastic components to become brittle to the point of cracking. Unless replaced, these cracked lines could leak fuel, increasing the potential for hazardous situations.
- When alcohol blended fuels become contaminated with water, the water combines with the alcohol then settles to the bottom of the tank. This leaves the gasoline on a top layer.

■ **Modern fuel lines and plastic fuel system components have been specially formulated to resist alcohol leaching effects.**

HIGH ALTITUDE OPERATION

At elevated altitudes there is less oxygen in the atmosphere than at sea level. Less oxygen means lower combustion efficiency and less power output. As a general rule, power output is reduced three percent for every thousand feet above sea level.

On carbureted engines, re-jetting for high altitude does not restore lost power, it simply corrects the air-fuel ratio for the reduced air density and makes the most of the remaining available power. The most important thing to remember when re-jetting for high altitude is to reverse the jetting when returning to sea level. If the jetting is left lean when you return to sea level conditions, the correct air/fuel ratio will not be achieved and possible engine damage may occur.

RECOMMENDATIONS

According to the fuel recommendations that come with your engine, there are only a few engines in the product line that requires more than 87 octane. Most MerCruiser engines need only 87 octane or less. An 89 octane rating generally means middle grade unleaded. Premium unleaded is more stable under severe conditions but also produces more combustion products. Therefore, when using premium unleaded, more frequent de-carboning is necessary.

Combustion

In a high heat environment like an modern engine, the fuel must be very stable to avoid detonation. If any parameters affecting combustion change suddenly (the engine runs lean for example), uncontrolled heat buildup will occurs very rapidly.

The combustion process is affected by several interrelated factors. This means that when one factor is changed, the other factors also must be changed to maintain the same controlled burn and level of combustion stability.

- Compression - determines the level of heat buildup in the cylinder when the air-fuel mixture is compressed. As compression increases, so does the potential for heat buildup
- Ignition Timing - determines when the gasses will start to expand in relation to the motion of the piston. If the ignition timing is too advanced, gasses will be ignited and begin to expand too soon, such as they would during pre-ignition. The motion of the piston opposes the expansion of the gasses, resulting in extremely high combustion chamber pressures and heat. If the ignition timing is retarded, the gases are ignited later in relation to piston position. This means that the piston has already traveled back down the bore toward the bottom of the cylinder, resulting in less usable power.
- Fuel Mixture - determines how efficient the burn will be. A rich mixture burns slower than a lean one. If the mixture is too lean, it can't become explosive. The slower the burn, the cooler the combustion chamber, because pressure buildup is gradual.
- Fuel Quality (Octane Rating) - determines how much heat is necessary to ignite the mixture. Once the burn is in progress, heat is on the rise. The unburned poor quality fuel is ignited all at once by the rising heat instead of burning gradually as a flame front of the burn passing by. This action results in detonation (pinging).

FUEL SYSTEM - CARBURETORS

There are two types of abnormal combustion - pre-ignition and detonation.
- Pre-ignition - occurs when the air-fuel mixture is ignited by some incandescent source other than the correctly timed spark from the spark plug.
- Detonation - occurs when excessive heat and or pressure ignites the air/fuel mixture rather than the spark plug. The burn becomes explosive.

In general, anything that can cause abnormal heat buildup can be enough to push an engine over the edge to abnormal combustion, if any of the four basic factors previously discussed are already near the danger point, for example, excessive carbon buildup raises the compression and retains heat as glowing embers.

CARBURETED FUEL SYSTEM

◆ See Figure 1

Troubleshooting

Troubleshooting fuel systems requires the same techniques used in other areas. A thorough, systematic approach to troubleshooting will pay big rewards. Build your troubleshooting checklist, with the most likely offenders at the top. Use your experience to adjust your list for local conditions. Everyone has been tempted to jump into the carburetor on a vague hunch. Pause a moment and review the facts when this urge occurs.

In order to accurately troubleshoot a carburetor or fuel system problem, you must first verify that the problem is fuel related. Many symptoms can have several different possible causes. Be sure to eliminate mechanical and electrical systems as the potential fault. Carburetion is the number one cause of most engine problems but there are other possibilities.

One of the toughest tasks with a fuel system is the actual troubleshooting. Several tools are at your disposal for making this process very simple. A timing light works well for observing carburetor spray patterns. Look for the proper amount of fuel and for proper atomization in the two fuel outlet areas (main nozzle and bypass holes). The strobe effect of the lights helps you see in detail the fuel being drawn through the throat of the carburetor. On multiple carburetor engines, always attach the timing light to the cylinder you are observing so the strobe doesn't change the appearance of the patterns. If you need to compare two cylinders, change the timing light hookup each time you observe a different cylinder.

Pressure testing fuel pump output can determine whether the fuel spray is adequate and if the fuel pump diaphragms are functioning correctly. A pressure gauge placed between the fuel pump(s) and the carburetor(s) will test the entire fuel delivery system. Normally a fuel system problem will show up at high speed where the fuel demand is the greatest. A common symptom of a fuel pump output problem is surging at wide open throttle but normal operation at slower speeds. To check the fuel pump output, install the pressure gauge and accelerate the engine to wide-open throttle. Observe the pressure gauge needle. It should always swing up the value indicated in the specification charts and remain steady. This reading would indicate a system that is functioning properly.

If the needle gradually swings down toward zero, fuel demand is greater than the fuel system can supply. This reading isolates the problem to the fuel delivery system (fuel tank or line). To confirm this, an auxiliary tank should be installed and the engine re-tested. Be aware that a bad anti-siphon valve on a built-in tank can create enough restriction to cause a lean condition and serious engine damage.

If the needle movement becomes erratic, suspect a ruptured diaphragm in the fuel pump.

Carburetor Applications

Year	Engine	Displacement L/Cu. In.	Engine Type	Carburetor Type	Fuel Delivery
2001	3.0L	3.0/181	L4	MerCarb	2 bbl
	4.3L	4.3/262	V6	MerCarb	2 bbl
		4.3/262	V6	Weber	4 bbl
	5.0L	5.0/305	V8	MerCarb	2 bbl
	5.7L	5.7/350	V8	MerCarb	2 bbl
2002	3.0L	3.0/181	L4	MerCarb	2 bbl
	4.3L	4.3/262	V6	MerCarb	2 bbl
	5.0L	5.0/305	V8	MerCarb	2 bbl
	5.7L	5.7/350	V8	MerCarb	2 bbl
2003	3.0L	3.0/181	L4	MerCarb	2 bbl
	4.3L	4.3/262	V6	MerCarb	2 bbl
	5.0L	5.0/305	V8	MerCarb	2 bbl
	5.7L	5.7/350	V8	MerCarb	2 bbl
2004	3.0L	3.0/181	L4	MerCarb	2 bbl
	4.3L	4.3/262	V6	MerCarb	2 bbl
	5.0L	5.0/305	V8	MerCarb	2 bbl
	5.7L	5.7/350	V8	MerCarb	2 bbl
2005	3.0L	3.0/181	L4	MerCarb	2 bbl
	4.3L	4.3/262	V6	MerCarb	2 bbl
	5.0L	5.0/305	V8	MerCarb	2 bbl
	5.7L	5.7/350	V8	MerCarb	2 bbl
2006-08	3.0L	3.0/181	L4	MerCarb	2 bbl
	4.3L	4.3/262	V6	MerCarb	2 bbl
	5.0L	5.0/305	V8	MerCarb	2 bbl
	5.7L	5.7/350	V8	MerCarb	2 bbl

Fig. 1 Carburetor Applications

6-4 FUEL SYSTEM - CARBURETORS

To check for air entering the fuel system, install a clear fuel hose between the fuel screen and fuel pump. If air is in the line, check all fittings back to the boat's fuel tank.

Spark plug tip appearance is a good indication of combustion efficiency. The tip should be a light tan. A White insulator or small beads on the insulator indicate too much heat. A dark or oil fouled insulator indicates incomplete combustion. To properly read spark plug tip appearance, run the engine at the RPM you are testing for about 15 seconds and then immediately turn the engine OFF without changing the throttle position.

Reading spark plug tip appearance is also the proper way to test jet verifications in high altitude.

COMMON PROBLEMS - W/O TKS

Fuel Delivery

Many times fuel system troubles are caused by a plugged fuel filter, a defective fuel pump or by a leak in the line from the fuel tank to the fuel pump. A defective choke may also cause problems. Would you believe, an majority of starting troubles which are traced to the fuel system are the result of an empty fuel tank or aged sour fuel.

Sour Fuel

Under average conditions (temperate climates), fuel will begin to break down in about four months. A gummy substance forms in the bottom of the fuel tank and in other areas. The filter screen between the tank and the carburetor and small passages in the carburetor will become clogged. The gasoline will begin to give off an odor similar to rotten eggs. Such a condition can cause the owner much frustration, time in cleaning components and the expense of replacement or overhaul parts for the carburetor.

Even with the high price of fuel, removing gasoline that has been standing unused over a long period of time is still the easiest and least expensive preventative maintenance possible. In most cases, this old gas can be used without harmful effects in an automobile using regular gasoline.

A gasoline preservative will keep the fuel fresh for up to twelve months. These products are available in most areas under various trade names.

Choke Problems

When the engine is hot, the fuel system can cause starting problems. After a hot engine is shut down, the temperature inside the fuel bowl may rise to 200°F and cause the fuel to actually boil. All carburetors are vented to allow this pressure to escape to the atmosphere. However, some of the fuel may percolate over the high-speed nozzle.

If the choke should stick in the open position, the engine will be hard to start. If the choke should stick in the closed position, the engine will flood, making it very difficult to start.

In order for this raw fuel to vaporize enough to burn, considerable air must be added to lean out the mixture. Therefore, the only remedy is to remove the spark plugs, ground the leads, crank the engine through about ten revolutions, clean the plugs, reinstall the plugs and start the engine.

If the needle valve and seat assembly is leaking, an excessive amount of fuel may enter the reed housing in the following manner. After the engine is shut down, the pressure left in the fuel line will force fuel past the leaking needle valve. This extra fuel will raise the level in the fuel bowl and cause fuel to overflow into the reed housing.

A continuous overflow of fuel into the reed housing may be due to a sticking inlet needle or to a defective float, which would cause an extra high level of fuel in the bowl and overflow into the reed housing.

Rough Engine Idle

If an engine does not idle smoothly, the most reasonable approach to the problem is to perform a tune-up to eliminate such areas as:
- Defective points
- Faulty spark plugs
- Timing out of adjustment

Other problems that can prevent an engine from running smoothly include:
- An air leak in the intake manifold
- Uneven compression between the cylinders

Of course any problem in the carburetor affecting the air/fuel mixture will also prevent the engine from operating smoothly at idle speed. These problems usually include:
- Too high a fuel level in the bowl
- A heavy float
- Leaking needle valve and seat
- Defective automatic choke
- Improper adjustments for idle mixture or idle speed

Excessive Fuel Consumption

Excessive fuel consumption can be the result of any one of four conditions or a combination of all.
- Inefficient engine operation.
- Faulty condition of the hull, including excessive marine growth.
- Poor boating habits of the operator.
- Leaking or out-of-tune carburetor.

If the fuel consumption suddenly increases over what could be considered normal, then the cause can probably be attributed to the engine or boat and not the operator.

Marine growth on the hull can have a very marked effect on boat performance. This is why sailboats always try to have a haul-out as close to race time as possible.

While you are checking the bottom, take note of the propeller condition. A bent blade or other damage will definitely cause poor boat performance.

If the hull and propeller are in good shape, then check the fuel system for possible leaks. Check the line between the fuel pump and the carburetor while the engine is running and the line between the fuel tank and the pump when the engine is not running. A leak between the tank and the pump many times will not appear when the engine is operating, because the suction created by the pump drawing fuel will not allow the fuel to leak. Once the engine is turned off and the suction no longer exists, fuel may begin to leak.

If a minor tune-up has been performed and the spark plugs, points (if equipped) and timing are properly adjusted, then the problem most likely is in the carburetor and an overhaul is in order.

Check the needle valve and seat for leaking. Use extra care when making any adjustments affecting the fuel consumption, such as the float level or automatic choke.

Engine Surge

If the engine operates as if the load on the boat is being constantly increased and decreased, even though an attempt is being made to hold a constant engine speed, the problem can most likely be attributed to the fuel pump or a restriction in the fuel line between the tank and the carburetor.

COMBUSTION RELATED PISTON FAILURES

When an engine has a piston failure due to abnormal combustion, fixing the mechanical portion of the engine is the easiest part of the equation. The hard part is determining what caused the problem in order to prevent a repeat failure. Think back to the four basic areas that affect combustion to find the cause of the failure.

Since you probably removed the cylinder head, inspect the failed piston and look for excessive deposit buildup that could raise compression or retain heat in the combustion chamber. Statically check the wide open throttle timing. Be sure that the timing is not over advanced. It is a good idea to seal these adjustments with paint to detect tampering.

Look for a fuel restriction that could cause the engine to run lean. Don't forget to check the fuel pump, fuel tank and lines, especially if a built in tank is used. Be sure to check the anti-siphon valve on built in tanks. If everything else looks good, the final possibility is poor quality fuel.

TKS DIAGNOSTICS

Normal Starting Procedure

1. Place the remote control handle in the Neutral position and start the engine.

2. Allow the engine to reach normal operating temperature - this should take about 6-10 minutes for the first start of the day. The engine should idle initially idle at 650-900 rpm and then return to normal idle speed when warm.

3. If the engine will not start after 3 attempts:

a. Push in the throttle only button and move the remote control to the 1/4 throttle position.

b. Start the engine.

4. If the engine still does not start:

a. Move the remote control to the full throttle position and then bring it back to the 1/4 position.

b. Start the engine.

FUEL SYSTEM - CARBURETORS

Diagnostic Charts

◆ See Figures 2 thru 5

Operation

BASIC FUNCTIONS

◆ See Figure 6

Traditional carburetor theory often involves a number of laws and principles. The diagram illustrates several carburetor basics. If you blow air across a straw inserted into a container of liquid, a pressure drop is created in the straw column. As the liquid in the column is expelled, an atomized mixture (air and fuel droplets) is created. In a carburetor this is mostly air and a little fuel.

The actual ratio of air to fuel differs with engine conditions but is usually from 15 parts air to one part fuel at optimum cruise to as little as 7 parts air to one part fuel at full choke.

Using our example, what if the top of the container is covered and sealed around the straw, what will happen? No flow. This is typical of a clogged carburetor bowl vent. If the base of the straw is clogged or restricted what will happen? No flow or low flow. This represents a clogged main jet. If the liquid in the glass is lowered and you blow through the straw with the same force what will happen? Not as much fuel will flow. A lean condition occurs. If the fuel level is raised and you blow again at the same velocity what happens? The result is a richer mixture.

FUEL & AIR METERING

The carburetor is merely a metering device for mixing fuel and air in the proper proportions for efficient engine operation.

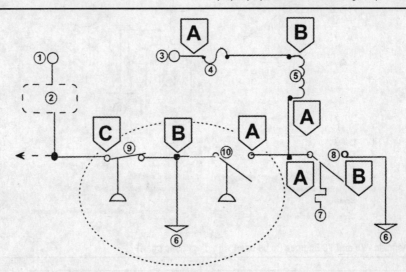

1	12 volts switched	7	Engine coolant
2	Alarm horn	8	Engine coolant temperature switch (normally open)
3	12 volts	9	Engine oil pressure switch (normally closed)
4	Fuse (20 amp)	10	Engine oil pressure switch (normally open)
5	TKS Module (on carburetor)	11	Additional alarm switches
6	Ground	A, B, C	Connector terminal identification (See wiring diagram for wire colors)

Fig. 2 TKS Diagnostic Schematic - 3.0L Engines (to be used with diagnostic chart)

Component Characteristics

Component	Characteristic	Function
Engine oil pressure switch (normally closed)	Closed 0 to 4 PSI Open above 4 PSI.	Turns on alarm when oil pressure is below 4 PSI.
Engine oil pressure switch (normally open)	Open 0 to 4 PSI Closed above 4 PSI.	Turns on TKS heater when oil pressure is above 4 PSI.
Engine coolant temperature switch (normally open)	Open at room temp. Closes at 130°F Re-opens at 110°F.	Used to keep the TKS module energized when the engine is warm.

General Diagnostics Tests

3.0L Voltage Checks - With key switch OFF, and engine stopped.			
Remove connector to	Test harness connector identification	Normal voltage	What to check if the voltage is not correct
20A Fuse	Fuse "A"	12	Battery connection, Battery Switch set to on
TKS connector	TKS "B"	12	Fuse
N.O. Temperature Switch and Dual Oil Pressure Switch	Temperature Switch "A"	12	TKS Module
	Oil Pressuure Switch "A"	12	TKS Module

Fig. 3 TKS Diagnostic Chart - 3.0L Engines

6-6 FUEL SYSTEM - CARBURETORS

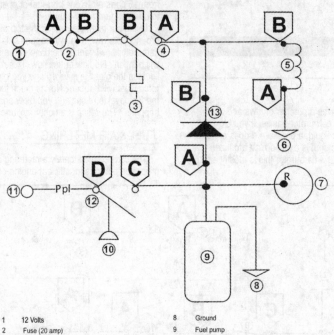

1. 12 Volts
2. Fuse (20 amp)
3. Engine coolant
4. Engine coolant temperature switch (normally open)
5. TKS Module (on carburetor)
6. Ground
7. Starter engagement solenoid
8. Ground
9. Fuel pump
10. Oil pressure
11. 12 Volts, switched
12. Engine oil pressure switch (normally open)
13. Diode

A, B, C Connector terminal identification (See wiring diagram for wire colors)

Fig. 4 TKS Diagnostic Schematic - V6 and V8 Engines (to be used with diagnostic chart)

Component Characteristics

Component	Characteristic	Function
Engine oil pressure switch (normally open)	Open 0 to 4 PSI Closed above 4 PSI.	Turns on TKS heater when oil pressure is above 4 PSI.
Engine coolant temperature switch (normally open)	Open at room temp. Closes at 130°F, Re-opens at 110°F.	Used to keep the TKS module energized when the engine is warm.
Diode	Passes current in one direction	Allows the engine coolant temperature switch to energize the TKS module without energizing the fuel pump, ignition, gauges, etc.

General Diagnostics Tests

V6 - V8 Voltage Checks - With key switch OFF, and engine stopped, and engine cold			
Remove connector to	Test harness connector identification	Normal voltage	What to check if the voltage is not correct
20A Fuse	Fuse "A"	12	Battery connection, Battery Switch set to on
N.O. Temperature Switch	N.O. Temperature Switch "B"	12	Fuse
TKS Module	TKS Coil "B"	0	N.O. Temp Switch

V6 - V8 Voltage Checks - With key switch ON, and engine stopped, and engine cold			
Remove connector to	Test harness connector identification	Normal voltage	What to check if the voltage is not correct
N.O. Oil Press Sw	N.O. Oil Pressure Switch "D"	12	Battery connection, Battery Switch set to on, Key Switch

V6 - V8 Voltage Checks - With key switch ON, and engine operating			
Remove connector to	Test harness connector identification	Normal voltage	What to check if the voltage is not correct
Diode	Diode "A"	12	N.O. Oil Pressure Switch
20A Fuse and TKS module	TKS Coil "B"	12	Diode

Fig. 5 TKS Diagnostic Chart - V6 and V8 Engines

FUEL SYSTEM - CARBURETORS

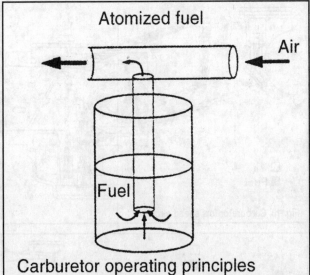

Fig. 6 If you blow air across a straw inserted into a container of liquid, a pressure drop is created in the straw column. As the liquid in the column is expelled, an atomized mixture (air and fuel droplets) is created

Float Systems

◆ See Figure 7

A small chamber in the carburetor serves as a fuel reservoir. A float valve admits fuel into the reservoir to replace the fuel consumed by the engine. If the carburetor has more than one reservoir, the fuel level in each reservoir (chamber) is controlled by identical float systems.

Fuel level in each chamber is extremely critical and must be maintained accurately. Accuracy is obtained through proper adjustment of the floats. This adjustment will provide a balanced metering of fuel to each cylinder at all speeds.

Following the fuel through its course, from the fuel tank to the combustion chamber of the cylinder, will provide an appreciation of exactly what is taking place. In order to start the engine, the fuel must be moved from the tank to the carburetor by a fuel pump installed in the fuel line.

After the engine starts, the fuel passes through the pump to the carburetor. All systems have some type of filter installed somewhere in the line between the tank and the carburetor. Most engines also have a filter as an integral part of the carburetor.

At the carburetor, the fuel passes through the inlet passage to the needle and seat and then into the float chamber (reservoir). A float in the chamber rides up and down on the surface of the fuel. After fuel enters the chamber and the level rises to a predetermined point, a tang on the float closes the inlet needle and the flow entering the chamber is cut off. When fuel leaves the chamber as the engine operates, the fuel level drops and the float tang allows the inlet needle to move off its seat and fuel once again enters the chamber. In this manner, a constant reservoir of fuel is maintained in the chamber to satisfy the demands of the engine at all speeds.

A fuel chamber vent hole is located near the top of the carburetor body to permit atmospheric pressure to act against the fuel in each chamber. This pressure assures an adequate fuel supply to the various operating systems.

Air/Fuel Mixture

◆ See Figure 8

A suction effect is created each time the piston moves upward in the cylinder. This suction draws air through the throat of the carburetor. A restriction in the throat, called a venturi, controls air velocity and has the effect of reducing air pressure at this point.

The difference in air pressures at the throat and in the fuel chamber, causes the fuel to be pushed out of metering jets extending down into the fuel chamber. When the fuel leaves the jets, it mixes with the air passing through the venturi. This fuel/air mixture should then be in the proper proportion for burning in the cylinders for maximum engine performance.

In order to obtain the proper air/fuel mixture for all engine speeds, some models have high and low speed jets. These jets have adjustable needle valves that are used to compensate for changing atmospheric conditions. In almost all cases, the high-speed circuit has fixed high-speed jets and is not adjustable.

A throttle valve controls the flow of air/fuel mixture drawn into the combustion chambers. A cold engine requires a richer fuel mixture to start and during the brief period it is warming to normal operating temperature. A choke valve is placed ahead of the metering jets and venturi. As this valve begins to close, the volume of air intake is reduced, thus enriching the mixture entering the cylinders. When this choke valve is fully closed, a very rich fuel mixture is drawn into the cylinders.

The throat of the carburetor is usually referred to as the barrel. Carburetors with single, double or four barrels have individual metering jets, needle valves, throttle and choke plates for each barrel. Single and two barrel carburetors are fed by a single float and chamber.

CARBURETOR CIRCUITS

The following section illustrates the circuit functions and locations of a typical marine carburetor.

Starting Circuit

◆ See Figure 9

The choke plate is closed, creating a partial vacuum in the venturi. As the piston rises, negative pressure in the crankcase draws the rich air-fuel mixture from the float bowl into the venturi and on into the engine.

Fig. 7 Fuel flow through a venturi, showing principal and related parts controlling intake and outflow

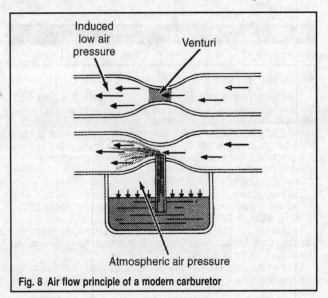

Fig. 8 Air flow principle of a modern carburetor

6-8 FUEL SYSTEM - CARBURETORS

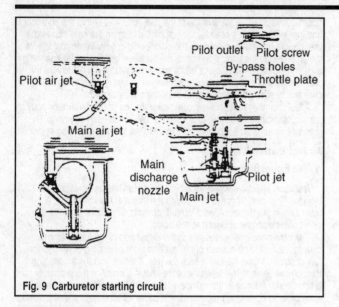

Fig. 9 Carburetor starting circuit

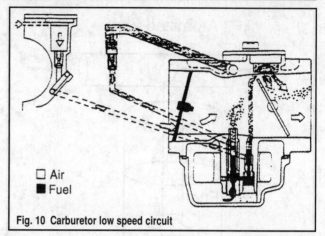

Fig. 10 Carburetor low speed circuit

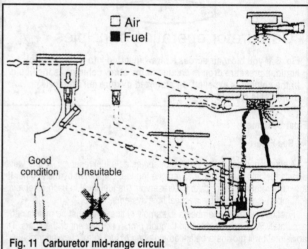

Fig. 11 Carburetor mid-range circuit

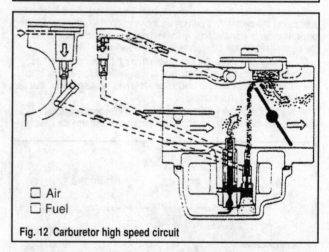

Fig. 12 Carburetor high speed circuit

Low Speed Circuit

◆ See Figure 10

Zero to one-eighth throttle, when the pressure in the crankcase is lowered, the air-fuel mixture is discharged into the venturi through the pilot outlet because the throttle plate is closed. No other outlets are exposed to low venturi pressure. The fuel is metered by the pilot jet. The air is metered by the pilot air jet. The combined air-fuel mixture is regulated by the pilot air screw.

Mid-Range Circuit

◆ See Figure 11

One-eighth to three-eighths throttle, as the throttle plate continues to open, the air-fuel mixture is discharged into the venturi through the bypass holes. As the throttle plate uncovers more bypass holes, increased fuel flow results because of the low pressure in the venturi. Depending on the model, there could be two, three or four bypass holes.

High Speed Circuit

◆ See Figure 12

Three-eighths to wide-open throttle, as the throttle plate moves toward wide open, we have maximum air flow and very low pressure. The fuel is metered through the main jet and is drawn into the main discharge nozzle. Air is metered by the main air jet and enters the discharge nozzle, where it combines with fuel. The mixture atomizes, enters the venturi and is drawn into the engine.

MerCarb 2bbl Carburetor

DESCRIPTION

◆ See Figure 13

The MerCarb carburetor used on 3.0L, 4.3L, 5.0L and 5.7L engines is a two barrel carburetor with a separate fuel feed for each venturi. It has one idle adjustment screw and is equipped with an electric choke. A removable venturi cluster - attached to the float bowl assembly - has the calibrated main well tubes and pump jets built into it. The venturi cluster is serviced as a unit.

A part number and a date code are embossed on the carburetor, as shown in the accompanying illustration.

The upper number is the Part Number.

The lower number is the date code and is interpreted as follows:

First digit is the year of manufacture. In the illustration "2" equals 2002; "6" would designate 2006.

The second digit is the month with "3" designating March, "9" designating September, etc. " X" designates October, "Y" designates November and, you guessed it, "Z" is for December.

The third and fourth digits indicate the day of the month, with the first nine days preceded by a "0". Thus "01" would be the first day; "08" the eighth day, etc.

Complete carburetor adjustment specifications can be found at the end of this section.

Beginning in 2004, most models will be equipped with a new Turn Key Start (TKS) carburetor. At its simplest, this new system accomplishes two things. . .provide additional fuel to air mixture when starting a cold engine, and prevent additional enrichment when starting a warm engine.

A casting in the carburetor allows for fuel to be drawn from the float bowl and then mixed with air in a chamber. The enriched mixture is then drawn into the engine via an opening in the carburetor body, just below the throttle plate.

A TKS module is mounted on the side of the carburetor body. When voltage (12V) is applied to the module, it will warm up and a plunger extends from the module to close the enrichment passage in the carburetor. On 3.0L engines, an oil pressure switch provides a ground path for the module while

FUEL SYSTEM - CARBURETORS   6-9

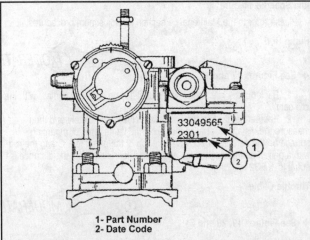

Fig. 13 Simple line drawing to indicate location of the I.D. numbers on the MerCarb 2bbl

Fig. 14 A good look at the mounting nuts

power is continuously supplied via a 20 amp fuse inline with the engine circuit breaker. On other engines, the existing oil pressure switch is what supplies the power and the module is continuously grounded.

■ On all models utilizing a TKS system, if the engine has not been run for a long period of time (Mercruiser does not quantify how long a period this is), it may require a few attempts to start until the fuel bowl fills up. In all cases, the engine will run at a higher than normal idle speed until it warms up sufficiently.

There are two versions of this system: Gen I (early V6/V8), and Gen II (all other engines). The easiest way to tell them apart is to find the enrichment reservoir on the side of the carburetor body - Gen I versions will have it about halfway up the side of the body, while Gen II versions will have it at the bottom, by the mounting surface.

✱✱ WARNING

Always disconnect the battery cables before attempting to work on the fuel system. Never smoke or allow open flame near the engine - this sounds like an obvious precaution, but you'd be surprised at how many people forget!

REMOVAL & INSTALLATION

◆ See Figure 14    MODERATE

■ No matter how much they look alike, marine carburetors are completely different from automotive carburetors. Never substitute an automotive carburetor for the one on your engine! Venting procedures are not the same and an automotive carburetor could allow dangerous vapors to escape into the bilge. Don't even think about it.

1. Open the engine hatch or remove the covers and then disconnect the battery cables. Turn the fuel petcock OFF and/or shut down the fuel supply at the tank.
2. Remove the flame arrestor after disconnecting the vent hose. It's always a good idea to plug the throttle bores with a clean, lint-free cloth.
3. Disconnect the throttle cable at the carburetor and carefully move it out of the way.
4. Using two open-end wrenches, hold the fuel inlet nut at the carburetor securely and loosen the fuel line nut. Disconnect the two and carefully move the line out of the way. Plug both the inlet and line open ends to prevent fuel seepage.
5. Disconnect the fuel pump sight tube at the unit (3.0L engines only).
6. Tag and disconnect the electric choke lead.
7. Tag and disconnect the TKS connector lead if so equipped.
8. Loosen and remove the carburetor mounting nuts/washers and lift the unit off the manifold. Plug the opening with a clean lint-free cloth.

■ On V6 and V8 models with TKS, the circuit breaker bracket is attached to the rear of the carburetor. As you lift off the carburetor, disconnect the bracket and carefully lay it aside.

To install:

9. Clean the mating surfaces thoroughly of all residual gasket material, position a new gasket and then install the carburetor. Tighten the nuts to 20 ft. lbs. (27 Nm) on the 3.0L and all TKS units or 132 inch lbs. (15 Nm) on the other engines. Hopefully you remembered to remove the rag!

■ On V6 and V8 models with TKS, don't forget to position the circuit breaker bracket on the rear stud before installing the nut.

10. Reconnect the fuel line to the inlet line after removing the plugs and tighten it to 18 ft. lbs. (24 Nm). Don't forget to use two wrenches.
11. Connect the plastic sight tube line (on the 3.0L) and then connect the choke lead.
12. Install and adjust the throttle cable as detailed later in this section.
13. Install the flame arrestor and reconnect the vent line.
14. Reconnect the choke or TKS leads.
15. Connect the battery cables. Start the engine and ensure there are no fuel leaks; shut down the engine immediately if there are. Check and adjust the idle speed and mixture.

ADJUSTMENT

Accelerator Pump Lever

◆ See Figures 15 and 16    MODERATE

The accelerator lever is equipped with 3 separate holes, allowing you to change the amount of fuel being delivered to the engine via the accelerator pump. The hole closest to the pump lever's shaft will deliver the full amount of fuel. The center hole will provide approximately 0.5cc less fuel with each stroke, while the hole furthest from the shaft will provide a flow of approximately 1.0cc less per stroke.

On TKS units, the hole closest to the shaft will be a richer mixture, the center hole be the center mix and the last hole will be a leaner mixture. Beginning with the hole closest to the shaft, each hole will deliver approximately 0.8cc less as you get further away from the shaft.

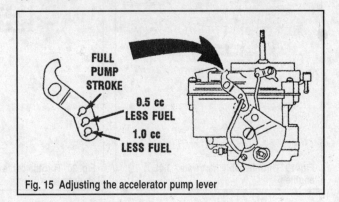

Fig. 15 Adjusting the accelerator pump lever

6-10 FUEL SYSTEM - CARBURETORS

Fig. 16 Most models will use the center hole

Idle Speed & Mixture

Please refer to the Maintenance section for adjustment procedures.

Pump Rod

◆ See Figures 17 and 18

1. Back out the idle speed screw until it is no longer in contact with the idle cam.
2. Ensure that the throttle valves are closed completely and then measure the distance between the top of the flame arrestor mounting surface on the carburetor body and the top of the pump rod. Grasp the rod with a pair of needle nose pliers and bend it (carefully) until the distance is equal to 1 5/32 in. (29mm).

Throttle Cable

◆ See Figures 19, 20 and 21

1. With the remote control lever in Neutral (idle), install the throttle cable end guide onto the lever and then push the barrel slightly toward the throttle lever end so as to preload the cable slightly and remove any slack. Adjust the barrel so that it aligns with the anchor stud.

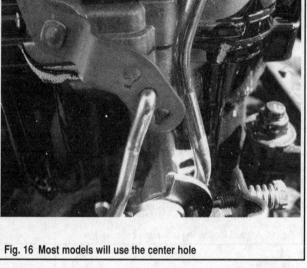

Fig. 17 Measure the distance between the flame arrestor mounting surface and the top of the pump rod

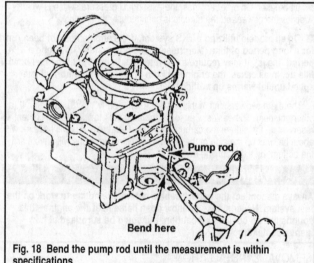

Fig. 18 Bend the pump rod until the measurement is within specifications

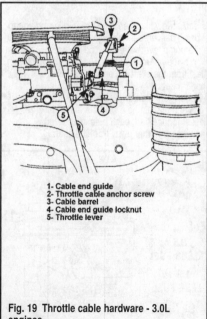

1- Cable end guide
2- Throttle cable anchor screw
3- Cable barrel
4- Cable end guide locknut
5- Throttle lever

Fig. 19 Throttle cable hardware - 3.0L engines

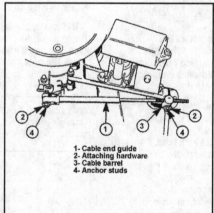

1- Cable end guide
2- Attaching hardware
3- Cable barrel
4- Anchor studs

Fig. 20 Throttle cable hardware - V6 and V8 engines

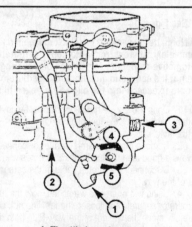

1- Throttle lever tang
2- Carburetor body
3- Idle speed adjustment screw
4- Wide open throttle
5- Idle

Fig. 21 Adjusting the throttle cable - 3.0L engines shown, but on all engines the lever should just be touching the idle screw end when the remote control handle is in the Neutral position

FUEL SYSTEM - CARBURETORS

2. Secure the throttle cable with its hardware and then tighten the locknut on the cable end guide until it just touches the cable, and then back it out one full turn on the 3.0L and 1/2 turn on 4.3L, 5.0L and 5.7L engines.

3. Tighten the throttle cable anchor screw securely - do not over-tighten.

4. Move the remote control lever to full throttle position. The throttle valves should be open all the way and the throttle shaft lever should be in contact with the body of the carburetor.

5. Move the lever back to the Neutral position; the throttle lever should touch the idle speed adjustment screw.

Choke Setting

◆ See Figures 22 and 23

Loosen the three choke housing cover screws and rotate the cover 2 index marks clockwise past the larger center index mark. Tighten the housing cover screws securely. On early 3.0L engines, the scribed mark on the housing cover should be aligned with the long case mark on the housing.

Choke Unloader

◆ See Figures 24 and 25

1. Remove the flame arrestor.
2. Hold the throttle valves so they are open fully.
3. Press down gently on the choke plate and then slide a 0.080 in. (5/64 in.0.2mm) drill bit between the upper edge of the choke plate and the air horn assembly - the bit should just fit through. If not, bend the tang on the throttle lever until it does.

Float Level And Drop

◆ See Figures 26 thru 30

1. Following the disassembly procedure detailed later, remove the air horn from the carburetor.
2. Turn the air horn assembly upside down and check that the float pivots freely on the pin. Raise the float and let it fall - do not force it!
3. Using a standard carburetor gauge (#91-36392 or an equivalent), measure the distance between the air horn gasket and the toe of the float. This is the level and it should be within the specifications given in the Carburetor Specifications chart - 3/8 in. (10mm) for models with a 2-pc solid inlet needle, and 9/16 in. (14mm) for 3.0L engines with spring loaded inlet needles or 11/32 in. (9mm) on other engines with the spring loaded needle (including TKS). Carefully bend the float arm with needle nose pliers to achieve the correct measurement. Make sure the float stays in alignment.
4. Its usually a good idea to check the float drop now as well. Turn the air horn right side up and allow the float to hang down freely.
5. Using the same gauge, measure the distance between the gasket and toe once again. If not 1 3/32 in. (27mm) on the 3.0L, or 15/16 in. (24mm) on the others, bend the float tang with needle nose pliers to achieve the correct measurement.
6. Recheck both measurements one more time.
7. Reinstall the air horn and carburetor.

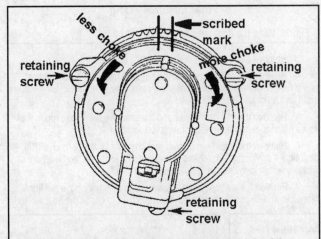

Fig. 22 The choke setting should be 2 marks to the right of the center index mark

Fig. 23 Many housing covers will have a dimple in place of the mark

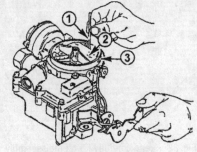

1- .08 in. (.2mm) drill bit
2- Choke pin
3- Air horn

Fig. 24 Insert a drill bit between the choke plate and the air horn to check the choke unloader

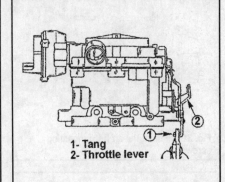

1- Tang
2- Throttle lever

Fig. 25 Bend the throttle lever tang to adjust the unloader if the bit does not fit properly

Fig. 26 When measuring from the float toe, use this example

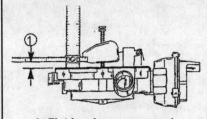

1- Flat level measurement

Fig. 27 Measuring the float level

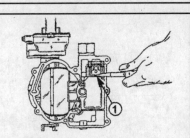

1- Bend float arm at this point

Fig. 28 Bend the float arm to adjust the float level

6-12 FUEL SYSTEM - CARBURETORS

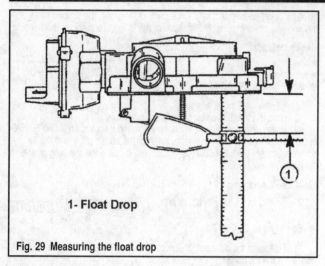

Fig. 29 Measuring the float drop

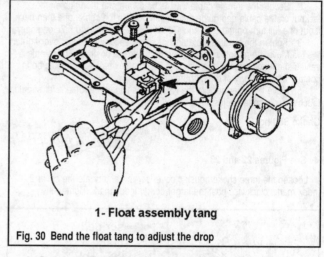

Fig. 30 Bend the float tang to adjust the drop

DISASSEMBLY

Models W/O TKS

◆ See Figures 31 thru 46   **DIFFICULT**

■ Always be certain that your carburetor rebuild kit is for marine applications.

1. Remove the carburetor from the manifold and mount it securely on a workbench.
2. Remove the choke cover, the lever and then the housing.
3. Remove the fuel inlet nut and then pull out the filter.
4. Remove the retaining clip at the bottom of the of the accelerator pump rod. Pivot the rod until its retaining ear aligns with the slot on the pump shaft and lever and then pull off the rod.
5. Remove the idle cam screw and then remove the choke rod in the same manner as the accelerator pump rod above.
6. Remove the 8 air horn screws and carefully lift the horn off of the float bowl - there should be 7 short screws and 1 long one, so take note of which hole the long one came out of.
7. Position the air horn upside down on the bench, remove the float hinge pin and lift out the float. Now is a good time to check the float weight!

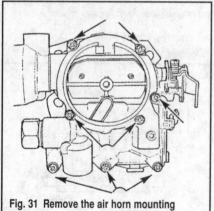

Fig. 31 Remove the air horn mounting screws...

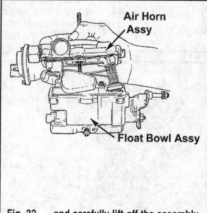

Fig. 32 ...and carefully lift off the assembly

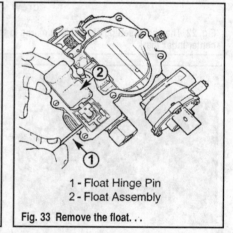

Fig. 33 Remove the float...

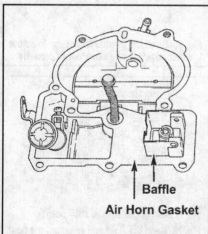

Fig. 34 ...and then the gasket and baffle

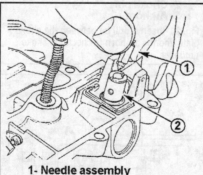

Fig. 35 Carefully remove the needle assembly from the seat...

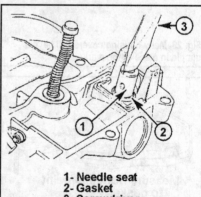

Fig. 36 ...and then use the slots to screw out the needle sea

FUEL SYSTEM - CARBURETORS 6-13

8. Remove the air horn gasket and baffle. Lift out the needle assembly from the seat. Insert a screwdriver into the seat slots and screw out the seat.

9. Loosen the accelerator pump set screw, slide the pump shaft and lever assembly out of the air horn and then remove the entire accelerator pump. Slide the retaining clip and washer from the pump shaft and disconnect the pump from the lever.

10. Now, move on to the float bowl and lift the accelerator pump return spring out of the pump well.

11. Remove the power valve assembly and then pull out the main metering jets.

12. Remove the mounting screws (should be 3) for the venturi cluster and lift the cluster and gasket straight up and out of the housing. Be very careful not to damage the brass tubes attached to the bottom of the cluster; they are permanently pressed into the cluster and are not replaceable.

13. Reach into the housing with needle nose pliers and pop out the check ball spring retainer for the accelerator pump. Turn the assembly over and allow the check ball and spring to fall out.

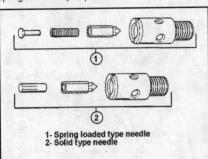

1- Spring loaded type needle
2- Solid type needle

Fig. 37 Your carburetor may come equipped with either of these needle assemblies

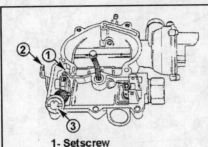

1- Setscrew
2- Lever Assembly
3- Pump Assembly

Fig. 38 Remove the entire accelerator pump assembly

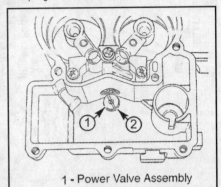

1- Power Valve Assembly
2- Gasket

Fig. 39 Remove the power valve...

1 - Main Metering Jets
2 - Gaskets

Fig. 40 ...and then thread out the main metering jets

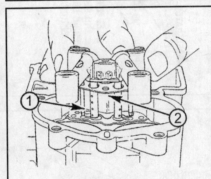

1 - Venturi Cluster Brass Tubes
2 - Gasket

Fig. 41 Lift the venturi cluster straight out, being very careful of the brass tubes

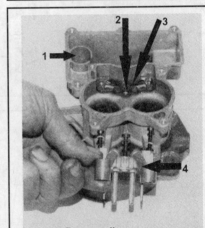

1- Pump well
2- Power valve
3- Main metering jet
4- Venturi cluster

Fig. 42 A close up view of the venturi cluster

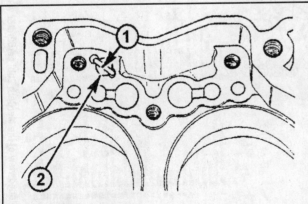

1- Spring retainer
2- Spring and check ball (not shown)

Fig. 43 Finally, find the check ball retainer, spring and ball...

Fig. 44 ...and remove the retainer

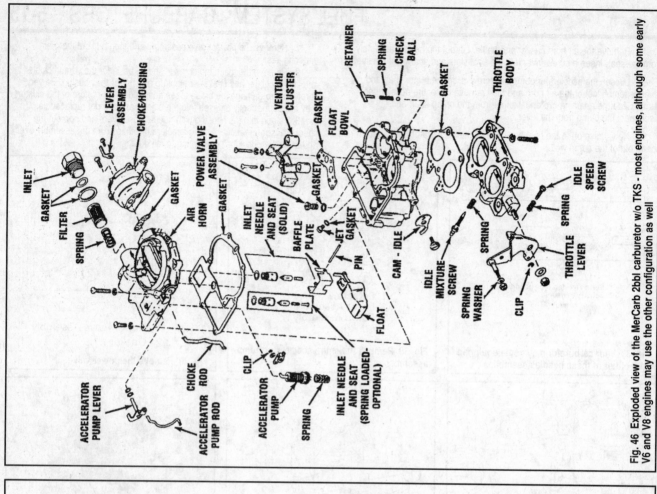

Fig. 46 Exploded view of the MerCarb 2bbl carburetor w/o TKS - most engines, although some early V6 and V8 engines may use the other configuration as well

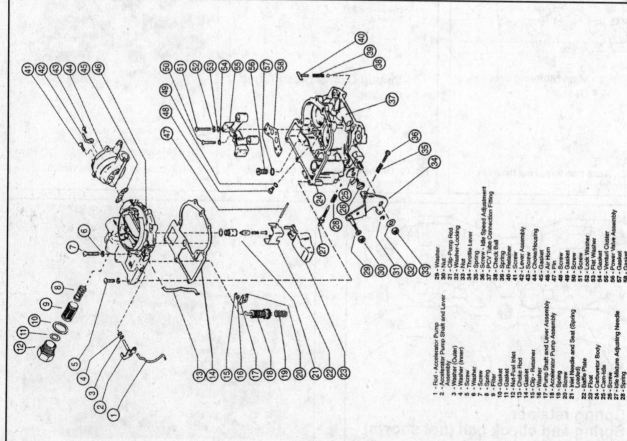

Fig. 45 Exploded view of the MerCarb 2bbl carburetor w/o TKS - certain early V6 and V8 engines

FUEL SYSTEM - CARBURETORS 6-15

Models W/TKS

◆ See Figures 43, 44 and 47 thru 49

■ Always be certain that your carburetor rebuild kit is for marine applications.

1. Remove the carburetor from the manifold and mount it securely on a workbench.

2. Remove the 2 mounting screws and the C-clip and then pull out the TKS module.

3. Remove the fuel inlet nut and then pull out the filter.

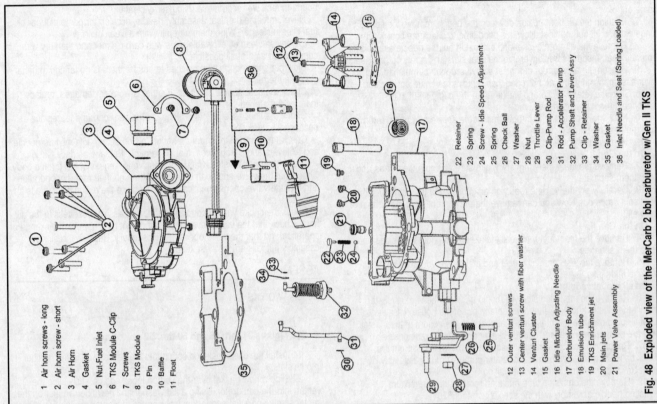

Fig. 48 Exploded view of the MerCarb 2 bbl carburetor w/Gen II TKS

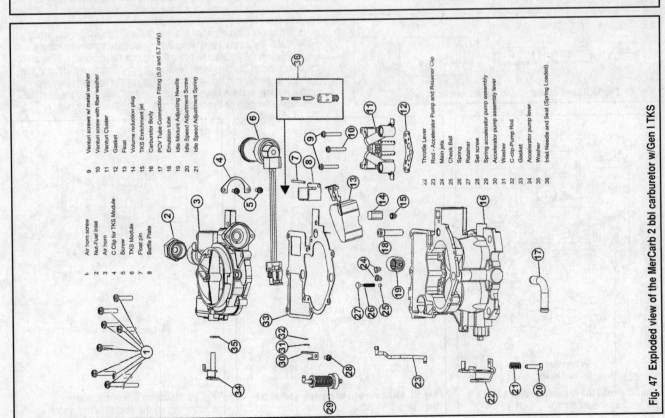

Fig. 47 Exploded view of the MerCarb 2 bbl carburetor w/Gen I TKS

6-16 FUEL SYSTEM - CARBURETORS

4. Remove the retaining clip at the bottom of the of the accelerator pump rod. Pivot the rod until its retaining ear aligns with the slot on the pump shaft and lever and then pull off the rod.

5. Remove the idle cam screw and then remove the choke rod in the same manner as the accelerator pump rod above.

6. Remove the 9 air horn screws and carefully lift the horn off of the float bowl - there should be 1 short screw on the Gen I and 1 long one with the others being the same length, so take note of which hole the long one came out of.

7. Position the air horn upside down on the bench, remove the float hinge pin and lift out the float. Now is a good time to check the float weight!

8. Remove the air horn gasket and baffle. Lift out the needle assembly from the seat. Insert a screwdriver into the seat slots and screw out the seat.

9. Loosen the accelerator pump set screw, slide the pump shaft and lever assembly out of the air horn and then remove the entire accelerator pump. Slide the retaining clip and washer from the pump shaft and disconnect the pump from the lever.

10. Now, move on to the float bowl and lift the accelerator pump return spring out of the pump well.

11. Remove the power valve assembly.

On the Gen I:

12. Remove the main metering jets. Make sure you note their sizes and keep them separate.

13. Carefully pry the volume reduction plug loose and then turn over the bowl so it drops out. Now you can remove the enrichment jet in the bottom of the well.

On the Gen II:

14. Remove the main metering jets. Make sure you note their sizes and keep them separate.

15. Remove the TKS enrichment jet and make sure you don't mix it up with the main jets.

On all models again:

16. Remove the mounting screws (should be 3) for the venturi cluster and lift the cluster and gasket straight up and out of the housing. Be very careful not to damage the brass tubes attached to the bottom of the cluster; they are permanently pressed into the cluster and are not replaceable.

17. Reach into the housing with needle nose pliers and pop out the check ball spring retainer for the accelerator pump. Turn the assembly over and allow the check ball and spring to fall out.

18. Remove the emulsion tube and then remove the idle mixture adjusting needle assembly with the special tool (#866201).

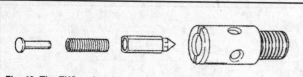

Fig. 49 The TKS carb uses a spring loaded needle

CLEANING & INSPECTION

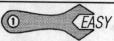

Never use a wire brush or drill to clean jet passages or tubes in the carburetor.

Never allow the carburetor to soak in a cleaner bath for more than two hours. In fact, we recommend using spray cleaner.

Never immerse the float assembly, needle, accelerator pump plunger or fuel filter in cleaner. Wipe them carefully with a clean cloth.

Otherwise clean all allowable parts with carb cleaner and then dry with compressed air if at all possible.

Blow out and through all passages to ensure there is no foreign material clogging them.

Check the float needle and seat, if either is worn or damaged, replace them as a matched set.

Check the float assembly and hinge pin for wear or damage, replace as necessary.

Check the pump plunger, return spring, piston spring, idle mixture needle and all levers and linkages for wear or damage. Replace as necessary.

Check the throttle valve shaft for excessive looseness in the throttle body. Check that the valve and shaft do not bind through their range of operation and that the valve opens and closes fully. Replace the assembly if it fails any of these tests.

Check the choke valve lever and shaft for excessive looseness in the air horn. Check that the valve, lever and shaft do not bind through their range of operation and that the valve opens and closes fully. Replace the air horn assembly if it fails any of these tests.

ASSEMBLY

Models W/O TKS

◆ See Figures 31 thru 46 and 50 thru 52

1. Install the check ball and spring into the passage. Clip the retainer into its slots.

2. Slide a new gasket over the venturi tubes and carefully slide the venturi into the carburetor. Using a flat washer and new fiber washer on the center screw and a flat and lock washer on the two outer screws, install the mounting screws and tighten them securely.

3. Install the main metering jets and gaskets, tightening them securely.

4. Install the power valve with a new gasket and tighten it securely.

5. Position the accelerator pump spring into the pump well. If the pump was disconnected from the lever, re-attach it with the retaining clip and then insert the assembly into the air horn (don't forget the washer). Align the pump lever hole with the shaft assembly and then slide it into the lever until its shoulder is resting on the lever. Tighten the set-screw securely.

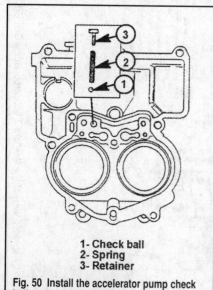

1- Check ball
2- Spring
3- Retainer

Fig. 50 Install the accelerator pump check ball and spring as shown

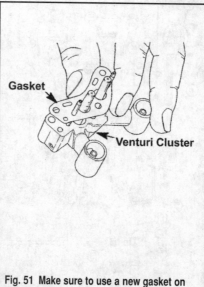

Fig. 51 Make sure to use a new gasket on the venturi

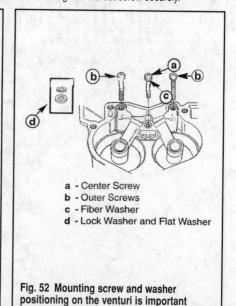

a - Center Screw
b - Outer Screws
c - Fiber Washer
d - Lock Washer and Flat Washer

Fig. 52 Mounting screw and washer positioning on the venturi is important

FUEL SYSTEM - CARBURETORS

6. Drop the needle seat and gasket into position and tighten it securely. Position the needle into the seat and then install the baffle and gasket.

7. Install the float assembly and slide in the hinge pin, making sure the float pivots smoothly through its range. Check the float level and drop as detailed in the Adjustment section.

8. Carefully drop the air horn assembly into the float bowl, making sure that accelerator pump slides correctly into the fuel well.

9. Insert the 8 mounting screws and tighten them securely, a little at a time.

10. Slide the end of the choke rod into the choke lever assembly.

11. Pop the idle cam onto the choke rod and attach it to the float bowl assembly using the screw so that the cam moves freely without binding.

12. Pop the ear end of the accelerator pump rod into the pump shaft and then insert the other end of the rod into the hole in the throttle lever. Slide the retaining clip over the end.

13. Position the choke housing onto the air horn and tighten the two screws securely. Install the choke lever and tighten the screw.

14. Install the choke cover so that the hook on the choke coil is engaged with the choke lever. Rotate the cover until the index marks align and then rotate the cover clockwise two marks. Tighten the three screws securely.

15. Install the carburetor.

Models W/TKS

◆ See Figures 43, 44 and 47 thru 52

1. Use the special tool (#91-866201) and thread the idle mixture needle and spring into the throttle body until it just seats itself, and then back it out the correct number of turns (as detailed in the Tune-Up Specifications chart).

2. Thread in the idle speed adjustment screw and spring.

3. Install the emulsion tube into the bowl.

4. Install the check ball and spring into the passage. Clip the retainer into its slots.

5. Slide a new gasket over the venturi tubes and carefully slide the venturi into the carburetor. Using a flat washer and new fiber washer on the center screw and a flat and lock washer on the two outer screws, install the mounting screws and tighten them securely.

On Gen I models:

6. Install the main metering jets and gaskets.

7. Install the TKS fuel enrichment jet into the bottom of the well. Press in the volume reduction plug so that the indent on the side slides around the screw hole.

On Gen II models:

8. Install the main metering jets and then the TKS enrichment jet.

On all models again:

9. Install the power valve with a new gasket and tighten it securely.

10. Position the accelerator pump spring into the pump well. If the pump was disconnected from the lever, re-attach it with the retaining clip and then insert the assembly into the air horn (don't forget the washer). Align the pump lever hole with the shaft assembly and then slide it into the lever until its shoulder is resting on the lever. Tighten the set-screw.

11. Drop the needle seat and gasket into position and tighten it securely. Position the needle into the seat and then install the baffle and gasket.

12. Install the float assembly and slide in the hinge pin, making sure the float pivots smoothly through its range. Check the float level and drop.

13. Carefully drop the air horn assembly into the float bowl, making sure that accelerator pump slides correctly into the fuel well.

14. Insert the 9 mounting screws and tighten them to 27-51 inch lbs. (3-6 Nm), a little at a time.

15. Slide the end of the choke rod into the choke lever assembly.

16. Pop the idle cam onto the choke rod and attach it to the float bowl assembly using the screw so that the cam moves freely without binding.

17. Pop the ear end of the accelerator pump rod into the pump shaft and then insert the other end of the rod into the hole in the throttle lever. Slide the retaining clip over the end.

18. Position the TKS module into the carburetor body with a new O-ring. Slide the C-clip into place and install the screws, tightening them to 17.5 inch lbs. (20 Nm). Make sure to coat the threads of the screws with Loctite 242 Threadlocker.

19. Install the carburetor.

Weber WFB 4bbl Carburetor

DESCRIPTION

◆ See Figure 53

The Weber 4bbl carburetor used on some early 4.3L engines has its serial number embossed on the lower mounting flange of the carburetor.

The Weber WFB is unique in that the main body and flange are cast as one piece. There are two separate float circuits.

Certain 4.3L models may come equipped with a special emissions carburetor, SAV1 Emissions. The big difference with the emissions version is that it has a PCV circuit, a ported vacuum switch (PVS) circuit and the idle mixture screws are sealed.

Mechanical procedures following are for both versions of the Weber carburetor.

※※ WARNING

Always disconnect the battery cables before attempting to work on the fuel system. Never smoke or allow open flame near the engine - this sounds like an obvious precaution, but you'd be surprised at how many people forget!

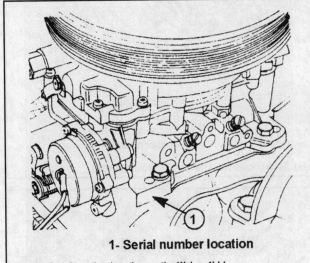

Fig. 53 Serial number location on the Weber 4bbl

REMOVAL & INSTALLATION

◆ See Figures 54 and 55

■ No matter how much they look alike, marine carburetors are completely different from automotive carburetors. Never substitute an automotive carburetor for the one on your engine! Venting procedures are not the same and an automotive carburetor could allow dangerous vapors to escape into the bilge. Don't even think about it.

1. Open the engine hatch or remove the covers and then disconnect the battery cables. Turn the fuel petcock OFF and/or shut down the fuel supply at the tank.

2. Remove the flame arrestor after disconnecting the vent hose. It's always a good idea to plug the throttle bores with a clean, lint-free cloth.

3. Disconnect the throttle cable hardware from the throttle bracket and lever anchor studs. Remove the cable and carefully move it out of the way.

4. Using two open-end wrenches, hold the fuel inlet nut at the carburetor securely and loosen the fuel line nut. Disconnect the two and carefully move the line out of the way. Plug both the inlet and line open ends to prevent fuel seepage.

5. Tag and disconnect the electric choke lead.

6. Tag and disconnect the two vacuum lines at the carburetor body on Emissions versions.

7. Loosen and remove the carburetor mounting nuts/washers and lift the unit off the manifold. Plug the opening with a clean lint-free cloth. Remove the throttle bracket.

6-18 FUEL SYSTEM - CARBURETORS

To install:

8. Clean the mating surfaces thoroughly of all residual gasket material, position a new gasket and then install the carburetor. Certain models may use an adaptor or wedge plate; install this first. Tighten the nuts to 132 inch lbs. (15 Nm). Hopefully you remembered to remove the rag!

9. Reconnect the fuel line to the inlet line after removing the plugs and tighten it to 18 ft. lbs. (24 Nm). Don't forget to use two wrenches.

10. Connect the two vacuum leads.
11. Connect the choke lead.
12. Install and adjust the throttle cable as detailed later in this section.
13. Install the flame arrestor and reconnect the vent line.
14. Connect the battery cables. Start the engine (remember to turn the fuel supply on) and ensure there are no fuel leaks; shut down the engine immediately if there are. Check and adjust the idle speed and mixture.

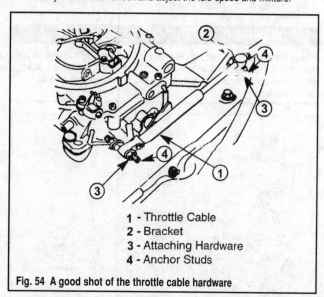

1 - Throttle Cable
2 - Bracket
3 - Attaching Hardware
4 - Anchor Studs

Fig. 54 A good shot of the throttle cable hardware

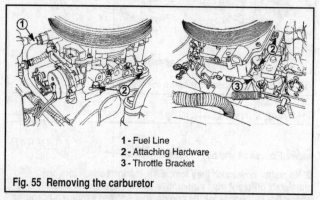

1 - Fuel Line
2 - Attaching Hardware
3 - Throttle Bracket

Fig. 55 Removing the carburetor

ADJUSTMENT

Accelerator Pump Lever

♦ See Figure 56

The accelerator lever is equipped with 3 separate holes, allowing you to change the amount of fuel being delivered to the engine via the accelerator pump. The hole closest to the pump lever's pivot point will deliver the full amount of fuel. The center hole will provide approximately 0.5cc less fuel with each stroke, while the hole furthest from the shaft will provide a flow of approximately 1.0cc less per stroke.

Accelerator Pump

♦ See Figures 57 and 58

1. Back out the idle speed screw until it is no longer in contact with the throttle lever.
2. Ensure that the throttle valves are closed fully and then measure the distance between the top of the carburetor and the bottom of the S-link on the plunger. Grasp the pump linkage with a pair of needle nose pliers and bend it (carefully) until the distance is equal to 7/16 in. (11mm).

Electric Choke

♦ See Figure 59

Loosen the three choke housing screws and rotate the housing until the center mark on the housing and the mark on the carburetor body are in alignment.

Float Level And Drop

♦ See Figures 60 thru 63

1. Following the disassembly procedure detailed later, remove the bowl cover from the carburetor.
2. Turn the cover assembly upside down and check that the float pivots freely on the pin. Raise the float and let it fall - do not force it!
3. Using a standard carburetor gauge, measure the distance between the bottom of the bowl cover gasket (remember its upside down, so this will be the top) and the toe of the float it should be within the specification given at the end of this section - 1-9/32 in. (33mm). Carefully bend the float arm with needle nose pliers to achieve the correct measurement. Make sure the float stays in alignment.
4. Its usually a good idea to check the float drop now. Turn the bowl cover assembly right side up and allow the float to hang down freely.
5. Using the same gauge, measure the distance between the bottom of the bowl cover and toe once again. If not 2 in. (51mm), bend the float tab with needle nose pliers to achieve the correct measurement.
6. Recheck both measurements one more time.
7. Reinstall the bowl cover and arrestor.

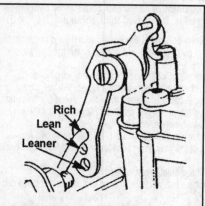

Fig. 56 Adjusting the accelerator pump lever

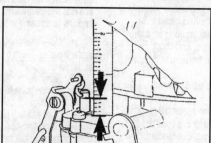

Fig. 57 Measure the distance between the top of the carburetor and the bottom of the S-link

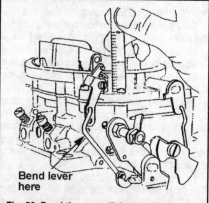

Fig. 58 Bend the pump linkage until the measurement is within specifications

FUEL SYSTEM - CARBURETORS 6-19

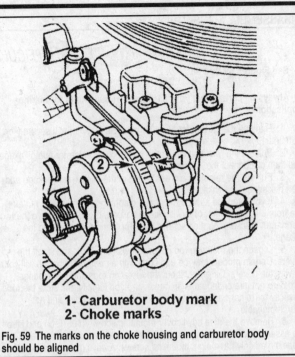

1- Carburetor body mark
2- Choke marks

Fig. 59 The marks on the choke housing and carburetor body should be aligned

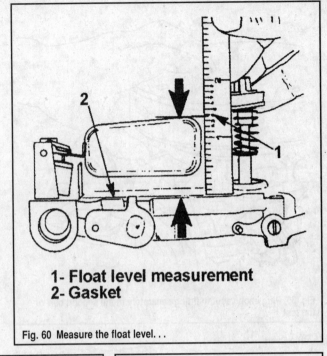

1- Float level measurement
2- Gasket

Fig. 60 Measure the float level...

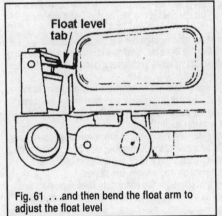

Fig. 61 ...and then bend the float arm to adjust the float level

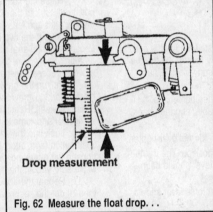

Fig. 62 Measure the float drop...

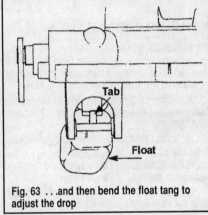

Fig. 63 ...and then bend the float tang to adjust the drop

Idle Speed & Mixture

Please refer to the Maintenance section for adjustment procedures.

Ported Vacuum Switch (PVS)

EMISSIONS VERSIONS ONLY

◆ See Figures 64 and 65

1. With the engine cold, start it and disconnect the vacuum line at the carburetor as shown. Cover the hose end with your finger and confirm that you feel no vacuum.
2. Connect the hose and allow the engine to run until it reaches normal operating temperature. Disconnect the hose again, cover the end with your finger and check that there is now vacuum. If so, the PVS is operating properly; if not, check that all hoses are properly connected and not plugged or cracked.
3. Reconnect the hose and make sure the engine is still running and at normal operating temperature. Disconnect the two PVS lines at the carburetor.
4. Connect a tachometer as per the manufacturer's instructions. The engine should be running at no more than 775 rpm. Plug one of the two vacuum line fittings at the carburetor and check that the engine speed increases. Now block the other fitting and look for the same results.
5. If vacuum is not found on either fitting, you may have a plugged port. Clean all ports and repeat the test.

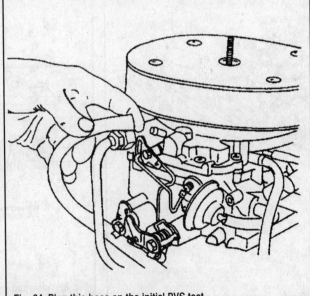

Fig. 64 Plug this hose on the initial PVS test

6-20 FUEL SYSTEM - CARBURETORS

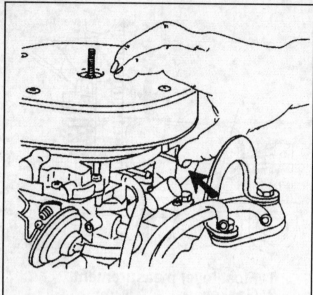

Fig. 65 Plug each vacuum fitting separately in the second part of the test

Throttle Cable

 MODERATE

◆ See Figures 66, 67 and 68

1. With the remote control lever in Neutral (idle), install the throttle cable end guide onto the lever and then push the barrel slightly toward the throttle lever end so as to preload the cable slightly and remove any slack. Adjust the barrel so that it aligns with the anchor stud.

2. Secure the throttle cable with its hardware and then tighten the locknut on the cable end guide until it just touches the cable, and then back it out 1/2 turn.

3. Tighten the throttle cable anchor screw securely - do not over-tighten.

4. Move the remote control lever to the full throttle position. The throttle valves should be open all the way and the throttle shaft lever should be in contact with the body of the carburetor.

5. Move the lever back to the Neutral position; the throttle lever should touch the idle speed adjustment screw.

DISASSEMBLY & ASSEMBLY

 DIFFICULT

◆ See Figures 69 thru 88

■ Always be certain that your carburetor rebuild kit is for marine applications.

1. Remove the carburetor from the manifold and mount it securely on a workbench.

2. Pull out the retaining clip at the end of the accelerator pump linkage and then disconnect the linkage.

3. Pull out the retaining clip at the end of the choke plate linkage and then disconnect the linkage.

4. If you intend to service the choke diaphragm, remove the vacuum line from the diaphragm, otherwise leave it alone. Likewise, remove the two diaphragm bracket mounting screws (#25 Torx) and remove it and the linkage.

5. Loosen the metering rod cover screws (#15 Torx) and lift off the covers. Certain models may be equipped with air deflectors, which will come off with the covers. Carefully lift out the metering rods and their springs. Rods are not interchangeable so make sure you identify and store the rods correctly for reinstallation. Also, the springs are color-coded and not interchangeable.

6. Remove the nine bowl cover mounting screws (#25 Torx) and lift off the assembly, disconnecting the choke linkage at the same time. Certain models may only have eight mounting screws. Always make sure to pull the top half straight up so as not to damage the accelerator pump or floats.

7. Raise the float slightly, grasp the float pin with needle nose pliers and pull it out. Lift out the float. Do not mix up the two floats - use a marker to identify right and left.

8. Lift out the two inlet needles from under the float attachments. Loosen the inlet seat and then lift the seat, gasket and filter from the housing as a unit. Make sure to keep the needles and seats together and marked as to which recess they came out of.

9. Pull up the gasket that should still be attached to the bottom of the bowl cover. Unscrew the accelerator pump lever and then lift out the pump assembly itself. Reach into the pump housing in the lower half of the carburetor and remove the pump spring.

10. Look into the carburetor and take note of which venturi clusters have 'distribution tags' attached and also of the ID number stamped on each one. Loosen the mounting screws (#25 Torx) and lift out each set of clusters. Make sure you lift them straight up and remove the gasket.

11. Repeat the above procedure for the secondary venturi clusters and then lift out the secondary air valve and weight assembly from underneath the clusters.

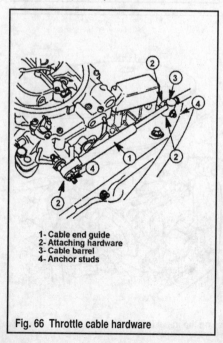

1- Cable end guide
2- Attaching hardware
3- Cable barrel
4- Anchor studs

Fig. 66 Throttle cable hardware

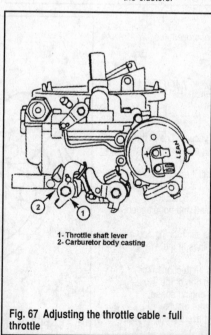

1- Throttle shaft lever
2- Carburetor body casting

Fig. 67 Adjusting the throttle cable - full throttle

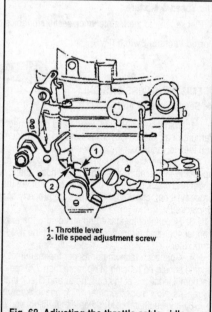

1- Throttle lever
2- Idle speed adjustment screw

Fig. 68 Adjusting the throttle cable - idle

FUEL SYSTEM - CARBURETORS

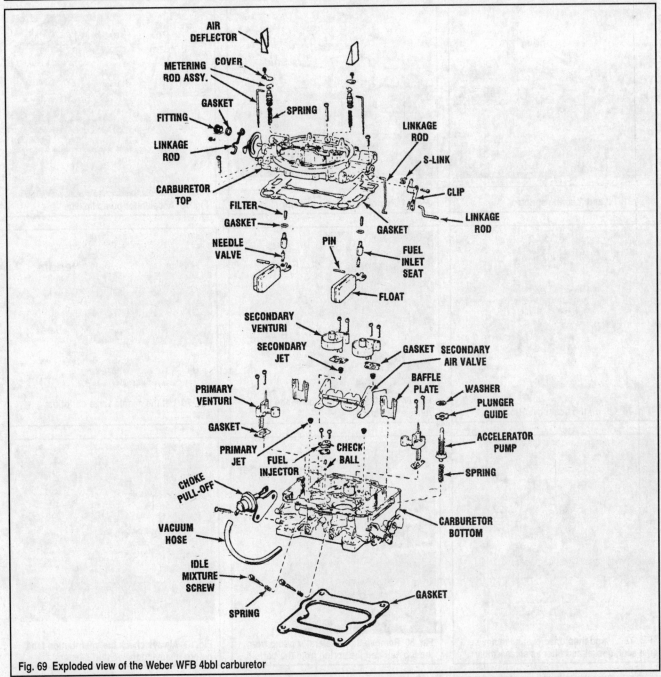

Fig. 69 Exploded view of the Weber WFB 4bbl carburetor

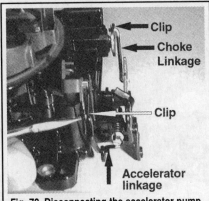

Fig. 70 Disconnecting the accelerator pump and choke plate linkages

Fig. 71 Unscrew the fuel metering rod cover...

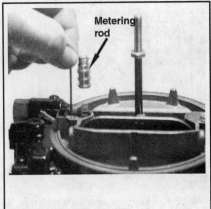

Fig. 72 Lift out the metering rod...

6-22 FUEL SYSTEM - CARBURETORS

Fig. 73 ...and then the spring

Fig. 74 Remove the flame arrestor stud

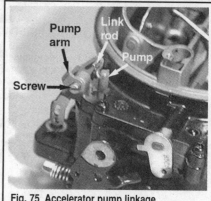
Fig. 75 Accelerator pump linkage

Fig. 76 Lift off the float cover (top piece)

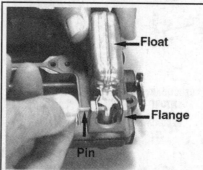
Fig. 77 Pull the pin out of the flange to release the float

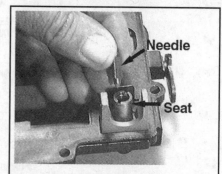
Fig. 78 Lift the needle valve out of the seat . . .

Fig. 79 ...and then unscrew and remove the seat, gasket and filter as an assembly

Fig. 80 Remove the accelerator pump from the top half and the spring from the bottom

Fig. 81 Always check the 'distribution tabs' before removing the venturi clusters

Fig. 82 Removing the secondary venturi cluster

Fig. 83 Lift out the secondary air valve and weight

Fig. 84 Remove the pump jet housing. . .

FUEL SYSTEM - CARBURETORS

Fig. 85 ...and then lift out the check ball

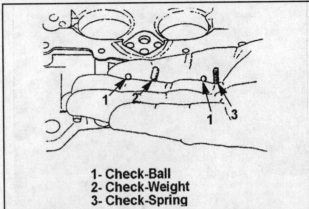

1- Check-Ball
2- Check-Weight
3- Check-Spring

Fig. 86 Some models may have a check ball and weight, while others may have a ball and spring

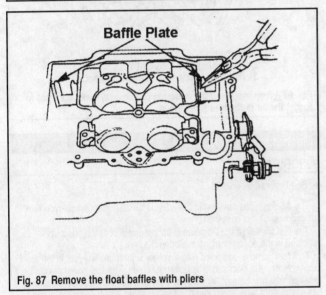

Fig. 87 Remove the float baffles with pliers

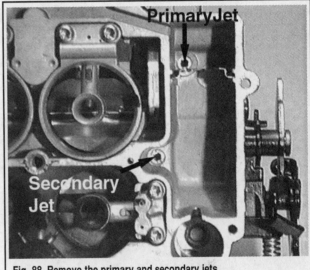

Fig. 88 Remove the primary and secondary jets

12. Loosen the two pump jet housing screws and pull out the housing and its gasket. At the bottom of the housing recess you should see a check ball and weight, or a check ball and spring - whichever it is, remove it(them).

13. Grasp the float bowl baffle plates with needle nose pliers and slide them up and out.

14. Remove the primary and secondary jets (#25 Torx). Make sure you note their location in the carb body and their sizes PRIOR to removal!

15. Turn each mixture screw in until it seats itself and not the number of turns (for installation); now you can remove the screws but be careful to note which one came from which hole.

To assemble:

◆ See Figures 89, 90 and 91

Never use a wire brush or drill to clean jet passages or tubes in the carburetor.

Never allow the carburetor to soak in a cleaner bath for more than two hours. In fact, we recommend using spray cleaner.

Never immerse the float assembly, needle, accelerator pump plunger or fuel filter in cleaner. Wipe them carefully with a clean cloth.

Otherwise clean all allowable parts with carb cleaner and then dry with compressed air if at all possible.

Blow out and through all passages to ensure there is no foreign material clogging them.

Check the float needle and seat, if either is worn or damaged, replace them as a matched set.

Check the float assembly and hinge pin for wear or damage, replace as necessary.

Check the pump plunger, return spring, piston spring, idle mixture needle and all levers and linkages for wear or damage. Replace as necessary.

Check the throttle valve shaft for excessive looseness in the throttle body. Check that the valve and shaft do not bind through their range of operation and that the valve opens and closes fully. Replace the assembly if it fails any of these tests.

Check the choke valve lever and shaft for excessive looseness in the air horn. Check that the valve, lever and shaft do not bind through their range of operation and that the valve opens and closes fully. Replace the air horn assembly if it fails any of these tests.

16. Install the idle mixture screws and tighten them until they just seat themselves. Now back them out to the number of turns you noted in the disassembly section.

17. Install the primary and secondary jets. Make sure you put them back into the same positions that you noted during removal.

18. Slide the float bowl baffle plates into position.

19. Install the check ball into the recess at the bottom of the pump jet housing. Slide the check weight of spring in on top of the ball. Never mix and match - if a spring came out, don't pop in a weight, or vice versa. Position a pump housing gasket and drop in the housing. Tighten the two screws securely.

20. Position the secondary air valve and weight assembly into the recess and then install the secondary venturi clusters with a new gasket. Tighten the two mounting screws on each cluster securely.

21. Repeat the procedure above and install the primaries.

22. Slip the accelerator pump spring into the housing on the lower carburetor body and then install the pump itself into the bowl cover. Remember to get the washer and guide into place first. Slip the end of the S-link on the accelerator pump lever into the hole on the end of the pump plunger and then install and tighten the lever pivot screw. Before you move on, check that the lever moves the pump plunger freely with no binding.

6-24 FUEL SYSTEM - CARBURETORS

Fig. 89 Install the idle mixture screw and spring

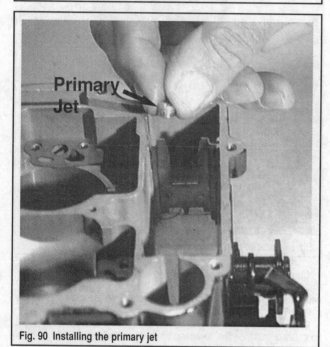

Fig. 90 Installing the primary jet

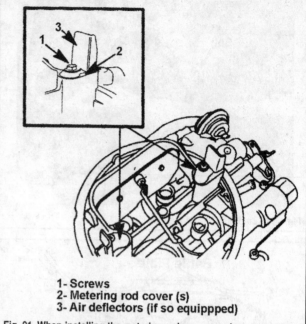

1- Screws
2- Metering rod cover (s)
3- Air deflectors (if so equippped)

Fig. 91 When installing the metering rod covers, make sure not to forget the air deflectors if equipped.

23. Position a new gasket onto the mating surface of the bowl cover. Press a new inlet filter into the bottom of the seat, install the gasket and press the assembly into the recess. Install the correct needles into their seats.

24. Position the floats (on the sides if reusing the old ones) and slide in the hinge pins. Carefully turn over the bowl cover and install it onto the lower half of the carburetor. Check that the gasket is still in the correct position and then tighten the nine mounting screws securely.

25. Install the metering rod springs into their holes on top of the carburetor - remember to check the specifications for the correct color. Drop the rods themselves into their original holes on top of the springs. Push down on them lightly to make sure the plunger is working properly and then install the covers and air deflectors (if equipped).

26. Reconnect the free end of the choke pull-off linkage and then install the diaphragm assembly. Reconnect the vacuum line to the rear of the diaphragm. Connect the choke plate linkage and press in the retaining clip.

27. Install the accelerator pump linkage end into the hole it came out of. If you've forgotten, refer to the correct procedure in Adjustments.

28. Install the carburetor.

Mechanical Fuel Pump

DESCRIPTION

◆ See Figures 92 and 93

The 3.0L engines covered in this service utilize a mechanical-type fuel pump driven off the camshaft.

The fuel filter on the 3.0L pump is an integral part of the pump as indicated in the accompanying exploded diagram.

This fuel pump is equipped with a yellow sight tube. Any fuel in the sight tube indicates the diaphragm has been ruptured. The fuel pump cannot be repaired with any type of "kit". Due to the hazardous condition indicated by fuel in the sight tube, the pump must be replaced with a complete pump assembly, at once.

The fuel pump sucks gasoline from the fuel tank and delivers it to the carburetor in sufficient quantities, under pressure, to satisfy engine demands under all operating conditions.

The pump is operated by a two-part rocker arm. The outer part rides on an eccentric on the camshaft and is held in constant contact with the camshaft by a strong return spring. The inner part is connected to the fuel pump diaphragm by a short connecting rod. As the camshaft rotates, the rocker arm moves up and down. As the outer part of the rocker arm moves downward, the inner part moves upward, pulling the fuel diaphragm upward. This upward movement compresses the diaphragm spring and creates a vacuum in the fuel chamber below the diaphragm. The vacuum causes the outlet valve to close and permits fuel from the gas tank to enter the chamber by way of the fuel filter and the inlet valve.

Now, as the eccentric on the camshaft allows the outer part of the rocker arm to move upward, the inner part moves downward, releasing its hold on the connecting rod. The compressed diaphragm spring then exerts pressure on the diaphragm, which closes the inlet valve and forces fuel out through the outlet valve to the carburetor.

Because the fuel pump diaphragm(s) is(are) moved downward only by the diaphragm spring, the pump delivers fuel to the carburetor only when the pressure in the outlet line is less than the pressure exerted by the diaphragm spring. This lower pressure condition exists when the carburetor float needle valve is unseated and the fuel passages from the pump into the carburetor float chamber are open.

When the needle valve is closed and held in place by the pressure of the fuel on the float, the pump builds up pressure in the fuel chamber until it overcomes the pressure of the diaphragm spring. This pressure almost stops movement of the diaphragm until more fuel is needed in the carburetor float bowl.

FUEL SYSTEM - CARBURETORS

REMOVAL & INSTALLATION

◆ See Figures 92 and 93

■ Always have a fire extinguisher handy when working on any part of the fuel system. Remember, a very small amount of fuel vapor in the bilge, has the tremendous potential explosive power.

1. Open or remove the engine hatch/covers. Disconnect the battery cables.
2. Position a container under the pump and remove the fuel inlet and outlet lines. It is important to use two open-end wrenches; one to hold the fitting nut and the other to loosen the line nut. Plug the line with a golf tee or something similar to prevent additional fuel spillage. Take care not to spill fuel on a hot engine, because such fuel may ignite.
3. Disconnect the sight tube if so equipped.
4. Loosen the two pump mounting screws and carefully pull the pump out. Scrape any old gasket material from the pump and block mating surfaces. The pump pushrod may fall out of position when removing the pump; to prevent this, slip a small screwdriver in behind the pump and support the pushrod. Check the pump pushrod for damage or wear.
5. Coat BOTH sides of a new gasket with Perfect Seal and position it on the pump. Swab the end of the pushrod with grease and insert the pump into the cylinder block so that the rod is riding on the camshaft eccentric. Coat the mounting bolts with Perfect Seal and then tighten them to 20 ft. lbs. (27 Nm).
6. Coat the threads of the fuel line pump fittings with #592 Loctite Pipe Sealant, thread them into the pump base until they are finger tight, and then tighten an additional 1 3/4 to 2 1/4 turns with a wrench. Do not over-tighten and do not use Teflon tape.
7. Install the fuel lines and tighten securely while holding the pump fittings with another wrench.
8. Install the sight tube, remove the container and connect the battery cables. Start the engine and check for fuel leaks.

PRESSURE TEST

◆ See Figure 94

1. Open or remove the engine hatch/covers. Disconnect the battery cables.
2. Position a container under the fuel line connection at the carburetor and remove the fuel inlet line. It is important to use two open-end wrenches; one to hold the fitting nut and the other to loosen the line nut. Plug the line with a golf tee or something similar to prevent additional fuel spillage.
3. Thread a 'tee' or a fuel pressure connector (#91-18078) into the fitting nut on the carburetor and then reconnect the fuel line to the other end.
4. Connect a fuel pressure test gauge to the remaining fitting on the connector.
5. Connect the battery cables, start the engine and let it idle. The fuel pressure should be 5 1/4 to 6 1/4 psi (36-44 kPa).
6. Slowly run the engine up to 1,800 rpm and check that the pressure remains the same.
7. If pressure varies, replace the pump.
8. Turn off the engine, remove the pressure gauge and fuel line and then remove the connector. Reinstall the fuel line to the fitting and tighten to 18 ft. lbs. (24 Nm).

FUEL LINE TEST

The fuel line, from the tank to the fuel pump, can be quickly tested by disconnecting the existing fuel line at the fuel pump and connecting a six-gallon portable tank and fuel line. This simple substitution eliminates the fuel tank and fuel lines in the boat. Now, start the engine and check the performance.

If the problem has been corrected, the fuel system between the fuel pump inlet and the fuel tank is at fault. This area includes the fuel line, the fuel pickup in the tank, the fuel filter, anti-siphon valve, the fuel tank vent, and

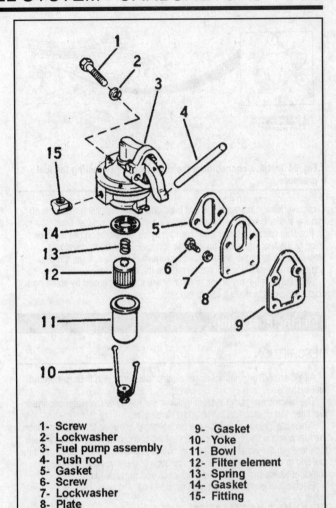

1- Screw
2- Lockwasher
3- Fuel pump assembly
4- Push rod
5- Gasket
6- Screw
7- Lockwasher
8- Plate
9- Gasket
10- Yoke
11- Bowl
12- Filter element
13- Spring
14- Gasket
15- Fitting

Fig. 92 Exploded view of the fuel pump - 3.0L engines

Fig. 93 A good look at the pump

6-26 FUEL SYSTEM - CARBURETORS

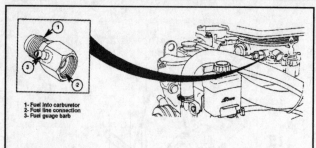

Fig. 94 Install a connector at the carburetor when testing the fuel pressure

excessive foreign matter in the fuel tank, and loose fuel fittings sucking air into the system. Improper size fuel fittings can also restrict fuel flow.

Possible cause of fuel line problems may be deterioration of the inside lining of the fuel line which may cause some of the lining to develop a blockage similar to the action of a check valve. Therefore, if the fuel line appears the least bit questionable, replace the entire line.

Another possible restriction in the fuel line may be caused by some heavy object lying on the line - a tackle box, etc.

Electric Fuel Pump

DESCRIPTION

All V6 and V8 engines utilize an electric fuel pump with their carbureted fuel systems.

The electric fuel pump system includes the fuel tank/s, a water separator fuel filter, the electric fuel pump and a carburetor.

When the ignition switch on the control panel is turned to the **ON** position, the fuel pump is energized. Operation of the pump draws fuel from the fuel tank through the water separator fuel filter and is then pushed on through the fuel line to the carburetor. Efficient operation of the carburetor supplies sufficient fuel to the engine under all loads and rpm speeds.

Electric fuel pumps are not repairable.

REMOVAL & INSTALLATION

◆ See Figures 95 and 96

■ **Always have a fire extinguisher handy when working on any part of the fuel system. Remember, a very small amount of fuel vapor in the bilge, has the tremendous potential explosive power.**

1. Open or remove the engine hatch/covers. Disconnect the battery cables.
2. Position a container under the pump and remove the fuel inlet (large) and outlet (small) lines. It is important to use two open-end wrenches; one to hold the fuel pump fitting nut and the other to loosen the line nut. Plug the line with a golf tee or something similar to prevent additional fuel spillage. Take care not to spill fuel on a hot engine, because such fuel may ignite.
3. Unplug the electrical harness at the pump connector and move it out of the way.
4. Grasp the pump in the center of its body and pull it out of its holding bracket. Remove the two inlet and outlet fittings with a wrench on the fitting nut and one on the pump flat.
5. Install the small rubber grommet onto the outlet side of the pump. Slide a new O-ring over the fitting and screw it into the pump. Install the large rubber grommet onto the inlet side of the pump. Slide a new O-ring over the fitting and screw it into the pump. Coat the threads of the lower fuel line pump fittings with #592 Loctite Pipe Sealant, thread them into the pump base until they are finger tight, and then tighten an additional 1-3/4 to 2-1/4 turns with a wrench. Do not over-tighten and do not use Teflon tape. Repeat this procedure for the upper line.
6. Press the pump into the bracket making sure the large grommet is at the bottom. Connect the wiring harness.
7. Connect the fuel lines and tighten them to 18 ft. lbs. (24 Nm).
8. Remove the container and connect the battery cables. Start the engine and check for fuel leaks.

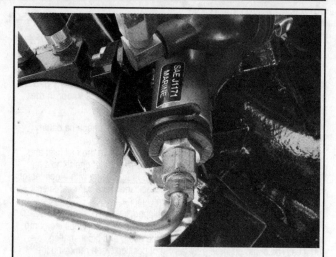

Fig. 95 The electric fuel pump is located on the starboard front side of the engine, near the water separating fuel filter

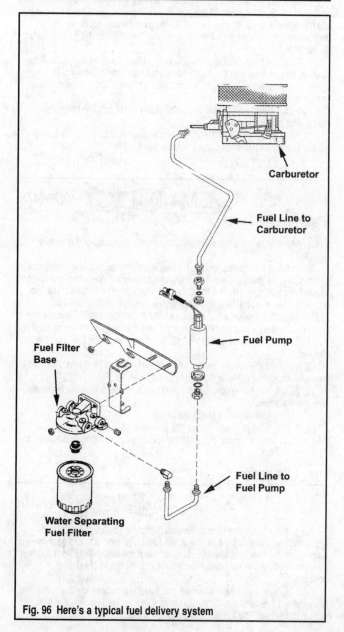

Fig. 96 Here's a typical fuel delivery system

FUEL SYSTEM - CARBURETORS   6-27

PRESSURE TEST

1. Open or remove the engine hatch/covers. Disconnect the battery cables.
2. Position a container under the fuel line connection at the carburetor (or the outlet side of the fuel pump) and remove the fuel inlet line. It is important to use two open-end wrenches; one to hold the fitting nut and the other to loosen the line nut. Plug the line with a golf tee or something similar to prevent additional fuel spillage.
3. Thread a 'tee' or a fuel pressure connector (#91-18078) into the fitting nut on the carburetor and then reconnect the fuel line to the other end.
4. Connect a fuel pressure test gauge to the remaining fitting on the connector.
5. Connect the battery cables, start the engine and let it idle. The fuel pressure should be 3-7 psi (21-49 kPa).
6. If pressure varies, replace the pump.
7. Turn off the engine, remove the pressure gauge and fuel line and then remove the connector. Reinstall the fuel line to the fitting and tighten to 18 ft. lbs. (24 Nm).

WIRING DIAGRAMS

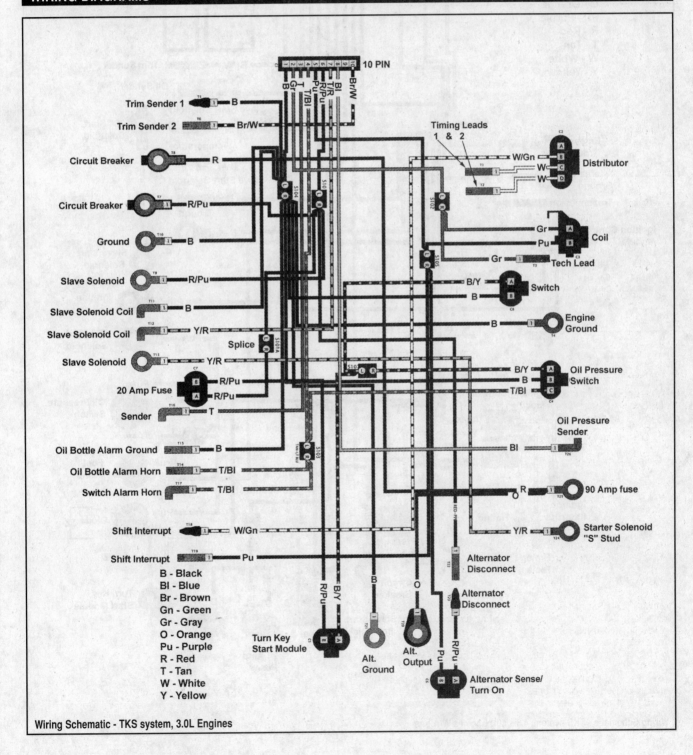

Wiring Schematic - TKS system, 3.0L Engines

6-28 FUEL SYSTEM - CARBURETORS

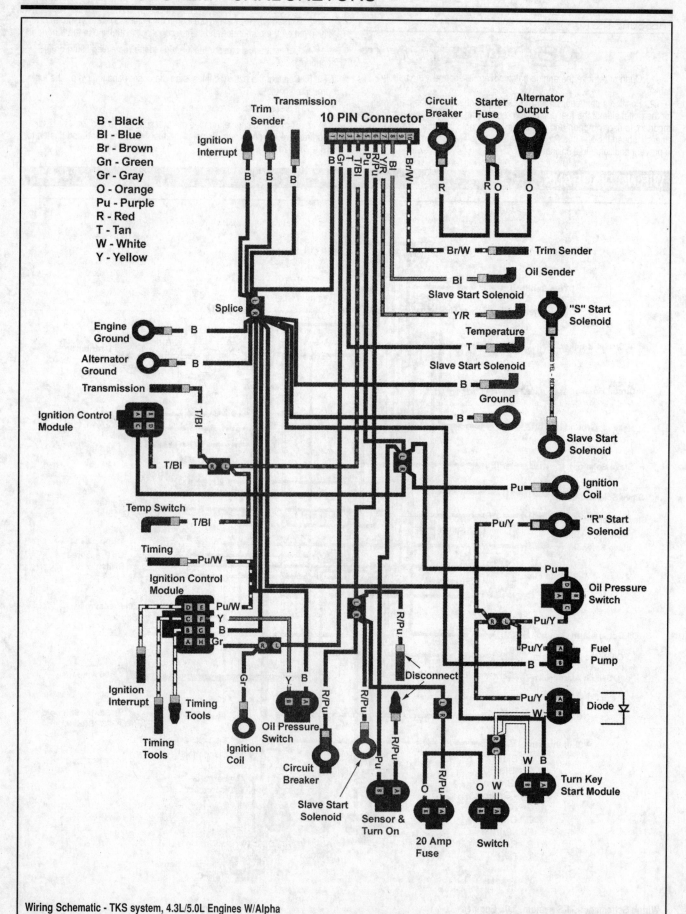

Wiring Schematic - TKS system, 4.3L/5.0L Engines W/Alpha

FUEL SYSTEM - CARBURETORS 6-29

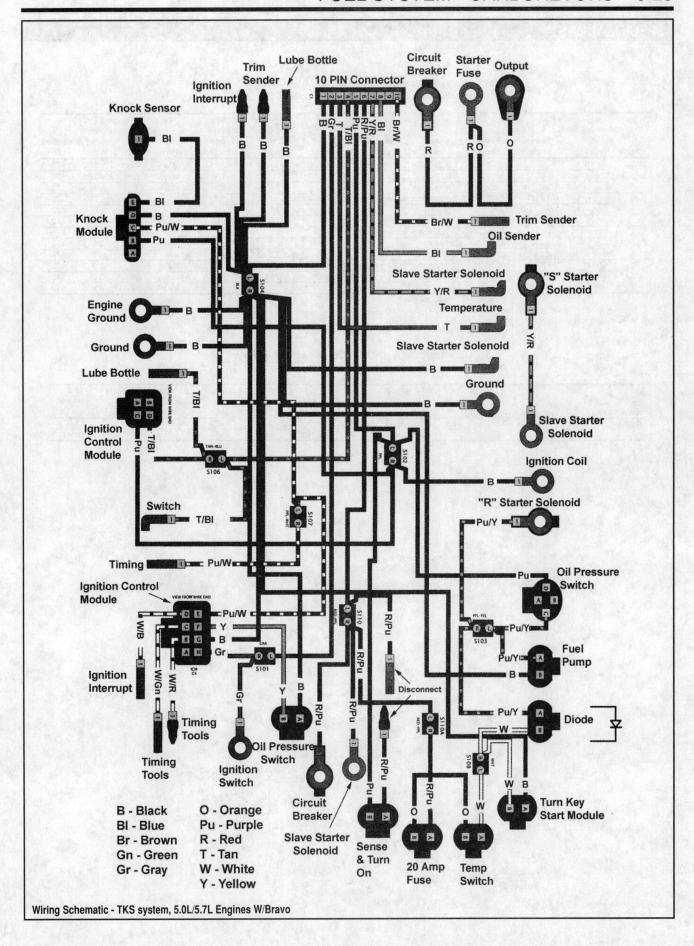

Wiring Schematic - TKS system, 5.0L/5.7L Engines W/Bravo

SPECIFICATIONS

Carburetor Specifications - 2 BBL MerCarb
Measurements are in.(mm) unless otherwise noted

Engine	Type	Float Level	Float Drop	Pump Rod	Choke Setting	Choke Unloader	TKS Enrichment Jet (mm)	Idle Mixture Screw (Turns Out)	Float Weight	Main Jet (mm)	Power Valve (mm)	Venturi I.D.
3.0L	36mm	①	1 3/32 (27)	1 5/32 (29)	2 marks lean	0.080 (0.2)	-	Factory	9g Max	1.45	2 x 0.60	470
	43mm	①	1 3/32 (27)	1 5/32 (29)	2 marks lean	0.080 (0.2)	-	1 1/4	9g Max	1.55	4 x 0.65	171
	Gen II TKS	0.550 (14)	1.080 (27)	Middle hole	-	-	70	3 1/8	9g Max	1.7	0.64	171
4.3L	43mm	9/16 (14)	1 3/32 (27)	Middle hole	2 marks lean	0.080 (0.2)	-	1 1/4	9g Max	1.55	0.74	472
	Gen I TKS	0.550 (14)	1.080 (27)	Middle hole	-	-	70	2 5/8	9g Max	1.55	0.74	472
	Gen II TKS	0.550 (14)	1.080 (27)	Middle hole	-	-	70	3 1/16	9g Max	1.55	0.74	472
5.0L	43mm	11/32 (9)	15/16 (24)	Middle hole	2 marks lean	0.080 (0.2)	-	1 1/2	9g Max	1.65	0.9	476
	Gen I TKS	0.350 (8.9)	15/16 (24)	Middle hole	-	-	70	2 1/2	9g Max	1.65	0.9	476
	Gen II TKS	0.350 (8.9)	0.937 (23.8)	Middle hole	-	-	70	2 1/4	9g Max	1.65	0.9	476
5.7L	43mm	11/32 (9)	15/16 (24)	Middle hole	2 marks lean	0.080 (0.2)	-	1 1/2	9g Max	1.65	0.9	475
	Gen I TKS	0.350 (8.9)	15/16 (24)	Middle hole	-	-	70	2 7/8	9g Max	1.65	0.9	475
	Gen II TKS	0.350 (8.9)	0.937 (23.8)	Middle hole	-	-	70	3	9g Max	1.65	0.9	475

① Solid needle: 3/8(10); spring loaded needle: 9/16(14)
② Measurement is in mm

Carburetor Specifications - 4bbl Weber
All measurements in.(mm) unless other wise noted

Engine	Type	Float Level	Float Drop	Pump Rod	Choke Pull-Off	Accel. Pump	Primary Jet (in.)	Secondary Jet (in.)	Metering Rod Number	Metering Rod Spring	Idle Mixture (Turns Out)
4.3L	WFB	1 9/32 (33)	2.0 (51)	#3 from end	1.28 (32.5)	7/16 (11)	0.092	0.089	16-686647	Green	1 1/4
	WFB SAV1	1 9/32 (33)	2.0 (51)	#3 from end	NA	7/16 (11)	0.087	0.086	16-656457	Natural	Sealed

COOL FUEL MODULE (GEN III) 7-12	FUEL RAIL & INJECTORS. 7-15
DISASSEMBLY & ASSEMBLY. 7-12	BALANCE TEST 7-17
REMOVAL & INSTALLATION 7-12	REMOVAL & INSTALLATION 7-15
COOL FUEL SYSTEM (GEN II). 7-10	FUEL SYSTEM APPLICATIONS 7-3
REMOVAL, DISASSEMBLY & INSTALLATION 7-10	GENERAL DIAGNOSTIC TEST SCHEMATICS - TBI 7-28
DESCRIPTION & OPERATION 7-4	GENERAL DIAGNOSTIC TESTS
MODES OF OPERATION 7-4	MPI EXC 8.1 7-52
MULTI-POINT INJECTION (MPI) 7-4	8.1L MPI. 7-56
SUBSYSTEMS. 7-7	TBI 7-27
THROTTLE BODY INJECTION (TBI) 7-4	IDLE AIR CONTROL VALVE (IAC) 7-23
DIAGNOSTIC TROUBLE CODE (DTC) CHART - MPI. 7-31	REMOVAL & INSTALLATION 7-23
CLEARING CODES. 7-31	INTERMITTENT FAULTS
READING CODES. 7-30	MPI. 7-48
DIAGNOSTIC TROUBLE CODE TEST SCHEMATICS 7-38	TBI 7-26
DIAGNOSTIC TROUBLE CODES (DTC) - MPI. 7-30	KNOCK SENSOR (KS) 7-24
ECM PIN LOCATIONS	REMOVAL & INSTALLATION 7-24
MPI. 7-59	KNOCK SENSOR MODULE 7-24
TBI 7-43	REMOVAL & INSTALLATION 7-24
ELECTRONIC CONTROL MODULE (ECM) . . 7-22	MANIFOLD ABSOLUTE PRESSURE SENSOR (MAP/MAPT) 7-25
REMOVAL & INSTALLATION 7-22	REMOVAL & INSTALLATION 7-25
ELECTRONIC FUEL INJECTION 7-4	OIL PRESSURE SWITCH/SENSOR 7-25
COOL FUEL MODULE (GEN III) 7-12	REMOVAL & INSTALLATION 7-25
COOL FUEL SYSTEM (GEN II). 7-10	PRECAUTIONS
DESCRIPTION & OPERATION 7-4	MPI. 7-48
ELECTRONIC CONTROL MODULE (ECM) . 7-22	TBI 7-26
ENGINE COOLANT TEMPERATURE SENSOR (ECT) 7-22	PROPULSION CONTROL MODULE (PCM) . . 7-22
FUEL BOOST PUMP (ELECTRIC) 7-13	REMOVAL & INSTALLATION 7-22
FUEL INJECTORS. 7-14	RELIEVING FUEL PRESSURE 7-9
FUEL METER COVER. 7-14	MPI ENGINES 7-9
FUEL PRESSURE REGULATOR 7-14	TBI ENGINES 7-9
FUEL PUMP - ELECTRIC 7-13	**SPECIFICATIONS 7-70**
FUEL PUMP RELAY 7-13	TORQUE - EFI ENGINES 7-70
FUEL RAIL & INJECTORS. 7-15	**SYSTEM DIAGNOSIS -**
IDLE AIR CONTROL VALVE (IAC) 7-23	**TBI ENGINES & EARLY 2001 MPI ENGINES 7-26**
KNOCK SENSOR (KS) 7-24	DIAGNOSTIC TROUBLE CODE (DTC) CHART. 7-31
KNOCK SENSOR MODULE 7-24	DIAGNOSTIC TROUBLE CODE TEST SCHEMATICS 7-38
MANIFOLD ABSOLUTE PRESSURE SENSOR (MAP/MAPT) 7-25	DIAGNOSTIC TROUBLE CODES (DTC) . . . 7-30
OIL PRESSURE SWITCH/SENSOR 7-25	ECM PIN LOCATIONS & SYMPTOMS CHARTS 7-43
PROPULSION CONTROL MODULE (PCM) . 7-22	GENERAL DIAGNOSTIC TEST SCHEMATICS 7-28
RELIEVING FUEL PRESSURE. 7-9	GENERAL DIAGNOSTIC TESTS 7-27
THROTTLE BODY. 7-17	INTERMITTENT FAULTS. 7-26
THROTTLE BODY ADAPTER PLATE 7-21	PRECAUTIONS 7-26
THROTTLE POSITION SENSOR (TP) 7-25	VACUUM DIAGRAMS 7-41
ENGINE COOLANT TEMPERATURE SENSOR (ECT) 7-22	WIRING SCHEMATICS 7-41
REMOVAL & INSTALLATION 7-22	**SYSTEM DIAGNOSIS -**
FUEL 7-2	**2001-08 MPI ENGINES 7-47**
ALCOHOL-BLENDED FUELS. 7-2	ECM PIN LOCATIONS. 7-59
OCTANE RATING 7-2	GENERAL DIAGNOSTIC TESTS
RECOMMENDATIONS 7-2	2001-08 4.3L, 5.0L, 5.7L & 6.2L MPI ENGINES. 7-52
VAPOR PRESSURE 7-2	8.1L MPI ENGINES 7-56
FUEL & COMBUSTION. 7-2	INTERMITTENT FAULTS. 7-48
APPLICATIONS 7-3	PRECAUTIONS 7-48
COMBUSTION. 7-2	TROUBLESHOOTING. 7-49
FUEL 7-2	WIRING SCHEMATICS 7-61
FUEL BOOST PUMP (ELECTRIC) 7-13	THROTTLE BODY. 7-17
REMOVAL & INSTALLATION 7-13	REMOVAL & INSTALLATION 7-17
FUEL INJECTORS. 7-14	THROTTLE CABLE ADJUSTMENT 7-20
REMOVAL & INSTALLATION 7-14	THROTTLE BODY ADAPTER PLATE 7-21
MPI ENGINES 7-15	REMOVAL & INSTALLATION 7-21
TBI ENGINES. 7-14	THROTTLE POSITION SENSOR (TP) 7-25
FUEL METER COVER. 7-14	REMOVAL & INSTALLATION 7-25
REMOVAL & INSTALLATION 7-14	TROUBLESHOOTING - MPI 7-49
FUEL PRESSURE REGULATOR 7-14	VACUUM DIAGRAMS - TBI 7-41
REMOVAL & INSTALLATION 7-14	WIRING SCHEMATICS
FUEL PUMP - ELECTRIC 7-13	MPI. 7-61
REMOVAL & INSTALLATION 7-13	TBI 7-41
FUEL PUMP RELAY 7-13	
REMOVAL & INSTALLATION 7-13	

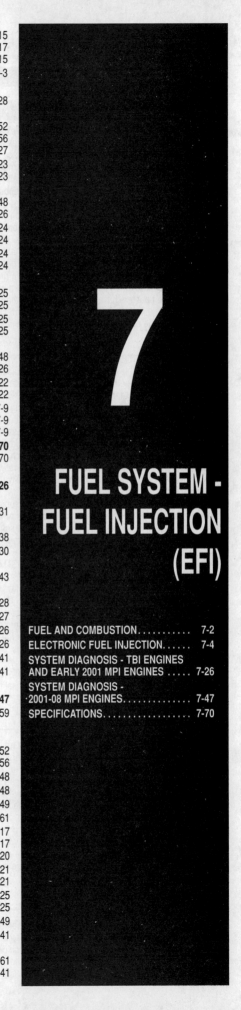

7

FUEL SYSTEM - FUEL INJECTION (EFI)

FUEL AND COMBUSTION.	7-2
ELECTRONIC FUEL INJECTION.	7-4
SYSTEM DIAGNOSIS - TBI ENGINES AND EARLY 2001 MPI ENGINES	7-26
SYSTEM DIAGNOSIS - 2001-08 MPI ENGINES.	7-47
SPECIFICATIONS.	7-70

7-2 FUEL SYSTEM - FUEL INJECTION (EFI)

FUEL & COMBUSTION

Fuel

Fuel recommendations have become more complex as the chemistry of modern gasoline changes. The major driving force behind the changes in gasoline chemistry is the search for additives to replace lead as an octane booster and lubricant. These new additives are governed by the types of emissions they produce in the combustion process. Also, the replacement additives do not always provide the same level of combustion stability, making a fuel's octane rating less meaningful.

In the 1960's and 1970's, leaded fuel was common. The lead served two functions. First, it served as an octane booster (combustion stabilizer) and second, in 4-stroke engines, it served as a valve seat lubricant. For 2-stroke engines, the primary benefit of lead was to serve as a combustion stabilizer. Lead served very well for this purpose, even in high heat applications.

Today, all lead has been removed from the refining process. This means that the benefit of lead as an octane booster has been eliminated. Several substitute octane boosters have been introduced in the place of lead. While many are adequate in an engine, most do not perform nearly as well as lead did, even though the octane rating of the fuel is the same.

OCTANE RATING

A fuel's octane rating is a measurement of how stable the fuel is when heat is introduced. Octane rating is a major consideration when deciding whether a fuel is suitable for a particular application. For example, in an engine, we want the fuel to ignite when the spark plug fires and not before, even under high pressure and temperatures. Once the fuel is ignited, it must burn slowly and smoothly, even though heat and pressure are building up while the burn occurs. The unburned fuel should be ignited by the traveling flame front, not by some other source of ignition, such as carbon deposits or the heat from the expanding gasses. A fuel's octane rating is known as a measurement of the fuel's anti-knock properties (ability to burn without exploding).

Usually a fuel with a higher octane rating can be subjected to a more severe combustion environment before spontaneous or abnormal combustion occurs. To understand how two gasoline samples can be different, even though they have the same octane rating, we need to know how octane rating is determined.

The American Society of Testing and Materials (ASTM) has developed a universal method of determining the octane rating of a fuel sample. The octane rating you see on the pump at a fuel dock is known as the pump octane number. Look at the small print on the pump. The rating has a formula. The rating is determined by the R+M/2 method. This number is the average of the research octane reading and the motor octane rating.

• The Research Octane Rating is a measure of a fuel's anti-knock properties under a light load or part throttle conditions. During this test, combustion heat is easily dissipated.

• The Motor Octane Rating is a measure of a fuel's anti-knock properties under a heavy load or full throttle conditions, when heat buildup is at maximum.

VAPOR PRESSURE

Fuel vapor pressure is a measure of how easily a fuel sample evaporates. Many additives used in gasoline contain aromatics. Aromatics are light hydrocarbons distilled off the top of a crude oil sample. They are effective at increasing the research octane of a fuel sample but can cause vapor lock (bubbles in the fuel line) on a very hot day. If you have an inconsistent running engine and you suspect vapor lock, use a piece of clear fuel line to look for bubbles, indicating that the fuel is vaporizing.

One negative side effect of aromatics is that they create additional combustion products such as carbon and varnish. If your engine requires high-octane fuel to prevent detonation, de-carbon the engine more frequently with an internal engine cleaner to prevent ring sticking due to excessive varnish buildup.

ALCOHOL-BLENDED FUELS

When the Environmental Protection Agency mandated a phase-out of the leaded fuels in January of 1986, fuel suppliers needed an additive to improve the octane rating of their fuels. Although there are multiple methods currently employed, the addition of alcohol to gasoline seems to be favored because of its favorable results and low cost. Two types of alcohol are used in fuel today as octane boosters, methanol (wood alcohol) or ethanol (grain alcohol).

When used as a fuel additive, alcohol tends to raise the research octane of the fuel. There are, however, some special considerations due to the effects of alcohol in fuel.

• Since alcohol contains oxygen, it replaces gasoline without oxygen content and tends to cause the air/fuel mixture to become leaner.

• On older engines, the leaching affect of alcohol may, in time, cause fuel lines and plastic components to become brittle to the point of cracking. Unless replaced, these cracked lines could leak fuel, increasing the potential for hazardous situations.

• When alcohol blended fuels become contaminated with water, the water combines with the alcohol then settles to the bottom of the tank. This leaves the gasoline on a top layer.

■ **Modern fuel lines and plastic fuel system components have been specially formulated to resist alcohol leaching effects.**

RECOMMENDATIONS

According to the fuel recommendations that come with your engine, there are only a few engines in the product line that requires more than 87 octane. Most MerCruiser engines need only 87 octane or less. An 89 octane rating generally means middle grade unleaded. Premium unleaded is more stable under severe conditions but also produces more combustion products. Therefore, when using premium unleaded, more frequent de-carboning is necessary.

Combustion

In a high heat environment like an modern engine, the fuel must be very stable to avoid detonation. If any parameters affecting combustion change suddenly (the engine runs lean for example), uncontrolled heat buildup will occurs very rapidly.

The combustion process is affected by several interrelated factors. This means that when one factor is changed, the other factors also must be changed to maintain the same controlled burn and level of combustion stability.

• Compression - determines the level of heat buildup in the cylinder when the air-fuel mixture is compressed. As compression increases, so does the potential for heat buildup

• Ignition Timing - determines when the gasses will start to expand in relation to the motion of the piston. If the ignition timing is too advanced, gasses will be ignited and begin to expand too soon, such as they would during pre-ignition. The motion of the piston opposes the expansion of the gasses, resulting in extremely high combustion chamber pressures and heat. If the ignition timing is retarded, the gases are ignited later in relation to piston position. This means that the piston has already traveled back down the bore toward the bottom of the cylinder, resulting in less usable power.

• Fuel Mixture - determines how efficient the burn will be. A rich mixture burns slower than a lean one. If the mixture is too lean, it can't become explosive. The slower the burn, the cooler the combustion chamber, because pressure buildup is gradual.

• Fuel Quality (Octane Rating) - determines how much heat is necessary to ignite the mixture. Once the burn is in progress, heat is on the rise. The unburned poor quality fuel is ignited all at once by the rising heat instead of burning gradually as a flame front of the burn passing by. This action results in detonation (pinging).

There are two types of abnormal combustion - pre-ignition and detonation.

• Pre-ignition - occurs when the air-fuel mixture is ignited by some incandescent source other than the correctly timed spark from the spark plug.

• Detonation - occurs when excessive heat and or pressure ignites the air/fuel mixture rather than the spark plug. The burn becomes explosive.

In general, anything that can cause abnormal heat buildup can be enough to push an engine over the edge to abnormal combustion, if any of the four basic factors previously discussed are already near the danger point, for example, excessive carbon buildup raises the compression and retains heat as glowing embers.

FUEL SYSTEM - FUEL INJECTION (EFI)

Fuel System Applications

Year		Engine Model	Displacement L/Cu. In.	Engine Type	Fuel Delivery
2004 (cont'd)		8.1S Horizon	8.1/496	V8	MPI
		496 Mag	8.1/496	V8	MPI
		496 Mag HO	8.1/496	V8	MPI
2005	4.3L	4.3L MPI	4.3/262	V6	MPI
	5.0L	5.0L MPI	5.0/305	V8	MPI
	5.7L	350 Mag MPI	5.7/350	V8	MPI
		350 Mag MPI/Horizon	5.7/350	V8	MPI
	6.2L	Black Scorpion	6.2/377	V8	MPI
		MX6.2 Black Scorpion	6.2/377	V8	MPI
		MX6.2 MPI	6.2/377	V8	MPI
		MX6.2 MPI/Horizon	6.2/377	V8	MPI
	8.1L	8.1S HO	8.1/496	V8	MPI
		8.1S Horizon	8.1/496	V8	MPI
		496 Mag	8.1/496	V8	MPI
		496 Mag HO	8.1/496	V8	MPI
2006	4.3L	4.3L MPI	4.3/262	V6	MPI
	5.0L	5.0L MPI	5.0/305	V8	MPI
	5.7L	350 Mag MPI	5.7/350	V8	MPI
		350 Mag MPI/Horizon	5.7/350	V8	MPI
	6.2L	Black Scorpion	6.2/377	V8	MPI
		MX6.2 Black Scorpion	6.2/377	V8	MPI
		MX6.2 MPI	6.2/377	V8	MPI
		MX6.2 MPI/Horizon	6.2/377	V8	MPI
	8.1L	8.1S HO	8.1/496	V8	MPI
		8.1S Horizon	8.1/496	V8	MPI
		496 Mag HO	8.1/496	V8	MPI
2007-08	3.0L	3.0L MPI ①	3.0/181	I4	MPI
	4.3L	4.3L MPI	4.3/262	V6	MPI
	5.0L	5.0L MPI	5.0/305	V8	MPI
	5.7L	5.7L MPI/Horizon	5.7/350	V8	MPI
		350 Mag MPI/Horizon	5.7/350	V8	MPI
		Scorpion 350	5.7/350	V8	MPI
		Tow Sport 350	5.7/350	V8	MPI
	6.2L	377 Mag	6.2/377	V8	MPI
		MX6.2 MPI/Horizon	6.2/377	V8	MPI
		Scorpion 377	6.2/377	V8	MPI
	8.1L	8.1S/Horizon	8.1/496	V8	MPI
		8.1S HO/Horizon	8.1/496	V8	MPI
		496 Mag MPI	8.1/496	V8	MPI
		496 Mag HO MPI	8.1/496	V8	MPI

MPI   Multi-Port Injection
TBI   Throttle Body Injection
①   Calif. Emissions only; no info available at publication

Fuel System Applications

Year		Engine Model	Displacement L/Cu. In.	Engine Type	Fuel Delivery
2001	4.3L	4.3L EFI	4.3/262	V6	TBI or MPI
	5.0L	5.0L EFI	5.0/305	V8	TBI or MPI
	5.7L	5.7L EFI	5.7/350	V8	TBI or MPI
		350 Mag MPI	5.7/350	V8	MPI
		350 Mag MPI/Horizon	5.7/350	V8	MPI
	6.2L	Black Scorpion	6.2/377	V8	MPI
		MX6.2 MPI	6.2/377	V8	MPI
	8.1	8.1S HO	8.1/496	V8	MPI
		8.1S Horizon	8.1/496	V8	MPI
		496 Mag	8.1/496	V8	MPI
		496 Mag HO	8.1/496	V8	MPI
2002	4.3L	4.3L EFI	4.3/262	V6	TBI
		4.3L MPI	4.3/262	V6	MPI
	5.0L	5.0L EFI	5.0/305	V8	TBI
		5.0L MPI	5.0/305	V8	MPI
	5.7L	5.7L EFI	5.7/350	V8	TBI or MPI
		350 Mag MPI	5.7/350	V8	MPI
		350 Mag MPI/Horizon	5.7/350	V8	MPI
	6.2L	Black Scorpion	6.2/377	V8	MPI
		MX6.2 Black Scorpion	6.2/377	V8	MPI
		MX6.2 MPI	6.2/377	V8	MPI
		MX6.2 MPI/Horizon	6.2/377	V8	MPI
	8.1L	8.1S HO	8.1/496	V8	MPI
		8.1S Horizon	8.1/496	V8	MPI
		496 Mag	8.1/496	V8	MPI
		496 Mag HO	8.1/496	V8	MPI
2003	4.3L	4.3L MPI	4.3/262	V6	MPI
	5.0L	5.0L MPI	5.0/305	V8	MPI
	5.7L	350 Mag MPI	5.7/350	V8	MPI
		350 Mag MPI/Horizon	5.7/350	V8	MPI
	6.2L	Black Scorpion	6.2/377	V8	MPI
		MX6.2 Black Scorpion	6.2/377	V8	MPI
		MX6.2 MPI	6.2/377	V8	MPI
		MX6.2 MPI/Horizon	6.2/377	V8	MPI
	8.1L	8.1S HO	8.1/496	V8	MPI
		8.1S Horizon	8.1/496	V8	MPI
		496 Mag	8.1/496	V8	MPI
		496 Mag HO	8.1/496	V8	MPI
2004	4.3L	4.3L MPI	4.3/262	V6	MPI
	5.0L	5.0L MPI	5.0/305	V8	MPI
	5.7L	350 Mag MPI	5.7/350	V8	MPI
		350 Mag MPI/Horizon	5.7/350	V8	MPI
	6.2L	Black Scorpion	6.2/377	V8	MPI
		MX6.2 Black Scorpion	6.2/377	V8	MPI
		MX6.2 MPI	6.2/377	V8	MPI
		MX6.2 MPI/Horizon	6.2/377	V8	MPI
	8.1L	8.1S HO	8.1/496	V8	MPI

7-4 FUEL SYSTEM - FUEL INJECTION (EFI)

ELECTRONIC FUEL INJECTION

Description & Operation

THROTTLE BODY INJECTION (TBI)

◆ See Figure 1

The purpose of this section is to describe - in layman's terms whenever possible, the Throttle Body Fuel Injection (TBI) system installed on many engines offered by Mercruiser.

Visual inspections, and simple tests that may be performed using only basic shop test equipment. Again, we emphasize: specialized test equipment, hours of training and considerable experience is required to perform detailed service on a fuel injection system which is beyond the scope of this manual.

The first fuel injection system was introduced in Europe over 60 years ago - in 1932 on diesel truck engines. In the beginning, the system and individual components were quite expensive. Over the years, state-of-the-art microprocessors (commonly referred to as "computer chips"), have lowered the cost of electronically controlled fuel injection systems. Today, the price of EFI is getting close to the cost of modern carbureted systems.

An electronic fuel injection system is quite different from a carburetor system - even though they appear similar (particularly TBI). The fuel injection system has a more efficient delivery of fuel to the cylinders than can be obtained with standard carburetor operation.

The EFI system provides a means of fuel distribution by precisely controlling the air/fuel mixture and under all operating conditions for, as near as possible, complete combustion.

This is accomplished by using and Electronic Control Module (ECM), a small 'on-board' microcomputer that receives electrical inputs from various sensors about engine operating conditions. The ECM uses these inputs to modify fuel delivery to achieve, as near as possible, an ideal air/fuel ratio of 14.7:1.

The ECM program automatically signals the fuel injectors in the throttle body assembly to provide the correct quantity of fuel for a wide range of operating conditions. Several sensors are used to determine existing operating conditions and the ECM then signals the injectors to provide the precise amount of fuel.

The ECM has a "learning" capability. That is, if the battery is disconnected for any reason, the learning process has to begin all over again.

The TBI assembly is located on the intake manifold, much like a carburetor, where air and fuel are distributed through a bore in the throttle body. Air for combustion is controlled by a throttle valve connected to the throttle linkage. Fuel is supplied by two injectors mounted in the TBI assembly, their metering tips are located directly above the throttle valve.

MULTI-POINT INJECTION (MPI)

◆ See Figures 2 thru 9

The purpose of this section is to describe - in layman's terms whenever possible, the Multi-Point Fuel Injection (MPI) systems installed on many engines offered by Mercruiser.

Visual inspections, and simple tests that may be performed using only basic shop test equipment. Again, we emphasize: specialized test equipment, hours of training and considerable experience is required to perform detailed service on a fuel injection system which is beyond the scope of this manual.

The first fuel injection system was introduced in Europe over 60 years ago - in 1932 on diesel truck engines. In the beginning, the system and individual components were quite expensive. Over the years, state-of-the-art microprocessors (commonly referred to as "computer chips"), have lowered the cost of electronically controlled fuel injection systems. Today, the price of EFI is getting close to the cost of modern carbureted systems.

An electronic fuel injection system is quite different from a carburetor system - even though they appear similar. The fuel injection system has a more efficient delivery of fuel to the cylinders than can be obtained with standard carburetor operation.

The EFI system provides a means of fuel distribution by precisely controlling the air/fuel mixture and under all operating conditions for, as near as possible, complete combustion.

This is accomplished by using and Electronic Control Module (ECM), a small 'on-board' microcomputer that receives electrical inputs from various sensors about engine operating conditions. The ECM uses these inputs to modify fuel delivery to achieve, as near as possible, an ideal air/fuel ratio of 14.7:1.

The ECM program automatically signals the fuel injectors in the throttle body assembly to provide the correct quantity of fuel for a wide range of operating conditions. Several sensors are used to determine existing operating conditions and the ECM then signals the injectors to provide the precise amount of fuel.

The ECM has a "learning" capability. That is, if the battery is disconnected for any reason, the learning process has to begin all over again.

On all MPI systems, there is a throttle body located on the intake manifold, much like a carburetor except that only air is distributed and metered through the bores on the throttle body Air for combustion is controlled by a throttle valve connected to the throttle linkage. Fuel is supplied by individual injectors mounted in the intake manifold and attached to a fuel rail assembly.

Each injector is "pulsed" or "timed" to open or close by an electronic signal from the ECM. While constantly receiving input from various sensors, the ECM performs high speed calculations of engine fuel requirements and then "pulses the injectors open or closed.

MODES OF OPERATION

The ECM receives signals from several sensors, and then responds by delivering the proper amount of fuel to the cylinders under one of several conditions. These conditions are labeled "modes" and are controlled by the ECM, as described in the following short sections.

Starting Mode

When the ignition switch is rotated to the cranking position, the ECM energizes the fuel pump relay and the pump builds up pressure. The ECM then checks the Engine Coolant Temperature (ECT) sensor and the Throttle Position (TP) sensor. From these incoming signals, the ECM determines the correct air/fuel ratio for starting the engine. The ECM controls the amount of fuel delivered in the starting mode by changing the length of time the injectors are turned on and off. This is accomplished by "pulsing" the injectors briefly.

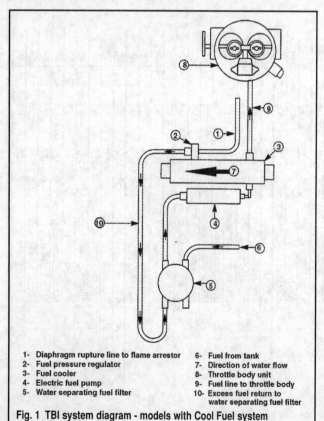

1- Diaphragm rupture line to flame arrestor
2- Fuel pressure regulator
3- Fuel cooler
4- Electric fuel pump
5- Water separating fuel filter
6- Fuel from tank
7- Direction of water flow
8- Throttle body unit
9- Fuel line to throttle body
10- Excess fuel return to water separating fuel filter

Fig. 1 TBI system diagram - models with Cool Fuel system

FUEL SYSTEM - FUEL INJECTION (EFI)

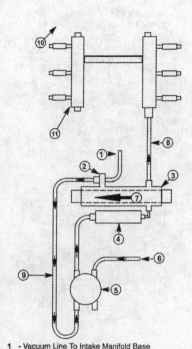

1 - Vacuum Line To Intake Manifold Base
2 - Fuel Pressure Regulator
3 - Fuel Cooler
4 - Electric Fuel Pump
5 - Water Separating Fuel Filter
6 - Fuel From Tank
7 - Water Flow
8 - Fuel Line To Fuel Rail
9 - Excess Fuel Return To Water Separating Fuel Filter
10 - Fuel Injectors (6)
11 - Fuel Rail

Fig. 2 MPI system diagram - 4.3L V6 models

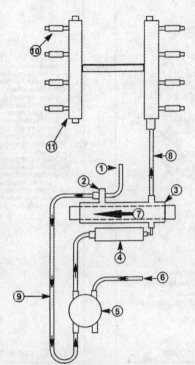

1 - Vacuum Line To Intake Manifold Base
2 - Fuel Pressure Regulator
3 - Fuel Cooler
4 - Electric Fuel Pump
5 - Water Separating Fuel Filter
6 - Fuel From Tank
7 - Water Flow
8 - Fuel Line to Fuel Rail
9 - Excess Fuel Return to Water Separating Fuel Filter
10 - Fuel Injectors (8)
11 - Fuel Rail

Fig. 3 MPI system diagram - 5.0L/5.7L/6.2L V8 models

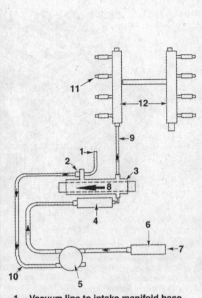

1 - Vacuum line to intake manifold base
2 - Fuel pressure regulator
3 - Fuel Cooler
4 - Electric fuel pump
5 - Water separating fuel filter
6 - Fuel boost pump
7 - Fuel from tank
8 - Direction of water flow
9 - Fuel line to fuel rail
10- Excess fuel return to water separating fuel filter
11- Fuel injectors
12- Fuel rail

Fig. 4 MPI system diagram - 8.1L V8 models

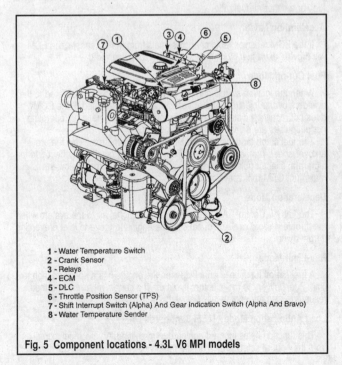

1 - Water Temperature Switch
2 - Crank Sensor
3 - Relays
4 - ECM
5 - DLC
6 - Throttle Position Sensor (TPS)
7 - Shift Interrupt Switch (Alpha) And Gear Indication Switch (Alpha And Bravo)
8 - Water Temperature Sender

Fig. 5 Component locations - 4.3L V6 MPI models

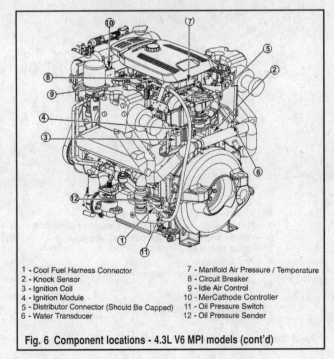

1 - Cool Fuel Harness Connector
2 - Knock Sensor
3 - Ignition Coil
4 - Ignition Module
5 - Distributor Connector (Should Be Capped)
6 - Water Transducer
7 - Manifold Air Pressure / Temperature
8 - Circuit Breaker
9 - Idle Air Control
10 - MerCathode Controller
11 - Oil Pressure Switch
12 - Oil Pressure Sender

Fig. 6 Component locations - 4.3L V6 MPI models (cont'd)

7-6 FUEL SYSTEM - FUEL INJECTION (EFI)

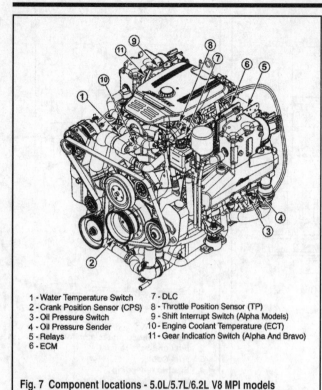

1 - Water Temperature Switch
2 - Crank Position Sensor (CPS)
3 - Oil Pressure Switch
4 - Oil Pressure Sender
5 - Relays
6 - ECM
7 - DLC
8 - Throttle Position Sensor (TP)
9 - Shift Interrupt Switch (Alpha Models)
10 - Engine Coolant Temperature (ECT)
11 - Gear Indication Switch (Alpha And Bravo)

Fig. 7 Component locations - 5.0L/5.7L/6.2L V8 MPI models

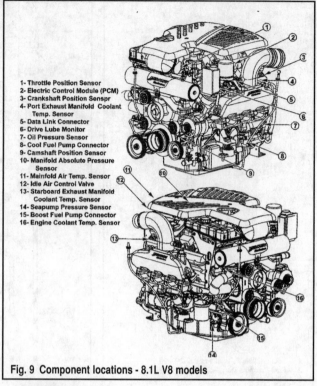

1- Throttle Position Sensor
2- Electric Control Module (PCM)
3- Crankshaft Position Sensor
4- Port Exhaust Manifold Coolant Temp. Sensor
5- Data Link Connector
6- Drive Lube Monitor
7- Oil Pressure Sensor
8- Cool Fuel Pump Connector
9- Camshaft Position Sensor
10- Manifold Absolute Pressure Sensor
11- Mainfold Air Temp. Sensor
12- Idle Air Control Valve
13- Starboard Exhaust Manifold Coolant Temp. Sensor
14- Seapump Pressure Sensor
15- Boost Fuel Pump Connector
16- Engine Coolant Temp. Sensor

Fig. 9 Component locations - 8.1L V8 models

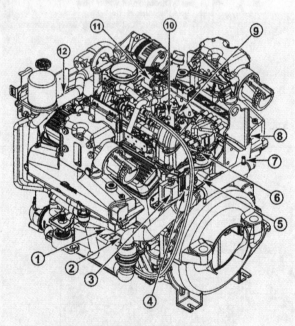

1 - Cool Fuel Harness Connector
2 - Knock Sensor (KS)
3 - Ignition Coil
4 - Ignition Module
5 - Transmission Temperature Connectors (If Equipped)
6 - Distributor Connector (Should Be Capped)
7 - Water Pressure Sender
8 - Knock Sensor
9 - Manifold Absolute Pressure / Temperature (MAPT)
10 - Circuit Breaker
11 - Idle Air Control (IAC)
12 - MerCathode Controller

Fig. 8 Component locations - 5.0L/5.7L/6.2L V8 MPI models (cont'd)

Clear Flood Mode

A flooded engine can be cleared by opening the throttle between 50% of its travel. Once this is done, the ECM shuts down the fuel injectors and no fuel is delivered to the cylinders. The ECM will hold this injector rate as long as the throttle remains 50-75% open and engine speed is below 300 rpm. If the throttle position changes to greater than 75% or slightly less than 50%, the ECM will return to the starting mode.

Run Mode

When the engine is first started and operating above 300 rpm, the system operates in the Run Mode. The ECM will calculate the desired air/fuel ratio based on rpm and input from the MAP and ECT sensors. Higher engine load (from the MAP) and colder engine temperature (from the ECT) requires more fuel, or a richer air/fuel ratio.

Acceleration Mode

If the ECM receives rapid change signals from the TP sensor, the ECM will provide extra fuel by increasing the injector pulse width.

Fuel Cut-off Mode

When the ignition switch is in the **OFF** position, no fuel is delivered to the cylinders by the injectors. Therefore, "dieseling" is prevented. If the ECM does not receive a distributor reference pulse - the engine is not operating - no fuel pulses are delivered to the injectors.

The fuel cutoff mode is also enabled at high engine rpm. This feature is an over-speed protection for the engine. Now, when cutoff is in effect due to high rpm, injector pulses will resume after engine rpm drops below the maximum OEM rpm specification.

Deceleration Mode

The Idle Air Control (IAC) valve provides additional air to the system when the throttle is rapidly released, causing the engine to move to idle, preventing it from dying.

Rev-Limit Mode

A fuel cut-off function is enabled when the engine hits a specified high rev limit. Fuel delivery to the injectors is cut off at a certain high rpm and then resumed when the engine falls below that rpm again.

Load Anticipation Mode - MEFI-3 Inboards Only

This function helps inboard engine during shifting. An electrical signal from the neutral safety switch on the transmission notifies the ECM if the

FUEL SYSTEM - FUEL INJECTION (EFI)

switch is open or closed - in Neutral the switch is closed (grounded), and when the shift lever is moved, putting the transmission into gear the switch opens. As the boat is shifted into gear the signal will cause the ECM to add bypass air mixture with the IAC valve, while removing the extra air mixture when the boat is shifted back into Neutral.

Moving Desired RPM Mode - MEFI-3 Inboards Only

This mode will increase the desired rpm at idle to a specified point according to the throttle position. When the TP sensor is showing a closed throttle setting, the ECM will utilize the IAC valve and the Ignition Control (IC) to maintain the specified rpm. This will smooth out the transition from idle to full throttle and also help maintain constant engine speeds between 600 and 1200 rpm.

SUBSYSTEMS

Fuel Metering System

As the name suggests, the fuel metering system "meters" the correct amount of fuel delivered to the cylinders through the intake manifold under all engine operating conditions. Fuel is delivered by the throttle body (TBI) or the individual injector (MPI) and is controlled by the ECM.

Fuel Supply

Naturally, the fuel supply will begin at the boat's fuel tank. From the fuel tank, the fuel is drawn through a water separating fuel filter and then moved by the fuel pump to the Cool Fuel (CF) System. All models utilize an electric fuel pump found in the CF. An additional electric fuel boost pump is used on 8.1L engine. A pressure regulator is standard on all models, while certain early V6 engines engine may have two.

Water Separating Fuel Filter

The water separator is a "typical" unit designed to prevent moisture from continuing on through the fuel lines and eventually through the injectors into the cylinders.

Electric Fuel Pump

When the ignition switch is moved to the **ON** position, the ECM will energize the fuel pump relay to ON, but only for a couple seconds. The fuel pump almost instantly pressurizes the fuel system. As soon as the ignition switch is moved to the **START** position, the ECM energizes the fuel pump relay again and the fuel pump begins to operate.

Now, if the ECM fails to receive ignition reference pulses - indicating the engine is either cranking or actually operating - the ECM de-energizes the fuel pump relay and the fuel pump will stop.

An inoperative fuel pump relay can cause an "Engine Cranks, But Fails To Operate" condition.

The pump is capable of providing more fuel than is required by the injectors at WOT (wide open throttle). A pressure regulator is an integral part of the system maintaining fuel to the injectors at a predetermined regulated pressure. Excess fuel not required by the injectors is returned to the fuel separator tank by a separate fuel line.

Cool Fuel System

This system consists of a fuel cooler, pressure regulator and a fuel pump inside a box located on the lower side of the engine.

Throttle Body - TBI

The throttle body consists of the following assemblies:
- Fuel Meter Cover - also houses the pressure regulator on models with vapor separator tank or the fuel damper on models with cool fuel.
- Fuel Meter Body - two fuel injectors.
- Throttle Body.
- Throttle Valves - two throttle valves controlling air flow into the cylinders.
- IAC (Idle Air Control) valve.
- TP (Throttle Position) sensor.

From the above list of components, it can easily be appreciated why the throttle body is considered one of the most critical items in the injection system. If the proper amount of air and fuel are not injected into the cylinders, the engine will not operate efficiently or may even fail to start.

Throttle Body - MPI

The throttle body assembly is attached to the intake manifold or plenum and controls air flow to the engine, thus controlling engine output. Throttle valves within the assembly are controlled by the throttle cables. At idle, the valves are almost completely closed, while they open wider in proportion to the amount of throttle applied by the operator.

Fuel Rail - MPI Only

The fuel rail positions the injectors in the intake manifold, distributes fuel evenly to each injector and integrates the pressure regulator into the entire metering system.

Fuel Pressure Regulator

◆ See Figure 10

The pressure regulator is a diaphragm operated relief valve with fuel pump pressure on one side, and regulator spring pressure on the other side. The purpose of the regulator is to maintain a constant pressure differential across the injectors at all times under all conditions.

When the ignition is ON, and the engine is not operating, fuel pressure is about 30 psi (207 kPa). A low/hi pressure condition would result in poor engine performance.

The regulator is located in the cool fuel assembly or on the fuel rail. On models using the Gen III cool fuel module, the regulator is located on the top of the module.

Fuel Damper - TBI

◆ See Figure 11

This unit acts as an equalization device to reduce and stabilize the fuel pressure spikes created by the fuel injectors on TBI engines.

Fuel Injectors - TBI

◆ See Figure 12

Each injector on TBI engines is a solenoid-operated device, controlled by the ECM, that meters pressurized fuel into the intake manifold. When the injector's solenoid is energized by the ECM, a pintle valve opens allowing fuel to flow past the valve and through the injector tip. The tip has holes that control the fuel flow, generating a conical spray pattern of atomized fuel at the injector tip. Fuel is directed at the intake valve, becoming further atomized and then vaporized prior to entering the combustion chamber.

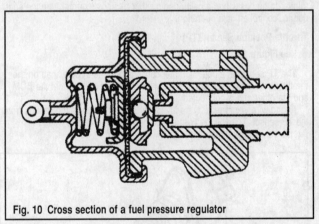

Fig. 10 Cross section of a fuel pressure regulator

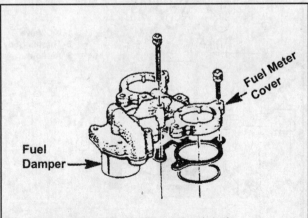

Fig. 11 The fuel damper is found in the throttle body

7-8 FUEL SYSTEM - FUEL INJECTION (EFI)

Fuel Injectors - MPI
◆ See Figure 13

On MPI engines, each injector is a solenoid-operated device, controlled by the ECM, that meters pressurized fuel into an individual engine cylinder. When the injector's solenoid is grounded by the ECM, a ball valve opens allowing fuel to flow past the valve and through a flow director plate. The director plate has six holes that control the fuel flow, generating a conical spray pattern of atomized fuel at the injector tip.

Idle Air Control Valve (IAC)
◆ See Figure 14

The IAC valve assembly is mounted in the throttle body and controls engine idle speed. At the same time, the assembly prevents stalls due to changes in engine load.

Operation of the IAC can best be described as a device to control bypass air around the throttle valves. This feature is accomplished by moving a conical valve know as a "pintle" inward towards a seat - to decrease air flow or outward, away from the seat - to increase air flow. In this manner, a controlled amount of air moves around the throttle valve.

If rpm is too low, more air is bypassed around the throttle valve to increase rpm. If rpm is too high, less air is bypassed around the throttle valve to decrease rpm.

The ECM moves the IAC valve in small increments. These increments can only be measured using expensive special scan tool test equipment plugged into a DLC (Data Link Connector).

During engine idle speed, the proper position of the IAC valve is calculated by the ECM, and is based on coolant temperature and engine rpm. If the rpm drops below specification and the throttle valve is closed, the ECM senses a near stall condition and calculates a new valve position to prevent the stall.

Understand - engine idle speed is a function of total air flow into the cylinders based on IAC valve "pintle" position to maintain the desired idle speed for all engine operating conditions and loads.

The minimum idle air rate is set at the factory with a stop screw. This setting allows sufficient air flow by the throttle valves to cause the IAC valve "pintle" to be positioned a calibrated number of steps (counts) from the seat during "controlled" idle operation.

Throttle Position Sensor (TP)
◆ See Figure 15

The TP sensor is a potentiometer connected to the throttle shaft on the throttle body. One end of the sensor is connected to 5-volts from the ECM and the other end is connected to ECM ground.

A third wire is connected directly to the ECM to measure the voltage from the TP sensor. The voltage output of the TP sensor changes when the throttle valve angle is changed.

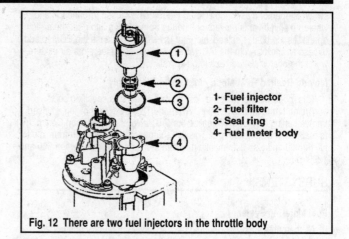

1- Fuel injector
2- Fuel filter
3- Seal ring
4- Fuel meter body

Fig. 12 There are two fuel injectors in the throttle body

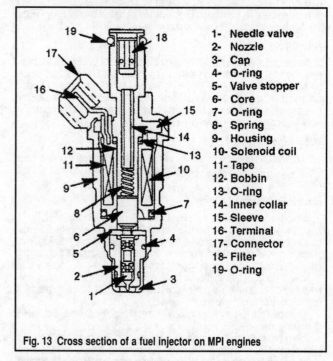

1- Needle valve
2- Nozzle
3- Cap
4- O-ring
5- Valve stopper
6- Core
7- O-ring
8- Spring
9- Housing
10- Solenoid coil
11- Tape
12- Bobbin
13- O-ring
14- Inner collar
15- Sleeve
16- Terminal
17- Connector
18- Filter
19- O-ring

Fig. 13 Cross section of a fuel injector on MPI engines

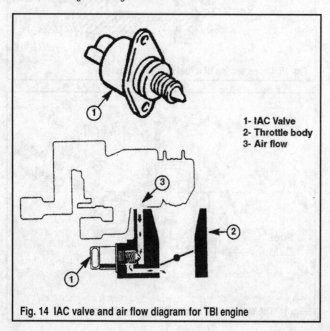

1- IAC Valve
2- Throttle body
3- Air flow

Fig. 14 IAC valve and air flow diagram for TBI engine

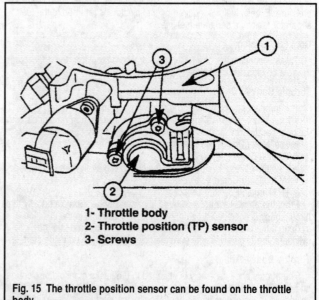

1- Throttle body
2- Throttle position (TP) sensor
3- Screws

Fig. 15 The throttle position sensor can be found on the throttle body

FUEL SYSTEM - FUEL INJECTION (EFI)

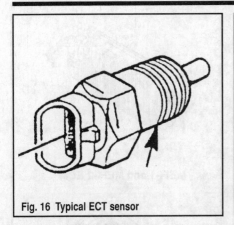

Fig. 16 Typical ECT sensor

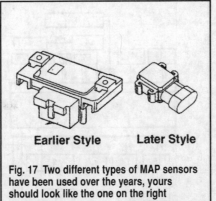

Fig. 17 Two different types of MAP sensors have been used over the years, yours should look like the one on the right

Fig. 18 Knock sensor

When the throttle position is closed, the voltage output of the TP sensor is low - about 1/2 volt. When the throttle valve is opened, the output increases and at WOT the output voltage should be close to 5 volts.

Therefore, the ECM can determine fuel delivery requirements based on throttle valve angle - operator demand.

Engine Coolant Temperature Sensor (ECT)

◆ See Figure 16

This sensor is a thermistor immersed in the engine coolant passageway. A thermistor is a resistor capable of changing value based on temperature. Low coolant temperature produces a high resistance and a high temperature causes low resistance.

Manifold Absolute Pressure Sensor (MAP)

◆ See Figure 17

The MAP sensor is a pressure transducer capable of measuring the changes in the intake manifold pressure. Pressure changes are the result of engine load and speed changes. MAP is the opposite of what is measured with a vacuum gauge.

When manifold pressure is high, vacuum is low - requires more fuel and of course the opposite is true. A low pressure - higher vacuum requires less fuel. The MAP sensor is also used to measure barometric pressure under certain conditions. This feature permits the ECM to automatically adjust for changes in operating altitude.

The ECM uses the MAP sensor to control fuel delivery and ignition timing.

Knock Sensor (KS)

◆ See Figure 18

The KS is mounted on the engine block drain "Y" fitting located on the lower starboard side of the engine block.

The ECM uses this signal in calculating ignition timing.

On TBI engines, the KS is mounted on the engine block drain "Y" fitting located on the lower starboard side of the engine block.

On a MPI engines, the KS sensor is mounted on the block, next to the starter motor.

Some early V6 andV8 models with MPI using the MEFI-3 ECM do not have a KS on the block, instead it is incorporated into the ECM.

The ECM uses this signal in calculating ignition timing.

Knock Sensor Module

◆ See Figure 19

The KS module contains solid-state circuitry monitoring the KS AC voltage signal. If no spark knock is present an 8-10 volt signal is sent to the ECM. If a spark knock is present, the module will remove the 8-10 volt signal to the ECM.

Distributor Reference Signal (DIST REF)

The Hi Reference signal is supplied to the ECM by way of the "REF" line from the High Energy Ignition (HEI). The Hi Reference signal is used by the ECM to determine engine speed. This pulse type signal input creates the timing signal for pulsing of the fuel injections as well as the Ignition Control (IC) functions.

Discrete Switch Inputs

Discrete switch inputs are utilized by the EFI system to identify abnormal conditions that could affect engine and stern drive operation. Normally these switches are at rest in an open position. When one of the discrete switches changes state from open to closed, the ECM will detect a change in the voltage value and responds by placing the engine into the power reduction mode.

The power reduction mode is an engine protection feature that allows the operator limited engine power up to 2800 rpm. Should the operator attempt to surpass this rpm limit, the ECM will reduce fuel and spark timing until the engine speed drops to approximately 1200 rpm. The power reduction mode will allow the operator maneuvering power but eliminates the possibility of high rpm engine damage, until the discrete switch fault condition is corrected.

Discrete switches are used to detect the critical engine and stern drive operations - namely Engine Oil Pressure and Stern Drive Fluid Level. Engine Coolant Temperature (ECT) sensor is part of the engine protection mode, but the sensor is not a discrete switch input to the ECM.

Engine Control Module

◆ See Figures 20 and 21

The ECM is the control center of the fuel injection system, monitoring input information from all sensors and then turning this information into system commands that affect engine performance.

Relieving Fuel Pressure

TBI ENGINES

※※ **CAUTION**

To reduce the risk of fire and personal injury, it is necessary to relieve the fuel system pressure before servicing any fuel system component. If this procedure is not performed, fuel may be sprayed out of a connection under extreme pressure. Always keep a dry chemical fire extinguisher near the work area when serving the fuel system.

Disconnect the electrical connection at the electric fuel pump and move the harness out of the way. Turn the ignition key and crank the engine for ten seconds to relieve any residual pressure in the system. If the engine starts, don't worry, allow it to run until it dies out.

MPI ENGINES

◆ See Figures 22, 23 and 24

※※ **CAUTION**

To reduce the risk of fire and personal injury, it is necessary to relieve the fuel system pressure before servicing any fuel system component. If this procedure is not performed, fuel may be sprayed out of a connection under extreme pressure. Always keep a dry chemical fire extinguisher near the work area when serving the fuel system.

7-10 FUEL SYSTEM - FUEL INJECTION (EFI)

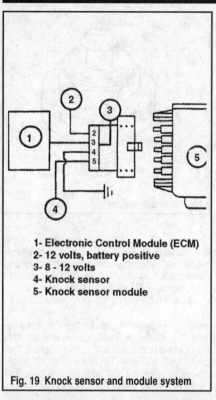

1- Electronic Control Module (ECM)
2- 12 volts, battery positive
3- 8 - 12 volts
4- Knock sensor
5- Knock sensor module

Fig. 19 Knock sensor and module system

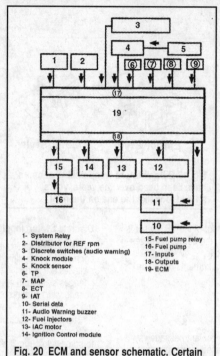

1- System Relay
2- Distributor for REF rpm
3- Discrete switches (audio warning)
4- Knock module
5- Knock sensor
6- TP
7- MAP
8- ECT
9- IAT
10- Serial data
11- Audio Warning buzzer
12- Fuel Injectors
13- IAC motor
14- Ignition Control module
15- Fuel pump relay
16- Fuel pump
17- Inputs
18- Outputs
19- ECM

Fig. 20 ECM and sensor schematic. Certain V8 engines with may not have the IAT sensor

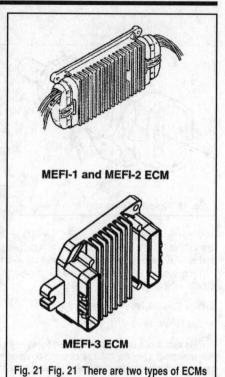

MEFI-1 and MEFI-2 ECM

MEFI-3 ECM

Fig. 21 Fig. 21 There are two types of ECMs used

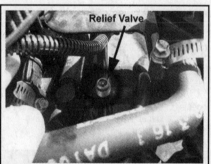

Fig. 22 MPI engine use a pressure relief valve

Fig. 23 Sometimes it's on the starboard fuel rail (8.1L shown)...

Fig. 24 ...but sometimes it's on the port rail (6.2L shown)

■ On the 2001 350 Mag, 6.2L models up to serial #299999 and 8.1L engines with the Gen II Cool Fuel Module; please use the TBI fuel pressure relief procedures.

Close the fuel supply valve if equipped.
Locate and activate the fuel pressure relief Schrader valve on the fuel rail. Tag and disconnect the lead connectors at each injector. Connect a fuel pressure gauge to the valve, position the relief line in a suitable container and then open the relief valve on the gauge. Remember that the valve can be on either rail, or also on the cross bar - looks just like an air valve on your tire, so you shouldn't have much problem finding it.

Cool Fuel System (Gen II)

REMOVAL, DISASSEMBLY & INSTALLATION

◆ See Figures 25, 26 and 27

1. Remove the engine compartment hatch and disconnect the battery cables.
2. If your boat has a fuel tank shut-off valve, close it and remove the inlet line at the water separating fuel filter or boost pump. If not, remove the line and quickly plug it. Make sure you've got a container and plenty of rags available.

3. If the boat is in the water, close the seacock, if equipped; or remove and plug the seawater inlet line.
4. Drain the seawater system as detailed in the Maintenance or Cooling System section.
5. Loosen the hose clamps and disconnect the seawater hoses at each end of the cool fuel assembly.
6. Relieve the fuel system pressure.
7. Disconnect and plug the fuel lines at the water separating filter and the throttle body/fuel rail. On 8.1L engines, disconnect and plug the fuel inlet line at the boost pump if you haven't already done so.
8. Remove the cover from the cool fuel assembly.
9. Tag and disconnect the electrical lead at the fuel pump.
10. Tag and disconnect the vacuum hose at the pressure regulator.
11. Remove the 2 upper engine mount bracket nuts that attach the assembly bracket to the engine and lift out the entire assembly.
12. Remove the 2 nuts from the cool fuel retaining bracket and lift the bracket and cooler assembly (with pump) out of the cover base. You may need to unclip the wiring harness from the cover base.
13. Remove the throttle body/fuel rail fuel line from the fuel cooler and then remove the return fuel line at the pressure regulator. The cooler-to-throttle body/fuel rail line utilizes a stepped screw to retain the line that is held in position by a retainer ring - do not remove the ring or the screw, loosen only!

FUEL SYSTEM - FUEL INJECTION (EFI)   7-11

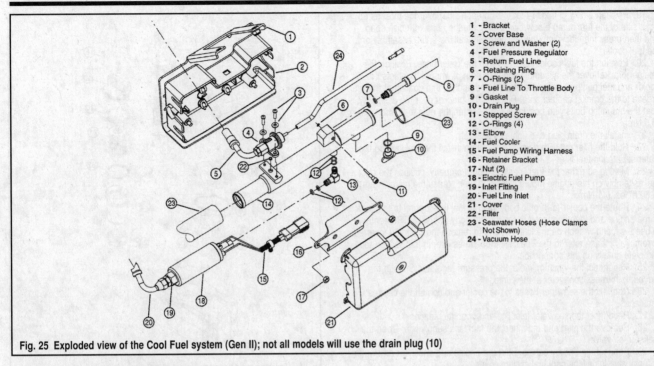

Fig. 25 Exploded view of the Cool Fuel system (Gen II); not all models will use the drain plug (10)

1 - Bracket
2 - Cover Base
3 - Screw and Washer (2)
4 - Fuel Pressure Regulator
5 - Return Fuel Line
6 - Retaining Ring
7 - O-Rings (2)
8 - Fuel Line To Throttle Body
9 - Gasket
10 - Drain Plug
11 - Stepped Screw
12 - O-Rings (4)
13 - Elbow
14 - Fuel Cooler
15 - Fuel Pump Wiring Harness
16 - Retainer Bracket
17 - Nut (2)
18 - Electric Fuel Pump
19 - Inlet Fitting
20 - Fuel Line Inlet
21 - Cover
22 - Filter
23 - Seawater Hoses (Hose Clamps Not Shown)
24 - Vacuum Hose

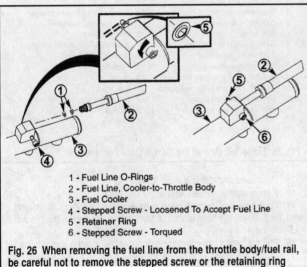

1 - Fuel Line O-Rings
2 - Fuel Line, Cooler-to-Throttle Body
3 - Fuel Cooler
4 - Stepped Screw - Loosened To Accept Fuel Line
5 - Retainer Ring
6 - Stepped Screw - Torqued

Fig. 26 When removing the fuel line from the throttle body/fuel rail, be careful not to remove the stepped screw or the retaining ring (Gen II)

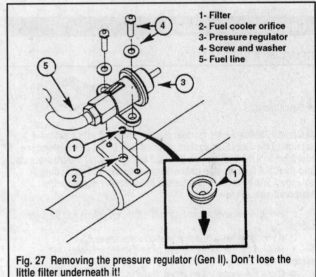

1 - Filter
2 - Fuel cooler orifice
3 - Pressure regulator
4 - Screw and washer
5 - Fuel line

Fig. 27 Removing the pressure regulator (Gen II). Don't lose the little filter underneath it!

14. Disconnect the fuel pump elbow fitting (with pump still attached) from the bottom of the fuel cooler. Remove the pump/elbow and then remove the elbow from the pump. Pull on the elbow to pop it out of the cooler and pump.
15. Remove the 2 mounting screws and detach the pressure regulator from the fuel cooler tube (don't lose the small filter in the tube recess under the regulator!).
16. Remove the seawater drain plug (with seal) from the cooler tube.

To assemble and install:
◆ See Figure 28

17. Install the small filter into the recess under the pressure regulator on the cooler tube so that the conical side is facing into the hole. Position the regulator on the tube and tighten the 2 bolts and washers to 53 inch lbs. (5.8 Nm). Hold the fitting on the regulator with a wrench and connect the fuel outlet line to the regulator, tightening it securely.
18. Install 2 new O-rings onto the cool fuel tube end of the throttle body/fuel rail fuel line. Loosen the stepped screw completely (be careful not to remove it), coat the O-rings with dish soap and insert the end of the fuel line into the cooler orifice. Hand-tighten the screw and then tighten to 81 inch lbs. (9 Nm). Do not over-tighten this screw!

1 - O-Rings (4)
2 - Elbow Fitting
3 - Fuel Pump
4 - Fuel Cooler

Fig. 28 Use new O-rings when installing the elbow

7-12 FUEL SYSTEM - FUEL INJECTION (EFI)

19. Position 2 new O-rings (4 total, 2 on each side) over the threads on each side of the fuel pump elbow fitting, lubricate the rings with dish soap and then press the elbow into the fuel pump. Install the entire assembly onto the fuel cooler.

20. Position the fuel cooler assembly into the cover base. Position the retainer bracket over the assembly and take note of where it contacts the cooler and fuel pump, remove the bracket and apply a thin coating of thermal grease to the bracket contact area. Reinstall the bracket over the assembly, coat the mounting bolts with Loctite 242 and tighten the nuts to 50 inch lbs. (5.6 Nm).

21. Install the drain plug and tighten it securely.

22. Hold the filter nut with a wrench and then install the fuel lines to the water separating fuel filter.

23. Making sure that the fuel lines are routed properly, position the cool fuel assembly on the engine mount studs and then tighten the mounting nuts to 30 ft. lbs. (41 Nm).

24. Route the cooler-to-throttle body/fuel rail fuel line around to the back of the engine and then connect it to the throttle body or fuel rail. Remember to use a second wrench on the filter nut and to remove the plug you inserted at removal. Please refer to the Fuel Rail procedures later in this section for complete details on this connection.

25. Reconnect the vacuum line at the pressure regulator. Snap in the fuel pump harness connector at the pump.

26. Connect the seawater hoses to the cooler and tighten the clamps securely.

27. Reconnect the seawater inlet line and open the seacock.

28. Remove the plug and reconnect the fuel tank supply line. Open the fuel shut-off valve.

29. Connect the battery cables. Start the engine and check for any water or fuel leaks. Install the cool fuel system cover.

Cool Fuel Module (Gen III)

REMOVAL & INSTALLATION

◆ See Figure 29

■ Certain 2004 and later models may be equipped with a different version of the Cool Fuel system, dubbed Gen III. On this system, the fuel filter and two fuel pumps (low and high pressure) are incorporated into the Cool Fuel module unlike on other versions. The module, if equipped, will be located just behind the seawater pump on the starboard side of the engine.

1. Remove the engine compartment hatch and disconnect the battery cables.
2. If your boat has a fuel tank shut-off valve, close it.
3. Relieve the fuel system pressure as detailed previously in this section.
4. Close the seacock if equipped. Drain the seawater system as detailed in the Maintenance or Cooling System section.
5. Loosen the hose clamps and disconnect the seawater hoses at each end of the cool fuel assembly.
6. Tag and disconnect the 2-pin harness connector at the module.
7. Disconnect and plug the fuel inlet line at the module.
8. Tag and disconnect the vacuum line at the pressure regulator on top of the module.
9. Loosen the nut on the cooling hose bracket and then remove the cooling hoses.
10. Remove the fuel outlet line retainer screw from the side of the fitting and then pull the line straight out of the fitting. Plug the line.
11. Remove the mounting bracket bolts and lift off the assembly.
12. If your model utilizes an J-clip to secure the fuel line, remove it and make sure you keep it.

To install:

13. Position the assembly on the engine and tighten the mounting bracket bolts to 17 ft. lbs. (23 Nm), the nuts to 35 ft. lbs. (47 Nm) and the support bracket not to 23 ft. lbs. (31 Nm).
14. Coat a new O-ring with a bit of oil, install on the fuel outlet line and then carefully press the line into the module. Thread in the retainer screw and tighten it to 80 inch lbs. (9 Nm).

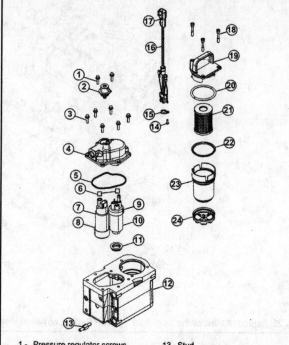

1 - Pressure regulator screws
2 - Pressure regulator
3 - Screws, Top cover
4 - Top cover
5 - Seal, Top cover
6 - Fuel pump outlet seal
7 - High-pressure fuel pump
8 - Isolator, High-pressure fuel pump
9 - Isolator, Low-pressure fuel pump
10 - Low-pressure fuel pump
11 - Inlet seal, Low-pressure fuel pump
12 - Cool fuel module housing
13 - Stud
14 - Screw, Wire harness retainer
15 - Wire harness retainer
16 - Wire harness
17 - 2-pin electrical connector (plug)
18 - Screws, Filter cap
19 - Filter cap
20 - O-ring
21 - Fuel filter
22 - Seal, Filter cup
23 - Filter cup
24 - Filter disc

Fig. 29 Exploded view of the Gen III Cool Fuel module

15. Reconnect the fuel line J-clip if used.
16. Reconnect the vacuum line at the regulator.
17. Coat new coolant hose quad rings with Parker Super Lube(r) and install them. Press the hose ends into the module and tighten the bracket nut to 13 inch lbs. (1.5 Nm).
18. Pull the heat sleeve over the fuel outlet line and the vacuum line.
19. Coat the fuel line inlet connector with Loctite 567 Pipe Sealant and install the line until it is hand-tight. Now hold the inlet fitting with a wrench and tighten line connector 1 3/4 to 2 1/2 turns with another wrench. Do not over-tighten!
20. Reconnect the harness connector and the battery cables.
21. Open the seacock and fuel shut-off valves if equipped. Start the engine and check for leaks.

DISASSEMBLY & ASSEMBLY

◆ See Figure 29

1. Remove the module from the engine and then remove the mounting and support brackets (2 bolts each).
2. Loosen the 3 mounting screws and then remove the filter assembly as detailed in the Maintenance section. Don't forget to relieve the fuel system pressure.
3. Remove the 2 mounting bolts and lift out the pressure regulator.
4. Remove the 6 mounting screws and lift the module top cover straight up and off. Note that the fuel pumps will come with it. There will still be fuel in the housing, so carefully drain into a suitable container.
5. Remove the O-ring from the cover.

FUEL SYSTEM - FUEL INJECTION (EFI)

■ Take a look at the positioning of the wires before removing the pumps, draw a little picture of their positions relative to the cover and installation will be easier; or take a picture with a digital camera.

  6. Tag and disconnect the electrical leads and then remove the fuel pumps.
  7. Remove the seal and isolator from the top of each pump and the seal from the bottom. The inlet seal (bottom) on the low pressure pump may not come out with the pump, so check in the module housing.

■ The low pressure pump is normally the only one to use an isolator, but the high pressure pump may also utilize one on later models as well.

  8. Drain any remaining fuel in the module into a suitable container.

To assemble:

  9. Install the outlet (upper) seals and isolator onto each pump. Install the inlet (lower) seal on the high pressure pump. Although not necessary, we would recommend using new ones.
  10. Position the lower inlet seal for the low pressure pump into the cavity in the module housing. Do not install it on the pump!
  11. Reconnect the electrical leads to each pump. On the low pressure pump, the red wire goes to the positive (+) terminal and the black to the negative (-) terminal, but the spade connectors will only fit one way.
  12. Wire routing is important.
  • The ground wire terminal at the harness end must be positioned against the outer wall of the casing
  • Wiring for the low pressure pump must be routed through the slot in the cover and above the pump itself
  • Wiring for the high pressure pump must also be above the pump itself.
  • Make sure that the wiring leads are not bound, pinched or otherwise compromised

■ You already knew how to handle the pump wiring because you drew a picture or took a picture as we suggested. . .right?

  13. Install the low pressure pump into the cover. Now do the same with the high pressure pump.
  14. Install a new cover seal/O-ring into the groove in the cover and carefully lower the assembly into the housing. Install the mounting bolts finger-tight.
  15. Make sure that nothing is binding and then tighten the cover in a criss-cross pattern, a little at a time, to 124 inch lbs. (14 Nm).
  16. Position the pressure regulator into the cover so that the hose barb is pointing toward the outside of the module, that is to say, away from the engine. Install the mounting screws and tighten to 124 inch lbs. (14 Nm).
  17. Install the filter assembly.
  18. If you removed the fuel inlet adapter fitting, coat the threads with Loctite 567 Pipe Sealant and thread it into the module. Tighten it carefully to 16 ft. lbs. (22 Nm).
  19. Connect the mounting and support brackets to the module and tighten them both to 17 ft. lbs. (23 Nm).
  20. Install the module.

Fuel Boost Pump (Electric)

REMOVAL & INSTALLATION

◆ See Figure 30

■ All 8.1L V8 engines and all 2002 and later 5.0L, 5.7L & 6.2L V8 engines (serial # OM336102 and higher) that DO NOT use a Gen III cool fuel module, utilize a boost pump in addition to the one in the cool fuel system housing. Models with the Gen III system, utilize a boost (low pressure) pump as well, but it is covered in the Module disassembly procedures.

  1. Open or remove the engine hatch/cover and disconnect the battery cables.
  2. Relieve the fuel system pressure as detailed previously in this section.
  3. Loosen the line nut and disconnect the fuel tank inlet line at the pump. Plug the line.
  4. Tag and disconnect the electrical harness.

Fig. 30 A good look at the pump. Certain models will have the pump on the other side of the bracket, behind the filter

  5. Disconnect the fuel line between the pump and the water separating filter base.
  6. Press the pump toward the engine and pop it out of the bracket. Remove the fuel line.

To install:

  7. Position the pump so that the grommets are lined up with the bracket and then press it into the bracket arms.
  8. Attach the fuel lines and plug in the electrical harness.
  9. Connect the battery cables, start the engine and check for fuel leaks.

Fuel Pump - Electric

REMOVAL & INSTALLATION

■ Most V8 MPI engines utilize an additional electric boost pump, this pump is detailed in the Fuel Boost Pump procedure earlier in this section.

■ Certain Tow Sports inboards that experience hard starts after a vapor locking problem may be eligible for a fuel boost pump retrofit (#861155A12) according to Mercruiser Service Bulletin #2002-03. Although this will not correct the vapor lock problem, it will allow the engine to restart quicker. Models affected are the 350 MPI (#OM310010 thru OM312122), 5.7L Black Scorpion (#OL678037 thru OL679432) and the 6.2L Black Scorpion (#OL679125 thru OL679176). Not all models were affected as many were retrofitted at the factory.

Please refer to the Cool Fuel System/Modules procedures in this section for fuel pump removal. The pump is incorporated into the assembly and removal is detailed in the overall procedure.

Fuel Pump Relay

REMOVAL & INSTALLATION

◆ See Figure 31

Locate the relay in the engine electrical box or on the bracket. Tag and disconnect the electrical lead. Remove the relay from the bracket. Do not soak the relay in any solvent of cleaner.

7-14 FUEL SYSTEM - FUEL INJECTION (EFI)

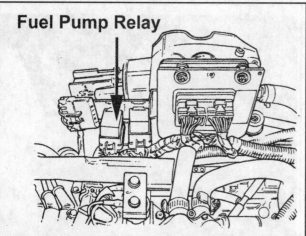

Fig. 31 The fuel pump relay is located on the engine electrical bracket, near the ECM

Fuel Pressure Regulator

REMOVAL & INSTALLATION

On most engines, the regulator is located in the Cool Fuel case/module, attached to the fuel cooler (Gen II) or module cover (Gen III). Please refer to the Cool Fuel System/Module procedures in this section for pressure regulator removal. The regulator is incorporated into the assembly and removal is detailed in the overall procedure. In almost all instances, there will also be another regulator attached to the fuel rail as well, and those procedures will be in the Fuel Rail & Injectors section.

Fuel Meter Cover

REMOVAL & INSTALLATION

TBI Engines Only

◆ See Figure 32

■ Do not remove the 4 mounting screws for the fuel damper installed in the fuel meter cover.

1. Open or remove the engine hatch/cover and relieve the fuel system pressure as detailed previously in this section. Disconnect the battery cable.

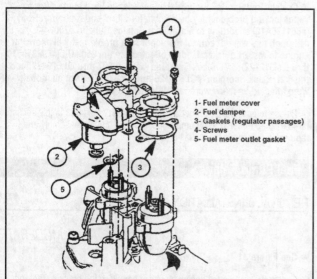

Fig. 32 When removing the fuel meter cover, keep track of the gaskets and seals

2. Remove the flame arrestor and then squeeze the plastic tabs on the injector electrical connectors and pull straight up until they come apart. Position the leads out of the way and make sure you tag them for correct reassembly.

3. Loosen and remove the meter cover mounting screws and carefully lift the cover straight up and off of the throttle body assembly. Watch for the damper seal and the outlet passage and cover gaskets - they may be hanging from the bottom of the cover.

4. Never immerse the cover in solvents or cleaner as damage to the damper diaphragm could occur. Inspect the mating surfaces of the throttle body assembly and the damper and cover for damage - use a magnifying glass if necessary! If any damage or scoring is found, replace the assembly.

5. Position a new damper seal in the throttle body cavity and then position the outlet passage and cover gaskets on the assembly.

6. Drop the meter cover into position and then slip in the mounting screws. Make sure they are coated with locking compound and that the short screws go into the holes next to the injectors. Tighten all screws to 28 inch lbs. (3 Nm).

7. Pop the injector lead connectors back on and connect the battery cables. Turn the ignition key to the **ON** position so that the fuel pump is energized and check for any fuel leaks. Do not actually start the engine until confirming that there are no leaks at the throttle body or fuel lines. Install the flame arrestor.

Fuel Injectors

REMOVAL & INSTALLATION

TBI Engines

◆ See Figures 33, 34 and 35

1. Open or remove the engine hatch/cover and relieve the fuel system pressure as detailed previously in this section. Disconnect the battery cable.

2. Remove the flame arrestor and then squeeze the plastic tabs on the injector electrical connectors and pull straight up until they come apart. Position the leads out of the way and make sure you tag them for correct reassembly.

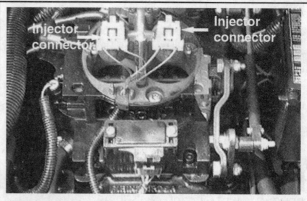

Fig. 33 Remove the connectors...

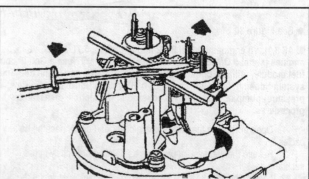

Fig. 34 ...and then carefully pry the injector out of the fuel meter body

FUEL SYSTEM - FUEL INJECTION (EFI)   7-15

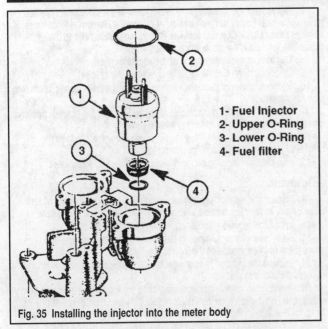

1- Fuel Injector
2- Upper O-Ring
3- Lower O-Ring
4- Fuel filter

Fig. 35 Installing the injector into the meter body

3. Loosen and remove the meter cover mounting screws and carefully lift the cover straight up and off of the throttle body assembly. Watch for the damper/regulator seal and the outlet passage and cover gaskets - they may be hanging from the bottom of the cover.

4. Make sure that the old meter cover gasket is still in place. Lay a small metal rod across the throttle body and then insert a small pry bar under the injector lip and carefully pry the injector out and up using the metal rod as a fulcrum.

To install:

5. Inspect the injectors for damage or clogging. MerCruiser recommends replacing damaged injectors. Always replace injectors with new ones that are the identical part number (as shown on the injector itself). All injectors are uniquely calibrated for flow rates so check the part number on the one being replaced and replace it with the same.

6. Slide the filter screen over the bottom of the injector and then install a new lower O-ring after soaking it in soapy water.

7. Soak a new upper O-ring in soapy water and install it into the fuel meter body.

8. Align the raised lug on the injector body with the notch in the meter cavity and install the injector. The injector terminals should now be parallel with the throttle shaft.

9. Install the fuel meter cover.

MPI Engines

Fuel injectors on MPI engines are an integral part of the fuel rail, please refer to the Fuel Rail & Injectors procedure detailed later in this section.

Fuel Rail & Injectors

REMOVAL & INSTALLATION

2001 350 Mag/6.2L MPI Engines (Up to Serial #299999)

◆ See Figures 36 and 37

1. Open or remove the engine hatch/cover and disconnect the battery cable.

2. Relieve the fuel system pressure as detailed previously in this section.

3. Remove the flame arrestor and throttle body.

4. Remove any lines or leads connected to the intake plenum. Remove the 6 retaining bolts and 2 nuts (under the front side) and then lift off the intake plenum.

5. Remove the distributor cap and position it out of the way with all wires still attached.

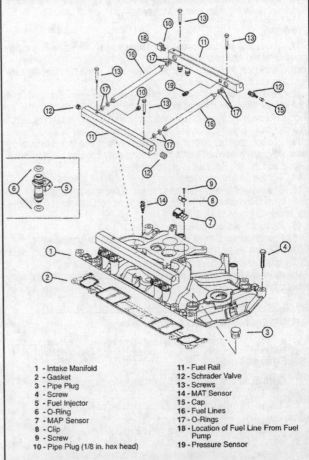

1 - Intake Manifold
2 - Gasket
3 - Pipe Plug
4 - Screw
5 - Fuel Injector
6 - O-Ring
7 - MAP Sensor
8 - Clip
9 - Screw
10 - Pipe Plug (1/8 in. hex head)
11 - Fuel Rail
12 - Schrader Valve
13 - Screws
14 - MAT Sensor
15 - Cap
16 - Fuel Lines
17 - O-Rings
18 - Location of Fuel Line From Fuel Pump
19 - Pressure Sensor

Fig. 36 Exploded view of the fuel rail assembly - 2001 350 Mag/6.2L MPI engines (up to serial #299999)

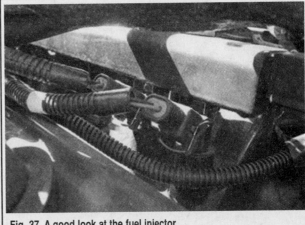

Fig. 37 A good look at the fuel injector

6. Carefully loosen and remove the fuel lines at the aft end of each rail. There should be a rail-to-rail fuel line on each side and a cool fuel-to-rail line, additionally, on the starboard side. Make sure you have a suitable container and plenty of rags available for the inevitable spills. Plug the line endings.

7. Loosen the connector fittings and remove the 2 rail-to-rail fuel lines at the front of the engine.

8. Tag and disconnect the electrical lead at each harness. Move them out of the way.

9. Remove the rail mounting screws. Grasp a rail at each end and then slowly rock it back and forth while pulling up on it until it (and the injectors) comes free of the manifold. Don't be surprised if some injectors stay in the manifold.

7-16 FUEL SYSTEM - FUEL INJECTION (EFI)

10. Carefully pop the injectors out of their fittings in the rail or manifold. If you are not replacing them, please be sure to keep them in order.

To install:

11. Inspect the injectors for damage. Clean the rails with solvent and dry with compressed air if possible; if not, make sure they are completely dry prior to installation. If you are replacing the injectors, make sure that the new ones have identical part numbers.

12. Slide the upper and lower seals (preferably new ones) over each end of the injector. Soak new O-rings in soapy water and position them on the injectors. Position the injectors and then pop them into their respective seats in the manifold - or you can do the same except with the fuel rail.

13. Position each rail (with or without the injectors) and rock it back and forth while pushing down on it until the injectors are all seated properly in the manifold and the rail. Repeat this on the other side.

14. Tighten the mounting screws to 105 inch lbs. (12 Nm).

15. Check that the rail-to-rail lines are still routed correctly and then hand tighten them to the connector fittings. Install the aft cool fuel-to-rail line and then tighten all fittings to 18 ft. lbs. (24 Nm).

16. Replace the distributor cap.

17. Position a new plenum gasket on the manifold so that the word **FRONT** is facing up and then install the plenum. Tighten the screws and nuts in a diagonal pattern to 150 inch lbs. (17 Nm).

18. Install the throttle body and arrestor. Connect the battery cables and check for fuel leaks.

2001-08 4.3L (Serial #OM322781 and later) And 5.0L, 5.7L/350 Mag & 6.2L MPI Engines (Serial #OM300000 and Later)

◆ See Figures 38 and 39

1. Open or remove the engine hatch/cover and disconnect the battery cable.
2. Relieve the fuel system pressure as detailed previously in this section.
3. Remove the cover, flame arrestor and throttle body.
4. Remove the IAC valve and its bracket.
5. Remove the upper engine wiring harness bracket studs and then move the harness itself out of the way.
6. Tag each injector harness and then disconnect from the injector. Repeat this procedure for each injector.
7. Disconnect the fuel inlet and return lines at the rail. Make sure you have some rags handy to catch the inevitable spills. Remember that these engines utilize Quick Connect fuel line fittings, so please refer to the procedure for details on disconnecting them.
8. Disconnect the vacuum line at the regulator.
9. Disconnect the throttle cable and remove the bracket.
10. Remove or disconnect any remaining lines, leads or hoses interfering with rail removal.
11. Remove the 4 rail mounting studs/nuts. Remove the bracket bolt and carefully lift off the fuel rail assembly with the injectors.
12. Remove the injectors from the rail being careful not to lose the retainers. Remove the O-rings from each injector.
13. Pop out the retaining clip and remove the pressure regulator.

To install:

14. Inspect the injector seating recess in the fuel rail for pitting, nicks, burrs or any other irregularities. Replace the rail assembly if necessary.
15. Install the regulator and snap in the retaining clip.
16. Soak new injector O-rings in clean engine oil for a minute and then press them onto each side of the injector.
17. Press each injector (with a new retaining clip) into its recess on the rail until it clicks into place.
18. Position the rail/injector assembly onto the intake manifold and gently press down on the rail until the injectors seat fully into their bores. If the studs came out with the rail, coat the threads with a 0.020 in. (5mm) bead of thread-locking compound and then tighten the stud/nuts to 27 inch lbs. (3 Nm) in a criss-cross pattern. Tighten the bracket bolt to 53 inch lbs. (6 Nm).
19. Reconnect the vacuum and fuel lines.
20. Install the injector connectors and the upper wiring harness.
21. Install the IAC bracket and valve. Tighten the bracket nuts to 132 inch lbs. (15 Nm), the valve-to-brackets bolts to 168 inch lbs. (19 Nm), the plug-to-throttle body screws to 24 inch lbs. (3 Nm) and the valve to 20 inch lbs. (2 Nm).
22. Install the throttle body and flame arrestor.
23. Install the throttle cable bracket and cable. Adjust the cable.
24. Install or connect any remaining lines, leads, or hoses you may have removed.
25. Connect the battery cables. Turn the ignition switch to the **ON** position for 2 seconds and then turn it OFF for 10 seconds. Now turn it back to the **ON** position and check for any fuel leaks before starting the engine.

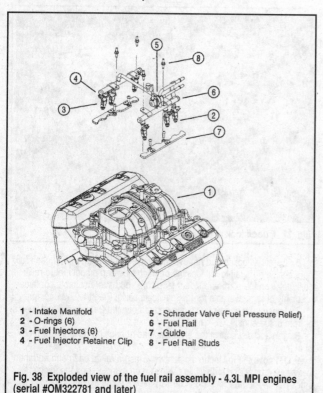

Fig. 38 Exploded view of the fuel rail assembly - 4.3L MPI engines (serial #OM322781 and later)

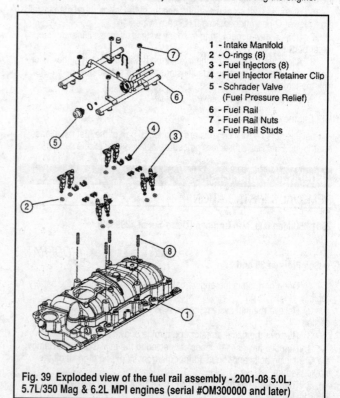

Fig. 39 Exploded view of the fuel rail assembly - 2001-08 5.0L, 5.7L/350 Mag & 6.2L MPI engines (serial #OM300000 and later)

FUEL SYSTEM - FUEL INJECTION (EFI)

8.1L V8 Engines

◆ See Figure 40

1. Open or remove the engine hatch/cover and disconnect the battery cable.
2. Relieve the fuel system pressure as detailed previously in this section.
3. Disconnect the throttle cable and remove the bracket
4. Remove the flame arrestor and throttle body.
5. Disconnect the fuel lines at the fuel rail.
6. Tag and disconnect the electrical lead at the injector harness. Do the same at each injector.
7. Although not strictly necessary, it's a good idea to remove the shift control bracket.
8. Remove or disconnect any remaining lines, leads or hoses interfering with rail removal.
9. Remove the screws, nut, MAP sensor retaining clip and stud from the rail. Lift out the rail/injector assembly from the manifold.
10. Pop off the injector retaining clip and remove each injector from the fuel rail.

To install:

11. Inspect the injector seating recess in the fuel rail for pitting, nicks, burrs or any other irregularities. Replace the rail assembly if necessary.
12. Soak new injector O-rings in soapy water for a minute and then press the injectors into the rail.
13. Pop the retaining clip into place over each injector and then connect the electrical lead.
14. Lubricate the remaining O-rings, position them in the manifold and then install the rail/injector assembly. Install the screws, nut and stud. Tighten to 105 inch lbs. (12 Nm).
15. Install the throttle body and arrestor.
16. Install the throttle cable bracket and cable. Adjust the cable.
17. Install or connect any remaining lines, leads, or hoses you may have removed.
18. Connect the battery cables. Turn the ignition switch to the **ON** position for 2 seconds and then turn it OFF for 10 seconds. Now turn it back to the **ON** position and check for any fuel leaks before starting the engine.

INJECTOR BALANCE TEST

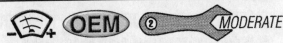

All injectors should have the same amount of pressure drop after being turned on and off for a controlled amount of time. Any injector with a drop of 1.5 psi or more, above or below the average drop across all injectors should be replaced. This procedure will require the use of an injector balance tester, available through your Mercury dealer or selected aftermarket suppliers. Be sure the tester has a pulse width of at least 200-400 milliseconds. At the very least, the tester must have a pulse width that is capable of dropping the fuel rail pressure to half of the normal operating pressure.

1. The engine should be cool, or at least have been shut down for 10 minutes prior to testing.
2. Relieve the pressure in the fuel system.
3. Turn OFF the ignition and connect a fuel pressure gauge to the tap on the fuel rail.
4. Tag and disconnect the electrical lead at each injector and move them out of the way.
5. Connect the injector tester (or scan tool) to an injector with the adaptor harness that should have come with the tool.
6. Make sure that the ignition has been OFF for at least 10 seconds and then turn it ON. The fuel pump should run for about 2 seconds.
7. Connect a length of tubing to the vent valve on the pressure gauge and put the other end into a suitable container. Open the valve to bleed off any air in the system and repeat until you are confident that there is no longer any air in the gauge.
8. Turn the ignition OFF for 10 seconds and then turn it ON. Repeat this procedure several times to get the fuel pressure up to its maximum. Record this reading. Energize the tester and record the pressure at its lowest point - you may notice a slight increase in pressure after the low point, don't worry about this as its normal.
9. Subtract the second reading from the first to get your pressure drop figure. Repeat the procedure for each injector and compare the readings. Good injectors will have almost identical readings. Retest any injector that deviates from the average more than 1.5 psi (up or down). If it still fails on the retest, replace the injector.
10. Disconnect the tester and the pressure gauge.
11. Connect all the leads at the injectors and reinstall the plenum.

Throttle Body

REMOVAL & INSTALLATION

TBI Engines

◆ See Figures 41 and 42

1. Open or remove the engine hatch/cover and disconnect the battery cable.
2. Relieve the fuel system pressure as detailed previously in this section.
3. Remove the flame arrestor and then squeeze the plastic tabs on the injector electrical connectors and pull straight up until they come apart. Position the leads out of the way and make sure you tag them for correct reassembly.
4. Disconnect the throttle cable and carefully move it out of the way.
5. Unplug the electrical connectors at the TP sensor and the IAC valve. Tag each lead and move them out of the way.
6. Hold a container under the connections and then remove the fuel inlet and outlet lines at the throttle body. Have some rags available to wipe up any spills. Plug the fuel lines to prevent any further spillage.
7. Remove the mounting screws and lift the throttle body off of the adapter plate. Cover the manifold opening with a rag.
8. Remove the TP sensor, IAC valve, fuel injectors and the meter cover and then thoroughly clean the throttle body assembly in an immersion-type cleaner. Dry with compressed air if available and make certain all passages are clean.

To install:

9. Inspect all mating and casting surfaces for damage or cracks and scoring.
10. Install the sensor, valve, injectors and meter cover as detailed elsewhere in this section.
11. Position a new gasket on the adapter plate, install the throttle body and tighten the mounting screws to 30 ft. lbs. (40 Nm).
12. Connect and adjust the throttle linkage.

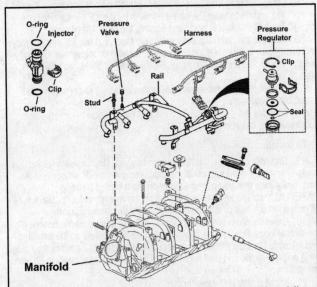

Fig. 40 Exploded view of the fuel rail assembly - 8.1L engines (all years similar)

7-18 FUEL SYSTEM - FUEL INJECTION (EFI)

13. Connect the fuel lines and tighten the flange nuts to 23 ft. lbs. (31 Nm).
14. Connect the TP sensor and IAC valve electrical leads.
15. Pop the injector lead connectors back on and connect the battery cables. Turn the ignition key to **ON** so that the fuel pump is energized and check for any fuel leaks. Do not actually start the engine until confirming that there are no leaks at the throttle body or fuel lines. Install the flame arrestor.

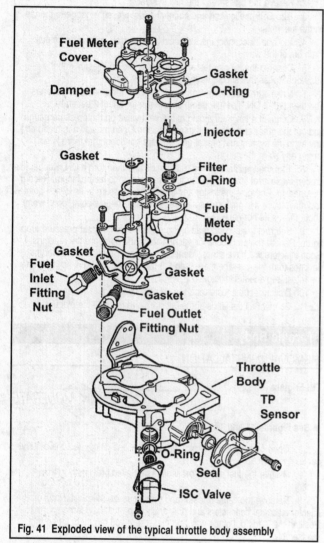

Fig. 41 Exploded view of the typical throttle body assembly

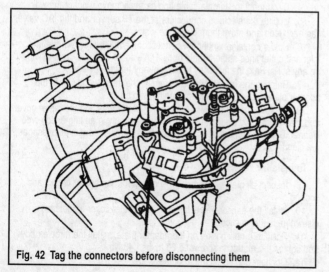

Fig. 42 Tag the connectors before disconnecting them

2001 350 Mag/6.2L MPI Engines (Up to Serial #299999)

◆ See Figures 43 and 44

1. Open or remove the engine hatch/cover and disconnect the battery cable.
2. Relieve the fuel system pressure as detailed previously in this section.
3. Remove mounting nut(s) and then remove the flame arrestor. Certain models will have a bracket holding the arrestor on.
4. Disconnect the throttle linkage.
5. Tag and disconnect the wiring harness for the IAC valve and TP sensor.
6. Remove the 3 throttle body mounting bolts (later model Black Scorpions may have a bolt and nuts). Lift out the throttle body and set it down in a holding fixture to avoid damage to the valves. Stuff a clean, lint-free rag into the plenum opening.
7. Remove the TP sensor and IAC valve with their O-rings from the assembly if necessary.

To install:

8. Carefully clean the throttle bore and valves. Do not use anything containing methyl ethyl ketone! If you didn't remove the TP sensor and IAC valve, make sure that you get no solvent on them during cleaning. CAREFULLY scrape any gasket material off the mating surfaces. Make sure all passages are free of dirt and completely dry before installation.
9. Install the sensor and valve. Position the throttle body and new gasket on the adaptor (350 Mag/6.2 MPI) or plenum and tighten the bolts to 75 inch lbs. (8.5 Nm). Later model Black Scorpions will have 2 gaskets and also a plate - the plate goes against the plenum.
10. Connect the IAC and TP leads.
11. Connect the throttle cable.
12. Install the flame arrestor.
13. Reconnect the battery cables.

2001-08 4.3L (Serial # OM322781 & Later) And 5.0L, 5.7L/350 Mag & 6.2L MPI Engines (Serial #OM300000 & Later)

◆ See Figures 45, 46 and 47

1. Open or remove the engine hatch/cover and disconnect the battery cable.
2. Relieve the fuel system pressure as detailed previously in this section.
3. Remove mounting nut(s) and then remove the flame arrestor. Certain models will have a bracket holding the arrestor on.
4. Disconnect the throttle linkage at the throttle body and move it aside.
5. Tag and disconnect the wiring harness for the TP sensor. Remove the IAC hose at the throttle body and remove the plug if necessary.
6. Remove the 3 throttle body mounting studs. Lift out the throttle body and set it down in a holding fixture to avoid damage to the valves. Carefully pull out the throttle body sealing ring. Stuff a clean, lint-free rag into the plenum opening.
7. Remove the TP sensor and IAC valve with their O-rings from the assembly if necessary.
8. Remove the small muffler for the IAC valve inside the throttle body - it looks like a small sponge.

To install:

9. Carefully clean the throttle bore and valves. Do not use anything containing methyl ethyl ketone! If you didn't remove the TP sensor and IAC plug, make sure that you get no solvent on them during cleaning. CAREFULLY scrape any gasket material off the mating surfaces. Make sure all passages are free of dirt and completely dry before installation.
10. Carefully press a new sealing ring into the groove on the bottom of the throttle body. Position the throttle body over the dowels on the manifold and install it. Coat the threads of the studs with Loctite 242 (unless they are new) and screw them in until finger-tight. Tighten them alternately to 80 inch lbs. (9 Nm).
11. Install a new IAC valve muffler.
12. Install the TP sensor and IAC plug. Connect the IAC hose and TP sensor lead.

FUEL SYSTEM - FUEL INJECTION (EFI) 7-19

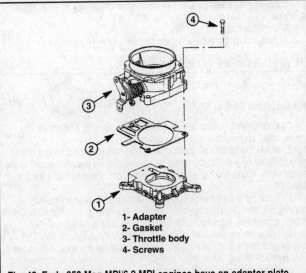

Fig. 43 Early 350 Mag MPI/6.2 MPI engines have an adaptor plate...

1- Adapter
2- Gasket
3- Throttle body
4- Screws

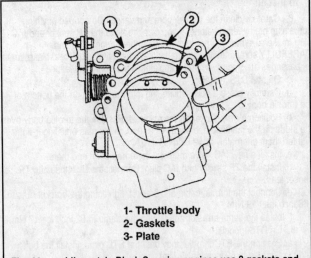

1- Throttle body
2- Gaskets
3- Plate

Fig. 44 ...while certain Black Scorpion engines use 2 gaskets and a plate

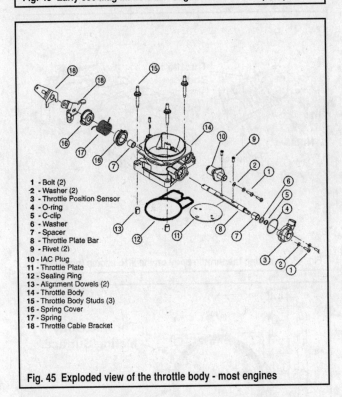

1 - Bolt (2)
2 - Washer (2)
3 - Throttle Position Sensor
4 - O-ring
5 - C-clip
6 - Washer
7 - Spacer
8 - Throttle Plate Bar
9 - Rivet (2)
10 - IAC Plug
11 - Throttle Plate
12 - Sealing Ring
13 - Alignment Dowels (2)
14 - Throttle Body
15 - Throttle Body Studs (3)
16 - Spring Cover
17 - Spring
18 - Throttle Cable Bracket

Fig. 45 Exploded view of the throttle body - most engines

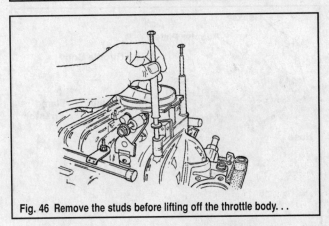

Fig. 46 Remove the studs before lifting off the throttle body...

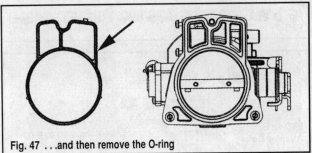

Fig. 47 ...and then remove the O-ring

13. Connect the throttle cable and bracket, tightening the bracket nuts to 168 inch lbs. (19 Nm).
14. Install the flame arrestor.
15. Reconnect the battery cables.

8.1L Engines

 MODERATE

◆ See Figures 45 and 48 thru 53

1. Open or remove the engine hatch/cover and disconnect the battery cables.
2. Relieve the fuel system pressure as detailed previously in this section.
3. Loosen the retaining clamp and then remove the flame arrestor.
4. On non-DTS models:
   a. Disconnect the throttle linkage at the throttle body and move it aside.
   b. Tag and disconnect the wiring harness for the TP sensor. Remove the IAC hose at the throttle body and remove the plug if necessary.

   c. Remove the 3 throttle body mounting bolt nuts. Lift out the throttle body and set it down in a holding fixture to avoid damage to the valves. Carefully pull out the throttle body sealing ring. Stuff a clean, lint-free rag into the plenum opening.

   d. Remove the TP sensor and IAC valve with their O-rings from the assembly if necessary.

   e. Remove the small muffler for the IAC valve inside the throttle body if equipped - it looks like a small sponge.

5. On DTS models with Electronic Throttle Control (ETC):
   a. Tag and disconnect the electrical harness at the ETC and move it out of the way.
   b. Remove the 4 mounting bolts and lift off the ETC (throttle body) from the adapter. Carefully pull out the throttle body sealing ring. Set it down in a holding fixture to avoid damage. Stuff a clean, lint-free rag into the plenum opening.

7-20 FUEL SYSTEM - FUEL INJECTION (EFI)

To install:

6. Carefully clean the throttle bore and valves. Do not use anything containing methyl ethyl ketone! If you didn't remove the TP sensor and IAC plug, make sure that you get no solvent on them during cleaning. CAREFULLY scrape any gasket material off the mating surfaces. Make sure all passages are free of dirt and completely dry before installation.

7. On non-DTS models:

   a. Carefully press a new sealing ring into the groove on the bottom of the throttle body.

   b. Position a new gasket on the manifold. Position the throttle body over the studs on the manifold and install it. Thread on the nuts until finger-tight. Tighten them alternately to 89 inch lbs. (10 Nm).

   c. Install a new IAC valve muffler if there had been one there.

   d. Install the TP sensor and IAC plug. Connect the IAC hose and TP sensor lead.

   e. Connect the throttle cable and bracket, tightening the bracket nuts to 168 inch lbs. (19 Nm).

   f. Install the flame arrester and tighten the clamp to 62 inch lbs. (7 Nm).

8. On DTS models:

   a. Position the ETC/throttle body onto the ETC plate, install the bolts and then tighten to 62 inch lbs. (7 Nm).

   b. Reconnect the electrical harness.

   c. Install the flame arrester and tighten the clamp ring to 42 inch lbs. (4.7 Nm).

9. On all models:

   a. Reconnect the battery cables and install the engine cover.

THROTTLE CABLE ADJUSTMENT

◆ See Figures 54 thru 57

1. Position the remote control handle in the Neutral. Make sure the engine cover and flame arrester are removed.

2. Install the cable end guide onto the throttle lever and then pre-load the shift cable by pushing the cable barrel end slightly toward the end of the throttle lever. Adjust the barrel so that it aligns with the hole in the anchor plate.

3. Secure the throttle cable with its hardware and tighten the nut securely. Tighten the locknut until it contacts the other nut on the 8.1L. On all others, tighten until it contacts the nut and then back it off 1/2 turn.

4. Place the remote control lever in the wide open throttle (WOT) position and check that the throttle plates are open all the way. If so, move the lever back to the idle position and confirm that the plates are closed fully.

5. Install the flame arrester and engine cover.

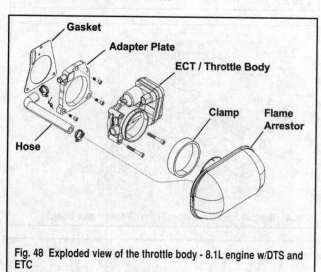

Fig. 48 Exploded view of the throttle body - 8.1L engine w/DTS and ETC

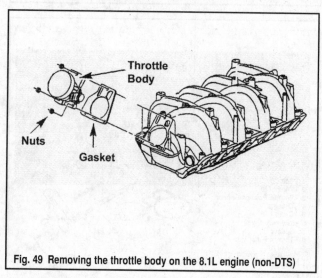

Fig. 49 Removing the throttle body on the 8.1L engine (non-DTS)

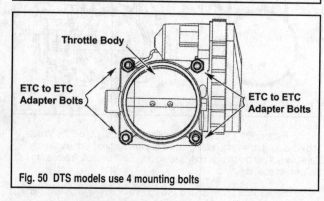

Fig. 50 DTS models use 4 mounting bolts

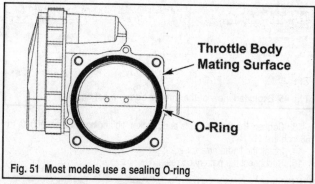

Fig. 51 Most models use a sealing O-ring

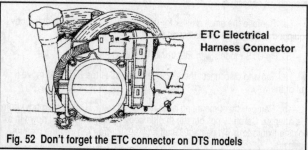

Fig. 52 Don't forget the ETC connector on DTS models

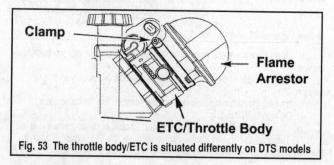

Fig. 53 The throttle body/ETC is situated differently on DTS models

FUEL SYSTEM - FUEL INJECTION (EFI)

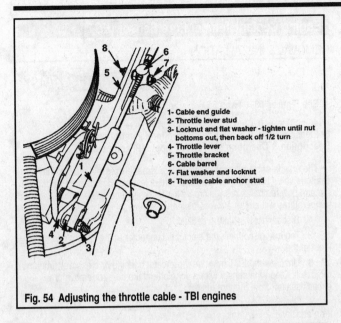

Fig. 54 Adjusting the throttle cable - TBI engines

1- Cable end guide
2- Throttle lever stud
3- Locknut and flat washer - tighten until nut bottoms out, then back off 1/2 turn
4- Throttle lever
5- Throttle bracket
6- Cable barrel
7- Flat washer and locknut
8- Throttle cable anchor stud

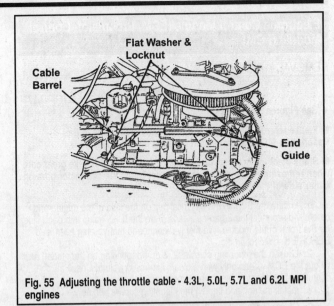

Fig. 55 Adjusting the throttle cable - 4.3L, 5.0L, 5.7L and 6.2L MPI engines

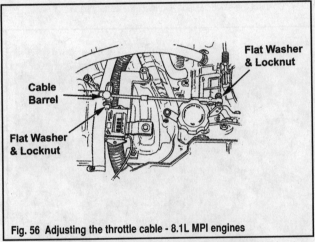

Fig. 56 Adjusting the throttle cable - 8.1L MPI engines

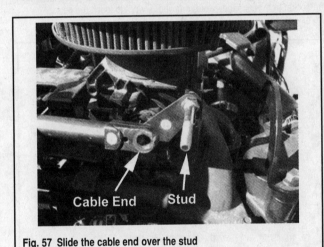

Fig. 57 Slide the cable end over the stud

Throttle Body Adapter Plate

REMOVAL & INSTALLATION

TBI Engines & 8.1L MPI Engines w/DTS

◆ See Figures 48 and 58

1. Open or remove the engine hatch/cover and relieve the fuel system pressure. Disconnect the battery cable.

2. On TBI models, remove the flame arrestor and then squeeze the plastic tabs on the injector electrical connectors and pull straight up until they come apart. Position the leads out of the way and make sure you tag them for correct reassembly.

3. Remove the throttle body or ECT. Lift the gasket off the adapter if it hasn't already come up with the throttle body on TBI models.

4. Remove the 4 mounting screws (3 on MPI engines) and lift off the adapter plate. Plug the intake manifold with a clean rag.

5. Install the adapter plate with a new gasket and tighten the mounting screws to 15 ft. lbs (19 Nm) on TBI engines, or securely on MPI engines.

6. Install the throttle body/ECT and flame arrester.

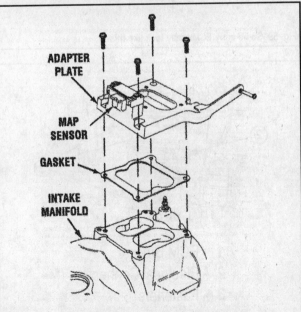

Fig. 58 Removing the throttle body adapter plate on TBI engines. Not all models will have the MAP sensor on the plate

7-22 FUEL SYSTEM - FUEL INJECTION (EFI)

Electronic Control Module (ECM)/Propulsion Control Module (PCM)

REMOVAL & INSTALLATION

◆ See Figures 59 thru 62

■ Please refer to the component locations illustrations in the Description & Operation section for most MPI engines.

■ Static electricity can severely damage the ECM/PCM! Take great care when removing the module that you never touch the connector pins on either side of the casing.

1. Locate the ECM in the electrical box on the side of the engine and carefully disconnect the 2 harness leads (3 on the 8.1L) going into each side (or the front) of the module. We highly recommend that you tag each lead BEFORE disconnecting it!
2. Loosen the mounting screws (2, 3 or 4 depending on the model) and lift out the ECM. Clean only with a clean, lint-free dry cloth. Check all connector pins for straightness and corrosion.
3. Install the unit and tighten the mounting screws securely.
4. Reconnect the electrical leads making sure they are going back into the socket from which they were removed. Try not to touch the pins themselves.

Engine Coolant Temperature Sensor (ECT)

REMOVAL & INSTALLATION

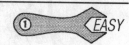

◆ See Figures 63, 64 and 65

■ Please refer to the component locations illustrations in the Description & Operation section for most MPI engines.

The engine coolant temperature sensor is a thermistor, a resister that changes value based on temperature, that is immersed in the engine coolant stream. Low temperatures will create higher resistance, while high temperatures will create a lower resistance.

1. Disconnect the battery cables.
2. Tag and disconnect the electrical connector at the sensor and move it out of the way.
3. Unscrew the ECT from the thermostat housing or crossover tube and remove it. Clean the sensor with a dry cloth; make sure to remove any excess sealant from the threads.
4. Screw the sensor into the housing until it is hand tight and then turn it an additional 2 1/2 turns.
5. Connect the electrical lead. Connect the battery cables.

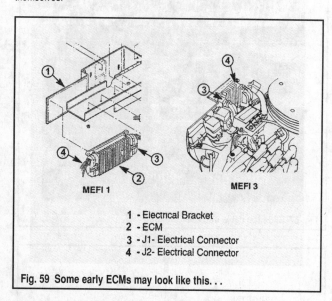

1 - Electrical Bracket
2 - ECM
3 - J1- Electrical Connector
4 - J2- Electrical Connector

Fig. 59 Some early ECMs may look like this. . .

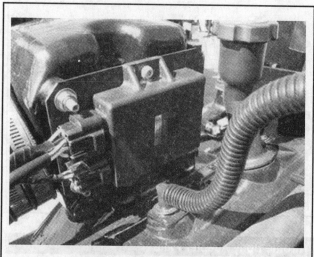

Fig. 60 . . .but most carb/TBI units will look like this. . .

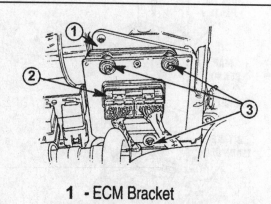

1 - ECM Bracket
2 - Electrical Connectors
3 - Fasteners

Fig. 61 . . .or this on V6 and 5.0L/5.7L/6.2L MPI engines. . .

Fig. 62 . . .and 8.1L engines

FUEL SYSTEM - FUEL INJECTION (EFI) 7-23

Fig. 63 The ECT is usually found on the side of the thermostat housing...

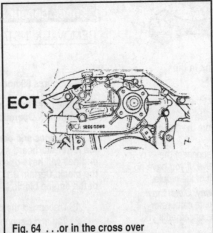

Fig. 64 ...or in the cross over

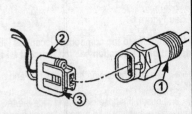

1 - Engine Coolant Temperature (ECT) Sensor
2 - Harness Connector
3 - Locking Tab

Fig. 65 A good look at the typical sensor and connector

Idle Air Control Valve (IAC)
REMOVAL & INSTALLATION

Except 8.1L Engines & Later MPI Engines

◆ See Figure 66

■ Please refer to the component locations illustrations in the Description & Operation section for most MPI engines.

1. Disconnect the battery cables.
2. Remove the flame arrestor and the throttle body.
3. Tag and disconnect the electrical connector at the valve and move it out of the way.
4. Unscrew the mounting screws and remove the valve. Remove the O-ring from the valve and throw it away. Clean the valve mating surfaces, pintle valve seat and air passage with carburetor cleaner - do not push or pull on the valve pintle. Don't be concerned if you notice shiny spots on the pintle or seat.

■ If replacing the IAC valve, make sure that the new one has the same pintle shape and diameter.

5. Install a new O-ring onto the valve.
6. Install the valve and tighten the screws to 20 inch lbs. (2 Nm).
7. Install the throttle body
8. Connect the electrical lead. Connect the battery cables. Turn the ignition key to the **ON** position for 10 seconds, turn the key to the **OFF** position for 10 seconds. Start the engine and check that the idle is OK.

Later MPI Engines Exc. 8.1L

◆ See Figure 67

■ Please refer to the component locations illustrations in the Description & Operation section for most MPI engines.

1. Remove the engine hatch or cover and disconnect the battery cables.
2. Remove the engine cover and the flame arrestor.
3. Disconnect the 2 hoses from the side of the valve.
4. Tag and disconnect the electrical connector.
5. Remove the IAV bracket nuts and lift off the bracket with the valve.
6. Remove the 2 mounting screws and lift off the valve.

To install:

7. Clean the valve thoroughly and check for wear or other visible signs of damage.
8. Install the valve onto the bracket and tighten the screws to 168 inch lbs. (19 Nm).
9. Position the bracket and valve on the manifold and tighten the nuts to 132 inch lbs. (15 Nm).
10. Connect the electrical lead and the two hoses. Tighten the hose clamps securely, but not so tight as to pinch the hose.
11. Install the flame arrestor and cover. Connect the battery cables and start the engine to check for any problems.

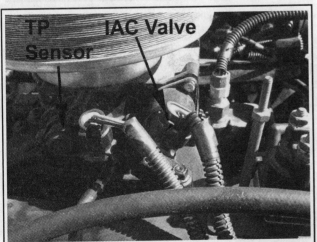

Fig. 66 A good look at the IAC valve on TBI and early MPI engines. Some engines are set up next to the TPS, others have them both on the other side, while others have one on each side

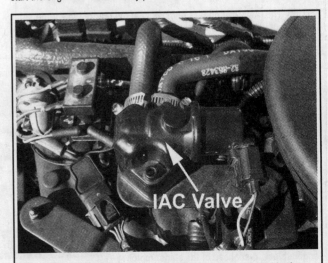

Fig. 67 On later MPI engines, the IAC valve is on a bracket in the top center of the engine

7-24 FUEL SYSTEM - FUEL INJECTION (EFI)

8.1L Engines

◆ See Figure 68

■ Please refer to the component locations illustrations in the Description & Operation section for most MPI engines.

1. Remove the engine hatch or cover and disconnect the battery cables.
2. Loosen the hose clamp and slide the IAC hose from the fitting on the throttle body if you plan to remove the valve and hose.
3. Locate the valve at the back of the engine and disconnect the electrical lead. Unscrew the 2 hex bolts and lift out the valve. If you have no need for the hose, simply loosen the clamp and slide it off of the nipple.
4. Remove the gasket from the valve and throw it away. Clean the valve mating surfaces, pintle valve seat and air passage with carburetor cleaner - do not push or pull on the valve pintle. Don't be concerned if you notice shiny spots on the pintle or seat.

■ If replacing the IAC valve, make sure that the new one has the same pintle shape and diameter.

5. Position a new gasket onto the valve.
6. Install the valve and tighten the screws to 20 inch lbs. (2 Nm). Connect the electrical lead.
7. Reconnect the vacuum hose, connect the battery cables, start the engine and ensure proper idle.

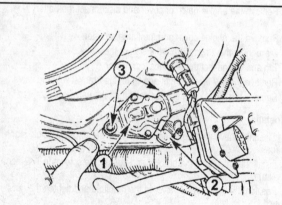

1 - IAC Valve
2 - Hose Connection
3 - Hex Bolts (1 Hidden)

Fig. 68 The IAC valve on 8.1L engines is at the rear of the engine, near the engine harness connector

Knock Sensor (KS)
REMOVAL & INSTALLATION

◆ See Figures 69 and 70

■ Please refer to the component locations illustrations in the Description & Operation section for most MPI engines.

■ TBI engines and early MPI engines will usually have the sensor on the lower starboard side of the engine block, while most other MPI engines will have one there and also another on the lower port side of the block. Certain 8.1L engines may also have another one on the rear of the engine block.

1. Disconnect the battery cables.
2. Tag and disconnect the electrical connector at the sensor and move it out of the way.
3. Unscrew the sensor from the block or Y-connector and remove the sensor. Clean the sensor with a dry cloth; especially the threads. If replacing the sensor, make sure the new one is the identical part number.
4. Repeat this procedure on engines with multiple sensors.
5. Install the sensor on the upper side of the Y-fitting on engines so equipped or into the block and tighten it to 12-16 ft. lbs. (16-22 Nm) on early 2001 models, or 15 ft. lbs. (20 Nm) on the 4.3L, 5.0L, 5.7L and 6.2L engines; or 14 ft. lbs. (19 Nm) on the 8.1L. Do not use any thread sealer or locking solution. Proper torque is imperative to ensure precise performance
6. Connect the electrical lead.

Knock Sensor Module
REMOVAL & INSTALLATION

◆ See Figure 71

■ Please refer to the component locations illustrations in the Description & Operation section for most MPI engines.

1. Disconnect the battery cables.
2. Tag and disconnect the electrical connector at the module and move it out of the way.
3. Remove the 2 (or 4) module mounting bolts and lift off the module.

✱✱ WARNING

Never soak the module in liquid cleaner or solvent.

4. Install the module and tighten the mounting bolts securely.
5. Connect the electrical lead and the battery cables.

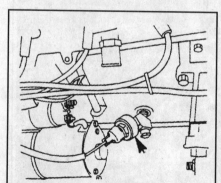

Fig. 69 The KS sensor is located on the engine block, just ahead of the starter motor on most engines. Some models use a "Y" fitting...

Fig. 70 ...while others do not

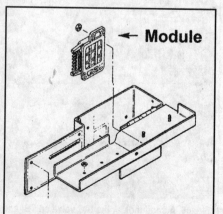

Fig. 71 The knock sensor module is usually found in or on the electrical box/bracket

FUEL SYSTEM - FUEL INJECTION (EFI) 7-25

Manifold Absolute Pressure Sensor (MAP/MAPT)

REMOVAL & INSTALLATION

◆ See Figures 72, 73 and 74

■ Please refer to the component locations illustrations in the Description & Operation section for MPI engines.

■ Most MPI engine utilize a pressure and temperature sensor, thus, MAPT.

1. Disconnect the battery cables.
2. Tag and disconnect the electrical connector at the sensor and move it out of the way.
3. Unscrew the mounting screw(s) and remove the sensor. Clean the sensor with a dry cloth; make sure that the sensor seal is in good condition; if not, replace it.
4. Install the sensor seal (coat it with a drop of clean oil first) and then install the sensor and tighten the screw to 44-62 inch lbs. (5-7 Nm) on TBI engines and early 2001 MPI engines; or 53 inch lbs. (6 Nm) on all other engines.
5. Connect the electrical lead and the battery cables.

Throttle Position Sensor (TP)

REMOVAL & INSTALLATION

◆ See Figures 75 and 76

■ Please refer to the component locations illustrations in the Description & Operation section for most MPI engines.

1. Disconnect the battery cables and remove the engine cover.
2. Remove the flame arrestor. Although not absolutely necessary, you may find it beneficial to remove the throttle body for better access to the sensor.
3. Tag and disconnect the electrical connector at the sensor and move it out of the way.
4. Unscrew the mounting screws and remove the sensor. Clean the sensor with a dry cloth; make sure that the sensor is in good condition (no wear, cracks or damage), if not replace it.

■ If replacing the sensor with a new one, make sure to use the 2 new screws that came with the package.

5. Install the sensor with a new seal and tighten the screws to 20 inch lbs. (2 Nm). Coat the threads with Loctite 242 and don't forget the washers!
6. Connect the electrical lead and install the throttle body and flame arrestor if removed. Start the engine and check the sensor output voltage. It should be approximately 0.7V at idle and 4.5V at wide open throttle.

Oil Pressure Switch/Sensor

REMOVAL & INSTALLATION

◆ See Figure 77

1. Open the engine hatch/covers and disconnect the battery cables.
2. Tag and disconnect the electrical connector at the switch.
3. Remove the switch.
4. Clean the switch thoroughly with a dry cloth and check for any obvious signs of wear or damage.
5. Install the switch and tighten it to 15 ft. lbs. (20 Nm). Connect the electrical lead and the battery cables.

Fig. 72 TBI engines and early 2001 MPI engines will have the MAP sensor in front of the lower edge of the throttle body

Fig. 73 Later V6 and 5.0/5.7/6.2L MPI engines will have it on the intake manifold, toward the back of the engine...

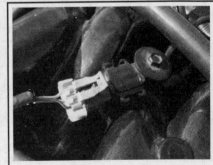

Fig. 74 ...while 8.1L engines are in the same place, but look like this

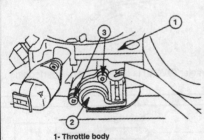

1- Throttle body
2- Throttle position (TP) sensor
3- Screws

Fig. 75 The TP sensor is always located on the throttle body and will look like this on TBI engines...

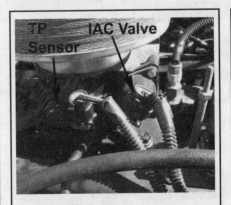

Fig. 76 ...or this, on MPI engines

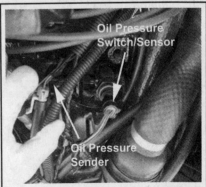

Fig. 77 The oil pressure switch is usually located on the port side of the engine block, toward the rear

FUEL SYSTEM - FUEL INJECTION (EFI)

SYSTEM DIAGNOSIS - TBI ENGINES & EARLY 2001 MPI ENGINES

◆ See Figure 78

■ The procedures contained within this section are valid for all TBI engines. It is also valid for early 2001 4.3L MPI engines (serial #322780 and below) and early 2001 5.0L, 5.7L and 6.2L V8 MPI engines (serial #299999 and below).

Scan Position		Units	Data Value
BARO		Volts	3-5
Battery		Volts	12-14.5
Coolant Temperature		Deg. F/C	150-170 (66-77)
Engine Overtemp		OK/Overheat	OK
Fuel Consumption		GPH	1 to 2
Idle Air Control Valve			
	Follower	Counts (Steps)	0
	Minimum	Counts (Steps)	0-40
	Normal	Counts (Steps)	0-40
Injector			
	On Time Cranking	Msec	2.5-3.5
	Pulse Width	Msec	2-3
Knock Retard		Deg.	0
Lanyard Stop Mode		OFF/ON	OFF
Manifold Air Temperature		Deg. F/C	Varies w/ambient
MAP		Volts	1-3
Memory Calibration Check		Sum of Check	Varies
Oil Pressure			
	I/O	OK/LOW	OK
	Transmission	OK/LOW	OK
Rpm			
	Desired	Rpm	600-700
	Normal	Rpm	600-650
Spark Advance		Deg.	10-30
Throttle			
	Angle	0-100%	0-1
	Position	Volts	0.4-0.8

Fig. 78 Normal specifications for a scan tool

Prior to performing any diagnostics on the EFI system, a diagnostic circuit check must first be completed. This is an organized approach to identifying a problem created by a malfunction in the electronic engine control management system and can only be performed with the correct scan tool - a Rinda Scan Tool or a Quicksilver Digital Diagnostic Tool. After hooking the tool up properly and finding that the on-board diagnostic system if functioning correctly and there are no codes displayed, the accompanying table should be used as a reference for what a normally functioning system should display. Use only the parameters listed in the chart.

If codes are displayed, move on to the next section.

Precautions

◆ See Figures 79 and 80

Good shop practice, as well as good judgment, requires all the following practices be observed:
• The negative battery cable must be disconnected before any ECM system component is removed
• Always check to be sure the battery cables are securely connected before starting the engine
• Never separate the battery from the on-board electrical system while the engine is operating
• Never separate the battery feed wire from the charging system while the engine is operating
• Disconnect the battery from the boat electrical system before starting to charge the battery
• Check to be sure all cable harnesses are securely connected and the battery terminals are clean and the cables securely connected
• Never connect or disconnect the wiring harness at the ECM when the ignition switch is in the ON position
• Before any electric arc welding is attempted, disconnect the battery leads and the ECM connector/s
• Never direct a steam-cleaning nozzle at ECM components. Such action will cause damage or corrosion the component terminals
• Do not use any test equipment not specified in the diagnostic charts. Such equipment may give an incorrect reading and/or may actually damage good components
• A digital voltmeter with a rating of 10 meg-ohms input impedance must be used when taking any voltage measurements.
• A "low-amphere rated" test light must be used when a test light is specified for a test. Never use a high-amphere rated test light. Power the setup with the boat battery. If the ammeter indicates less than 3/10 amp current flow, the test light is safe to use.

A final word here. If a test light with 100 mA or less is used, a faint glow may show, when the test actually states "no light".

Intermittent Faults

■ A specific problem may or may not set a code and turn on the MIL. An actual fault must be recorded before you can use the DTC charts and tests.

Most intermittent problems are caused by bad electrical connections and/or wiring problems. Always check for the following:
• Poor mating of connector halves
• Terminal connectors not fully seated on the connector body
• Damaged, worn or corroded terminals

All terminals, wiring and/or connections in a circuit with a suspected problem should be checked carefully prior to proceeding with any other test procedures.

Running the vessel with a DVOM connected to a suspicious circuit may work, as abnormal voltage when the fault occurs is a good indicator that there is a fault in the circuit being checked.

A scan tool can also be used as it has several features built into it to aid in detecting intermittents.

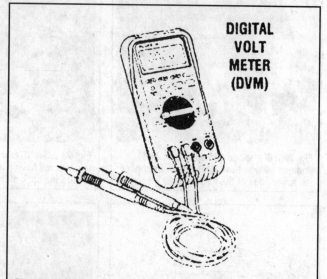

Fig. 79 A quality multi-meter (DVOM) is necessary when performing diagnostic tests. . .

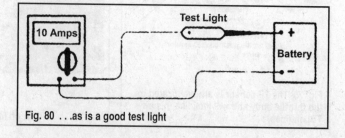

Fig. 80 . . .as is a good test light

FUEL SYSTEM - FUEL INJECTION (EFI)   7-27

An intermittent fault (MIL) which is not causing a stored DTC can be caused by the following:
- Ignition coil shorted to ground
- Arcing at the coil wire or spark plug wires
- MIL wire to the ECM is shorted to ground
- Poor ECM grounds
- Electrical system interference caused by a sharp surge - normally occurs when the bad components is being operated
- Incorrect installation of electrical accessories (GPS, radio, lights, etc.)
- Knock sensor electrical lead routed near or touching spark plug wires or ignition/charging system components
- Secondary ignition system components shorted to ground
- Starters, alternators and relays, etc. shorted to ground
- All IC module wiring should be routed away from the alternator.

General Diagnostic Tests

◆ See Figure 78

A scan tool or CodeMate Tester is a must to perform these tests.

ON-BOARD DIAGNOSTIC SYSTEM CHECK

■ **Please refer to the General Diagnostic Test Schematics and/or ECM Pin Locations in this section for correct pin and circuit locations**

1. With the ignition switch in the ON position and the engine OFF, connect the tester to the diagnostic link connector (DLC) in the electrical box and switch it to the Normal mode. The malfunction indicator lamp (MIL) should come on, if not please refer to the following procedure for "A1 - No MIL or DLC Data".

2. If the tester flashes a Code 12, check circuit 451 for a short to ground. Otherwise, switch the tester to the Service mode and confirm that it flashes Code 12. If it doesn't, check the "A2 - MIL on Steady. . ." procedure following.

3. Once you've gotten your Code 12, switch the tester back to the Normal mode and start the engine. If the engine continues to run, move to the next step; if it won't run, move to the "A3 - Engine Cranks But Won't Run" procedure.

4. Turn off the engine but leave the ignition switch in the ON position. Switch the tester back to Service and check for any stored codes. Refer to the DTC charts at the end of this section.

A1 - NO MIL OR NO DLC DATA

◆ See Figure 81

■ **Please refer to the General Diagnostic Test Schematics and/or ECM Pin Locations in this section for correct pin and circuit locations**

When a tester is connected to the DLC it should receive voltage via circuit 440 (terminal **F**) and be grounded through circuit 419 (**E**). There should always be a steady MIL when the ignition is ON (engine stopped). Obviously, always check for bad connections or frayed wires. It's also a good idea to check that the indicator bulb is not burned out.

1. Remove the CodeMate tester or scan tool and turn the ignition switch to the ON position, engine OFF. Connect a test light to ground and touch the probe to terminal **F** of the DLC. If the light goes on move to Step 2; if it does not come on, circuit 440 is open or has a short to ground.

2. Now connect the test light to the battery positive (**B+**) terminal and touch the probe to terminal **E** of the DLC. If the light goes on, your tester is bad; if it does not come on, go to Step 3.

3. With the ignition OFF, disconnect the front and rear (J1 and J2) connectors at the ECM. Connect a multi-meter (DVOM) and measure the resistance between ECM circuit 419 and terminal **E** on the DLC. If less than an ohm, go to Step 4; if more than an ohm, circuit 419 is open.

4. If you have a scan tool, go to Step 5; if you don't have access to a scan tool, replace your Codemate tester.

5. Check the ECM/DLC fuse. If it's OK, go to Step 6; if it's blown, you have a short to ground somewhere in the battery feed circuit.

6. With the ignition switch in the OFF position and the ECM still disconnected, connect a test light to ground and touch the other probe to ECM circuit 440. If the light comes on, go to Step 7; if it doesn't, circuit 440 is open or shorted to ground.

7. Turn the ignition switch back to the ON position (engine OFF) and now touch the probe to ECM circuit 439. If the light comes on and your initial problem was no DLC data (while using the scan tool), circuit 461 is open or shorted to ground, or you've got a bad ground on the ECM; if the light doesn't come on, go to Step 8.

8. Check the ignition/injection fuses(s). If OK, perform the "A6 - EFI System/Ignition Relay Check" test following; if the fuse is blown, there is a short to ground in circuit 439.

A2 - MIL ON STEADY - WILL NOT FLASH DTC 12

◆ See Figure 81

■ **Please refer to the General Diagnostic Test Schematics and/or ECM Pin Locations in this section for correct pin and circuit locations**

When the CodeMate tester is installed, it receives voltage through circuit 440 (terminal **F**) and is grounded through circuit 419 (**E**). When the ignition is ON and the engine OFF there should always be a steady MIL. Obviously, always check for bad connections or frayed wires. It's also a good idea to check that the indicator bulb is not burned out.

1. With the ignition switch in the ON position and the engine OFF, switch the CodeMate tester to Normal mode. If the MLC flashes DTC 12, circuit 451 has a short to ground; if no DTC 12, go to Step 2.

2. With the ignition switch still in the ON position, switch the tester to the Service mode. If the MIL flashes DTC 12, look for bad connections in the circuit; if DTC 12 does not flash, go to Step 3.

3. Turn the ignition switch to the OFF position and disconnect the two connectors at the ECM. Turn the ignition back to the ON position. If the MIL comes ON, circuit 419 has a short to ground; if the light does not come on, go to Step 4.

4. Turn the ignition switch back to the OFF position and connect a jumper wire between terminals **A** and **B** at the DLC. Connect a test light between circuit 451 and battery positive (**B+**). If the light goes ON, you'll need to verify that the Codemate tester is working properly by hooking it up to a known good system - we know its unlikely that you have another engine available, so all we can suggest is running down to your local dealer and asking them to confirm that its OK. If the light does not go on, circuits 450 and/or 451 are open.

A3 - ENGINE CRANKS BUT WILL NOT RUN

◆ See Figure 82

■ **Please refer to the General Diagnostic Test Schematics and/or ECM Pin Locations in this section for correct pin and circuit locations**

The distributor ignition (DI) system and the fuel injector circuit are both supplied voltage via the EFI system relay. From the relay, circuit 902 delivers voltage to the injector/ECM fuse and to the coil. The following test assumes a properly functioning battery and adequate fuel supply. Is also a good idea to check all connections and wires before hand.

1. Ensure that the ignition switch has been in the OFF position for at least 10 seconds and then switch it ON. If the fuel pump runs for 2 seconds, go to Step 2; if not, move on to the "A5 - Fuel System Electrical Test" procedure following.

2. Check for secondary ignition spark. If adequate, go to Step 3; if not, move on to the "A7 - Ignition System Check" procedure following.

3. Install a fuel pressure gauge as per the manufacturer's instructions or as detailed in elsewhere in this manual. Turn the ignition OFF for 10 seconds and then turn the ignition ON and let the pump run for about 2 seconds. Check the fuel pressure while the pump is running. If above 25 psi, go to Step 4; if below 25 psi, move on to the "A4 - Fuel System Diagnosis" procedure.

4. Turn the ignition OFF and disconnect the two ECM connectors. Connect a multi-meter (DVOM) and measure the resistance between ECM circuits 467 and 468. If more than an ohm, go to Step 5; if less than an ohm, circuits 467 or 468 have a short to ground.

7-28 FUEL SYSTEM - FUEL INJECTION (EFI)

5. Check the resistance across each injector. If higher than 1 ohm on TBI systems or 10 ohms on MPI systems, go to Step 6; if less than 1 ohm on TBI systems or 10 ohms on MPI systems, look for a short to ground at the injector, if none replace the injector.

6. Reconnect the injectors, turn the ignition OFF and disconnect the ECM. Connect a test light to ground and touch the probes to ECM circuits 467 and 468. If the light comes on, go to Step 7; if it doesn't, circuits 467 or 468 have an open connection.

7. Disconnect the injectors and turn the ignition ON. Connect a test light to ground and touch the probe to circuits 467 and 468 on the ECM side of the injector harness. Test the harness on each side of the engine, if the light goes on you have a short.

A4 - FUEL SYSTEM DIAGNOSIS

1. Install a fuel pressure gauge. Turn the ignition switch OFF and then ON. The fuel pump should run for about 2 seconds. If the fuel pressure is less than 25 psi, go to Step 2; if higher than 25 psi, go to Step 4.

2. Try and start and idle the engine. If it starts, go to Step 3; if it won't start, go to Step 5.

3. With the engine still idling, connect a vacuum gauge to the pressure regulator and apply 10 in. of vacuum. If the pressure drops by about 5 psi, the fuel supply is probably restricted, check the pump and lines; if the pressure doesn't drop to the above spec, replace the fuel pressure regulator.

4. Was the any fuel pressure at all? If there was, go to Step 5; if no pressure, move to the "A5 - Fuel System Electrical Test" procedure.

5. Did the system register pressure and then drop off to no pressure? If it does, go to Step 6; if it doesn't, check for a restriction in the fuel lines.

6. Turn the ignition OFF, disconnect the fuel line between the pump and the throttle body or fuel rail, and plug it. Turn the ignition On and check to see if pressure holds. If it does, check for leaking injectors of fuel line connectors. If it does not, go to the next step.

7. With the ignition still in the OFF position, plug the fuel return line. If the pressure holds, replace the fuel pressure regulator; if it doesn't hold, check the pump, fuel lines and inlet filter for leaks. Check that the battery isn't low. If everything is OK, replace the fuel pump.

A5 - FUEL SYSTEM ELECTRICAL TEST

◆ See Figure 83

■ Please refer to the General Diagnostic Test Schematics and/or ECM Pin Locations in this section for correct pin and circuit locations

The fuel system receives voltage via a system relay on circuit 902. It is protected by an inline 15 amp fuse that then sends the voltage on to the pump relay via circuit 339 (terminal 30 on the relay). The ECM turns the pump relay on via circuit 465; it will remain on as long as the engine is cranking or running, and the ECM continues to receive a reference pulse. If no pulse is received, the ECM will de-energize the pump within 2 seconds after the ignition switch is turned ON or the engine is stopped.

Always check for bad connections and frayed wires. The following test also assumes an uncontaminated fuel supply.

1. Turn the ignition OFF, remove the fuel pump relay and then turn the ignition back ON. Connect a test light to ground and touch the probe to the pump relay harness connector terminal 30. If the light goes on, move to Step 2; if it doesn't, go to Step 5.

2. Turn the ignition OFF again and connect a fused jumper wire between terminals 30 and 87 on the pump relay connector. Turn the ignition ON. If the fuel pump runs, go to Step 3; if it doesn't energize, check that circuits 120 or 150 may be open. If they're OK, replace the fuel pump.

3. Turn the ignition off and disconnect the jumper wire. Connect a test light to battery positive terminal (B+) and touch the probe to terminal 86 on the pump relay connector. If the light comes on, go to Step 4; if not, circuit 450 is open.

4. Now touch the probe to terminal 85 on the relay connector and turn the ignition ON. If the light goes on for 2 seconds and then goes out, check the fuel system for a blocked filter, vapor lock, plugged or disconnected lines or hoses, and make sure there is fuel in the tank; if it doesn't, you've either got a bad connection at the ECM J2 terminal 9 or circuit 465 is open.

5. Check the fuse between the system relay and pump relay. If it's OK, circuits 339 or 902 are open; if blown, look for a short to ground in circuit 339 or 120. Also check for contaminated fuel. If everything looks good, replace the fuel pump. And also the bad fuse!

A6 - EFI SYSTEM/IGNITION RELAY CHECK

◆ See Figure 84

■ Please refer to the General Diagnostic Test Schematics and/or ECM Pin Locations in this section for correct pin and circuit locations

The battery supplies voltage to terminal 30 of the system relay. When the ignition switch is turned on and moved to the RUN position, battery voltage is also supplied, through the ignition, to terminal 86 on the system relay. The pull-in coil is then energized which creates a magnetic field and then closes the system relay contacts. Voltage and current are then supplied to the ignition control module, injectors, ECM and pump relay via terminal 87 on circuit 902.

Always check for bad connections and frayed wires. The following test also assumes an uncontaminated fuel supply.

1. Turn the ignition OFF and remove the EFI system relay connector. Turn the ignition ON, connect a test light to ground and touch the probe to terminals 30 and 86 on the relay harness connector. If the light goes on at both terminals, move to the next step; if it doesn't, check for an open or short to ground on the circuit where it didn't light (circuit 2 or 3).

2. Now connect the test light to battery positive (B+) and touch the probe to terminal 85. If it lights, check the relay connector for a poor contact otherwise replace the relay; if it doesn't light, circuit 150 is open to ground.

A7 - IGNITION SYSTEM CHECK

Please refer to the Ignition & Electrical section for complete testing on the ignition system.

A8 - IDLE AIR CONTROL VALVE FUNCTIONAL TEST

◆ See Figure 79

■ Please refer to the General Diagnostic Test Schematics and/or ECM Pin Locations in this section for correct pin and circuit locations

The ECM controls idle speed to a pre-set, desired rpm based on input from sensors and the actual engine rpm. Four separate circuits are used to move the IAC valve; said movement varying the amount of air allowed to bypass the throttle plates. Idle speed then is controlled, via the ECM, by the position of the IAC valve.

1. With the engine at normal operating temperature, turn it off, connect a tachometer and then restart it and allow the idle to stabilize. Record the idle speed and turn off the engine for 10 seconds. Disconnect the IAC harness connector, restart the engine and check the rpm. If the second recorded idle speed is higher than the first by 200 rpm or more, go to Step 2; if not, go to Step 3.

2. Reconnect the IAC harness. If the idle speed returns to within 75 rpm of the originally recorded speed in Step 1 within 30 seconds, the IAC circuit is functioning properly; if it does not, go to the next step.

3. Turn off the ignition for 10 seconds. Disconnect the IAC again and restart the engine. Connect a test light to ground and then touch the probe to each of the four IAC harness terminals. If the light blinks on each terminal, check for a bad IAC connection somewhere in the circuit, otherwise replace the valve; if it doesn't blink, look for an open or shorted circuit on the terminal(s) that didn't blink. Also, check for any bad connections at the ECM.

General Diagnostic Test Schematics

◆ See Figures 81 thru 85

FUEL SYSTEM - FUEL INJECTION (EFI) 7-29

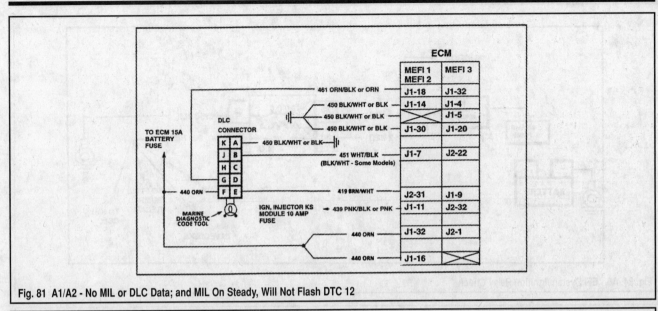

Fig. 81 A1/A2 - No MIL or DLC Data; and MIL On Steady, Will Not Flash DTC 12

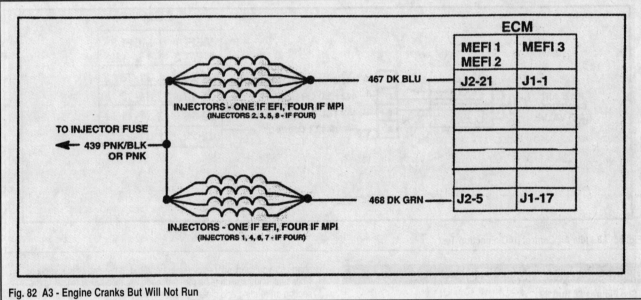

Fig. 82 A3 - Engine Cranks But Will Not Run

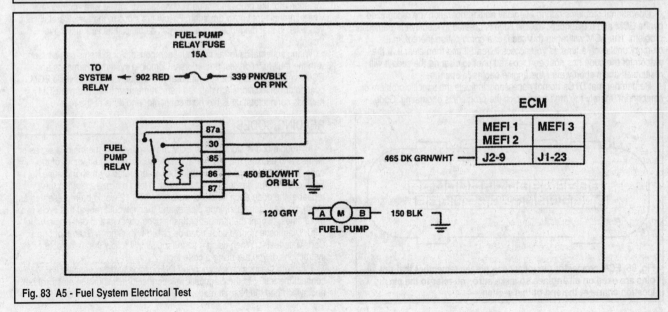

Fig. 83 A5 - Fuel System Electrical Test

7-30 FUEL SYSTEM - FUEL INJECTION (EFI)

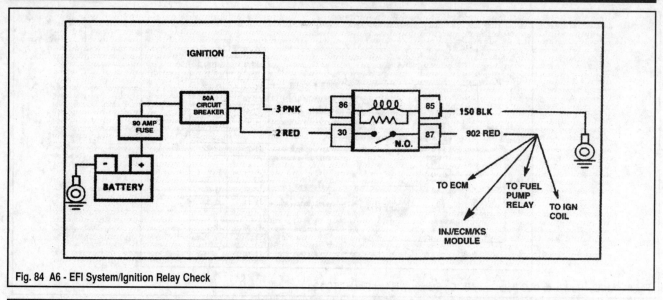

Fig. 84 A6 - EFI System/Ignition Relay Check

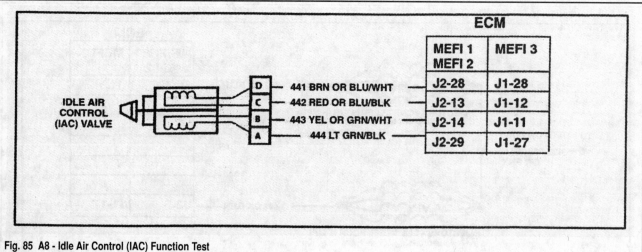

Fig. 85 A8 - Idle Air Control (IAC) Function Test

Diagnostic Trouble Codes (DTC)

◆ See Figures 86 thru 89

Operational problems during everyday engine operation are recognized by the ECM and a diagnostic trouble code is created to identify the particular problem. The ECM will retain this code, or a combination thereof, in its memory until such a time as you access it, read it and then clear it. If the reason for the code has not been repaired prior to clearing the code it will reset itself again shortly after the engine begins to operate.

Remember that DTCs do not necessarily indicate the specific problem or component, merely the area from which the problem is originating. Code charts for your particular engine can be found at the end of this section; while fault and function tests for each code are also found following the section on reading and clearing codes. You will find that more times than not, the source of the problem is a bad connection or frayed wire - particularly where intermittent codes are seen. Follow the steps in the tests carefully and you will find that diagnostics is not always as difficult as you may have thought.

Wiring schematics for all individual components or systems can be found with the fault/function test procedures. Complete system schematics and ECM symptoms charts can be found at the end of this section, while ECM connector pin locations are detailed here; remember there are two ECM harness connectors, **J1** is the front connector and **J2** is at the rear.

READING CODES

All trouble codes stored in the ECM are displayed by means of a series of flashes and pauses; the number of flashes represents the number of the code, with the first and second digits being separated by a short pause. Each code will be flashed three times with a long pause separating the repeat of the code each time. For instance, DTC 12 would be represented as "flash, pause, flash-flash, long pause; and then it would repeat this cycle two more times. Count the number of flashes you observe and determine your DTC. Diagnostic trouble codes can be pulled with either a Scan tool or a CodeMate tester. When using a scan tool, simply follow the manufacturer's instructions. If you are using a code tool:

1. With the ignition switch in the **ON** position and the engine OFF, connect the tester to the diagnostic link connector (DLC) in the electrical box and switch it to the Normal mode. The malfunction indicator lamp (MIL)

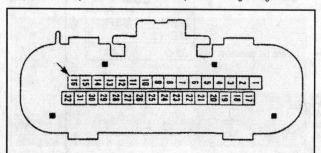

Fig. 86 ECM connector pin locations - please remember that not all pins are used on all engines so make sure you refer to the pin location charts at the end of this section

FUEL SYSTEM - FUEL INJECTION (EFI)

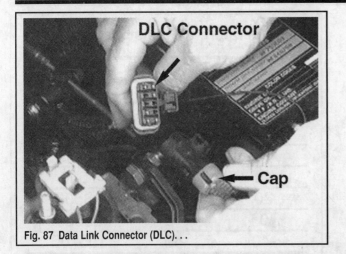

Fig. 87 Data Link Connector (DLC)...

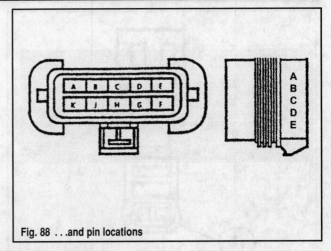

Fig. 88 ...and pin locations

Diagnostic Trouble Codes (DTC)

Code	Description	MEFI-1	MEFI-2	MEFI-3
14	Engine coolant temperature sensor (low temp. indicated)	X	X	X
15	Engine coolant temperature sensor (high temp. indicated)		X	X
21	Throttle position sensor (signal voltage high)	X	X	X
22	Throttle position sensor (signal voltage low)		X	X
23	Intake air temperature (low temp. indicated)			X
25	Intake air temperature (high temp. indicated)			X
33	Manifold absolute pressure (signal voltage high)	X	X	X
34	Manifold absolute pressure (signal voltage low)		X	X
41	Ignition control (open circuit)		X	X
42	Ignition control (gounded circuit, open or grounded bypass)	X	X	X
43	Knock sensor (continuous knock detected)	X	X	X
44	Knock sensor (no knock detected)		X	X
45	Coil driver			X
51	ECM calibration memory failure	X	X	X
52	ECM EEProm failure			X
61	Fuel pressure high			X
62	Fuel pressure low			X

Fig. 89 Diagnostic Trouble Code (DTC) Chart

should come on and flash a Code 12 (3 times), indicating the diagnostic system is operating properly. If Code 12 is not displayed, please refer to the following procedure for "No MIL or DLC Data".

2. Shortly after displaying Code 12, the system will flash any stored codes from the system; three times each and in ascending order if more than one code has been stored. If Code 12 continues being displayed then you have no stored codes (or an intermittent one).

CLEARING CODES

Once again, if using a scan tool, follow the tool manufacturer's instructions. If using a CodeMate tester:
1. Ensure that the battery is fully charged.
2. Connect the tester to the DLC in the electrical box and switch the ignition key to the **ON** position.
3. Select the Service mode on the tester and then move the throttle through its full range, from idle to full throttle and then back to idle.
4. Switch out of Service mode on the tester, start the engine and let it run for fifteen seconds.
5. Turn the ignition switch to OFF for five seconds, switch the tester back to Service mode and then turn the ignition switch ON again.
6. There should no longer be any codes present. If there are, check the battery again and then perform the procedure one more time. If codes are still apparent at the end of the second go-around (we're assuming that you fixed the code's problem before attempting to clear it!), refer to the appropriate troubleshooting or diagnostic charts. If the battery is not at full charge, you should hear the audio warning buzzer come on after engine start-up.

CODE 14 - ENGINE COOLANT TEMPERATURE SENSOR (ECT)

◆ See Figures 90 and 91

The ECT sensor utilizes a thermistor to control the signal voltage being sent to the ECM. The ECM then applies specified voltage back through the 410 circuit to the sensor. When the coolant is cold, the sensor resistance is high. As the coolant warms, resistance lessens and voltage drops.

With A Scan Tool

◆ See Figures 90 and 91

1. Perform the On-Board Diagnostic System check in the General Diagnostic Test section.
2. Hook up the scan tool as per the manufacturer's instructions.

FUEL SYSTEM - FUEL INJECTION (EFI)

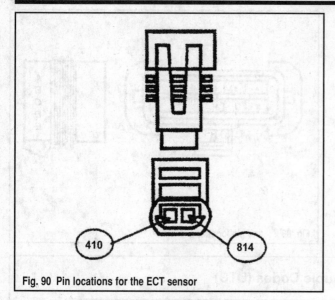

Fig. 90 Pin locations for the ECT sensor

°F	°C	Ohms
210	100	185
160	70	450
100	38	1800
70	20	3400
40	4	7500
20	-7	13500
0	-18	25000
-40	-40	100700

Temperature-to-Resistance Values

Fig. 91 ECT sensor values chart

3. Turn the ignition switch to the **ON** position. If the tool displays a coolant temperature higher than 266°F (130°C) or less than -22°F (-30°C), move to the next step. If not, Code 14 is intermittent.

4. If the scan tool showed a temperature higher than 266°F (130°C) in the previous step, turn OFF the ignition and disconnect the ECT harness. Turn the ignition back ON. If the tool displays a temperature less than -22°F (-30°C), look for a bad electrical connection or replace the sensor. If the temperature shown is not less than -22°F (-30°C), the 410 circuit lead has a short and it must be found and repaired.

5. If the temperature shown in Step 4 is not higher than 266°F, turn OFF the ignition and disconnect the ECT harness. Connect terminals **A** and **B** with a jumper wire and then turn the ignition back ON. If the tool displays a temperature higher than 266°F (130°C), look for a bad electrical connection or replace the sensor. If the temperature shown is less than 266°F (130°C), the 410 circuit lead or the sensor has a short to ground and it must be found and repaired.

Without A Scan Tool

◆ See Figures 90 and 91

1. Perform the On-Board Diagnostic System check in the General Diagnostic Test section.
2. With the ignition OFF, disconnect the ECT sensor harness connector. Turn the ignition switch to the **ON** position (but don't start the engine) and then connect a multi-meter (DVOM) across the **A** and **B** sensor harness terminals.
3. If the voltage is above 4 volts, look for a bad electrical connection at the ECM or an open sensor ground in the 814 circuit.
4. If the voltage is below 4 volts, disconnect the DVOM and then connect a test light to battery positive (**B+**).
5. Now touch terminal **B** of the harness connector (circuit 410). If the light does not go ON, check for an open or bad connection in the 410 circuit or a bad connection at the ECM.
6. If the light does go ON, disconnect the J-1 connector at the ECM on MEFI-1/MEFI-2 engines or the J-2 connector on MEFI-3 engines. If the light goes ON again, you've got a bad ground in the 410 circuit. If it does not go ON, you've either got a bad ground in the 410 circuit or it's shorted to sensor ground.

CODE 15 - ENGINE COOLANT TEMPERATURE SENSOR (ECT)

◆ See Figures 90 and 91

The ECT sensor utilizes a thermistor to control the signal voltage being sent to the ECM. The ECM then applies specified voltage back through the 410 circuit to the sensor. When the coolant is cold, the sensor resistance is high. As the coolant warms, resistance lessens and voltage drops.

With A Scan Tool

◆ See Figures 90 and 91

1. Perform the On-Board Diagnostic System check in the General Diagnostic Test section.
2. Hook up the scan tool as per the manufacturer's instructions.
3. Turn the ignition switch to the **ON** position. If the tool displays a coolant temperature higher than 266°F (130°C), turn OFF the ignition and disconnect the ECT harness. Turn the ignition back ON (engine still OFF).
4. If the tool displays a temperature less than -22°F (-30°C), look for a bad electrical connection or replace the sensor. If the temperature shown is not less than the above, the 410 circuit lead has a short to ground and it must be found and repaired. Also check for bad ECM connections.
5. If the temperature shown in Step 3 is not higher than 266°F, look for a bad electrical connection or replace the sensor.

Without A Scan Tool

◆ See Figures 90 and 91

1. Perform the On-Board Diagnostic System check in the General Diagnostic Test section.
2. With the ignition OFF, disconnect the ECT sensor harness connector. Turn the ignition ON (but don't start the engine) and then connect a multi-meter (DVOM) across the **A** and **B** terminals at the sensor.
3. If the voltage is above 4 volts, look for a bad electrical connection or replace the sensor.
4. If the reading is below 4 volts, the 410 circuit lead has a short and it must be found and repaired.

CODE 21 - THROTTLE POSITION SENSOR (TP)

◆ See Figure 86

The TP sensor provides voltage signal to the ECM that varies with the opening and closing of the throttles. Signal voltage will vary from approximately 0.5 volts at idle to slightly more than 4 volts at wide open throttle.

With A Scan Tool

◆ See Figure 92

1. Perform the On-Board Diagnostic System check in the General Diagnostic Test section.
2. Connect a scan tool as per the manufacturer's instructions.

FUEL SYSTEM - FUEL INJECTION (EFI) 7-33

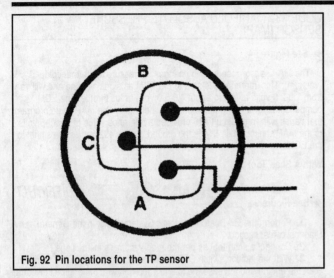

Fig. 92 Pin locations for the TP sensor

3. With the throttle closed and the ignition switch in the **ON** position, check the reading on the tool. If over 4 volts go to Step 5; if in between 4 volts and 0.3 volts, the fault is intermittent. If under 0.3 volts, go to Step 4.

4. Turn the ignition OFF, disconnect the TP sensor electrical lead and run a jumper wire between harness terminals **A** and **C**. Turn the ignition ON. If the tool reads higher than 4 volts, replace the TP sensor; if it reads less than 4 volts, go to Step 6.

5. Turn the ignition OFF, disconnect the TP sensor connector and then turn the ignition back to the **ON** position. If the tool reads over 4 volts, 417 circuit is shorted to voltage; if the tool reads less than 4 volts, go to Step 6.

6. Turn the ignition switch OFF, disconnect the TP sensor harness connector and then turn the ignition back to the **ON** position. Connect a multi-meter (DVOM) between harness terminal **A** and **B**. If the reading is over 4 volts, go to Step 7; if the reading is less than 4 volts, go to Step 10.

7. Connect the multi-meter between harness terminals **A** (circuit 416) and **C** (circuit 417). If the reading is higher than 4 volts, go to Step 8; if the reading is less than 4 volts, go to Step 9.

8. Turn the ignition switch OFF. Connect a test light to the positive battery terminal (**B+**). Touch the harness terminal **C**. If the light goes ON, go to Step 11; if the light does not come on, replace the TP sensor.

9. Connect the multi-meter between harness terminal **C** and an engine ground. If the reading is higher than 4 volts, check 417 circuit for a short to voltage; if the reading is less than 4 volts, 417 circuit is open or you have a bad connection at the ECM.

10. Connect the multi-meter between harness terminal **A** and an engine ground. If the reading is higher than 4 volts, 813 circuit is open or you have a bad connection at the ECM; if the reading is less than 4 volts, 416 circuit is open or shorted to ground, or you have a bad connection at the ECM.

11. Disconnect the ECM and touch the test light to harness terminal **C**. If the light comes ON, 417 circuit is shorted to ground; if the light doesn't go ON, 417 circuit is open or you have a bad connection at the ECM.

Without A Scan Tool

◆ See Figure 92

1. Perform the On-Board Diagnostic System check in the General Diagnostic Test section.

2. With the ignition switch OFF, disconnect the TP sensor harness connector and then turn the ignition switch to the **ON** position. Connect a multi-meter (DVOM) between the **A** and **B** harness terminals.

3. If the reading is over 4 volts, go to Step 3. If the reading is under 4 volts, go to Step 6.

4. Connect a DVOM between the harness terminals **A** on circuit 416 and terminal **C** on circuit 417. l

5. If the reading is over 4 volts, turn the ignition OFF and connect a test light to battery positive (**B**). Touch the probe to the **C** harness terminal. If the light doesn't come ON, replace the TP sensor. If the light comes on, disconnect the ECM and touch the harness terminal **C** with the probe. If the light comes ON, 417 circuit is shorted to ground; if the light doesn't come ON, 417 is an open circuit or you have a bad connection at the ECM.

6. If the reading in Step 3 is under 4 volts, connect the multi-meter between terminal **C** and engine ground. If the reading is over 4 volts, 417 circuit is shorted to voltage; if the reading is less than 4 volts, 417 is an open circuit or you have a bad connection at the ECM.

7. If the reading in Step 2 is less than 4 volts, connect a multi-meter between harness connector terminal **A** and a good engine ground. If the reading is over 4 volts, 813 circuit is open or you have a bad connection at the ECM; if less than 4 volts, 416 circuit is open or shorted to ground, or you have a bad connection at the ECM.

CODE 22 - THROTTLE POSITION SENSOR (TP)

◆ See Figure 92

The TP sensor provides voltage signal to the ECM that varies with the opening and closing of the throttles. Signal voltage will vary from approximately 0.5 volts at idle to slightly more than 4 volts at wide open throttle.

With A Scan Tool

◆ See Figure 92

1. Perform the On-Board Diagnostic System check in the General Diagnostic Test section.

2. Connect a scan tool as per the manufacturer's instructions.

3. With the throttle closed and the ignition switch in the **ON** position, check the reading on the tool. If less than 0.3 volts, go to Step 4. If not less than 0.3 volts, the fault is intermittent. Look for bad wiring and/or connections.

4. With the ignition OFF, disconnect the TP sensor harness connector and connect a jumper wire between terminal **A** (circuit 416) and **B** (circuit 417). Turn the ignition ON, but the engine OFF. If the reading is greater than 4 volts, check for bad connections or replace the TP sensor; if less than 4 volts, go to Step 5.

5. With the ignition OFF, connect a multi-meter (DVOM) between terminal **A** (circuit 416) and a good ground. If the tool indicates sensor voltage greater than the specified value, 417 circuit has is open or shorted to ground, or you have a bad connection at the ECM; if less than the correct value, 416 circuit has is open or shorted to ground, or you have a bad connection at the ECM.

Without A Scan Tool

◆ See Figure 92

1. Perform the On-Board Diagnostic System check in the General Diagnostic Test section.

2. Connect a CodeMate Tester and switch it to the Normal mode.

3. Turn the ignition switch to the **OFF** position and disconnect the TP sensor harness connector.

4. Connect a jumper wire between terminals **A** (416 circuit) and **C** (417 circuit). Start the engine and allow it to idle for two minutes or until the tester indicates a stored code, whichever comes first.

5. Turn the engine OFF, but leave the ignition switch in the **ON** position. Switch the tester to the Service mode and note the trouble code. If DTC 21, check all connections and/or replace the TP sensor; if no DTC 21, 417 circuit is open or shorted to ground; if everything is OK, you have a bad connection at the ECM.

CODE 23 - INTAKE AIR TEMPERATURE SENSOR (IAT)

◆ See Figure 93

The IAT sensor utilizes a thermistor to control a voltage signal to the ECM. Resistance is high when intake air is cold so the ECM will see a high voltage. As the air warms, resistance lessens and voltage drops. At normal operating temperature, voltage should be 1.5-2.0 volts.

Only engines using MEFI-3 ECMs use this component.

1. Perform the On-Board Diagnostic System check in the General Diagnostic Test section.

7-34 FUEL SYSTEM - FUEL INJECTION (EFI)

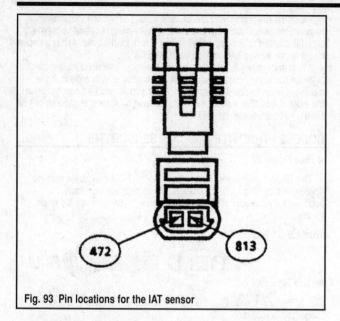

Fig. 93 Pin locations for the IAT sensor

2. If you are using a scan tool, go to Step 3; if not, go to Step 5.

3. Connect a scan tool as per the manufacturer's instructions. Turn the ignition switch to the **ON** position. If the tool displays a temperature less than -22°F (-30°C), go to Step 4; if the temperature is higher than specified, the fault is intermittent and you should check all system circuits for bad connections or frayed wires.

4. Turn the ignition OFF and disconnect the harness at the IAT sensor. Connect a jumper wire between harness terminals **A** (472 circuit) and **B** (813 circuit). Turn the ignition ON but leave the engine OFF. If the tool shows a temperature above 266°F (130°C), check all system circuits for bad connections or frayed wires, if OK, replace the sensor. If the temperature is below 266°, circuits 472 or 813 are open.

5. Turn the ignition OFF and disconnect the harness at the sensor. Turn the ignition switch to the **ON** position and connect a multi-meter (DVOM) across the harness terminals. If voltage is higher than 4 volts, check all system circuits for bad connections or frayed wire, otherwise replace the sensor; if lower than 4 volts, go to Step 6.

6. Connect the positive meter lead to harness terminal **B** and the negative lead to a good ground. If the voltage is over 4 volts, circuit 813 is open to ground or you have a bad connection at the ECM; if under 4 volts, circuit 472 is open or you have a bad connection at the ECM.

CODE 25 - INTAKE AIR TEMPERATURE SENSOR (IAT)

◆ See Figure 93

The IAT sensor utilizes a thermistor to control a voltage signal to the ECM. Resistance is high when intake air is cold so the ECM will see a high voltage. As the air warms, resistance lessens and voltage drops. At normal operating temperature, voltage should be 1.5-2.0 volts.

1. Perform the On-Board Diagnostic System check in the General Diagnostic Test section.

2. If you are using a scan tool, go to Step 3; if not, go to Step 5.

3. Connect a scan tool as per the manufacturer's instructions. Turn the ignition switch to the **ON** position. If the tool displays a temperature greater than 266°F (130°C), go to Step 4; if the temperature is less than specified, check all system circuits for bad connections or frayed wires.

4. Turn the ignition OFF and disconnect the harness at the IAT sensor. Turn the ignition ON, but leave the engine OFF. If the tool shows a temperature below -22°F (-30°C), check all system circuits for bad connections or frayed wires, if OK, replace the sensor. If the temperature is above -22°, circuit 472 is open and will require repair, if not open you have a bad connection at the ECM.

5. Turn the ignition OFF and disconnect the harness at the sensor. Turn the ignition switch to the **ON** position and connect a multi-meter (DVOM) across the harness terminals. If voltage is higher than 4 volts, check all system circuits for bad connections or frayed wire, otherwise replace the sensor; if lower than 4 volts, circuit 472 is open or you have a bad connection at the ECM

CODE 33 - MANIFOLD ABSOLUTE PRESSURE SENSOR (MAP)

◆ See Figure 94

The MAP sensor responds to changes in pressure in the manifold (vacuum). This information is sent to the ECM as signal voltage and will vary from 1.0-2.0 volts at idle to about 4.0-5.0 volts at full throttle. The ECM compensates for a failing Map sensor by defaulting to a MAP value program that varies with rpm. Circuit 416 provides a reference value of about 5 volts to the MAP sensor, while 814 is the ground. Circuit 432 carries the signal to the ECM.

With A Scan Tool

◆ See Figure 94

1. Perform the On-Board Diagnostic System check in the General Diagnostic Test section.

2. Connect a scan tool as per the manufacturer's instructions.

3. With the ignition switch OFF, install a vacuum gauge to the manifold. Start the engine and increase the idle to 1000 rpm with the engine in Neutral. If the reading is steady and above 14 in. Hg., go to Step 4; if the reading fluctuates or is below 14 in. Hg., look for and repair a vacuum leak.

4. With the engine idling, check the voltage reading on the scan tool. If sensor voltage is greater than 4 volts, go to Step 5; if less than 4 volts, the fault is intermittent and you should check all sensor circuits for bad connections or scraped wires.

5. With the ignition switch OFF, disconnect the sensor wiring harness. Turn the ignition switch to the **ON** position (engine OFF). If the tool reads less than 1 volt, go to Step 6; if the tool reads more than 1 volt, you have a short to voltage in circuit 432 circuit or a bad connection at the ECM.

6. With the ignition OFF, connect a multi-meter (DVOM) between the harness terminal **A** (circuit 814) and **C** (circuit 416). Turn the ignition switch to the **ON** position. If sensor voltage is higher than 4 volts, check for a leaky or damaged vacuum fitting on the sensor, otherwise replace the sensor; if less than 4 volts, 814 circuit is open or you have a bad connection at the ECM.

Without A Scan Tool

◆ See Figure 94

1. Perform the On-Board Diagnostic System check in the General Diagnostic Test section.

2. With the ignition OFF, install a vacuum gauge to the manifold, start the engine and increase the idle to 1000 rpm. If the gauge reads steady at 14 in. Hg. or higher, go to Step 3; if it fluctuates or is lower, you have vacuum leak somewhere.

3. Connect a CodeMate Tester and switch it to the Normal mode. Turn the ignition switch OFF and disconnect the MAP sensor harness connector. Start the engine and allow it to idle for two minutes or until the tester indicates a stored code, whichever comes first. Turn the engine OFF, but leave the ignition switch in the **ON** position. Switch the tester to the Service mode and note the trouble code. If DTC 34, check for a leaking or damaged vacuum fitting on the sensor or replace the MAP sensor; if no DTC 34, go to Step 4.

4. With the Map sensor harness disconnected, turn the ignition switch ON. Connect a multi-meter (DVOM) between the harness terminals **A** (circuit 814) and **C** (circuit 416). If the reading is higher than 4 volts, you have a short to voltage in 432 circuit or a bad connection at the ECM; if lower than 4 volts, 814 circuit is open or you have a bad ECM connection.

CODE 34 - MANIFOLD ABSOLUTE PRESSURE SENSOR (MAP)

◆ See Figure 94

The MAP sensor responds to changes in pressure in the manifold (vacuum). This information is sent to the ECM as signal voltage and will vary from 1.0-2.0 volts at idle to about 4.0-5.0 volts at full throttle. The ECM compensates for a failing Map sensor by defaulting to a MAP value program that varies with rpm. Circuit 416 provides a reference value of about 5 volts to the MAP sensor, while 814 is the ground. Circuit 432 carries the signal to the ECM.

FUEL SYSTEM - FUEL INJECTION (EFI) 7-35

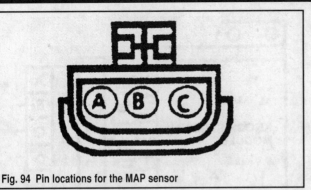

Fig. 94 Pin locations for the MAP sensor

MEFI-2/3 Engines With A Scan Tool

◆ See Figure 94

1. Perform the On-Board Diagnostic System check in the General Diagnostic Test section.
2. Connect a scan tool as per the manufacturer's instructions.
3. With the ignition switch OFF, install a vacuum gauge to the manifold. Start the engine an increase the idle to 1000 rpm with the engine in Neutral. If the reading is steady and above 14 in. Hg., go to Step 4; if the reading fluctuates or is below 14 in. Hg., look for and repair a vacuum leak.
4. With the engine idling, check the voltage reading on the scan tool. If sensor voltage is less than 1 volt, go to Step 5; if greater than 1 volt, check all sensor circuits for bad connections or scraped wires, if OK replace the MAP sensor.
5. With the ignition switch OFF, disconnect the sensor wiring harness. Connect a jumper wire between harness terminals **B** (circuit 432) and **C** (circuit 416). Turn the ignition ON (engine OFF). If the tool reads less than 4 volts, go to Step 6; if the tool reads more than 4 volts, check for a leaking or damaged vacuum fitting on the sensor, otherwise replace the sensor.
6. With the ignition OFF, connect a multi-meter (DVOM) between the harness terminal **C** and a known good ground. Turn the ignition switch to the ON position. If sensor voltage is higher than 4 volts, 432 circuit is open or shorted to voltage or you have a bad connection at the ECM; if less than 4 volts, 416 circuit is open or shorted to ground, or you have a bad connection at the ECM. You should also the TP sensor circuit 416 for a short to ground.

MEFI-2/3 Engines With A Scan Tool

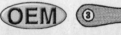 

◆ See Figure 94

1. Perform the On-Board Diagnostic System check in the General Diagnostic Test section.
2. Connect a CodeMate Tester and switch it to the Normal mode. Turn the ignition switch OFF and disconnect the MAP sensor harness connector. Connect a jumper wire between the harness terminals **B** (circuit 432) and **C** (circuit 416). Start the engine and increase the idle to 1000 rpm. Turn the engine OFF, but leave the ignition switch in the **ON** position and switch the tester to the Service mode and note the trouble code. If DTC 33, check the system circuits for bad connections or damaged wires, if OK replace the sensor; if no DTC 33, go to Step 3.
3. Remove the jumper wire and connect a multi-meter (DVOM) between the harness terminals **A** (circuit 814) and **C**. If the reading is higher than 4 volts, 432 circuit is open or you have a short to voltage, or a bad connection at the ECM; if lower than 4 volts, 614 circuit is open or shorted to ground, or you have a bad ECM connection. You should also check the TP sensor circuit 416 for a short to ground.

CODE 41 - IGNITION CONTROL CIRCUIT (IC)

◆ See Figure 95

When the system is running on the ignition module or in cranking mode there is no voltage through circuit 424 and the module grounds circuit 423. The ECM is programmed for low voltage in this condition so if it detects voltage, Code 41 is set and it will not move into the IC mode.
1. Perform the On-Board Diagnostic System check in the General Diagnostic Test section.
2. Clear the trouble code as detailed in this section. Start the engine and allow it to idle for two minutes or until the code sets itself again, whichever comes first. If DTC 41 sets, go to Step 3; if not, the fault is intermittent and you should check the system harness and connections for a bad connection.
3. With the ignition OFF, disconnect the ECM harness connectors. Connect a multi-meter (DVOM), set it to Ohms and probe the ECM circuit 423 to ground. If the resistance is over 3000 ohms, go to Step 4; if under 3000 ohms, go to Step 5.
4. Reconnect the ECM, start the engine and allow it to idle for two minutes or until DTC 41 is set, which ever comes first. If the code sets, move to Step 5; if not, check the system harness and wiring for a bad connection on circuit 423.
5. Look for circuit 423 to be open or a bad ECM connection. If you found either of these problems, fix it; if no problems were found, replace the IC module.

CODE 42 - IGNITION CONTROL CIRCUIT (IC)

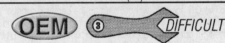

◆ See Figure 95

When the idle speed reaches the necessary rpm for IC and voltage is applied to circuit 424, circuit 423 should no longer be grounded. If 424 is open or grounded the module will not switch to IC mode, 423 voltage will be low and code 42 will then be set.
1. Perform the On-Board Diagnostic System check in the General Diagnostic Test section.
2. Clear the code as detailed in this section. Start the engine and allow it to idle for two minutes or until the code sets itself again. If DTC 42 sets itself again, go to Step 3; if not, circuit 424 is open or shorted to ground, or circuit 423 has a short.

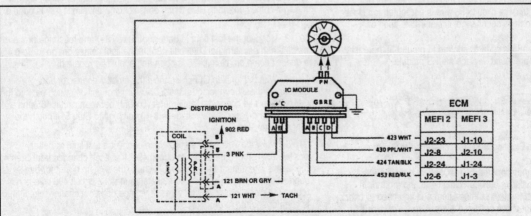

Fig. 95 Pin locations for the IC circuit

7-36 FUEL SYSTEM - FUEL INJECTION (EFI)

3. With the ignition OFF, disconnect the ECM harnesses. Connect a multi-meter, set it to Ohms and then ground circuit 423. If resistance is over 3000 ohms, go to Step 4; if under 3000 ohms, circuit 423 has a short to ground or there is a bad connection at the ECM.

4. With the DVOM still connected and grounding 423, connect a test light to the positive battery and touch the probe to circuit 424. The resistance should switch from over 3000 ohms to under 1000 ohms. If it does, circuit 424 is open or shorted to ground, or circuit 423 has a short; if it doesn't drop, go to Step 5.

5. Touch the test light probe to circuit 424. If it lights, go to Step 6; if not, circuit 424 is open.

6. With the light still connected from Step 5, disconnect the 4-wire connector at the module. If the light goes on, circuit 424 is shorted to ground or there is a bad connection at the ECM; if not, replace the IC module.

CODE 43 - KNOCK SENSOR (KS)

◆ See Figure 96

On models with MEFI-1/MEFI-2 ECM's, detonation or spark knock is sensed by the module which sends a voltage signal to the ECM. As the sensor detects knock, the voltage drops, signaling the ECM to begin retarding timing. The ECM will retard timing whenever knock is detected, and rpm and idle speed are above a specified level.

On models with MEFI-3 ECM's, detonation or spark knock is sensed by the sensor which sends a voltage signal to the ECM. As the sensor detects knock, the voltage increases, signaling the ECM to begin retarding timing. The ECM will retard timing whenever knock is detected, and rpm and idle speed are above a specified level. MEFI-3 ECM's are not equipped with a module.

Octane ratings that are to high may trip either of these codes.

Engines With MEFI-1/2 ECM

◆ See Figure 96

1. Perform the On-Board Diagnostic System check in the General Diagnostic Test section.

2. Disconnect the 5-wire KS module connector. Turn the ignition switch to the **ON** position (without starting the engine), connect a test light to ground and then touch the probe to the module harness terminal **B** (circuit 439). If the light comes on, go to Step 3; if it doesn't come on, circuit 439 is open or shorted to ground.

3. Connect the test light to battery positive (**B+**) and touch the probe to module harness terminal **D** (circuit 486). If the light comes on, go to Step 4; if it doesn't come on, circuit 486 is open and should be repaired.

4. With the ignition OFF, reconnect the module connector and disconnect the **J1** connector at the ECM. Turn the ignition switch back to the **ON** position and connect a multi-meter (DVOM) between circuit 485 and a known ground. If the voltage is 8-10 volts, go to Step 5; if not in the range, go to Step 6.

5. Allow the voltage to stabilize and connect a test light to battery positive (**B+**). Touch the probe to circuit 496. If voltage changes from Step 4, the fault is intermittent, check the system circuit for bad connections or frayed wires. Also make sure that the circuit 496 wire is not routed too close to any ignition wires. If voltage does not change, circuit 496 is open or shorted to ground.

6. Check to see if circuit 485 is open or shorted to ground. If there is a problem, fix it; if no problems are found, replace the KS module.

Engines W/MEFI-3 ECM

◆ See Figure 96

1. Perform the On-Board Diagnostic System check in the General Diagnostic Test section.

2. Disconnect the **J1** connector at the ECM. Connect a multi-meter (DVOM) and measure the resistance between circuit 496 and ground. If resistance is 3000-5000 ohms, the fault is intermittent and you need to check the system circuit for bad connections or frayed wires. Also make sure that the circuit 496 wire is not routed too close to any ignition wires. If voltage does not change, go to Step 3.

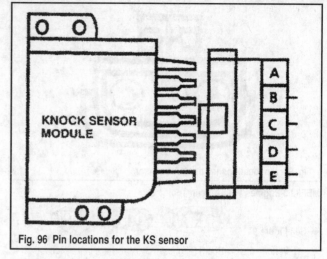

Fig. 96 Pin locations for the KS sensor

3. Disconnect the KS connector and measure the resistance between the sensor and ground. If resistance is 3000-5000 ohms, circuit 496 is open or shorted to ground; if resistance is not as specified, replace the KS sensor.

CODE 44 - KNOCK SENSOR (KS)

◆ See Figure 96

On models with MEFI-2 ECM's, detonation or spark knock is sensed by the module which sends a voltage signal to the ECM. As the sensor detects knock, the voltage drops, signaling the ECM to begin retarding timing. The ECM will retard timing whenever knock is detected, and rpm and idle speed are above a specified level.

On models with MEFI-3 ECM's, detonation or spark knock is sensed by the sensor which sends a voltage signal to the ECM. As the sensor detects knock, the voltage increases, signaling the ECM to begin retarding timing. The ECM will retard timing whenever knock is detected, and rpm and idle speed are above a specified level. MEFI-3 ECM's are not equipped with a module.

Octane ratings that are too high may trip either of these codes.

Engines W/MEFI-2 ECM

◆ See Figure 96

1. Perform the On-Board Diagnostic System check in the General Diagnostic Test section.

2. Disconnect the 5-wire KS module connector. Turn the ignition switch to the **ON** position (without starting the engine), connect a test light to ground and then touch the probe to the module harness terminal **B** (circuit 439). If the light comes on, go to Step 3; if it doesn't come on, circuit 439 is open or shorted to ground.

3. Connect the test light to battery positive (**B+**) and touch the probe to module harness terminal **D** (circuit 486). If the light comes on, go to Step 4; if it doesn't come on, circuit 486 is open and should be repaired.

4. With the ignition OFF, reconnect the module connector and disconnect the **J1** connector at the ECM. Turn the ignition switch back to the **ON** position and connect a multi-meter (DVOM) between circuit 485 and a known good ground. If the voltage is 8-10 volts, go to Step 5; if not in the range, go to Step 6.

5. Allow the voltage to stabilize and connect a test light to battery positive (**B+**). Touch the probe to circuit 496. If voltage changes from Step 4, the fault is intermittent and you should check the system circuit for bad connections or frayed wires. Also make sure that the circuit 496 wire is not routed too close to any ignition wires. If voltage does not change, circuit 496 is open or shorted to ground.

6. Check to see if circuit 483 is open or shorted to ground. If there is a problem, fix it; if no problems are found, replace the KS module.

FUEL SYSTEM - FUEL INJECTION (EFI)

Engines W/MEFI-3 ECM

◆ See Figure 96

1. Perform the On-Board Diagnostic System check in the General Diagnostic Test section.

2. Disconnect the **J1** connector at the ECM. Connect a multi-meter (DVOM) and measure the resistance between circuit 496 and ground. If resistance is 3000-5000 ohms, the fault is intermittent and you should check the system circuit for bad connections or frayed wires. Also make sure that the circuit 496 wire is not routed too close to any ignition wires. If resistance is not within specifications, go to Step 3.

3. Disconnect the KS connector and measure the resistance between the sensor and ground. If resistance is 3000-5000 ohms, circuit 496 is open or shorted to ground; if resistance is not as specified, replace the KS sensor.

CODE 45 - IGNITION COIL DRIVER

On models with an MEFI-3 ECM, coil driver circuitry is integrated into the control module. Whenever the engine is running, a diagnostic check of the driver circuit is ongoing. If the system detects excessive current to the coil via the 121 circuit, it will set a code 45, but allow the engine to continue running.

1. Perform the On-Board Diagnostic System check in the General Diagnostic Test section.

2. With the ignition switch in the **ON** position and the engine OFF, connect a multi-meter (DVOM) at the positive (+) terminal on the coil. If voltage is greater than 12 volts, go to Step 3; if less than 12 volts, circuit 902 is open or shorted to ground.

3. With the ignition switch OFF, disconnect the **J1** connector at the ECM. Check the continuity between the ECM circuit 121 and the negative (-) terminal on the coil. If resistance is less than an ohm, go to Step 4; if more than an ohm, circuit 121 is open and should be repaired.

4. With the ignition switch OFF and the **J1** terminal still disconnected, check the resistance between circuit 121 and ground. If resistance is less than an ohm, circuit 121 is shorted to ground. If more than an ohm, go to Step 5.

5. With the ignition switch OFF and the **J1** terminal still disconnected, check the resistance between the positive (+) and negative (-) terminals on the coil. If less than 2 ohms, go to Step 6; if more than 2 ohms, replace the ignition coil.

6. With the ignition switch OFF and the **J1** terminal still disconnected, remove the secondary ignition wire at the coil and check the resistance between the secondary coil tower and the negative coil terminal. Resistance should be between 5 ohms and 15 ohms. If it is, the fault is intermittent and you should check the system circuits for bad connections or frayed wires. If it is not within the range, replace the ignition coil.

CODE 51 - ECM CALIBRATION MEMORY FAILURE

■ **If DTC 51 has shown more than once but is intermittent, replace the ECM.**

Perform the On-Board Diagnostic System check in the General Diagnostic Test section.

Turn the ignition switch to the **ON** position and clear the code as detailed elsewhere in this section. If the code resets itself, replace the ECM; if not, the fault is intermittent and you should check all system circuits for bad connections and/or frayed wires.

CODE 52 - ECM EEPROM FAILURE

■ **If DTC 52 has shown more than once but is intermittent, replace the ECM.**

Perform the On-Board Diagnostic System check in the General Diagnostic Test section.

Clear the code as detailed elsewhere in this section and turn the ignition switch to the **ON** position. If the code resets itself, replace the ECM; if not, check all system circuits for bad connections and/or frayed wires.

CODE 61 - FUEL PRESSURE (FP) CIRCUIT

General Information

◆ See Figure 97

The FP sensor sends a voltage signal that changes as fuel pressure does. Signal voltage should always be 2.5-3.5 volts.

With A Scan Tool

◆ See Figure 97

1. Perform the On-Board Diagnostic System check in the General Diagnostic Test section.

2. Connect a scan tool as per the manufacturer's instructions and then turn the ignition switch to the **ON** position. If sensor voltage is higher than 4 volts, go to Step 4; if lower than 4 volts, but higher than 0.3 volts, the fault is intermittent and you should check the system circuit for bad connections or frayed wires. If sensor voltage is lower than 0.3 volts, go to Step 3; if not, the fault is intermittent and you should check the system circuit for bad connections or frayed wires.

3. Turn the ignition OFF and disconnect the connector at the pressure sensor. Connect a jumper wire between terminals **B** and **C**, then turn the ignition ON. If the voltage reads higher than 4 volts, replace the sensor; if lower than 4 volts, go to Step 5.

4. Turn the ignition OFF and disconnect the connector at the sensor. Turn the ignition switch to the **ON** position. If the reading is higher than 4 volts, check if circuit 475 is shorted to voltage; if less than 4 volts, go to Step 5.

5. Turn the ignition ON and connect a multi-meter (DVOM) between sensor harness terminals **A** (ground) and **B** (5 V reference). If the reading is over 4 volts, go to Step 6; if lower than 4 volts, go to Step 9.

6. Connect the meter between harness terminals **B** (circuit 416) and **C** (circuit 475). If the reading is over 4 volts, go to Step 7; if lower than 4 volts, go to Step 8.

7. Turn the ignition OFF and connect a test light to the battery positive (**B+**). Touch the probe to terminal **C**. If the light goes on, move to Step 10; if it does not light, replace the sensor.

8. Connect the multi-meter between harness terminal **C** and an engine ground. If the reading is over 4 volts, circuit 475 is shorted to voltage. If under 4 volts, circuit 475 is open or you have a bad connection at the ECM.

9. Now move the meter probe to terminal **B**. If over 4 volts, circuit 813 is open or you have a bad connection at the ECM. If under 4 volts, circuit 416 is open or shorted to ground, or you have a bad connection at the ECM.

10. Disconnect the ECM and then touch the test light probe to sensor terminal **C**. If it lights, circuit 475 is shorted to ground; if it doesn't light, circuit 475 is open or you have a bad ECM connection.

Without A Scan Tool

◆ See Figure 97

1. Perform the On-Board Diagnostic System check in the General Diagnostic Test section.

2. Turn the ignition switch to the **ON** position and connect a multi-meter (DVOM) between sensor harness terminals **A** and **B**. If the reading is over 4 volts, go to Step 3; if lower than 4 volts, go to Step 5.

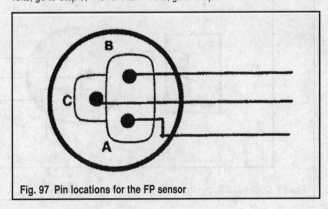

Fig. 97 Pin locations for the FP sensor

7-38 FUEL SYSTEM - FUEL INJECTION (EFI)

3. Connect the meter between harness terminals **B** and **C**. If the reading is over 4 volts, go to Step 4; if lower than 4 volts, go to Step 7.

4. Turn the ignition OFF and connect a test light to the battery positive (**B+**). Touch the probe to terminal **C**. If the light goes on, move to Step 6; if it does not light, replace the sensor.

5. Now move the meter probe to terminal **B**. If over 4 volts, circuit 813 is open or you have a bad connection at the ECM; if under 4 volts, circuit 416 is open or shorted to ground, or you have a bad connection at the ECM.

6. Disconnect the ECM and then touch the test light probe to sensor terminal **C**. If it lights, circuit 475 is shorted to ground; if it doesn't light, circuit 475 is open or you have a bad ECM connection.

CODE 62 - FUEL PRESSURE (FP) SENSOR CIRCUIT

◆ See Figure 97

The FP sensor sends a signal that changes as fuel pressure does. Signal voltage should always be 2.5-3.5 volts.

With A Scan Tool

◆ See Figure 97

1. Perform the On-Board Diagnostic System check in the General Diagnostic Test section.

2. Connect a scan tool as per the manufacturer's instructions. Turn the ignition OFF. If the sensor voltage is less than 0.3 volts, go to Step 3; if greater than 0.3 volts, the fault is intermittent and you should check the system circuit for bad connections or frayed wires.

3. With the ignition still OFF, disconnect the sensor harness connector. Connect a jumper wire between harness terminals **B** (circuit 416) and **C** (circuit 475). Turn the ignition switch to the **ON** position without starting the engine. If the tool shows voltage greater than 4 volts, check the system circuit for bad connections or frayed wires, if OK replace the sensor. if less than 4 volts, go to Step 4.

4. Turn the ignition OFF again and connect a multi-meter (DVOM) between terminal **B** and a known ground. If voltage is greater than 4 volts, circuit 475 is open or shorted to ground, or you have a bad connection at the ECM. If less than 4 volts, circuit 416 is open or shorted to ground, or you have a bad connection at the ECM.

Without A Scan Tool

◆ See Figure 97

1. Perform the On-Board Diagnostic System check in the General Diagnostic Test section.

2. Connect a CodeMate tester and set it in the Normal mode. Turn the ignition OFF and disconnect the harness at the sensor.

3. Connect a jumper wire between terminals **B** (circuit 416) and **C** (circuit 475), then start the engine and let it idle for two minutes or until a stored code comes up; whichever comes first.

4. Turn the engine OFF but leave the ignition switch in the **ON** position. Switch the tester to the Service mode. If DTC 61 shows, check the system circuit for bad connections or frayed wires, if OK replace the sensor. If no DTC 61, circuit 475 is open or shorted to ground, or you have a bad connection at the ECM.

Diagnostic Trouble Code Test Schematics

■ Please use these schematics when performing the individual fault tests under each code.

◆ See Figures 98 thru 105

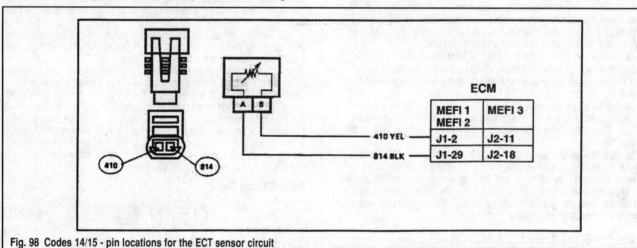

Fig. 98 Codes 14/15 - pin locations for the ECT sensor circuit

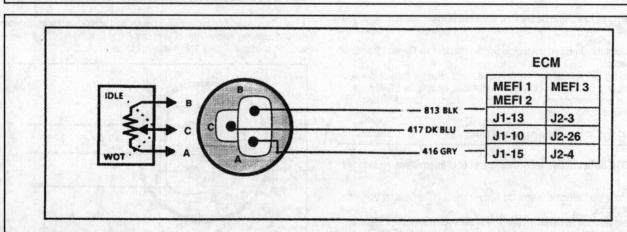

Fig. 99 Codes 21/22 - pin locations for the TP sensor circuit

FUEL SYSTEM - FUEL INJECTION (EFI) 7-39

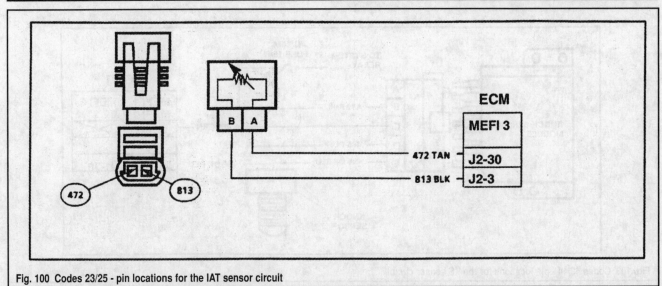

Fig. 100 Codes 23/25 - pin locations for the IAT sensor circuit

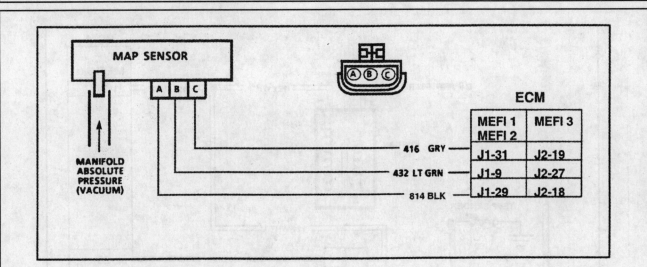

Fig. 101 Codes 33/34 - pin locations for the MAP sensor circuit

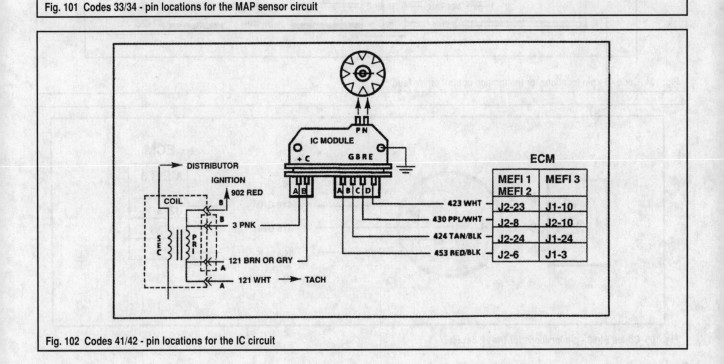

Fig. 102 Codes 41/42 - pin locations for the IC circuit

7-40 FUEL SYSTEM - FUEL INJECTION (EFI)

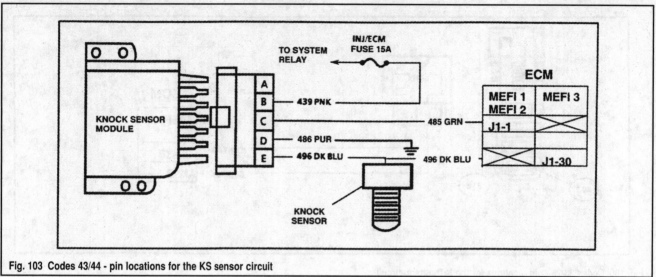

Fig. 103 Codes 43/44 - pin locations for the KS sensor circuit

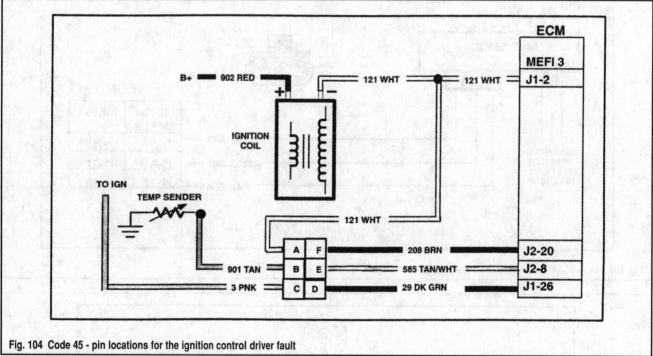

Fig. 104 Code 45 - pin locations for the ignition control driver fault

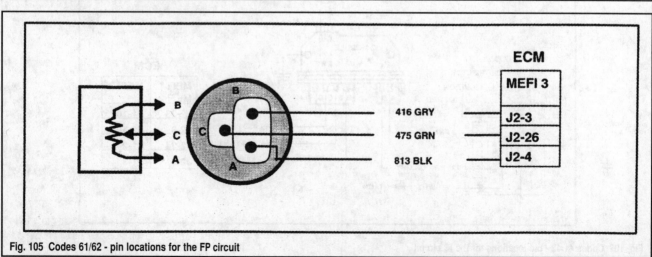

Fig. 105 Codes 61/62 - pin locations for the FP circuit

FUEL SYSTEM - FUEL INJECTION (EFI) 7-41

Vacuum Diagrams

◆ See Figures 106, 107 and 108

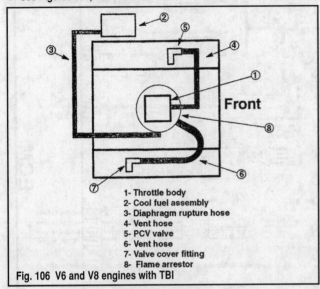

1- Throttle body
2- Cool fuel assembly
3- Diaphragm rupture hose
4- Vent hose
5- PCV valve
6- Vent hose
7- Valve cover fitting
8- Flame arrestor

Fig. 106 V6 and V8 engines with TBI

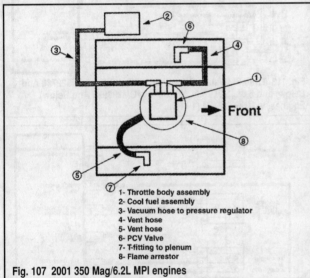

1- Throttle body assembly
2- Cool fuel assembly
3- Vacuum hose to pressure regulator
4- Vent hose
5- Vent hose
6- PCV Valve
7- T-fitting to plenum
8- Flame arrestor

Fig. 107 2001 350 Mag/6.2L MPI engines

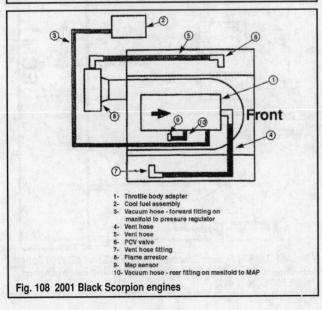

1- Throttle body adapter
2- Cool fuel assembly
3- Vacuum hose - forward fitting on manifold to pressure regulator
4- Vent hose
5- Vent hose
6- PCV valve
7- Vent hose fitting
8- Flame arrestor
9- Map sensor
10- Vacuum hose - rear fitting on manifold to MAP

Fig. 108 2001 Black Scorpion engines

Wiring Schematics

◆ See Figures 109 thru 116

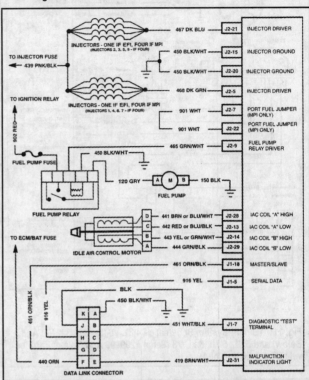

Fig. 109 TBI/MPI System Schematic - 4.3L V6 (Serial #322780 And Below) and 5.0L, 5.7L, 6.2L V8 (Serial #299999 And Below) Engines, w/MEFI-1/2 ECM (1 of 4)

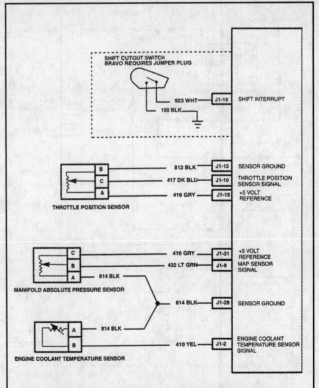

Fig. 110 TBI/MPI System Schematic - 4.3L V6 (Serial #322780 And Below) and 5.0L, 5.7L, 6.2L V8 (Serial #299999 And Below) Engines, w/MEFI-1/2 ECM (2 of 4)

7-42 FUEL SYSTEM - FUEL INJECTION (EFI)

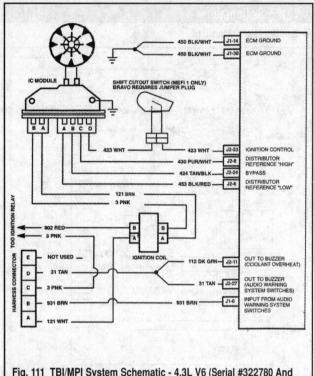

Fig. 111 TBI/MPI System Schematic - 4.3L V6 (Serial #322780 And Below) and 5.0L, 5.7L, 6.2L V8 (Serial #299999 And Below) Engines, w/MEFI-1/2 ECM (3 of 4)

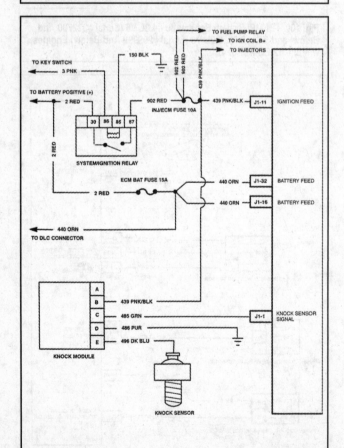

Fig. 112 TBI/MPI System Schematic - 4.3L V6 (Serial #322780 And Below) and 5.0L, 5.7L, 6.2L V8 (Serial #299999 And Below) Engines, w/MEFI-1/2 ECM (4 of 4)

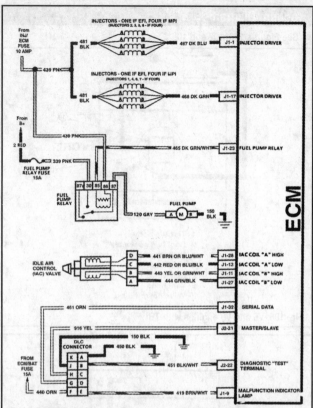

Fig. 113 TBI/MPI System Schematic - 4.3L V6 (Serial #322780 And Below) and 5.0L, 5.7L, 6.2L V8 (Serial #OM299999 And Below) Engines, w/MEFI-3 ECM (1 of 4)

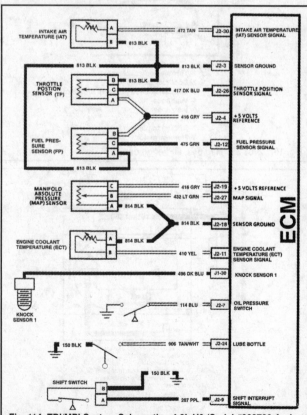

Fig. 114 TBI/MPI System Schematic - 4.3L V6 (Serial #322780 And Below) and 5.0L, 5.7L, 6.2L V8 (Serial #299999 And Below) Engines, w/MEFI-3 ECM (2 of 4)

FUEL SYSTEM - FUEL INJECTION (EFI)

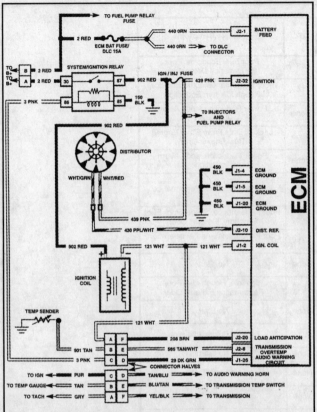

Fig. 115 TBI/MPI System Schematic - 4.3L V6 (Serial #322780 And Below) and 5.0L, 5.7L, 6.2L V8 (Serial #299999 And Below) Engines, w/MEFI-3 ECM, Mercury Distributor (3 of 4)

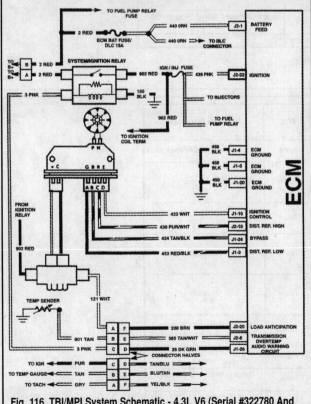

Fig. 116 TBI/MPI System Schematic - 4.3L V6 (Serial #322780 And Below) and 5.0L, 5.7L, 6.2L V8 (Serial #299999 And Below) Engines, w/MEFI-3 ECM, GM EST Distributor (4 of 4)

ECM Pin Locations & Symptoms Charts

◆ See Figures 117 thru 125

The accompanying charts should help you in the diagnosis of symptoms related to the fuel injection system on your engine. All voltages have been derived from a known-good engine with the electrical system intact and operating.

✳✳ CAUTION

Never attempt to obtain these voltages by probing wires or connectors - they are at the pin connector only. Voltage may also vary depending on battery condition, so make sure that the battery is fully charged.

■ Only pins used are shown, if a pin is not listed, then it's not used in these systems. Not all pins are used on all models, and they should be noted as such.

The following conditions must be met prior to beginning any testing:
- The engine should be at normal operating temperature
- The ignition must be ON or the engine must be running
- There should be no scan tools connected

The following Notes apply to all charts, where called for. When the **B+** symbol is shown in the charts, it means normal battery voltage.
- NOTE 1: Battery voltage for 2 seconds and then 0 volts
- NOTE 2: Varies with temperature
- NOTE 3: Varies with manifold vacuum
- NOTE 4: Varies with throttle movement
- NOTE 5: Less than 0.5 volts (500 mV).

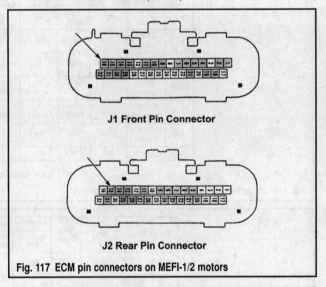

Fig. 117 ECM pin connectors on MEFI-1/2 motors

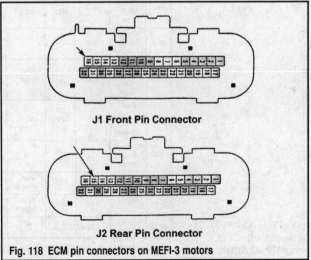

Fig. 118 ECM pin connectors on MEFI-3 motors

7-44 FUEL SYSTEM - FUEL INJECTION (EFI)

Pin	Pin Function	CKT	Wire Color	Normal Voltage Ignition ON	Normal Voltage Engine Running	DTC	Possible Symptoms
J1-1	Knock Sensor	485	GRN	9.5V	9.5V	43	Poor Fuel Economy, Poor Performance Detonation
J1-2	ECT Signal	410	YEL	1.95V (NOTE 2)	1.95V (NOTE 2)	14	Poor Performance, Exhaust Odor, Rough Idle rpm Reduction
J1-4	Discrete Switch	931	BRN	-	-	None	Power Reduction Mode Alarm Activation
J1-5	Master/Slave	916	YEL	B+	B+	None	Lack Of Data From Other Engine (Dual Engine Only)
J1-6	Discrete Switch	931	BRN	-	-	None	Power Reduction Mode Alarm Activation
J1-7	Diagnostic Test	451	BLK/WHT	B+	B+	None	Incorrect Idle, Poor Performance
J1-9	Map Signal	432	LT GRN	4.9V	1.46V (NOTE 3)	33	Poor Performance, Surge, Poor Fuel Economy, Exhaust Odor
J1-10	TP Signal	417	DK BLU	.62V (NOTE 4)	.62V (NOTE 4)	21	Poor Performance And Acceleration, Incorrect Idle
J1-11	Ignition Fused	439	PNK/BLK	B+	B+	None	No Start
J1-13	Sensor Ground	813	BLK	0 (NOTE 5)	0 (NOTE 5)	21,23	High Idle, Rough Idle, Poor Performance Exhaust Odor
J1-14	ECM Ground	450	BLK/WHT	0 (NOTE 5)	0 (NOTE 5)	None	No Start
J1-15	TP 5V Power	416	GRY	5V	5V	21	Lack Of Power, Idle High
J1-16	Battery	440	ORN	B+	B+	None	No Start
J1-18	Serial Data	461	ORN/BLK	5V	5V	None	No Serial Data
J1-19	Shift Switch	923	WHT	0	0	None	Incorrect Idle
J1-21	Lanyard Stop	942	PNK	0	0	None	No Start
J1-29	MAP Ground	814	BLK	0 (NOTE 5)	0 (NOTE 5)	33	Lack Of Performance, Exhaust Odor, Stall
J1-30	ECM Ground	450	BLK/WHT	0 (NOTE 5)	0 (NOTE 5)	None	No Start
J1-31	MAP 5V Power	416	GRY	5V	5V	33	Lack Of Power, Surge, Rough Idle, Exhaust Odor
J1-32	Battery	440	ORN	B+	B+	None	No Start

See Text for Notes

Fig. 119 J1 connector circuit tests on MEFI-1/2 motors

Pin	Pin Function	CKT	Wire Color	Normal Voltage Ignition ON	Normal Voltage Engine Running	DTC	Possible Symptoms
J2-5	Injector Driver	468	LT GRN	B+	B+	None	Rough Idle, Lack Of Power, Stall
J2-6	Ignition Control Ref. Low	463	RED/BLK	0 (NOTE 5)	0 (NOTE 5)	None	Poor Performance
J2-7	Port Fuel Jumper	901	WHT	-	-	None	-
J2-8	Ignition Control Ref. High	430	PUR/WHT	5V	1.6V	None	No Restart
J2-9	Fuel Pump Relay Driver	465	DK GRN/WHT	0 (NOTE 1&5)	B+	None	No Start
J2-11	Coolant Over temp.	112	DK GRN	0	0	None	Power Reduction Mode or Improper Audio Warning
J2-13	IAC "A" Low	442	RED	Not Usable	Not Usable	None	Rough Unstable or Incorrect Idle
J2-14	IAC "B" Low	443	YEL	Not Usable	Not Usable	None	Rough Unstable or Incorrect Idle
J2-15	Injector Ground	450	BLK/WHT	0 (NOTE 5)	0 (NOTE 5)	None	Rough Running, Lack Of Power, Poor Performance

Fig. 120 J2 connector circuit tests on MEFI-1/2 motors   See Text for Notes

FUEL SYSTEM - FUEL INJECTION (EFI)

Pin	Pin Function	CKT	Wire Color	Normal Voltage Ignition ON	Normal Voltage Engine Running	DTC	Possible Symptoms
J2-20	Fuel Injector Ground	450	BLK/WHT	0 (NOTE 5)	0 (NOTE 5)	None	Rough Running, Poor Idle, Lack Of Performance
J2-21	Injector Driver	467	DK BLU	B+	B+	None	Rough Idle, Lack Of Power, Stalling
J2-22	Port Fuel Jumper	901	WHT	–	–	–	–
J2-23	Ignition Control Signal	423	WHT	0 (NOTE 5)	1.2V	42	Stall, Will Restart In Bypass Mode, Lack Of Power
J2-24	Ignition Control Bypass	424	TAN/BLK	0 (NOTE 5)	4.5V	42	Lack Of Power, Fixed Timing
J2-26	Discrete Switch	31	TAN	–	–	–	Audio Warning System Activation
J2-27	Discrete Switch	31	TAN	–	–	–	Audio Warning System Activation
J2-28	IAC "A" High	441	BRN	Not Usable	Not Usable	None	Rough, Unstable or Incorrect Idle
J2-29	IAC "B" Low	444	GRN/BLK	Not Usable	Not Usable	None	Rough, Unstable or Incorrect Idle
J2-31	MIL Lamp	419	BRN/WHT	0 (NOTE 5)	0 (NOTE 5)	None	Lamp Inoperative

NOTE: J2-22 is not used on the Throttle Body Injection system.

See Text For Notes

Fig. 121 J2 connector circuit tests on MEFI-1/2 motors (cont'd)

Pin	Pin Function	CKT	Wire Color	Normal Voltage Ignition ON	Normal Voltage Engine Running	DTC	Possible Symptoms
J1-1	Injector Driver	467	DK BLU	B+	B+	None	Rough Idle, Lack of Power, Stalling
J1-2	Ignition Coil	121	WHT	Not Usable	Not Usable	45	Rough Running, Poor Idle, Lack of Performance
J1-3	Ignition Control Ref. Low	453	RED/BLK	0 (NOTE 5)	0 (NOTE 5)	None	Poor Performance
J1-4	ECM Ground	450	BLK	0 (NOTE 5)	0 (NOTE 5)	None	No Start
J1-5	ECM Ground	450	BLK	0 (NOTE 5)	0 (NOTE 5)	None	No Start
J1-9	MIL Lamp	419	BRN/WHT	0 (NOTE 5)	0 (NOTE 5)	None	Lamp Inoperative
J1-10	Ignition Control Signal	423	WHT	0 (NOTE 5)	1.2V	42	Stall, Will Restart In Bypass Mode, Lack Of Power
J1-11	IAC "B" Low	443	GRN/WHT	Not Usable	Not Usable	None	Rough Unstable or Incorrect Idle
J1-12	IAC "A" Low	442	BLU/BLK	Not Usable	Not Usable	None	Rough Unstable or Incorrect Idle
J1-17	Injector Driver	468	DK GRN	B+	B+	None	Rough Idle, Lack Of Power, Stall
J1-20	ECM Ground	450	BLK	0 (NOTE 5)	0 (NOTE 5)	None	Rough Running, Poor Idle, Lack Of Performance

See Text For Notes

Fig. 122 J1 connector circuit tests on MEFI-3 motors

7-46 FUEL SYSTEM - FUEL INJECTION (EFI)

Pin	Pin Function	CKT	Wire Color	Normal Voltage Ignition ON	Normal Voltage Engine Running	DTC	Possible Symptoms
J1-23	Fuel Pump Relay Driver	465	DK GRN/WHT	0 (NOTES 1&5)	B+	None	No Start
J1-24	Ignition Control Bypass	424	TAN/BLK	0 (NOTE 5)	4.5V	42	Lack Of Power, Fixed Timing
J1-26	Audio Warning Horn	29	DK GRN	–	–	None	–
J1-27	IAC "B" Low	444	GRN/BLK	Not Usable	Not Usable	None	Rough Unstable or Incorrect Idle
J1-28	IAC "A" High	441	BLU/WHT	Not Usable	Not Usable	None	Rough Unstable or Incorrect Idle
J1-30	Knock Sensor Signal	496	BLU	–	–	43, 44	Poor Fuel Economy, Poor Performance Detonation
J1-32	Serial Data	461	ORN	5V	5V	None	No Serial Data

See Text for Notes

Fig. 123 J1 connector circuit tests on MEFI-3 motors (cont'd)

Pin	Pin Function	CKT	Wire Color	Normal Voltage Ignition ON	Normal Voltage Engine Running	DTC	Possible Symptoms
J2-1	Battery	440	ORN	B+	B+	None	No Start
J2-3	Sensor Ground	813	BLK	0 (NOTE 5)	0 (NOTE 5)	21, 23	High Idle, Rough Idle, Poor Performance Exhaust Odor
J2-4	TP 5V Power	416	GRY	5V	5V	21	Lack Of Power, Idle High
J2-7	Discrete Switch	114	BLU	–	–	None	–
J2-8	Discrete Switch	585	TAN/WHT	–	–	None	–
J2-9	Shift Switch	923	WHT	0	0	None	Incorrect Idle
J2-10	Ignition Control Ref. High	430	PUR/WHT	5V	1.6V	None	No Restart
J2-11	ECT Signal	410	YEL	1.95V (NOTE 2)	1.95V (NOTE 2)	14	Poor Performance, Exhaust Odor, Rough Idle rpm Reduction
J2-12	Fuel Pressure	475	GRN	3V	3V	61, 62	
J2-18	MAP Ground	814	BLK	0 (NOTE 5)	0 (NOTE 5)	33	Lack Of Performance, Exhaust Odor, Stall
J2-19	MAP 5V Reference	416	GRY	5V	5V	33	Lack Of Power, Surge, Rough Idle, Exhaust Odor
J2-20	Discrete Switch	208	BRN	–	–	–	–

See Text for Notes

Fig. 124 J2 connector circuit tests on MEFI-3 motors

Pin	Pin Function	CKT	Wire Color	Normal Voltage Ignition ON	Normal Voltage Engine Running	DTC	Possible Symptoms
J2-21	Master/Slave	916	YEL	B+	B+	None	Lack Of Data From Other Engine (Dual Engine Only)
J2-22	Diagnostic Test	451	BLK/WHT	B+	B+	None	Incorrect Idle, Poor Performance
J2-24	Discrete Switch	906	TAN/WHT	–	–	None	
J2-26	TP Signal	417	DK BLU	.62V (NOTE 4)	.62V (NOTE 4)	21	Poor Performance And Acceleration, Incorrect Idle
J2-27	Map Signal	432	LT GRN	4.9V	1.46V (NOTE 3)	33	Poor Performance, Surge, Poor Fuel Economy, Exhaust Odor
J2-30	IAT Sensor	472	TAN	5V	(NOTE 2)	23	Poor Fuel Economy, Exhaust Odor
J2-32	Ignition Fused	439	PNK	B+	B+	None	No Start

See Text for Notes

Fig. 125 J2 connector circuit tests on MEFI-3 motors (cont'd)

FUEL SYSTEM - FUEL INJECTION (EFI) 7-47

SYSTEM DIAGNOSIS - 2001-08 MPI ENGINES

◆ See Figures 126 thru 130

■ The procedures contained in this section are valid for all 4.3L V6 MPI engine above serial #OM322781, also for all 5.0L, 5.7L and 6.2L V8 MPI engines above serial #OM300000 and for all 8.1L V8 engines.

Electronic fuel injection systems on these engines are equipped with an ECM/PCM (555) that provides state of the art control of fuel and spark delivery throughout all engine operating conditions.

These ECMs are the control center of the systems, constantly monitoring input from all sensors and system controls affecting engine performance throughout it's range of operation.

The ECM performs a constant and ongoing diagnostic function check of the systems; recognizing operating problems and storing fault codes identifying all problem areas.

■ These are not the numerical or flash codes that you are used to with previous systems.

Almost all sensors and switches are supplied 5 or 12 volts by the ECM; consequently, a 10 meg-ohm input impedance DVOM is necessary for all diagnostic work.

Coupled with this system is an Engine Guardian system which is the mainstay of the self-diagnostic system on all engines covered here. Guardian protects the engine from possible damage resulting from faulty conditions - the system monitors the sensors and if a malfunction occurs, the fault description will then be stored in the ECM and available power will usually be reduced.

Additionally, there is an Audio Warning System that will sound an alarm whenever a malfunction fault is stored. Whenever the ignition switch is turned to the **ON** position, the alarm will momentarily activate and test the warning system. The alarm will sound once if the system is functioning correctly. Please refer to the accompanying charts for deciphering any additional alarms.

Fault	Smartcraft Gauges	Audio Alarm	Available Power	Description
Main Power Relay Output	Yes	No	N/A	Engine will not start
Main Power Relay Backfeed	Yes	No	N/A	Engine will not start
MAP Sensor 1 Input High	Yes	2 Bp/min	90%	High voltage or short
MAP Sensor 1 Input Low	Yes	2 Bp/min	90%	Open, no visual on SC1000
Oil PSI CKT Hi	Yes	2 Bp/min	90%	Open, defaults to 50.7 psi
Oil PSI CKT Lo	Yes	2 Bp/min	90%	Short, defaults to 50.7 psi
Overspeed	Yes	Constant	rpm limit	Engine over rpm limit
Pitot CKT Hi	No	No	100%	Short or high voltage
Pitot CKT Lo	No	No	100%	Open
Sea Pump PSI Lo	Yes	Constant	6-100%	Guardian Strategy
Sea Pump CKT Hi	Yes	2 Bp/min	90%	Open - 0 psi reading
Sea Pump CKT Lo	Yes	2 Bp/min	90%	Voltage high or short
Sea Water Temp	No	No	N/A	Defaults to -31 degrees C
Fuel Level #1	No	No	N/A	Only if turned on
STB EMCT CKT Hi	N/A	N/A	N/A	N/A
STB EMCT CKT Lo	N/A	N/A	N/A	N/A
STB EMCT CKT Overheat	N/A	N/A	N/A	N/A

NOTE: *If any 5v sensor becomes shorted to ground the engine will not start. If the engine is operating when the short occurs the engine may stop operating and will not start.*

Fig. 127 Engine Guardian audible warning fault codes - V6 and V8 MPI engines (exc. 8.1L) (cont'd)

Fault	Smartcraft Gauges	Audio Alarm	Available Power	Description
ECT CKT HI	Yes	2 Bp/min	90%	Open
ECT CKT LO	Yes	2 Bp/min	90%	Short
ECT Coolant Overheat	Yes	Constant	6 - 100 %	Engine guardian overheat condition
EST 1 Open ①	Yes	2 Bp/min	100%	Coil harness wire open
EST 1 Short ①	Yes	2 Bp/min	100%	Coil harness wire short
Fuel Injector 1-7-4-6 Open	Yes	2 Bp/min	100%	Fuel injector wire open
Fuel Injector 3-5-2-8 Open	Yes	2 Bp/min	100%	Fuel injector wire open
Guardian Strategy	Yes	Constant	0% - 100%	Protection Strategy
IAC Output LO/HI ②	Yes	2 Bp/min	90%	Open
Knock Sensor 1 Lo	Yes	2 Bp/min	90%	Open
Knock Sensor 1 Hi	Yes	2 Bp/min	90%	Short
Low Drive Lube Strategy	Yes	Constant	100%	Low oil in sterndrive
Low Oil Pressure Strategy	Yes	Constant	0 - 100%	Low oil pressure strategy

NOTE: *If any 5v sensor becomes shorted to ground the engine will not start. If the engine is operating when the short occurs the engine may stop operating and will not start.*

① *GM EFI ignition system failure open or shorted, driver will flag EST 1 fault.*
② *Tps must see 5 % throttle then back to 0 % to flag IAC fault.*

Fig.126 Engine Guardian audible warning fault codes - V6 and V8 MPI engines (exc. 8.1L)

Fault	Smartcraft Gauges	Audio Alarm	Available Power	Description
Steer CKT Hi	No	No	100%	Short or high voltage
Steer CKT Lo	No	No	100%	Open
TPS1 CKT or Range Hi	Yes	2 Bp/min	90%	Short or high voltage
TPS1 CKT or Range Lo	Yes	2 Bp/min	90%	Open or low voltage
Trim CKT or Range Hi	Yes	No	100%	Open or high voltage
Trim CKT or Range Lo	Yes	No	100%	Short
5 VDC PWR Low ①	Yes	2 Bp/min	90%	Short or low - engine may not start
MAT Sensor Hi	Yes	2 Bp/min	90%	Open - default to -32 degrees F
MAT Sensor Lo	Yes	2 Bp/min	90%	Short - default to -32 degrees F
Shift Switch ②	Yes	2 Bp/min	90%	Open Circuit

NOTE: *If any 5v sensor becomes shorted to ground the engine will not start. If the engine is operating when the short occurs the engine may stop operating and will not start.*

① *VDC PWR Low - if shorted or no voltage engine will not start; if VDC PWR voltage falls velow 22 volts will set sensor faults.*
② *Shift Switch - will activate code when engine rpm are above 3500 rpm and 40 % load.*

Fig. 128 Engine Guardian audible warning fault codes - V6 and V8 MPI engines (exc. 8.1L) (cont'd)

7-48 FUEL SYSTEM - FUEL INJECTION (EFI)

Fault	SC1000	Audio Alarm	Available Power	Description
Cam Sensor	Yes	2 Bp/min	90%	Open or short, engine must be cranking to set this fault code.
ECT CKT HI	Yes	2 Bp/min	90%	Open
ECT CKT LO	Yes	2 Bp/min	90%	Short
ECT Coolant Overheat	Yes	Constant	6-100%	Engine guardian overheat condition
EST 1-8 Open	Yes	2 Bp/min	NA	Coil harness wire open
EST 1-8 Short	Yes	2 Bp/min	NA	Coil harness wire short
Fuel Injector 1-8 Open	Yes	2 Bp/min	NA	Fuel injector wire open
Fuel Injector 1-8 Short	Yes	2 Bp/min	NA	Fuel injector wire short
IAC Output	Yes	2 Bp/min	90%	Only with rpm
Knock Sensor 1	Yes	2 Bp/min	90%	Alarm sounds for 20 seconds in NEUTRAL and indefinitely in gear.
Knock Sensor 2	Yes	2 Bp/min	90%	Alarm sounds for 20 seconds in NEUTRAL and indefinitely in gear.
Low Drive Lube Strategy	Yes	Steady Bp	0-100%	Low oil in sterndrive
Low Oil Pressure Strategy	Yes	Constant	0-100%	Low oil pressure strategy
MAP Sensor 1 Input High	No	2 Bp/min	90%	Short, no visual on SC1000
MAP Sensor 1 Input Low	No	2 Bp/min	90%	Open, no visual on SC1000
MAT Sensor	Yes	2 Bp/min	90%	Open or short in MAT circuit

NOTE: If any 5v sensor becomes shorted to ground the engine will not start. If the engine is operating when the short occurs the engine may stop operating and will not start.

Fig. 129 Engine Guardian audible warning fault codes - 8.1L V8 engines

Fault	SC1000	Audio Alarm	Available Power	Description
Oil PSI CKT Hi	Yes	2 Bp/min	90%	Short, defaults to 51.7 psi
Oil PSI CKT Lo	Yes	2 Bp/min	90%	Open, zero oil pressure
Overspeed	Yes	Constant	RPM Limit	Engine over rpm limit
Port EMCT CKT Hi	Yes	2 Bp/min	90%	Open, defaults to 32 degrees F (0 degree C)
Port EMCT CKT Lo	Yes	2 Bp/min	90%	Short, defaults to 32 degrees F (0 degree C)
Port EMCT CKT Overheat	Yes	Constant	6-100%	Overheat condition, 212 degrees F (100 degrees C) limit
Sea Pump PSI Lo	Yes	Constant	6-100%	Low water pressure strategy, defaults to 43.4 psi
Sea Pump CKT Hi	Yes	2 Bp/min	90%	Open
Sea Pump CKT Lo	Yes	2 Bp/min	90%	Short
STB EMCT CKT Hi	Yes	2 Bp/min	90%	Open, defaults to 32 degrees F (0 degrees C)
STB EMCT CKT Lo	Yes	2 Bp/min	90%	Short, defaults to 32 degrees F (0 degrees C)
STB EMCT CKT Overheat	Yes	Constant	6-100%	Overheat condition, 212 degrees F (100 degrees C) limit
Steer CKT Hi	Yes	No	No	Open and short
TPS1 CKT Hi	Yes	2 Bp/min	90%	Short, signal to 5v+, engine will not start. Refer to data monitor screen.
TPS1 CKT Lo	Yes	2 Bp/min	90%	Open
TPS 1 Range Hi	Yes	2 Bp/min	90%	Above 4.8v, 994 counts
TPS 1 Range Lo	Yes	2 Bp/min	90%	Below 0.5v, 35 counts
Trim CKT Hi	Yes	No	No	Short, high range, visual warning on SC1000 only.
Trim CKT Lo	Yes	No	No	Open, low range, visual warning on SC1000 only.
5 VDC PWR Low	Yes	2 Bp/min	varies	Short any 5v+ to ground

NOTE: If any 5v sensor becomes shorted to ground the engine will not start. If the engine is operating when the short occurs the engine may stop operating and will not start.

Fig. 130 Engine Guardian audible warning fault codes - 8.1L V8 engines (cont'd)

Precautions

◆ See Figures 79 and 80

Good shop practice, as well as good judgment, requires all the following practices be observed:
- The negative battery cable must be disconnected before any ECM system component is removed
- Always check to be sure the battery cables are securely connected before starting the engine
- Never separate the battery from the on-board electrical system while the engine is operating
- Never separate the battery feed wire from the charging system while the engine is operating
- Disconnect the battery from the boat electrical system before starting to charge the battery
- Check to be sure all cable harnesses are securely connected and the battery terminals are clean and the cables securely connected
- Never connect or disconnect the wiring harness at the ECM when the ignition switch is in the **ON** position
- Before any electric arc welding is attempted, disconnect the battery leads and the ECM connector/s
- Never direct a steam-cleaning nozzle at ECM components. Such action will cause damage or corrosion the component terminals
- Do not use any test equipment not specified in the diagnostic charts. Such equipment may give an incorrect reading and/or may actually damage good components
- A digital voltmeter with a rating of 10 meg-ohms input impedance must be used when taking any voltage measurements. Always make sure that the DVOM is switched to the **OFF** position prior to connecting it to any circuit being tested.
- A "low-amphere rated" test light must be used when a test light is specified for a test. Never use a high-amphere rated test light. Power the setup with the boat battery. If the ammeter indicates less than 3/10 amp current flow, the test light is safe to use.

A final word here. If a test light with 100 mA or less is used, a faint glow may show, when the test actually states "no light".

Intermittent Faults

■ A specific problem may or may not set/store a fault. An actual fault must be recorded before you can locate the problem.

Most intermittent problems are caused by bad electrical connections and/or wiring problems. Always check for the following:
- Poor mating of connector halves
- Terminal connectors not fully seated on the connector body
- Damaged, worn or corroded terminals

All terminals, wiring and/or connections in a circuit with a suspected problem should be checked carefully prior to proceeding with any other test procedures.

Running the vessel with a DVOM connected to a suspicious circuit may work as abnormal voltage when the fault occurs is a good indicator that there is a fault in the circuit being checked.

A diagnostic tool (DDT) can also be used as it has several features built into it to aid in detecting intermittents. The tool allows for manipulation of harnesses and components when the engine is not running. It can also be used while running the vessel.

When the problem seems to be related to parameters that can be checked on the tool, they should be checked with the vessel running if at all possible. When there seems to be no correlation between the problem and a specific circuit, use the tool data to check for any change in readings that may indicate an intermittent.

The DDT is also a great way to compare the operating characteristics of a bad engine with that of a known-good one. A sensor may shift value, but not enough to actually set a fault - comparing the sensor's reading with those of typical tool data reading may actually help you uncover the root of the problem.

The tool will also allow you to save diagnosis time and, quite possibly, prevent accidental replacement of good parts. Knowledge and understanding of the systems being diagnosed are the key to successful diagnostic work!

FUEL SYSTEM - FUEL INJECTION (EFI)

To check for loss of fault memory:
1. Disconnect the TP sensor and allow the engine to idle at normal operating temperature.
2. Attach the digital diagnostic tool as per the manufacturer's instructions.
3. Turn the ignition switch the **OFF** position and check for the fault **TPS1 CKT Lo**. If this is not stored in the system's memory, the ECM is bad and must be replaced.
4. Clear the fault.

An intermittent fault which is not causing a stored fault can be caused by the following:
- Ignition coil shorted to ground
- Arcing at the coil wire or spark plug wires
- Poor ECM grounds
- Electrical system interference caused by a sharp surge - normally occurs when the bad component is being operated
- Incorrect installation of electrical accessories (GPS, radio, lights, etc.)
- Knock sensor electrical lead routed near or touching spark plug wires or ignition/charging system components
- Secondary ignition system components shorted to ground
- Starters, alternators and relays, etc. shorted to ground

Troubleshooting

◆ See Figures 131 thru 136

WITHOUT A DIAGNOSTIC/SCAN TOOL

Troubleshooting without a digital diagnostic or scan tool is essentially limited to checking sensor resistance with a DVOM. Typical failures do not usually involve the ECM as loose connections or wear are generally at fault.
1. Confirm the engine is in good mechanical condition.
2. Check that the ECM grounds and all sensor connections are clean, tight and properly located.
3. Check all vacuum lines for splits, kinks and good connections. Check for leaks or restrictions.
4. Check for any air leaks at the throttle body and intake manifold sealing surfaces.
5. Check the ignition wires.
6. Inspect all wiring for cracks, routing and proper connections.
7. Make sure there is no moisture in the primary or secondary ignition circuits.
8. Check the fuel pump and pressure.
9. Make sure that the throttle cable is adjusted correctly for the TP sensor at 0 degrees.

WITH A DIAGNOSTIC/SCAN TOOL

The Quicksilver digital diagnostic terminal (DDT) and the Mercury scan tool have been developed specifically for these engines and we highly recommend that you do not undertake any form of diagnostic work without one.

Symptom	Possible Cause	Action
1. Engine cranks but will not start	1.0 Lanyard stop switch in wrong position	1.0 Reset lanyard stop switch
	1.1 Weak battery or bad starter motor. Battery voltage drops below 8 volts while cranking.	1.1 Replace or recharge battery. Inspect condition of starter motor. Inspect condition of battery connections.
	1.2 No fuel	1.2 Key ON engine to verify fuel pump operates for 3 seconds. Check fuel tank for fuel. Verify fuel pressure is 43 psi. Listen for fuel pump relay to click.
	1.3 Blown fuse	1.3 Inspect engine harness and electrical components. Replace fuse.
	1.4 Main power relay (MPR) malfunction	1.4 Listen for MPR to click when the key switch is turned ON.
	1.5 Crankshaft or camshaft sensor defective	1.5 Inspect for loose connection or corrosion. Check for tachometer signal while cranking engine. If no signal, faulty crankshaft sensor. Inspect continuity between sensors and PCM.
	1.6 ECM malfunction	1.6 Listen for fuel injector ticking when cranking the engine. Check battery voltage. Check for blown fuse. Check battery voltage to the fuse from the MPR. Inspect harness connections. Replace PCM.
2. Engine overheat	2.0 Reduced or no water flow	2.0 Verify water inlet valve is open. Inspect seawater strainer for debris. Clogged water hose.
	2.1 Faulty seapump impeller	2.1 Replace impeller.
	2.2 Faulty thermostat	2.2 Replace thermostat.
	2.3 Faulty water pump	2.3 Replace water pump.

Symptom	Fault Cause	Action
3. Engine cranks, starts and stalls	3.0 Low fuel pressure	3.0 Key on engine to verify fuel pump operates for 3 seconds. Check fuel tank for fuel. Check fuel pressure is 43 psi. Listen for fuel pump relay to click.
	3.1 Contaminated fuel	3.1 Change water separating fuel filter.
	3.2 TPS sensor range	3.2 Inspect throttle linkage for wear and binding. Verify TPS is in range.
	3.3 Engine mechanical malfunction	3.3 Check for low compression, cylinder head gasket leaks, worn camshaft, valve train problem or restricted exhaust system.
4. Engine lacks power, sluggish	4.0 Extremely dirty flame arrestor	4.0 Clean or replace flame arrestor.
	4.1 Contaminated fuel	4.1 Change water separating fuel filter.
	4.2 Improper ignition voltage	4.2 Check ignition voltage.
	4.3 Fouled spark plugs	4.3 Change spark plugs.
	4.4 Engine mechanical problems	4.4 Check for low compression, cylinder head gasket leaks, worn camshaft, valve train problem or restricted exhaust system.
	4.5 Engine Guardian	4.5 Read fault descriptions.
5. Engine idle is rough	5.0 Fouled spark plugs	5.0 Check ignition voltage.
	5.1 Weak spark	5.1 Inspect coils, spark plug wires and harness connections.
	5.2 IAC faulty	5.2 Listen for IAC motor upon key ON. Read fault descriptions.
	5.3 Faulty injectors	5.3 Perform Injector Balance Test.
	5.4 Engine mechanical malfunction	5.4 Check for low compression, cylinder head gasket leaks, worn camshaft, valve train problem or restricted exhaust system.
	5.5 Faulty motor mounts	5.5 Inspect motor mounts.
	5.6 Vacuum leak	5.6 Check vacuum lines and gaskets for leaks and wear. Replace.
	5.7 Throttle cable not adjusted properly	5.7 Adjust throttle cable.
6. Detonation or spark knock	6.0 Faulty knock sensor circuit	6.0 Inspect both knock sensor circuits.
	6.1 Poor ignition system ground	6.1 Inspect ignition system connections.
	6.2 Contaminated fuel	6.2 Replace fuel with known high quality fuel.

Fig. 131 Quick reference symptom chart - 2001-08 MPI engines

7-50 FUEL SYSTEM - FUEL INJECTION (EFI)

Faults	Possible Causes	Action
1. ECT CKT Hi or Lo	1.0 Open(Hi) or Short (Lo) in harness wiring, faulty connection	1.0 Repair harness connection or cut in wire.
	1.1 Open(Hi) or Short (Lo) sensor	1.1 Replace sensor.
	1.2 Water in the connector	1.2 Dry connector and inspect for cracks or wear. Replace.
2. ECT Coolant Overheat	2.0 Coolant leak	2.0 Inspect closed cooling system.
	2.1 Restricted waterflow	2.1 Check for blockage in inlet and outlet water hoses.
	2.2 Faulty seawater pump	2.2 Inspect seawater pump water ports, impeller and seals for damage. Replace damaged parts.
	2.3 Faulty thermostat	2.3 Replace thermostat.
	2.4 Faulty seawater pump	2.4 Replace seawater pump.
	2.5 Worn or broken drive belt	2.5 Replace drive belt.
3. EST 1 Open or Short	3.0 Loose spark plug wire connection.	3.0 Verify spark plug boot firmly connected.
	3.1 Broken spark plug	3.1 Inspect spark plug for damage. Replace.
	3.2 Open or Short in harness wiring, bad harness connection	3.2 Inspect coil harness. Repair or replace.
	3.3 Faulty coil	3.3 Replace coil.
	3.4 Water in connection	3.4 Dry connector and inspect for cracks or wear. Replace.
4. Fuel Injector 1-8 Open or Short	4.0 Open or short in harness wire, bad harness connection or corroded terminals	4.0 Inspect fuel injector harness. Repair or replace.
	4.1 Faulty fuel injector	4.1 Replace fuel injector.
	4.2 12 volt fuel injector wire shorted to ground	4.2 Inspect fuel injector harness. Repair or replace.
	4.3 Corroded harness terminals	4.3 Inspect fuel injector harness. Repair or replace.
5. Guardian Strategy	5.0 Engine block pressure, map sensor, oil pressure, starboard exhaust temperature, engine coolant temperature or overspeed readings are out of normal ranges	5.0 Other fault codes will appear on the diagnostic tool. Verify repairs associated with the other faults. Scan for faults again.
6. IAC Output Lo or Hi	6.0 Cut harness wire, bad harness connection, short in harness	6.0 Inspect IAC wiring circuit. Repair.
	6.1 Faulty IAC, pindle stuck	6.1 Replace IAC.
7. Knock Sensor 1 Hi or Lo	7.0 Corrosion or wear on the sensor	7.0 Replace sensor.
	7.1 Open or short in circuit	7.1 Inspect harness. Repair or replace harness.
8. Low Drive Lube Strategy MCM engines only.	8.0 Improper hose routing	8.0 Route hose as shown in installation manual.
	8.1 Float in bottle stuck	8.1 Replace bottle.
	8.2 Short in circuit	8.2 Inspect circuit. Repair or replace harness.
	8.3 Incorrect gear lube level	8.3 Fill drive lube monitor. Refer to appropriate service manual for filling instructions.
	8.4 System leak	8.4 Inspect drive lube system. Repair or replace any worn parts.
9. Low Oil Pressure Strategy	9.0 Faulty sensor readings	9.0 Replace sensor with a known good sensor.
	9.1 Faulty oil pump	9.1 Replace oil pump.
	9.2 Low oil level	9.2 Check oil level, add oil.
10. Main Power Relay Output	10.0 Short in MPR circuit	10.0 Inspect MPR circuit. Repair or replace harness.
	10.1 Low battery voltage	10.1 Charge battery or replace battery.
	10.2 Faulty MPR	10.2 Replace MPR.
11. Main Power Relay Backfeed	11.0 Driver power from some other source	11.0 Test for voltage at 87 of the MPR with key OFF.
	11.1 Short in circuit at splice 105	11.1 Inspect wiring harness. Repair or replace harness.

Faults	Possible Causes	Action
12. MAP Sensor 1 Input Hi or Lo	12.0 Loose connection, corrosion, open (Lo) or short (Hi) in circuit	12.0 Inspect circuit. Repair or replace harness.
	12.1 Faulty sensor	12.1 Replace sensor.
13. MAT Sensor Hi or Lo	13.0 Loose connection, corrosion, open (Lo) or short (Hi) in circuit	13.0 Inspect circuit. Repair or replace harness.
	13.1 Faulty sensor	13.1 Replace sensor.
14. Oil PSI CKT Hi or Lo	14.0 Loose connection, corrosion, open (Lo) or short (Hi) in circuit	14.0 Inspect circuit. Repair or replace harness.
	14.1 Faulty sensor	14.1 Replace sensor.
15. Overspeed	15.0 Underpropped	15.0 Change propeller.
	15.1 Over trimmed condition	15.1 Trim drive properly.
	15.2 Rev-limit out of range	15.2 Check rev-limit.
16. Pitot CKT Hi or Lo	16.0 Loose connection, open (Lo) or short (Hi) in wiring circuit	16.0 Inspect wiring harness. Repair or replace harness.
	16.1 Corroded or faulty sensor	16.1 Check sensor for damage. Replace seals or sensor.
	16.2 Loose hose connection	16.2 Tighten hose connection.
17. seawater pump PSI Lo	17.0 Restricted waterflow	17.0 Check for blockage in inlet and outlet water hoses.
	17.1 Faulty seawater pump	17.1 Inspect seawater pump water ports, impeller and seals for damage. Replace damaged parts.
	17.2 Faulty thermostat	17.2 Replace thermostat.
	17.3 Faulty seawater pump	17.3 Replace seawater pump.
	17.4 Worn or broken drive belt	17.4 Replace drive belt.
18. seawater pump CKT Hi or Lo	18.0 Loose connection, corrosion, open (Hi) or short (Lo) in circuit	18.0 Inspect circuit. Repair or replace harness.
	18.1 Faulty sensor	18.1 Replace sensor.
19. Steer CKT Hi or Lo	19.0 Loose connection, corrosion, open (Hi) or short (Lo) in circuit	19.0 Inspect circuit. Repair or replace harness.
	19.1 Faulty sensor	19.1 Replace sensor.
20. TPS1 CKT Hi or Lo	20.0 Loose connection, corrosion, open (Hi) or short (Lo) in circuit	20.0 Inspect circuit. Repair or replace harness.
	20.1 Faulty sensor	20.1 Replace sensor.
21. TPS1 Range Hi or Lo	21.0 Worn or damaged sensor, count reading over 990 for Hi, under 45 for Lo	21.0 Inspect sensor for damage. Replace TPS.
	21.1 Short in transducer ground circuit	21.1 Inspect harness for short to ground. Repair or replace harness.
	21.2 Worn, bent or corroded throttle lever	21.2 Inspect throttle lever. Repair any damage.
22. Transmission Overtemp	22.0 Loose connection, corrosion, open (Lo) or short (Hi) in circuit	22.0 Inspect circuit. Repair or replace harness.
	22.1 Faulty sensor	22.1 Inspect circuit. Repair or replace harness.
23. Trim CKT Hi or Lo	23.0 Loose connection, corrosion, open (Hi) or short (Lo) in circuit	23.0 Inspect circuit, repair or replace harness.
	23.1 Faulty sensor	23.1 Replace sensor.
24. VDC PWR Lo	24.0 Short to ground in the 5 volt system, harness or 3-wire sensor	24.0 Read other faults for a starting point for finding the short. Repair or replace harness or faulty sensor.

Fig. 132 Engine fault reference chart - 2001-08 MPI engines

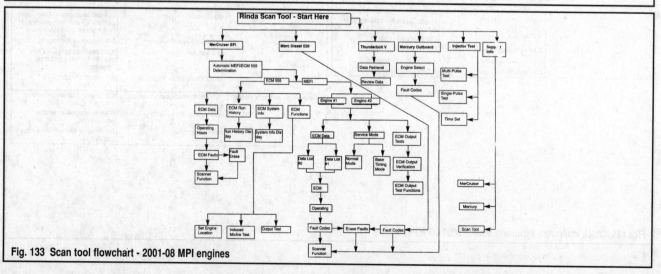

Fig. 133 Scan tool flowchart - 2001-08 MPI engines

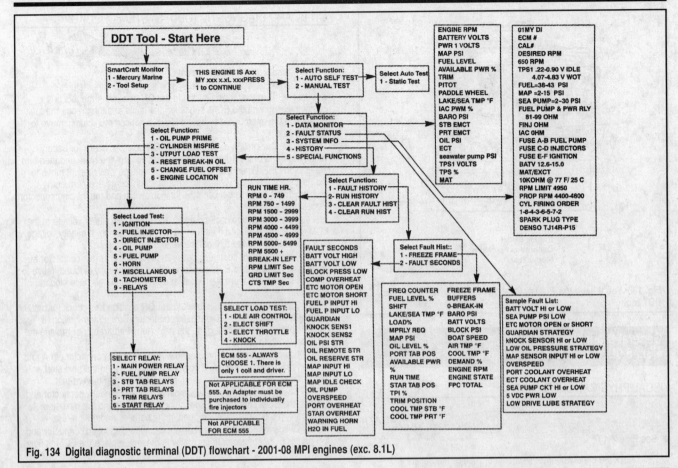

Fig. 134 Digital diagnostic terminal (DDT) flowchart - 2001-08 MPI engines (exc. 8.1L)

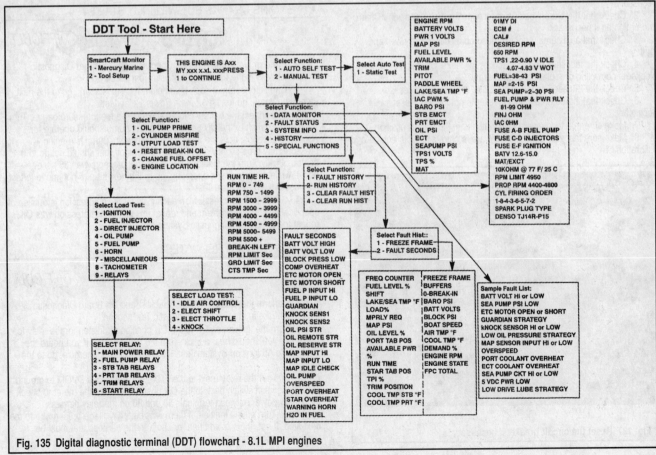

Fig. 135 Digital diagnostic terminal (DDT) flowchart - 8.1L MPI engines

7-52 FUEL SYSTEM - FUEL INJECTION (EFI)

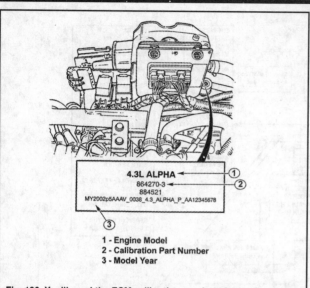

Fig. 136 You'll need the ECM calibration number when using the DDT tool

1 - Engine Model
2 - Calibration Part Number
3 - Model Year

General Diagnostic Tests - 2001-08 4.3L, 5.0L, 5.7L & 6.2L MPI Engines

VISUAL/PHYSICAL CHECKLIST

◆ See Figure 137

1. Check the battery charge. Recharge or replace the battery if necessary.
2. Check that the battery cable connections are clean and tight. Clean and/or tighten if necessary.
3. Check that all external engine ground connections are clean and tight.
4. Check the fuel lines and connections for leaks, corrosion or blockage, correcting or replacing as necessary.
5. Check the 3 fuses located at the ECM, replacing if necessary.
6. Check that the 50 amp circuit breaker is not tripped. Reset if necessary.
7. Make sure that the lanyard stop switch is in the correct position.
8. Make sure that your scan tool software is correct for your engine and calibration.

Fig. 137 Reset the circuit breaker if necessary

ON-BOARD DIAGNOSTIC (OBD) SYSTEM CHECKS

1. Complete the Visual/Physical Checklist.
2. Connect a scan tool to the engine as per the manufacturer's instructions. Turn the ignition switch to the **ON** position. If the tool is communicating with the ECM, check for any faults and then repair the fault or move to the appropriate symptom chart. If not communicating, move to the next step.
3. With the ignition switch still in the **ON** position, check for battery voltage at Pin **5** of the 10-pin connector coming from the helm and then turn the ignition OFF. When the switch was on, if voltage was present move to the next step. If battery voltage was not present, locate and correct the problem in the harness between the ignition switch and the 10-pin connector. Repeat this test.
4. Turn the ignition back on again and check for battery voltage at the ECM connector **B-18** and then turn off the ignition switch. If battery voltage was present, go to Step 6. If not, move to the next step.
5. Now check for continuity between pin **5** on the 10-pin connector at the engine harness and ECM connector **B-18**. If continuity is found, move to the next step; if not, there is an open somewhere in the harness - locate, repair and re-test.
6. Turn the ignition switch back to the **ON** position and check for battery voltage at the **D** pin on the Diagnostic connector. Turn off the ignition. If battery voltage was present, go to the next step; if not, repair or replace the harness and then retest.
7. Check continuity of the ground wire, Diagnostic connector pin **A** and pin **1** on the 10-pin connector. If continuity is found, go to the next step; otherwise locate and repair the open in the harness and then re-test.
8. Check for continuity between pin **B** on the diagnostic connector and **A-12** on the ECM connector. If continuity is present, go to the next step; if not, locate, repair or replace the open in the harness and then retest.
9. Check for continuity between pin **C** on the diagnostic connector and **A-5** on the ECM connector. If continuity is present, replace the ECM; if not, locate, repair or replace the open in the harness and then retest.

A1 - ENGINE CRANKS BUT WILL NOT START

1. Perform the Visual/Physical Checklist and On-Board Diagnostic System checks in the General Diagnostic Test section.
2. Check for spark at each spark plug. If there is spark, go to the next step. If no spark, go to test A2 main Power Relay test.
3. Disconnect the battery cables and install a fuel pressure gauge. Turn the ignition switch to the **ON** position. The fuel pump should operate for about 3-5 seconds; note the fuel pressure while the pump is running - pressure may drop slightly after the pump turns off, but should not go to 0 psi too quickly. Turn the ignition OFF. If the fuel pressure was 35 psi or above while the pump was operating, go to the next step. If not, move to test A3 Fuel System Electrical Test.
4. Check the compression on each cylinder. If compression is outside of specifications, fix the problem and repeat this test. If compression was OK, repeat the On-Board Diagnostic System checks.

A2 - MAIN POWER RELAY TEST

1. Perform the Visual/Physical Checklist and On-Board Diagnostic System checks in the General Diagnostic Test section.
2. Turn the ignition switch to the **ON** position and listen for the relay to energize - you should hear a click. Turn the ignition OFF. If you heard the relay, go to A5 Ignition System Test. If the relay did not energize, go to the next step.
3. Pull out the relay (next to the ECM) and connect a DVOM to ground. Turn the ignition switch back to the **ON** position and use the meter to check for battery power between terminals **30** and **86** on the relay harness connector. Turn off the ignition. If there was battery voltage present, go to the next step. If not, there is an open, or short, in the harness that must be repaired - repair and re-test.

FUEL SYSTEM - FUEL INJECTION (EFI) 7-53

4. Check for continuity between terminal **85** of the relay harness connector and **A-22** of the ECM harness connector. If continuity was present, install a known-good relay and re-test. If no continuity, there is a short, or open, in the harness - repair and re-test.

A3 - FUEL SYSTEM ELECTRICAL TEST

1. Perform the Visual/Physical Checklist and On-Board Diagnostic System checks in the General Diagnostic Test section.
2. Turn the ignition switch to the **ON** position and listen for the fuel pump to operate. Turn off the ignition off. If the fuel pump operated for 3-5 seconds, go to A4 Fuel System Diagnosis. If it didn't operate, go to the next step.
3. Connect a DVOM and then turn the ignition switch to the **ON** position. Check for battery voltage at **A** on the pump harness connector. Turn off the ignition. If voltage was present, install a known-good pump and re-test the system. If battery voltage was not present, go to the next step.
4. Remove the fuel pump relay (next to the ECM) and then turn the ignition switch back on. Check for battery power at terminal **30** of the relay harness connector and then turn the ignition off again. If battery voltage was present, go to the next step. If not, there is a short, or open, in the harness that will require fixing.
5. Check for continuity between terminal **86** in the harness connector and **A-19** in the ECM harness connector. If there was continuity present, install a known good pump relay and re-test. If none present, there is a short, or open, in the harness that will require fixing.

A4 - FUEL SYSTEM DIAGNOSIS

1. Perform the Visual/Physical Checklist and On-Board Diagnostic System checks in the General Diagnostic Test section.
2. Disconnect the battery cables and install a fuel pressure gauge as per the manufacturer's instruction. Turn the ignition switch to the **ON** position. The fuel pump should operate for about 3-5 seconds; note the fuel pressure while the pump is running - pressure may drop slightly after the pump turns off, but should not go to 0 psi too quickly. Turn the ignition OFF. If the fuel pressure was 35 psi or above while the pump was operating, go to the next step. If there was pressure, but not above the 35 psi, go to Step 5. If there was no pressure at all, go to A3 Fuel System Electrical test.
3. Attempt to start the engine and allow it to idle until it reaches normal operating temperature. If it started, go to the next step. If not, or if it wouldn't stay running, go to Step 5.
4. With the engine still idling, connect an external vacuum source to the fuel pressure regulator and apply 10 in. vacuum. If the pressure decreased by approximately 5 psi, either the problem is intermittent or you have a low or restricted fuel supply. If the pressure did not decrease, replace the pressure regulator and re-test.
5. If the gauge showed the correct pressure and then immediately fell to 0 psi, go to next step. Otherwise, re-test the entire system.
6. With the ignition off, plug the fuel pressure line between the pump and the rail. Now turn the ignition on and see if pressure remains steady. If so, replace the fuel pressure regulator; if not install a known-good fuel pump and re-test the system.

A5 - IGNITION SYSTEM TEST

1. Perform the Visual/Physical Checklist and On-Board Diagnostic System checks in the General Diagnostic Test section.
2. Install an analog tachometer to the auxiliary tach lead located near the ECM. Crank the engine and then turn the ignition off. If there was a tach signal while cranking, go to the next step. If not, there is a mechanical problem.
3. Check the plug wires for open circuits, cracks or bad seating on the plugs, cap or coil. Fix anything you find; if nothing was evident, go to the next step.
4. Check that all spark plugs have sufficient spark. If they do, go to the next step; if not, go to Step 6.
5. Check the spark plugs themselves. Gap or replace if necessary, otherwise go to Step 10.
6. Turn the ignition switch to the **ON** position and check for battery voltage at the **A** terminal on the ignition coil. Turn off the ignition. If voltage was present, go to the next step; if not look for an open, or ground, in the harness. Re-test the system.
7. Turn on the ignition switch again and check for battery voltage at the **A** terminal on the ignition coil driver harness connector. Turn off the ignition. If voltage was present, go to the next step; if not look for and repair an open, or ground, in the harness. Re-test the system.
8. Now check for continuity between terminal **C** on the coil driver harness connector and the engine ground. If continuity was present, go to the next step; if not, look for and repair an open in the harness. Re-test the system.
9. Once again, check for continuity between terminal **B** on the coil driver harness connector and the **B-23** terminal on the ECM connector. If continuity is found, replace the coil and driver and then re-test the system. If not, look for and repair an open in the harness. Re-test the system.
10. Disconnect the crankshaft position sensor harness and turn the ignition switch to the **ON** position. Check for 5 volt power at the harness connector terminal **A** and then turn off the ignition. If you found power, go to the next step; otherwise, look for and repair an open in the harness. Re-test the system.
11. Now check for continuity between the **B** terminal on the crankshaft position sensor harness connector and the engine ground. If continuity is found, go to the next step; otherwise, look for and repair an open in the harness. Re-test the system.
12. Check for continuity between the **C** terminal in the harness connector and **B-10** on the ECM harness connector. If continuity is found, go to the next step; otherwise, look for and repair an open in the harness. Re-test the system.
13. Check that the sensor resistance values are 23.3 megohms between terminals **A** and **B** and 23.21 between terminals **B** and **C**. If they are correct, go back to the Visual/Physical Checklist; if they are not, replace the crankshaft position sensor.

A6 - HARD START SYMPTOMS

◆ See Figure 138

■ **A hard start is defined as an engine that cranks fine, but takes a longer time than normal to actually start.**

1. Perform the Visual/Physical Checklist and On-Board Diagnostic System checks in the General Diagnostic Test section.
2. Check for contaminated or poor quality fuel. Check any and all fuel filters for blockage. If the fuel is bad or the filters are clogged, replace them; otherwise go to the next step.
3. Check that each spark plug has sufficient spark. If not, go back to A2 Main Power Relay Test; otherwise go to the next step.
4. Disconnect the battery cables, install a fuel pressure gauge as per the manufacturer's instructions and then reconnect the cables. Turn the ignition switch to the **ON** position. The fuel pump should operate for about 3-5 seconds; note the fuel pressure while the pump is running - pressure may drop slightly after the pump turns off, but should not go to 0 psi too quickly.

ECT Sensor Approx. Temperature-to-Resistance Values		
Deg. (°) F	Deg. (°) C	Ohms
210	100	185
160	70	450
100	38	1,800
70	20	3,400
40	4	7,500
20	-7	13,500
0	-18	25,000
-40	-40	100,700

Fig. 138 ECT sensor values

7-54 FUEL SYSTEM - FUEL INJECTION (EFI)

Turn the ignition OFF. If the fuel pressure was 35 psi or above while the pump was operating, go to the next step. If pressure was not above the 35 psi, go to A3 Fuel System Electrical test.

5. If you are using a scan tool, go to the next step. If you are not using a scan tool, go to Step 12.

6. Check to see if the ECT sensor has shifted in value. With the engine completely cool, use the tool to compare the sensor temperature with the ambient outside air temperature. If the temperatures are with 10° F (5.5° C) of each other, go to the next step; otherwise replace the sensor.

7. Use the scan tool to display the temperature again and take note of the value. Check the resistance of the sensor. Now compare the approximate temperature of the sensor to the ambient outdoor temperature. If the ECT sensor temperature is near the resistance temperature (in the chart), go to the next step; if not, find and repair high resistance or a bad connection in the ECT signal circuit or at the sensor ground.

8. Check for intermittent opens or shorts to ground in the MAP sensor circuit. If you find a problem, repair the open in the harness and then re-test. If not, go to the next step.

9. Use the scan tool to check that the TP sensor is operating properly. Check the throttle linkage for sticking, binding or wear. If a problem was found with either, find and repair it and then re-test. If you found no problems, go to the next step.

10. Check for the following:
- Low compression
- Leaking cylinder head gaskets
- Worn camshaft
- Bad valve timing or another problem with the valve train
- Restricted exhaust system

Fix any problems you may find, otherwise go to the next step.

11. Review everything you've done so far. If no problem has been found, redo the Visual/Physical test, check your scan tool and/or check all electrical connections in the suspected circuit.

12. Check to see if the ECT sensor has shifted in value. With the engine completely cool, measure the resistance at the sensor and jot it down. Next, compare the approximate sensor temperature with the ambient outside air temperature. If the readings are similar, go back to Step 8; otherwise, replace the sensor.

A7 - ENGINE SURGE SYMPTOMS

■ An engine surge is defined as a power variation while under steady throttle.

1. Perform the Visual/Physical Checklist and On-Board Diagnostic System checks in the General Diagnostic Test section.

2. Check for contaminated or poor quality fuel. Check any and all fuel filters for blockage. If the fuel is bad or the filters are clogged, replace them; otherwise go to the next step.

3. Install fuel pressure gauge and check the fuel pressure during the condition (A4 Fuel System Diagnosis). If there is a problem with pressure, go to A3 Fuel System Electrical Test; otherwise, go to the next step.

4. Remove the spark plugs and check them thoroughly as detailed in the Maintenance section. Fix any problems or replace the plugs. If no problems are found, go to the next step.

5. Check the ignition coil for cracks or carbon tracks. If anything is found, repair of replace the coil and re-test the system; if nothing is found, go to the next step.

6. Check all primary and secondary wires. Check the distributor, cap and wires, including where they are routed. If something is found, repair of replace the problem; if not, go to the next step.

7. Check all vacuum lines for cracks, kinks, splits or bad connections. Repair of replace any hoses with a problem and then go to the next step.

8. Check the fuel injector wiring harness for proper connections and any intermittent opens or shorts. If something is found, repair or replace and then re-test the system; otherwise go to the next step.

9. Do the same with the ECM harness and then move to the next step.

10. Check that the voltage output at the alternator is 13.9-14.7 volts. If it is, move to the next step; otherwise, refer to the Charging System section for repairs.

11. Review everything you've done so far. If no problem has been found, redo the Visual/Physical test, check your scan tool and/or check all electrical connections in the suspected circuit.

A8 - LACK OF POWER, SLUGGISH OR SPONGY SYMPTOMS

■ These symptoms are defined as less than expected power, little or no increase in speed when the throttle is advanced part of the way.

1. Perform the Visual/Physical Checklist and On-Board Diagnostic System checks in the General Diagnostic Test section.

2. Has the engine gone into Guardian strategy? If so, confirm the fault and repair it; otherwise, go to the next step.

3. Check the flame arrestor for dirt, damage or some other restriction. Clean or replace if necessary; otherwise, go to the next step.

4. Check for contaminated or poor quality fuel. Check any and all fuel filters for blockage. If the fuel is bad or the filters are clogged, replace them; otherwise go to the next step.

5. Install a fuel pressure gauge and check the fuel pressure during the condition (A4 Fuel System Diagnosis). If there is a problem with pressure, go to A3 Fuel System Electrical Test; otherwise, go to the next step.

6. Remove the spark plugs and check them thoroughly as detailed in the Maintenance section. Fix any problems or replace the plugs. If no problems are found, go to the next step.

7. Check the ignition coil for cracks or carbon tracks. If anything is found, repair of replace the coil and re-test the system; if nothing is found, go to the next step.

8. Check the ECT, MAP, TP and KS sensor circuits for any shorts or opens. If you find a problem, fix it (or the harness); otherwise go to the next step.

9. Do the same with the ECM harness and then move to the next step.

10. Check that the voltage output at the alternator is 13.9-14.7 volts. If it is, move to the next step; otherwise, refer to the Charging System section for repairs.

11. Check for the following:
- Low compression
- Leaking cylinder head gaskets
- Worn camshaft
- Bad valve timing or another problem with the valve train
- Restricted exhaust system

Fix any problems you may find, otherwise go to the next step.

12. Check the hull and/or keel for dirt or barnacles. Confirm that the propeller is of the correct size and pitch. Fix any problems and re-test the system; otherwise go to the next step.

13. Review everything you've done so far. If no problem has been found, redo the Visual/Physical test, check your scan tool and/or check all electrical connections in the suspected circuit.

A9 - DETONATION/SPARK KNOCK SYMPTOMS

♦ See Figure 138

■ These symptoms are defined as a mild to severe pinging, usually worse under hard acceleration.

1. Perform the Visual/Physical Checklist and On-Board Diagnostic System checks in the General Diagnostic Test section.

2. Check that the engine has the proper propeller for the boat and operating range - if so, go to the next step.

3. Confirm that the engine is running the correct spark plugs and that they are clean and gapped correctly. Do the same with the spark plug wires. Replace if necessary, or go to the next step.

4. Check for cracks, breaks or other damage at the distributor, cap and rotor. If something is found, repair or replace it and then re-test the system. If not, go to the next step.

5. Check for contaminated or poor quality fuel. Check any and all fuel filters for blockage. If the fuel is bad or the filters are clogged, replace them; otherwise go to the next step.

6. Install a fuel pressure gauge and check the fuel pressure during the condition (A4 Fuel System Diagnosis). If there is a problem with pressure, go to A3 Fuel System Electrical Test; otherwise, go to the next step.

7. Is the engine operating above the normal operating temperature range? If it is, go to the next step. If not, go to Step 9.

8. Check for obvious overheating issues:

FUEL SYSTEM - FUEL INJECTION (EFI)

- Loose serpentine belt
- Bad seawater pump
- Cooling system restriction
- Bad thermostat

If anything is wrong, repair or replace it and then re-test the system. If not, go to the next step.

9. If you are using a scan tool, go to the next step; if you're not, go to Step 14.

10. Check to see if the ECT sensor has shifted in value. With the engine completely cool, use the tool to compare the sensor temperature with the ambient outside air temperature. If the temperatures are with 10° F (5.5° C) of each other, go to the next step; otherwise replace the sensor.

11. Check for the following:
- Low compression
- Leaking cylinder head gaskets
- Worn camshaft
- Bad valve timing or another problem with the valve train
- Restricted exhaust system

Fix any problems you may find, otherwise go to the next step.

12. Use a good engine cleaner to remove excessive carbon build-up from the combustion chambers and then re-test the system, If the symptoms are still present, go to the next step.

13. Review everything you've done so far. If no problem has been found, redo the Visual/Physical test, check your scan tool and/or check all electrical connections in the suspected circuit.

14. Check to see if the ECT sensor has shifted in value. With the engine completely cool, measure the resistance at the sensor and jot it down. Next, compare the approximate sensor temperature with the ambient outside air temperature. If the readings are similar, go back to Step 11; otherwise replace the ECT sensor.

A10 - HESITATION, SAG OR STUMBLE SYMPTOMS

■ These symptoms are defined as a momentary lack of response as the throttle is increased. They can occur at all engine speeds, but are usually worst when just starting out. The engine may stall sometimes.

1. Perform the Visual/Physical Checklist and On-Board Diagnostic System checks in the General Diagnostic Test section.

2. Check the flame arrestor for dirt, damage or some other restriction. Clean or replace if necessary; otherwise go to the next step.

3. Check for intermittent opens or shorts to ground in the MAP sensor circuit. If anything is found, correct the problem and then re-test the system; otherwise go to the next step.

4. Using the scan tool, check that the TP sensor is operating correctly. Check the throttle linkage for sticking, binding or general wear. Correct any problems an then re-test the system; otherwise go to the next step.

5. Check for contaminated or poor quality fuel. Check any and all fuel filters for blockage. If the fuel is bad or the filters are clogged, replace them; otherwise go to the next step.

6. Install a fuel pressure gauge and check the fuel pressure during the condition (A4 Fuel System Diagnosis). If there is a problem with pressure, go to A3 Fuel System Electrical Test; otherwise, go to the next step.

7. Check the fuel injectors and perform the Injector Balance test. Repair or replace any problems found; otherwise go to the next step.

8. Remove the spark plugs and check them thoroughly as detailed in the Maintenance section. Fix any problems or replace the plugs. If no problems are found, go to the next step.

9. Check that the voltage output at the alternator is 13.9-14.7 volts. If it is, move to the next step; otherwise, refer to the Charging System section for repairs.

10. Check for obvious overheating issues:
- Loose serpentine belt
- Bad seawater pump
- Cooling system restriction
- Bad thermostat

If anything is wrong, repair or replace it and then re-test the system. If not, go to the next step.

11. Check for the following:
- Low compression
- Deposits on the intake valves

Fix any problems you may find, otherwise go to the next step.

12. Review everything you've done so far. If no problem has been found, redo the Visual/Physical test, check your scan tool and/or check all electrical connections in the suspected circuit.

A11 - CUTS OUT OR MISSES SYMPTOMS

■ These symptoms are defined as a steady pulsation or jerking that changes with engine speed and is generally more pronounced as load increases. The exhaust will have a steady 'spitting' sound at idle, low speeds and/or on hard acceleration. Fuel starvation can cause the engine to cut out.

1. Perform the Visual/Physical Checklist and On-Board Diagnostic System checks in the General Diagnostic Test section.

2. Check that the high voltage switch (distributor) is properly aligned as detailed in the Ignition section. If not, align it and re-test the system. If its alright, go to the next step.

3. Check for contaminated or poor quality fuel. Check any and all fuel filters for blockage. If the fuel is bad or the filters are clogged, replace them; otherwise go to the next step.

4. Install a fuel pressure gauge and check the fuel pressure during the condition (A4 Fuel System Diagnosis). If there is a problem with pressure, go to A3 Fuel System Electrical Test; otherwise, go to the next step.

5. Check the fuel injectors and perform the Injector Balance test. Repair or replace any problems found; otherwise, go to the next step.

6. Check that all spark plugs are receiving adequate spark. If not, move to the next step; if they are, go to A2 Main Power Relay test.

7. Remove the spark plugs and check them thoroughly as detailed in the Maintenance section. Fix any problems or replace the plugs. If no problems are found, go to the next step.

8. Check for the following:
- Low compression
- Sticking, or leaking valves
- Bent pushrods
- Worn rocker arms
- Broken valve springs
- Worn camshaft
- Bad valve timing or some other valvetrain problem
- Restricted exhaust system

Fix any problems you find; otherwise, go to the next step.

9. Check the intake or exhaust manifolds for casting flash. Repair any problems or replace the manifolds and then re-test the system. If the manifolds are ok, go to the next step.

10. Check for electromagnetic interference by monitoring engine rpm with a scan tool or tachometer. A sudden increase in rpm with little change in the actual engine rpm will indicate interference. If present, locate and correct the source of interference; otherwise, go to the next step.

11. Review everything you've done so far. If no problem has been found, redo the Visual/Physical test, check your scan tool and/or check all electrical connections in the suspected circuit.

A12 - ROUGH, UNSTABLE OR INCORRECT IDLE AND STALLING SYMPTOMS

■ These symptoms are defined as the engine operating unevenly at idle. When severe, the engine or vessel may actually shake. The engine speed may vary with rpm. Each of these conditions may prove severe enough to stall the engine.

1. Perform the Visual/Physical Checklist and On-Board Diagnostic System checks in the General Diagnostic Test section.

2. Check for contaminated or poor quality fuel. Check any and all fuel filters for blockage. If the fuel is bad or the filters are clogged, replace them; otherwise go to the next step.

3. Install a fuel pressure gauge and check the fuel pressure during the condition (A4 Fuel System Diagnosis). If there is a problem with pressure, go to A3 Fuel System Electrical Test; otherwise, go to the next step.

4. Check the fuel injectors and perform the Injector Balance test. Repair or replace any problems found; otherwise, go to the next step.

5. Check that all spark plugs are receiving adequate spark. If not, move to the next step; if they are, go to A2 Main Power Relay test.

6. Check for cracks or other damage to the distributor, cap and rotor. Check for proper alignment of the distributor (Ignition System section). If anything is found, repair and/or replace the problem and the re-test the system. If everything is ok, go to the next step.
7. Remove the spark plugs and check them thoroughly as detailed in the Maintenance section. Fix any problems or replace the plugs. If no problems are found, go to the next step.
8. Check for the following:
- Low compression
- Sticking, or leaking valves
- Bent pushrods
- Worn rocker arms
- Broken valve springs
- Worn camshaft
- Bad valve timing or some other valvetrain problem
- Restricted exhaust system

Fix any problems you find; otherwise, go to the next step.
9. Review everything you've done so far. If no problem has been found, redo the Visual/Physical test, check your scan tool and/or check all electrical connections in the suspected circuit.

A13 - POOR FUEL ECONOMY SYMPTOMS

■ These symptoms are defined as fuel economy being noticeably than expected; or now consistently lower than it once was.

1. Perform the Visual/Physical Checklist and On-Board Diagnostic System checks in the General Diagnostic Test section.
2. Check or consider driving habits...excessive loads being carried? Regular hard acceleration? If so, there's no problem; otherwise, go to the next step.
3. Check all fuel lines and their connections for leaks. If something is found, fix it. If nothing, go to the next step.
4. Check the hull and keel for excessive dirt and/or barnacles. Check that the propeller is the correct size and pitch for the vessel. If a problem is found, clean the hull or replace the propeller; if not, go to the next step.
5. Check the flame arrestor for dirt, damage or some other restriction. Clean or replace if necessary; otherwise, go to the next step.
6. Check for contaminated or poor quality fuel. Check any and all fuel filters for blockage. If the fuel is bad or the filters are clogged, replace them; otherwise go to the next step.
7. Install a fuel pressure gauge and check the fuel pressure during the condition (A4 Fuel System Diagnosis). If there is a problem with pressure, go to A3 Fuel System Electrical Test; otherwise, go to the next step.
8. Check the fuel injectors and perform the Injector Balance test. Repair or replace any problems found; otherwise, go to the next step.
9. Check that all spark plugs are receiving adequate spark. If not, move to the next step; if they are, go to A2 Main Power Relay test.
10. Remove the spark plugs and check them thoroughly as detailed in the Maintenance section. Fix any problems or replace the plugs. If no problems are found, go to the next step.
11. Check all vacuum lines for splits, kinks or loose connections, fixing any problems. If nothing is found, go to the next step.
12. Check the engine compression. If low, repair the problem; if ok. Go to the next step.
13. Check the exhaust system for possible restrictions and/or damaged pipes. Repair or replace any problems found; if all is ok, go to the next step.
14. Review everything you've done so far. If no problem has been found, redo the Visual/Physical test, check your scan tool and/or check all electrical connections in the suspected circuit.

A14 - DIESELING OR RUN-ON SYMPTOMS

■ These symptoms are defined as the engine continuing to operate after the key is turned to the OFF position. If the engine itself operates smoothly during normal running though, check the ignition switch and adjustment.

1. Perform the Visual/Physical Checklist and On-Board Diagnostic System checks in the General Diagnostic Test section.
2. Install a fuel pressure gauge and check the fuel pressure during the condition (A4 Fuel System Diagnosis). If there is a problem with pressure, go to A3 Fuel System Electrical Test; otherwise, go to the next step.
3. Check the fuel injectors and perform the Injector Balance test. Repair or replace any problems found; otherwise, go to the next step.
4. Check for obvious overheating issues:
- Loose serpentine belt
- Bad seawater pump
- Cooling system restriction
- Bad thermostat

If anything is wrong, repair or replace it and then re-test the system. If not, go to the next step.
5. Check the fuel pump relay as detailed in A3 Fuel System Electrical Test; otherwise, go to the next step.
6. Review everything you've done so far. If no problem has been found, redo the Visual/Physical test, check your scan tool and/or check all electrical connections in the suspected circuit.

A15 - BACKFIRE SYMPTOMS

1. Perform the Visual/Physical Checklist and On-Board Diagnostic System checks in the General Diagnostic Test section.
2. Check the flame arrestor for dirt, damage or some other restriction. Clean or replace if necessary; otherwise, go to the next step.
3. Check for contaminated or poor quality fuel. Check any and all fuel filters for blockage. If the fuel is bad or the filters are clogged, replace them; otherwise go to the next step.
4. Install a fuel pressure gauge and check the fuel pressure during the condition (A4 Fuel System Diagnosis). If there is a problem with pressure, go to A3 Fuel System Electrical Test; otherwise, go to the next step.
5. Check the fuel injectors and perform the Injector Balance test. Repair or replace any problems found; otherwise, go to the next step.
6. Inspect the spark plug wires for open circuits, cracks in the insulation of loose connections. Find, repair or replace any problems and then re-test the system. If nothing is found, go to the next step.
7. Check that all spark plugs are receiving adequate spark. If not, move to the next step; if they are, go to A2 Main Power Relay test.
8. Remove the spark plugs and check them thoroughly as detailed in the Maintenance section. Fix any problems or replace the plugs. If no problems are found, go to the next step.
9. Check the MAP sensor circuit for opens or shorts to ground. If a problem is found, repair it and then re-test the system. If nothing is found, go to the next step.
10. Check that the TP sensor is operating properly. Check to see if the throttle linkage is sticking, binding or worn, which could cause the sensor voltage to be higher than normal. If the sensor has a problem or is higher voltage than normal, locate and repair the problem, or replace the sensor; and then retest the system. If everything checks out, go to the next step.
11. Check for the following:
- Low compression
- Sticking, or leaking valves
- Bent pushrods
- Worn rocker arms
- Broken valve springs
- Worn camshaft
- Bad valve timing or some other valvetrain problem
- Restricted exhaust system

Fix any problems you find; otherwise, go to the next step.
12. Review everything you've done so far. If no problem has been found, redo the Visual/Physical test, check your scan tool and/or check all electrical connections in the suspected circuit.

General Diagnostic Tests - 8.1L MPI Engines

ENGINE WILL NOT CRANK

1. Turn the ignition switch to the **ON** position. You should hear a click as the main power relay turns on and the fuel pump should energize and run for about 10 seconds. If this happens, go to Step 3; if not, go to the next step.

FUEL SYSTEM - FUEL INJECTION (EFI)

2. Check the battery for 12 volts. Ensure that the battery switch is in the **ON** position and that the engine-to-battery wiring is in good condition with sound connections. If there are problems found, repair them and recheck the system. If not problems are found, check for the following:
- Remote control lever not in Neutral
- Circuit breaker tripped
- Blown fuse
- Bad ignition switch
- Bad slave solenoid
- Bad or incorrectly adjusted neutral safety switch
- Bad starter or solenoid

3. With the ignition off, connect an analog tachometer to the grey auxiliary tach lead found under the PCM and then turn the ignition switch to the **ON** position. If there is no tach signal evident, go to the next step; otherwise:
- check the ignition system and there is spark at the plugs
- check for a clogged or dirty flame arrestor, too much fuel pump pressure or a bad pressure regulator
- check that there is fuel in the tank and the shut-off is open
- check for bad fuel, water in the fuel or a bad IAC valve

4. Disconnect the crankshaft position sensor and hook up a DVOM. Turn the ignition switch to the **ON** position and make sure there is 5 volts between the Black/Pink wire and the Grey wire. If there is proper voltage, go to the next step. If no voltage, or too low voltage, go to Step 11.

5. Check for continuity on the Tan wire (PCM pin **B14**). If continuity is found, go to the next step, if not, repair or replace the harness.

6. Now check for continuity between PCM pin **A23** and the 5 volt sensors (Grey wire) in the harness splice 101 (you can find this in the 5 Volt Sensor schematic). If continuity is found, move to the next step; otherwise, repair or replace the harness.

7. Use the scan tool and perform the Auto Self Test function. Disconnect the fuel pump and remove the fuse. Perform the test with spark plugs. If any problems were found in the self test, repair them; if no problems were found, the coils, plugs and harness from the PCM are good and you can move to the next step.

88. Remove the crankshaft position sensor and check the resistance across the sensor. Terminals **A - B** should be 23.30 meg-ohms and terminals **B - C** should be 23.21 meg-ohms. If the sensor values were correct, go to the next step; if not, replace the crankshaft position sensor.

9. Unplug the **A** and **C** PCM connectors and check for continuity between pins **A22** and **C15**, **A22** and **C16** and finally, **A22** and **C24** on the PCM. If there was continuity present in each instance, go to the next step; otherwise, replace the PCM.

10. Inspect the reluctor wheels. If they are damaged, replace them. If no damage is evident, check for the following:
- Remote control lever not in Neutral
- Circuit breaker tripped
- Blown fuse
- Bad ignition switch
- Bad slave solenoid
- Bad or incorrectly adjusted neutral safety switch
- Bad starter or solenoid

11. Connect the scan tool to the engine. If faults are present, repair them and recheck the system. If no faults have been stored, go back to Step 6.

ELECTRONIC SPARK TIMING (EST)

1. Connect the scan tool and choose Manual Test, Special Functions, Output Load Test, Ignition. Choose NO when prompted to perform test with any spark plugs. Select the cylinder number you wish to test.

2. Install a spark gap tester (#91-63998A-1) between plug and the plug wire end on the cylinder you've chosen. Make sure that the tester ground is connected to a know-good ground.

3. Repeat the plug test on each cylinder, verifying that spark is present. If spark is present on each cylinder, the ignition coils are all good. If not, go to the next step.

4. Check for continuity between the ignition coil and each coil harness (see the coil harness circuit in the schematic section). If you find a problem, repair it; if nothing is found, go to the next step.

5. Now check for continuity from the coil connector to the PCM (see the Ignition circuit schematic). If you discover a problem, fix it; if nothing shows up, substitute a known good coil for the suspect one and re-test.

ECM Pin Locations

◆ See Figures 139 thru 142

Connector A	Connector B
1 - Empty	1 - Splice 100
2 - Fuel Injector Circuit, Injectors 2, 3, 5, 8	2 - MAPT Connector Pin 2
3 - Empty	3 - MAPT Connector Pin 4
4 - CAN Line Connector Pin J	4 - Fuel Level Connector Pin C
5 - Diagnostic Connector Pin C	5 - Empty
6 - Odd Knock Connector Pin B	6 - Empty
7 - Even Knock Connector Pin B	7 - Oil Presure Connector Pin C
8 - Splice109	8 - Transom Connector Pin E
9 - !0-Pin Connector Pin 4	9 - Paddle Wheel Connector Pin C
10 - 10-Pin Connector 2	10 - Crankshaft Connector Pin C
11 - CAN Line Connector Pin K	11 - Seapump Pressure Connector Pin C
12 - Diagnostic Connector Pin B	12 - Transom Connector Pin D
13 - Odd Knock Connector Pin A	13 - Digital Trim
14 - Even Knock Connector Pin A	14 - Coolant Connector Pin B
15 - CAN Line Connector E	15 - Empty
16 - Splice 104	16 - Empty
17 - Fuel Injector Circuit, Injectors 1, 4 6, 7	17 - Empty
18 - Empty	18 - Splice 102
19 - Fuel Pump Relay Pin 86	19 - Shift Interupt Connector Pin C
20 - IAC Connector Pin 1	20 - Throttle Position Connector Pin C
21 - Gear Indicator Connector Pin B	21 - Splice 101
22 - Main Power Relay Pin 85	22 - Paddlwe Wheel Connector Pin D
23 - Splice 106	23 - Coil Driver Connector Pin B
24 - Splice 104	24 - Empty

Fig. 139 ECM pin-outs - 2001-08 V6 and V8 MPI engines (exc. 8.1L)

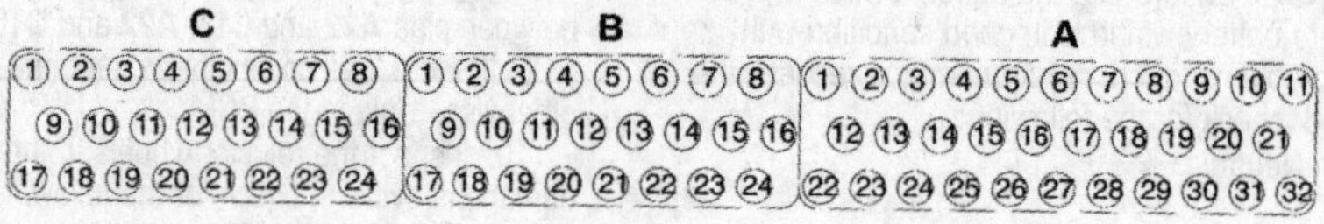

7-58 FUEL SYSTEM - FUEL INJECTION (EFI)

Connector A

1. Key-on Power
2. Empty
3. Map Sensor Connector Pin B
4. Oil Pressure Sensor Connector Pin C
5. Pitot Pressure Smart Transom Connector Pin D
6. Throttle Position Sensor Connector Pin C
7. Empty
8. Trim Position Smart Transom Connector Pin C
9. Starboard Tab Position Tab Connector Pin A
10. Port Tab Position Tab Connector Pin B
11. CAN Line Pos (+) Connector Pin J
12. Steering Position Smart Transom Connector Pin E
13. Seawater Temperature Paddle Wheel Connector Pin D
14. MAT Sensor Connector Pin B
15. Coolant Temperature Sensor Connector Pin B
16. Port Exhaust Water Temperature Connector Pin B
17. Starboard Exhaust Temperature Connector Pin B
18. Data Link Connector Pin C
19. Port Knock Sensor Connector Pin B
20. Starboard Knock Sensor Connector Pin B
21. CAN Line Neg (–) Connector Pin K
22. Splice 100
23. Splice 101
24. Seapump Pressure Connector Pin C
25. Fuel Level Connector Pin C
26. Fuel Level Connector Pin B
27. Empty
28. Data Link Connector Pin B
29. Port Knock Sensor Connector Pin A
30. Starboard Knock Connector Pin B
31. CAN2 Line Pos (+) Connector Pin G
32. CAN2 Line Neg (–) Connector Pin H

Fig. 140 ECM pin-outs - 8.1L V8 MPI engines, connector A

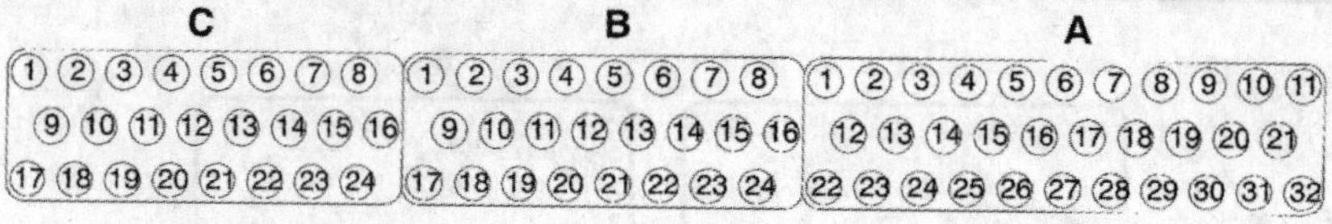

Connector B

1. Splice 106 Coil Return
2. Port Coil 1 Connector Pin G
3. Paddle Wheel Connector Pin C
4. Main Power Relay Pin 85
5. Empty
6. CAM Sensor Connector Pin C
7. IAC Connector Pin 1
8. Transmission Temperature Ground
9. Port Coil 5 Connector Pin C
10. Port Coil 3 Conector Pin F
11. Fuel Pump Relay Pin 85
12. Tachometer
13. Empty
14. Crankshaft Position Sensor Connector Pin C
15. Charging Harness Connector Pin F
16. Drive Lube Monitor Or Transmission Oil Temperature Switch
17. Splice 107
18. Splice 107
19. Empty
20. Fuel Injector 4 Connector Pin B
21. Charging Harness Connector Pin D
22. Fuel Injector 3 Connector Pin A
23. CAN Line Connector Pin E
24. Empty

Fig. 141 ECM pin-outs - 8.1L V8 MPI engines, connector B

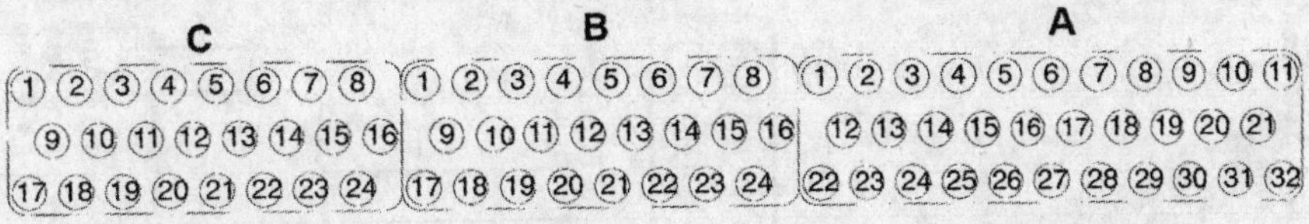

FUEL SYSTEM - FUEL INJECTION (EFI)

Connector C

1 - Port TAB Up Solenoid Connector Pin G
2 - Empty
3 - Fuel Injector 8 Connector Pin E
4 - Empty
5 - Fuel Injector 7 Connector Pin G
6 - Fuel Injector 1 Connector Pin H
7 - Starboard Coil 4 Connector Pin C
8 - Starboard Coil 2 Connector Pin B
9 - Starboard TAB Up Solenoid Connector Pin C
10 - Starboard TAB Down Solenoid Connector Pin D
11 - Fuel Injector 2 Connector Pin F
12 - Starboard Coil 6 Connector Pin F
13 - Port Coil 7 Connector Pin B
14 - Starboard Coil 8 Connector Pin G
15 - Splice 104
16 - Splice 104
17 - Empty
18 - Empty
19 - Port TAB Down Solenoid Connector Pin H
20 - Smart Transom Connector Pin G
21 - Fuel Injector 6 Connector Pin D
22 - Trim Up Relay Pin 85
23 - Fuel Injector 5 Connector Pin C
24 - Splice 104

Fig. 142 ECM pin-outs - 8.1L V8 MPI engines, connector C

Wiring Schematics

Wire Splice Descriptions 7-60
Fuses, IAC & Relays - 2001-08 V6 & V8 MPI (Exc. 8.1L) 7-60
MAP/T, CPS & TP Sensors - 2001-08 V6 & V8 MPI (Exc. 8.1L) 7-61
ECT, Seawater Pump & Oil Pressure Sensors -
2001-08 V6 & V8 MPI (Exc. 8.1L) 7-61
Fuel Injector Control & Diagnostic Circuits -
2001-08 V6 & V8 MPI (Exc. 8.1L) 7-61
CAN, Fuel Level, Paddle Wheel & Temp Circuits -
2001-08 V6 & V8 MPI (Exc. 8.1L) 7-62
Engine 12 Volt Ground Circuit - 2001-08 V6 & V8 MPI (Exc. 8.1L) .. 7-62
Diagnostics Circuit - 2001-08 V6 & V8 MPI (Exc. 8.1L) 7-62
ECT Circuit - 2001-08 V6 & V8 MPI (Exc. 8.1L) 7-63
IAC Circuit - 2001-08 V6 & V8 MPI (Exc. 8.1L) 7-63
MAP/T Circuit - 2001-08 V6 & V8 MPI (Exc. 8.1L) 7-63
TPS Circuit - 2001-08 V6 & V8 MPI (Exc. 8.1L) 7-63
Knock Sensor Circuit - 2001-08 V6 & V8 MPI (Exc. 8.1L) 7-64
Harness-To-Paddle Wheel Connector Circuit -
2001-08 V6 & V8 MPI (Exc. 8.1L) 7-64
Fuel Level Circuit - 2001-08 V6 & V8 MPI (Exc. 8.1L) 7-64
Fuel Pump Relay Circuit - 2001-08 V6 & V8 MPI (Exc. 8.1L) 7-64
Control Area Network (CAN) Circuit - 2001-08 V6 & V8 MPI (Exc. 8.1L) . 7-65
5 Volt Sensor Circuit - 8.1L MPI 7-65
Fuel Injector Harness - 8.1L MPI 7-65
Main Power Relay Circuit - 8.1L MPI 7-65
Diagnostics Circuit - 8.1L MPI 7-66
ECT Circuit - 8.1L MPI 7-66
IAC Circuit - 8.1L MPI 7-66
MAT Circuit - 8.1L MPI 7-67
MAP Circuit - 8.1L MPI 7-67
Crank & Cam Sensor Circuit - 8.1L MPI 7-67
TP Sensor Circuit - 8.1L MPI 7-67
10-Pin Harness Circuit - 8.1L MPI 7-68
Knock Sensor Circuits - 8.1L MPI 7-68
Harness-To-Paddle Wheel Harness Circuit - 8.1L MPI 7-68
Fuel Level Sensor Circuit - 8.1L MPI 7-68
Fuel Pump Relay Circuit - 8.1L MPI 7-69
Control Area Network (CAN) Circuit - 8.1L MPI 7-69

7-60 FUEL SYSTEM - FUEL INJECTION (EFI)

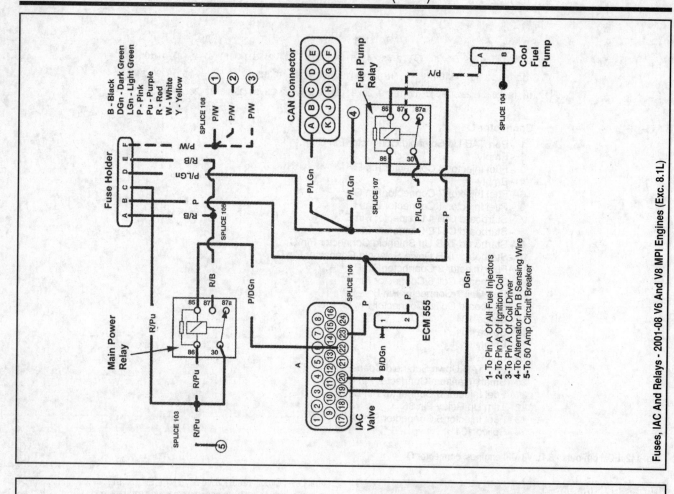

Fuses, IAC And Relays - 2001-08 V6 And V8 MPI Engines (Exc. 8.1L)

4.3 L, 5.0 L, 5.7 L & 6.2 L MPI

Splice Number	Description
100	5 Volt Transducer Ground
101	5 Volt Transducer Power
102	Wake Line
103	12 Volt 50 amp Protected
104	12 Volt Engine Ground
105	12 Volt From MPR
106	Switched 12 Volt Fused
107	12 Volt Fused
108	12 Volt Fused to All Injectors
109	Transmission and Drive Lube
110	Injectors 1, 4, 6, 7
111	Injectors 2, 3, 5, 8
113	Tachometer Lead
114	Ignition Coil and Coil Driver

8.1 L MPI

Splice Number	Description
100	5 Volt Transducer Ground
101	5 Volt Transducer Power For Sensors
102	Fused 20 AMP 12 Volt Power, Key ON Only
103	Continuous 12 Volt Battery Power
104	Battery Ground
105	Main Power Relay, Key ON Only
106	Coil Return
107	Fused 12 Volt Power to the PCM
108	Fused 12 Volt Power to the Fuel Pump Relay and the Injectors
110	Continuous 12 Volt Battery Power to the Main Power Relay
111	Fused 12 Volt Power to the Fuel Pumps

Wire Splice Descriptions

FUEL SYSTEM - FUEL INJECTION (EFI) 7-61

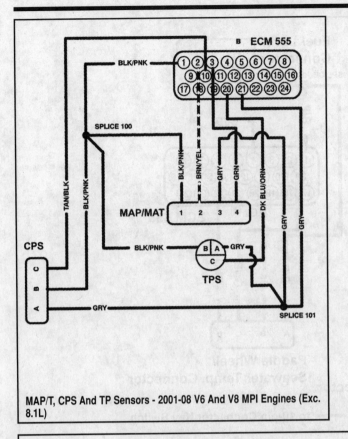

MAP/T, CPS And TP Sensors - 2001-08 V6 And V8 MPI Engines (Exc. 8.1L)

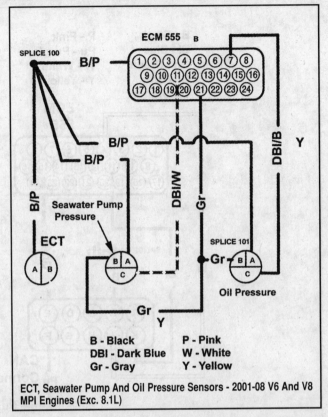

ECT, Seawater Pump And Oil Pressure Sensors - 2001-08 V6 And V8 MPI Engines (Exc. 8.1L)

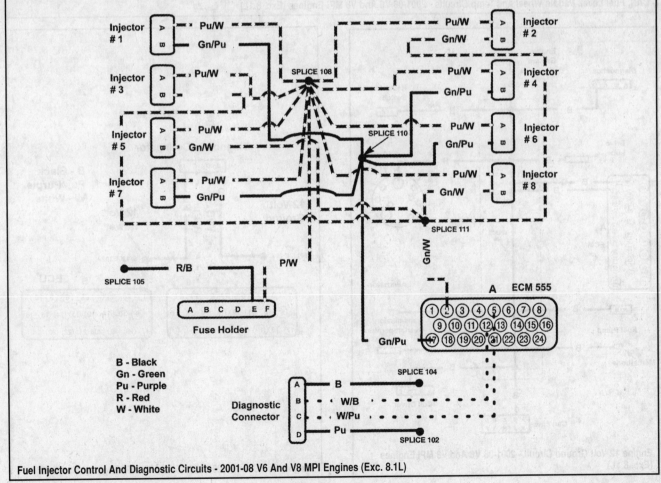

Fuel Injector Control And Diagnostic Circuits - 2001-08 V6 And V8 MPI Engines (Exc. 8.1L)

7-62 FUEL SYSTEM - FUEL INJECTION (EFI)

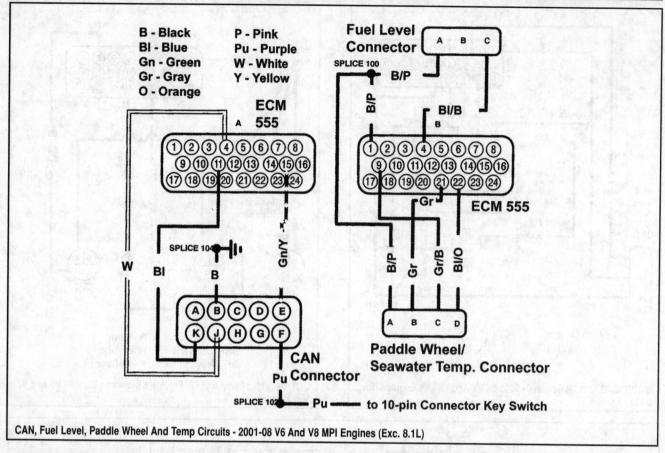

CAN, Fuel Level, Paddle Wheel And Temp Circuits - 2001-08 V6 And V8 MPI Engines (Exc. 8.1L)

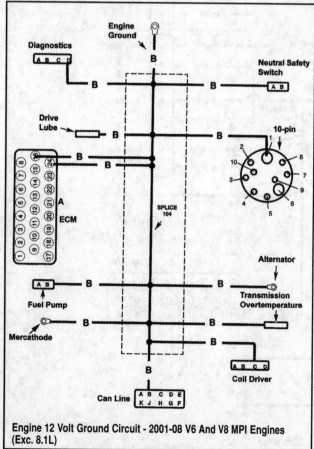

Engine 12 Volt Ground Circuit - 2001-08 V6 And V8 MPI Engines (Exc. 8.1L)

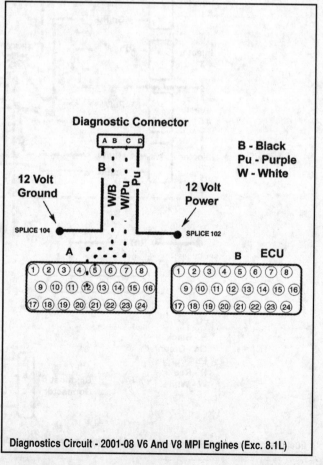

Diagnostics Circuit - 2001-08 V6 And V8 MPI Engines (Exc. 8.1L)

FUEL SYSTEM - FUEL INJECTION (EFI) 7-63

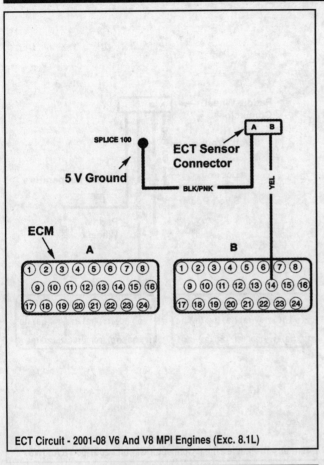

ECT Circuit - 2001-08 V6 And V8 MPI Engines (Exc. 8.1L)

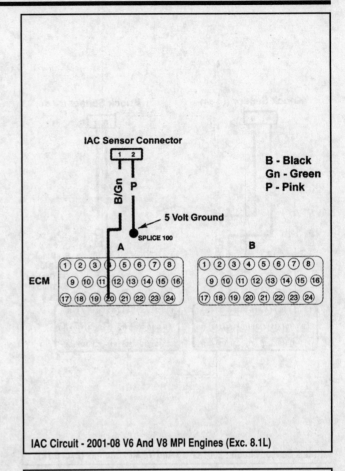

IAC Circuit - 2001-08 V6 And V8 MPI Engines (Exc. 8.1L)

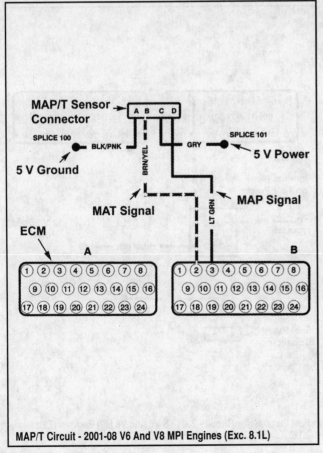

MAP/T Circuit - 2001-08 V6 And V8 MPI Engines (Exc. 8.1L)

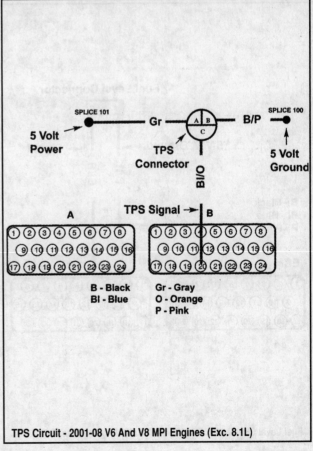

TPS Circuit - 2001-08 V6 And V8 MPI Engines (Exc. 8.1L)

7-64 FUEL SYSTEM - FUEL INJECTION (EFI)

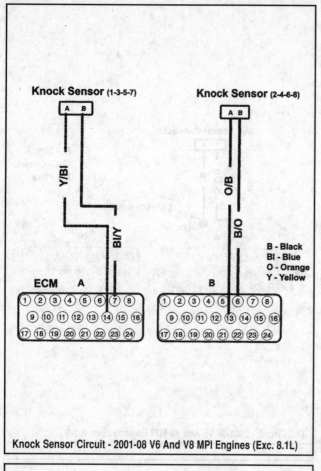

Knock Sensor Circuit - 2001-08 V6 And V8 MPI Engines (Exc. 8.1L)

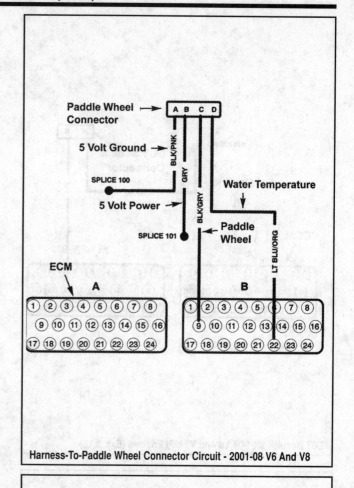

Harness-To-Paddle Wheel Connector Circuit - 2001-08 V6 And V8

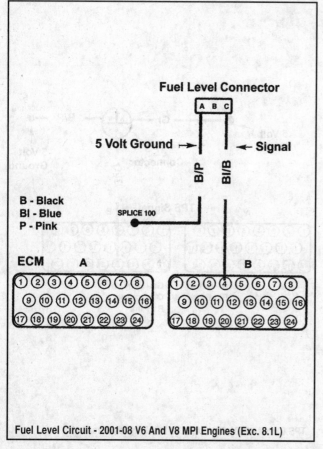

Fuel Level Circuit - 2001-08 V6 And V8 MPI Engines (Exc. 8.1L)

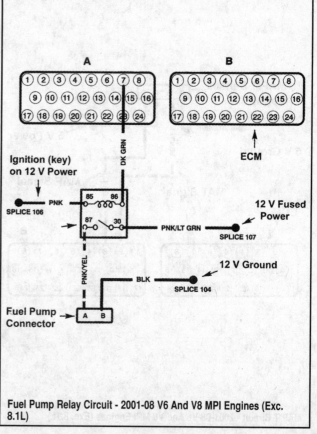

Fuel Pump Relay Circuit - 2001-08 V6 And V8 MPI Engines (Exc. 8.1L)

FUEL SYSTEM - FUEL INJECTION (EFI) 7-65

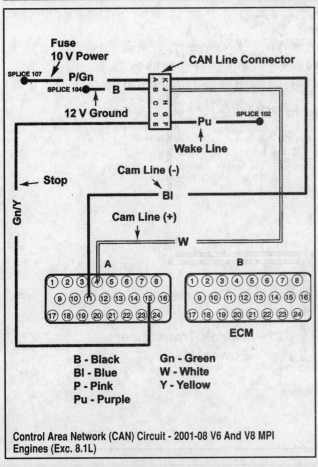

Control Area Network (CAN) Circuit - 2001-08 V6 And V8 MPI Engines (Exc. 8.1L)

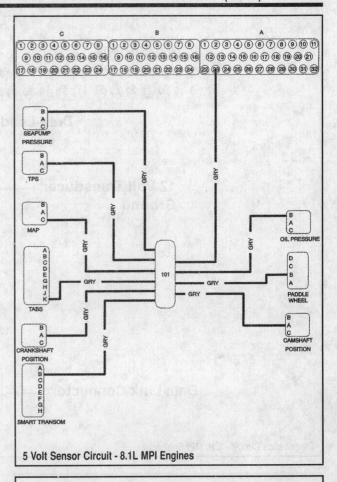

5 Volt Sensor Circuit - 8.1L MPI Engines

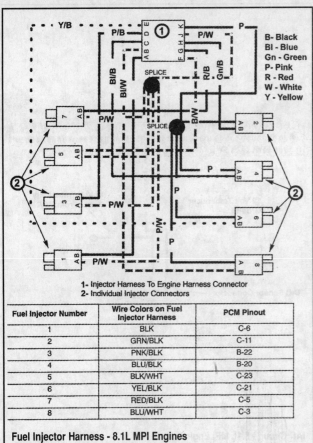

Fuel Injector Number	Wire Colors on Fuel Injector Harness	PCM Pinout
1	BLK	C-6
2	GRN/BLK	C-11
3	PNK/BLK	B-22
4	BLU/BLK	B-20
5	BLK/WHT	C-23
6	YEL/BLK	C-21
7	RED/BLK	C-5
8	BLU/WHT	C-3

Fuel Injector Harness - 8.1L MPI Engines

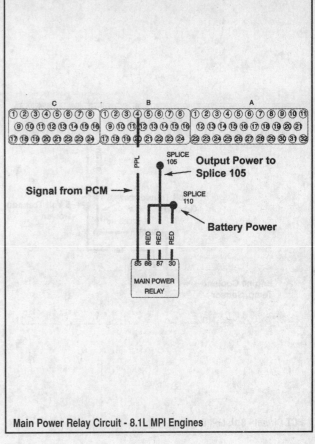

Main Power Relay Circuit - 8.1L MPI Engines

7-66 FUEL SYSTEM - FUEL INJECTION (EFI)

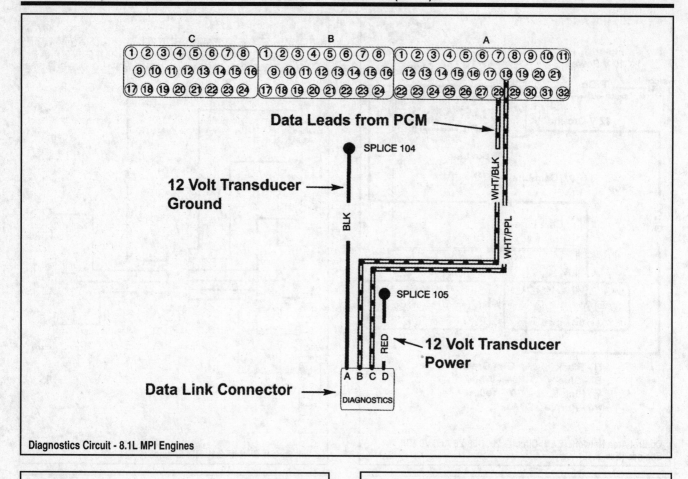

Diagnostics Circuit - 8.1L MPI Engines

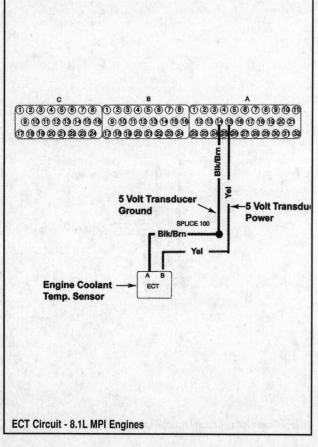

ECT Circuit - 8.1L MPI Engines

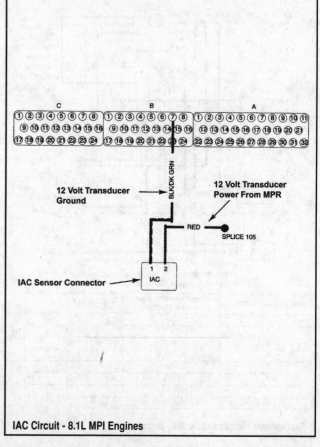

IAC Circuit - 8.1L MPI Engines

FUEL SYSTEM - FUEL INJECTION (EFI) 7-67

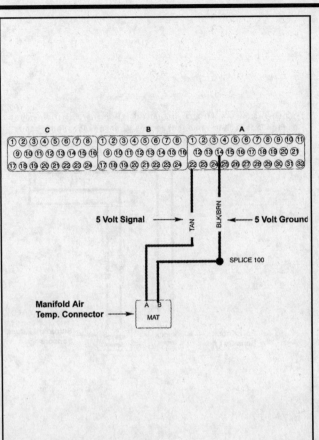

MAT Circuit - 8.1L MPI Engines

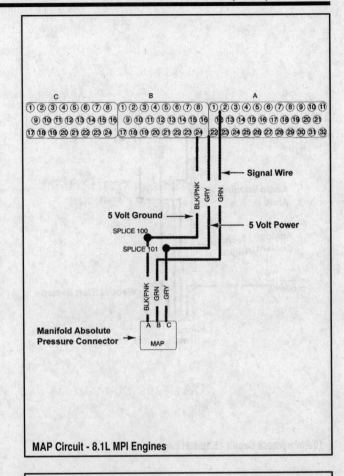

MAP Circuit - 8.1L MPI Engines

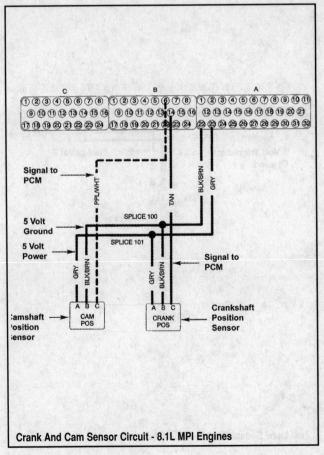

Crank And Cam Sensor Circuit - 8.1L MPI Engines

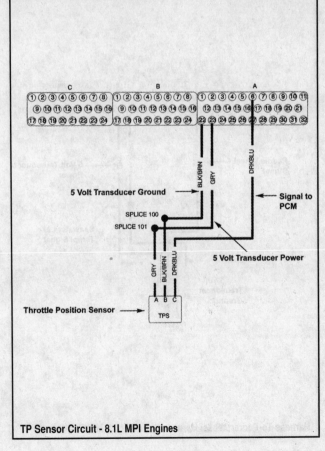

TP Sensor Circuit - 8.1L MPI Engines

7-68 FUEL SYSTEM - FUEL INJECTION (EFI)

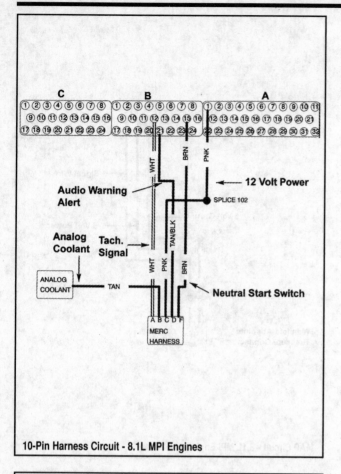

10-Pin Harness Circuit - 8.1L MPI Engines

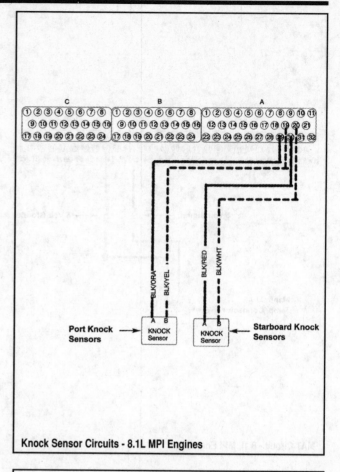

Knock Sensor Circuits - 8.1L MPI Engines

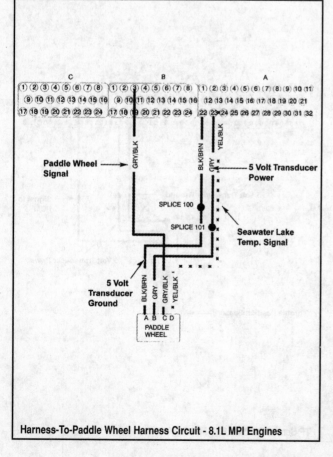

Harness-To-Paddle Wheel Harness Circuit - 8.1L MPI Engines

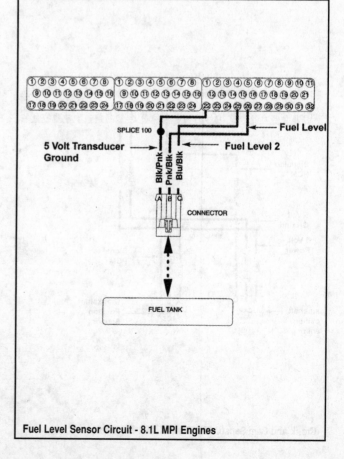

Fuel Level Sensor Circuit - 8.1L MPI Engines

FUEL SYSTEM - FUEL INJECTION (EFI) 7-69

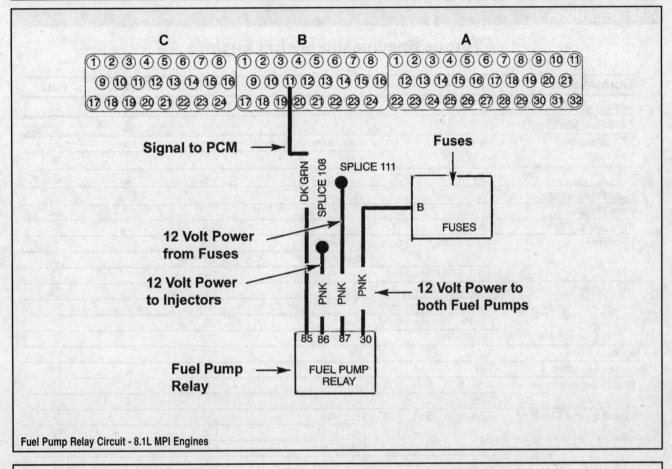

Fuel Pump Relay Circuit - 8.1L MPI Engines

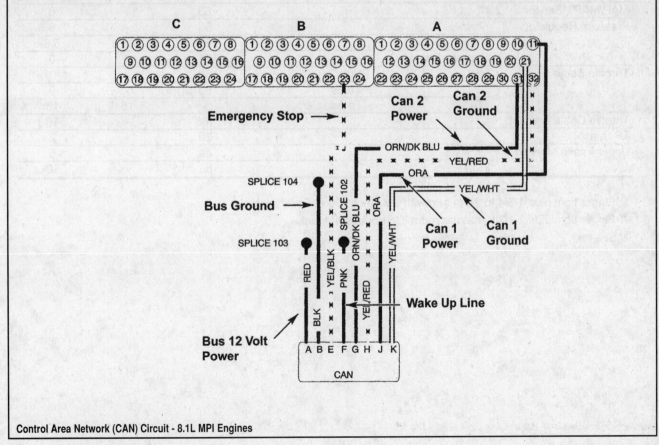

Control Area Network (CAN) Circuit - 8.1L MPI Engines

FUEL SYSTEM - FUEL INJECTION (EFI)

SPECIFICATIONS

Torque Specifications - EFI Engines

Component			Inch lbs.	ft. lbs.	Nm
Cool Fuel Assy (Gen II)			-	30	41
Cool Fuel Module (Gen III)			-	③	③
ECT Sensor	MPI	4.3L, 5.0L, 5.7L, 6.2L	-	15	20
		8.1L	-	①	①
	TBI		-	①	①
Fuel Inlet Fitting			-	②	②
Fuel Lines			-	19	24
Fuel Meter Cover	TBI		28	-	3
Fuel Rail	4.3L, 5.0L, 5.7L, 6.2L	Bracket Bolt	53	-	6
		Retainer Nut	27	-	3
	8.1L, 2001 350 Mag		105	-	12
IAC Valve	Bracket Nut		132	-	15
	Valve to Bracket Bolt		168	-	19
	Plug to Throttle Body		24	-	3
	IAC Valve		20	-	2
Ignition Coil	Studs		106	-	12
Intake Manifold	TBI		150	-	17
Knock Sensor	4.3L, 5.0L, 5.7L, 6.2L		-	15	20
	8.1L		-	14	19
Lower Intake Manifold	4.3L, 5.0L, 5.7L, 6.2L	1st Pass	27	-	3
		2nd Pass	106	-	12
		Final Pass	132	-	15
	8.1L		150	-	17
MAP/MAPT Sensor			53	-	6
Pressure Regulator	Screws		53	-	6
	Retainer Nuts		50	-	5.6
	Stepped Screw		81	-	9
Throttle Body	MPI	Nuts	88	-	10
		Studs	80	-	9
	TBI	To Adapter	-	15	19
Throttle Cable Bracket			168	-	19
TP Sensor			20	-	2
Upper Intake Manifild	1st Pass		44	-	5
	Final Pass		89	-	10

① Finger tight plus 1-3/4 to 2-1/4 turn with a wrench. Do not overtighten!
② Finger tight plus 2-1/2 turns maximum. Do not overtighten!
③ See text

8 COOLING SYSTEM

COOLING SYSTEMS 8-2
MAINTENANCE AND TESTING 8-2
FLOW DIAGRAMS 8-8

Index

- COOLING SYSTEM 8-2
  - DRAINING, FLUSHING & FILLING 8-2
  - DRIVE BELTS 8-3
  - FLUID LEVEL CHECK 8-3
  - PRESSURE CAP TEST 8-3
  - PRESSURE TEST 8-3
  - RAW WATER COOLING TEST 8-3
  - RAW WATER FLOW TEST 8-3
- COOLING SYSTEMS 8-2
  - DESCRIPTION & OPERATION 8-2
  - TROUBLESHOOTING 8-2
- CROSSOVER 8-3
  - REMOVAL & INSTALLATION 8-3
- FLOW DIAGRAMS 8-8
  - INDEX 8-8
  - 3.0L W/SEAWATER 8-9
  - 3.0L W/CLOSED 8-9
  - CARB/TBI V6 - SEAWATER (ALPHA) 8-10
  - CARB/TBI V6 - SEAWATER (BRAVO) 8-11
  - CARB/TBI V6 - CLOSED (ALPHA) 8-12
  - CARB/TBI V6 - CLOSED (BRAVO) 8-13
  - MPI V6 - SEAWATER (ALPHA) 8-13
  - MPI V6 - SEAWATER (BRAVO) 8-14
  - MPI V6 – CLOSED 8-14
  - CARB/TBI V8 - SEAWATER (ALPHA) 8-15
  - CARB/TBI V8, ALSO INC. 2001 MPI UP TO SERIAL #OM299999 - SEAWATER (BRAVO, INBOARD & SKI) 8-15
  - CARB/TBI V8, ALSO INC. 2001 MPI UP TO SERIAL #OM299999 - CLOSED 8-16
  - 5.0L, 5.7L & 6.2L MPI V8 (EXC. DRY JOINT) - SEAWATER (ALPHA) .. 8-16
  - 5.0L, 5.7L & 6.2L MPI V8 (EXC. DRY JOINT) - SEAWATER (BRAVO, INBOARD & TOW SPORT) 8-17
  - 5.0L, 5.7L & 6.2L MPI V8 (EXC. DRY JOINT) - CLOSED 8-17
  - 8.1L V8 (EXC. DRY JOINT) - CLOSED 8-18
  - V8 W/ALPHA & DRY JOINT (EXC. MPD MODELS) - SEAWATER 8-18
  - V8 W/ALPHA & DRY JOINT (MPD MODELS) - SEAWATER 8-19
  - V8 W/BRAVO & DRY JOINT (EXC. MPD MODELS) - SEAWATER 8-19
  - V8 W/BRAVO & DRY JOINT (MPD MODELS) - SEAWATER 8-20
  - V8 W/STERNDRIVES & DRY JOINT - CLOSED 8-20
  - V8 W/INLINE INBOARDS & DRY JOINT - SEAWATER 8-21
  - V8 W/V-DRIVE INBOARDS & DRY JOINT - SEAWATER 8-21
  - V8 W/INLINE INBOARDS & DRY JOINT - CLOSED 8-22
  - V8 W/V-DRIVE INBOARDS & DRY JOINT - CLOSED 8-22
  - V8 W/BRAVO OR INLINE INBOARDS, COLD RISERS & DRY JOINT - SEAWATER . 8-23
  - V8 W/V-DRIVE INBOARDS, COLD RISERS & DRY JOINT - SEAWATER 8-23
  - V8 W/STERN DRIVE OR INLINE INBOARDS, COLD RISERS & DRY JOINT - CLOSED 8-24
  - V8 W/V-DRIVE INBOARDS, COLD RISERS & DRY JOINT - CLOSED 8-24
  - V8 W/BRAVO OR INLINE INBOARDS, WARM RISERS & DRY JOINT - SEAWATER 8-25
  - V8 W/V-DRIVE INBOARDS, WARM RISERS & DRY JOINT - SEAWATER 8-25
  - V8 W/BRAVO OR INLINE INBOARDS, WARM RISERS & DRY JOINT - CLOSED ... 8-26
  - V8 W/V-DRIVE INBOARDS, WARM RISERS & DRY JOINT - CLOSED 8-26
- HEAT EXCHANGER 8-4
  - CLEANING 8-4
  - REMOVAL, DISASSEMBLY & INSTALLATION 8-4
  - TESTING 8-5
- IMPELLER 8-5
  - DISASSEMBLY & ASSEMBLY 8-5
    - ALPHA DRIVES 8-5
    - BRAVO/INBOARD - CARB & TBI 8-5
    - BRAVO/INBOARD - MPI 8-6
  - OUTPUT TEST 8-7
  - REMOVAL & INSTALLATION 8-5
- MAINTENANCE AND TESTING 8-2
  - COOLING SYSTEM 8-2
  - CROSSOVER 8-3
  - GENERAL INFORMATION 8-2
  - HEAT EXCHANGER 8-4
  - SEA STRAINER 8-5
  - SEAWATER PUMP/IMPELLER 8-5
  - THERMOSTAT 8-7
  - WATER (ENGINE) CIRCULATING PUMP 8-8
  - WATER DISTRIBUTION HOUSING 8-7
- SEA STRAINER 8-5
  - REMOVAL & INSTALLATION 8-5
- SEAWATER PUMP/IMPELLER 8-5
  - DISASSEMBLY & ASSEMBLY 8-5
    - ALPHA DRIVES 8-5
    - BRAVO/INBOARD - CARB & TBI 8-5
    - BRAVO/INBOARD - MPI 8-6
  - OUTPUT TEST 8-7
  - REMOVAL & INSTALLATION 8-5
- THERMOSTAT 8-7
- TROUBLESHOOTING 8-2
  - ENGINES OVERHEATS - CLOSED SYSTEM . 8-2
  - ENGINE OVERHEATS - MECHANICAL 8-2
  - ENGINE OVERHEATS - SEAWATER SYSTEM 8-2
  - WATER IN CYLINDERS 8-2
  - WATER IN OIL 8-2
- WATER (ENGINE) CIRCULATING PUMP 8-8
  - REMOVAL & INSTALLATION 8-8
- WATER DISTRIBUTION HOUSING 8-7
  - REMOVAL & INSTALLATION 8-7

8-2 COOLING SYSTEM

COOLING SYSTEMS

Description & Operation

All MerCruiser engines are cooled by means of one of two systems: an external-water, Seawater system; or a Closed system, which actually incorporates the Seawater system into a closed automotive-style anti-freeze system. All engines may come equipped with either of the two systems. Please refer to the flow diagrams at the end of this section.

SEAWATER SYSTEM

As implied by the name, this system utilizes water from outside the boat to cool the engine and certain related components. All versions of this system utilize two water pumps, hoses and a thermostat.

Models with an Alpha One stern drive unit use a pump mounted on top of the lower unit which draws water in through the drive unit and sends it on to an engine circulating pump attached to the front of the cylinder block - although different in construction, this pump is quite similar to a typical water pump you would find on your car or truck. From here the water is circulated through the engine block, cylinder heads and exhaust manifold(s); and then expelled back into the body of water where it originated.

Models with a Bravo stern drive or inboard use a belt-driven seawater pump mounted on the lower end of the engine to draw the water up through the drive unit and then on to the engine circulating pump just like on Alpha models.

CLOSED SYSTEM

This system is actually two systems in one - a closed freshwater system and a seawater system. All versions of this system utilize two water pumps, hoses, a thermostat and a heat exchanger.

In the closed portion of the system, a mixture of freshwater and anti-freeze is circulated thru-out the engine block, cylinder head and a heat exchanger (similar to your car's radiator). On certain V8 engines, the exhaust manifolds are also cooled. The pressurized water/coolant never leaves the system and is thermostatically controlled.

Lacking a radiator and fan like an automobile, it is necessary to find another means of keeping the fluid in the closed system from boiling and this is where the second portion of this system comes into play. Seawater from outside the vessel is drawn in by means of a pick-up pump in the stern drive unit on Alpha models, or by a belt-driven engine-mounted seawater pump on all other models. Unlike true Seawater models as described previously, the seawater is pumped into the heat exchanger rather than through the engine. In fact, it bypasses the engine circulating pump altogether. Once in the exchanger, the seawater is routed through tubes surrounding the closed system tubes, thus cooling the fresh water/anti-freeze mix as it passes through the exchanger. The heated seawater is then routed through the exhaust manifold(s) and then back overboard.

MAINTENANCE AND TESTING

General Information

First, the most important words in this section: The engine cannot be operated for even five seconds without water moving through the water pick-up/seawater pump or the pump impeller will be damaged. Therefore, never start the engine, even for testing purposes, without the boat being in the water, or provision having been made for water to pass through the seawater pick-up pump.

■ Marine thermostats are rated at 143°-160°. An automotive-type thermostat must never be used because of the higher temperature ratings. Such a high rating would cause the engine to run much hotter than normal.

Troubleshooting

ENGINE OVERHEATS - SEAWATER SYSTEM

- Loose or broken circulating pump belt or pick-up water pump belt.
- Inaccurate temperature gauge or sender.
- An accessory or barnacles in front of water pickup causing turbulence.
- Defective water pick-up pump or seawater pump.
- Loose hose connections between pick-up and pump - sucking air.
- Pump fails to hold prime due to air leaks.
- Ice in water passages.
- Defective engine circulating pump.
- Defective thermostat.

ENGINES OVERHEATS - CLOSED SYSTEM, IN ADDITION TO ABOVE

- Closed cooling system reservoir level low.
- Plugged heat exchanger cores.
- Too much anti-freeze in closed system.
- Fresh water kit not installed properly.
- Exhaust elbow dump fittings "bottomed" on inner water jacket.

ENGINE OVERHEATS - MECHANICAL

- Incorrect ignition timing.
- Spark plug wires crossed.
- Lean air/fuel mixture.
- Pre-ignition - spark plugs are wrong heat range.
- Engine laboring - engine rpm below specifications at WOT.
- Poor lubrication.
- Water in cylinders - due to warped cylinder head.
- Distributor not functioning properly.
- Clogged exhaust elbows.
- Exhaust flappers stuck closed.
- Cabin hot water heater incorrectly connected to engine.

WATER IN CYLINDERS

- Backwash through exhaust system.
- Loose cylinder head bolts.
- Blown cylinder head gasket.
- Warped cylinder head.
- Cracked or corroded exhaust manifold.
- Cracked block in valve lifter area.
- Improper engine or exhaust hose installation.
- Cracked cylinder wall.

WATER IN OIL

- Backwash through exhaust system.
- Water seeping past piston rings from flooded combustion chamber.
- Thermostat stuck open or missing - condensation forms because engine is operating too cool.
- Cracked cylinder block.
- Intake manifold water passage leak.

Cooling System

The cooling system, raw water or closed, should be cleaned and flushed at least once every two years; more often if possible. Please refer to the Maintenance section for detailed flushing procedures.

DRAINING, FLUSHING & FILLING

The cooling system should be drained, cleaned, and refilled each season, although Mercruiser's recommendations are for every two years on normal anti-freeze systems. We think its cheap insurance to do it every season, but you certainly can't go wrong by following the factory's suggestion. The bow of the boat must be higher than the stern to properly drain the cooling system. If the bow is not higher than the stern, water will remain in the

COOLING SYSTEM

cylinder block and in the exhaust manifold. Insert a piece of wire into the drain holes, but not in the petcock, to ensure sand, silt, or other foreign material is not blocking the drain opening.

If the engine is not completely drained for winter storage, trapped water can freeze and cause severe damage. The water in the oil cooler - if so equipped - must also be drained.

■ For complete details, procedures and illustrations on draining, filling and/or flushing of the cooling system, please refer to the Maintenance section.

DRIVE BELTS

■ For complete details, procedures and illustrations on drive belt removal and adjustment, please refer to the Maintenance section.

FLUID LEVEL CHECK

■ For complete details, procedures and illustrations on checking the fluid level in the closed cooling system, please refer to the Maintenance section.

PRESSURE TEST

■ For complete details, procedures and illustrations covering pressure testing the cooling system, please refer to the Maintenance section.

PRESSURE CAP TEST

■ For complete details, procedures and illustrations covering testing the closed cooling system pressure cap, please refer to the Maintenance section.

RAW WATER COOLING TEST

If you suspect that the raw water system is taking in air, perform the following test.

✱✱ WARNING

This test MUST be performed with the vessel in the water. DO NOT perform this test with the vessel out of water and a flushing attachment attached.

1. Disconnect the water hose at the raw water pump outlet and then disconnect the same hose at the inlet on the thermostat housing.
2. Connect a clear vinyl hose of the same diameter between the pump and the housing, Tighten the clamps securely.
3. Start the engine and drive the boat at the lowest rpm at which the overheating problem has been observed. Have an assistant (do not do this yourself - stay at the helm!) check the clear hose for evidence of air bubbles being drawn through the system.
4. If air bubbles are present, air is being sucked into the system somewhere between the pump and the inlet holes on the lower unit. Check all hoses, clamps and fittings for damage or other signs of deterioration. Check the lower unit water tube, guide, seal, grommet and water passage cover gasket for damage, signs of deterioration or leakage. Check the pump impeller plate gasket and housing O-ring for damage and/or leakage.
5. If no bubbles are observed in Step 3, remove the clear hose and reconnect the original water hose.

RAW WATER FLOW TEST

If you suspect that a low cooling water supply is causing an overheat condition, perform the following test.

✱✱ WARNING

This test MUST be performed with the vessel in the water. DO NOT perform this test with the vessel out of water and a flushing attachment attached.

1. Disconnect the water hose at the thermostat housing inlet. Hold the hose vertically so that the open end is not higher than the top of the engine.
2. Start the engine and allow it to idle. A column of water should rise from the hose approximately 2-4 in. high. Less than 2 in. indicates a likely restriction in the supply line, or the raw water pump is defective. Check for a blocked inlet screen at the drive or a crimped/broken pivot housing at the gimbal housing/transom plate. Also, confirm that the pump impeller is not worn or broken.

✱✱ CAUTION

Don't forget to shut the engine down immediately after observing the height of water coming out of the hose!

3. Reconnect the hose to the housing and tighten the clamp securely.

Crossover

REMOVAL & INSTALLATION

8.1L Engines Only

◆ See Figures 1 and 2

1. Drain the cooling system.
2. Disconnect the coolant lines at the heat exchanger if you haven't already. Disconnect the hose leading to the overflow tank.
3. Remove the coolant reservoir.
4. Loosen the clamps on the water inlet seal and hose.
5. Loosen the heat exchanger mounting bolts and then loosen the hex bolt on the retaining strap.
6. Remove the retaining strap, take the mounting bolts out and then lift off the exchanger.
7. Remove the thermostat retainer and pull out the thermostat.
8. Tag and disconnect any electrical connections at the crossover.
9. Disconnect any remaining coolant lines at the crossover.
10. Remove the 2 mounting bolts on each side of the crossover at the cylinder heads and remove the crossover.

To install:

11. Coat the inside rubber portion of the bypass seal on the bottom of the crossover with a little soapy water. Position the hose clamp around the rubber portion of the seal.
12. Position the crossover so that the rubber portion of the bypass seal is directly over the fitting at the top of the water pump. Press the unit down until the end of the seal is resting against the top of the pump.
13. Position a new flange gasket between the flange and cylinder head at each end. Make sure all gaskets and lines are in alignment and then tighten the mounting bolts to 37 ft. lbs. (50 Nm).
14. Tighten the clamp on the bypass seal to 16 ft. lbs. (21 Nm).

Fig. 1 The crossover is mounted underneath the heat exchanger

8-4 COOLING SYSTEM

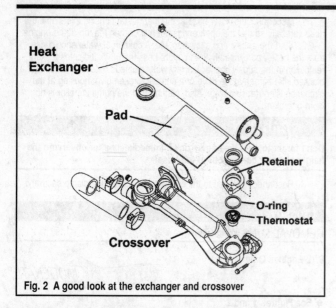

Fig. 2 A good look at the exchanger and crossover

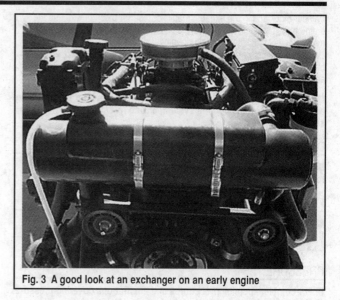

Fig. 3 A good look at an exchanger on an early engine

15. If you removed any of the fittings or pipe plugs, install them and tighten to 17 ft. lbs. (23 Nm).
16. Reconnect all electrical leads and coolant lines. Tighten the hose clamps securely.
17. Insert the thermostat and install the retainer.
18. Carefully position the heat exchanger and install the retaining strap. Do not tighten to tight or you could damage the holding slots.
19. Tighten the retainer bolts securely. Refill the cooling system, start the engine and check the system for leaks.

Heat Exchanger - Closed Cooling System Only

REMOVAL, DISASSEMBLY & INSTALLATION

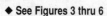 

◆ See Figures 3 thru 6

■ Heat exchangers are used on closed cooling system engines only.

1. Make sure that the engine is cold, and then drain the seawater and closed cooling systems.
2. Remove the flame arrestor if necessary.
3. Disconnect all hoses at the exchanger and position them out of the way. You may want to tag each hose for ease of installation later on.
4. On models with an air actuator, push in on the bottom of the air manifold and pull out the air hoses.
5. Remove the 2 large hose clamps and lift off the exchanger.
6. If you haven't unclipped the air pump (on models so equipped), do it now and then remove the bracket and air manifold.
7. Loosen the end cap bolts and remove them along with the gaskets and sealing washers.

To assemble and install:

8. Coat new end cap gaskets with Perfect Seal and then install the end caps with new sealing washers. Tighten the bolts to 36-72 inch lbs. (4-8 Nm) on carbureted and TBI engines; on MPI engines they should be tightened to 54 inch lbs. (6 Nm).

■ Make sure the O-ring is installed between the end cover and the gasket.

9. If equipped, install the air pump bracket and the air manifold, tightening the mounting bolts securely. Snap in the air pump.
10. Lower the exchanger into position on the bracket and connect all water/coolant hoses so that they are aligned and fully seated on the fittings. Tighten each of their hose clamps securely (without pinching the hose) and then install the 2 large clamps holding the exchanger to the bracket.

■ Certain newer models may utilize a rubber hose over the clamp - make sure this hose is between the clamp and the mounting bracket.

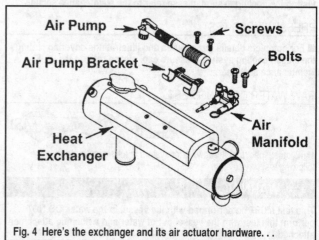

Fig. 4 Here's the exchanger and its air actuator hardware...

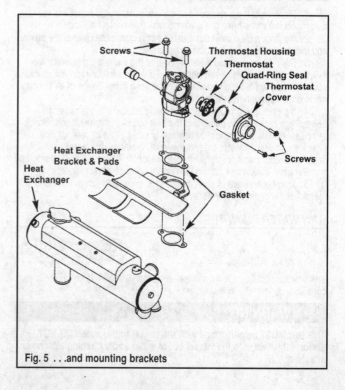

Fig. 5 ...and mounting brackets

COOLING SYSTEM 8-5

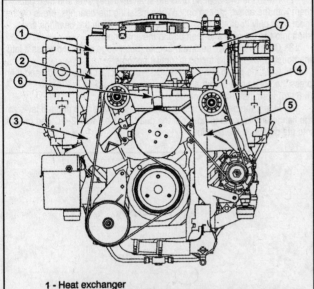

1 - Heat exchanger
2 - Starboard heat exchanger-to-exhaust manifold hose
3 - Water circulating pump hose
4 - Port heat exchanger-to-exhaust manifold hose (MCM)
5 - Cool fuel system-to-heat exchanger hose
6 - Heat exchanger bypass hose
7 - Coolant identification decal

Fig. 6 Hose routing on mid-2001-06 V6 and V8 engines (exc. 8.1L)

11. If your engine had the air actuator, connect the air lines to the manifold. Install the air pump over the fitting and pull the locking lever up. Fill the system with air until the green indicators extend fully - if they fail to extend, the lines are not connected properly.
12. Install the flame arrestor if removed.
13. Fill the closed system.

CLEANING

♦ See Figure 7

■ On some models it may be necessary to remove the exchanger prior to cleaning it. If you need to do this, make sure to refill the closed portion of the system again.

1. Remove the drain plug at the bottom of the exchanger making sure you have a suitable container underneath it. When the water has drained completely, coat the threads of the drain plug with Perfect Seal and tighten the plug securely.
2. Remove the end plate retaining bolts on each side of the exchanger and then pry off the plates, seals and gaskets. Carefully scrape any residual gasket material from the mating surfaces.
3. Clean each water passage in the unit with a small wire brush (carefully!) and then blow out each passage with compressed air if available.
4. Coat both side of new gaskets and seals with Perfect Seal. Install the gasket, seal and then end plate. Tighten the bolt to 36-72 inch lbs. (4-8 Nm) on carbureted and TBI engines; on MPI engines they should be tightened to 54 inch lbs. (6 Nm). Fill the system with coolant if necessary.

TESTING

1. An internal leak will cause coolant to mix with the seawater system when pressurized. Remove the seawater hose from the exchanger, but do not drain it. Install a cooling system tester and pressurize the circuit to 16-20 psi (110-138 kPa). If seawater begins to seep out of the fitting, there is a leak and the exchanger will require replacement.

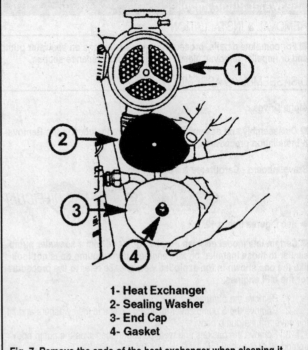

1- Heat Exchanger
2- Sealing Washer
3- End Cap
4- Gasket

Fig. 7 Remove the ends of the heat exchanger when cleaning it

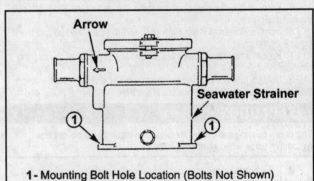

1 - Mounting Bolt Hole Location (Bolts Not Shown)

Fig. 8 The arrow indicates the direction of water flow and must point toward the water pump

2. To check for blockage, remove the end caps and inspect for any visible blockage. Remove the coolant hoses and inspect the tubes just inside the fittings. Although you can plug the coolant lines here, you may wish to drain the system instead. If any blockage is apparent, replace the exchanger.

Sea Strainer

REMOVAL & INSTALLATION

♦ See Figure 8

If performing the following while boat is in the water, close the seacock. If your boat is not equipped with a seacock, disconnect and plug the seawater inlet line to prevent water from entering the system.

■ The engine should be OFF and cool.

1. Disconnect the inlet and outlet hoses at the strainer.
2. Remove the mounting bolts and lift off the strainer. Clean the strainer and inspect it as detailed in the Maintenance section.
3. Mount the strainer so that the arrow on the casing points toward the seawater pump and tighten the mounting bolts securely.
4. Install the inlet and outlet lines with two hose clamps on each side.

8-6 COOLING SYSTEM

Seawater Pump/Impeller

REMOVAL & INSTALLATION

■ For complete details, procedures and illustrations on seawater pump and/or impeller removal, please refer to the Maintenance section.

DISASSEMBLY & ASSEMBLY

Alpha Drives

■ Disassembly and assembly procedures are included in the Removal & Installation procedures.

Bravo/Inboard - Carbureted & TBI Engines

◆ See Figures 9 thru 12

■ Certain late model engines may be equipped with a seawater pump similar to those installed on MPI engines. If your pump does not look like the one shown in the exploded view, please refer to the procedures for the MPI engines.

1. Remove the pump.
2. Remove the 5 pump cover mounting bolts and their washers and lift the cover off the pump body.
3. Remove the gaskets and wear plate and then slide the pump body up and off of the shaft.
4. Pull out the rubber plug and then lift out the impeller.
5. Lift or scrape the gasket off the lower mating surface and pinch out the quad ring seal.
6. Install a Universal Puller Plate (# 91-37241) over the hub and then press it off the shaft.
7. Puncture the front oil seal with an awl and pry it out of the bearing housing.
8. Using a pair of snap-ring pliers, remove the ring from the bore and then press the shaft and bearings out of the bore through the pulley end.

✳✳ CAUTION

The pump bearings are slip-fitted into the housing so do not use excessive force when removing them.

9. If the bearing need replacement, press them off the shaft with the Puller tool. Remember that if you remove them, you must replace them.
10. Press out the rear seals.

To assemble:

11. Coat two new rear housing seals with Loctite and then install them so that the lips face the impeller. Press in the lower seal until it bottoms and then the upper so that it is flush with the housing. Pack the cavity between the seals with Shell Alvania No. 2 grease (in a pinch, you can use Quicksilver 2-4-C).

12. Press the two bearing onto the shaft until they seat and then pack them with the above grease only. When installing the bearings, press on the inner race only in order to avoid damaging them. Slide the bearings and shaft into the housing and then pop in the snap-ring.
13. Coat a new front seal with Loctite (outer edge only) and press it into the housing, with the lip facing in, until it bottoms.
14. Coat the pump shaft with Quicksilver Special Lubricant 101 and then press the pulley hub onto the shaft until it's outer surface is 17/64 in. (6.6mm) from the outer end of the shaft.
15. Mount the bearing housing carefully into a vise with padded jaws so the flange end is pointing up. Coat the quad ring seal with 2-4-C and press it into the groove in the housing.

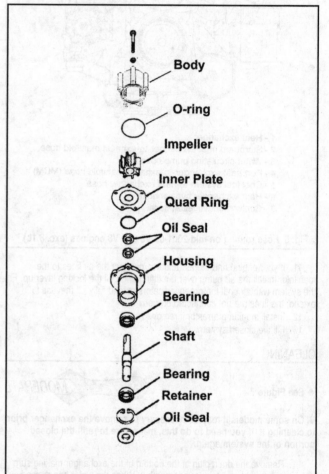

Fig. 9 Exploded view of the seawater pump - most carb and TBI engines

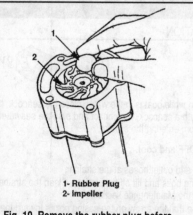

Fig. 10 Remove the rubber plug before lifting out the impeller

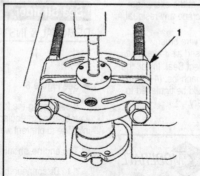

Fig. 11 A special puller tool and an arbor press are needed to remove the hub and bearings

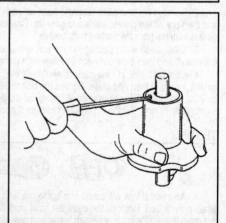

Fig. 12 Pry out the oil seal with an awl

COOLING SYSTEM

16. Position the wear plate onto the housing flange so the holes are in alignment. Coat both sides of a new gasket with Perfect Seal and lay it on top of the plate.
17. Position the impeller over the bore in the pump body and then turn it slowly in its normal direction of rotation while pressing down on it until it seats into the bore. Install the pump assembly onto the shaft so that all holes line up with those in the flange and gasket. Press in the rubber plug.
18. Coat both side of the two cover gaskets with Perfect Seal and position them onto the pump body so that all holes align correctly. The gasket with the single large hole goes down first, then the wear plate and finally the gasket with the two holes. Drop the cover into place and tighten the bolts to 120 inch lbs. (14 Nm).
19. Install the pump.

Bravo/Inboard - MPI Engines

◆ See Figure 13

■ Certain early 350 Mag and 6.2L engines may be equipped with a seawater pump similar to those installed on TBI engines. If your pump does not look like the one shown in the exploded view, please refer to the procedures for the TBI engines.

1. Remove the pump and mounting bracket. Press off the pulley with an appropriate puller.
2. Remove the 6 mounting bolts and their washers from the rear of the pump.
3. Separate the halves and pull out the O-ring from the back half.
4. Reach into the housing and remove the impeller.
5. Remove the retainer/seal from the front half of the pump and then remove the tolerance ring.
6. From inside the housing (impeller side), press out the pump shaft.
7. Remove the rear oil seal from the front housing.

To assemble:

8. Clean all metal components in solvent and then dry thoroughly.
9. Lightly coat the shaft and bearings with clean engine oil.
10. Clean all gasket material and sealer from the pump mating surfaces and then inspect the 2 pump halves for cracks or other damage.
11. Inspect the impeller, but we recommend replacing it regardless of its condition...why take the chance.
12. Position the rear seal into the bearing housing so that the side with the spring is toward the main pump body.
13. Slide the pump shaft assembly into the housing and then install the tolerance ring.
14. Coat the outer edges of the retainer/seal with Loctite 609 adhesive and then press it into place over the shaft. Do not get the adhesive onto the seal portion.
15. Lubricate the impeller in warm soapy water and twist it into the housing.
16. Install a new O-ring and then align the pump halves correctly. Install the retaining bolts and tighten to 88 inch lbs. (9.9 Nm).
17. Install the pump.

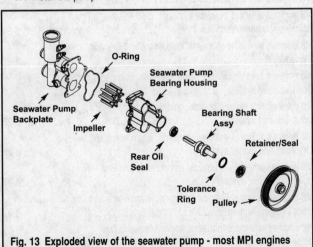

Fig. 13 Exploded view of the seawater pump - most MPI engines

OUTPUT TEST

If an overheating problem exists, this test can be used to confirm that a sufficient flow of cooling water is being supplied to the engine.

■ The vessel MUST be in the water for this test...never attempt with a flushing device installed.

The ability of this test to accurately detect a problem is entirely dependent upon the accuracy with which the test is performed! Make sure you use a shop tachometer and not the tach on your boat.

1. On models with Alpha drives, disconnect and remove the water inlet line at the gimbal housing water tube (the line between the transom and engine). Replace this line with a hose of the same diameter that is about 3 ft. (1 m) longer - clamp the hose at the gimbal housing water tube only, do not clamp it at the other end. Make sure that the new hose has wall thickness sufficient to resist kinking.
2. On models with Bravo drives, disconnect and remove the water line at the outlet on the seawater pump (the line between the pump and engine). Replace this line with a hose of the same diameter that is about 3 ft. (1 m) longer - clamp the hose at the pump outlet only, do not clamp it at the other end. Make sure that the new hose has wall thickness sufficient to resist kinking.
3. Position a container that is at least 10 qts. (9.5L) near the end of the new hose.
4. Slide the unclamped end of the new hose onto the engine fitting; hold it there, but do not secure it with the clamp. Start the engine and move the throttle until you can achieve an idle speed of exactly 1000 rpm on the tach (NOT the one on the helm!). Pop the hose off the connection and immediately feed it into the empty container for EXACTLY 15 seconds - use a stop watch for this, not your wrist watch! As soon as the 15 seconds is up, direct the end of the hose overboard and turn Off the engine.
5. Reconnect the open end of the hose to the engine. Measure the water in the container and compare it with the quantities as shown.

On V8 models with dry joint exhaust:

- Alpha drive w/2.40:1 ratio - 1.4 qts. (1.3L) minimum
- Alpha drive w/2.00:1 ratio - 1.7 qts. (1.6L) minimum
- Alpha drive w/1.81:1 ratio - 1.5 qts. (1.4L) minimum
- Alpha drive w/1.62:1 ratio - 1.7 qts. (1.6L) minimum
- Alpha drive w/1.47:1 ratio - 1.9 qts. (1.8L) minimum
- Bravo drives w/seawater - 8.0 qts. (7.6L) minimum
- Bravo drives w/closed - 10.0 qts. (9.5L) minimum
- Inboard & Tow Sport models - 10.0 qts. (9.5L) minimum

On all other models:

- Alpha drive w/2.40:1 ratio - 1.4 qts. (1.3L) minimum
- Alpha drive w/2.00:1 ratio - 1.7 qts. (1.6L) minimum
- Alpha drive w/1.98:1 ratio - 3.0 qts. (2.8L) minimum
- Alpha drive w/1.84:1 ratio - 3.3 qts. (3.1L) minimum
- Alpha drive w/1.81:1 ratio - 1.5 qts. (1.4L) minimum
- Alpha drive w/1.65:1 ratio - 3.6 qts. (3.4L) minimum
- Alpha drive w/1.62:1 ratio - 1.7 qts. (1.6L) minimum
- Alpha drive w/1.50:1 ratio - 4.0 qts. (3.8L) minimum
- Alpha drive w/1.47:1 ratio - 1.9 qts. (1.8L) minimum
- Alpha drive w/1.32:1 ratio - 4.5 qts. (4.3L) minimum
- Bravo drives - 7.5 qts. (7.1L) minimum
- Inboard & Tow Sport models - 10.0 qts. (9.5L) minimum

6. Repeat this test 4 more times to ensure that you are getting essentially the same quantity of water in the container each time. If not, you have a restriction or a problem with the pump.

8-8 COOLING SYSTEM

SEAWATER PUMP PRESSURE SENSOR

8.1L Sterndrives (serial # 0W060000 & below) and inboards (serial #0W090000 & below) utilize a 0-100 psi pump pressure sensor mounted in the pump's impeller cover. Later models use a 0-50 psi sensor in either the power steering cooler or the transmission cooler (back of engine, over the flywheel housing).

1. Disconnect the electrical harness and unscrew the sensor.
2. Thread the sensor in and tighten it to 115 inch lbs. (13 Nm). Connect the electrical lead.

Thermostat

■ For all removal and testing procedures, please refer to the Engine & Drive Maintenance section.

Water Distribution Housing

REMOVAL & INSTALLATION

◆ See Figures 14 and 15

1. Drain the seawater section of the cooling system.
2. On air-actuated single point drain systems, remove each air line from the distribution housing by pressing on the fitting release while pulling on the line.
3. On manual single point drain systems, remove the C-clip from the drain rod (blue handle), turn the rod counterclockwise until the rod threads come clear of the alignment bracket and then pull it straight up and out.
4. Tag and disconnect all hoses at the distribution housing (front of the engine, port side).

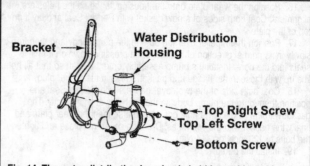

Fig. 14 The water distribution housing is held in position with a bracket

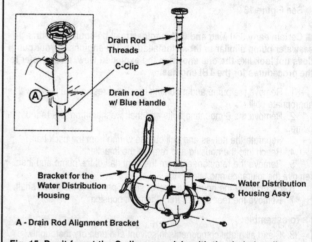

Fig. 15 Don't forget the C-clip on models with the drain handle

FLOW DIAGRAMS

■ Certain components in the following diagrams may be different on your specific application, but water and coolant flow paths will be similar.

3.0L engines w/seawater system	8-9
3.0L engine w/closed system	8-10
Carbureted and TBI V6 engines - seawater system (Alpha)	8-11
Carbureted and TBI V6 engines - seawater system (Bravo)	8-12
Carbureted and TBI V6 engines - closed system (Alpha)	8-13
Carbureted and TBI V6 engines - closed system (Bravo)	8-14
MPI V6 engines - seawater system (Alpha)	8-14
MPI V6 engines - seawater system (Bravo)	8-15
MPI V6 engines - closed system	8-15
Carbureted and TBI V8 engines - seawater system (Alpha)	8-16
Carbureted and TBI V8 engines, also includes 2001 MPI engines up to serial #0M299999 - seawater system (Bravo, Inboard & Ski)	8-16
Carbureted and TBI V8 engines, also includes 2001 MPI engines up to serial #0M299999 - closed system	8-17
5.0L, 5.7L & 6.2L MPI V8 engines (exc. Dry Joint) - seawater system (Alpha)	8-17
5.0L, 5.7L & 6.2L MPI V8 engines (exc. Dry Joint) - seawater system (Bravo, Inboard & Tow Sport)	8-18
5.0L, 5.7L & 6.2L MPI V8 engines (exc. Dry Joint) - closed system	8-18
8.1L V8 engines (exc. Dry Joint) - closed system	8-19
8.1L V8 engines (serial # 0M000000 & above) - closed system	8-19
V8 engines w/Alpha and Dry Joint (exc. MPD models) - seawater system	8-20
V8 engines w/Alpha and Dry Joint (MPD models) - seawater system	8-20
V8 engines w/Bravo and Dry Joint (exc. MPD models) - seawater system	8-21
V8 engines w/Bravo and Dry Joint (MPD models) - seawater system	8-21
V8 engines w/stern drives and Dry Joint - closed system	8-22
V8 engines w/inline inboards and Dry Joint - seawater system	8-22
V8 engines w/V-drive inboards and Dry Joint - seawater system	8-23
V8 engines w/inline inboards and Dry Joint - closed system	8-23
V8 engines w/V-drive inboards and Dry Joint - closed system	8-24
V8 engines w/Bravo or inline inboards, cold risers and Dry Joint - seawater system	8-24
V8 engines w/V-drive inboards, cold risers and Dry Joint - seawater system	8-25
V8 engines w/stern drive or inline inboards, cold risers and Dry Joint - closed system	8-25
V8 engines w/V-drive inboards, cold risers and Dry Joint - closed system	8-26
V8 engines w/Bravo or inline inboards, warm risers and Dry Joint - seawater system	8-26
V8 engines w/V-drive inboards, warm risers and Dry Joint - seawater system	8-27
V8 engines w/Bravo or inline inboards, warm risers and Dry Joint - closed system	8-27
V8 engines w/V-drive inboards, warm risers and Dry Joint - closed system	8-28

COOLING SYSTEM 8-9

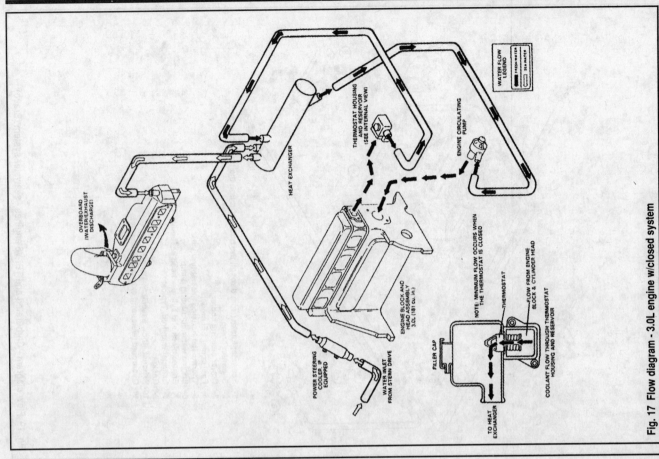

Fig. 17 Flow diagram - 3.0L engine w/closed system

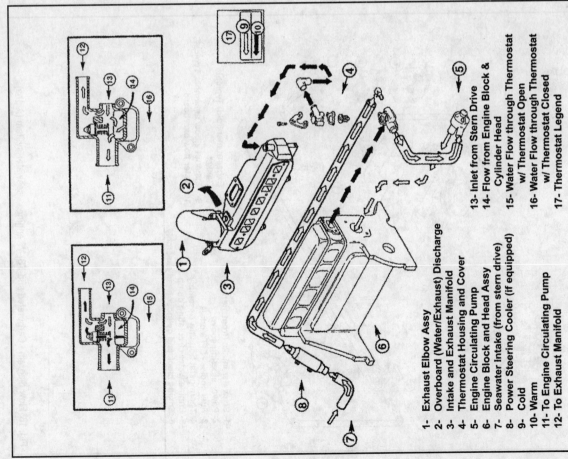

Fig. 16 Flow diagram - 3.0L engines w/seawater system

1- Exhaust Elbow Assy
2- Overboard (Water/Exhaust) Discharge
3- Intake and Exhaust Manifold
4- Thermostat Housing and Cover
5- Engine Circulating Pump
6- Engine Block and Head Assy
7- Seawater Intake (from stern drive)
8- Power Steering Cooler (if equipped)
9- Cold
10- Warm
11- To Engine Circulating Pump
12- To Exhaust Manifold
13- Inlet from Stern Drive
14- Flow from Engine Block & Cylinder Head
15- Water Flow through Thermostat w/ Thermostat Open
16- Water Flow through Thermostat w/ Thermostat Closed
17- Thermostat Legend

8-10 COOLING SYSTEM

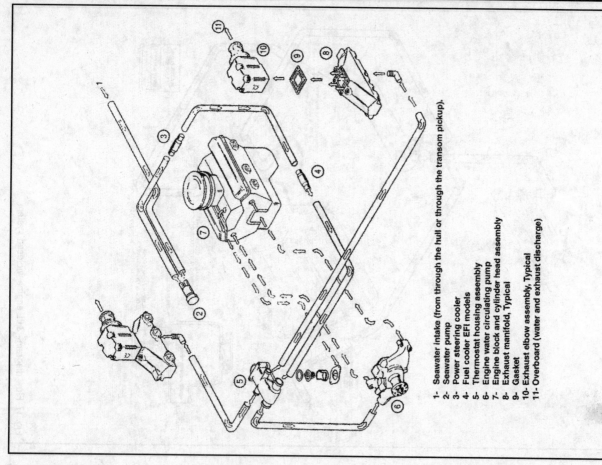

1- Seawater intake (from through the hull or through the transom pickup).
2- Seawater pump
3- Power steering cooler
4- Fuel cooler EFI models
5- Thermostat housing assembly
6- Engine water circulating pump
7- Engine block and cylinder head assembly
8- Exhaust manifold, Typical
9- Gasket
10- Exhaust elbow assembly, Typical
11- Overboard (water and exhaust discharge)

Fig. 19 Flow diagram - Carbureted and TBI V6 engines - seawater system (Bravo)

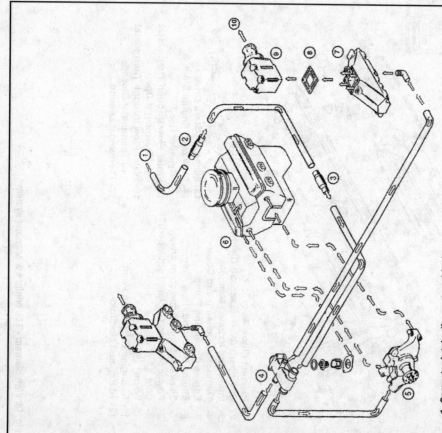

1- Seawater Intake (from sterndrive)
2- Power steering cooler
3- Fuel cooler - EFI Models
4- Thermostat housing Assembly
5- Engine water circulating pump
6- Engine block and cylinder head assembly
7- Exhasut manifold, Typical
8- Gasket
9- Exhaust elbow assembly, Typical
10- Water flow overboard

Fig. 18 Flow diagram - Carbureted and TBI V6 engines - seawater system (Alpha)

COOLING SYSTEM 8-11

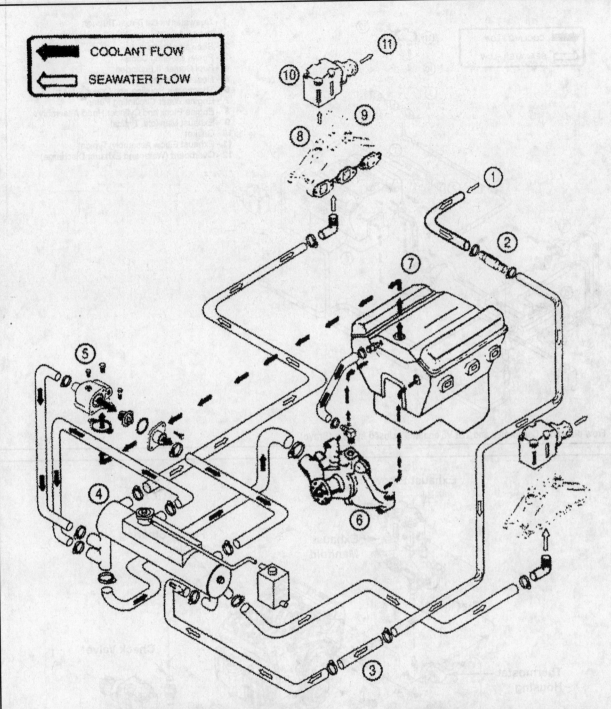

1 - Seawater Intake (From Sterndrive)
2 - Power Steering Cooler
3 - Fuel Cooler - EFI Models
4 - Heat Exchanger, Typical
5 - Thermostat Housing Assembly
6 - Engine Water Circulating Pump
7 - Engine Block And Cylinder Head Assembly
8 - Exhaust Manifold, Typical
9 - Gasket
10 - Exhaust Elbow Assembly, Typical
11 - Overboard (Water and Exhaust Discharge)

Fig. 20 Flow diagram - Carbureted and TBI V6 engines - closed system (Alpha)

8-12 COOLING SYSTEM

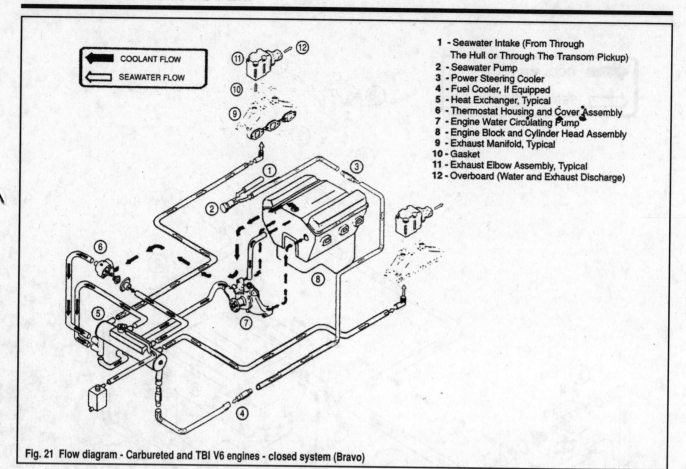

Fig. 21 Flow diagram - Carbureted and TBI V6 engines - closed system (Bravo)

1 - Seawater Intake (From Through The Hull or Through The Transom Pickup)
2 - Seawater Pump
3 - Power Steering Cooler
4 - Fuel Cooler, If Equipped
5 - Heat Exchanger, Typical
6 - Thermostat Housing and Cover Assembly
7 - Engine Water Circulating Pump
8 - Engine Block and Cylinder Head Assembly
9 - Exhaust Manifold, Typical
10 - Gasket
11 - Exhaust Elbow Assembly, Typical
12 - Overboard (Water and Exhaust Discharge)

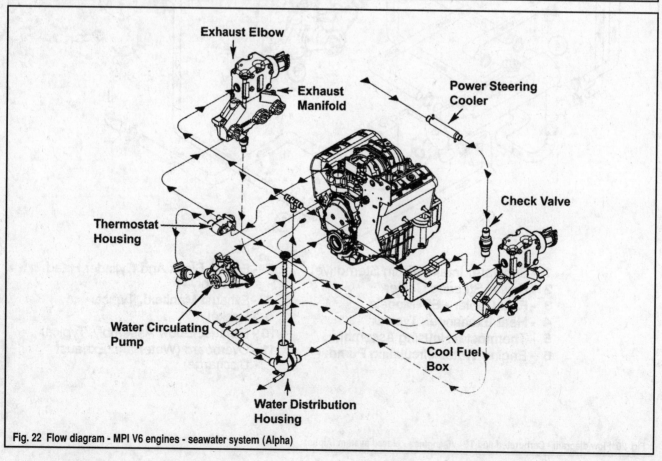

Fig. 22 Flow diagram - MPI V6 engines - seawater system (Alpha)

COOLING SYSTEM   8-13

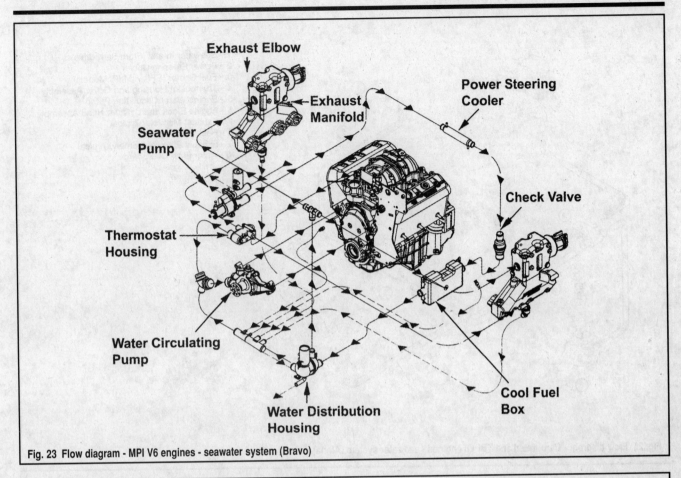

Fig. 23 Flow diagram - MPI V6 engines - seawater system (Bravo)

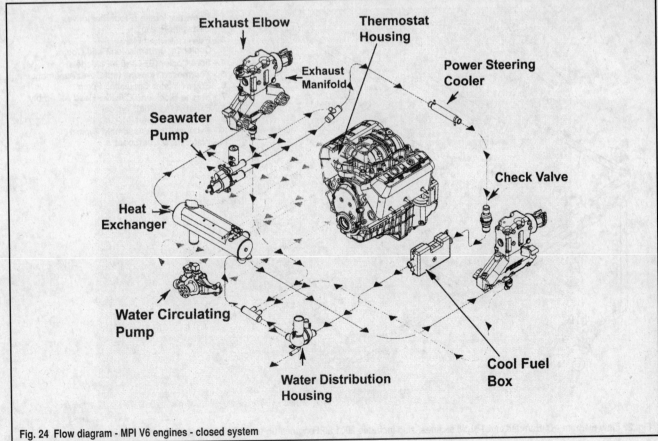

Fig. 24 Flow diagram - MPI V6 engines - closed system

8-14 COOLING SYSTEM

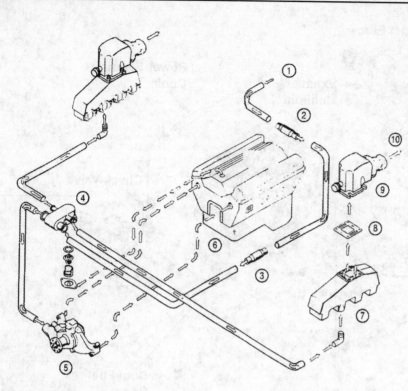

1 - Seawater Intake (From Sterndrive)
2 - Power Steering Cooler
3 - Fuel Cooler (EFI and MPI Models)
4 - Thermostat Housing and Cover Assembly
5 - Engine Water Circulating Pump
6 - Engine Block and Cylinder Head Assembly
7 - Exhaust Manifold, Typical
8 - Restrictor Gasket
9 - Exhaust Elbow Assembly, Typical
10 - Water Flow Overboard

Fig. 25 Flow diagram - Carbureted and TBI V8 engines - seawater system (Alpha)

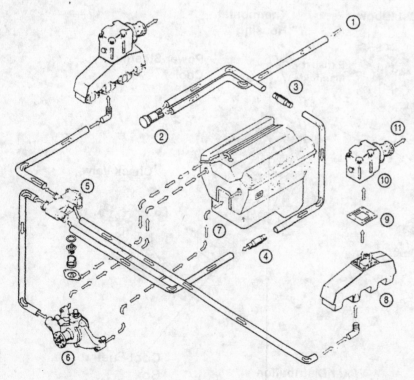

1 - Seawater Intake (From Sterndrive)
2 - Seawater Pump
3 - Power Steering Fluid Cooler Or Transmission Fluid Cooler
4 - Fuel Cooler (EFI and MPI Models)
5 - Thermostat Housing and Cover Assembly
6 - Engine Water Circulating Pump
7 - Engine Block and Cylinder Head Assembly
8 - Exhaust Manifold, Typical
9 - Restrictor Gasket
10 - Exhaust Elbow Assembly, Typical
11 - Water Flow Overboard

Fig. 26 Flow diagram - Carbureted and TBI V8 engines, also includes 2001 MPI engines up to serial #OM299999 - closed system

COOLING SYSTEM 8-15

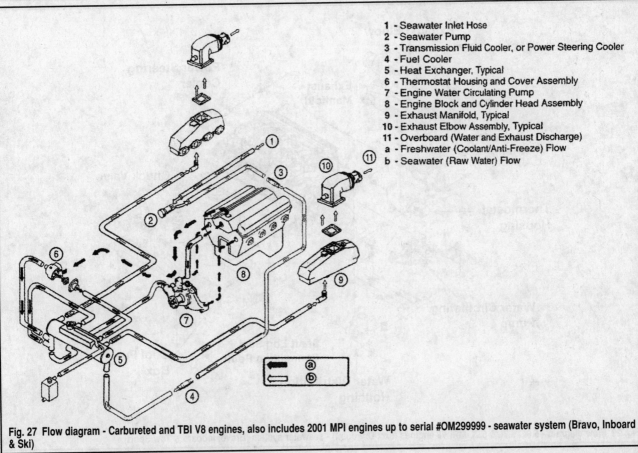

1 - Seawater Inlet Hose
2 - Seawater Pump
3 - Transmission Fluid Cooler, or Power Steering Cooler
4 - Fuel Cooler
5 - Heat Exchanger, Typical
6 - Thermostat Housing and Cover Assembly
7 - Engine Water Circulating Pump
8 - Engine Block and Cylinder Head Assembly
9 - Exhaust Manifold, Typical
10 - Exhaust Elbow Assembly, Typical
11 - Overboard (Water and Exhaust Discharge)
a - Freshwater (Coolant/Anti-Freeze) Flow
b - Seawater (Raw Water) Flow

Fig. 27 Flow diagram - Carbureted and TBI V8 engines, also includes 2001 MPI engines up to serial #OM299999 - seawater system (Bravo, Inboard & Ski)

Fig. 28 Flow diagram - 5.0L, 5.7L & 6.2L MPI V8 engines (exc. Dry Joint) - seawater system (Alpha)

8-16 COOLING SYSTEM

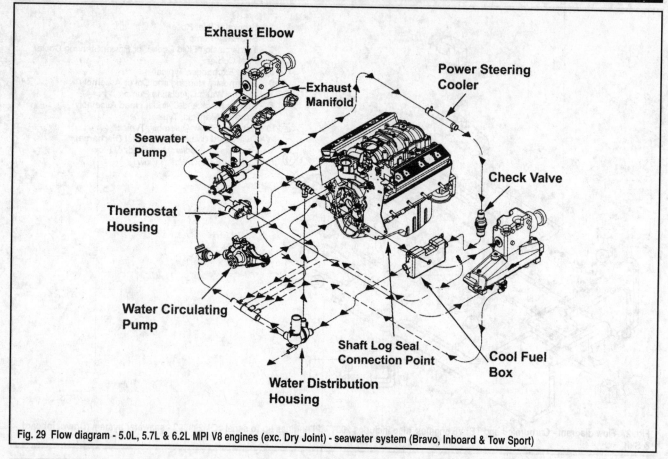

Fig. 29 Flow diagram - 5.0L, 5.7L & 6.2L MPI V8 engines (exc. Dry Joint) - seawater system (Bravo, Inboard & Tow Sport)

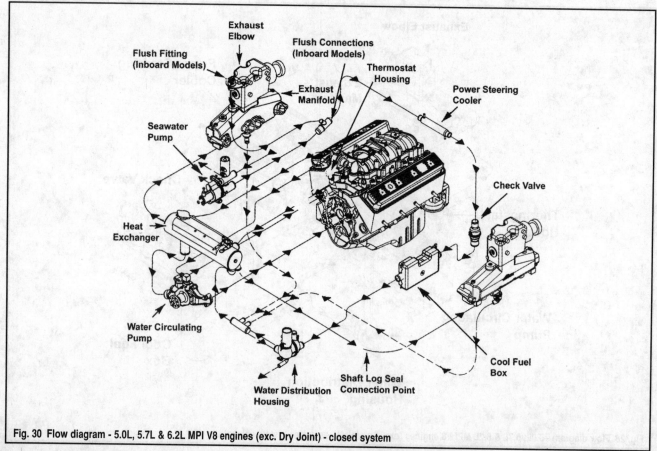

Fig. 30 Flow diagram - 5.0L, 5.7L & 6.2L MPI V8 engines (exc. Dry Joint) - closed system

COOLING SYSTEM 8-17

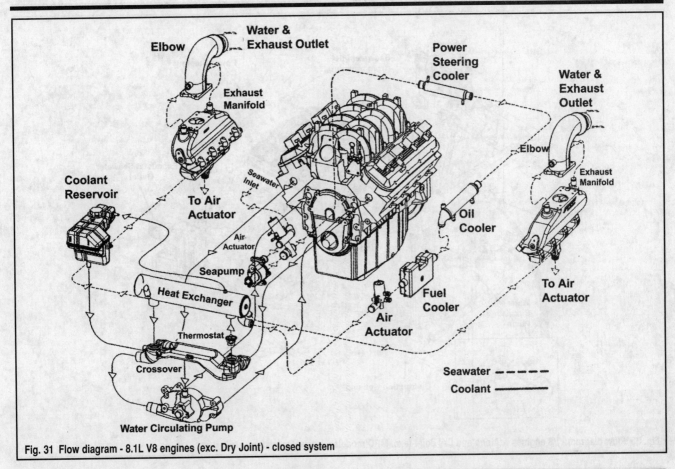

Fig. 31 Flow diagram - 8.1L V8 engines (exc. Dry Joint) - closed system

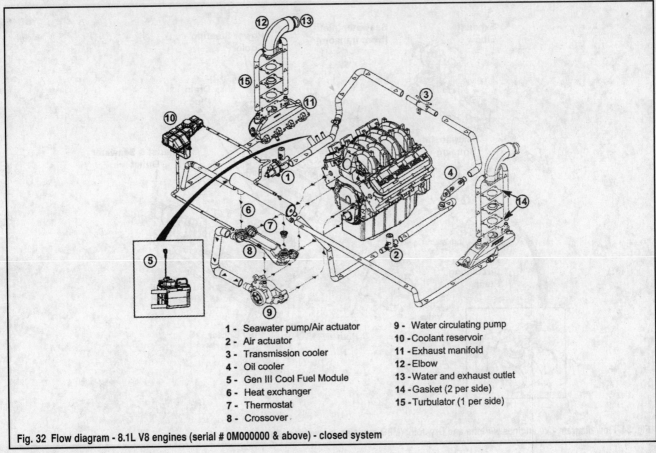

1 - Seawater pump/Air actuator
2 - Air actuator
3 - Transmission cooler
4 - Oil cooler
5 - Gen III Cool Fuel Module
6 - Heat exchanger
7 - Thermostat
8 - Crossover
9 - Water circulating pump
10 - Coolant reservoir
11 - Exhaust manifold
12 - Elbow
13 - Water and exhaust outlet
14 - Gasket (2 per side)
15 - Turbulator (1 per side)

Fig. 32 Flow diagram - 8.1L V8 engines (serial # 0M000000 & above) - closed system

8-18 COOLING SYSTEM

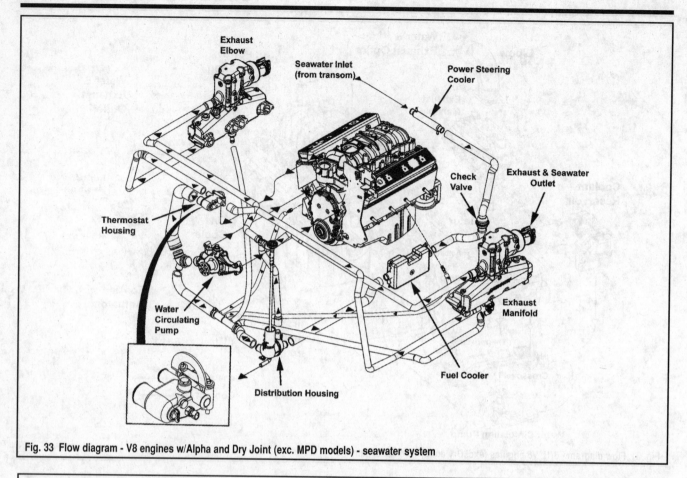

Fig. 33 Flow diagram - V8 engines w/Alpha and Dry Joint (exc. MPD models) - seawater system

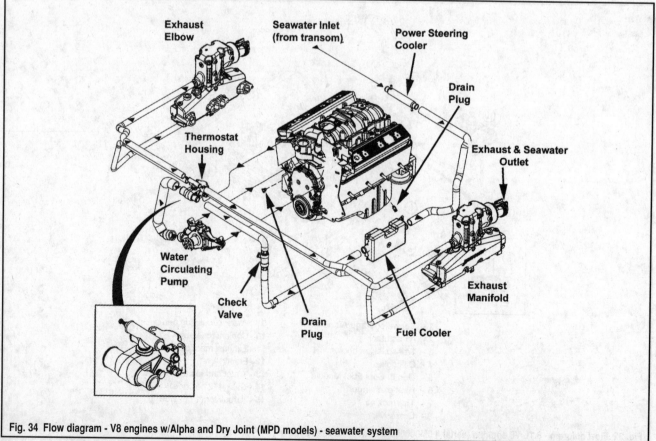

Fig. 34 Flow diagram - V8 engines w/Alpha and Dry Joint (MPD models) - seawater system

COOLING SYSTEM 8-19

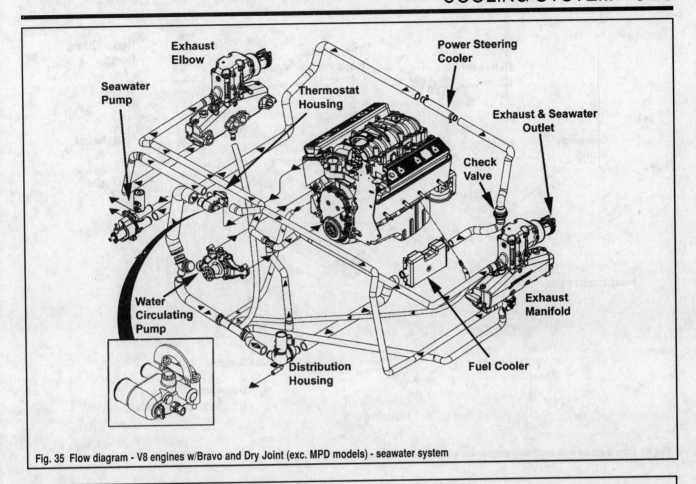

Fig. 35 Flow diagram - V8 engines w/Bravo and Dry Joint (exc. MPD models) - seawater system

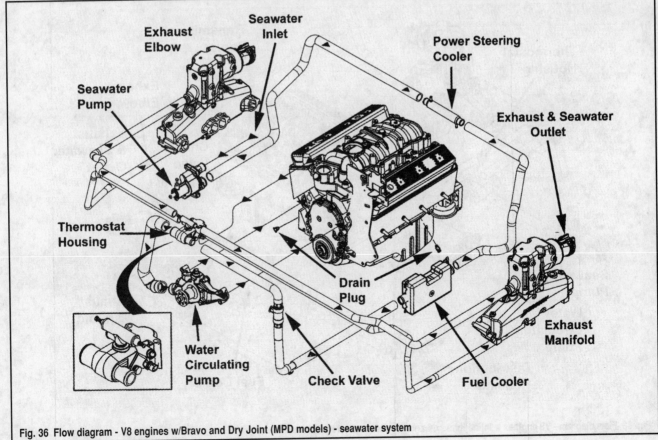

Fig. 36 Flow diagram - V8 engines w/Bravo and Dry Joint (MPD models) - seawater system

8-20 COOLING SYSTEM

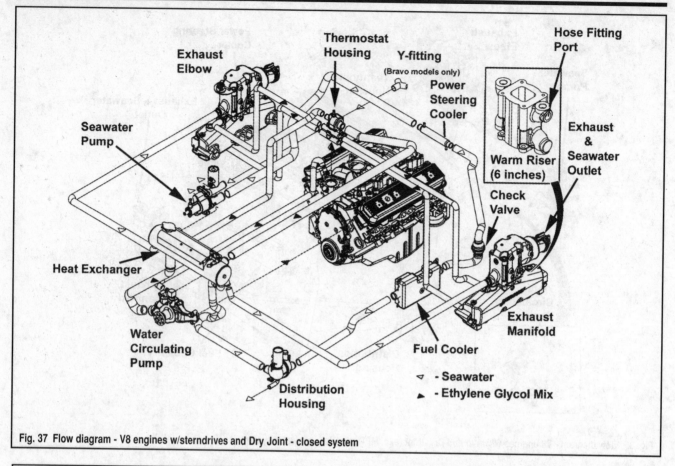

Fig. 37 Flow diagram - V8 engines w/sterndrives and Dry Joint - closed system

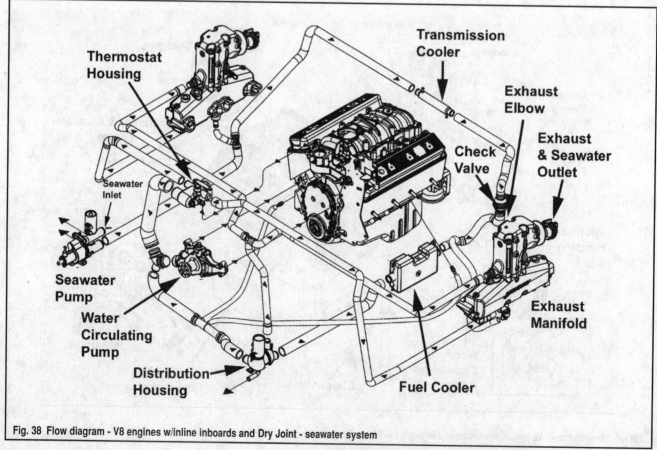

Fig. 38 Flow diagram - V8 engines w/inline inboards and Dry Joint - seawater system

COOLING SYSTEM 8-21

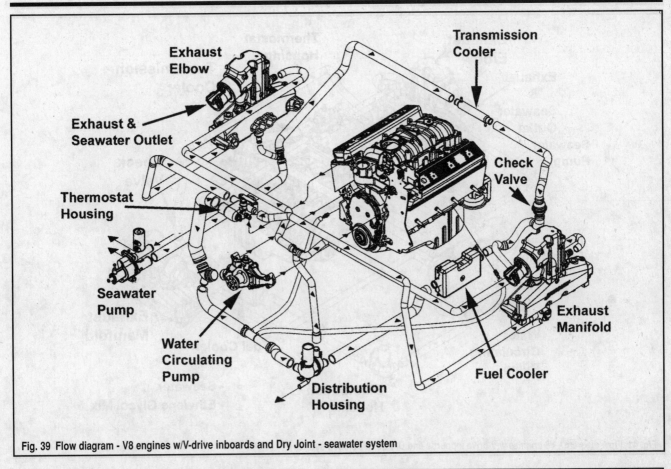

Fig. 39 Flow diagram - V8 engines w/V-drive inboards and Dry Joint - seawater system

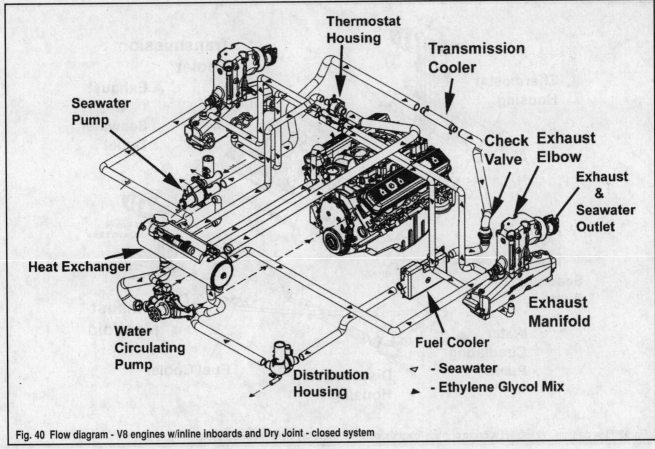

Fig. 40 Flow diagram - V8 engines w/inline inboards and Dry Joint - closed system

8-22 COOLING SYSTEM

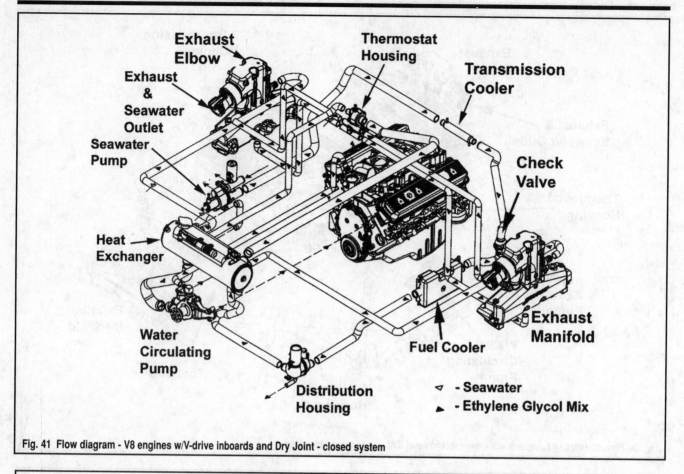

Fig. 41 Flow diagram - V8 engines w/V-drive inboards and Dry Joint - closed system

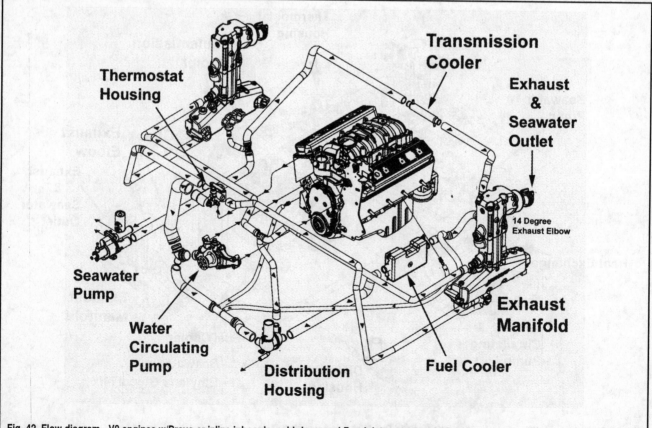

Fig. 42 Flow diagram - V8 engines w/Bravo or inline inboards, cold risers and Dry Joint - seawater system

COOLING SYSTEM 8-23

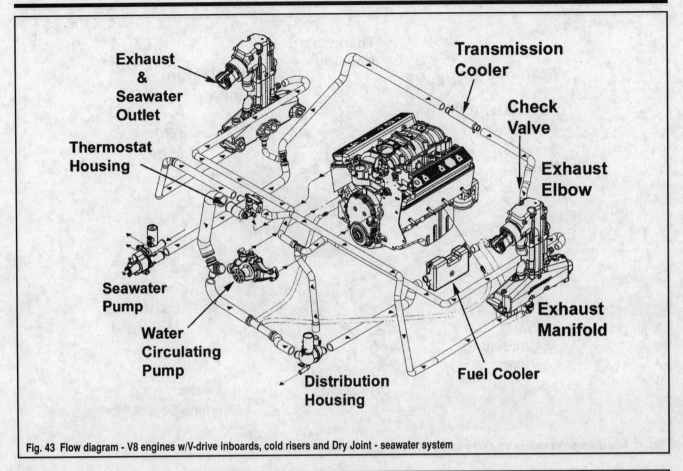

Fig. 43 Flow diagram - V8 engines w/V-drive inboards, cold risers and Dry Joint - seawater system

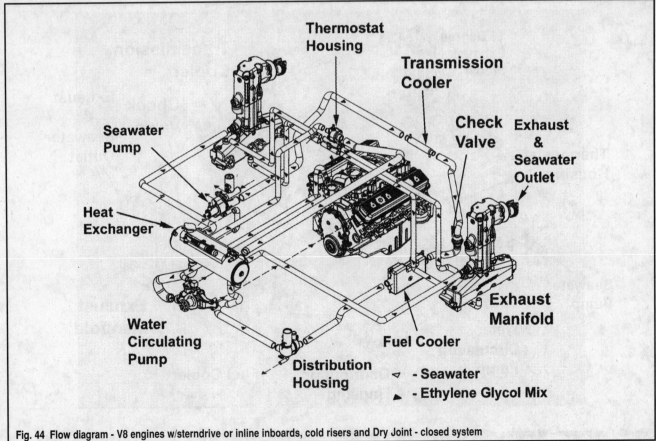

Fig. 44 Flow diagram - V8 engines w/sterndrive or inline inboards, cold risers and Dry Joint - closed system

8-24 COOLING SYSTEM

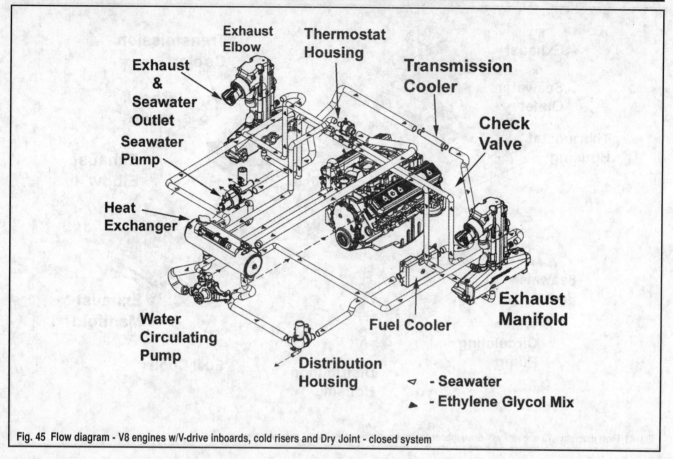

Fig. 45 Flow diagram - V8 engines w/V-drive inboards, cold risers and Dry Joint - closed system

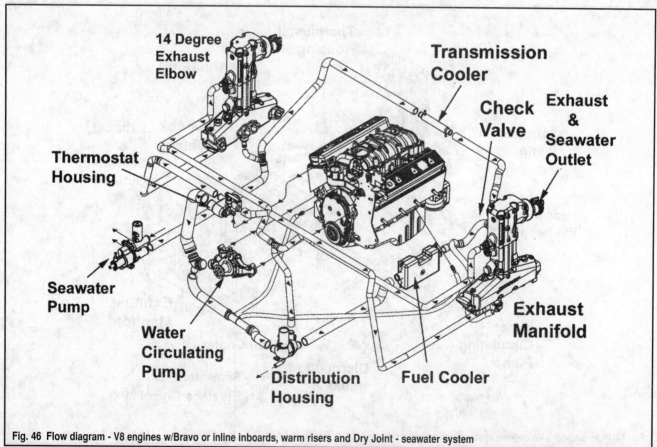

Fig. 46 Flow diagram - V8 engines w/Bravo or inline inboards, warm risers and Dry Joint - seawater system

COOLING SYSTEM 8-25

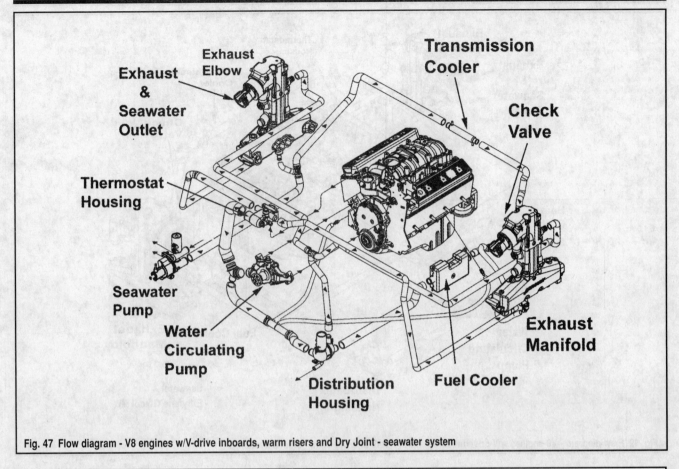

Fig. 47 Flow diagram - V8 engines w/V-drive inboards, warm risers and Dry Joint - seawater system

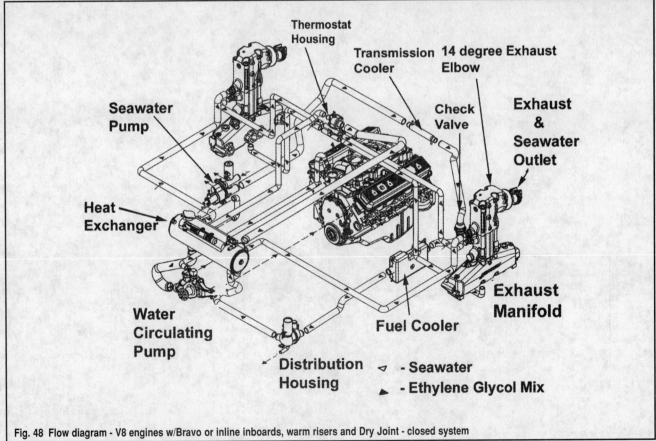

Fig. 48 Flow diagram - V8 engines w/Bravo or inline inboards, warm risers and Dry Joint - closed system

8-26 COOLING SYSTEM

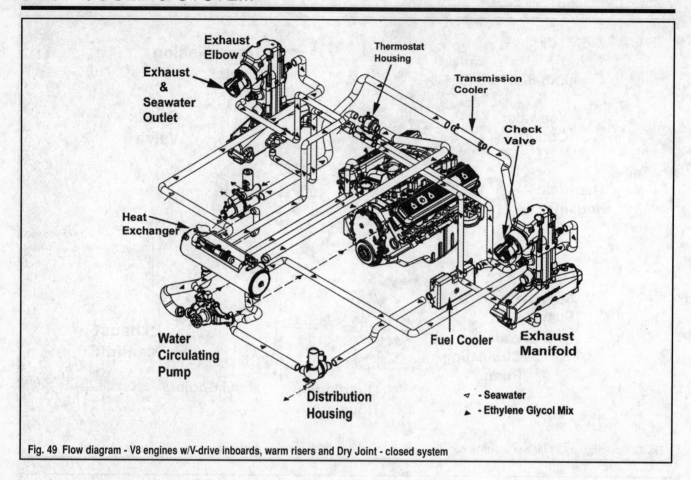

Fig. 49 Flow diagram - V8 engines w/V-drive inboards, warm risers and Dry Joint - closed system

ALTERNATOR 9-12
  CIRCUIT TESTING 9-13
  COMPONENT TESTING 9-14
  DISASSEMBLY & ASSEMBLY 9-16
  INSPECTION 9-12
  PRECAUTIONS 9-12
  REMOVAL & INSTALLATION 9-12
AUDIO WARNING SYSTEM 9-43
  TESTING 9-43
BATTERY **9-8**
  BATTERY GAUGE 9-38
  REMOVAL & INSTALLATION 9-38
  TESTING 9-38
CHARGING SYSTEM **9-10**
  ALTERNATOR 9-12
  GENERAL INFORMATION 9-10
  TROUBLESHOOTING 9-11
CRUISELOG METER 9-39
  REMOVAL & INSTALLATION 9-39
  TESTING 9-39
DISTRIBUTOR
  DEI 9-35
  EST 9-27
  THUNDERBOLT 9-33
DISTRIBUTOR CAP
  DEI 9-34
  EST 9-27
  THUNDERBOLT 9-32
DISTRIBUTOR ELECTRONIC IGNITION
(DEI) SYSTEM - ECM 555 .... **9-34**
  DISTRIBUTOR 9-35
  DISTRIBUTOR CAP 9-34
  GENERAL INFORMATION 9-34
  IGNITION COIL 9-36
DISTRIBUTORLESS ELECTRONIC
IGNITION (DIS) SYSTEM - PCM 555 .... **9-36**
  GENERAL INFORMATION 9-36
  IGNITION COIL 9-36
ELECTRICAL COMPONENTS 9-2
ELECTRONIC SPARK TIMING SYSTEM (EST) .. **9-24**
  DESCRIPTION 9-24
  DISTRIBUTOR 9-27
  DISTRIBUTOR CAP 9-27
  IGNITION COIL 9-30
  TROUBLESHOOTING 9-26
EMERGENCY (LANYARD) STOP SWITCH 9-43
  TESTING 9-43
FUEL GAUGE 9-37
  REMOVAL & INSTALLATION 9-38
  TESTING 9-38
  TROUBLESHOOTING 9-37
FUEL TANK SENDING UNIT 9-41
  TESTING 9-41
IGNITION COIL
  DEI 9-36
  DIS 9-36
  EST 9-30
IGNITION OR KNOCK CONTROL MODULES ... 9-33
  REMOVAL & INSTALLATION 9-33
IGNITION SWITCH 9-42
  TESTING 9-42
IGNITION SYSTEMS **9-24**
  APPLICATIONS 9-24
  GENERAL INFORMATION 9-24
INSTRUMENTS & GAUGES **9-37**
  BATTERY GAUGE 9-38
  CRUISELOG METER 9-39
  FUEL GAUGE 9-37
  OIL & TEMPERATURE GAUGES 9-37
  SPEEDOMETER 9-39
  TACHOMETER 9-39
  VACUUM GAUGE 9-40
OIL & TEMPERATURE GAUGES 9-37
  REMOVAL & INSTALLATION 9-37
  TESTING 9-37
  TROUBLESHOOTING 9-37
OIL PRESSURE SENDING UNIT 9-40
  GENERAL INFORMATION 9-40
  TESTING 9-40
OIL PRESSURE SWITCH 9-43
  TESTING 9-43
ROTOR/SENSOR WHEEL 9-32
  REMOVAL & INSTALLATION 9-32
SENDING UNITS & SWITCHES **9-40**
  AUDIO WARNING SYSTEM 9-43
  EMERGENCY (LANYARD) STOP SWITCH 9-43
  FUEL TANK SENDING UNIT 9-41
  IGNITION SWITCH 9-42
  OIL PRESSURE SENDING UNIT 9-40
  OIL PRESSURE SWITCH 9-43
  START/STOP SWITCH 9-43
  STERN DRIVE GEAR LUBE MONITOR SWITCH . 9-44
  TRANSMISSION FLUID
   TEMPERATURE SWITCH 9-44
  WATER TEMPERATURE SENDER 9-40
  WATER TEMPERATURE SWITCH 9-44
SENSOR 9-33
  REMOVAL & INSTALLATION 9-33
  TESTING 9-33
SPECIFICATIONS **9-74**
  ALTERNATOR 9-74
  STARTER 9-74
SPEEDOMETER 9-39
  REMOVAL & INSTALLATION 9-39
  TESTING 9-39
START/STOP SWITCH 9-43
  TESTING 9-43
STARTER CIRCUIT **9-18**
  DESCRIPTION & OPERATION 9-18
  STARTER MOTOR 9-18
  TROUBLESHOOTING 9-18
STARTER MOTOR 9-18
  DESCRIPTION & OPERATION 9-18
  DISASSEMBLY & ASSEMBLY 9-21
  PRECAUTIONS 9-19
  REMOVAL & INSTALLATION 9-21
  TESTING 9-20
  TROUBLESHOOTING 9-19
TACHOMETER 9-39
  REMOVAL & INSTALLATION 9-39
  TESTING 9-39
TEST EQUIPMENT 9-4
THUNDERBOLT V IGNITION SYSTEM **9-30**
  DESCRIPTION 9-30
  DISTRIBUTOR 9-33
  DISTRIBUTOR CAP 9-32
  IGNITION OR KNOCK CONTROL MODULES .. 9-33
  ROTOR/SENSOR WHEEL 9-32
  SENSOR 9-33
  TROUBLESHOOTING 9-31
TRANSMISSION FLUID TEMPERATURE SWITCH . 9-44
  TESTING 9-44
TROUBLESHOOTING
  CHARGING SYSTEM 9-11
  EST 9-26
  STARTING SYSTEM 9-18
  THUNDERBOLT 9-31
UNDERSTANDING AND
TROUBLESHOOTING ELECTRICAL SYSTEMS .. 9-2
  BASIC ELECTRICAL THEORY 9-2
  ELECTRICAL COMPONENTS 9-2
  TEST EQUIPMENT 9-4
  TROUBLESHOOTING THE
   ELECTRICAL SYSTEM 9-6
  WIRE AND CONNECTOR REPAIR 9-8
VACUUM GAUGE 9-40
  REMOVAL & INSTALLATION 9-40
  TESTING 9-40
WATER TEMPERATURE SENDER 9-40
  GENERAL INFORMATION 9-40
  TESTING 9-41
WATER TEMPERATURE SWITCH 9-44
  TESTING 9-44
WIRE AND CONNECTOR REPAIR 9-8
WIRING DIAGRAMS **9-45**
  INDEX 9-45

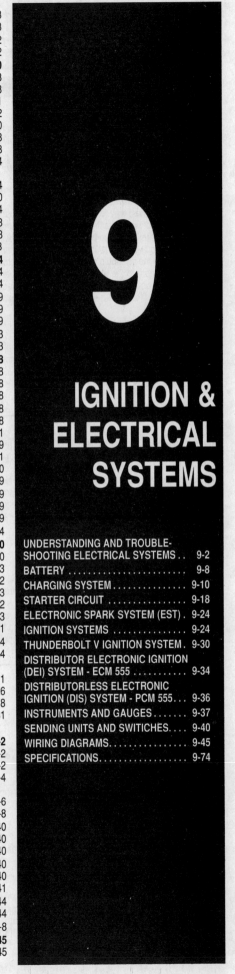

9

IGNITION & ELECTRICAL SYSTEMS

UNDERSTANDING AND TROUBLE-SHOOTING ELECTRICAL SYSTEMS ..	9-2
BATTERY	9-8
CHARGING SYSTEM	9-10
STARTER CIRCUIT	9-18
ELECTRONIC SPARK SYSTEM (EST) .	9-24
IGNITION SYSTEMS	9-24
THUNDERBOLT V IGNITION SYSTEM .	9-30
DISTRIBUTOR ELECTRONIC IGNITION (DEI) SYSTEM - ECM 555	9-34
DISTRIBUTORLESS ELECTRONIC IGNITION (DIS) SYSTEM - PCM 555	9-36
INSTRUMENTS AND GAUGES	9-37
SENDING UNITS AND SWITCHES....	9-40
WIRING DIAGRAMS	9-45
SPECIFICATIONS	9-74

IGNITION & ELECTRICAL SYSTEMS

UNDERSTANDING AND TROUBLESHOOTING ELECTRICAL SYSTEMS

Basic Electrical Theory

◆ See Figure 1

For any 12 volt, negative ground, electrical system to operate, the electricity must travel in a complete circuit. This simply means that current (power) from the positive terminal (+) of the battery must eventually return to the negative terminal (-) of the battery. Along the way, this current will travel through wires, fuses, switches and components. If for any reason the flow of current through the circuit is interrupted, the component(s) fed by that circuit will cease to function properly.

Perhaps the easiest way to visualize a circuit is to think of connecting a light bulb (with two wires attached to it) to the battery - one wire attached to the negative (-) terminal of the battery and the other wire to the positive (+) terminal. With the two wires touching the battery terminals, the circuit would be complete and the light bulb would illuminate. Electricity would follow a path from the battery to the bulb and back to the battery. It's easy to see that with longer wires on our light bulb, it could be mounted anywhere. Further, one wire could be fitted with a switch so that the light could be turned on and off.

The normal marine circuit differs from this simple example in two ways. First, instead of having a return wire from each bulb to the battery, the current travels through a single ground wire, which handles all the grounds for a specific circuit. Secondly, most marine circuits contain multiple components, which receive power from a single circuit. This lessens the amount of wire needed to power components.

HOW ELECTRICITY WORKS: THE WATER ANALOGY

Electricity is the flow of electrons - the sub-atomic particles that constitute the outer shell of an atom. Electrons spin in an orbit around the center core of an atom. The center core is comprised of protons (positive charge) and neutrons (neutral charge). Electrons have a negative charge and balance out the positive charge of the protons. When an outside force causes the number of electrons to unbalance the charge of the protons, the electrons will split off the atom and look for another atom to balance out. If this imbalance is kept up, electrons will continue to move and an electrical flow will exist.

Many people have been taught electrical theory using an analogy with water. In a comparison with water flowing through a pipe, the electrons would be the water and the wire is the pipe.

The flow of electricity can be measured much like the flow of water through a pipe. The unit of measurement used is amps, frequently abbreviated as amps (a). You can compare amperage to the volume of water flowing through a pipe. When connected to a circuit, an ammeter will measure the actual amount of current flowing through the circuit. When relatively few electrons flow through a circuit, the amperage is low. When many electrons flow, the amperage is high.

Water pressure is measured in units such as pounds per square inch (psi), electrical pressure is measured in units called volts (V). When a voltmeter is connected to a circuit, it is measuring the electrical pressure. When electrical pressure is low, then voltage is considered to be low. When electrical pressure is high, then voltage is considered to be high.

The actual flow of electricity depends not only on voltage and amperage but also on the resistance of the circuit. The higher the resistance, the higher the force necessary to push the current through the circuit. The standard unit for measuring resistance is an ohm (Ω). Resistance in a circuit varies depending on the amount and type of components used in the circuit and the overall condition of the components and wires. If we assume that everything in our circuit is new, then, the main factors which determine resistance are:

• Material - some materials have more resistance than others. Those with high resistance are said to be insulators. Rubber materials (or rubber-like plastics) are some of the most common insulators used, as they have a very high resistance to electricity. Very low resistance materials are said to be conductors. Copper wire is among the best conductors. Silver is actually a superior conductor to copper and is used in some relay contacts but its high cost prohibits its use as common wiring. Most marine wiring is made of copper.

• Size - the larger the wire size being used, the less resistance the wire will have. This is why components that use large amounts of electricity usually have large wires supplying current to them.

• Length - for a given thickness of wire, the longer the wire, the greater the resistance. The shorter the wire, the less the resistance. When determining the proper wire for a circuit, both size and length must be considered to design a circuit that can handle the current needs of the component.

• Temperature - with many materials, the higher the temperature, the greater the resistance (positive temperature coefficient). Some materials exhibit the opposite trait of lower resistance with higher temperatures (negative temperature coefficient). These principles are used in many of the sensors on the engine.

OHM'S LAW

There is a direct relationship between current, voltage and resistance that can be summed up by a statement known as Ohm's law.

• Voltage (E) is equal to amperage (I) times resistance (R): $E = I \times R$
• Other forms of the formula are $R = E/I$ and $I = E/R$

In each of these formulas, **E** is the voltage in volts, **I** is the current in amps and **R** is the resistance in ohms. The basic point to remember is that as the resistance of a circuit goes up, the amount of current that flows in the circuit will go down, if voltage remains the same.

The amount of work that the electricity can perform is expressed as power. A unit of power is known as a watt (W). There is a direct relationship between power, voltage and current that can be summed up by the following formula:

• Power (W) is equal to amperage (I) times voltage (E): $W = I \times E$

■ This formula is only true for direct current (DC) circuits. The alternating current formula is a tad different but since the electrical circuits in most boats are DC type, we need not get into AC circuit theory.

Electrical Components

POWER SOURCE

◆ See Figure 2

Power is supplied to the boat by two devices: The battery and the alternator. The battery supplies electrical power during starting or during periods when the current demand of the boat's electrical system exceeds the output capacity of the alternator. The alternator supplies electrical current when the engine is running. The alternator does not just supply the current needs of the boat but it also recharges the battery.

In most modern boats, the battery is a lead/acid electrochemical device consisting of six 2 volt subsections (cells) connected in series, so that the unit is capable of producing approximately 12 volts of electrical pressure. Each subsection consists of a series of positive and negative plates held a short distance apart in a solution of sulfuric acid and water.

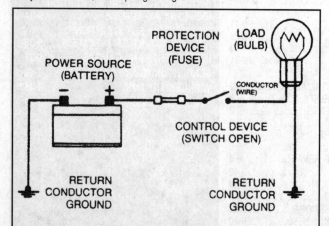

Fig. 1 This example illustrates a simple circuit. When the switch is closed, power from the positive (+) battery terminal flows through the fuse and the switch and then to the light bulb. The light illuminates and the circuit is completed through the ground wire back to the negative (-) battery terminal.

IGNITION & ELECTRICAL SYSTEMS

The two types of plates are of dissimilar metals, thus setting up a chemical reaction inside the battery case. It is this reaction that produces current flow from the battery when its positive and negative terminals are connected to an electrical load. The alternator, restoring the battery to its original chemical state replaces the power removed from the battery.

The following is a brief description of how the system works:

The battery stores electricity and acts as a "sponge" for the whole system. It mops up generated current until it's fully charged and it releases energy on demand.

Permanent magnets that create the moving magnetic field. If your engine has good spark, you can take it for granted that the magnets are in working order because the ignition and charging systems share the same magnets.

The alternator windings are the stationary coils of wire that the magnets rotate around. They produce the electrical charge. Simply put, the more windings in your stator, the greater the potential output in amps your charging system you'll have.

The rectifier consists of a series of diodes or electrical one-way valves. The rectifier overcomes one of the disadvantages of a current-generating system using permanent magnets and stator windings, which is that the current produced within the windings is alternating current (AC). You can't use AC to charge batteries. They accept only direct current (DC). So the rectifier is designed to convert AC current to a usable form of DC current simply called "rectified AC."

All engines covered here utilize a voltage regulator; either combined with the rectifier or standing alone. The regulator automatically reduces the output of generated current as the battery becomes fully charged.

GROUND

All boats use some sort of a ground return circuit. Direct ground components are grounded to an electrically conductive metal component through their mounting points. These electrically conductive metal components are then grounded to the battery.

All other components use some sort of ground wire which leads directly back to the battery. The electrical current runs through the ground wire and returns to the battery through the ground (-) cable. If you look, you'll see that the battery ground cable connects between the battery and a heavy gauge ground wire.

■ It should be noted that a good percentage of electrical problems can be traced to bad grounds.

PROTECTIVE DEVICES

◆ See Figure 3

It is possible for large surges of current to pass through the electrical system of your boat. If this surge of current were to reach components in the circuit, the surge could burn them out or severely damage them. Surges can also overload the wiring, causing the harness to get hot and melt the insulation. To prevent this, fuses, circuit breakers and/or fusible links are connected into the supply wires of the electrical system. These items are nothing more than a built-in weak spot in the system. When an abnormal amount of current flows through the system, these protective devices work as follows to protect the circuit:

• Fuse - when an excessive electrical current passes through a fuse, the fuse "blows" (the conductor melts) and opens the circuit, preventing the passage of current.

• Circuit Breaker - a circuit breaker is basically a self-repairing fuse. It will open the circuit in the same fashion as a fuse but when the surge subsides, the circuit breaker can be reset and does not need replacement.

• Fusible Link - a fusible link (fuse link or main link) is a short length of special, high temperature insulated wire that acts as a fuse. When an excessive electrical current passes through a fusible link, the thin gauge wire inside the link melts, creating an intentional open to protect the circuit.

To repair the circuit, the link must be replaced. Some newer type fusible links are housed in plug-in modules, which are simply replaced like a fuse, while older type fusible links must be cut and spliced if they melt. Since this link is very early in the electrical path, it's the first place to look if nothing on the boat works, yet the battery seems to be charged and is properly connected.

✱✱CAUTION

Always replace fuses, circuit breakers and fusible links with identically rated components. Under no circumstances should a component of higher or lower amperage rating be substituted.

SWITCHES & RELAYS

◆ See Figure 4

Switches are used in electrical circuits to control the passage of current. The most common use is to open and close circuits between the battery and the various electric devices in the system. Switches are rated according to the amount of amperage they can handle. If a switch rated for the sufficient amperage is not used in a circuit, the switch could overload and cause damage.

Some electrical components which require a large amount of current to operate use a special switch called a relay. Since these circuits carry a large amount of current, the thickness of the wire in the circuit is also greater. If this large wire were connected from the load to the control switch, the switch would have to carry the high amperage load and the space needed for wiring in the boat would be twice as big to accommodate the increased size of the wiring harness. To prevent these problems, a relay is used.

Relays are composed of a coil and a set of contacts. When the coil has a current passed though it, a magnetic field is formed and this field causes the contacts to move together, completing the circuit. Most relays are normally open, preventing current from passing through the circuit but they can take any electrical form depending on the job they are intended to do. Relays can be considered "remote control switches." They allow a smaller current to operate devices that require higher amperages. When a small current

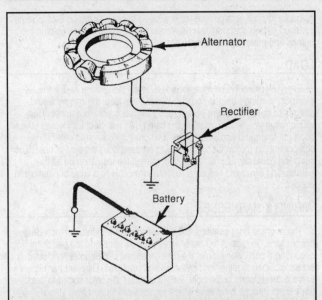

Fig. 2 Functional diagram of a typical charging circuit showing the relationship of the stator, solid-state rectifier and the batter

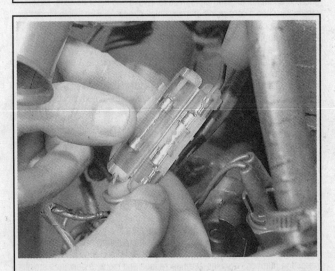

Fig. 3 Fuses protect the vessel's electrical system from abnormally high amounts of current flow

9-4 IGNITION & ELECTRICAL SYSTEMS

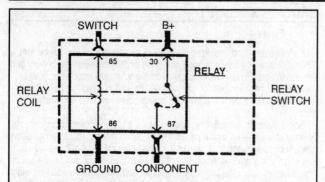

Fig. 4 Relays are composed of a coil and a switch. These two components are linked together so that when one operates, the other operates at the same time. The large wires in the circuit are connected from the battery to one side of the relay switch (B+) and from the opposite side of the relay switch to the load (component). Smaller wires are connected from the relay coil to the control switch for the circuit and from the opposite side of the relay coil to ground

operates the coil, a larger current is allowed to pass by the contacts. Some common circuits that may use relays are horns, lights, starter, electric fuel pumps and other high draw circuits.

LOAD

Every electrical circuit must include a "load" (something to use the electricity coming from the source). Without this load, the battery would attempt to deliver its entire power supply from one pole to another. This would result in a "short circuit" of the battery. All this electricity would take a short cut to ground and cause a great amount of damage to other components in the circuit by developing a tremendous amount of heat. This condition could develop sufficient heat to melt the insulation on all the surrounding wires and reduce a multiple wire cable to a lump of plastic and copper.

WIRING & HARNESSES

The average boat contains miles of wiring, with hundreds of individual connections. To protect the many wires from damage and to keep them from becoming a confusing tangle, they are organized into bundles, enclosed in plastic or taped together and called wiring harnesses. Different harnesses serve different parts of the boat. Individual wires are color coded to help trace them through a harness where sections are hidden from view.

Marine wiring can be either single strand wire, multi-strand wire or printed circuitry. Single strand wire has a solid metal core and is usually used inside such components as alternators, motors, relays and other devices. Multi-strand wire has a core made of many small strands of wire twisted together into a single conductor. Most of the wiring in a marine electrical system is made up of multi-strand wire, either as a single conductor or grouped together in a harness. All wiring is color coded on the insulator, either as a solid color or as a colored wire with an identification stripe. A printed circuit is a thin film of copper or other conductor that is printed on an insulator backing. Occasionally, a printed circuit is sandwiched between two sheets of plastic for more protection and flexibility. A complete printed circuit, consisting of conductors, insulating material and connectors is called a printed circuit board. Printed circuitry is used in place of individual wires or harnesses in places where space is limited, such as behind instrument panels.

Since marine electrical systems are very sensitive to changes in resistance, the selection of properly sized wires is critical when systems are repaired. A loose or corroded connection or a replacement wire that is too small for the circuit will add extra resistance and an additional voltage drop to the circuit.

The wire gauge number is an expression of the cross-section area of the conductor. Boats from countries that use the metric system will typically describe the wire size as its cross-sectional area in square millimeters. In this method, the larger the wire, the greater the number. Another common system for expressing wire size is the American Wire Gauge (AWG) system. As gauge number increases, area decreases and the wire becomes smaller. An 18 gauge wire is smaller than a 4 gauge wire. A wire with a higher gauge number will carry less current than a wire with a lower gauge number. Gauge wire size refers to the size of the strands of the conductor, not the size of the complete wire with insulator. It is possible, therefore, to have two wires of the same gauge with different diameters because one may have thicker insulation than the other.

It is essential to understand how a circuit works before trying to figure out why it doesn't. An electrical schematic shows the electrical current paths when a circuit is operating properly. Schematics break the entire electrical system down into individual circuits. In a schematic, usually no attempt is made to represent wiring and components as they physically appear on the boat, switches and other components are shown as simply as possible. Face views of harness connectors show the cavity or terminal locations in all multi-pin connectors to help locate test points.

CONNECTORS

◆ See Figures 5 thru 8

Weatherproof connectors are most commonly used where the connector is exposed to the elements. Terminals are protected against moisture and dirt by sealing rings that provide a weather tight seal. All repairs require the use of a special terminal and the tool required to service it.

Unlike standard blade type terminals, these weatherproof terminals cannot be straightened once they are bent. Make certain that the connectors are properly seated and all of the sealing rings are in place when connecting leads.

Test Equipment

Pinpointing the exact cause of trouble in an electrical circuit is most times accomplished by the use of special test equipment. The following sections describe different types of commonly used test equipment and briefly explain how to use them in diagnosis. In addition to the information covered below, the tool manufacturer's instruction manual (provided with most tools) should be read and clearly understood before attempting any test procedures.

JUMPER WIRES

◆ See Figure 9

※※CAUTION

Never use jumper wires made from a thinner gauge wire than the circuit being tested. If the jumper wire is of too small a gauge, it may overheat and possibly melt. Never use jumpers to bypass high resistance loads in a circuit. Bypassing resistances, in effect, creates a short circuit. This may, in turn, cause damage and fire. Jumper wires should only be used to bypass lengths of wire or to simulate switches.

Jumper wires are simple, yet extremely valuable, pieces of test equipment. They are basically test wires that are used to bypass sections of a circuit. Although jumper wires can be purchased, they are usually fabricated from lengths of standard marine wire and whatever type of connector (alligator clip, spade connector or pin connector) that is required for the particular application being tested. In cramped, hard-to-reach areas, it is advisable to have insulated boots over the jumper wire terminals in order to prevent accidental grounding.

It is also advisable to include a standard marine fuse in any jumper wire. This is commonly referred to as a "fused jumper". By inserting an in-line fuse holder between a set of test leads, a fused jumper wire can be used for bypassing open circuits. Use a 5-amp fuse to provide protection against voltage spikes.

Jumper wires are used primarily to locate open electrical circuits. If an electrical component fails to operate, connect the jumper wire between the component and a good ground. If the component operates only with the jumper installed, the ground circuit is open.

If the ground circuit is good but the component does not operate, the circuit between the power feed and component may be open. By moving the jumper wire successively back from the component toward the power source, you can isolate the area of the circuit where the open is located. When the component stops functioning or the power is cut off, the open is in the segment of wire between the jumper and the point previously tested.

You can sometimes connect the jumper wire directly from the battery to the "hot" terminal of the component but first make sure the component uses 12 volts in operation. Some electrical components, such as fuel injectors or sensors are designed to operate on about 4 to 5 volts and running 12 volts directly to these components will cause damage.

IGNITION & ELECTRICAL SYSTEMS

TEST LIGHTS

◆ See Figure 10

The test light is used to check circuits and components while electrical current is flowing through them. It is used for voltage and ground tests. To use a 12-volt test light, connect the ground clip to a good ground and probe wherever necessary with the pick. The test light will illuminate when voltage is detected. This does not necessarily mean that 12 volts (or any particular amount of voltage) is present, it only means that some voltage is present.

■ It is advisable before using the test light to touch its ground clip and probe across the battery posts or terminals to make sure the light is operating properly.

✳✳WARNING

Do not use a test light to probe electronic ignition, spark plug or coil wires. Never use a pick-type test light to probe wiring on electronically controlled systems unless specifically instructed to do so. Any wire insulation that is pierced by the test light probe should be taped and sealed with silicone after testing.

Like the jumper wire, the 12-volt test light is used to isolate opens in circuits. But, whereas the jumper wire is used to bypass the open to operate the load, the 12-volt test light is used to locate the presence of voltage in a circuit. If the test light illuminates, there is power up to that point in the circuit, if the test light does not illuminate, there is an open circuit (no power). Move the test light in successive steps back toward the power source until the light in the handle illuminates. The open is between the probe and a point that was previously probed.

The self-powered test light is similar in design to the 12-volt test light but contains a 1.5-volt penlight battery in the handle. It is most often used in place of a multi-meter to check for open or short circuits when power is isolated from the circuit (continuity test).

The battery in a self-powered test light does not provide much current. A weak battery may not provide enough power to illuminate the test light even when a complete circuit is made (especially if there is high resistance in the circuit). Always make sure that the test battery is strong. To check the battery, briefly touch the ground clip to the probe, if the light glows brightly, the battery is strong enough for testing.

✳✳WARNING

A self-powered test light should not be used on any electronically controlled system or component. The small amount of electricity transmitted by the test light is enough to damage many electronic marine components.

MULTI-METERS

◆ See Figure 11

Multi-meters are an extremely useful tool for troubleshooting electrical problems. They can be purchased in either analog or digital form and have a price range to suit any budget. A multi-meter is a voltmeter, ammeter and multi-meter (along with other features) combined into one instrument. It is often used when testing solid-state circuits because of its high input impedance (usually 10 mega-ohms or more). A brief description of the multi-meter main test functions follows:

• Voltmeter - the voltmeter is used to measure voltage at any point in a circuit or to measure the voltage drop across any part of a circuit. Voltmeters usually have various scales and a selector switch to allow the reading of different voltage ranges.

The voltmeter has a positive and a negative lead. To avoid damage to the meter, always connect the negative lead to the negative (-) side of the circuit (to ground or nearest the ground side of the circuit) and connect the positive lead to the positive (+) side of the circuit (to the power source or the nearest power source).

■ The negative voltmeter lead will always be Black and the positive voltmeter will always be some color other than Black (usually Red).

• Multi-meter - the multi-meter is designed to read resistance (measured in ohms) in a circuit or component. Most multi-meters will have a selector switch which permits the measurement of different ranges of resistance (usually the selector switch allows the multiplication of the meter reading by 10, 100, 1,000 and 10,000). Some multi-meters are "auto-ranging" which means the meter itself will determine which scale to use.

Since an internal battery powers the meters, the multi-meter can be used like a self-powered test light. When the multi-meter is connected, current from the multi-meter flows through the circuit or component being tested. Since the multi-meter's internal resistance and voltage are known values, the

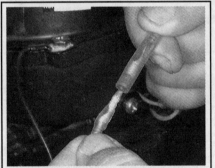
Fig. 5 Bullet connectors are some of the more common electrical connectors found on an engine

Fig. 6 A typical weatherproof electrical connector

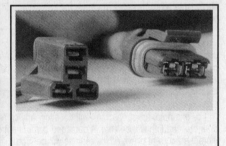
Fig. 7 Hard shell (left) and weatherproof (right) connectors have replaceable terminals

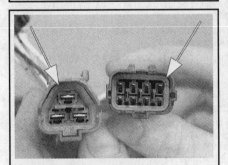
Fig. 8 The seals on weatherproof connectors must be kept in good condition to prevent the terminals from corroding

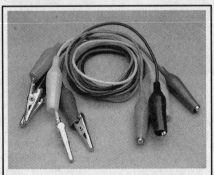
Fig. 9 Jumper wires are simple, yet extremely valuable, pieces of test equipment

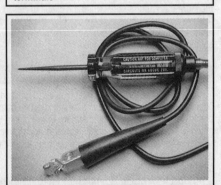
Fig. 10 A test light is used to detect the presence of voltage in a circuit

9-6 IGNITION & ELECTRICAL SYSTEMS

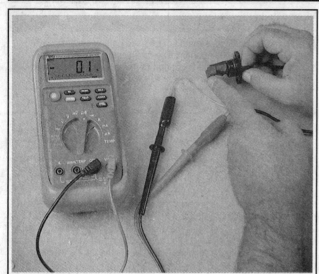

Fig. 11 Multi-meters are essential for diagnosing faulty wires, switches and other electrical components

amount of current flow through the meter depends on the resistance of the circuit or component being tested.

The multi-meter can also be used to perform a continuity test for suspected open circuits. In using the meter for making continuity checks, do not be concerned with the actual resistance readings. Zero resistance (or any ohm reading) indicates continuity in the circuit. Infinite resistance indicates an opening in the circuit. A high resistance reading where there should be none indicates a problem in the circuit.

Checks for short circuits are made in the same manner as checks for open circuits, except that the circuit must be isolated from both power and normal ground. Infinite resistance indicates no continuity, while zero resistance indicates a dead short.

✱✱WARNING

Never use a multi-meter to check the resistance of a component or wire while there is voltage applied to the circuit.

- **Ammeter** - an ammeter measures the amount of current flowing through a circuit in units called amps (amps). At normal operating voltage, most circuits have a characteristic amount of amps, called "current draw" which can be measured using an ammeter. By referring to a specified current draw rating, then measuring the amps and comparing the two values, one can determine what is happening within the circuit to aid in diagnosis.

For example, an open circuit will not allow any current to flow, so the ammeter reading will be zero. A damaged component or circuit will have an increased current draw, so the reading will be high.

The ammeter is always connected in series with the circuit being tested. All of the current that normally flows through the circuit must also flow through the ammeter; if there is any other path for the current to follow, the ammeter reading will not be accurate. The ammeter itself has very little resistance to current flow and, therefore, it will not affect the circuit but it will measure current draw only when the circuit is closed and electricity is flowing. Excessive current draw can blow fuses and drain the battery, while a reduced current draw can cause motors to run slowly, lights to dim and other components to not operate properly.

Troubleshooting the Electrical System

When diagnosing any electrical problem organized troubleshooting is a must. The complexity of electrical systems on modern boats and their power plants demands that you approach any problem in a logical organized manner. There are certain troubleshooting techniques, which are standard:
- **Establish when the problem occurs** - Does the problem appear only under certain conditions? Were there any noises, odors or other unusual symptoms?
- **Check for obvious problems** - Problems such as broken wires and loose or dirty connections can cause major problems. Always check the obvious before assuming something complicated (or expensive) is the cause.

■ Experience has shown that most problems tend to be the result of a fairly simple and obvious cause, such as loose or corroded connectors, bad grounds or damaged wire insulation, which causes a short. This makes careful visual inspection of components during testing essential to quick and accurate troubleshooting.

- **Isolate the problem area** - Make some simple tests and observations, then eliminate the systems that are working properly. Test for problems systematically to determine the cause once the problem area is isolated. Are all the components functioning properly? Is there power going to electrical switches and motors? Performing careful, systematic checks will often turn up most causes on the first inspection, without wasting time checking components that have little or no relationship to the problem.

- **Verify all systems after repairs are completed** - Some causes can be traced to more than one component, so a careful verification of repair work is important in order to pick up additional malfunctions that may cause a problem to reappear or a different problem to arise. A blown fuse, for example, is a simple problem that may require more than another fuse to repair. If you don't look for a problem that caused a fuse to blow, a shorted wire (for example) may go undetected.

VOLTAGE

◆ See Figure 12

This test determines voltage available from the battery and should be the first step in any electrical troubleshooting procedure after visual inspection. Many electrical problems, especially on electronically controlled systems, can be caused by a low state of charge in the battery. Excessive corrosion at the battery cable terminals can cause poor contact that will prevent proper charging and full battery current flow.

1. Set the voltmeter selector switch to the 10-volt position.
2. Connect the multi-meter negative lead to the battery's negative (-) terminal and the positive lead to the battery's positive (+) terminal.
3. Turn the battery (ignition) switch **ON** to provide a load.
4. A well-charged battery should register over 12 volts. If the meter reads below 11.5 volts, the battery power may be insufficient to operate the electrical system properly.
5. Charge the battery and retest.

VOLTAGE DROP

◆ See Figure 13

When current flows through a load, the voltage beyond the load drops. This voltage drop is due to the resistance created by the load and also by small resistances created by corrosion at the connectors and damaged insulation on the wires. The maximum allowable voltage drop under load is critical, especially if there is more than one load in the circuit, since all voltage drops are cumulative.

1. Set the voltmeter selector switch to the 10-volt position.
2. Connect the multi-meter negative lead to the battery negative (-) terminal or another good ground.
3. Touch the multi-meter positive lead to the battery's positive (+) terminal to determine battery voltage.
4. Operate the circuit and check the voltage prior to the first component (load).
5. There should be little or no voltage drop (from battery voltage) in the circuit prior to the first component. If a voltage drop exists, the wire or connectors in the circuit are suspect.
6. While operating the first component in the circuit, probe the ground-side of the component with the positive (+) meter lead and observe the voltage readings. A small voltage drop should be noticed. The resistance of the component causes this voltage drop.
7. Repeat the test for each component (load) down the circuit.
8. If a large voltage drop is noticed, the preceding component, wire or connector is suspect.

IGNITION & ELECTRICAL SYSTEMS 9-7

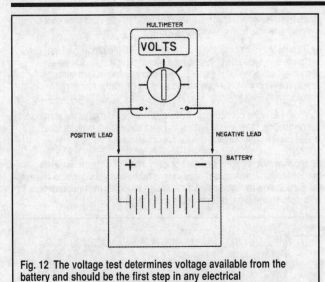

Fig. 12 The voltage test determines voltage available from the battery and should be the first step in any electrical troubleshooting procedure after visual inspection

RESISTANCE

 MODERATE

◆ See Figures 14 and 15

✱✱WARNING

Never use a multi-meter with power applied to the circuit. The multi-meter is designed to operate on it's own power supply. The normal 12-volt electrical system voltage can damage the meter!

1. Isolate the circuit from the boat's power source.
2. Ensure that the battery (ignition) switch is **OFF**.
3. Isolate at least one side of the circuit to be checked, in order to avoid reading parallel resistances. Parallel circuit resistances will always give a lower reading than the actual resistance of either of the branches.
4. Connect the meter leads to both sides of the circuit (wire or component) and read the actual resistance measured in ohms on the meter scale. Make sure the selector switch is set to the proper ohm scale for the circuit being tested, to avoid misreading the multi-meter test value.
5. Compare this reading to the resistance specification for the component or formulate the theoretical resistance using Ohms Law.

Fig. 14 Using a multi-meter to check resistance on the secondary side of the ignition coil

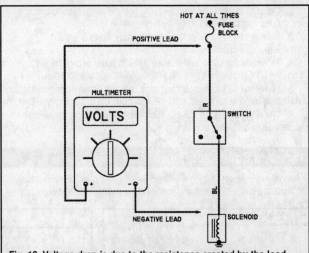

Fig. 13 Voltage drop is due to the resistance created by the load and also by small resistances created by corrosion at the connectors and damaged insulation on the wires

OPEN CIRCUITS

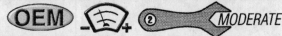 MODERATE

◆ See Figures 16 and 17

This test already assumes the existence of an open in the circuit and it is used to help locate the open portion.
1. Isolate the circuit from power and ground.
2. Connect the self-powered test light or multi-meter ground clip to the ground-side of the circuit and probe sections of the circuit sequentially.
3. If the light is out or there is infinite resistance, the open is between the probe and the circuit ground.
4. If the light is on or the meter shows continuity, the open is between the probe and the end of the circuit toward the power source.

SHORT CIRCUITS

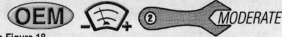 MODERATE

◆ See Figure 18

■ Never use a self-powered test light to perform checks for opens or shorts when power is applied to the circuit under test. The test light can be damaged by outside power.

1. Isolate the circuit from power and ground.

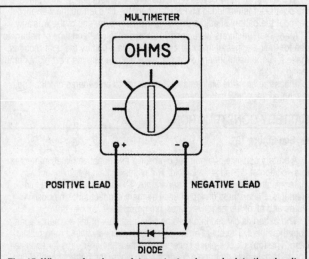

Fig. 15 When performing resistance tests, always isolate the circuit from power

9-8 IGNITION & ELECTRICAL SYSTEMS

2. Connect the self-powered test light or multi-meter ground clip to a good ground and probe any easy-to-reach point in the circuit.

3. If the light comes on or there is continuity, there is a short somewhere in the circuit.

4. To isolate the short, probe a test point at either end of the isolated circuit (the light should be on or the meter should indicate continuity).

5. Leave the test light probe engaged and sequentially open connectors or switches, remove parts, etc. until the light goes out or continuity is broken.

6. When the light goes out, the short is between the last two circuit components, which were opened.

Wire and Connector Repair

Almost anyone can replace damaged wires, as long as the proper tools and parts are available. Wire and terminals are available to fit almost any need. Even the specialized weatherproof, molded and hard shell connectors are now available from aftermarket suppliers.

Be sure the ends of all the wires are fitted with the proper terminal hardware and connectors. Wrapping a wire around a stud is never a permanent solution and will only cause trouble later. Replace wires one at a time to avoid confusion. Always route wires in the same manner of the manufacturer.

When replacing connections, make absolutely certain that the connectors are certified for marine use. Automotive wire connectors may not meet United States Coast Guard (USCG) specifications.

■ **If connector repair is necessary, only attempt it if you have the proper tools. Weatherproof connectors require special tools to release the pins inside the connector. Attempting to repair these connectors with conventional hand tools will damage them.**

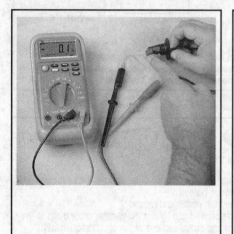

Fig. 16 The infinite display on this multimeter (0.1) indicates that the circuit is open

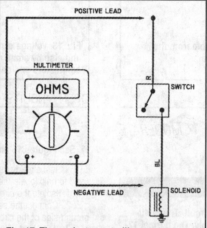

Fig. 17 The easiest way to illustrate an open circuit is to consider an example circuit with a switch. When the switch is turned OFF, power does not flow through the circuit to the load. Thus, the circuit is open

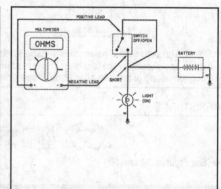

Fig. 18 In this illustration, the circuit between the battery and light should be open because the switch is turned OFF. However, battery voltage is reaching the light at the point of the short; this could possibly be caused by chaffed wires

BATTERY

The battery is one of the most important parts of the electrical system. In addition to providing electrical power to start the engine, it also provides power for operation of the running lights, radio and electrical accessories.

Because of its job and the consequences (failure to perform in an emergency), the best advice is to purchase a well-known brand, with an extended warranty period, from a reputable dealer.

The usual warranty covers a pro-rated replacement policy, which means the purchaser is entitled to consideration for the time left on the warranty period if the battery should prove defective before the end of the warranty.

Many manufacturers have specifications on the size and type of battery to use for their engines. If in doubt as to how large a battery the boat requires, make a liberal estimate and then purchase the one with the next higher amp rating.

Please refer to the Maintenance section for all procedures on cleaning, testing, storage and maintenance.

BATTERY CONSTRUCTION

◆ See Figure 19

A battery consists of a number of positive and negative plates immersed in a solution of diluted sulfuric acid. The plates contain dissimilar active materials and are kept apart by separators. The plates are grouped into elements. Plate straps on top of each element connect all of the positive plates and all of the negative plates into groups.

The battery is divided into cells holding a number of the elements apart from the others. The entire arrangement is contained within a hard plastic case. The top is a one-piece cover and contains the filler caps for each cell. The terminal posts protrude through the top where the battery connections for the boat are made. Each of the cells is connected to its neighbor in a positive-to-negative manner with a heavy strap called the cell connector.

MARINE BATTERIES

◆ See Figure 20

Because marine batteries are required to perform under much more rigorous conditions than automotive batteries, they are constructed differently than those used in automobiles or trucks. Therefore, a marine battery should always be the No. 1 unit for the boat and other types of batteries used only in an emergency.

Marine batteries have a much heavier exterior case to withstand the violent pounding and shocks imposed on it as the boat moves through rough water and in extremely tight turns. The plates are thicker and each plate is securely anchored within the battery case to ensure extended life. The caps are spill proof to prevent acid from spilling into the bilge when the boat heels to one side in a tight turn or is moving through rough water. Because of these features, the marine battery will recover from a low charge condition and give satisfactory service over a much longer period of time than any type intended for automotive use.

✷✷ WARNING

Never use a Maintenance-free battery with an engine that is not voltage regulated. The charging system will continue to charge as long as the engine is running and it is possible that the electrolyte could boil out if periodic checks of the cell electrolyte level are not done.

BATTERY RATINGS

◆ See Figure 21

Three different methods are used to measure and indicate battery electrical capacity:

IGNITION & ELECTRICAL SYSTEMS

- Amp/hour rating
- Cold cranking performance
- Reserve capacity

The amp/hour rating of a battery refers to the battery's ability to provide a set amount of amps for a given amount of time under test conditions at a constant temperature. Therefore, if the battery is capable of supplying 4 amps of current for 20 consecutive hours, the battery is rated as an 80 amp/hour battery. The amp/hour rating is useful for some service operations, such as slow charging or battery testing.

Cold cranking performance is measured by cooling a fully charged battery to 0°F (-17°C) and then testing it for 30 seconds to determine the maximum current flow. In this manner the cold cranking amp rating is the number of amps available to be drawn from the battery before the voltage drops below 7.2 volts.

The illustration depicts the amount of power in watts available from a battery at different temperatures and the amount of power in watts required of the engine at the same temperature. It becomes quite obvious - the colder the climate, the more necessary for the battery to be fully charged.

Reserve capacity of a battery is considered the length of time, in minutes, at 80°F (27°C), a 25 amp current can be maintained before the voltage drops below 10.5 volts. This test is intended to provide an approximation of how long the engine, including electrical accessories, could operate satisfactorily if the stator assembly or lighting coil did not produce sufficient current. A typical rating is 100 minutes.

■ **If possible, the new battery should have a power rating equal to or higher than the unit it is replacing.**

BATTERY LOCATION

◆ See Figure 22

Every battery installed in a boat must be secured in a well protected, ventilated area. If the battery area lacks adequate ventilation, hydrogen gas, which is given off during charging, is very explosive. This is especially true if the gas is concentrated and confined.

DUAL BATTERY INSTALLATION

◆ See Figures 23 thru 26

Three methods are available for utilizing a dual-battery hook-up:

1. A high-capacity switch can be used to connect the two batteries. The accompanying illustration details the connections for installation of such a switch. This type of switch installation has the advantage of being simple, inexpensive, and easy to mount and hookup. However, if the switch is forgotten in the closed position, it will let the convenience loads run down both batteries and the advantage of the dual installation is lost. However, the switch may be closed intentionally to take advantage of the extra capacity of the two batteries, or it may be temporarily closed to help start the engine under adverse conditions.

2. A relay, can be connected into the ignition circuit to enable both batteries to be automatically put in parallel for charging or to isolate them for ignition use during engine cranking and start. By connecting the relay coil to the ignition terminal of the ignition-starting switch, the relay will close during the start to aid the starting battery. If the second battery is allowed to run down, this arrangement can be a disadvantage since it will draw a load from the starting battery while cranking the engine. One way to avoid such a condition is to connect the relay coil to the ignition switch accessory terminal. When connected in this manner, while the engine is being cranked, the relay is open, but when the engine is running with the ignition switch in the normal position, the relay is closed, and the second battery is being charged at the same time as the starting battery.

3. A heavy duty switch installed as close to the batteries as possible can be connected between them. If such an arrangement is used it must meet the standards of the American Boat and Yacht Council, or the Fire Protection Standard for Motor Craft, N.F.P.A. No. 302.

BATTERY CHARGERS

◆ See Figure 27

Before using any battery charger, consult the manufacturer's instructions for its use. Battery chargers are electrical devices that change Alternating

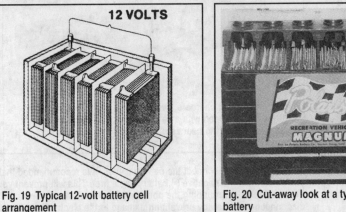

Fig. 19 Typical 12-volt battery cell arrangement

Fig. 20 Cut-away look at a typical marine battery

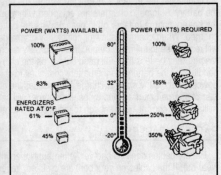

Fig. 21 Comparison of battery efficiency and engine demands at various temperatures

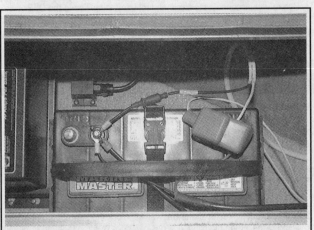

Fig. 22 Comparison of battery efficiency and engine demands at various temperatures

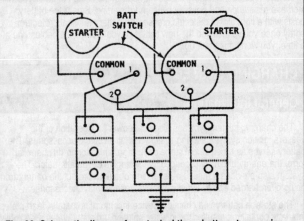

Fig. 23 Schematic diagram for a typical three battery, two engine hookup.

9-10 IGNITION & ELECTRICAL SYSTEMS

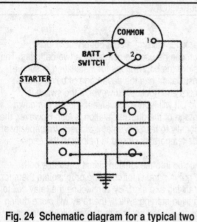

Fig. 24 Schematic diagram for a typical two battery, one engine hookup.

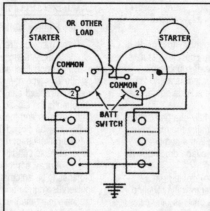

Fig. 25 Schematic diagram for a typical two battery, two engine hookup.

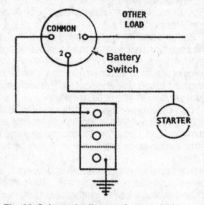

Fig. 26 Schematic diagram for a typical single battery, one engine hookup.

Current (AC) to a lower voltage of Direct Current (DC) that can be used to charge a marine battery. There are two types of battery chargers - manual and automatic.

A manual battery charger must be physically disconnected when the battery has come to a full charge. If not, the battery can be overcharged and possibly fail. Excess charging current at the end of the charging cycle will heat the electrolyte, resulting in loss of water and active material, substantially reducing battery life.

■ **As a rule, on manual chargers, when the ammeter on the charger registers half the rated amperage of the charger, the battery is fully charged. This can vary and it is recommended to use a hydrometer to accurately measure state of charge.**

Automatic battery chargers have an important advantage - they can be left connected (for instance, overnight) without the possibility of overcharging the battery. Automatic chargers are equipped with a sensing device to allow the battery charge to taper off to near zero as the battery becomes fully charged. When charging a low or completely discharged battery, the meter will read close to full rated output. If only partially discharged, the initial reading may be less than full rated output, as the charger responds to the condition of the battery. As the battery continues to charge, the sensing device monitors the state of charge and reduces the charging rate. As the rate of charge tapers to zero amps, the charger will continue to supply a few milliamps of current - just enough to maintain a charged condition.

BATTERY CABLES

Battery cables don't go bad very often but like anything else, they can wear out. If the cables on your boat are cracked, frayed or broken, they should be replaced.

When working on any electrical component, it is always a good idea to disconnect the negative (-) battery cable. This will prevent potential damage to many sensitive electrical components

Always replace the battery cables with one of the same length or you will increase resistance and possibly cause hard starting. Smear the battery posts with a light film of dielectric grease or a battery terminal protectant-spray once you've installed the new cables. If you replace the cables one at a time, you won't mix them up.

Fig. 27 Automatic chargers, such as the Battery Tender(r) from Deltran, are equipped with a sensing device to allow the battery charge to taper off to near zero as the battery becomes fully charged

■ **Any time you disconnect the battery cables, it is recommended that you disconnect the negative (-) battery cable first. This will prevent you from accidentally grounding the positive (+) terminal when disconnecting it, thereby preventing damage to the electrical system.**

Before you disconnect the cable(s), first turn the ignition to the **OFF** position. This will prevent a draw on the battery that could cause arcing. When the battery cable(s) are reconnected (negative cable last), be sure to check all electrical accessories are all working correctly.

CHARGING SYSTEM

General Information

The charging system provides electrical power for operation of the vessel's ignition system, starting system and all electrical accessories. The battery serves as an electrical surge or storage tank, storing (in chemical form) the energy originally produced by the engine driven alternator. The system also provides a means of regulating output to protect the battery from being overcharged and to avoid excessive voltage to the accessories.

The storage battery is a chemical device incorporating parallel lead plates in a tank containing a sulfuric acid/water solution. Adjacent plates are slightly dissimilar, and the chemical reaction of the two dissimilar plates produces electrical energy when the battery is connected to a load such as the starter motor. The chemical reaction is reversible, so that when the alternator is producing a voltage (electrical pressure) greater than that produced by the battery, electricity is forced into the battery, and the battery is returned to its fully charged state.

Newer engines use alternating current alternators, because they are more efficient, can be rotated at higher speeds, and have fewer brush problems. In an alternator, the field usually rotates while all the current produced passes only through the stator winding. The brushes bear against continuous slip rings. This causes the current produced to periodically reverse the direction of its flow. Diodes (electrical one way valves) block the flow of current from traveling in the wrong direction. A series of diodes is wired together to permit the alternating flow of the stator to be rectified back to 12 volts DC for use by the vessel's electrical system.

IGNITION & ELECTRICAL SYSTEMS

The voltage regulating function is performed by a regulator. The regulator is often built into the alternator; this system is termed an integrated or internal regulator.

An alternator differs from a DC shunt generator in that the armature is stationary, and is called the stator, while the field rotates and is called the rotor. The higher current values in the alternator's stator are conducted to the external circuit through fixed leads and connections, rather than through a rotating commutator and brushes as in a DC generator. This eliminates a major point of maintenance.

The rotor assembly is supported in the drive end frame by a ball bearing and at the other end by a roller bearing. These bearings are lubricated during assembly and require no maintenance. There are six diodes in the end frame assembly. These diodes are electrical check valves that also change the alternating current developed within the stator windings to a Direct Current (DC) at the output (**BAT**) terminal. Three of these diodes are negative and are mounted flush with the end frame while the other three are positive and are mounted into a strip called a heat sink. The positive diodes are easily identified as the ones within small cavities or depressions.

The alternator charging system is a negative (-) ground system which consists of an alternator, a regulator, a charge indicator, a storage battery and wiring connecting the components, and fuse link wire.

The alternator is belt-driven from the engine. Energy is supplied from the alternator/regulator system to the rotating field through two brushes to two slip-rings. The slip-rings are mounted on the rotor shaft and are connected to the field coil. This energy supplied to the rotating field from the battery is called excitation current and is used to initially energize the field to begin the generation of electricity. Once the alternator starts to generate electricity, the excitation current comes from its own output rather than the battery.

The alternator produces power in the form of alternating current. The alternating current is rectified by 6 diodes into direct current. The direct current is used to charge the battery and power the rest of the electrical system.

When the ignition key is turned **ON**, current flows from the battery, through the charging system indicator light, to the voltage regulator, and to the alternator. When the engine is started, the alternator begins to produce current. As the alternator turns and produces current, the current is divided in two ways: part to the battery (to charge the battery and power the electrical components of the vessel), and part is returned to the alternator (to enable it to increase its output). In this situation, the alternator is receiving current from the battery and from itself. A voltage regulator is wired into the current supply to the alternator to prevent it from receiving too much current, which would cause it to put out too much current. Conversely, if the voltage regulator does not allow the alternator to receive enough current, the battery will not be fully charged and will eventually go dead.

The battery is connected to the alternator at all times, whether the ignition key is turned **ON** or not. If the battery were shorted to ground, the alternator would also be shorted. This would damage the alternator. To prevent this, a fuse link is installed in the wiring between the battery and the alternator. If the battery is shorted, the fuse link melts, protecting the alternator.

An alternator is better that a conventional, DC shunt generator because it is lighter and more compact, because it is designed to supply the battery and accessory circuits through a wide range of engine speeds, and because it eliminates the necessary maintenance of replacing brushes and servicing commutators.

General System Troubleshooting

ALTERNATOR FAILS TO CHARGE

Drive belt loose or broken. Replace and/or adjust drive belt.
Corroded or loose wires or connection in the charging circuit. Inspect, clean, and tighten.
Open charging circuit. Trace and repair.

ALTERNATOR CHARGES LOW OR UNSTEADY

Drive belt loose or broken. Replace and/or adjust drive belt.
Battery charge too low. Charge or replace the battery.
High resistance at the battery terminals. Remove the cables, clean the connectors and battery posts, replace and tighten.
High resistance in the charging circuit. Trace and repair.
High resistance in the ground circuit. Trace and repair.

ALTERNATOR OUTPUT TOO HIGH - BATTERY OVERCHARGED

Regulator base not grounded properly. Correct condition to make good ground.
Faulty ignition switch. Replace switch.

ALTERNATOR TOO NOISY

Worn, loose, or frayed drive belt. Replace belt and adjust properly.
Alternator mounting loose. Tighten all mounting hardware securely.
Worn alternator bearings. Replace.
Interference between rotor and stator leads or rectifier leads. Check and correct.
Rotor or fan damaged. Replace.
Open or shorted rectifier. Replace.
Open or shorted winding in stator. Replace.

AMMETER FLUCTUATES CONSTANTLY

High resistance connection in the alternator or voltage regulator circuit. Trace and repair.

TESTING

Many times the alternator is suspected of being defective when the battery is not receiving a charge and is constantly being depleted of its energy. Most of the time a heavily corroded wire terminal, broken wire or worn out battery is the actual problem. Perform the preliminary checks listed above to eliminate any problem areas in the charging circuitry before performing the output tests.

If the battery is constantly undercharged, verify all accessories are being switched off when the engine is not running. Check to see if a new accessory (fish finder, live bait tank, etc.) has been added which will place a heavy ampere draw on the battery when operating at low speeds. The battery may be drained when operating at slow speeds for long periods of time.

Check the physical condition and charge state of the battery. The battery should be 75% (1.230 specific gravity) of a full charge. If one or more cells in the battery are defective, the battery should be replaced.

Inspect the alternator system wiring for corroded or loose terminals, damaged or frayed wiring and/or loose wire harness connectors. Check the drive belt for physical condition and proper tension.

If the charging system has passed all the above visual checks, perform the output tests to determine if the alternator is defective.

The following tests require the use of a voltmeter/multi-meter capable of reading 0-20 volts DC. These tests will determine if the alternator and other components within the alternator circuit are in satisfactory working condition.

CHARGING SYSTEM

◆ See Figure 28

1. Check the drive belt and battery condition.
2. With a fully charged battery, connect a DVOM directly to the battery terminals. Start the engine and run it at 1300-1500 rpm; confirm that the meter reads 13.8-14.8 volts.
3. If the voltage reading is within specifications, switch the meter to the AC Volt position and check that there is no more than a 0.250 AC voltage reading while the engine is still running. If the reading is higher, the alternator has bad diodes.
4. If the reading in Step 2 is below 13.5 V, connect the positive lead of the meter to the output terminal on the back of the alternator and connect the negative lead to the negative terminal on the alternator. Start the engine and run it at 1300-1500 rpm; confirm that the meter reads 13.8-14.8 volts.
5. If the reading is now within specification, there is too much resistance between the battery and the alternator.
6. If the reading is below 12.5 V, the alternator may not be charging - check all wiring leading to the alternator.

9-12 IGNITION & ELECTRICAL SYSTEMS

CHARGING SYSTEM RESISTANCE

◆ See Figures 29 and 30

1. Remove the high tension line at the ignition coil and ground it so the engine will not start. Crank the engine over on the starter for about 15 seconds so that the battery discharges slightly.
2. Reconnect the high tension lead at the coil and make sure that all accessories are turned OFF.
3. Connect the positive lead of a DVOM to the alternator output terminal and the negative lead to the positive battery terminal - use the post, not the cable end.
4. Start the engine and run it at 1300-1500 rpm. Any reading higher than 0.5 V indicates too much resistance in the circuit wiring.
5. Now connect the negative lead on the meter to the ground terminal on the alternator and the positive lead to the negative battery terminal (not the cable end).
6. Start the engine and run it at 1300-1500 rpm. Any reading higher than 0.5 V indicates too much resistance in the circuit wiring.

Alternator

PRECAUTIONS

Several precautions must be observed when performing work on alternator equipment.
- If the battery is removed for any reason, make sure that it is reconnected with the correct polarity. Reversing the battery connections may result in damage to the one-way rectifiers.
- Never short across or ground any of the alternator terminals unless specifically mentioned in a particular test procedure.
- Never operate the alternator with the main circuit broken. Make sure that the battery, alternator, and regulator leads are not disconnected while the engine is running.
- Never attempt to polarize an alternator.
- When charging a battery that is installed in the vessel, disconnect the battery cables.
- When arc (electric) welding is to be performed on any part of the vessel, disconnect the negative battery cable and alternator leads.
- Never disconnect the battery cables while the engine is running

CHARGING SYSTEM INSPECTION

1. Make sure the battery connections are clean and tight and that the battery is in good condition and fully charged.
2. Check the drive belt for damage or looseness.
3. Check the wiring harness at the alternator. The harness connector should be tight and latched. Make sure that the output terminal of the alternator is connected to the vessel battery positive lead.
4. Verify that all charging system related fuses and electrical connections are tight and free of damage.
5. Check the mounting bolts for proper torque.

The alternator does not require periodic lubrication. The rotor shaft is mounted on bearings at the drive end and the slip ring end. Each bearing contains its own permanent grease supply.

REMOVAL & INSTALLATION

All Exc. 8.1L Engines

◆ See Figure 31

■ All engines covered here use either a Mando or Delco alternator.

1. Disconnect both the negative and positive battery cables from the battery terminal posts. On the back of the alternator, disconnect the Orange, Purple, and Red/Purple wire leads from the alternator terminals. Disconnect the excitation wire.
2. Loosen the serpentine belt with the adjustment pulley and then remove the belt from the pulley.
3. Remove the alternator mounting bolts and then lift the alternator free of the engine/bracket.

To install:

4. Install the alternator into the mounting bracket with the bolt, washers, spacer and nut. If your engine uses washers, make sure they go on either side of the spacer. Do not tighten the bolt yet.
5. Attach the brace to the alternator if equipped.
6. Install the serpentine belt and adjust it.
7. Reconnect the wiring and tighten the alternator-to-brace bolt to 8 ft. lbs. (11 Nm), the alternator-to-bracket bolt to 35 ft. lbs. (48 Nm) and the brace/bracket-to-engine bolt(s) to 30 ft. lbs. (41 Nm).

8.1L Engines

◆ See Figure 32

1. Disconnect both the negative and positive battery cables from the battery terminal posts.
2. On the back of the alternator, disconnect the Orange, Purple, and Red/Purple wire leads from the alternator terminals. Disconnect the excitation wire.
3. Loosen the serpentine belt with the adjustment/idler pulley and then remove the belt from the pulley.
4. If your engine is equipped with a belt adjustment assembly, remove the assembly (bottom of alternator) and then remove the upper alternator mounting bolt. Remove the alternator.
5. If your engine is equipped with an automatic belt tensioner, remove the alternator mounting bolts and lift out the alternator.

To install:

6. Install the alternator into the mounting bracket with the bolt, washers, spacer and nut. If your engine uses washers, make sure they go on either side of the spacer. Do not tighten the bolt yet.
7. Attach the brace to the alternator if equipped.
8. If your engine has used the belt adjustment assembly, install the upper bolt finger-tight. Install the adjustment assembly and tighten the bolt to 26 ft. lbs. (35 Nm). Install the serpentine belt over the idler pulley and alternator pulley and then tighten the upper bolt to 19 ft. lbs. (25 Nm).
9. If your engine used the automatic belt tensioner, install the upper and lower bolts/nut and tighten the upper bolt/lock nut to 27 ft. lbs. (36 Nm) and the lower bolt to 72 ft. lbs. (91 Nm). Install the serpentine belt.
10. Reconnect the wiring harness and battery cables.

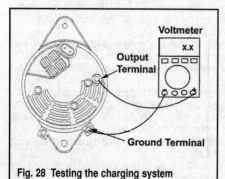

Fig. 28 Testing the charging system

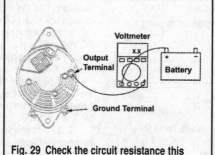

Fig. 29 Check the circuit resistance this way...

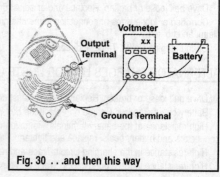

Fig. 30 ...and then this way

IGNITION & ELECTRICAL SYSTEMS

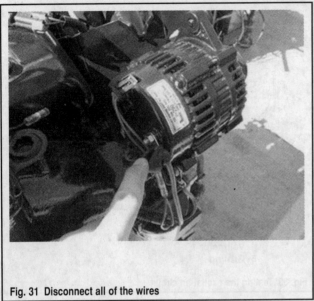

Fig. 31 Disconnect all of the wires

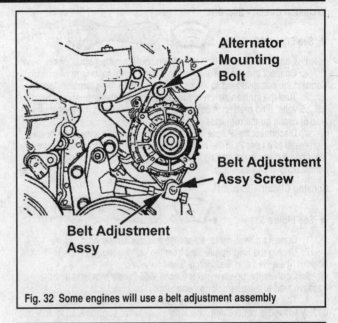

Fig. 32 Some engines will use a belt adjustment assembly

CIRCUIT TESTING

Output Circuit - Mando

 DIFFICULT

◆ See Figure 33

1. Connect a multi-meter as per the manufacturer's instructions.
2. Connect the positive lead to the alternator output wire terminal (Orange). Connect the negative lead to the ground terminal on the alternator.
3. Start the engine and increase the idle speed to 1500 rpm.
4. The meter should indicate approximate battery voltage (13.8-14.2 V).
5. If the reading is below 13.8 V, wiggle the engine wire harness and look for any drop in the meter reading that would indicate a loose or broken wire. If no reading is obtained or the reading varies, perform the Charging System Resistance test.
6. Disconnect the multi-meter leads from the alternator.

Output Circuit - Delco

DIFFICULT

◆ See Figure 34

■ Early 2001 350 Mag and 6.2L engines (up to serial #299999) should use the Mando test.

1. Connect a multi-meter as per the manufacturer's instructions.
2. Start the engine and increase the idle speed to 1300 rpm. Observe the meter reading.
3. If the reading is 13.5-14.2 volts, switch the meter to the AC volt position and check the reading. If 0.25 V or less, the diodes are functioning properly. Anything above this indicates that the diodes are bad and the alternator must be replaced.
4. If the reading in Step 2 was below 13.5 V, connect the positive lead to the alternator output wire terminal (Orange). Connect the negative lead to the ground terminal on the alternator.
5. Start the engine and increase the idle speed to 1300 rpm. Wiggle the engine wire harness and look for any drop in the meter reading that would indicate a loose or broken wire. If no reading is obtained or the reading varies, perform the Charging System Resistance test.
6. If the reading is above 15 V, the alternator is overcharging and will require replacement.
7. Disconnect the multi-meter leads from the alternator.

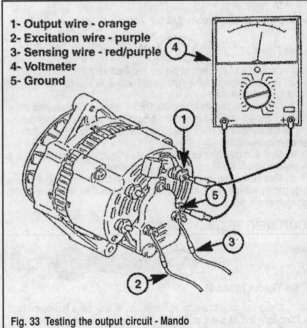

1- Output wire - orange
2- Excitation wire - purple
3- Sensing wire - red/purple
4- Voltmeter
5- Ground

Fig. 33 Testing the output circuit - Mando

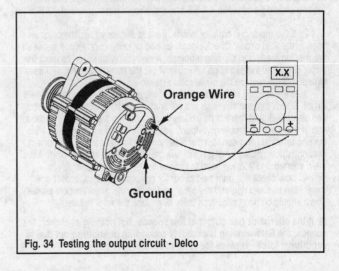

Fig. 34 Testing the output circuit - Delco

9-14 IGNITION & ELECTRICAL SYSTEMS

Excitation Circuit - Exc. 8.1L

◆ See Figure 35

1. Connect a multi-meter as per the manufacturer's instructions.
2. Connect the positive lead to the tie strap terminal on the alternator. Connect the negative lead to the ground terminal on the alternator (test 1).
3. Turn the ignition switch to the **ON** position. The meter should read 1.3-2.5 volts. If no reading is observed, there is an open in the excitation lead or circuit on the regulator.
4. Disconnect the Purple lead at the regulator and connect the positive meter lead to it (test 2). If the meter indicates battery voltage, replace the regulator. If there is no reading, check the excitation circuit for bad connections or frayed wiring.

Sensing Circuit - Exc. 8.1L

◆ See Figure 36

1. Connect a multi-meter as per the manufacturer's instructions.
2. Unplug the Red/Purple lead from the voltage regulator. On MPI engines, it's a Red and Red/Purple wire connector.
3. Connect the positive meter lead to Red/Purple wire lead and the negative lead to the alternator ground terminal.
4. The voltmeter should indicate approximately battery voltage. If battery voltage is not present, check the Red/Purple sensing lead for loose dirty or damaged connections.

Sensing & Excitation Circuit - 8.1L

◆ See Figure 37

1. Connect a multi-meter as per the manufacturer's instructions.
2. Tag and disconnect the Purple and Red/Purple wire connector at the rear of the alternator.
3. Connect the positive lead of the DVOM to the Red/Purple lead in the connector and the negative lead to the ground terminal. The meter should indicate battery voltage. If not, check the sensing circuit for loose and/or damaged connections.
4. Now, with the negative meter lead still connected to ground, connect the positive lead to the Purple lead in the connector.
5. Turn the ignition switch to the **ON** position and check that the meter reads battery voltage. If not, check the excitation circuit for loose and/or damaged connections.

COMPONENT TESTING - MANDO

Rotor

◆ See Figures 38 and 39

1. Check the rotor circuit for opens, shorts or excessive resistance (test 1). Connect a DVOM (set it on the Rx1 scale) and check the rotor windings for continuity. Connect the ohmmeter between the two slip rings, and the reading should be between 4.2-5.5 ohms at a room temperature of 70°F to 80°F (21-26°C).
2. If the reading is high, or infinite, there is excessive resistance or an open in the field circuit. Check closely for bad or loose connections between the winding leads and slip ring terminals. If everything looks ok, connect the meter directly to the slip rings - if you now get the correct reading, or if its still high or infinite, replace the entire rotor assembly.
3. If the reading in Step 1 was low, there is probably a short in the field circuit. Check that the slip rings are not bent or touching the outer ring. You can also check that there is no excess solder shorting the terminals to the aft ring. If there are no problems evident, unsolder the field winding leads from the ring terminals and connect the meter directly to the leads. If the correct reading is now shown, or of it is still low, the slip rings or field windings have been shorted and you will need to replace the entire rotor assembly.
4. Now check the rotor field circuit for grounds (test 2). Connect one meter lead to a slip ring and the other lead to the rotor shaft (or pole pieces). There should be no continuity; if there is, replace the rotor assembly.

■ **If the alternator has output at low speeds, but nothing at higher speeds, the field winding is probably grounding or shorting out due to centrifugal force - replace the rotor assembly**

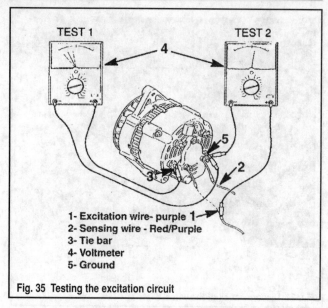

1- Excitation wire- purple
2- Sensing wire - Red/Purple
3- Tie bar
4- Voltmeter
5- Ground

Fig. 35 Testing the excitation circuit

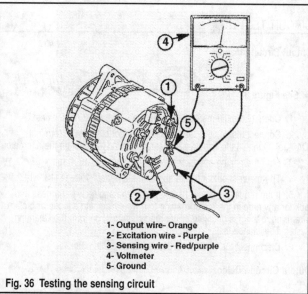

1- Output wire- Orange
2- Excitation wire - Purple
3- Sensing wire - Red/purple
4- Voltmeter
5- Ground

Fig. 36 Testing the sensing circuit

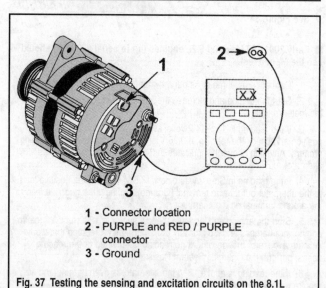

1 - Connector location
2 - PURPLE and RED / PURPLE connector
3 - Ground

Fig. 37 Testing the sensing and excitation circuits on the 8.1L engine

IGNITION & ELECTRICAL SYSTEMS 9-15

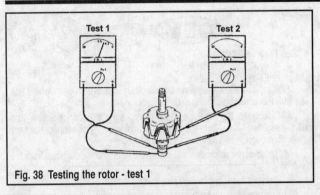

Fig. 38 Testing the rotor - test 1

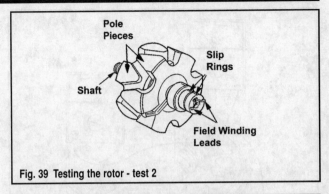

Fig. 39 Testing the rotor - test 2

Stator

◆ See Figure 40

1. The stator is not checked for shorts due to the very low resistance of the windings and the need for specialized test equipment. However, tests are made to check the stator for an open or grounded circuit. Use the following procedures to check the stator for possible defects.

■ **The stator leads must be disconnected from the rectifier for the following test.**

2. Test the stator for ground circuits by obtaining an ohmmeter and setting the switches for Rx1 scale (test 1). Connect one meter lead to one of the stator leads. Make a good positive contact with the other meter lead to the stator frame.
3. The meter should indicate no continuity (no movement). If continuity exists, the stator is grounded and must be replaced.
4. Test the stator for open circuits by obtaining an ohmmeter and setting the switches for Rx1 scale (test 2). Connect the meter leads to each pair of the stator windings (3 different ways). The meter should indicate continuity.
5. If no continuity exists on any stator lead, the stator circuit is open and must be replaced.
6. If all of the tests check out satisfactorily, but the alternator still fails to meet its rated output, the rotor field windings may be shorting or grounding out, due to centrifugal force.

Rectifier Negative (-)

◆ See Figure 41

■ **The rectifier must be disconnected from the stator to perform the following test. Each rectifier must be tested to be sure it is not open or shorted. Check the three rectifiers using the following procedures.**

1. Obtain an ohmmeter and set the switches to Rx1 scale. Connect one lead of the meter to the negative rectifier heat sink. Connect the other meter lead to one of the rectifier terminals and note the reading.
2. Reverse the meter leads and again note the meter reading. The meter should indicate a high or infinite resistance (no movement) in one direction and a low resistance indication when the leads are reversed.
3. If both readings are high (or infinite), indicating the rectifier is open, or if both readings are low, indicating the rectifier is shorted, the rectifier assembly must be replaced. Repeat this step on the other two rectifiers.

Rectifier Positive (+) & Diodes

◆ See Figures 42 and 43

■ **The rectifier must be disconnected from the stator to perform the following test. Each rectifier must be tested to be sure it is not open or shorted. Check the three rectifiers using the following procedures.**

1. Obtain an ohmmeter and set the switches to the Rx1 scale. Connect one meter lead to the 1/4 in. stud on the positive rectifier heat sink. Connect the other meter lead to one of the rectifier terminals and note the reading.
2. Reverse the meter leads and again note the meter reading. The meter should indicate a high or infinite resistance (no movement) in one direction and a low resistance indication when the leads are reversed.

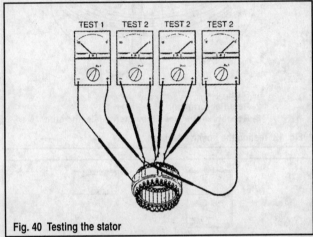

Fig. 40 Testing the stator

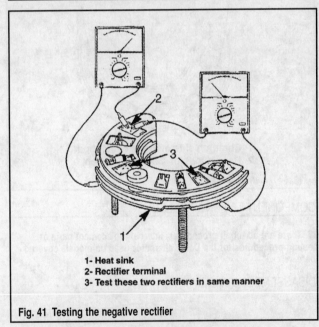

1- Heat sink
2- Rectifier terminal
3- Test these two rectifiers in same manner

Fig. 41 Testing the negative rectifier

3. If both readings are high, indicating the rectifier is open, or if both readings are low indicating the rectifier is shorted, the rectifier assembly must be replaced. Repeat this step on the other two rectifiers.
4. Now connect one meter lead to the common side of the diode and the other lead to the other side of one of the 3 diodes. Note the reading. Now reverse the leads and note the reading again.
5. The meter should indicate a high, or infinite, resistance (no movement) in one direction and a low resistance indication when the leads are reversed.
6. If both readings are high, indicating the diode is open, or if both readings are low indicating the diode is shorted, the rectifier assembly must be replaced. Repeat this step on the other two diodes.

9-16 IGNITION & ELECTRICAL SYSTEMS

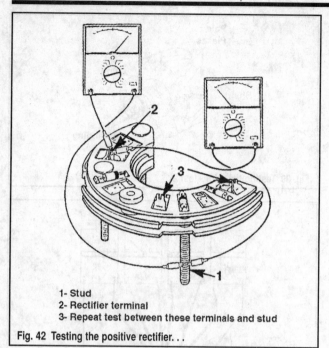

1- Stud
2- Rectifier terminal
3- Repeat test between these terminals and stud

Fig. 42 Testing the positive rectifier...

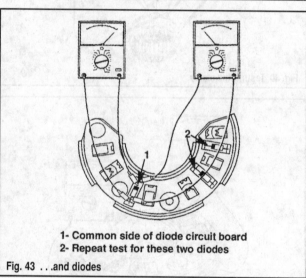

1- Common side of diode circuit board
2- Repeat test for these two diodes

Fig. 43 ...and diodes

COMPONENT TESTING - DELCO

■ There are no repair procedures, internal replacement parts or testing procedures for the Delco alternator used on models covered here.

DISASSEMBLY & ASSEMBLY

■ Some times the equipment, small parts and experience required to repair the alternator can be more cost effective if the alternator is taken to a shop which specializes in alternator rebuilding. Be sure to inform the shop the alternator is for marine application, to ensure the terminal ends and covers are installed properly.

If a specialty shop is not available, use the following instructions and exploded diagram to disassemble the alternator.

Delco

There are no repair procedures or internal replacement parts for this model alternator. If it fails any of the preceding tests, replace the entire unit.

Mando

◆ See Figures 44 thru 48

1. Remove the alternator.
2. Place the alternator on a suitable clean work bench or mounted in a vise with the end frame facing up.
3. Disconnect the regulator leads. Remove the 4 mounting screws and pull the regulator away from the alternator until you can disconnect the field leads. Remove the regulator and its rubber seal.
4. Remove the stud cover insulator, two nuts and the tie strap from the assembly.
5. Remove the 2 brush/regulator screws and lift out the assembly.
6. Scribe a mark on both end frames and matching marks on the stator, as an aid to properly assemble the alternator frame.
7. Remove the four thru-bolts. Separate the end frame from the stator by carefully inserting 2 screwdrivers no more than 1/16 in. into the slots on opposite sides of the stator frame. Pry the end frame and then the rear end frame from the stator. Never pry anywhere except at the slot or the castings will be damaged.
8. Place the rear end frame on the work bench with the stator pointing down. Remove the nuts, washers, insulators and condenser from the studs.
9. Turn the end frame over and remove the Phillips head screw securing the rectifier assembly to the end frame.
10. Insert screwdrivers into the slots again and separate the stator and rectifier.
11. Unsolder the three stator leads at the rectifier heat sink. Place the tips of needle nose pliers onto the diode terminal between the solder joints and diode body. This action will serve as a heat sink, while soldering the stator terminals and prevent damaging the diodes

■ Perform the next step only if the pulley is damaged, front bearing is defective or if the pulley must be transferred to another alternator rotor shaft.

12. Place an oversized V-belt around the pulley and clamp the pulley in a vise. Make sure the vise is clamping on the V-belt only. Place a 7/8 in. socket or wrench over the pulley nut and remove the nut, lockwasher, pulley, fan spacer, and fan from the end of the rotor shaft. Slide the front end frame off the rotor shaft.
13. Remove the three Phillips head screws and washers securing the front bearing retaining plate to the front end frame and lift off the retaining plate. Place the end frame in an arbor press with the face pointing up. Use a suitable size mandrel and press the bearing out of the front end frame and discard the bearing.

To assemble:

The following components are to be visually inspected for serviceability. Electrical checks and tests are detailed previously in this section.

Use a clean soft and wipe any dirt or debris off the parts. Do not clean electrical components with solvent, because such action may damage the item.

Inspect the brush casing for cracks or damaged brush leads. Check the brush leads for poor solder connections and damaged leads. Check for broken springs and excessive brush wear. If the brushes are worn to less than a 1/4 (6.35mm) long the brush set must be replaced.

Inspect the end of the rotor shaft for stripped or damaged threads. Check the pole piece fingers for damage caused by worn bearings. Inspect the rear bearing for smooth quiet rolling action. If the threads cannot be repaired or the bearing is defective, the entire rotor assembly must be replaced.

Clean the rotor slip rings with 400 grit sand paper. Blow off any dust using compressed air and inspect the rotor slip rings for grooves, pits, flat spots or out-of-round more than 0.002 in. (0.051mm). If any of the above conditions are found, the rotor assembly must be replaced.

Inspect the insulating enamel for heat discoloration because discoloration is an indication of a shorted diode or grounded winding. Check the stator for damaged insulation and wires from contact with the rotor. Replace the stator if any of the above defects are obvious.

Inspect the end frames for cracks, distortion, stripped threads or worn bearing bore (bearing seized on shaft and spinning in bore). If any of the above damage is obvious the end frame must be replaced.

IGNITION & ELECTRICAL SYSTEMS

Inspect the fan for bent or cracked fins, broken welds or worn mounting hole. Check the pulley sheaves for trueness, excessive wear, grooves, pits, nicks and corrosion. If the pulley sheave damage cannot be repaired with a file or wire brush, the pulley must be replaced or drive belt wear will be accelerated.

14. Position a new bearing into the front end frame bearing bore. Position the front end frame into an arbor press. Using the appropriate size mandrel, press the bearing into the bore until the bearing contacts the end frame. Place the bearing retainer plate over the bearing and secure with three Phillips head screws and washers.

15. Slide the threaded end of the rotor shaft through the front end frame. Slide the fan spacer, fan, pulley spacer, pulley, lockwasher and nut onto the end of the shaft. Place an oversize V-belt around the pulley and tighten the pulley in a vise. Using a 7/8 in. socket, tighten the pulley nut to 42 ft. lbs. (58 Nm).

■ The insulating washers must be installed correctly or you will damage the alternator.

16. Assemble the stator and rectifier and then solder the three leads. Position into the rear frame and install the mounting screw.
17. Position the rear end frame over the front and align the scribe marks made previously. Press the frames together and then install the screws, tightening them to 55 inch lbs. (5.5 Nm).
18. Push the brushes down so they are flush with the top of the holder and then insert a #54 (0.050 in./1mm) drill bit into the hole in the holder so they stay compressed. Position the brush/regulator in the cavity on the back of the rear end frame and tighten the two mounting screws to 42 inch lbs. (4.2 Nm). Pull out the drill bit so that the brushes release against the slip rings.
19. Install the tie strap between the two studs and tighten the nuts securely.
20. Install the cover. Install the screw, connect the two leads and then install the two nuts; tighten all three securely. Pop on the insulator caps.
21. Install the alternator.

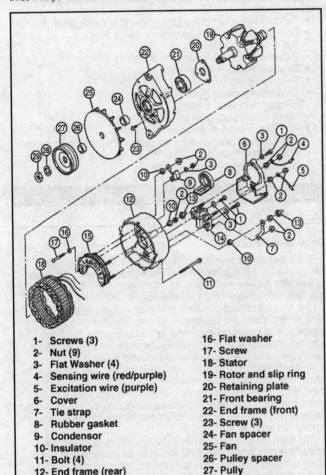

1- Screws (3)
2- Nut (9)
3- Flat Washer (4)
4- Sensing wire (red/purple)
5- Excitation wire (purple)
6- Cover
7- Tie strap
8- Rubber gasket
9- Condensor
10- Insulator
11- Bolt (4)
12- End frame (rear)
13- Cap (2)
14- Brush / regulator assembly
15- Rectifier assembly
16- Flat washer
17- Screw
18- Stator
19- Rotor and slip ring
20- Retaining plate
21- Front bearing
22- End frame (front)
23- Screw (3)
24- Fan spacer
25- Fan
26- Pulley spacer
27- Pully
28- Lockwasher
29- Nut

Fig. 44 Exploded view of the Mando alternator

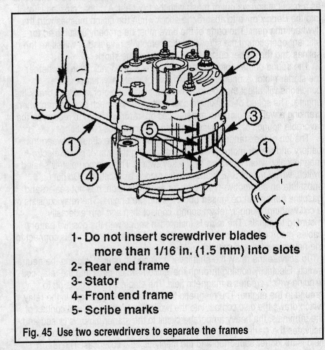

1- Do not insert screwdriver blades more than 1/16 in. (1.5 mm) into slots
2- Rear end frame
3- Stator
4- Front end frame
5- Scribe marks

Fig. 45 Use two screwdrivers to separate the frames

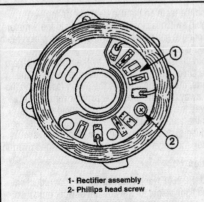

1- Rectifier assembly
2- Phillips head screw

Fig. 46 A single screw holds the rectifier assembly

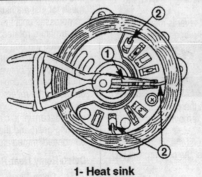

1- Heat sink
2- Stator leads

Fig. 47 Use needle nose pliers to dissipate heat when using a solder gun

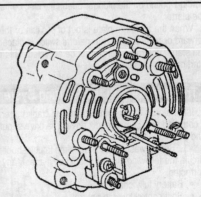

Fig. 48 Use a drill bit to compress the brushes

9-18 IGNITION & ELECTRICAL SYSTEMS

STARTER CIRCUIT

Description and Operation

All engines utilize an electric starter motor coupled with a mechanical gear mesh between the starter motor and the flywheel, similar to the method used to crank an automobile engine.

As the name implies, the sole purpose of the starter motor circuit is to control operation of the starter motor to crank the engine until it is operating. The circuit includes a relay or magnetic switch (solenoid) to connect or disconnect the motor from the battery. The operator controls the switch with a key switch.

A neutral safety switch is installed into the circuit to permit operation of the starter motor only if the shift control lever is in neutral. This switch is a safety device to prevent accidental engine start when the engine is in gear.

The starter motor is a series-wound electric motor that draws a heavy current from the battery. It is designed to be used only for short periods of time to crank the engine for starting. To prevent overheating the motor, cranking should not be continued for more than 30-seconds without allowing the motor to cool for at least three minutes. Actually, this time can be spent in making preliminary checks to determine why the engine fails to start.

Power is transmitted from the starter motor to the flywheel through a Bendix drive. This drive has a pinion gear mounted on screw threads. When the motor is operated, the pinion gear moves upward and meshes with the teeth on the flywheel ring gear.

When the engine starts, the pinion gear is driven faster than the shaft and as a result, it screws out of mesh with the flywheel. A rubber cushion is built into the Bendix drive to absorb the shock when the pinion meshes with the flywheel ring gear. The parts of the drive must be properly assembled for efficient operation. If the screw shaft assembly is reversed, it will strike the splines and the rubber cushion will not absorb the shock.

The sound of the motor during cranking is a good indication of whether the starter motor is operating properly or not. Naturally, temperature conditions will affect the speed at which the starter motor is able to crank the engine. The speed of cranking a cold engine will be much slower than when cranking a warm engine. An experienced operator will learn to recognize the favorable sounds of the engine cranking under various conditions.

The job of the starter motor relay is to complete the circuit between the battery and starter motor. It does this by closing the starter circuit electromagnetically, when activated by the key switch. This is a completely sealed switch, which meets SAE standards for marine applications. Do not substitute an automotive-type relay for this application. It is not sealed and gasoline fumes can be ignited upon starting the engine. The relay consists of a coil winding, plunger, return spring, contact disc and four externally mounted terminals. The relay is installed in series with the positive battery cables mounted to the two larger terminals. The smaller terminals connect to the neutral switch and ground.

To activate the relay, the shift lever is placed in neutral, closing the neutral switch. Electricity coming through the ignition switch goes into the relay coil winding which creates a magnetic field. The electricity then goes on to ground in the engine. The magnetic field surrounds the plunger in the relay, which draws the disc contact into the two larger terminals. Upon contact of the terminals, the heavy amperage circuit to the starter motor is closed and activates the starter motor. When the key switch is released, the magnetic field is no longer supported and the magnetic field collapses. The return spring working on the plunger opens the disc contact, opening the circuit to the starter.

When the armature plate is out of position or the shift lever is moved into forward or reverse gear, the neutral switch is placed in the open position and the starter control circuit cannot be activated. This prevents the engine from starting while in gear.

Troubleshooting the Starting System

If the starter motor spins but fails to crank the engine, the cause is usually a corroded or gummy Bendix drive. The drive should be removed, cleaned and given an inspection.

1. Before wasting too much time troubleshooting the starter motor circuit, the following checks should be made. Many times, the problem will be corrected.
- Battery fully charged.
- Shift control lever in neutral.
- Main 20-amp fuse is good (not blown).
- All electrical connections clean and tight.
- Wiring in good condition, insulation not worn or frayed.

2. Starter motor cranks slowly or not at all.
- Faulty wiring connection
- Short-circuited lead wire
- Shift control not engaging neutral (not activating neutral start switch)
- Defective neutral start switch
- Starter motor not properly grounded
- Faulty contact point inside ignition switch
- Bad connections on negative battery cable to ground (at battery side and engine side)
- Bad connections on positive battery cable to magnetic switch terminal
- Open circuit in the coil of the magnetic switch (relay)
- Bad or run-down battery
- Excessively worn down starter motor brushes
- Burnt commutator in starter motor
- Brush spring tension slack
- Short circuit in starter motor armature

3. Starter motor keeps running.
- Melted contact plate inside the magnetic switch
- Poor ignition switch return action

4. Starter motor picks up speed, put pinion will not mesh with ring gear.
- Worn down teeth on clutch pinion
- Worn down teeth on flywheel ring gear

Starter Motor

DESCRIPTION & OPERATION

Delco-Remy Direct Drive

Delco-Remy direct drive starters (14MT) used on certain V8 motors consist of a set of field coils positioned over pole pieces, which are attached to the inside of a heavy iron frame. An armature, an over-running clutch drive mechanism, and a solenoid are included inside the iron frame.

The armature consists of a series of iron laminations placed over a steel shaft, a commutator, and the armature winding. The windings are heavy copper ribbons assembled into slots in the iron laminations. The ends of the windings are soldered or welded to the commutator bars. These bars are electrically insulated from each other and from the iron shaft.

An overrunning clutch drive arrangement is installed near one end of the starter shaft. This clutch drive assembly contains a pinion that is made to move along the shaft by means of a shift lever to engage the engine ring gear for cranking. The relationship between the pinion gear and the ring gear on the engine flywheel provides sufficient gear reduction to meet cranking requirement speed for starting.

The overrunning clutch drive has a shell and sleeve assembly, which is splined internally to match the spiral splines on the armature shaft. The pinion is located inside the shell. Spring-loaded rollers are also inside the shell and they are wedged against the pinion and a taper inside the shell. Some starters use helical springs and others use accordion type springs. Four rollers are used. A collar and spring, located over a sleeve completes the major parts of the clutch mechanism.

When the solenoid is energized and the shift lever operates, it moves the collar endwise along the shaft. The spring assists movement of the pinion into mesh with the ring gear on the flywheel. If the teeth on the pinion fail to mesh for just an instant with the teeth on the ring gear, the spring compresses until the solenoid switch is closed; current flows to the armature; the armature rotates; the spring is still pushing on the pinion; the pinion teeth mesh with the ring gear; and cranking begins.

Torque is transferred from the shell to the pinion by the rollers, which are wedged tightly between the pinion and the taper cut into the inside of the shell. When the engine starts, the ring gear drives the pinion faster than the armature; the rollers move away from the taper; the pinion overruns the shell; the return spring moves the shift lever back; the solenoid switch is opened; current is cutoff to the armature; the pinion moves out of mesh with the ring gear; and the cranking cycle is completed. The start switch should be opened immediately when the engine starts to prevent prolonged overrun.

Delco-Remy Gear Reduction

The Delco-Remy PG260, PG260F1 and PG260L series gear reduction starter motors are small and light weight for the amount of work produced. These starters have small permanent magnets mounted inside the field frame. The placement and strength of these magnets differ between models. Therefore the field frames are not interchangeable.

IGNITION & ELECTRICAL SYSTEMS

The permanent magnets take the place of the large current-carrying iron core field-coil magnets previously used on the larger direct drive starter motors. The motor armature is supported on both ends by a permanently lubricated roller or ball bearing assembly to reduce the drag and friction created by the motor speed of approximately 7,000 rpm.

A planetary gear reduction unit is mated between the drive motor armature and the Bendix drive gear. This planetary gear drive results in an over all gear reduction of approximately 4:1. Through the gear reduction, the conversion of motor high speed, low torque is converted to a high torque, low speed gear drive output. The gear reduction results in a final engine cranking speed of approximately 1,750 rpm.

The starter is designed to operate under heavy loads and produce high power for only short periods of time. Therefore, never crank the engine for more than 30 seconds without allowing a minimum cooling off time of two minutes, before attempting to crank the engine again.

As with all marine installations, a safety switch is installed in the remote control shift box. This switch is designed to open the starter circuit to prevent the engine from starting, if the shift lever is in any position other than the neutral position.

The unit has a Bendix follow-thru type drive designed to overcome disengagement of the flywheel ring gear when engine speed exceeds cranking motor speed.

A helical cut shaft is designed to quickly engage and disengage the Bendix drive. An internal one-way clutch in the Bendix allows the motor to drive the Bendix. If the engine should start, and the flywheel begin to drive the Bendix, the one-way clutch will release and allow the drive to overrun and release the cranking motor armature. The helical splined shaft will then disengage the Bendix from the flywheel.

On the 260-F1 and some 260 series, the solenoid and lever are replaceable components. Any other differences between these models are minor and will be identified in the text.

PRECAUTIONS

- Always make sure that each battery cable is connected to the correct terminal on the battery.
- Never disconnect the battery while the engine is running.
- When using a battery charger or booster, always make sure that the positive battery cable on the charger is connected to the positive terminal on the battery. The same goes for the negative side.
- Always make sure that both battery cables are disconnected prior to connecting a charger or booster.
- Always make sure the battery is in good operating condition.
- Always make sure the battery leads and terminals are clean.

TROUBLESHOOTING

◆ See Figures 49 and 50

Regardless of how or where the solenoid is mounted, the basic circuits of the starting system on all makes of cranking motors are the same and similar tests apply. In the following testing and troubleshooting procedures, the differences are noted.

✶✶ CAUTION

Always take time to vent the bilge when making any of the tests as a prevention against igniting any fumes accumulated in that area. As a further precaution, remove the high-tension wire from the center of the distributor cap and ground it securely to prevent sparks.

All cranking motor problems fall into one of three areas:
1. The cranking motor fails to rotate.
2. The cranking motor spins rapidly, but does not crank the engine.
3. The cranking motor cranks the engine, but too slowly to affect engine start.

The following paragraphs provide a logical sequence of tests designed to isolate a problem in the cranking system.

Battery

1. Turn on several of the cabin lights (or any accessories). Turn the ignition switch to the **ON** position and note the effect on the brightness of the lights. With a properly functioning electrical system, the lights will dim slightly and the starter will crank the engine at a reasonable rate. If the lights dim considerably and the engine does not turn over, one of several causes may be at fault.

2. If the lights go out completely, or dim considerably, the battery charge is low or almost dead. The obvious remedy is to charge the battery; switch over to a secondary battery if one is available; or to replace it with a known fully charged one.

3. If the starting relay clicks, sounding similar to a machine gun firing, the battery charge is too low to keep the starting relay engaged when the starter load is brought into the circuit.

4. If the starter spins without cranking the engine, the drive is broken. The starter will have to be removed for repairs.

5. If the lights do not dim, and the starter does not operate, then there is an open circuit. Proceed to the Cable Connection Test.

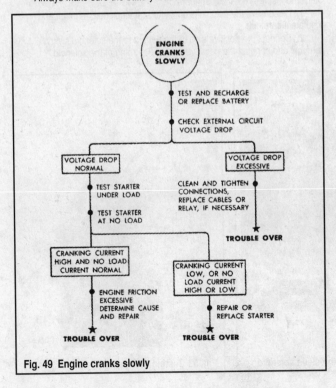

Fig. 49 Engine cranks slowly

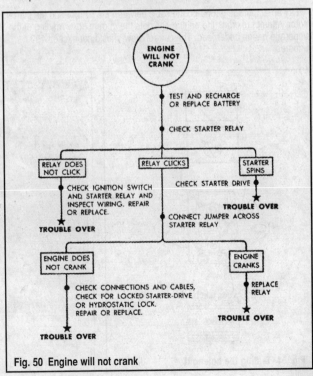

Fig. 50 Engine will not crank

9-20 IGNITION & ELECTRICAL SYSTEMS

Cable Connection

 EASY

1. If the starter fails to operate and the lights do not dim when the ignition switch is turned to start, the first area to check is the connections at the battery, starting relay, starter and neutral-safety switch.
2. First, remove the cables at the battery; clean the connectors and posts; replace the cables; and tighten them securely.
3. Now, try the starter. If it still fails to crank the engine, try moving the shift box selector lever from neutral to forward to determine if the neutral-safety switch is out of adjustment or the electrical connections need attention.
4. Sometimes, after working the shift lever back-and-forth and perhaps a bit sideways, the neutral-switch connections may be temporarily restored and the engine can be started. Disconnect the leads; clean the connectors and terminals on the switch; replace the leads; and tighten them securely at the first opportunity.
5. If the starter still fails to crank the engine, move on to the Solenoid Test.

Slave Solenoid

 OEM MODERATE

◆ See Figures 51 and 52

1. The solenoid, commonly called the starter relay, is checked by directly bridging between the terminal from the battery (the large heavy one) to the terminal from the ignition switch.

** CAUTION

Take every precaution to ensure there is no gasoline fumes in the bilge before making these tests.

2. Connect the battery and DVOM as shown in the illustration. If there is no meter movement, replace the solenoid.

Current Draw

 MODERATE

◆ See Figure 53

Lay an amperage gauge on the cable between the battery and the starter motor. Attempt to crank the engine and note the current draw reading of the amperage gauge under load. The current draw should not exceed 190 amperes.

TESTING

Voltage Drop

 MODERATE

■ This test must be performed with a DVOM and with the starter still installed in the engine.

1. Make sure the DVOM is set to **0** and that the battery is fully charged.
2. Disconnect the coil wire at the distributor cap and ground it to ensure that the engine will not start.
3. Connect the positive lead on the meter to the large threaded terminal on the solenoid and the negative lead to an unpainted metal surface on the starter housing.
4. Crank the engine over for 10-15 seconds and watch the meter.
5. If the meter is reading 9.5 V or more, then the starter is getting sufficient voltage from the battery. Check the starter and/or engine.
6. If the meter is reading less than 9.5 V, there is a voltage loss somewhere between the battery and starter. Check the battery cables and/or connections, and the battery switch.

■ The minimum voltage drop allowed is 0.5 V.

PG260 Starter

 MODERATE

◆ See Figures 54 and 55

1. Check the armature for a short circuit by placing it on a growler and holding a hack saw blade over the armature core while the armature is rotated. If the saw blade vibrates, the armature is shorted. Clean between the commutator bars, and then check again on the growler. If the saw blade still vibrates, the armature must be replaced.
2. Connect one probe of the test light to the armature core or shaft and the other probe to the commutator. If the light comes on, the armature is grounded and must be replaced.

Pinion Clearance

 MODERATE

◆ See Figures 56

Pinion clearance should always be checked after reassembly of the starter to insure proper adjustment.

1. Disconnect the field coil connector at the solenoid terminal and insulate it carefully.
2. Connect the positive side of the battery to the solenoid switch terminal and connect the negative side to the frame of the solenoid.

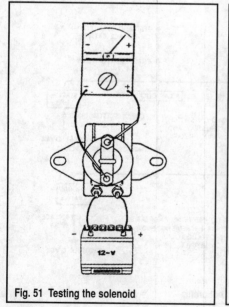

Fig. 51 Testing the solenoid

Fig. 52 A good look at the slave solenoid

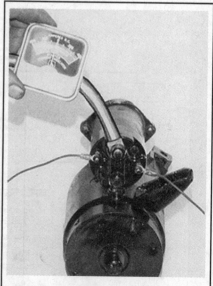

Fig. 53 Testing the current draw

IGNITION & ELECTRICAL SYSTEMS 9-21

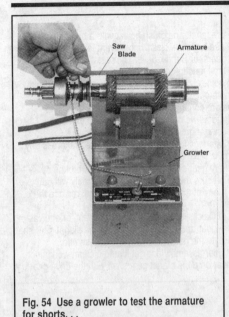

Fig. 54 Use a growler to test the armature for shorts...

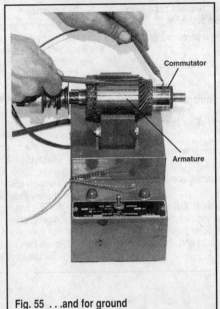

Fig. 55 ...and for ground

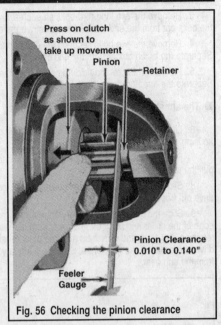

Fig. 56 Checking the pinion clearance

3. Connect a jumper wire to the solenoid motor terminal and quickly touch the other end to the body of the starter so the pinion moves into the cranking position.

4. Push the pinion back toward the commutator end until there is no slack. Measure the gap between the pinion and the retainer. If not 0.010-0.140 in. (0.25-3.5mm) on direct drive models (14MT) or 0.010-0.160 in. (0.25-4.06mm) on gear reduction models (PG260 series), the solenoid linkage or shift lever yoke buttons are worn and require replacement; or the shift lever mechanism has been assembled improperly.

REMOVAL & INSTALLATION

 MODERATE

◆ See Figures 57, 58 and 59

1. Disconnect the battery cables.
2. Tag and disconnect all wires at the starter solenoid. Move them out of the way.
3. If your engine uses a brace at the forward end of the starter, remove the mounting bolt.
4. Loosen the two mounting bolts and then pull them out of the starter housing. Carefully lift out the starter. Certain versions of the PG260 starter may have three bolts used for mounting the starter to the block.

■ Be aware that many applications will utilize a shim(s) between the motor and cylinder block. Do not lose this shim!

To install:

5. Position the shim (if equipped) and then install the starter. Slide in the bolts and tighten them to 50 ft. lbs. (68 Nm) on V8 direct drives (14MT); or 30 ft. lbs. (41 Nm) on gear reduction starters (PG260 series).

6. Install the brace mounting bolt (if equipped).
7. Connect all wires to the solenoid and then connect the battery cables.

DISASSEMBLY & ASSEMBLY

■ Before disassembling begins, read these few words first on a time and money saving idea. Most dealers do not find it economical to stock all the small parts or spend the time to service a starter motor. In most cases the labor cost alone will exceed the cost of a new or rebuilt unit.

Throughout the country, specialty shops may be found specializing in and servicing starter motors, alternators and other electrical assemblies. A customer can usually obtain a new or rebuilt starter for a modest cost usually on the spot or same day service. Some marine dealers even stock new and rebuilt cranking motors ready for installation. This means the repair can be completed and the boat back in service before the end of the day.

If the decision is made to disassemble the unit, check with the marine dealer first for parts availability (because there's a good chance you can't get them!), order time, and cost compared to a new or rebuilt unit off the shelf. These units usually come with some type of warranty. Therefore, if it fails within the warranty period, additional cost is not involved.

Delco-Remy - PG260

 DIFFICULT

◆ See Figures 60 thru 63

1. Remove the starter.
2. Disconnect the wire brush lead from the brush block to the solenoid terminal. Place a scribe mark or a strip of tape on the drive housing, field frame and end cap. The tape or scribe marks will assist in the alignment of the drive housing and field frame during the assembling procedures.

Fig. 57 Tag and disconnect all wiring leads at the solenoid switch

Fig. 58 Some models may use an additional brace...

Fig. 59 ...but they all have these mounting bolts

9-22 IGNITION & ELECTRICAL SYSTEMS

3. Remove the two long thru-bolts on each side of the motor case. Slide the bolts out from the end cap and then remove the cap assembly.

4. Remove the two bolts securing the end cap to the brush holder. Separate the brush holder from the end cap.

5. Pull the armature and field frame out from the drive housing. Separate the armature from the field frame.

■ The armature will be held in the field frame by permanent magnets.

6. Lift out the shield and washer from the open end of the drive housing.

7. Remove the 3 screws securing the solenoid to the drive housing and pull the solenoid free of the housing.

8. Slide the planetary gear set, solenoid lever, grommet, and Bendix drive out from the drive housing.

9. Slide the thrust collar washer off the end of the shaft. Tap on the edge of the collar and pry it off and away from the retaining ring. Spread the retaining ring open and slide it up and free of the shaft. Slide the Bendix drive and collar off the gear shaft. Slide the planetary gears and shaft out from the sun gear.

To assemble:

10. Clean the armature, shaft and brush holder with a brush and compressed air. Do not use a grease dissolving solvent to clean electrical components, planetary gears, or Bendix drive. Solvent will damage wiring insulation and washout the lubricant in bearings and drive gears.

11. Verify the brush holder is not damaged and the brushes are held firmly against the commutator.

12. Inspect the armature commutator for grooves or out of round condition. If the commutator is worn it can be turned down on a lathe. Check the armature for shorts using a growler or test light.

13. Check the roller bearings for wear and rough spots. If there is any roughness or the bearing is stiff, replace the bearing.

14. Check the planetary gear set and the gear teeth for broken or missing teeth. Check to be sure the gears mesh and roll freely without binding or rough spots. If any of these conditions are found, replace the entire planetary gear set.

15. Apply a coating of Quicksilver 2-4-C Marine Lubricant, or equivalent, to the sun gear, planetary gear and roller bearing. Slide the planet gear shaft through the sun gear. Rotate the gears until they are fully seated.

16. Invert the planet gear assembly and set it onto a clean workbench. Slide the Bendix drive gear onto the planet gear shaft with the drive gear

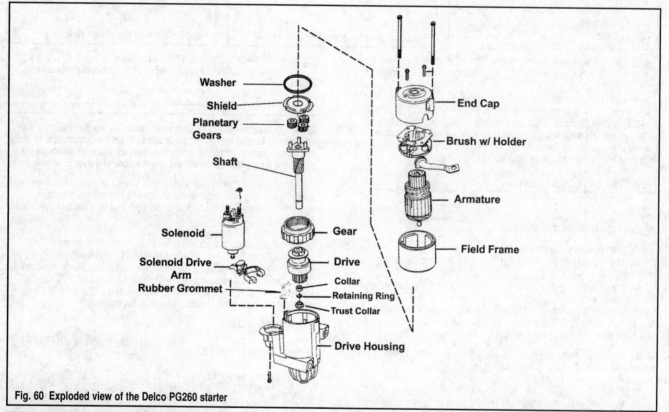

Fig. 60 Exploded view of the Delco PG260 starter

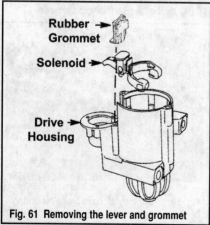

Fig. 61 Removing the lever and grommet

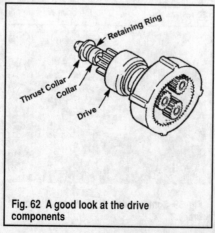

Fig. 62 A good look at the drive components

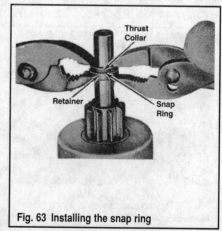

Fig. 63 Installing the snap ring

IGNITION & ELECTRICAL SYSTEMS   9-23

facing up. Rotate the drive and index the pins in the drive with the helical grooves in the shaft. Slide the collar over the end of the shaft with the open side facing up. Place the retaining ring over the end of the shaft. Tap the ring onto the shaft with a wood block and hammer or spread the ring open with a pair of pliers.

17. Slide the retainer ring down and into the groove on the shaft. Squeeze the ring closed around the groove on the shaft.

18. Slide the thrust collar onto the shaft against the retaining ring on one side and the collar up against the retaining ring on the opposite side. Use a pair of pliers and squeeze the collar evenly around until the collar is fully seated over the retaining ring.

19. Spread the ends of the solenoid lever open and insert the tabs of the Bendix drive into the lever. Place a dab of grease on the thrust collar to hold it in place on the shaft. Insert the Bendix drive, planet gear set, and rubber grommet into the drive housing.

20. Insert the solenoid lever return spring and solenoid into the drive housing opening. Position the solenoid with the brush block terminal at the 6 o'clock position, or next to the rubber grommet. Secure the solenoid with three screws and lock washer.

21. Place the shield over the end of the planet gear assembly. Place the large washer against the shield.

22. Slide the armature into the field frame assembly. Align the scribe marks or tape on the outside of the motor field frame assembly with the tape or marks on the drive housing. Guide the end of the armature into the center of the planetary gear set while aligning the marks on the motor case.

23. Position the end frame over the brush block and align the tape or scribe marks with the motor case. Guide the wire lead from the brush block and rubber grommet into place in the end frame. Secure the end frame to the brush block with two short bolts. Tighten the bolts securely.

24. Insert two long bolts down through the end frame. Align the bolts with the threaded openings in the drive housing. Tighten the bolts alternately and evenly until they are secure.

Delco-Remy - PG260F1

◆ See Figure 64

■ Although Mercury does not provide disassembly procedures for this unit (at least not after the late 1990's), if you absolutely must take it apart, follow the procedures for the PG260 starter.

Delco-Remy - PG260L

■ The exploded view is provided for reference purposes only, there are no overhaul procedures available for this model, as there are no serviceable parts.

Delco-Remy 14MT

◆ See Figure 65

■ The exploded view is provided for reference purposes only, there are no overhaul procedures available for this model, as there are no serviceable parts.

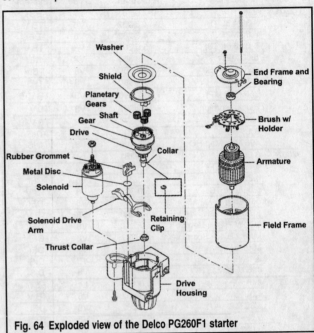

Fig. 64 Exploded view of the Delco PG260F1 starter

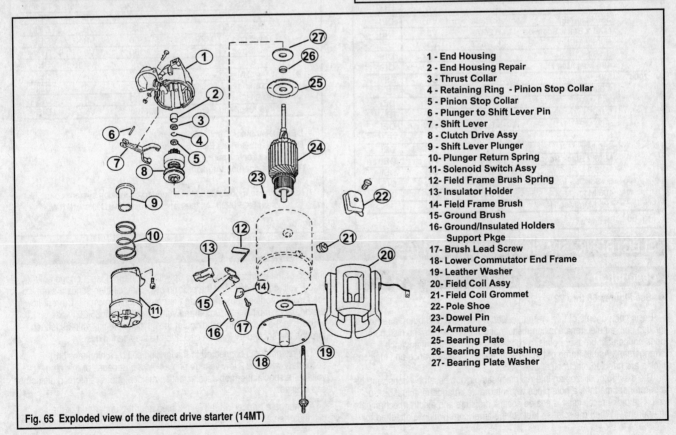

1 - End Housing
2 - End Housing Repair
3 - Thrust Collar
4 - Retaining Ring - Pinion Stop Collar
5 - Pinion Stop Collar
6 - Plunger to Shift Lever Pin
7 - Shift Lever
8 - Clutch Drive Assy
9 - Shift Lever Plunger
10- Plunger Return Spring
11- Solenoid Switch Assy
12- Field Frame Brush Spring
13- Insulator Holder
14- Field Frame Brush
15- Ground Brush
16- Ground/Insulated Holders Support Pkge
17- Brush Lead Screw
18- Lower Commutator End Frame
19- Leather Washer
20- Field Coil Assy
21- Field Coil Grommet
22- Pole Shoe
23- Dowel Pin
24- Armature
25- Bearing Plate
26- Bearing Plate Bushing
27- Bearing Plate Washer

Fig. 65 Exploded view of the direct drive starter (14MT)

9-24 IGNITION & ELECTRICAL SYSTEMS

IGNITION SYSTEMS

General Information

♦ See Figures 66 and 67

Mercruiser engine covered in this manual utilize one of three different ignition systems: Thunderbolt V Electronic Ignition, Electronic Spark Timing System (EST), Distributor Electronic Ignition (DEI) and Distributorless Electronic Ignition (DIS).

Descriptions of each system, and complete repair procedures follow.

Ignition System Applications

Year	Model	Displacement L/Cu. In.	Engine Type	Ignition System
2001	3.0L	3.0/181	L4	EST
	4.3L/4.3L TBI	4.3/262	V6	TB V
	4.3L MPI	4.3/262	V6	DEI
	5.0L/5.0L TBI	5.0/305	V8	TB V
	5.0L MPI	5.0/305	V8	DEI
	5.7L/5.7L TBI	5.7/350	V8	TB V
	5.7L MPI	5.7/350	V8	DEI
	350 Mag MPI	5.7/350	V8	TB V ①
	Black Scorpion	6.2/377	V8	TB V ①
	MX6.2 MPI	6.2/377	V8	TB V ①
	8.1S HO	8.1/496	V8	DIS
	496 Mag/Mag HO	8.1/496	V8	DIS
2002	3.0L	3.0/181	L4	EST
	4.3L/4.3L EFI	4.3/262	V6	TB V
	4.3L MPI	4.3/262	V6	DEI
	5.0L/5.0L TBI	5.0/305	V8	TB V
	5.0L MPI	5.0/305	V8	DEI
	5.7L/5.7L TBI	5.7/350	V8	TB V
	5.7L MPI/350 Mag MPI	5.7/350	V8	DEI
	MX6.2 Black Scorpion	6.2/377	V8	DEI
	MX6.2 MPI	6.2/377	V8	DEI
	8.1S HO	8.1/496	V8	DIS
	496 Mag/Mag HO	8.1/496	V8	DIS
2003	3.0L	3.0/181	L4	EST
	4.3L	4.3/262	V6	TB V
	4.3L MPI	4.3/262	V6	DEI
	5.0L	5.0/305	V8	TB V
	5.0L MPI	5.0/305	V8	DEI
	5.7L	5.7/350	V8	TB V
	350 Mag MPI	5.7/350	V8	DEI
	MX6.2 Black Scorpion	6.2/377	V8	DEI
	MX6.2 MPI	6.2/377	V8	DEI
	8.1S HO	8.1/496	V8	DIS
	496 Mag/Mag HO	8.1/496	V8	DIS
2004	3.0L	3.0/181	L4	EST
	4.3L	4.3/262	V6	TB V
	4.3L MPI	4.3/262	V6	DEI
	5.0L	5.0/305	V8	TB V
	5.0L MPI	5.0/305	V8	DEI
	5.7L	5.7/350	V8	TB V
	350 Mag MPI	5.7/350	V8	DEI
	MX6.2 Black Scorpion	6.2/377	V8	DEI
	MX6.2 MPI	6.2/377	V8	DEI
	8.1S HO	8.1/496	V8	DIS
	496 Mag	8.1/496	V8	DIS
	496 Mag/Mag HO	8.1/496	V8	DIS

Fig. 66

Ignition System Applications

Year	Model	Displacement L/Cu. In.	Engine Type	Ignition System
2005	3.0L	3.0/181	L4	EST
	4.3L	4.3/262	V6	TB V
	4.3L MPI	4.3/262	V6	DEI
	5.0L	5.0/305	V8	TB V
	5.0L MPI	5.0/305	V8	DEI
	5.7L	5.7/350	V8	TB V
	350 Mag MPI	5.7/350	V8	DEI
	MX6.2 Black Scorpion	6.2/377	V8	DEI
	MX6.2 MPI	6.2/377	V8	DEI
	8.1S HO	8.1/496	V8	DIS
	496 Mag/Mag HO	8.1/496	V8	DIS
2006	3.0L	3.0/181	L4	EST
	4.3L	4.3/262	V6	TB V
	4.3L MPI	4.3/262	V6	DEI
	5.0L	5.0/305	V8	TB V
	5.0L MPI	5.0/305	V8	DEI
	5.7L	5.7/350	V8	TB V
	350 Mag MPI	5.7/350	V8	DEI
	MX6.2 Black Scorpion	6.2/377	V8	DEI
	MX6.2 MPI	6.2/377	V8	DEI
	8.1S HO	8.1/496	V8	DIS
	496 Mag/Mag HO	8.1/496	V8	DIS
2007-08	3.0L TKS	3.0/181	L4	EST
	4.3L TKS	4.3/262	V6	TB V
	4.3L MPI	4.3/262	V6	DEI
	5.0L TKS	5.0/305	V8	TB V
	5.0L MPI	5.0/305	V8	DEI
	5.7L TKS	5.7/350	V8	TB V
	350 Mag MPI	5.7/350	V8	DEI
	Scorpion 350	5.7/350	V8	DEI
	Tow Sport 5.7 MPI	5.7/350	V8	DEI
	Tow Sport 5.7L TKS	5.7/350	V8	TB V
	377 Mag	6.2/377	V8	DEI
	MX6.2 MPI	6.2/377	V8	DEI
	Scorpion 377	6.2/377	V8	DEI
	8.1S HO	8.1/496	V8	DIS
	496 Mag/Mag HO	8.1/496	V8	DIS

DEI	Distributor Electronic Ignition
DIS	Distributorless Electronic Ignition
EST	Electronic Spark Timing
TB V	Thunderbolt V Ignition

① These models used the Thunderbolt ignition system up to serial #299999. All later engines uses the DEI system

Fig. 67

ELECTRONIC SPARK TIMING SYSTEM (EST)

Description

♦ See Figures 68 thru 72

Basic components of this system include a pointless distributor, a remote ignition coil, an electronic ignition module and a pick-up coil. The distributor does not contain breaker points, a condenser or a centrifugal advance arrangement. As the name of the system implies, spark advance and ignition timing are handled electronically.

The distributor utilizes an internal magnetic pick-up assembly consisting of a permanent magnet, a pole piece with internal teeth and the actual pick-up coil. The pick-up coil itself is sealed against moisture and electro-mechanical interference. When the rotating teeth of the timer core (which is attached to the top of the distributor shaft) line up with the internal teeth on the pole piece, voltage is induced in the pick-up coil. The coil then sends a signal to the ignition module; thus triggering the primary ignition circuit. Current flow in the primary circuit is interrupted and voltage as high as 35,000 volts is induced in the secondary windings of the ignition coil; this voltage is then through the secondary ignition circuit to fire the spark plugs.

The distributor is connected to the ignition coil by a high tension secondary wire and 2 low voltage primary wires. Because of this high voltage, a special thermoplastic, injection molded, glass-reinforced distributor cap is used.

A timer core (trigger wheel) is mounted near the upper end of the distributor shaft. This "wheel" has "teeth" - the number corresponding to the

IGNITION & ELECTRICAL SYSTEMS   9-25

number of engine cylinders. An ignition module is mounted close to the wheel.

The ignition coil produces greater spark voltage, longer spark, and operates at a much higher rpm than normal. The secondary windings are wrapped around the primary windings, which are in turn wrapped around an iron core. The coil generates a very high secondary voltage of up to 35,000 V when the primary circuit has been broken. There are a pair of 2-wire connectors used for:
- Battery voltage input
- Primary voltage to the ignition module
- Ignition module trigger signal input
- Tachometer output signal

The ignition module is located inside the distributor and is a solid state unit that uses transistorized relays and switches to control any associated circuits. The module serves the following functions:
- Changing the voltage signal from the pick-up coil to a "square" digital signal
- Sending the digital signal as a reference signal (**REF HI**) for ignition control
- Providing a ground reference signal (**REF LO**)
- Providing a means to control spark advance (Ignition mode and/or Module mode)
- Providing a trigger signal for the ignition coil

The pick-up coil assembly consists of the stationary pole piece and a pick-up coil and magnet; which rides inside the pole piece. The coil produces an alternating voltage signal as the teeth pass the magnet. A signal is produced for each cylinder of the engine during one complete revolution of the engine. The coil assembly is connected to the ignition module with a 2-wire connector.

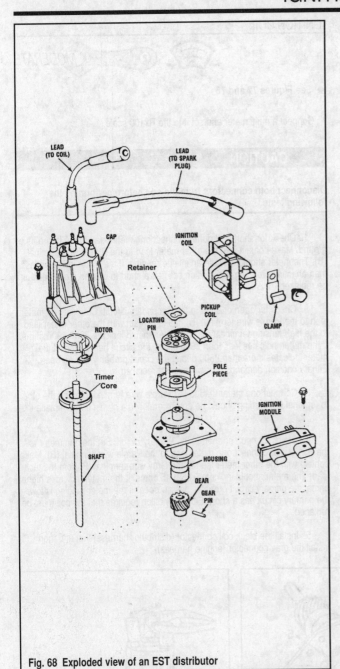

Fig. 68 Exploded view of an EST distributor

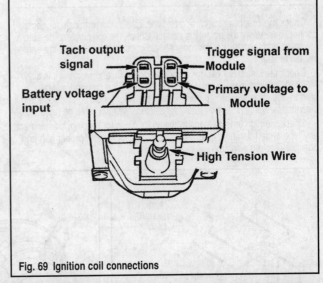

Fig. 69 Ignition coil connections

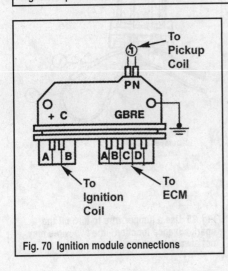

Fig. 70 Ignition module connections

Fig. 71 A good look at the distributor...

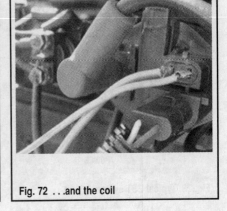

Fig. 72 ...and the coil

9-26 IGNITION & ELECTRICAL SYSTEMS

Troubleshooting

◆ See Figures 69 thru 75

✱✱ WARNING

This Delco EST breakerless ignition system requires the use of a jumper wire (#91-818812A1) or one may be fabricated using a six inch piece of 16AWG wire and a male bullet terminal on each end. This jumper is required to shunt (turn off), the electronic spark advance function in the module while setting initial timing.

✱✱ CAUTION

Always ground the high tension of the coil any time it is disconnected from the distributor. Failure to disconnect the lead could result in fire or an explosion, if gas vapors are present. Reason: A high voltage discharge in the secondary circuit of the coil may occur when the ignition switch is turned to the on and off positions. This high voltage discharge may occur even if the engine is not cranked.

Before spending too much time and money attempting to trace a problem to the ignition system, a compression check of the cylinders should be made. If the cylinders do not have adequate compression, troubleshooting and attempted service of the ignition or fuel system will fail to give the desired results of satisfactory engine performance.

VISUAL CHECKS

Check the coil tower for carbon tracking. Check the terminals for secure connections. Verify the polarity is correct. Check the coil nipple to be sure it is sealed and insulated. If flashover should occur at this location, the engine will fail to start.

Check the distributor cap for carbon tracking. Clean the cap if it is dirty. Moisture and dirt make a good path for flashover. If a carbon track has started, the cap must be replaced. Check the rotor for carbon tracking and cleanliness. Again, if carbon track has started, the rotor must be replaced.

Check the high tension leads to the spark plugs for burning, cracking and any type of deterioration. Check the spark plug for fouling, proper gap and possible cracked insulator.

IGNITION COIL

◆ See Figures 72 and 76

Connect a multi-meter and set it to the Rx100 scale.

✱✱ CAUTION

Disconnect both connectors from the coil before performing the following tests.

1. Check for short to ground: Connect one meter lead to the frame as a ground. Make contact with the other meter lead to the Purple wire terminal (**B**). The meter should indicate an infinite reading. If the meter indicates less than an infinite reading, the ignition coil has a short to ground. The coil must be replaced.

2. Check for open or short in the Secondary circuit. Connect one meter lead to the Purple wire terminal (**B**). Make contact with the other meter lead to the high tension terminal (**A**). The meter should indicate 7800-8800 ohms. If the meter indicates less than 7800 ohms, the circuit has a short. If the meter indicates more than 8800 ohms, the circuit probably has an open. Either condition demands the coil must be replaced.

3. Connect one meter lead to the Purple wire terminal (**B**) and the other to terminal **C**. Reading should be approximately 0.4 ohms; if not, replace the coil.

4. Check for open or short in the Primary circuit: Set the ohmmeter to the Rx1 scale. Connect one meter lead to the Purple wire terminal (**B**). Make contact with the other meter lead to the Gray wire terminal (**F**) from the tach lead. The meter should indicate 0.35-0.45 ohms. If the meter indicates higher ohms, the primary circuit probably has an open. If the meter reading is low, the primary circuit has a short. Either condition requires that the coil must be replaced.

5. Install the black coil connector (distributor harness) first and then install the gray connector (engine harness).

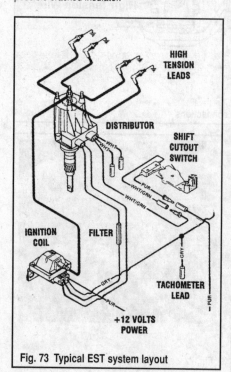

Fig. 73 Typical EST system layout

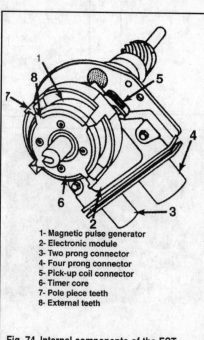

Fig. 74 Internal components of the EST distributor

1- Magnetic pulse generator
2- Electronic module
3- Two prong connector
4- Four prong connector
5- Pick-up coil connector
6- Timer core
7- Pole piece teeth
8- External teeth

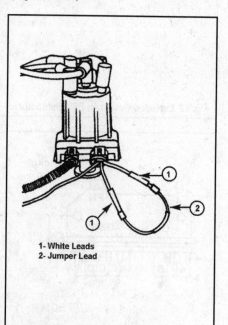

Fig. 75 Use a jumper wire to turn off the spark advance function - the two wires may not always be white

1- White Leads
2- Jumper Lead

IGNITION & ELECTRICAL SYSTEMS 9-27

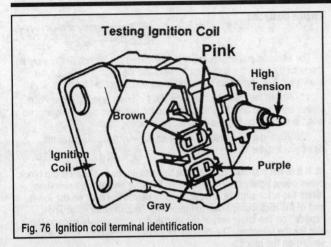

Fig. 76 Ignition coil terminal identification

PICK-UP COIL

◆ See Figure 77

1. Connect a multi-meter as per the manufacturers instructions.
2. Remove the screws securing the distributor cap, and then remove the cap and the rotor. Release the locking tab and unplug the pickup coil connector. On almost all models these two wires should be white and green.
3. Check for short to ground: Set the ohmmeter to the Rx10 scale. Connect one meter lead to the distributor body. Make contact with the other meter lead to the white wire terminal (**C**) of the pick-up coil. The meter should indicate an infinite reading. A reading of less than infinite, indicates the coil has a shorted circuit. The coil must be replaced.
4. Check for short to ground: Set the ohmmeter to the Rx10 scale. Connect one meter lead to the distributor body. Make contact with the other meter lead to the green wire terminal (**B**) of the pick-up coil. The meter should indicate an infinite reading. A reading of less than infinite, indicates the coil has a shorted circuit. The coil must be replaced.
5. Check for open or shorted coil: Set the ohmmeter to the Rx100 scale. Connect one meter lead to the green wire pickup coil terminal (**B**). Make contact with the other meter lead to the white wire coil terminal (**C**). The meter should have a constant indication of 500-1500 ohms. If the meter indicates more than 1500 ohms, the coil probably has an open circuit and the coil must be replaced. If the meter indicates less than 500 ohms the coil probably has a shorted circuit and the coil must be replaced. Make sure that you bend the leads slightly while testing to determine any intermittent open circuits.

IGNITION MODULE

The ignition module in the distributor either produces a spark or it fails to produce an adequate spark. If all other tests have been performed with satisfactory results and the problem still exists - replace the ignition module.

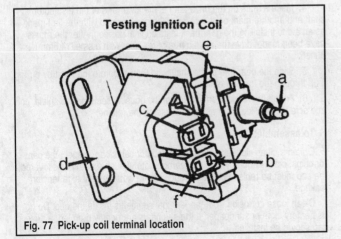

Fig. 77 Pick-up coil terminal location

Distributor Cap

REMOVAL & INSTALLATION

◆ See Figure 78

■ It is not necessary to remove the spark plug leads at the cap when removing it. If you do remove them, be sure to tag each of them to ensure correct installation.

1. Disconnect the battery cables.
2. Remove the distributor cap mounting screws and lift off the cap. Some caps may have a vent connection; if yours does, unplug the hose.
3. Clean the cap with warm soapy water - if you have not removed the spark plug wires, you must, obviously, do it at this time. Blow it dry with compressed air; or if not available, make sure it air dries completely prior to installation.
4. Check the cap contacts for excessive burning, wear or corrosion. Check the cap for cracks or wear.
5. Install the cap by fitting the tab into the notch in the distributor housing and tighten the screws securely. Reconnect the plug wires if disconnected.

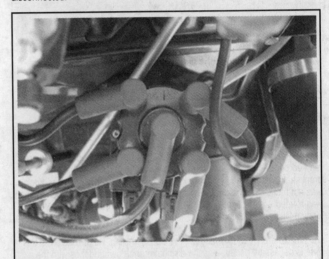

Fig. 78 Remove the cap with the wires attached

Distributor

REMOVAL

◆ See Figure 79

1. Begin by disconnecting the high tension leads from the distributor. Take time to identify and tag each lead as an aid during installation to ensure they are properly installed for correct cylinder firing.
2. Lift the locking tabs and disconnect the 2-and 4-terminal connectors. Rotate the crankshaft until No. 1 cylinder is in the firing position - both valves for No. 1 are closed and the timing mark on the harmonic balancer is aligned with the 0° mark on the grid attached to the front chain cover.

■ The 4 terminal connector may have 3 or 4 wires coming from it.

3. Remove the two screws securing the distributor cap in place, and then remove the cap.
4. Now, note the position of the rotor tip. Take time to make a reference mark on the distributor housing to enable the rotor and the housing to be properly aligned during installation.
5. Make a mark on the distributor base and a matching mark on the engine as an aid during installation to ensure the distributor will be installed back in its original position.

9-28 IGNITION & ELECTRICAL SYSTEMS

■ Take care to prevent the crankshaft from being rotated - even slightly - while the distributor is out of the block. If the crankshaft should be rotated - follow the procedures listed under Engine Disturbed later in this section.

6. Remove the distributor clamp bolt and lift the distributor straight up and clear of the engine. The gasket should be discarded.

Fig. 79 A good look at the distributor clamp bolt

INSTALLATION

Engine Not Disturbed

1. Perform the following procedures if the distributor and engine were marked prior to removal and there is no evidence to indicate that the crankshaft was rotated while the distributor was removed. If the distributor was not marked, as instructed in the removal procedures, or if the crankshaft was rotated - even slightly - while the distributor was out, proceed to the Engine Disturbed section.
2. Install the rotor and rotate it approximately 1/8 turn counterclockwise from the reference mark made prior to disassembling. Rotor rotation at this time is necessary because as the distributor shaft gear indexes with the camshaft gear, the shaft will rotate clockwise about 1/8 turn - the rotor will then be back where it should be for the No. 1 cylinder to fire.
3. Position a new distributor gasket in place on the engine. A reference mark should have been made on the distributor and the block prior to removal, as instructed in the removal procedures. Install the distributor into the block with the mark on the distributor roughly aligned with the mark on the block. Push the distributor fully into the block until the housing is seated.

※※ WARNING

If necessary, rotate the rotor slightly to permit the gear on the lower end of the distributor shaft to index with the gear on the camshaft. The "spade" on the lower end of the distributor shaft must engage the slot for the oil pump. Failure to index would result in no engine oil circulation - disaster. The distributor shaft collar should now be fully seated on the block. However, once the distributor is in place the rotor reference marks and the distributor housing marks should both be aligned.

4. Secure the distributor in place with the hold down clamp and bolt. Tighten the bolt to 20 ft. lbs (27 Nm).
5. Connect the 2-and 3-wire leads into the distributor.

■ The 4 terminal connector may have 3 or 4 wires coming from it.

6. Install the cap and tighten the screws securely. The cap can only be installed properly - one way. Lubricate the sockets in the distributor cap with Quicksilver 2-4-C lubricant, or the equivalent. Install the spark plug high tension leads (if they were removed) using the identification made during removal to ensure proper cylinder firing.
7. Check the ignition timing.

Engine Disturbed

The following procedures are to be performed if reference marks were not made for the rotor, the distributor housing and the engine block or if the crankshaft was rotated while the distributor was out of the block.

1. Rotate the crankshaft until the No. 1 cylinder is ready to fire - both valves are closed and the timing mark on the harmonic balancer is aligned with the 0° mark on the grid attached to the timing cover.
2. Position a new gasket in place on the engine block. Install the distributor into the block.

■ If it is not possible to fully seat the distributor in place on the block, press down lightly on the distributor housing and at the same time rotate the rotor slightly. This action will permit the gear on the lower end of the distributor shaft to index with the camshaft gear. The "spade" on the lower end of the distributor shaft should engage the slot for the oil pump. The distributor shaft collar should now be fully seated on the block.

3. Once the distributor is fully seated, install the clamp and bolt, but leave it just loose enough to permit rotating the distributor with strong hand pressure. At this point, the rotor must be in position to fire No. 1 cylinder.
4. Just place the cap in position on the distributor. Scribe a mark on the distributor housing aligned with the No. 1 spark plug terminal. Now, remove the cap and verify the rotor is aligned with the mark just made for the No. 1 spark plug terminal. If the rotor does not align with the mark, rotate the distributor housing - with difficulty, because the clamp was not tightened - right or left to align the rotor with the mark for the No. 1 terminal. If it is not possible to align the rotor with the No. 1 mark, the distributor must be removed the entire procedure started over again.
5. Tighten the distributor clamp bolt to 20 ft. lbs (27 Nm).
6. Check all high tension leads and connect them in the proper sequence for correct cylinder firing.
7. Connect the 2-wire engine harness connector to the distributor.
8. Check the ignition timing.

DISASSEMBLY & ASSEMBLY

◆ See Figures 68 and 80 thru 87

Remove the distributor from the engine, as outlined in the previous section. The following procedures pick up the work after the distributor has been removed and is on a suitable work surface.

1. Pull the rotor free from the upper end of the distributor shaft.
2. Disconnect the leads from the ignition module, and then pry the retainer free. Remove the pick-up coil.
3. Disconnect the ignition module lead. Remove the two screws securing the module to the distributor, and then remove the module. It may be necessary to carefully pry the module loose from the distributor.
4. Make a mark on the drive tang at the lower end of the distributor shaft and another mark on the collar of the gear aligned with the first mark as an aid to installing the gear back in its original position. After the marks have been made, drive the roll pin free of the gear with a small (4.5mm) punch.
5. Slide the gear free of the shaft, and then separate the shaft by pulling it up through the top of the pick-up coil.
6. Pry off the retainer on the end of the housing securing the shield, coil and pole piece. Slide them all off and discard the retainer.

To assemble:

Obtain a clean cloth and wipe the distributor cap clean. Inspect the cap for chips, cracks and any sign of a carbon path. If such a path is discovered, the cap must be replaced because such a path would permit high tension leakage.

Clean loose corrosion from the terminal segments inside the cap. Do not use emery cloth or sandpaper. If the segments are deeply grooved, a new cap should be installed.

IGNITION & ELECTRICAL SYSTEMS 9-29

Inspect the terminal sockets for corrosion. Clean the sockets using a stiff wire brush to loosen the corrosion. After the sockets are clean, lubricate them with Quicksilver 2-4-C lubricant, or equivalent.

Inspect the rotor for cracks. Check to be sure the tip of the rotor is not badly burned. Such a condition demands the rotor be replaced.

Inspect the trigger wheel for any sign of contact with the sensor. If the distributor has been removed from the block, make an attempt to check the distributor shaft for excessive wear between the shaft and the bushings in the housing.

The following procedures pickup the work after disassembled parts have been cleaned and replacement items have been obtained and are on hand.

7. Align the pole piece with the housing and slide it onto the end of the housing. Align the tab on the coil with the hole in the base of the housing.

8. Install the shield and a new retainer over the end of the housing. Position a 5/8 in. socket over the retainer and lightly tap it into place over the end of the housing - make sure it is seated firmly into the groove.

9. Coat the distributor shaft with oil and slide into and through the center of the housing. Spin it several times to lubricate it and the housing.

Fig. 80 A good view of the EST distributor

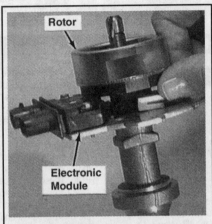
Fig. 81 Lift off the rotor

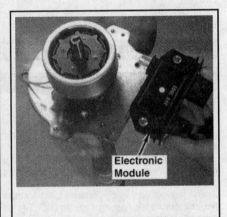
Fig. 82 Removing the ignition module

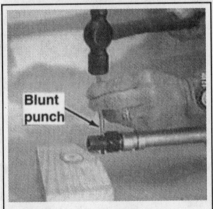
Fig. 83 Driving out the roll pin

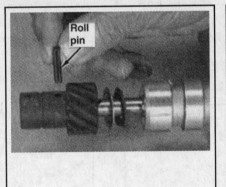
Fig. 84 Close up of the pin and the gear

Fig. 85 Removing the gear

Fig. 86 Pull the shaft up through the coil

Fig. 87 Coat the plate with grease

9-30 IGNITION & ELECTRICAL SYSTEMS

10. Slide the star washer onto the shaft so that the teeth are pointing up toward the top of the housing. Slide the flat washer and the gear onto the shaft so that the alignment marks made earlier line up. Insert the roll pin and drive it down until it is flush.

11. Make sure the surface of the distributor plate is clean and then spread an even coat of Thermal Conductive grease onto the bottom of the module. Position the module on the plate and tighten the two screws securely; clean any excess grease from the plate and module.

12. Connect the lead from the coil to the back of the module and lock it in place.

13. Align the rotor with the notch in the distributor shaft, and then press it to secure it in place.

Ignition Coil

REMOVAL & INSTALLATION

◆ See Figure 72

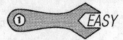

1. Disconnect the negative battery cable.
2. Disconnect the high tension lead coming from the distributor.
3. Tag and disconnect the coil electrical leads (two).
4. Remove the two coil mounting bolts and lift out the coil with the bracket attached.
5. Install the coil and tighten the mounting hardware securely.
6. Connect the electrical leads and the high tension lead.
7. Connect the battery cable.

THUNDERBOLT V IGNITION SYSTEM

Description

◆ See Figures 88, 89 and 90

The Thunderbolt V is an electronic transistorized High Energy Ignition (HEI) system. This system is quite different from the standard point-type ignition systems.

The system does not utilize breaker points or a mechanical advance mechanism. It was designed to be almost maintenance free without the requirement for periodic adjustment.

The Thunderbolt V ignition systems look very much like a conventional type distributor shaft, housing, cap, and rotor. The ignition coil is mounted externally in close proximity to the distributor. The ignition module is mounted on the exhaust elbow for early models or on the side of the distributor housing and is known as the Digital Electronic Spark Advance (D.E.S.A.). A second module for spark Knock Control is mounted on top of the D.E.S.A. module.

Internally, a sensor and sensor wheel are mounted near the top of the distributor shaft under a conventional-type rotor. The ignition sensor performs similar to a switch, when subjected to a magnetic field. The distributor electrical operation uses a Hall effect vane switch assembly, causing the ignition coil to be switched ON and OFF by the ECM. The sensor has a Hall switch on one side and a permanent magnet on the opposite side. The rotary sensor wheel is made of a ferrous metal and triggers the on and off signal. When a window opening of the sensor wheel cup is in front of the magnet and the Hall effect switch, a magnetic flux field is completed from the magnet through the Hall effect switch and back to the magnet.

As the sensor wheel cup rotates and closes the window opening, the flux lines are shunted through the vane and back to the magnet. During this time, a voltage is produced as the vane passes through the window opening. When the vane clears the opening, the window edge causes the signal to go to 0 volts. This pulsed signal provides a **REF HI** and **REF LO** signal to the ECM. These two signals from the distributor to the ECM provide a precise indication of engine speed to the ECM for spark and timing control. Two additional lines called **IC** and **BYPASS** between the distributor and the ECM are used to sense and control ignition module functions.

Ignition timing advancement is normally controlled by the ECM. Should a failure occur or a fault be detected within the ECM, the ignition module will take control of the ignition timing causing the engine to operate at a reduced power output.

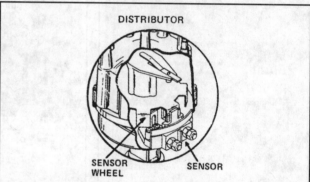

Fig. 89 Thunderbolt systems use a sensor and wheel in place of points

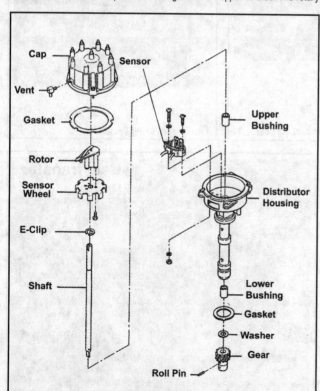

Fig. 88 Exploded view of the Thunderbolt V distributor (V8 shown)

Fig. 90 A good look at a Thunderbolt distributor

IGNITION & ELECTRICAL SYSTEMS 9-31

To change the base timing on engines equipped with the Thunderbolt V ignition system and ECM, requires the use of a scan tool or code tool. Once the tool is connected into the DTC connector and set to on, the ECM will shift into Ignition Control Mode and the distributor ignition module will shift to base timing. The initial timing can now be adjusted by rotating the distributor, as with other ignition systems.

The ignition module will withstand a reverse battery connection for only about a minute, before it is severely damaged.

The ignition coil appears to be a standard size and shape. However, the HEI coil has a special winding and core. If a standard coil is used, the ignition module will not be damaged, but it will supply a low output and will overheat.

Standard tachometers and most synchronizers monitoring ignition impulses at the negative terminal of the coil will operate satisfactorily with the system.

Troubleshooting

GENERAL SYSTEM

The only equipment needed to troubleshoot the Thunderbolt V ignition system are an analog or digital voltmeter (multi-meter) and a spark gap tester.

✱✱ WARNING

Check to be sure the engine compartment is well ventilated and free of any gasoline vapors before starting any of the following tests. A spark will be generated creating a potential fire hazard if fuel vapors are present.

1. Check to be sure the battery is up to full charge. If not, correct the condition by charging the battery or making a substitution. Check to be sure the distributor, and clamp bolt are tight. If loose, rotate the distributor and check the timing.
2. Check all terminal connections on the distributor, ignition module and ignition coil. Verify the Gray Tachometer lead from the coil (-) side, to the tachometer on the control panel is not shorted to ground. Temporarily disconnect the tachometer lead if it is suspected of having a short.
3. Set the ignition key switch to the **RUN** position. Connect a multi-meter and set the switches for 12V DC reading. Attach the Red meter lead to the positive terminal of the ignition coil and the Black meter lead to a good engine ground. The meter should indicate 12 volts. If no voltage is present, check the engine and instrument wiring harness, battery cables, and the key switch for damaged wiring or unplugged harness connectors. If 12 volts is indicated on the meter, proceed to the next step.
4. Disconnect the White/Red wire lead from the distributor bullet connector. Connect the Red meter lead to the ignition module side of the White/Red lead. Connect the Black meter lead to a good engine ground. The meter should indicate 12 volts. If no voltage is present, replace the ignition module. If 12 volts is present, proceed to the next step.
5. Connect the White/Red bullet connectors at the distributor, if previously disconnected. Remove the high tension lead from the distributor to the coil. Insert a spark gap tester from the coil tower to ground. Disconnect the White/Green lead from the distributor terminal. Turn the ignition switch to the **RUN** position. Now, with a rapid motion, strike the White/Green lead coming from the distributor 2-3 times per second against a good engine ground. If there is spark at the coil, replace the defective ignition sensor in the distributor. If there is no spark at the coil, proceed with the next step.
6. Substitute a new or known good ignition coil and repeat the previous step. If there is now spark at the coil, install a new ignition coil. If there is no spark at the coil, replace the defective ignition module.

KNOCK CONTROL MODULE

The knock control module contains the circuits that monitor AC voltage signal output from the knock sensor. If no spark knock is present, the knock control module supplies 8-10 volts to the ignition control module and normal timing advancement will occur. However, if spark knock is detected by the sensor, the knock control module will remove the 8-10 volt signal to the ignition control module and the ignition timing will then be retarded to eliminate the spark knock.

Inspect the routing of the Blue wire lead from the knock sensor to the module. If this wire is too close to a high tension lead, the knock module may interpret this interference as a knock signal, which could result in a false retard timing.

1. Obtain a multi-meter and an un-powered test light. The test light should be low power and use less than 300mA of current draw. Set the switches on the meter to auto range or 12 volt DC reading.

✱✱ CAUTION

Water must circulate through the lower unit to the engine any time the engine is run to prevent damage to the water pump mounted on the engine. Just a few seconds without water will damage the water pump.

2. Start the engine and allow it to warm to normal operating temperature.
3. Connect the Red lead from the DVOM to the Purple/White bullet connector between the ignition control and knock control modules. Connect the Black lead from the DVOM to a good ground. The meter should indicate 8-10 volts. If there is no voltage, move the Red meter lead to terminal **B** on the knock control module connector plug, or probe the Purple wire to the connector. The meter should indicate 12 volts power to the knock control module. If the meter indicates 12 volts to the module, but no output from the module when connected to the Purple/White lead, then the knock control module is probably defective. Continue with the remainder of the test to eliminate all probable failures before replacing the knock control module.
4. Connect the Red meter lead back to the Purple/White bullet connector. If there is no voltage indicated on the Purple lead, troubleshoot the engine harness for an open circuit or damaged wiring.
5. Disconnect the Blue wire connector from the knock sensor on the side of the engine block. Connect the test light to a 12 volt power source. Move the throttle remote control until engine speed is slightly above 1500 rpm. Now, simulate an AC voltage signal from the knock sensor by tapping on the knock sensor harness terminal with the test light while observing the meter. If a voltage drop is observed, the wiring from the sensor to the knock control module is good but the knock sensor could be defective. If no voltage drop is observed, check the Blue wire from the sensor to the knock module for a short or open circuit.
6. Connect the Blue wire terminal onto the knock sensor. Using a small hammer, tap rapidly on the side of the engine block near the knock sensor. If a voltage drop is observed on the meter, the knock sensor and wiring is functioning properly. If no voltage drop is observed on the meter, then the knock sensor is defective and should be replaced.
7. Place the remote control throttle in the idle position. Shut down the engine and disconnect the DVOM meter and test light.

PRECAUTIONS

The following safety precautions should always be practiced when servicing or testing any portion of the ignition system.

1. Check to be sure the engine compartment is well ventilated. After the engine hood has been removed, allow some time for any existing fumes to dissipate and be replaced with fresh air. Service work should not be started if any gasoline vapors are present to prevent the possibility of fire.
2. Work slowly and deliberately - thinking of the next movement to made before proceeding with the work.
3. Take extra precautions to keep hands, feet and clothing completely clear of any moving part.
4. Be especially careful not to touch or disconnect any ignition component while the engine is operating.
5. Never reverse the battery connections for any reason. The electrical and ignition systems comprise a negative (-) ground.
6. Never disconnect either end of the battery cables while the engine is operating.

■ **A careful inspection should be made of the wiring during troubleshooting and before expensive parts are purchased for replacement. The manufacturer uses specially designed features and environmental protection extensively to protect ignition contacts. The following areas could be considered likely causes of an electrical or ignition problem:**

9-32 IGNITION & ELECTRICAL SYSTEMS

7. Damaged contacts caused by improper engagement.
8. The mating surfaces of contacts contaminated by corrosion, sealer or other foreign material.
9. Terminals not fully seated.
10. Microscopic damage or holes in an electrical lead allowing water to enter and cause corrosion, or actual circuit failure.

Distributor Cap

REMOVAL & INSTALLATION

◆ See Figures 91 and 92

■ It is not necessary to remove the spark plug leads at the cap when removing it. If you do remove them, be sure to tag each of them to ensure correct installation.

1. Disconnect the battery cables.
2. Remove the four distributor cap mounting screws and lift off the cap. Some caps may have a vent connection; if yours does, unplug the hose.
3. Clean the cap with warm soapy water - if you have not removed the spark plug wires, you must, obviously, do it at this time. Blow it dry with compressed air; or if not available, make sure it air dries completely prior to installation.
4. Check the cap contacts for excessive burning, wear or corrosion. Check the cap for cracks or wear.
5. Install the cap by fitting the tab into the notch in the distributor housing and tighten the screws securely. Reconnect the plug wires if disconnected.

Rotor/Sensor Wheel

REMOVAL & INSTALLATION

◆ See Figures 93 thru 95

1. Remove the distributor cap.

■ The rotor and sensor wheel are mounted to the distributor shaft with Loctite; you may wish to carefully use a heat gun or hair drier before prying the assembly off of the shaft

2. Position two small flat blade screwdrivers underneath the sensor wheel so the end of the blade is against the distributor shaft. The screwdrivers should be 180° apart. Press down on each screwdriver (using the housing as a fulcrum) and pry the assembly up and off the shaft.

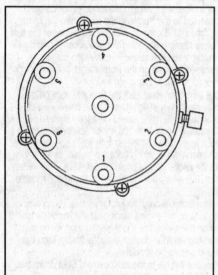

Fig. 91 Distributor cap terminal locations - V6 engines

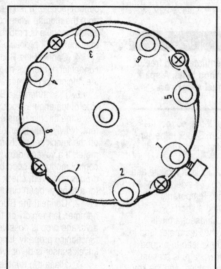

Fig. 92 Distributor cap terminal locations - V8 engines

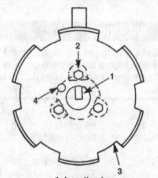

1- Locating key
2- Bolts (hex head)
3- Sensor Wheel
4- Locating pin

Fig. 93 Check the sensor wheel indexing by laying it upside down (teeth pointing up) on top of this illustration so that the bolt holes and locating pin hole are aligned. All teeth should be in alignment with those in the illustration if indexed correctly - V6 engines

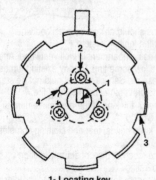

1- Locating key
2- Screws
3- Sensor wheel
4- Locating pin

Fig. 94 Check the sensor wheel indexing by laying it upside down (teeth pointing up) on top of this illustration so that the bolt holes and locating pin hole are aligned. All teeth should be in alignment with those in the illustration if indexed correctly - V8 engines

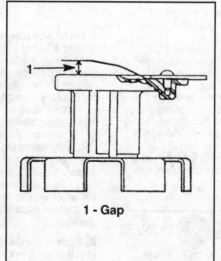

1 - Gap

Fig. 95 Adjusting the carbon brush gap

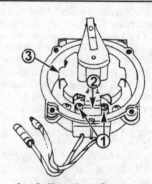

1 - Adjusting Screws
2 - Sensor
3 - Sensor Wheel Fingers

Fig. 96 You may have to adjust the sensor if the wheel is coming in contact with it

IGNITION & ELECTRICAL SYSTEMS

3. Inspect the locating key found in the ramp at the bottom of the splined hole in the assembly. It should be about 1/8 in. (3mm) wide and should not be shaved or burred in any way.

4. Check the rotor for burns, corrosion, cracks and carbon tracks. Replace the rotor if necessary - just remove the three small hex bolts securing it to the sensor wheel. When installing the new rotor to the wheel, make sure that the pin on the rotor fits into the locating hole on the wheel and then tighten the three bolts securely.

5. Check that the carbon brush tang gap between it and the rotor is 1/4 in. (6.4mm). It can be adjusted by carefully bending it until the correct gap is achieved.

6. Squeeze two drops of Loctite 271 onto the locating key in the rotor and then squeeze two more drops into the keyway on the upper edge of the distributor shaft and then immediately install the rotor assembly onto the shaft. Ensure that the key is positioned fully in the keyway and then press the assembly completely down with the palm of your hand until it seats fully.

■ Let the Loctite cure overnight before operating the engine. It is a good idea to allow it to cure with the distributor upside down so no sealant can run into the housing bushing. Yes, this means you should remove the distributor.

7. Make sure that the fingers on the sensor wheel do not touch the sensor when the wheel is rotated; if so, loosen the sensor screws and adjust its positioning slightly.

8. Install the distributor cap and connect the battery cables.

Sensor

TESTING

1. Tag and disconnect the two sensor leads at the harness.
2. Connect a DVOM and check for resistance across the two leads. If anything less than 100 ohms, replace the sensor.

REMOVAL & INSTALLATION

◆ See Figure 97

1. Disconnect the battery cables, remove the distributor cap and the rotor/sensor wheel assembly.
2. Tag and disconnect the sensor leads.
3. Remove the two mounting screws and lift the sensor out of the housing.
4. Inspect the two jumper leads hanging from the inside edge of the sensor with a magnifying glass. If either is cracked, replace the sensor.
5. Install the sensor into the distributor housing and tighten the two screws securely.

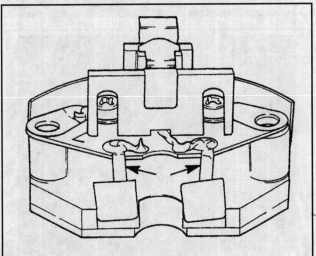

Fig. 97 Check the two jumper leads. Note the mounting screw holes on each side of the sensor

6. Install the rotor assembly and cap.
7. Connect the battery cables.

Ignition Or Knock Control Modules

REMOVAL & INSTALLATION

◆ See Figures 98

1. Disconnect the battery cables.
2. Unsnap the wiring harness at the modules.
3. Almost all others have it mounted onto the side of the distributor housing. Remove the mounting bolts and pull off the modules.
4. Coat the seating base of the module with thermal-conductive grease and install the module. Tighten the mounting screws to 10 inch lbs. (1.1 Nm).
5. Reconnect the wiring harness and the battery cables.

Distributor

REMOVAL

◆ See Figures 79 and 98

1. Disconnect the battery cables.
2. Disconnect the high tension leads from the distributor cap and remove the cap. Take time to identify each lead as an aid during the installation to ensure they are properly installed for correct cylinder firing. Also, it's a good idea to paint a little mark on the housing even with where the No. 1 plug wire terminal is. Although this is not necessary, it's a good idea anyway.
3. Lift the locking tabs on the module wire connectors and disconnect the wire harness from the distributor on the ignition and/or knock modules.
4. Rotate the crankshaft until the No. 1 cylinder is in the firing position - both valves for the No. 1 cylinder are closed and the timing mark on the harmonic balancer is aligned with the 0° mark on the timing grid attached to the front cover. Now, note the position of the rotor tip - it should be pointing toward the spot where the No. 1 cylinder terminal was in the distributor cap.
5. Make a mark on the distributor base and a matching mark on the engine block as an aid during installation to ensure the distributor will be installed back into its original position.

■ Take care to prevent the crankshaft from rotating - even slightly - while the distributor is out of the block. If the crankshaft should be rotated - follow the procedures listed under Engine Disturbed following the assembly instructions.

6. Loosen and remove the distributor clamp bolt and clamp. Lift the distributor straight up and clear of the engine. Remove the gasket from the base of the distributor and discard the gasket.

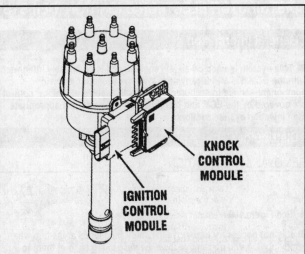

Fig. 98 Most engines have both an ignition module and a knock control module mounted on the side of the housing

9-34 IGNITION & ELECTRICAL SYSTEMS

INSTALLATION

Engine Not Disturbed

 MODERATE

1. Install a new gasket over the distributor shaft.
2. Turn the rotor approximately 1/8 of a turn counterclockwise past the mark painted on the housing and then install the distributor slowly into the cylinder block. You may have to wiggle the rotor slightly until it meshes with the pump gear to get the unit fully seated. The key is that the rotor ends up pointing to the mark on the housing made during removal.
3. Install the clamp and clamp bolt and tighten it to 20 ft. lbs. (27 Nm).
4. Connect all harness leads and install the cap. Connect the spark plug leads if removed.
5. Connect the battery cables.

Engine Disturbed

 MODERATE

1. Set the No. 1 piston to TDC. Remove the No. 1 spark plug, put your thumb over the hole and slowly crank the engine until you feel compression on the No. 1 cylinder. Crank the engine a bit more until the pointer lines up with the mark on the scale.
2. Install a new gasket on the housing.
3. Grasp the rotor and spin the shaft about 1/8 of a turn, counterclockwise, past the alignment mark. Position the distributor over the hole in the block so that the alignment marks match and then work it down into the hole until it engages completely with the gear.

■ It is OK to wiggle the rotor slightly during installation so that the distributor gear meshes properly with the oil pump. It is imperative though that the rotor ends up pointing to the alignment mark made during removal when installation is complete.

4. Install the clamp bolt and tighten it finger-tight. Rotate the distributor carefully until the marks all line up and then tighten the clamp bolt to 20 ft. lbs. (27 Nm).
5. Install the distributor cap into position and confirm that the rotor points to the terminal for the No. 1 spark plug lead. If it does, go to the next step; if not, repeat the installation procedure.
6. Install the cap and coil/plug leads.
7. Check the ignition timing.

DISASSEMBLY & ASSEMBLY

 DIFFICULT

◆ See Figures 88 and 99

1. Remove the distributor, cap, rotor/sensor wheel assembly and sensor as detailed previously.
2. Carefully clamp the distributor in a vise and drive out the roll pin at the base of the shaft with a small drift.
3. Slide the gear and washer off the end of the distributor shaft. Check for side play between the distributor shaft and the distributor housing bushings by pulling and pushing the shaft sideways in the distributor housing. If the shaft moves sideways in the housing in excess of 0.002 in. (0.05 mm), the bushings are worn and the distributor should be replaced.
4. Pull the distributor shaft out of the distributor housing from the top. Check the distributor shaft for being trueness by placing it in a pair of V-blocks. Place a dial indicator gauge against the shaft and rotate the distributor shaft. If the shaft maximum run-out exceeds 0.002 in. (0.5mm), the shaft is bent and must be replaced.

To assemble:

5. Apply a coating of engine oil to the distributor driveshaft. Install the E-clip onto the shaft, if it was removed during disassembling. Slide the shaft down through the center of the distributor housing until the clip contacts the distributor housing.

■ If a new driven gear is being installed, the new gear will come with only one hole drilled in it. Slide the new gear onto the driveshaft, aligning the hole in the gear with the hole in the distributor shaft. This hole is offset and the gear will only fit on the shaft in one direction. Use the two holes as a pilot guide and insert a 3/16 in. carbide-tipped drill bit through the hole in the gear and shaft. Now, drill the second hole through the other side of the gear.

6. Slide the washer and driven gear onto the distributor shaft aligning the hole in the gear with the hole in the driveshaft. This hole is offset and the gear will only fit in one direction onto the shaft. With the gear and shaft holes aligned, insert the roll pin. Using a hammer and blunt end punch, tap the roll pin until it is flush with the gear surface.
7. Install the sensor, rotor/wheel, cap and distributor as detailed previously.

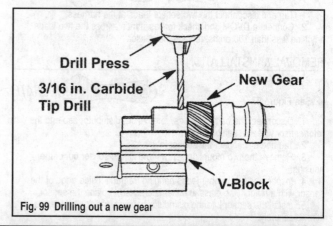

Fig. 99 Drilling out a new gear

DISTRIBUTOR ELECTRONIC IGNITION (DEI) SYSTEM - ECM 555

General Information

■ This system is used on all MPI engines and is completely integrated with the ECM 555 MPI fuel system. With the exception of main component service (distributor and coil), all aspects of ignition control are covered by the ECM and therefore detailed within the appropriate Fuel Injection system sections.

Distributor Cap

REMOVAL & INSTALLATION

 EASY

◆ See Figure 100

■ It is not necessary to remove the spark plug leads at the cap when removing it. If you do remove them, be sure to tag each of them to ensure correct installation.

1. Disconnect the battery cables.

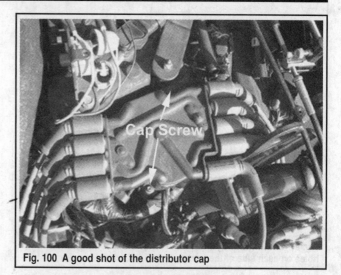

Fig. 100 A good shot of the distributor cap

IGNITION & ELECTRICAL SYSTEMS 9-35

2. Remove the 2 distributor cap mounting screws and lift off the cap.
3. Clean the cap with warm soapy water - if you have not removed the spark plug wires, you must, obviously, do it at this time. Blow it dry with compressed air; or if not available, make sure it air dries completely prior to installation.
4. Check the cap contacts for excessive burning, wear or corrosion. Check the cap for cracks or wear.
5. Install the cap by fitting it over the distributor housing and tighten the screws securely. Reconnect the plug wires if disconnected.

Distributor

REMOVAL

◆ See Figure 101

1. Disconnect the battery cables.
2. Disconnect the high tension leads from the distributor cap and remove the cap. Take time to identify each lead as an aid during the installation to ensure they are properly installed for correct cylinder firing. Also, it's a good idea to paint or scribe a little mark on the housing even with where the No. 1 plug wire terminal is. Although this is not necessary, it's a good idea anyway
3. Rotate the crankshaft until the No. 1 cylinder is in the firing position - both valves for the No. 1 cylinder are closed and the timing mark on the harmonic balancer is aligned with the 0° mark on the timing grid attached to the front cover. Now, note the position of the rotor tip - it should be pointing toward the spot where the No. 1 cylinder terminal was in the distributor cap.
4. Make a mark on the distributor base and a matching mark on the manifold as an aid during installation to ensure the distributor will be installed back into its original position.

■ Take care to prevent the crankshaft from rotating - even slightly - while the distributor is out of the block. If the crankshaft should be rotated - follow the procedures listed under Engine Disturbed following the assembly instructions.

5. Loosen and remove the distributor clamp bolt and clamp. Lift the distributor straight up and clear of the engine. Remove the gasket from the base of the distributor and discard the gasket.

INSTALLATION

Engine Not Disturbed

1. Install a new gasket over the distributor shaft.
2. Install the distributor slowly into the cylinder block. You may have to wiggle the rotor slightly until it meshes with the gear to get the unit fully seated. The key is that the rotor ends up pointing to the mark on the housing made during removal.
3. Install the clamp and clamp bolt and tighten it to 30 ft. lbs. (40 Nm).
4. Install the cap. Connect the spark plug leads if removed.
5. Connect the battery cables.

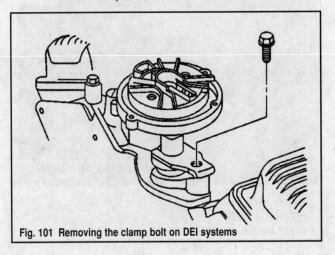

Fig. 101 Removing the clamp bolt on DEI systems

Engine Disturbed

1. Set the No. 1 piston to TDC. Remove the No. 1 spark plug, put your thumb over the hole and slowly crank the engine until you feel compression on the No. 1 cylinder. Crank the engine a bit more until the pointer lines up with the mark on the scale.
2. Install a new gasket on the housing.
3. Position the distributor over the hole in the block so that the alignment marks match and then work it down into the hole until it engages completely with the gear.

■ It is OK to wiggle the rotor slightly during installation so that the distributor gear meshes properly with the gear. It is imperative though that the rotor ends up pointing to the alignment mark made during removal when installation is complete.

4. Install the clamp bolt and tighten it finger-tight. Rotate the distributor carefully until the marks line up and then tighten the clamp bolt to 30 ft. lbs. (40 Nm).
5. Install the distributor cap into position and confirm that the rotor points to the terminal for the No. 1 spark plug lead. If it does, go to the next step; if not, repeat the installation procedure.
6. Install the cap and coil/plug leads.
7. Check the ignition timing.

DISASSEMBLY & ASSEMBLY

◆ See Figures 102 thru 105

1. Remove the distributor, cap, rotor and sensor (if equipped) as detailed previously.
2. Carefully clamp the distributor in a vise and drive out the roll pin at the base of the shaft with a small drift.
3. Slide the gear, washer and tang washer off the end of the distributor shaft.
4. Pull the distributor shaft out of the distributor housing from the top.

To assemble:

5. Check the distributor cap for cracks, carbon tracks or other signs of visible damage. Check the metal terminals for any signs of corrosion.

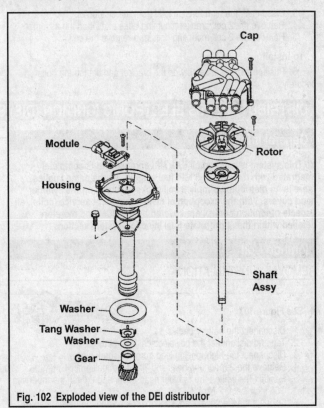

Fig. 102 Exploded view of the DEI distributor

9-36 IGNITION & ELECTRICAL SYSTEMS

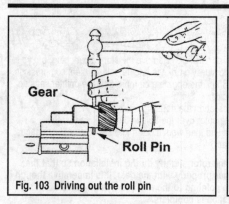
Fig. 103 Driving out the roll pin

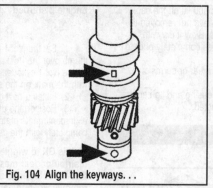
Fig. 104 Align the keyways...

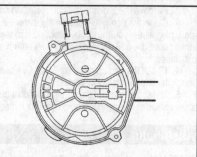
Fig. 105 ...and then make sure the tang is between the marks

6. Inspect the outer terminal of the rotor for any sign of wear or burning.
7. Inspect the distributor shaft-to-bushing for any sign of looseness - insert into the housing and see if it wobbles.
8. Clean all metal components in solvent and allow to dry thoroughly; using compressed air is recommended.
9. Apply a light coating of grease to the distributor driveshaft and slide the shaft down through the center of the distributor housing.
10. Slide the tang washer, washer and driven gear onto the distributor shaft.
11. Now install the rotor temporarily and align any and all scribe or paint marks. Once in alignment, install the roll pin through the gear and shaft. Spin the shaft to ensure that it is unimpeded.
12. Align the keyways in the shaft and then reinstall the rotor. If the engine is at the No. 1 cylinder, the tang must be aligned between the two marks as shown in the illustration.
13. Install the sensor, cap and distributor as detailed previously.

Ignition Coil

REMOVAL & INSTALLATION

◆ See Figure 106

1. Disconnect the battery cables.
2. Tag and disconnect the coil and ignition module wiring harnesses at the coil.
3. Disconnect the high tension lead from the distributor.
4. Remove the 2 coil bracket mounting bolts and lift off the assembly.
5. Remove the 2 coil mounting bolts and remove the coil.

To install:

6. Position the coil assembly on the bracket and tighten the bolts securely.

7. Position the bracket and coil on the engine and tighten the bolts securely.
8. Coat the outside of the coil lead tower with silicone dielectric compound and then press the high tension lead onto the coil.

** CAUTION

Do not allow any dielectric compound to get into the center of the coil tower.

9. Reconnect the 2 electrical connectors and the battery cables.

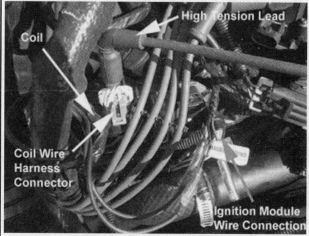
Fig. 106 A nice shot of the coil and its various connections

DISTRIBUTORLESS ELECTRONIC IGNITION (DIS) SYSTEM - PCM 555

General Information

■ This system is used on all 8.1L MPI engines and is completely integrated with the PCM 555 MPI fuel system. As the name implies, there is no distributor, simply 8 coil packs mounted on the cylinder head covers. With the exception of main component service (coils), all aspects of ignition control are covered by the PCM and therefore detailed within the appropriate Fuel Injection system sections.

Ignition Coil

REMOVAL & INSTALLATION

◆ See Figure 107

1. Disconnect the battery cables.
2. Tag and disconnect the coil electrical lead at the coil.
3. Disconnect the high tension lead from the spark plug.
4. Remove the 2 mounting bolts and lift off the ignition coil.
5. Position the coil on the cylinder head cover and tighten the mounting bolts to 106 inch lbs. (12 Nm).
6. Connect the electrical lead, spark plug lead and the battery cables.

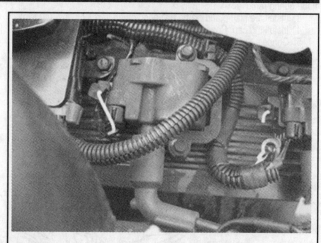
Fig. 107 The 8.1L engine uses coil packs in place of a distributor

IGNITION & ELECTRICAL SYSTEMS

INSTRUMENTS AND GAUGES

Oil And Temperature Gauges

TROUBLESHOOTING

◆ See Figures 108 and 109

The body of oil and temperature gauges must be grounded and they must be supplied with 12 volts. Many gauges have a terminal on the mounting bracket for attaching a ground wire. A tang from the mounting bracket makes contact with the gauge. Check to be sure the tang does make good contact with the gauge.

Ground the wire to the sending unit and the needle of the gauge should move to the full right position indicating the gauge is in serviceable condition.

Check the sender unit for a defective temperature warning system.

If a problem arises on a boat equipped with water, temperature, and oil pressure lights, the first area to check is the light assembly for loose wires or burned-out bulbs.

When the ignition key is turned to the **ON** position, the light assembly is supplied with 12 volts and grounded through the sending unit mounted on the engine. When the sending unit makes contact because the water temperature is too hot or the oil pressure is too low, the circuit to ground is completed and the lamp should light.

To check the bulb: turn the ignition switch to the **ON** position. Disconnect the wire at the sending unit, and then ground the wire. The lamp on the dash should light. If it does not light, check for a burned-out bulb or a break in the wiring to the light.

■ Certain Commodore and International Series gauges can supply power to the bulb via the ignition switch or and alternate lighting switch. These gauges can be identified by the lack of a contact strip on the left side of the rear of the gauge.

TESTING

◆ See Figure 110

1. With the ignition switch in the **OFF** position, remove the wire to the sending unit (**S** or **SEND**) terminal.
2. Turn the ignition switch to the **RUN** position and confirm that the gauge needle is seated on the post at the left side; or all the way to the left of the scale if there is no post.
3. Turn the ignition switch OFF and connect a jumper wire between the ground terminal (**GND, G, or -**) and the sending unit terminal.
4. Turn the ignition switch back to the **RUN** position and confirm that the gauge needle is seated on the post at the right side or all the way to the right of the scale if there is no post.
5. Replace the gauge if anything fails.

REMOVAL & INSTALLATION

1. Disconnect the battery cables.
2. Tag and disconnect the electrical leads at the back of the gauge.
3. Disconnect the light socket, remove the holding strap and lift out the gauge.

To install:

4. Position the gauge into the mounting hole, install the strap and tighten the nuts securely.

■ Be careful not to tighten the holding strap too tightly or you risk distorting the gauge casing.

5. Connect the ground wire and then connect the remaining leads.
6. Install the light socket and then coat all terminal connections with liquid neoprene or equivalent.
7. Connect the battery cables.

Fuel Gauge

TROUBLESHOOTING

In order for the fuel gauge to operate properly the sending unit and the receiving unit must be of the same type and preferably of the same make. The following symptoms and possible corrective actions will be helpful in restoring a faulty fuel gauge circuit to proper operation.

If you suspect the gauge is not operating properly, the first area to check is all electrical connections from one end to the other. Be sure they are clean and tight. Next, check the common ground wire between the negative side of the battery, the fuel tank, and the gauge on the dash.

If all wires and connections in the circuit are in good condition, check the sending unit.

If the pointer does not move from the empty position one of four faults could be to blame:

1. The dash receiving unit is not properly grounded.
2. No voltage at the dash receiving unit.
3. Negative meter connections are on a positive grounded system.
4. Positive meter connections are on a negative grounded system.

If the pointer fails to move from the full position, the problem could be one of three faults.

5. The tank sending unit is not properly grounded.
6. Improper connection between the tank sending unit and the receiving unit on the dash.
7. The wire from the gauge to the ignition switch is connected at the wrong terminal.

If the pointer remains at the 3/4 full mark, it indicates a six-volt gauge is installed in a 12-volt system.

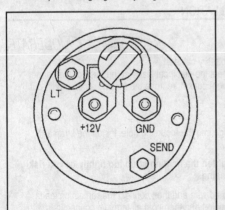

Fig. 108 Rear of a typical QSI Series gauge

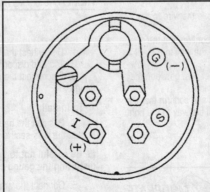

Fig. 109 Commodore and International Series gauges are slightly different\

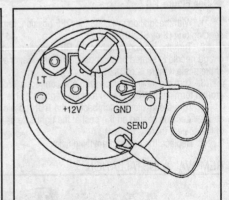

Fig. 110 Testing the oil, fuel and temperature gauges

9-38 IGNITION & ELECTRICAL SYSTEMS

If the pointer remains at about 3/8 full, it indicates a 12-volt gauge is installed in a six-volt system.

Erratic Fuel Gauge Readings

Inspect all of the wiring in the circuit for possible damage to the insulation or conductor. Carefully check:
1. Ground connections at the receiving unit on the dash.
2. Harness connector to the dash unit.
3. Body harness connector to the chassis harness.
4. Ground connection from the fuel tank to the trunk floor pan.
5. Feed wire connection at the tank sending unit.

Gauge Always Reads Full - When Ignition Switch Is In ON Position

1. Check the electrical connections at the receiving unit on the dash; the body harness connector to chassis harness connector; and the tank unit connector in the tank.
2. Make a continuity check of the ground wire from the tank to the tank floor pan.
3. Connect a known good tank unit to the tank feed wire and the ground lead. Raise and lower the float and observe the receiving unit on the dash. If the dash unit follows the arm movement, replace the tank sending unit.

Gauge Always Reads Empty - When Ignition Switch Is ON

Disconnect the tank unit feed wire and do not allow the wire terminal to ground. The gauge on the dash should read full.

If Gauge Reads Empty

1. Connect a spare control unit into the control unit harness connector and ground the unit. If the spare unit reads full, the original unit is shorted and must be replaced.
2. A reading of empty indicates a short in the harness between the tank sending unit and the gauge on the control panel.

If Gauge Reads Full

1. Connect a known good tank sending unit to the tank feed wire and the ground lead.
2. Raise and lower the float while observing the gauge on the control panel. If the control panel gauge follows movement of the float, replace the tank sending unit.

Gauge Never Reads Full

This test requires shop test equipment.
1. Disconnect the feed wire to the tank unit and connect the wire to ground thru a variable resistor or thru a spare tank unit.
2. Observe the control panel gauge reading. The reading should be full when resistance is increased to about 90 ohms. This resistance would simulate a full tank.
3. If the check indicates the control panel gauge is operating properly, the trouble is either in the tank sending unit rheostat being shorter, or the float is binding. The arm could be bent, or the tank may be deformed. Inspect and correct the problem.

TESTING

◆ See Figure 110

1. With the ignition switch in the **OFF** position, remove the wire to the sending unit (**S** or **SEND**) terminal.
2. Turn the ignition switch to the **RUN** position and confirm that the gauge needle is seated on the post at the left side; or all the way to the left of the scale if there is no post.
3. Turn the ignition switch OFF and connect a jumper wire between the ground terminal (**GND, G, or -**) and the sending unit terminal.
4. Turn the ignition switch back to the **RUN** position and confirm that the gauge needle is seated on the post at the right side or all the way to the right of the scale if there is no post.
5. Replace the gauge if anything fails.

REMOVAL & INSTALLATION

1. Disconnect the battery cables.
2. Tag and disconnect the electrical leads at the back of the gauge.
3. Disconnect the light socket, remove the holding strap and lift out the gauge.

To install:

4. Position the gauge into the mounting hole, install the strap and tighten the nuts securely.

■ **Be careful not to tighten the holding strap too tightly or you risk distorting the gauge casing.**

5. Connect the ground wire and then connect the remaining leads.
6. Install the light socket and then coat all terminal connections with liquid neoprene or equivalent.
7. Connect the battery cables.

Battery Gauge

TESTING

◆ See Figure 111

1. Make sure that the battery is fully charged and then disconnect the cables.
2. Turn the ignition switch to the **RUN** position. The gauge should read battery voltage, otherwise replace it.
3. If the gauge is out of the dash, turn the ignition switch OFF and connect a jumper wire between the ground terminal (**GND, G, or -**) and the negative battery terminal. Connect another jumper between the power terminal (**I or +12V**) and the positive battery terminal. The gauge should register battery voltage, otherwise replace it.

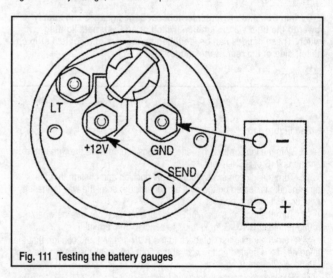

Fig. 111 Testing the battery gauges

REMOVAL & INSTALLATION

1. Disconnect the battery cables.
2. Tag and disconnect the electrical leads at the back of the gauge.
3. Disconnect the light socket, remove the holding strap and lift out the gauge.

To install:

4. Position the gauge into the mounting hole, install the strap and tighten the nuts securely.

■ **Be careful not to tighten the holding strap too tightly or you risk distorting the gauge casing.**

5. Connect the ground wire and then connect the remaining leads.
6. Install the light socket and then coat all terminal connections with liquid neoprene or equivalent.
7. Connect the battery cables.

IGNITION & ELECTRICAL SYSTEMS   9-39

CruiseLog Meter

TESTING

◆ See Figure 112

Turn the ignition switch to the **RUN** position. The run indicator dial should be turning; if not, replace the gauge.

If the gauge is out of the dash, connect a jumper wire between terminal **A** and the positive battery cable. Connect another jumper between terminal **B** and the negative battery cable. The run indicator dial should be turning; if not, replace the gauge.

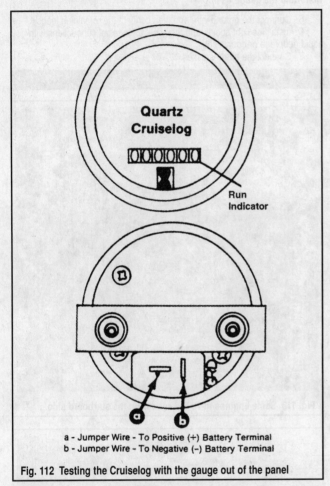

a - Jumper Wire - To Positive (+) Battery Terminal
b - Jumper Wire - To Negative (-) Battery Terminal

Fig. 112 Testing the Cruiselog with the gauge out of the panel

REMOVAL & INSTALLATION

1. Disconnect the battery cables.
2. Tag and disconnect the electrical leads at the back of the gauge.
3. Disconnect the light socket, remove the holding strap and lift out the gauge.

To install:

4. Position the gauge into the mounting hole, install the strap and tighten the nuts securely.

■ Be careful not to tighten the holding strap too tightly or you risk distorting the gauge casing.

5. Connect the ground wire and then connect the remaining leads.
6. Install the light socket and then coat all terminal connections with liquid neoprene or equivalent.
7. Connect the battery cables.

Speedometer

TESTING

■ An air compressor is necessary for this procedure. It is imperative that its pressure gauge is extremely accurate.

Disconnect the hose at the back of the gauge and then connect a line from an air compressor. Apply 5.3 psi of air pressure to the speedometer and check that the gauge reads 20 +/- 1 mph (32 +/- 1.6 kph). Now apply 27.8 psi pressure and check that the gauge reads 45 +/- 1 mph (72 +/- 1 kph). If either of the readings is not as specified, replace the speedometer.

REMOVAL & INSTALLATION

1. Disconnect the battery cables.
2. Tag and disconnect the electrical leads and the hose at the back of the gauge.
3. Disconnect the light socket, remove the holding strap and lift out the gauge.

To install:

4. Position the gauge into the mounting hole, install the strap and tighten the
nuts securely.

■ Be careful not to tighten the holding strap too tightly or you risk distorting the gauge casing.

5. Connect the ground wire and then connect the remaining leads. Connect the hose.
6. Install the light socket and then coat all terminal connections with liquid neoprene or equivalent.
7. Connect the battery cables.

Tachometer

TESTING

Connect a tachometer as per the manufacturer's instructions. Start the engine and check a few different engine speeds on the boat's tachometer against the service tach. If using a 6000 rpm service tach, variations of +/- 150 rpm are acceptable; if using an 8000 rpm tach, variations of +/- 200 rpm are acceptable. Replace the tachometer if not within specifications.

REMOVAL & INSTALLATION

8. Disconnect the battery cables.
9. Tag and disconnect the electrical leads at the back of the gauge.
10. Disconnect the light socket, remove the holding strap and lift out the gauge.

To install:

11. Position the gauge into the mounting hole, install the strap and tighten the nuts securely.

■ Be careful not to tighten the holding strap too tightly or you risk distorting the gauge casing.

12. Connect the ground wire and then connect the remaining leads. .
13. Install the light socket and then coat all terminal connections with liquid neoprene or equivalent.
14. Connect the battery cables.

9-40 IGNITION & ELECTRICAL SYSTEMS

Vacuum Gauge

TESTING

1. Disconnect the vacuum hose from the engine.

2. Connect a service vacuum gauge to the engine. Start the engine and then record the vacuum readings at 1000, 2000 and 3000 rpm. Shut off the engine and remove the gauge.

3. Reconnect the engine vacuum line and restart the engine. Compare the readings on the helm gauge at each of the engine speeds with those taken with the external gauge. If any of the readings are not within 3 inch Hg. of the recorded readings, then you will have to replace the gauge.

■ Obviously, before replacing the gauge, make sure that the vacuum line connections are secure and that the line has no leaks.

SENDING UNITS AND SWITCHES

■ Mercury does not provide service or testing information on sending units or switches for any MPI engines covered within except the 2001 350 Mag and 6.2L. That being said, we believe the following procedures are correct for all models, including the MPI engines.

Oil Pressure Sending Unit

GENERAL INFORMATION

◆ See Figures 113 and 114

The oil pressure sending unit is located on the starboard side of early 3.0L engine blocks, above the starter motor; and on the lower rear port side of the block on most other models. It is readily identifiable, so please refer to the accompanying illustrations if you are not certain where it is on your engine.

TESTING

◆ See Figures 115 and 116

■ This test should be performed only after confirming that the oil pressure gauge is operating properly.

Connect a multi-meter as per the manufacturer's instructions.

Disconnect the electrical lead at the sender terminal. Connect the positive lead of the meter to the sender terminal and the negative lead to the hex nut on the back of the sender housing.

1. With the engine not running, the meter should register continuity as shown in the accompanying chart.

Start the engine and check the sender at the different oil pressure readings as shown in the accompanying chart. If any readings vary, replace the unit.

■ Dual station units can be identified by 353-AM stamped on the hex nut of the sender.

Water Temperature Sender

GENERAL INFORMATION

◆ See Figures 117 and 118

The water temperature sending unit is located in the thermostat housing. It may be on the front of the housing (3.0L), or on either side (all other engines), but it should always be identifiable by the tan wire connected to it.

REMOVAL & INSTALLATION

1. Disconnect the battery cables.
2. Tag and disconnect the electrical leads at the back of the gauge.
3. Disconnect the light socket, remove the holding strap and lift out the gauge.

To install:

4. Position the gauge into the mounting hole, install the strap and tighten the nuts securely.

■ Be careful not to tighten the holding strap too tightly or you risk distorting the gauge casing.

5. Connect the ground wire and then connect the remaining leads.
6. Install the light socket and then coat all terminal connections with liquid neoprene or equivalent.
7. Connect the battery cables.

Fig. 113 Some engines have the sender on the starboard side...

Fig. 114 ...but on most it will be on the port side

IGNITION & ELECTRICAL SYSTEMS   9-41

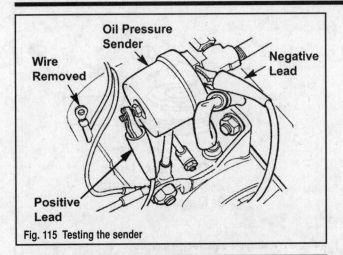

Fig. 115 Testing the sender

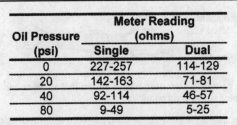

Oil Pressure (psi)	Meter Reading (ohms)	
	Single	Dual
0	227-257	114-129
20	142-163	71-81
40	92-114	46-57
80	9-49	5-25

Fig. 116 Oil pressure sender unit test specifications

TESTING

◆ See Figure 119

1. Drain the cooling system. Disconnect the electrical lead and remove the switch from the thermostat housing.
2. Connect a multi-meter to the switch with the positive lead on the terminal and the negative lead on the hex. Carefully immerse it a pan of water. Heat the water and observe that the readings are as follows with an accurate thermometer:
   - 140°F (60°C): 121-147 ohms
   - 194°F (90°C): 47-55 ohms
   - 212°F (100°C): 36-41 ohms
3. Turn the heat off and allow the water to cool, checking the readings again as it cools. If outside the parameters, replace the sensor.
4. Install the sender after coating the threads with Loctite Pipe Sealant and then tighten it securely.

■ Dual station units can be identified by 362-BC stamped on the hex nut of the sender.

Fuel Tank Sending Unit

TESTING

Flange Type

◆ See Figure 120

1. Disconnect the electrical lead at the sender (in the tank) and loosen the mounting screw with the ground wire attached; remove the ground wire.
2. Remove the remaining mounting screws and lift the sender out of the fuel tank - be careful not to damage the float arm.
3. Connect a multi-meter as per the manufacturer's instructions. Connect the positive lead to the terminal on the sender and the negative lead to the flange on the housing. Move the float arm to the full position (arm is horizontal) and check that the meter reads 30 ohms (+/- 5).
4. Move the float arm to the empty position (arm is vertical) and check that the meter reads 240 ohms (+/- 5).
5. Replace the sender unit if it is outside either of the specifications.

Fig. 117 The sender will either be on the front of the thermostat housing. . .

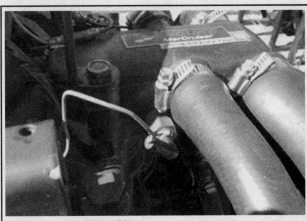

Fig. 118 . . .or on either side

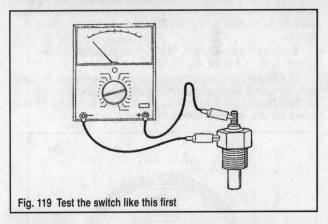

Fig. 119 Test the switch like this first

Capsule Type

◆ See Figure 121

1. Disconnect the electrical lead at the sender unit.
2. Remove the two mounting screws and carefully lift the unit from the fuel tank.
3. Position a large magnet under the bottom side of the unit.
4. Connect a multi-meter as per the manufacturer's instructions. Connect the positive lead to the sender terminal and the negative lead to the metal portion of the unit housing. Rotate the magnet counterclockwise until the gauge reads empty. The meter should show 240 ohms (+/- 5).
5. Rotate the magnet the other way until the gauge reads FULL. The meter should read 30 ohms (+/- 5).
6. If the unit is not within specifications on either test, replace it.

9-42 IGNITION & ELECTRICAL SYSTEMS

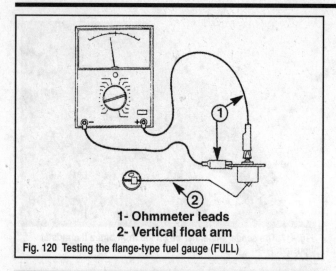

Fig. 120 Testing the flange-type fuel gauge (FULL)

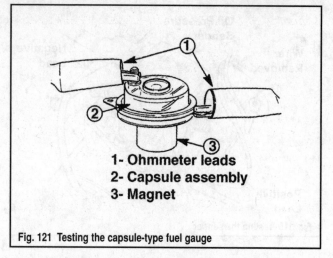

Fig. 121 Testing the capsule-type fuel gauge

Ignition Switch

TESTING

3.0L Engines

◆ See Figure 122

■ Before performing this test, please ensure that all fuses and the starter motor are in good condition.

1. Disconnect the battery cables and connect a multi-meter as per the manufacturer's instructions.
2. With the switch in the **OFF** position, test for continuity across all of the terminals on the back of the switch - there should be none.
3. Move the ignition switch to the **RUN** position. Check that there is continuity between terminals **A** and **B** and between terminals **B** and **I**. There should be no continuity between terminal **C** and any of the other three terminals.
4. Move the ignition switch to **START**. There should be continuity between terminals **A** and **B**, **B** and **I** and between terminals **B** and **C**.
5. If the switch fails any of the tests, unsolder the wire connections and remove the actual switch and repeat Steps 2-4. There should be no continuity between any of the terminals with the switch removed. If the switch passes this time, you've got a bad wiring harness.

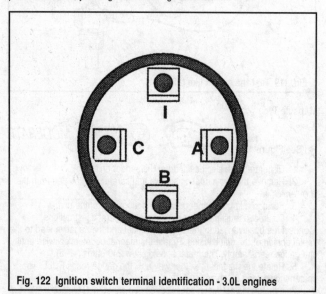

Fig. 122 Ignition switch terminal identification - 3.0L engines

All Other Engines

◆ See Figure 123

■ Before performing this test, please ensure that all fuses and the starter motor are in good condition.

1. Disconnect the battery cables and connect a multi-meter as per the manufacturer's instructions.
2. With the switch in the **OFF** position, test for continuity across all of the terminals on the back of the switch - there should be none.
3. Move the ignition switch to the **RUN** position. Check that there is continuity between terminals **B** and **I**. There should be no continuity between terminal **S** and any of the other three terminals.
4. Move the ignition switch to **START**. There should be continuity between terminals **I** and **B** and between terminals **B** and **S**.
5. The terminal should all make contact at the angle as shown in the accompanying illustration and then stay in contact as the switch is moved toward the **START** position.
6. If the switch fails any of the tests, unsolder the wire connections and remove the actual switch and repeat Steps 2-4. There should be no continuity between any of the terminals with the switch removed. If the switch passes this time, you've got a bad wiring harness.

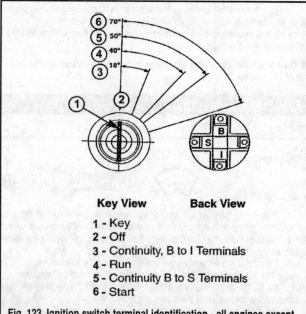

Fig. 123 Ignition switch terminal identification - all engines except the 3.0L

IGNITION & ELECTRICAL SYSTEMS 9-43

Emergency (Lanyard) Stop Switch

TESTING

These switches can be either panel-mounted or remote control-mounted, but the test is the same.

Disconnect the switch leads and connect a multi-meter across each lead (Black/Yellow on 4.3L and 5.0/5.7/6.2L EFI; Purple on 5.0/5.7L w/Carb). There should be continuity with the switch lanyard connected and no continuity with it disconnected.

Start/Stop Switch

TESTING

◆ See Figure 124

1. Disconnect the battery cables and connect a multi-meter as per the manufacturer's instructions.

2. Check for continuity between the start switch terminals - there should be none. Press the button and you should see continuity.

3. Now check the stop switch in the same manner. If either switch fails either test, replace the switch.

Audio Warning System

TESTING

◆ See Figure 125

1. Turn the ignition switch to the **RUN** position. The buzzer should sound and then turn off.

2. If the buzzer does not sound, disconnect the Tan/Blue wire at the back of the system and touch it to a known ground. If the buzzer sounds, you have a problem in the wire; if not, replace the unit.

Oil Pressure Switch

TESTING

Non-TKS Models

◆ See Figure 126

■ This test should be performed only after confirming that the oil pressure gauge is operating properly.

Connect a multi-meter as per the manufacturer's instructions. Connect the positive lead of the meter to the sender terminal and the negative lead to the hex nut on the back of the sender housing. With the engine Off, the meter should indicate full continuity. With the engine running and the oil pressure above 6 psi, the meter should register no continuity. If it does, replace the switch.

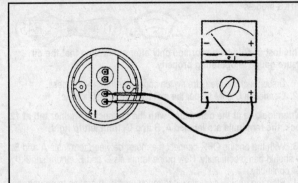

Fig. 124 Ignition switch terminal identification - all engines except the 3.0L

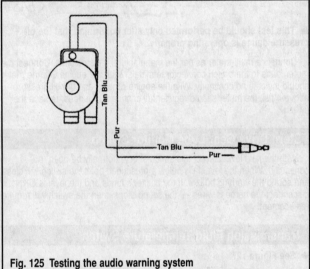

Fig. 125 Testing the audio warning system

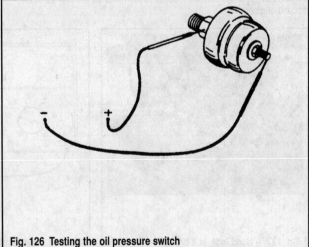

Fig. 126 Testing the oil pressure switch

9-44 IGNITION & ELECTRICAL SYSTEMS

3.0L TKS Models

■ This test should be performed only after confirming that the oil pressure gauge is operating properly.

1. Disconnect the pressure switch connector from the harness.
2. Connect a DVOM as per the manufacturer's instructions.

■ When looking at the connector, with the connector retainer tab at 12 o'clock, the terminals are labeled A, B and C, from left to right.

3. With the engine OFF, connect the meter between terminals **A** and **B** - there should be no continuity. Now probe terminals **C** and **B** - there should be full continuity.
4. Start the engine and allow it to reach normal operating temperature. With the oil pressure above 4 psi, the meter should show full continuity when across terminals **A** and **B**. When probing terminals **C** and **B**, there should be no continuity.
5. Replace the switch if any readings are bad.

V6 And V8 TKS Models

■ This test should be performed only after confirming that the oil pressure gauge is operating properly.

Connect a multi-meter as per the manufacturer's instructions. Connect the meter leads to the switch connector terminals. With the engine Off, the meter should indicate no continuity. With the engine running and the oil pressure above 4 psi, the meter should register full continuity. If it does, replace the switch.

Stern Drive Gear Lube Monitor Switch

With the fluid level correct this switch should normally be open (no continuity). When the level falls below a prescribed point the switch will close and sound the warning buzzer. If the buzzer sounds and the level is OK, disconnect the harness wires - if the sound stops, then the switch will require replacement.

Transmission Fluid Temperature Switch

◆ See Figure 127

These switches are normally open until the engine reaches a certain temperature, at which time they close. These switches are only used on inboard engines.

TESTING

◆ See Figures 128 and 129

1. Disconnect the electrical lead and remove the switch from the transmission.
2. Connect a multi-meter to the switch (each terminal on the top) and check that there is no continuity. If there is, replace the switch; if not, proceed to the next step.
3. With the meter still attached, carefully immerse the switch in a pan of water. Heat the water and verify with an accurate thermometer that the switch closes (continuity present) at 220°-240°F (104°-116°C).
4. Turn off the heating element and allow the container to cool. The switch should open again when it reaches 180°-200°F (82°-93°C).
5. Replace the switch if it fails either test.

Water Temperature Switch

◆ See Figures 117 and 118

This switches is normally open until the engine reaches a certain temperature, at which time it closes.

The water temperature switch is located in the thermostat housing. It may be on the front of the housing (3.0L), or on either side (all other engines), but it should always be identifiable by the tan wire connected to it.

TESTING

◆ See Figure 119

1. Disconnect the electrical lead and remove the switch from the thermostat housing.
2. Connect a multi-meter to the switch and carefully immerse it a pan of water. Heat the water and use an accurate thermometer to verify that the switch closes at 190°-200°F (88°-93°C) on #48592 switches (red sleeve); or at 215°-225°F (102°-107°C) on #87-86080 switches (black sleeve).
3. Turn off the heating element and allow the container to cool. The switch should open again when it reaches 150°-170°F (66°-77°C) on #48952 switches; or at 175°-195°F (79°-91°C) on #87-86080 switches (black sleeve).
4. Replace the switch if it fails either test.

Fig. 127 A good look at a typical transmission fluid temperature switch

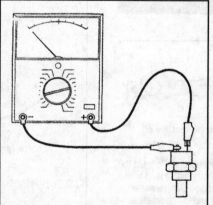

Fig. 128 First test the transmission fluid switch like this...

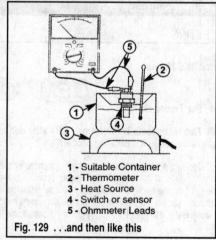

1 - Suitable Container
2 - Thermometer
3 - Heat Source
4 - Switch or sensor
5 - Ohmmeter Leads

Fig. 129 ...and then like this

IGNITION & ELECTRICAL SYSTEMS

WIRING DIAGRAMS

3.0L ENGINES

- ENGINE - W/CIRCUIT BREAKER 9-46
- ENGINE - W/O CIRCUIT BREAKER 9-46
- INSTRUMENTATION - SINGLE STATION 9-47
- INSTRUMENTATION - DUAL STATION
- USING NEUTRAL SAFETY SWITCH IN ONE REMOTE CONTROL ... 9-48
- INSTRUMENTATION - DUAL STATION
- USING NEUTRAL SAFETY SWITCH
  IN BOTH REMOTE CONTROLS 9-48

4.3L V6 ENGINES

- ENGINE - W/CARBURETOR, ALPHA 9-49
- ENGINE - W/CARBURETOR, BRAVO 9-49
- ENGINE - W/TBI (MEFI 1) 9-50
- ENGINE - W/TBI (MEFI 3) 9-50
- ENGINE - W/MPI 9-51
- INSTRUMENTATION - W/CARB OR TBI, SINGLE STATION 9-52
- INSTRUMENTATION - W/MPI, SINGLE STATION 9-52
- INSTRUMENTATION - W/MPI, DUAL STATION USING
  NEUTRAL SAFETY SWITCH IN ONE REMOTE CONTROL 9-53
- INSTRUMENTATION - W/MPI, DUAL STATION USING
  NEUTRAL SAFETY SWITCH IN BOTH REMOTE CONTROLS 9-53
- INSTRUMENTATION - W/MPI, DUAL STATION USING
  NEUTRAL SAFETY SWITCH IN ENGINE WIRING HARNESS 9-54
- MERCATHODE SYSTEM - W/MPI 9-54
- STARTING CIRCUIT - W/MPI (ECM/PCM 555) 9-63
- WAKE, HORN & TACHOMETER - W/MPI (ECM 555) 9-64
- GEAR INDICATOR & SHIFT INTERRUPT CIRCUIT -
  W/MPI (ECM 555) 9-64
- TRANSOM HARNESS - W/MPI (ECM 555) 9-65
- TRANSOM CONNECTOR (ENG. SIDE) - W/MPI (ECM 555) 9-65
- TRANSOM HARNESS (TRAN. SIDE) - W/MPI (ECM 555) 9-66
- SLAVE SOLENOID - W/MPI (ECM 555) 9-66
- ALTERNATOR OUTPUT CIRCUIT - W/MPI (ECM 555) 9-66
- SEAWATER PUMP CIRCUIT - W/MPI (ECM 555) 9-66
- OIL PRESSURE CIRCUIT - W/MPI (ECM 555) 9-67
- IGNITION SYSTEM - W/MPI (ECM 555) 9-67

5.0L, 5.7L & 6.2L V8 ENGINES

- ENGINE - W/CARBURETOR, ALPHA & BRAVO 9-55
- ENGINE - W/CARBURETOR, INBOARD 9-55
- ENGINE - 2001 W/EFI (MEFI 1/2), UP TO SERIAL #299999 9-56
- ENGINE - 2001 W/EFI (MEFI 3), UP TO SERIAL #299999,
  STERNDRIVE 9-56
- ENGINE - 2001 W/EFI (MEFI 3), UP TO SERIAL #299999,
  INBOARD 9-57
- ENGINE - 2001 AND LATER W/MPI, SERIAL #300000 AND ABOVE,
  STERNDRIVE 9-58
- ENGINE - 2001 AND LATER W/MPI, SERIAL #300000 AND ABOVE,
  INBOARD 9-59
- INSTRUMENTATION - W/CARB OR TBI, AND 2001 MPI
  (UP TO SERIAL #299999), SINGLE STATION 9-62
- INSTRUMENTATION - W/MPI (SERIAL #300000 AND ABOVE),
  SINGLE STATION 9-62
- MERCATHODE SYSTEM WIRING SCHEMATIC - W/MPI 9-63
- STARTING CIRCUIT - W/MPI (ECM/PCM 555) 9-63
- WAKE, HORN & TACHOMETER - W/MPI (ECM 555) 9-64
- GEAR INDICATOR & SHIFT INTERRUPT CIRCUIT - W/MPI (ECM 555) .. 9-64
- TRANSOM HARNESS - W/MPI (ECM 555) 9-65
- TRANSOM CONNECTOR (ENG. SIDE) - W/MPI (ECM 555) 9-65
- TRANSOM HARNESS (TRAN. SIDE) - W/MPI (ECM 555) 9-66
- SLAVE SOLENOID - W/MPI (ECM 555) 9-66
- ALTERNATOR OUTPUT CIRCUIT - W/MPI (ECM 555) 9-66
- SEAWATER PUMP CIRCUIT - W/MPI (ECM 555) 9-66
- OIL PRESSURE CIRCUIT - W/MPI (ECM 555) 9-67
- IGNITION SYSTEM - W/MPI (ECM 555) 9-67

8.1L V8 ENGINES

- ENGINE - W/MPI, MECHANICAL HARNESS 9-60
- ENGINE - W/MPI, INTEGRATED MECHANICAL HARNESS 9-61
- INSTRUMENTATION - W/MPI (SERIAL #300000 AND ABOVE),
  SINGLE STATION 9-62
- MERCATHODE SYSTEM WIRING SCHEMATIC - W/MPI 9-63
- CHARGING HARNESS (PCM 555) 9-68
- COIL HARNESS (PCM 555) 9-69
- IGNITION CIRCUIT (PCM 555) 9-69
- SEAWATER PUMP (PCM 555) 9-70
- OIL PRESSURE CIRCUIT (PCM 555) 9-70
- EXHAUST MANIFOLD COOLANT TEMP CIRCUIT (PCM 555) 9-71
- GEAR LUBE MONITOR/TRANSMISSION
  OVERTEMP CIRCUIT (PCM 555) 9-71
- TRANSOM HARNESS (PCM 555) 9-72
- FUSE CIRCUIT (PCM 555) 9-73
- SMART TRANSOM CIRCUIT (PCM 555) 9-73

9-46 IGNITION & ELECTRICAL SYSTEMS

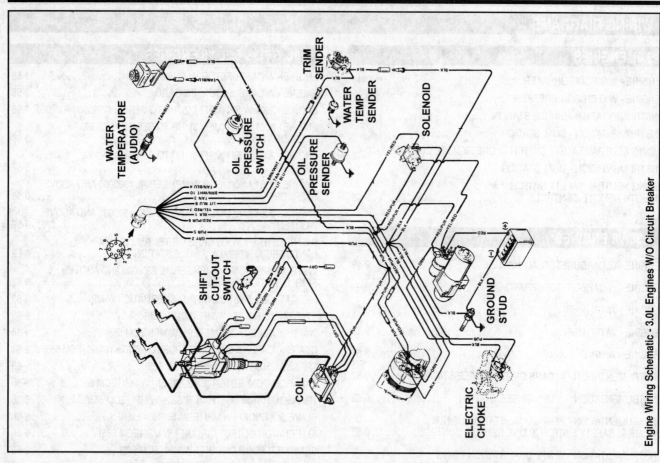

Engine Wiring Schematic - 3.0L Engines W/O Circuit Breaker

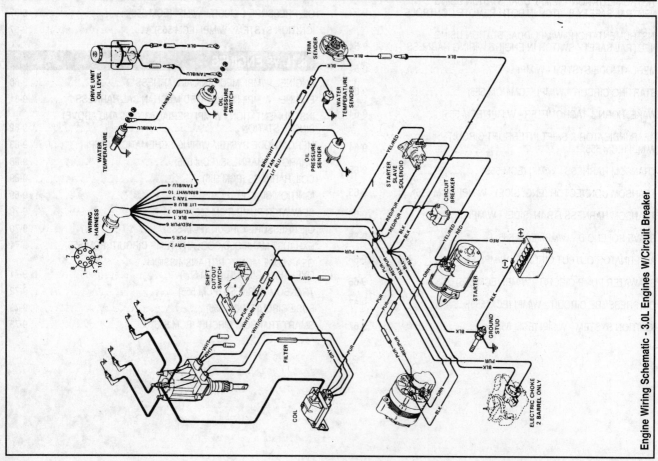

Engine Wiring Schematic - 3.0L Engines W/Circuit Breaker

IGNITION & ELECTRICAL SYSTEMS 9-47

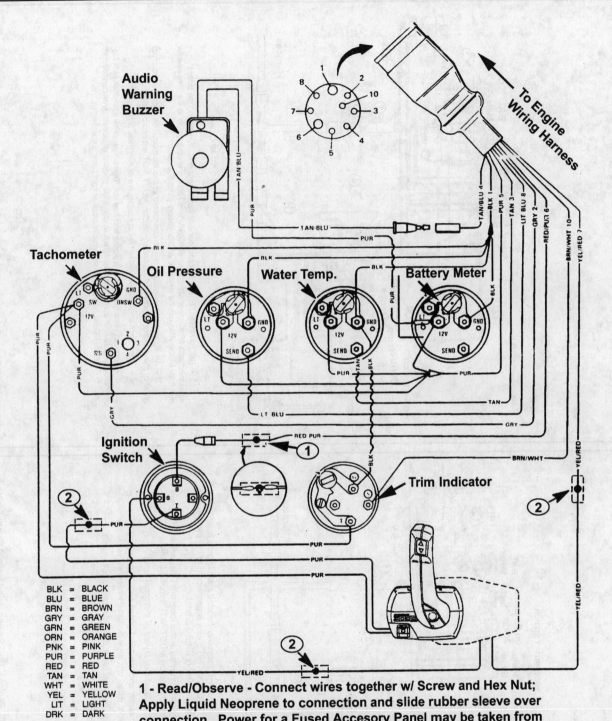

1 - Read/Observe - Connect wires together w/ Screw and Hex Nut; Apply Liquid Neoprene to connection and slide rubber sleeve over connection. Power for a Fused Accesory Panel may be taken from the connection. Load must not exceed 40 Amps. Panel ground wire must be connected to Instrument Terminal that has an 8-gauge Black (Ground) Harness wire conected to it.

2 - Lanyard Top Switch Lead and Neutral Safety Switch Leads must be soldered and covered with shrink tube for a water proof connection. If an alternate method of connection is made, verify connection is secure and sealed for a water proof connection.

Instrumentation Wiring Schematic - 3.0L Engines, Single Station

9-48 IGNITION & ELECTRICAL SYSTEMS

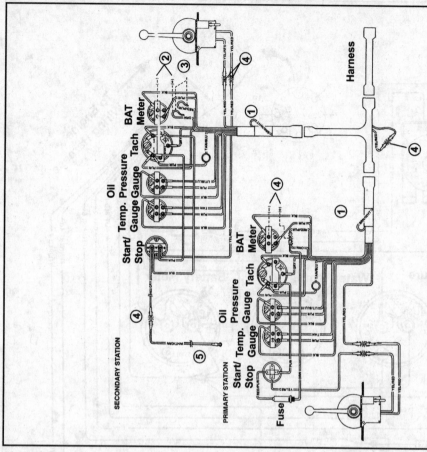

Instrumentation Wiring Schematic - 3.0L Engines, Dual Station Using Neutral Safety Switch in Both Remote Controls

Instrumentation Wiring Schematic - 3.0L Engines, Dual Station Using Neutral Safety Switch in One Remote Control

IGNITION & ELECTRICAL SYSTEMS

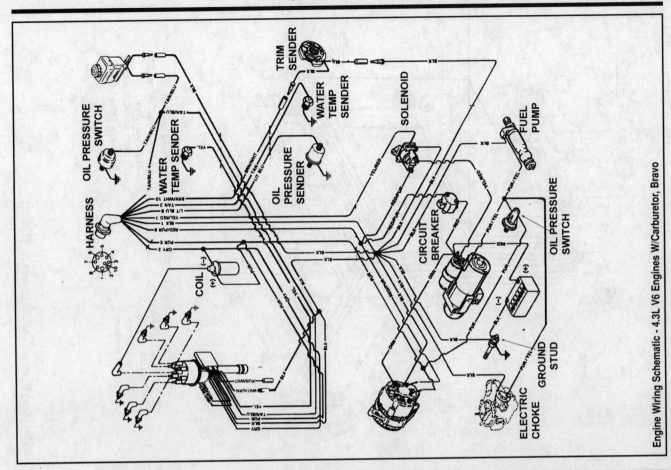

Engine Wiring Schematic - 4.3L V6 Engines W/Carburetor, Bravo

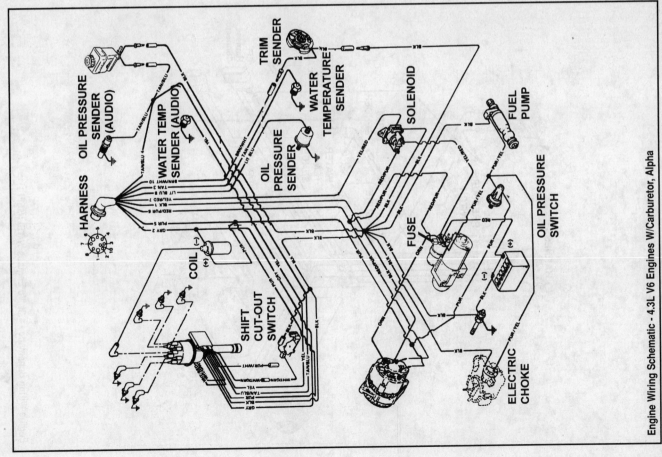

Engine Wiring Schematic - 4.3L V6 Engines W/Carburetor, Alpha

9-50 IGNITION & ELECTRICAL SYSTEMS

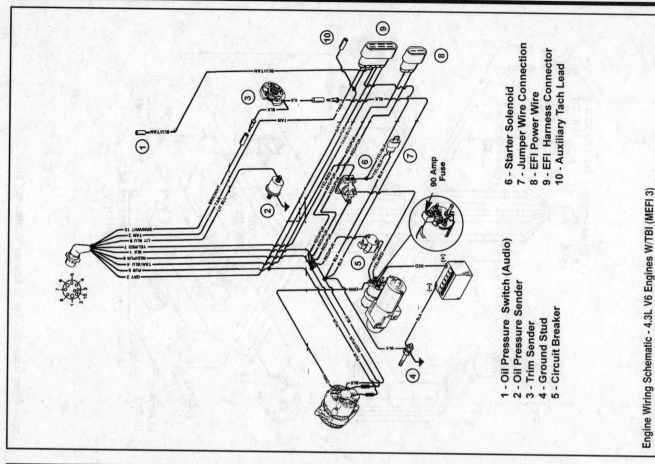

Engine Wiring Schematic - 4.3L V6 Engines W/TBI (MEFI 3)

1 - Oil Pressure Switch (Audio)
2 - Oil Pressure Sender
3 - Trim Sender
4 - Ground Stud
5 - Circuit Breaker
6 - Starter Solenoid
7 - Jumper Wire Connection
8 - EFI Power Wire
9 - EFI Harness Connector
10 - Auxiliary Tach Lead

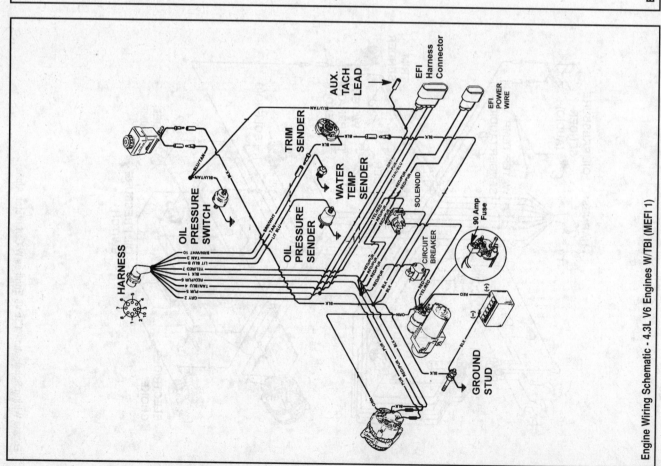

Engine Wiring Schematic - 4.3L V6 Engines W/TBI (MEFI 1)

IGNITION & ELECTRICAL SYSTEMS 9-51

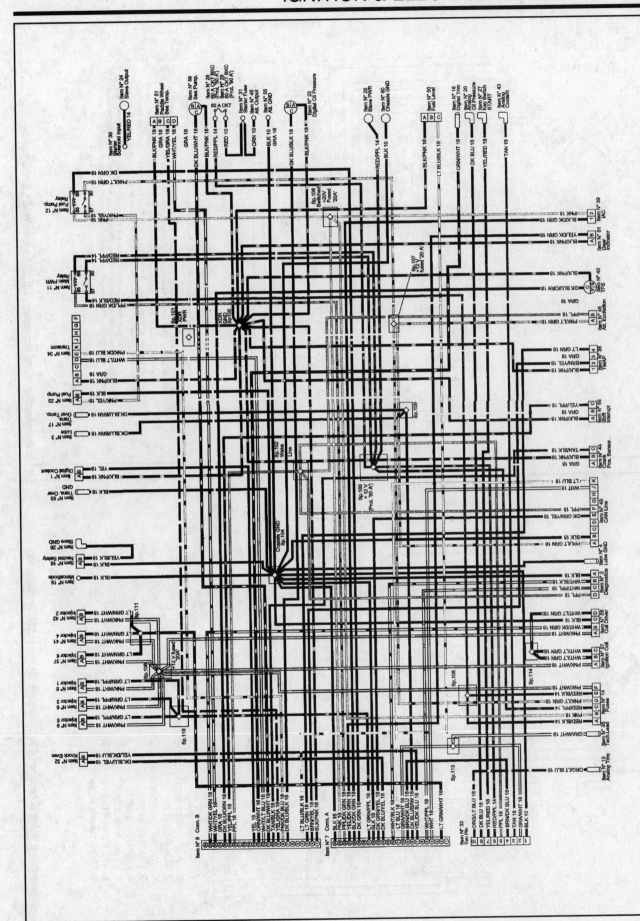

Engine Wiring Schematic - 4.3L V6 Engines W/MPI

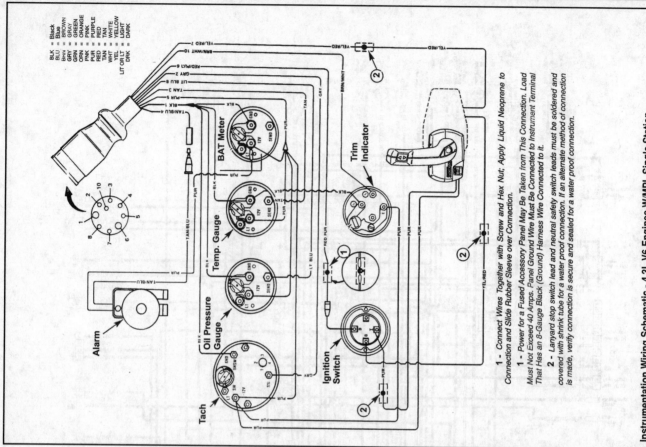

Instrumentation Wiring Schematic - 4.3L V6 Engines W/MPI, Single Station

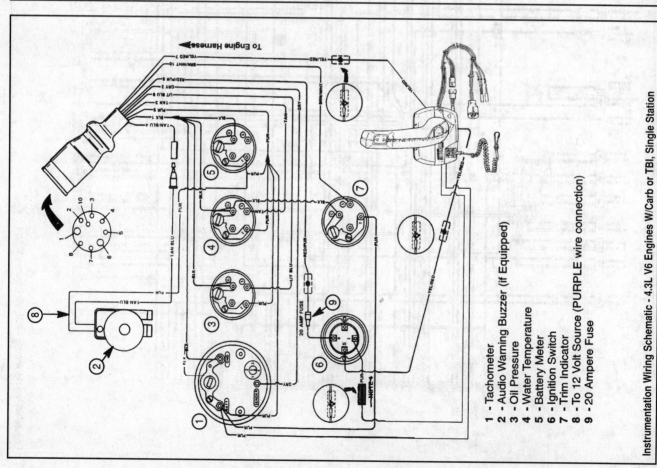

1 - Tachometer
2 - Audio Warning Buzzer (if Equipped)
3 - Oil Pressure
4 - Water Temperature
5 - Battery Meter
6 - Ignition Switch
7 - Trim Indicator
8 - To 12 Volt Source (PURPLE wire connection)
9 - 20 Ampere Fuse

Instrumentation Wiring Schematic - 4.3L V6 Engines W/Carb or TBI, Single Station

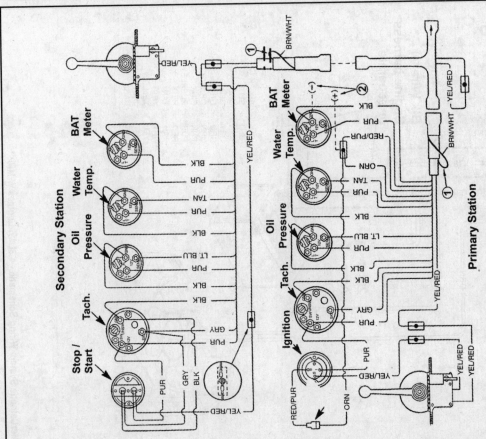

9-54 IGNITION & ELECTRICAL SYSTEMS

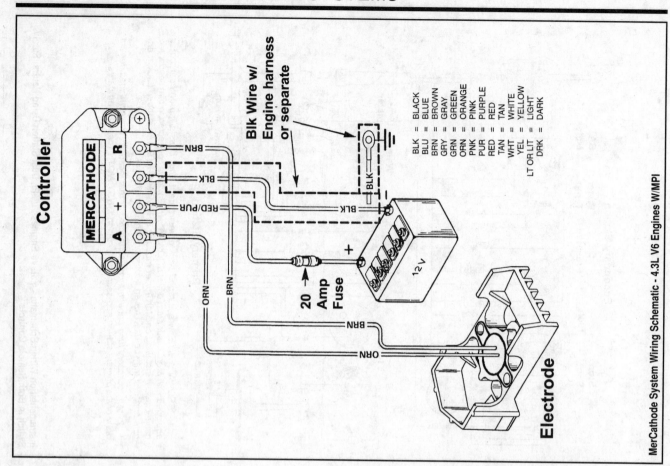

MerCathode System Wiring Schematic - 4.3L V6 Engines W/MPI

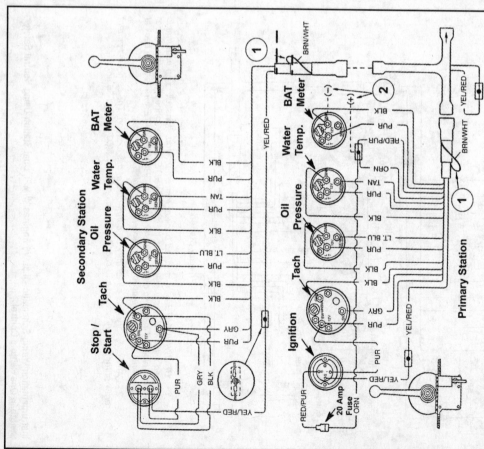

Instrumentation Wiring Schematic - 4.3L V6 Engines W/MPI, Dual Station Using Neutral Safety Switch in Engine Wiring Harness

IGNITION & ELECTRICAL SYSTEMS 9-55

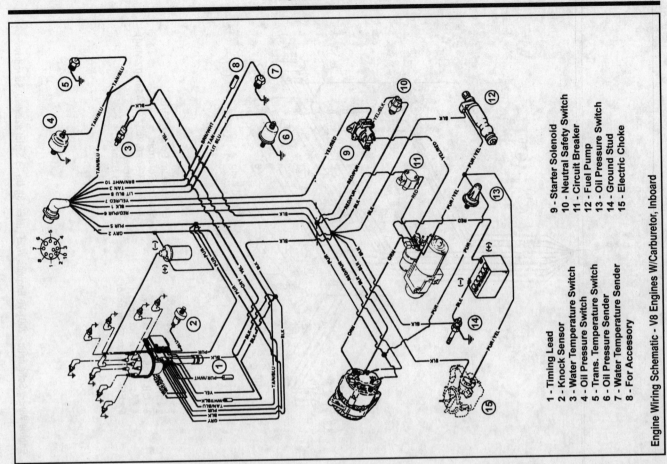

Engine Wiring Schematic - V8 Engines W/Carburetor, Inboard

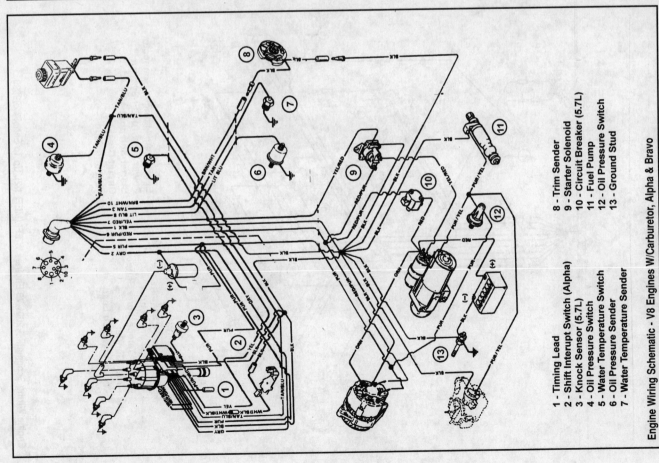

Engine Wiring Schematic - V8 Engines W/Carburetor, Alpha & Bravo

9-56 IGNITION & ELECTRICAL SYSTEMS

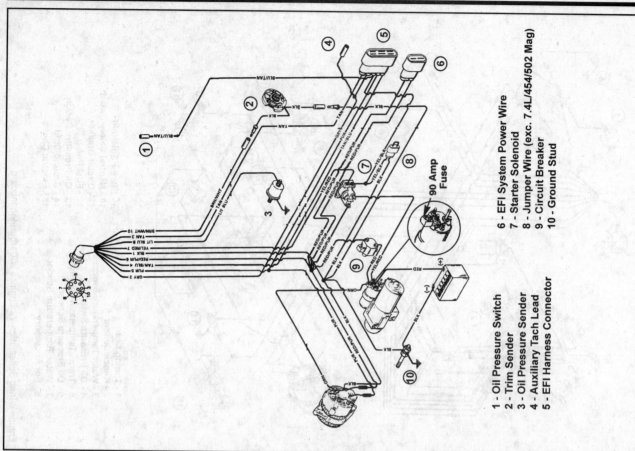

Engine Wiring Schematic - 2001 V8 Engines W/EFI (MEFI 3), Up To Serial #299999, Sterndrive

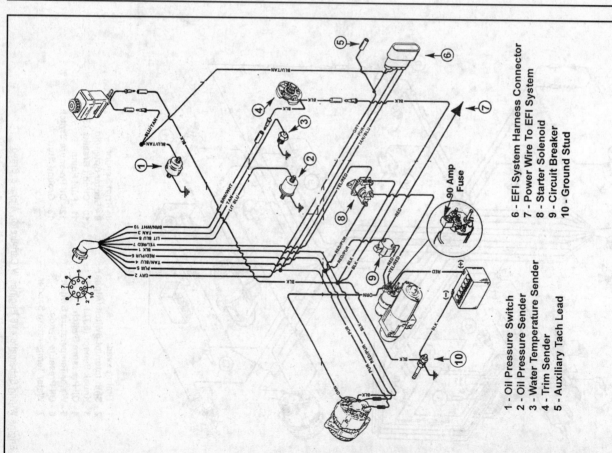

Engine Wiring Schematic - 2001 V8 Engines W/EFI (MEFI 1/2), Up To Serial #299999

IGNITION & ELECTRICAL SYSTEMS 9-57

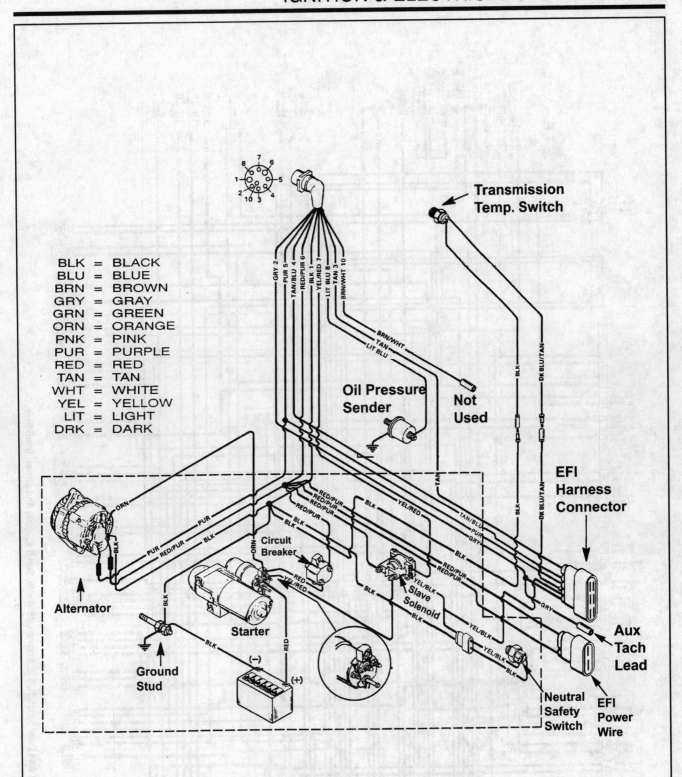

Engine Wiring Schematic - 2001 V8 Engines W/EFI (MEFI 3), Up To Serial #299999, Inboard

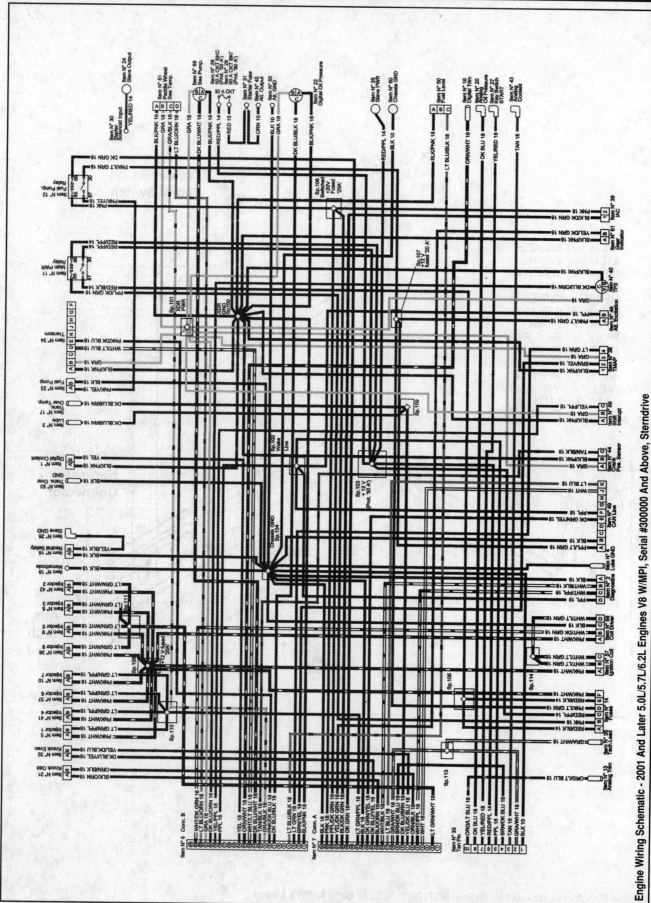

Engine Wiring Schematic - 2001 And Later 5.0L/5.7L/6.2L Engines V8 W/MPI, Serial #300000 And Above, Sterndrive

IGNITION & ELECTRICAL SYSTEMS 9-59

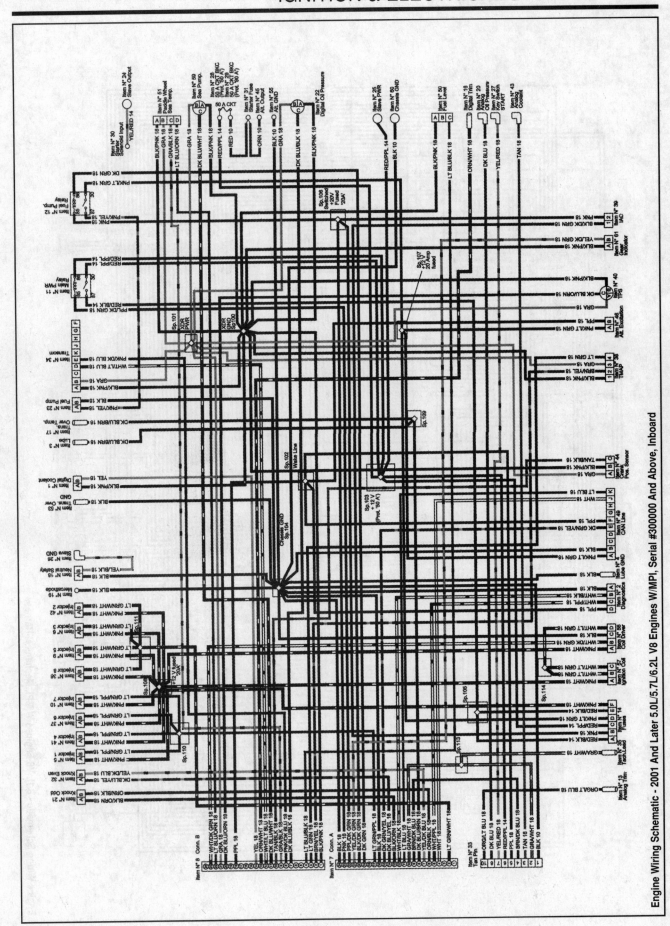

Engine Wiring Schematic - 2001 And Later 5.0L/5.7L/6.2L V8 Engines W/MPI, Serial #300000 And Above, Inboard

9-60 IGNITION & ELECTRICAL SYSTEMS

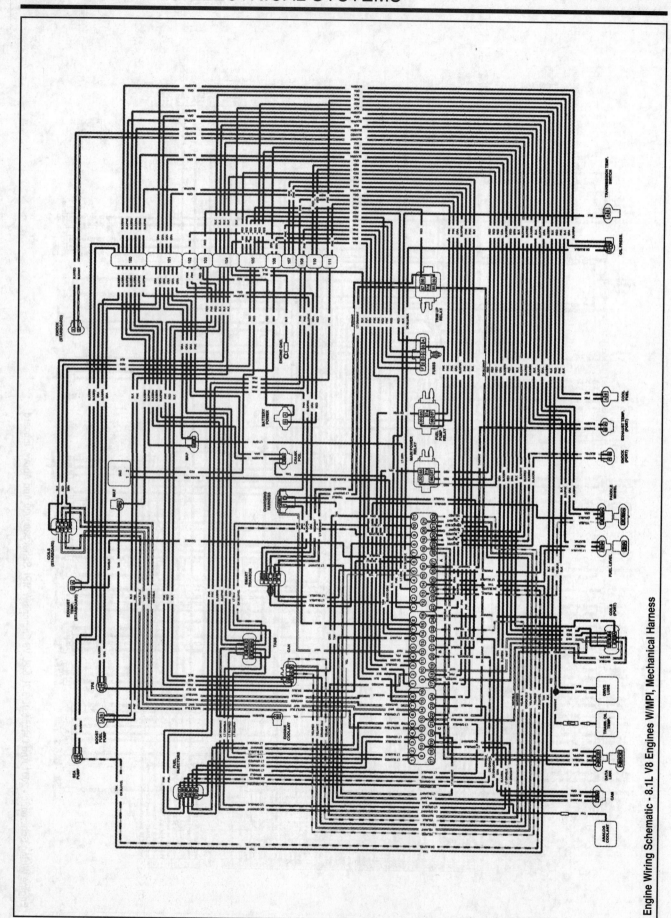

Engine Wiring Schematic - 8.1L V8 Engines W/MPI, Mechanical Harness

IGNITION & ELECTRICAL SYSTEMS 9-61

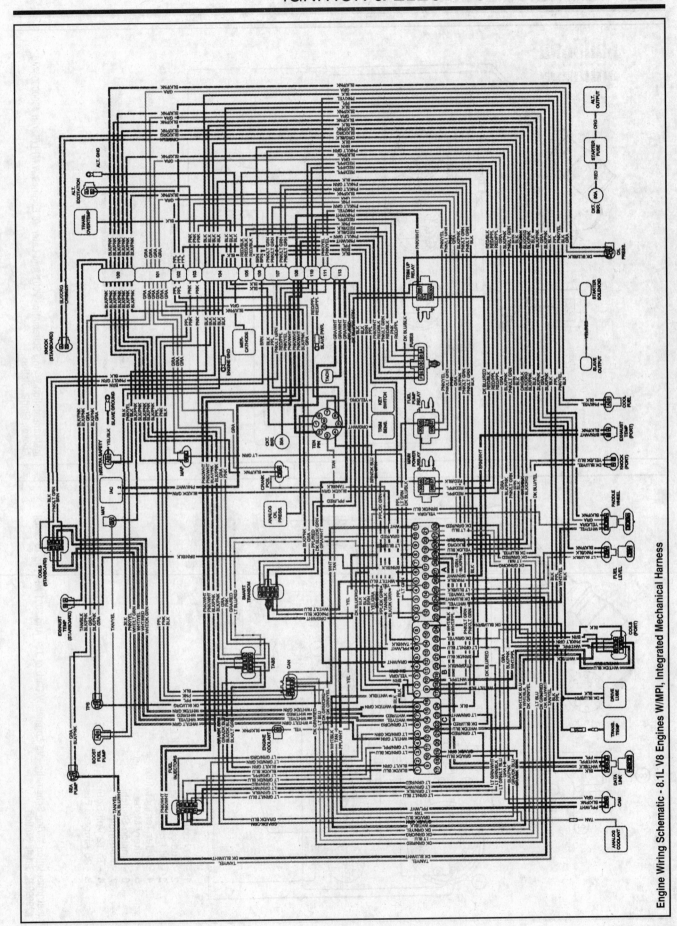

Engine Wiring Schematic - 8.1L V8 Engines W/MPI, Integrated Mechanical Harness

9-62 IGNITION & ELECTRICAL SYSTEMS

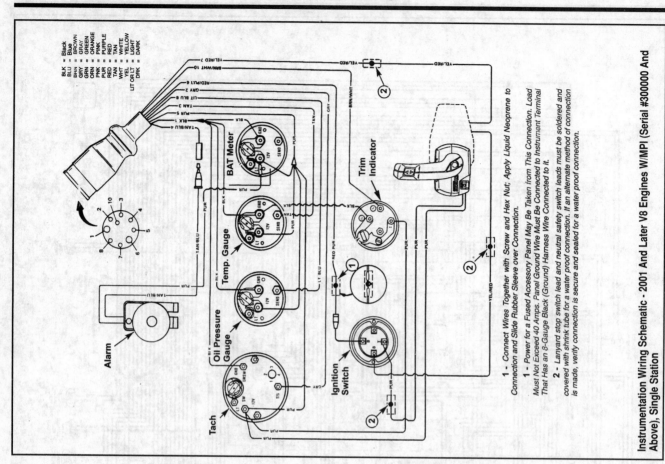

Instrumentation Wiring Schematic - 2001 And Later V8 Engines W/MPI (Serial #300000 And Above), Single Station

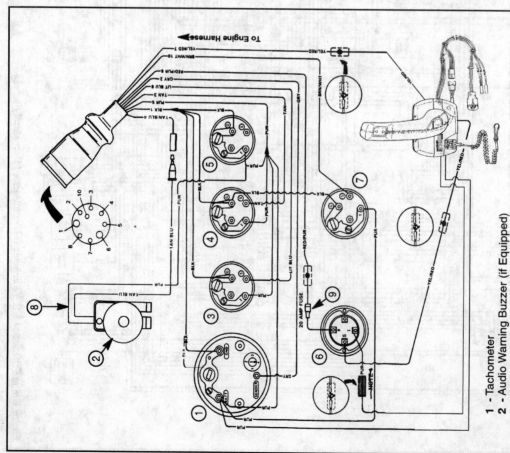

1 - Tachometer
2 - Audio Warning Buzzer (if Equipped)
3 - Oil Pressure
4 - Water Temperature
5 - Battery Meter
6 - Ignition Switch
7 - Trim Indicator
8 - To 12 Volt Source (PURPLE wire connection)
9 - 20 Ampere Fuse

Instrumentation Wiring Schematic - V8 Engines W/Carb or TBI, And 2001 MPI (Up To Serial #299999), Single Station

IGNITION & ELECTRICAL SYSTEMS

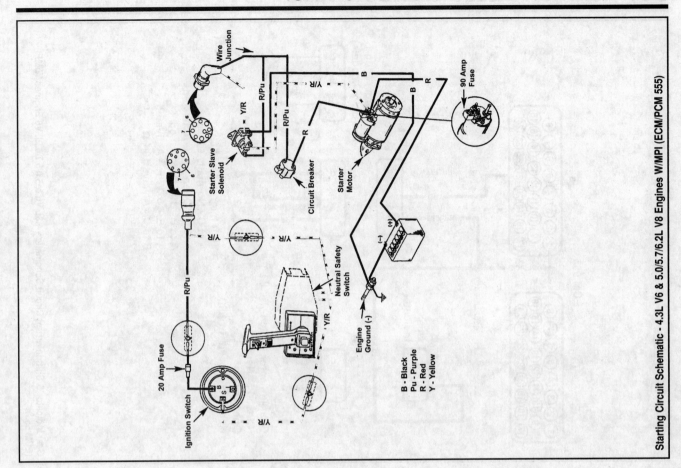

Starting Circuit Schematic - 4.3L V6 & 5.0/5.7/6.2L V8 Engines W/MPI (ECM/PCM 555)

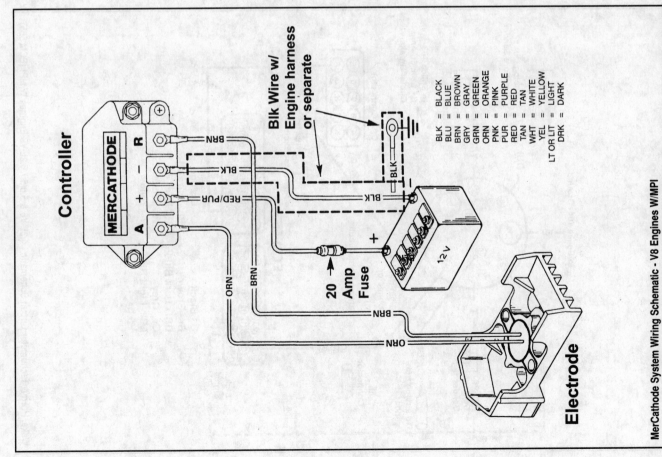

MerCathode System Wiring Schematic - V8 Engines W/MPI

9-64 IGNITION & ELECTRICAL SYSTEMS

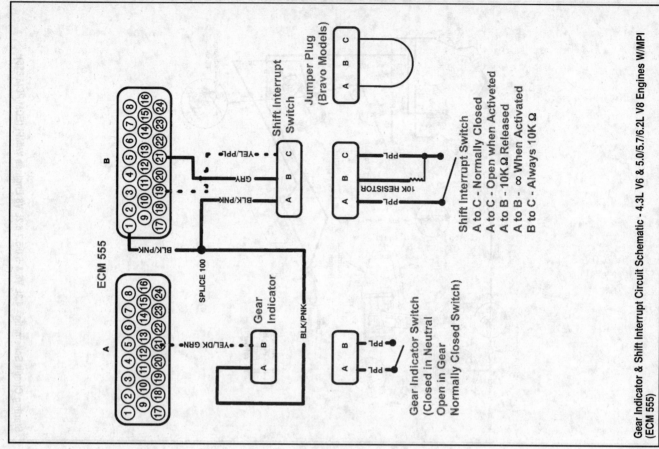

Gear Indicator & Shift Interrupt Circuit Schematic - 4.3L V6 & 5.0/5.7/6.2L V8 Engines W/MPI (ECM 555)

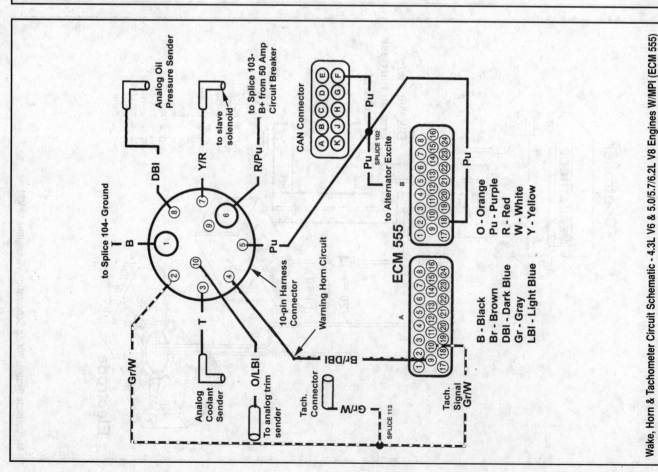

Wake, Horn & Tachometer Circuit Schematic - 4.3L V6 & 5.0/5.7/6.2L V8 Engines W/MPI (ECM 555)

IGNITION & ELECTRICAL SYSTEMS 9-65

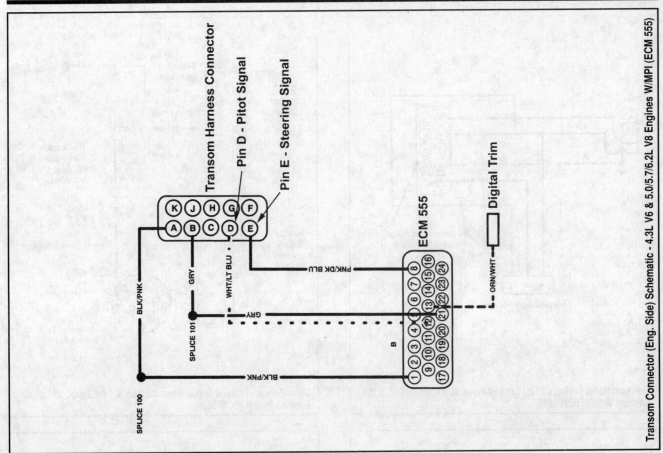

Transom Connector (Eng. Side) Schematic - 4.3L V6 & 5.0/5.7/6.2L V8 Engines W/MPI (ECM 555)

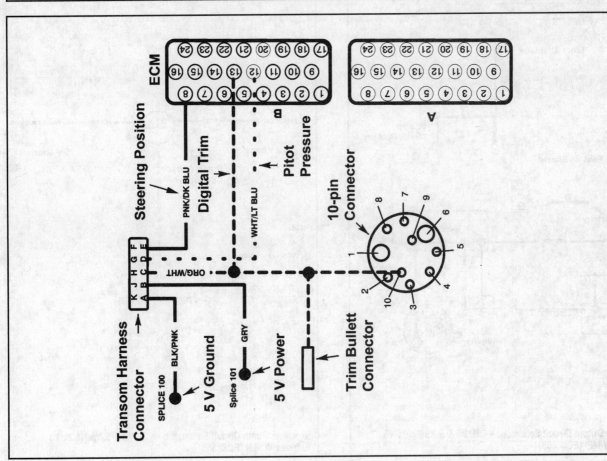

Transom Harness Schematic - 4.3L V6 & 5.0/5.7/6.2L V8 Engines W/MPI (ECM 555)

9-66 IGNITION & ELECTRICAL SYSTEMS

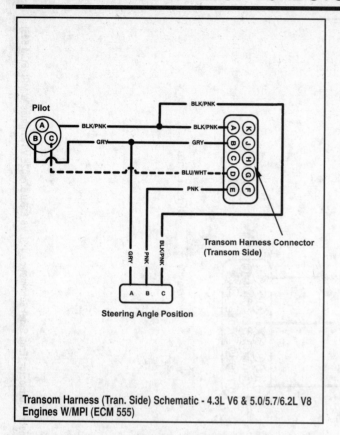

Transom Harness (Tran. Side) Schematic - 4.3L V6 & 5.0/5.7/6.2L V8 Engines W/MPI (ECM 555)

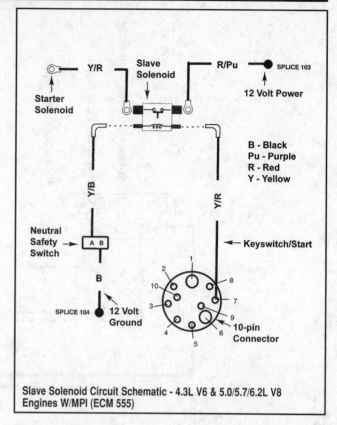

Slave Solenoid Circuit Schematic - 4.3L V6 & 5.0/5.7/6.2L V8 Engines W/MPI (ECM 555)

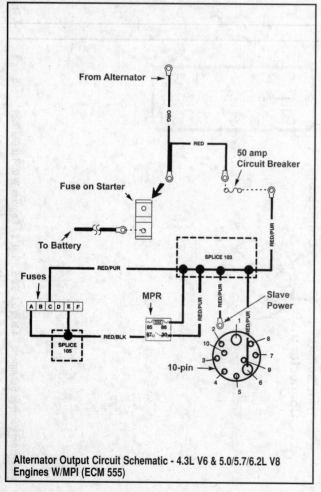

Alternator Output Circuit Schematic - 4.3L V6 & 5.0/5.7/6.2L V8 Engines W/MPI (ECM 555)

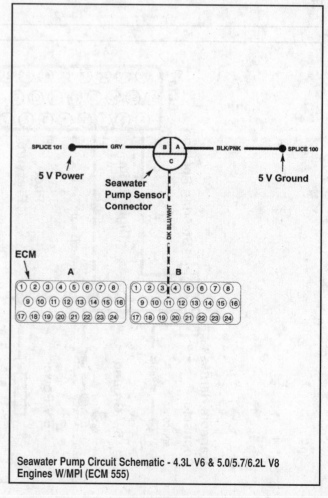

Seawater Pump Circuit Schematic - 4.3L V6 & 5.0/5.7/6.2L V8 Engines W/MPI (ECM 555)

IGNITION & ELECTRICAL SYSTEMS 9-67

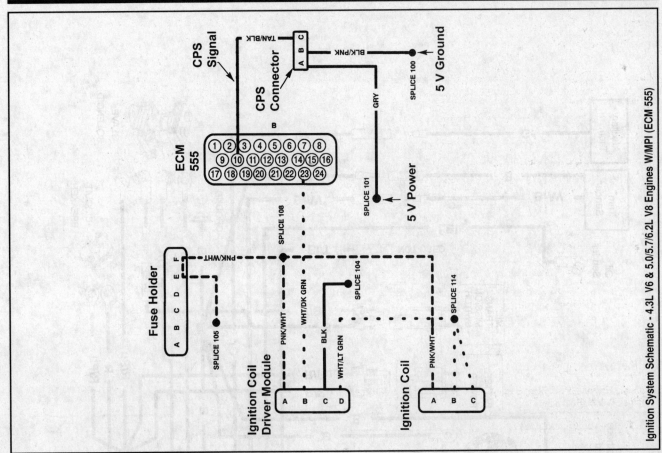

Ignition System Schematic - 4.3L V6 & 5.0/5.7/6.2L V8 Engines W/MPI (ECM 555)

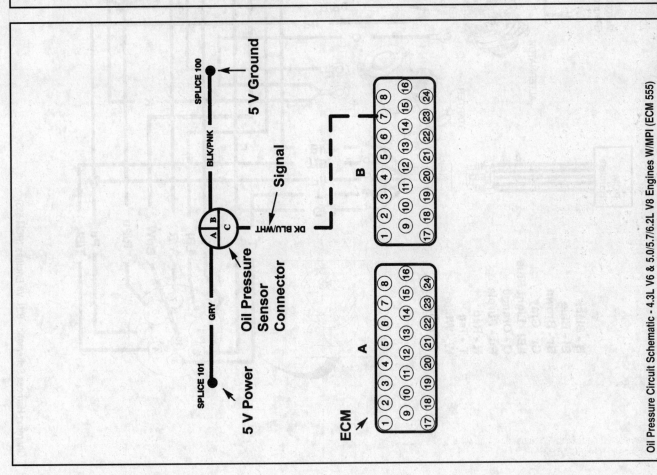

Oil Pressure Circuit Schematic - 4.3L V6 & 5.0/5.7/6.2L V8 Engines W/MPI (ECM 555)

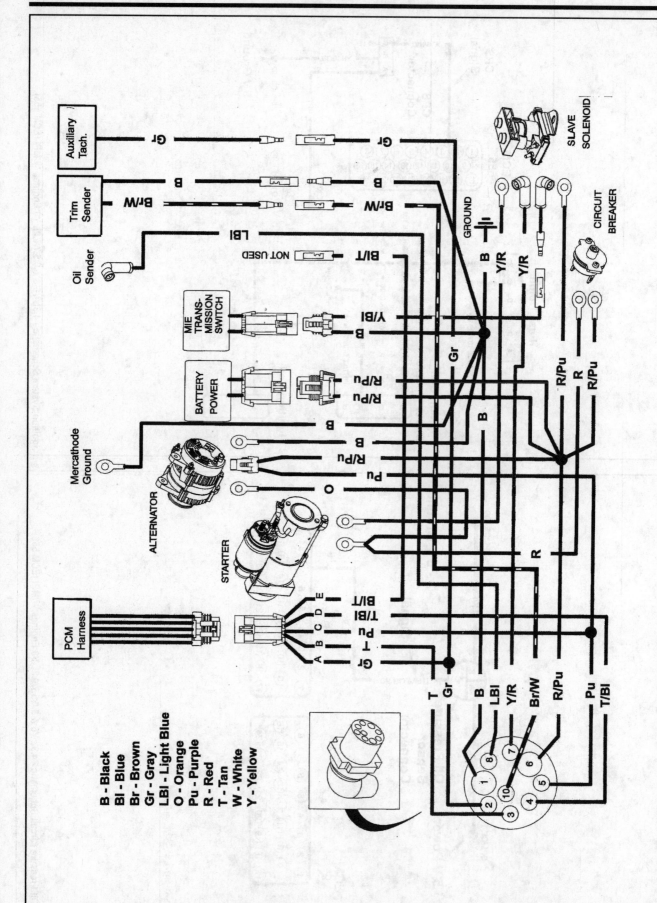

IGNITION & ELECTRICAL SYSTEMS

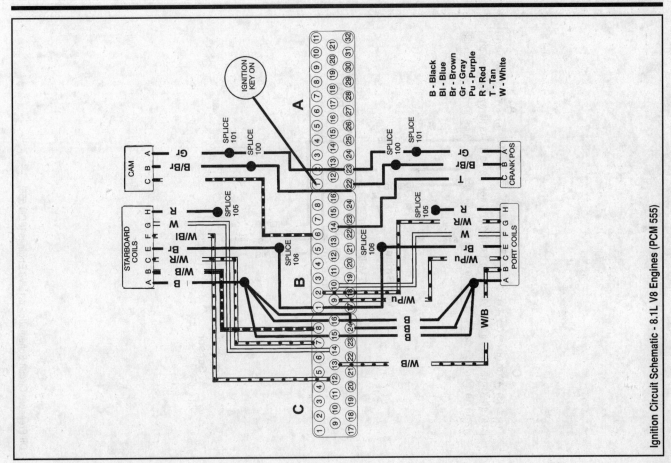

Ignition Circuit Schematic - 8.1L V8 Engines (PCM 555)

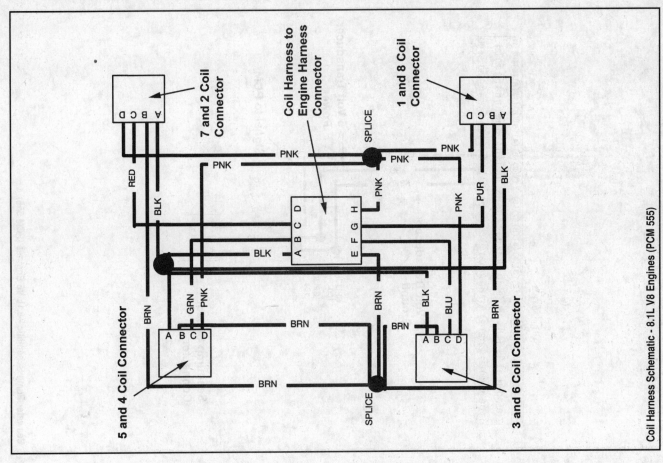

Coil Harness Schematic - 8.1L V8 Engines (PCM 555)

9-70 IGNITION & ELECTRICAL SYSTEMS

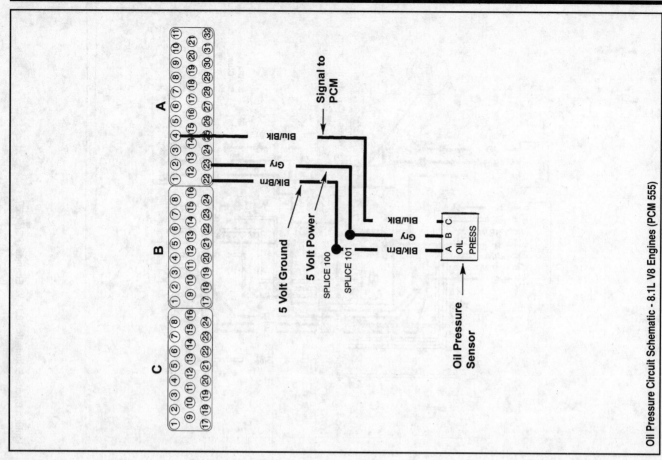

Oil Pressure Circuit Schematic - 8.1L V8 Engines (PCM 555)

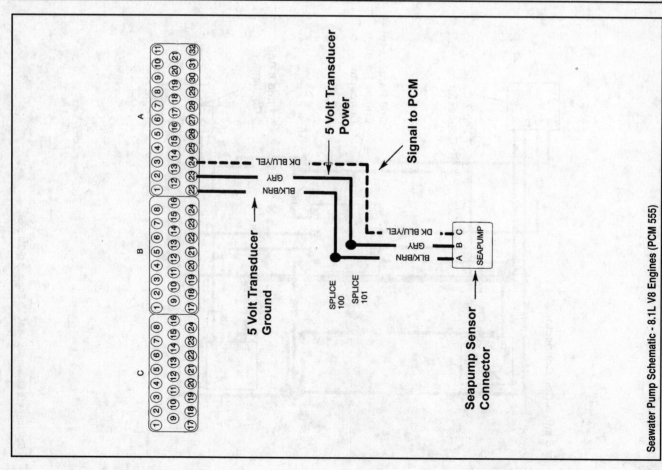

Seawater Pump Schematic - 8.1L V8 Engines (PCM 555)

IGNITION & ELECTRICAL SYSTEMS 9-71

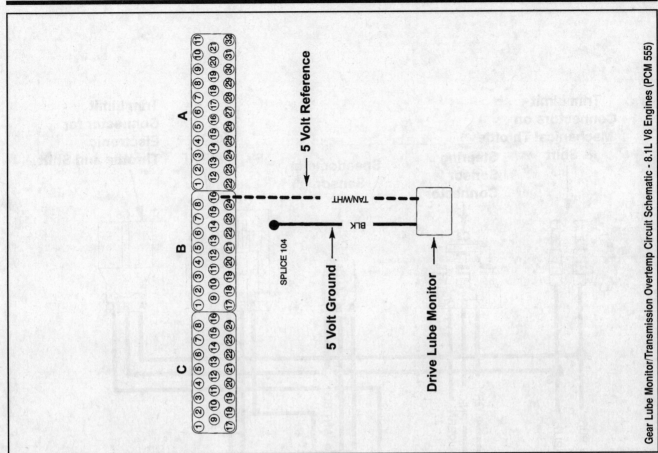

Gear Lube Monitor/Transmission Overtemp Circuit Schematic - 8.1L V8 Engines (PCM 555)

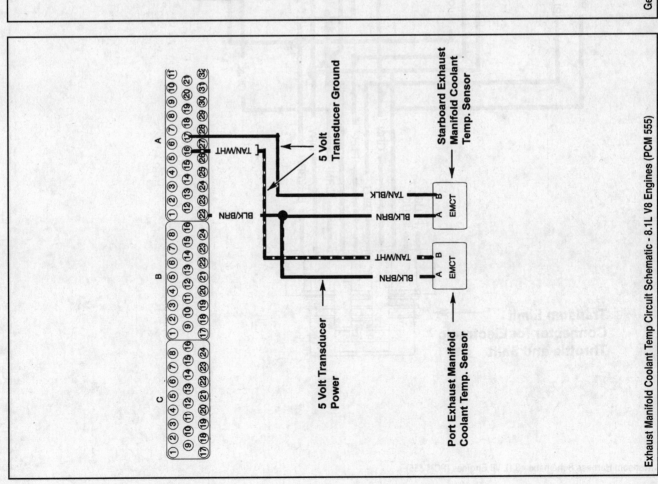

Exhaust Manifold Coolant Temp Circuit Schematic - 8.1L V8 Engines (PCM 555)

9-72 IGNITION & ELECTRICAL SYSTEMS

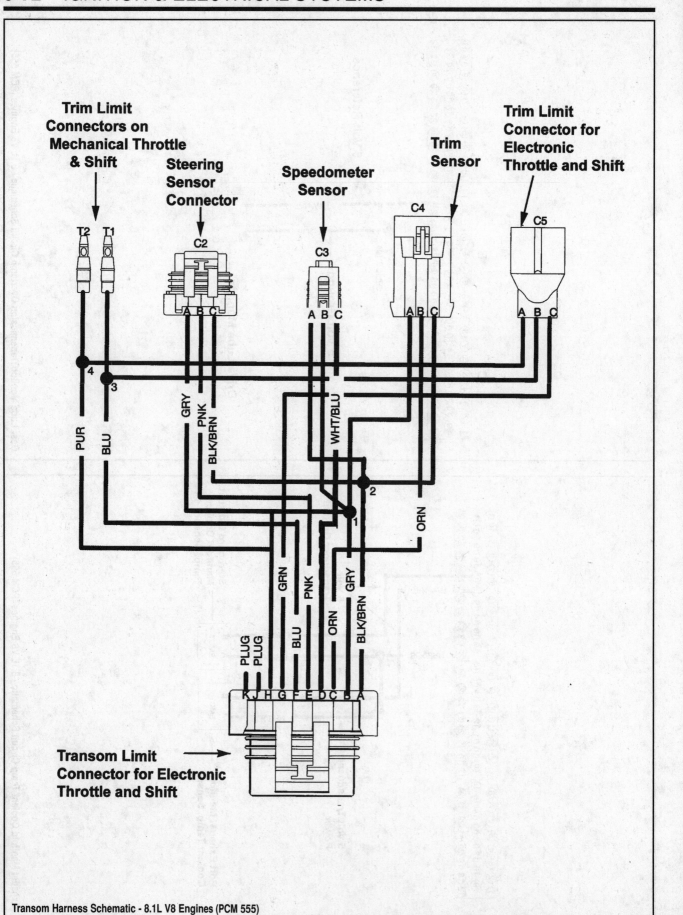

Transom Harness Schematic - 8.1L V8 Engines (PCM 555)

IGNITION & ELECTRICAL SYSTEMS 9-73

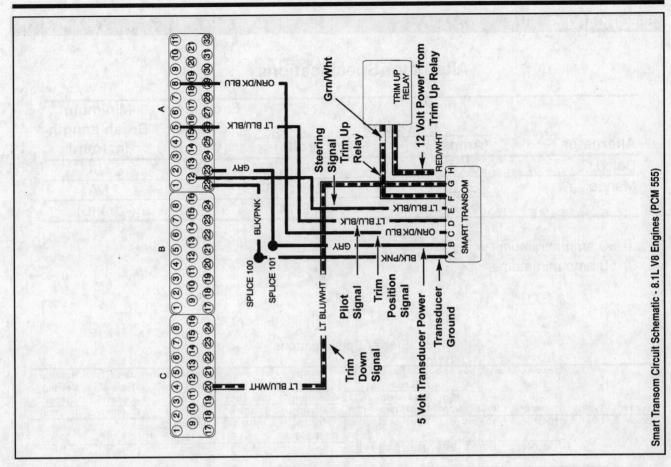

Smart Transom Circuit Schematic - 8.1L V8 Engines (PCM 555)

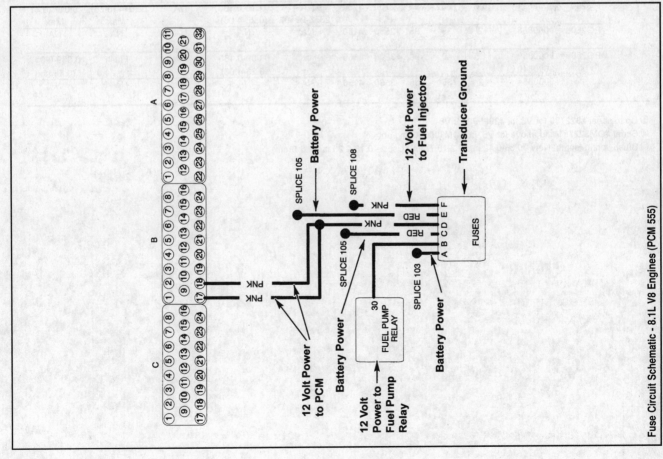

Fuse Circuit Schematic - 8.1L V8 Engines (PCM 555)

SPECIFICATIONS

Alternator Specifications

Alternator	Output (Amps)	(Volts)	Excitation Circuit (Volts)	Minimum Brush Length In. (mm)
Delco	65 ①	13.8-14.8	NA	NA
Mando	65 ①	13.8-14.8	NA	NA
	55 ②	13.9-14.7	1.3-2.5	1/4 (6)

① 60 amp minimum
② 50 amp minimum

Starter Specifications

Brand	Type	Delco ID Number	Volts	No Load Min Amps	No Load Max Amps	No Load Min rpm	No Load Max rpm	Brush Spring Tension Oz. (g)	Pinion Clearance In. (mm)	Commutator End Frame Gap In. (mm)	Gear Bearing Depth In. (mm)	Housing Bearing Depth In. (mm)
Delco	14MT	10455602 ①	10.6	70	120	5400	10800	56-105 (1588-2976)	0.010-0.140 (0.25-3.5)	0.025 (0.6) Max	NA	NA
		19010615 ②	12	60	100	6000	9200	NA	NA	NA	NA	NA
	PG260	9000821	10.6	60	95	2750	3250	83-104 (2353-2948)	0.010-0.160 (0.25-4.06)	NA	0.014 (0.38) Max	0.017 (0.4) Max
	PG260L	9000888	12	35	85	2550	4150	NA	NA	NA	NA	NA
	PG260F1	9000839-9000840 ①	11.5	40	90	3200	4800	③	0.009-0.160 (0.23-4.06)	NA	Flush	0.009-0.017 (0.4 Max)
		9000884 ②	12	35	85	2550	4150	NA	NA	NA	NA	NA

① Up to serial #322780 for V6, or #299999 for V8
② Serial #OM322781 and above for V6, or #OM300000 and above
③ Brush spring length: New - 0.36-0.42 in. (9.2-10.7mm); Used - 0.18-0.23 in. (4.6-6mm)

10 DRIVE SYSTEMS - ALPHA

STERN DRIVE UNIT - ALPHA	10-2
GEAR HOUSING (LOWER UNIT)	10-5
DRIVESHAFT HOUSING (UPPER UNIT)	10-27
TRANSOM ASSEMBLY	10-36
SPECIFICATIONS	10-52

BELL HOUSING 10-44
   REMOVAL & INSTALLATION 10-44
DRIVESHAFT HOUSING 10-28
   DISASSEMBLY & ASSEMBLY 10-28
     DRIVEN GEAR/DRIVESHAFT ASSEMBLY 10-32
     EXPLODED VIEWS 10-28
     U-JOINT/DRIVE GEAR ASSEMBLY 10-28
   PRE-LOAD AND SHIM ADJUSTMENTS ... 10-34
     DRIVE (PINION) GEAR SHIMMING 10-35
     DRIVEN GEAR SHIMMING 10-34
     GENERAL INFORMATION............ 10-34
     UPPER DRIVESHAFT BEARING PRELOAD 10-34
   REMOVAL & INSTALLATION 10-28
DRIVESHAFT HOUSING (UPPER UNIT) .. 10-27
   DESCRIPTION 10-27
   DRIVESHAFT HOUSING (UPPER UNIT) .. 10-28
   GEAR RATIO IDENTIFICATION 10-27
   DRIVESHAFT OIL SEAL CARRIER 10-26
   REMOVAL & INSTALLATION 10-26
EXHAUST BELLOWS 10-42
   REMOVAL & INSTALLATION 10-42
GEAR HOUSING 10-5
   ASSEMBLY 10-15
     COUNTER ROTATION 10-19
     STANDARD ROTATION 10-15
   CLEANING & INSPECTION 10-14
   DISASSEMBLY - STANDARD ROTATION .. 10-7
     BEARING CARRIER & REVERSE GEAR . 10-10
     MAIN COMPONENTS 10-7
     PROPELLER SHAFT 10-11
   DISASSEMBLY - COUNTER ROTATION .. 10-12
   REMOVAL & INSTALLATION 10-5
GEAR HOUSING (LOWER UNIT) 10-5
   DESCRIPTION 10-5
   DRIVESHAFT OIL SEAL CARRIER 10-26
   GEAR HOUSING (LOWER UNIT) 10-5
   PROPELLER 10-23
   SEA WATER PUMP & IMPELLER 10-26
   GEAR RATIO IDENTIFICATION 10-27
GIMBAL BEARING 10-36
   REMOVAL & INSTALLATION 10-36
GIMBAL HOUSING/TRANSOM PLATE 10-49
   REMOVAL & INSTALLATION 10-49
GIMBAL RING 10-45
   REMOVAL & INSTALLATION 10-45
     ENGINE/TRANSOM
     ASSEMBLY INSTALLED 10-45
     ENGINE/TRANSOM
     ASSEMBLY REMOVED 10-48
LOWER UNIT 10-5
   ASSEMBLY 10-15
     COUNTER ROTATION 10-19
     STANDARD ROTATION 10-15
   CLEANING & INSPECTION 10-14
   DISASSEMBLY - STANDARD ROTATION .. 10-7
     BEARING CARRIER & REVERSE GEAR . 10-10
     MAIN COMPONENTS 10-7
     PROPELLER SHAFT 10-11
   DISASSEMBLY - COUNTER ROTATION .. 10-12
   REMOVAL & INSTALLATION 10-5
PROPELLER........................... 10-23
   GENERAL INFORMATION 10-23
     CAVITATION 10-24
     CUPPING 10-24
     DIAMETER AND PITCH.............. 10-23
     PITCH 10-24

RAKE 10-24
ROTATION 10-25
SELECTION 10-23
SHOCK ABSORBERS 10-24
VIBRATION 10-24
INSPECTION 10-25
REMOVAL & INSTALLATION 10-26
SEA WATER PUMP & IMPELLER 10-26
   REMOVAL & INSTALLATION 10-26
SHIFT CABLE 10-37
   ADJUSTMENT 10-38
     REMOTE CONTROL CABLE......... 10-38
     REMOTE CONTROL CABLE -
     CHECKING OUTPUT............... 10-40
     TRANSOM CABLE - CHECKING PLAY .. 10-40
     TRANSOM CABLE - ISOLATING PLAY. . 10-41
   REMOVAL & INSTALLATION 10-37
     REMOTE CONTROL CABLE.......... 10-37
     TRANSOM CABLE................. 10-37
SHIFT CUT-OUT SWITCH 10-41
   ADJUSTMENT 10-41
     MODELS W/PLUNGER SWITCH 10-42
     MODELS W/ROLLER SWITCH 10-41
SPECIFICATIONS 10-52
   APPLICATIONS 10-52
   TORQUE............................ 10-52
STERN DRIVE UNIT 10-3
   REMOVAL & INSTALLATION 10-3
STERN DRIVE UNIT - ALPHA.............. 10-2
   DESCRIPTION 10-2
   STERN DRIVE UNIT 10-3
   TROUBLESHOOTING 10-2
TRANSOM ASSEMBLY 10-36
   BELL HOUSING 10-44
   DESCRIPTION 10-36
   EXHAUST BELLOWS 10-42
   GIMBAL BEARING 10-36
   GIMBAL HOUSING/TRANSOM PLATE 10-49
   GIMBAL RING 10-45
   SHIFT CABLE....................... 10-37
   SHIFT CUT-OUT SWITCH 10-41
   U-JOINT BELLOWS 10-43
TROUBLESHOOTING 10-2
   DRIVE UNIT WILL NOT SHIFT -
   SHIFT HANDLE DOES NOT MOVE 10-3
   DRIVE UNIT WILL NOT SHIFT -
   SHIFT HANDLE MOVES 10-2
   DRIVE UNIT WILL NOT SLIDE INTO
   BELL HOUSING 10-2
   DRIVESHAFT HOUSING NOISE 10-2
   GEAR HOUSING NOISE 10-2
   HARD SHIFTING 10-3
   JUMPS OUT OF GEAR 10-3
U-JOINT BELLOWS 10-43
   REMOVAL & INSTALLATION 10-43
UPPER UNIT 10-28
   DISASSEMBLY & ASSEMBLY 10-28
     DRIVEN GEAR/DRIVESHAFT ASSEMBLY 10-32
     EXPLODED VIEWS 10-28
     U-JOINT/DRIVE GEAR ASSEMBLY 10-28
   PRE-LOAD AND SHIM ADJUSTMENTS ... 10-34
     DRIVE (PINION) GEAR SHIMMING 10-35
     DRIVEN GEAR SHIMMING 10-34
     GENERAL INFORMATION 10-34
     UPPER DRIVESHAFT BEARING PRELOAD 10-34
   REMOVAL & INSTALLATION 10-28

DRIVE SYSTEMS - ALPHA

STERN DRIVE UNIT - ALPHA

Description

◆ See Figures 1 and 2

That which we refer to as the stern drive is actually a number of individual components attached and working together to transfer the power of the engine into a viable propulsion system for your boat. All stern drive units can be broken down into their component assemblies.

The transom assembly, consisting of an inner transom plate, a gimbal housing and gimbal plate, and a bell housing is just what it sounds like - the unit attached to the transom of the vessel. The inner transom plate is, obviously, attached to the inner side of the transom and actually makes up the rear engine mounts.

On the other side of the transom, and attached to the transom plate, are the gimbal housing, gimbal plate (or ring) and bell housing. The bell housing is attached to the gimbal plate via roller bearings and is what allows for the up and down (trim) movement of the stern drive unit itself. The gimbal plate is also attached to the gimbal housing via roller bearings and is what allows for side-to-side movement of the unit, or, steering.

The stern drive unit, or at least that thing that is most visible when viewing the stern of the boat, is made of two component assemblies: the driveshaft housing (upper gear housing) and the lower gear housing.

The driveshaft housing, frequently called the upper gear housing or simply the upper unit is attached to the bell housing at the top and the gear housing at the bottom. Power from the engine, brought through the transom assembly via the driveshaft is transferred to a vertical shaft leading to the lower unit by means of a set of drive and driven gears.

The gear housing, or lower unit, is attached to the bottom of the driveshaft housing. Power, or propulsion, comes through the vertical shaft from the upper unit, is transferred to the propeller shaft via a pinion gear and causes the propeller to rotate.

Output power from the engine is connected to the stern drive through a horizontal driveshaft. A coupler is bolted to the flywheel and has a splined hub in the center. The end of the horizontal driveshaft indexes with and slides into the center of the hub. Power from the engine is then transmitted through the horizontal driveshaft to a pinion gear set where power direction is changed from horizontal to vertical.

The upper driven gear is pressed onto the outside diameter of the upper driveshaft. The upper driveshaft is splined on the lower end. When the upper gear housing is mated to the lower unit, the end of the lower unit driveshaft indexes into the splined end of the upper driveshaft. Engine power is then transferred down into the lower gear unit.

The horizontal and vertical driveshafts are both mechanically connected. Therefore, anytime the engine is operating, the horizontal and vertical driveshafts are constantly rotating with engine rpm. A double yoke universal joint assembly in the horizontal driveshaft allows the stern drive to be raised or lowered to a required trim/tilt position (within limits), while the engine is operating.

Troubleshooting

The following are a list of potential drive unit problems and their possible causes:

GEAR HOUSING NOISE

- Metal particles in the unit oil supply
- Propeller incorrectly installed
- Propeller or propeller shaft bent
- Incorrect drive gear shimming - gear housing back-lash or pinion gear height
- Worn or damaged gears or bearings

DRIVESHAFT HOUSING NOISE

- Steering lever may be contacting the transom cut-out edges when turning
- Flywheel housing on the engine coming in contact with the inner transom plate or the exhaust pipe
- Bad propeller
- U-joint cross and bearing assembly O-rings of incorrect size or installed wrong
- U-joint cross and bearing assemblies have excessive side-play

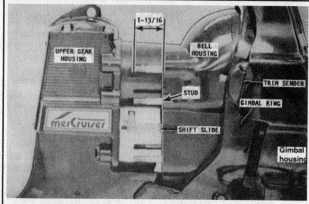

Fig. 1 A good look at the entire drive system

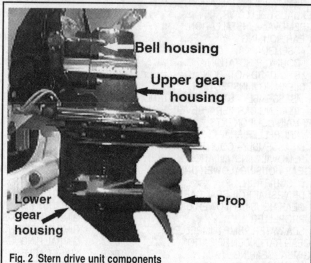

Fig. 2 Stern drive unit components

- Bearing caps on the U-joint are coming in contact with the center socket or the driveshaft housing bearing retainer
- Rough or scored U-joint cross and bearings
- Worn or missing O-rings on the U-joint shaft rattling against the gimbal bearing
- Worn or damaged splines on the driveshafts and/or couplers
- Incorrect engine alignment
- Damaged, rough, worn or loose gimbal bearing
- Gimbal bearing seated incorrectly
- Incorrect clearance between the gimbal housing and plate
- Bell housing and gimbal housing incorrectly aligned
- Weak or flexing transom
- Weak, missing or mis-aligned rear engine mount

DRIVE UNIT WILL NOT SLIDE INTO BELL HOUSING

- U-joint and engine coupler splines not aligned
- Unit not in Forward gear
- Incorrectly aligned shift shaft coupler
- Engine out of alignment
- Incorrectly installed gimbal bearing
- Damaged or worn splines on the driveshaft or coupler

DRIVE UNIT WILL NOT SHIFT - SHIFT HANDLE MOVES

- Shift cables not adjusted properly
- Shift cables not connected
- Inner wire on cable broken
- Gear housing crank installed incorrectly

DRIVE SYSTEMS - ALPHA

DRIVE UNIT WILL NOT SHIFT - SHIFT HANDLE DOES NOT MOVE

- Remote control box/assembly installed incorrectly
- Broken, worn or damaged linkage
- Shift shaft or lever stuck
- Controls or cables installed or adjusted incorrectly

HARD SHIFTING

- Shift cables out of adjustment
- Shift cut-out switch broken or improperly adjusted
- Shift cable too short, or too long
- Corroded cables
- Shift shaft bushings corroded or damaged
- Shift crank or clutch actuating spool worn or damaged
- Shaft shaft damaged, worn or broken
- Shift cable attaching nuts too tight

JUMPS OUT OF GEAR

- Incorrectly adjusted shift cables
- Worn or damaged clutch or gears

Stern Drive Unit

REMOVAL & INSTALLATION

◆ See Figures 3 thru 10

1. Drain the stern drive unit oil. Check the drained lubricant carefully to determine if it contains any water or metal particles. Rub some of the lubricant between your fingers; any metal particles in the lubricant will thus be evident. Do not be mislead if the color of the lubricant is metal colored.

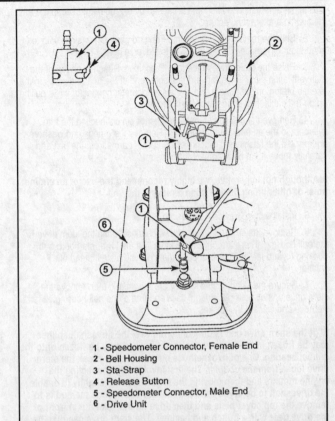

1 - Speedometer Connector, Female End
2 - Bell Housing
3 - Sta-Strap
4 - Release Button
5 - Speedometer Connector, Male End
6 - Drive Unit

Fig. 3 Disconnecting the speedometer

Fig. 4 There are six mounting bolts...

Fig. 5 ...remove them in a criss-cross pattern

Fig. 6 You can try it this way, but we recommend a lifting hoist

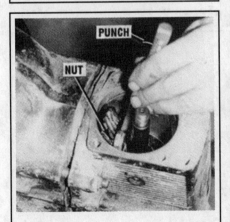

Fig. 7 Removing the yoke nut with a punch

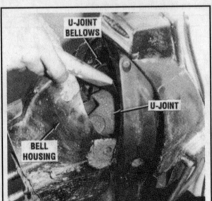

Fig. 8 Cut back the bellows after removing the bell housing

Fig. 9 Use the pliers to remove the gimbal bearing snap-ring

10-4 DRIVE SYSTEMS - ALPHA

This is not a harmful condition and is caused by the lubricant used in the unit the first time after manufacture.

2. Shift the remote control unit into Forward gear. On engines with left hand (counter) rotation, move the lever to Reverse gear.

3. Raise the stern drive to the full UP position. Reach in on top of the anti-ventilation plate at the forward (transom) end of the unit and press in the release button on the female end of the speedometer connector while pulling up on the cable. Move the cable out of the way.

4. Remove the plastic cap over the anchor pin connecting the trim cylinders to the aft end of the upper unit. Remove the e-rings and washers and pry out the bushings. Disconnect the cylinder arms from the unit and carefully move them back against the transom.

■Although not imperative, we highly recommend the use of an engine hoist or other lifting device when removing the unit.

5. Install a lifting device if so desired.

6. Remove the six 5/8 in. nuts and washers securing the stern drive to the bell housing. There are only five washers as the lower starboard bolt (looking toward the bow) uses a permanent ground plate instead of a washer.

7. Secure the lifting chains so they are taught and then remove the stern drive unit by pulling it straight back and free of the bell housing. Remove and discard the gasket.

■ If the stern drive unit is difficult to remove, the driveshaft splines may be frozen in the engine coupler or the shaft may be frozen onto the gimbal bearing. One of two methods may be used to break the stern drive loose from the coupler. The first involves disconnecting the engine mounts and then moving the engine forward slightly to enable the driveshaft to be pryed from the splines. The second method is to remove the top cover plate and then drive off the yoke nut in front of the drive gear with a punch and hammer. The stern drive can then be removed, leaving the frozen shaft in the bell housing.

8. If you were unable to remove the drive unit with the driveshaft attached, remove the bell housing as detailed later in this section and then cut or remove the exhaust and U-joint bellows.

9. Remove the snap-ring securing the gimbal bearing with a pair of tru-arc pliers - yes the U-joints are in the way, but be patient and you'll get it out.

10. Use a slide hammer and pull out the shaft assembly. If it is frozen to the bearing, or the bearing is frozen, apply a little heat from a torch while working the slide hammer. See the Transom Assembly procedures later in this section for complete details on the gimbal bearing.

To install:
◆ See Figures 11 thru 18

The engine must be properly aligned or the female splines in the engine coupler and the male splines of the driveshaft will be destroyed after a short time of engine operation.

If troubleshooting indicates the coupler is damaged and requires replacement, remove the engine and replace the coupler. The coupler is simply secured to the flywheel with attaching bolts. If the necessary tools are not available for proper engine alignment, the boat should be taken to an authorized marine dealer.

11. Install a new O-ring in the water passageway between the upper unit and the bell housings and the rubber sealing ring inside the bell housing. Install a new gasket on the face of the bell housing.

12. Coat all 6 studs and their threads with Quicksilver 2-3-C marine lubricant.

13. Check the bell housing bearing to be sure it turns freely without binding or rough spots.

14. Oil the shaft to prevent damage to the seals, and then slide two new O-ring seals over the splines of the universal joint shaft, and into the grooves provided.

15. Coat the splines of the driveshaft and the driveshaft housing pilot, with Multi-purpose Lubricant. As an aid to installation, apply a light coating of Multi-purpose Lubricant to the inside surfaces of the bell housing bore.

16. Snap a shift shaft slide stabilizer tool (sorry, no part number available) onto the stud directly below the shift slide and position it as shown in the illustration.

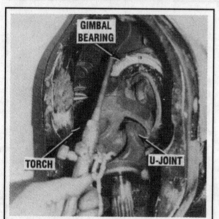

Fig. 10 You will probably need a torch to free up the frozen bearing

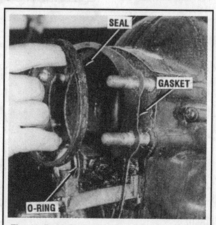

Fig. 11 Don't forget the O-ring and seal when installing the gasket

Fig. 12 Make sure to install two new O-rings

Fig. 13 The drive should be in the correct gear for installation

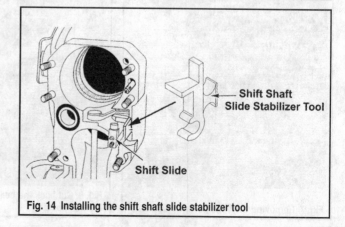

Fig. 14 Installing the shift shaft slide stabilizer tool

DRIVE SYSTEMS - ALPHA

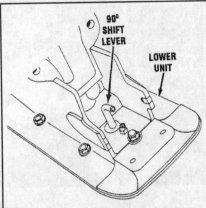

Fig. 15 Correct positioning of the shift lever shaft...

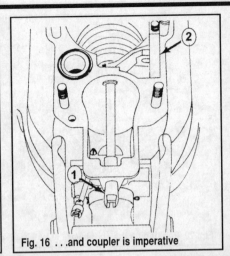

Fig. 16 ...and coupler is imperative

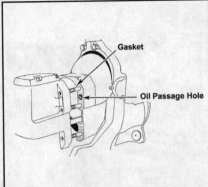

Fig. 17 Make sure the hole in the gasket is aligned correctly if your unit uses a lube monitor

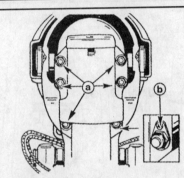

- a - Locknut and Flat Washers
- b - Locknut and Continuity Circuit Washer (No Flat Washer at this Location)

Fig. 18 Remember that the lower starboard nut does not use a washer

17. On standard rotation models, move the shift lever to the full forward position. On counter rotation models, move the lever to the full reverse position. This will position the bell housing shift slide so that the slot in the intermediate shaft coupler is positioned straight ahead (fore and aft).

■ Remember that the shift slide assembly rotates freely on the core wire - be very careful that the slide remains in the upright position and is engaged with the shift shaft lever.

18. Apply a coating of Multi-purpose Lubricant onto the shift shaft coupler, and then align the slot straight by moving the lever to the right (clockwise); while turning the propeller shaft counterclockwise at the same time. Lubricate the slot and the reverse lock roller.

19. As a double check, be sure the stern drive unit is in Forward gear. Rotate the propeller shaft to align the intermediate shift shaft coupling with the upper shift shaft slot. Bad News: If the coupler of the intermediate shift shaft in the driveshaft housing and the locating slot of the shift lever in the bell housing are not aligned properly, the bell housing and shift shaft couplers will be damaged when the drive unit is installed. Remember that on engines with left hand rotation, the unit must be in Reverse.

20. Install a new gasket on the bell housing. If your boat utilizes a gear lube monitor, be sure that the housing gasket has a hole where the passage is (starboard side of the bell housing flange, between the top and center mounting bolt holes).

21. Coat the U-joint O-rings and gimbal housing ball bearing with Multi-purpose Lubricant, to assist installation. Install the drive unit into the bell housing bore. If the driveshaft splines do not index with the splines of the engine coupler, slide the propeller onto the propeller shaft, and then slowly rotate the shaft counterclockwise until the drive unit can be pushed completely into position. While performing this installation maneuver, make sure the shifting slide remains upright and engaged. If the stern drive will not move completely into place, check underneath and observe if the shift cam is properly aligned in the groove on the stern drive.

■ Once the U-joint shaft has begun to engage with the coupler splines, remove the stabilizer tool and install the unit fully. Make sure that the stud the tool was attached to is thoroughly coated with Quicksilver 2-4-C.

22. Secure the driveshaft housing to the bell housing with the six elastic-stop nuts with flat washers. Tighten the nuts alternately and evenly to a torque value of 50 ft. lbs. (68 Nm). If you removed the top cover because the driveshaft was stuck, tighten the screws to 20 ft. lbs. (27 Nm).
23. Return the remote control lever to the Neutral position.
24. Install the trim cylinders.
25. Connect the speedometer cable in the opposite manner that you disconnected it - press it in until it snaps into place.
26. Fill the unit with oil.

GEAR HOUSING (LOWER UNIT)

Description

The upper gear housing transfers engine power to the lower unit through a driveshaft. Splines on the forward end of the drive shaft mate with matching splines of a coupler attached to the engine flywheel. The aft end of the driveshaft is connected to a universal joint and the upper gear assembly. The pinion gear and drive gear changes the direction of the power train from horizontal to vertical. The power is transferred from the upper gear housing down to the lower unit through a vertical driveshaft. A water pump installed in (on top of) the lower unit is constantly driven by the vertical driveshaft. This pump supplies cooling water directly to the engine, or indirectly through a heat exchanger.

The forward and reverse gears are contained within the lower unit. A sliding clutch, actuated by the shift linkage, engages a forward gear and creates a direct coupling for transmitting power through the pinion gear and forward gear to the propeller shaft. The reverse gear utilizes the same mechanics as the forward gear, except that the clutch is coupled with the reverse gear.

Gear Housing (Lower Unit)

REMOVAL & INSTALLATION

◆ See Figure 19

The lower unit may be separated and removed from the upper gear housing while the stern drive remains attached to the boat or after the stern drive has been removed.

1. If you plan on disassembling the unit, remove the propeller from the lower gear housing.
2. If the unit is still on the boat, drain the drive oil. Allow the lubricant to drain into the container. As the lubricant drains, check the color. Dark brown to black indicates normal old lubricant. A chalky white to cream color indicates the presence of water. The presence of any water in the gear

10-6 DRIVE SYSTEMS - ALPHA

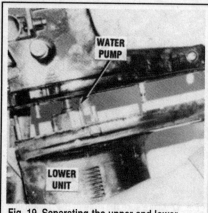

Fig. 19 Separating the upper and lower units

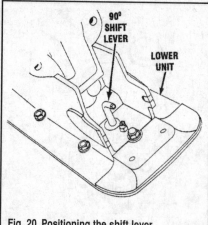

Fig. 20 Positioning the shift lever

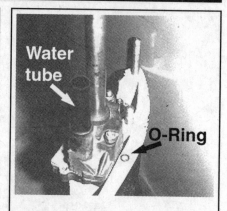

Fig. 21 A view of the water pump and tube, don't forget the O-ring

lubricant is bad news. The unit must be completely disassembled, inspected, the cause of the problem determined and corrected. Close attention should be given to the back-to-back seals on the propeller shaft, driveshaft and the bearing carrier O-ring. Examine the magnet on the end of the fill/drain plug for evidence of metal particles. The presence of tiny small metal dust like shavings indicates normal wear of the gears, bearings and shafts within the lower unit. Large metal chips or heavy fillings indicate extensive internal damage is taking place and the lower unit must be completely disassembled and inspected. All worn and/or damaged components must be replaced.

3. If the stern drive is installed on the boat, place the remote control shift lever in the Forward gear position.

4. If the unit is still on the boat, disconnect the speedometer fitting as detailed in the drive unit removal and installation procedures. If the unit has been removed, then you've already done this.

5. Place an alignment mark on the trim tab trailing edge and the anti-cavitation plate as an aid during assembling. Remove the plastic plug from the top of the anti-cavitation plate. Remove the bolt (1/2 in.) directly above the trim tab and then lift off the trim tab. Now remove the aft mounting bolt in the cavity.

6. Remove the 4 hex nuts and washers (2 on each side), from the sides of the anti-cavitation plate.

7. Separate the two housings by tapping downward on the lower unit with a soft face mallet and a wooden block. In some cases the housing may be difficult to separate because the driveshaft may be corroded and "frozen" in the upper gear housing vertical shaft. Lower the gear housing away from the upper gear housing.

To install:
◆ See Figures 20 thru 25

8. Verify that the remote control shift handle is in the Forward gear position for standard right hand stern drives. The 90° shift lever should be facing forward and in the center of the lower unit. Rotate the propeller shaft clockwise to verify gear engagement.

■ **If the unit being serviced is a counter-rotation (left hand drive) the gear shift lever should be in the reverse gear position.**

9. Verify that the trim tab bolt is in place in the aft section of the lower unit.

10. Check to be sure the O-ring is in place on the passageway for the gear oil.

11. Apply a coating of Quicksilver 2-4-C Lubricant, or equivalent, to the splines on the end of the driveshaft.

12. Some early lower unit housings may have an aluminum dam behind the water pump; although it is doubtful that any units covered here will still have this. If the aluminum dam is corroded or damaged it must be replaced with a rubber filler unit. If the aluminum dam is serviceable, check the drain hole on the starboard side of the dam to verify it is not plugged. If this drain hole is plugged, severe damage to the gear housing will result.

13. Check to be sure the plastic guide tube is installed in the water pump housing. Verify the water pump face seal is positioned on top of the water pump cover. This seal prevents exhaust gases from entering the water cooling system.

If the unit has an aluminum dam and it has not been replaced with the rubber plug, place a bead of Permatex Ultra Blue Silicone Sealant, or

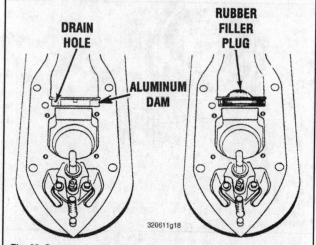

Fig. 22 Some early model Alpha One stern drives were equipped with an aluminum dam. Later models replace the aluminum dam with a rubber filler plug. Aluminum dams require a bead of silicone sealant, as explained in the text.

equivalent, across the top of the dam. If a rubber dam is installed, no sealant is required.

14. Now, raise the lower unit (or lower the upper unit) to mate with the upper gear housing, and at the same time check to be sure the water tube starts into the water pump guide tube.

15. It may be necessary to slowly rotate the propeller shaft clockwise until the splines on the upper and lower drive shafts index.

16. Screw the nut onto the stud at the front of the unit. Install the aft screw into the forward hole in the trim tab well.

17. Install the two port and two starboard bolts and their nuts and washers. Tighten the four 5/8 in. nuts alternately, and a little at-a-time, until the lower unit is tight against the upper gear housing.

18. Now tighten the front nut and the four side nuts to 35 ft. lbs. (47.5 Nm). Tighten the aft screw to 28 ft. lbs. (41 Nm).

19. Install the trim tab with the marks made during disassembly aligned. Tighten the screw(s) to 23 ft. lbs. (31 Nm).

■ **The trim tab performs two very important jobs, one of which you may not realize. First, the tab compensates for steering torque. If the boat continually seems to move to port or starboard while the helmsman is on a straight course, the trim tab can be adjusted to the side of the pull. The tab also prevents electrolysis from damaging expensive parts. The tab is not expensive; it should show signs of electrolytic action; and should always be replaced after some of the material has been eaten away or is pitted. Now, if the tab shows no signs of electrolytic action after the boat has been in use over a period of time, the grounding should be checked to ensure more expensive parts are not being damaged. Install the trim tab plastic cover plug.**

20. Install the propeller if it was removed and refill the drive unit with oil.

DRIVE SYSTEMS - ALPHA   10-7

Fig. 23 Make sure that the water tube slides into the guide

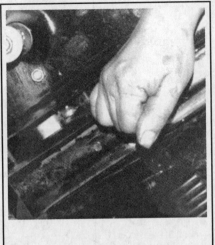

Fig. 24 Tightening the port attaching nuts

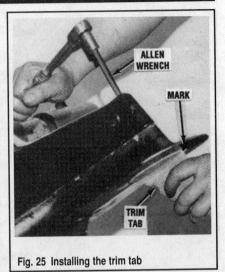

Fig. 25 Installing the trim tab

DISASSEMBLY - STANDARD ROTATION

Main Components

◆ See Figures 26 thru 41

■ The following procedures detail disassembly of the main components of the lower unit; further disassembly procedures for the bearing carrier and propeller shaft components (once they have been removed) are shown under a separate set of procedures.

The following procedures assume that the lower unit, water pump, seal carrier and propeller have already been removed.

1. To remove the bearing carrier, first straighten the locking tab on the tab washer. Loosen the cover (retainer) nut turning it counterclockwise with special wrench (#91-61069). Remove the nut, tab washer and thrust washer.

■ If the retainer (cover nut) is corroded in place, drill 4 holes in the nut and then break the retainer into pieces with a chisel. Pry out the pieces.

2. Pull out the bearing carrier assembly using a bearing carrier puller (2-armed puller) on the outer rings of the carrier. The thrust hub for the propeller, which was removed previously, can be used to keep the puller jaws in place. On early models, install the puller arms on the inner bosses (center area near the oil seals. On later models, install the puller arms on the outer ring of the carrier.

■ If the bearing carrier is corroded in place, apply heat to the housing to loosen the corrosion. While the housing is still hot, use a bearing carrier puller to remove the assembly do not overheat the housing or it will become distorted and useless.

3. Reach in and remove the thrust washer-to-gear housing shim material. Tag and wire them together, as an aid during assembling.

4. Remove the driveshaft pinion nut by sliding the pinion nut holding tool (#91-61067A3) over the propeller shaft so the slot is facing the pinion

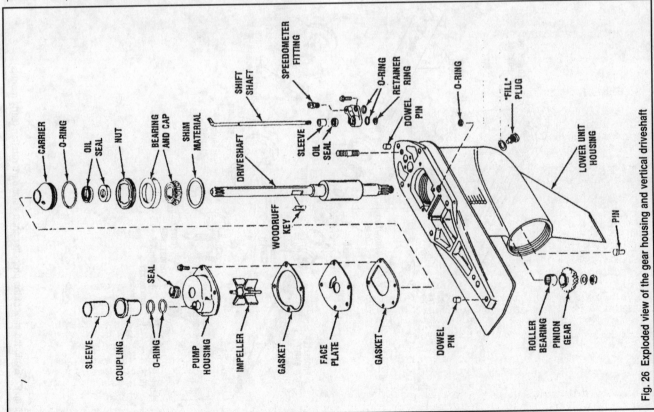

Fig. 26 Exploded view of the gear housing and vertical driveshaft

10-8 DRIVE SYSTEMS - ALPHA

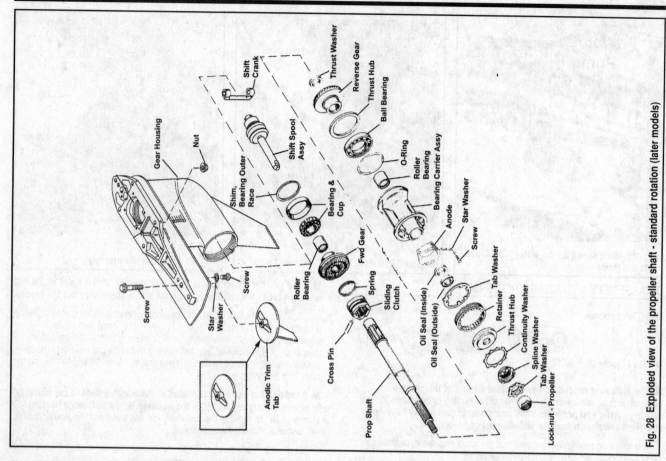

Fig. 28 Exploded view of the propeller shaft - standard rotation (later models)

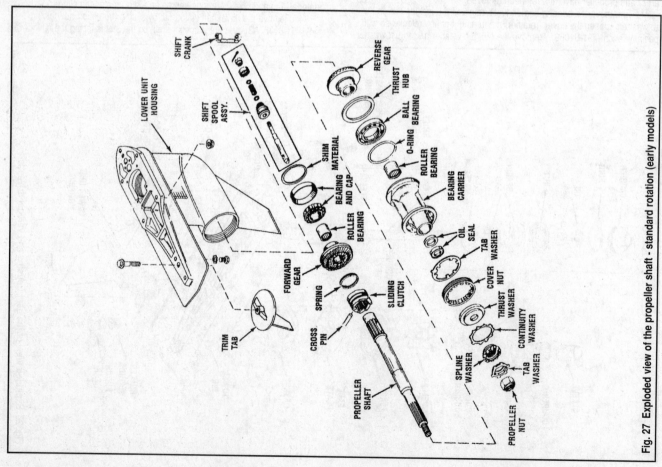

Fig. 27 Exploded view of the propeller shaft - standard rotation (early models)

DRIVE SYSTEMS - ALPHA

gear. You may need to lift the driveshaft slightly and wiggle it to get the tool to slide over the pinion nut.

5. Once the tool is properly positioned on the nut, slide the bearing carrier into the bore, BACKWARDS, so that it will support the propeller shaft and keep the tool aligned properly.

6. Turn the driveshaft counterclockwise with the special splined wrench adapter (#91-56775) and a break-over handle. Loosen the pinion nut, but do not completely unscrew it.

■ If the pinion nut holding tool is not available, a substitute can be made from a box-end wrench of the proper size for the nut. Grind the wrench down on both sides until it will clear the clutch dog and the pinion gear. Place the wrench on the nut and proceed with disassembling. Remove the nut.

7. Remove the driveshaft bearing retainer in order for the driveshaft to clear the housing. To remove the retainer, obtain driveshaft bearing retainer tool (#91-43506). Slide the tool down the driveshaft until the tool is indexed over the retainer. Now, use a breaker bar on the special tool and loosen the nut. Slide the special tool free of the driveshaft, and then back the nut off and clear of the driveshaft.

8. Rotate the driveshaft until the pinion nut is free and then remove all tools.

9. Pull the driveshaft straight up and set it aside. Remove the bearing carrier and the pinion nut tool and then remove the nut, washer and gear from the bore.

■ The pinion bearing rollers may drop out of the bearing once the shaft is removed - be careful that you collect all 18 rollers.

Fig. 29 Bend out the tab on the washer

Fig. 30 If the nut is corroded, drill some holes and then break it with a chisel

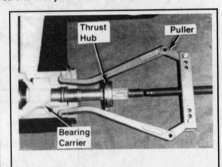
Fig. 31 Pulling out the bearing carrier

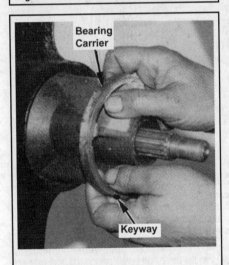
Fig. 32 Removing the bearing carrier after disconnecting the puller

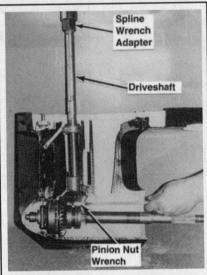
Fig. 33 Removing the driveshaft pinion nut

Fig. 34 Bearing nut retainer tool in position

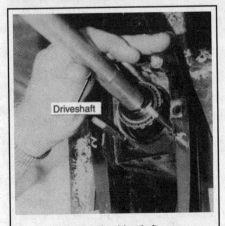
Fig. 35 Removing the driveshaft

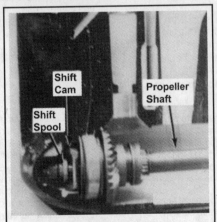
Fig. 36 Removing the propeller shaft

Fig. 37 Shim material locations

10-10 DRIVE SYSTEMS - ALPHA

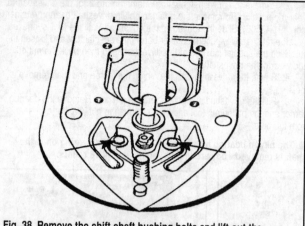

Fig. 38 Remove the shift shaft bushing bolts and lift out the assembly...

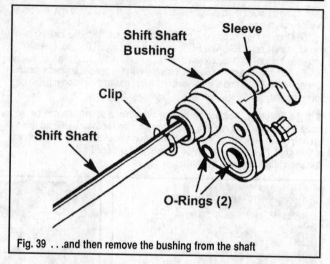

Fig. 39 ...and then remove the bushing from the shaft

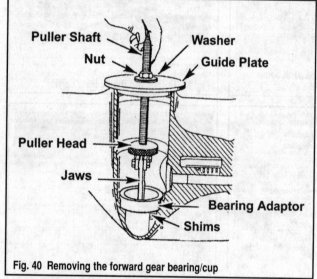

Fig. 40 Removing the forward gear bearing/cup

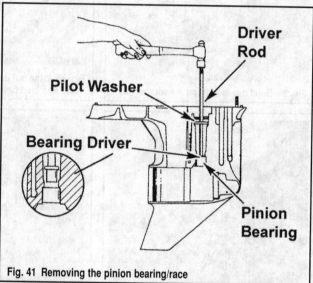

Fig. 41 Removing the pinion bearing/race

10. On the top of the housing where the driveshaft bearing was positioned, reach inside the bearing bore and remove the shim material for the bearing. Identify and save the shim material for later measurement and count.

11. Install the lower unit in a stand so that the propeller shaft bore is facing up.

12. Move the propeller shaft to the port side of the lower unit slightly to clear the shift crank. Pull out the propeller shaft assembly and forward gear.

If further disassembly of the shaft assembly is required, please refer to the separate Propeller Shaft procedures.

13. Remove the two bolts securing the shift shaft bushing to the lower unit housing. Grasp the shift shaft bushing and pull it straight up and out of the lower unit housing.

14. Rotate the lower unit housing or reach inside through the propeller shaft opening and retrieve the shift crank from the gear housing.

15. Remove the clip from the shift shaft and then slide the bushing assembly and sleeve off of the straight end of the shaft.

■ The next step is not required, unless the bearings are to be replaced. Any time the bearings are replaced, they must be adjusted properly with shim material or their service life will be drastically reduced. You will need a puller shaft (#91-31229), nut (#11-24156), guide plate (#91-816243), washer (#91-34961) and a puller head and jaws from #91-34589A1 to perform the next step.

16. Install the special puller tools as shown in the illustration, remove the propeller shaft forward gear roller bearing cup. Tag and wire the shim material pieces from behind the bearing cup together, as an aid during assembly (but you should not reuse them!).

■ The forward gear bearing bore on early units (prior to serial #OF680000) is larger than that on later units (#OF680000 and above). The later style bearing cup is thinner and cannot be interchanged with and earlier unit, or vice versa.

17. Install a pinion bearing driver (#91-36569), pilot washer (#91-36571) and driver rod (#91-37323) into the driveshaft bore from the top, and over the pinion bearing. Give the rod a sharp tap with a small mallet and force out the pinion bearing (race and rollers).

■ If any of the 18 needles had fallen out during another stage of the procedure, make sure that they are placed back into the race prior to pressing it out. Once removed, the bearing race and 18 rollers will require replacement.

Bearing Carrier And Reverse Gear

◆ See Figures 27, 28 and 42 thru 45

1. Certain later models may be equipped with a thrust spacer at the reverse gear - if so, remove it.

2. Clamp the bearing carrier in a vise with soft jaws. Be sure to clamp on the bearing carrier at the reinforcement ribs.

3. Position the jaws of a universal slide hammer puller down through the center of the reverse gear and bearing. Pull the reverse gear, thrust hub/ring and bearing out of the bearing carrier. Check the bearing for roughness, catches, and overheating. If the bearing is removed from the reverse gear, it must be replaced with a new item. If the bearing is found to be defective, press the bearing from the reverse gear using a universal bearing plate (#91-37241) and an arbor press.

DRIVE SYSTEMS - ALPHA

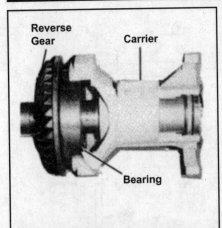

Fig. 42 Cross-section of the assembly

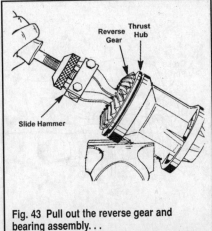

Fig. 43 Pull out the reverse gear and bearing assembly...

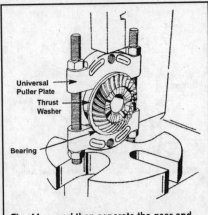

Fig. 44 ...and then separate the gear and bearing

Fig. 45 Removing the oil seal

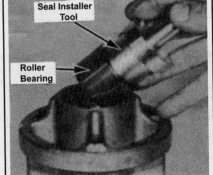

Fig. 46 Installing the roller bearing

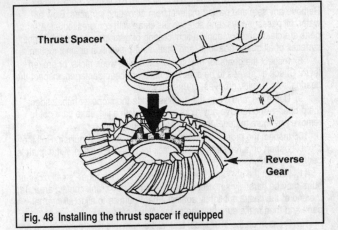

Fig. 47 Installing the oil seals

4. Remove the two propeller shaft oil seals from the bearing carrier. Press the needle bearing free of the carrier, using an arbor press.

To assemble:

◆ See Figures 46, 47 and 48

5. Wash all parts in solvent and blow them dry with compressed air. Remove any corrosion from the outside surface of the bearing carrier. Remove any seal and gasket material from all mating surfaces. Blow all water, oil passageways, and screw holes clean with compressed air. After all parts are clean and dry, apply a light coating of gear lubricant to the bright surfaces of all gears, bearings, and shaft, as a prevention against rusting.

6. Inspect the gears for any sign of excessive wear, nicks, or broken teeth. Check all splines to be sure they have not been damaged. Inspect the bearing sets for pits, grooves, or uneven wear. If the bearing carrier is worn or damaged, it may be replaced with a new assembly at the same time the bearings and seals are installed.

7. Install the new bearing into the bearing carrier with the numbers on the bearing facing the installation tool (#91-15755), as shown. Coat the bearing bore with Quicksilver 2-4-C lubricant.

8. Press two new oil seals into the front of the bearing carrier using the driver tool (#91-31108). These seals are installed back-to-back. Seat the first one with the lip facing in to prevent the lubricant from escaping. Seat the second seal with the lip facing out to prevent water from entering. Make sure you coat the bore for the 2nd seal with a thin film of Loctite(r) 271.

9. Take time to obtain the proper mandrel (#91-36571) for the ball bearing to ensure the force will be applied to the inner race of the bearing. Press the reverse gear ball bearing and thrust washer into place with the washer facing the reverse gear.

10. Turn the bearing carrier over with the seal side down and then press the reverse gear and bearing assembly into the carrier. Always use an adapter to protect the reverse gear. Set the assembly aside for later installation.

11. Install a new O-ring into the groove on the carrier.

12. If your carrier utilized a thrust spacer, install it into the gear recess.

Fig. 48 Installing the thrust spacer if equipped

Propeller Shaft

◆ See Figures 27, 28, 49 and 50

■ Forward gear bearing cup removal is detailed in the Main Components procedures.

1. With a small screwdriver or prybar, unwind the spring from around the sliding clutch. Take care not to damage the spring by bending it out of shape - if it does not coil back to its original position, it will require replacement.

2. Remove the cross-pin and then slide the shift-spool and the shift-actuating shaft off of the shaft.

3. Remove the cotter pin and nut, and then pull off the spool and clutch assemblies.

DRIVE SYSTEMS - ALPHA

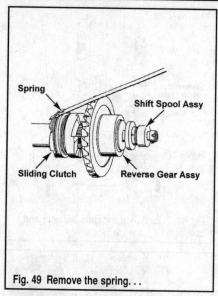

Fig. 49 Remove the spring...

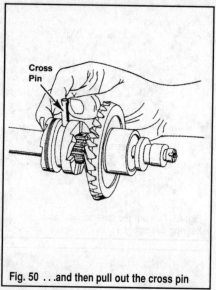

Fig. 50 ...and then pull out the cross pin

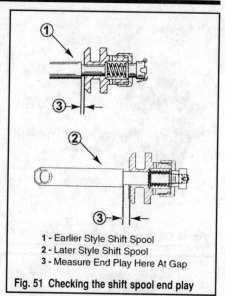

1 - Earlier Style Shift Spool
2 - Later Style Shift Spool
3 - Measure End Play Here At Gap

Fig. 51 Checking the shift spool end play

4. Remove the forward gear and bearing assembly.

■ Separating the forward gear bearing from the gear should only be performed if you intend to replace the bearings.

5. Remove the needle bearings from the forward gear by first splitting the needle bearing inside the gear with a chisel and hammer. Remove the bearings. The bearing case is high tensile steel and may be difficult to cut, if so you can use a grinder to make a notch in the casing.

6. Use a universal puller plate to remove the tapered roller bearing. Press the gear from the bearing.

To assemble:
◆ See Figure 51

7. Wash all parts in solvent and blow them dry with compressed air. Remove any seal and gasket material from all mating surfaces. Blow all water, oil passageways, and screw holes clean with compressed air. After all parts are clean and dry, apply a light coating of gear lubricant to the bright surfaces of all gears, bearings, and shaft, as a prevention against rusting.

8. Inspect the gears for any sign of excessive wear, nicks, or broken teeth. Check all splines to be sure they have not been damaged. Inspect the bearing sets for pits, grooves, or uneven wear.

9. Use a file, and clean the grooves inside the propeller hub. Inspect the propeller and remove any nicks and burrs with a file. Take care not to remove any more material than is absolutely necessary.

10. Inspect the propeller for cracks, damage, or bent condition. Roll the propeller shaft on a flat surface and check for a bent condition. If the shaft is bent over 0.007 in. (0.178mm), it must be replaced.

11. Inspect the shift spool assembly for signs of damage - small nicks or burrs may be carefully smoothed out. Check the shift crank contact area. Tap the end of the castle nut lightly against a hard surface to align all internal parts and then make sure that the spool spins freely.

12. Press in on the shift spool and take a measurement between the end of the shift shaft and the bottom of the spool. Now pull out on the spool and take the measurement again. The difference between the two measurements is your end play and should be 0.002-0.010 in. (0.051-0.254mm).

■ There is a latter style shift spool that has a larger gap than earlier units. It began at serial #0K041000 and may be used when replacing earlier spools. Although the gap is larger, the endplay measurement is still the same as that for earlier models.

13. Slide the clutch assembly onto the propeller shaft so that the cross pin holes line up with the slot in the propeller shaft. The grooved end of the clutch should be facing the propeller end of the propeller shaft.

14. Use a suitable adapter and press the tapered roller bearing onto the forward gear if it was removed. The force must be applied only against the center race. Seat the bearing against the shoulder of the forward gear. Use the proper adapter and press the forward gear needle bearing into the gear from the numbered side. Seat the bearing against the inner gear shoulder.

15. Install the forward gear-and-bearing assembly, with the gear facing the sliding clutch.

16. Now slide on the shift spool so that the cross pin hole lines up with the hole in the clutch assembly.

17. Press in the cross pin and then carefully position the retaining spring over the cross pin.

DISASSEMBLY - COUNTER ROTATION

◆ See Figures 26 and 52 thru 59

The following procedures assume that the lower unit, water pump, seal carrier and propeller have already been removed.

1. To remove the bearing carrier, first straighten the locking tab on the tab washer. Loosen the cover nut (retainer), turning it counterclockwise with the special wrench (#91-61069). Remove the cover nut and tab washer.

■ If the retainer (cover nut) is corroded in place, drill 4 holes in the nut and then break the retainer into pieces with a chisel. Pry out the pieces.

2. Pull out the bearing carrier assembly using a bearing carrier puller (2-armed puller) on the outer rings of the carrier. The thrust hub for the propeller, which was removed previously, can be used to keep the puller jaws in place.

■ If the bearing carrier is corroded in place, apply heat to the housing to loosen the corrosion. While the housing is still hot, use a bearing carrier puller to remove the assembly do not overheat the housing or it will become distorted and useless.

3. The thrust washer may come off with the carrier assembly, so make sure that you reach in and pull it out of the carrier recess.

4. If there is no need to remove the needle bearing, carefully pry out each of the oil seals from the aft end of the carrier - throw them away.

5. If both the seals and the needle bearing are to be removed, insert a driver head (#91-36569) into the carrier with a rod (#91-37323) and drive all three out of the housing. Discard the seals and the bearing.

6. Reach in and remove the aft thrust washer and thrust collar.

7. Pull upward (or outward) on the propeller shaft while pressing down on the forward thrust collar. Doing so should expose the 2 keepers; remove them and then lift out the thrust collar and bearing.

8. Fabricate a removal tool out of a wire hanger and pull out the forward gear bearing adaptor as shown in the illustration.

9. Align the pins on a bearing removal tool (#91-816245) with the holes in the adaptor and apply pressure to the tool so it is evenly distributed across the pins until the adaptor bearing comes free. Get rid of the bearing.

10. Reach in and remove the O-ring. Now install the forward gear installation tool (#91-815850) onto the gear and remove the forward gear, thrust race and thrust bearing at the same time. Remove the forward gear shim.

■ The thrust race is a very tight fit in the propeller shaft bore. If the preceding step does not succeed in removing all of the components, fabricate a wire hanger so there is a small hook on the end and use it to pull out the race.

11. Slide a driveshaft bearing retainer wrench (#91-43506) over the shaft and into position, but do not loosen the retainer yet.

12. Remove the driveshaft pinion nut by sliding the pinion nut holding tool (#91-61067A1) over the propeller shaft so the MR slot is facing the pinion gear. You may need to lift the driveshaft slightly and wiggle it to get the tool to slide over the pinion nut.

13. Once the tool is properly positioned on the nut, slide the bearing carrier into the bore, BACKWARDS, so that it will support the propeller shaft and keep the tool aligned properly.

14. Turn the driveshaft counterclockwise with the special splined wrench adapter (#91-56775) and a break-over handle. Loosen the nut, but do not completely unscrew it.

■ If the pinion nut holding tool is not available, a substitute can be made from a box-end wrench of the proper size for the nut. Grind the wrench down on both sides until it will clear the clutch dog and the pinion gear. Place the wrench on the nut and proceed with disassembling. Remove the nut.

15. Now you can remove the driveshaft bearing retainer in order for the driveshaft to clear the housing. Use a breaker bar on the special tool installed previously and loosen the retainer. Slide the special tool free of the driveshaft, and then back the retainer nut off and clear of the driveshaft.

16. Rotate the driveshaft until the pinion nut is free and then remove all tools.

■ The pinion bearing rollers may drop out of the bearing once the shaft is removed - be careful that you collect all 18 rollers.

17. Pull the driveshaft straight up and set it aside.

18. Remove the bearing carrier and the pinion nut tool and then remove the nut and washer from the bore. Fabricate a retrieval tool by bending a small hook onto the end of a wire hanger and inserting it into the bore. Snag the pinion gear with the hook and then move the propeller shaft around to allow enough room to pull out the gear.

19. On the top of the housing where the driveshaft bearing was positioned, reach inside the bearing bore and remove the shim material for the bearing. Identify and save the shim material for later measurement and count.

20. Install the lower unit in a stand so that the propeller shaft bore is facing down.

21. Grasp the propeller shaft and move it to the port side of the lower unit slightly to clear the shift crank. Pull out the propeller shaft assembly and reverse gear.

■ The rollers in the reverse gear bearing adaptor may come loose while removing the shaft. If so, check the bearing cage carefully to ensure there has been no damage and then snap them back into position. If you notice damage, replace the entire bearing.

22. The thrust race and bearing should come out with the shaft - if so, take them off of the top of the reverse gear. If they do not, they're probably stuck to the adaptor, so pry them off before going further.

23. With a small screwdriver or prybar, unwind the spring from around the sliding clutch. Take care not to damage the spring by bending it out of shape. If it does not spring back to its normal position on removal it will require replacement.

24. Remove the cross-pin and then slide the shift-spool, clutch and reverse gear off of the propeller shaft.

25. If necessary, pry out the needle bearings with a suitable mandrel.

26. If deemed necessary to remove the reverse gear bearing adaptor, you'll need a number of special tools: bolt (#91-31229), nut (#11-24156),

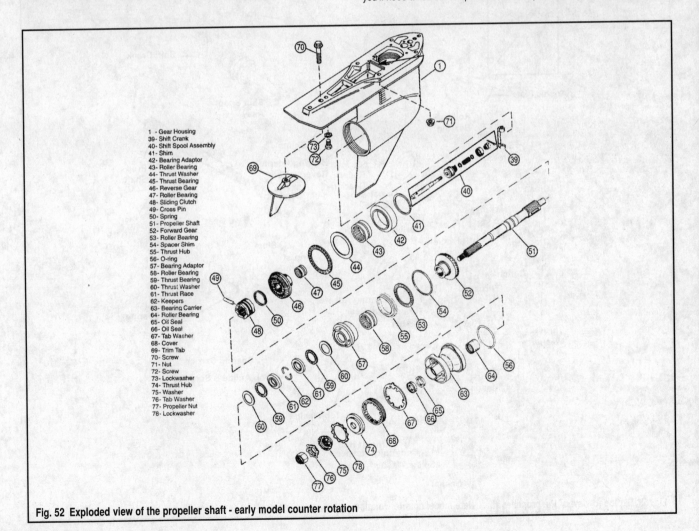

Fig. 52 Exploded view of the propeller shaft - early model counter rotation

10-14 DRIVE SYSTEMS - ALPHA

guide plate (#91-816243), washer (#91-34961), puller head (#90-34569A1) and jaws (#91-816242). We suspect this is all available as a kit, but are unable to find a number for said kit. Assemble the tool as per the instructions, insert it into the bore (and the adaptor) and remove the assembly. Reach into the bore and remove the old shim material; save it, or mark down the total thickness for use later.

27. Remove the two bolts securing the shift shaft bushing to the lower unit housing. Grasp the shift shaft bushing and pull it straight up and out of the lower unit housing.

28. Rotate the lower unit housing or reach inside through the propeller shaft opening and retrieve the shift crank from the gear housing.

29. Remove the clip from the shift shaft and then slide the bushing assembly and sleeve off of the straight end of the shaft.

30. Install a pinion bearing driver (#91-36569), pilot washer (#91-36571) and driver rod (#91-37323) into the driveshaft bore from the top, and over the pinion bearing. Give the rod a sharp tap with a small mallet and force out the pinion bearing (race and rollers).

■ If any of the 18 needles had fallen out during another stage of the procedure, make sure that they are placed back into the race prior to pressing it out. Once removed, the bearing race and 18 rollers will require replacement.

CLEANING AND INSPECTION

1. Wash all parts in solvent and blow them dry with compressed air. Remove all traces of seal and gasket material from all mating surfaces. Blow all water, oil passageways and screw holes clean with compressed air. After cleaning, apply a light coating of gear lubricant to the bright surfaces of all gears, bearings, and shafts as prevention against rusting and corrosion. Check the shift crank to be sure it is not bent.

2. Inspect the water pump impeller plate for wear and corrosion. Replace any worn, corroded, or damaged parts. Use a fine file to remove burrs. Replace all O-rings, gaskets, and seals to ensure satisfactory service from the unit. Clean the corrosion from inside the housing where the bearing carrier was removed.

3. Check to be sure the water intake is clean and free of any foreign material.

4. Inspect the gear case, housings, and covers inside and out for cracks. Check carefully around screw and shaft holes. Check for burrs around machined faces and holes. Check for stripped threads in screw holes and traces of gasket material remaining on mating surfaces.

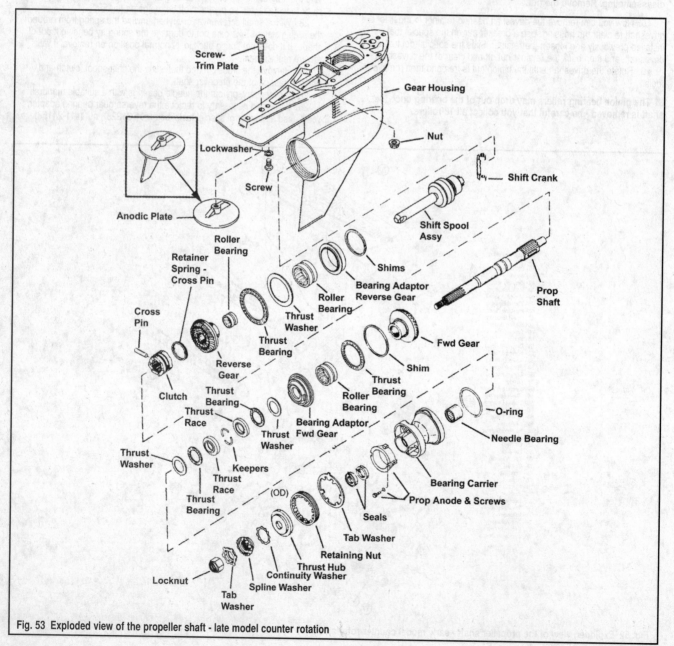

Fig. 53 Exploded view of the propeller shaft - late model counter rotation

DRIVE SYSTEMS - ALPHA   10-15

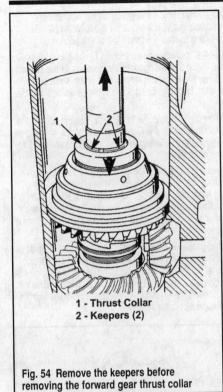

Fig. 54 Remove the keepers before removing the forward gear thrust collar

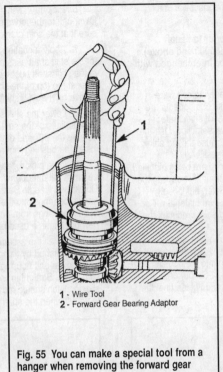

Fig. 55 You can make a special tool from a hanger when removing the forward gear adaptor

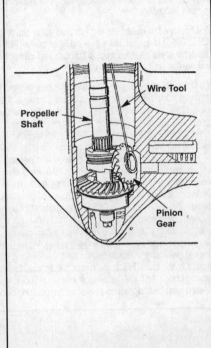

Fig. 56 You can also make your own tool when removing the pinion gear

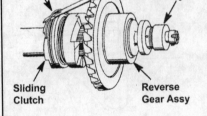

Fig. 57 Remove the spring from the propeller shaft assembly...

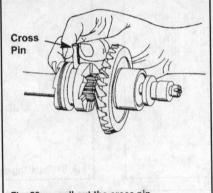

Fig. 58 ...pull out the cross pin...

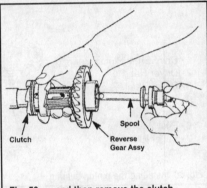

Fig. 59 ...and then remove the clutch, spool and reverse gear

5. Check O-ring seal grooves for sharp edges, which could cut a new seal. Check all oil holes.

6. Inspect the bearing surfaces of the shafts, splines, and keyways for wear and burrs. Look for evidence of an inner bearing race turning on the shaft. Check for damaged threads. Measure the run-out on all shafts to detect any bent condition. If possible, check the shafts in a lathe for out-of-roundness.

7. Inspect the gear teeth and shaft holes for wear and burrs. Hold the center race of each bearing and turn the outer race. The bearing must turn freely without binding or evidence of rough spots. Never spin a ball bearing with compressed air or it will be ruined. Inspect the outside diameter of the outer races and the inside diameter of the inner races for evidence of turning in the housing or on a shaft. Deep discoloration and scores are evidence of overheating.

8. Inspect the thrust washers for wear and distortion. Measure the washers for uniform thickness and flatness.

9. Inspect the forward and reverse gears for distortion, burrs, and cracks. Check for any sign of discoloration, which means the gears are running too hot.

10. Replace all seals, O-rings, and gaskets to ensure maximum service after the work is completed.

ASSEMBLY

Standard Rotation

♦ See Figures 22, 27, 28 and 60 thru 79

※※ CAUTION

The pinion bearing has a slight taper; therefore, if the roller bearing is reversed during installation, the driveshaft will fail. The only way the bearing can be installed properly, is with the numbered (and lettered) end of the bearing case facing up and away from the pinion gear and toward the anti-cavitation plate.

1. Coat the pinion bearing bore in the housing with a thin layer of Quicksilver High Performance Gear Lube.

2. Position the bearing (right side up!) with the cardboard shipping sleeve still in place on the driver head (#91-38628) and then insert the two through the propeller bore and into position in the driveshaft bore.

3. Use the rest of the Bearing Installation and Removal Kit (#91-31229A5), and install the driveshaft roller bearing, pay attention to the

DRIVE SYSTEMS - ALPHA

previous Caution. Pull the bearing up until it bottoms on the lower unit housing shoulder.

■ **The forward gear bearing bore on early units (prior to serial #OF680000) is larger than that on later units (#OF680000 and above). The later style bearing cup is thinner and cannot be interchanged with and earlier unit, or vice versa.**

4. If the forward gear, forward gear bearing and bearing cup were removed from the lower unit housing, and none of these components are being replaced with new items, install the same amount of shim material behind the bearing cup which was removed during disassembling. Use the same thickness and quantity of NEW shim material because the old shim material will have been distorted and give a false thickness.

5. If a new forward gear, or bearing and bearing cup are being installed, place a nominal thickness of 0.020 in. (0.51mm) shim material between the bearing cup and the housing. You'll come back to this later, but loosely speaking, adding or subtracting 0.001 in. (0.25mm) of shim material will change the backlash by an equivalent amount - adding shims reduces the backlash, removing shims increases the lash.

Lubricate the bearing cup bore with a light coating of gear lubricant (Quicksilver High Performance Gear Lube). Place the shim material into the forward gear bore in the housing. Slide the bearing cup into the housing and aligned in the bore. Place the driver cup special tool (#91-31106) onto the bearing cup. Install the installation tool (#91-18605A1) and tighten until the forward gear bearing cup is seated in the housing bore.

6. Pick up the shift actuating crank. Now, reach all the way into the lower unit to the forward end and install the crank on the locating pin. Be sure that the shift crank is facing toward the port side of the housing.

7. Apply a coating of 2-4-C marine lubricant to the O-rings and oil seal for the shift shaft assembly. Place the O-rings into the grooves in the bottom of the shift shaft bushing. Slide the oil seal over the end of the shaft and against the top of the shift shaft bushing. Ensure the E-clip is seated into the groove on the shift shaft.

8. Lower the shift shaft assembly into the lower unit housing. When the male splines of the shift shaft index with the female splines of the shift crank, be sure the 90° end of the shift shaft is facing forward. It is not necessary to install the bushing mounting bolts yet.

9. If the pinion gear bearing cage was not removed, but the needle bearings fell out upon disassembly, coat the 18 needle bearing with Quicksilver Needle Bearing Assembly Lubricant, or equivalent. Place the 18 needle bearings into the bearing cage. Do not mix the needle bearings with needle bearings from another bearing and never use needle bearings from a new bearing in a used bearing cage.

10. Install the assembled propeller shaft into the lower unit housing. This is accomplished by tilting the propeller end of the shaft toward the port side of the lower unit housing to allow the actuating shaft and spool to engage the shift crank. Straighten the propeller shaft and operate the shift shaft. The sliding clutch should move forward when the shaft is turned clockwise and move aft when the shaft is turned counterclockwise.

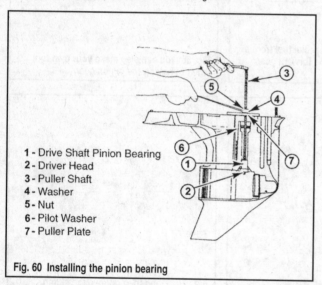

1 - Drive Shaft Pinion Bearing
2 - Driver Head
3 - Puller Shaft
4 - Washer
5 - Nut
6 - Pilot Washer
7 - Puller Plate

Fig. 60 Installing the pinion bearing

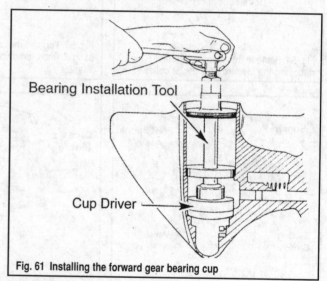

Fig. 61 Installing the forward gear bearing cup

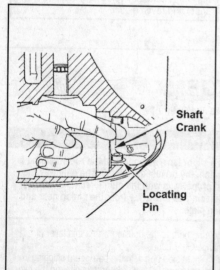

Fig. 62 Make sure the shift crank faces the port side of the housing

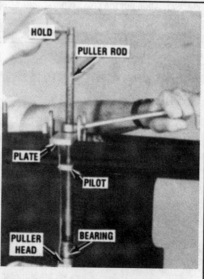

Fig. 63 Use the special tool to install the driveshaft roller bearing

Fig. 64 The shift cam should be facing the port side

DRIVE SYSTEMS - ALPHA   10-17

■ **The shift shaft must point forward when rotated clockwise. If not, pull up on it slightly to re-align the lower splines with the shift crank.**

11. Install the shift shaft bushing bolts and tighten them to 60 inch lbs. (6.8 Nm). Recheck the shaft alignment and then slide the rubber sleeve over the top of the shaft until it is sitting on the oil seal in the bushing.

12. If the tapered roller bearing was pressed off the driveshaft, install the bearing onto the driveshaft. Insert the driveshaft into a suitable adapter, against the bearing shoulder on the shaft, and then press against the inner bearing race to seat the bearing. Use a wrench adapter on the upper end of the driveshaft to protect the splines.

13. If the driveshaft, driveshaft bearing and cup, or gear housing were not replaced and you saved the original driveshaft bearing shim material, position an equivalent amount of new shim material into the bore (but do not use the old shims). If any of the mentioned components were replaced, or you neglected to save the old shim material like we suggested, position 0.030 in. (0.76mm) of shim material into the bore.

■ **A thin washer is used to more evenly distribute the load of the pinion nut against the pinion gear. When the washer is new, both sides are flat. However, after use, one side becomes slightly concave, the other side will be slightly raised. When installing a used washer, the raised side must face toward the pinion gear. We recommend replacing the washer, but it is not absolutely necessary.**

14. Glue the pinion gear washer to the gear with 3M adhesive or Bellows adhesive so you needn't worry about it falling off during installation. Coat the threads of the pinion nut with Loctite 271 and then position it in the MR slot on the end of the special tool (#91-61067A3) - a dab of grease in the slot will help hold the nut in position. Now position the gear and washer into the bore and then slide in the tool with the pinion nut.

15. Install the driveshaft, threaded end first, through the two bearings and into the pinion gear. You may need to wiggle the shaft a bit to get the splines to align with those in the pinion gear, but once in alignment, rotate the driveshaft until the pinion nut is just snug.

16. After the driveshaft is in place and the pinion nut is snugged up, slide the outer bearing race/cup and retainer down the driveshaft and over the bearing and secure it hand tight.

17. Install the bearing carrier backwards into the propshaft bore temporarily to hold the shaft and pinion nut tool in position. Install a driveshaft spline protector tool (#91-56775) onto the top of the shaft and then turn the shaft to tighten the pinion nut to 70 ft. lbs. (95 Nm).

18. Remove the bearing carrier. Now you can secure the retainer by tightening it with the special tool (#91-43506) to 100 ft. lbs. (135 Nm).

19. Obtain the special pre-load tool kit (#91-14311A2). This tool consists of a number of components and should be installed in the order shown in the illustration.

20. Once all components are in position, pull up on the driveshaft and then tighten the 2 Allen screws in the top nut evenly and alternately. Now, screw the bottom nut down until the top of the nut is 1 in. below the bottom of the top nut. Rotate the driveshaft for 3 complete revolutions.

21. Insert a pinion gear shimming tool (#91-56084. Mercury is a little confusing here, because they also call it #56048; in either case, you need to order the pinion gear shimming tool), over the propeller shaft and into the housing. The design of this tool will allow the flat portion of the front of the gauge to clear the pinion gear. After the tool is inserted all the way past the pinion gear, turn the tool until the flat on the tool is away from the pinion gear. Now, with the tool held in this position, insert a 0.025 in. (0.64mm) feeler gauge between the bottom face of the pinion gear and the rounded part of the gauge. Rotate the driveshaft clockwise 120° and take another reading. Do this in at least 3 places.

22. Add the 3 readings together and then divide the sum by 3 for the average pinion gear height. Write it down.

■ **Most feeler gauges are not long enough to make this measurement properly. However, a suitable gauge can be made by grinding the teeth off a hacksaw blade and then using it to take the clearance measurement.**

23. If the average height is not within specifications (0.25 in.), remove the preload tool, the driveshaft retainer and the bearing cup. Add or remove shim material to achieve the correct average gear height. Reinstall everything and recheck the gear height. Continue this until correct. Leave the preload shimming tool installed for forward and reverse gear shimming procedures. Remember, adding 0.001 in. of shim material will increase the gear location by 0.001 in., while subtracting 0.001 in. of shim material will decrease the gear location by 0.001 in.

24. Install the bearing carrier assembly temporarily - you may have to rotate the driveshaft slightly to align the pinion and reverse gear teeth. Align the V-notch in the carrier with the alignment hole in the gear housing.

25. Install the tab washer so the external tab inserts into the hole in the gear housing. Make sure the V-tab is aligned with the V-notch in the carrier.

26. Coat the threads of the gear housing cover (retainer nut) with Quicksilver Lubricant 101 and then thread it into place by hand to prevent cross-threading. Tighten it a couple of turns by hand, and then use bearing carrier wrench (#C-91-61069) and tighten the cover nut to 210 ft. lbs. (285 Nm). Do not secure the tabbed washer at this time. The tab is not secured until after the forward and reverse gear backlash has been adjusted.

■ **The driveshaft pre-load tool will remain installed for both the forward and reverse gear shimming procedures.**

27. Install bearing carrier puller with the arms of the puller on the carrier bosses (or on the carrier outer ring) and the center bolt on the end of the propeller shaft. Tighten the puller center bolt to 45 inch lbs. (5 Nm). This action places a preload on the forward gear, pushing the forward gear into the forward gear bearing. Rotate the driveshaft about 3 full revolutions, and then recheck the torque value on the center puller bolt.

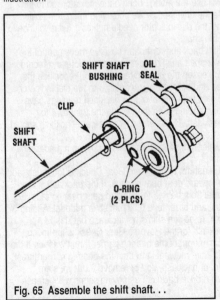

Fig. 65 Assemble the shift shaft...

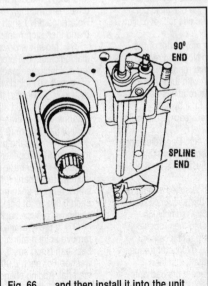

Fig. 66 ...and then install it into the unit

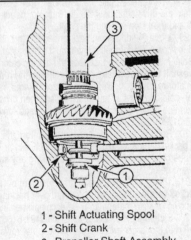

1 - Shift Actuating Spool
2 - Shift Crank
3 - Propeller Shaft Assembly

Fig. 67 Proper propeller shaft to shift crank engagement is important

10-18 DRIVE SYSTEMS - ALPHA

Fig. 68 Pressing the bearing onto the driveshaft

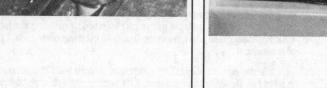
Fig. 69 Use the tool to tighten the driveshaft retaining nut

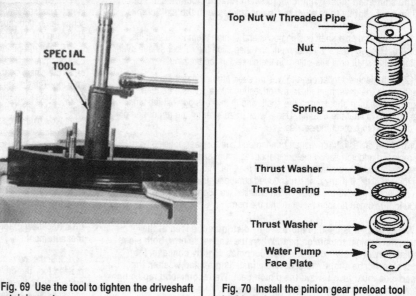
Fig. 70 Install the pinion gear preload tool in this order...

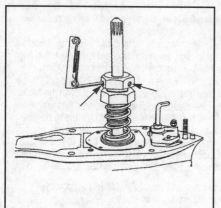
Fig. 71 ...and then tighten the Allen screws to secure it to the driveshaft

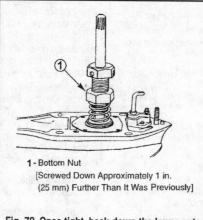
1 - Bottom Nut
[Screwed Down Approximately 1 in. (25 mm) Further Than It Was Previously]
Fig. 72 Once tight, back down the lower nut

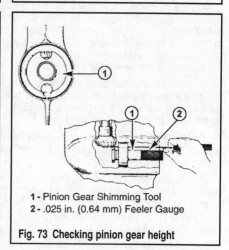
1 - Pinion Gear Shimming Tool
2 - .025 in. (0.64 mm) Feeler Gauge
Fig. 73 Checking pinion gear height

28. Install the backlash indicator (#91-58222A1, except for models with 14:28 gears, which use #91-78473). Position the dial indicator shaft so that it is perpendicular to and touching the line marked **I** on the dial indicator rod (models w/14:28 gears will use a **2** mark). Tighten the indicator rod to the driveshaft and then rotate the driveshaft while observing the dial indicator. Rotate the drive shaft until the dial indicator needle makes one full revolution to **0** and then stop rotating the shaft.

29. Move the driveshaft back-and-forth, and observe movement of the dial indicator. The propeller shaft must remain stationary. Note and record the dial indicator reading. Loosen the dial indicator rod and reposition the driveshaft 90° (clockwise), and then tighten the rod again with the dial indicator resting on **0**. Repeat this step until four backlash readings have been obtained. Add all four readings and then divide the total by 4 to determine the average backlash. The average forward gear backlash should be 0.017-0.028 in. (0.43-0.71mm).

30. Remove the bearing carrier puller from the lower unit. Regardless of the backlash measurements, if they are correct or incorrect, perform the reverse gear backlash check before making any shim adjustments on the forward gear.

31. The reverse gear backlash adjustment outlined in this step should be checked before correcting the forward gear backlash. In this sequence, the changing of shim material may be accomplished in one disassembly job.

32. Slide the pinion nut adapter tool (#91-61061A2) onto the propeller shaft with the larger end of the pinion adapter toward the bearing carrier. Install the special washer (#12-54048) over the end of the exposed threads of the propeller shaft. Thread the propeller nut onto the shaft, and then tighten the nut to 45 inch lbs (5.0 Nm). This action will place a preload onto the reverse gear - pushing it into the reverse gear bearing.

33. Install the backlash indicator (#91-58222A1, except for models with 14:28 gears, which use #91-78473). Position the dial indicator shaft so that it is perpendicular to and touching the line marked **I** on the dial indicator rod (models w/14:28 gears will use a **2** mark). Tighten the indicator rod to the driveshaft and then rotate the driveshaft while observing the dial indicator. Rotate the drive shaft until the dial indicator needle makes 1 full revolution to **0** and then stop rotating the shaft.

34. Move the driveshaft back-and-forth, and observe movement of the dial indicator. The propeller shaft must remain stationary. Note and record the dial indicator reading. Loosen the dial indicator rod and reposition the driveshaft 90° (clockwise), and tighten the rod again with the dial indicator resting on **0**. Repeat this step until a total of four backlash readings have been obtained. Add all four readings and then divide the total by 4 to determine the average backlash. The average reverse gear backlash should be 0.028-0.052 in. (0.71-1.32mm).

35. Remove the propeller nut, special washer, and pinion nut adapter tool from the propeller shaft.

36. If the average backlash for the Forward Gear Check is too much, add shim material behind the forward gear bearing race. If the backlash is too small, remove shim material from behind the forward gear bearing race. For each 0.001 in. of backlash, add or remove 0.001 in. shim material. Adding shims will reduce backlash, removing shims will increase backlash.

37. If the average backlash for the Reverse Gear Check is too much, remove shim material from in front of the bearing carrier thrust washer. If the backlash is too small, add shim material in front of the bearing carrier thrust washer. For each 0.001 in. of backlash, add or remove 0.001 in. shim material. Adding shims will reduce backlash, removing shims will increase backlash.

DRIVE SYSTEMS - ALPHA 10-19

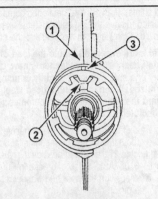

1 - Gear Case Alignment Hole
2 - "V" Shaped Notch In Bearing Carrier
3 - Alignment Tab Of Tab Washer

Fig. 74 Make sure that everything lines up correctly

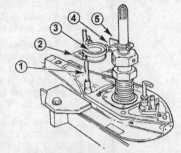

1 - Threaded Rod [3/8 in. (9.5 mm) Obtain Locally]
2 - Dial Indicator Holding Tool
3 - Backlash Indicator
4 - Indicator Pointer
5 - Backlash Indicator Rod

Fig. 75 Checking backlash on the forward gear with the special tool; reverse gear similar

Fig. 76 Cut-away of the lower unit showing the clutch dog engaged with the forward gear. . .

Fig. 77 . . .and the reverse gear

Fig. 78 The pinion gear is properly engaged with the teeth of the reverse gear

Fig. 79 A good shot of the shim material locations on the propeller shaft

38. After correcting the forward gear backlash by changing the thickness of the shim material behind the forward gear cup, or the reverse gear backlash by changing the thickness of shim material between the reverse gear and the reverse gear bearing carrier assembly, proceed with the next step.

39. Remove the bearing preload tool.

40. Remove the bearing carrier assembly and clean off all lubricant used during assembling. After the assembly is clean, install a new O-ring onto the carrier. Apply a liberal coating of Perfect Seal Sealant, or equivalent, to the outer diameter of the bearing carrier where it contacts the lower unit housing (both forward and aft). Do not allow any sealant to enter the ball bearings or the reverse gear. Fill the space below the oil seals with Quicksilver 2-4-C lubricant. Coat the threads of the bearing carrier nut with sealant and install the tab washer and nut as detailed previously. Using a Bearing Carrier Nut special tool (#91-61069), tighten the bearing carrier retainer nut to 210 ft. lbs. (285 Nm). Bend one of the tabs on the locking washer into one of the slots in the cover; bend all other tabs outward.

41. Install the anode plate if your unit was equipped

42. Install the water pump and propellers. Install the lower unit to the upper unit.

Counter Rotation

◆ See Figures 26, 52, 53, 63, 64, 65, 66, 68, 69, 70, 71, 72 and 80 thru 89

✱✱ CAUTION

The pinion bearing has a slight taper; therefore, if the roller bearing is reversed during installation, the driveshaft will fail. The only way the bearing can be installed properly, is with the numbered (and lettered) end of the bearing case facing up and away from the pinion gear and toward the anti-cavitation plate.

1. Coat the pinion bearing bore in the housing with a thin layer of Quicksilver High Performance Gear Lube.

2. Position the bearing (right side up!) with the cardboard shipping sleeve still in place on the driver head (#91-38628) and then insert the two through the propeller bore and into position in the driveshaft bore.

3. Use the rest of the Bearing Installation and Removal Kit (#91-31229A5), and install the driveshaft roller bearing, pay attention to the previous Caution. Pull the bearing up until it bottoms on the lower unit housing shoulder.

4. If the reverse gear, reverse gear adaptor, thrust bearing or bearing race were removed from the lower unit housing, and none of these components are being replaced with new items, install the same amount of shim material behind the bearing cup that was removed during disassembling. Use the same thickness and quantity of NEW shim material because the old shim material will have been distorted and give a false thickness.

If a new reverse gear, reverse gear adaptor, thrust bearing or bearing race are being installed, place a nominal thickness of 0.020 (0.51mm) shim material between the bearing cup and the housing. You'll come back to this later, but loosely speaking, adding or subtracting 0.001 in. (0.25mm) of shim material will change the backlash by an equivalent amount - adding shims reduces the backlash, removing shims increases the lash.

■ The forward and reverse gears look similar. . .if you have mixed them up, the reverse gear has a shorter hub and also has a groove scribed in the flat surface on the back of the gear.

■ Mercury has changed the bearing adaptor style - early models (prior to serial #OF680000) and later models (#OF680000 and above) may not interchange the adaptors.

5. Lubricate the bearing adaptor bore with a light coating of gear lubricant (Quicksilver Special Lubricant 101). Place the shim material into the reverse gear bore in the housing. Slide the bearing adaptor into the housing (without the thrust bearing or race at this time) and align in the bore. Place

10-20 DRIVE SYSTEMS - ALPHA

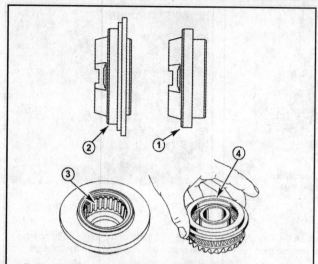

1 - Earlier Style Forward Gear Bearing Adaptor (Prior To S/N 0F680000)
2 - Later Style Forward Gear Bearing Adaptor (S/N 0F680000 And Above)
3 - Later Style Forward Gear Bearing Adaptor Needle Bearing
4 - Earlier Style Thrust Washer

Fig. 80 A good look at the different types of forward gear bearing adaptors that have been used

the adaptor installation tool (#91-18605A1) into the bore so it is resting on the reverse gear and then turn the tool bolt until the adaptor bottoms out on the shoulder in the bore. You will notice that the resistance on the bolt increases dramatically when the adaptor is seated, do not tighten it any further. Remove the installation tool and lift out the reverse gear.

6. Pick up the shift actuating crank. Now, reach all the way into the lower unit to the forward end and install the crank on the locating pin. Be sure that the shift crank is facing toward the port side of the housing.

7. Apply a coating of 2-4-C marine lubricant to the O-rings and oil seal for the shift shaft assembly. Place the O-rings into the grooves in the bottom of the shift shaft bushing. Slide the oil seal over the end of the shaft and against the top of the shift shaft bushing. Ensure the E-clip is seated into the groove on the shift shaft.

8. Lower the shift shaft assembly into the lower unit housing. When the male splines of the shift shaft index with the female splines of the shift crank, be sure the 90° end of the shift shaft is facing forward. It is not necessary to install the bushing mounting bolts yet, but we'd at least hand tighten them.

9. If the pinion gear bearing cage was not removed, but the needle bearings fell out upon disassembly, coat the 18 needle bearing with Quicksilver Needle Bearing Assembly Lubricant, or equivalent. Place the 18 needle bearings into the bearing cage. Do not mix the needle bearings with needle bearings from another bearing and never use needle bearings from a new bearing in a used bearing cage.

10. Coat the large reverse gear thrust bearing with Quicksilver High Performance Gear Lube and then position the thrust race and bearing into the bore and on the adaptor. Once in position, you can install the reverse gear again.

11. Install the assembled propeller shaft into the lower unit housing temporarily. This is accomplished by tilting the propeller end of the shaft toward the port side of the lower unit housing to allow the actuating shaft and spool to engage the shift crank. Straighten the propeller shaft and operate the shift shaft. The sliding clutch should move forward when the shaft is turned clockwise and move aft when the shaft is turned counterclockwise.

■ The shift shaft must point forward when rotated clockwise. If not, pull up on it slightly to re-align the lower splines with the shift crank.

12. If the tapered roller bearing was pressed off the driveshaft, install the bearing onto the driveshaft. Insert the driveshaft into a suitable adapter, against the bearing shoulder on the shaft, and then press against the inner bearing race to seat the bearing. Use a wrench adapter on the upper end of the driveshaft to protect the splines.

13. If the pinion gear, driveshaft, driveshaft bearing and cup, or gear housing were not replaced and you saved the original driveshaft bearing shim material, position an equivalent amount of NEW shim material into the bore (but do not use the old shims). If any of the mentioned components were replaced, or you neglected to save the old shim material like we suggested, position 0.038 in. (0.96mm) of shim material into the bore.

■ A thin washer is used to more evenly distribute the load of the pinion nut against the pinion gear. When the washer is new, both sides are flat. However, after use, one side becomes slightly concave, the other side will be slightly raised. When installing a used washer, the raised side must face toward the pinion gear. We recommend replacing the washer, but it is not absolutely necessary.

14. Glue the pinion gear washer to the gear with 3M adhesive or Bellows adhesive so you needn't worry about it falling off during installation. Coat the threads of the pinion nut with Loctite 271 and then position it in the slot on the end of the special tool (#91-61067A3) - a dab of grease in the slot will help hold the nut in position. Now position the gear and washer into the bore and then slide in the tool with the pinion nut.

15. Install the driveshaft, threaded end first, through the two bearings and into the pinion gear. You may need to wiggle the shaft a bit to get the splines to align with those in the pinion gear, but once in alignment, rotate the driveshaft until the pinion nut is just snug,

16. After the driveshaft is in place and the pinion nut is snugged up, slide the outer bearing race/cup and retainer down the driveshaft and over the bearing. Secure the retainer by tightening it with the special tool (#91-43506) to 100 ft. lbs. (130 Nm).

17. Install the bearing carrier backwards into the propshaft bore temporarily to hold the shaft and pinion nut tool in position. Install a driveshaft spline protector tool (#91-56775) onto the top of the shaft and then turn the shaft to tighten the pinion nut to 70 ft. lbs. (95 Nm).

18. Remove the bearing carrier and any special tools.

19. Obtain the special pre-load tool kit (#91-14311A1). This tool consists of a number of components and should be installed in the order shown in the illustration.

20. Once all components are in position, pull up on the driveshaft and then tighten the 2 Allen screws in the top nut evenly and alternately. Now, screw the bottom nut down until the top of the nut is 1 in. below the bottom of the top nut. Rotate the driveshaft for 3 complete revolutions.

21. Insert pinion gear shimming tool (#91-56084. Mercury is a little confusing here, because they also call it #56048; in either case, you need to order the pinion gear shimming tool), over the propeller shaft and into the housing. The design of this tool will allow the flat portion of the front of the gauge to clear the pinion gear. After the tool is inserted all the way past the pinion gear, turn the tool until the flat on the tool is away from the pinion gear. Now, with the tool held in this position, insert a 0.025 in. (0.64mm) feeler gauge between the bottom face of the pinion gear and the rounded part of the gauge. Rotate the driveshaft clockwise 120° and take another reading. Do this in at least 3 places.

22. Add the 3 readings together and then divide the sum by 3 for the average pinion gear height. Write it down.

■ Most feeler gauges are not long enough to make this measurement properly. However, a suitable gauge can be made by grinding the teeth off a hacksaw blade and then using it to take the clearance measurement.

23. If the average height is not within specifications (0.25 in.), loosen the preload tool, and remove the driveshaft retainer and the bearing cup. Add or remove shim material to achieve the correct average gear height. Reinstall everything and recheck the gear height. Continue this until correct. Leave the preload shimming tool installed for forward and reverse gear shimming procedures. Remember, adding 0.001 in. of shim material will increase the gear location by 0.001 in., while subtracting 0.001 in. of shim material will decrease the gear location by 0.001 in.

24. Install the reverse gear bearing adaptor tool (#91-18605A1) into the propeller shaft bore and onto the reverse gear. Tighten the driver bolt to 45 inch lbs. (5 Nm).

25. Attach the backlash indicator tool (#91-58222 A1, except on models with 14:28 gears, which use #91-78473) to the driveshaft. Position the dial indicator pointer so that it is perpendicular to and touching the line marked I (2 on models with 14:28 gears) on the dial indicator rod. Tighten the indicator rod to the driveshaft and then rotate the driveshaft while observing the dial indicator. Rotate the driveshaft until the dial indicator needle makes 1 full revolution to 0 and then stop rotating the shaft.

26. Move the driveshaft back-and-forth, and observe movement of the dial indicator. The propeller shaft must remain stationary. Note and record the dial indicator reading. Loosen the dial indicator rod and reposition the driveshaft 90° (clockwise), and then tighten the rod again with the dial

DRIVE SYSTEMS - ALPHA　　10-21

indicator resting on **0**. Repeat this step until four backlash readings have been obtained. Add all four readings and then divide the total by 4 to determine the average backlash. The average reverse gear backlash should be 0.040-0.050 in. (1.02-1.27mm).

27. If backlash is within specifications, move to the next step. If the average backlash for the Reverse Gear Check is too much, add shim material behind the reverse gear bearing adaptor. If the backlash is too small, remove shim material from behind the adaptor. For each 0.001 in. of backlash, add or remove 0.001 in. shim material. Adding shims will reduce backlash, removing shims will increase backlash.

28. Remove the installer tool and the reverse gear from the bore.

29. Install the spacer shim into bore and then insert the propeller shaft (with one thrust collar) until it sits on the splined area.

30. Coat the forward gear thrust bearing with Quicksilver High Performance Gear Lube and position it onto the back of the gear. Position the thrust race (if equipped) over the bearing so that the stepped side is up, that is, facing away from the gear.

■ The thrust race has been replaced with a thrust washer on all models above #OF680000.

31. Install the forward gear assembly into an installation tool (#91-815850) and then slide them down and over the propeller shaft, making sure the thrust race seats evenly onto the shim. Tap the race into position lightly with a soft mallet, making sure not to damage the surface of the race.

32. Install the forward gear bearing adaptor so that it seats evenly on the thrust race - you can use the tool you made from the coat hanger during disassembly for this. Remove the hanger tool and the installation tool.

■ Please refer to the illustration regarding different style adaptors.

33. Slide the bearing carrier down and over the propeller shaft until it is fully seated and the V-notch in the carrier is lined up with the alignment hole in the gear housing.

34. Install the tab washer so the external tab inserts into the hole in the gear housing. Make sure the V-tab is aligned with the V-notch in the carrier.

35. Coat the threads of the carrier retainer with Quicksilver Lubricant 101 and then thread it into place by hand to prevent cross-threading. Tighten it a couple of turns by hand, and then use bearing carrier wrench (#91-61069) and tighten the cover nut to 210 ft. lbs. (285 Nm). Do not secure the tabbed washer at this time. The tab is not secured until after the backlash has been adjusted.

36. Install the pinion nut adaptor (#91-61067A3) over the propeller shaft, followed by the washer and propeller nut. Insert a small screwdriver through the access hole in the tools to hold the shaft in position and then tighten the prop nut to 45 inch lbs. (5 Nm). Rotate the driveshaft 3 complete revolutions and then re-torque the nut.

37. Attach the bearing preload tool (#91-89897) to the driveshaft if you had removed it an earlier step. Position the dial indicator pointer so that it is perpendicular to and touching the line marked **I** on the dial indicator rod. Tighten the indicator rod to the driveshaft and then rotate the driveshaft while observing the dial indicator. Rotate the driveshaft until the dial indicator needle makes one full revolution to **0** and then stop rotating the shaft.

38. Move the driveshaft back-and-forth, and observe movement of the dial indicator. The propeller shaft must remain stationary. Note and record the dial indicator reading. Loosen the dial indicator rod and reposition the driveshaft 90° (clockwise), and then tighten the rod again with the dial indicator resting on **0**. Repeat this step until four backlash readings have been obtained. Add all four readings and then divide the total by 4 to determine the average backlash. The average forward gear backlash should

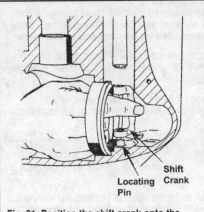

Fig. 81 Position the shift crank onto the locating pin

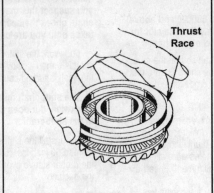

Fig. 82 The stepped side of the thrust race should be facing away from the forward gear

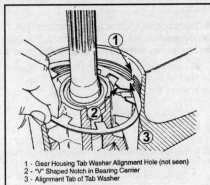

1 - Gear Housing Tab Washer Alignment Hole (not seen)
2 - "V" Shaped Notch in Bearing Carrier
3 - Alignment Tab of Tab Washer

Fig. 83 The V-notch in the carrier should align with the alignment hole in the housing...

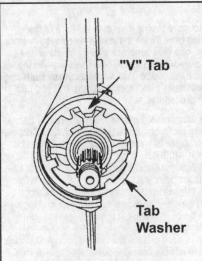

Fig. 84 ...and the V-tab in the washer should align with the V-notch in the carrier

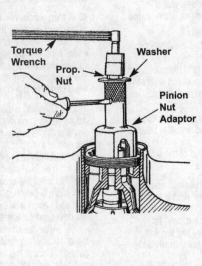

Fig. 85 Use the pinion nut adaptor tool...

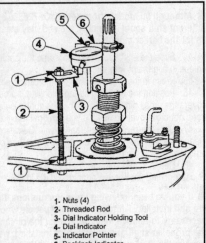

1- Nuts (4)
2- Threaded Rod
3- Dial Indicator Holding Tool
4- Dial Indicator
5- Indicator Pointer
6- Backlash Indicator

Fig. 86 ...and a dial indicator to check the forward gear backlash

10-22 DRIVE SYSTEMS - ALPHA

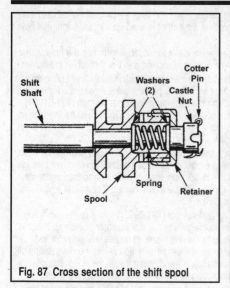

Fig. 87 Cross section of the shift spool

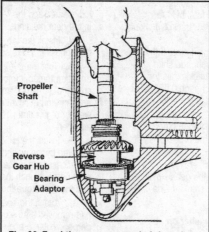

Fig. 88 Feed the reverse gear hub into the adaptor with the propeller shaft tilted to port...

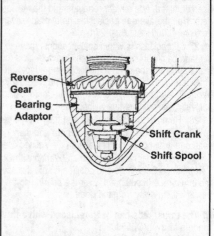

Fig. 89 ...and then straighten the shaft as the gear seats in the adaptor

be 0.017-0.028 in. (0.43-0.71mm).

39. If backlash is within specifications, move to the next step. If the average backlash for the Forward Gear Check is too much, add shim material behind the reverse gear bearing adaptor. If the backlash is too small, remove shim material from behind the adaptor. For each 0.001 in. of backlash, add or remove 0.001 in. shim material. Adding shims will reduce backlash, removing shims will increase backlash. Remove the propeller assembly.

■ The propeller shaft on all later models (serial #OF680000 and above) has a groove on the shaft where the clutch slides on. Also, the slot for the keeper has been moved closer to the forward end of the shaft,

40. Assemble the sliding clutch onto the propeller shaft so that the cross pin holes in the clutch are in alignment with the slot in the shaft. Ensure that the grooved end of the clutch is facing aft, or toward the propeller end of the shaft.

41. Slide the reverse gear onto the shaft.

42. If the shift spool assembly has not been disassembled, remove the cotter pin and throw it away. Thread the castle nut down until it touches the washer and you feel a slight resistance and then loosen it until the slot for the cotter pin is aligned with the hole in the shaft (if the holes were aligned when the nut was snugged down, loosen until the next alignment). Insert a new pin and bend over the ends. Move to the next step.

43. If the spool assembly was disassembled, or if you are continuing from the previous step, tap the forward (castle nut) end of the spool shaft on a hard surface and then make sure that the spool spins freely. Confirm that the spool has no more than 0.002-0.010 in. (0.051-0.254mm) of end play - if it does, go back and repeat the previous step.

■ Although all models from serial #OF726586 and above use a different shift spool assembly, the end play measurement is the same as for the earlier models.

44. Slide the shift spool assembly onto the shaft so that the cross pin hole is in alignment with the clutch slot.

45. Press in the cross pin through the clutch, shaft and shift spool shaft.

46. Position the retaining spring and wind it around the clutch so that it covers the cross pin. Make sure that the spring is flat against the clutch and does not cross over on itself. If you can't get it to lie flat, replace it.

47. Coat the reverse gear bearing with Quicksilver High Performance Gear Lube and position the bearing race and then bearing onto the bearing adaptor.

48. Reach in and spin the shift crank toward the aft end of the unit until it is just touching the bearing adaptor. Hold it in this position.

■ If you've been following these instructions, the driveshaft and pinion are still installed. Although we know people who swear that it is not necessary to remove them (or at least the pinion gear) in order to get the propeller shaft and reverse gear installed, we think it best to remove the pinion gear and the driveshaft. You decide, but we will proceed assuming that they are not installed; if they are, simply skip those particular procedures.

49. Carefully insert the propeller shaft assembly into the housing bore.

Tilt the propeller end of the shaft to the port side of the housing to allow the shift spool to properly engage the shift crank. As soon as the gear hub comes into contact with the adaptor, slowly move the shaft to the center of the bore while inserting it into the adaptor until the gear seats on the adaptor and the spool engages the crank.

■ There is a good chance that one of the 18 needle bearing in the adaptor could become dislodged during propeller shaft installation. If you suspect this may have happened, please be sure to remove the shaft, check the bearings and start over. Do this however many times it takes until you are comfortable that the bearings were not dislodged.

50. Work the shift shaft to make sure it is working correctly - when the shaft is turned clockwise, the clutch should move forward or backward when the shaft is turned counterclockwise.

■ The shift shaft must point forward when the shaft is rotated clockwise. If it does not, lift up on it slightly to realign the shaft splines in the shift crank.

51. Install the shift shaft bushing bolts (if you haven't already) and tighten them to 60 inch lbs. (6.8 Nm). Recheck the shaft alignment and then slide the rubber sleeve over the top of the shaft until it is sitting on the oil seal in the bushing.

■ If you left the driveshaft and pinion gear in position and were still able to get the propeller shaft into position with the reverse gear attached, you can skip these next few steps.

52. Confirm that you have the correct thickness of shims installed in the driveshaft bore (as per the backlash procedures previously).

■ A thin washer is used to more evenly distribute the load of the pinion nut against the pinion gear. When the washer is new, both sides are flat. However, after use, one side becomes slightly concave, the other side will be slightly raised. When installing a used washer, the raised side must face toward the pinion gear. We recommend replacing the washer, but it is not absolutely necessary, particularly if you just replaced it in the backlash procedures.

53. Glue the pinion gear washer to the gear with 3M adhesive or Bellows adhesive so you needn't worry about it falling off during installation. Coat the threads of the pinion nut with Loctite 271 and then position it in the slot on the end of the special tool (#91-61067A3) - a dab of grease in the slot will help hold the nut in position. Now position the gear and washer into the bore and then slide in the tool with the pinion nut.

54. Install the driveshaft, threaded end first, through the two bearings and into the pinion gear. You may need to wiggle the shaft a bit to get the splines to align with those in the pinion gear, but once in alignment, rotate the driveshaft until the pinion nut is just snug,

55. After the driveshaft is in place and the pinion nut is snugged up, slide the outer bearing race/cup and retainer down the driveshaft and over the bearing. Secure the retainer by tightening it with the special tool (#91-43506) to 100 ft. lbs. (135 Nm).

56. Install the bearing carrier backwards into the propshaft bore

temporarily to hold the shaft and pinion nut tool in position. Install a driveshaft spline protector tool (#91-56775) onto the top of the shaft and then turn the shaft to tighten the pinion nut to 70 ft. lbs. (95 Nm). Yes, we know that you already done the last three steps earlier!

57. Remove the carrier and the special tools.
58. Install the appropriate size spacer shim into the propeller shaft bore so that it rests on the shoulder.
59. Coat the thrust bearing with Quicksilver High Performance Gear Lube and position it on the back of the forward gear. Place the thrust race on top of the bearing so the stepped side is facing up - remember that later model adaptors utilize a thrust washer instead of the race.
60. Install the forward gear assembly into an installation tool (#91-815850) and then slide them down and over the propeller shaft, making sure the thrust race seats evenly onto the shim. Tap the race into position lightly with a soft mallet, making sure not to damage the surface of the race.
61. Install the forward gear bearing adaptor so that it seats evenly on the thrust race - you can use the tool you made from the coat hanger during disassembly for this. Remove the hanger tool and the installation tool.
62. Make sure that the top of the adaptor is clean and slide the small thrust race over the shaft so it seats on the adaptor. Coat the small thrust bearing with Quicksilver High Performance Gear Lube and position it on the race.
63. Slide the thrust collar over the shaft so that it rests on the bearing with the stepped side. Pull up on the propeller shaft a little to expose the groove in the shaft and install the 2 keepers. Ease off on the shaft.

■ **Models after serial #OF680000 use a different (thinner) thrust collar. They are not interchangeable with the style used on the earlier models.**

64. Now install the 2nd thrust collar over the shaft so the stepped side is facing up (toward the propeller). Coat the 2nd small thrust bearing with Quicksilver High Performance Gear Lube and position it on the collar.
65. Coat the 2nd small thrust bearing race with Quicksilver High Performance Gear Lube and position it on the inside surface of the bearing carrier (the side opposite the propeller).
66. Coat the large O-ring with Quicksilver High Performance Gear Lube and position it into the bore.
67. Coat the outer surfaces of the bearing carrier with Perfect Seal where they come in contact with the gear housing. Fill the space between the carrier oil seals with Quicksilver 2-4-C marine lubricant. Lubricate the needle bearing with Quicksilver High Performance Gear Lube. Slide the assembly down and over the propeller shaft until it is fully seated and the V-notch in the carrier is lined up with the alignment hole in the gear housing.
68. Install the tab washer so the external tab inserts into the hole in the gear housing. Make sure the V-tab is aligned with the V-notch in the carrier.
69. Coat the threads of the carrier retainer with Quicksilver Lubricant 101 and then thread it into place by hand to prevent cross-threading. Tighten it a couple of turns by hand, and then use bearing carrier wrench (#91-61069) and tighten the cover to 210 ft. lbs. (285 Nm). Bend one of the tabs outward between two of the notches, bend all other tabs inward.
70. Install the water pump and propellers. Install the lower unit to the upper unit.

Propeller

GENERAL INFORMATION

Diameter and Pitch

◆ See Figures 90 and 91

Only two dimensions of the propeller are of real interest to the boat owner: diameter and pitch. These two dimensions are stamped on the propeller hub and always appear in the same order, the diameter first and then the pitch.

The diameter is the measured distance from the tip of one blade to the tip of the other.

The pitch of a propeller is the angle at which the blades are attached to the hub. This figure is expressed in inches of water travel for each revolution of the propeller. In our example of a 9 7/8 in. x 10 1/2 in., the propeller should travel 10 1/2 inches through the water each time it revolves. If the propeller action was perfect and there was no slippage, then the pitch multiplied by the propeller rpm would be the boat speed.

Most manufacturers equip their units with a standard propeller, having a diameter and pitch they consider to be best suited to the engine and boat. Such a propeller allows the engine to run as near to the rated rpm and

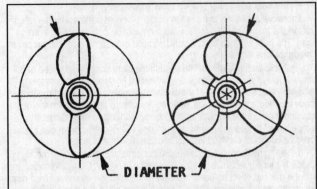

Fig. 90 Diameter and pitch are the two basic dimensions of a propeller. Diameter is measured across the circumference of a circle scribed by the propeller blades

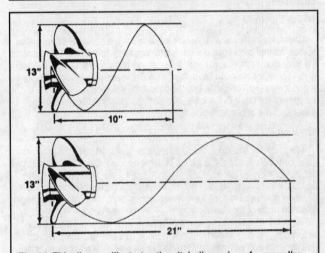

Fig. 91 This diagram illustrates the pitch dimension of a propeller. The pitch is the theoretical distance a propeller would travel through water if there were no friction

horsepower (at full throttle) as possible for the boat design.

The blade area of the propeller determines its load-carrying capacity. A two-blade propeller is used for high-speed running under very light loads.

A four-blade propeller is installed in boats intended to operate at low speeds under very heavy loads such as tugs, barges or large houseboats. The three-blade propeller is the happy medium covering the wide range between high performance units and load carrying workhorses.

Propeller Selection

There is no one propeller that will do the proper job in all cases. The list of sizes and weights of boats is almost endless. This fact, coupled with the many boat-engine combinations, makes the propeller selection for a specific purpose a difficult task. Actually, in many cases the propeller may be changed after a few test runs. Proper selection is aided through the use of charts set up for various engines and boats. These charts should be studied and understood when buying a propeller. However, bear in mind that the charts are based on average boats with average loads; therefore, it may be necessary to make a change in size or pitch, in order to obtain the desired results for the hull design or load condition.

Propellers are available with a wide range of pitch. Remember, a low pitch design takes a smaller bite of water than a high pitch propeller. This means the low pitch propeller will travel less distance through the water per revolution. However, the low pitch will require less horsepower and will allow the engine to run faster.

All engine manufacturers design their units to operate with full throttle at, or in the upper end of the specified operating rpm. If the powerhead is operated at the rated rpm, several positive advantages will be gained.

- Spark plug life will be increased.
- Better fuel economy will be realized.
- Steering effort is often reduced.

DRIVE SYSTEMS - ALPHA

- The boat and power unit will provide best performance.

Therefore, take time to make the proper propeller selection for the rated rpm of the engine at full throttle with what might be considered an average load. The boat will then be correctly balanced between engine and propeller throughout the entire speed range.

A reliable tachometer must be used to measure powerhead speed at full throttle, to ensure that the engine achieves full horsepower and operates efficiently and safely. To test for the correct propeller, make a test run in a body of smooth water with the lower unit in forward gear at full throttle. If the reading is above the manufacturer's recommended operating range, try propellers of greater pitch, until one is found allowing the powerhead to operate continually within the recommended full throttle range.

If the engine is unable to deliver top performance and the powerhead is properly tuned, then the propeller may not be to blame. Operating conditions have a marked effect on performance. For instance, an engine will lose rpm when run in very cold water. It will also lose rpm when run in salt water, as compared with fresh water. A hot, low-barometer day will also cause the engine to lose power.

Cavitation

◆ See Figure 92

Cavitation is the forming of voids in the water just ahead of the propeller blades. Marine propulsion designers are constantly fighting the battle against the formation of these voids, due to excessive blade tip speed and engine wear. The voids may be filled with air or water vapor, or they may actually be a partial vacuum. Cavitation may be caused by installing a piece of equipment too close to the lower unit, such as the knot indicator pickup, depth sounder or bait tank pickup.

Vibration

The propeller should be checked regularly to ensure that all blades are in good condition. If any of the blades become bent or nicked, this condition will set up vibrations in the drive unit and motor. If the vibration becomes very serious, it will cause a loss of power, efficiency, and boat performance. If the vibration is allowed to continue over a period of time, it can have a damaging effect on many of the operating parts.

Vibration in boats can never be completely eliminated, but it can be reduced by keeping all parts in good working condition and through proper maintenance and lubrication. Vibration can also be reduced in some cases by increasing the number of blades. For this reason, many racers use two-blade propellers, while luxury cruisers have four- and five-blade propellers installed.

Shock Absorbers

◆ See Figure 93

The shock absorber in the propeller plays a very important role in protecting the shafting, gears and engine against the shock of a blow, should the propeller strike an underwater object. The shock absorber allows the propeller to stop rotating at the instant of impact, while the power train continues turning.

How much impact the propeller is able to withstand, before causing the shock absorber to slip, is calculated to be more than the force needed to propel the boat, but less than the amount that could damage any part of the power train. Under normal propulsion loads of moving the boat through the water, the hub will not slip. However, it will slip if the propeller strikes an object with a force that would be great enough to stop any part of the power train.

If the power train was to absorb an impact great enough to stop rotation, even for an instant, something would have to give, resulting in severe damage. If a propeller is subjected to repeated striking of underwater objects, it will eventually slip on its clutch hub under normal loads. If the propeller should start to slip, a new shock absorber/cushion hub will have to be installed by a propeller repair shop.

Propeller Rake

◆ See Figure 94

If a propeller blade is examined on a cut extending directly through the center of the hub, and if the blade is set vertical to the propeller hub, the propeller is said to have a zero degree (0°) rake. As the blade slants back, the rake increases. Standard propellers have a rake angle from 0° to 15°.

A higher rake angle generally improves propeller performance in a cavitating or ventilating situation. On lighter, faster boats, a higher rake often will increase performance by holding the bow of the boat higher.

Progressive Pitch

◆ See Figure 95

Progressive pitch is a blade design innovation that improves performance when forward and rotational speed is high and/or the propeller breaks the surface of the water.

Progressive pitch starts low at the leading edge and progressively increases to the trailing edge. The average pitch over the entire blade is the number assigned to that propeller. In the illustration of the progressive pitch, the average pitch assigned to the propeller would be 21.

Cupping

◆ See Figure 96

If the propeller is cast with an edge curl inward on the trailing edge, the blade is said to have a cup. In most cases, cupped blades improve performance. The cup helps the blades to HOLD and not break loose, when operating in a cavitating or ventilating situation.

A cup has the effect of adding to the propeller pitch. Cupping usually will reduce full-throttle engine speed about 150-300 rpm below that of the engine equipped with the same pitch propeller without a cup to the blade. A propeller repair shop is able to increase or decrease the cup on the blades. This change, as explained, will alter powerhead rpm to meet specific operating demands. Cups are rapidly becoming standard on propellers.

In order for a cup to be the most effective, the cup should be completely concave (hollowed) and finished with a sharp corner. If the cup has any convex rounding, the effectiveness of the cup will be reduced.

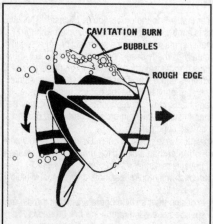

Fig. 92 Cavitation (air bubbles) can damage a prop

Fig. 93 A damaged rubber hub could cause the propeller to slip

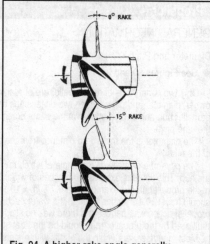

Fig. 94 A higher rake angle generally improves propeller performance

DRIVE SYSTEMS - ALPHA

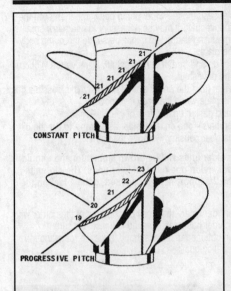

Fig. 95 Comparison of a constant and progressive pitch propeller. Notice how the pitch of the progressive propeller (right) changes to give the blade more thrust

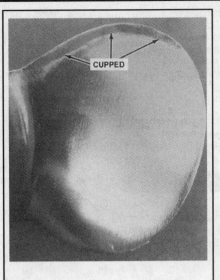

Fig. 96 Propeller with a cupped leading edge. Cupping gives the propeller a better hold in the water

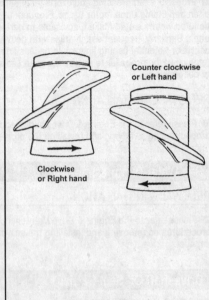

Fig. 97 Note the blade angle is reversed on right and left-hand propellers

Rotation

◆ See Figure 97

Propellers are manufactured as right-hand (RH) rotation or left-hand (LH) rotation. The standard propeller for most units is RH rotation.

A right-hand propeller can easily be identified by observing it. Observe how the blade of the right-hand propeller slants from the lower left to upper right. The left-hand propeller slants in the opposite direction, from lower right to upper left.

When the RH propeller is observed rotating from astern the boat, it will be rotating clockwise when the outboard unit is in forward gear. The left-hand propeller will rotate counterclockwise.

INSPECTION

◆ See Figures 98, 99 and 100

The propeller should be inspected before and after each use to be sure the blades are in good condition. If any of the blades become bent or nicked, this condition will set up vibrations in the motor. Remove and inspect the propeller. Use a file to trim nicks and burrs. Take care not to remove any more material than is absolutely necessary.

** CAUTION

Never run the engine with serious propeller damage, as it can allow for excessive engine speed and/or vibration that can damage the motor. Also, a damaged propeller will cause a reduction in boat performance and handling.

Also, check the rubber and splines inside the propeller hub for damage. If there is damage to either of these, take the propeller to your local marine dealer or a "prop shop". They can evaluate the damaged propeller and determine if it can be saved by re-hubbing.

Additionally, the propeller should be removed AT LEAST every 100 hours of operation or at the end of each season, whichever comes first for cleaning, greasing and inspection. Whenever the propeller is removed, apply a fresh coating of an all-purpose water-resistant marine grade grease to the propeller shaft and the inner diameter of the propeller hub. This is necessary to prevent possible propeller seizure onto the shaft that could lead to costly or troublesome repairs. Also, whenever the propeller is removed, any material entangled behind the propeller should be removed before any damage to the shaft and seals can occur. This may seem like a waste of time at first, but the small amount of time involved in removing the propeller is returned many times by reduced maintenance and repair, including the replacement of expensive parts.

Fig. 98 This propeller is long overdue for repair or replacement

Fig. 99 Although minor damage can be dressed with a file...

Fig. 100 ...a propeller specialist should repair large nicks or damage

10-26 DRIVE SYSTEMS - ALPHA

■ Propeller shaft greasing and debris inspection should occur more often depending upon motor usage. Frequent use in salt, brackish or polluted waters would make it advisable to perform greasing more often. Similarly, frequent use in areas with heavy marine vegetation, debris or potential fishing line would necessitate more frequent removal of the propeller to ensure the gearcase seals are not in danger of becoming cut.

REMOVAL & INSTALLATION

■ Please refer to the Engine & Drive Maintenance section for all procedures on removing and installing the propeller.

Sea Water Pump & Impeller

REMOVAL & INSTALLATION

■ Please refer to the Engine & Drive Maintenance section for all procedures on removing and installing the seawater pump and impeller.

Driveshaft Oil Seal Carrier

REMOVAL & INSTALLATION

◆ See Figures 101 thru 108

1. Remove the lower unit.
2. Pull the water tube and coupling (water) seal out of the water pump body. Slide the water seal up and free of the driveshaft. Discard the seal.
3. Remove the bolts securing the water pump body, and then slide the body up and free of the driveshaft. It may be necessary to use a few small prybars - one on each mounting flange - to gently persuade the pump body to "break" loose from the plate.
4. Slide the impeller up and free of the driveshaft. Remove the Woodruff key from the cutout in the driveshaft.
5. Remove the gasket from the under side of the body. It is possible this gasket will remain on the plate when the body is removed.
6. Slide the plate and gasket up and free of the driveshaft.
7. Use two small prybars - one on each side, as shown - to pry the oil seal carrier free from the gear housing. Remove the carrier.

■ The seal carrier on older units will be brown aluminum and should be replaced by the black nylon one used on later units. The old units can be identified by the 2 support feet as opposed to the 4 feet on the newer style.

8. If the oil seals are damaged, they may be pried out of the carrier with a screwdriver after the carrier has been clamped in a vise. Be sure replacements have been obtained and are at hand, because removing the seals will destroy their sealing qualities.

To install:

** WARNING

The water pump impeller must be in very good condition for satisfactory service. The pump performs an extremely important function by supplying sufficient water to properly cool the stern drive and the engine. Therefore, good shop practice dictates - replace the water pump impeller whenever the unit is disassembled.

9. Inspect the water tube coupling for wear or damage. Its always a good idea to replace the two O-rings.
10. Inspect the impeller for any wear or damage. Replace as necessary.

Fig. 101 Slide the water seal up and off the shaft

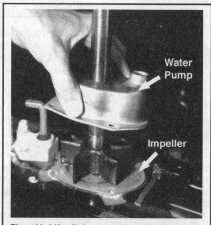

Fig. 102 Lift off the pump housing...

Fig. 103 ...and then remove the old gasket

Fig. 104 Lift off the face plate and gasket...

Fig. 105 ...and then pry off the seal carrier

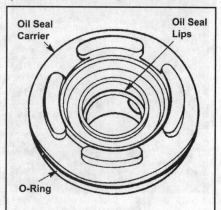

Fig. 106 A good shot of the oil seal and carrier

DRIVE SYSTEMS - ALPHA

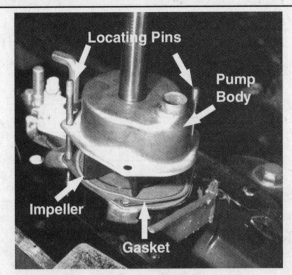

Fig. 107 Use alignment pins when installing the pump...

Fig. 108 ...and then tighten the bolts

11. If the seals in the seal carrier were removed, apply a thin coating of Quicksilver Perfect Seal to the oil seal bore, and then press the seals in back-to-back with the small seal going in first, with the lip facing down - away from seal driver (#91-817569). The seal is properly installed when the driver bottoms against the carrier. Do not press further or the carrier may be damaged. Press the large seal in with the lip facing up, until the driver bottoms on the carrier. This places the seals in the carrier back-to-back. Install a new O-ring around the perimeter of the carrier.

12. Apply a light coating of Quicksilver 2-4-C Marine Lubricant, or equivalent, to the lips of the oil seals and to the O-ring. Slide the carrier down the driveshaft and into the lower unit opening. Do not use a hammer to seat the carrier. Use only hand pressure.

13. Slide the small hole gasket down the driveshaft, followed by the face plate and the large hole gasket. Holes in the gaskets and plate will only align with each other and the holes in the lower unit one way. If the holes do not align one or more of the items is upside down. Correct the situation by turning one or more of the items over.

14. Apply just a "dab" of grease to the Wood-ruff key, and then place it in the driveshaft keyway. Slide the impeller down the driveshaft and onto the face plate with the cutout in the impeller indexed over the Woodruff key. If an old impeller with a "set" to the blades is being installed, face the curl of the blades in a counterclockwise direction. If the direction is reversed, premature impeller failure will surely occur.

15. Slide the pump body down the driveshaft and just to the top of the impeller. Keeping the gasket, impeller, and body mounting holes aligned is not an easy task. However, if a couple of pins are inserted down through just two opposite holes, as shown, the holes will stay aligned while the body is worked down over the impeller. The pins can be old drill bits, small diameter bolts, rod, whatever is handy. Exert some downward pressure on the pump body and at the same time rotate the driveshaft clockwise and the impeller blades will "set" in the proper direction.

16. Start a couple of the body mounting bolts, and then remove the two pins. Install the remaining mounting bolts, and tighten them all to 60 inch lbs. (6.8 Nm).

■ A special water pump seal seating tool is required to properly seat the seal on top of the water pump case. This tool is only available from MerCruiser in a kit (#26-81657A2).

17. Apply a light coating of lubricant to the driveshaft. Slide the water pump face seal down the driveshaft until it is one inch above the pump body. Obtain special seal seating tool from the kit identified above. Slide the tool down the driveshaft and onto the seal. Push the seal down with the tool until the tool makes contact with the pump body. If the tool is not available, a large washer with an inside diameter slightly larger than the driveshaft can be used. Push the seal down evenly onto the water pump body until the seal face makes light contact with the pump body.

DRIVESHAFT HOUSING (UPPER UNIT)

Description

Output power from the engine is connected to the stern drive through a horizontal driveshaft. A coupler is bolted to the flywheel and has a splined hub in the center. The end of the horizontal driveshaft indexes with, and slides into, the center of the hub. Power from the engine is then transmitted through the horizontal driveshaft to a pinion gear set where power direction is changed from horizontal to vertical.

The upper driven gear is pressed onto the outside diameter of the upper driveshaft. The upper driveshaft is splined on the lower end. When the upper gear housing is mated to the lower unit, the end of the lower unit driveshaft indexes into the splined end of the upper driveshaft. Engine power is then transferred down into the lower gear unit.

The horizontal and vertical driveshafts are both mechanically connected. Therefore, anytime the engine is operating, the horizontal and vertical driveshafts are constantly rotating with engine rpm. A double yoke universal joint assembly in the horizontal driveshaft allows the stern drive to be raised or lowered to a required trim/tilt position (within limits), while the engine is operating.

Gear Ratio Identification

◆ See Figures 109 and 110

The gear ratio for the stern drive must be known to permit the proper special tool selection to be made in order to properly position the gears in the housing. The gear ratio for all Alpha stern drives is identified in two places - a decal is affixed to the port side of the upper gear housing. A number such as **1.50R** is followed by the serial number.

The second location of the gear ratio is on the universal joint splined yoke. The identification mark here will be a letter such as **F**.

This letter represents the gear ratio in the upper gear housing. Use the accompanying chart to determine the gear ratio when the letter is known.

If the gears inside the housing have previously been changed from the factory markings on the housing, another method may be used to determine the gear ratio. This method is the least desired because the unit has to be disassembled before the ratio is known. With the gears removed from the housing count the number of teeth on the drive and driven gears. Compare the gear teeth number count to the chart below for the gear ratio of the unit being serviced.

10-28 DRIVE SYSTEMS - ALPHA

Gear Ratio	
Letter	Ratio
A	2.00:1
B	1.98:1
C	1.62:1, 1.65:1
D	1.81:1, 1.84:1
F	1.47:1, 1.50:1
H	1.32:1
K	2.40:1
M	1.50:1 Mag

Fig. 109 Gear ratio identifier

Gear Ratio			
Tooth Count	Drive Gear	Driven Gear	Gear Ratio
14-28	20	24	2.40:1
14-28	24	24	2.00:1
17-28	20	24	1.98:1
17-28	17	19	1.81:1, 1.84:1
17-28	24	24	1.62:1, 1.65:1
17-28	22	20	1.47:1, 1.50:1
17-28	20	16	1.32:1

Fig. 110 Gear tooth chart

Driveshaft Housing (Upper Unit)

REMOVAL & INSTALLATION

◆ See Figures 19 thru 25

1. Remove the stern drive unit and support it in a stand.
2. If you haven't already done so, drain the drive oil. Allow the lubricant to drain into the container. As the lubricant drains, check the color. Dark brown to black indicates normal old lubricant. A chalky white to cream color indicates the presence of water. The presence of any water in the gear lubricant is bad news. The unit must be completely disassembled, inspected, the cause of the problem determined and corrected. Close attention should be given to the back-to-back seals on the propeller shaft, driveshaft and the bearing carrier O-ring. Examine the magnet on the end of the fill/drain plug for evidence of metal particles. The presence of tiny small metal dust like shavings indicates normal wear of the gears, bearings and shafts within the lower unit. Large metal chips or heavy fillings indicate extensive internal damage is taking place and the lower unit must be completely disassembled and inspected. All worn and/or damaged components must be replaced.
3. Place an alignment mark on the trim tab trailing edge and the anti-cavitation plate as an aid during assembling. Remove the plastic plug from the top of the anti-cavitation plate. Remove the bolt directly above the trim tab and lift out the trim tab.
4. Remove the 4 hex nuts and washers (2 on each side), from the sides of the anti-cavitation plate.
5. Remove the nut and washer in front of the speedometer cable fitting. Remove the bolt up and inside the cavity for the trim tab.
6. Separate the two housings by tapping downward on the lower unit with a soft face mallet and a wooden block. In some cases the housing may be difficult to separate because the driveshaft may be corroded and "frozen" in the upper gear housing vertical shaft. Lower the gear housing away from the upper gear housing.

To install:

7. Verify that the remote control shift handle is in the Forward gear position for standard right hand stern drives. The 90° shift lever should be facing forward and in the center of the lower unit. Rotate the propeller shaft clockwise to verify gear engagement.

■ If the unit being serviced is a counter-rotation - left hand drive - the gear shift lever should be in the reverse gear position.

8. Verify that the trim tab bolt is in place in the aft section of the lower unit.
9. Check to be sure the O-ring is in place on the passageway for the gear oil.
10. Apply a coating of Quicksilver 2-4-C Lubricant, or equivalent, to the splines on the end of the driveshaft.
11. Some early lower unit housings may have an aluminum dam behind the water pump; although it is doubtful that any units covered here will still have this. If the aluminum dam is corroded or damaged it must be replaced with a rubber filler unit. If the aluminum dam is serviceable, check the drain hole on the starboard side of the dam to verify it is not plugged. If this drain hole is plugged, severe damage to the gear housing will result.
12. Check to be sure the plastic guide tube is installed in the water pump housing. Verify the water pump face seal is positioned on top of the water pump cover. This seal prevents exhaust gases from entering the water cooling system.

If the unit has an aluminum dam and it has not been replaced with the rubber plug, place a bead of Permatex Ultra Blue Silicone Sealant, or equivalent, across the top of the dam. If a rubber dam is installed, no sealant is required.

13. Now, lower the upper unit (or raise the lower unit) to mate with the lower gear housing, and at the same time check to be sure the water tube starts into the water pump guide tube.
14. It may be necessary to slowly rotate the propeller shaft clockwise until the splines on the upper and lower driveshafts index.
15. Screw the nut onto the stud at the front of the unit. Install the aft screw into the forward hole in the trim tab well.
16. Install the 2 port and 2 starboard bolts and their nuts and washers. Tighten the four 5/8 in. nuts alternately, and a little at-a-time, until the upper unit is tight against the lower gear housing.
17. Now tighten the front nut and the four side nuts to 35 ft. lbs. (47.5 Nm). Tighten the aft screw to 28 ft. lbs. (41 Nm).
18. Install the trim tab with the marks made during disassembly aligned. Tighten the screw(s) to 23 ft. lbs. (31 Nm).

■ The trim tab performs two very important jobs, one of which you may not realize. First, the tab compensates for steering torque. If the boat continually seems to move to port or starboard while the helmsman is on a straight course, the trim tab can be adjusted to the side of the pull. The tab also prevents electrolysis from damaging expensive parts. The tab is not expensive; it should show signs of electrolytic action; and should always be replaced after some of the material has been eaten away or is pitted. Now, if the tab shows no signs of electrolytic action after the boat has been in use over a period of time, the grounding should be checked to ensure more expensive parts are not being damaged. Install the trim tab plastic cover plug.

19. Install the drive unit. Install the propeller if it was removed and refill the drive unit with oil.

DISASSEMBLY & ASSEMBLY

Exploded Views

◆ See Figures 111 and 112

U-Joint/Drive Gear Assembly

◆ See Figures 111 thru 119

Remove the drive unit and separate the upper and lower units. Clamp the upper gear housing into a large shop vise or place the unit into a shop holding fixture designed for the upper gear housing.

1. Remove the dipstick and gasket from the center of the top cover (if equipped). Next, remove the four bolts securing the cover to the housing and lift off the top cover. If the cover is difficult to remove, insert a small pry bar into the cut-out on each side of the housing and pry up on the cover. Check the O-ring around the bearing race inside the cover for damage and deterioration.

DRIVE SYSTEMS - ALPHA 10-29

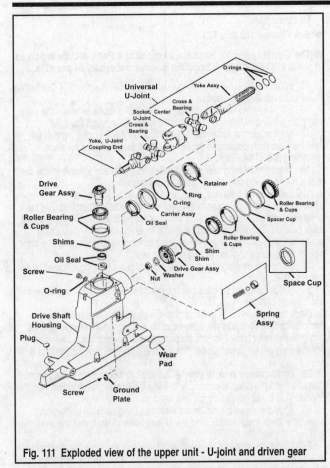

Fig. 111 Exploded view of the upper unit - U-joint and driven gear

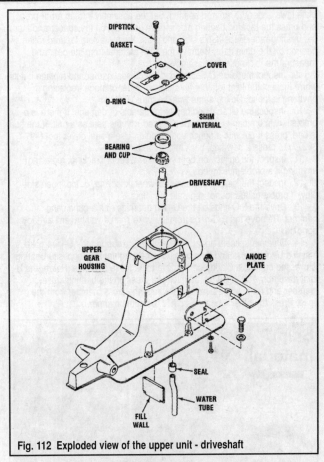

Fig. 112 Exploded view of the upper unit - driveshaft

2. If necessary, position a slide hammer into the bearing race/cup in the underside of the top cover and pull out the race. Carefully remove any shim material under the race and save it if its in good condition; if you're not going to reuse it, make sure that you mark down the total thickness that you removed.

3. Scribe a mark on the bearing retainer and a matching mark on the housing, as an aid during assembling. Loosen the U-joint roller bearing retainer with special wrench (#91-17256). Work the tool counterclockwise to remove the retainer. Pull the U-joint/shaft assembly from the housing. Leave the tool attached to the retainer.

4. Reach inside the housing bore and remove the shim material and spacer ring (certain models). Save and tag the shim material, as an aid during assembling.

5. Clean the assembly in solvent and allow it to dry thoroughly. Inspect the drive (pinion) gear and the driven gear for broken teeth, pitting, etc. Make sure all teeth are worn evenly

6. Slide the two small O-rings off the end of the coupling yoke (the long shaft that goes toward the engine) and then clamp the retainer tool in a vise as a holding fixture. Insert the U-joint assembly, gear side facing up, into the tool.

7. Slide a breaker bar through the joint so that it is levered against the side of the vise. Remove the locknut and washer from the end of the gear yoke and pull off the driven gear/bearing. Set it aside for now.

■ **Removing the locknut will cause the drive gear and bearing assembly to separate from the yoke and u-joint assembly. Make certain that both are supported securely during the separation process or they will fall; causing damage to themselves, or your toes.**

8. If the drive or driven gears are found to be defective, both gears must be replaced as a set. Obtain a Universal Bearing Puller Plate (#91-37241) or equivalent tool. Place the tool under the tapered bearing and tighten the

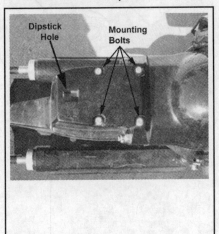

Fig. 113 A good look at the top cover...

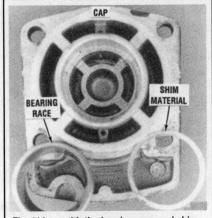

Fig. 114 ...with the bearing race and shim material

Fig. 115 Loosen the retainer to remove the joint assembly...

10-30 DRIVE SYSTEMS - ALPHA

plate jaws under the tapered bearing. Set the assembly into an arbor press and press the tapered bearing off the drive gear. Lift off the spacer and then the bearing race. Repeat this step and press the remaining tapered roller bearing off the drive gear. Remove the shim material from the gear and measure the thickness, for assembling.

9. Inspect the large O-ring around the oil seal carrier and replace it if its worn. Inspect the joint retainer for cracks or other damage, replacing if anything is found. Do the same with the thrust ring.

10. Inspect the oil seal and carrier for defects or damage. If there is a problem with the carrier it must be replaced with the seal as a unit. If the seal is bad, press it out with a punch and hammer. Use a seal driver tool (#91-36577) to press a new seal back into the retainer.

11. Inspect the splines on both yokes for wear or cracking; replacing anything if a problem is found.

12. Inspect the joints themselves for wear, knocking, or too much side play. If problems are found, separate the joints.

13. Drive the 8 C-rings off the U-joint bearings with a punch and hammer. Remove the 2 O-rings from the yoke shaft if you haven't already done so.

14. Obtain a suitable adaptor (#91-38756) to support the U-joint yoke. Install a U-joint press and then press one bearing until the opposite bearing is pressed out into the adaptor. Remove the 2 loose bearings. Rotate the U-joint assembly 180°. Use the adaptor and press on the bearing cross-member and remove the bearings. Remove the cross-member from the yoke. Press the other 4 bearings out in the same manner.

Fig. 116 ...and then pull out the shims

To assemble:
◆ See Figures 120 thru 127

■ The U-joint assembly has been changed to a Perm-A-Lube set-up as of serial #OD899000, no lubrication is either necessary or possible.

15. Lubricate the U-joint bearing cups with a liberal amount of Quicksilver U-Joint/Gimbal Bearing Grease.

16. Still using the adaptor and press, position a cross between the yoke and then press one bearing cup in until it is almost through the yoke.

17. Now rotate the assembly 180° and press another bearing cup on until they are both positioned properly. Repeat these 2 steps for each remaining pair of bearings.

18. Drive the bearing cup retaining C-rings into place (groove in the cap) with a hammer and punch until each is fully seated.

19. Lubricate all bearings, cups, and gears with Quicksilver U-Joint/Gimbal Bearing Grease or an equivalent, before assembling. Set the drive gear into an arbor press with the gear facing down. Place the tapered roller bearing onto the drive gear with the tapered side facing up. Using driver tool (#91-90774), press the bearing down onto the gear until the aft side of the bearing makes contact with the gear.

20. Place the bearing cup over the tapered roller bearing followed by the large spacer.

21. Set the second tapered bearing cup onto the large spacer ring with the tapered side facing up.

Place the tapered roller bearing onto the shaft. Obtain a suitable size mandrel and slowly press the tapered roller bearing down into the bearing cup until the roller barely makes contact with the bearing cup - then stop. The bearing will be drawn closer into the cup when the bearing preload is set.

22. On all models prior to serial #OL10009, temporarily install a hose clamp around the bearing assembly to keep the bearing cups in line with the spacer. This is not necessary on later models.

23. Slide the retainer ring, thrust washer, O-ring, oil seal carrier, and drive gear bearing assembly onto the U-joint yoke shaft. Install the washer and nut - finger-tight.

24. Clamp the unit in a vise with soft face jaws or clamp the retainer tool in a vise and set the U-joint into the retainer tool, as shown. Insert a breaker bar into the lower U-joint to prevent the yoke from rotating.

25. Tighten the nut on the end of the gear yoke until the preload on the bearing begins to increase slightly. Remove the hose clamp from around the bearing (on early models). Pull the bar out of the lower U-joint to permit the shaft to rotate.

26. Attach a socket and an inch-lb torque wrench to the nut on the end of the gear yoke. Hold the bearings stationary and at the same time, rotate the torque wrench and nut 2 full turns. Now turn the wrench and nut a third

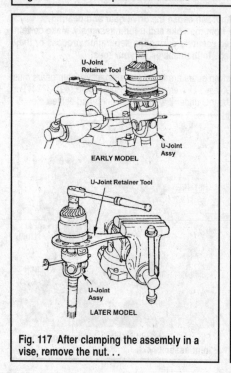

Fig. 117 After clamping the assembly in a vise, remove the nut...

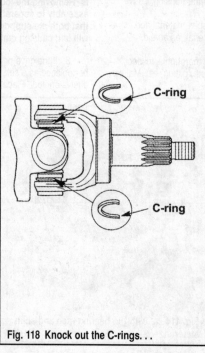

Fig. 118 Knock out the C-rings...

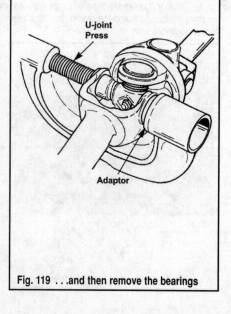

Fig. 119 ...and then remove the bearings

DRIVE SYSTEMS - ALPHA 10-31

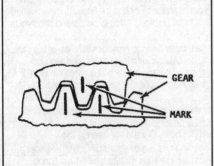

Fig. 120 Most drive and driven gears will have matchmarks on them

Fig. 121 Press the taper bearing onto the drive gear...

Fig. 122 ...and then slide on the bearing cup and the large spacer

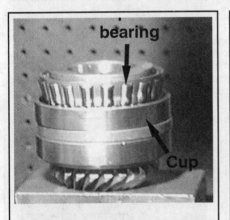

Fig. 123 Install the second bearing cup and then press in the bearing

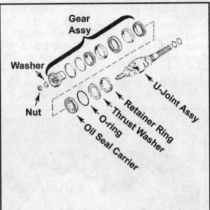

Fig. 124 A good look at the order of assembly

Fig. 125 Use a torque wrench to adjust the preload

revolution and observe the torque wrench indication for a bearing pre-load torque of 8 inch lbs. (0.9 Nm) on new bearings, or 5.25 inch lbs. (0.55 Nm) for used bearings. If the bearing preload torque is under specifications, repeat the procedure, tightening the nut slightly each time until the required bearing preload is obtained.

■ If the bearing preload torque exceeds the specified limit, the bearings must be removed from the gear, inspected and then reassembled, as previously described in this section. If the bearing is damaged it must be replaced with a new bearing and cup. Failure to follow these procedures will place excessive loads on the bearing and gear, resulting in early bearing failure and possible gear damage.

27. Install the same number of bearing-to-driveshaft housing shims in the top cover that were removed and tagged during disassembling - we'd use new ones.

28. Most MerCruiser drive and driven gears have matching marks. These marks must be aligned when the gears are meshed in the gear housing. However, on some models, the gears do not have these marks. In this case, the gears may be installed with any gear mesh. Position the driven and drive gears to mesh properly before installing the U-joint shaft assembly.

29. Install the spacer ring and shims. Insert the assembled shaft, with the gears properly meshed into the upper gear housing. Tighten the bearing retainer nut to 200 ft. lbs. (271 Nm). It is very difficult to tighten the retainer nut to the proper torque value due to the retainer wrench taking the torque.

On beam-type wrenches, measure from the square drive to the fulcrum (pivot) point of the handle. On click-stop or dial-type wrenches, measure from the square drive to the reference mark on the handle (2 bands, a line, etc.). Find your length in the accompanying chart and then use the suggested torque figure.

An alternate method is to tighten the nut securely and then to bring it around to the mark you made during disassembling. From this point tighten it another 1/4 in. past the mark.

Torque Wrench Length in Inches (cm)	Torque Wrench Reading in Lb. Ft. (N·m)
15 (38)	111 (151)
16 (41)	114 (155)
17 (43)	117 (159)
18 (46)	120 (163)
19 (48)	123 (167)
20 (51)	125 (170)
21 (53)	127 (172)
22 (56)	129 (175)
23 (58)	131 (178)
24 (61)	133 (180)
25 (64)	135 (183)
26 (66)	136 (184)
27 (69)	138 (187)
28 (71)	140 (190)
29 (74)	141 (191)
30 (76)	143 (194)
31 (79)	144 (195)
32 (81)	145 (197)
33 (84)	147 (200)
34 (86)	148 (201)
35 (89)	149 (202)
36 (91)	150 (203)

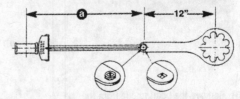

Fig. 126 Use this chart when tightening the bearing retainer

DRIVE SYSTEMS - ALPHA

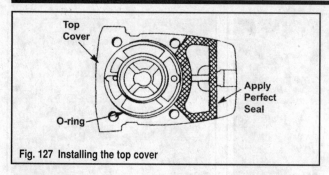

Fig. 127 Installing the top cover

30. Install a new O-ring into the top cover and then coat the area shown in the illustration with Perfect Seal. Install the top cover and tighten the screws to 20 ft. lbs. (27 Nm) in a criss-cross pattern.
31. Connect the upper and lower units and then install the drive.

Driven Gear/Driveshaft Assembly

◆ See Figures 111, 112 and 128 thru 136

If you've gotten here, we're sure that you've already removed the drive, separated the units and pulled out the U-joint and drive gear assembly; but, in the event that you haven't for some reason - do it now.

1. Reach into the top of the unit and withdraw the driven gear assembly, straight up and out of the housing bore.
2. If the driven gear large bearing race is in good condition, the shim material under the race providing bearing clearance is also in good condition and probably does not have to be changed - skip this step and proceed to Step 6.
3. Check the driven gear taper bearing race for evidence of the race spinning inside the housing bore. Look for pits, grooves, scoring and discoloration of the bearing race from overheating and contamination. If any of these conditions exist, the bearing race and shim material must be replaced.
4. Place a slide hammer tool (#91-34569A1) into the housing and expand the jaws out under the bearing race. Using the slide hammer, pull the bearing race out of the housing bore. Discard the bearing race.
5. Reach inside the housing and remove the shim material under the taper bearing race. Save and tag the shim material, as an aid during assembling.

■ The top cover should only be disassembled if the bearing cup is damaged, or the shim material under the cup needs to be changed for gear location or bearing pre-load. These procedures are identified and given in detail in the U-Joint/Drive Gear section.

6. Inspect the bearing cup in the top cover for signs of the cup spinning inside the bore of the cover. Look for pits, grooves, scoring and discoloration on the bearing cup from overheating and contamination. If any of these conditions exist, the cover, bearing cup, shims, and small bearing on the end of the upper drive gear assembly must all be replaced as one unit.
7. Reach into the housing from the top and remove the water tube and seal from inside the housing. Discard the seal. If the water tube is damaged, replace the tube.

Fig. 128 Lift out the driven gear assembly...

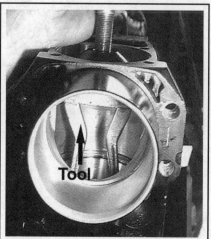

Fig. 129 ...and then pull the bearing race out with a slide hammer

Fig. 130 Removing the shims from the bore

Fig. 131 Remove the bearing cup from the cover and lift out the shims

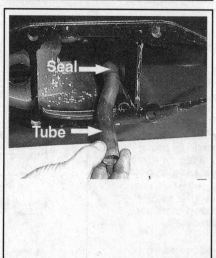

Fig. 132 Remove the water tube and seal...

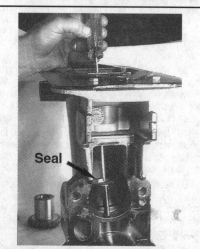

Fig. 133 ...and then pry out the two oil seals

DRIVE SYSTEMS - ALPHA

8. Turn the housing upside down and insert a long shank common screwdriver down through the driveshaft bore in the housing. Tap the two oil seals free of the driveshaft bore. Discard the seals.

9. Clean the driven gear and tapered bearings in solvent and blow them dry with clean compressed air. Do not spin the bearing with compressed air because such action will damage the bearing. Visually examine the gear for pitting, chipped, broken or damaged teeth. Examine both the large and small taper bearings on the shaft. Check for smooth rolling action without any rough spots or dragging. Check to be sure the bearing race inside the top cover and the gear housing are free of pits, grooves, scores, uneven wear and discoloration from overheating. If any of the above damage is found on the bearings or gear, replace the bearing or gear.

10. Place the Universal Puller plate (#91-37241) between the large tapered roller bearing and the driven gear. Tighten the puller plate securely between the driven gear and the large tapered bearing. Press the large tapered bearing free of the driveshaft.

11. Position the puller plate under the small tapered bearing, and then press the bearing free of the shaft.

12. Removal of the gear from the driveshaft is not necessary unless it is damaged and unfit for further service. If the gear is damaged, position the gear under a suitable size mandrel and press the gear free of the driveshaft.

To assemble:

◆ See Figures 137 and 138

13. Press the driveshaft into the driven gear using an arbor press and suitable mandrel until the shoulder on the shaft seats against the gear collar.

14. Position the tapered roller bearing over the end of the shaft with the number side of the bearing facing towards the gear teeth. Using a suitable size mandrel, press the bearing onto the shaft until the inner race of the bearing contacts the shoulder on the shaft. Remove the unit from the arbor press.

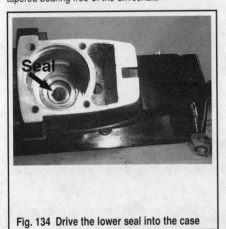

Fig. 134 Drive the lower seal into the case

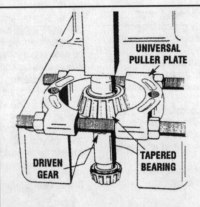

Fig. 135 Remove the large tapered bearing...

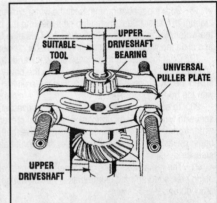

Fig. 136 ...and then remove the small one

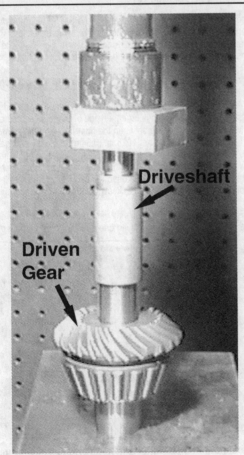

Fig. 137 Press the driveshaft into the driven gear...

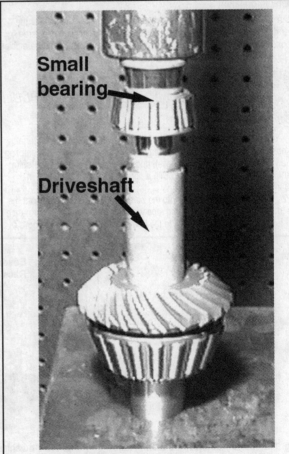

Fig. 138 ...and then press on the bearing

DRIVE SYSTEMS - ALPHA

15. Inspect the housing and cover inside and out for cracks. Check carefully around screw and shaft holes. Inspect machined faces and holes for burrs. Verify all old gasket material has been removed. Check O-ring seal grooves for sharp edges, which could cut a new seal. Inspect all oil passages to ensure they are clear and free of obstruction.

16. Inspect the bearing surfaces of the shafts, splines, and keyways for wear and burrs. Check for evidence of an inner bearing race rotating on the shaft. Measure the run-out on all shafts to detect any bent condition. If possible, check the shafts in a lathe for out-of-roundness.

17. Inspect the gear teeth and shaft holes for wear and burrs. Hold the center race of each bearing and turn the outer race. The bearing must turn freely without any evidence of binding or rough spots. Never spin a bearing with compressed air or the bearing will be ruined. Inspect the balls and rollers for pitting and flat spots. Inspect the outside diameter of the outer race and the inside diameter of the inner races for evidence of turning in the housing or on a shaft. Dark discoloration and/or deep scores are evidence the bearing has overheated.

18. Coat the rubber grommet for the water pick-up tube with a small amount of Perfect Seal. Install the grommet into the housing. Coat the end of the water pickup tube with Quicksilver 2-4-C Marine Lubricant or equivalent and slide the water pick-up tube into the water pick-up.

19. Coat the outside metal case of the lower driveshaft oil seal with a small amount of Loctite. Drive the seal into the cavity with the lip of the seal facing the driven gear. Continue to drive the seal into place until it is flush with the housing.

20. To ease the installation of the upper driveshaft, lubricate the lip of the seal with Multi-Purpose Lubricant. This oil seal prevents exhaust gases and water from entering the splined area of the driveshaft. Do not install the upper driveshaft oil seal until the upper driveshaft bearing preload has been measured.

21. The remainder of the components are installed while the bearing is being preloaded and measurements are taken during the assembling procedures.

PRE-LOAD AND SHIM ADJUSTMENTS

General Information

The following procedures cover preload and shim adjustments for the upper gear housing. The sequence of instructions are divided into three sections and they all must be performed in the order given to ensure efficient and long life operation of the upper gear housing.

Upper Driveshaft Bearing Preload

◆ See Figures 139, 140 and 141

1. If the lower bearing race was removed and the old shim material thickness is known, install the shim material removed during disassembling (or new shims of the same thickness), under the bearing race. If the shim thickness is unknown, then begin with a nominal thickness of 0.015 in. (0.038mm).

2. Place the shim material into the bore of the gear housing. Coat the bore in the housing for the bearing race with Quicksilver High Performance Gear lube or equivalent. Now, slide the bearing race down into the bore of the housing. Place the correct size mandrel (# 91-33493) onto the bearing race and drive the race into the bore until it contacts the shim material on the shoulder in the housing.

3. If the upper oil seal is in place, it must be removed before replacing the driveshaft. This seal must be removed in order to obtain an accurate bearing pre-load measurement. The friction a seal exerts on the shaft would give a false measurement.

4. Coat the gear, bearings, seals, and O-rings with Quicksilver High Performance Gear lube or equivalent. Install the upper driveshaft into the gear housing.

5. If the upper bearing race was removed from the top cover, install the same amount of shim material under the bearing race of the top cover as was removed during disassembling - remember? We told you to either save it or write it down! Press in the top tapered bearing race. Install the top cover temporarily, without the O-ring seal in place. A gasket is not used under the top cover. Now, tighten the 4 bolts, securing the top cover in place, to 20 ft. lbs. (27 Nm).

6. With the unit turned upside down, insert an old gear housing driveshaft, with a pinion gear retaining nut in place, into the upper driveshaft splines. Determine the effort required to turn the upper driveshaft. This may be done with a torque wrench calibrated in inch pounds. The torque value for new or used bearings should be as follows;
• Used Bearings - 3.0-7.5 inch lbs. (0.3-0.8 Nm)
• New Bearings - 6.0-10 inch lbs. (0.7-1.7 Nm)

7. If the reading is too high, remove shim material from under the bearing cup in the top cover. If the torque value is less than the minimum listed, add shim material under the bearing cup in the housing bore. If the shim material pack under the top cover is changed, the bolts securing the cover in place must be tightened again to the proper torque value.

8. Repeat this procedure until the required bearing preload is obtained.

Driven Gear Shimming

◆ See Figures 142, 143 and 144

These instructions pickup the work after the driveshaft bearing preload has been properly adjusted, as described in the previous section. The measurements in this section must be done slowly and precisely, to avoid problems surely to develop if inaccurate measurements are taken.

1. Begin by obtaining a shimming gauge tool (#91-854377), and then install the tool into the drive gear cavity, with the proper tool position (A, B or C), according to the accompanying table.

2. Align the gauge tool with at least two full teeth of the driven gear, making sure that one full tooth is on either side of the tool centerline.

3. Measure the clearance between the gear face and the shimming gauge. This clearance should be 0.025 in. (0.64mm). Most feeler gauges do not have a blade of sufficient length to make this clearance measurement. A hacksaw blade is usually about 0.023 in. thick. Therefore, if the teeth of the blade are ground off, it may then be used to check the clearance. It would be best to check the thickness of the hacksaw blade with a micrometer before grinding the teeth down.

4. Rotate the shimming tool slightly each way to obtain a slight drag on the feeler gauge when it is aligned with one outside tooth. Without moving

Fig. 139 Add shims under the bearing race in the top cover

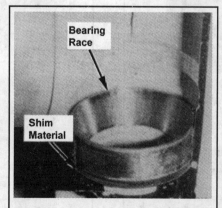

Fig. 140 Add shims under the bearing cup race in the bore

Fig. 141 Insert an old shaft into the unit from the bottom

DRIVE SYSTEMS – ALPHA

Driven Gear Shim Tool Position

Tool Position	Gear Ratio
C	1.47:1
A	1.62:1
B	1.81:1
B	1.94:1
A	2.0:1
B	2.40:1

Fig. 142 Shim tool position

Fig. 143 Use a feeler gauge and the special tool to shim the driven gear...

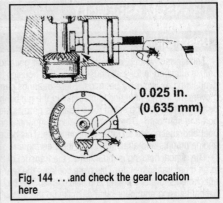

Fig. 144 ...and check the gear location here

the shimming tool, remove the gauge and insert it between the shimming tool and the other outside tooth. This procedure is necessary to align the face of the gauge parallel to the driven gear teeth.

　5. Accurately determine the clearance. If the feeler gauge can be inserted with only a slight drag, then the shimming is correct.

　If the measured clearance is less than 0.025 in. (0.64mm) - the gauge didn't fit; determine the actual clearance and subtract it from the specified 0.025 in. (0.64mm). Write it down! Remove shim material of that same thickness from under the driven gear tapered roller bearing race in the bore. Add an equivalent amount of shim material under the upper driveshaft tapered roller bearing race (in the top cover) to maintain the previously corrected bearing preload.

　If the measured clearance is greater than 0.025 in. (0.64mm) - the gauge was too loose; determine the actual clearance and subtract 0.025 in. (0.64mm) from that figure. Write it down! Add shim material of that same thickness beneath the driven gear tapered roller bearing race in the bore. Remove an equivalent amount of shim material under the upper driveshaft tapered roller bearing race (in the top cover) to maintain the previously corrected bearing preload.

　6. Check the clearance again once you think you've got it right. If the clearance is correct, coat the outside metal case of the upper driveshaft oil seal with a thin layer of Loctite. Apply a thin coating of Multi-Purpose Lubricant to the lip of the seal to ease installation of the upper driveshaft.

■ **It was necessary to remove this seal when measuring the driveshaft bearing preload because of the extra drag a seal places on a rotating shaft. This seal prevents lubricants from working from the driveshaft housing into the exhaust chamber.**

Drive (Pinion) Gear Shimming

◆ See Figures 145 and 146

　This procedure must be followed precisely as described in order to obtain a correct reading of the clearance between the drive gear and the shimming tool. The measurement is needed to ensure the shimming tool face is parallel with the face of the gear.

　1. Insert the shimming tool (#91-60523) into the driveshaft housing top cover cavity. Align the proper tool face (X, Y or Z) with two full teeth of the drive gear. One full tooth should be on each side of the tool centerline.

　2. Insert a 0.025 in. feeler gauge blade between one of the outside teeth and the shimming tool. Rotate the tool to obtain a slight drag on the feeler gauge blade, and then, without moving the shimming tool, remove the gauge and insert it between the tool and the other outside tooth.

　3. If the gauge can be inserted with only a slight drag then the shimming and gear location is correct and you're done.

　If the clearance is greater than the gauge thickness (there is no drag on the feeler gauge), repeat the measuring procedure with increasingly thicker gauges until the same clearance is obtained between both outside gear teeth. Write the actual clearance down and then subtract 0.025 in. (0.635mm) from the measurement and write this figure down. Remove the U-joints and subtract an amount of shim material that equals the last figure.

　If the clearance is less than specified (the gauge would not fit), use a thinner thickness gauge blade and repeat the measuring procedures described until you get an accurate clearance measurement. Subtract the actual clearance from 0.025 in. (0.635mm) and write it down. Remove the U-joints and add an amount of shim material that equals the last figure.

　4. Check the clearance a final time after making changes to the shims. Remove the shimming tool.

　5. After the shimming procedure is complete, position a new O-ring seal on the upper cover. Apply a coating of Quicksilver Perfect Seal, or equivalent, to the area shown on the cover. Install the cover and tighten the four bolts to 20 ft. lbs. (27 Nm).

　The upper driveshaft housing assembly is now completely rebuilt and ready to be assembled to the lower unit, if the lower unit does not require service.

Drive Gear Shim Tool Position

Tool Position	Gear Ratio
Z	1.47:1, 1.50:1
Y	1.62:1, 1.65:1
Y	1.81:1, 1.84:1
Y	1.94:1, 1.98:1
Y	2.0:1
Y	2.40:1

Fig. 145 Shim tool position

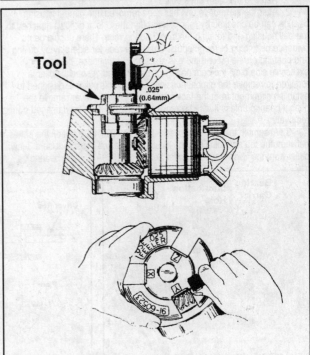

Fig. 146 Use a feeler gauge and the special tool to shim the drive gear

DRIVE SYSTEMS - ALPHA

TRANSOM ASSEMBLY

Description

The stern drive unit, consisting of the upper gear housing and the lower unit, is attached to the bell housing. The bell housing is secured to the gimbal ring by two roller bearings. These bearings permit up-and-down trim movement. The gimbal ring is mounted to the gimbal housing by two roller bearings. This second set of roller bearings permits steering movement to port and starboard. The gimbal housing is mounted on the transom of the boat and attached to an inner transom plate. This plate contains the rear engine mounts and attachments for the steering and shift cables.

The gimbal housing is mounted on the transom of the boat and attached to an inner transom plate. This plate contains the rear engine mounts and attachments for the steering and shift cables. The bell housing, gimbal ring, gimbal housing and inner plate all make up the transom assembly.

The bell housing has extended flanges to connect the exhaust, universal and shift cable bellows, and water hoses between the stern drive unit and the gimbal housing.

The bell housing is held in place by the gimbal ring.

Gimbal housing/transom plate service can generally be performed without removing the gimbal housing. However, in those cases when a part of the housing has been broken, or if the inner transom plate must be removed, the gimbal housing must be removed before the transom plate.

Before the gimbal housing can be removed, several tasks involving considerable work and time must be performed in the following order:
- Stern drive removed.
- Engine removed.
- Bell housing and gimbal ring removed.

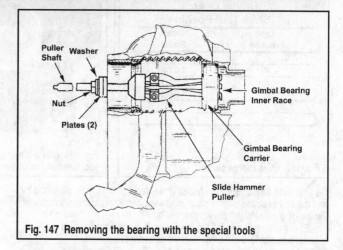

Fig. 147 Removing the bearing with the special tools

Gimbal Bearing

REMOVAL & INSTALLATION

◆ See Figures 147 thru 150

■ The gimbal bearing and carrier are a matched set and must be replaced as an assembly; further, the tolerance ring must be replaced whenever the bearing is replaced.

1. Remove the stern drive unit.
2. After the stern drive unit has been removed, the gimbal housing bearing can be checked by reaching through the U-joint bellows attached to the bell housing and rotating the inner bearing race. There should be no evidence of binding or rough spots. Check the race for side play by pulling and pushing on the race. If there is any sign of roughness, binding, or excessive side play, the bearing should be replaced. A special puller is required to remove the gimbal housing bearing. This puller is designed to establish alignment from the face of the bell housing. Never remove the gimbal bearing unless it is to be replaced, because it will be damaged during removal.
3. Assemble the plates of special tool (#91-29310). Position the plates between the top and middle studs located on the bell housing. Use a 3-jaw puller from the Slide Hammer Puller Set (#91-34569A1). If the bearing assembly is tight, tap the end of the Puller Shaft (#91-31229), with a mallet while attempting to turn the nut (#11-24156).
4. Remove the tolerance ring from the carrier.
5. Reach into the housing and remove the grease seal.

To install:

6. Clean all metal parts in solvent and dry them with compressed air. Never spin ball bearings with compressed air, because such action will ruin the bearing.
7. Inspect the bellows carefully for cracks, cuts, and punctures. Verify that the bellows are still flexible. If the condition of the bellows is doubtful, replace them. In most cases, if the gimbal bearing is damaged due to water, the water has entered through, or around, the bellows.
8. The gimbal housing bearing can only be installed with the bell housing in place in order to establish an alignment reference, so if you have removed the housing install it now.
9. Use a suitable mandrel and pres the grease seal into position until it is seated on the shoulder.
10. Lubricate the outside of a new gimbal bearing carrier assembly with Multi-Purpose Lubricant, or equivalent.
11. Install the tolerance ring over the carrier. The opening in the tolerance band must align with the lubrication opening in the bearing carrier before the carrier is installed. The opening must also align with the lubrication opening in the gimbal housing after the carrier is installed.

■ Observe the notches (could be called "cutouts), in the bearing carrier, indicated in the accompanying illustration. These notches must face forward when the carrier is installed.

12. Install the assembled bearing carrier using the tools indicated in the cutaway line drawing. Insert the Driver Head (#91-32325), through the Mandrel (#91-30366-1), and into the inside diameter of the new bearing. Align the bearing with the gimbal housing by positioning the Plate (#91-29310) between the top and middle studs on the bell housing. Positioning

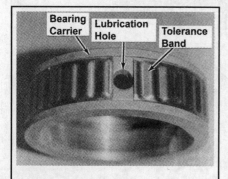

Fig. 148 It is important to install the tolerance ring correctly

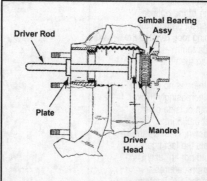

Fig. 149 Installing the bearing with special tools

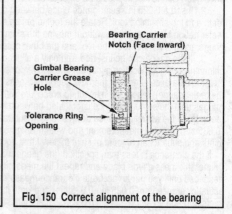

Fig. 150 Correct alignment of the bearing

DRIVE SYSTEMS - ALPHA

the plate as described will ensure the driver rod will remain at right angles to the bearing carrier bore.

13. Now, drive the bearing into the gimbal housing cavity with a hammer. If you can see about 1/8 in. of exposed bore, than you know the bearing is seated fully. Lubricate the bearing using only Quicksilver Multi-Purpose Lubricant through the grease fitting. To lubricate the bearing properly, pump 40 full strokes, to deliver about one full ounce of lubricant.

14. Install the stern drive.

Shift Cable

REMOVAL & INSTALLATION

Remote Control Cable

1. Remove the control box and disconnect the cables. Remove the cotter pin and cable connector securing the cable to the throttle lever. Remove the screw and nut securing the anchor bracket to the housing and then remove the bracket.

2. Trace the cable back to the shift bracket on the engine and disconnect the cable guide from the shift lever pin.

■ **Take note of the positioning of the cable pin in the slot for ease of installation.**

3. Now loosen the anchor block retainer screw (upper on standard rotation props and lower on counter rotation props) and pivot the retainer away from the trunnion. Remove the shift cable.

To install:

4. Lubricate the control unit end of the cable with grease.
5. Install the cable trunnion into the anchor and then install the anchor on the trunnion. Install the screw and locknut, tightening securely.
6. Pull the cable forward so it aligns with the pin.
7. Connect the other end at the remote control and install the unit.

Transom Cable

♦ See Figures 151 thru 157

1. Remove the stern drive.
2. Disconnect the stern drive unit shift cable from the shift plate mounted on the engine. Remove the nut and washers at the forward end of the cable and slide the cable end off of the shift lever pin. Remove the cotter pin at the rear attaching point and slide the trunnion out of the retainer.
3. Loosen the 2 set screws from the cable end guide, and then remove the end guide. If only the inner cable is to be replaced, burrs made by the set screws must be removed from the core wire, to prevent the inner lining of the shift cable from being damaged when the wire is removed. Loosen the jam nut on the metal cable end, and then turn the metal end out of the cable.
4. Pull the threaded support tube from the end of the inner core wire.
5. Remove the protective wrapping from the shift cable in the area where the cable passes through the inner transom plate.
6. Working at the bell housing now, cut the safety wire holding the set screw in the end of the shift slide. Pull out the inner core wire and the slide.
7. Obtain the special shift cable removal and installation tool (#91-12037). Reach into the cable bore in the bell housing and remove the shift cable bellows clamp at the small end of the bellows. Pull the shift cable

Fig. 151 Disconnect the shift cable at the engine

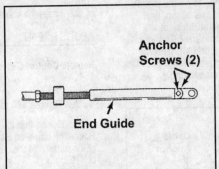

Fig. 152 Loosen the core wire set screws and remove the cable end guide...

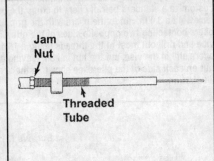

Fig. 153 ...and then remove the threaded support tube

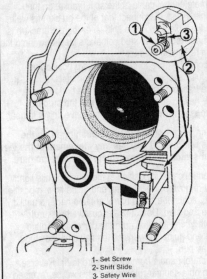

Fig. 154 The screw on the cable slide will be safety wired

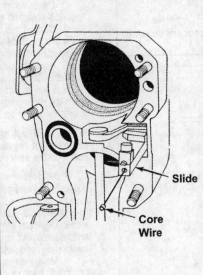

Fig. 155 Remove the core wire and the shift slide

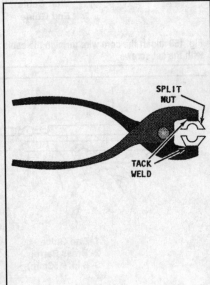

Fig. 156 Fabricate a crimping tool with a pair of pliers and a nut

10-38 DRIVE SYSTEMS - ALPHA

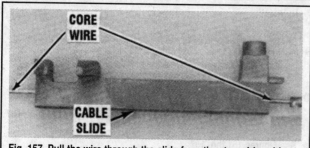

Fig. 157 Pull the wire through the slide from the stern drive side

through the bellows. Using the special tool, completely loosen the locking nut and remove the cable from the bell housing. The locking nut is located on the aft (stud) end of the bell housing in the bore.

To install:
◆ See Figures 158 and 159

8. Inspect the shift bellows for cracks, cuts, and punctures. Verify the bellows are still flexible. If there is the least doubt about the condition of the bellows, they should be replaced. If the bellows are defective and leak, water will enter the boat.

9. Inspect the cable locking nut threads for any damage such as cross-threading and check for any indication of the locking nut separating from the outer casing. Check the length of the shift cable for kinks, cuts or chafing and the inner core wire for signs of unraveling.

■ The manufacturer strongly recommends that the small shift cable bellows clamp be the crimp type, not a worm type clamp and not a tie wrap. To crimp such a clamp, a special tool is required. However, a pair of common pliers may be easily and quickly modified to do the job. To customize a standard pair of pliers to crimp the bellows clamp, first tack weld a 3/4 in. nut to the pliers with the gripping surfaces of the pliers contacting two opposite sides of the nut. Clamp the pliers in a vice and drill out most of the threads using a 1/2 in. drill bit. With the pliers still in the vise, cut the nut in half, leaving an equal amount of the nut on each side of the pliers, as shown in the accompanying illustration.

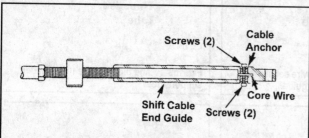

Fig. 158 Insert the core wire through the cable anchor and secure it with the set screws

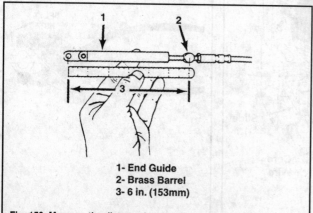

Fig. 159 Measure the distance between the barrel and the end guide

10. Insert the shift cable through the bell housing and bellows.
11. Apply a coating of Perfect Seal to the locking nut threads. Secure the shift cable to the bell housing and tighten the nut securely using the special cable tool. No more than 2 threads on the retainer nut should be visible.
12. Install the bellows clamp on the small end of the shift bellows with the end of the bellows 2 in. from the cable locking nut. Use the modified pliers, described in the previous Note, and squeeze the bellows clamp securely around the bellows. After the clamp is properly installed, water will not be able to enter the bellows and find its way into the boat. Do not allow the bellows to flatten.
13. Install the inner core wire through the shift slide from the stern drive side. Install the inner core wire through the shift cable.
14. Install the Allen head screw securely, and then back it off about 1/8 turn. This adjustment will permit the inner wire to rotate freely. Install the safety wire, twist it tightly and cut off any excess.
15. Coat the inner shift wire with light weight oil, and then feed it into the shift cable until the cable slide enters the bell housing. Now, from inside the boat, carefully pull on the inner wire and be sure it is fully extended.
16. Install the protective wrapping around the cable.
17. Install the threaded tube on the shift cable end and tighten it until it just bottoms; now tighten the jam nut against the cable end.
18. Install the stern drive.
19. Push in on the drive unit shift cable while having someone slowly spin the propeller counterclockwise until it stops. Stretch a bungee cord over the prop to maintain pressure and keep the clutch engaged.

■ Remember this would be opposite on a counter revolution drive.

20. Install the cable end guide over the core wire and then insert the core wire through the anchor. Tighten the 2 set screws securely.
21. Rotate the barrel on the cable threads until the distance between the center of the barrel and the center of the hole in the end guide is 6 in. (153mm).
22. Install the cable to the shift bracket and adjust the cable.

ADJUSTMENT

Remote Control Cable

◆ See Figures 160 thru 165

1. Ensure that the shift bracket end of the transom cable is set correctly. Rotate the barrel on the cable threads until the distance between the center of the barrel and the center of the hole in the end guide is approximately 6 in. (153mm).
2. Reconnect the cable. The brass barrel should be in the retainer with a cotter pin holding it in position. Make sure that the end guide has washers on both sides and that the locknut has been backed out from bottom 1/4 - 1/2 turns.
3. Disconnect the remote control cable at the shift bracket on the engine. If equipped with shift assist, disconnect that at the bracket as well.
4. Confirm that the adjustable stud in the shift lever (on the bracket) is at the bottom of the slot. If not, loosen the stud bolt and slide it down into the bottom of the slot and then re-tighten the nut.
5. Move the remote control handle to the full Forward position on standard rotation units; or full Reverse on counter-rotation units.
6. Push inward on the transom shift cable (this is the bottom cable that's still connected) while having someone slowly rotate the propeller counterclockwise until it stops - this will confirm that the drive unit is in gear.

■ You need only maintain a light pressure while pushing in on the cable so that any slack is removed. If you see any movement at the v-notch on the shift cut-out switch, you know you're pressing too hard.

7. Hold the remote control cable in position at the bracket while lightly pulling on the end guide to remove slack. With the end guide hole in position over the stud (but not on it), adjust the brass barrel at the other end until it lines up with clevis pin hole and stud. Do not use excessive pressure!
8. Once you've got everything lined up, turn the barrel 4 complete revolutions AWAY from the end guide. Now install the cable on the stud (temporarily) and push in the pin.
9. Move the remote control handle from full Forward to full Reverse. Obviously, if you're working on a counter-rotation unit, this will be reversed (Reverse to Forward). At the same time you are doing this have an assistant rotate the propeller shaft clockwise to make sure the clutch is fully engaged.

DRIVE SYSTEMS - ALPHA   10-39

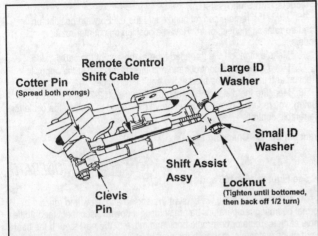

Fig. 160 A good look at the attaching hardware on models with shift assist

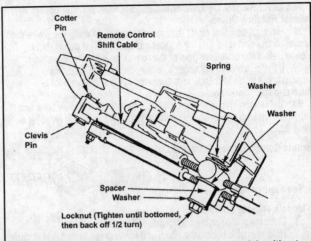

Fig. 161 A good look at the attaching hardware on models without shift assist

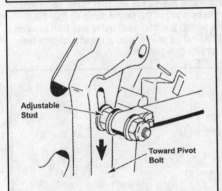

Fig. 162 Make sure the stud is at the bottom of the slot

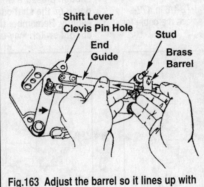

Fig. 163 Adjust the barrel so it lines up with the stud...

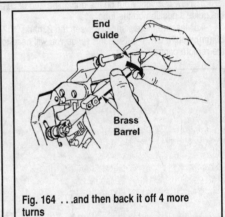

Fig. 164 ...and then back it off 4 more turns

10. Check the position of the shift cut-out switch - early models use a roller switch while later models use a plunger. Whichever your application has, the roller or plunger must be centered in the depression.

11. If not, check that the adjustable stud is still at the bottom of the slot in the bracket shift lever. Check the remote control for correct shift cable output (3 +/- 1/8 in. {76 +/- 3mm}). If both of these situations are OK, check the cable run for kinking or binding.

12. Now you can do a final installation on the shift cable (and shift assist, if equipped).

■ If you can not line up the shift assist attachment points, press in or pull out on the assist assembly to get it right - do not attempt to mess with the remote control cable adjustment!

■ If an extra long remote control cable is used on the vessel, or if there a number of bends in the cable, or if there is insufficient output travel, you may need to perform further adjustment.

On Models With Single Lever Shift/Throttle Remotes And Standard Rotation Drives

13. Shift the remote control lever into the full throttle reverse position while an assistant rotates the propeller shaft clockwise until the clutch engages and the prop stops moving.

14. If the clutch does not engage, loosen the forward stud on the transom cable guide end and move it upward in the shift lever slot until the clutch engages reverse gear.

15. Tighten the stud in the slot. Shift the remote control lever back and forth and then stop it in reverse; confirm that the shift cut-out lever roller is centered.

On Models With Single Lever Shift/Throttle Remotes And Counter Rotation Drives

16. Shift the remote control lever into the full throttle forward position while an assistant rotates the propeller shaft clockwise until the clutch engages and the prop stops moving.

17. If the clutch does not engage, loosen the forward stud on the transom cable guide end and move it upward in the shift lever slot until the clutch engages forward gear.

18. Tighten the stud in the slot. Shift the remote control lever back and forth and then stop it in forward; confirm that the shift cut-out lever roller is centered.

Two Lever Remote Controls With Separate Shift/Throttle Levers And Standard Rotation Drives

19. Have an assistant rotate the propeller clockwise while you move the remote control shift lever into the full reverse position. The clutch should engage before the shift lever gets all the way to the full reverse position.

20. If the clutch does not engage, loosen the forward stud on the transom cable guide end and move it upward in the shift lever slot until the clutch engages reverse gear.

21. Tighten the stud in the slot. Shift the remote control lever back and forth and then stop it in reverse; confirm that the shift cut-out lever roller is centered.

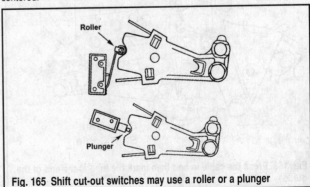

Fig. 165 Shift cut-out switches may use a roller or a plunger

DRIVE SYSTEMS - ALPHA

Two Lever Remote Controls With Separate Shift/Throttle Levers And Counter Rotation Drives

22. Have an assistant rotate the propeller clockwise while you move the remote control shift lever into the full forward position. The clutch should engage before the shift lever gets all the way to the full reverse position.

23. If the clutch does not engage, loosen the forward stud on the transom cable guide end and move it upward in the shift lever slot until the clutch engages forward gear.

24. Tighten the stud in the slot. Shift the remote control lever back and forth and then stop it in forward; confirm that the shift cut-out lever roller is centered.

Remote Control Cable - Checking Output

◆ See Figures 166 and 167

Models Without Shift Assist

1. Disconnect the remote control cable at the shift bracket.
2. Hold the cable behind the barrel and lightly press in on the shift cable end guide. Make a mark on the threaded tube at the end of the cable end guide.
3. Now pull lightly out on the end guide and once again mark the end of the guide's position on the tube.
4. Measure the distance between the 2 marks. If 2-7/8 - 3-1/8 in. (73-80mm), you're OK. If outside this range you will need to replace the cable (or the remote control).

Models With Shift Assist

5. Move the remote control handle into the full Forward position on standard rotation models, or full Reverse position on counter-rotation models.
6. Make a mark on the threaded tube at the end of the cable guide.
7. Move the remote control handle to the full position in the opposite direction and then mark the tube at the end of the guide.
8. Measure the distance between the 2 marks. If 2-7/8 - 3-1/8 in. (73-80mm), you're OK. If outside this range you will need to replace the cable (or the remote control).

Transom Cable - Checking Play

◆ See Figures 165, 168 and 169

1. If the boat is in the water, start the engine and allow it to reach normal operating temperature before shutting it down. Disconnect the throttle cable at the carburetor or throttle body and move to the next step. If the boat is out of the water, just start with the next step.
2. Disconnect the remote control shift cable at the engine bracket.

■ When checking for excessive play in the transom cable, apply just enough pressure when pushing or pulling on the cable so that the V-notch on the shift cut-out switch just begins to move and then ease off slightly. Remember that depending on the year of the application, the cut-out switch may use a roller or a plunger.

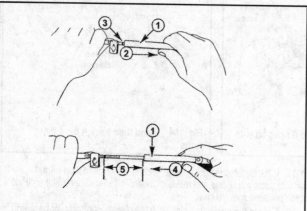

1 - Shift Cable end Guide
2 - Remote Control Shift Cable - Lightly Push On End Guide
3 - Place A Mark On Tube Against Edge Of Cable End Guide
4 - Remote Control Shift Cable - Lightly Pull In On End Guide
5 - Measurement Taken From Mark To Edge Of Cable End Guide

Fig. 166 Checking the output on models without shift assist

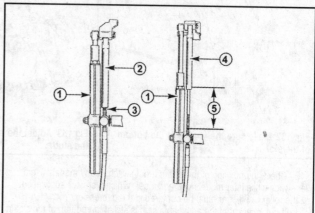

1 - Shift Assist Assembly
2 - Remote Control Shift Cable - Retracted
3 - Place a Mark On Tube Against Edge Of Cable End Guide
4 - Remote Control Shift Cable - Extended
5 - Measurement Taken from Mark To Edge Of Cable End Guide

Fig. 167 Checking the output on models with shift assist

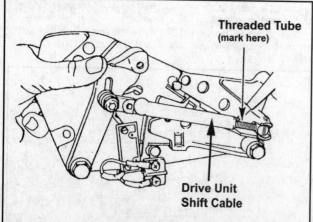

Fig. 168 Press the cable in and then mark the tube at the end of the guide

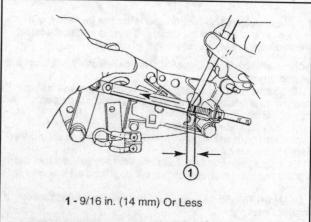

1 - 9/16 in. (14 mm) Or Less

Fig. 169 Now pull it out and mark the end of the guide again - measure the difference

DRIVE SYSTEMS - ALPHA

3. Lightly press in on the transom cable shift lever while having an assistant rotate the propeller/shaft counterclockwise until it stops - this will ensure full clutch engagement with the unit in gear. Use a marker to mark the threaded tube at the end of the cable end guide.

4. Now, with your assistant still maintaining pressure on the propeller to keep the clutch locked with the gear, lightly pull the cable end outward and then place another mark on the threaded tube where the end guide stops.

5. Measure the distance between the 2 marks. If cable play is 9/16 in. (14mm) or less, you're in good shape. If play is greater than the figure, you will need to remove the drive unit and recheck the cable set-up.

6. Reconnect the remote control cable.

7. If your boat was in the water, reconnect the throttle cable.

Transom Cable - Isolating Play

◆ See Figures 170, 171 and 172

■ You will need to create a pointer and scale as shown in the accompanying illustration in order to complete this procedure.

1. Install a scale and pointer to the shift shaft at the front of the drive unit as shown.

2. Have an assistant maintain pressure on the propeller/shaft so the clutch is locked with the gear. Now turn the shift shaft coupler (lightly!) counterclockwise and note the position on the degree scale.

3. Now have the assistant rotate the propeller counterclockwise until it locks with the gear while you turn the coupler clockwise at the same time - note the pointer location on the scale.

4. Still maintaining pressure on the propeller, now rotate the coupler counterclockwise and note the final position.

5. Determine the breadth of the pointers movement through each mark. If the total travel is 12° or less, the problem is with the transom cable, the upper shift shaft assembly and/or the lever. If travel is more than 12°, the problem is in the shift spool assembly on the propeller shaft.

Shift Cut-Out Switch

ADJUSTMENT

Models W/Roller Switch

◆ See Figures 173 and 174

■ The cut-out switch is found on the shift plate.

1. Disconnect the switch wire (White/Green) at the terminal block.

2. Connect an ohmmeter as per the manufacturer's instructions. Connect the positive lead to the White/Green wire and the negative lead to the switch's black wire at the terminal block. Set the meter to the RX1 scale.

3. Slowly move the switch roller off of its seat. The circuit should close (showing continuity) when the roller has been moved 1/8 in. (3mm). If the switch closes before being moved as specified, bend the roller away from its seat. If it closes after the specification, bend it toward the seat.

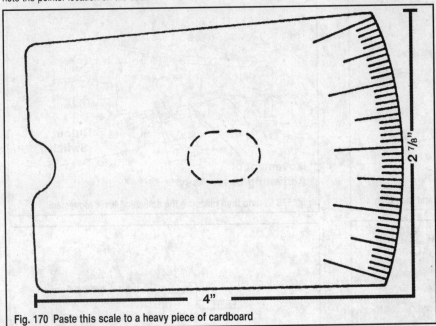

Fig. 170 Paste this scale to a heavy piece of cardboard

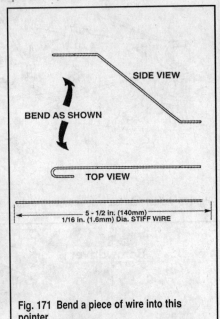

Fig. 171 Bend a piece of wire into this pointer

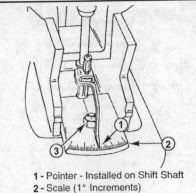

1 - Pointer - Installed on Shift Shaft
2 - Scale (1° Increments)
3 - Nut - Hand Tight Only

Fig. 172 Set up the scale and pointer to check for play

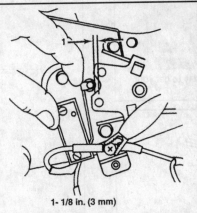

1- 1/8 in. (3 mm)

Fig. 173 Move the roller off its seat until the circuit closes

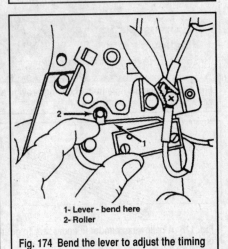

1- Lever - bend here
2- Roller

Fig. 174 Bend the lever to adjust the timing

10-42 DRIVE SYSTEMS - ALPHA

■ Although Mercury provides a special tool for this adjustment, we think it is easily adjusted without the tool.

4. Reconnect the wires at the block and cover them with Liquid Neoprene.

Models W/Plunger Switch

◆ See Figures 175 and 176

1. Loosen the switch mounting screws while holding the nuts on the back of the shift plate.
2. Slowly move the switch until you can achieve a 1/32 in. (0.8mm) clearance between the end of the plunger and the bottom of the V-notch.
3. Now slide a thin ruler behind the activating lever assembly and push up or down on it until the switch opens or closes. The measurement of lever movement must be 3/16 +/- 1/32 in. (4.8 +/- 0.8mm).
4. If outside the measurement range, move the switch ever so slightly and then try again. Repeat until within range.
5. Tighten the switch mounting screws and recheck the plunger-to-notch clearance.

Exhaust Bellows

REMOVAL & INSTALLATION

◆ See Figures 177, 178 and 179

■ Although in the past we have provided instructions for removing the bellows without removing the drive unit, we now believe the best way to replace the bellows is by removing the Stern drive.

1. Remove the stern drive.
2. Tilt the bell housing slightly up and move it to the port side. Access to the aft clamp on the exhaust bellows is from the bottom of the bell housing. Remove the clamp.
3. Access to the forward clamp on the exhaust bellows is through the access hole on the port side of the gimbal housing. Insert a screwdriver through the access hole and remove the forward exhaust bellows clamp.
4. Pull the exhaust bellows from the gimbal housing and bell housing exhaust flanges. It may be necessary to exert considerable force to pull the bellows loose because of the adhesive used during installation.

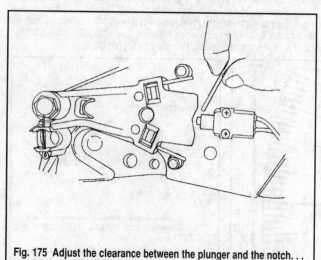

Fig. 175 Adjust the clearance between the plunger and the notch...

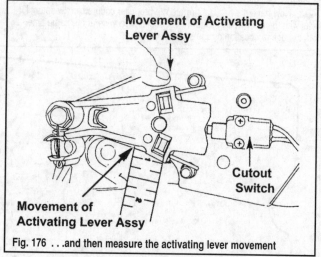

Fig. 176 ...and then measure the activating lever movement

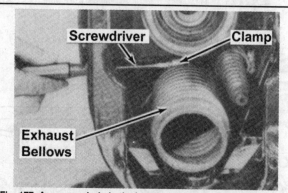

Fig. 177 An access hole in the housing provide a way to get your screwdriver on the clamp screw

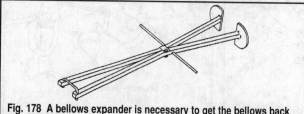

Fig. 178 A bellows expander is necessary to get the bellows back onto the flange

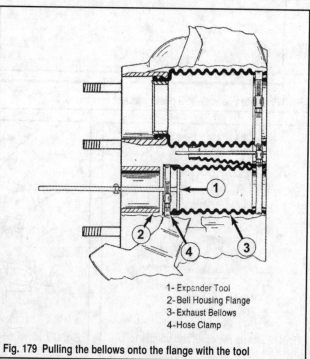

Fig. 179 Pulling the bellows onto the flange with the tool

DRIVE SYSTEMS - ALPHA

To install:

5. Clean the bellows mounting flanges of the upper bell housing with a wire brush or sandpaper, and then wipe the surface clean with lacquer thinner to remove any old glue.

6. If new bellows are being installed (and if you've removed it, you should install a new one), be sure to clean the powder residue from the bellows surface with warm soapy water and dry thoroughly. The powdery substance is a mold release agent and the adhesive will not bond if this powder is left on the sealing surfaces.

7. Check the flanges and housing for cracks, nicks, or corrosion. Clean the bellows clamps thoroughly to ensure a good ground. New bellows have a clip on each end of the bellows to ground the clamp. Check the clamps for cracks or nicks. Replace the clamp if in doubt as to their condition. Always use stainless steel clamps for satisfactory service.

✳✳ WARNING

Bellows Adhesive must be used to ensure a satisfactory installation. This adhesive is extremely toxic and flammable. Therefore, make every effort to ensure adequate ventilation in the work area during its use. Work in the outdoors, if at all possible. Vapors from the adhesive may cause a flash fire or ignite explosively. Keep the adhesive away from heat, sparks, and open flame. DO NOT smoke. Extinguish all flames and pilot lights in the area. Turn off stoves, heaters, electric motors, and all other possible sources of ignition while using the adhesive and until there is no doubt but what all vapors have left the area. Close the container immediately after use. The adhesive is harmful or fatal if swallowed. Avoid prolonged contact with the skin or breathing of the vapors. If swallowed, do not induce vomiting. Call a physician immediately. Keep the adhesive out-of-reach of children.

8. Apply a coating of Bellows Adhesive to the inside of each end of the bellows and around the mounting flanges. Allow the adhesive to dry approximately 10 minutes, or until the material is no longer tacky. Install the clamp on one end of the bellows, and then install it to the gimbal housing flange so that the clamp screw is accessible through the access hole. Tighten the clamp screw to 35 inch lbs. (4 Nm).

9. Now the hard part. Install the bellow clamp on the other end of the bellows.

■ A special bellows tool is almost a necessity to "pull" the bellows over the flange of the stern drive. Therefore, obtain expander tool (#91-45497A1) and place the tool into the first bellows convolution. Pull on the tool until the tool touches the flange on the bell housing - the hose starts to slip onto the flange - then release the tool. The accompanying cross-section line drawing illustrates the position of the tool in the first convolution of the bellows.

10. Move the tool into the third bellows convolution and pull the bellows onto the bell housing flange. Tighten the hose clamp to 35 inch lbs. (4 Nm).

11. Install the drive, start the engine and check for leaks.

U-Joint Bellows

REMOVAL & INSTALLATION

◆ See Figures 180 thru 183

■ Although not necessary, we strongly recommend removing the stern drive. The bell housing does not have to be removed in order to replace the U-joint bellows. However, this is not an easy task. Therefore, work slowly and carefully. If the job proves too difficult, the decision may be made to remove the exhaust bellows and the water intake hose first.

1. Remove the single hose clamp securing the bellows to the gimbal housing using a long shank screwdriver. Work the bellows off the gimbal housing flange. Considerable force may be required to remove the bellows due to the adhesive used during installation.

2. There is no hose clamp securing the other end of the bellows to the bell housing. Instead, a sleeve fits tightly inside the bellows to hold it against the bell housing flange. The sleeve must be pried free of the bellows. Obtain a can of aerosol cleaner such as "Gunk" or WD-40. Spray the cleaner around the edge of the sleeve to help loosen it. Pry the sleeve free using a thin blade screwdriver. Push the bellows back from the bell housing flange until it is free. No adhesive is used at this end of the bellows, therefore, just push.

3. If you were smart and removed the drive, insert a sleeve removal tool (#91-818169) through the housing and pull out the sleeve. Now you simply need to pry the bellows off of the flange

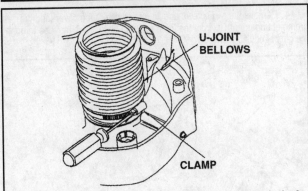

Fig. 180 The U-joint bellows clamp is placed in approximately the 3 o'clock position and tighten. A sleeve is wedged into the bell housing end as explained in the text

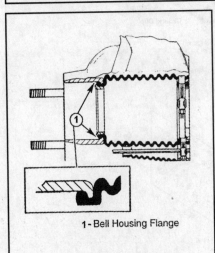

1 - Bell Housing Flange

Fig. 181 Make sure that the bell housing flange rests in the groove at the end of the bellows

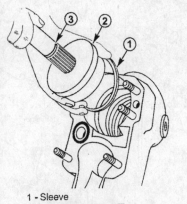

1 - Sleeve
2 - Sleeve Installation Tool
3 - Suitable Driving Rod

Fig. 182 You'll need the special tool to install the sleeve

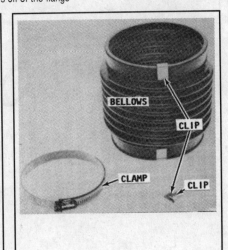

Fig. 183 Don't forget the ground clips

10-44 DRIVE SYSTEMS - ALPHA

To install:

4. Remove all bellow adhesive from the inside diameter of the bellows, if it is to be reused. Inspect the bellows for cracks, cuts, punctures and to be sure it is still flexible. If the least bit of doubt exists concerning the condition of the bellow, install new bellows. If you are considering re-using the bellows, stop and think - why not replace it while you have everything apart? You may wish you had down the road!

5. Clean the bellows mounting flanges of the bell housing with a wire brush or sandpaper, and then wipe the surface clean with lacquer thinner.

6. Check the flanges and housing for cracks, nicks, or corrosion. Clean the bellows clamp thoroughly. Check the clamp for cracks or nicks. Replace the clamp if in doubt as to their condition. Always use stainless steel clamps as a replacement.

✱✱ WARNING

Bellows Adhesive must be used to ensure a satisfactory installation. This adhesive is extremely toxic and flammable. Therefore, make every effort to ensure adequate ventilation in the work area during its use. Work in the outdoors, if at all possible. Vapors from the adhesive may cause flash fire or ignite explosively. Keep the adhesive away from heat, sparks, and open flame. Observe no smoking. Extinguish all flames and pilot lights in the area. Turn off stoves, heaters, electric motors, and all other possible sources of ignition while using the adhesive and until there is no doubt but what all vapors have left the area. Close the container immediately after use. The adhesive is harmful or fatal if swallowed. Avoid prolonged contact with the skin or breathing of the vapors. If swallowed, do not induce vomiting. Call a physician immediately. Keep the adhesive out-of-reach of children.

7. Apply a coating of Bellows Adhesive to the inside diameter of the gimbal housing bellows end. Do not apply the adhesive to the bell housing end. Allow the adhesive to dry approximately 10 minutes, or until the material is no longer tacky. Install a grounding clip over the forward edge of the bellows with the shorter side of the clip on the inside of the bellows.

8. Find the word **TOP** embossed on the bellows. This word must face up after installation. Be sure to position the bead on the inner diameter of the bellows in the groove on the gimbal housing flange. Install the U-joint bellows over the flange and position the clamp tightening screw at the 3 o'clock position. Tighten the screw 35 inch lbs. (4 Nm). Don't forget the ground clip.

9. Position the U-joint bellows on the bell housing. Ensure the bell housing flange is indexed in the groove at the end of the bellows.

10. Obtain a Sleeve Installation Tool (#91-818162). Lubricate the outside surface of the sleeve with soapy water or engine cleaner, and then install the sleeve tool. Use a suitable driving rod and move the sleeve in place on the bell housing - this is why we recommend removing the drive!

11. If the water hose and the exhaust bellows were removed, install them to the gimbal housing.

12. Replace the stern drive.

13. Start the engine and check the completed work.

Bell Housing

REMOVAL & INSTALLATION

◆ See Figures 184, 185 and 186

1. Remove the stern drive.

2. Loosen the mounting screws, bend back the retainer clip and remove the trim position sender from the starboard side and the trim limit switch from the port side. Let them dangle temporarily.

3. Disconnect and remove the shift cable as detailed in the Transom Cable section.

4. Remove the two screws and lift off the water tube cover (and the grommet underneath it) from the inner transom plate. Push the tube through the gimbal housing.

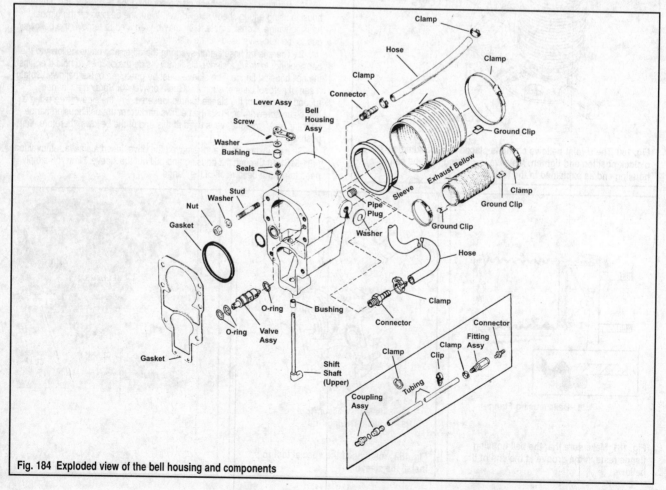

Fig. 184 Exploded view of the bell housing and components

DRIVE SYSTEMS - ALPHA   10-45

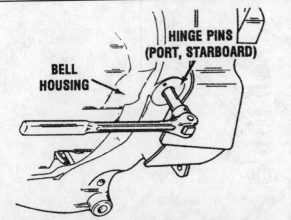

Fig. 185 The hinge pins are removed and installed with a special tool. Without this tool, removal of the hinge pins will almost be an impossible task

5. Insert a sleeve removal tool (#91-818169) into the U-joint bellows hole, tighten the nut and remove the sleeve.
6. Remove the exhaust bellows.
7. Pop out the speedometer tubing clip at the bottom of the housing (forward edge).
8. Attach a hinge pin tool (#91-78310) to a socket wrench. Swivel the housing to the port side and remove the starboard hinge pin. Now swivel it over to starboard and remove the port hinge pin. Carefully remove the bell housing.
9. If you need to replace the trim limit or position switches, loosen the screw and remove the retaining clamp. Disconnect the wire leads at the engine harness and pull them through the hole in the gimbal housing.

To install:
◆ See Figures 187, 188 and 189

10. Install the bellows to the gimbal housing.
11. Lubricate the end of the shift cable with 2-4-C lubricant and insert the shaft cable into the shift cable bellows.
12. Insert the water tube through the gimbal housing. Install a new grommet and then install the plate cover.
13. If you disconnected the trim switches, route the trim limit switch and trim position sender leads, as indicated in the accompanying illustration. Install a Sta-strap around both sets of leads approximately 5 in. (12cm) from the retaining cover. Tighten the retainer to 90-100 inch lbs. (10-11 Nm).
14. Bring the bell housing together with the gimbal housing. Push evenly on the perimeter of the bell housing until the U-joint bellows slides into the flange on the gimbal housing. Check to be sure the exhaust bellows aligns with the flange on the bell housing. Refer to the appropriate procedures in this section for pertinent details on final bellows connections.
15. Coat the hinge pin threads (inside the housing and on the bolts) with Locquic Primer "T" and allow the material to dry. Next, apply a coating of Loctite 271 to the housing threads. Install and tighten the hinge pins to 100 ft. lbs. (136 Nm).
16. Check final bellows installation against the individual procedures given earlier.
17. Install and adjust the shift cable.
18. Install and adjust the trim switches on either side of the gimbal ring
19. Install the stern drive.

Gimbal Ring

REMOVAL & INSTALLATION

Engine/Transom Assembly Installed

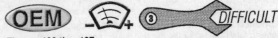

◆ See Figures 190 thru 197

■ Unless a previous owner has already drilled an access hole in the gimbal housing, you will need to drill (and plug) your own hole in order to remove the steering lever.

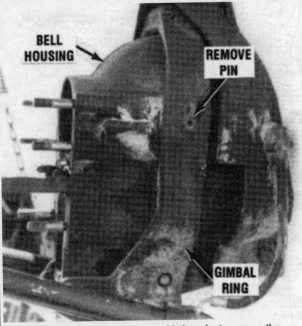

Fig. 186 Swivel the housing to one side in order to remove the opposite side's hinge pin

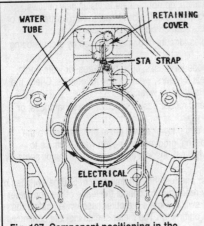

Fig. 187 Component positioning in the gimbal housing before installing the bell housing

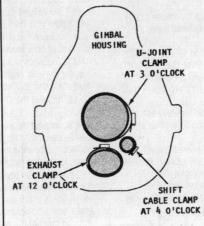

Fig. 188 Bellows clamp locations - gimbal housing

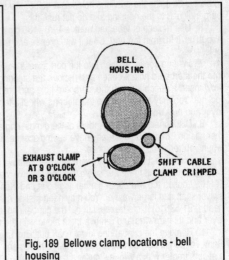

Fig. 189 Bellows clamp locations - bell housing

10-46 DRIVE SYSTEMS - ALPHA

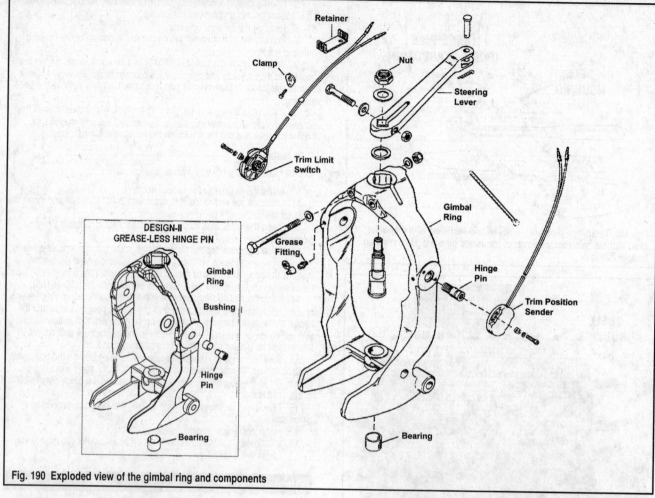

Fig. 190 Exploded view of the gimbal ring and components

1. Remove the stern drive and the bell housing.
2. Pump a liberal amount of grease into the fitting at the top of the gimbal housing. Disregard this step if you have a Type II greaseless hinge pin.
3. You'll need Access Plug Kit (#22-88847A-1). Use the template provided and position it on the housing using the dimple in the front of the housing - it will be under the decal in the top center of the flat.
4. Use a center punch and the template to mark the location of the holes on each side of the gimbal housing and then drill a 1/4 in. (6mm) hole through each side of the housing.
5. Using a 1-1/8 in. hole saw with a 1/4 in. (6mm) pilot rod, drill a hole in each side using the previous holes as a guide. Make sure the drill is perpendicular to the housing and do not rush this.
6. Use a piece of tape and mark a 1 in. #180 pipe tap (hardware store) about 1-1/8 in. from the end. Coat it with grease and then thread it into the hole.
7. Insert a socket wrench into the port access hole and an open end into the starboard hole. Loosen and remove the steering lever clamping bolt. Now insert a punch through the starboard hole and unthread the elastic locknut. You may have to move the steering wheel a few times to jockey the lever and nuts into position.
8. Disconnect the trim cylinders at the gimbal ring and carefully set them aside - they should already have been disconnected on the other end when you removed the stern drive.
9. Disconnect the continuity wire and then remove the lower swivel cotter pin. Drive out the lower swivel pin and remove the anti-gauling washer.
10. Loosen the two upper gimbal ring bolts and then remove the upper swivel shaft and large washer. You may need a slide hammer and puller head (#91-63616) if it is frozen. Lift out the gimbal ring. Look for the steering lever and its hardware and remove them along with the ring.
11. Remove the gimbal ring lower shaft bushing, using a suitable driver. Remove the gimbal ring upper swivel shaft bushing and single oil seal, using a slide hammer with a two jaw puller attachment.

12. Remove the gimbal ring hinge pins, using a Hinge Pin tool (#91-78310) to remove the pins, if they are threaded into the housing.
13. Obtain a suitable driver and drive out both hinge pin bushings.

To install:

14. Clean all metal parts in solvent and blow them dry with compressed air. Never spin ball bearings with compressed air because such action will ruin the bearings.
15. Remove all bellows adhesive from the inside diameter of the bellows, if the bellows are to be reused. Inspect the bellows for cracks, cuts, punctures and to be sure they are still flexible. If there is the least bit of doubt concerning their condition, do not hesitate to install new bellows. Clean the bellows mounting flanges of the bell housing with a wire brush or sandpaper and then wipe the surface clean with lacquer thinner.
16. Check the shift cables for cuts or damage caused by a cable being pinched or bent too short.
17. Inspect the water hose for cracks, cuts, punctures, or worn spots. Inspect the shift shaft oil seal for tears, wear, or any roughness. Check the shift shaft and shift shaft bushings for wear.
18. Inspect the O-ring and rubber gasket for cuts, nicks, hardness, or cracks.
19. Inspect the surface of the lower swivel pin in the area where the needle bearing rides. Any pitting, grooves, or uneven wear is cause to replace the bearing and swivel pin. Inspect the lip surface of the oil seals for wear, tears, and roughness.
20. Obtain an ohmmeter and test the trim limit switch. Set the switches of the ohmmeter for continuity reading and connect the leads of the ohmmeter to the leads of the trim limit switch. Rotate the outer case while holding the inner rotor and observe the ohmmeter. The meter should indicate continuity through most of the movement and then no continuity. If the meter indicates continuity all the time or no continuity, the switch is defective and must be replaced.

DRIVE SYSTEMS - ALPHA   10-47

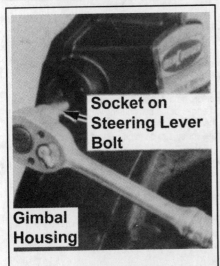

Fig. 191 Use a socket on the port side when removing the steering lever bolt

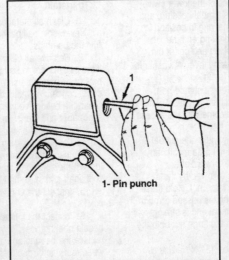

Fig. 192 Use a punch to drive off the elastic nut

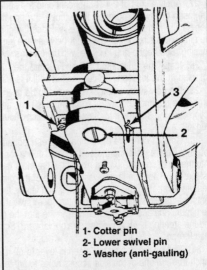

1- Cotter pin
2- Lower swivel pin
3- Washer (anti-gauling)

Fig. 193 Removing the lower swivel pin

Fig. 194 You may need a slide hammer to remove the upper swivel shaft

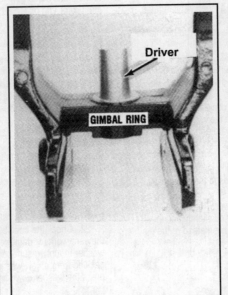

Fig. 195 Removing the lower bushing...

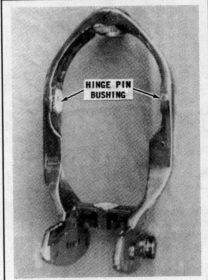

Fig. 196 ...and then remove the hinge pin bushings

21. Inspect the long steering lever retaining bolt. Any grooves found on the bolt are a result of friction against the upper swivel shaft. If grooves are discovered, both the steering lever and the bolt must be replaced.

22. Obtain Resiweld Sealer and apply a coating of this sealer to the outer diameter of the lower bushing. Using a suitable driver, tap the bushing into place.

23. To install the small upper swivel shaft bushing, first obtain a Bearing and Seal Driver tool (#91-43578). Next, place the bushing on the tool, and then use a hammer and tap the bushing into place. Install the larger bushing in a similar manner.

24. Pack the single oil seal lip with Multi-Purpose lubricant and place the seal on the driver with the lip facing the smaller diameter of the tool. Install the oil seal into the gimbal ring until the seal seats against the larger bushing.

25. Check the condition of the synthane washers on the gimbal ring. If they are worn, or damaged, remove and discard the washers. Clean the washer mounting surface free of all adhesives, dirt, grease and oil. Peel the backing from a new washer and place it on the mounting surface, and then press the washer down firmly. The adhesive backing will hold the washer in place.

26. Apply a coating of Resiweld to the outer diameter of both hinge pin bushings. Using a suitable driver, tap the bushings into place on the port and starboard sides of the gimbal ring.

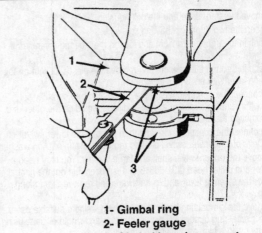

1- Gimbal ring
2- Feeler gauge
3- Gimbal housing mount

Fig. 197 Use a feeler gauge to check the lower swivel

10-48 DRIVE SYSTEMS - ALPHA

27. Apply a coat of 2-4-C Lubricant onto the surface of the lower swivel pin. Insert the switch harness, if equipped, through the gimbal housing, and then secure it in place with the screws. Place the gimbal ring in position in the gimbal housing. Place a washer between the gimbal ring and the housing. Secure the ring in place with the lower swivel pin and cotter pin.

28. Install the long clamp bolt and nut into the steering lever. A washer is not used at this location. Tighten the nut until it is just "snug".

29. If a new swivel shaft elastic nut is to be used (and it should be!), thread the nut onto the swivel shaft until it is tight, and then remove it. This action cuts the threads into the new nut (which should come with the access plug kit).

30. Check to be sure the gimbal ring rests squarely within the gimbal housing. Before installing the swivel shaft, check for a flat area machined onto the shaft splines. If a flat exists, the flat must face forward after installation. Place the larger washer on top of the ring, followed by the steering lever, then the smaller washer and finally, the large nut. Hold these components aligned together while the upper swivel shaft is passed through the gimbal ring and these parts. Make sure that the gimbal ring is straight and the steering lever is pointed straight forward. Start the large nut onto the shaft threads.

31. Tighten the bolt on the steering lever to 60 ft. lbs. (81 Nm). Use a punch through one of the side access holes to tighten the nut, until a clearance of 0.002-0.010 in. (0.05-0.25mm) exists between the lower swivel pin washer and the gimbal housing.

32. Use a rawhide mallet and strike the top of the gimbal ring flanges a couple times each to seat both pins. Recheck the clearance and adjust the large nut (using the punch and hammer method), as necessary to maintain the required clearance.

33. Tighten the two gimbal ring bolts to 55 ft. lbs. (74 Nm). Apply a coating of Perfect Seal to the threads or the sealing surface of the plugs and close the two large openings drilled earlier on the sides of the gimbal housing for access to the steering lever.

34. Install the steering lever ground wire so that it comes off the transom plate to port and then back in to the connector on the steering arm - envision a 'U' laying on its side with the open end facing starboard and the closed end facing to port. Don't forget to re-install the wire running from the gimbal housing to the gimbal ring on the lower side.

35. Install the trim cylinders by first attaching them in place with the forward anchor pin. Next, install the two hydraulic hoses to the connector. Finally, install the connector to the gimbal housing.

36. Connect the steering link rod to the steering lever. Tighten the castellated nut to 24 inch lbs. (3 Nm), and then back it off slightly and insert the cotter pin. If the assembly is secured with a Nylok nut instead of the cotter pin, tighten the nut to 5 ft. lbs. (7 Nm).

37. Install the bell housing and stern drive.

Engine/Transom Assembly Removed

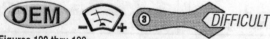

◆ See Figures 190 thru 198

1. Remove the stern drive and the bell housing.
2. Loosen the two upper gimbal ring screws.
3. Reach in through the transom and loosen the steering lever clamp bolt.
4. Reach in through the transom and remove the elastic locknut on top of the steering lever with an 1-1/16 in. wrench.
5. Disconnect the trim cylinders at the gimbal ring and carefully set them aside - they should already have been disconnected on the other end when you removed the stern drive.
6. Disconnect the continuity wire and then remove the lower swivel cotter pin. Drive out the lower swivel pin and remove the anti-gauling washer.
7. Remove the upper swivel shaft and large washer. You may need a slide hammer and puller head (#91-63616) if it is frozen. Lift out the gimbal ring. Look for the steering lever and its hardware and remove them along with the ring.
8. Remove the gimbal ring lower shaft bushing, using a suitable driver. Remove the gimbal ring upper swivel shaft bushing and single oil seal, using a slide hammer with a two jaw puller attachment.
9. Remove the gimbal ring hinge pins, using a Hinge Pin tool (#91-78310) to remove the pins, if they are threaded into the housing.
10. Obtain a suitable driver and drive out both hinge pin bushings.

To install:

11. Clean all metal parts in solvent and blow them dry with compressed air. Never, spin ball bearings with compressed air, because such action will ruin the bearings.

12. Remove all bellow adhesive from the inside diameter of the bellows, if the bellows are to be reused. Inspect the bellows for cracks, cuts, punctures and to be sure they are still flexible. If there is the least bit of doubt concerning their condition, do not hesitate to install new bellows. Clean the bellow mounting flanges of the bell housing with a wire brush or sandpaper and then wipe the surface clean with lacquer thinner.

13. Check the shift cables for cuts or damage caused by a cable being pinched or bent too short.

14. Inspect the water hose for cracks, cuts, punctures, or worn spots. Inspect the shift shaft oil seal for tears, wear, or any roughness. Check the shift shaft and shift shaft bushings for wear.

15. Inspect the O-ring and rubber gasket for cuts, nicks, hardness, or cracks.

16. Inspect the surface of the lower swivel pin in the area where the needle bearing rides. Any pitting, grooves, or uneven wear is cause to replace the bearing and swivel pin. Inspect the lip surface of the oil seals for wear, tears, and roughness.

17. Obtain an ohmmeter and test the trim limit switch. Set the switches of the ohmmeter for continuity reading and connect the leads of the ohmmeter to the leads of the trim limit switch. Rotate the outer case while holding the inner rotor and observe the ohmmeter. The meter should indicate continuity through most of the movement and then no continuity. If the meter indicates continuity all the time or no continuity, the switch is defective and must be replaced.

18. Inspect the long steering lever retaining bolt. Any grooves found on the bolt are a result of friction against the upper swivel shaft. If grooves are discovered, both the steering lever and the bolt must be replaced.

19. Obtain Resiweld Sealer. Apply a coating of this sealer to the outer diameter of the lower bushing. Using a suitable driver, tap the bushing into place.

20. To install the small upper swivel shaft bushing, first obtain a Bearing and Seal Driver tool (#91-43578). Next, place the bushing on the tool, and then use a hammer and tap the bushing into place. Install the larger bushing in a similar manner.

21. Pack the single oil seal lip with Multi-Purpose lubricant and place the seal on the driver with the lip facing the smaller diameter of the tool. Install the oil seal into the gimbal ring until the seal seats against the larger bushing.

22. Check the condition of the synthane washers on the gimbal ring. If they are worn, or damaged, remove and discard the washers. Clean the washer mounting surface free of all adhesives, dirt, grease and oil. Peel the backing from a new washer and place it on the mounting surface, and then press the washer down firmly. The adhesive backing will hold the washer in place.

23. Apply a coating of Resiweld to the outer diameter of both hinge pin bushings. Using a suitable driver, tap the bushings into place on the port and starboard sides of the gimbal ring.

24. Apply a coat of 2-4-C Lubricant onto the surface of the lower swivel pin. Insert the switch harness, if equipped, through the gimbal housing, and then secure it in place with the screws. Place the gimbal ring in position in the gimbal housing. Place a washer between the gimbal ring and the housing. Secure the ring in place with the lower swivel pin and cotter pin.

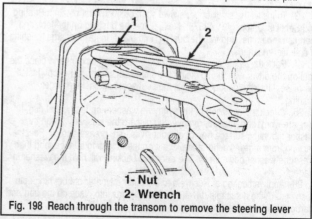

1- Nut
2- Wrench
Fig. 198 Reach through the transom to remove the steering lever

DRIVE SYSTEMS - ALPHA

25. Install the long bolt and nut into the steering lever. A washer is not used at this location. Tighten the nut until it is just "snug". If a new swivel shaft elastic nut is to be used, thread the nut onto the swivel shaft until it is tight, and then remove it. This action cuts the threads into the new nut (which should come with the access plug kit).

26. Check to be sure the gimbal ring rests squarely within the gimbal housing. Before installing the swivel shaft, check for a flat area machined onto the shaft splines. If a flat exists, the flat must face forward after installation. Place the larger washer on top of the ring, followed by the steering lever, then the smaller washer and finally, the large nut. Hold these components aligned together while the upper swivel shaft is passed through the gimbal ring and these parts. Start the large nut onto the shaft threads.

27. Tighten the bolt on the steering lever to 60 ft. lbs. (81 Nm). Now tighten it further until a clearance of 0.002-0.010 in. (0.05-0.25mm) exists between the lower swivel pin washer and the gimbal housing.

28. Use a rawhide mallet and strike the top of the gimbal ring flanges a couple times each to seat both pins. Recheck the clearance and adjust the large nut (using the punch and hammer method), as necessary to maintain the required clearance.

29. Tighten the two gimbal ring bolts to 55 ft. lbs. (74 Nm).

30. Install the trim cylinders by first attaching them in place with the forward anchor pin. Next, install the two hydraulic hoses to the connector. Finally, install the connector to the gimbal housing.

31. Connect the steering link rod to the steering lever. Tighten the castellated nut to 24 inch lbs. (3 Nm), and then back it off slightly and insert the cotter pin. If the assembly is secured with a Nylok nut instead of the cotter pin, tighten the nut to 5 ft. lbs. (7 Nm).

32. Install the bell housing and stern drive.

Gimbal Housing/Transom Plate

REMOVAL & INSTALLATION

◆ See Figures 199 thru 203

1. If you are removing both the inner transom plate and the gimbal housing, remove the stern drive, bell housing, gimbal ring and bushing and then skip to Step 3.

2. If you are only intending on removing the inner transom plate, remove the engine - although not always necessary in all applications, removing the engine makes this procedure much simpler.

■ Many of the following steps may have already been performed in the process of removing the engine, drive, etc; but since there are a number of routes to follow to get either of these two components off we have chosen to detail each step in its order. If you have previously removed a component detailed in a step, please bear with us and simply skip to the next step.

3. If equipped with power steering, disconnect the clevis from the steering lever and then disconnect the cable at the clevis. Loosen the pivot bolts and then remove the power steering unit. More information on this can be found in the Power Steering section.

4. If equipped with manual steering, disconnect the steering cable and clevis at the steering lever. Loosen the pivot bolts and remove the steering swivel ring.

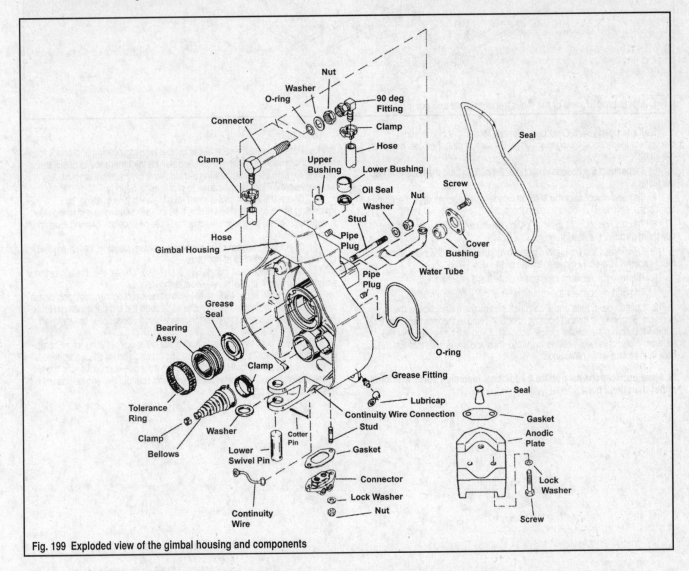

Fig. 199 Exploded view of the gimbal housing and components

10-50 DRIVE SYSTEMS - ALPHA

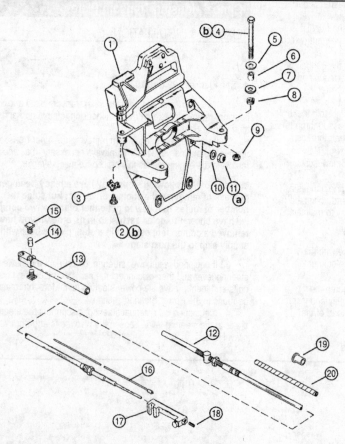

1 - Transom Plate Assembly
2 - Pivot Bolts
3 - Tab Washers
4 - Screw Engine Mounting
5 - Washer
6 - Spacer
7 - Washer - Fiber
8 - Lockwasher - Double Wound
9 - Locknut
10 - Washer
11 - Locknut
12 - Shift Cable Outer Casing
13 - End Guide
14 - Core Wire Anchor
15 - Anchor Screws
16 - Core Wire
17 - Shift Slide
18 - Screw - Core Wire Cavity
19 - Gimbal Housing Eyelet
20 - Cable Wrapping

Torque Specifications
a - 25 lb-ft (34 Nm)
b - 37 lb-ft (50 Nm)

Fig. 200 Exploded view of the inner transom plate and components

5. If equipped with a Gear Lube Monitor, loosen the clamp and disconnect the hose at the transom plate. Be sure to plug the hose and cap the fitting.

6. Disconnect the speedometer tube. Be sure to plug the hose and cap the fitting.

7. Tag and disconnect the MerCathode wires (if equipped) at the control unit on the engine. Pull them through the plate.

8. Tag and disconnect the trim switch/sender wires at the engine and pull them through the assembly.

9. Disconnect the power trim pump hydraulic lines at the plate and plug them securely. Carefully move them out of the way.

10. Remove the exhaust pipe at the plate if still attached.

11. Tag and disconnect the ground wire at the steering lever.

12. Loosen the 6 nuts and 2 bolts that secure the inner plate to the transom. Support the gimbal housing and then remove the retainers and lift out the transom plate and/or gimbal housing. If you are removing only the transom plate, carefully feed the hydraulic lines and shift cable through the hole in the plate while removing it.

■ Many applications may utilize 2 additional mounting studs with nuts rather than the 2 bolts.

To install:

13. Rotate the inside race of the gimbal housing bearing and check for rough or any sign of binding. Push and pull on the inner race to check the bearing for side play. The bearing should move only a slight amount. Any excessive movement is just cause to replace the bearing.

14. Check the upper swivel shaft roller bearing and bushing, by inspecting the area of the swivel shaft, where the bearing and bushing ride. Any pits, grooves, or uneven wear is cause to replace the bearing, bushing, or swivel shaft.

15. Check the transom plate for pitting or worn housing. Clean any old sealer from the back side of the plate.

16. Clean the exhaust outlet. Use a tap, and "chase" the threads of the screw holes where the exhaust elbow is mounted.

17. Secure the water tube to the water hose and tighten the hose clamp securely. Install the water hose and tube through the gimbal housing. Route the trim limit switch and trim position sender leads down over the installed tube.

18. Secure both sets of leads with a Sta-strap 5 in. (12cm) away from their retaining bracket. Tilt the bell housing up and secure the water hose over the flange on the bell housing. Tighten the hose clamp over the tube. Install the grommet and water tube cover and connect the engine water inlet hose to the water tube.

DRIVE SYSTEMS - ALPHA   10-51

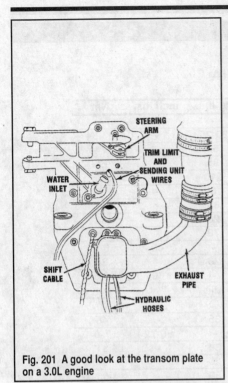

Fig. 201 A good look at the transom plate on a 3.0L engine

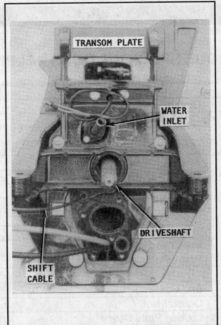

Fig. 202 Another good look with the exhaust pipe(s) removed

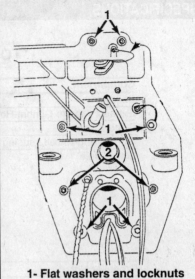

1- Flat washers and locknuts
2- Long screws, lock washers and square flat washers

Fig. 203 Remove these nuts and bolts to pull of the plate

19. Insert the shift cable, the trim limit switch lead, and the hydraulic hoses through the appropriate opening in the transom plate, and then move the gimbal housing into position. The hydraulic hose must be positioned on the port side of the exhaust elbow on in-line engines and on the starboard side for V6 or V8 engines.

■ Do not hold onto the trim limit switch wires to support the gimbal housing while installing the unit. The trim limit switch or the switch leads will be damaged, if the gimbal housing is supported by the switch leads while fastening the inner transom plate.

20. Hold the gimbal housing in position, and at the same time insert the shift cable and hydraulic hose through the large opening; insert the trim limit switch leads through the center hole of the three small holes in the inner transom plate; and then set the plate in position.

21. Install the steering lever ground wire so that the crimps are leading the wire to the right side of the connections as you face the back of the vessel.

22. Thread the two short cap screws, with lockwashers, through the top two holes of the inner transom plate and into the gimbal housing. Insert the two special anode head bolt assemblies, with NEW rubber seals between the bolt and gimbal housing, through the bottom two holes of the gimbal housing, the boat transom, and the inner transom plate.

23. Insert a flat washer and elastic stop nut on each bolt or stud, but do not tighten them at this time. Install a flat washer and elastic stop nut onto each of the two studs protruding through the inner transom plate.

24. Do not attempt to drive the cap screws through the transom, because the threads in the gimbal housing will be damaged. Install the two long cap screws; with a flat washer and lockwasher, through the remaining holes in the inner transom plate and into the gimbal housing. The square flat washer must be against the transom plate.

25. Now, tighten the transom plate cap screws and elastic stop nuts evenly. Work from the center up, and then down. Tighten the screws and nuts to 22.5 ft. lbs. (30.5 Nm).

26. Check to be sure the mating surfaces on the exhaust elbow and the gimbal housing are clean. Place a new O-ring seal(s) into the groove in the gimbal housing opening. Position the lower exhaust pipe and/or separator onto the gimbal housing and install the bolts with their lockwashers. Tighten the screws to 22.5 ft. lbs. (30.5 Nm), and at the same time check to be sure the O-ring remains properly seated in the groove. If equipped with a thru-transom exhaust system, don't forget the block-off plate and seal if it was removed earlier.

■ Move quickly while working with the hydraulic hoses to prevent spilling any more hydraulic fluid than is necessary. Follow the sequence and take care to route the hoses properly, or the hoses may be damaged and the system become inoperative.

27. Connect the hydraulic lines and tighten the connections to 125 inch lbs. (14.1 Nm).

28. Install the steering gear and tighten the pivot bolts on either system to 25 ft. lbs. (34 Nm). Bend the washer tabs up and against the bolt head.

29. Connect the Gear Lube line, speedometer tube and MerCathode wires.

30. Install the engine if removed.

31. Install the bell housing and stern drive if removed.

32. Bleed the hydraulic system. Adjust the shift cable.

DRIVE SYSTEMS - ALPHA

SPECIFICATIONS

TORQUE SPECIFICATIONS - ALPHA

Component			ft. lbs.	inch lbs.	Nm
Bellows	Clamp		-	35	4
Exhaust Pipe-to-Gimbal Housing			20-25	-	27-34
Gimbal ring			55	-	74
Gimbal ring-to-bell housing	Hinge pins		95	-	129
Lower Unit	Bearing Carrier Retainer		210	-	285
	Driveshaft Retainer		100	-	136
	Lower-to-Upper Unit	Nuts	35	-	47
		Screw	28	-	38
	Pinion Gear Nut		70	-	95
	Shift Shaft Bushing		-	60	6.8
	Trim Tab		23	-	31
	Water Pump		-	60	6.8
Propeller Nut			55 Min. ①	-	75 Min. ①
Shift Cable	End Guide		②	-	②
Steering Cable	Coupler Nut		35	-	48
	Locking Plate Screw		-	60-72	7-8
Steering Lever			60	-	81
Steering Systrm Pivot Bolts			25	-	34
Stern Drive-to-Bell Housing			50	-	68
Transom Plate			20-25	-	27-34
Trim Limit Switch			-	90-100	10-11
Trim Position Sender			-	90-100	10-11
Upper Unit	Top Cover Screws		17-23	-	23-31
	U-Joint Drive Gear Nut		①	-	①
	U-Joint Retainer Nut		①	-	①
	U-Joint Bearing Preload	New	-	6-10	0.7-1.0
		Used	-	3-7.5	0.3-0.8
	Upper Shaft Bearing Preload	New	-	6-15	0.7-1.7
		Used	-	3-7.5	0.3-0.8

① See text
② Tighten until nut bottoms on washer and then back off 1/2 turn

STERN DRIVE DRIVE MODELS - ALPHA

Year	Model	Gear Ratio	U-Joint Shaft Marking	Tooth Count	Number Of Teeth Per Gear Driveshaft Housing	
					Drive	Driven
2001-08	Alpha	2.40:1	K	14-28	20	24
		2.00:1	A	14-28	24	24
		1.98:1	B	17-28	20	24
		1.94:1	B	17-28	20	24
		1.84:1	D	17-28	17	19
		1.81:1	D	17-28	17	19
		1.65:1	C	17-28	24	24
		1.62:1	C	17-28	24	24
		1.50:1	F, M	17-28	22	20
		1.47:1	F	17-28	22	20

BELL HOUSING	11-40
REMOVAL & INSTALLATION	11-40
DRIVESHAFT HOUSING (UPPER UNIT)	**11-18**
DESCRIPTION	11-18
DRIVESHAFT HOUSING	11-19
GEAR RATIO IDENTIFICATION	11-18
EXHAUST BELLOWS	11-36
REMOVAL & INSTALLATION	11-36
EXHAUST TUBE	11-36
REMOVAL & INSTALLATION	11-36
EXPLODED VIEWS	11-38
GEAR HOUSING (LOWER UNIT)	**11-4**
DESCRIPTION	11-4
GEAR HOUSING - BRAVO I/II/X/XR/XZ	11-4
GEAR HOUSING - BRAVO III	11-11
PROPELLER(S)	11-18
GIMBAL BEARING	11-32
REMOVAL & INSTALLATION	11-32
GIMBAL HOUSING/TRANSOM PLATE	11-43
REMOVAL & INSTALLATION	11-43
GIMBAL RING	11-41
ACCESS PLUG INSTALLATION	11-43
REMOVAL & INSTALLATION	11-41
LOWER UNIT - BRAVO I/II/X/XR/XZ	11-4
ASSEMBLY	11-10
CLEANING AND INSPECTION	11-9
DISASSEMBLY	11-5
REMOVAL & INSTALLATION	11-4
LOWER UNIT - BRAVO III	11-11
ASSEMBLY	11-16
CLEANING AND INSPECTION	11-16
DISASSEMBLY	11-12
REMOVAL & INSTALLATION	11-11
PROPELLER(S)	11-18
GENERAL INFORMATION	11-18
REMOVAL & INSTALLATION	11-18
SHIFT CABLE	11-33
ADJUSTMENT	11-35
REMOVAL & INSTALLATION	11-33
REMOTE CONTROL CABLE	11-33
TRANSOM CABLE	11-33
SPECIFICATIONS	**11-46**
APPLICATIONS - BRAVO	11-47
TORQUE - BRAVO	11-46
STERN DRIVE UNIT	11-2
REMOVAL & INSTALLATION	11-2
STERN DRIVE UNIT - BRAVO	**11-2**
DESCRIPTION	11-2
STERN DRIVE UNIT	11-2
TROUBLESHOOTING	11-2
TRANSOM ASSEMBLY	**11-32**
DESCRIPTION	11-32
EXHAUST BELLOWS	11-36
EXHAUST TUBE	11-36
EXPLODED VIEWS	11-38
GIMBAL BEARING	11-32
GIMBAL HOUSING/TRANSOM PLATE	11-43
GIMBAL RING	11-41
SHIFT CABLE	11-33
U-JOINT BELLOWS	11-37
WATER HOSE & FITTING	11-38
U-JOINT BELLOWS	11-37
REMOVAL & INSTALLATION	11-37
UPPER UNIT	11-19
DISASSEMBLY & ASSEMBLY	11-23
EXPLODED VIEWS	11-19
REMOVAL & INSTALLATION	11-19
BRAVO I/II/Z/XR/XZ	11-19
BRAVO III/X	11-22
WATER HOSE & FITTING	11-38
REMOVAL & INSTALLATION	11-38

11

DRIVE SYSTEMS - BRAVO

STERN DRIVE UNIT - BRAVO	11-2
GEAR HOUSING (LOWER UNIT)	11-4
DRIVESHAFT HOUSING (UPPER UNIT)	11-18
TRANSOM ASSEMBLY	11-32
SPECIFICATIONS	11-46

11-2 DRIVE SYSTEMS - BRAVO

STERN DRIVE UNIT - BRAVO

Description

That which we refer to as the stern drive is actually a number of individual components attached and working together to transfer the power of the engine into a viable propulsion system for your boat. All stern drive units can be broken down into their components assemblies.

The transom assembly, consisting of an inner transom plate, a gimbal housing and gimbal plate, and a bell housing is just what it sounds like - the unit attached to the transom of the vessel. The inner transom plate is, obviously, attached to the inner side of the transom and actually makes up the rear engine mounts.

On the other side of the transom, and attached to the transom plate, are the gimbal housing, gimbal plate (or ring) and bell housing. The bell housing is attached to the gimbal plate via roller bearings and is what allows for the up and down (trim) movement of the stern drive unit itself. The gimbal plate is also attached to the gimbal housing via roller bearings and is what allows for side-to-side movement of the unit, or, steering.

The stern drive unit, or at least that thing that is most visible when viewing the stern of the boat, is made of two component assemblies: the driveshaft housing (upper gear housing) and the lower gear housing.

The driveshaft housing, frequently called the upper gear housing, or simply the upper unit, is attached to the bell housing at the top and the gear housing at the bottom. Power from the engine, brought through the transom assembly via the driveshaft is transferred to a vertical shaft leading to the lower unit by means of a set of drive and driven gears.

The gear housing, or lower unit, is attached to the bottom of the driveshaft housing. Power, or propulsion, comes through the vertical shaft from the upper unit, is transferred to the propeller shaft(s) via a pinion gear and causes the propeller(s) to rotate.

Output power from the engine is connected to the stern drive through a horizontal driveshaft. A coupler is bolted to the flywheel and has a splined hub in the center. The end of the horizontal driveshaft indexes with and slides into the center of the hub. Power from the engine is then transmitted through the horizontal driveshaft to a pinion gear set where power direction is changed from horizontal to vertical.

The upper driven gear is pressed onto the outside diameter of the upper driveshaft. The upper driveshaft is splined on the lower end. When the upper gear housing is mated to the lower unit, the end of the lower unit driveshaft indexes into the splined end of the upper driveshaft. Engine power is then transferred down into the lower gear unit.

All shifting is accomplished inside the upper gear housing. The shift cable actuates a yoke and cam assembly via linkage. The yoke rides in the center of a clutch spool between two brass collars of the clutch gears. The contact surfaces of the collars are not perfectly horizontal. By pivoting left or right, the yoke is able to block one of the clutch gears and permit the clutch shaft to rotate in only one direction.

The lower splines of the short clutch shaft are coupled with the upper splines of the driveshaft.

The pinion gear and single driven gears are contained within the lower unit. Unlike many other stern drive units, the lower unit is free of any shifting mechanisms and water pump components.

The horizontal and vertical driveshafts are both mechanically connected. Therefore, anytime the engine is operating, the horizontal and vertical driveshafts are constantly rotating with engine rpm. A double yoke universal joint assembly in the horizontal driveshaft allows the stern drive to be raised or lowered to a required trim/tilt position (within limits), while the engine is operating.

Troubleshooting

Troubleshooting must be done before the unit is removed from the boat to permit isolating the problem to one area.

1. Check the propeller and the rubber hub for shredding. If the propeller has been subjected to many strikes with underwater objects, it could slip on its hub. If the hub appears to be damaged, replace it with a new hub. Replacement of the hub must be done by a propeller rebuilding shop equipped with the proper tools and experience for such work.

2. **Shift Mechanism Check**: Verify the ignition switch is in the off position to prevent possible injury, should the engine start. Shift the unit into Reverse and at the same time have an assistant turn the propeller shaft to ensure the clutch is fully engaged. If the shift handle is hard to move, the trouble may be in the stern drive unit, transom shift cable, remote control cable, or the shift box.

3. **Isolate the Problem**: Disconnect the remote control cable at the transom plate, by first removing the two nuts, and then lifting off the remote control shift cable. Operate the shift lever. If shifting is still hard, the problem is in the shift cable or control box. If the shifting feels normal with the remote control cable disconnected, the problem must be in the stern drive. To verify the problem is in the stern drive unit, have an assistant turn the propeller and at the same time move the shift cable between the transom plate and stern drive back-and-forth. Determine if the clutch engages properly. Most of hard shifting problems are caused because this cable is not moved during the off-season.

Water entering the cable, especially salt water, will cause rapid corrosion and hard shifting. If the cable moves freely, then reconnect the control cables at the transom plate.

4. **Stern Drive Noise Check**: First, a word about stern drive noise. When the stern drive is positioned for dead-ahead operation, the U-joints will make very little noise. However, as the stern drive is moved to port or starboard, the noise will increase, due to the working of the U-joints. This test and procedure is for abnormal noises. Attach a Flush-Test device to the stern drive and turn on the water. Start the engine and shift into gear.

✱✱ CAUTION

Water must circulate through the lower unit to the engine any time the engine is run to prevent damage to the water pump mounted on the engine. Just a few seconds without water will damage the water pump.

Operate the engine at idle speed. Turn the stern drive slightly to port and then slightly to starboard, and at the same time, listen at the upper gear housing. An unusual noise is an indication the U-joints are worn or the gimbal housing bearing is defective. If the bearing is defective, refer to Bell Housing. Shut down the engine. Turn off the water and disconnect the Flush-Test device.

5. **Tilt/Trim System Check**: Operate the control buttons on the control panel or on the shift control lever. Check to determine if the stern drive unit moves to the full up and full down position. If the drive unit fails to move, the U-joints may be "frozen" due to corrosion from water entering through the U-joint bellow; lack of lubrication; or from non-operation over a long period of time. If the stern drive does not move properly, the trim/tilt mechanism may need service, see the section on Trim/Tilt.

6. **Noise in Upper Gear Housing**:
- Oil level low
- Worn U-joints
- Worn bearings
- Incorrect thrust washer placement
- Gear timing incorrect
- Engine not aligned properly
- Worn engine coupler
- Transom too light for stern drive
- Worn, damaged, or loose internal parts

Stern Drive Unit

REMOVAL & INSTALLATION

OEM ③ DIFFICULT

◆ See Figures 1 thru 7

1. Drain the stern drive unit oil as detailed in the Maintenance section. Check the drained lubricant carefully to determine if it contains any water or metal particles. Rub some of the lubricant between your fingers. Any metal particles in the lubricant will thus be evident. Do not be mislead if the color of the lubricant is metal colored. This is not a harmful condition and is caused by the lubricant used in the unit the first time after manufacture.

2. Shift the unit into Neutral and then move it to the full UP/OUT position.

3. Reach around to the rear of the anti-ventilation plate and move the speedometer hose fitting release lever over and up. Pull out the hose and move it out of the way.

4. Move the drive unit back to the full DOWN/IN position. Remove the plastic cap over the anchor pin connecting the trim cylinders to the aft end of the upper unit. Remove the nuts and washers and pry out the bushings. Disconnect the cylinder arms from the unit and carefully move them back against the transom. It's a good idea to tie them up and out of the way to avoid damage during drive removal.

DRIVE SYSTEMS - BRAVO

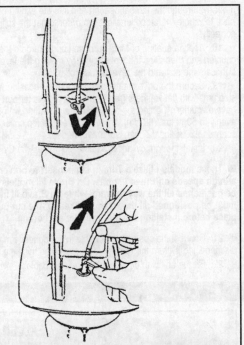

Fig. 1 Move the speedometer lever counterclockwise to release the tube

■ Although not imperative, we highly recommend the use of an engine hoist or other lifting device when removing the unit.

5. Install a lifting device if so desired.

6. Remove the six 5/8 in. nuts and washers securing the stern drive to the bell housing. There are only five washers as the center, port bolt (looking toward the bow) uses a permanent ground plate instead of a continuity washer.

7. Secure the lifting chains so they are taught and then remove the stern drive unit by pulling it straight back and free of the bell housing. Remove and discard the gasket. Make sure that the jaws on the shift cable linkage open properly and release the cable end as you are pulling the unit back and out.

To install:

The engine must be properly aligned or the female splines in the engine coupler and the male splines of the driveshaft will be destroyed after a short time of engine operation.

If troubleshooting indicates the coupler is damaged and requires replacement, remove the engine and replace the coupler. The coupler is simply secured to the flywheel with attaching bolts. If the necessary tools are not available for proper engine alignment, the boat should be taken to an authorized marine dealer.

8. Ensure that the remote control shift lever on the drive is in Neutral.

9. Lubricate and install new O-rings into the two passageways just under the driveshaft on the forward side of the housing. Install a new gasket on the face of the bell housing.

10. Coat all six studs and their threads with Quicksilver 2-4-C marine lubricant (w/Teflon).

11. Check the bell housing bearing to be sure it turns freely without binding or rough spots.

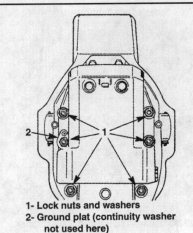

1- Lock nuts and washers
2- Ground plat (continuity washer not used here)

Fig. 2 Remove the six locknuts

Fig. 3 You can try it this way, but we recommend a lifting hoist

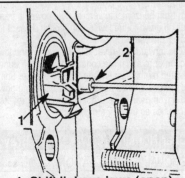

1- Shift linkage jaws (open)
2- Shift cable end (released from jaws)

Fig. 4 Make sure the shift cable comes out of the assembly jaws

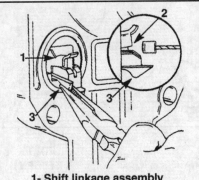

1- Shift linkage assembly
2- Jaws - open
3- Underside of lower lip

Fig. 5 Pull the linkage assembly out until the jaws open all the way

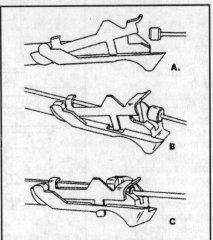

Fig. 6 Shift cable linkage assembly

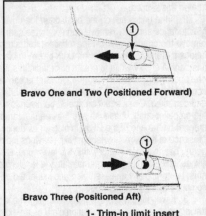

Bravo One and Two (Positioned Forward)

Bravo Three (Positioned Aft)
1- Trim-in limit insert

Fig. 7 It is very important that the trim-in insert is positioned correctly

11-4 DRIVE SYSTEMS - BRAVO

12. Oil the shaft with coupler grease to prevent damage to the seals, and then slide three new O-ring seals over the splines of the universal joint shaft, and into the grooves provided.
13. Coat the splines of the driveshaft and the driveshaft housing pilot, with Multi-purpose Lubricant. As an aid to installation, apply a light coating of Multi-purpose Lubricant to the inside surfaces of the bell housing bore.
14. Check the U-joint bellows for cracks wear or other damage. Replace it if there is any doubt as to its condition whatsoever as it acts as a seal between the bell housing and the upper unit.
15. Check the driveshaft bellows for cracks wear or other damage. Replace it if there is any doubt as to its condition whatsoever.
16. Grab the lower lip of the shift linkage assembly with needle nose pliers and pull it out until the jaws open all the way. Coat the underside of the assembly with Special Lubricant 101. While moving the drive unit into position with the bell housing, make sure that the shift cable enters the jaws and drops down into the slot in the lower lip. Once it is in the slot it should force the entire assembly back into the upper unit as you move the unit further into position, thus closing the jaws. You may want to guide the cable into the jaws by hand to make sure this goes smoothly.
17. Position the drive unit and slowly move it into the bell housing. The trim cylinders should be untied and positioned so they are pointing straight backwards. Make sure the driveshaft slips into the bell housing bore properly, through the gimbal bearing and into the engine coupler. You may have to wiggle the propeller slightly in order to get the splines to line up correctly.
18. Apply a coating of Multi-purpose Lubricant onto the shift lever coupler, and then align the slot straight by moving the lever to the right. Lubricate the slot and the reverse lock roller.
19. Secure the driveshaft housing to the bell housing with the six elastic-stop nuts with flat washers (remember that the center port stud does not have a washer). Tighten the nuts alternately (starting with the center) and evenly to 50 ft. lbs. (68 Nm). If you removed the top cover, tighten the screws to 20 ft. lbs. (27 Nm) on Bravo models.
20. Install the trim cylinders as detailed in the Trim Cylinder section.

■ These models utilize a Trim-In Limit insert to prevent over trimming at high speeds on Bravo III units. On Bravo I/II models the insert must be positioned all the way forward, while on Bravo III models the insert must be positioned all the way to the rear of the cut-out. This must take place before installing the trim cylinder anchor pin.

21. Connect the speedometer cable in the opposite manner that you disconnected it - turn the release lever clockwise this time.
22. Fill the unit with oil and adjust the shift cable.

GEAR HOUSING (LOWER UNIT)

Description

The upper gear housing transfers engine power to the lower unit through a driveshaft. Splines on the forward end of the horizontal driveshaft mate with matching splines of a coupler attached to the engine flywheel. The aft end of this driveshaft is connected to a universal joint and the pinion (drive) gear assembly. The pinion (drive) gear changes the direction of the power train from horizontal to vertical. The power is transferred from the upper gear housing down to the lower unit through a set of upper and lower gears and a vertical driveshaft.

The forward and reverse gears (upper and lower depending on standard or counter rotation) are contained within the upper unit. A sliding clutch, actuated by the shift linkage (all in the upper unit), engages a forward gear and creates a direct coupling for transmitting power through the pinion gear and driven gear to the propeller shaft. The reverse gear utilizes the same mechanics as the forward gear.

Gear Housing (Lower Unit) - Bravo I/II/X/XR/XZ

REMOVAL & INSTALLATION

◆ See Figures 8 and 9

The lower unit may be separated and removed from the upper gear housing while the stern drive remains attached to the boat or after the stern drive has been removed.

1. Remove the propeller(s) from the lower gear housing as detailed later in this section.
2. If the unit is still on the boat, drain the drive oil as detailed in the Maintenance section - remove and empty the gear lube monitor and also remove the drain plug. Allow the lubricant to drain into the container. As the lubricant drains, check the color. Dark brown to black indicates normal old lubricant. A chalky white to cream color indicates the presence of water. The presence of any water in the gear lubricant is bad news. The unit must be completely disassembled, inspected, the cause of the problem determined and corrected. Close attention should be given to the back-to-back seals on the propeller shaft, driveshaft and the bearing carrier O-ring. Examine the magnet on the end of the fill/drain plug for evidence of metal particles. The presence of tiny small metal dust like shavings indicates normal wear of the gears, bearings and shafts within the lower unit. Large metal chips or heavy fillings indicate extensive internal damage is taking place and the lower unit must be completely disassembled and inspected. All worn and/or damaged components must be replaced.
3. If working on a Bravo I, place an alignment mark on the trim tab trailing edge and the anti-cavitation plate as an aid during assembling. Remove the plastic (or rubber) plug from the top of the anti-cavitation plate. Remove the 1/2 in. bolt directly above the trim tab, lift off the trim tab. You will need a long extension to your socket wrench for this one.

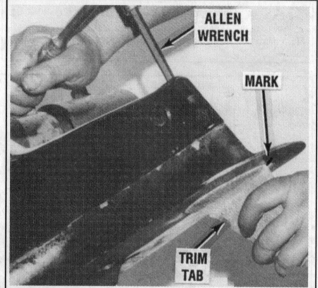

Fig. 8 Removing the trim tab bolt will require a socket extension - Bravo I

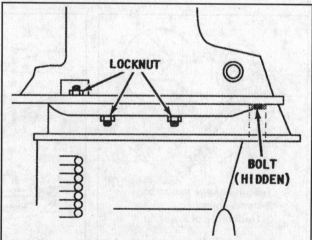

Fig. 9 Remove the six nuts and bolt to separate the two drive halves

DRIVE SYSTEMS - BRAVO

4. Remove the six hex nuts and washers (three on each side), from the sides of the anti-cavitation plate. Remove the single bolt hidden inside the trim tab cavity on the Bravo I; on the Bravo II this bolt is in a recess, just forward of the trim tab.

5. Separate the two housings by tapping downward on the lower unit with a soft face mallet and a wooden block. In some cases the housing may be difficult to separate because the driveshaft may be corroded and "frozen" in the upper gear housing vertical shaft. Lower the gear housing away from the upper gear housing.

To install:

6. Coat a new O-ring with 3-M Adhesive and install it on the upper face of the lower unit at the water passage opening. No gasket is used between the two housings. All sealing is accomplished with two O-rings: The water passageway O-ring and the O-ring installed around the driveshaft spacer. The spacer has two vertical holes drilled through it to allow oil to pass through the spacer and lubricate not only the double roller bearing assembly on the driveshaft but the entire lower unit. Oil passes freely through the double bearing arrangement. Therefore, the oil passageway O-ring is perhaps the most crucial component in a Bravo stern drive.

✱✱ WARNING

In the Bravo stern drive, no locating pins are utilized in aligning the lower unit with the upper gearcase housing. Therefore, if the bolt holes show serious signs of elongation, or the securing bolts show signs of stress or "necking", the lower unit, the upper gearcase housing and all securing bolts should be replaced. This is not as drastic or expensive as it sounds. All internal components are removed and only the housing castings are replaced.

7. Install the coupler and retainer onto the upper end of the driveshaft. Raise the lower unit to mate with the upper gearcase housing. The lower end of the short clutch shaft must slide into the coupler at the upper end of the lower driveshaft. It may be necessary to slowly rotate the propeller shaft counterclockwise until the splines index.

8. Screw the aft bolt into the hole in the trim tab cavity on Bravo I or into the hole just forward of the trim tab on Bravo II.

9. Install the three port and three starboard nuts and their washers. Tighten the six nuts alternately, and a little at-a-time, until the lower unit is tight against the upper gear housing.

10. Now tighten the rear bolt and the six side nuts to 35 ft. lbs. (48 Nm).

11. Install the trim tab (Bravo I) with the marks made during disassembly aligned. Tighten the screw(s) to 23 ft. lbs. (32 Nm) on the Bravo I, or 20 ft. lbs. (27 Nm) on the Bravo II.

■ The trim tab performs two very important jobs, one of which you may not realize. First, the tab compensates for steering torque. If the boat continually seems to move to port or starboard while the helmsman is on a straight course, the trim tab can be adjusted to the side of the pull. The tab also prevents electrolysis from damaging expensive parts. The tab is not expensive; it should show signs of electrolytic action; and should always be replaced after some of the material has been eaten away or is pitted. Now, if the tab shows no signs of electrolytic action after the boat has been in use over a period of time, the grounding should be checked to ensure more expensive parts are not being damaged. Install the trim tab plastic cover plug.

12. Install the propeller(s) if it was removed and refill the drive unit and lube monitor with oil.

DISASSEMBLY

♦ See Figures 10 thru 23

The following procedures assume that the lower unit and propeller have already been removed.

1. Install a clamp plate (#91-43559T) over the driveshaft and tighten the nuts onto the studs securely.

2. Use a puller tool and carefully remove the propeller hub on the XR and XZ models.

3. Straighten the locking tabs on the bearing carrier retainer tab washer. Loosen the bearing carrier retainer by turning it counterclockwise with the special wrench (#91-61069 on the Bravo I/XR/XZ, or #91-17257 on the Bravo II). Remove the retainer and tab washer.

4. Pull out the bearing carrier assembly using a bearing carrier puller (#91-90338A1 - Bravo I/X/XR/XZ; #91-34569A1 & #91-46086 A1 - Bravo II).

■ If the bearing carrier is corroded in place, apply heat to the housing to loosen the corrosion. While the housing is still hot, use a bearing carrier puller to remove the assembly do not overheat the housing or it will become distorted and useless.

5. Reach in and remove the O-ring, thrust washer and load ring. Tag and wire them together, as an aid during assembling - you may want to do this after removing the propeller shaft.

6. Grab the end of the propeller shaft and pull it out of the housing (leaving the driven gear still in the housing).

7. Temporarily reinstall the bearing retainer by hand and then install a driveshaft adapter tool (#91-61077) on the top of the vertical shaft. Insert a long breaker bar (with socket) into the propeller shaft bore and remove the pinion gear bolt (nut on Bravo II) while holding the adaptor tool so the driveshaft does not rotate. Lift out the bolt and washer.

8. Remove the clamp plate installed previously. Lift the O-ring, spacer, shim material and tab washer up and off of the driveshaft. Tie the shims together and save them for assembly.

9. Grab the top of the driveshaft and pull it out of the housing. Pull out the pinion gear.

10. Reach into the propeller shaft bore and pull out the driven gear/bearing and cup.

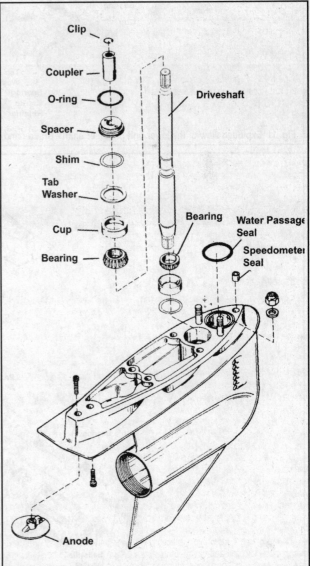

Fig. 10 Exploded view of the lower unit driveshaft - Bravo I, X, XR and XZ

11-6 DRIVE SYSTEMS - BRAVO

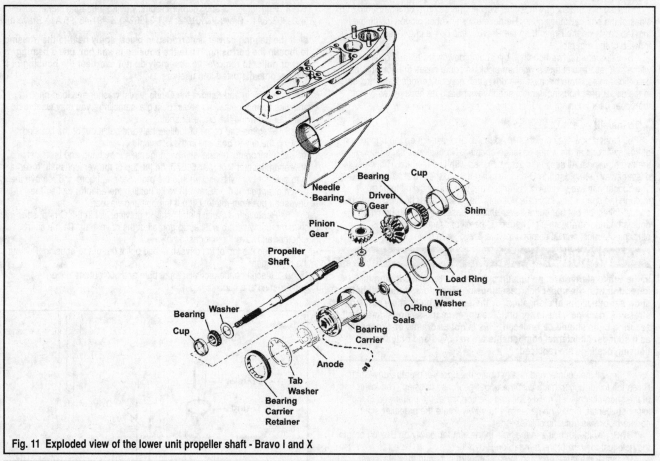

Fig. 11 Exploded view of the lower unit propeller shaft - Bravo I and X

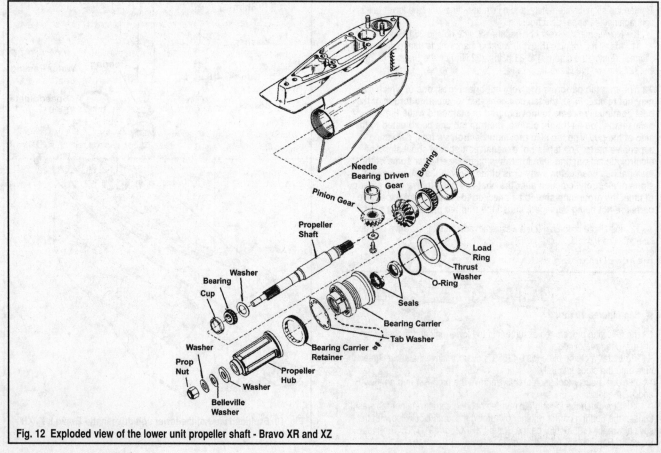

Fig. 12 Exploded view of the lower unit propeller shaft - Bravo XR and XZ

DRIVE SYSTEMS - BRAVO   11-7

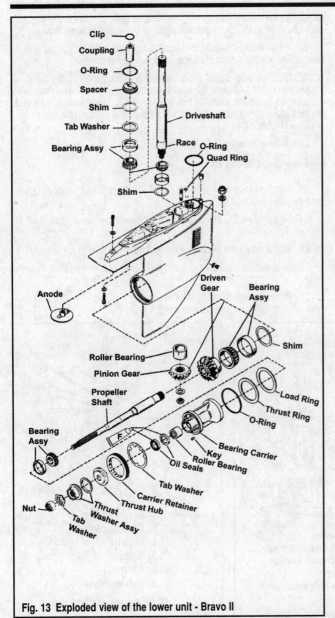

Fig. 13 Exploded view of the lower unit - Bravo II

■ The next step is not required, unless the bearings are to be replaced. Any time the bearings are replaced, they must be adjusted properly with shim material or their service life will be drastically reduced.

11. Using a slide hammer with a two-armed puller, remove the driveshaft tapered roller bearing cup. Tag and wire the shim material pieces from underneath the bearing cup together, as an aid during assembly.

12. Remove the needle bearings from the driveshaft needle bearing race unless you intend to remove the pinion bearing.

13. Pull the O-ring out of the water passage on top of the housing and remove the oil passage quad ring.

Driveshaft And Propeller Shaft Bearings

◆ See Figure 24

■ Bearing assemblies will be damaged on removal; always replace the roller bearings if they have been removed.

14. Install a Universal Puller Plate (#91-37241) to the shaft to support the bearing.

15. Install the assembly into an arbor press and press off the bearing.

16. Repeat this procedure for the second tapered roller bearing on the driveshaft.

17. Install a suitable mandrel and press the bearings onto the shaft. When installing the bearings on the driveshaft, first press the smaller bearing onto the shaft so that the smaller O.D. is facing the pinion end of the shaft and then press on the larger bearing making sure the larger O.D. faces the pinion end.

Bearing Carrier

◆ See Figures 25, 26 and 27

■ The oil seals or the bearing cup may be replaced individually. It is not necessary to remove one in order to remove the other.

18. Clamp the bearing carrier in a vise with soft jaws. Be sure to clamp on the bearing carrier at the reinforcement ribs. Place the jaws of a universal slide hammer puller down through the center of the carrier until the teeth on the puller are through the bearing cup. Pull the cup out of the bearing carrier.

19. Remove the two propeller shaft oil seals from the bearing carrier with a hammer and punch.

To assemble:

20. Wash all parts in solvent and blow them dry with compressed air. Remove any corrosion from the outside surface of the bearing carrier. Remove any seal and gasket material from all mating surfaces. Blow all water, oil passageways, and screw holes clean with compressed air. After all

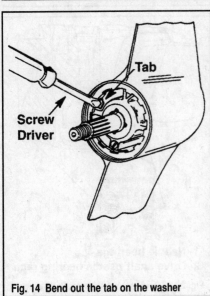

Fig. 14 Bend out the tab on the washer

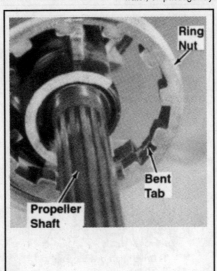

Fig. 15 A better view of the tab

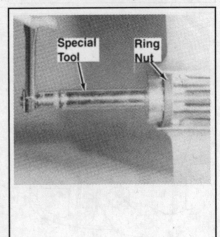

Fig. 16 Use a special tool to loosen the bearing carrier retainer. . .

11-8 DRIVE SYSTEMS - BRAVO

Fig. 17 ...and then spin out the nut and remove the washer

parts are clean and dry, apply a light coating of gear lubricant to the bright surfaces of all gears, bearings, and shaft, as prevention against rusting.

21. Install the new bearing cup into the bearing carrier with a cup and seal driver, making sure the bearing surface is facing inward.

22. Coat the outer edges of two new seals with Loctite 271 and then press two new oil seals into the front of the bearing carrier. These seals are installed back-to-back. Seat the first one with the lip facing inward. Seat the second seal with the lip facing outward. Fill the area between the two seals with 2-4-C lubricant.

Driven Gear Bearing

◆ See Figures 28 and 29

23. Install a Universal Puller Plate (#91-37241) between the driven gear and the roller bearing.

24. Sit the assembly on two pieces of wood and press the bearing off with an arbor press.

25. Sit the gear on a block of wood with the teeth toward the wood.

26. Coat the inside of a new bearing with gear lube and position over the gear.

27. Place a suitable mandrel (an old bearing race works well) over the new bearing race and then press it into the gear.

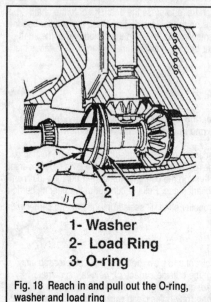

1- Washer
2- Load Ring
3- O-ring

Fig. 18 Reach in and pull out the O-ring, washer and load ring

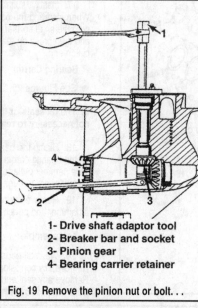

1- Drive shaft adaptor tool
2- Breaker bar and socket
3- Pinion gear
4- Bearing carrier retainer

Fig. 19 Remove the pinion nut or bolt...

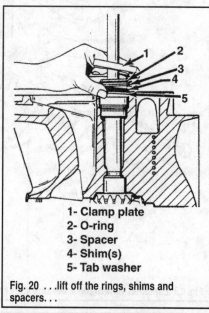

1- Clamp plate
2- O-ring
3- Spacer
4- Shim(s)
5- Tab washer

Fig. 20 ...lift off the rings, shims and spacers...

1- Drive shaft - Pull up
2- Pinion gear

Fig. 21 ...and then pull out the driveshaft

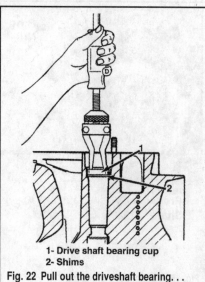

1- Drive shaft bearing cup
2- Shims

Fig. 22 Pull out the driveshaft bearing...

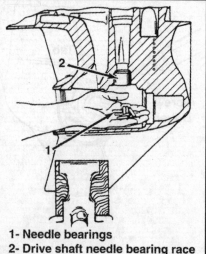

1- Needle bearings
2- Drive shaft needle bearing race

Fig. 23 ...and remove the needle bearings

DRIVE SYSTEMS - BRAVO

CLEANING AND INSPECTION

◆ See Figures 10 thru 13

1. Wash all parts in solvent and blow them dry with compressed air. Remove all traces of seal and gasket material from all mating surfaces. Blow all water, oil passageways and screw holes clean with compressed air. After cleaning, apply a light coating of gear lubricant to the bright surfaces of all gears, bearings, and shafts as prevention against rusting and corrosion. Check the shift crank to be sure it is not bent.

2. Check to be sure the water intake is clean and free of any foreign material.

3. Inspect the gear case, housings, and covers inside and out for cracks. Check carefully around screw and shaft holes. Check for burrs around machined faces and holes. Check for stripped threads in screw holes and traces of gasket material remaining on mating surfaces.

4. Check O-ring seal grooves for sharp edges, which could cut a new seal. Check all oil holes.

5. Inspect the bearing surfaces of the shafts, splines, and keyways for wear and burrs. Look for evidence of an inner bearing race turning on the shaft. Check for damaged threads. Measure the run-out on all shafts to detect any bent condition. If possible, check the shafts in a lathe for out-of-roundness.

6. Inspect the gear teeth and shaft holes for wear and burrs. Hold the center race of each bearing and turn the outer race. The bearing must turn freely without binding or evidence of rough spots. Never spin a ball bearing with compressed air or it will be ruined. Inspect the outside diameter of the outer races and the inside diameter of the inner races for evidence of turning in the housing or on a shaft. Deep discoloration and scores are evidence of overheating.

7. Inspect the thrust washers for wear and distortion. Measure the washers for uniform thickness and flatness.

8. Inspect the gears for distortion, burrs, and cracks. Check for any sign of discoloration, which means the gears are running too hot.

9. Replace all seals, O-rings, and gaskets to ensure maximum service after the work is completed.

10. Inspect the pilot tube opening on the fore edge for the housing for any obstructions and clean thoroughly with a stiff piece of wire. If you are unable to dislodge an obstruction, use a 5/64 in. (2mm) drill bit, but do not go deeper then 2 in.

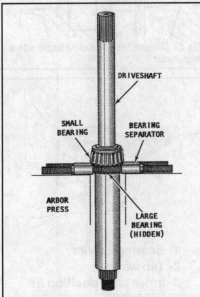

Fig. 24 Use an arbor press and the special tool to remove the roller bearings on either of the shafts

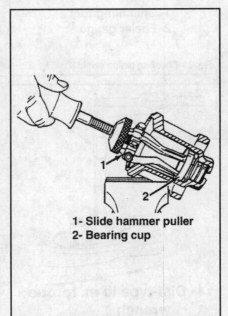

Fig. 25 Using a puller to remove the bearing cup

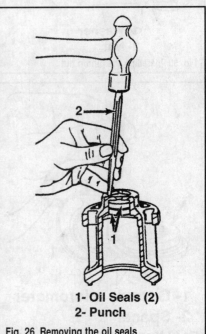

Fig. 26 Removing the oil seals

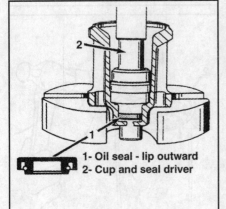

Fig. 27 Installing the oil seals (outer seal shown); bearing cup similar

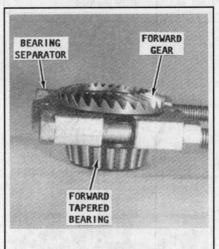

Fig. 28 Use the special tool to separate the bearing from the driven (forward) gear

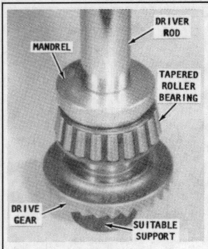

Fig. 29 Press the roller bearing into the driven gear

DRIVE SYSTEMS - BRAVO

ASSEMBLY

◆ See Figures 10 thru 13 and 30 thru 38

1. Lubricate all gears, splines and bearings with gear lube so as to get accurate pre-load readings.

2. Install a new speedometer water passage seal with 3-M Adhesive. Make sure that the top edge of the seal is flush with the housing surface. Coat a new water passage O-ring and quad ring with 3-M Adhesive and install them into the recess.

3. Coat the needle bearings with Quicksilver Needle Bearing lubricant and install them into the lower driveshaft bearing casing - the grease will hold them in place.

4. Position the driven gear and bearing assembly into the propeller bore.

5. Drop the original shims into the driveshaft bore and drive the lower cup into position with the driver tool (#91-67443). If the original shims were damaged or misplaced, install new ones with a thickness of 0.050 in. (1.27mm) as a starting point.

6. Reach into the propeller shaft bore and position the pinion gear while an assistant slides the driveshaft into the vertical bore. Coat the threads of a new pinion bolt or nut with Loctite® 271, install the washer and then screw the bolt/nut in finger-tight.

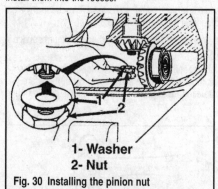

1- Washer
2- Nut

Fig. 30 Installing the pinion nut

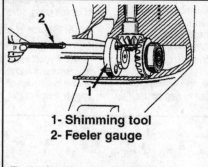

1- Shimming tool
2- Feeler gauge

Fig. 31 Checking pinion height

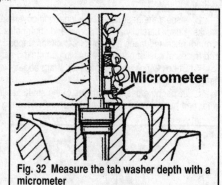

Fig. 32 Measure the tab washer depth with a micrometer

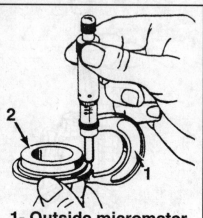

1- Outside micrometer
2- Spacer

Fig. 33 Measure the spacer thickness with an outside micrometer

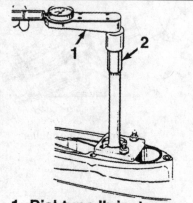

1- Dial-type lb in. torque wrench
2- Drive shaft adaptor

Fig. 34 Checking the pre-load

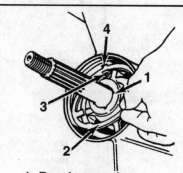

1- Bearing carrier
2- Tab washer
3- Inner tab - position in V-notch
4- Outer tab

Fig. 35 Position the inner tab washer in the V-notch

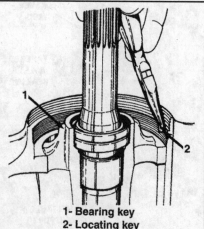

1- Bearing key
2- Locating key

Fig. 36 Make sure you don't forget the locating key on Bravo II units

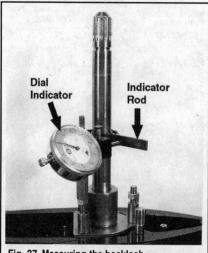

Fig. 37 Measuring the backlash

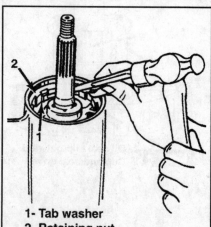

1- Tab washer
2- Retaining nut

Fig. 38 Bend in the tabs once the pre-load is within specifications

DRIVE SYSTEMS - BRAVO

7. Thread in the bearing carrier retainer again and then install the driveshaft adaptor tool (#91-61077) to the shaft. With a long handled socket on the pinion bolt/nut and a torque wrench on the adaptor, tighten the pinion bolt to 45 ft. lbs. (61 Nm) on all but the Bravo II where it should be 100 ft. lbs. (136 Nm). Remove the bearing retainer and adaptor tool.

8. Install the upper driveshaft bearing cup, tab washer, original shims, spacer and the O-ring over the shaft. If the original shims were damaged or misplaced, install new ones with a thickness of 0.050 in. (1.27mm).

Or, use a micrometer to measure the depth from the top surface of the gear housing to the top of the tab washer. Next, measure the thickness of the spacer from the top to the bottom of the machined surface. Now, subtract the thickness of the spacer from the tab washer depth in the bore and add 0.001 in. - this can be the starting shim thickness.

9. Install the clamp plate (#91-43559) back over the driveshaft and onto the top of the gear housing with 2 nuts and 4 washers. Tighten the nuts to 35 ft. lbs. (47 Nm).

10. Position an adaptor over the upper splines on the driveshaft and attach a torque wrench. Slowly turn the wrench and check that the rolling pre-load is 3-5 inch lbs. (0.3-0.6 Nm). If not within specifications, you can adjust the pre-load by adding or subtracting shim material at the upper bearing cup. Once you get the load within the range, make sure that you jot it down somewhere.

11. Rotate the driveshaft at least 3 times to seat the bearings and then insert a shimming tool (#91-42840 for all but the Bravo II which uses #91-96512) into the propeller shaft bore. Check the clearance between the tool and the gear with a feeler gauge (you're going to need a long one!) - check this in three places, 120° apart. This is known as pinion height and the clearance must be 0.025 in. (0.635mm). If less than spec, add shims of the appropriate thickness under the lower bearing cup to bring it into specification. If greater than spec, remove shim material from under the bearing cup. Recheck the clearance again until you're sure it is correct.

■ Any thickness of shims added to the lower cup, must have an equivalent amount subtracted from the shim thickness at the upper bearing cup. And also, any thing removed from the lower cup must be added to the upper.

12. Once the pinion height is correct, recheck the rolling pre-load again.
13. Install the original load ring, thrust ring and O-ring into the propeller shaft bore. If you tied them all together in disassembly, make sure you remove the tie. Slide the propeller shaft through the bore and into the driven gear.
14. Install the bearing carrier and tab washer so the inner tab on the washer lines up with the V-notch. Now thread in the bearing carrier retainer and tighten it with the retainer wrench (#91-61069 or #91-17257) until resistance to the propeller shaft rotation can felt - you have pre-loaded the bearings.
15. Install a dial indicator adaptor (#91-83155), backlash indicator rod (#91-53459) and dial indicator onto the driveshaft so that the dial rod is aligned with the I on the indicator rod (II on Bravo II). Rotate the driveshaft back and forth slightly without turning the propeller shaft; the dial indicator should read 0.012-0.015 in. (0.28-0.38mm) on all except the Bravo II which should read 0.009-0.015 (0.23-0.38mm). If backlash is less than specified, remove shims from behind the driven gear bearing cup; if greater, add shims. Continue this process until determining that backlash is correct.

16. Remove the bearing carrier, O-ring, thrust washer and load ring. Reinstall everything using a new load ring. Lubricate the space between the carrier oil seals with Quicksilver 2-4-C lubricant and coat the mounting surfaces on the carrier with Perfect Seal. Lubricate the threads on the carrier retainer with Quicksilver Special Lubricant 101. On the Bravo II, insert the locating key with a pair of needle nose pliers.
17. Install the bearing carrier and tab washer. Align the inner tab on the washer with the V-notch in the retainer and the outer tab with the hole in the housing bore.

■ On XR and XZ models, you must use a carrier installation tool (#91-840388) to center the carrier and get it into position.

18. Lubricate the threads of the retainer with Special Lubricant 101 and tighten it until resistance to the propeller shaft rotation can just be felt.
19. Determine what the overall gearcase pre-load should be by adding the driveshaft pre-load you recorded previously to the propeller shaft bearing preload (which should be 8-12 inch lbs. (0.9-1.4 Nm) for new bearings, or 5-8 inch lbs. (0.6-0.9 Nm) on used bearings) - this is your final gearcase pre-load.
20. Install the retainer wrench (#91-61069 on the Bravo I and Z, #91-840393 on the XR and XZ; #91-17257 on the Bravo II) and tighten the retainer slowly and in small increments. Each time you stop, spin on the propeller nut and check the gearcase pre-load with a torque wrench by rotating the propeller shaft in its normal direction of rotation with a slow and steady motion. Continue tightening and then checking until you arrive at the correct gearcase pre-load figure and then bend one tab from the washer out and into the retainer and then the remaining tabs down into the housing.
21. Reconnect the upper and lower cases.

Gear Housing (Lower Unit) - Bravo III

REMOVAL & INSTALLATION

 MODERATE

◆ See Figures 39, 40 and 41

The lower unit may be separated and removed from the upper gear housing while the stern drive remains attached to the boat or after the stern drive has been removed.

1. Remove the propeller(s) from the lower gear housing as detailed elsewhere in this section.
2. If the unit is still on the boat, drain the drive oil as detailed in the Maintenance section. Allow the lubricant to drain into the container. As the lubricant drains, check the color. Dark brown to black indicates normal old lubricant. A chalky white to cream color indicates the presence of water. The presence of any water in the gear lubricant is bad news. The unit must be completely disassembled, inspected, the cause of the problem determined and corrected. Close attention should be given to the back-to-back seals on the propeller shaft, driveshaft and the bearing carrier O-ring. Examine the magnet on the end of the fill/drain plug for evidence of metal particles. The presence of tiny small metal dust like shavings indicates normal wear of the gears, bearings and shafts within the lower unit. Large metal chips or heavy fillings indicate extensive internal damage is taking place and the lower unit

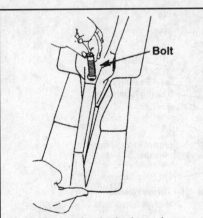

Fig. 39 Pop out the plastic plug and remove the anode mounting bolt

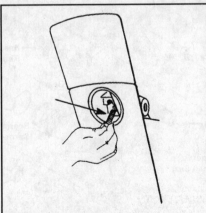

Fig. 40 After removing the anode, remove the bolt in the cavity...

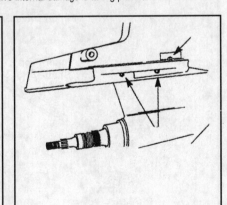

Fig. 41 ...remove the six nuts to separate the two drive halves

11-12 DRIVE SYSTEMS - BRAVO

must be completely disassembled and inspected. All worn and/or damaged components must be replaced.

3. Remove the plastic plug from the top of the gear housing. Remove the bolt directly above the anode, lift off the anode. You may need an extension to your socket wrench for this one.
4. Look inside the anode cavity and remove the mounting bolt.
5. Remove the six hex nuts and washers (three on each side), from the sides of the housing.
6. Separate the two housings by tapping downward on the lower unit with a soft face mallet and a wooden block. In some cases the housing may be difficult to separate because the driveshaft may be corroded and "frozen" in the upper gear housing vertical shaft. Lower the gear housing away from the upper gear housing and place it in a holding device.

To install:

7. Install the coupler and retainer onto the upper end of the driveshaft. Raise the lower unit to mate with the upper gearcase housing. The lower end of the short clutch shaft must slide into the coupler at the upper end of the lower driveshaft. It may be necessary to slowly rotate the propeller shaft counterclockwise until the splines index.
8. Install the three port and three starboard nuts and their washers. Tighten the six nuts alternately, and a little at-a-time, until the lower unit is tight against the upper gear housing. Now tighten all six side nuts to 35 ft. lbs. (47.5 Nm).
9. Install the mounting bolt in the anode cavity and tighten to 35 ft. lbs. (47.5 Nm).
10. Install the anode into the extension and tighten the bolt to 20 ft. lbs. (27 Nm).
11. Install the drive unit if removed.
12. Install the propellers and refill the drive unit with oil.

DISASSEMBLY

◆ See Figures 42 thru 57

The following procedures assume that the lower unit and propeller have already been removed.

1. Install a clamp plate (#91-43559) over the driveshaft and tighten the 2 nuts and 4 washers onto the studs securely.
2. Install a bearing carrier removal tool (#91-805374-1) over the propeller shaft until it seats onto the leading edge of the carrier.
3. Heat the end of the housing around the carrier (with a gun or lamp, no torches) and then turn the carrier off the propeller shaft with a breaker bar. Remember that this is a left hand thread so you will loosen it by turning it CLOCKWISE - opposite of what you're used to.
4. Slide a bearing retainer tool (#91-805302) and a bearing carrier tool (#91-805374) over the propeller shaft and into the bore until it fits into the bearing retainer. Install the carrier removal tool over the retainer tool and loosen the retainer by turning the tools CLOCKWISE also. Unscrew the retainer and remove it from the bore.
5. Grab the inner and outer propeller shaft assembly and pull it out of the housing bore, leaving the front driven gear in the bore. Reach back into the housing bore and remove the shim(s).
6. Install a driveshaft adapter tool (#91-61077) on the top of the vertical shaft (with 1 1/4 in. socket). Insert a long breaker bar (with 9/16 in. socket) into the propeller shaft bore and remove the pinion gear bolt. Lift out the bolt and washer.
7. Remove the clamp plate installed previously. Lift the O-ring, spacer, shim material and tab washer up and off of the driveshaft.

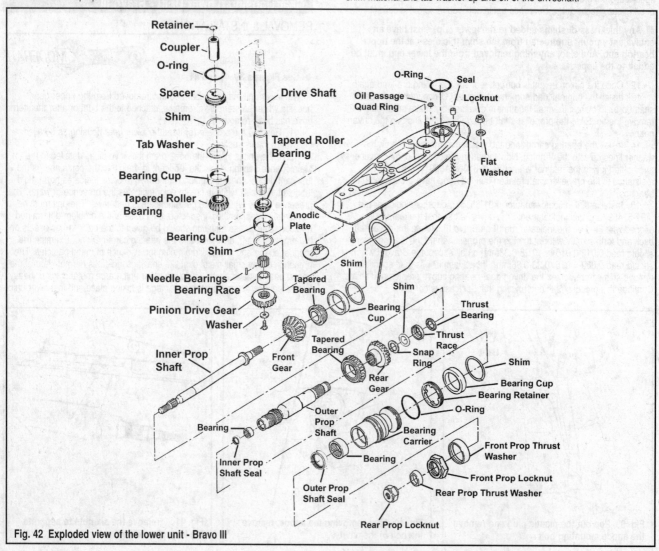

Fig. 42 Exploded view of the lower unit - Bravo III

8. Grab the top of the driveshaft and pull it out of the housing.
9. Reach into the propeller shaft bore and pull out the pinion gear.
10. Reach into the propeller shaft bore and pull out the driven gear and bearing/cup.

■ The next step is not required, unless the bearings are to be replaced. Any time the bearings are replaced, they must be adjusted properly with shim material or their service life will be drastically reduced.

11. Using a slide hammer with a two-armed puller, remove the driveshaft tapered roller bearing cup. Tag and wire the shim material pieces from underneath the bearing cup together, as an aid during assembly.

12. Remove the needle bearings from the driveshaft needle bearing race unless you intend to remove the pinion bearing.
13. Pull the O-ring out of the water passage on top of the housing, and also the oil passage quad ring.

Driveshaft

◆ See Figure 24

■ Bearing assemblies will be damaged on removal; always replace the roller bearings if they have been removed.

14. Install a Universal Puller Plate (#91-37241) to the shaft to support the bearing.

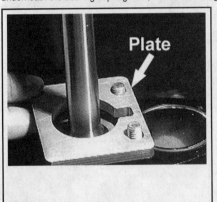
Fig. 43 Install the clamp plate over the shaft

Fig. 44 Unscrew the bearing carrier with the special tool...

Fig. 45 ...and remove it from the housing

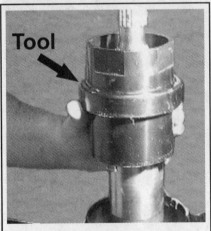
Fig. 46 Slide the retainer nut tool over the propeller shaft...

Fig. 47 ...and then install the carrier tool to remove the nut

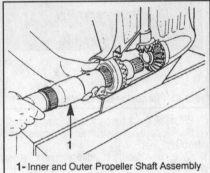
Fig. 48 Pull out the propeller shaft assembly...

Fig. 49 ...and then pull out the shims

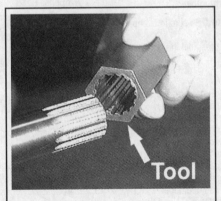
Fig. 50 Install an adaptor tool to the top of the shaft...

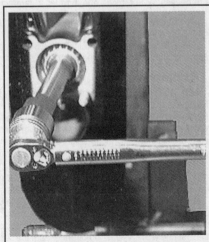
Fig. 51 ...and then loosen the pinion bolt

11-14 DRIVE SYSTEMS - BRAVO

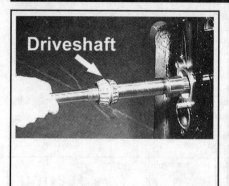
Fig. 52 Pull out the driveshaft

Fig. 53 Pull out the pinion gear

Fig. 54 Pull out the forward driven gear and bearing

Fig. 55 Pull out the bearing race...

Fig. 56 ...and the shim material

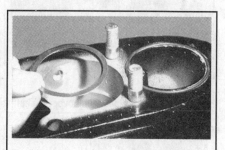
Fig. 57 Remove the shim material from the vertical bore

15. Install the assembly into and arbor press and press off the bearing.
16. Repeat this procedure for the second tapered roller bearing on the driveshaft.
17. Install a suitable mandrel and press the bearings onto the shaft. When installing the bearings on the driveshaft, first press the smaller bearing onto the shaft so that the smaller O.D. is facing the pinion end of the shaft and then press on the larger bearing making sure the larger O.D. faces the pinion end.

Bearing Carrier
◆ See Figure 58

■ The oil seals or the bearing cup may be replaced individually. It is not necessary to remove one in order to remove the other.

18. Clamp the bearing carrier in a vise with soft jaws. Be sure to clamp on the bearing carrier at the reinforcement ribs.
19. Use a seal removal tool (#91-862064A1) and press out the two propeller shaft oil seals and the bearing.

To assemble:
20. Wash all parts in solvent and blow them dry with compressed air. Remove any corrosion from the outside surface of the bearing carrier. Remove any seal and gasket material from all mating surfaces. Blow all water, oil passageways, and screw holes clean with compressed air. After all parts are clean and dry, apply a light coating of gear lubricant to the bright surfaces of all gears, bearings, and shaft, as a prevention against rusting.
21. Use and installation tool (#91-805356) and install the new bearing into the bearing carrier with the bearing surface facing inward.
Make sure you coat the outer edge of the bearing with gear lube.
22. Coat the outer edges of two new seals with Loctite® 271 and then press two new oil seals into the front of the bearing carrier with a seal installation tool (#91-805372). These seals are installed back-to-back. Seat the first one with the lip facing outward. Seat the second seal with the lip facing inward. Fill the area between the two seals with Special Lubricant 101.

Driven Gear Bearing
◆ See Figures 28 and 29

23. Install a Universal Puller Plate (#91-37241) between the driven gear and the roller bearing.
24. Sit the assembly on two pieces of wood and press the bearing off with an arbor press.
25. Sit the gear on a block of wood with the teeth toward the wood.
26. Coat the inside of a new bearing with gear lube and position over the gear.
27. Place a suitable mandrel (an old bearing race works well) over the new bearing race and then press it into the gear.

Propeller Shaft
◆ See Figures 59 thru 63

28. Grasp the outer shaft and pull the inner shaft and thrust bearing out of the assembly.
29. Remove the bearing cup from the outer shaft.
30. Position a punch on the thrust cap so it is lined up with the gap in the snap-ring and then press off the thrust cap. Remove the shim.
31. Using snap-ring pliers, remove the snap-ring from the end of the shaft and then slide off the rear driven gear and bearing assembly.
32. Insert a slide hammer with a three-armed puller attachment into the outer shaft and pull out the bearing and two oil seals.

To assemble:
33. Press a new bearing into the outer shaft with the appropriate installation tool (#91-805352).
34. Coat the edges of the two new oil seals with Loctite® 271 and press the seals into the outer shaft so the seal lips face away from each other. The inner seal lips should be facing inward and the outer seal lips should be facing out. Press the seals in until the tool (#91-805358) bottoms out against the shaft and then lubricate the lips and the area between the seals with Quicksilver Special Lubricant 101.

DRIVE SYSTEMS - BRAVO 11-15

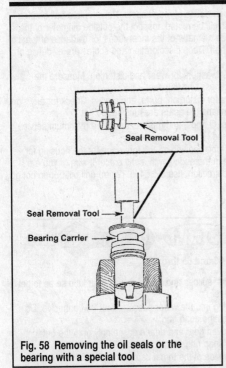

Fig. 58 Removing the oil seals or the bearing with a special tool

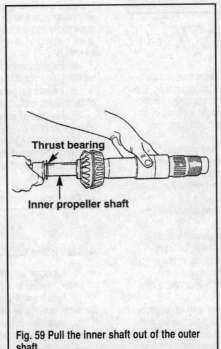

Fig. 59 Pull the inner shaft out of the outer shaft...

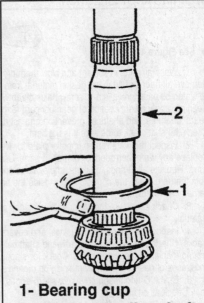

1- Bearing cup
2- Outer propeller shaft

Fig. 60 ...and then remove the bearing cup from the outer shaft

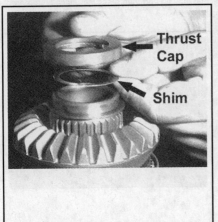

Fig. 61 Remove the thrust cap and shim...

Fig. 62 ...and then remove the snap-ring...

Fig. 63 ...and gear/bearing assembly

35. Press the driven gear and bearing onto the end of the outer shaft and install a new snap-ring.
36. Coat the inside ring of the thrust cap lightly with Loctite® 242 and then position the shim into the cap. Position the cap and shim so it is squarely over the shaft and then press it on. If the original shim material was damaged or misplaced, use a 0.015 in. (0.38mm) shim pack as the starting point for installation and then adjust it after checking end-play.
37. Do not install the inner shaft at this time, see the Assembly procedure for this.

Spline Lash Check
◆ See Figure 64

■ This test is necessary to complete the final backlash checks upon final reassembly of the unit.

38. Position the inner propeller shaft on two V-blocks.
39. Install a dial indicator rod on the shaft just behind the shaft splines.
40. Slide the front driven gear assembly onto the splines and then install a dial indicator so its tip lines up with the mark on the indicator rod.
41. Grasp the gear assembly and wiggle it back and forth a bit while observing the indicator reading.
42. Repeat this procedure for the outer shaft and record both readings for use later on in assembly when checking backlash.

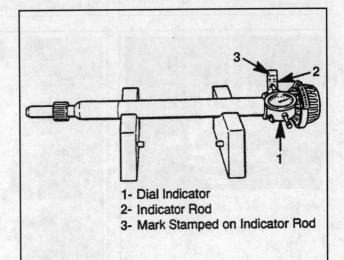

1- Dial Indicator
2- Indicator Rod
3- Mark Stamped on Indicator Rod

Fig. 64 Checking the spline lash on the propeller shafts

DRIVE SYSTEMS - BRAVO

CLEANING AND INSPECTION

◆ See Figure 42

1. Wash all parts in solvent and blow them dry with compressed air. Remove all traces of seal and gasket material from all mating surfaces. Blow all water, oil passageways and screw holes clean with compressed air. After cleaning, apply a light coating of gear lubricant to the bright surfaces of all gears, bearings, and shafts as prevention against rusting and corrosion. Check the shift crank to be sure it is not bent.

2. Inspect the water pump impeller plate for wear and corrosion. Replace any worn, corroded, or damaged parts. Use a fine file to remove burrs. Replace all O-rings, gaskets, and seals to ensure satisfactory service from the unit. Clean the corrosion from inside the housing where the bearing carrier was removed.

3. Check to be sure the water intake is clean and free of any foreign material.

4. Inspect the gear case, housings, and covers inside and out for cracks. Check carefully around screw and shaft holes. Check for burrs around machined faces and holes. Check for stripped threads in screw holes and traces of gasket material remaining on mating surfaces.

5. Check O-ring seal grooves for sharp edges, which could cut a new seal. Check all oil holes.

6. Inspect the bearing surfaces of the shafts, splines, and keyways for wear and burrs. Look for evidence of an inner bearing race turning on the shaft. Check for damaged threads. Measure the run-out on all shafts to detect any bent condition. If possible, check the shafts in a lathe for out-of-roundness.

7. Inspect the gear teeth and shaft holes for wear and burrs. Hold the center race of each bearing and turn the outer race. The bearing must turn freely without binding or evidence of rough spots. Never spin a ball bearing with compressed air or it will be ruined. Inspect the outside diameter of the outer races and the inside diameter of the inner races for evidence of turning in the housing or on a shaft. Deep discoloration and scores are evidence of overheating.

8. Inspect the thrust washers for wear and distortion. Measure the washers for uniform thickness and flatness.

9. Inspect the gears for distortion, burrs, and cracks. Check for any sign of discoloration, which means the gears are running too hot.

10. Replace all seals, O-rings, and gaskets to ensure maximum service after the work is completed.

11. Inspect the pilot tube opening on the fore edge for the housing for any obstructions and clean thoroughly with a stiff piece of wire. If you are unable to dislodge an obstruction, use a 5/64 in. (2mm) drill bit, but do not go deeper then 2 in.

ASSEMBLY

◆ See Figures 42, 31, 33 and 65 thru 75

1. Lubricate all gears, splines and bearings with gear lube so as to get accurate pre-load readings.

2. Install a new O-ring into the water passage with 3-M adhesive. Do the same with the oil passage quad ring.

3. Install the rear driven gear and bearing assembly onto the outer propeller shaft using the snap ring.

4. Lightly coat the inside of the thrust cap with Loctite® 242, but make sure that it does not come in contact with the shim mating surface. Position the original shim material into the cap and then press into place on the end of the outer propeller shaft.

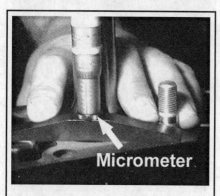

Fig. 65 Measure the tab washer depth with a micrometer

Fig. 66 Slide the tab washer over the shaft...

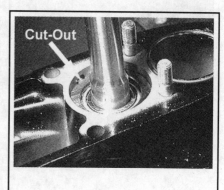

Fig. 67 ...so it fits into the cutout

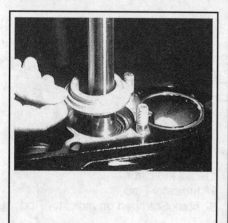

Fig. 68 Slide the spacer over the shaft

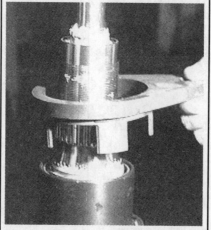

Fig. 69 Install the bearing carrier tool

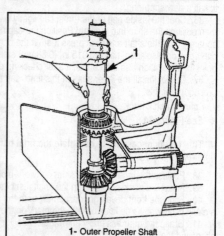

1- Outer Propeller Shaft

Fig. 70 Installing the outer shaft over the inner shaft

DRIVE SYSTEMS - BRAVO

■ Using the original shim material is as a starting point only. If you have replaced any shafts, gears or bearing assemblies do not use the original shim material; instead, use new shims to the thickness of 0,020 in.

5. Coat the needle bearings with bearing assembly lubricant and install them into the lower driveshaft bearing casing.

6. Position the forward driven gear and bearing into the propeller bore.

7. Drop the original shims into the driveshaft bore and drive the lower cup into position with a bearing cup driver (#91-67443) or an old propeller shaft. If the original shims were damaged or misplaced, install new ones with a thickness of 0.050 in. (1.27mm).

8. Slide the driveshaft into the bore being careful not to knock loose any needles. Position the pinion gear on the bottom of the shaft, coat the bolt threads with Loctite 271 and then install the pinion bolt and washer. Tighten it to 45 ft. lbs. (61 Nm). Remember, you hold the pinion nut and tighten it by turning the top of the driveshaft (don't forget the adaptor).

9. Install the upper bearing cup and tab washer over the top of the shaft. If the original shims were damaged or misplaced, install new ones with a nominal thickness of 0.050 in. (1.27mm).

The best way to do this though is to measure the distance between the top of the gear housing and the top of the tab washer with a micrometer. Now measure the thickness of the machined side of the spacer from the top side to the bottom side. Subtract this figure from the first measurement and then add 0.001 in. (0.025mm) - this should be the thickness of the shims you need to start with.

10. Either way, install the correct shim(s), the spacer and a new O-ring.

11. Install the clamp plate back onto the top of the gear housing.

12. With the driveshaft adapter still installed, check the rolling pre-load with a dial-type torque wrench. It should be 3-5 inch lbs. (0.3-0.6 Nm) while using a slow, steady force on the wrench. If not within specifications, you'll need to remove everything and add or remove shim material from under the spacer. Continue doing this until the pre-load falls within spec.

13. Now its time to check the pinion height. Insert a shimming tool (#91-805462) into the propeller shaft bore and check the clearance between the tool and the gear with a feeler gauge (you'll need a long one). Check this in three places, 120° apart. Clearance must be 0.023-0.028 in. (0.575-0.700mm). If less than spec, add shims of the appropriate thickness under the lower bearing cup to bring it into specification. If greater than spec, remove shim material from under the bearing cup. Recheck the clearance again until you're sure it is correct.

■ When setting up the tool, use position 15:19 for units with 1.50:1 or 1.36:1 gear ratios; use tool position 16:27 for all other gear ratios.

■ Any thickness of shims added to the lower bearing must be subtracted from the shim thickness at the upper bearing. And of course, if you remove shims at the bottom you'll need to add them at the top.

14. Install the inner propeller shaft and thrust bearing into the housing and then slide on the outer shaft. Make sure you fill the area between the outer shaft oil seal lips with Quicksilver Special Lubricant 101.

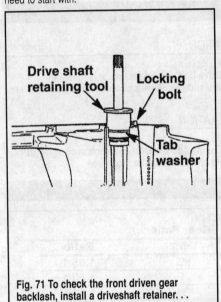

Fig. 71 To check the front driven gear backlash, install a driveshaft retainer...

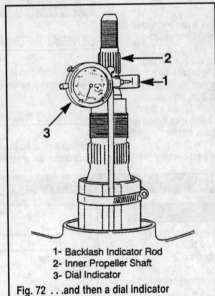

1- Backlash Indicator Rod
2- Inner Propeller Shaft
3- Dial Indicator

Fig. 72 ...and then a dial indicator

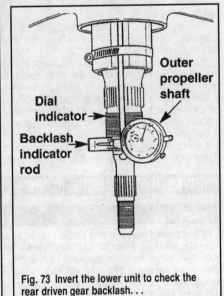

Fig. 73 Invert the lower unit to check the rear driven gear backlash...

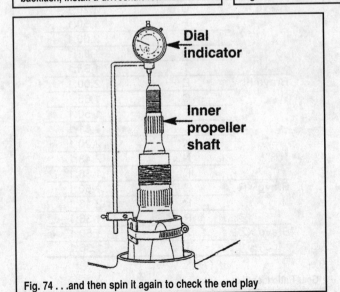

Fig. 74 ...and then spin it again to check the end play

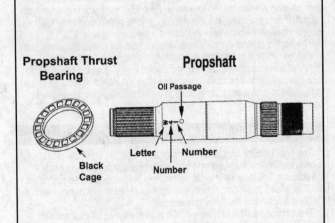

Fig. 75 Sometime in 2004, a new propshaft/thrust bearing was introduced. They are a matched set and cannot be used on earlier versions of the Bravo III

11-18 DRIVE SYSTEMS - BRAVO

■ Sometime in 2004, the propeller shaft was shortened to accommodate the thickness of the propeller shaft thrust bearing. Newer models (serial #OW100000 and above may be identified by the 3 stampings (a letter and 2 numbers) on the exterior of the shaft, near the oil passage and also by the black composite bearing cage. Previous models (serial #OM99999 and below) will have 2 stampings (a letter and a number) and a silver colored steel bearing cage. The shaft and bearing are a matched set and cannot be cross-matched; nor can a new set be used in an early drive, or vice versa.

15. Install a new shim pack of the same thickness over the shafts as that removed, or start with a nominal thickness of 0.050 in. (1.3mm).

16. Install the bearing cup over the shaft assembly and seat it against the shim pack.

17. Coat the threads of the bearing retainer with lubricant and screw it onto the shaft assembly. Install the retainer tools (#91-805382, #805374) and tighten it to 200 ft. lbs. (271 Nm). Remember this is a left hand thread, so tighten it counterclockwise.

■ In order to make sure the threads are engaged correctly, start the retainer by hand. Rotate the retainer clockwise until you feel the threads engage and then tighten it counterclockwise.

18. Coat the threads of the bearing carrier and the O-ring surface with Special Lubricant 101. Fill the area in between the oil seals with this as well. Coat the tapered surface of the carrier with Perfect Seal and then screw the carrier onto the shaft and into the housing. Install the special tool (#91-805374) and tighten the carrier to 150 ft. lbs. (203 Nm) - remember this is counterclockwise also!

19. Remove the clamp plate, O-ring, spacer and shims again and install a driveshaft retaining tool (# 91-805381) over the shaft and onto the tab washer. Do not tighten the lock bolt yet.

20. Position the unit so the propeller shafts are facing up, in a vertical position and then wiggle the propeller shafts several times to seat the bearing. Press down lightly on the retainer tool while wiggling the driveshaft to seat the bearings in the cups and then tighten the lock bolt. Now install a backlash indicator rod to the end of the inner propeller shaft. Mount a dial indicator to the bearing carrier with a hose clamp and position it so the tip is at the mark on the indicator rod. Wiggle the inner shaft back and forth slightly and record the reading. Subtract the reading observed in the Spline Lash Check detailed during Disassembly to arrive at your total gear backlash. It should be 0.012-0.016 in. (0.3-0.4mm). If the reading is too high, disassemble everything and add shims behind the front driven gear bearing cup; if too low, remove shims. Recheck the backlash reading again after reassembly.

21. Spin the unit 180° so that the propeller shaft is now facing down in the vertical position. Loosen the lock bolt on the driveshaft retainer, wiggle the propeller shaft a few times to seat the bearings and then press down on the retainer again and tighten the lock bolt. With the dial indicator and rod still installed from the previous step, wiggle the outer propeller shaft back and forth while observing the indicator. Record the reading. Subtract the reading observed in the Spline Lash Check detailed previously to arrive at your total gear backlash. It should be 0.012-0.016 in. (0.3-0.4mm). If the reading is too high, disassemble everything and remove shims from in front of the rear driven gear bearing cup; if too low, add shims. Recheck rear driven gear backlash again after reassembly.

22. Swivel the unit back around so the propeller shaft is again pointing upward and install the dial indicator as illustrated. Move the inner propeller shaft up and down while observing the indicator (the outer shaft should be lifted by the inner shaft at the same time). End play should be 0.001-0.050 in. (0.025-1.27mm). If you reading is outside of this range, add or subtract the appropriate number of thrust race shims to the outer propeller shaft until the reading comes into range.

23. Remove the retainer tool and reinstall the O-ring, spacer and shims.

24. Reconnect the housing halves.

Propeller(s)

GENERAL INFORMATION

■ Please refer to the Drive Unit - Alpha section or all general information on propellers such as diameter, pitch, rake, inspection, etc.

REMOVAL & INSTALLATION

■ Please refer to the Engine And Drive Maintenance section for all procedures on removing and installing the propeller.

DRIVESHAFT HOUSING (UPPER UNIT)

Description

Output power from the engine is connected to the stern drive through a horizontal driveshaft. A coupler is bolted to the flywheel and has a splined hub in the center. The end of the horizontal driveshaft indexes with and slides into the center of the hub. Power from the engine is then transmitted through the horizontal driveshaft to a pinion gear set where power direction is changed from horizontal to vertical.

The upper driven gear is pressed onto the outside diameter of the upper driveshaft. The upper driveshaft is splined on the lower end. When the upper gear housing is mated to the lower unit, the end of the lower unit driveshaft (or clutch assembly) indexes into the splined end of the upper driveshaft. Engine power is then transferred down into the lower gear unit.

The horizontal and vertical driveshafts are both mechanically connected. Therefore, anytime the engine is operating, the horizontal and vertical driveshafts are constantly rotating with engine rpm. A double yoke universal joint assembly in the horizontal driveshaft allows the stern drive to be raised or lowered to a required trim/tilt position (within limits), while the engine is operating.

Gear Ratio Identification

The gear ratio for the stern drive must be known to permit the proper special tool selection to be made in order to properly position the gears in the housing. The gear ratio for all stern drives is identified in two places - a decal is affixed to the port side of the upper gear housing. A number such as 1.50R is followed by the serial number.

The second location of the gear ratio is on the universal joint splined yoke. The identification mark here will be a letter such as "C".

This letter represents the gear ratio in the upper gear housing. Use the accompanying chart to determine the gear ratio when the letter is known.

Model	Gear Ratio Letter	Ratio
Bravo I	C	1.65:1
	F	1.50:1
	H	1.36:1
Bravo II	C	2.20:1
	F	1.50:1
	F	2.00:1
	H	1.81:1
	T	1.65:1
Bravo III	B	2.00:1
	C	1.65:1
	F	1.50:1
	G	1.81:1
	K	2.20:1
	N	2.43:1
	P	1.36:1
Bravo XR	R	1.50:1
	R	2.43:1
	R	1.36:1
Bravo XZ	Z	1.50:1
	T	1.36:1

Gear Ratio Identifier

DRIVE SYSTEMS - BRAVO

Driveshaft Housing (Upper Unit)

EXPLODED VIEWS

◆ See Figures 78 thru 84

REMOVAL & INSTALLATION

Bravo I/II/Z/XR/XZ

◆ See Figures 8 and 9

1. Remove the stern drive unit.
2. Remove the propeller(s) from the lower gear housing as detailed later in this section.
3. If you haven't already done so, drain the drive oil as detailed in the Maintenance section - remove and empty the gear lube monitor and also remove the drain plug. Allow the lubricant to drain into the container. As the lubricant drains, check the color. Dark brown to black indicates normal old lubricant. A chalky white to cream color indicates the presence of water. The presence of any water in the gear lubricant is bad news. The unit must be completely disassembled, inspected, the cause of the problem determined and corrected. Close attention should be given to the back-to-back seals on the propeller shaft, driveshaft and the bearing carrier O-ring. Examine the magnet on the end of the fill/drain plug for evidence of metal particles. The presence of tiny small metal dust like shavings indicates normal wear of the gears, bearings and shafts within the lower unit. Large metal chips or heavy fillings indicate extensive internal damage is taking place and the lower unit must be completely disassembled and inspected. All worn and/or damaged components must be replaced.
4. If working on a Bravo I/X/XR/XZ, place an alignment mark on the trim tab trailing edge and the anti-cavitation plate as an aid during assembling. Remove the plastic (or rubber) plug from the top of the anti-cavitation plate. Remove the 1/2 in. bolt directly above the trim tab, lift off the trim tab. You will need a long extension to your socket wrench for this one.

If the gears inside the housing have previously been changed from the factory markings on the housing, another method may be used to determine the gear ratio. This method is the least desired because the unit has to be disassembled before the ratio is known. With the gears removed from the housing count the number of teeth on the drive and driven gears. Compare the gear teeth number count to the chart below for the gear ratio of the unit being serviced.

Model	Gear Ratio		
	Drive Gear	Driven Gear	Gear Ratio
Bravo I	23	30	1.65:1
	27	32	1.50:1
	27	29	1.36:1
Bravo II	23	30	2.20:1
	27	32	2.00:1
	27	29	1.81:1
	27	32	1.65:1
	27	29	1.50:1
Bravo III	23	30	2.43:1
	23	30	2.20:1
	27	32	2.00:1
	27	29	1.81:1
	27	32	1.65:1
	27	32	1.50:1
	27	29	1.36:1
Bravo XR	16	19	1.50:1
	16	19	1.35:1
	16	19	1.26:1
Bravo XZ	27	32	1.50:1
	27	29	1.36:1

Gear tooth chart

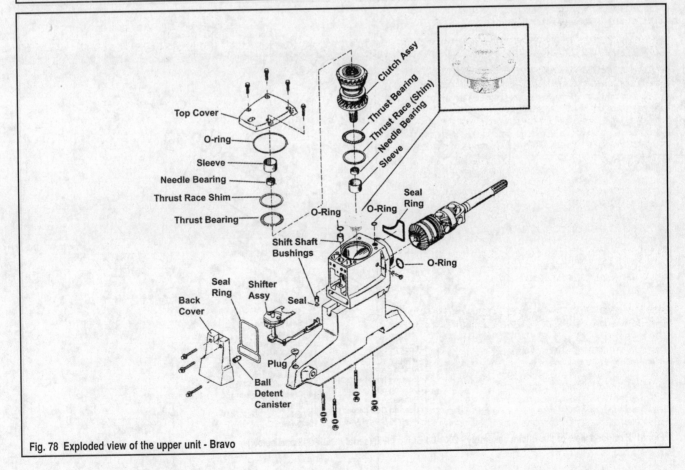

Fig. 78 Exploded view of the upper unit - Bravo

11-20 DRIVE SYSTEMS - BRAVO

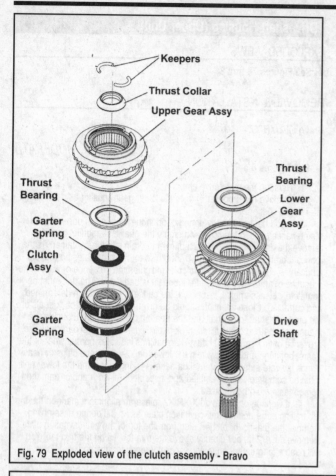

Fig. 79 Exploded view of the clutch assembly - Bravo

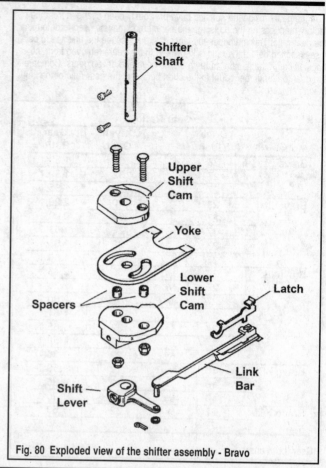

Fig. 80 Exploded view of the shifter assembly - Bravo

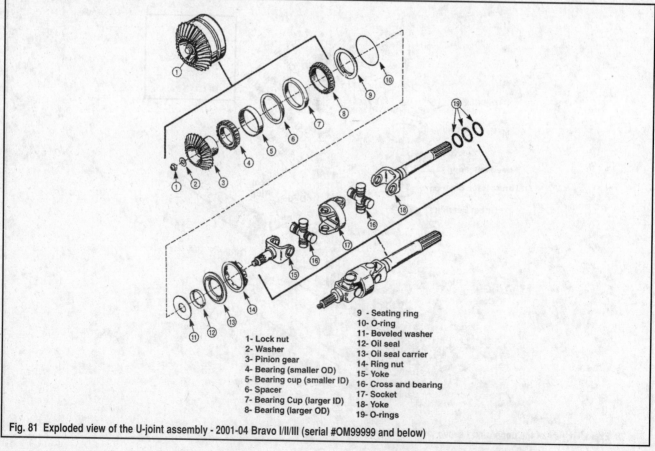

1- Lock nut
2- Washer
3- Pinion gear
4- Bearing (smaller OD)
5- Bearing cup (smaller ID)
6- Spacer
7- Bearing Cup (larger ID)
8- Bearing (larger OD)
9 - Seating ring
10- O-ring
11- Beveled washer
12- Oil seal
13- Oil seal carrier
14- Ring nut
15- Yoke
16- Cross and bearing
17- Socket
18- Yoke
19- O-rings

Fig. 81 Exploded view of the U-joint assembly - 2001-04 Bravo I/II/III (serial #OM99999 and below)

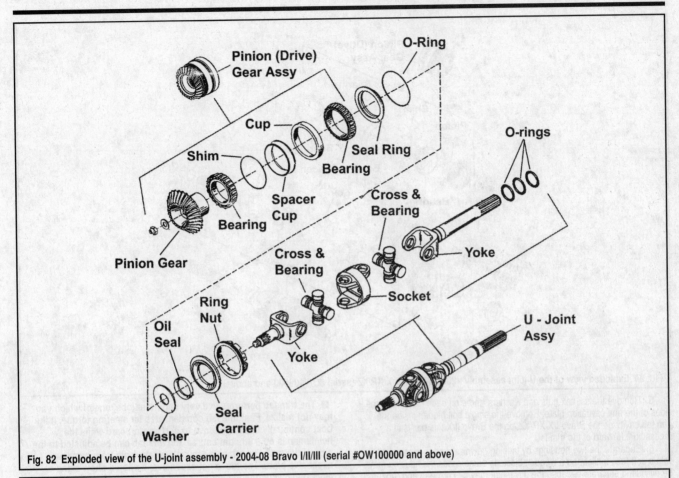

Fig. 82 Exploded view of the U-joint assembly - 2004-08 Bravo I/II/III (serial #OW100000 and above)

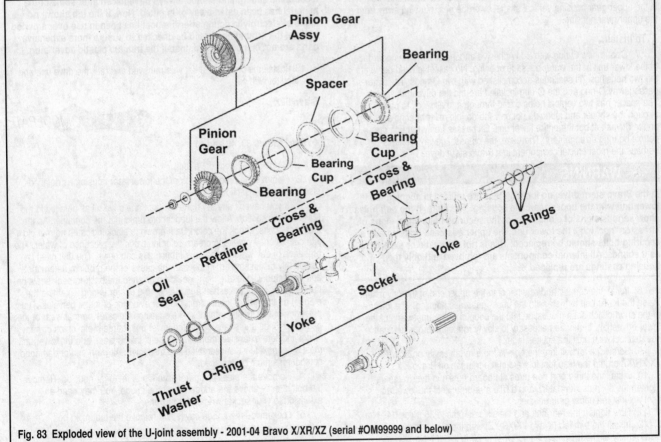

Fig. 83 Exploded view of the U-joint assembly - 2001-04 Bravo X/XR/XZ (serial #OM99999 and below)

11-22 DRIVE SYSTEMS - BRAVO

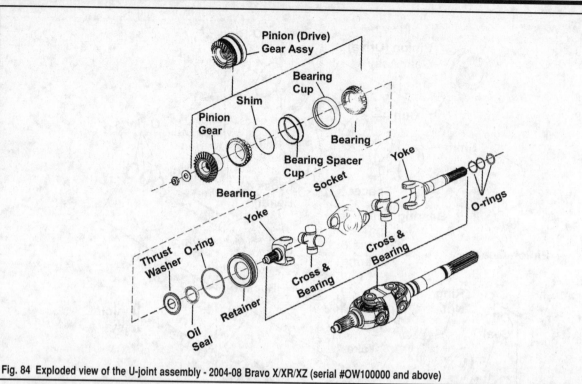

Fig. 84 Exploded view of the U-joint assembly - 2004-08 Bravo X/XR/XZ (serial #OW100000 and above)

5. Remove the six hex nuts and washers (three on each side), from the sides of the anti-cavitation plate. Remove the single bolt hidden inside the trim tab cavity on the Bravo I/X/XR/XZ; on the Bravo II/X this bolt is in a recess, just forward of the trim tab.

6. Separate the two housings by tapping downward on the lower unit with a soft face mallet and a wooden block. In some cases the housing may be difficult to separate because the driveshaft may be corroded and "frozen" in the upper gear housing vertical shaft. Lower the gear housing away from the upper gear housing.

To install:

7. Coat a new O-ring with 3-M Adhesive and install it on the upper face of the lower unit at the water passage opening. No gasket is used between the two housings. All sealing is accomplished with two O-rings: The water passageway O-ring and the O-ring installed around the driveshaft spacer. The spacer has two vertical holes drilled through it to allow oil to pass through the spacer and lubricate not only the double roller bearing assembly on the driveshaft but the entire lower unit. Oil passes freely through the double bearing arrangement. Therefore, the oil passageway O-ring is perhaps the most crucial component in a Bravo stern drive.

✱✱ WARNING

In the Bravo stern drive, no locating pins are utilized in aligning the lower unit with the upper gearcase housing. Therefore, if the bolt holes show serious signs of elongation, or the securing bolts show signs of stress or "necking", the lower unit, the upper gearcase housing and all securing bolts should be replaced. This is not as drastic or expensive as it sounds. All internal components are removed and only the housing castings are replaced.

8. Install the coupler and retainer onto the upper end of the driveshaft. Raise the lower unit to mate with the upper gearcase housing. The lower end of the short clutch shaft must slide into the coupler at the upper end of the lower driveshaft. It may be necessary to slowly rotate the propeller shaft counterclockwise until the splines index.

9. Screw the aft bolt into the hole in the trim tab cavity on Bravo I/X/XR/XZ or into the hole just forward of the trim tab on Bravo II/X.

10. Install the three port and three starboard nuts and their washers. Tighten the six nuts alternately, and a little at-a-time, until the lower unit is tight against the upper gear housing.

11. Now tighten the rear bolt and the six side nuts to 35 ft. lbs. (48 Nm).

12. Install the trim tab (Bravo I/X/XR/XZ) with the marks made during disassembly aligned. Tighten the screw(s) to 20 ft. lbs. (27 Nm).

■ The trim tab performs two very important jobs, one of which you may not realize. First, the tab compensates for steering torque. If the boat continually seems to move to port or starboard while the helmsman is on a straight course, the trim tab can be adjusted to the side of the pull. The tab also prevents electrolysis from damaging expensive parts. The tab is not expensive; it should show signs of electrolytic action; and should always be replaced after some of the material has been eaten away or is pitted. Now, if the tab shows no signs of electrolytic action after the boat has been in use over a period of time, the grounding should be checked to ensure more expensive parts are not being damaged. Install the trim tab plastic cover plug.

13. Install the propeller(s) if it was removed and refill the drive unit and lube monitor with oil.

Bravo III/X

◆ See Figures 39, 40 and 41

1. Remove the stern drive unit.

2. Remove the propeller(s) from the lower gear housing as detailed elsewhere in this section.

3. If you haven't already done so, drain the drive oil as detailed in the Maintenance section. Allow the lubricant to drain into the container. As the lubricant drains, check the color. Dark brown to black indicates normal old lubricant. A chalky white to cream color indicates the presence of water. The presence of any water in the gear lubricant is bad news. The unit must be completely disassembled, inspected, the cause of the problem determined and corrected. Close attention should be given to the back-to-back seals on the propeller shaft, driveshaft and the bearing carrier O-ring. Examine the magnet on the end of the fill/drain plug for evidence of metal particles. The presence of tiny small metal dust like shavings indicates normal wear of the gears, bearings and shafts within the lower unit. Large metal chips or heavy fillings indicate extensive internal damage is taking place and the lower unit must be completely disassembled and inspected. All worn and/or damaged components must be replaced.

4. Remove the plastic plug from the top of the gear housing. Remove the bolt directly above the anode, lift off the anode. You may need an extension to your socket wrench for this one.

5. Look inside the anode cavity and remove the mounting bolt.

6. Remove the six hex nuts and washers (three on each side), from the sides of the housing.

DRIVE SYSTEMS - BRAVO

7. Separate the two housings by tapping downward on the lower unit with a soft face mallet and a wooden block. In some cases the housing may be difficult to separate because the driveshaft may be corroded and "frozen" in the upper gear housing vertical shaft. Lower the gear housing away from the upper gear housing and place it in a holding device.

To install:

8. Install the coupler and retainer onto the upper end of the driveshaft. Raise the lower unit to mate with the upper gearcase housing. The lower end of the short clutch shaft must slide into the coupler at the upper end of the lower driveshaft. It may be necessary to slowly rotate the propeller shaft counterclockwise until the splines index.

9. Install the three port and three starboard nuts and their washers. Tighten the six nuts alternately, and a little at-a-time, until the lower unit is tight against the upper gear housing. Now tighten all six side nuts to 35 ft. lbs. (47.5 Nm).

10. Install the mounting bolt in the anode cavity and tighten to 35 ft. lbs. (47.5 Nm).

11. Install the anode and tighten the bolt to 20 ft. lbs. (27 Nm).

12. Install the drive unit if removed.

13. Install the propellers and refill the drive unit with oil.

DISASSEMBLY & ASSEMBLY

Upper Unit Components - Units W/Serial #OM99999 And Below

◆ See Figures 78 and 85 thru 96

■ As of mid-2004, Mercury has added rather extensive pinion gear shimming and driven gear backlash procedures for units with serial #OW100000 and above. Please be sure that you use the accompanying procedures for these drives. If your unit is serial #OM99999 and below, proceed through the remainder of this procedure. If you have a later drive though, we'll let you know now that disassembly procedures are the same and we're going to refer you right back here!

1. Make sure that the drive unit is in the Neutral detent position.

2. Loosen the three rear cover bolts evenly and remove the cover (the Blackhawk uses two Allen screws). Unscrew the vent screw on the forward starboard side of the housing (if equipped).

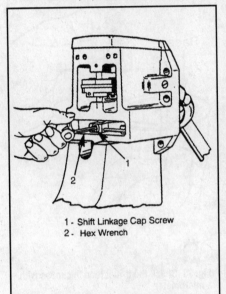

1 - Shift Linkage Cap Screw
2 - Hex Wrench

Fig. 85 Loosen the shift linkage cap screw

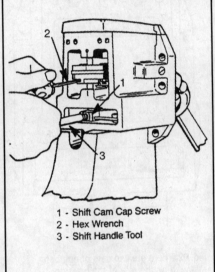

1 - Shift Cam Cap Screw
2 - Hex Wrench
3 - Shift Handle Tool

Fig. 86 Loosen the shift cam cap screw

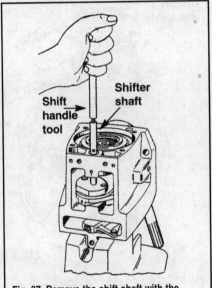

Fig. 87 Remove the shift shaft with the special tool

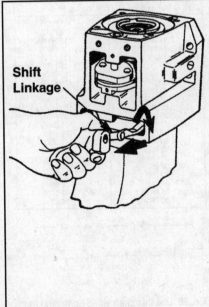

Fig. 88 Pull out the shift linkage

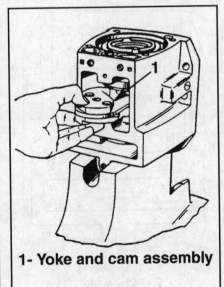

1- Yoke and cam assembly

Fig. 89 Pull out the shift cams

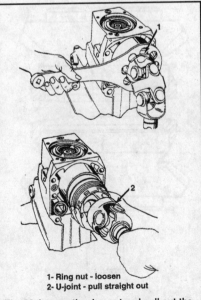

1- Ring nut - loosen
2- U-joint - pull straight out

Fig. 90 Loosen the ring nut and pull out the driveshaft assembly

11-24 DRIVE SYSTEMS - BRAVO

3. Insert an Allen wrench into the shift linkage cap screw (bottom slot in the back housing), loosen it and remove the screw. Now insert a shift handle tool (#91-17302) into the linkage cap screw hole to hold it steady and then remove the shift cam cap screw (larger upper opening) with an Allen wrench.

4. Loosen and remove the four top cover mounting bolts and lift off the cover. There may be a small Allen screw at the forward edge of the cover on early models - leave it there. Throw away the O-ring.

5. Screw the same shift handle tool onto the top of the shifter shaft from the top of the housing and then pull the shaft up and out of the housing.

6. Grab the shift linkage assembly from the rear (bottom slot), turn it so the link bar rotates about 1/4 turn clockwise and pull the entire assembly out of the housing. You may have to wiggle it a little as you're pulling it out. This can be tricky, but just examine the situation and you'll get it out. In fact, unless something is broken, there's really no reason to remove it in the first place other than to inspect the components.

7. Reach in and grab the shifter yoke and cam assembly from the upper opening.

8. Install a U-joint retainer wrench (#91-17256) over the yoke shaft and joints so that it indexes with the notches in the retaining ring nut. Loosen the ring nut and then pull the shaft assembly straight out of the housing. Do not twist or turn it in any manner, just pull it straight back. If you're working on one of the mid-2004 and later models, make sure you carefully remove the shim material resting against the bearing cup spacer shoulder and lay it aside.

■ **Bravo Z/XR/XZ models will need a different version of the retainer wrench (#91-862219) that will have small pins which index with the indents in the nut.**

9. Lift the upper thrust race (shim) and bearing out of the top of the housing. Put them aside and keep them together in the exact position that they were removed.

10. Reach in and carefully lift the clutch/gear assembly straight up and out of the housing. Reach into the bore and lift out the lower thrust bearing and race. Store them as you did the upper set.

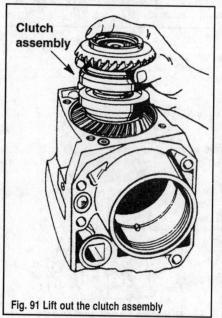

Fig. 91 Lift out the clutch assembly

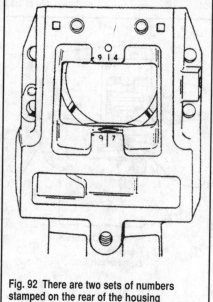

Fig. 92 There are two sets of numbers stamped on the rear of the housing

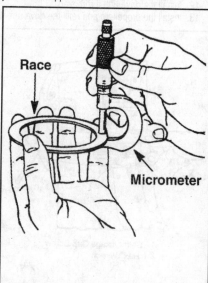

Fig. 93 Check the thrust race thickness with a micrometer

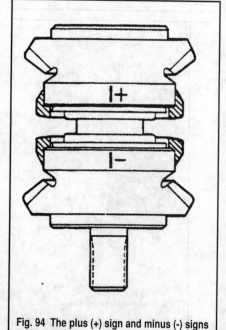

Fig. 94 The plus (+) sign and minus (-) signs must always be opposite each other

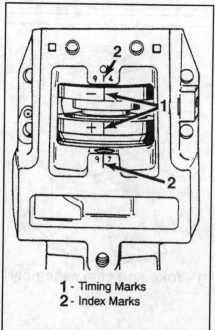

1 - Timing Marks
2 - Index Marks

Fig. 95 All four index marks must line up

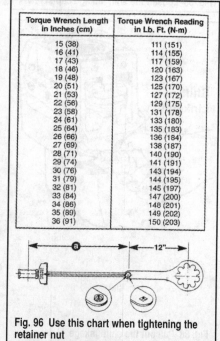

Torque Wrench Length in Inches (cm)	Torque Wrench Reading in Lb. Ft. (N-m)
15 (38)	111 (151)
16 (41)	114 (155)
17 (43)	117 (159)
18 (46)	120 (163)
19 (48)	123 (167)
20 (51)	125 (170)
21 (53)	127 (172)
22 (56)	129 (175)
23 (58)	131 (178)
24 (61)	133 (180)
25 (64)	135 (183)
26 (66)	136 (184)
27 (69)	138 (187)
28 (71)	140 (190)
29 (74)	141 (191)
30 (76)	143 (194)
31 (79)	144 (195)
32 (81)	145 (197)
33 (84)	147 (200)
34 (86)	148 (201)
35 (89)	149 (202)
36 (91)	150 (203)

Fig. 96 Use this chart when tightening the retainer nut

DRIVE SYSTEMS - BRAVO 11-25

11. Inspect and disassemble the individual components as detailed in the accompanying procedures.

To assemble:

12. There are two sets of numbers stamped on the rear of the housing, above and below the shifter cavity that designate the upper and lower thrust bearing race thickness. You guessed it. . .the top number is for the upper race and the lower number is for the lower one! These are guides to what the actual thickness should be - "94" equates to 0.094 in., etc. Measure the thickness of your races with a micrometer to ensure that they are within specification. If using the original races, make sure they are installed exactly as they were removed; do not flip them so the contact sides are different.

13. Position the correct lower race into the clutch bore after coating it thoroughly with High Performance Gear Lube.

14. Coat the lower face of the bottom clutch assembly gear and the thrust bearing with 60% mixture of High Performance Gear lube and 40% Special Lubricant 101 (have fun with this!). Stick the bearing onto the bottom of the gear and insert the assembly into the housing bore.

15. Repeat this procedure with the top gear and the upper bearing and attach the bearing to the gear set. Position the race on top of the bearing.

16. Each side of each clutch gear has an index mark and a plus (+) and minus (-) sign, align the upper and lower gears so that the index marks line up AND there is a plus (+) on one and a minus (-) on the other. It doesn't matter whether the plus or minus sign is on the top or bottom, just that they are different. There should never be two plus (+) signs or two minus (-) signs on top of each other. You will also notice that there is an index mark stamped above and below the shifter cavity (in between the number discussed earlier) - make sure the aligned index marks on the two clutch gears are in line with each of these.

17. Coat the threads of the retaining ring nut with Special Lubricant 101 and carefully guide the driveshaft assembly back into the housing. Tighten the ring nut to 200 ft. lbs. (271 Nm) with the special tool. Check the index marks on the rear of the housing and the clutch gears are still in alignment; if not, remove the shaft assembly and try it again. It is imperative that the nut is not cross-threaded - turn the nut counterclockwise until you can feel that the threads have engaged and then turn it clockwise.

■ You must use the retainer wrench when starting the nut on Z/XR and XZ drives.

■ It is very difficult to tighten the retainer nut to the proper torque value due to the retainer wrench taking the torque. On beam-type wrenches, measure from the square drive to the fulcrum (pivot) point of the handle. On click-stop or dial-type wrenches, measure from the square drive to the reference mark on the handle (2 bands, a line, etc.). Find your length in the accompanying chart and then use the suggested torque figure.

18. Insert the shift cam assembly into the cavity in the rear of the housing so that the boss and nuts are on the bottom side, facing the bottom of the housing.

19. Slide the shifter linkage assembly into the cavity while twisting it side-to-side. Once in, rotate it 1/4 turn counterclockwise.

20. Reinstall the shift handle tool to the shift shaft and press the shaft down through the housing and shifter assemblies. Now take the shift handle tool and insert it through the shift linkage and into the shift shaft so you can swivel the shaft until the hole aligns with the one in the lower shift cam. Coat the 1st few threads of the cam cap screw with Loctite® 271 and screw it into the cam. Tighten the screw to 100-120 inch lbs. (12-13 Nm).

21. Remove the tool and then install the linkage cap screw in the same way and to the same torque. Coat the inner diameter of the recess liberally with Special Lubricant 101.

22. Move the linkage into the Neutral detent position.

23. Coat two new O-rings with 3-M adhesive and install them into the shift linkage and water passages in the front of the housing.

24. Coat two new O-rings with 3-M adhesive and install them into the shifter shaft and oil passages in the top of the housing.

25. Coat a new rear cover O-ring with adhesive and install it into the groove in the cover. Do the same with the top cover.

26. Install the top cover and tighten the bolts, in a criss-cross pattern, to 18-22 ft. lbs. (25-29 Nm). Repeat this procedure for the rear cover, to the same torque.

27. Connect the upper and lower units and install the stern drive.

Upper Unit Components - Units W/Serial #OW100000 And Above

■ As of mid-2004, Mercury has added rather extensive pinion gear shimming and driven gear backlash procedures for units with serial #OW100000 and above. Although the assembly procedures are significantly different, the disassembly procedures are identical, so please refer back to the disassembly procedures for models with serial #OM99999 and below.

To assemble:
◆ See Figures 97 thru 102

1. OK, now that you've finished disassembling and assembling all the various components, its time to have some fun!

2. Make sure that pinion gear/bearing pre-load has been set as detailed in the U-Joint/Pinion (Drive) Gear Assembly procedures, and then disassemble the pinion gear/bearing assembly again. Remove the rings, washers, seals/carriers and retainer nut.

3. Now re-install them all in the correct order on a stub shaft tool (#91-865084 for I/II/III drives and #91-865083 for the X-series drives). Install the washer and nut and tighten it until the washer just seats itself against the gear.

4. Slide the old shim material over the gear end and position it up against the shoulder on the bearing spacer cup. If you've lost the old shim or it was destroyed during disassembly, use a new one and start with a thickness of 0.035 in. (0.09mm).

5. Install the gear assembly along with the tool into the upper unit. Coat the threads of the retainer nut with Special Lubricant 101 and thread it in by hand until you are sure that the threads are engaged properly. It is imperative that the nut is not cross-threaded - turn the nut counterclockwise until you can feel that the threads have engaged and then turn it clockwise. Tighten the ring nut to 200 ft. lbs. (271 Nm) with the special tool.

■ You must use the retainer wrench starting the nut on Z/XR and XZ drives.

■ It is very difficult to tighten the retainer nut to the proper torque value due to the retainer wrench taking the torque. On beam-type wrenches, measure from the square drive to the fulcrum (pivot) point of the handle. On click-stop or dial-type wrenches, measure from the square drive to the reference mark on the handle (2 bands, a line, etc.). Find your length in the accompanying chart and then use the suggested torque figure.

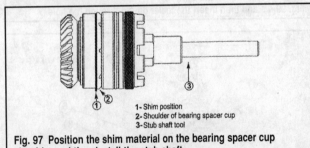

1 - Shim position
2 - Shoulder of bearing spacer cup
3 - Stub shaft tool

Fig. 97 Position the shim material on the bearing spacer cup shoulder and then install the stub shaft

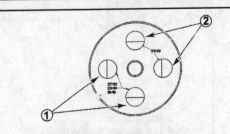

1 - Access holes for gear tooth counts 27/32, 23/30, 16/19
2 - Access holes for gear tooth counts 27/29

Fig. 98 Install the shimming tool so the correct hole is facing the pinion gear. . .

11-26 DRIVE SYSTEMS - BRAVO

6. Insert the pinion gear shimming tool (#91-865114) into the top of the unit and rotate it until the driven gear tooth count marking is lined up with the pinion drive gear. Position the pinion gear so there are 2 full teeth centered on the gauging surface with one tooth on each side of the gauge centerline. Insert a feeler gauge and measure the clearance between the pinion gear tooth and the tool. Clearance should be 0.025 in. (0.64mm). Rotate the tool until one side of the gauging surface comes into contact with the feeler gauge and there is a slight drag on the gauge. Remove the feeler gauge and insert it between the other tooth and the shimming tool without moving the tool. If there is only a slight drag, then the shimming is correct. If there is no drag, continue using progressively thicker gauges until you feel a slight drag and take note of the clearance. If you can't fit the spec gauge, use a smaller thickness feeler until it fits with a slight drag.

■ Please refer to the charts in the Gear Ratio Information section and confirm that the number of teeth on the driven gear match that listed for your gear ratio.

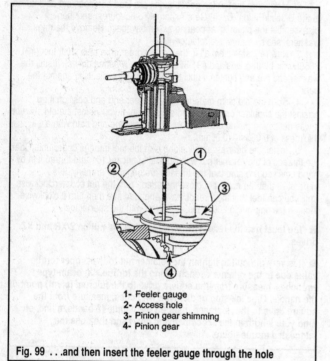

1- Feeler gauge
2- Access hole
3- Pinion gear shimming
4- Pinion gear

Fig. 99 ...and then insert the feeler gauge through the hole

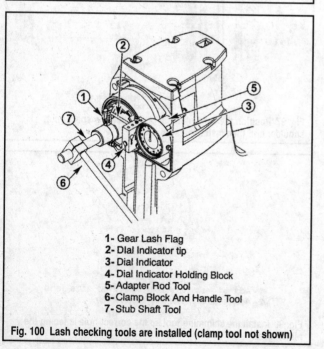

1- Gear Lash Flag
2- Dial Indicator tip
3- Dial Indicator
4- Dial Indicator Holding Block
5- Adapter Rod Tool
6- Clamp Block And Handle Tool
7- Stub Shaft Tool

Fig. 100 Lash checking tools are installed (clamp tool not shown)

7. If the gear location was in spec, remove the pinion and stub shaft assembly and the tool and proceed to the driven gear lash procedures. If the clearance is not within specifications, calculate the difference between the measured clearance and the correct clearance (0.025 in.).

If the measured clearance is too great, subtract the specified clearance from it and then remove the resulting shim thickness figure from the existing shim material on the gear/bearing assembly. For instance, if your measurement was 0.045 in., subtract 0.025 in. (the correct clearance) from it and the result is 0.020 in. This is the figure that will need to be subtracted from the actual shim material on the gear/bearing assembly - you started with 0.035 in. so it now must be changed to 0.015 in.

Now if the measured clearance was too tight, you need to subtract that figure from the correct clearance and then add that figure to the existing shim material on the gear/bearing assembly. For instance, if your measurement was 0.020 in., subtract this figure from 0.025 in. (the correct clearance) and the result is 0.005 in. This is the figure that will need to be added to the actual shim material on the gear/bearing assembly - you started with 0.035 in. so it now must be increased to 0.040 in.

Get it? Of course, you need to recheck everything once you think you've got it right.

8. Loosen the retainer nut and remove the assembly from the upper unit. Set it aside, but leave it attached to the stub shaft.

9. Now its time for the driven gear lash measurement. Make sure that the pinion gear/bearing shimming has been completed correctly and then measure the upper and lower driven gear thrust races (shims) with a micrometer. Record this measurement. If you are unable to use the original races, use a new 0.064 in. (1.63mm) thick race as your starting point.

10. Coat the thrust races and bearings lightly with High Performance Gear Lube. Position the lower race into the housing bore and then sit the bearing on top of it so their centers are perfectly aligned. If you are re-using the old race, make sure the original contact side is against the bearing.

11. Coat the clutch and driven gear assembly lightly with High Performance Gear Lube and then carefully install the assembly into the housing bore. Make sure that the top of the assembly is below the top surface of the housing - if not, you'll need to pull everything out and reposition the thrust bearing and race.

12. Install the upper thrust bearing and race.

13. Each side of each clutch gear has an index mark and a plus (+) and minus (-) sign, align the upper and lower gears so that the index marks line up AND there is a plus (+) on one and a minus (-) on the other. It doesn't matter whether the plus or minus sign is on the top or bottom, just that they are different. There should never be two plus (+) signs or two minus (-) signs on top of each other. You will also notice that there is an index mark stamped above and below the shifter cavity - make sure the aligned index marks on the two clutch gears are in line with each of these.

14. Position the shift shaft and top cover O-rings and then install the top cover. Tighten the 4 bolts to 20 ft. lbs. (27 Nm).

15. Coat the threads of the retaining ring nut with Special Lubricant 101 and carefully guide the driveshaft assembly back into the housing. Tighten the ring nut to 200 ft. lbs. (271 Nm) with the special tool. Check that the index marks on the rear of the housing and the clutch gears are still in alignment; if not, remove the shaft assembly and try it again. It is imperative that the nut is not cross-threaded - turn the nut counterclockwise until you can feel that the threads have engaged and then turn it clockwise.

■ You must use the retainer wrench when starting the nut on Z/XR and XZ drives.

■ It is very difficult to tighten the retainer nut to the proper torque value due to the retainer wrench taking the torque. On beam-type wrenches, measure from the square drive to the fulcrum (pivot) point of the handle. On click-stop or dial-type wrenches, measure from the square drive to the reference mark on the handle (2 bands, a line, etc.). Find your length in the accompanying chart and then use the suggested torque figure.

16. Install an adaptor rod tool (#91-865086) over the stub shaft assembly and into the top stud on the housing. Install a gear lash flag onto the shaft and make sure that the drive tooth count matches the numbers on the flag.

17. Line up the indicator mark on the flag toward the stub shaft and install a clamp block and handle (#91-865085) on the end of the shaft so it is just snug.

18. Install a dial indicator holding block (#91-865097) to the adaptor rod and then install a dial indicator with a short extension tip. Move the block so that the indicator tip is parallel and at 90 degrees to the flag. The flag must be vertical.

DRIVE SYSTEMS - BRAVO 11-27

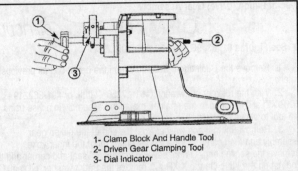

Fig. 101 Lash checking tools are installed (clamp tool not shown)

1- Clamp Block And Handle Tool
2- Driven Gear Clamping Tool
3- Dial Indicator

Driven Gear Lash

Drive Gear Tooth	Driven Gear Tooth	Gear Lash in. (mm)
27	32	0.011-0.016 (0.279-0.406)
27	29	0.013-0.018 (0.330-0.457)
23	30	0.011-0.016 (0.279-0.406)
16	19	0.009-0.015 (0.229-0.381)

Fig. 102 Driven gear lash table

19. Insert a small pry bar through the rear of the unit and lever the clutch upward so you have access to the upper driven gear. Install a driven gear clamping tool (#91-865115) to the upper gear torus ring in a position that you will be able to hold the tool while and the same time reading the dial indicator.

20. Adjust the clamping tool and handle so that you san hold them both at the same time while observing the dial indicator. While holding the driven gear locked with the clamp, move the stub shaft handle back and forth while watching the dial indicator. Jot down the figure and call it upper driven gear measurement.

21. Remove the clamping tool, push down on the clutch and reinstall the tool so it locks the lower gear at the torus ring. Repeat the previous step and mark down the lower gear measurement.

22. If lash measurements are within the specifications as shown in the accompanying table, remove all the tools, remove the clutch and driven gears assembly, remove the pinion gear and bearing assembly from the stub shaft and you're ready to continue with assembly.

If lash was correct on one gear and bad on another, replace only the thrust race for that particular gear.

If lash on one or both gears was out of spec, calculate the difference between the measured lash and the correct lash as per the table.

If the lash was to high, subtract the correct measurement and add that figure to the existing measurement of the thrust race for that gear. In example, if the measured lash for the upper gear was 0.018 in. and the correct lash should have been 0.013 in., subtract 0.013 in. from 0.018 in. and get 0.004 in. This figure then needs to be added to the thickness of the existing race to determine what thickness race to replace it with.

If the lash was too low, subtract the measured lash from the correct lash and remove the difference from the existing measurement of the thrust race for that gear. In example, If the measured lash 0.013 in., subtract that figure from the correct lash (say, 0.015 in.) and get 0.002 in. This figure then needs to be subtracted from the thickness of the existing race to determine what thickness race to replace it with.

Reducing the thickness of a thrust race moves the driven gear away from the pinion, thus increasing lash. Increasing the thickness of a thrust race moves the gear closer to the pinion, thus reducing lash.

Thrust bearing races come color-coded:
- Brown: 0.058 in. (1.47mm)
- White: 0.059 in. (1.50mm)
- Orange: 0.060 in. (1.52mm)
- Green: 0.061 in. (1.55mm)
- Yellow: 0.062 in. (1.57mm)
- Red: 0.063 in. (1.60mm)
- Lt. Blue: 0.064 in. (1.63mm)
- Black: 0.065 in. (1.65mm)
- Pink: 0.066 in. (1.68mm)
- Purple: 0.067 in. (1.70mm)

23. Recheck the lash for a final time. If correct finally, remove all the tools, remove the clutch and driven gears assembly, remove the pinion gear and bearing assembly from the stub shaft and you're ready to continue with assembly.

24. Reset the pinion gear/bearing pre-load as detailed in the U-Joint/Pinion (Drive) Gear Assembly procedures. Yes, we know you've already done this, but you must do it again after having set the pinion gear shims and driven gear lash.

25. Coat the thrust races and bearings lightly with High Performance Gear Lube. Position the lower race into the housing bore and then sit the bearing on top of it so their centers are perfectly aligned. If you are re-using the old race, make sure the original contact side is against the bearing.

26. Coat the clutch and driven gear assembly lightly with High Performance Gear Lube and then carefully install the assembly into the housing bore. Make sure that the top of the assembly is below the top surface of the housing - if not, you'll need to pull everything out and reposition the thrust bearing and race.

27. Install the upper thrust bearing and race.

28. Each side of each clutch gear has an index mark and a plus (+) and minus (-) sign, align the upper and lower gears so that the index marks line up AND there is a plus (+) on one and a minus (-) on the other. It doesn't matter whether the plus or minus sign is on the top or bottom, just that they are different. There should never be two plus (+) signs or two minus (-) signs on top of each other. You will also notice that there is an index mark stamped above and below the shifter cavity - make sure the aligned index marks on the two clutch gears are in line with each of these.

29. Coat the threads of the pinion gear/bearing retaining ring nut with Special Lubricant 101 and carefully guide the driveshaft assembly back into the housing. Tighten the ring nut to 200 ft. lbs. (271 Nm) with the special tool. Check the index marks on the rear of the housing and the clutch gears are still in alignment; if not, remove the shaft assembly and try it again. It is imperative that the nut is not cross-threaded - turn the nut counterclockwise until you can feel that the threads have engaged and then turn it clockwise.

■ **You must use the retainer wrench when starting the nut on Z/XR and XZ drives.**

■ **It is very difficult to tighten the retainer nut to the proper torque value due to the retainer wrench taking the torque. On beam-type wrenches, measure from the square drive to the fulcrum (pivot) point of the handle. On click-stop or dial-type wrenches, measure from the square drive to the reference mark on the handle (2 bands, a line, etc.). Find your length in the accompanying chart and then use the suggested torque figure.**

30. Insert the shift cam assembly into the cavity in the rear of the housing so that the boss and nuts are on the bottom side, facing the bottom of the housing.

31. Slide the shifter linkage assembly into the cavity while twisting it side-to-side. Once in, rotate it 1/4 turn counterclockwise.

32. Reinstall the shift handle tool to the shift shaft and press the shaft down through the housing and shifter assemblies. Now take the shift handle tool and insert it through the shift linkage and into the shift shaft so you can swivel the shaft until the hole aligns with the one in the lower shift cam. Coat the 1st few threads of the cam cap screw with Loctite® 271 and screw it into the cam. Tighten the screw to 100-120 inch lbs. (12-13 Nm).

33. Remove the tool and then install the linkage cap screw in the same way and to the same torque. Coat the inner diameter of the recess liberally with Special Lubricant 101.

34. Move the linkage into the Neutral detent position.

35. Coat two new O-rings with 3-M adhesive and install them into the shift linkage and water passages in the front of the housing.

36. Coat two new O-rings with 3-M adhesive and install them into the shifter shaft and oil passages in the top of the housing.

37. Coat a new rear cover O-ring with adhesive and install it into the groove in the cover. Do the same with the top cover.

38. Install the top cover and tighten the bolts, in a criss-cross pattern, to 18-22 ft. lbs. (25-29 Nm). Repeat this procedure for the rear cover, to the same torque.

39. Connect the upper and lower units and install the stern drive.

11-28 DRIVE SYSTEMS - BRAVO

Shifter Assembly

 MODERATE

◆ See Figures 80 and 103 thru 105

1. Remove the shifter assembly from the upper unit as detailed previously.
2. Disconnect the latch at the link bar. Remove the link bar from the shift lever and throw away the cotter pin.
3. Pull the ball detent canister out of the rear cover; being careful not to lose the small spring.
4. Remove the locknuts and then separate the upper and lower cams from the yoke. Don't lose the 2 spacers.
5. Pry the upper O-ring out of the shifter shaft recess in the aft side of the top of the upper unit. Install a bushing removal tool (#91-17273) and press out the upper bushings. Do the same with the lower oil seal and bushing.

To assemble:

6. Press the lower bushing in from the top until it is about 2/3 of the way into the bore. Slide the lower seal onto the bolt of a seal driver (#91-17275A1) so that the lip is facing the bolt head and then coat the outer diameter of the seal with Loctite 271. Now, insert the driver into the bottom of the shaft bore and the pilot through the bushing from the top. Install the bolt (with seal) through the pilot and then tighten the bolt until the tool bottoms out on the top and bottom surfaces of the casting.
7. Drive the upper bushing into the shaft bore until the bottom of the bushing is flush with the bottom of the bore.
8. Coat all contact surfaces of the cam and yoke assembly with gear lube and assemble the upper and lower cams to the yoke without forgetting the 2 spacers. Tighten the locknuts to 80 inch lbs. (9 Nm).
9. Place a small dab of Special Lubricant 101 on the compression spring and detent canister before installing them into the rear cover.
10. Feed the pin on the link bar through the hole in the shift lever (from the top), install the washer and then install a new cotter pin.
11. Position the latch in the bar.
12. Install the assembly into the upper unit.

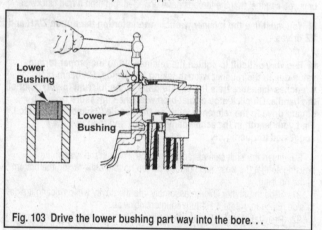

Fig. 103 Drive the lower bushing part way into the bore...

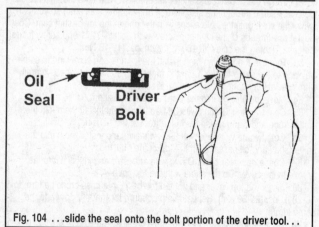

Fig. 104 ...slide the seal onto the bolt portion of the driver tool...

U-Joint/Pinion (Drive) Gear Assembly

 DIFFICULT

◆ See Figures 81, 83, 96 and 106 thru 109

1. Remove the U-joint/gear assembly from the upper unit as previously detailed.
2. With the retainer nut wrench (#91-17256 - I/II/III, or #91-862219 - X/XR/XZ) still attached to the retainer, mount the assembly in a vise using the tool to clamp down on.
3. Remove the nut and washer from the end of the pinion gear assembly and then carefully slide the assembly off of the inner yoke. Remove the sealing ring, O-ring, beveled washer, oil seal and carrier and the ring nut on the I/II/II. On the X/XR/XZ, remove the thrust washer, oil seal, O-ring and retainer.

■ On 2004 and later models (serial #OW100000 and above), a change has been made to the gear/bearing assembly. Remove the shim material from the assembly where it rests against the shoulder Tie the shim material together for later use.

4. If gear is in good shape but the bearings are in need of replacement, both bearings must be replaced as a set. Obtain a Universal Bearing Puller Plate (#91-37241) or equivalent tool. Place the tool under the tapered bearing and tighten the plate jaws under the tapered bearing. Set the assembly into an arbor press and press the tapered bearing off the gear. Lift off the spacer and then the bearing cup. Repeat this step and press the remaining tapered roller bearing off the gear.

■ On 2004 and later models (serial #OW100000 and above), a change has been made to the gear/bearing assembly. After pressing off the bearing, remove the spacer and the spacer cup. Tie the shim material together for later use.

5. If the oil seal is bad, pry or press the seal out of the carrier on I/II/II drives or out of the bearing retainer on X/XR/XZ drives.
6. Clean the assembly in solvent and allow it to dry thoroughly. Inspect the pinion (drive) gear and the driven gear for broken teeth, pitting, etc. Make sure all teeth are worn evenly
7. Slide the three small O-rings off the end of the coupling yoke.
8. Inspect the large O-ring around the oil seal carrier and replace it if its worn. Inspect the joint retainer for cracks or other damage, replacing if anything is found. Do the same with the thrust ring.
9. Inspect the oil seal and carrier (I/II/II only) for defects or damage. If there is a problem with the carrier it must be replaced with the seal as a unit

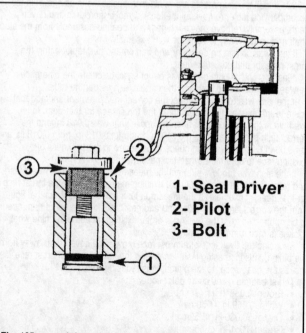

1- Seal Driver
2- Pilot
3- Bolt

Fig. 105 ...and then pull the seal and bushing into position with the other parts of the tool

- the same goes for X/XR/XZ models, only the seal is matched to the bearing retainer. If the seal is bad, press it out with a punch and hammer. Use a seal driver tool (#91-36577) to press a new seal back into the carrier or retainer.

10. Inspect the splines on both yokes for wear or cracking, replacing anything if a problem is found.

11. Inspect the joints themselves for wear, knocking or too much side play. If problems are found, separate the joints.

12. Drive the eight C-rings off the U-joint bearings with a punch and hammer.

13. Obtain a suitable adaptor (#91-38756) to support the U-joint yoke. Press one bearing until the opposite bearing is pressed into the adaptor. Remove the two loose bearings. Rotate the U-joint assembly 180°. Use the adaptor and press on the bearing cross-member and remove the bearings. Remove the cross-member from the yoke. Press the other four bearings out in the same manner, as described in the previous step and in the first part of this step.

To assemble:

14. Lubricate the U-joint bearing cups with a liberal amount of Quicksilver 2-4-C lubricant. Place the lubricated cups in the yoke and start them on the bearing cross-members. Press the bearings through the yoke onto the cross-members, as shown. Make sure that the grease fittings on the crosses are facing toward the coupler yoke (the long one).

15. Drive the bearing cup retaining C-rings into place with a hammer and punch. Lubricate and install the other sets of bearings in the same manner.

16. Lubricate all bearings, cups, and gears with Quicksilver High Performance Gear lube or an equivalent, before assembling. Set the drive gear into an arbor press with the gear facing down. Place the tapered roller bearing onto the drive gear with the tapered side facing up. Using driver tool (#91-90774), press the bearing down onto the gear until the aft side of the bearing makes contact with the gear.

17. Place the bearing cup over the tapered roller bearing followed by the large spacer so the flat face is toward the gear. On 2004 and later models (see earlier Note), reposition the original shim material and then the bearing spacer cup.

18. Set the second tapered bearing cup onto the large spacer ring (or spacer cup) with the tapered side facing up.

Place the tapered roller bearing onto the shaft. Obtain a suitable size mandrel and slowly press the tapered roller bearing down into the bearing cup until the roller barely makes contact with the bearing cup - then stop. The bearing will be drawn closer into the cup when the bearing preload is set. If you press the bearing on too tightly, so the spacer is unable to move freely, reinstall the puller plate and tap (lightly!) the end of the gear to loosen up the bearing.

19. On X/XR/XZ drives, press a new oil seal into the retainer nut, with the lip of the seal facing the pinion gear. Coat the lip of the seal with Multi-Purpose lubricant, or equivalent. Install the O-ring.

20. On I/II/III drives, press a new oil seal into the oil seal carrier, with the lip of the seal facing the concave side of the carrier (toward the gear), until the seal is flush with the carrier. Coat the lip of the seal with Multi-Purpose lubricant, or equivalent. Install the oil seal carrier, with the lip of the oil seal facing down, toward the drive gear. The seal prevents lubricant from working its way out of the upper housing and onto the U-joints.

21. Slide the retainer nut onto the short shaft yoke, followed by the oil seal and carrier on I/II/III drives and the beveled or thrust washer.

22. Slide the gear/bearing assembly onto the shaft, making sure everything lines up and then install the washer and nut. Tighten the nut until the washer just comes into contact with the gear.

23. Slide on the retainer nut wrench and clamp the handle in a vise so the shaft assembly is hanging straight down with the gear on top. Slowly tighten the pinion nut with a torque wrench, 1/16 of a turn at a time, until the pre-load is 6-10 inch lbs. (0.7-1.1 Nm). If the nut becomes too tight before reaching the proper pre-load figure, loosen it a few turns and then give the end of the shaft a few light hits with a bar and hammer. Now attempt to set the pre-load again.

Although we recommend the above, an alternate means is to temporarily install the assembly in the housing, turn it sideways so the assembly is hanging straight down with the gear up and then follow the remainder of the procedure detailed above.

24. Install the assembly in the upper unit.

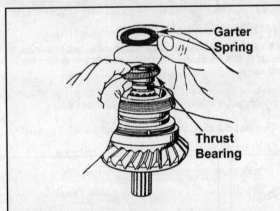

Fig. 106 After clamping the assembly in a vise, remove the nut and then slide off the drive gear/bearing assembly

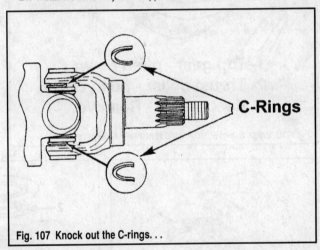

Fig. 107 Knock out the C-rings...

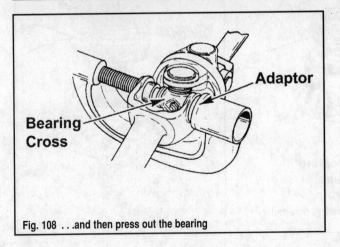

Fig. 108 ...and then press out the bearing

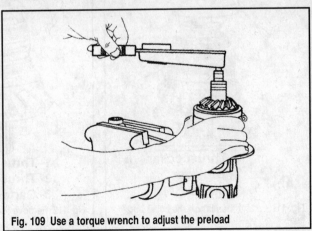

Fig. 109 Use a torque wrench to adjust the preload

DRIVE SYSTEMS - BRAVO

Clutch/Gear Assembly

◆ See Figures 79 and 110 thru 113

1. Remove the clutch assembly as previously detailed.
2. Obtain a holding fixture (#91-17301) and slide the clutch shaft splined end into the holding fixture, or clamp the splined end in a vise equipped with soft jaws.
3. Push down on the upper gear and upper thrust collar until the collar just clears the two keepers. Remove the two keepers from the groove in the clutch shaft.
4. Slide the thrust collar and upper gear free of the shaft.

■ If either the gear or the internal needle bearing is unfit for further service, both are replaced as a set. Both gears, as removed, are not serviceable, even though it appears a circlip holds the needle bearing within the gear. The gear and the needle bearing are purchased together under the same part number. Therefore, separating them would be a pointless task.

5. Lift out the upper thrust bearing and garter spring. Stack these items closely together if they are to be reused after inspection. Identify the stack as the upper set.

6. Rotate the clutch spool counterclockwise, up over the spiral worm gear on the clutch shaft and remove it.
7. Remove the lower garter spring, thrust bearing. Stack these items closely together, if they are to be reused after inspection. Identify the stack as the lower set.
8. Lift the lower gear from the clutch shaft.

To assemble:

9. Apply a good grade of gear oil to the splines. Slide the lower gear assembly down over the splines with the brass collar facing upward. The gear assembly will rotate clockwise as it follows the curved splines going down the shaft and will seat against the lower collar.
10. Slide the lower thrust bearing down over the shaft so the silver side is facing up. Stretch the lower garter spring over the shaft and down onto the bearing. The garter spring will not lie flat. However if it does, the spring has lost too much tension and needs to be replaced.
11. Lower the clutch spool down onto the shaft. The spool will rotate clockwise as it follows the curved splines downward to rest against the lower garter spring.
12. Slide the upper garter spring and upper thrust bearing over the shaft, making sure that the silver side of the bearing faces the spring. At this stage, these items will not appear to be seated. Fear not, for this is a normal condition.
13. Position the upper gear over the shaft with the brass collar facing downward. Install the upper thrust collar, with the taper (smaller diameter), facing upward.
14. Snap the gear down sharply to spread and seat both garter springs around each end of the clutch spool. This action will expose the upper groove in the shaft. Hold the gear down and insert the two keepers into the groove in the shaft. Release pressure on the gear allowing the upper thrust collar and gear to snap up and retain the split ring in the groove.
15. Install the assembly in the upper unit.

Top Cover Bearing And Sleeve

◆ See Figures 78 and 114 thru 119

■ On the X/XR/XZ, simply install a slide hammer attached to a Snap-On adaptor (#CG-40-4) and puller head (#CG-40-A10) and remove the bearing and sleeve.

1. Remove the top cover and place it upside down on a clean surface.
2. Using a puller set (#92-90244), position 2 puller jaws around the bearing sleeve.
3. Position the puller guide over the jaws and install the bolt supplied with the tool.

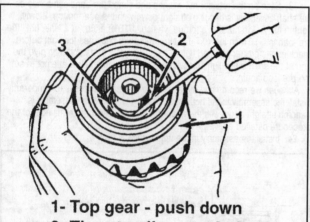

1- Top gear - push down
2- Thrust collar - push down
3- Keepers (2) - Remove

Fig. 110 Press the gear down and remove the keepers...

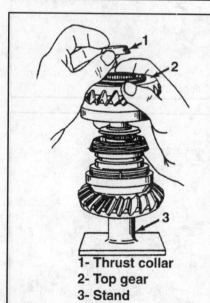

1- Thrust collar
2- Top gear
3- Stand

Fig. 111 ...and then lift off the collar and gear

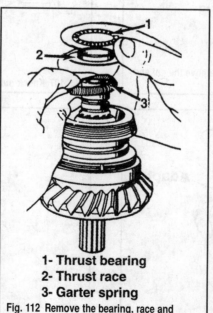

1- Thrust bearing
2- Thrust race
3- Garter spring

Fig. 112 Remove the bearing, race and spring...

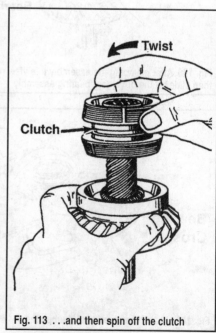

Fig. 113 ...and then spin off the clutch

DRIVE SYSTEMS - BRAVO

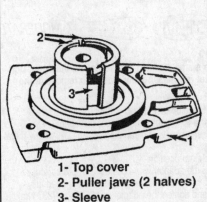

Fig. 114 Install the puller jaws...
1- Top cover
2- Puller jaws (2 halves)
3- Sleeve

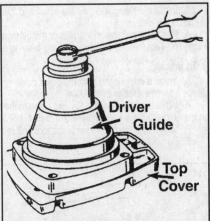

Fig. 115 ...and then slide on the puller guide
1- Puller guide
2- Puller bolt
3- Top cover
4- Suitable spacers

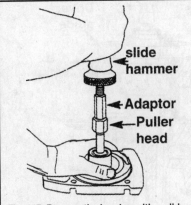

Fig. 116 Slide the driver guide over the puller and turn the bolt

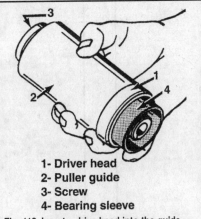

Fig. 117 Remove the bearing with a slide hammer

Fig. 118 Insert a drive head into the guide...
1- Driver head
2- Puller guide
3- Screw
4- Bearing sleeve

Fig. 119 ...and then slide it into the driver guide before tapping the bolt

4. Install a driver guide over the puller and then turn the bolt clockwise to pull off the sleeve.

5. Clamp the cover in a vise (carefully!) and pull out the bearing with a slide hammer if necessary.

To install:

6. Install a driver head (#91-862530) onto the puller guide (#91-90774) and then position the new sleeve against the edge of the driver.

7. Move the sleeve and puller into place on the top cover and slide on the guide. Tap the bolt lightly with a mallet until the tool bottoms out.

8. Now install the bearing onto the tool and repeat the previous step.

9. Remove the tools, install a new O-ring and install the cover. Tighten the four bolts to 20 ft. lbs. (27 Nm).

Driveshaft Housing Bearing And Sleeve

♦ See Figures 78 and 120 thru 124

■ On the X/XR/XZ, simply install a slide hammer attached to a Snap-On adaptor (#CG-40-4) and puller head (#CG-40-A10) and remove the bearing and sleeve.

1. Using a puller set (#92-90244), position the puller jaws around the sleeve in the housing bore.

2. Position the puller guide over the jaws and install the bolt supplied with the tool.

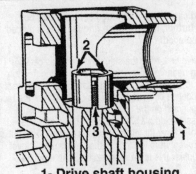

Fig. 120 Install the puller jaws...
1- Drive shaft housing
2- Pull jaws (2 halves)
3- Sleeve

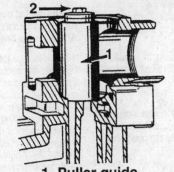

Fig. 121 ...and then slide on the puller guide
1- Puller guide
2- Puller bolt

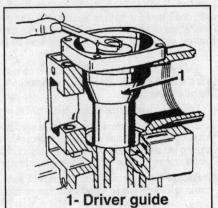

Fig. 122 Slide the driver guide over the puller and turn the bolt
1- Driver guide

3. Install a driver guide over the puller and then turn the bolt clockwise to pull off the sleeve.

4. Position a suitable mandrel over the bearing in the clutch bore. Place a long bar over the mandrel and drive the bearing down into the oil cavity

To install:

5. Install a driver head onto the puller guide and then position the new sleeve against the edge of the driver.

6. Move the sleeve and puller into place in the housing bore and slide on the guide. Tap the bolt lightly with a mallet until the tool bottoms out.

7. Now install the bearing onto the tool and repeat the previous step.

8. Remove the tools.

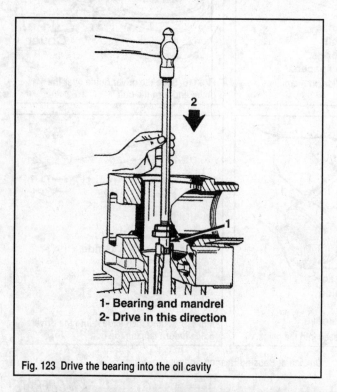

Fig. 123 Drive the bearing into the oil cavity

TRANSOM ASSEMBLY

Description

The stern drive unit, consisting of the upper gear housing and the lower unit, is attached to the bell housing. The bell housing is secured to the gimbal ring by two roller bearings. These bearings permit up-and-down trim movement. The gimbal ring is mounted to the gimbal housing by two roller bearings. This second set of roller bearings permits steering movement to port and starboard. The gimbal housing is mounted on the transom of the boat and attached to an inner transom plate. This plate contains the rear engine mounts and attachments for the steering and shift cables. The bell housing, gimbal ring, gimbal housing and inner plate all make up the transom assembly.

The bell housing has extended flanges to connect the exhaust, universal and shift cable bellows, and water hoses between the stern drive unit and the gimbal housing.

The bell housing is held in place by the gimbal ring.

Gimbal housing/transom plate service can generally be performed without removing the gimbal housing. However, in those cases when a part of the housing has been broken, or if the inner transom plate must be removed, the gimbal housing must be removed before the transom plate.

Before the gimbal housing can be removed, several tasks involving considerable work and time must be performed in the following order:

Stern drive removed.

Engine removed.

Bell housing and gimbal ring removed.

Bearing Adaptor

◆ See Figure 78

■ X/XR/XZ drives utilize a steel bearing adaptor inside the upper unit. Although there should not normally be any reason to remove it, the procedures follow.

1. Position a bearing adaptor socket (#91-862531) into the steel adaptor. Heat the area with a torch lamp and remove the adaptor.

2. Pry out the oil passage plug and clean the passageway of any metal chips or contamination.

3. Press the plug back into position.

4. Coat the threads of the adaptor with Loctite® 277 and carefully thread it back into the bore by hand until the threads are engaged fully. Install the adaptor socket and tighten the adaptor to 175 ft. lbs. (237 Nm).

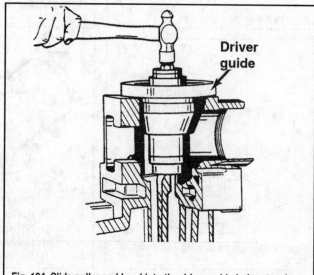

Fig. 124 Slide puller and head into the driver guide before tapping the bolt

Gimbal Bearing

REMOVAL & INSTALLATION

◆ See Figures 125 thru 128

■ The gimbal bearing and carrier are a matched set and must be replaced as an assembly; further, the tolerance ring must be replaced whenever the bearing is replaced.

1. Remove the stern drive unit. After the stern drive unit has been removed, the gimbal housing bearing can be checked by reaching through the U-joint bellows attached to the bell housing and rotating the inner bearing race. There should be no evidence of binding or rough spots. Check the race for side play by pulling and pushing on the race. If there is any sign of roughness, binding, or excessive side play, the bearing should be replaced. A special puller is required to remove the gimbal housing bearing. This puller is designed to establish alignment from the face of the bell housing. Never remove the gimbal bearing unless it is to be replaced because it will be damaged during removal.

2. Assemble the plates of special tool (#91-29310). Position the plates between the top and middle studs located on the bell housing. Use a 3-jaw puller from the Slide Hammer Puller Set (#91-34569A1). If the bearing assembly is tight, tap the end of the Puller Shaft (#91-31229), with a mallet while attempting to turn the nut (#11-24156).

DRIVE SYSTEMS - BRAVO

3. Remove the tolerance ring from the carrier.
4. Reach into the housing and remove the grease seal - you may need a slide hammer for this.

To install:

5. Clean all metal parts in solvent and dry them with compressed air. Never spin ball bearings with compressed air, because such action will ruin the bearing.
6. Inspect the bellows carefully for cracks, cuts, and punctures. Verify the bellows are still flexible. If the condition of the bellows is doubtful, replace them. In most cases, if the gimbal bearing is damaged due to water, the water has entered through, or around, the bellows.
7. The gimbal housing bearing can only be installed with the bell housing in place in order to establish an alignment reference, so if you have removed the housing, install it now.
8. Use a suitable mandrel and press the grease seal into position in the housing bore.
9. Lubricate the outside of a new gimbal bearing carrier assembly with Multi-Purpose Lubricant, or equivalent.
10. Install the tolerance ring over the carrier. The opening in the tolerance band must align with the lubrication opening in the bearing carrier before the carrier is installed. The opening must also align with the lubrication opening in the gimbal housing after the carrier is installed.

■ Observe the notches (could be called "cutouts), in the bearing carrier, indicated in the accompanying illustration. These notches must face forward when the carrier is installed.

11. Install the assembled bearing carrier using the special tools. Insert the Driver Head (#91-32325), through the Mandrel (#91-30366), and into the inside diameter of the new bearing. Align the bearing with the gimbal housing by positioning Plate (#91-29310) between the top and middle studs on the bell housing. Positioning the plate as described will ensure the driver rod (#91-37323) will remain at right angles to the bearing carrier bore.
12. Now, drive the bearing into the gimbal housing cavity with a hammer. Lubricate the bearing using only Quicksilver Multi-Purpose Lubricant through the grease fitting. To lubricate the bearing properly, pump 40 full strokes, to deliver about one full ounce of lubricant.
13. Install the stern drive.

Shift Cable

REMOVAL & INSTALLATION

Remote Control Cable

◆ See Figure 129

1. Remove the control box and disconnect the cables. Remove the cotter pin and cable connector securing the cable to the lever. Remove the screw and nut securing the anchor bracket to the housing and then remove the bracket.
2. Trace the cable back to the shift bracket on the engine and disconnect the cable guide from the shift lever pin.

■ Take note of the positioning of the cable pin in the slot for ease of installation.

3. Now loosen the anchor block retainer screw (upper on standard rotation props and lower on counter rotation props) and pivot the retainer away from the trunnion. Remove the shift cable.

To install:

4. Lubricate the control unit end of the cable with grease.
5. Install the cable trunnion into the anchor and then install the anchor on the trunnion. Install the screw and locknut, tightening securely.
6. Pull the cable forward so it aligns with the pin.
7. Connect the other end at the remote control and install the unit.

Transom Cable

◆ See Figures 129 thru 130

1. Remove the stern drive.
2. Disconnect the stern drive unit shift cable from the shift plate mounted on the engine. Remove the nut and washers at the forward end of the cable and slide the cable end off of the shift lever pin. Remove the cotter pin at the rear attaching point and slide the trunnion out of the retainer.

■ In mid-2004, Mercruiser made a change to models with serial #OW100000 and above - both the cable end guide and the barrel trunnion are secured to the shift plate with cotter pins installed from the top down. 2004 and earlier models with serial #OM99999 and below use the not on the cable end and insert the barrel cotter pin from the bottom.

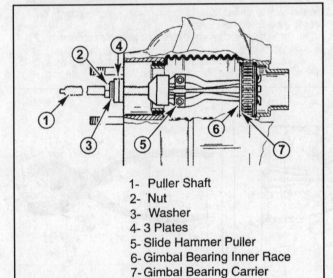

1- Puller Shaft
2- Nut
3- Washer
4- 3 Plates
5- Slide Hammer Puller
6- Gimbal Bearing Inner Race
7- Gimbal Bearing Carrier

Fig. 125 Removing the bearing with the special tools

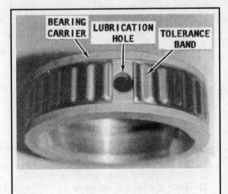

Fig. 126 It is important to install the tolerance ring correctly

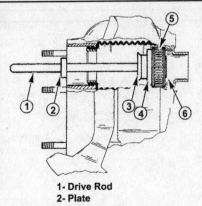

1- Drive Rod
2- Plate

Fig. 127 Installing the bearing with special tools

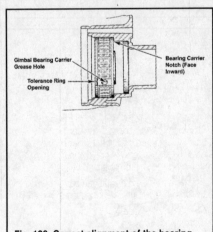

Fig. 128 Correct alignment of the bearing

11-34 DRIVE SYSTEMS - BRAVO

3. Loosen the set screws from the cable end guide, and then remove the end guide. If only the inner cable is to be replaced, burrs made by the set screws must be removed from the core wire to prevent the inner lining of the shift cable from being damaged when the wire is removed. Loosen the jam nut on the metal cable end, and then turn the metal end out of the cable.

4. Pull the threaded support tube from the end of the inner core wire.

5. If the inner wire is to be replaced, hold the cable slide on the drive end and pull the inner wire out of the shift cable.

6. If removing the entire cable, hold the shift cable retaining nut on the transom side of the housing and remove the flanged nut on the other side.

7. Remove the protective wrapping from the shift cable in the area where the cable passes into and through the transom.

8. Loosen or remove the shift cable bellows crimp clamp at the small end of the bellows. Pull the shift cable through the bellows.

To install:

9. Inspect the shift bellows for cracks, cuts and punctures. Verify the bellows are still flexible. If there is the least doubt about the condition of the bellows, they should be replaced. If the bellows are defective and leak, water will enter the boat.

10. Inspect the cable locking nut threads for any damage such as cross-threading and check for any indication of the locking nut separating from the outer casing. Check the length of the shift cable for kinks, cuts or chafing and the inner core wire for signs of unraveling.

■ The manufacturer strongly recommends that the small shift cable bellows clamp be the crimp type, not a worm type clamp and not a tie wrap. To crimp such a clamp, a special tool is required. However, a pair of common pliers may be easily and quickly modified to do the job. To customize a standard pair of pliers to crimp the bellows clamp, first tack weld a 3/4 in. nut to the pliers with the gripping surfaces of the pliers contacting two opposite sides of the nut. Clamp the pliers in a vice and drill out most of the threads using a 1/2 in. drill bit. With the pliers still in the vise, cut the nut in half, leaving an equal amount of the nut on each side of the pliers, as shown in the accompanying illustration.

11. If you only removed the core wire, slide it through the bellows and cable. If you removed the entire cable, insert the forward (control) end of the shift cable into and through the shift cable bellows.

12. Apply a coating of Perfect Seal to the flanged nut threads. Secure the shift cable to the bell housing, hold the retaining nut and tighten the flanged nut to 65 inch lbs. (7 Nm).

13. Install the protective wrapping around the cable approximately 2 in. out from the housing.

14. Install the shift bellows crimp clamp on the small end of the shift bellows. Use the modified pliers, described in the previous Note, and

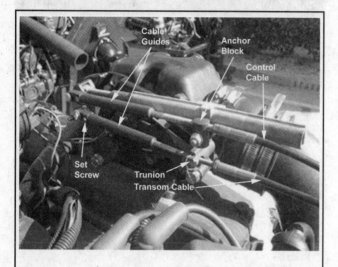

Fig. 129 A good shot of the shift cable bracket (standard rotation shown)

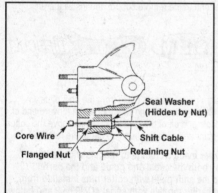

Fig. 130 Although you can remove just the core wire, you'll generally remove the entire cable

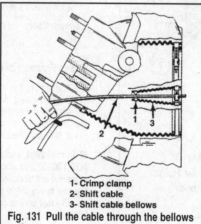

1- Crimp clamp
2- Shift cable
3- Shift cable bellows

Fig. 131 Pull the cable through the bellows and out

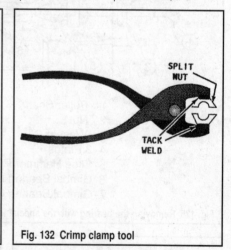

Fig. 132 Crimp clamp tool

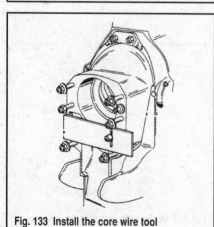

Fig. 133 Install the core wire tool

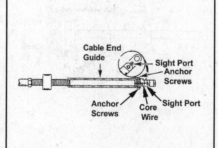

Fig. 134 Ensure that the core wire is visible through the sight port before tightening the screws

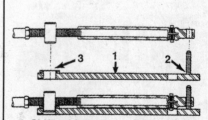

1- Shift cable anchor adjustment tool
2- Stud- placed thru hole in end guide
3- Hole- barrel placed here

Fig. 135 Install the shift cable anchor adjustment tool

DRIVE SYSTEMS - BRAVO   11-35

squeeze the bellows clamp securely around the bellows so that you maintain an O. D. of 1/2 in. and the clamp is round. After the clamp is properly installed, water will not be able to enter the bellows and find its way into the boat.

15. Install a core wire locating tool (#91-17263) onto the aft end of the bell housing.

16. Install the threaded tube on the shift cable end and tighten it until it just bottoms; now tighten the jam nut against the cable end.

17. Install the cable end guide over the core wire and then insert the core wire through the anchor. Make sure that the core wire is visible through the small sight port and then tighten the set screws to 20 inch lbs. (2.3 Nm).

18. Install the shift cable anchor adjustment tool (#91-17262) onto the end of the cable so the stud is through the hole in the guide. Make sure that the end of the core wire is pulled tight against the core wire locating tool.

19. Rotate the barrel on the cable threads until it lines up with the hole in the tool and seats itself in the recess. Remove the tools and install the cable on the shift plate.

20. Adjust the cable.

ADJUSTMENT

◆ See Figures 136 thru 141

■ The front propeller on Bravo III drives is always left hand rotation and the rear propeller is always right hand rotation, so the shift cable should always move as shown in the illustration. On Bravo I/II drives, if the cable moves in the direction of A when the control lever is moved to Forward, it is set up for RH propeller rotation. If it moves in the direction of B when the lever is moved to the Forward position, it is set up for LH propeller rotation.

■ When installing shift cables, it is imperative that both cables are routed in such a manner that there are no kinks or sharp bends, and out of contact with any moving parts.

1. Install the forward cable into the remote control.
2. Loosen the upper stud on the shift plate and move it within the slot until it is exactly 3 in. from the center of the lower pivot bolt and then install the transom shift cable guide over the stud. There should be a washer on each side of the guide, tighten the lock nut until it contacts the washer and then back it off 1 full turn. Feed the cable barrel/trunnion into the retainer and install a new cotter pin (from the bottom).

■ In mid-2004, Mercruiser made a change to models with serial #OW100000 and above - both the cable end guide and the barrel trunnion are secured to the shift plate with cotter pins installed from the top down. 2004 and earlier models with serial #OM99999 and below use the not on the cable end and insert the barrel cotter pin from the bottom.

3. Place the adjustment tool (#91-12427) over the shift cable as shown and then tape it in place over the barrel retainer.
4. Shift the remote control lever to Neutral. Push in on the remote control cable end just enough to relieve the play and mark the position on the tube. Now pull out on the cable end until all play is removed and mark that position. Mark the center point of these two positions.
5. Install the control cable end guide into the shift lever and insert the anchor pin. Adjust the cable barrel so that the hole in the barrel is centered with the vertical centerline of the stud. Make sure that the center mark you made in the last step is in alignment with the edge of the cable end guide. Do not insert the barrel yet or you risk bending the cable end.
6. Remove the cable end guide from the shift lever by pulling out the pin, and then install the barrel onto the stud. Tighten the barrel locknut until it bottoms out on the barrel and then install a new cotter pin (if equipped). Now reinstall the other end into the shift lever.

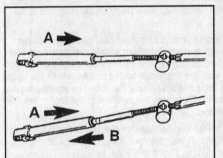

Fig. 136 Cable direction tells you propeller rotation. Upper image is Bravo III, lower image is Bravo I/II

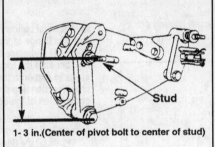

Fig. 137 Set the stud dimension...

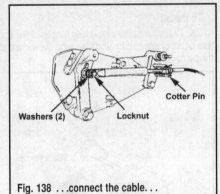

Fig. 138 ...connect the cable...

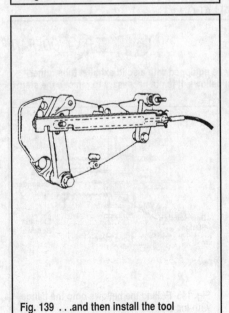

Fig. 139 ...and then install the tool

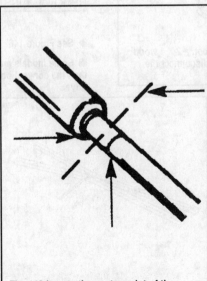

Fig. 140 Locate the center point of the remote control cable backlash

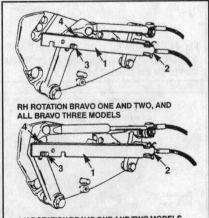

Fig. 141 Fit the stud into the slot on the tool for final adjustment

DRIVE SYSTEMS - BRAVO

7. Pop the front of the adjustment tool off of the stud, but keep it taped over the barrel retainer.

8. Shift the control lever into the full Forward position - the rear slot in the tool should fit over the shift lever stud on Bravo I/II RH rotation models and all Bravo III models; on Bravo I/II LH rotation models it should be the forward slot. If the slot does not fit over the stud, loosen the stud and slide it up or down until it fits. Tighten the stud.

9. Now you can remove the tool. Lubricate the shift cable pivot points as detailed in the Engine & Drive Maintenance section.

Exhaust Bellows

REMOVAL & INSTALLATION

◆ See Figures 142 thru 145

■ Although in the past we have provided instructions for removing the bellows without removing the drive unit, we now believe the best way to replace the bellows is by removing the Stern drive.

1. Remove the stern drive.
2. Tilt the bell housing slightly up and move it to the port side. Access to the aft clamp on the exhaust bellows is from the bottom of the bell housing. Remove the clamp.
3. Access to the forward clamp on the exhaust bellows is through the access hole on the port side of the gimbal housing. Insert a screwdriver through the access hole and remove the forward exhaust bellows clamp.
4. Pull the exhaust bellows from the gimbal housing and bell housing exhaust flanges. It may be necessary to exert considerable force to pull the bellows loose because of the adhesive used during installation.

To install:

5. Clean the bellows mounting flanges of the upper bell housing with a wire brush or sandpaper, and then wipe the surface clean with lacquer thinner to remove any old glue.

6. If new bellows are being installed (and if you've removed it, you should install a new one), be sure to clean the powder residue from the bellows surface with warm soapy water and dry thoroughly. The powdery substance is a mold release agent and the adhesive will not bond if this powder is left on the sealing surfaces.

7. Check the flanges and housing for cracks, nicks, or corrosion. Clean the bellows clamps thoroughly to ensure a good ground. New bellows have a clip on each end of the bellows to ground the clamp. Check the clamps for cracks or nicks. Replace the clamp if in doubt as to their condition. Always use stainless steel clamps for satisfactory service.

✷✷ WARNING

Bellows Adhesive must be used to ensure a satisfactory installation. This adhesive is extremely toxic and flammable. Therefore, make every effort to ensure adequate ventilation in the work area during its use. Work in the outdoors, if at all possible. Vapors from the adhesive may cause a flash fire or ignite explosively. Keep the adhesive away from heat, sparks, and open flame. DO NOT smoke. Extinguish all flames and pilot lights in the area. Turn off stoves, heaters, electric motors, and all other possible sources of ignition while using the adhesive and until there is no doubt but what all vapors have left the area. Close the container immediately after use. The adhesive is harmful or fatal if swallowed. Avoid prolonged contact with the skin or breathing of the vapors. If swallowed, do not induce vomiting. Call a physician immediately. Keep the adhesive out-of-reach of children.

8. Apply a coating of Bellows Adhesive to the inside of each end of the bellows and around the mounting flanges. Allow the adhesive to dry approximately 10 minutes, or until the material is no longer tacky. Install the clamp on one end of the bellows, and then install it to the gimbal housing flange so that the clamp screw is accessible through the access hole. Tighten the clamp screw to 35 inch lbs. (4 Nm).

9. Now the hard part. Install the bellow clamp on the other end of the bellows.

■ A special bellows tool is almost a necessity to "pull" the bellows over the flange of the stern drive. Therefore, obtain expander tool (#91-45497A1) and place the tool into the first bellows convolution. Pull on the tool until the tool touches the flange on the bell housing - the hose starts to slip onto the flange - then release the tool. The accompanying cross-section line drawing illustrates the position of the tool in the first convolution of the bellows.

10. Move the tool into the third bellows convolution and pull the bellows onto the bell housing flange. Tighten the hose clamp to 35 inch lbs. (4 Nm).
11. Install the drive, start the engine and check for leaks.

Exhaust Tube

REMOVAL & INSTALLATION

◆ See Figure 146

■ Some models may be equipped with a solid exhaust tube rather than the conventional bellows. It is not necessary to remove the stern drive.

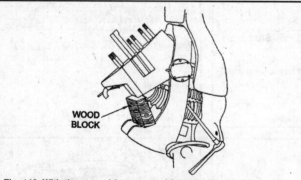

Fig. 142 With the stern drive removed from the boat, a 2 X 4 wood support under the bell housing will be helpful to disconnect the clamps on the exhaust bellows

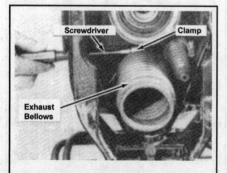

Fig. 143 An access hole in the housing provides a way to get your screwdriver on the clamp screw

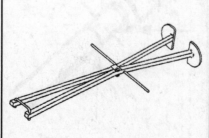

Fig. 144 A bellows expander is necessary to get the bellows back onto the flange

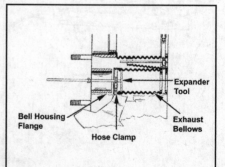

Fig. 145 Pulling the bellows onto the flange with the tool

DRIVE SYSTEMS - BRAVO

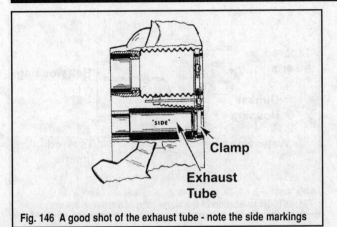

Fig. 146 A good shot of the exhaust tube - note the side markings

1. Raise the drive unit to the full UP position.
2. Reach into the bell housing and loosen the clamp and then pull out the tube.
3. Inspect the tube for charring, cracks, hardening or other visible signs of damage.
4. Using fine grit sand paper, carefully rough up the mating surfaces on each end of the tube and then clean them with a small amount of lacquer thinner.
5. Position a grounding clip (we recommend replacing the old one) over the lip on the forward edge of the tube.
6. Slide the clamp over the flange on the forward end of the tube so that the screw is on the port side and aligned with the access hole in the housing. Move the tube into position so that the **SIDE** markings are, you guessed it, on each side and then tighten the clamp securely (or to 35 inch lbs. [4 Nm]).

U-Joint Bellows

REMOVAL & INSTALLATION

◆ See Figures 147 thru 150

■ Although not necessary, we strongly recommend removing the stern drive. The bell housing does not have to be removed in order to replace the U-joint bellows. However, this is not an easy task. Therefore, work slowly and carefully. If the job proves too difficult, the decision may be made to remove the exhaust bellows and the water intake hose first.

1. Remove the single hose clamp securing the bellows to the gimbal housing using a long shank screwdriver. Work the bellows off the gimbal housing flange. Considerable force may be required to remove the bellows due to the adhesive used during installation.

2. There is no hose clamp securing the other end of the bellows to the bell housing. Instead, a sleeve fits tightly inside the bellows to hold it against the bell housing flange. The sleeve must be pried free of the bellows. Obtain a can of aerosol cleaner such as Gunk® or WD-40®. Spray the cleaner around the edge of the sleeve to help loosen it. Pry the sleeve free using a thin blade screwdriver. Push the bellows back from the bell housing flange until it is free. No adhesive is used at this end of the bellows, therefore, just push.

3. If you were smart and removed the drive, insert a sleeve removal tool (#91-818169) through the housing and pull out the sleeve. Now you simply need to pry the bellows off of the flange

To install:

4. Remove all bellow adhesive from the inside diameter of the bellows if it is to be reused. Inspect the bellows for cracks, cuts, punctures and to be sure it is still flexible. If the least bit of doubt exists concerning the condition of the bellow, install new bellows. If you are considering re-using the bellows, stop and think - why not replace it while you have everything apart? You may wish you had down the road!

5. Clean the bellows mounting flanges of the bell housing with a wire brush or sandpaper, and then wipe the surface clean with lacquer thinner.

6. Check the flanges and housing for cracks, nicks, or corrosion. Clean the bellows clamp thoroughly. Check the clamp for cracks or nicks. Replace the clamp if in doubt as to their condition. Always use stainless steel clamps as a replacement.

✶✶ WARNING

Bellows Adhesive must be used to ensure a satisfactory installation. This adhesive is extremely toxic and flammable. Therefore, make every effort to ensure adequate ventilation in the work area during its use. Work in the outdoors, if at all possible. Vapors from the adhesive may cause flash fire or ignite explosively. Keep the adhesive away from heat, sparks, and open flame. Observe no smoking. Extinguish all flames and pilot lights in the area. Turn off stoves, heaters, electric motors, and all other possible sources of ignition while using the adhesive and until there is no doubt but what all vapors have left the area. Close the container immediately after use. The adhesive is harmful or fatal if swallowed. Avoid prolonged contact with the skin or breathing of the vapors. If swallowed, do not induce vomiting. Call a physician immediately. Keep the adhesive out-of-reach of children.

7. Apply a coating of Bellows Adhesive to the inside diameter of the gimbal housing bellows end. Do not apply the adhesive to the bell housing end. Allow the adhesive to dry approximately 10 minutes, or until the material is no longer tacky. Install a grounding clip over the forward edge of the bellows with the shorter side of the clip on the inside of the bellows.

8. Find the word **TOP** embossed on the bellows. This word must face up after installation. Be sure to position the bead on the inner diameter of the bellows in the groove on the gimbal housing flange. Install the U-joint bellows over the flange and position the clamp tightening screw at the 3 o'clock position. Tighten the screw securely, or to 35 inch lbs. (4 Nm). Don't forget the ground clip.

9. Position the U-joint bellows on the bell housing. Ensure the bell housing flange is indexed in the second groove from the end of the bellows.

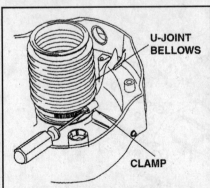

Fig. 147 The U-joint bellows clamp is placed in approximately the 3 o'clock position. A sleeve is wedged into the bell housing end as explained in the text

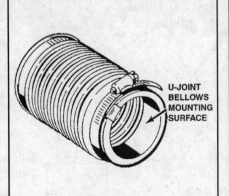

Fig. 148 Coat the gimbal housing end of the bellows with adhesive

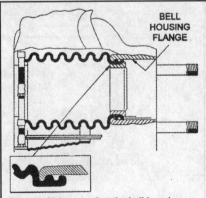

Fig. 149 Make sure that the bell housing flange rests in the second groove on the bellows

11-38 DRIVE SYSTEMS - BRAVO

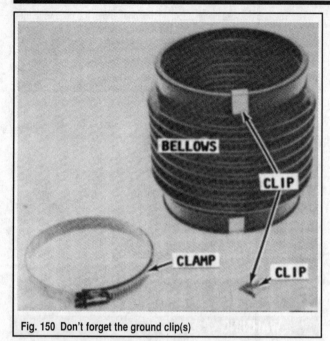

Fig. 150 Don't forget the ground clip(s)

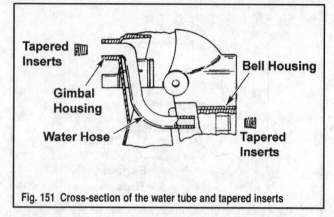

Fig. 151 Cross-section of the water tube and tapered inserts

get to this without removing the engine, but some applications will require it. Do the same with the insert on the bell housing.

3. Carefully pull out the water hose from the back of the drive.

To install:

※※ **CAUTION**

The water hose is preformed and must always be installed with the short end going into the gimbal housing.

4. Feed the new hose, short end first, into the unit from the bell housing side until the forward end of the hose is flush with the mounting surface on the engine side of the inner transom plate. The other end of the hose should be maneuvered until it is protruding 1/8 in. out of the hole in the back of the bell housing.

5. Coat the threads of an insert with Quicksilver 2-4-C Lubricant. Install the insert into the tool and thread it into the hose on the inner plate while holding the hose securely so it stays flush with the mating surface and does not slide back into the transom assembly.

6. Do the same thing with the other end of the hose, making sure that it is still protruding from the bell housing.

Exploded Views

◆ See Figures 152 thru 155

10. Obtain a Sleeve Installation Tool (#91-818162). Lubricate the outside surface of the sleeve with soapy water or engine cleaner, and then install the sleeve tool. Use a suitable driving rod and move the sleeve in place on the bell housing - this is why we recommend removing the drive!

11. If the water hose and the exhaust bellows were removed, install them to the gimbal housing.

12. Replace the stern drive if you removed it.

13. Start the engine and check the completed work.

Water Hose & Fitting

REMOVAL & INSTALLATION

◆ See Figure 151

1. Remove the stern drive unit.
2. Using a tapered insert tool (#91-43579), remove the tapered insert from the water line bore on the inner transom plate - you should be able to

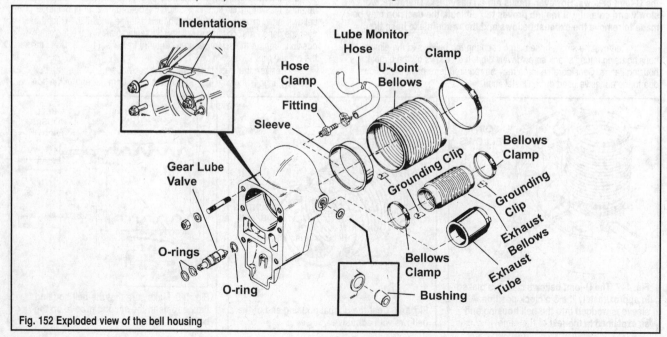

Fig. 152 Exploded view of the bell housing

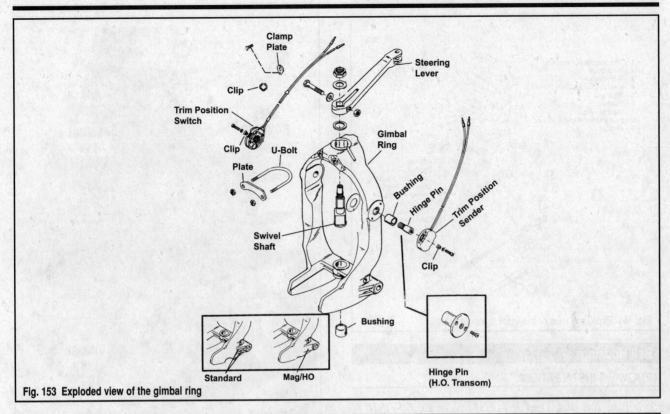

Fig. 153 Exploded view of the gimbal ring

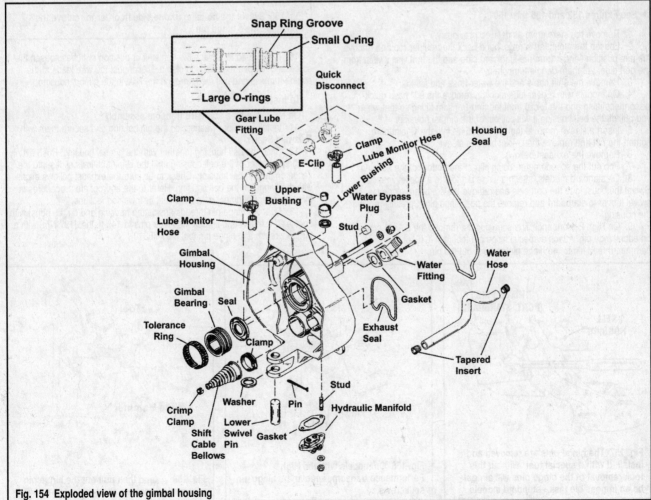

Fig. 154 Exploded view of the gimbal housing

11-40 DRIVE SYSTEMS - BRAVO

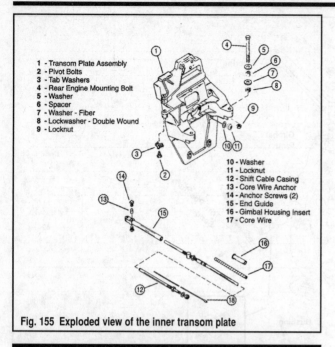

1 - Transom Plate Assembly
2 - Pivot Bolts
3 - Tab Washers
4 - Rear Engine Mounting Bolt
5 - Washer
6 - Spacer
7 - Washer - Fiber
8 - Lockwasher - Double Wound
9 - Locknut
10 - Washer
11 - Locknut
12 - Shift Cable Casing
13 - Core Wire Anchor
14 - Anchor Screws (2)
15 - End Guide
16 - Gimbal Housing Insert
17 - Core Wire

Fig. 155 Exploded view of the inner transom plate

Bell Housing

REMOVAL & INSTALLATION

◆ See Figures 152 and 156 thru 160

1. Remove the stern drive as detailed previously.
2. Loosen the mounting screws, bend back the retainer clip and remove the trim position sender from the starboard side and the trim limit switch from the port side. Let them dangle temporarily.
3. Remove the shift cable and the water hose and fittings.
4. On models with a gear lube monitor, remove the 90° hose quick-disconnect fitting and the E-clip from the thru-transom fitting so that when you pull off the bell housing it will pull out of the gimbal housing.
5. Insert a sleeve removal tool (#91-862546) into the U-joint bore, tighten the nut and remove the U-joint bellows sleeve.
6. Remove the exhaust bellows.
7. Pop out the speedometer tubing clip at the bottom of the housing.
8. On standard models, attach a hinge pin tool to a socket wrench. Swivel the housing to the port side and remove the starboard hinge pin. Now swivel it over to starboard and remove the port hinge pin. Carefully remove the housing.
9. On High Performance Transom models, remove the 2 Allen screws on each hinge pin. Attach a special hinge pin tool (#91-63616) to a slide hammer, thread it into the hinge pin and pull out the pin.

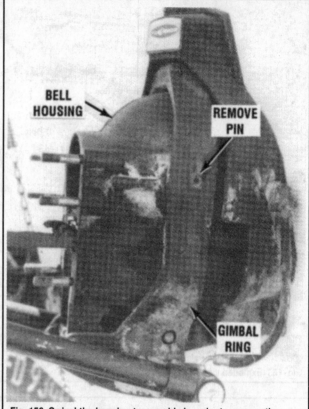

Fig. 156 Swivel the housing to one side in order to remove the opposite side's hinge pin

10. If you need to replace the trim limit or position switches, loosen the screw and remove the retaining clamp. Disconnect the wire leads at the engine harness and pull them through the hole in the gimbal housing.

To install:

On models with a standard transom assembly:

11. Install new fiber washers on the gimbal ring and secure them with Reiswald Sealer.
12. Bring the bell housing together with the gimbal housing. Push evenly on the perimeter of the bell housing until the U-joint bellows slides into the flange on the gimbal housing. Check to be sure the exhaust bellows aligns with the flange on the bell housing. Refer to the appropriate procedures in this section for pertinent details on final bellows connections.
13. Coat the hinge pin threads (inside the housing and on the bolts) with Loctite 271. Install and tighten the hinge pins to 145 ft. lbs. (197 Nm) with a torque wrench and hinge pin tool (#91-78310).

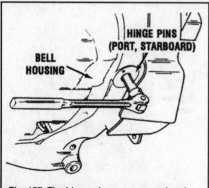

Fig. 157 The hinge pins are removed and installed with a special tool. Without this tool, removal of the hinge pins will almost be an impossible task - standard models

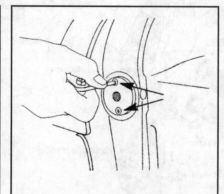

Fig. 158 On models with the High Performance transom, remove the hinge pin set screws...

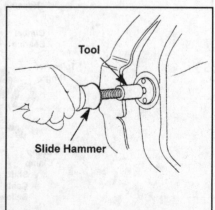

Fig. 159 ...and then pull out the hinge pin with a slide hammer and tool

DRIVE SYSTEMS - BRAVO 11-41

Gimbal Ring

REMOVAL & INSTALLATION

♦ See Figures 153 and 161 thru 167

■ Unless a previous owner has already drilled an access hole in the gimbal housing, you will need to drill (and plug) your own hole in order to remove the steering lever.

1. Remove the stern drive and bell housing.
2. Remove the trim position sender and switch.
3. If the engine and inner transom plate are installed:
  a. Remove the two access plugs on either side of the top of the gimbal housing.
  b. Insert a socket wrench into the port access hole and an open end into the starboard hole. Loosen and remove the steering lever clamping bolt.
  c. Now insert a punch through the starboard hole and unthread the elastic locknut. You may have to move the steering wheel a few times to jockey the lever and nuts into position.

Fig. 161 Use a socket on the port side when removing the steering lever bolt

1- Bell housing
2- Tapered insert -turn clockwise
3- Tapered insert tool
4- Water hose

Fig. 160 Use the special tool to remove the tapered insert from the water tube

On models with a high performance transom:

14. Coat the hinge pins with Quicksilver Special Lubricant and thread a pin onto the installation tool (#91-63616). Slide it through the gimbal ring and bell housing so the holes in the pin flange line up with those in the ring.

15. Tap the pin into place (lightly!) with a slide hammer. Coat the threads of the 2 screws with Loctite 242, thread them in and tighten to 25-30 inch lbs. (2.8-3.3 Nm). Repeat for the other pin.

On all models:

16. Install the speedometer hose clip.

17. Install the exhaust and U-joint bellows to the gimbal housing as detailed in the appropriate section.

18. Install the water hose as detailed in the appropriate section.

19. Install and adjust the shift cable as detailed in the appropriate section.

20. If you disconnected the trim switches, route the trim limit switch and trim position sender leads, as indicated in the accompanying illustration. Install a Sta-strap around both sets of leads approximately 5 in. (12cm) from the retaining cover.

21. Install the trim switches on either side of the gimbal ring

22. Install the stern drive.

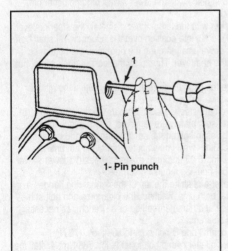

1- Pin punch

Fig. 162 Use a punch to drive off the elastic nut (engine not removed)

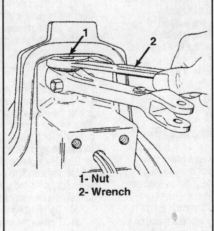

1- Nut
2- Wrench

Fig. 163 Reach through the transom to remove the steering lever (engine removed)

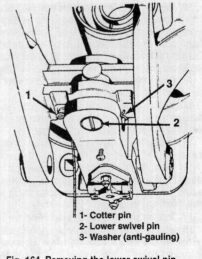

1- Cotter pin
2- Lower swivel pin
3- Washer (anti-gauling)

Fig. 164 Removing the lower swivel pin

11-42 DRIVE SYSTEMS - BRAVO

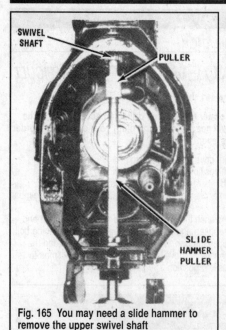

Fig. 165 You may need a slide hammer to remove the upper swivel shaft

Fig. 166 Removing the lower bushing...

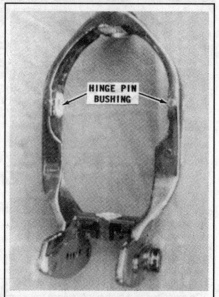

Fig. 167 ...and then remove the hinge pin bushings

4. If the engine and inner transom plate have been removed:
   a. Reach into the gimbal housing from the inside and use 2 open end wrenches to loosen the steering lever clamp bolt.
   b. Insert a 1-1/16 in. open end into the gimbal housing from the inside and then remove the swivel shaft lock nut atop the steering lever.
5. They've probably already been removed when you pulled off the drive unit, but if not, disconnect the trim cylinders at the gimbal ring and carefully set them aside - tie them off with a piece of rope.
6. Disconnect the continuity wire and then remove the lower swivel cotter pin. Drive out the lower swivel pin and remove the anti-gauling washer.
7. Loosen the two upper gimbal ring bolts and then remove the steering lever with the two washers and locknut. Don't forget to remove the ground wire from the lever/transom.
8. Remove the upper swivel shaft - you may need a slide hammer if it is frozen. Lift out the gimbal ring.
9. Remove the gimbal ring lower swivel shaft bushing, using a suitable mandrel. Remove the gimbal housing upper swivel shaft bushing and single oil seal, using a slide hammer with a two jaw puller attachment.
10. Obtain a suitable driver and drive out both hinge pin bushings.

To install:

11. Clean all metal parts in solvent and blow them dry with compressed air. Never, spin ball bearings with compressed air, because such action will ruin the bearings.
12. Remove all bellows adhesive from the inside diameter of the bellows, if the bellows are to be reused. Inspect the bellows for cracks, cuts, punctures and to be sure they are still flexible. If there is the least bit of doubt concerning their condition, do not hesitate to install new bellows. Clean the bellows mounting flanges of the bell housing with a wire brush or sandpaper and then wipe the surface clean with lacquer thinner.
13. Check the shift cables for cuts or damage caused by a cable being pinched or bent too short.
14. Inspect the water hose for cracks, cuts, punctures, or worn spots. Inspect the shift shaft oil seal for tears, wear, or any roughness. Check the shift shaft and shift shaft bushings for wear.
15. Inspect the O-ring and rubber gasket for cuts, nicks, hardness, or cracks.
16. Inspect the surface of the lower swivel pin in the area where the needle bearing rides. Any pitting, grooves, or uneven wear is cause to replace the bearing and swivel pin. Inspect the lip surface of the oil seals for wear, tears, and roughness.
17. Test the trim limit switch as detailed in the Trim/Tilt section.
18. Inspect the long steering lever retaining bolt. Any grooves found on the bolt are a result of friction against the upper swivel shaft. If grooves are discovered, both the steering lever and the bolt must be replaced.
19. Obtain Resiweld Sealer. Apply a coating of this sealer to the outer diameter of the lower bushing. Using a suitable driver, tap/press the bushing into place.
20. To install the small upper swivel shaft bushing, first obtain a Bearing and Seal Driver tool (#91-43578). Next, place the bushing on the tool, and then use a hammer and tap the bushing into place. Install the larger bushing in a similar manner.
21. Pack the single oil seal lip with Multi-Purpose lubricant and place the seal on the driver with the lip facing the smaller diameter of the tool. Install the oil seal into the gimbal ring until the seal seats against the larger bushing.
22. Check the condition of the synthane washers on the gimbal ring. If they are worn, or damaged, remove and discard the washers. Clean the washer mounting surface free of all adhesives, dirt, grease and oil. Peel the backing from a new washer and place it on the mounting surface, and then press the washer down firmly. The adhesive backing will hold the washer in place.
23. Apply a coating of Resiweld to the outer diameter of both hinge pin bushings. Using a suitable driver, tap the bushings into place on the port and starboard sides of the gimbal ring.
24. Apply a coat of 2-4-C Lubricant onto the surface of the lower swivel shaft. Insert the switch harness, if equipped, through the gimbal housing, and then secure it in place with the screws. Place the gimbal ring in position in the gimbal housing. Place a washer between the gimbal ring and the housing. Secure the ring in place with the lower swivel pin and cotter pin. Install the ground wire.
25. Position the washer with the larger inner diameter, the steering lever, the smaller washer and the nut into position over the upper swivel shaft bore.
26. Insert the upper swivel shaft through the bore from the bottom so it feeds through the washers and lever. Thread on the locknut a few turns, but do not yet tighten it.
27. Check to be sure the gimbal ring rests squarely within the gimbal housing.
28. Tighten the bolt on the steering lever to approximately 60 ft. lbs. (81 Nm) if the engine and transom plate were removed. Use a punch through one of the side access holes to tighten the nut if the engine and transom plate are still in position. Either way, you want to tighten the nut until a clearance of 0.002-0.010 in. (0.05-0.25mm) exists between the lower swivel pin washer and the gimbal housing mounting surface.
29. Use a plastic mallet and strike the top of the gimbal ring flanges a couple times each to seat both pins. Recheck the clearance and adjust the large nut (using the punch and hammer method or a wrench), as necessary to maintain the required clearance.
30. Tighten the two gimbal ring U-bolt nuts to 53 ft. lbs. (72 Nm).
31. Tighten the steering lever clamp bolt to 50 ft. lbs. (68 Nm). Install the steering lever ground wire so that the loop is on the port side of the lever.
32. If the engine and transom plate were not removed, apply a coating of Perfect Seal to the threads or the sealing surface of the 2 upper gimbal

DRIVE SYSTEMS - BRAVO

housing access plugs and thread them into the housing securely until they are flush with the surface of the housing. They take a 5/8 in. Allen socket/wrench.

33. Install the trim cylinders by first attaching them in place with the forward anchor pin. Next, install the two hydraulic hoses to the connector. Finally, install the connector to the gimbal housing.

34. Install the bell housing and stern drive.

ACCESS PLUG INSTALLATION

◆ See Figure 168

1. Remove the stern drive and the bell housing.
2. Pump a liberal amount of grease into the fitting at the top of the gimbal housing.
3. You'll need an Access Plug Kit (#22-88847A-1). Use the template provided and position it on the top of the gimbal housing using the dimple in the front of the housing - it will be under the decal.
4. Use a center punch and the template to mark the location of the holes on each side of the gimbal housing and then drill a 1/4 in. (6mm) hole through each side of the housing.
5. Using a 1 1/8 in. hole saw with a 1/4 in. (6mm) pilot rod, drill a hole in each side using the previous holes as a guide. Make sure the drill is perpendicular to the housing and do not rush this.
6. Use a piece of tape and mark a 1 in. #180 pipe tap (hardware store) about 1 1/8 in. (28mm) from the end. Coat it with grease and then thread it into the holes until you reach the line marked with the tape.
7. Thread in the two plugs supplied with the kit.

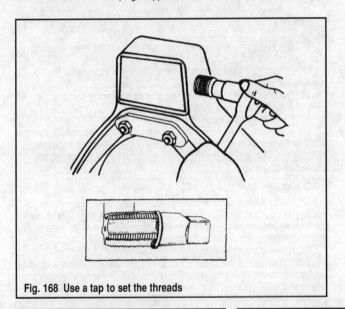

Fig. 168 Use a tap to set the threads

Gimbal Housing/Transom Plate

REMOVAL & INSTALLATION

2001-04 Models (Serial #OM99999 And Below)

◆ See Figures 154, 155 and 169 thru 171

1. Remove the stern drive, bell housing, gimbal ring and bushing.
2. Remove the engine.

If equipped with power steering:

3. Disconnect the rear clevis from the steering lever and then disconnect the cable at the front clevis.
4. Place a wrench on the cable guide flats and move it so they are vertical. Loosen the coupler nut and then remove the steering cable.
5. Bend the tabs on the washers down and away from the pivot bolts and then remove the bolts and power steering unit.

If equipped with manual steering:

6. Disconnect the steering cable at the lever. Loosen the pivot bolts and remove the steering swivel ring.

On all models now:

7. Tag and disconnect the trim switch wires at the engine and pull them through the assembly.
8. If equipped with a Gear Lube Monitor, disconnect the hose at the transom plate.
9. Disconnect the trim pump hydraulic lines and plug them securely. Carefully move them out of the way.
10. Remove the exhaust pipe at the plate. If your's is a thru-transom model, remove the exhaust block-off plate.
11. Disconnect the speedometer tube.
12. Tag and disconnect the MerCathode wires.
13. Disconnect the ground wire at the steering lever.
14. Disconnect the seawater intake hose at the pick-up outlet and then remove the outlet.
15. Loosen the 8 locknuts (some early models may use 2 bolts on the top) that secure the inner plate to the transom/gimbal housing. Support the gimbal housing and then remove the retainers and lift out the transom plate and gimbal housing.

To install:

16. Rotate the inside race of the gimbal housing bearing and check for roughness or any signs of binding. Push and pull on the inner race to check the bearing for side play. The bearing should move only a slight amount. Any excessive movement is just cause to replace the bearing.
17. Check the upper swivel shaft roller bearing and bushing, by inspecting the area of the swivel shaft, where the bearing and bushing ride. Any pits, grooves, or uneven wear is cause to replace the bearing, bushing, or swivel shaft.
18. Check the transom plate for pitting or worn housing. Clean any old sealer from the back side of the plate. Clean the exhaust outlet. Use a tap, and "chase" the threads of the screw holes where the exhaust elbow is mounted.

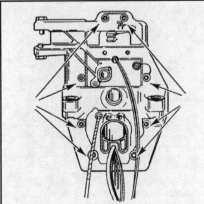

Fig. 169 Remove these nuts and bolts to pull off the plate

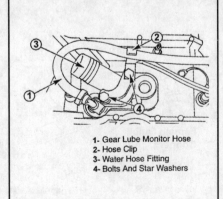

1- Gear Lube Monitor Hose
2- Hose Clip
3- Water Hose Fitting
4- Bolts And Star Washers

Fig. 170 Make sure you position the gear lube hose correctly

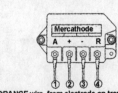

1- ORANGE wire- from electrode on transom assy
2- RED/PURPLE wire- connect other end to Pos (+) battery
3- BLACK wire- from engine harness
4- BROWN Wire- from electrode on transom assy

Fig. 171 Make sure you make the proper connection on the MerCathode controller

11-44 DRIVE SYSTEMS - BRAVO

19. Secure the water tube to the water hose and tighten the hose clamp securely. Install the water hose and tube through the gimbal housing. Route the trim limit switch and trim position sender leads down over the installed tube.

20. Secure both sets of leads with a Sta-strap 5 in. (12cm) away from their retaining bracket. Tighten the hose clamp over the tube. Install the grommet and water tube cover and connect the engine water inlet hose to the water tube.

21. Insert the shift cable, the trim limit switch lead, and the hydraulic hose through the opening in the transom, and then move the gimbal housing into position. The hydraulic hose must be positioned on the starboard side.

■ Do not hold onto the trim limit switch wires to support the gimbal housing while installing the unit. The trim limit switch or the switch leads will be damaged, if the gimbal housing is supported by the switch leads while fastening the inner transom plate.

22. Hold the gimbal housing in position, and at the same time insert the shift cable and hydraulic hose through the large opening; insert the trim limit switch leads through the center hole of the three small holes in the inner transom plate; and then set the plate in position.

23. Insert a flat washer and locknut on each bolt and then tighten the nuts to 23 ft. lbs. (31 Nm), starting with the center nuts and working your way outward.

24. Reconnect the ground wire to the steering lever, making sure that the 'U' that it creates is on the port side of the lever.

25. Position the water inlet with a new gasket and tighten the bolts to 45 inch lbs. (5 Nm). Make sure that the gear lube monitor hose is positioned as shown in the illustration.

26. Connect any remaining water hoses and tighten their clamps securely.

27. Connect the speedometer tube and MerCathode wires.

28. Check to be sure the mating surfaces on the exhaust separator and the gimbal housing are clean. Place new O-ring seals into the grooves in the gimbal housing. Install the exhaust separator, with the exhaust elbows and exhaust bellows attached, to the gimbal housing. Thread the bolts, with lockwashers, into the gimbal housing. Tighten the screws to 23 ft. lbs. (31 Nm) and at the same time check to be sure the O-ring seals remain properly seated in the grooves. The exhaust block-off plate (if equipped) would be tightened to the same torque.

■ Move quickly while working with the hydraulic hoses to prevent spilling any more hydraulic fluid than is necessary. Follow the sequence and take care to route the hoses properly, or the hoses may be damaged and the system become inoperative.

29. Connect the hydraulic lines and tighten the connections to 125 inch lbs. (14 Nm).

30. Install the steering gear and tighten the pivot bolts to 25 ft. lbs. (34 Nm). Bend the washer tabs up and against the bolt head. Tighten the coupler nut to 35 ft. lbs. (48 Nm). We recommend that you refer to the appropriate procedures in the Steering section when installing the control unit...we know your memory is good, but suggest you check the procedures anyway just to make sure the hydraulic lines are routed correctly and the cable is set.

31. Install the engine.
32. Install the bell housing and stern drive.
33. Bleed the hydraulic system. Adjust the shift cable.

2004 And Later Models (Serial #OW100000 And Above)

◆ See Figures 154, 155 and 172 thru 178

1. Disconnect the seawater inlet hose. Insert a small prybar under the retainer clip for the inlet hose assembly, pop it to the open position and remove the assembly. Remove the extension hose if equipped.
2. Remove the stern drive, bell housing, gimbal ring and bushing.
3. Remove the engine.
4. Remove the 2 bolts and lift off the water inlet fitting. One bolt may contain a J-clip for the gear lube monitor hose - unclip the hose and don't lose the clip.

If equipped with power steering:

5. Disconnect the rear clevis from the steering lever and then disconnect the cable at the front clevis.
6. Place a wrench on the cable guide flats and move it so they are vertical. Loosen the coupler nut and then remove the steering cable.
7. Bend the tabs on the washers down and away from the pivot bolts and then remove the bolts and power steering unit.

If equipped with manual steering:

8. Disconnect the steering cable at the lever. Loosen the pivot bolts and remove the steering swivel ring.

On all models now:

9. Tag and disconnect the trim switch wires at the engine and pull them through the assembly.
10. If equipped with a Gear Lube Monitor, disconnect the hose at the transom plate by pressing the quick release button in while pulling out on the line.
11. Disconnect the trim pump hydraulic lines and plug them securely. Carefully move them out of the way.
12. Remove the exhaust pipe at the plate. If your model uses thru-transom exhaust, remove the exhaust block-off plate.
13. Disconnect the speedometer tube quick-connect.
14. Tag and disconnect the MerCathode wires.
15. Disconnect the ground wire at the steering lever.
16. Loosen the 8 locknuts (some early models may also use 2 bolts) that secure the inner plate to the transom/gimbal housing. Support the gimbal housing and then remove the retainers and lift out the transom plate and gimbal housing.

To install:

17. Rotate the inside race of the gimbal housing bearing and check for roughness or any signs of binding. Push and pull on the inner race to check the bearing for side play. The bearing should move only a slight amount. Any excessive movement is just cause to replace the bearing.
18. Check the upper swivel shaft roller bearing and bushing, by inspecting the area of the swivel shaft, where the bearing and bushing ride. Any pits, grooves, or uneven wear is cause to replace the bearing, bushing, or swivel shaft.

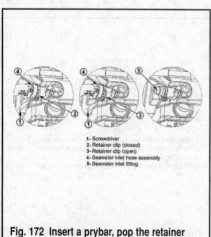

Fig. 172 Insert a prybar, pop the retainer open and remove the inlet hose assembly

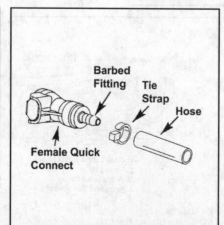

Fig. 173 These models use a quick connect fitting on the speedometer...

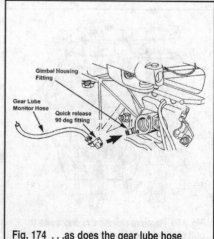

Fig. 174 ...as does the gear lube hose

DRIVE SYSTEMS - BRAVO   11-45

19. Check the transom plate for pitting or worn housing. Clean any old sealer from the back side of the plate. Clean the exhaust outlet. Use a tap, and "chase" the threads of the screw holes where the exhaust elbow is mounted.

20. Position the gimbal housing and transom assembly and feed all wires, hoses and cables through the appropriate holes in the transom and plate.

21. Move the housing and plate into their final position and install the mounting nuts. Tighten them to 25 ft. lbs. (34 Nm) in the sequence shown in the accompanying illustration.

22. Reconnect the ground wire to the steering lever, making sure that the 'U' that it creates is on the port side of the lever.

23. Position the water inlet with a new gasket and tighten the bolts to 45 inch lbs. (5 Nm).

24. Connect the gear lube hose so that the quick release button is positioned away from the water inlet fitting. Make sure that the gear lube monitor hose is positioned as shown in the illustration.

25. Connect the speedometer tube and MerCathode wires.

26. Connect the seawater hose to the hose assembly and then pop the retainer in so it's in the closed position. Position the hose assembly near the fitting so that the decal on the hose, and the retainer clip, are facing the engine. Line up the slots on the quick connect assembly with the tabs on the inlet fitting and press the assembly onto the fitting until the retainer clip snaps into place.

✴✴ WARNING

Grab the seawater hose close to the quick connect assembly and pull outward on it with approximately 25 ft. lbs. of force (this is a pretty good pull).

The connection should not release, but if it does, you'll need to repeat the installation process again. If it still comes apart, replace the quick connect assembly.

■ If your application is using the extension hose assembly, make sure the center of the retainer clip on the quick connect is facing AWAY from the engine. You will also need to perform the proceeding pull test at this connection as well; just make sure that you secure the side of the hose that you are pulling against.

27. Connect any remaining water hoses and tighten their clamps securely.

28. Check to be sure the mating surfaces on the exhaust separator and the gimbal housing are clean. Place new O-ring seals into the grooves in the gimbal housing. Install the exhaust separator, with the exhaust elbows and exhaust bellows attached, to the gimbal housing. Thread the bolts, with lockwashers, into the gimbal housing. Tighten the screws to 23 ft. lbs. (31 Nm) and at the same time check to be sure the O-ring seals remain properly seated in the grooves. If your engine has thru-transom exhaust, the block-off plate should be tightened to the same torque.

■ Move quickly while working with the hydraulic hoses to prevent spilling any more hydraulic fluid than is necessary. Follow the sequence and take care to route the hoses properly, or the hoses may be damaged and the system become inoperative.

29. Connect the hydraulic lines and tighten the connections to 125 inch lbs. (14 Nm).

30. Install the steering gear and tighten the pivot bolts to 25 ft. lbs. (34 Nm). Bend the washer tabs up and against the bolt head. Tighten the coupler nut to 35 ft. lbs. (48 Nm). We recommend that you refer to the appropriate procedures in the Steering section when installing the control unit...we know your memory is good, but suggest you check the procedures anyway just to make sure the hydraulic lines are routed correctly and the cable is set.

31. Install the engine.
32. Install the bell housing and stern drive.
33. Bleed the hydraulic system. Adjust the shift cable.

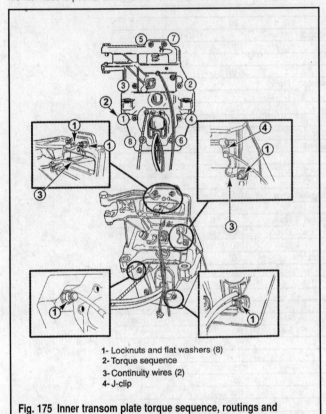

1- Locknuts and flat washers (8)
2- Torque sequence
3- Continuity wires (2)
4- J-clip

Fig. 175 Inner transom plate torque sequence, routings and connections

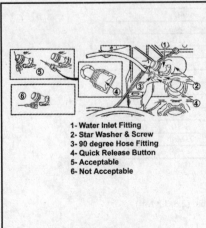

1- Water Inlet Fitting
2- Star Washer & Screw
3- 90 degree Hose Fitting
4- Quick Release Button
5- Acceptable
6- Not Acceptable

Fig. 176 Gear lube hose quick connect positioning is very important

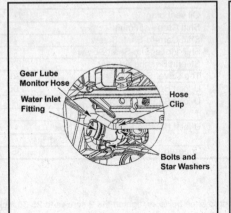

Fig. 177 Make sure you position the gear lube hose connector correctly

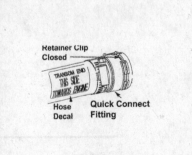

Fig. 178 The seawater hose and quick connect assembly must be positioned correctly!

SPECIFICATIONS

TORQUE SPECIFICATIONS - BRAVO

Component			ft. lbs.	inch lbs.	Nm
Exhaust Bellows			-	35	4
Exhaust Back-Off Plate-to-Gimbal Housing			23 ④	-	31 ④
Exhaust Pipe-to-Gimbal Housing			23	-	31
Lower Unit - Bravo I	Anode		23	-	32
	Bearing Carrier Retainer		①	-	①
	Driveshaft Bearing Preload		-	3-5	0.3-0.8
	Lower-to-Upper Unit		35	-	48
	Oil Fill/Drain Plug		-	40	5
	Pinion Gear Bolt		45	-	61
	Prop Shaft Bearing Preload	New	-	8-12	0.9-1.4
		Used	-	5-8	0.6-0.9
	Trim Tab		23	-	32
Lower Unit - Bravo II	Anode		20	-	27
	Driveshaft Bearing Preload		-	3-5	0.3-0.6
	Lower-to-Upper Unit		35	-	48
	Oil Fill/Drain Plug		-	40	4.5
	Pinion Gear Nut		100	-	136
	Prop Shaft Bearing Preload	New	-	8-12	0.9-1.4
		Used	-	5-8	0.6-0.9
	Trim tab		20	-	27
Lower Unit - Bravo III	Anode		20	-	27
	Bearing Carrier Retainer		150 ②	-	203 ②
	Driveshaft Bearing Preload		-	3-5	0.3-0.6
	Lower-to-Upper Unit		35	-	48
	Oil Fill/Drain Plug		-	40	5
	Oil Vent Plug		-	40	4.5
	Outer Prop Shaft Bearing Retainer		200 ②	-	271 ②
	Pinion Gear Bolt		45	-	61
	Prop Shaft Bearing Preload	New	-	8-18	0.9-2
		Used	-	5-15	0.6-1.7
Gimbal Ring	Hinge Pins		150 ③	-	203 ③
	Steering Lever Clamp Bolt		50	-	68
	Trim Wire Retainer		-	95	11
	U-Bolt Nuts		53	-	72
Propeller Nut	Bravo I		55	-	75
	Bravo II		60	-	82
	Bravo III	Front	100	-	136
		Rear	60	-	81
Shift Cable	Bellows Clamp		-	35	4
	Core Wire		-	20	2.3
Steering Unit	Cable Coupler		35	-	48
	Pivot Bolts		25	-	34
Stern Drive-to-Bell Housing			50	-	68
Transom Assembly			23 ④	-	31 ④
U-Joint Bellows			-	35	4
Upper Unit	Anode		23	-	32
	Back Cover Screws		20	-	27
	Oil vent plug		-	40	4
	Shift Cam Assembly		-	80	9
	Shift Cam-to-Shaft Screw		-	110	13
	Shift Linkage-to-Shaft Screw		-	110	13
	Steerin Bearing Adaptor		175	-	237
	Top Cover Screws		20	-	27
	U-joint Bearing Retainer Nut		200	-	271
	U-joint Bearing Preload ⑤	New	-	8	0.9
		Used	-	5	0.6
	U-joint Bearing Preload ⑥	New	-	6-10	0.7-1.0
		Used	-	7-Mar	0.3-0.8
	Upper-to-Lower Unit		35	-	48

① See text
② Left hand thread
③ On models with a high performance transom, tighten the 2 screws to 25-30 inch lbs. (2.8-3.3 Nm)
④ 2004 and later models w/serial #OW100000 and above: 25 ft. lbs. (34 Nm)
⑤ 2001-04 models w/serial # OM99999 and below
⑥ 2004 and later models w/serial # OW100000 and above

STERN DRIVE DRIVE MODELS - BRAVO

Year	Model	Gear Ratio	U-Joint Shaft Marking	Number Of Teeth Per Gear Driveshaft Housing	
				Drive	Driven
2001	Bravo One	1.65:1	C	23	30
		1.50:1	F	27	32
		1.36:1	H	27	29
	Bravo Two	2.20:1	C	23	30
		2.00:1	F	27	32
		1.81:1	H	27	29
	Bravo Three	2.43:1	N	23	30
		2.20:1	K	23	30
		2.00:1	B	27	32
		1.81:1	G	27	29
		1.65:1	C	27	32
		1.50:1	F	27	32
		1.36:1	P	27	29
	Bravo XR	1.50:1	R	16	19
	Bravo XZ	1.50:1	Z	27	32
		1.36:1	T	27	29
2002	Bravo One	1.65:1	C	23	30
		1.50:1	F	27	32
		1.36:1	H	27	29
	Bravo Two	2.20:1	C	23	30
		2.00:1	F	27	32
		1.81:1	H	27	29
	Bravo Three	2.43:1	N	23	30
		2.20:1	K	23	30
		2.00:1	B	27	32
		1.81:1	G	27	29
		1.65:1	C	27	32
		1.50:1	F	27	32
		1.36:1	P	27	29
	Bravo XR	1.50:1	R	16	19
	Bravo XZ	1.50:1	Z	27	32
		1.36:1	T	27	29
2003	Bravo One	1.65:1	C	23	30
		1.50:1	F	27	32
		1.36:1	H	27	29
	Bravo Two	2.20:1	C	23	30
		2.00:1	F	27	32
		1.81:1	H	27	29
	Bravo Three	2.43:1	N	23	30
		2.20:1	K	23	30
		2.00:1	B	27	32
		1.81:1	G	27	29
		1.65:1	C	27	32
		1.50:1	F	27	32
		1.36:1	P	27	29
	Bravo XR	1.50:1	R	16	19
	Bravo XZ	1.50:1	Z	27	32
		1.36:1	T	27	29
2004	Bravo One, One X	1.65:1	C	23	30
		1.50:1	F	27	32
		1.36:1	H	27	29
	Bravo Two, Two X	2.20:1	C	23	30
		2.00:1	F	27	32
		1.81:1	H	27	29
		1.65:1	T	27	32

STERN DRIVE DRIVE MODELS - BRAVO

Year	Model	Gear Ratio	U-Joint Shaft Marking	Number Of Teeth Per Gear Driveshaft Housing	
				Drive	Driven
2004 (cont'd)		1.50:1	F	27	29
	Bravo Three, Three X	2.43:1	N	23	30
		2.20:1	K	23	30
		2.00:1	B	27	32
		1.81:1	G	27	29
		1.65:1	C	27	32
		1.50:1	F	27	32
		1.36:1	P	27	29
	Bravo XR	1.50:1	R	16	19
		1.35:1	R	16	19
		1.26:1	R	16	19
	Bravo XZ	1.50:1	Z	27	32
		1.36:1	T	27	29
2005	Bravo One, One X	1.65:1	C	23	30
		1.50:1	F	27	32
		1.36:1	H	27	29
	Bravo Two, Two X	2.20:1	C	23	30
		2.00:1	F	27	32
		1.81:1	H	27	29
		1.65:1	T	27	32
		1.50:1	F	27	29
	Bravo Three, Three X	2.43:1	N	23	30
		2.20:1	K	23	30
		2.00:1	B	27	32
		1.81:1	G	27	29
		1.65:1	C	27	32
		1.50:1	F	27	32
		1.36:1	P	27	29
	Bravo XR	1.50:1	R	16	19
		1.35:1	R	16	19
		1.26:1	R	16	19
	Bravo XZ	1.50:1	Z	27	32
		1.36:1	T	27	29
2006-08	Bravo One, One X	1.65:1	C	23	30
		1.50:1	F	27	32
		1.36:1	H	27	29
	Bravo Two, Two X	2.20:1	C	23	30
		2.00:1	F	27	32
		1.81:1	H	27	29
		1.65:1	T	27	32
		1.50:1	F	27	29
	Bravo Three, Three X	2.43:1	N	23	30
		2.20:1	K	23	30
		2.00:1	B	27	32
		1.81:1	G	27	29
		1.65:1	C	27	32
		1.50:1	F	27	32
		1.36:1	P	27	29
	Bravo XR	1.50:1	R	16	19
		1.35:1	R	16	19
		1.26:1	R	16	19
	Bravo XZ	1.50:1	Z	27	32
		1.36:1	T	27	29
		2.00:1	F	27	32
		1.81:1	H	27	29
		1.65:1	T	27	32
		1.50:1	F	27	29
	Bravo Three, Three X	2.43:1	N	23	30
		2.20:1	K	23	30
		2.00:1	B	27	32
		1.81:1	G	27	29
		1.65:1	C	27	32
		1.50:1	F	27	32
		1.36:1	P	27	29
	Bravo XR	1.50:1	R	16	19
		1.35:1	R	16	19
		1.26:1	R	16	19
	Bravo XZ	1.50:1	Z	27	32
		1.36:1	T	27	29

GENERAL INFORMATION	
VELVET	12-2
ZF/HURTH	12-6
IDENTIFICATION	
VELVET	12-2
ZF/HURTH	12-6
PRESSURE TEST	
VELVET	12-6
ZF/HURTH	12-9
SHIFT CABLE - VELVET	
ADJUSTMENT	12-2
EXCEPT 5000 SERIES	12-2
5000 SERIES	12-3
SHIFT CABLE - ZF/HURTH	
ADJUSTMENT	12-7
GENERAL INFORMATION	12-6
SPECIFICATIONS	**12-10**
PRESSURE - VELVET	12-12
TORQUE - INBOARDS	12-11
IDENTIFICATION	12-10
TRANSMISSION - VELVET	
PRESSURE TEST	12-6
REMOVAL & INSTALLATION	12-5
EXCEPT 5000 SERIES	12-5
5000 SERIES	12-5
TRANSMISSION - ZF/HURTH	
PRESSURE TEST	12-9
REMOVAL & INSTALLATION	12-9
TRANSMISSION FLUID	
VELVET	12-2
ZF/HURTH	12-6
VELVET TRANSMISSIONS	**12-2**
GENERAL INFORMATION	12-2
IDENTIFICATION	12-2
SHIFT CABLE	12-2
TRANSMISSION	12-5
TRANSMISSION FLUID	12-2
ZF/HURTH TRANSMISSIONS	**12-6**
GENERAL INFORMATION	12-6
IDENTIFICATION	12-6
SHIFT CABLE	12-6
TRANSMISSION	12-9
TRANSMISSION FLUID	12-6

12

DRIVE SYSTEMS - TRANSMISSIONS

VELVET TRANSMISSIONS	12-2
ZF/HURTH TRANSMISSIONS	12-6
SPECIFICATIONS	12-10

12-2 DRIVE SYSTEMS - TRANSMISSIONS

VELVET TRANSMISSIONS

General Information

This section contains service procedures for all Velvet in-line and V-drive transmissions. Disassembly and assembly procedures are not available, or recommended; nor does Mercury stock or sell replacement parts. Internal transmission problems should be referred to a local Velvet repair facility.

Identification

■ Transmission identification details can be found in the Engine & Drive Maintenance section.

Transmission Fluid

■ Transmission fluid level checking procedures and also drain/fill procedures are detailed in the Engine & Drive Maintenance section. Please refer there for all appropriate procedures relating to fluids.

Shift Cable

ADJUSTMENT

Except Velvet 5000 Series

◆ See Figures 1 thru 9

■ The following procedures are for Quicksilver cables only. If using cables from another manufacturer, please refer to the instructions provided with the cables.

■ We suspect that there may be late model carbureted models that should utilize the adjustment procedures detailed under the MPI section. Please compare your shift lever to the illustration shown here and in the MPI section and use the procedure that it matches.

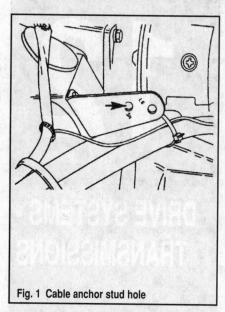

Fig. 1 Cable anchor stud hole

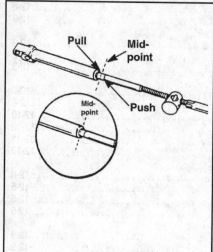

Fig. 2 When checking end-play (backlash), adjust the guide to the mid-point

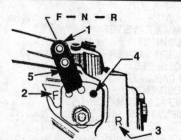

1- Transmission shift lever
2- Shift lever must be over this letter when propelling boat forward
3- Shift lever must over this letter when propelling boat in reverse
4- Proppet ball must be centered in detent hole for each F-N-R position (forward gear shown)
5- Install shift lever stud in this hole, if necessary, to center poppet ball in forward or reverse detent holes

Fig. 3 Shift lever positioning - carbureted, TBI and 2001 350 Mag MPI

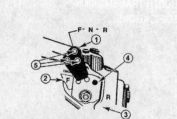

1- Transmission Shift Lever
2- Shift Lever Must Be Over This Letter When Propelling Boat FORWARD
3- Shift Lever Must Be Over This Letter When Propelling Boat In REVERSE
4- Poppet Ball Must Be Centered In Detent Hole For Each F-N-R Position (Forward Gear Shown)
5- Shift Lever Stud Holes

Fig. 4 Shift lever positioning - MPI (exc. 2001 350 Mag MPI)

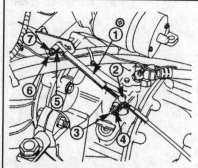

1- Cable end guide
2- Cable barrel
3- Cable barrel stud
4- Elastic stop nut and washer
5- Spacer
6- Cable end guide stud
7- Elastic stop nut and washer

Fig. 5 Shift cable set-up - single cable, rear entry

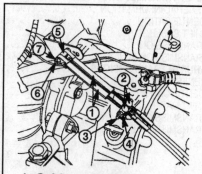

1- Cable end guide
2- Cable barrel
3- Cable barrel stud
4- Elastic stop nut and washer
5- Spacer
6- Cable end guide stud
7- Elastic stop nut and washer

Fig. 6 Shift cable set-up - dual cable, rear entry

DRIVE SYSTEMS - TRANSMISSIONS

✱✱ WARNING
Removing the shift lever poppet ball or spring, or repositioning the shift lever will void the manufacturer's warranty.

1. Confirm that the shift cable stud is in the correct stud hole, tighten the elastic stop nut securely and the place the remote control lever in the Neutral position.
2. Disconnect the cable from the attaching studs on the shift plate.
3. Push in (toward the rest of the cable) on the cable end guide just enough to remove any play and then mark the inner end of the guide on the tube.
4. Now pull out on the cable end guide just enough to remove any play and mark the inner end of the guide once again.
5. Measure and then mark the halfway point between the first and second marks on the tube and move the end guide so its inner edge is now lined up with the mid-point mark on the tube.
6. Loosen the screw and adjust the cable barrel so that the mounting holes in it and the end guide line up with the mounting studs. Install the cable on the studs, but do not secure it.
7. Move the remote control shift lever to the full Forward position and check that the shift lever on the transmission is exactly as depicted in the illustration. The lever must be over the **F** stamped in the housing and the poppet must be in the last detent hole.
8. Move the remote control shift lever to the full Reverse position and check that the shift lever on the transmission is exactly as depicted in the illustration. The lever must be over the **R** stamped in the housing and the poppet must be in the first detent hole.
9. If the lever does not position correctly in either gear, move the shift lever stud from the top hole in the lever to the bottom hole and then recheck positioning. If it still won't position properly you'll need to replace the remote control. If the lever positions correctly in one position but not the other, recheck the adjustment made earlier in this procedure.
10. Install the nut and washer on the cable end guide stud, tighten it until its snug and then back it off 1/2 turn.
11. Install the nut and washer to the barrel stud and tighten it until it just bottoms out - do not over-tighten it.

Velvet 5000 Series

 MODERATE

◆ See Figures 2, 9 thru 19

Propeller rotation is determined by the cable installation on the transmission remote control. For standard left hand propeller rotation, the cable hook-up must cause the shift lever to move forward when the remote control lever is moved to the Forward position. For right hand propeller rotation, the cable hook-up must cause the shift lever to move backwards when the remote control lever is moved to the Forward position.

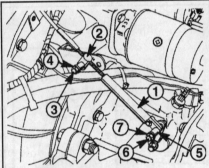

1- Cable end guide
2- Cable barrel
3- Cable barrel stud
4- Elastic stop nut and washer
5- Spacer
6- Cable end guide stud
7- Eleastic stop nut and washer

Fig. 7 Shift cable set-up - single cable, forward entry

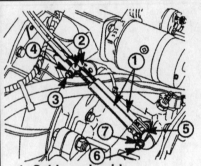

1- Cable end guide
2- Cable barrel
3- Cable barrel stud
4- Elastic stop nut and washer
5- Spacer
6- Cable end guide stud
7- Elastic stop nut and washer

Fig. 8 Shift cable set-up - dual cable, forward entry

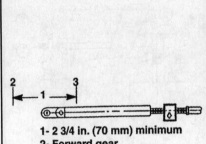

1- 2 3/4 in. (70 mm) minimum
2- Forward gear
3- Reverse gear

Fig. 9 Total shift cable travel at the transmission must be at least 2 3/4 in. (70mm)

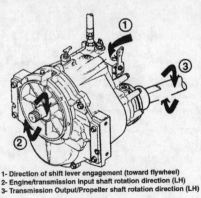

1- Direction of shift lever engagement (toward flywheel)
2- Engine/transmission input shaft rotation direction (LH)
3- Transmission Output/Propeller shaft rotation direction (LH)

Fig. 10 Shift lever engagement (LH rotation)

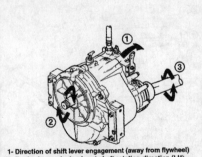

1- Direction of shift lever engagement (away from flywheel)
2- Engine/transmission input shaft rotation direction (LH)
3- Transmission Output/Propeller shaft rotation direction (RH)

Fig. 11 Shift lever engagement (LH rotation)

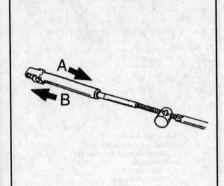

Fig. 12 Cable direction is different for left hand and right hand propeller rotation

12-4 DRIVE SYSTEMS - TRANSMISSIONS

※※ CAUTION

All 5000 series transmissions are full power reversing and designed to work in conjunction with standard left hand rotation engines. Never install a 5000 transmission to an engine with right hand rotation.

On left hand propeller rotation units, the shift cable must remove in direction **A** when the remote control handle is moved to the Forward position.

On right hand propeller rotation units, the shift cable must remove in direction **B** when the remote control handle is moved to the Forward position.

The anchor stud must be installed in the forward hole on the shift cable bracket and there must be at least 2 3/4 in. (70mm) of total shift cable travel so that the remote control will be able to fully position the shift lever in the forward and reverse positions.

Confirm that the distance between the two studs (on the shift lever and the shift bracket) is 7-1/8 in. (318mm). If not, loosen the clamping bolt at the bottom of the lever and move it slowly until it meets the dimension.

1. Place the remote control lever and the shift lever (on the transmission) in the Neutral position.
2. Confirm that the anchor stud is installed in the correct hole on the shift lever and bracket.
3. Disconnect the cable from the attaching studs on the shift plate.
4. Push in (toward the rest of the cable) on the cable end just enough to remove any play and then mark the inner end of the guide on the tube.
5. Now pull out on the cable end just enough to remove any play and mark the inner end of the guide once again.
6. Mark the halfway point between the first and second marks on the tube and move the end guide so its inner edge is now lined up with the mid-point mark on the tube.
7. Loosen the screw and adjust the cable barrel so that the mounting holes in it and the end guide line up with the mounting studs. Install the cable on the studs, but do not secure it.
8. Move the remote control shift lever to the full Forward position. Hold the lever in position and then confirm that the poppet ball is in the **2** detent hole.
9. Move the remote control shift lever to the full Reverse and confirm that the poppet ball is in the **3** detent hole.
10. If the lever does not position correctly in one or both gears, recheck the adjustment and travel. If it still won't position properly you'll need to replace the remote control.
11. Install the nut and washer on the cable end guide stud, tighten it until snug and then back it off one (1) full turn on carb, TBI and 2001 MPI models. On all others, back off the nut 1/2 turn.
12. Install the nut and washer to the barrel stud and tighten it until it just bottoms out - do not over-tighten it.
13. Refer to the accompanying illustrations for cable hardware layouts.

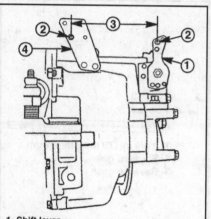

1- Shift lever
2- Anchor stud
3- Dimension between stud (7 1/8 in. (181mm)
4- Shift cable bracket

Fig. 13 The distance between studs on the bracket and the shift lever should be 7 1/8 in. (181mm)

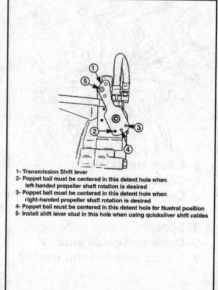

1- Transmission Shift lever
2- Poppet ball must be centered in this detent hole when left-handed propeller shaft rotation is desired
3- Poppet ball must be centered in this detent hole when right-handed propeller shaft rotation is desired
4- Poppet ball must be centered in this detent hole for Nuetral position
5- Install shift lever stud in this hole when using quicksilver shift cables

Fig. 14 Proper shift lever positioning

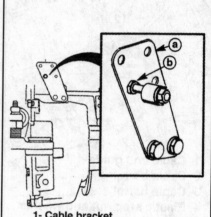

1- Cable bracket
2- Shift cable anchor stud location

Fig. 15 Anchor stud positioning in the bracket

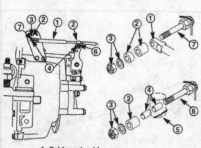

1- Cable end guide
2- Spacer (as required)
3- Elastic stop nut and washer
4- Bushing(s)
5- Cable barrel(s)
6- Cable barrel stud
7- Cable end guide stud

Fig. 16 Shift cable hardware - single cable, rear entry

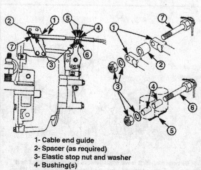

1- Cable end guide
2- Spacer (as required)
3- Elastic stop nut and washer
4- Bushing(s)
5- Cable barrel(s)
6- Cable barrel stud
7- Cable end guide stud

Fig. 17 Shift cable hardware - dual cable, rear entry

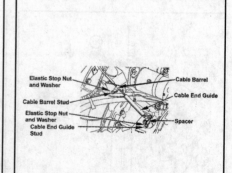

Fig. 18 Shift cable hardware - single cable, front entry

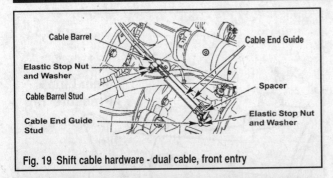

Fig. 19 Shift cable hardware - dual cable, front entry

Transmission

REMOVAL & INSTALLATION

Except Velvet 5000 Series

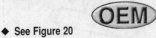

◆ See Figure 20

This procedure details removal of the transmission without removing the engine. Engine removal procedures can be found in the proper section.

1. Disconnect the battery cables.
2. Drain the transmission fluid as detailed previously.
3. Disconnect the fluid cooler hoses and the shift cable (at the transmission).

■ **The cooler should be removed with the transmission.**

4. Tag and disconnect the electrical leads at the neutral safety and oil temperature switches. Move the wires out of the way.
5. Unscrew the flange bolts and disconnect the propeller/driveshaft at the transmission. Make sure you support the shaft(s) with a block of wood.
6. Loosen and remove the four rear mount-to-engine bed bolts.
7. Support the rear of the engine. You can wedge some 2 x 4s underneath the flywheel housing, but we recommend using an engine hoist. It's a bit more trouble, but it's certainly worth it from a safety standpoint!
8. Remove the two center transmission-to-flywheel housing bolts and insert two long non-threaded studs in their place to help support the transmission.
9. Remove the remaining bolts and then pull the transmission straight back and off of the engine, sliding on the studs. Once again, we recommend using a suitable lifting device; although it's not absolutely necessary.

To install:

10. On carb, TBI and 2001 350 Mag MPI models, check the output shaft rolling torque with a torque wrench and socket on the coupling nut. Refer to the chart.

Output Shaft Rolling Torque ①

Ratio In Forward Gear	Torque ft. lbs. (Nm)
1:1	50 (68)
1.52:1	55 (75)
1.88:1	60 (81)
2.57:1	65 (88)
2.91:1	70 (95)

① Transmission not installed and no fluid

Fig. 20 Output shaft rolling torque

11. Coat the input and drive plate shaft splines with Engine Coupler Spline grease.
12. Check that the rear engine mount brackets on the transmission housing are tightened to 45 ft. lbs. (61 Nm).
13. Position the transmission so the input shaft splines are aligned with the drive plate and then slide it forward and into place. Tighten all mounting bolts to 50 ft. lbs. (68 Nm).
14. Lower the engine back into place with the hoist or remove the wooden blocks. Tighten the engine mounts-to-bed bolts securely.
15. Connect the electrical leads to the two switches. Connect the cooler hoses.
16. Connect and adjust the shift cable(s).
17. Check engine alignment as detailed in the Engine Removal & Installation section and then install the propeller shaft to the transmission output shaft. Install the washers, nuts and bolts and tighten to 50 ft. lbs. (68 Nm).
18. Refill the transmission with fluid and connect the battery cables.

Velvet 5000 Series

◆ See Figure 20

This procedure details removal of the transmission without removing the engine. Engine removal procedures can be found in the proper section.

1. Disconnect the battery cables.
2. Drain the transmission fluid as detailed previously.
3. Disconnect the fluid cooler hoses and the shift cable (at the transmission).

■ **The cooler should be removed with the transmission.**

4. Tag and disconnect the electrical leads at the neutral safety and oil temperature switches. Move the wires out of the way.
5. Loosen the trunnion clamp fasteners on the port and starboard engine mounts.
6. Unscrew the coupling nuts and bolts at the output flange and disconnect the propeller/driveshaft at the transmission. Make sure you support the shaft(s) with a block of wood.
7. Loosen and remove the four rear mount-to-engine bed bolts.
8. Support the rear of the engine. You can wedge some 2 X 4s underneath the flywheel housing, but we recommend using an engine hoist. It's a bit more trouble, but it's certainly worth it from a safety standpoint!
9. Attach another hoist to the transmission and remove the rear mount brackets (base and trunnion) from the transmission.
10. Remove the transmission-to-engine mounting bolts and then pull the transmission straight back and off of the engine. Once again, we recommend using a suitable lifting device; although it's not absolutely necessary.

To install:

11. Check the output shaft rolling torque as listed in the accompanying chart.
12. Coat the input and drive plate shaft splines with Engine Coupler Spline grease.
13. Position the transmission so the input shaft splines are aligned with the drive plate and then slide it forward and into place. Tighten all mounting bolts to 55 ft. lbs. (75 Nm).
14. Install the rear engine mount brackets on the transmission housing and tighten to 45 ft. lbs. (61 Nm).
15. Lower the engine back into place with the hoist or remove the wooden blocks. Tighten the engine mounts-to-bed bolts securely.
16. Connect the electrical leads to the two switches.
17. Connect the cooler hoses and tighten the fittings to 25 ft. lbs. (34 Nm).
18. Connect and adjust the shift cable(s).
19. Check engine alignment as detailed in the Engine Removal & Installation section and then install the propeller shaft to the transmission output shaft. Install the washers, nuts and bolts and tighten to 50 ft. lbs. (68 Nm).
20. Refill the transmission with fluid and connect the battery cables.

12-6 DRIVE SYSTEMS - TRANSMISSIONS

PRESSURE TEST

◆ See Figures 21 and 22

■ This test should be conducted with the vessel in the water.

1. Remove the temperature switch from the transmission.
2. Install a pressure gauge into the main line pressure tap.
3. Start the engine and allow it to idle until it reaches normal operating temperature.
4. Confirm that all readings are as shown in the Pressure Specifications chart

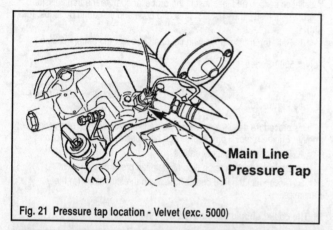

Fig. 21 Pressure tap location - Velvet (exc. 5000)

Fig. 22 Pressure tap location - Velvet 5000

ZF/HURTH TRANSMISSIONS

General Information

This section contains service procedures for all Hurth and ZF down angle and V-drive transmissions. Disassembly and assembly procedures are not available, or recommended. Internal transmission problems should be referred to a local Hurth/ZF repair facility.

Identification

■ Transmission identification details can be found in the Engine & Drive Maintenance section.

Transmission Fluid

■ Transmission fluid level checking procedures and also drain/fill procedures are detailed in the Engine & Drive Maintenance section. Please refer there for all appropriate procedures relating to fluids.

Shift Cable

GENERAL INFORMATION

◆ See Figure 23

Propeller rotation is determined by the cable installation on the transmission remote control. For standard left hand propeller rotation, the cable hook-up must cause the shift lever to move backwards (4) when the remote control lever is moved to the Forward position. For right hand propeller rotation, the cable hook-up must cause the shift lever to move forward (1) when the remote control lever is moved to the Forward position.

※※ CAUTION

All ZF/Hurth transmissions are built to work in conjunction with standard left hand rotation engines. Never install a ZF/Hurth transmission to an engine with right hand rotation.

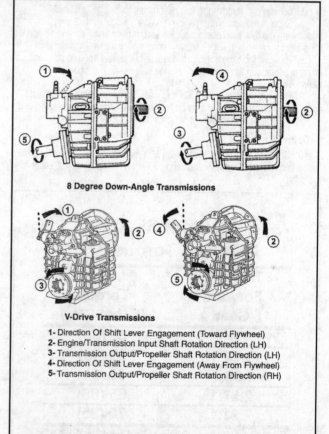

1- Direction Of Shift Lever Engagement (Toward Flywheel)
2- Engine/Transmission Input Shaft Rotation Direction (LH)
3- Transmission Output/Propeller Shaft Rotation Direction (LH)
4- Direction Of Shift Lever Engagement (Away From Flywheel)
5- Transmission Output/Propeller Shaft Rotation Direction (RH)

Fig. 23 Make sure the transmission shift lever moves in the right direction for intended propeller rotation

DRIVE SYSTEMS - TRANSMISSIONS

ADJUSTMENT

◆ See Figures 24 thru 33

1. Place the remote control lever in the Neutral position.
2. Confirm that the anchor stud is installed in the correct hole for your transmission. The holes should be marked **630** and **800** on the shift cable bracket; while on the lever itself, quicksilver cables should always be in the top hole and other manufacturer's cables in the hole that they suggest in their instructions.
3. Confirm that the shift lever on the transmission is positioned approximately 10° aft of vertical and that the distance between the 2 studs is 7-1/8 in. (318mm). If not, loosen the clamping bolt at the bottom of the lever and move it slowly until it meets the dimension.
4. Disconnect the cable from the attaching studs on the shift plate.
5. Push in (toward the rest of the cable) on the cable end just enough to remove any play and then mark the inner end of the guide on the tube.
6. Now pull out on the cable end just enough to remove any play and mark the inner end of the guide once again.
7. Mark the halfway point between the 1st and 2nd marks on the tube and move the end guide so its inner edge is now lined up with the mid-point mark on the tube.
8. Loosen the screw and adjust the cable barrel so that the mounting holes in it and the end guide line up with the mounting studs. Install the cable on the studs, but do not secure it.
9. Move the remote control shift lever to the full Forward position. Hold the lever in position and then have a friend carefully slide the cable off the anchor points on the transmission. See if you can move the lever any farther forward.
10. Reconnect the shift cable at the transmission.
11. Move the remote control shift lever to the full Reverse. Hold the lever in position and then have a friend carefully slide the cable off the anchor points on the transmission. See if you can move the lever any farther backward.
12. If the lever does not position correctly in either gear, move the shift lever stud from the top hole in the lever to the bottom hole and then recheck positioning. If it still won't position properly you'll need to replace the remote control. If the lever positions correctly in one position but not the other, recheck the adjustment made earlier in this procedure.
13. Install the nut and washer on the cable end guide stud, tighten it until snug and then back it off one full turn on carb, TBI and 2001 350 Mag models; or 1/2 turn on all other models.
14. Install the nut and washer to the barrel stud and tighten it until it just bottoms out - do not over-tighten it; and then back the nut off 1/2 turn.
15. Refer to the accompanying illustrations for cable hardware layouts.

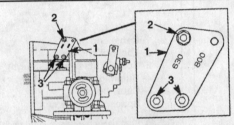

1- Shift cable bracket
2- Anchor stud location
3- Bracket mounting bolts

Fig. 24 The anchor stud must be in the correct hole

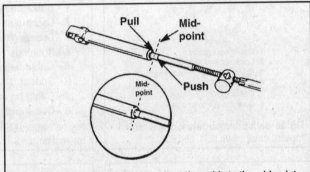

Fig. 26 When checking end-play, adjust the guide to the mid-point

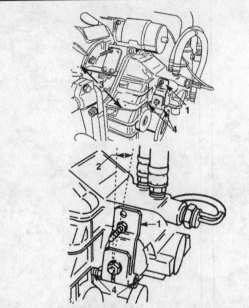

1- Shift lever
2- Lever, in neutral detent, must be approximately 10 degrees Aft
3- Dimension between studs - 7 1/8 in. (318mm)
4- Clamping bolt

Fig. 25 Shift lever positioning

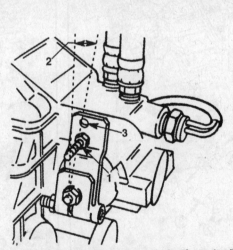

1- Shift lever stud (in bottom hole, if required)
2- Lever, in neutral detent, must be approximately 10 Aft of vertical
3- Shift lever top hole

Fig. 27 Moving the lever stud to the other hole

12-8 DRIVE SYSTEMS - TRANSMISSIONS

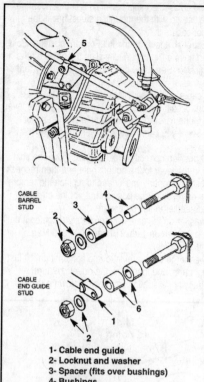

1- Cable end guide
2- Locknut and washer
3- Spacer (fits over bushings)
4- Bushings
5- Cable barrel
6- Spacers (fit over studs)

Fig. 28 Shift cable hardware; single cable, forward entry - 5.0L, 5.7L and 6.2L

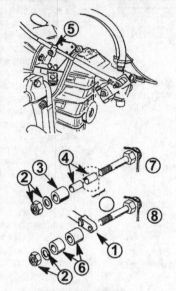

1 - Cable end guide
2 - Locknut and washer
3 - Spacer (fits over bushings)
4 - Bushings
5 - Cable barrel location
6 - Spacer (fits over stud)
7 - Cable barrel stud
8 - Cable end guide stud

Fig. 29 Shift cable hardware; single cable, forward entry - 8.1L

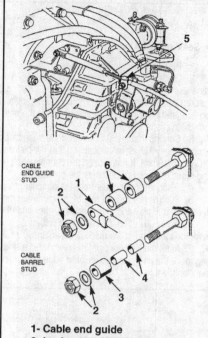

1- Cable end guide
2- Locknut and washer
3- Spacer (fits over bushings)
4- Bushings
5- Cable barrel
6- Spacers (fit over stud)

Fig. 30 Shift cable hardware; single cable, rear entry - 5.0L, 5.7L and 6.2L

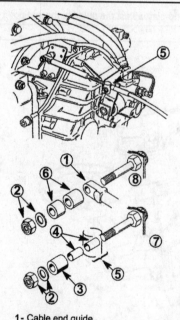

1- Cable end guide
2- Locknut and washer
3- Spacer (fits over bushings)
4- Bushings
5- Cable barrel location
6- Spacer (fits over stud)
7- Cable barrel stud
8- Cable end guide stud

Fig. 31 Shift cable hardware; single cable, rear entry - 8.1L

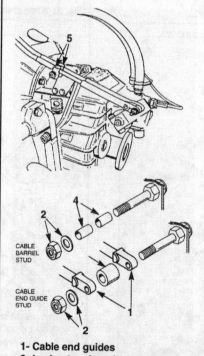

1- Cable end guides
2- Locknut and washer
3- Spacer (fits over stud)
4- Bushings
5- Cable barrels

Fig. 32 Shift cable hardware - dual cable, forward entry

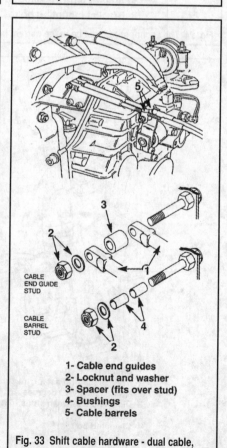

1- Cable end guides
2- Locknut and washer
3- Spacer (fits over stud)
4- Bushings
5- Cable barrels

Fig. 33 Shift cable hardware - dual cable, rear entry

DRIVE SYSTEMS - TRANSMISSIONS

Transmission

REMOVAL & INSTALLATION

◆ See Figure 34

This procedure details removal of the transmission without removing the engine. Engine removal procedures can be found in the proper section.

1. Disconnect the battery cables.
2. Drain the transmission fluid as detailed previously in the Engine & Drive Maintenance section.
3. Disconnect the seawater hoses from the transmission fluid cooler - the cooler should come out with the transmission.
4. Disconnect the shift cable on non-DTS models.
5. Tag and disconnect the electrical leads at the neutral safety and audio warning temperature switches. Move the wires out of the way.
6. On DTS models, tag and disconnect the harness connectors at the shift solenoids and pressure transducers.
7. Loosen the trunnion clamp bolts on the port and starboard rear mounts.
8. Remove the nuts/bolts from the coupling and then pull the propeller shaft coupler from the transmission output flange.
9. Loosen and remove the 4 rear mount-to-engine bed bolts.
10. Support the rear of the engine. You can wedge some 2 x 4s underneath the flywheel housing, but we recommend using an engine hoist. It's a bit more trouble, but it's certainly worth it from a safety standpoint!
11. Support the transmission with a hoist and then remove the rear mount brackets (base and trunnion) from each side.
12. Unscrew the flange bolts and disconnect the propeller/driveshaft at the transmission. Make sure you support the shaft(s) with a block of wood.
13. Remove the 2 mounting bolts and 4 locknuts, and then pull the transmission straight back and off of the engine; and then out of the boat. Once again, we recommend using a suitable lifting device; although it's not absolutely necessary.

To install:

14. Coat the input and drive plate shaft splines with Engine Coupler Spline grease.
15. Position the transmission so the input shaft splines are aligned with the drive plate and then slide it forward and into place. Tighten all mounting bolts and nuts to 45 ft. lbs. (61 Nm).
16. Lower the engine back into place with the hoist or remove the wooden blocks. Tighten the engine mount brackets to 45 ft. lbs. (61 Nm). Now you can tighten the 4 engine mount-to-engine bed bolts securely
17. Connect the electrical leads to all switches, sensors, solenoids and transducers. Coat the connections with Liquid Neoprene if necessary.
18. Connect the seawater hoses to the cooler and tighten the hose clamps securely.
19. Connect and adjust the shift cable(s) on non-DTS engines.
20. Check engine alignment as detailed in the appropriate Engine Removal & Installation section and then install the propeller shaft to the transmission output shaft. Install the washers, nuts and bolts and tighten to 50 ft. lbs. (68 Nm). Tighten the trunnion clamp bolts to the same torque.
21. Refill the transmission with fluid.

PRESSURE TEST

◆ See Figure 35

■ This test should be conducted with the vessel in the water.

1. Tag and disconnect the electrical connector, and then remove the temperature switch from the transmission (**B**). Install a thermocouple into the hole.
2. Install a pressure gauge into the main line pressure tap after removing the plug (**A**).
3. Start the engine and allow it to idle until it reaches normal operating temperature.
4. Confirm that the reading at both ports is 312-341 psi with the engine warm and at 2000 rpm.
5. Remove the gauge and thermocouple, coat the switch and plug with Loctite 567 pip sealant and then install each, tightening securely.
6. Check for leaks the next time you run the boat.

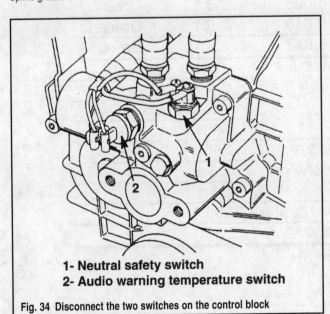

Fig. 34 Disconnect the two switches on the control block
1- Neutral safety switch
2- Audio warning temperature switch

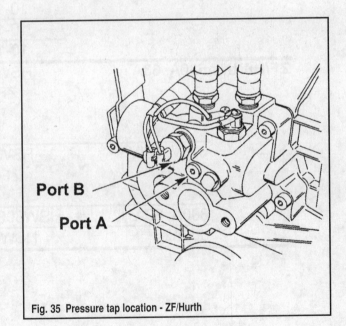

Fig. 35 Pressure tap location - ZF/Hurth

SPECIFICATIONS

Transmission Identification

Make	Model	Model Number	Ratio	ID Plate Color Code
Velvet	5000A Down Angle	20-01-002	1.25:1	Black
		20-01-003	1.5:1	Black
		20-01-004	2.1:1	Black
		20-01-005	2.5:1	Black
		20-01-006	2.8:1	Black
	5000V V-Drive	20-02-003	1.5:1	Blue
		20-02-004	2.0:1	Blue
		20-02-005	2.5:1	Blue
	71C Direct	10-17-004	1.0:1	Red
	71C In-Line	10-17-006	1.5:1	Red
		10-17-012	1.5:1	Red
	72C V-Drive	10-05-011	1.5:1	Green
		10-05-002	2.0:1	Green
		10-05-005 (Gear)	2.5:1	Green
		10-05-004 (Chain)	2.5:1	Green
	72C Direct	10-18-002	1.0:1	Green
	72C In-Line	10-18-004	1.5:1	Green
		10-18-006	2.0:1	Green
		10-18-010	2.5:1	Green
		10-18-012	3.0:1	Green
ZF/Hurth	630A, 63A	HSW630A, ZF63A	1.5:1	-
			2.0:1	-
			2.5:1	-
			2.7:1	-
	630V, 63IV	HSW630V, ZF63IV	1.55:1	-
			2.0:1	-
			2.5:1	-
	800A, 80A	HSW800A, ZF80A	2.85:1	-
	800A2	HSW800A2	2.85:1	-

DRIVE SYSTEMS - TRANSMISSIONS

TORQUE SPECIFICATIONS

Manufacturer	Model	Component	ft. lbs.	inch lbs.	Nm
Velvet	71, 72	Drain Plug (Bushing)	25	-	34
		Fluid Hose-to-Bushing	25	-	34
		Neutral Start Switch	-	111	12.5
		Prop Coupler-to-Output Flange	50	-	68
		Pump Housing-to-Adaptor	19	-	26
		Rear Mounts-to-Transmission	45	-	61
		Shift Lever-to-Valve	-	115	13
		Shift Lever Nut	-	115	13
		Trans-to-Flywheel Housing	50	-	68
	5000	Drain Plug	25	-	34
		Fluid Hose-to-Cooler	25	-	34
		Fluid Hose-to-Housing	25	-	34
		Neutral Start Switch	-	115①	13①
		Prop Coupler-to-Output Flange	50	-	68
		Pump Housing-to-Adaptor	20	-	27
		Rear Mounts-to-Transmission	45	-	61
		Shift Lever-to-Valve	-	115	13
		Trans-to-Flywheel	55②	-	75②
ZF/Hurth	630, 800	Coupler-to-Output Flange	50	-	68
		Fluid Cooler Hose Fittings	25	-	34
		Rear Mount Brackets	45	-	61
		Trans-to-Flywheel Housing	45	-	61

① 8.1L: 120 inch lbs. (14 Nm)
② 8.1L: 45 ft. lbs. (61 Nm) - 5000A; 50 ft. lbs. (68 Nm) - 5000V

Pressure Specifications - Velvet

Transmission	Model	Engine Speed (rpm)	Neutral psi (kPa) Min	Neutral psi (kPa) Max	Forward psi (kPa) Min	Forward psi (kPa) Max	Rear psi (kPa) Min	Rear psi (kPa) Max
Velvet	Direct, In-Line, V-drive	250	-	-	70 (483)	-	70 (483)	-
		600	115 (793)	135 (931)	115 (793)	140 (965)	120 (827)	140 (965)
		2000	-	-	125 (862)	160 (1103)	125 (862)	160 (1103)
		3000	-	-	135 (931)	180 (1241)	-	-
	5000 (early)	250	-	-	70 (483)	-	70 (483)	-
		600	115 (793)	135 (931)	115 (793)	140 (965)	120 (827)	140 (965)
		2000	-	-	125 (862)	160 (1103)	125 (862)	160 (1103)
		3000	-	-	135 (931)	180 (1241)	-	-
	5000 (later)	900	10 (69)	50 (345)	250 (1724)	400 (2758)	250 (1724)	400 (2758)
		2400	15 (103)①	70 (483)	250 (1724)	400 (2758)	250 (1724)	400 (2758)
		4500	-	-	250 (1724)	400 (2758)	250 (1724)	400 (2758)

① 8.1L: 50 (344)

13
TRIM & TILT

AUTO TRIM II SYSTEM 13-24
  CONTROL MODULE 13-24
    DESCRIPTION 13-24
    OPERATION 13-24
  CONTROL MODULE 13-24
    ADJUSTMENT 13-24
  DUAL POWER TRIM SYSTEM 13-22
    DIODE MODULE REPLACEMENT 13-23
    RELAY REPLACEMENT 13-22
    TESTING 13-22
      DIODE MODULE 13-23
      GENERAL 13-22
      PORT TRIM SWITCH 13-23
      RELAY NO. 1 13-22
      RELAY NO. 2 13-23
      STARBOARD TRIM SWITCH 13-23
      TRAILER SWITCH 13-23
    TRIM CONTROL SWITCH REPLACEMENT 13-24
  PUMP ASSEMBLY 13-9
    FLUID LEVEL 13-9
    DISASSEMBLY & ASSEMBLY 13-10
    GENERAL INFORMATION 13-9
    REMOVAL & INSTALLATION 13-9
STANDARD POWER TRIM AND TILT 13-2
  DESCRIPTION & OPERATION 13-2
  DUAL POWER TRIM SYSTEM 13-22
  PUMP ASSEMBLY 13-9
  SERVICE PRECAUTIONS 13-6
  SYSTEM BLEEDING 13-7
  SYSTEM TESTING 13-7
  TRIM CYLINDER 13-17
  TRIM LIMIT SWITCH/TRIM POSITION SENDER 13-8
  TROUBLESHOOTING 13-3
  SYSTEM BLEEDING 13-7
  TRIM CYLINDER 13-17
    DISASSEMBLY & ASSEMBLY 13-19
    GENERAL INFORMATION 13-17
    REMOVAL & INSTALLATION 13-18
  TRIM LIMIT SWITCH/TRIM POSITION SENDER 13-8
    ADJUSTMENT 13-8
      TRIM LIMIT SWITCH 13-8
      TRIM POSITION SENDER 13-8
    GENERAL INFORMATION 13-8
    REMOVAL & INSTALLATION 13-8
  TROUBLESHOOTING 13-3
    ELECTRICAL 13-3
      GENERAL 13-3
      PUMP MOTOR - IN BOAT 13-4
      PUMP MOTOR - OUT OF BOAT 13-4
      SOLENOID - PUMP IN BOAT 13-4
      SOLENOID - PUMP OUT OF BOAT 13-4
      110 AMP FUSE 13-4
      20 AMP FUSE 13-5
    GENERAL INFORMATION 13-3
    HYDRAULIC AND MECHANICAL PROBLEMS 13-6
    SYMPTOMS AND PROBLEMS 13-5
WIRING SCHEMATICS 13-25
  AUTO TRIM II SYSTEM 13-27
  DUAL POWER TRIM 13-26
  HANDLE AND SEPARATE TRAILER SWITCH 13-25
  SMARTCRAFT DTS (14-PIN) - TRIM CIRCUIT 13-29
  SMARTCRAFT DTS (14-PIN) - TRIM HARNESS 13-28
  TRIM UP RELAY CIRCUIT 13-30
  3 BUTTON TRIM/TRAILER PANEL 13-25

STANDARD POWER TRIM AND TILT.. 13-2
AUTO TRIM II SYSTEM 13-24
WIRING SCHEMATICS............... 13-25

13-2 TRIM & TILT

STANDARD POWER TRIM AND TILT

Description & Operation

GENERAL

◆ See Figures 1 and 2

The power trim/tilt system consists of a valve body, hydraulic reservoir, electric motor, up and down solenoids, trim/tilt switch, trim sending unit, trim gauge, trim limit switch, two hydraulic trim cylinders, and the necessary hydraulic lines and electrical harnesses for the system to function efficiently.

The valve body, hydraulic reservoir, electric motor, and up/down solenoids make up a single unit known as the pump assembly, which is mounted inside the boat. Two hydraulic trim cylinders are mounted, one on each side of the stern drive unit. One end of the cylinder is connected to the gimbal ring and the opposite end is attached to the upper gear housing.

The trim sender unit is mounted on the starboard side of the gimbal housing and is connected to the trim gauge on the control panel through the wiring harness. When the stern drive is trimmed **IN** or **OUT** the gauge moves to indicate the stern drives position.

A trim limit switch is mounted on the port side of the gimbal ring and is connected to the **UP** solenoid on the trim/tilt hydraulic pump. This limit switch allows the stern drive to be trimmed out only to a pre-set position. If the stern drive is moved out past the support brackets under high throttle setting could impose high stress loads on the boat transom and possibly loss of steering control.

The relationship of the various units in the trim/tilt system are discussed in the following paragraphs.

HYDRAULIC UP CIRCUIT

When the **UP/UP** buttons on the control panel, or the **Trailer** switch on the remote control throttle, is pressed and held, the trim/tilt electric motor operates. The hydraulic pump is splined to the electric motor and the pump begins to create hydraulic pressure. This pressure created in the UP side of the system moves past the UP pressure relief valve, through lines and hoses to the trim/tilt cylinder.

At this point, the pressure pushes the dual hydraulic pistons outward. Fluid in the DOWN side of the system is forced out of the cylinder by movement of the piston. Fluid movement continues back through connecting hoses and lines to the DOWN pressure relief valve. The valve is pushed open and the fluid returns to the pump reservoir.

The UP pressure relief valve regulates the system pressure between 2200-2600 psi (15173-17932 kPa). A thermal relief valve prevents hydraulic lock-up in the full tilt-up position caused by the thermal expansion of the hydraulic fluid. A check valve in the end of the UP pressure relief valve maintains pressure in the UP side of the system when the pump is not running and the unit is shifted into forward gear.

HYDRAULIC DOWN CIRCUIT

When the **IN** button is pushed on the control panel, or the throttle remote control lever, the electric motor and pump begin to operate. Hydraulic pressure is created in the DOWN circuit of the trim/tilt system. The pressure moves a slide valve in the valve body that unseats the poppet valve in the UP circuit. Hydraulic fluid moves past the DOWN pressure relief valve where the DOWN circuit pressure is regulated between 400-600 psi (2759-4183 kPa). Hydraulic fluid under pressure flows through lines, fittings and hoses to the DOWN side of the trim/tilt cylinder piston.

The hydraulic pressure moves the piston inward, forcing fluid out the UP side of the cylinder through hoses, fittings and check valves back to the reservoir. This fluid returning to the reservoir prevents the pump from operating under an out of fluid condition.

HYDRAULIC TRIM CIRCUITS

When the **OUT** or **IN** trim button on the control panel or the remote control throttle lever is pressed, the hydraulic trim/tilt motor begins to operate.

Trimming the stern drive OUT while at high engine power settings, could impose severe stress - possibly damaging the stern drive, boat transom and could cause loss of directional control. A trim limit switch is wired into the UP circuit to prevent such action. If the trim switch is held to the OUT position, the stern drive will raise or trim OUT. When the upper limit of the limit switch is reached, power to the UP solenoid on the pump is de-energized and the trim/tilt pump stops operating, preventing further trim movement of the stern drive. When the throttle lever is reduced to slightly above the idle position, the tilt **UP** switch is energized, allowing the stern drive to be raised further for beach and shallow water operation.

When the **IN** trim button on the control panel or the remote control throttle lever is pressed, the hydraulic trim/tilt motor begins to run and pressurizes the DOWN circuit. The stern drive moves inward until the gimbal housing contacts the transom stop pin.

A sliding check valve assembly and poppet valves mounted internally in the valve body seat to create hydraulic locks that prevent the stern drive from raising when the shift lever is moved to the **REVERSE** position. These same check valves create a hydraulic lock which holds the stern drive in a trim OUT position.

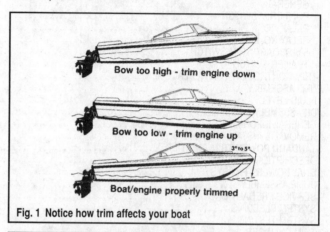

Fig. 1 Notice how trim affects your boat

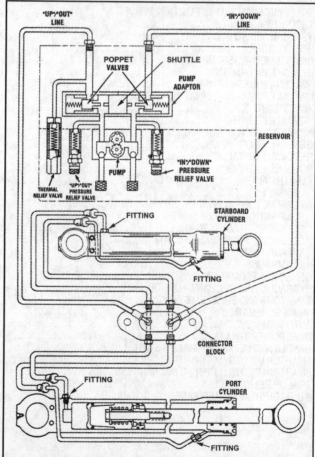

Fig. 2 A simplified functional diagram of the standard Oildyne trim/tilt hydraulic system

TRIM & TILT

If the stern drive should strike an object under water while moving, the rapid return of fluid from the cylinders being extended will cause the check valve in the valve body and the cylinder pistons to open and release the hydraulic lock. This will allow the stern drive to kick-up and away from the underwater object.

TRAILERING OR LAUNCHING

Both the UP buttons on the control panel must be pushed at the same time, or the **TRAILER** switch on the remote control throttle must be pressed, to raise the stern drive to the full UP position for trailering or launching the boat. By pushing the middle **UP/OUT** button, current is passed to the top **UP** switch. The current passing through the **UP** switch while the button is depressed, will by-pass the trim limit switch and permit the UP circuit to raise the stern drive to the full UP position. If the middle **UP/OUT** button is depressed during normal boat operation, a trim limit switch will prevent the stern drive unit from moving out beyond the gimbal ring support guides.

The **TRAILER** switch on the remote throttle lever will raise the stern drive to the full UP position when approaching the beach or for trailering the boat. Never operate the engine above idle speed with the stern drive up past the limit switch settings. Such action would cause severe damage to the stern drive gimbal housing and possibly a loss of steering control.

When the hydraulic cylinders have reached their full extended travel, the pump motor will begin to "labor" if the control buttons are not released. Therefore, to prevent damage to the system, a bi-metal switch in the pump motor will open the circuit to stop the pump motor and prevent the motor from overheating. The switch will automatically close after the motor has cooled allowing the motor to again be operated.

The following is a short description of some additional major components in the trim/tilt system and their function.

RESERVOIR

The reservoir is a translucent 1 1/2 quart reservoir that attaches to the pump valve body. A **MIN** and **MAX** mark cast into the reservoir makes it easy to check the fluid quantity. A large screw-on cap and opening provides for easy servicing of the reservoir. The cap vents the reservoir to atmosphere pressure when the fill neck seal is removed inside the cap.

HYDRAULIC PUMP

The hydraulic fluid pump is a gear-type pump, very similar to an engine oil pump. The pump is attached to the valve body with four fasteners and cannot be disassembled. If the pump fails, it must be replaced as an assembly. The reservoir must be drained and removed to gain access to the pump.

VALVE BODY

The valve body, or adapter, is the central mounting point for the pump, valves, reservoir and electric motor. The hydraulic pump is capable of creating pressure up to 3000 psi. Internal pressure relief valves set at the factory control the output pressure in the UP and DOWN circuits. A system thermal relief valve protects the pump, valves, lines, and cylinder packing from thermal expansion of the hydraulic fluid in the system when operating or sitting in the hot sun. The pump, pressure and thermal relief valves are all set at the factory and adjustments are not possible. If any of these components are found to be defective during troubleshooting or disassembly, they must be replaced with identical items. Each of the valves is color coded to ensure the correct valve is installed in the correct port. A replaceable filter is mounted on the inlet opening to the pump.

There are no provisions for a manual relief valve on the valve body. If the pump or motor fails, manually raising or lowering the stern drive would require the appropriate hydraulic line be loosened on the valve body external fitting.

Access to the valve body requires the reservoir to be drained and removed.

PRESSURE RELIEF VALVE - UP/OUT

The UP pressure relief valve is threaded onto the valve body. This valve regulates the maximum oil pressure in the hydraulic system during the UP cycle to between 2200-2600 psi (15173-17932 kPa). The valve is non-adjustable, and the replacement valve is usually color coded Blue.

PRESSURE RELIEF VALVE - DOWN/IN

The DOWN pressure relief valve is threaded onto the valve body. The valve regulates the hydraulic pressure between 400-600 psi (2759-4138 kPa) in the DOWN circuit. The valve opens under pressure and allows hydraulic fluid to flow from the pump to the aft chamber of the trim/tilt cylinders. The aft chamber of the cylinder is identified as the chamber behind the piston. The valve prevents the stern drive from raising during Reverse operation, but does allow fluid to flow from the aft chamber of the hydraulic cylinders to the reservoir during the UP cycle. The valve is non-adjustable, and the replacement valve is usually color coded Green.

THERMAL RELIEF VALVE

The thermal relief valve prevents hydraulic lock-up in the full tilt-up position caused by the thermal expansion of the fluid. The pressure setting of this valve is non-adjustable, and the replacement valve is usually color coded Gold.

Troubleshooting

GENERAL INFORMATION

The electric pump motor must operate to drive the mechanical hydraulic pump. The hydraulic pump will convert the electrical energy into mechanical energy. The mechanical energy moves the hydraulic fluid to extend and retract the trim cylinders. Use the following troubleshooting procedures to determine if the problem is electrical or mechanical.

Determine If Problem Is Electrical Or Mechanical

1. Verify that the battery is fully charged before continuing the troubleshooting procedure. If the battery is not fully charged, the system could give a false indication of a problem when none actually exists.
2. Verify that the fluid level in the reservoir is up to the **MAX** mark on the side of the reservoir.
3. Depress the **UP** or **DOWN** button on the control panel and listen for the pump motor to operate or the solenoid to click.
4. If the pump motor does not operate, and the solenoids do not "click", the problem is electrical. Refer to electrical troubleshooting.
5. If the pump motor operates, but the stern drive does not move, the problem is most likely a hydraulic or mechanical problem. Refer to mechanical troubleshooting.

ELECTRICAL

General

◆ See Figure 3

1. Obtain a DVOM and set the switches for 12-volt DC readings. Connect the meter Red lead to the 110 amp circuit breaker terminal. Connect the meter Black lead to the pump ground terminal. The meter should indicate 12-volts or battery voltage. If no voltage is indicated, check the battery cable connections to the pump 110 amp circuit breaker.
2. Set the DVOM switches for continuity reading. Connect the Red meter lead to the pump 110 amp circuit breaker terminal. Connect the Black meter lead to the solenoid buss bar. The meter should indicate no resistance and full continuity. If there is no continuity, or a high resistance is indicated, the 110 amp circuit breaker is defective and should be replaced with a new unit.
3. Depress and hold both **UP** switches on the control panel or the **TRAILER** switch on the remote control throttle. Listen for the UP solenoid to "click". If the solenoid fails to "click", connect the ohmmeter Red lead to the Blue/White wire terminal and the Black meter lead to the Black wire terminals on the UP solenoid. Again depress and hold both **UP** switches on the control panel or the **TRAILER** switch. If 12-volts is not indicated, the switch is defective or the 20 amp fuse is blown. Replace the blown fuse or replace the defective switch. If 12-volts is indicated on the meter, the UP solenoid is defective and must be replaced.
4. Repeat this test but hold in the switch for the **DOWN** position. Connect the ohmmeter to the Green/White wire on the DOWN solenoid terminal and the Black meter lead to the Black wire terminal on the DOWN solenoid terminals. Depress the **IN** or **DOWN** switch on the control panel. If 12-volts is not indicated, the switch or wiring is defective. Replace the defective switch or wiring. If 12-volts is indicated on the meter, the DOWN solenoid is defective and must be replaced.

13-4 TRIM & TILT

Pump Motor - In Boat

◆ See Figures 4 and 5

1. **Out/Up Operation** - Connect a jumper wire between the Blue/White lead on the motor terminal and the positive (+) solenoid terminal. The motor should run

2. **In/Down Operation** - Connect a jumper wire between the Green/White lead at the motor terminal and the positive (+) solenoid terminal. The motor should run.

3. If the motor does not run in either of these tests, rebuild the motor.

Pump Motor - Out Of Boat

◆ See Figures 6 and 7

■ Obviously, you must first remove the motor and drain the fluid!

1. **Out/Up Operation** - Connect the positive (+) lead from a 12 V power source to the Blue/White lead terminal at the motor. Connect the negative lead (-) to a good ground on the pump. The motor should run.

2. **In/Down Operation** - Connect the positive (+) lead from a 12 V power source to the Green/White lead terminal at the motor. Connect the negative lead (-) to a good ground on the pump. The motor should run.

3. If the motor does not run in either of these tests, rebuild the motor.

Solenoid - Pump In Boat

◆ See Figures 8 and 9

1. **Out/Up Solenoid** - Connect a jumper wire between the positive (+) terminal at the solenoid and the Blue/White lead terminal. The motor should run.

2. **In/Down Solenoid** - Connect a jumper wire between the positive (+) terminal at the solenoid and the Green/White lead terminal. The motor should run.

3. If the motor does not run in one or the other direction, replace the appropriate solenoid.

Solenoid - Pump Out Of Boat

◆ See Figures 10 and 11

■ Obviously, you must first remove the motor and drain the fluid!

1. **Out/Up Solenoid** - Connect the positive (+) lead from a 12 V power source to the Blue/White lead terminal at the motor. Connect the negative lead (-) to the ground terminal on the solenoid.

2. Now connect an ohmmeter to the two large (upper) terminals on the solenoid as shown in the accompanying illustration. If the meter shows full continuity (zero ohms), the solenoid is OK. If the meter shows no continuity (High ohms), replace it.

3. **In/Down Solenoid** - Connect the positive (+) lead from a 12 V power source to the Green/White lead terminal at the motor. Connect the negative lead (-) to the ground terminal on the solenoid.

4. Now connect an ohmmeter to the two large (lower) terminals on the solenoid as shown in the accompanying illustration. If the meter shows full continuity (zero ohms), the solenoid is OK. If the meter shows no continuity (High ohms), replace it.

110 Amp Fuse

◆ See Figures 12, 13 and 14

1. If the pump is in the boat, check for voltage at the battery power terminal by touching the positive (+) lead of a DVOM to the terminal and the negative (-) lead to ground. If there is battery voltage, move to the next step; if not check the battery and wiring.

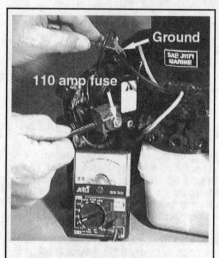

Fig. 3 Check that there is battery voltage

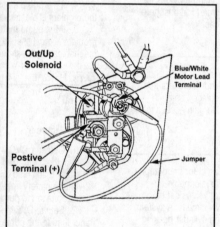

Fig. 4 Testing the pump for Out/Up operation (in boat)

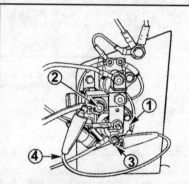

1- IN/DOWN Solenoid
2- Positive Terminal (+)
3- GREEN-WHITE Motor Lead Terminal
4- Jumper Wire

Fig. 5 Testing the pump for In/Down operation (in boat)

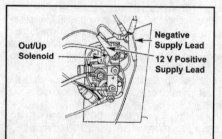

Fig. 6 Testing the pump for Out/Up operation (out of boat)

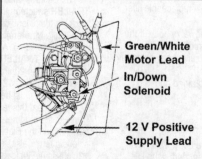

Fig. 7 Testing the pump for In/Down operation (out of boat)

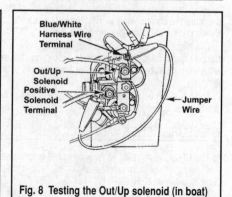

Fig. 8 Testing the Out/Up solenoid (in boat)

TRIM & TILT 13-5

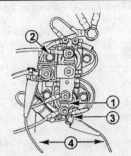

1- IN/DOWN Solenoid
2- Positive (+) Solenoid Terminal
3- GREEN/WHITE Harness Wire Terminal
4- Jumper Wire

Fig. 9 Testing the In/Down solenoid (in boat)

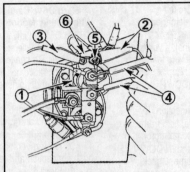

1- OUT/UP Solenoid
2- 12 volt Positive (+) Supply Lead
3- Negative (–) Supply Lead
4- Ohmmeter Leads
5- BLUE/WHITE Harness Wire Terminal
6- Solenoid Ground Terminal

Fig. 10 Testing the Out/Up solenoid (out of boat)

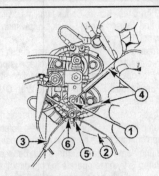

1- IN/DOWN Solenoid
2- 12 Volt Positive (+) Supply Lead
3- Negative (–) Supply Lead
4- Ohmmeter Leads
5- GREEN/WHITE Harness Wire Terminal
6- Solenoid Ground Terminal

Fig. 11 Testing the In/Down solenoid (out of boat)

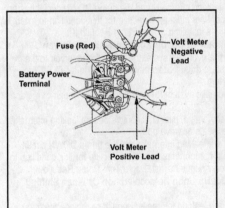

Fig. 12 Check for battery voltage...

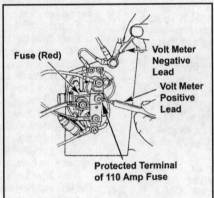

Fig. 13 ...and then check the fuse

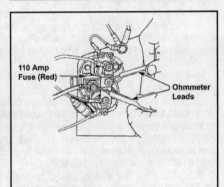

Fig. 14 If the pump is out of the boat, check for continuity

2. Now connect the positive lead of the meter to the protected terminal on the fuse while the other is still touching ground. If there is voltage, the fuse is OK; if not, replace the fuse.

3. If the pump is out of the boat, connect an ohmmeter between the two fuse terminals. If the meter shows continuity (zero ohms), the fuse is OK; if it shows no continuity, replace the fuse.

20 Amp Fuse

◆ See Figure 15

■ While 2001-03 models use a tube fuse, all 2004 and later models utilize a blade-type fuse. Procedures are the same for either.

Remove the fuse from the holder and connect an ohmmeter lead to each end. If the meter shows continuity (zero ohms), the fuse is OK; if it shows no continuity, replace the fuse.

■ Yeah, we know, you could just look at the fuse and see if it's blown!

SYMPTOMS AND PROBLEMS

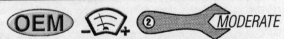

Troubleshooting some problems can be done more easily by identifying the symptom first. Below is a list of the most common symptoms found on the trim/tilt system and the areas to check for defective components.

Trim Pump Motor Fails To Operate With OUT/UP or IN/DOWN Switches Pressed - Solenoids Fail To "Click"

1. Thermal breaker in pump open. Allow motor to cool. Replace breaker switch in motor end plate if it does not reset.

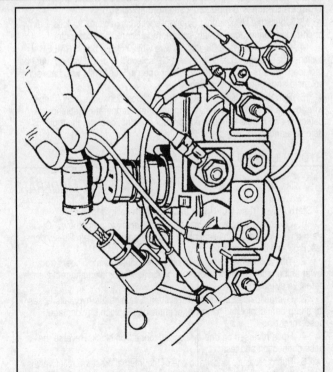

Fig. 15 Remove the 20 amp fuse from the holder

13-6 TRIM & TILT

2. 20 Amp fuse blown. Check trim limit switch wires for broken insulation and shorted wires. Check pump wiring harness for damaged insulation, open or shorted wiring.

3. Check for 12 V at pump 110 amp circuit breaker terminal. If voltage is present, replace 110 amp fuse. If no voltage is present, check wire lead from the battery for open, short or loose connection at the battery.

Trim Pump Motor Fails To Operate With OUT/UP Or IN/DOWN Switches Pressed - Solenoids "Click"

4. Check solenoid connections for corrosion or loose connection.

5. Pump motor has worn out brushes, defective armature, field frame, or has been contaminated by water or oil. Repair or replace pump motor.

6. Wiring harness has internal short, between OUT/UP and IN/DOWN circuits. Pump is trying to operate in both directions at same time. Check wire harness for broken, damaged, melted insulation. Check control panel switch assembly for damaged wiring and burned contacts in the switch assembly.

Trim Pump Motor Will Not Operate In The OUT/UP Direction - Solenoid Does Not "Click"

7. Check UP solenoid wire connections for loose, dirty, or corroded connections. Clean, tighten, repair or replace connector.

8. Check for 12 V on Blue/White wire while holding OUT/UP switch depressed. Voltage present - UP solenoid is defective and must be replaced. Voltage not present - check for open Blue/White wire between solenoid and control panel switch.

9. Possible bad field winding in pump motor. Connect a 12 V jumper wire to Blue/White wire terminal on motor. If motor operates, solenoid is defective and should be replaced. If motor does not run, the motor is defective and must be replaced.

Trim Pump Motor Will Not Operate In The IN/DOWN Direction - Solenoid "Clicks"

10. Check DOWN solenoid wire connections for loose, dirty, or corroded connections. Clean, tighten, repair or replace connector.

11. Check for 12 V on Green/White wire while holding the IN/DOWN switch depressed. If voltage is present - DOWN solenoid is defective and must be replaced. Voltage not present - check for open Green/White wire between solenoid and control panel switch.

12. Possible bad field winding in pump motor. Connect a 12 V jumper wire to Green/White wire terminal on motor. If motor operates - solenoid is defective and should be replaced. If motor does not operate - motor is defective and must be replaced.

Trim Pump Motor Will Not Operate In The Out/Up Direction With TRAILER Switches Depressed - Solenoid "Clicks"

13. Check the UP solenoid Blue/White wire connections for loose, dirty, or corroded connections. Clean, tighten, repair or replace connector.

14. Check for 12 V on Blue/White wire while holding both TRAILER switches depressed. Voltage present - UP solenoid is defective and must be replaced. Voltage not present - check for open in Blue/White wire between solenoid and control panel switch.

15. Possible bad field winding in pump motor. Connect a 12 V jumper wire to Blue/White wire terminal on motor. If motor operates - solenoid is defective and should be replaced. If motor does not operate, the motor is defective and must be replaced.

HYDRAULIC AND MECHANICAL PROBLEMS

Stern Drive Trims Out/Up Or In/Down With Jerky Motion

1. Oil level in reservoir is low. Service reservoir to proper level. Cycle stern drive Up and Down five times to remove any trapped air. Re-service reservoir and check for leaks.

2. Trim cylinders sticking or binding. Disconnect trim cylinders from lower unit end and cycle. Observe which cylinder is binding and replace the defective cylinder.

3. Hydraulic hose(s) pinched in system. Examine the hydraulic hoses for sharp bends, pinched hoses, or bent tubing. Replace any damaged hoses or tubing.

4. Hose reversed on one cylinder at connection block. Reverse the hose connection and test.

5. Pump pressure is low and one of the internal pressure relief valves has malfunctioned. Repair or replace the pressure relief valve(s) on the pump manifolds.

Stern Drive Cannot Be Lowered From Full Up Position Or Lowers With A Jerky Motion

6. Oil level low in reservoir or air in the system. Service reservoir to proper level. Cycle stern drive UP and DOWN five times to remove any trapped air. Re-service reservoir and check for leaks.

7. Hydraulic hose(s) pinched in system. Examine the hydraulic hoses for sharp bends, pinched hoses, or bent tubing. Replace any damaged hoses or tubing.

8. Trim cylinders sticking or binding. Disconnect trim cylinders from lower unit end and cycle. Observe which cylinder is binding and replace the defective cylinder.

9. Hose reversed on one cylinder at connection block. Reverse the hose connection and test.

10. Stern drive unit is binding on gimbal ring housing. Check the hinge pins and bushings in the gimbal ring for corrosion and binding. Check the driveshaft housing for misalignment and/or damage. Check the upper gear housing universal bearings for binding, damage and/or misalignment with the engine. Replace any defective components.

Stern Drive Slowly Drops From Up Position During Extended Storage

11. Check entire hydraulic system for external leaking. Repair or replace leaking components.

12. Pump thermal relief valve, pilot check valve or seal leaking internally. Replace defective valve, seals or valve body.

13. Trim cylinder(s) leaking internally and the pump DOWN circuit has an internal leak (both components must be defective for this condition to exist). The manufacturer suggests installing a Trim Pump Rebuild Kit.

Stern Drive Will Not Stay In Out/Up Position While Underway

14. Check for air in the hydraulic system. Service the reservoir to the proper level. Cycle the stern drive five times from full UP to full DOWN to remove air from the system.

15. Check entire hydraulic system for external leaking. Repair or replace leaking components.

16. Pump thermal relief valve, pilot check valve or seal leaking internally. Replace defective valve, seals or valve body.

17. Trim cylinder(s) are leaking internally and the pump DOWN circuit has an internal leak (both components must be defective for this condition to happen). The manufacturer suggests installing a Trim Pump Rebuild Kit.

Stern Drive Trails Out/Up Upon De-acceleration Or When Shifting The Unit Into Reverse

18. Trim cylinder(s) are leaking internally. Rebuild or replace the trim cylinders.

19. The pump DOWN circuit has an internal leak. The manufacturer suggests installing a Trim Pump Rebuild Kit.

Oil Foams Out The Vent Fill Cap On The Reservoir

20. The hydraulic system oil is contaminated. Drain and flush the hydraulic system with clean oil. Fill the reservoir and cycle the system four or five times to remove any air.

21. Fluid level is low which caused the fluid to foam. Service reservoir to the proper level. Cycle stern drive UP and DOWN five times to remove any trapped air. Re-service reservoir and check for leaks.

Service Precautions

The following points should always be observed when installing, testing, or servicing any part of the power trim/tilt system.

1. Coat the threads of all fittings and O-rings with Quicksilver Power Trim and Steering Fluid, or an equivalent product.

2. Use clean Power Trim and Steering fluid when filling the hydraulic fluid reservoir. If the Power Trim and Steering fluid is not available, substitute with SAE 10W-30 or 10W-40 motor oil in an emergency and then drain the system as soon as possible.

3. The stern drive must be in the full DOWN position with the hydraulic cylinders collapsed when checking the fluid level or adding fluid to the reservoir. Normal fluid level should be maintained between the **MIN** and **MAX** marks on the side of the reservoir. If the quantity is low, add fluid until the level is up to the **MAX** mark on the side of the reservoir.

4. If the pump stops during long use, allow the pump motor to cool at least five minutes before starting again. An internal thermal circuit breaker protects the pump motor from overheating. If the pump still will not operate, the 110 amp fuse may have blown.

5. Keep the work area clean when servicing or disassembling parts. The smallest amount of dirt or lint can cause failure of the pump to operate.

6. The valve body, pump assembly, pressure relief valves, and thermal

TRIM & TILT 13-7

relief valves must be replaced as individual assemblies. There are no piece parts available for these components except for the O-rings that are sold in a kit.

7. The motor is protected from internal fluid leaks and external moisture by seals, O-rings and a grommet on the field winding harness. If the motor fails due to hydraulic fluid, clean the motor thoroughly and replace the internal pump seal and O-ring. If the motor fails from moisture as evidenced by internal corrosion, the wire harness grommet is damaged or deteriorated, and the pump motor should be replaced. Sometimes liquid neoprene can be used on the wire harness grommet to seal minor cracks. Use liquid neoprene with care to prevent contaminating the electric motor or hydraulic valves.

8. Keep the trim cylinders attached to the forward anchor pin during servicing and repairs. Do not allow the trim cylinders to hang by their hydraulic hoses. Such practice may kink and damage the hoses. Use care when handling trim cylinders during removal/installation of the stern drive unit. Rough treatment of the hoses could result in a weak-ended hose, partial separation at the fittings, or bending of the metal tubing, any one of which could restrict the flow of fluid to the trim cylinders.

9. Always hold both fittings with a wrench when tightening or loosening hydraulic hoses and fittings. When installing flex hoses, be sure there are no twists or sharp bends in the flex hose. The fitting on the cylinder end of the hose must point approximately 45° from the stern drive centerline towards the transom. A twisted hose will cause severe loads and result in stress that could bend the tubing ends. On the stern drive models covered here, the hoses are routed directly from the cylinder to a block under the gimbal housing. Take extra care when attaching the hoses to the block. If the fittings are cross threaded, the block would have to be replaced. This job would entail the stern drive being removed and possibly the engine.

System Testing

High pressure testing of hydraulic components requires expensive special gauges, tools and highly trained personnel to accurately interpret test results. A danger to the operator and others in the area always exists during the testing process. Therefore, it is highly recommended that the boat, with the trim/tilt system installed, be taken to a qualified repair facility having the necessary equipment to conduct the testing properly and safely.

Do not attempt to work on hydraulic hose connections without the proper tools designed for that specific purpose. The use of a common box end wrench will quite likely result in "rounding off" the flats on a hydraulic fitting. It is imperative that you have a set of "flare" or "line" wrenches that will almost guarantee the fitting will not to be damaged.

System Bleeding

◆ See Figures 16 and 17

■ Please refer to the Maintenance section for procedures on filling the pump reservoir.

The power trim/tilt system is designed to be self-purging with nothing to be done other than raising and lowering the stern drive unit several times. If, during service work, components were installed already filled with hydraulic fluid, or a line was disconnected and then reconnected, the self-purging feature of the system will adequately eliminate the small amount of unwanted air. However, if components were removed and installed 'dry', the following procedures must be performed to adequately purge all air from the system.

✲✲ WARNING

Check to be sure the vent hole opening (if equipped) in the reservoir fill cap is free and clear. A small amount of air must be allowed to enter the reservoir tank as the fluid level drops or vacuum could develop in the tank making it difficult for the pump to draw the fluid from the reservoir.

1. Depress the **IN/DOWN** switch to operate the pump and lower the stern drive to the full DOWN position. Verify that the fluid level in the power trim reservoir is between the **MAX** and **MIN** marks cast into the side of the reservoir. If necessary, fill the reservoir with Quicksilver Power Trim and Steering fluid, or an equivalent product.

2. Begin by bleeding the OUT/UP side of the trim cylinder. If one or both cylinders were replaced without filling the cylinders with fluid, be sure to check the reservoir fluid level often while performing this procedure. Disconnect the OUT/UP hose from the end of the trim cylinder. If one cylinder was rebuilt, then disconnect the one hose. If both trim cylinders were rebuilt, disconnect the hose from both trim cylinders. Place the ends of the hose(s) into a suitable container. Have an assistant press and hold the **UP** button on the control panel. When a solid stream of fluid, free of any air bubbles, is observed escaping from the end of the hoses, release the **UP** control button. Connect the hoses to the hydraulic cylinders, and service the reservoir, if needed.

3. Bleed the IN/DOWN side of the trim/tilt cylinder. Again, check the fluid level in the power trim pump reservoir and fill to the **MAX** mark cast into the side of the reservoir. Be sure to check the fluid level often while performing this procedure.

4. Disconnect the **IN/DOWN** hose from the hydraulic connector block (NOT the cylinder as before!). If one cylinder was rebuilt, then disconnect the one hose. If both trim cylinders were rebuilt, disconnect the hoses for both cylinders. Install a special pipe thread plug (#22-38609), or equivalent into the opening in the hydraulic connector block.

5. Place the ends of the hose(s) into a suitable container. Stand clear of the stern drive because it will begin to move in this next step.

6. Have an assistant press and hold the **UP** button on the control panel. The cylinders will begin to extend raising the stern drive. At the same time, air and fluid will be forced out the end of the hoses. When the cylinders are fully extended - have your assistant release the control button.

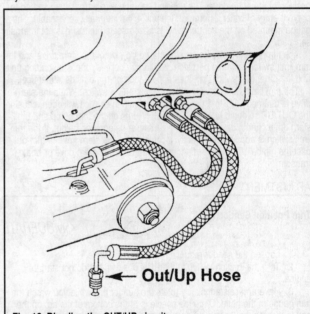

Fig. 16 Bleeding the OUT/UP circuit

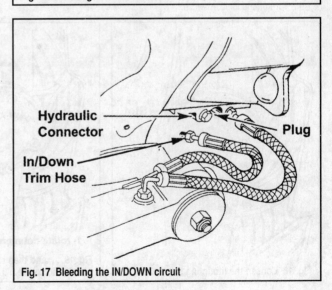

Fig. 17 Bleeding the IN/DOWN circuit

13-8 TRIM & TILT

✱✱ CAUTION

Be sure to check the power trim reservoir fluid level often during this procedure. If the reservoir fluid level drops to the MIN mark, stop and service the reservoir. Failing to maintain the proper fluid level will cause the pump to draw air into the system.

7. Remove the special pipe plug(s) from the connector block and just momentarily press the **DOWN** button on the control panel. This will remove any air from the DOWN circuit. Now, connect the trim cylinder hoses to the connector block and tighten them securely. Press and hold the **IN/DOWN** button on the control panel until the stern drive is in the full DOWN position. Check and service the reservoir fluid level, if needed. Raise and lower the stern drive four or five times to remove any trapped air in the trim system.

Trim Limit Switch/Trim Position Sender

GENERAL INFORMATION

The gauge mounted on the control panel indicates to the helmsperson the trim position of the stern drive at all times. An electrical sending unit on the starboard side of the stern drive is mechanically indexed to the hinge pin. When the stern drive is raised or lowered, the sending unit sends a variable voltage signal to the gauge unit on the control panel indicating the trim position of the stern drive. The sending unit and wire harness are sealed to prevent entry of water. Therefore, the unit is not internally serviceable. The gauge mounted on the control panel is also a sealed unit and is not internally serviced.

The trim sender is located on the starboard side of the stern drive and the trim limit switch is on the port side. Would you believe, one is a sending unit and the other is a switch, but their appearance on the outside is identical.

The trim limit switch on the other hand is only a switch. When the stern drive is trimmed OUT to a pre-determined limit, the switch will open, causing the OUT/UP solenoid to open. This action turns off the trim pump and the stern drive cannot be trimmed out past its pre-determined height. If the trim limit switch is adjusted incorrectly, it could cause the stern drive to trim out past the gimbal ring support flanges and damage the stern drive or boat transom.

ADJUSTMENT

Trim Position Sender

 MODERATE

1. Loosen the sender mounting screws at the drive.
2. Turn the ignition switch to the **RUN** position, but do not start the engine.
3. Have an assistant slowly rotate the sender unit while you watch the trim gauge in the helm. Once the needle is at the bottom of the arc on the gauge, tighten the 2 retaining screws and then recheck the gauge.
4. Turn off the ignition switch.

Trim Limit Switch

 MODERATE

◆ See Figures 18 thru 21

■ On models with the Auto Trim II system, make sure that the trim mode switch is set in the Manual position. Make sure that the stern drive unit is in the full DOWN/IN position.

1. Loosen the 2 screws on the slotted portion of the limit switch housing and turn it clockwise to the end of its slots. Don't confuse these screws with the 4 housing screws!
2. Have an assistant press and hold the **UP/OUT** trim switch - do not use the **TRAILER** switch. Slowly rotate the trim limit switch housing counterclockwise until the trim cylinders have extended to a distance of 20 3/4 in. (520mm), measured between the forward and aft cylinder anchor pins on Alphas; On Bravos, and Auto Trim II systems, the dimension should be 21 3/4 in. (552mm) maximum.
3. Tighten the screws on the switch securely.

REMOVAL & INSTALLATION

 MODERATE

◆ See Figures 22, 23 and 24

1. Remove the stern drive as detailed in the Drive System section.
2. Remove the two mounting/adjustment screws with their washers and retainers and lift off the switch or sender unit, allowing it to dangle by the electrical leads.
3. Remove the U-joint bellows sleeve.
4. Remove the gimbal ring hinge pins.
5. Grab the bell housing and pull backwards on it while rotating it about 90°, until you have access to the wire harness retainer.
6. Remove the harness retainer, disconnect the wires at the engine and pull them through.

To install:

7. Feed the new wires through the hole in the housing and then slide the rubber grommet halves together in position so that the straight edges are vertical - you will have to maintain a little pressure on the wires in order to hold the grommets in position.

■ You may wish to use the old wires to feed the new ones through the hole in the housing.

8. Install the retainer and tighten the screw to 95 inch lbs. (11 Nm).
9. Install the U-joint bellows, hinge pins and stern drive as detailed in the Drive System section.
10. With the drive unit in the full DOWN position, rotate the center rotor on the switch/sender until its index mark aligns with the index mark on the switch/sender body.
11. Install the cap and then install the switch/sender unit to the gimbal ring with the retainers and screws.
12. Adjust the units as detailed above.

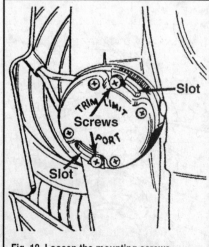

Fig. 18 Loosen the mounting screws...

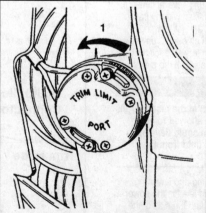

Fig. 19 ...and then rotate the switch counterclockwise...

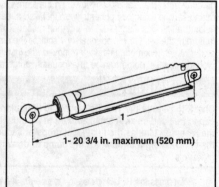

Fig. 20 ...until the cylinder meets this dimension (Alpha shown)

TRIM & TILT 13-9

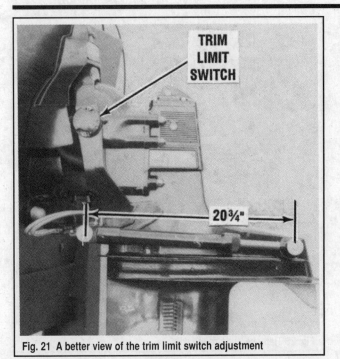

Fig. 21 A better view of the trim limit switch adjustment

13. Don't forget to attach the limit switch harness to the water hose with the plastic clip.
14. Reconnect the leads at the engine harness and connect the battery cables.

Pump Assembly

GENERAL INFORMATION

The pump motor is a 12 V DC bi-directional motor. The armature of the motor rests inside the field winding and is energized through brushes contacting the commutator on the end of the armature. When the up or down switches are closed, current is directed through the motor field in one of two directions causing the motor to run in one desired direction.

The opposite end of the armature is mechanically splined to the hydraulic pump. The gear rotor action of the pump causes the fluid to flow in either direction.

The motor case is sealed with O-rings and a shaft seal to prevent the entry of water or hydraulic fluid into the motor cavity.

Before going directly to the pump as a source of trouble, check the battery to be sure it is up to a full charge. Inspect the wiring for loose connections, corrosion and damaged wires. Take a good look at the control switches and connection for evidence of a problem.

The control switches may be quickly eliminated as a source of trouble by connecting the pump directly to the battery for testing purposes.

Testing the control panel switch must be done at the back of the switches because the wires are soldered to the switch terminals and cannot be disconnected. Unplug the trim harness from the trim pump.

Connect an ohmmeter, set for continuity reading, between the terminals on the back of the switch being tested. The meter must indicate an open circuit - no meter movement - with the button in the free position.

The meter must indicate continuity - full meter deflection - with the button(s) depressed. Remember to press both up buttons when checking the Trailering circuit on a 3-button control panel.

If the control panel switches are found to be satisfactory, the problem is with the solenoid(s), pump motor or the hydraulic pump. These components may be checked after the trim/tilt assembly is removed from the boat.

FLUID LEVEL

■ Please refer to the Engine & Drive Maintenance section for fluid level checking and filling procedures.

REMOVAL & INSTALLATION

◆ See Figures 25 and 26

1. Lower the stern drive to the full down position. If the stern drive will not lower electrically, place a suitable drain pan under the hydraulic lines. Now, loosen the UP/OUT line fittings on the hydraulic connector. Allow the fluid to flow into the drain pan until the stern drive is in the full down position and then tighten the fittings.
2. Disconnect the battery cables from the battery terminals.
3. Disconnect the positive and negative battery leads at the assembly.
4. Mark or tag the hydraulic hoses (right or left) before disconnecting them from the pump. The Black hydraulic hose on the left is for the UP circuit and the Gray hydraulic hose on the right is for the DOWN circuit. If possible, cap the hoses to prevent fluid leakage into the boat.
5. Pull to disconnect the 3-prong trim harness connector from the trim pump.
6. Remove the 4 lag bolts (and their washers) securing the trim/tilt pump assembly to the boat. Lift the assembly out and free of the boat.

To install:

7. Position the trim/tilt pump assembly in the boat. Install the four lag bolts through the pump support bracket into the floor. Tighten the bolts securely.
8. Remove the cap from the hose and connect the Gray (DOWN) hose to the right fitting as you face the pump (it may be marked **DN** on the adapter). Thread the hose fitting into the adapter several turns by hand to ensure it is not cross-threaded. Tighten the fitting to 70-150 inch lbs. (7.9-16.9 Nm) on the Alpha; or 110 inch lbs. (12 Nm) on the Bravo.
9. Remove the cap from the hose and connect the Black (UP) hose to the left fitting as you face the pump (it may be marked **UP** on the adapter. Thread the hose fitting into the adapter several turns by hand to ensure it is not cross-threaded. Tighten the fitting to 70-150 inch lbs. (7.9-16.9 Nm) on the Alpha; or 110 inch lbs. (12 Nm) on the Bravo.

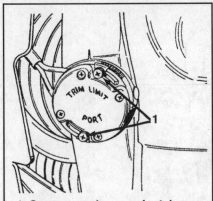

1- Screws, washers, and retainers

Fig. 22 Remove the trim limit switch. . .

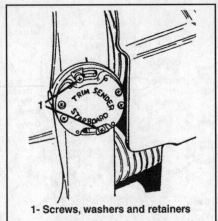

1- Screws, washers and retainers

Fig. 23 . . .and the trim position sensor

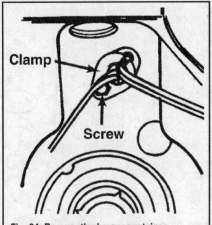

Fig. 24 Remove the harness retainer

13-10 TRIM & TILT

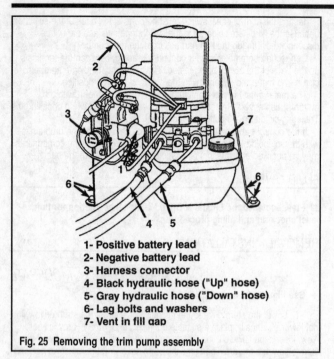

1- Positive battery lead
2- Negative battery lead
3- Harness connector
4- Black hydraulic hose ("Up" hose)
5- Gray hydraulic hose ("Down" hose)
6- Lag bolts and washers
7- Vent in fill gap

Fig. 25 Removing the trim pump assembly

Fig. 26 Remove the ground wire

10. Connect the 3-wire trim/tilt harness to the pump bracket. Connect the battery (Red) positive 12-volt cable to the 110 amp circuit breaker. Apply a coating of liquid neoprene to the wire and terminal. Connect the battery (Black) negative to the ground terminal lug on the side of the support bracket. Secure the cable with the bolt and nut.

11. Remove the cap from the reservoir filler opening. Verify the hole in the cap is free and clear (if yours has a hole). If this hole becomes clogged, it could cause the hydraulic pump to cavitate, and the stern drive to raise or lower improperly.

12. Fill the reservoir with Quicksilver Power Trim and Steering Fluid to the **MAX** line.

13. Cycle the stern drive Up and Down several times to bleed all air from the system. Check the reservoir level after each cycle and service, if needed.

DISASSEMBLY & ASSEMBLY

Support Bracket

◆ See Figures 27 thru 34

1. Remove the pump assembly.
2. Drain all the remaining fluid in the reservoir into a suitable container. Discard the fluid in accordance with any local hazardous waste ordinances.
3. Using a Phillips head screwdriver, loosen the screw securing the terminal cover (if equipped) on the solenoid buss. Slide the cover off the solenoid buss bar.

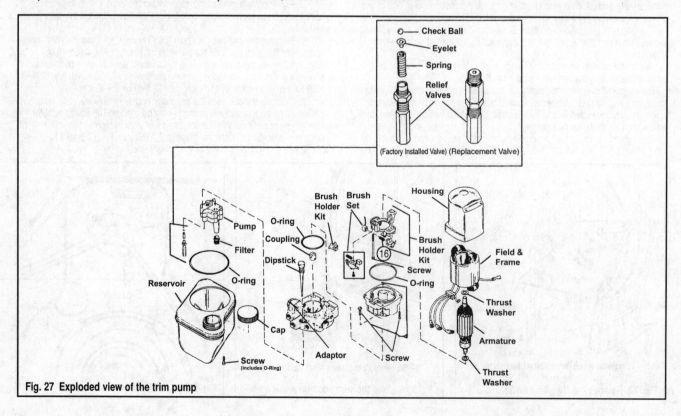

Fig. 27 Exploded view of the trim pump

TRIM & TILT   13-11

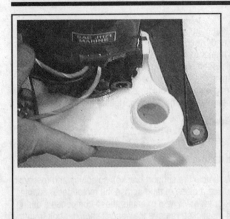
Fig. 28 Drain the fluid into a container

Fig. 29 Remove the cover...

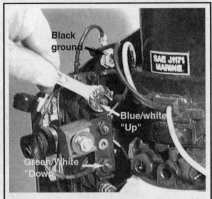
Fig. 30 ...and disconnect the trim motor wires

Fig. 31 Remove the mounting bolts...

Fig. 32 ...and lift out the pump assembly

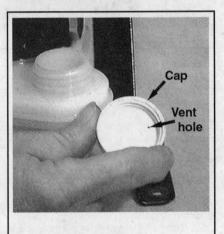
Fig. 33 Check the vent hole...

4. Remove the nuts and washers securing the Blue/White (Up) wire lead and Green/White (Down) wire leads from the pump motor to the solenoid terminals. Remember to tag the wires. Remove the ground wire.

5. On the back of the support bracket, remove the two bolts and lockwashers securing the adapter with motor and reservoir to the support bracket. Slide the motor, adapter and reservoir out of the support bracket.

To assemble:

6. Align the back of the adapter to the support bracket and install the bolt and lock washer through the support bracket into the adapter. Tighten the bolts securely.

7. Connect the Green/White wire lead from the motor harness to the DOWN solenoid terminal. Secure the wire to the terminal with a flat washer and nut. Apply a coating of liquid neoprene to the wire and terminal.

8. Connect the Blue/White wire lead from the motor harness to the UP solenoid terminal. Secure the wire to the terminal with a flat washer and nut. Apply a coating of liquid neoprene to the wire and terminal.

9. Install the terminal cover over the solenoid terminals and secure it in place with a Phillips head screw in the center.

10. Connect the pump motor ground wire terminal to the support bracket and temporarily secure it in place with a bolt nut.

Solenoids

◆ See Figures 27 and 35

Replacement of the UP or DOWN solenoid is easily performed with the support bracket and pump assembly separated.

1. Remove the pump assembly and support bracket.
2. Mark and disconnect the wires from the defective solenoid terminal. The UP solenoid is on top and the DOWN solenoid is on the bottom.
3. Remove the two bolts securing the solenoid to the support bracket.

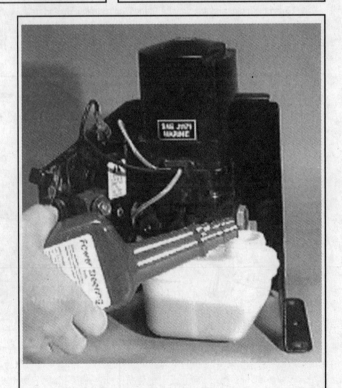
Fig. 34 ...and then fill the reservoir with fluid

13-12 TRIM & TILT

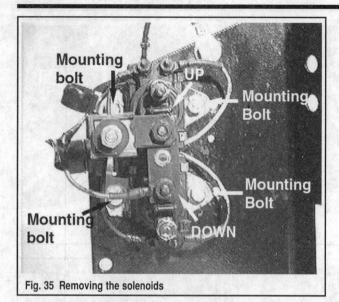

Fig. 35 Removing the solenoids

4. Install the new solenoid and secure it with two bolts. Connect the wires to the terminals as marked during disassembly. Refer to the wiring schematic for the correct color code. Coat all wire terminals with liquid neoprene after the system has been checked.

Pump Reservoir

◆ See Figures 27 and 36 thru 38

1. Remove the pump assembly, support bracket and solenoids.
2. Place a scribe mark or a piece of tape on the motor case, adapter, and reservoir. This mark or tape will help orientate these components during assembling.
3. Using a 1/4 in. socket, remove the special shoulder bolt from the bottom of the reservoir. Slide the O-ring off the shoulder bolt and discard the O-ring.
4. Grasp the motor and adapter and lift them both straight up and free of the reservoir.
5. Slide the O-ring off the shoulder on the adapter and discard the O-ring.

To assemble:

6. Place a new O-ring onto the adapter and lightly lubricate it with Quicksilver Power Trim and Steering Fluid, or regular old motor oil.
7. Align the scribe marks or tape strips placed on the adapter and reservoir tank prior to disassembly. Carefully lower the adapter onto the reservoir to avoid damaging the filters on the ends of the pickup tubes.
8. Place a new O-ring onto the shoulder bolt. Insert the bolt up through the bottom of the reservoir and into the adapter. Tighten the bolt securely.

Hydraulic Pump And Adaptor

◆ See Figures 27 and 39 thru 51

1. Remove or disassemble the support bracket, solenoids and reservoir.
2. Scribe mark or place a piece of tape across the motor case, adapter, and reservoir. This mark or tape will help orientate these components during assembling. Using a 1/4 in. socket, remove the special shoulder bolt from the bottom of the reservoir. Slide the O-ring off the shoulder bolt and discard the O-ring.
3. Set the adapter and motor assembly back onto the reservoir temporarily. Remove the two bolts securing the electric motor to the adapter. Lift the motor straight up and free of the adapter. Remove and discard the O-ring on the shoulder of the adapter. Reach in and lift out the motor shaft coupler and set it aside for safe keeping.

■ The UP, DOWN and THERMAL pressure relief valves are installed and preset at the factory. A jam nut at the base of the valve holds this factory setting. Therefore, if the jam nut is loosened, as in removal, the factory setting of the valve will be changed and the valve must be replaced with a new replacement valve. Do not remove these valves unless you intend to replace them. The replacement valves are set at the factory and are secured to an adapter fitting. The adapter fitting is then threaded into the original pressure relief valve port and tightened. The spring, eyelet and ball are contained in the adapter for the replacement valve. Remove the factory valve only if it is known to have failed and then it must be replaced with a new replacement valve assembly. All three valves look identical except the replacement valves are color-coded to prevent the wrong valve being installed. The colors for the replacement valves are as follows:

- UP Pressure Relief Valve - blue
- DOWN Pressure Relief Valve - green
- THERMAL Pressure Relief Valve - gold

Be sure the valve has failed before it is removed from the adapter. When installing a new replacement valve, make sure it is the correct color code and installed into the correct adapter position. Replacement valves may be

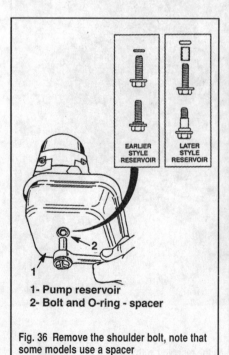

Fig. 36 Remove the shoulder bolt, note that some models use a spacer

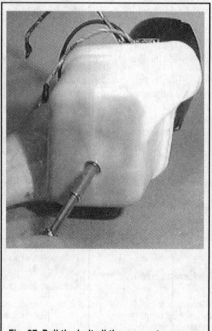

Fig. 37 Pull the bolt all the way out...

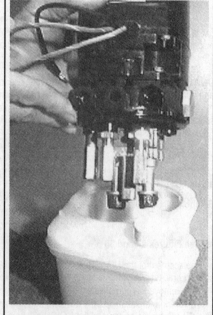

Fig. 38 ...and lift out the pump

TRIM & TILT

installed and removed provided a new O-ring is used each time the adapter fitting is threaded into the adapter. Place the wrench only on the adapter fitting. Never tighten the valve by the hex flats on the valve or the jam nut.

4. If the filter is damaged or clogged, twist and pull the screen mesh and retaining ring off the end of the adapter inlet pipe. A new filter will be installed during assembly.

5. Using the correct size wrench, loosen the jam nut on one, two, or all three pressure relief valves. Un-thread the valves from the adapter and remove the ball, eyelet, and spring. Discard the valve and small parts.

■ **The hydraulic pump is non-serviceable and is replaced only as a complete unit. The pump is secured to the adapter housing with two lobular head socket screws. If the special wrench for these lobular head screws is not available, use a standard 3/16 in. (5mm) six point socket.**

6. Examine the base of the pump where it contacts the adapter. Two screws go through the pump and into the adapter. Two other screws only secure the pump halves together. Remove the two screws securing the pump to the adapter.

7. Lift the pump straight up and out of the adapter housing. Remove the two O-rings on the bottom of the pump and discard the O-rings. If the pump is defective, discard the pump.

8. Insert a flat blade screwdriver through the center of the pump shaft oil seal and pry the seal from the adapter. Discard the oil seal.

9. Place the adapter in a vise with soft face jaws with the large hex plug retainers exposed. Remove the large hex plug (2), O-ring(s) and spring(s).

10. Remove the adapter from the vise and remove the loose poppet valve(s) from the adapter.

11. A rebuild kit is required to remove and replace the check valve bodies and spool valve in the adapter. Insert the 1/8 in. (Alpha) or 1/4 in. (Bravo) diameter rod which comes with the rebuild kit against the check valve. Using a soft face mallet, tap the check valves, and spool valve out the other side of the adapter. Discard the check valve bodies. Clean the hex plugs, retainers and spool valve.

To assemble:

12. Clean all components in cleaning solvent and dry with low pressure compressed air. If compressed air is not available, make sure they air-dry completely. Replace all O-rings, seals and gaskets.

13. Inspect the check valve bodies and spool valve for signs of scoring and scratches. Check the poppet valve and seat for signs of grooves or wear on the poppet valve taper seat. If the seat or valve is worn, both the valve and valve body should be replaced.

14. Good shop practice dictates that a service rebuild kit for the hydraulic pump be purchased and all components in the kit be installed, any time the pump is disassembled.

15. Lubricate all components during assembly with Quicksilver Power Trim and Steering Fluid or motor oil.

16. Lubricate the check valve body, O-ring, and spool valve with Quicksilver Power Trim and Steering Fluid or motor oil. Slide the O-ring onto the check valve bodies and insert the spool valve into the center of the adapter. Slide a new check valve body onto each side of the spool valve. Work slowly and carefully so as not to force the check valve bodies into the adapter which could damage the O-rings. Center the spool and check valve bodies in the adapter.

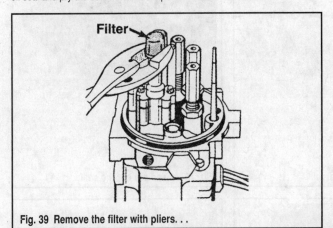

Fig. 39 Remove the filter with pliers...

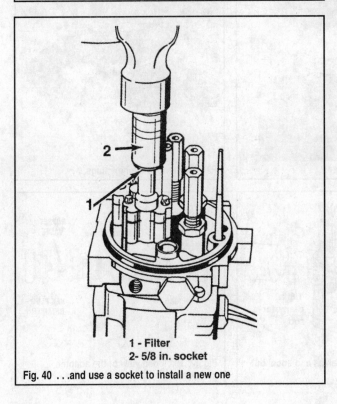

1- Filter
2- 5/8 in. socket

Fig. 40 ...and use a socket to install a new one

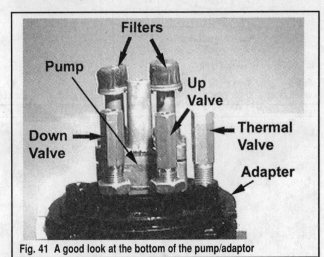

Fig. 41 A good look at the bottom of the pump/adaptor

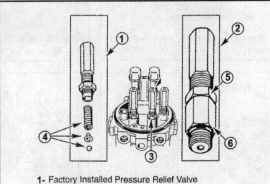

1- Factory Installed Pressure Relief Valve
2- Replacement Pressure Relief Valve
3- Jam Nut
4- Pump Body Components (Spring, Eyelet And Check Ball)
5- Jam Nut
6- O-ring

Fig. 42 Factory valve on left, replacement valve on right

13-14 TRIM & TILT

17. Insert the poppet valves and place them into the check valve bodies on both sides.

18. Lubricate the new hex plug O-rings and slide the O-ring onto the hex plugs. Insert the spring into the center of the hex plug, and then thread the hex plug into the adapter so that it is just finger-tight. Working on both sides equally, tighten the hex plugs approximately 1/4 turn, and then back them off 1/8 turn. Continue tightening both hex plugs equally in this manner until the plugs are tightened securely to 44 ft. lbs. (60 Nm). This procedure will keep the spool valve centered and will prevent damage to the check valve O-rings.

19. Place a new pump shaft oil seal into the adapter with the lips of the seal pointing towards the pump. Press the seal into place with your thumb. Lubricate the lips of the seal with oil.

20. Place two new O-rings into the base of the hydraulic pump. Lower the pump into position on the adapter.

21. Using a 3/16 in. socket, tighten the two hex lobular screws alternately and evenly to 70 inch lbs. (7.9 Nm).

22. If you removed any of the relief valves, place a new O-ring onto the replacement valve fitting. Lubricate the O-ring with power steering fluid or motor oil and thread the color-coded valve into the port in the adapter. Tighten the new valve by the hex flats on the adapter fitting to 70 inch lbs. (7.9 Nm).

23. Place a new filter and sleeve over the pump inlet. Using a 5/8 in. deep socket, tap the end of the socket and drive the sleeve onto the end of the pump inlet. Repeat this step for the other filter.

24. Place a new O-ring onto the adapter and lightly lubricate it with Quicksilver Power Trim and Steering Fluid or motor oil.

25. Align the scribe marks or tape strips placed on the adapter and reservoir tank prior to disassembly. Carefully lower the adapter onto the reservoir to avoid damaging the filters on the ends of the pick-up tubes.

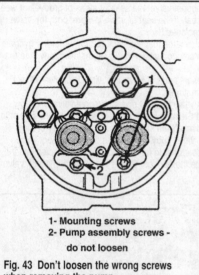

Fig. 43 Don't loosen the wrong screws when removing the pump

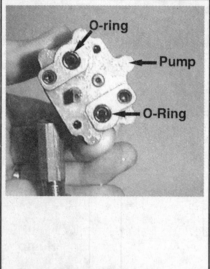

Fig. 44 Lift off the pump...

Fig. 45 ...and pry out the seal

Fig. 46 Don't forget to replace the O-ring on the adapter...

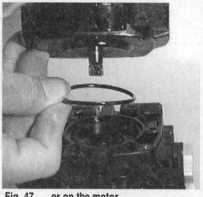

Fig. 47 ...or on the motor

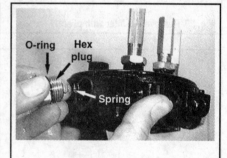

Fig. 48 Remove the hex plugs...

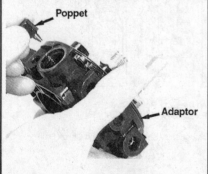

Fig. 49 ...and then the poppet valves

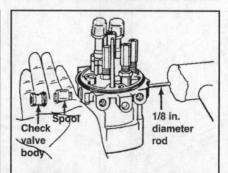

Fig. 50 Tap the check valves and spool out with a small rod

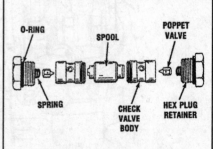

Fig. 51 Exploded view of the adapter components

TRIM & TILT

Pump Motor

◆ See Figures 27 and 52 thru 65

The following procedures pickup the work after the trim/tilt pump assembly has been removed from the boat and the unit has been separated from the support bracket, as previously described in this section.

Be sure to keep the work area, tools and hands as clean as possible while working on the trim/tilt motor.

1. Place a scribe mark or a piece of tape on both halves of the motor case and the hydraulic adapter. These marks and/or tape will be very helpful during assembling of these components. Remove the two bolts securing the electric motor to the adapter. Lift the electric motor straight up and free of the adapter. Remove and discard the O-ring on the shoulder of the adapter.

2. Reach in and lift out the motor shaft coupler. Set the coupler aside for safekeeping.

3. Remove the four Phillips head screws securing the end cover to the motor. Lift off the end cover. It may be necessary to pry between the cover and motor because a tight seal may have formed around the wire harness grommet. Gently pull and twist on the grommet to break the tight bond. Once the grommet is free, lift off the cover. Be sure to remove the washer from inside the cover or from the armature shaft.

4. Remove the O-ring from the motor housing and discard the O-ring.

5. Loosen the screw and remove the tab securing the brush holder to the brush frame.

6. Grasp the brush holder and lift it from the brush frame. A spring behind the brush will push the brush out of the holder when it clears the end of the commutator on the armature.

7. Remove the Phillips head screw securing the thermal switch to the brush frame. Disconnect the Black wire terminal from the thermal switch and lift the switch free of the motor.

8. Remove the two Phillips head screws securing the brush frame to the motor housing. Gently move the wires aside while lifting the brush frame from the casing.

9. Grasp the armature and lift it out of the motor housing. If the thrust washer is on the end of the armature shaft, remove the thrust washer.

10. Place a scribe mark on the motor field frame and the motor housing. Lift out the field frame from the housing.

To assemble:

Any sign of oil in the pump motor indicates either the pump shaft oil seal is damaged or the vent hole in the reservoir fill cap is plugged. If the vent hole is plugged, air in the reservoir tank may not escape and oil is forced into the pump motor.

Clean the motor case with warm soap and water and blow-dry with low pressure compressed air.

Clean the armature, and field with a spray electrical contact cleaner and blow-dry with low pressure compressed air.

11. Check the armature on a growler for shorts, open windings, or shorted windings. If the commutator is worn, true it on a lathe, and undercut the mica. If a growler is not available check the armature with an ohmmeter.

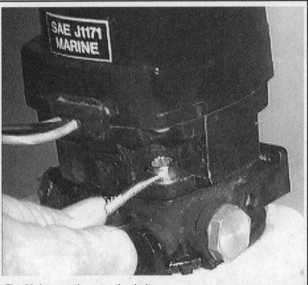

Fig. 52 Loosen the mounting bolts...

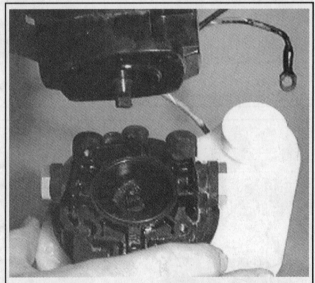

Fig. 53 ...and lift the motor off the adapter

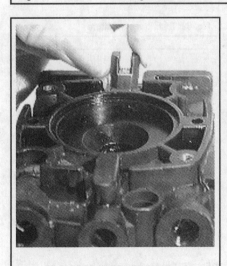

Fig. 54 Lift out the motor shaft coupler

Fig. 55 Remove the end cover screws...

Fig. 56 ...and lift off the cover

13-16 TRIM & TILT

Set an ohmmeter to the Rx1 scale. Connect the Black lead of the meter to the center of the armature shaft and connect the Red meter lead to each one of the commutator bars. If the meter indicates continuity between the commutator and the armature shaft, the armature is grounded and it must be replaced. If no continuity is indicated, the armature is good.

12. Check the thermal switch for continuity. Obtain an ohmmeter and set the switches for Rx1 scale. Connect the Black meter lead to the thermal switch spade terminal, connect the Red meter lead to the brush lead. If continuity is indicated, the switch is good. If no continuity is indicated the switch is defective and must be replaced. If the switch has high resistance, it must be replaced. Open the switch contacts and insert a piece of paper or other insulator. If continuity is indicated the switch is defective and must be replaced. If no continuity is indicated the switch is good.

13. Check the field frame for open circuits. Obtain an ohmmeter and set the switches for Rx1 scale. Connect the Red meter lead to the Blue/White wire lead on the field frame. Connect the Black meter Black lead to the brush lead. If zero ohms is indicated - full continuity - the field is good. If ohms are indicated or - no continuity - the field is open and must be replaced.

Move the Red meter Red lead over to the Green/White wire on the field. If zero ohms is indicated - full continuity - the field is good. If ohms are indicated or - no continuity - the field is open and must be replaced.

14. Check the field frame for a short. Obtain an ohmmeter and set the switches for Rx1 scale. Connect the Red meter lead to the brush lead. Connect the Black meter lead to the field metal frame. If zero ohms is indicated - full continuity - the field is defective, shorted out, and must be replaced. If zero ohms is not indicated or - no continuity - the field windings are good.

15. Check the positive brush lead on the field frame for damaged or broken insulation. Check the negative brush lead on the thermal switch for damage or frayed braided lead.

16. Check the amount of wear to the brushes. If they are worn to half their original length, approximately 3/8 in. or less, the brushes should be replaced. When replacing the brush connected to the field frame wires, cut the braided wire as close to the old brush as possible. Insert the new brush braided wire and the old field braided wire into a wire crimp provided in the

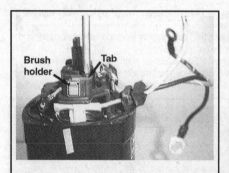

Fig. 57 Remove the brush hold down arms

Fig. 58 Remove the brush holder, spring and brush

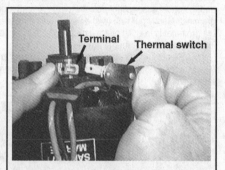

Fig. 59 Remove the thermal switch

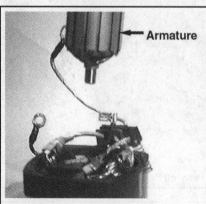

Fig. 60 Lift out the armature...

Fig. 61 ...and then the field frame

Fig. 62 Testing the armature

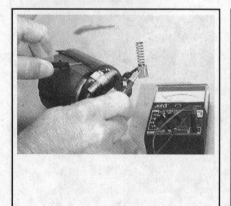

Fig. 63 Testing for a short

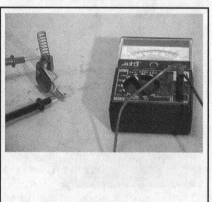

Fig. 64 Testing the thermal switch

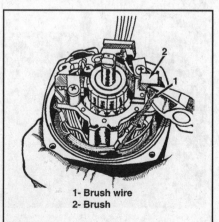

1- Brush wire
2- Brush

Fig. 65 Cut the brush wire as close as possible

TRIM & TILT

brush kit. Squeeze the crimp tightly around both wires. The thermal cut-out switch and brush are replaced as an assembly.

17. Align the marks on the field frame with the marks on the motor case. These marks should have been made during disassembly. If no marks were made, the wire harness from the field must point towards the front of the motor case. The two screw holes in the field frame should also be aligned with the field winding. Slide the field frame down into the case.

18. Place the bronze thrust washer onto the end of the armature. Lower the armature through the center of the field frame and at the same time, guide the end of the armature into the motor case.

19. Align the brush frame with the field screw holes. Note the location of the thermal switch mounting pad. The pad must be directly in front of the Black spade wire or switch, if already installed. Gently pull back the wires on the field and lower the brush frame onto the end of the field frame. Align the screw holes in the brush frame, field, and motor case. Install the two Phillips head screws and tighten them securely.

20. Connect the Black wire terminal to the spade on the thermal switch. Place the thermal switch onto the mounting pad of the brush frame and secure it with the Phillips head screw.

21. Slide the brush holder over the spring and brush. Compress the spring while pushing the brush into the holder. Align the end of the brush with the commutator on the armature. Align the tabs on the bottom of the brush holder with the slots in the brush frame. Insert the brush holder into the brush frame and hold in place with finger pressure. Verify the brush holder is fully seated in the frame and the brush is contacting the commutator end of the armature.

22. Place the lock tab over the brush holder and secure the brush holder in place with a Phillips head screw. Do the same for the other brush and brush holder.

23. Slide a new O-ring over the wire harness and grommet. Place the O-ring into the groove in the motor case.

24. Place the thrust washer over the end of the armature against the commutator. Align the cover with the wire harness grommet. Place a thin coating of liquid neoprene on the grommet to ensure a good seal.

25. Lower the cover over the end of the armature and align the tape marks and fastener holes. Check to be sure the O-ring remains in the groove and the grommet fits into place. Apply a drop of Loctite® to the screw threads and secure the cover in place with the four Phillips head screws. Tighten the screws alternately and evenly in a cross-sequence. Do not over-tighten the screws.

26. Apply a small amount of Quicksilver 2-4-C marine lubricant to the pump and motor coupler. Position the coupler with the narrow slot facing down towards the pump and the larger slot facing up. Place the coupler onto the end of the hydraulic pump shaft. Be sure the coupler is fully engaged with the slot on the end of the pump shaft.

27. Place a new O-ring into the groove in the adapter. Align the scribe marks or tape strip on the motor case with the adapter. The wire harness should be facing forward.

28. Slowly lower the motor onto the adapter. Be sure the flats on the end of the motor shaft engage the slot in the coupler. Secure the motor to the adapter with two hex head bolts. Tighten the bolts to 25 inch lbs. (2.9 Nm).

Install the pump and motor assembly into the support bracket.

Trim Cylinder

GENERAL INFORMATION

◆ See Figures 66 and 67

■ If troubleshooting procedures have isolated a problem to the trim cylinders, for example: a leaking oil scraper seal around the rod or a defective impact valve, it is strongly recommended that the cylinders be removed and replaced with new ones, rather than attempting to disassemble the cylinders. This recommendation is based on the fact considerable difficulty may be encountered in removing the end cap from the cylinder. Even with the aid of the special tool for this purpose, the task is most difficult, sometimes impossible.

■ Removing the end cap without the use of the special tool, could be termed "impossible". The risk of damaging the small holes in the end cap used to hold the tool is very high. Should these holes be even slightly damaged, not even the special tool will remove the end cap. The cylinder would be completely unserviceable.

Other than the end cap removal, the rest of the components are relatively easily serviced.

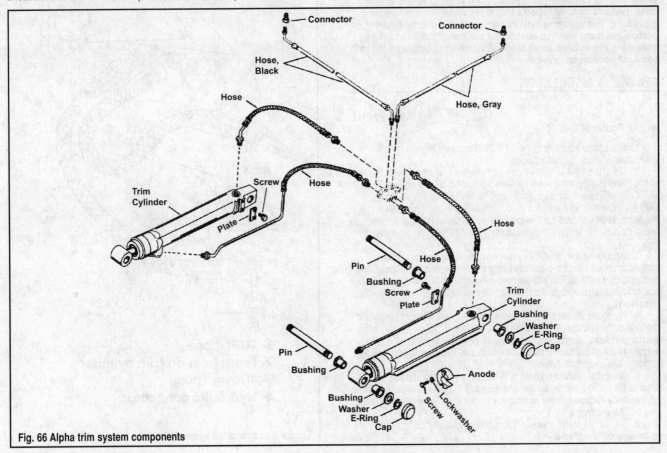

Fig. 66 Alpha trim system components

13-18 TRIM & TILT

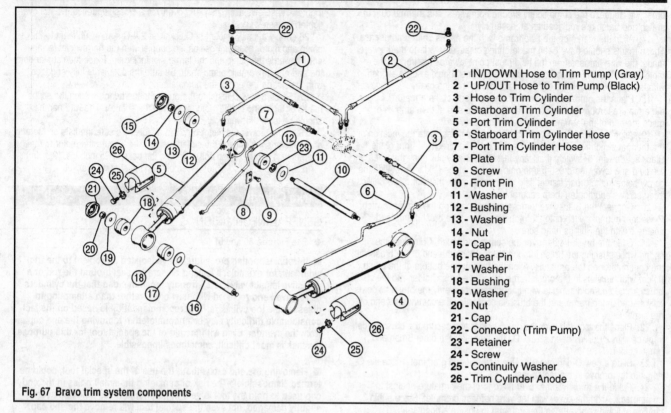

1 - IN/DOWN Hose to Trim Pump (Gray)
2 - UP/OUT Hose to Trim Pump (Black)
3 - Hose to Trim Cylinder
4 - Starboard Trim Cylinder
5 - Port Trim Cylinder
6 - Starboard Trim Cylinder Hose
7 - Port Trim Cylinder Hose
8 - Plate
9 - Screw
10 - Front Pin
11 - Washer
12 - Bushing
13 - Washer
14 - Nut
15 - Cap
16 - Rear Pin
17 - Washer
18 - Bushing
19 - Washer
20 - Nut
21 - Cap
22 - Connector (Trim Pump)
23 - Retainer
24 - Screw
25 - Continuity Washer
26 - Trim Cylinder Anode

Fig. 67 Bravo trim system components

The following procedures cover the removal and installation of a single cylinder. Simply repeat the procedures for the other side unless specific instructions are included.

■ Bravo units contain a Trim-In Limit insert in the drive unit to keep certain vessels from rolling onto their sides under extreme operating conditions. This system is detailed extensively in the Drive System section, but make sure you observe the position of the insert before attempting any work on these drive units - forward position on Bravo I and II, aft position on Bravo III

REMOVAL & INSTALLATION

MODERATE

◆ See Figures 66 thru 74

1. Move the stern drive to the full DOWN position. Place a suitable drain pan under the hydraulic hoses.
2. Disconnect the UP/OUT hydraulic hose from the front hole in the end of the cylinder using the correct size flare nut wrench or line wrench. These wrenches will prevent damaging the hex flats on the line or hose fittings, if they are extremely tight and/or a slight amount of corrosion has built up on the fitting. Most standard open-end wrenches will flex under high torque loads, causing the wrench to slip, damaging the hex fitting on the hydraulic line.
3. Disconnect the IN/DOWN hydraulic hose fitting at the hydraulic connector on the transom assembly (using the line wrench also). Install an hydraulic plug (#22-38609) into the hole in the connector, and plug the hoses and/or fittings to prevent draining the trim/tilt hydraulic system any more than necessary.
4. Pry off the plastic cover on the end of the forward anchor pin (this is the side of the cylinder closest to the boat). Pull the E-Clip off the end of the anchor pin (Alpha) or remove the locknut (Bravo) and then slide off the flat washer and then the bushing from the anchor pin. Repeat this step for the opposite end of the trim cylinder.
5. Grasp the cylinder on the inside of each anchor pin and pull both ends of the cylinder off the anchor pins. Remove the flat washer and bushing on the inside surface cylinder ends or the anchor pins. There is also a snapring on Bravo drives.
6. Repeat the above steps for the opposite cylinder if both cylinders are to be repaired or replaced.

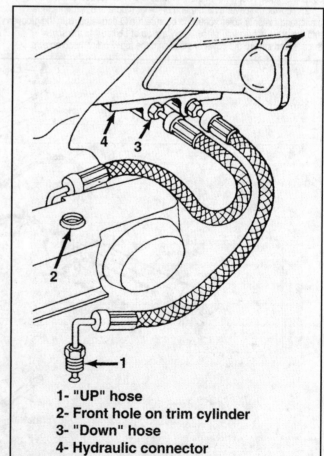

1- "UP" hose
2- Front hole on trim cylinder
3- "Down" hose
4- Hydraulic connector

Fig. 68 Disconnect the hydraulic hoses

TRIM & TILT 13-19

To install:

7. Position the port and starboard trim cylinders so the offset of the piston rod eyelets face the stern drive. The UP port for the cylinder hose connections should also be facing. . .up.

8. Install a washer and bushing onto the forward and aft anchor pin for the cylinder (Bravos will need the snapring as well). Place the ends of the cylinder over the bushings and push the cylinder ends onto the bushing and the anchor pins.

9. Install another set of bushings onto the forward and aft anchor pins for the cylinder. Push the bushing onto the anchor pin and into the end of the cylinder. Slide a flat washer onto each anchor stud and secure the cylinder to the anchor stud with an E-ring on the Alpha. On the Bravo, tighten the locknuts finger-tight and then tighten each one until the nut and washer just bottom on the shoulder of the anchor pin - no more.

■ Bravo units, ensure that the trim-in insert is in the same position it was on cylinder removal - forward for the Bravo I and II units and aft for Bravo III units.

10. Install a new plastic end cap over the ends of the anchor pins. Repeat the last three steps for the other trim cylinder if it was also removed.

11. Remove the plugs from the end of the hose fittings and cylinder port fittings. Bleed the system as detailed previously and then connect the hydraulic hoses to the ports. Carefully tighten the fittings to 70-150 inch lbs. (8-17 Nm) on the Alpha, or 110 inch lbs. (12 Nm) on the Bravo.

DISASSEMBLY & ASSEMBLY

◆ See Figures 75 thru 84

1. Remove the trim cylinder.
2. Hold the cylinder over the drain pan; extend, and then retract the cylinder 2-3 times to remove all fluid from the cylinder.
3. Remove the two bolts securing the retaining plate on the IN/DOWN hydraulic hose. Using the correct size line or flare nut wrench, disconnect the hose from the cylinder. Place the hose into a plastic bag to prevent contamination from entering the hose.

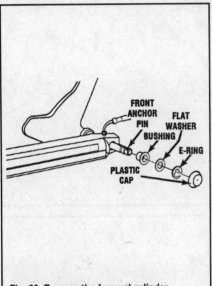

Fig. 69 Remove the forward cylinder hardware (Alpha). . .

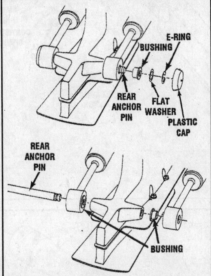

Fig. 70 . . .and then remove the aft hardware (Alpha)

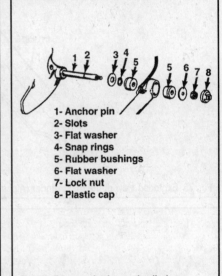

Fig. 71 Remove the forward cylinder hardware (Bravo). . .

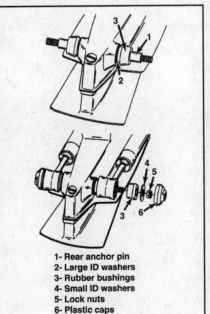

Fig. 72 . . .and then remove the aft hardware (Bravo)

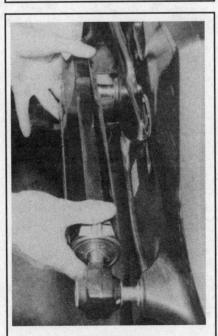

Fig. 73 Removing the cylinder

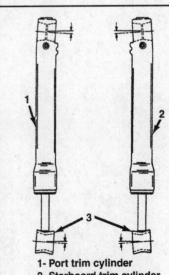

Fig. 74 Install the cylinder like this on the Bravo

13-20 TRIM & TILT

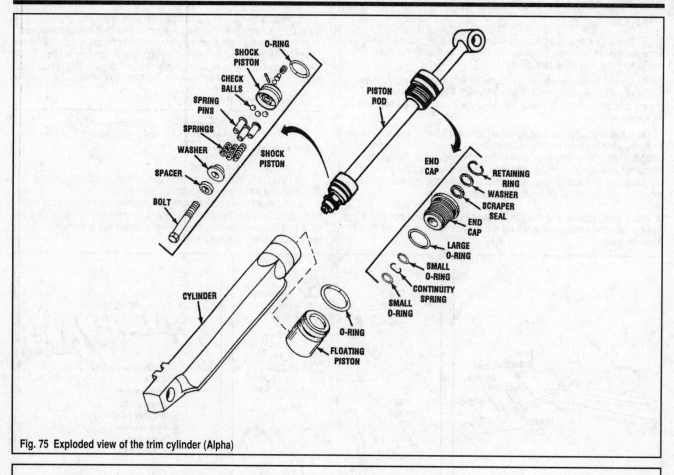

Fig. 75 Exploded view of the trim cylinder (Alpha)

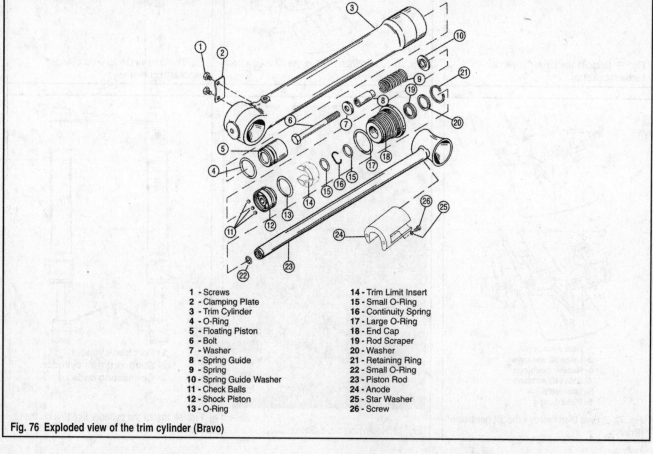

1 - Screws
2 - Clamping Plate
3 - Trim Cylinder
4 - O-Ring
5 - Floating Piston
6 - Bolt
7 - Washer
8 - Spring Guide
9 - Spring
10 - Spring Guide Washer
11 - Check Balls
12 - Shock Piston
13 - O-Ring
14 - Trim Limit Insert
15 - Small O-Ring
16 - Continuity Spring
17 - Large O-Ring
18 - End Cap
19 - Rod Scraper
20 - Washer
21 - Retaining Ring
22 - Small O-Ring
23 - Piston Rod
24 - Anode
25 - Star Washer
26 - Screw

Fig. 76 Exploded view of the trim cylinder (Bravo)

TRIM & TILT 13-21

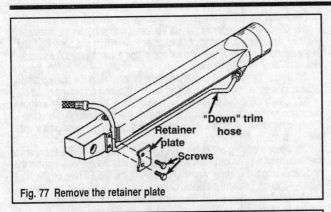
Fig. 77 Remove the retainer plate

4. Remove the trim cylinder anode from the end cap (two screws) on Bravo units.

5. Place the front mounting flange of the cylinder in a vise equipped with soft-face jaws and carefully tighten it.

6. Obtain a spanner wrench (#91-821709) or an equivalent tool. Removal of the end cap is difficult using the special tool: and could be termed "impossible" without the tool. Insert the tangs of the special tool into the holes in the end cap. If needed, slide a long breaker bar onto the tool so the bar is in the same plane as the tool. This position will provide maximum mechanical advantage. Remove the end cap. Tap the breaker bar, if necessary, but bear in mind - if the holes in the end of the cap become damaged (elongated), the cylinder might as well be given the "deep six". If the attempt to remove the cap is successful, congratulations! Continue to unscrew the end cap until it is held by a single thread. Extend the rod, and then continue to remove the end cap and piston assembly.

7. Pop the tilt limit insert off of the piston rod on Bravo units

8. Remove the cylinder from the vise and invert it to remove the floating piston from the cylinder. Note the orientation of the piston as it is removed from the cylinder. Remove and discard the O-ring from the floating piston.

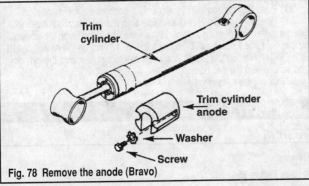
Fig. 78 Remove the anode (Bravo)

■ The factory, or boat builder, may have installed a spacer on the end of the piston rod to prevent full IN trim on certain Alpha drives. This spacer - if equipped - is located between the head of the piston bolt and the flat washer on the piston assembly. If this spacer is installed, it must be re-installed during assembling. Failure to ensure these parts are correctly installed could cause the bow to plow during full IN trim and possibly induce unwanted or dangerous steering changes. Replacement cylinders will usually have the capability for increased trim-in range (approx. 1 1/2° with spacer removed) which could improve acceleration on some boats, but we recommend discussing this with an authorized Mercury facility prior to deciding to remove the spacer.

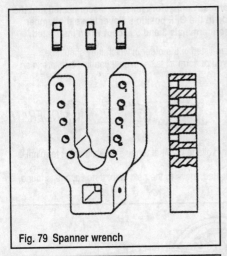
Fig. 79 Spanner wrench

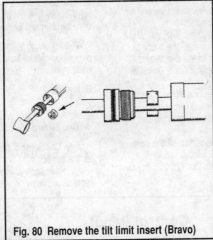
Fig. 80 Remove the tilt limit insert (Bravo)

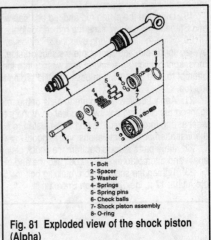
Fig. 81 Exploded view of the shock piston (Alpha)

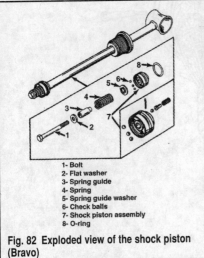
Fig. 82 Exploded view of the shock piston (Bravo)

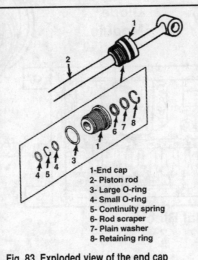

Fig. 83 Exploded view of the end cap

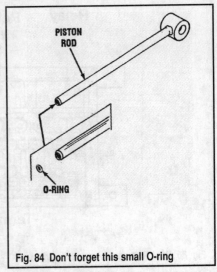
Fig. 84 Don't forget this small O-ring

13-22 TRIM & TILT

※※ CAUTION

Never attempt to substitute a trim cylinder from another style drive unit or application.

9. Place the eyelet end of the piston into a vise and clamp it tight. Slowly remove the bolt from the end of the piston. As the bolt loosens, the tension on the shock piston and springs will lessen. When all tension has been removed, unscrew the bolt and slide off the spacer, washer, springs, spring pins, check balls, and shock piston assembly from the end of the piston rod.
10. Grasp the end cap and slide it off the piston rod. Remove the retainer ring, washer, scraper, O-rings, and continuity spring from the end cap.
11. Remove the small O-ring from inside the end of the piston rod.

To assemble:

12. Make an effort to keep the work area clean, because any contamination on the piston could lead to a malfunction. Inspect the interior of the cylinder for any signs of scoring or roughness. Clean all surfaces with safety solvent and blow them dry with compressed air.
13. The internal check balls in the shock piston are an integral part of the piston, and should not be removed. If the shock piston has any signs of corrosion or damage it must be replaced with a new unit.
14. Clean the end cap threads with a wire brush to remove all traces of the old sealant.
15. Lubricate all O-rings and interior components with Power Trim and Steering Fluid.
16. Place a new small O-ring into the end of the piston rod.
17. Insert a small O-ring, continuity spring, and another small O-ring into the rod end cap.
18. Insert a new scraper seal, flat washer and retaining ring into the opposite end of the rod cap. Slide the large O-ring over the threads of the rod cap and into the groove.
19. Lubricate the piston rod and rod end cap with Quicksilver Power Trim and Steering Fluid. Now, slide the rod cap onto the piston rod.
20. Slide the large O-ring over the shock piston. Install a check ball and eyelet, spring pin, and spring into the shock piston. Now install the 3 check balls, spring guide(s), spring(s) and washer into the end of the piston and attempt to hold everything in position while sliding the piston into the end of the rod.
21. Apply Loctite® 271 to the threads of the retaining bolt. Now, attempt to insert the retaining bolt (with spacer on Alpha if equipped) through the washer and thread it into the end of the piston rod. If success was gained on the first attempt, congratulations, you're good!
22. If the parts slipped out from the shock piston, reposition all the parts again and attempt to thread the bolt into the end of the cylinder rod.
23. Tighten the shock piston retaining bolt to 15-20 ft. lbs. (20-27 Nm) on the Alpha; 17 ft. lbs. (23 Nm) on Bravo units.
24. Install the tilt limit insert onto the piston rod (Bravo).
25. Clamp the cylinder into a vise equipped with soft-jaws as detailed in the third step. Apply a coating of Quicksilver Power Trim and Steering Fluid to the interior surfaces of the cylinder and piston assembly - you can also use 10W-30 or 10W-40 motor oil.
26. Place a new O-ring on the floating piston. Insert the floating piston into the cylinder in the correct position, as noted during disassembly.
27. Slowly lower the piston rod into the cylinder. Work the loose end of the piston rod in a circular motion to help start the shock piston O-ring into the cylinder. Once the piston starts down into the cylinder, press gently until the cap contacts the cylinder.
28. Apply a coating of Special Lubricant 101 or 2-4-C Marine Lubricant to the threads of the end cap. Thread the end cap into the cylinder. Tighten the end cap, using the spanner wrench, to 40-50 ft. lbs. (55-68 Nm) on the Alpha; 45 ft. lbs. (61 Nm) on Bravo units.
29. On Bravo units, install the anode to the end cap and tighten the two screws to 30 inch lbs. (3.4 Nm).
30. Push the rod into the cylinder until it bottoms out for installation.
31. Position the DOWN hydraulic line onto the cylinder. Start the line fitting into the cylinder finger-tight. Tighten the line fitting using the correct size flare or line wrench to 70-150 inch lbs. (8-17 Nm) on the Alpha, or 110 inch lbs. (12 Nm) on the Bravo. Place the clamping plate onto the cylinder and secure with the two bolts.
32. Install the cylinder(s).

Dual Power Trim System

TESTING

General

◆ See Figure 85

■ The control box harness connectors must be disconnected and the ignition switch must be in the OFF position. Never leave the jumper wire connected between terminals 3 and 5 except when instructed.

The following tests are listed in the order in which they should be performed. Make sure you perform all tests, even if you find a failure in an early one.

Relay No. 1

◆ See Figure 85

1. Check for battery voltage (12V) at terminal **2** while using terminal **4** as the ground.
2. If voltage is indicated, move to the next step. If no voltage is found, replace the relay.

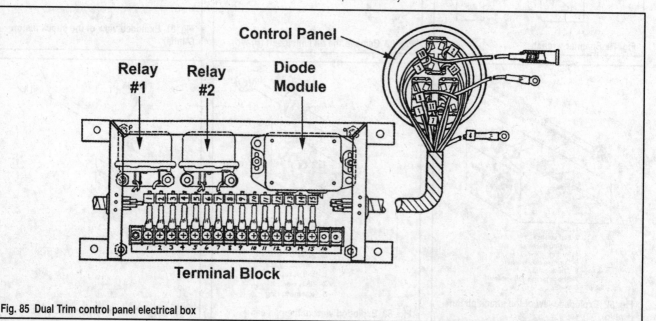

Fig. 85 Dual Trim control panel electrical box

TRIM & TILT 13-23

3. Connect a jumper wire between terminals **3** and **5** and then repeat Step 1 again.
4. If there is no voltage shown, the relay is OK; if voltage is indicated, replace the relay.

Relay No. 2

◆ See Figures 85

1. Check for continuity between terminals **13** and **9**.
2. If continuity is indicated, move to the next step. If no continuity is found, replace the relay.
3. Connect a jumper wire between terminals **3** and **5** and then repeat Step 1 again.
4. If there is no continuity shown, the relay is OK; if continuity is indicated, replace the relay.

Diode Module

◆ See Figure 85

Perform all test with an ohmmeter set on the RX1 scale. Take 2 readings for each test; note the 1st reading and then reverse the leads and note the 2nd reading.

A good diode will indicate a high or infinite resistance reading when tested one way and then upon reversing the meter leads, it should test as a low reading (under 60 ohms). Or vice versa, depending on which side you start with. If both readings are the same, replace the diode module.

1. **Diode 1**: Connect a jumper wire between terminals **3** and **5**. Now test the diode between terminals **9** and **10**.
2. **Diode 2**: Connect a jumper wire between terminals **3** and **5**. Now test the diode between terminals **10** and **13**.

■ Remove the fuse from the red/purple harness lead to ensure you can't short the control box or the meter.

3. **Diode 3**: Test the diode between terminals **6** and **12**.
4. **Diode 4**: Test the diode between terminals **12** and **7**.
5. **Diode 5**: Test the diode between terminals **8** and **11**.
6. **Diode 6**: Test the diode between terminals **14** and **15**.
7. **Diode 7**: Test the diode between terminals **8** and **5**.
8. **Diode 8**: Test the diode between terminals **5** and **15**.
9. Replace the fuse.

Trailer Switch

◆ See Figure 85

1. Remove the fuse from the red/purple harness lead.
2. Set your ohmmeter on the RX1 scale.
3. Have someone press down on the TRAILER switch and then check for continuity between terminals **10** and **3**. If there is no continuity indicated on the meter, replace the switch; otherwise move to the next step.
4. Have someone push up on the TRAILER switch and then check for continuity between terminals **2** and **12**. If there is continuity indicated on the meter, the switch is OK; otherwise replace the switch.

Starboard Trim Switch

◆ See Figure 85

1. Set your ohmmeter on the RX1 scale.
2. Have someone press down on the STARBOARD TRIM switch and then check for continuity between terminals **1** and **9**. If there is no continuity indicated on the meter, replace the switch; otherwise move to the next step.
3. Have someone push up on the STARBOARD TRIM switch and then check for continuity between terminals **11** and **6**. If there is no continuity indicated on the meter, the switch is OK; otherwise replace the switch.

Port Trim Switch

◆ See Figure 85

1. Set your ohmmeter on the RX1 scale.
2. Have someone press down on the PORT TRIM switch and then check for continuity between terminals **2** and **13**. If there is no continuity indicated on the meter, replace the switch; otherwise move to the next step.
3. Have someone push up on the PORT TRIM switch and then check for continuity between terminals **14** and **7**. If there is continuity indicated on the meter, the switch is OK; otherwise replace the switch.

RELAY REPLACEMENT

◆ See Figures 86 and 87

1. Disconnect the negative battery cables.
2. Remove the eight screws and lift off the control box cover.
3. Carefully cut off the relay electrical leads from the terminal block as close to the relay terminals as possible.
4. Remove the two mounting screws and nuts and lift out the relay.

To install:

5. Strip some insulation off the ends of each electrical lead and carefully re-solder each to its respective relay terminal.

■ Use only 63/67 (tin/lead) alloy solder when re-attaching the leads. Never use acid core solder or you will damage the relay. Always coat the terminal connections with Liquid Neoprene.

6. Reinstall the relay and tighten the mounting screws securely.
7. Install the control box cover and tighten the screws securely.
8. Connect the battery cable.

DIODE MODULE REPLACEMENT

◆ See Figure 88

1. Disconnect the negative battery cables.
2. Remove the eight screws and lift off the control box cover.
3. Tag and disconnect the electrical leads from the terminal block.

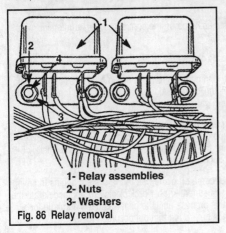

Fig. 86 Relay removal
1- Relay assemblies
2- Nuts
3- Washers

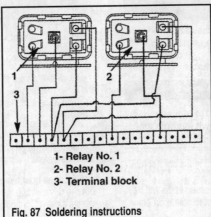

Fig. 87 Soldering instructions
1- Relay No. 1
2- Relay No. 2
3- Terminal block

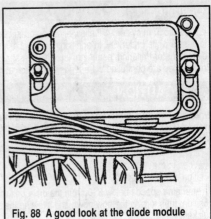

Fig. 88 A good look at the diode module

13-24 TRIM & TILT

4. Remove the two mounting screws and nuts and lift out the module.

To install:

5. Connect the electrical leads to their respective terminals on the module.
6. Install the module and tighten the mounting screws securely.
7. Install the control box cover and tighten the screws securely.
8. Connect the battery cable.

TRIM CONTROL SWITCH REPLACEMENT

◆ See Figure 89 and 90

1. Disconnect the battery cables.
2. Remove the trim control panel at the dash.
3. Carefully cut off the electrical leads as close to the switch terminal as possible.
4. Remove the bezel nut and pull out the switch.

To install:

5. Slide the switch back into the panel and thread on the bezel nut until it seats against the panel.
6. Loop the electrical leads through the respective terminal eyelet and solder them to the terminal.

■ Use only 63/67 (tin/lead) alloy solder when re-attaching the leads. Never use acid core solder or you will damage the relay. Always coat the terminal connections with Liquid Neoprene.

7. Install the panel and connect the battery cables.

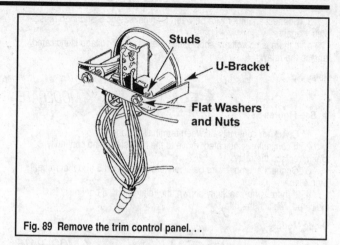

Fig. 89 Remove the trim control panel...

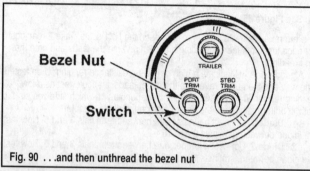

Fig. 90 ...and then unthread the bezel nut

AUTO TRIM II SYSTEM

Description

The auto trim system consists of the following components:
- Auto trim pump. This is the same pump used in the standard system and all service procedures are the same.
- Control module. An electronic device sensing engine speed which regulates the time at which the stern drive is trimmed In and Out while operating in the Auto mode. The module is protected by a 20 amp fuse.
- Mode switch. Allows the captain to select either auto or manual modes at the helm.
- Manual trim control. This will only work with the mode switch in the **Manual** mode and the ignition switch in the **RUN** position. It will allow the operator to raise and lower the drive for trailering, beaching, launching and shallow water operation. It will also allow the operator to manually adjust the trim angle while underway in conditions that may not warrant Auto operation.
- Trim limit switch. This switch controls the maximum trim out limit in both manual and auto modes. Service procedures are the same as detailed for the switch in the standard trim section.

Operation

The Auto Trim system will allow either automatic or manual trimming of the vessel while underway. In addition, the manual mode will allow drive adjustment for trailering, beaching, launching or shallow water operations. Both modes are controlled via a 2 position switch at the helm.

✽✽ CAUTION

When the ignition is turned to the RUN position and the trim switch is in the AUTO position, the drive unit will automatically lower from the raised position.

When in the Auto mode, the system will keep the drive in the correct trim position automatically.

At engine speeds below 3100 rpm the drive unit will remain in the full IN position, thus forcing the bow down and allowing the boat to get up on plane quicker.

Once the engine speed exceeds 3100 rpm, and after a short delay (5-6 seconds) the drive will be trimmed Out to a preset position. The Trim Out position can be adjusted to tune the trim angle for each particular vessel.

On deceleration, the unit will begin to trim In as the engine slows below 2600 rpm (most vessels will still be on plane at this time). Once the drive reaches the full In position, the trim pump motor will continue to run for about 5 seconds.

Moving the trim switch to the **Manual** mode will deactivate the Auto system and allow for manual trim adjustment. You will also need this mode to raise the unit for trailering, etc.

Control Module

ADJUSTMENT

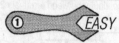

1. Rotate the adjustment knob on the control module counterclockwise until it stops - do not force it.
2. Turn the knob out (clockwise) a total of 4 clicks.
3. Make sure that the trim mode switch is in the Auto position and that the drive unit is in the full In/Down position. Slowly accelerate the boat while keeping an eye on the trim gauge.
4. If the boat is on plane before the system begins to trim out the drive, rotate the adjustment knob counterclockwise 1 click and accelerate slowly. Continue this process until the boat has been on plane for approximately 5-6 seconds before the trim out occurs.
5. If the boat has not come on plane before the system begins trimming out the drive, rotate the adjustment knob 1 click clockwise and accelerate the vessel slowly. Continue this process until the boat has been on plane for approximately 5-6 seconds before the trim out occurs.

TRIM & TILT

WIRING SCHEMATICS

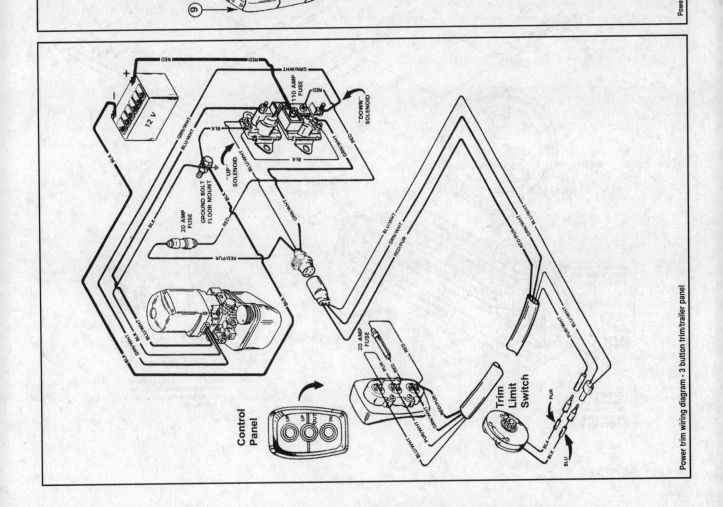

13-26 TRIM & TILT

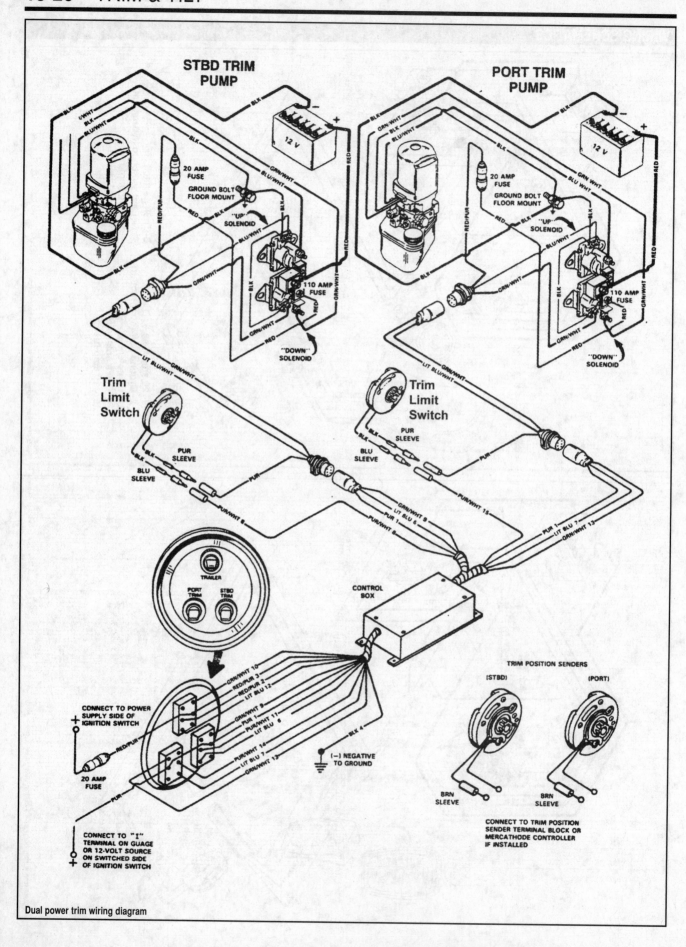

Dual power trim wiring diagram

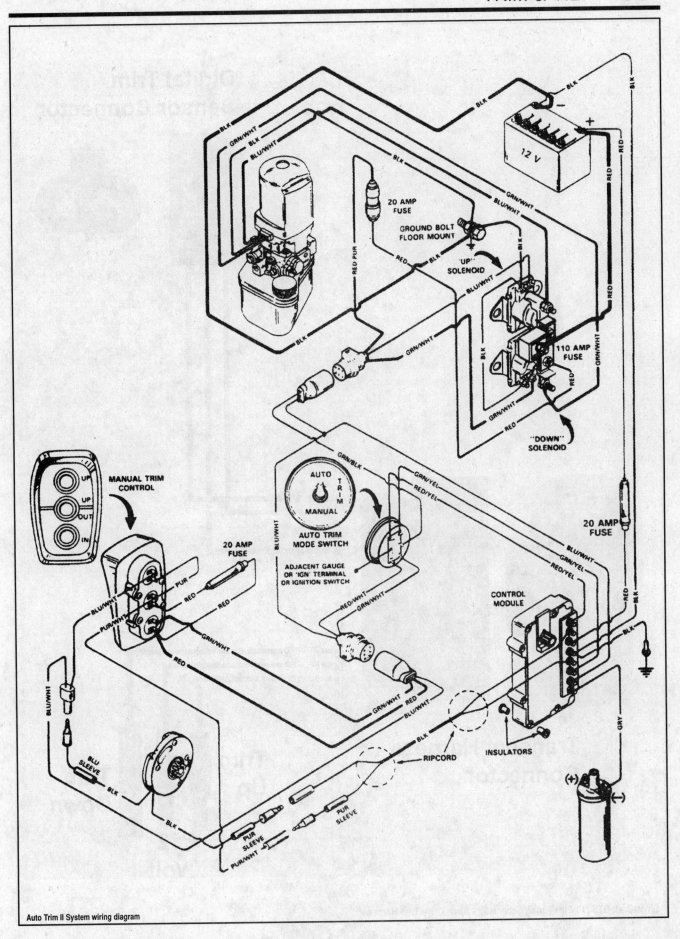

Auto Trim II System wiring diagram

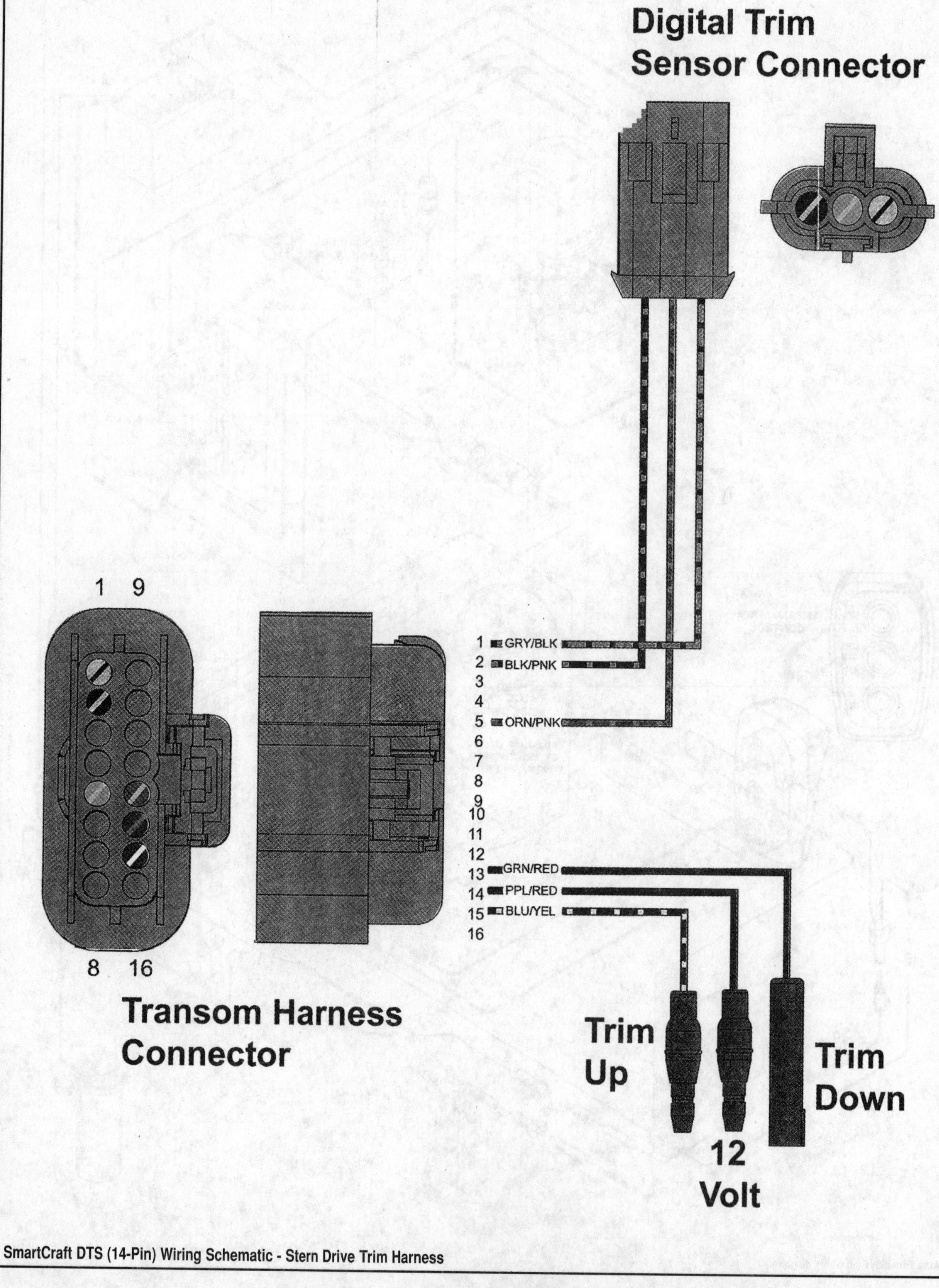

13-28 TRIM & TILT

Digital Trim Sensor Connector

1 — GRY/BLK
2 — BLK/PNK
3
4
5 — ORN/PNK
6
7
8
9
10
11
12
13 — GRN/RED
14 — PPL/RED
15 — BLU/YEL
16

Trim Up — 12 Volt — Trim Down

Transom Harness Connector

1 9

8 16

SmartCraft DTS (14-Pin) Wiring Schematic - Stern Drive Trim Harness

TRIM & TILT 13-29

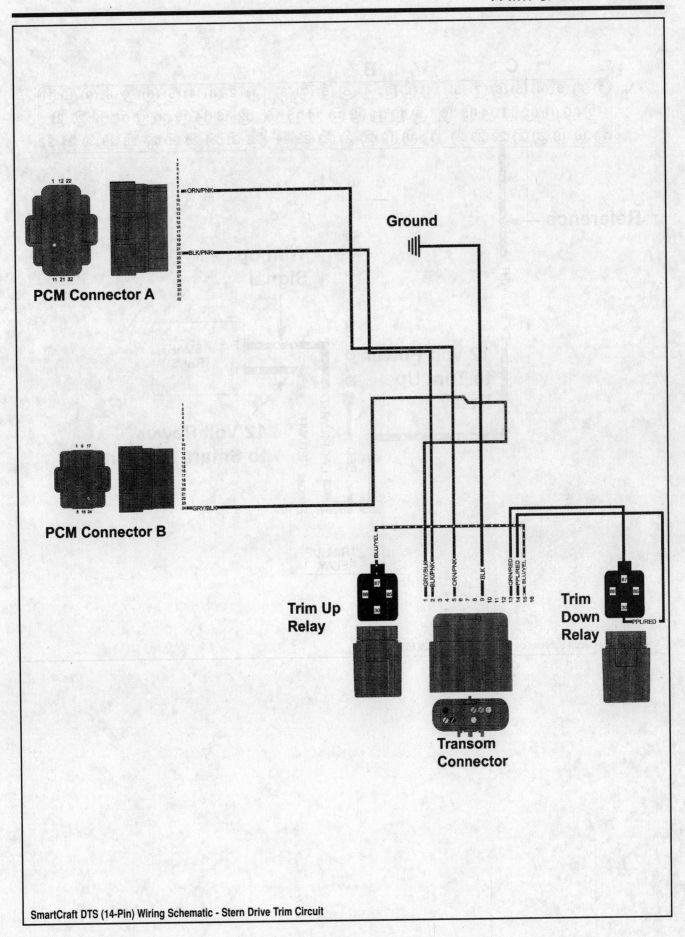

SmartCraft DTS (14-Pin) Wiring Schematic - Stern Drive Trim Circuit

13-30 TRIM & TILT

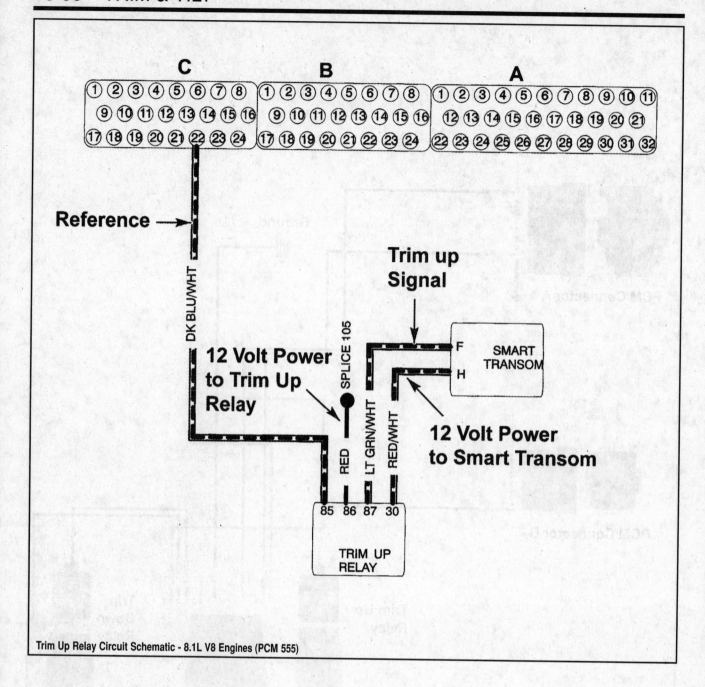

Trim Up Relay Circuit Schematic - 8.1L V8 Engines (PCM 555)

BOOSTER CYLINDER - POWER ... 14-12
   REMOVAL & INSTALLATION ... 14-12
   TESTING ... 14-13
COMPACT HYDRAULIC SYSTEM ... **14-19**
   DESCRIPTION ... 14-19
   HYDRAULIC FLUID ... 14-19
   STEERING UNIT ... 14-19
CONTROL VALVE - POWER ... 14-10
   BALANCING ... 14-12
   DISASSEMBLY & ASSEMBLY ... 14-10
   REMOVAL & INSTALLATION ... 14-10
MANUAL STEERING SYSTEM ... **14-2**
   GENERAL INFORMATION ... 14-2
   STEERING CABLE ... 14-2
   SWIVEL RING ... 14-2
POWER STEERING COOLER ... 14-16
   REMOVAL & INSTALLATION ... 14-16
POWER STEERING PUMP ... 14-13
   BLEEDING ... 14-13
   DISASSEMBLY & ASSEMBLY ... 14-15
   FLOW CONTROL VALVE SERVICE ... 14-14
   FLUID LEVEL ... 14-13
   OIL SEAL REPLACEMENT ... 14-15
   PUMP PULLEY ... 14-14
   REMOVAL & INSTALLATION ... 14-13
      4 CYLINDER ENGINES ... 14-13
      V6/V8 ENGINES (EXC. 8.1L) ... 14-13
      8.1L V8 ENGINES ... 14-14
POWER STEERING SYSTEM ... **14-4**
   BOOSTER CYLINDER ... 14-12
   CONTROL VALVE ... 14-10
   DESCRIPTION ... 14-4
   POWER STEERING COOLER ... 14-16
   POWER STEERING PUMP ... 14-13
   POWER STEERING UNIT ... 14-9
   STEERING CABLE ... 14-8
   STEERING CABLE GUIDE TUBE ... 14-8
   TESTING ... 14-4
   TIE BAR - MULTIPLE DRIVES ... 14-17
POWER STEERING UNIT ... 14-9
   GENERAL INFORMATION ... 14-9
   REMOVAL & INSTALLATION ... 14-9
STEERING CABLE - MANUAL ... 14-2
   REMOVAL & INSTALLATION ... 14-2
STEERING CABLE - POWER ... 14-8
   REMOVAL & INSTALLATION ... 14-8
STEERING CABLE GUIDE TUBE - POWER ... 14-8
   REMOVAL & INSTALLATION ... 14-8
STEERING UNIT - COMPACT ... 14-19
   REMOVAL & INSTALLATION ... 14-19
SWIVEL RING - MANUAL ... 14-2
   DISASSEMBLY & ASSEMBLY ... 14-3
   REMOVAL & INSTALLATION ... 14-2
TESTING - POWER ... 14-4
   PUMP LUG TEST ... 14-4
   PUMP PRESSURE TEST ... 14-7
   SYSTEM PRESSURE TEST ... 14-4
      ALPHA ... 14-4
      BRAVO ... 14-6
TIE BAR - MULTIPLE DRIVES ... 14-17
   INSTALLATION ... 14-18
      PORT CABLE ... 14-18
      STARBOARD CABLE ... 14-18
   TIE BAR LENGTH ... 14-17

14

STEERING

MANUAL STEERING SYSTEM ... 14-2
POWER STEERING SYSTEM ... 14-4
COMPACT HYDRAULIC SYSTEM ... 14-19

14-2 STEERING

MANUAL STEERING SYSTEM

General Information

The manual steering system on your boat is very simple - a steering wheel, a steering cable and a steering lever in the stern drive unit. Due to the variety of steering systems available, the following procedures are general in their scope. Please refer to the Maintenance section for Lubrication procedures and Drive Systems for Steering Lever procedures.

Steering Cable

REMOVAL & INSTALLATION

◆ See Figures 1 thru 5  MODERATE

1. Locate the end of the steering cable at the transom. Pull out the cotter pin securing the clevis pin to the cable and the steering lever. Pull out the clevis and disconnect the cable from the steering lever. On dual engine installations, support the tie bar and then repeat the procedure on the opposite side. You will also need to disconnect the clevis assembly from the cable.

■ All Quicksilver guides come standard with a self-locking coupler between the cable and the guide tube. On applications not coming with a self-locking coupler, and external locking device must be in place.

2. On self-locking applications, loosen the coupler. On others, remove the cotter pin and slide the locking sleeve off of the coupler nut, and then loosen the nut. On either application, carefully pull the cable through the guide tube and swivel ring after loosening the nut.
3. Carefully remove the steering cable and then disconnect it at the helm station.

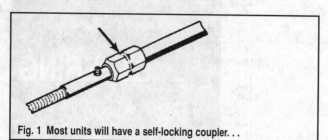

Fig. 1 Most units will have a self-locking coupler. . .

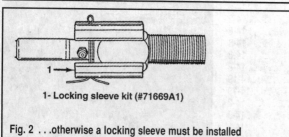

1- Locking sleeve kit (#71669A1)

Fig. 2 . . .otherwise a locking sleeve must be installed

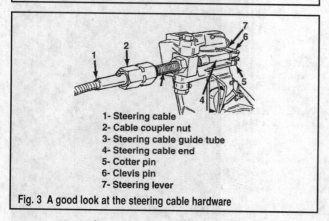

1- Steering cable
2- Cable coupler nut
3- Steering cable guide tube
4- Steering cable end
5- Cotter pin
6- Clevis pin
7- Steering lever

Fig. 3 A good look at the steering cable hardware

To install:

4. Clean the guide tube and steering lever thoroughly and inspect carefully for cracks, wear or other damage.
5. Connect the new cable to the helm as per the cable manufacturer's instructions and run it back through the boat.
6. Coat the end of the steering cable liberally with Special Lubricant 101 and feed it through the coupler and guide tube.
7. Attach the clevis assembly to the end of the cable (if equipped) and then align everything with the hole in the steering lever and insert the cotter pin.
8. Follow the manufacturer's instructions for final installation and adjustment. As a general rule of thumb though, turn the steering wheel all the way to starboard and check that the guide tube protrudes through the ring 3/4 in. (19mm). Thread the coupler nut onto the tube and tighten to 35 ft. lbs. (48 Nm). Check that the distance from the inner end of the nut to the centerline of the cable end hole is 21-3/8 in. (543mm). Mid-point of the cable travel should be 16-7/8 in. (429mm) with no more than 4 1/2 in. (114mm) travel in either direction.
9. Center the steering wheel and confirm that the steering lever is centered in the drive unit.

Swivel Ring

REMOVAL & INSTALLATION

◆ See Figures 6 and 7  MODERATE

1. Remove the steering cable.
2. Use a screwdriver or pair of pliers and bend the locking tabs away from the upper and lower pivot bolts. One side will be away from the bolt and the other will be away from the ridge in the transom assembly.
3. Remove the two pivot bolts and tab washers from the transom assembly and then pull out the swivel ring and guide tube assembly.
4. Lubricate the bushings inside the swivel ring with a good amount of Special Lubricant 101. Do the same to each pivot bolt.
5. Position the assembly into the transom assembly and then screw the bolts (don't forget the washers) in all the way by hand - no wrenches yet.
6. Make sure that the tab washers tangs are straddling the ridge in the transom plate on one side and then tighten them to 25 ft. lbs. (35 Nm) - you can use a wrench this time! Hopefully when you hit the correct torque figure, the remaining tab will align with a flat on the bolt. If so, bend it up and against the bolt head and then bend the other tab down over the transom assembly ridge. If it doesn't align, unscrew it and start over.
7. Move the swivel ring back and forth to ensure that it is pivoting correctly and then install the steering cable.

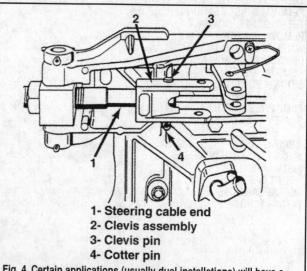

1- Steering cable end
2- Clevis assembly
3- Clevis pin
4- Cotter pin

Fig. 4 Certain applications (usually dual installations) will have a clevis assembly attached to the end of the cable

STEERING 14-3

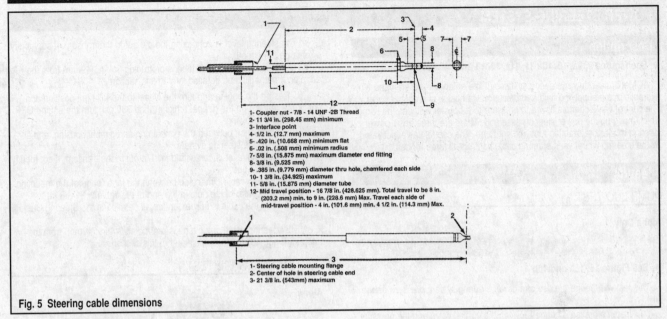

Fig. 5 Steering cable dimensions

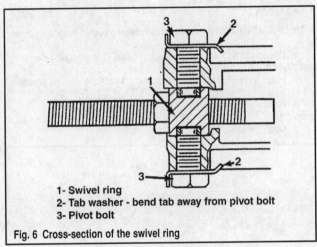

Fig. 6 Cross-section of the swivel ring

1- Swivel ring
2- Tab washer - bend tab away from pivot bolt
3- Pivot bolt

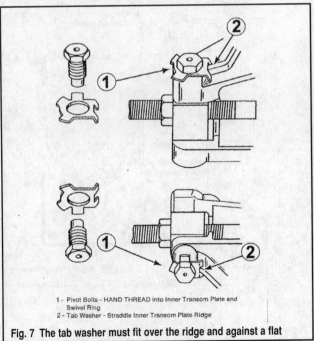

1 - Pivot Bolts - HAND THREAD into Inner Transom Plate and Swivel Ring
2 - Tab Washer - Straddle Inner Transom Plate Ridge

Fig. 7 The tab washer must fit over the ridge and against a flat

DISASSEMBLY & ASSEMBLY

◆ See Figure 8

1. Disconnect the steering cable and remove the swivel ring/guide tube assembly.
2. Mount the unthreaded end of the guide tube in a soft-jawed vise and then carefully apply heat to the locknut to loosen the Loctite.

✶✶ CAUTION

Be very careful when applying heat to the nut that you don't damage the bushings inside the swivel ring.

3. Remove the swivel nut and then thread the swivel itself off of the guide tube.
4. Pry out the two bushings.

To assemble:

5. Inspect the bushings for cracks, wear or other damage. It's not a bad idea to replace them anyway while they're out.
6. Clean the tube threads and the inside of the nut with a wire brush to remove any remaining Loctite and then reapply a good amount of Loctite® 271 to the threads on the area of the tube where the swivel and nut will go.
7. Thread the swivel ring onto the guide tube until the unthreaded end of the tube is protruding 2 1/2 in. (64mm). Thread the nut onto the tube until it seats against the swivel and then tighten it to 35-40 ft. lbs. (47-54 Nm).
8. Install the assembly into the transom assembly and connect the steering cable.

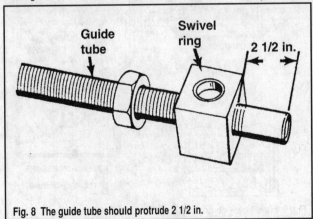

Fig. 8 The guide tube should protrude 2 1/2 in.

14-4 STEERING

POWER STEERING SYSTEM

Description

◆ See Figures 9, 9a, 10, 10a 11, 11a, 12, 13 and 14

All MerCruiser power steering systems utilize an engine-driven, vane-type hydraulic pump supplying fluid and pressure, via hoses, to a control valve. The control valve controls flow and pressure to a booster cylinder, the two of which make up the power steering assembly. There are three basic power steering modes: neutral, left turn and right turn. The control valve, activated by the steering wheel via a steering cable controls all three modes.

Testing

PUMP LUG TEST

Alpha Only

◆ See Figure 15, 15a and 15b

This test will require the use of a power steering system pressure gauge kit.

1. Install a pressure gauge between the control valve and the pressure line from the power steering pump - this is the line on the left as you are looking at the power steering unit from inside the boat.
2. Ensure that the power steering fluid is at the correct level in the reservoir(s).
3. Open the valve on the test gauge, start the engine and allow it to idle.
4. Turn the steering wheel to the full left position and check the reading on the gauge. If higher than 300 psi, turn the engine OFF and check the following:

• Any obstruction between the gimbal ring and gimbal housing, and all moving parts of the steering system.

• The steering lever is contacting the cut-out in the transom. If so, modify or enlarge the cut-out.

• The steering cable guide tube is protruding 3/4 in. (19mm) from the adapter. If not, adjust as detailed later in this section.

5. Restart the engine and turn the wheel to the full right position and check the reading on the gauge. If higher than 300 psi, turn the engine OFF and check the following:

• Any obstruction between the gimbal ring and gimbal housing, and all moving parts of the steering system.

• The steering lever is contacting the cut-out in the transom. If so, modify or enlarge the cut-out.

• The steering cable dimension between the inner edge of the mounting flange and the center of the hole in the end of the cable is 21-3/8 in. (543mm) when the cable is fully extended as detailed in the Steering Cable section.

• The steering cable guide tube is protruding 3/4 in. (19mm) from the adapter. If not, adjust as detailed later in this section.

SYSTEM PRESSURE TEST

Alpha

◆ See Figure 15

This test will require the use of a power steering system pressure gauge kit.

1. Disconnect and remove the steering cable. Disconnect the clevis assembly from the steering lever.
2. Install a pressure gauge between the control valve and the pressure line from the power steering pump.
3. Ensure the power steering fluid is at the correct level in the reservoir(s).
4. Open the valve on the test gauge.

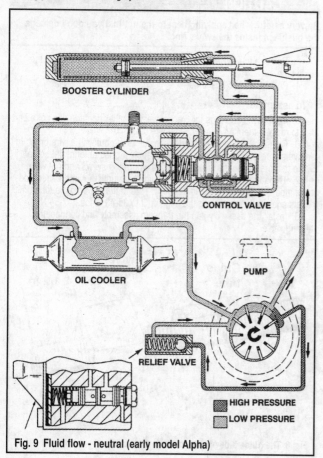

Fig. 9 Fluid flow - neutral (early model Alpha)

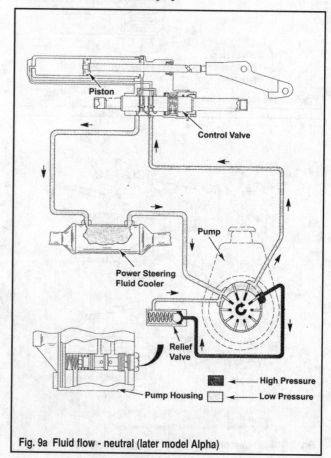

Fig. 9a Fluid flow - neutral (later model Alpha)

STEERING 14-5

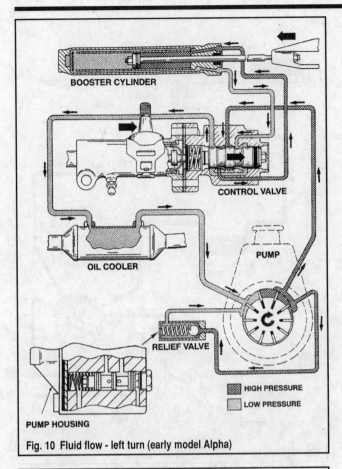

Fig. 10 Fluid flow - left turn (early model Alpha)

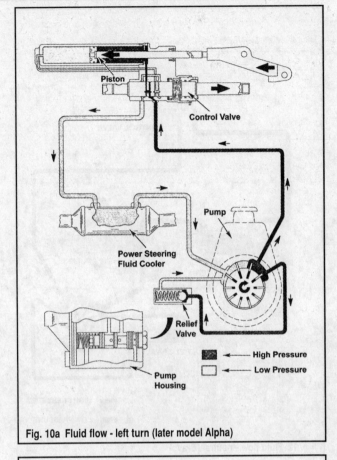

Fig. 10a Fluid flow - left turn (later model Alpha)

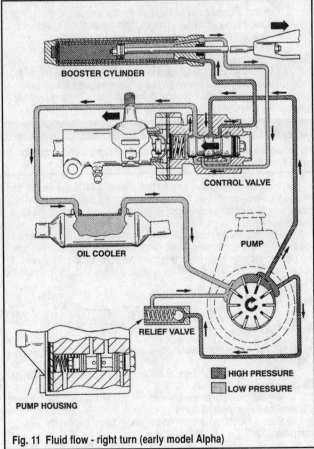

Fig. 11 Fluid flow - right turn (early model Alpha)

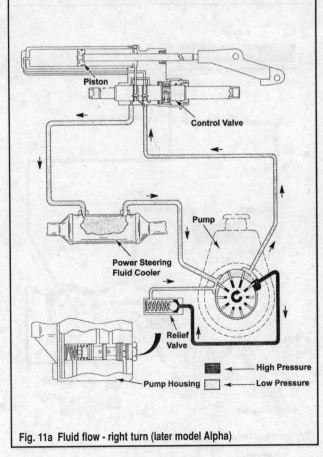

Fig. 11a Fluid flow - right turn (later model Alpha)

14-6 STEERING

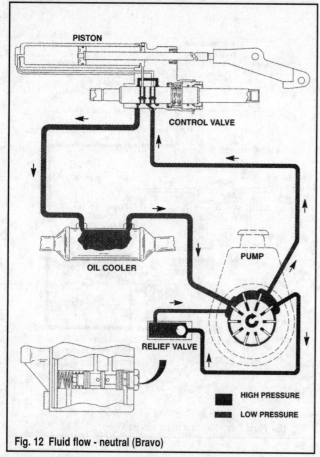

Fig. 12 Fluid flow - neutral (Bravo)

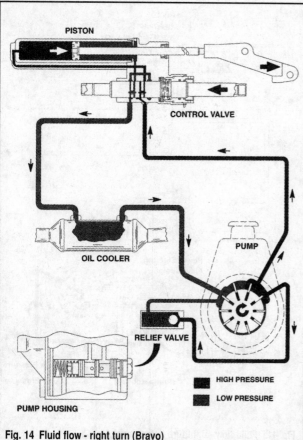

Fig. 14 Fluid flow - right turn (Bravo)

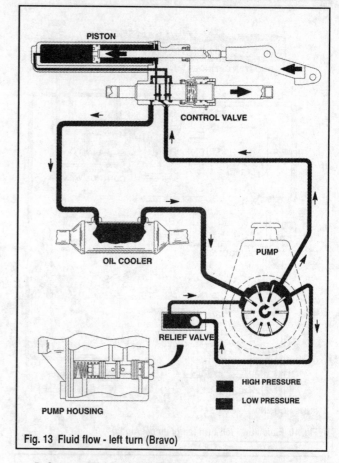

Fig. 13 Fluid flow - left turn (Bravo)

5. Connect a flushing device between a water source and the flush port on the starboard side of the drive. Turn on the water to about half of its maximum flow and fill the cooling system. Do not use full tap pressure.
6. Start the engine and run it at 1000-1500 rpm until it reaches normal operating temperature.
7. Allow the engine to drop down to idle and check the reading on the gauge. If lower than 70 psi, move on to the Pump Pressure Test. If higher than 125 psi, check for restrictions in the hydraulic lines. For any pressure in between the above readings, move to the next step.

✳✳ CAUTION

Never lug the pump at maximum pressure for more than 5 seconds when performing the next steps.

8. Push the control valve adapter block MOMENTARILY to the left and then to the right - do not hold the block over for more than 5 seconds. The gauge should show an instant increase in pressure when pushing the block in either direction.
9. Push the adapter block to the right until the booster cylinder piston rod is fully retracted into the cylinder. Once retracted, MOMENTARILY push the block to the right again until the highest pressure reading is obtained. If above 1000 psi, the system pressure is good. If below 1000 psi, move to the Pump Pressure Test.

Bravo

◆ See Figure 16

OEM ② MODERATE

This test will require the use of a power steering system pressure gauge kit.

1. Remove the front and rear clevis pins and then retract the cable into the guide tube.
2. Install a pressure gauge between the control valve and the pressure line from the power steering pump.
3. Ensure the power steering fluid is at the correct level in the reservoir(s).

STEERING 14-7

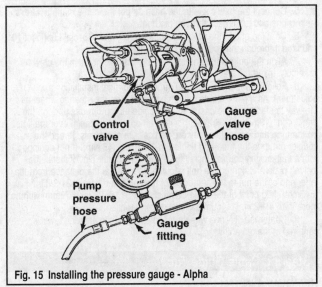

Fig. 15 Installing the pressure gauge - Alpha

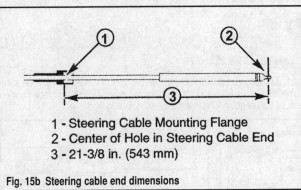

1 - Steering Cable Mounting Flange
2 - Center of Hole in Steering Cable End
3 - 21-3/8 in. (543 mm)

Fig. 15b Steering cable end dimensions

4. Open the valve on the test gauge.
5. Connect a flushing device between a water source and the flush port on the starboard side of the drive. Turn on the water to about half of its maximum flow and fill the cooling system. Do not use full tap pressure.
6. Start the engine and run it at 1000-1500 rpm until it reaches normal operating temperature.
7. Allow the engine to drop down to idle and check the reading on the gauge. If lower than 70 psi, move on to the Pump Pressure Test. If higher than 125 psi, check for restrictions in the hydraulic lines. For any pressure in between the above readings, move to the next step.

✱✱ CAUTION

Never lug the pump at maximum pressure for more than 5 seconds when performing the next steps.

8. Push in on the steering cable and then pull it out for a moment while observing the gauge - you should see an instant increase in pressure in both directions.
9. Now push the steering cable in again until the booster cylinder piston rod is fully retracted. Once the rod is fully retracted, push in on the cable momentarily until a maximum pressure reading is obtained. If above 1000 psi, the system pressure is good. If below 1000 psi, move to the Pump Pressure Test.

PUMP PRESSURE TEST

♦ See Figures 15 and 16

This test will require the use of a power steering system pressure gauge kit.

1. Install a pressure gauge between the control valve and the pressure line from the power steering pump.

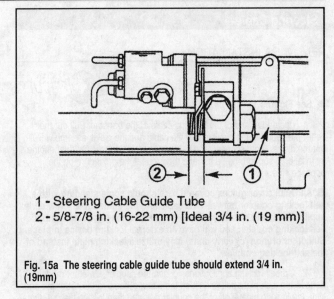

1 - Steering Cable Guide Tube
2 - 5/8-7/8 in. (16-22 mm) [Ideal 3/4 in. (19 mm)]

Fig. 15a The steering cable guide tube should extend 3/4 in. (19mm)

2. Ensure the power steering fluid is at the correct level in the reservoir(s).
3. Connect a flushing device between a water source and the flush port on the starboard side of the drive. Turn on the water to about half of its maximum flow and fill the cooling system. Do not use full tap pressure.
4. Start the engine and run it at 1000-1500 rpm until it reaches normal operating temperature.
5. Close the valve on the gauge just long enough to observe a maximum pressure reading.
6. Now close and open the valve 3 times in a row; observing the maximum pressure reading each time. Record each reading.
7. If the three readings were between 1150 psi and 1250 psi, and were within 50 psi of each other, the pump is OK, you're done. If the pump is OK, but you got low readings in the System Pressure Test, proceed to the Booster Cylinder Test later in this section.
8. If the three readings were between 1150 psi and 1250 psi, but were not within 50 psi of each other, the flow control valve in the power steering pump is sticking or you have clogged hydraulic lines.
9. If all three readings were constant, but under 1000 psi, replace the power steering pump.

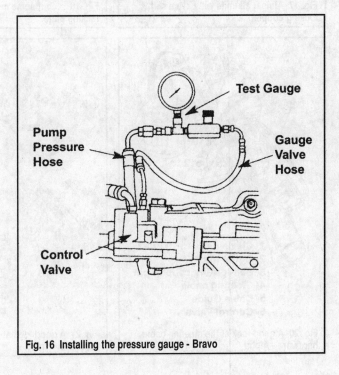

Fig. 16 Installing the pressure gauge - Bravo

14-8 STEERING

Steering Cable

REMOVAL & INSTALLATION

◆ See Figures 17 thru 23

1. Locate the end of the steering cable at the transom. Pull out the cotter pin securing the clevis pin to the cable and the clevis assembly. Pull out the clevis and disconnect the cable from the assembly. On dual engine installations, support the tie bar and then repeat the procedure on the opposite side.

■ All Quicksilver guides covered here should come standard with a self-locking coupler between the cable and the guide tube. A few early applications, and possibly other brand cables, may not come with a self-locking coupler, and will have an external locking device in place. Still other brands (or early units) may utilize a locking plate instead of the self-locking coupler.

2. On self-locking applications, loosen the coupler. On other models, remove the lock plate bolt and lift off the plate, or remove the cotter pin and slide the locking sleeve off of the coupler nut, and then loosen the nut. On all applications, carefully pull the cable through the guide tube after loosening the nut.

3. Carefully remove the steering cable and then disconnect it at the helm station.

To install:

4. Clean the guide tube and clevis assembly thoroughly and inspect carefully for cracks, wear or other damage.

5. Connect the new cable to the helm as per the cable manufacturer's instructions and run it back through the boat.

6. Coat the end of the steering cable liberally with Special Lubricant 101 and feed it through the coupler and guide tube.

7. Align the hole in the end of the cable with the holes in the clevis assembly, insert the clevis pin and then insert the cotter pin.

8. Follow the manufacturer's instructions for final installation and adjustment. As a general rule of thumb though, turn the steering wheel all the way to starboard and check that the guide tube protrudes through the ring 3/4 in. (19mm). On Bravos, position an open end wrench over the flats on the tube and rotate it so they are vertical. On all models, thread the coupler nut onto the tube and tighten to 35 ft. lbs. (48 Nm). If not equipped with a self-locking coupler, install the sleeve and cotter pin, or install the locking plate and tighten the bolt securely. Check that the distance from the inner end of the nut to the centerline of the cable end hole is 21-3/8 in. (543mm). Mid-point of the cable travel should be 16-7/8 in. (429mm) with no more than 4 1/2 in. (114mm) travel in either direction.

9. Center the steering wheel and confirm that the steering lever is centered in the drive unit.

Steering Cable Guide Tube

REMOVAL & INSTALLATION

Alpha Only

◆ See Figures 24, 25 and 26

1. Disconnect the hydraulic lines at the control valve assembly. Be sure to plug both of the lines and the inlets on the valve. Have some rags handy because you're going to spill some power steering fluid.

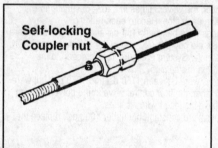

Fig. 17 Almost all units will have a self-locking coupler...

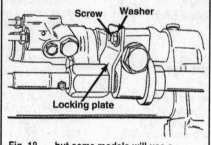

Fig. 18 ...but some models will use a locking plate...

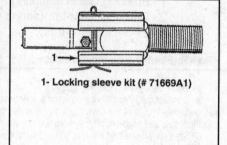

Fig. 19 ...or a locking sleeve instead

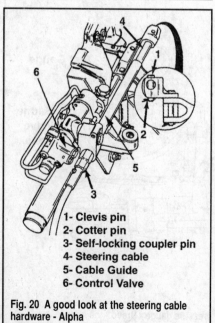

1- Clevis pin
2- Cotter pin
3- Self-locking coupler pin
4- Steering cable
5- Cable Guide
6- Control Valve

Fig. 20 A good look at the steering cable hardware - Alpha

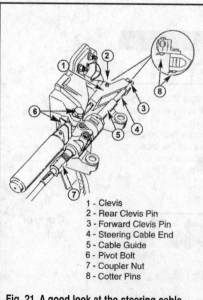

1 - Clevis
2 - Rear Clevis Pin
3 - Forward Clevis Pin
4 - Steering Cable End
5 - Cable Guide
6 - Pivot Bolt
7 - Coupler Nut
8 - Cotter Pins

Fig. 21 A good look at the steering cable hardware - Bravo

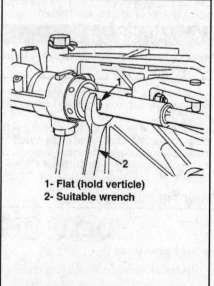

1- Flat (hold verticle)
2- Suitable wrench

Fig. 22 On the Bravo, the flats on the tube must be vertical when tightening the coupler

STEERING 14-9

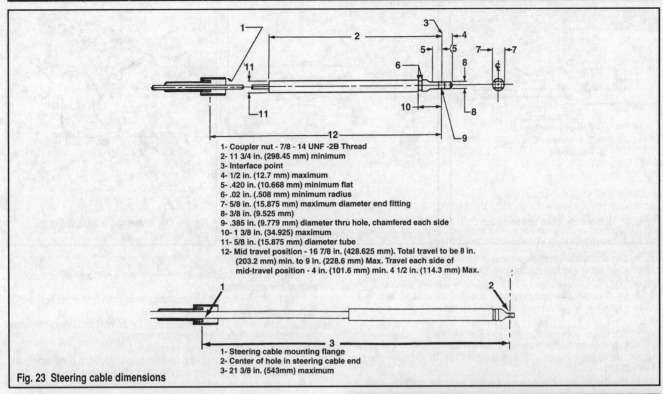

Fig. 23 Steering cable dimensions

2. Disconnect and remove the steering cable.

3. Loosen the mounting bolt on the adapter block and then tap the block lightly with a rubber mallet to break it loose from the assembly. Remove the bolt and then lift off the adapter block and guide tube.

4. Carefully mount the adaptor body in a soft-jawed vise. Apply heat, very carefully, to the inner tube nut to loosen the Loctite and then loosen the nut. Pull out the guide tube, nut, guide and bushing.

To install:

5. Check the tube, bushing and block for any cracks, wear or other signs of damage. Clean them all thoroughly.

6. Clean the threads of the guide tube with a wire brush to remove any residual Loctite and then coat the area of the threads where the adapter block and nut will rest with Loctite® 271.

7. Thread the nut onto the tube and slide on the guide and bushing. Install the tube into the adapter block until the threaded end protrudes 3/4 in. (19mm) through the block. Tighten the nut to 40 ft. lbs. (54 Nm).

8. Install the adapter block assembly and tighten the mounting bolt to 30-40 ft. lbs. (41-54 Nm).

9. Install and connect the steering cable.

10. Unplug and reconnect the two hydraulic lines. Tighten the large fitting nut to 20-25 ft. lbs. (27-34 Nm) and the small fitting to 96-108 inch lbs. (11-12 Nm).

Power Steering Unit

GENERAL INFORMATION

The power steering unit is made up of the control valve assembly, adapter and booster cylinder. It is attached to the inner transom assembly. On Alpha models, the control valve can be removed, while on the Bravo the unit is one piece and unserviceable.

REMOVAL & INSTALLATION

◆ See Figures 22, 27, 28 and 29

1. Disconnect the hydraulic lines at the control valve assembly. Be sure to plug both of the lines and the inlets on the valve. Have some rags handy because you're going to spill some power steering fluid.

2. Disconnect and remove the steering cable.

3. Remove the lower cotter pin and pull out the clevis connecting the clevis assembly and booster piston rod to the steering lever.

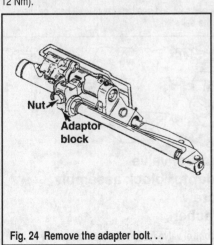

Fig. 24 Remove the adapter bolt...

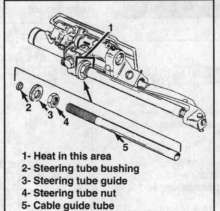

Fig. 25 ...and then the guide tube

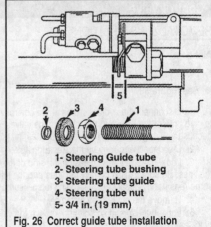

Fig. 26 Correct guide tube installation

14-10 STEERING

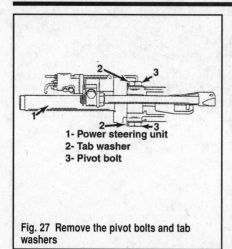

Fig. 27 Remove the pivot bolts and tab washers

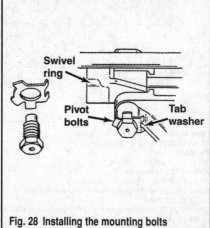

Fig. 28 Installing the mounting bolts

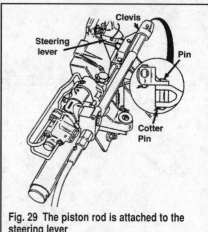

Fig. 29 The piston rod is attached to the steering lever

4. Use a screwdriver or pair of pliers and bend the locking tabs away from the two, upper and lower, pivot bolts. One side will be away from the bolt and the other will be away from the ridge in the transom assembly.

5. Remove the upper and lower pivot bolts and tab washers from the transom assembly and then pull out the power steering unit.

To install:

6. Lubricate the bushings inside the swivel ring (at the inner end of the booster) with a good amount of Special Lubricant 101. Do the same to each pivot bolt.

7. Position the assembly onto the transom assembly and then screw the bolts (don't forget the washers) in all the way by hand - no wrenches yet.

8. Make sure that the tab washer tangs are straddling the ridge in the transom plate on one side and then tighten them to 25 ft. lbs. (34 Nm) - you can use a wrench this time! Hopefully when you hit the correct torque figure, the remaining tab will align with a flat on the bolt. If so, bend it up and against the bolt head and then bend the other tab down over the transom assembly ridge. If it doesn't align, unscrew it and start over. If you need to tighten the bolt slightly to get the flat and tab to line up, it's Ok; do not though, loosen the bolt to line things up.

9. Connect the piston rod to the steering lever and slide in the clevis pin. Install the cotter pin.

10. Grab the booster cylinder and move it back and forth to ensure that it is pivoting correctly and then install the steering cable. Install the clevis pin and cotter pin and then tighten the coupler nut to 35 ft. lbs. (47 Nm).

■ **When tightening the coupler nut on the Bravo, remember to hold the flats on the cable guide tube in a vertical position with and open end wrench while tightening the coupler.**

11. Unplug and reconnect the two hydraulic lines. Tighten the large fitting nut to 20-25 ft. lbs. (27-34 Nm) and the small fitting to 96-108 inch lbs. (11-12 Nm). If you have a later model with the same size fittings (Bravo particularly), tighten them both to 23 ft. lbs. (31 Nm).

Control Valve

REMOVAL & INSTALLATION

Alpha

◆ See Figure 30

1. Disconnect the hydraulic lines at the control valve assembly. Be sure to plug both of the lines and the inlets on the valve. Have some rags handy because you're going to spill some power steering fluid.

2. Disconnect and remove the steering cable.

3. Loosen the two line fittings on the side of the valve and disconnect the lines. Be sure to plug both of the lines and the inlets on the valve. Have some rags handy because you're going to spill some power steering fluid.

4. Remove the guide tube assembly.

5. Loosen the mounting bolt and remove the valve from the adapter block assembly.

To install:

6. Clean the control valve thoroughly and check it carefully for cracks, wear or any other damage.

7. Mount the valve on the adapter assembly and tighten the bolt to 25-35 ft. lbs. (34-47 Nm).

8. Unplug the two hydraulic lines from the booster and thread them into the control valve. Tighten the fitting nuts securely.

9. Install and connect the steering cable.

10. Unplug and reconnect the two hydraulic lines. Tighten the large fitting nut to 20-25 ft. lbs. (27-34 Nm) and the small fitting to 96-108 inch lbs. (11-12 Nm).

Bravo

The power steering unit on these models is not serviceable and must be replaced as a unit.

DISASSEMBLY & ASSEMBLY

Alpha

◆ See Figures 31 thru 38

1. Remove the control valve and remove the adapter block if not already done.

2. Pop the dust cap from the starboard side of the valve housing and remove the adjusting nut.

3. Remove the two screws and washers and separate the control valve housing from the adapter.

4. Carefully remove all the internal valve components as shown in the accompanying exploded view.

5. Being very careful not to nick the top surface of the adapter housing, turn the valve shaft counterclockwise to remove the adjuster plug, spring and upper ball seat.

6. Pull out the lower ball seat and bearing sleeve. Pull back on the rubber boot and remove the ball stud.

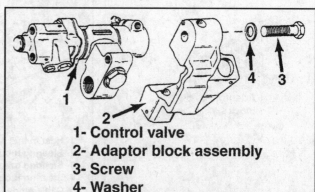

Fig. 30 Remove the control valve from the adapter assembly

STEERING 14-11

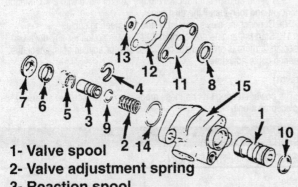

1- Valve spool
2- Valve adjustment spring
3- Reaction spool
4- Spring thrust washer
5- Valve spring
6- Spring retainer
7- Annulus seal
8- Large ID washer
9- Reaction spool o-ring
10- Valve spool "V" block seal
11- Annulus spacer
12- Gasket
13- Small washer
14- O-ring
15- Valve housing

Fig. 31 Exploded view of the control valve assembly

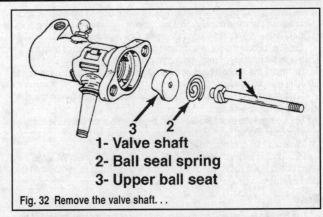

1- Valve shaft
2- Ball seal spring
3- Upper ball seat

Fig. 32 Remove the valve shaft...

To assemble:

7. Clean all parts thoroughly and inspect fro cracks, wear or other damage.
8. Press the lower seat into the bearing sleeve and then insert them into the bore. Press the ball stud through the boot and into the sleeve. Install the upper seat.
9. Install the ball seat spring with the small coil facing downward. Insert the valve shaft into the plug and thread it in until it's tight. Back off the plug until the slot lines up with the notches in the bearing sleeve. Install the key over the shaft so that the tangs in the key fit into the notches.
10. Slide the small washer over the shaft and then position the gasket and spacer. Slide on the large washer.
11. Install a new O-ring into the adapter side of the valve housing.
12. Install a V-block seal onto the valve spool so that the lip on the seal faces the lands on the spool. Insert the assembly into the adjusting nut side of the housing.
13. Assemble the reaction spool as shown in the illustration. Insert the valve adjustment spring into the inner opening on the valve and then press in the reaction spool.

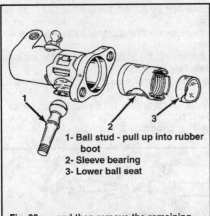

1- Ball stud - pull up into rubber boot
2- Sleeve bearing
3- Lower ball seat

Fig. 33 ...and then remove the remaining components from the adapter

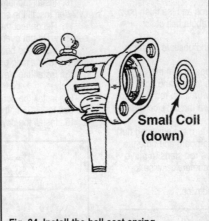

Small Coil (down)

Fig. 34 Install the ball seat spring

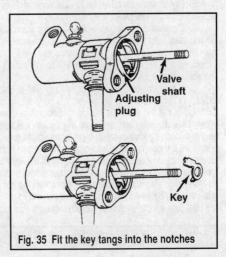

Valve shaft
Adjusting plug
Key

Fig. 35 Fit the key tangs into the notches

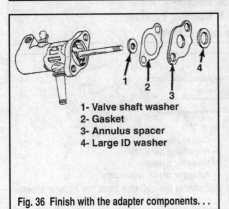

1- Valve shaft washer
2- Gasket
3- Annulus spacer
4- Large ID washer

Fig. 36 Finish with the adapter components...

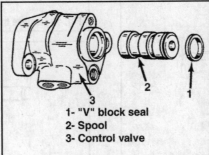

1- "V" block seal
2- Spool
3- Control valve

Fig. 37 ...and then insert the valve spool in the valve housing

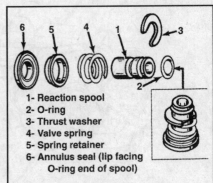

1- Reaction spool
2- O-ring
3- Thrust washer
4- Valve spring
5- Spring retainer
6- Annulus seal (lip facing O-ring end of spool)

Fig. 38 Assemble the reaction spool

14-12 STEERING

14. Line up the gasket and spacer holes with those on the adapter and then position the valve housing onto the adapter. Tighten the mounting bolts to 20-30 ft. lbs. (27-41 Nm). Don't forget the lock washers.

15. Press down the valve spool and thread the adjusting nut onto the shaft about 4 rotations. Press in the dust cover.

16. Install the adapter block to the valve assembly and tighten the bolt to 30-40 ft. lbs. (41-54 Nm).

17. Install the control valve to the adapter assembly and tighten the bolt to 25-35 ft. lbs. (34-47 Nm).

18. Connect the hydraulic lines from the booster and tighten the fitting nuts securely. Do not over-tighten.

19. Install the cable guide and steering cable.

20. Unplug and reconnect the two hydraulic lines. Tighten the large fitting nut to 20-25 ft. lbs. (27-34 Nm) and the small fitting to 96-108 inch lbs. (11-12 Nm).

Bravo

The power steering unit on these models is not serviceable and must be replaced as a unit.

BALANCING

Alpha

◆ See Figure 39

All control valves are balanced when delivered from the factory and should not, in regular operation, require further adjustment. However, occasionally a drive unit will develop a creeping problem - engine running, in Neutral, hands off the steering wheel, yet the drive moves slightly in one direction or another. If you are experiencing this, the control valve must be balanced.

1. With the engine off, disconnect the steering cable at the clevis assembly.
2. Disconnect the clevis assembly on the end of the piston rod from the steering lever.
3. Pry off the dust cover on the starboard side of the control valve assembly to expose the adjustment nut.
4. Connect a flushing device between a water source and the flush port on the starboard side of the drive. Turn on the water to about half of its maximum flow. Do not use full tap pressure.
5. Start the engine (make sure you're in Neutral) and observe what the booster cylinder piston rod does.
6. If the booster cylinder piston rod moves to Starboard, turn the adjusting nut clockwise until the rod just begins to move to Port and note the position of the nut. Turn the nut counterclockwise slowly until the rod starts to move to Starboard again and note this position. Now, turn the nut clockwise to the midpoint between the two positions you noted.
7. If the booster cylinder piston rod moves to Port, turn the adjusting nut counterclockwise until the rod just begins to move to Starboard and note the position of the nut. Turn the nut clockwise slowly until the rod starts to move to Port again and note this position. Now, turn the nut counterclockwise to the midpoint between the two positions you noted.
8. Turn off the engine and disconnect the flushing device.

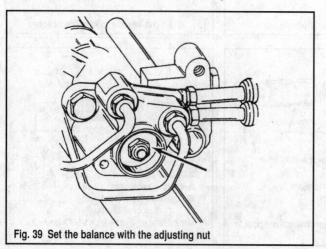

Fig. 39 Set the balance with the adjusting nut

9. Fill the adjustment nut cavity in the control valve with 2-4-C Marine Lubricant and then install the dust cover.
10. Connect the clevis assembly and the steering cable.
11. Start the engine and confirm that there is no longer any drive unit creeping. If there is and you are confident of your adjustments, check the steering cable adjustment.

Bravo

The power steering unit on these models is not serviceable and must be replaced as a unit.

Booster Cylinder

REMOVAL & INSTALLATION

Alpha

◆ See Figures 40 and 41

■ Although not absolutely necessary, we recommend removing the power steering unit before performing this procedure.

1. Remove the power steering unit.
2. Remove the control valve.
3. Loosen the nut on the end of the booster piston and pull off the clevis assembly.
4. Disconnect the booster cylinder from the adapter block assembly.
5. Remove the cotter pin and then thread a small screw into each of the two retaining pins and remove the pins from the adapter block assembly.
6. If necessary, remove the two hydraulic lines at the end of the booster. Mark which line went to which fitting.
7. Remove the snap-ring and then pry out the large and small oil seals.

To install:

8. Clean the assembly thoroughly and inspect for cracks, wear or any other damage.
9. Install the small oil seal over the piston rod with the lip facing toward the cylinder.
10. Install the large oil seal over the rod with the lip facing away from the cylinder. Install the snap-ring.
11. Reconnect the hydraulic lines to their appropriate fittings and tighten securely. Do not over-tighten.
12. Install the booster cylinder into the adapter block assembly and press in the two retaining pins. Install the cotter pin.
13. Install the clevis assembly over the piston rod and tighten the end nut securely.
14. Install the control valve into the adapter block assembly and tighten the bolt to 25-35 ft. lbs. (34-47 Nm).
15. Install the power steering assembly.

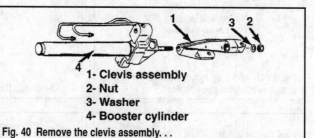

1- Clevis assembly
2- Nut
3- Washer
4- Booster cylinder

Fig. 40 Remove the clevis assembly...

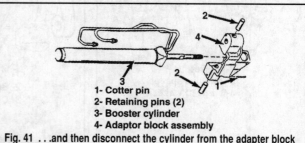

1- Cotter pin
2- Retaining pins (2)
3- Booster cylinder
4- Adaptor block assembly

Fig. 41 ...and then disconnect the cylinder from the adapter block assembly

STEERING 14-13

Bravo

The power steering unit on these models is not serviceable and must be replaced as a unit.

TESTING

1. Connect a flushing device between a water source and the flush port on the starboard side of the drive. Turn on the water to about half of its maximum flow. Do not use full tap pressure.
2. Start the engine and reach in and push the control valve adapter block to the right until the booster cylinder piston rod is fully retracted into the cylinder.
3. Turn the engine OFF and disconnect the upper hydraulic line (the metal one to the booster, not the line from the power steering pump) on the control valve. Have some rags handy to take care of the inevitable spilled fluid and make sure you plug both the valve fitting and the line end.
4. Start the engine again and MOMENTARILY push the valve adapter block to the right and observe what happens.

✳✳ WARNING

Do not lug the pump at maximum pressure for more than 5 seconds in the above step or you risk damaging the pump.

5. If the piston rod extends while pushing the adapter block, the booster cylinder has a leak and will require replacement. Repeat this test again after replacement; if pressure is still low (the rod moves) you will need to replace the control valve also.
6. If the rod does not move, but you are still experiencing low system pressure, replace the control valve.
7. Turn the engine OFF and connect the hydraulic line, clevis assembly and steering cable.

Power Steering Pump

FLUID LEVEL

■ Please refer to the Engine & Drive Maintenance section for fluid level checking procedures.

BLEEDING THE SYSTEM

■ For all system bleeding procedures, please refer to the Engine & Drive Maintenance section.

REMOVAL & INSTALLATION

4 Cylinder Engines

◆ See Figures 42 and 43

■ The power steering pump utilizes Metric fasteners.

1. Loosen the locknut on the idler pulley adjusting stud, turn the stud to release belt tension and remove the serpentine belt from around the pump.

■ On certain engines applications you may have to remove the pump before disconnecting the hydraulic lines.

2. Loosen the hose clamp on the return line (lower) and pull the line off the fitting on the pump housing. Make sure to have a suitable container and some rags available. Drain the fluid reservoir. Plug the line and secure it somewhere with the plugged end facing up.
3. Loosen the pressure line fitting on the rear of the pump housing and remove the line. Plug it and tie it up as you did with the return line.
4. Loosen the pump to brace bolt (if you haven't already) and disconnect the pump from the engine brace. If necessary, remove the brace from the engine.
5. Remove the two mounting bracket bolts and lift out the pump assembly.

■ On dual engine installations there may be a remote reservoir connected by hoses to the top of each pump reservoir. Remove the hose if necessary.

6. Remove the pump pulley and mounting bracket if necessary.

To install:

7. Install the mounting bracket and pulley if removed.

■ On certain engines applications you may have to connect the hydraulic lines prior to installing the pump.

8. Install the pump and bracket assembly onto the brace and tighten the bolts securely.
9. If removed, attach the pump brace to the engine and tighten it to 30 ft. lbs. (41 Nm).
10. Connect the pump to the engine brace and tighten the bolt securely.
11. Connect the pressure line to the flow control fitting and tighten it securely; be sure to use a new O-ring. Slide the return line over the pump fitting and tighten the hose clamp.
12. Install and adjust the drive/serpentine belt.
13. Fill the reservoir with power steering fluid and bleed the system.

V6/V8 Engines (Exc. 8.1L)

◆ See Figure 44

1. Loosen the locknut on the idler pulley adjusting stud, turn the stud to release belt tension and remove the serpentine belt from around the pump.
2. Loosen the hose clamp on the return line (low pressure) and pull the line off the fitting on the pump housing. Make sure to have a suitable container and some rags available. Drain the fluid reservoir. Plug the line and secure it somewhere with the plugged end facing up.
3. Loosen the pressure line fitting on the rear of the pump housing and remove the line. Plug it and tie it up as you did with the return line.
4. Loosen the pump to brace/bracket fasteners (bolts, nuts or both) and lift out the pump assembly.

■ On dual engine installations there may be a remote reservoir connected by hoses to the top of each pump reservoir. Remove the hose if necessary.

5. Remove the pump pulley and mounting bracket if necessary.

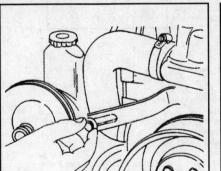

Fig. 42 Remove the brace bolt...

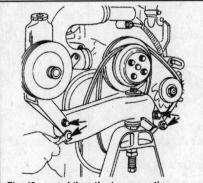

Fig. 43 ...and then the two mounting bracket bolts

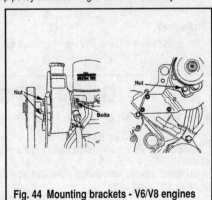

Fig. 44 Mounting brackets - V6/V8 engines (exc. 8.1L)

14-14 STEERING

To install:

6. Install the mounting bracket and pulley if removed.
7. Connect the pump to the engine brace/bracket and tighten the nuts and/or bolts to 30 ft. lbs. (41 Nm).
8. Connect the pressure line to the flow control fitting and tighten it securely; be sure to use a new O-ring. Slide the return line over the pump fitting and tighten the hose clamp.
9. Install and adjust the serpentine belt.
10. Fill the reservoir with power steering fluid and bleed the system.

8.1L V8 Engines

◆ See Figure 45

These engines utilize a remote power steering fluid reservoir.

1. Loosen the serpentine belt idler pulley and ease belt tension. Remove the serpentine belt.
2. Loosen the hose clamp on the return line (low pressure) and pull the line off the fitting on the pump housing. Make sure to have a suitable container and some rags available. Drain the fluid reservoir. Plug the line and secure it somewhere with the plugged end facing up.
3. Loosen the pressure line fitting on the front of the pump housing and remove the line. Plug it and tie it up as you did with the return line.
4. Loosen the three pump to brace/bracket bolts and lift out the pump assembly.

To install:

5. Connect the pump to the engine brace/bracket and tighten the bolts to 30 ft. lbs. (41 Nm) on all models but the 8.1L where it should be 19 ft. lbs. (26 Nm).
6. Connect the pressure line to the larger fitting and tighten it to 23 ft. lbs. (31 Nm); be sure to use a new O-ring. Slide the return line over the pump fitting and tighten the hose clamp.
7. Install and adjust the serpentine belt.
8. Fill the reservoir with power steering fluid and bleed the system.

FLOW CONTROL VALVE SERVICE

All Except 8.1L Engines

◆ See Figure 46

■ Although removal of the flow control valve may be possible without removing the power steering pump first on certain installations, we recommend removing the pump.

1. Loosen and remove the fitting nut. The outer O-ring will probably be attached; discard it.

■ If you were able to get to the valve without removing the pump, make sure that you drain the pump fluid before removing the valve.

2. Pull out the control valve assembly and spring. Pry out the inner O-ring and discard it.
3. Clean all components thoroughly. Inspect them for cracks, wear or other damage.
4. Install as they were removed, using new O-rings. Tighten the fitting nut to 35 ft. lbs. (47 Nm).

8.1L Engines

The power steering pump on these engines is not serviceable.

PUMP PULLEY

◆ See Figures 47 and 48

1. Remove the power steering pump.
2. Carefully lay the pump on a clean surface with the pulley facing upward.
3. Install a pulley removal tool (#J-25034) or equivalent onto the end of the pulley and pump shaft.
4. Hold the flats of the tool with an open end wrench while turning the tool bolt clockwise until the pulley breaks loose from the shaft.

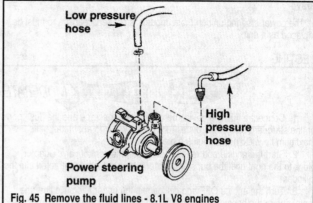

Fig. 45 Remove the fluid lines - 8.1L V8 engines

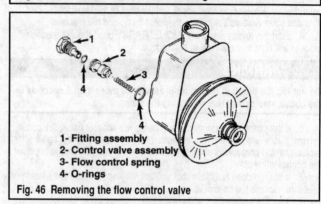

1- Fitting assembly
2- Control valve assembly
3- Flow control spring
4- O-rings

Fig. 46 Removing the flow control valve

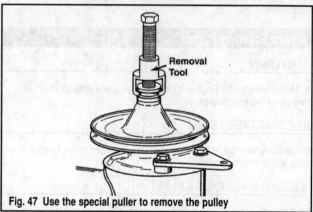

Fig. 47 Use the special puller to remove the pulley

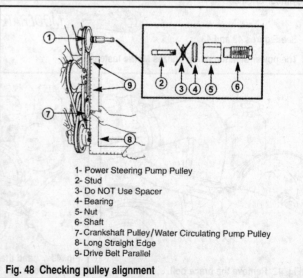

1- Power Steering Pump Pulley
2- Stud
3- Do NOT Use Spacer
4- Bearing
5- Nut
6- Shaft
7- Crankshaft Pulley/Water Circulating Pump Pulley
8- Long Straight Edge
9- Drive Belt Parallel

Fig. 48 Checking pulley alignment

STEERING 14-15

5. Press the pulley onto the shaft with a pulley installation tool (#91-93656A1). Thread the stud all the way into the shaft, place the bearing over the stud - don't use the spacer supplied with the kit. Thread the nut onto the shaft and then thread the shaft all the way onto the stud.
6. Install the pump (with the tool still attached). Install the drive belt and place a long straight edge across the pulley on the other side of the belt (crankshaft or circulating pump).
7. Turn the large pusher nut on the tool until the drive belt is parallel with the straightedge.
8. Remove the tool and adjust the belt tension.
9. Fill and bleed the system.

OIL SEAL REPLACEMENT

All Except 8.1L Engines

◆ See Figures 49 and 50

1. Remove the power steering pump and then remove the pulley.
2. Wrap a small piece of shim stock (0.005 in. {0.13mm}) around the shaft and feed inside the seal until it bottoms in the body of the pump; about 2 1/2 in. (63mm).
3. Use a small chisel to cut the seal and tear the metal body about 1 in.
4. Insert a small pry bar, or awl, between the seal edge and the housing and then pry out the seal. Remove the shim stock.
5. Install a new seal over the shaft with the metal side facing up.
6. Support the pump housing/reservoir on two pieces of wood, position a small mandrel (a 1 in. socket should work) over the seal and lightly tap it into place in the housing.
7. Install the pump and pulley. Check the pulley alignment.

8.1L Engines

The power steering pump on these engines is not serviceable.

DISASSEMBLY & ASSEMBLY

All Except 8.1L V8 Engines

◆ See Figures 51 thru 56

1. Remove the pump and pulley.
2. Remove the flow control valve assembly.
3. Loosen the studs and separate the pump housing from the reservoir. You may have to persuade the separation a little, but don't hit it too hard and use a rubber mallet.
4. Insert a small awl into the hole in the end plate and maneuver the retaining ring around in its recess until the end of the ring is approximately 1 in. (25mm) past the hole. Press in on the ring with the awl while prying up on it with a small screwdriver until it pops out of the recess.

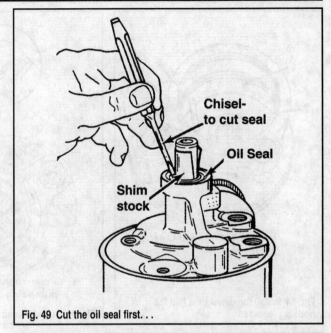

Fig. 49 Cut the oil seal first...

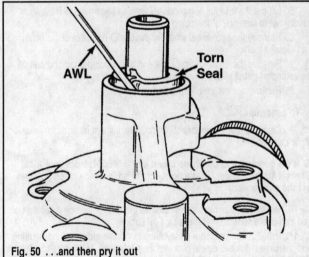

Fig. 50 ...and then pry it out

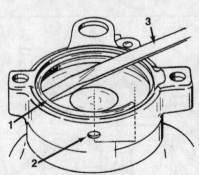

1- Retaining ring - position so that ring end is 1 in. (25mm) from end of hole in housing
2- Hole- insert into Awl in housing to push ring from recess
3- Screwdriver - Pry ring from pump body

Fig. 51 Pry out the retaining ring...

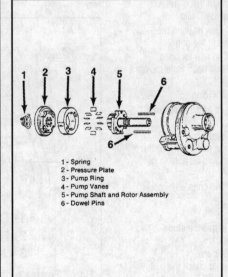

1 - Spring
2 - Pressure Plate
3 - Pump Ring
4 - Pump Vanes
5 - Pump Shaft and Rotor Assembly
6 - Dowel Pins

Fig. 52 ...and then remove the pump components

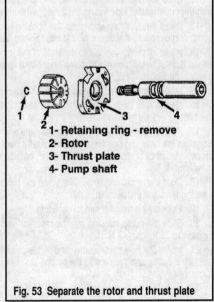

1- Retaining ring - remove
2- Rotor
3- Thrust plate
4- Pump shaft

Fig. 53 Separate the rotor and thrust plate

14-16 STEERING

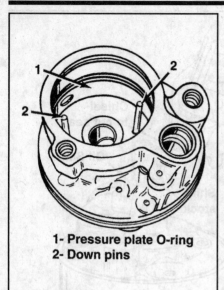

Fig. 54 Install the dowel pins into the housing recesses

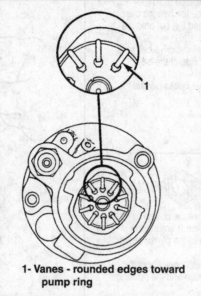

Fig. 55 Correct pump vane positioning

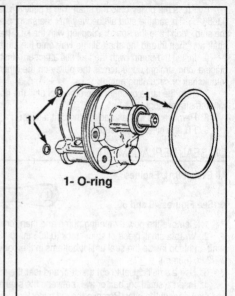

Fig. 56 Install the remaining O-rings

5. Lift out the internal pump components as shown in the illustration. Be careful not to lose any of the pump vanes.

6. Remove the two dowel pins. Pry the two O-rings out of the housing and discard them.

7. Remove the small C-clip on the end of the pump shaft and pull off the rotor and thrust plate.

8. Remove the magnet from inside the housing.

To assemble:

9. Clean all metal components thoroughly. Inspect all parts for wear or other damage.

■ It's a good idea to purchase a seal kit (#5688044) from your local General Motors dealer. That's right, this is an instance when you can get the car version!

10. Install a new pump shaft oil seal so that the metal side is facing up. You can use a suitable mandrel or a 1 in. socket.

11. Install a new pressure plate O-ring in the 3rd groove of the housing and then insert the two dowel pins into their recesses. Make sure to coat the O-ring with power steering fluid.

12. Slide the thrust plate onto the pump shaft and then install the rotor with the countersunk side facing the thrust plate. Install the C-clip.

13. Insert the shaft assembly into the housing and then position the pump ring over it so the dowel pins fit into the two holes in the ring.

14. Insert all pump vanes into the slots in the rotor so that rounded edges face outward toward the pump ring. Make sure that each vane moves freely in its slot.

15. Install the pressure plate so the spring groove is facing up.

16. Coat a new large O-ring with power steering fluid and install it into the second groove in the housing.

17. Position the spring on the pressure plate and install the endplate. Squeeze the retaining ring together and install it over the plate and into the groove. An arbor press will make this easier, but you should be able to get it in without it.

18. Coat the remaining large O-ring and the two small ones with power steering fluid and install them.

19. Position the magnet into the housing.

20. Slide the pump housing into the reservoir and install the studs. Tighten to 35 ft. lbs. (47 Nm).

21. Install the flow control valve and tighten it to 35 ft. lbs. (47 Nm). Install the pulley and then install the pump - don't forget to check the pulley alignment.

8.1L Engines

The power steering pump on these engines is not serviceable.

Power Steering Cooler

All engines covered here utilize a raw-water cooled steering cooler on the low pressure (return) side of the power steering system; located in-line between the control valve and the power steering pump.

REMOVAL & INSTALLATION

◆ See Figure 57

1. Locate the cooler. It can generally be found at the rear of the engine, although certain engines will have it tucked in at the front of the engine. If it's not readily visible, just follow the lines from the pump.

2. Drain the cooling system as detailed in the Maintenance section.

3. Loosen the hose clamps and wiggle off the cooling hoses on each end of the unit. Have some rags and a container handy as there will still be some residual water in the cooler and lines. Tie the hoses up and out of the way.

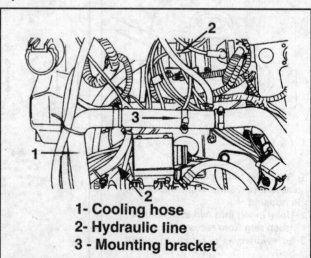

Fig. 57 Power steering cooler

STEERING 14-17

■ Although not absolutely necessary, we recommend draining the power steering system before attempting this procedure.

4. Loosen the hydraulic lines running into and out of the cooler unit. Most will simply use a hose clamp, but a few applications will use a screw-in fitting. Have some rags and a container handy as there will still be some residual fluid in the cooler and lines. Tie the hoses up and out of the way. Plug the openings to prevent contamination.

5. Loosen and remove the mounting brackets - usually just hose clamps, and lift out the cooler.

To install:

6. Clean the cooler thoroughly and inspect it for cracks, wear or other damage.

7. Position the cooler and tighten the mounting hardware.

8. Slide the cooling hoses over the end flanges and tighten the hose clamps securely. Check the hose ends for cracks, crimping or bulges.

9. Install the hydraulic lines onto their original fitting and tighten the hose clamps or nuts securely.

10. Refill the cooling and power steering systems.

Tie Bar - Multiple Drives

TIE BAR LENGTH

◆ See Figures 58 and 59

※※ WARNING

On dual installations that use a starboard tie bar lit, the steering cable MUST have a minimum radius of 8 in. (203mm) at the transom side of the cable. Anything less could possibly cause the cable to kink, affecting steering performance. If, due to the design of the vessel, this minimum radius is not possible, you MUST convert the system over to a port tie bar kit.

■ Although optimum handling and performance will result from parallel stern drives units, certain boats may require that the units be toed-in or out. If your drives are such, make sure that you take your measurements from the centerlines of the steering levers rather than the units themselves.

1. Rub your finger across the decal on the top rear facing portion of the drive unit until you feel the small dimple underneath it - this is the centerline of the drive.

2. Use a t-square or other tool to mark the centerline on the top forward surface and then measure between the two drives at this centerline.

3. Use the accompanying chart to determine tie bar size.

■ Remember if your drives are not parallel, be sure to measure from the centerlines of the steering levers.

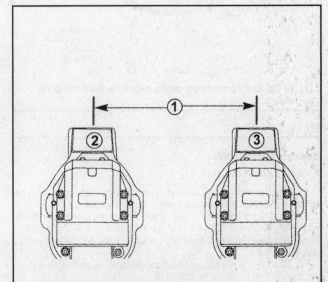

1- Distance Between Centerlines
2- Port Transom Assembly Centerline
3- Starboard Transom Assembly Centerline

Fig. 58 Determining tie bar length on the centerlines

Tie Bar Length

Steering Cable	Length (in.)	Part Number
Starboard Steering Cable	16-30	92020A1
	30-46	92020A2
	46-62	92020A3
Port Steering Cable	28-37 1/2	96708A4
	37 1/2-55	96708A5
	55-72	96708A6

* If length measurement matches max figure in a range, choose the next longer bar kit

Fig. 59 Tie bar chart

14-18 STEERING

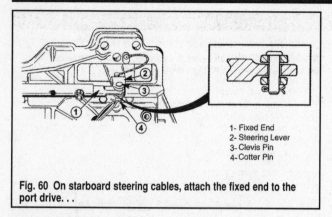

Fig. 60 On starboard steering cables, attach the fixed end to the port drive...

1- Fixed End
2- Steering Lever
3- Clevis Pin
4- Cotter Pin

INSTALLATION

Starboard Steering Cable

◆ See Figures 60 and 61

1. Connect the fixed end of the of the tie bar to the steering lever on the port drive unit. Slide the clevis pin in and install a new cotter pin.
2. Position the drives in the desired position (usually parallel) and then turn the adjustable end of the bar in or out until it lines up with the holes in the steering lever (starboard drive) and the clevis on the end of the steering valve piston rod.
3. Now turn the adjustable end out an additional 3-4 revolutions and coat the exposed threads with Loctite® 277.
4. Turn the adjustable end back in the same number of turns as you turned it out and then slide in the clevis pin and a new cotter pin.
5. Coat any remaining exposed threads on the tie bar with Loctite® 277 and then back down the adjusting nut and tighten it to 50 ft. lbs. (68 Nm).

Port Steering Cable

◆ See Figures 62 and 63

1. Connect the fixed end of the of the tie bar to the steering lever on the starboard drive unit. Slide the clevis pin in and install a new cotter pin.
2. Position the drives in the desired position (usually parallel) and then turn the adjustable end of the bar in or out until it lines up with the holes in the steering lever (port drive) and the clevis on the end of the steering valve piston rod.
3. Now turn the adjustable end out an additional 3-4 revolutions and coat the exposed threads with Loctite® 277.
4. Turn the adjustable end back in the same number of turns as you turned it out and then slide in the clevis pin and a new cotter pin.
5. Coat any remaining exposed threads on the tie bar with Loctite® 277 and then back down the adjusting nut and tighten it to 50 ft. lbs. (68 Nm).

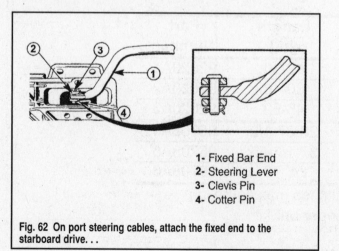

1- Fixed Bar End
2- Steering Lever
3- Clevis Pin
4- Cotter Pin

Fig. 62 On port steering cables, attach the fixed end to the starboard drive...

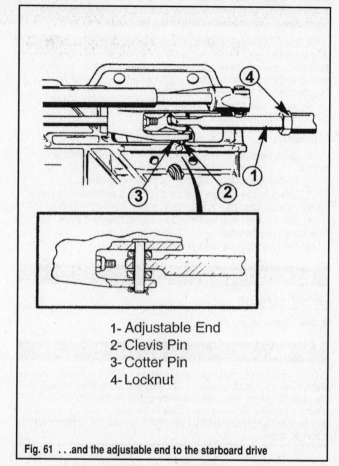

1- Adjustable End
2- Clevis Pin
3- Cotter Pin
4- Locknut

Fig. 61 ...and the adjustable end to the starboard drive

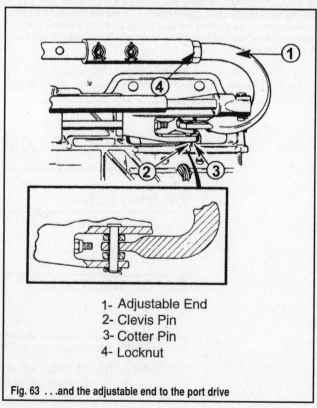

1- Adjustable End
2- Clevis Pin
3- Cotter Pin
4- Locknut

Fig. 63 ...and the adjustable end to the port drive

STEERING 14-19

COMPACT HYDRAULIC SYSTEM

Description

This system uses no conventional pump, cable or control valve. Steering is controlled hydraulically via the helm.

Hydraulic Fluid

FILLING/PURGING THE SYSTEM

◆ See Figures 64 and 65

■ The following procedures are for a single station, single cylinder system. If you have twin station, twin cylinder system, please refer to the instructions provided by SeaStar® that came with the vessel. To complete successful purging of this system you will need two people.

■ Hydraulic fluid must be visible in the filler tube during the entire procedure. Never allow the fluid container used for filling to empty, causing the filler tube to be empty and allowing air to enter the system. You will need at least 2 qts. of Quicksilver Hydraulic fluid to complete this procedure, but the length of the hydraulic line will cause the amount to vary.

1. Remove the vent/fill plug from above the helm.
2. Obtain a Filler Kit from your local Mercury representative and screw the filler tube into the vent/fill plug hole. Hand-tighten the tube and then screw in a container of hydraulic fluid to the top of the tube.
3. Turn the container upside down, support it and then carefully punch a small hole in its bottom.
4. You must always have fluid in the tube, so with fluid still visible in the fill tube and in the container, remove the container and install another one. Punch a hole in this one also...and you thought this was going to be neat!
5. When no air bubbles are visible in the filler tube any longer, the helm is full. Leave the fluid container connected, and securely supported.
6. Pop off the caps on the bleeder valves in the T-fittings on the steering cylinder (at the transom). Attach a sufficient length of clear plastic tubing from each bleeder and into a suitable container - sorta like when you used to bleed the brake on your car.
7. Have an assistant start to turn the steering wheel SLOWLY clockwise while you open the Starboard bleeder. Continue turning the wheel until you observe an air-free (no bubbles) stream of fluid coming through the tube and then close the bleeder. Continue turning the wheel until the piston rod is extended fully from the cylinder.
8. Now, have your assistant start to turn the steering wheel SLOWLY counterclockwise while you open the Port bleeder. Continue turning the wheel until you observe an air-free stream of fluid coming through the tube and then close the bleeder. Continue turning the wheel until the piston rod is fully retracted into the cylinder and the wheel comes to a stop.
9. Open the Starboard bleeder again. Hold the piston rod in the retracted position while continuing to turn the steering wheel counterclockwise until an air-free stream of fluid is visible. Close the bleeder while continuing to turn the wheel.
10. Check the fluid level once again.

CHECKING THE LEVEL

1. Turn the steering wheel until the steering cylinder piston rod is fully retracted into the cylinder.
2. Obtain a Filler Kit from your local Mercury representative and screw the filler tube into the vent/fill hole above the helm.
3. Pour a container of hydraulic fluid into the tube very slowly until it is about 1/2 full. Make sure there are no air bubbles in the tube.
4. Pop off the cap on the Starboard bleeder valve in the T-fitting on the steering cylinder. Attach a sufficient length of clear plastic tubing from each bleeder and into a suitable container.
5. Have an assistant start to turn the steering wheel SLOWLY clockwise while you open the Starboard bleeder. Continue turning the wheel until the fluid level in the tube drops to the top of the plastic fitting. Continue turning the wheel for another 1/4 of a turn and then stop. Close the bleeder.
6. Remove the filler tube and install the vent/fill plug. The fluid level should be at the bottom of the filler hole. Never allow the fluid level to drop more than 1/4 in. (6mm) below the hole.

Steering Unit

REMOVAL & INSTALLATION

◆ See Figures 66 thru 69

1. Disconnect the hydraulic lines at the steering cylinder assembly T-fittings. Be sure to plug both of the lines and the inlets on the cylinder. Have some rags handy because you're going to spill some power steering fluid. Plug the lines as quickly as possible and tie them up and out of the way.

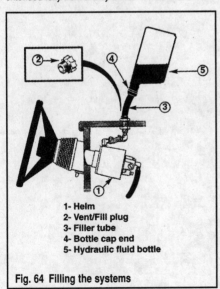

1- Helm
2- Vent/Fill plug
3- Filler tube
4- Bottle cap end
5- Hydraulic fluid bottle

Fig. 64 Filling the systems

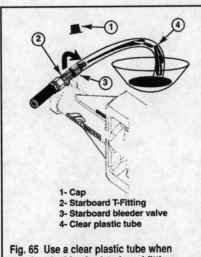

1- Cap
2- Starboard T-Fitting
3- Starboard bleeder valve
4- Clear plastic tube

Fig. 65 Use a clear plastic tube when opening the bleeder (starboard fitting shown)

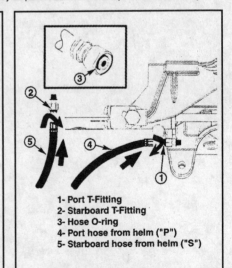

1- Port T-Fitting
2- Starboard T-Fitting
3- Hose O-ring
4- Port hose from helm ("P")
5- Starboard hose from helm ("S")

Fig. 66 Remove the hydraulic lines

14-20 STEERING

2. Remove the lower cotter pin and pull out the clevis connecting the clevis assembly and piston rod to the steering lever.

3. Use a screwdriver or pair of pliers and bend the locking tabs away from the upper and lower pivot bolts. One side will be away from the bolt and the other will be away from the ridge in the transom assembly.

4. Remove the two pivot bolts and tab washers from the transom assembly and then pull out the steering cylinder.

To install:

5. Lubricate the upper and lower assembly bushings with a good amount of Special Lubricant 101. Do the same to each pivot bolt.

6. Position the assembly onto the transom assembly and then screw the bolts (don't forget the washers) in all the way by hand - no wrenches yet.

7. Make sure that the tab washers tangs are straddling the ridge in the transom plate on one side and then tighten them to 25 ft. lbs. (34 Nm) - you can use a wrench this time! Hopefully when you hit the correct torque figure, the remaining tab will align with a flat on the bolt. If so, bend it up and against the bolt head and then bend the other tab down over the transom assembly ridge. If it doesn't align, unscrew it and start over. If you need to tighten the bolt slightly to get the flat and tab to line up, it's Ok; but do not, though, loosen the bolt to line things up.

8. Unplug and reconnect the two hydraulic lines. Lubricate a new O-ring with hydraulic fluid and press the hoses all the way into their fitting. Tighten the fitting nuts to 130 inch lbs. (15 Nm).

9. Fill and purge the hydraulic system.

10. Connect the cylinder piston rod to the steering lever and slide in the clevis pin. Install the cotter pin. Make sure that the angled notch in the clevis is facing aft as shown.

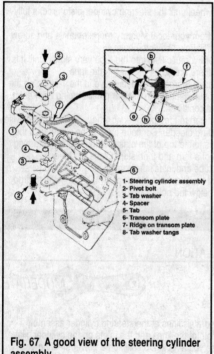

Fig. 67 A good view of the steering cylinder assembly

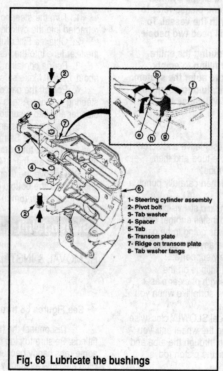

Fig. 68 Lubricate the bushings

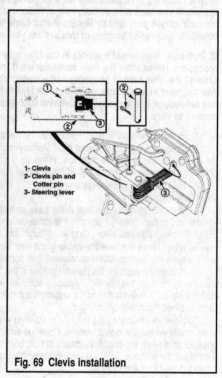

Fig. 69 Clevis installation

15 REMOTE CONTROLS

CABLE ADJUSTMENT - CONVENTIONAL ... 15-2
CONTROLLER ... 15-36
  REMOVAL & INSTALLATION ... 15-36
CONVENTIONAL REMOTE CONTROLS ... **15-2**
  CABLE ADJUSTMENT ... 15-2
  EXPLODED VIEWS ... 15-2
DESCRIPTION & OPERATION -
SMARTCRAFT ... 15-7
  CONTROL AREA NETWORK (CAN)
  DATA HARNESS ... 15-7
  DTS COMMAND MODULE ... 15-7
  ELECTRONIC REMOTE CONTROL (ERC) . 15-7
  ELECTRONIC SHIFT CONTROL (ESC) ... 15-7
  ELECTRONIC THROTTLE CONTROL (ETC) 15-7
  KEY SWITCH & START/STOP PANEL ... 15-8
  PROPULSION CONTROL MODULE (PCM) . 15-7
DUAL HANDLE CONSOLE MOUNT & SHADOW
MOUNT CONSOLE REMOTE CONTROLS 15-10
  REMOVAL, DISASSEMBLY &
  INSTALLATION ... 15-10
ELECTRONIC SHIFT CONTROL (ESC) ... 15-8
  ACTUATOR ADJUSTMENT ... 15-9
  SHIFT ACTUATOR REMOVAL &
  INSTALLATION ... 15-8
  SHIFT OVERRIDE PROCEDURE ... 15-9
ELECTRONIC THROTTLE CONTROL (ETC) . 15-9
  REMOVAL & INSTALLATION ... 15-9
EQUALIZER ... 15-36
  REMOVAL & INSTALLATION ... 15-36
EXPLODED VIEWS
  CONVENTIONAL ... 15-2
  PRECISION PILOT ... 15-35
FRONT SAFETY GUARD ... 15-36
  REMOVAL & INSTALLATION ... 15-36
IMPELLER ... 15-36
  REMOVAL & INSTALLATION ... 15-36
JOYSTICK ... 15-37
  REMOVAL & INSTALLATION ... 15-37
MOTORS ... 15-37
  REMOVAL & INSTALLATION ... 15-37
PANEL MOUNT REMOTE CONTROL ... 15-9
  REMOVAL, DISASSEMBLY &
  INSTALLATION ... 15-9
PRECISION PILOT ... **15-33**
  CONTROLLER ... 15-36

EQUALIZER ... 15-36
  EXPLODED VIEWS ... 15-35
  FRONT SAFETY GUARD ... 15-36
  IMPELLER ... 15-36
  JOYSTICK ... 15-37
  MOTORS ... 15-37
  REAR SAFETY GUARD ... 15-37
  TROUBLESHOOTING ... 15-33
  WIRING SCHEMATICS ... 15-37
REAR SAFETY GUARD ... 15-37
  REMOVAL & INSTALLATION ... 15-37
SINGLE HANDLE CONSOLE MOUNT
REMOTE CONTROL ... 15-10
  REMOVAL, DISASSEMBLY &
  INSTALLATION ... 15-10
SLIM BINNACLE SINGLE HANDLE
CONSOLE MOUNT REMOTE CONTROL ... 15-12
  ADJUSTMENT ... 15-22
  REMOVAL, DISASSEMBLY &
  INSTALLATION ... 15-12
**SMARTCRAFT DIGITAL THROTTLE &
SHIFT (DTS)** ... **15-7**
  DESCRIPTION & OPERATION ... 15-7
  DUAL HANDLE CONSOLE MOUNT REMOTE
  CONTROLS ... 15-10
  ELECTRONIC SHIFT CONTROL (ESC) ... 15-8
  ELECTRONIC THROTTLE CONTROL (ETC) 15-9
  PANEL MOUNT REMOTE CONTROL ... 15-9
  SHADOW MOUNT CONSOLE REMOTE
  CONTROLS ... 15-10
  SINGLE HANDLE CONSOLE MOUNT
  REMOTE CONTROL ... 15-10
  SLIM BINNACLE SINGLE HANDLE
  CONSOLE MOUNT REMOTE CONTROL 15-12
  WIRING SCHEMATICS ... 15-13
  ZERO EFFORT CONSOLE MOUNT
  REMOTE CONTROL ... 15-11
TROUBLESHOOTING -
PRECISION PILOT ... 15-33
WIRING SCHEMATICS
  PRECISION PILOT ... 15-37
  SMARTCRAFT DTS (10-PIN) ... 15-13
  SMARTCRAFT DTS (14-PIN) ... 15-16
ZERO EFFORT CONSOLE MOUNT
REMOTE CONTROL ... 15-11
  REMOVAL & INSTALLATION ... 15-11

CONVENTIONAL CONTROLS ... 15-2
SMARTCRAFT DIGITAL THROTTLE
& SHIFT (DTS) ... 15-7
PRECISION PILOT ... 15-33

15-2 REMOTE CONTROLS

CONVENTIONAL REMOTE CONTROLS

Cable Adjustment

■ For all conventional, cable operated remote control cable adjustment procedures, please refer to the appropriate Fuel System section for the throttle cable or the appropriate Drive System section for the shift cable.

Exploded Views

◆ See Figures 1 thru 19

■ As space does not permit actual disassembly procedures for the many versions of the Commander conventional remote controls, we have provided detailed exploded views for your convenience.

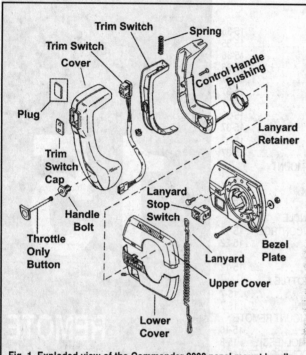

Fig. 1 Exploded view of the Commander 3000 panel mount handle and bezel

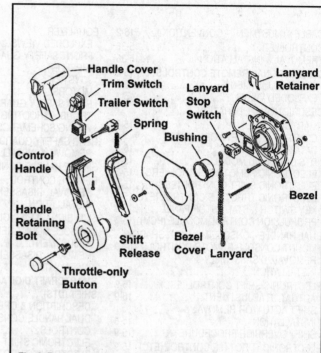

Fig. 2 Exploded view of the Commander 3000 Classic panel mount handle and bezel

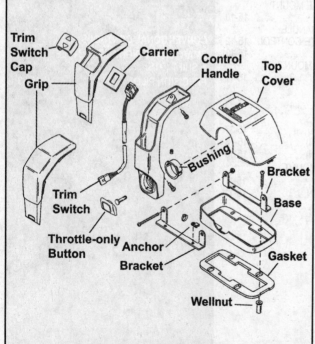

Fig. 3 Exploded view of the Commander 3000 console mount single handle

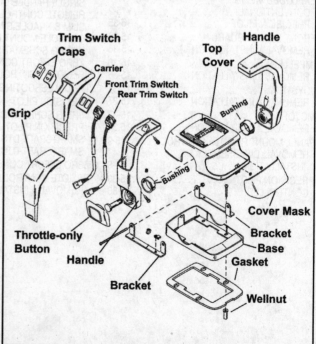

Fig. 4 Exploded view of the Commander 3000 console mount dual handle

REMOTE CONTROLS 15-3

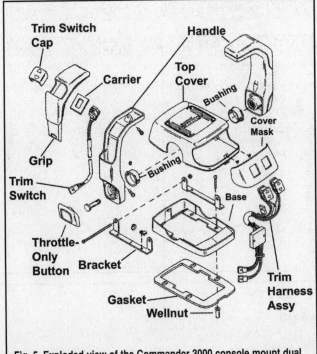

Fig. 5 Exploded view of the Commander 3000 console mount dual handle, 3 trim button

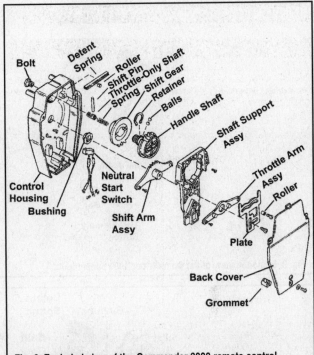

Fig. 6 Exploded view of the Commander 3000 remote control module (design II)

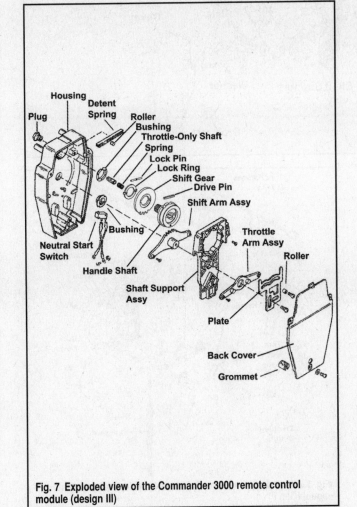

Fig. 7 Exploded view of the Commander 3000 remote control module (design III)

Fig. 8 Exploded view of the Commander 4000 side mount

15-4 REMOTE CONTROLS

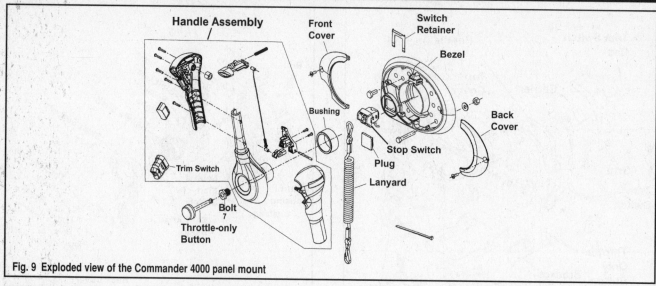

Fig. 9 Exploded view of the Commander 4000 panel mount

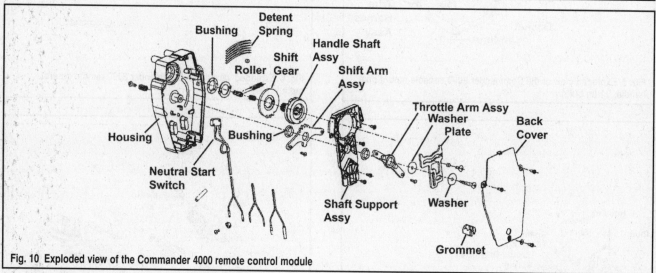

Fig. 10 Exploded view of the Commander 4000 remote control module

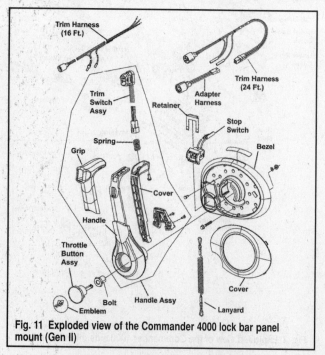

Fig. 11 Exploded view of the Commander 4000 lock bar panel mount (Gen II)

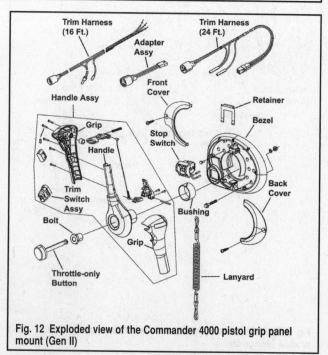

Fig. 12 Exploded view of the Commander 4000 pistol grip panel mount (Gen II)

REMOTE CONTROLS 15-5

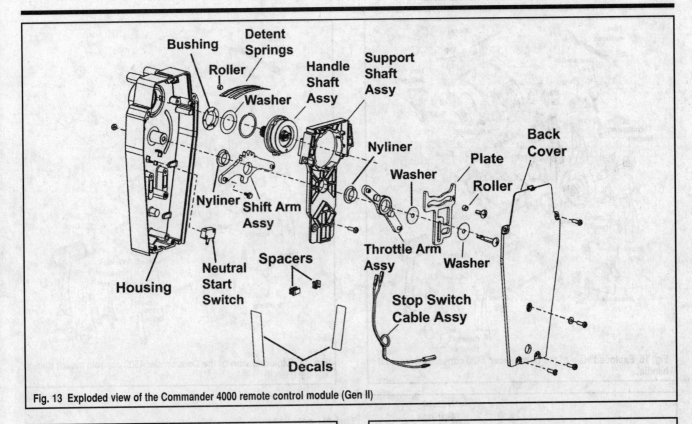

Fig. 13 Exploded view of the Commander 4000 remote control module (Gen II)

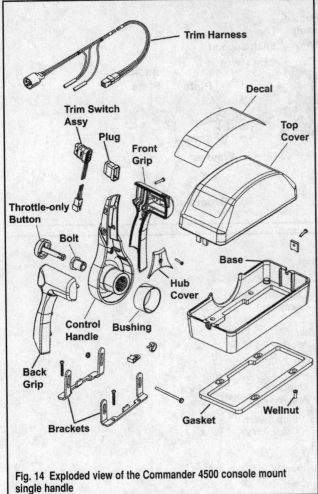

Fig. 14 Exploded view of the Commander 4500 console mount single handle

Fig. 15 Exploded view of the Commander 4500 console mount single handle (Gen II)

15-6 REMOTE CONTROLS

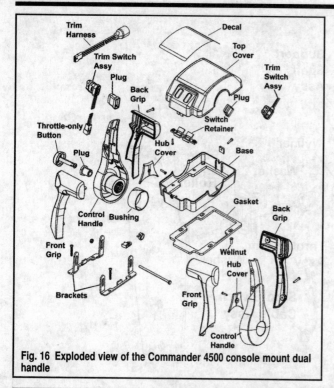

Fig. 16 Exploded view of the Commander 4500 console mount dual handle

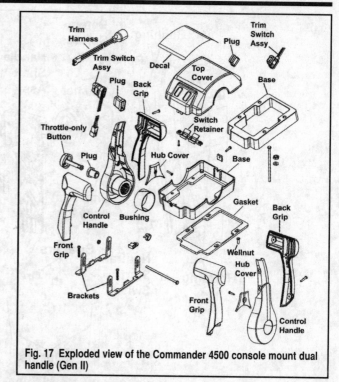

Fig. 17 Exploded view of the Commander 4500 console mount dual handle (Gen II)

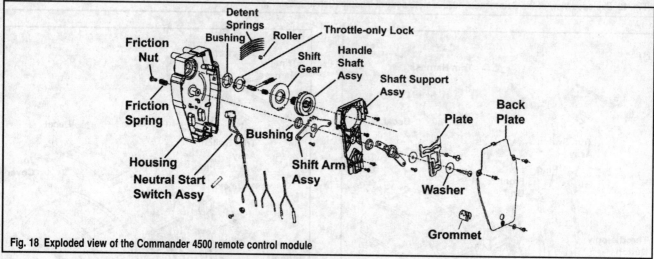

Fig. 18 Exploded view of the Commander 4500 remote control module

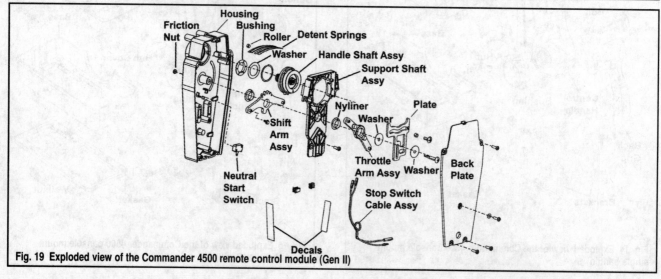

Fig. 19 Exploded view of the Commander 4500 remote control module (Gen II)

REMOTE CONTROLS

SMARTCRAFT DIGITAL THROTTLE & SHIFT (DTS)

Description & Operation

◆ See Figure 20

Beginning in 2004 Mercruiser offers the SmartCraft digital throttle and shift (DTS) control system as an option to conventional remote controls. The initial version used a 10-pin engine connection.

Shortly thereafter a new 14-pin version was introduced. Although most of the functions and procedures remain identical between the two, they are different systems and do not share parts. The biggest noticeable differences are the 14-pin panel mount remote control uses a shift lock on the handle; and the 14-pin console mount remote control uses a Throttle Only or Station Select button on the console.

DTS COMMAND MODULE

The command module is, very simply, a small computer that is meant to manage communications between the PCM and ERC. Once the key is turned to the Run position, a 12V signal is sent throughout the module harness, turning on the module. The power relay (if equipped), System View (if equipped) and any other gauges connected to the junction box will also be powered up. The voltage signal is also sent back down the data cable (via the purple wire) to the engine PCM.

CONTROL AREA NETWORK (CAN) DATA HARNESS

SmartCraft DTS utilizes 2 sets of CAN lines. CAN1 connects engine(s) and command modules to the gauges so that engine data such as temperature, pressure, depth, speed and rpm can be transmitted through the system and to the gauges at the helm.

CAN2 carries throttle and shift data only and is isolated between the engine and command module. The systems are redundant so as a back-up, if CAN2 loses connection than CAN1 will take over and transmit the throttle and shift data. When this happens, the system will move into Guardian mode and will automatically reduce power to the engine.

14-pin units utilize a DTS terminator resistor as a CAN line signal conditioner by placing a known load on the line to ensure proper communication between the command module and the PCM. There is generally a resistor at each end of the line (one near the helm and one near the engine).

PROPULSION CONTROL MODULE (PCM)

The PCM receives throttle shift, start, trim and shut off commands from the DTS command module via two control area network (CAN) signal and ground circuits. These commands initiate a software process which will control and perform all shift, throttle, start and stop functions normally controlled by the conventional remote control system. The PCM verifies and quantifies the shift and throttle commands from the DTS command module to ensure that the electronic shift controller (ESC) and throttle controller (ETC) are correctly positioning the throttle and shifters.

ELECTRONIC THROTTLE CONTROL (ETC)

The ETC is a single bore cylindrical valve and DC electric motor used to control the position of the throttle valve. Positive spring force is used to return the throttle valve to a preset position when there is no signal to the DC motor. Idle airflow is controlled by the throttle valve - there is no need for a conventional idle air control system.

There are two throttle position potentiometers (TPS), operating in opposite, that provide a positive feedback signal. As voltage increases in one potentiometer, it will decrease in the other. Interestingly though, the PCM processes the signals from both sensors in a manner that the digital diagnostic terminal display will show both TPS sensors increasing and decreasing in tandem.

ELECTRONIC SHIFT CONTROL (ESC)

The ESC, or shift actuator, is used as a means of shifting the drive unit into forward, neutral and reverse gears; without any conventional mechanical cables from the remote control at the helm. The actuator motor rotates a ball/screw assembly through reduction gears in the actuator. The screw shaft will then extend or retract the actuator shaft, while the gear set rotates a potentiometer in the actuator at the same time.

The PCM provides a 5 V reference signal to the potentiometer which will confirm the actual position of the actuator shaft.

ELECTRONIC REMOTE CONTROL (ERC)

Although it looks almost identical to a conventional remote control handle, the ERC handle position is what determines the throttle position and also the shift direction. The handle has a range of 78° (+/-) from the vertical, or neutral position; and is held in the neutral position by a spring detent mechanism which prevents any unintended movement.

There are 3 potentiometers in the EDC, arranged along the handle's axis of rotation, which communicate handle position to the DTS command module. The movement of the handle will vary the voltage output of the potentiometers.

Voltage from on potentiometer is used to set the engine speed, while the other two are used (in combination) to specify shift direction. Handle movement will determine, and activate, shift direction prior to increasing the engine speed - output from the 2 shift potentiometers is used as a redundancy check against the throttle command.

The shift potentiometer voltages are used by the DTS command module, which functions as follows:

• Reads the voltage from the two sensors and checks for consistency to accurately determine the position of the handle

• Creates a shift direction command

• Passes this command, with the voltages, on to the propulsion control module (PCM).

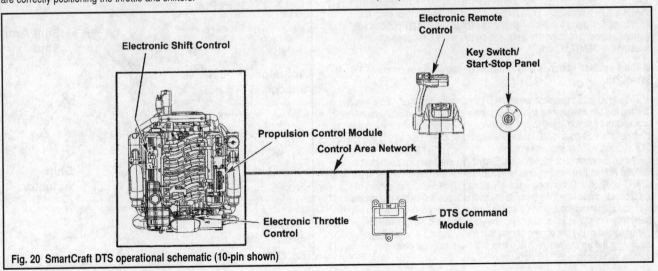

Fig. 20 SmartCraft DTS operational schematic (10-pin shown)

15-8 REMOTE CONTROLS

The PCM will then create its own shift command to be sent to the ESC once it verifies that the signal from the command module is valid. Once it receives the signal and command from the command module, the PCM will:
- Compares the voltage signals and command module shift command to verify consistency across the 3 signals/commands
- Creates its own shift command and then sends it back to the DTS command module, which then confirms that this new signal has the same values as that which it just sent to the PCM (and that the interpretation of them is accurate). Once done, the PCM then resend the signal back to the command module confirming the original signal
- Finally, the PCM forwards this command signal on to the ESC.

This shift command, essentially just voltage applied to the DC motor connections in the ESC/actuator, will determine the direction that the motor will rotate; thus determining the shift direction.

Throttle setting is changed via movement of the ERC handle. As mentioned, the handle position specifies both the throttle and shifter setting, which is then communicated to the DTS command module via the 3 potentiometers. The command module uses these potentiometer voltage signals to verify the throttle position as follows:
- It reads the signals from the ERC and checks for consistency by comparing the voltage from the throttle potentiometer with that from the 2 shift control potentiometers
- Upon verifying the consistency of the 3 voltage signals, it creates a command for the indicated throttle setting
- It then passes the throttle setting command (and voltages) on to the PCM.

The PCM verifies the command and then must set the commanded throttle position. It will do this as follows:
- Compare the command module throttle command with the ERC potentiometer voltages to confirm consistency
- Creates its own throttle setting command
- Once created, it will send this command, and the potentiometer voltages back to the command module. The command module will then confirm that the new PCM command and voltages match those that it originally sent to the PCM, and, if so, send them back to the PCM
- Once this is done, the PCM will send the command and voltages through to the electronic throttle control (ETC).

This command is basically a simple torque specification of the DC throttle motor. The throttle is opened and closed by the motor acting against the throttle spring via a train of spur gears. Once this torque is applied to the throttle plate, its resultant position, determined by the two potentiometers in the throttle body, is sent back to the PCM. The PCM then confirms that the throttle position is equal to the command that it sent originally.

As a back up, the PCM also monitors the manifold absolute pressure (MAP) and the DC motor duty cycle (via its pulse width modulation {PWM}).

KEY SWITCH & START/STOP PANEL

The key switch, obviously, supplies power to the DTS command module, which then powers on the remainder of the SmartCraft system (including the PCM and all gauges.

You may have a 3-position switch (OFF, ON, START), a 4-position switch (OFF, ACC, ON, START), or a 2-position switch (ON, OFF) coupled with an START/STOP push button. Dual engine set-ups will have one of each for each engine. Turning the 3- or 4-position switch to the **START** position will start the engine, or turning the 2-position switch to the **ON** position and pressing the **START** button.

■ The 4-position switch is only available on the 14-pin versions of SmartCraft.

The signal from the switch feeds the DTS command module - if the key is ON and the module senses a switch closure at the **START** position when the engine is not running, it will initiate cranking. If the engine is already running, it will send a stop command. Pulling the lanyard or turning the key to the **OFF** position will also stop the engine.

The DTS system also incorporates SmartStart - allowing for push button starting. Rather than holding down the Start button or holding the key in the **START** position until the engine actually starts, SmartStart takes control of the process immediately. As soon as the button is pushed or the key is held over, the command module will send a signal to the engine PCM, which will then close the starter slave solenoid. The slave solenoid then closes the starter solenoid and the engine begins to crank. Upon a 3-4 second time-out expiring or the engine reaching 400 rpm, the slave solenoid circuit will be opened and the cranking will stop.

Electronic Shift Control (ESC)
SHIFT ACTUATOR REMOVAL & INSTALLATION

◆ See Figures 20 and 21

1. Tag and disconnect the electrical harness at the actuator and move it out of the way.
2. Remove the nut and washer securing the actuator arm to the shift arm.
3. Remove the mounting bolt at the base of the actuator and lift the actuator off of the shift plate.

※※ **CAUTION**

Do not rotate the piston while the actuator is removed.

To install:

4. Coat the outer surface of the actuator spacer with Unirex-N grease.
5. Line up the actuator with the stud on the shift lever and the spacer on the shift plate and then slide it into position.
6. Coat the threads of the mounting bolt with Loctite® 271 and then tighten the bolt to 14 ft. lbs. (19 Nm).
7. Install the washer and nut onto the shift lever stud and tighten it to 72 inch lbs. (8 Nm). Check the adjustment.

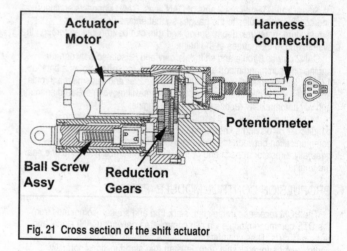

Fig. 21 Cross section of the shift actuator

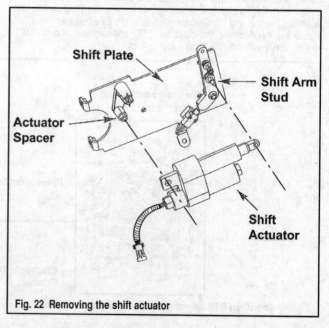

Fig. 22 Removing the shift actuator

REMOTE CONTROLS   15-9

ACTUATOR ADJUSTMENT

◆ See Figure 23

■ All new actuators are pre-set at the factory and equipped with plastic ties to ensure no inadvertent movement of the piston. If you are installing a new actuator, simply cut the ties and install the unit. . .no adjustment is necessary.

1. Turn the ignition key to the **ON** position without starting the engine and make sure that the remote control handle is in Neutral.
2. Measure from the center point of the actuator mounting bolt to the center of the shift lever stud and record the measurement.
3. If the measurement is not 7.782 in. +/- 0.40 (19.766 cm +/- 0.10mm), remove the unit and rotate the actuator lever piston clockwise to shorten the distance or counterclockwise to lengthen it.

■ The ESC adjustment length is monitored by the PCM and the engine will not start if the adjustment is out of specification.

SHIFT OVERRIDE PROCEDURE

◆ See Figure 24

If your vessel has the optional System View display installed and shows an error message reading **GEAR POS DIFF** and/or the engine will not start or shift into gear, there is a problem with the shift actuator.

The actuator can be disengaged so that the drive can be manually shifted into Neutral for starting and then into Forward gear. Engine speed will be limited to 1000-1200 rpm, but at least you'll be able to move the boat.

1. Turn the ignition switch to the **OFF** position.
2. Tag and disconnect the wiring harness at the actuator assembly and move it out of the way.
3. Make sure that the remote control handle is in Neutral and the shift lever is in the vertical position. The shift interrupt switch should be fully engaged and on the rounded portion of the bottom of the shift lever.
4. You should now be able to start the engine. Once done, manually move the shift lever into a gear position as shown.

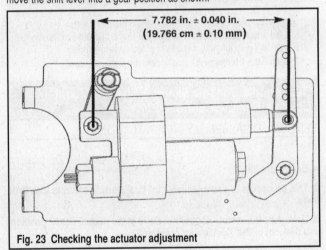

Fig. 23 Checking the actuator adjustment

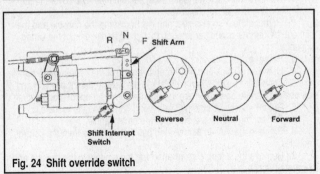

Fig. 24 Shift override switch

Electronic Throttle Control (ETC)

REMOVAL & INSTALLATION

◆ See Figure 25

1. Loosen the hose clamp and remove the flame arrestor.
2. Tag and disconnect the electrical harness at the ETC unit.
3. Remove the 4 mounting bolts and lift the ETC unit off of the adapter.
4. Inspect the adapter and throttle body mating surfaces for any kind of marring or degradation that might affect the seal.
5. Inspect the O-ring on the throttle body - we suggest simply replacing it.
6. Line up the ETC unit on the adapter plate and install the mounting bolts. Tighten them, alternately, to 62 inch lbs. (7 Nm).
7. Reconnect the harness and install the flame arrestor.

Panel Mount Remote Control

REMOVAL, DISASSEMBLY & INSTALLATION

◆ See Figure 26

■ Make sure the remote control handle is in the neutral detent position and take not of that position for reassembly.

1. Remove the 2 screws securing the side covers to the bezel and lift out the covers.
2. Remove the 3 screws securing the bezel to the side panel and then pull out the remote control far enough that you can disconnect the wiring connectors.
3. Remove the unit.
4. Use a small pick and pry the emblem out of the center of the handle.
5. Make sure that the handle is still in the neutral detent position and remove the handle bolt. Separate the handle from the control module shaft.
6. Remove the 4 Allen screws and pull the bezel from the control module.

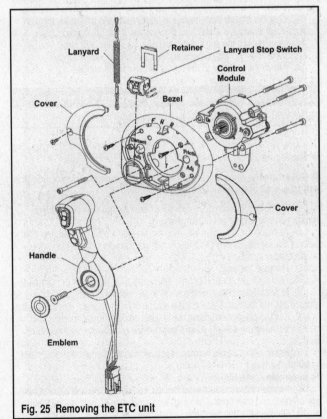

Fig. 25 Removing the ETC unit

15-10 REMOTE CONTROLS

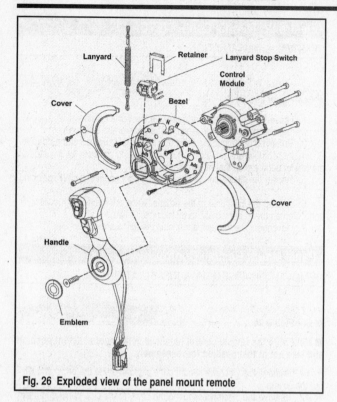

Fig. 26 Exploded view of the panel mount remote

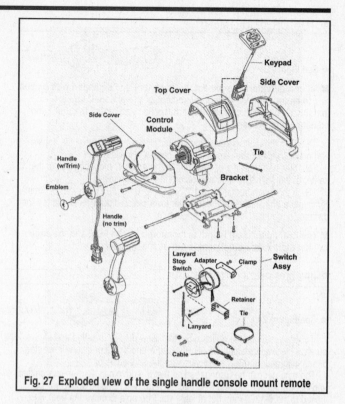

Fig. 27 Exploded view of the single handle console mount remote

To assemble & install:

7. Position the bezel on the module and tighten the mounting screws to 50 inch lbs. (5.6 Nm).
8. Connect the handle to the module shaft so that is still in the neutral detent position and also in the same position as when it was removed. Coat the handle bolt threads with Loctite®271 and screw it in finger-tight.
9. Pull the wiring harnesses/connectors out through the hole in the side panel and connect them to the remote control unit.
10. Install the unit into the side panel and tighten the screws securely.
11. Shift the handle from neutral to full forward and then back through neutral to full reverse to make sure that you have the full range of travel.
12. Now, tighten the control handle bolt to 20 ft. lbs. (27 Nm).
13. Install the two side covers and tighten the screws securely.
14. Attach the emblem to the center of the control handle.

Single Handle Console Mount Remote Control

REMOVAL, DISASSEMBLY & INSTALLATION

◆ See Figure 27

■ Make sure the remote control handle is in the neutral detent position and take note of that position for reassembly.

1. Remove the 2 screws (each side) securing the side covers to the control housing and lift out the covers.
2. Remove the 4 screws securing the bracket to the console and then pull out the remote control far enough that you can disconnect the wiring connectors.
3. Remove the unit.
4. Use a small pick and pry the emblem out of the center of the handle.
5. Make sure that the handle is still in the neutral detent position and remove the handle bolt. Separate the handle from the control module shaft.
6. Cut the plastic tie holding the keypad harness to the module.
7. Remove the 4 Allen screws and pull the top cover from the control module.
8. Remove the 2 long screws attaching the module to the bracket and separate the two.

To assemble & install:

9. Position the module on the bracket so the spacers and bracket holes are aligned, and then tighten the mounting screws to 50 inch lbs. (5.6 Nm).
10. Connect the handle to the module shaft so that it is still in the neutral detent position and also in the same position as when it was removed. Coat the handle bolt threads with Loctite® 271 and tighten the bolt to 20 ft. lbs. (27 Nm).
11. Pull the wiring harnesses/connectors out through the hole in the console and connect them to the remote control unit.
12. Install the unit into the side panel and tighten the screws securely.
13. Shift the handle from neutral to full forward and then back through neutral to full reverse to make sure that you have the full range of travel.
14. Install the top cover, tightening the screws securely. Make sure that the harness is routed between the bracket and mounting tabs.
15. Use a new plastic tie and secure the keypad harness to the module.
16. Attach the emblem to the center of the control handle.
17. Install the 2 side covers and tighten the screws securely.

Dual Handle Console Mount & Shadow Mount Console Remote Controls

REMOVAL, DISASSEMBLY & INSTALLATION

◆ See Figure 28

■ This procedure is also applicable for dual console controls with CAN pads.

■ Make sure the remote control handle is in the neutral detent position and take not of that position for reassembly.

1. Remove the 2 screws (each side) securing the side covers to the control housing and lift out the covers.
2. Remove the 4 screws securing the bracket to the console and then pull out the remote control far enough that you can disconnect the wiring connectors.
3. Remove the unit.
4. Use a small pick and pry the emblem out of the center of the handle.
5. Make sure that the handle is still in the neutral detent position and remove the handle bolt. Separate the handle from the control module shaft.
6. Cut the plastic tie holding the keypad harness to the module.
7. Repeat the previous 2 steps for the other handle.
8. Remove the 4 Allen screws and pull the top cover from the control module.
9. Remove the 2 long screws attaching the module to the bracket and separate the two.

REMOTE CONTROLS 15-11

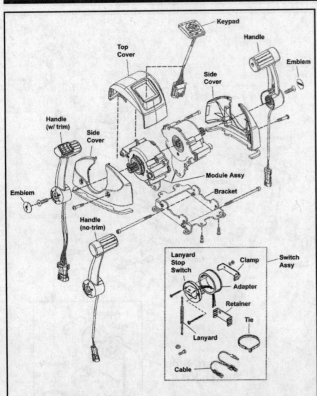

Fig. 28 Exploded view of the dual handle console mount remote, shadow mount similar

10. Remove the 2 screws connecting the two module halves and then separate them.

To assemble & install:

11. Position the module halves and tighten the 2 screws to 50 inch lbs. (5.6 Nm).
12. Position the modules on the bracket so the spacers and bracket holes are aligned, and then tighten the mounting screws to 50 inch lbs. (5.6 Nm).
13. Connect the handle to the module shaft so that it is still in the neutral detent position and also in the same position as when it was removed. Coat the handle bolt threads with Loctite® 271 and tighten the bolt to 20 ft. lbs. (27 Nm).
14. Repeat the previous step for the other handle.
15. Shift the handle from neutral to full forward and then back through neutral to full reverse to make sure that you have the full range of travel.
16. Pull the wiring harnesses/connectors out through the hole in the console and connect them to the remote control unit.
17. Install the unit into the console and tighten the screws securely.
18. Install the top cover, tightening the screws securely. Make sure that the harness is routed between the bracket and mounting tabs.
19. Use a new plastic tie and secure the keypad harness to the module.
20. Attach the emblem to the center of the control handle(s).
21. Install the 2 side covers and tighten the screws securely.

Zero Effort Console Mount Remote Control

REMOVAL & INSTALLATION

◆ See Figures 29 and 30

1. Move the control handle to the Neutral detent position and make sure to note the position for installation.
2. Remove the 4 screws and nylon washers, lift the unit out of the console and disconnect the electrical connection.

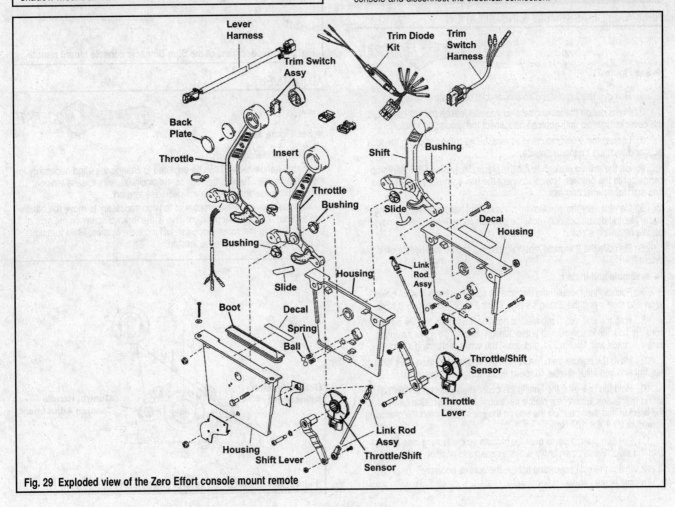

Fig. 29 Exploded view of the Zero Effort console mount remote

15-12 REMOTE CONTROLS

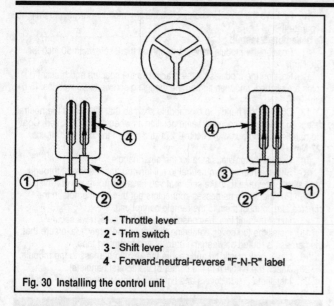

1 - Throttle lever
2 - Trim switch
3 - Shift lever
4 - Forward-neutral-reverse "F-N-R" label

Fig. 30 Installing the control unit

3. Make sure the levers are positioned as shown, position the unit over the console so you can connect the electrical leads and then position it onto the console.

4. Install the mounting screws with new nylon washers and tighten them securely.

Slim Binnacle Single Handle Console Mount Remote Control

REMOVAL, DISASSEMBLY & INSTALLATION

◆ See Figure 31

1. Remove the 2 mounting screws and lift off the side cover.
2. Press out on the locking tabs so you can loosen the top cover. Lift the cover straight up until you can disconnect the keypad connector.
3. Loosen the 4 unit mounting screws, being careful to catch the lock nuts and washers on the underside.
4. Lift the unit up enough to cut the plastic tie securing the start/stop connector to the bracket. Now disconnect the trim harness, lever harness and start/stop panel harness.
5. Carefully pry the emblem from the control handle. Make sure that it is in the Neutral detent position (note the position as well) and then remove the handle mounting screw.
6. Remove the 2 module mounting screws and lift the module off of the bracket.

To assemble and install:

7. Position the module into the bracket so the spacers and all holes align and then insert the 2 long screws. Tighten to 50 inch lbs. (5.6 Nm).
8. With the unit still in the Neutral position, install the handle and make sure the position is as you had noted during removal. Coat the mounting screw threads with Loctite 271 and install the screw, tightening it securely.
9. Move the handle from Neutral to full Forward and then back through Neutral and into full Reverse, checking its full range of movement.
10. Hold the unit over the hole in the console so you can connect all wiring harnesses, use a new plastic tie to connect the start/stop harness to the bracket and then position the unit on the console. Tighten the mounting screws to 20 ft. lbs. (27 Nm).
11. Position the top cover over the module so the locking tabs line up with the slots. Press down firmly until they snap into position in the slots.
12. Install the side cover and tighten the screws securely.
13. Press the emblem into the control handle.

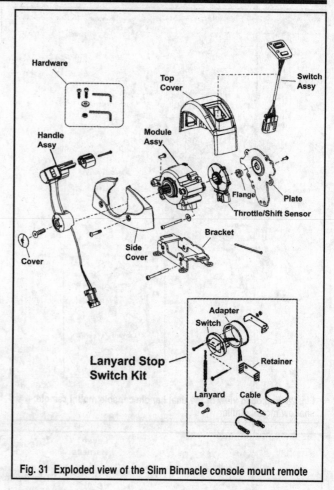

Fig. 31 Exploded view of the Slim Binnacle console mount remote

ADJUSTMENT

◆ See Figure 32

Remove the side cover.
Control handle tension can be adjusted to change the effort necessary to move the handle. Turning the rear screw clockwise will increase tension, while turning counterclockwise will decrease tension.

You can also adjust the amount of tension necessary to move the handle into and out of the detent position. This is the forward screw on the side of the unit. Turning the screw clockwise will increase tension, while turning counterclockwise will decrease tension.

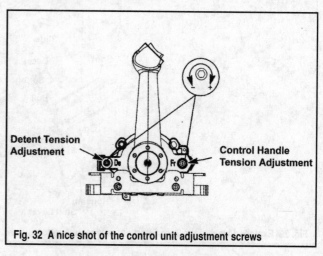

Fig. 32 A nice shot of the control unit adjustment screws

REMOTE CONTROLS 15-13

Wiring Schematics - SmartCraft DTS (10-Pin)

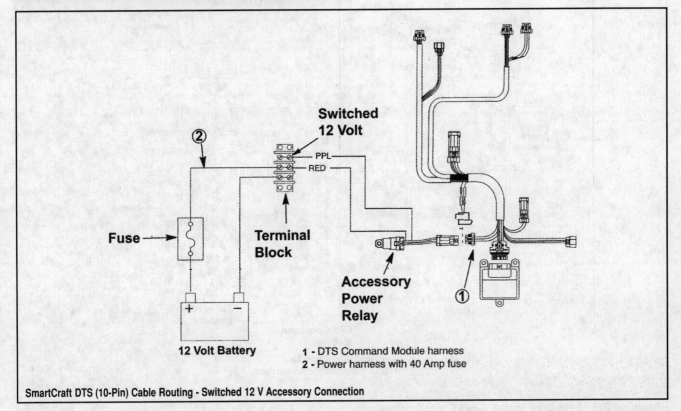

SmartCraft DTS (10-Pin) Cable Routing - Switched 12 V Accessory Connection

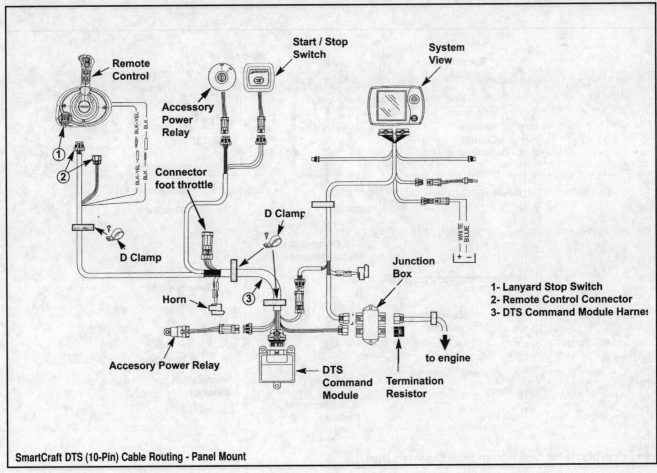

SmartCraft DTS (10-Pin) Cable Routing - Panel Mount

15-14 REMOTE CONTROLS

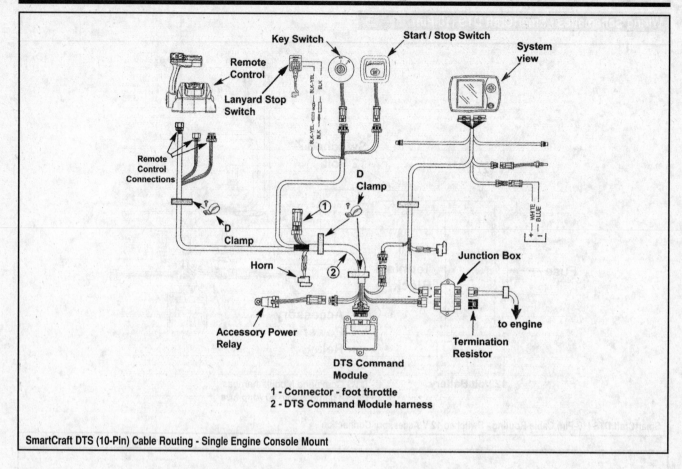

SmartCraft DTS (10-Pin) Cable Routing - Single Engine Console Mount

1 - Connector - foot throttle
2 - DTS Command Module harness

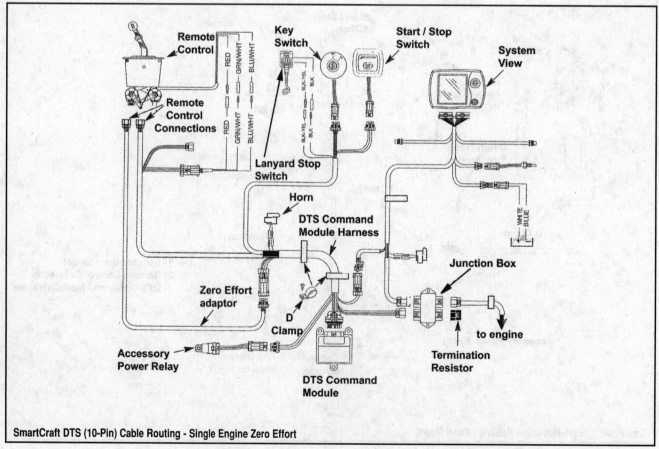

SmartCraft DTS (10-Pin) Cable Routing - Single Engine Zero Effort

REMOTE CONTROLS 15-15

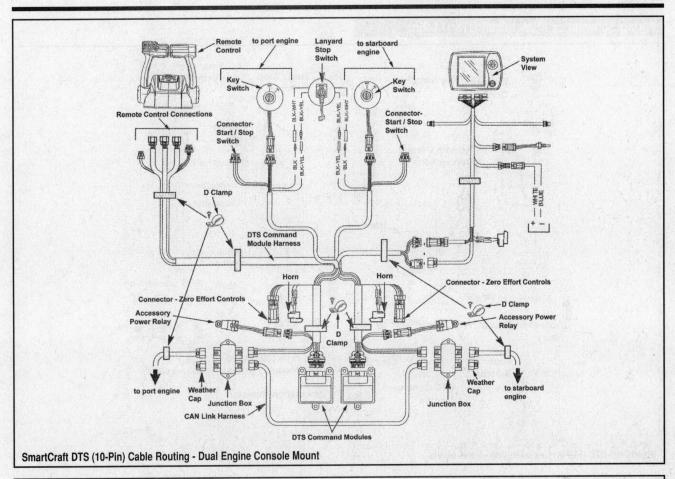

SmartCraft DTS (10-Pin) Cable Routing - Dual Engine Console Mount

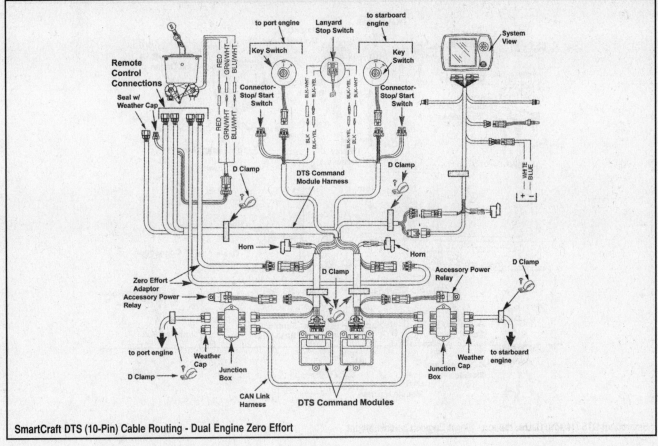

SmartCraft DTS (10-Pin) Cable Routing - Dual Engine Zero Effort

15-16 REMOTE CONTROLS

Wiring Schematics - SmartCraft DTS (14-Pin)

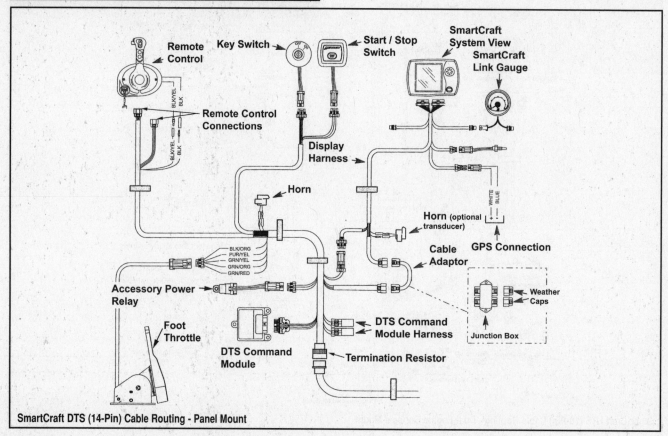

SmartCraft DTS (14-Pin) Cable Routing - Panel Mount

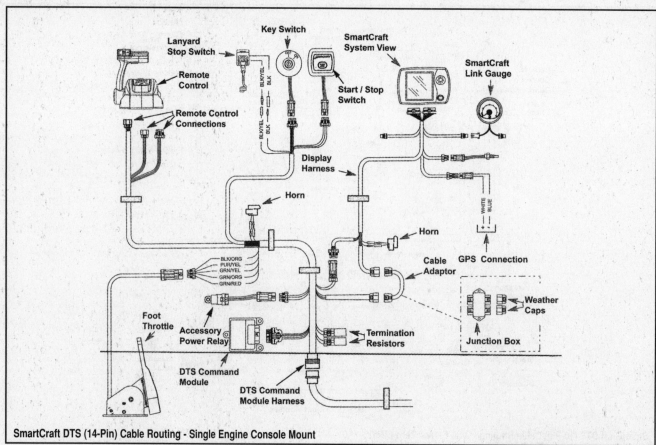

SmartCraft DTS (14-Pin) Cable Routing - Single Engine Console Mount

REMOTE CONTROLS 15-17

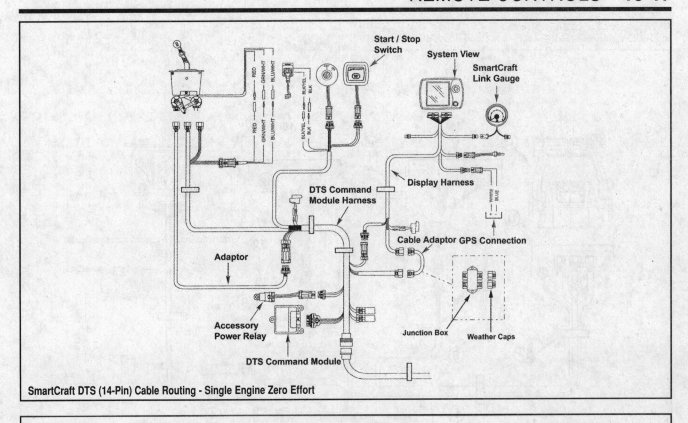

SmartCraft DTS (14-Pin) Cable Routing - Single Engine Zero Effort

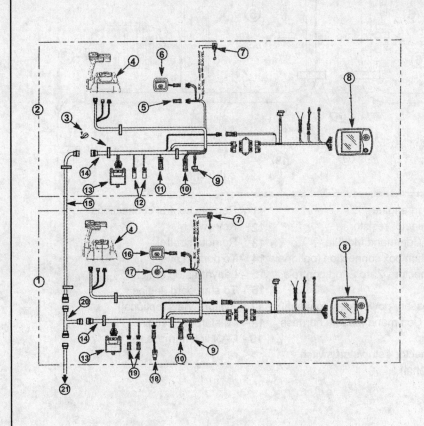

1 - Helm 1
2 - Helm 2
3 - Clamps (use as required)
4 - Remote control
5 - Key switch connector not used – seal with weather cap
6 - Start/stop switch (2 required at helm)
7 - Lanyard stop switch
8 - System View (optional)
9 - Horn
10 - Foot throttle connector
11 - Accessory power relay connector (optional, not used at second helm)
12 - Terminator resistors (helm 2 only)
13 - DTS Command Module
14 - DTS Command Module harness
15 - Data harness
16 - Start/stop switch (optional at helm 1)
17 - Key switch
18 - Accessory power relay
19 - Terminator resistors not used – seal connectors with weather caps
20 - Dual helm adaptor harness
21 - Data harness to engine

SmartCraft DTS (14-Pin) Cable Routing - Single Engine Console Mount, Dual Helm

15-18 REMOTE CONTROLS

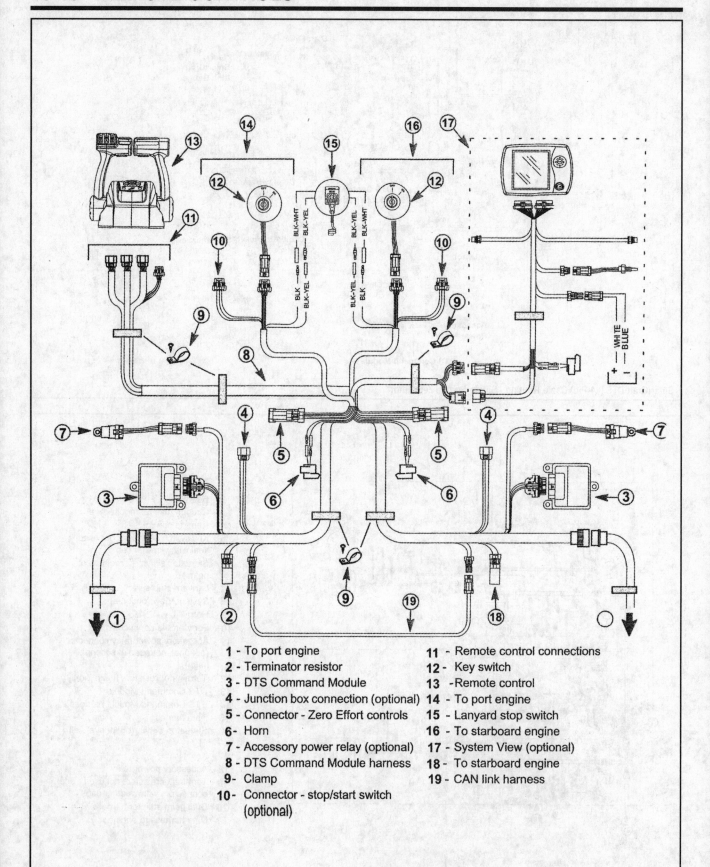

1 - To port engine
2 - Terminator resistor
3 - DTS Command Module
4 - Junction box connection (optional)
5 - Connector - Zero Effort controls
6 - Horn
7 - Accessory power relay (optional)
8 - DTS Command Module harness
9 - Clamp
10 - Connector - stop/start switch (optional)
11 - Remote control connections
12 - Key switch
13 - Remote control
14 - To port engine
15 - Lanyard stop switch
16 - To starboard engine
17 - System View (optional)
18 - To starboard engine
19 - CAN link harness

SmartCraft DTS (14-Pin) Cable Routing - Dual Engine Console Mount

REMOTE CONTROLS 15-19

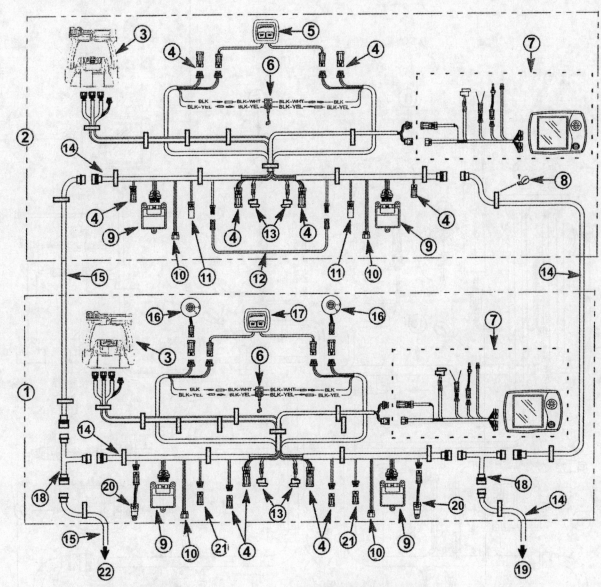

1 - Helm 1 (primary)
2 - Helm 2 (auxiliary)
3 - Remote control
4 - Connector not used – seal with weather cap
5 - Start/stop switch (2 required at helm)
6 - Lanyard stop switch
7 - System View (optional)
8 - Clamps (use as required)
9 - DTS Command Module
10 - Junction box connection (optional)
11 - Terminator resistor (2 installed at helm only)
12 - CAN link harness
13 - Horn
14 - DTS Command Module harness
15 - Data harness
16 - Key switch
17 - Start/stop switch (optional at helm 1)
18 - Dual helm adaptor harness
19 - Data harness to starboard engine
20 - Accessory power relay
21 - Remove terminator resistor seal with weather cap
22 - Data harness to port engine

SmartCraft DTS (14-Pin) Cable Routing - Dual Engine Console Mount, Dual Helm

15-20 REMOTE CONTROLS

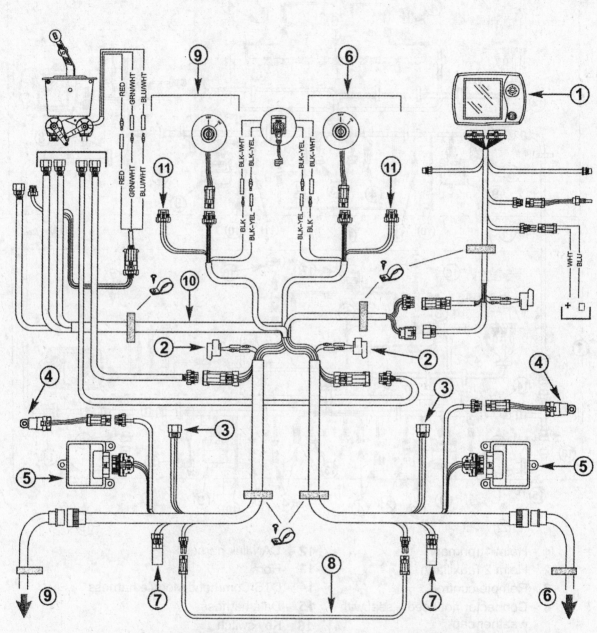

1 - System View (optional)
2 - Horn
3 - Junction box connection (optional)
4 - Accessory power relay (optional)
5 - DTS Command Module
6 - To starboard engine
7 - Terminator resistor
8 - CAN link harness
9 - To port engine
10 - DTS Command Module harness
11 - Connector - Start/stop switch (optional)

SmartCraft DTS (14-Pin) Cable Routing - Dual Engine Zero Effort

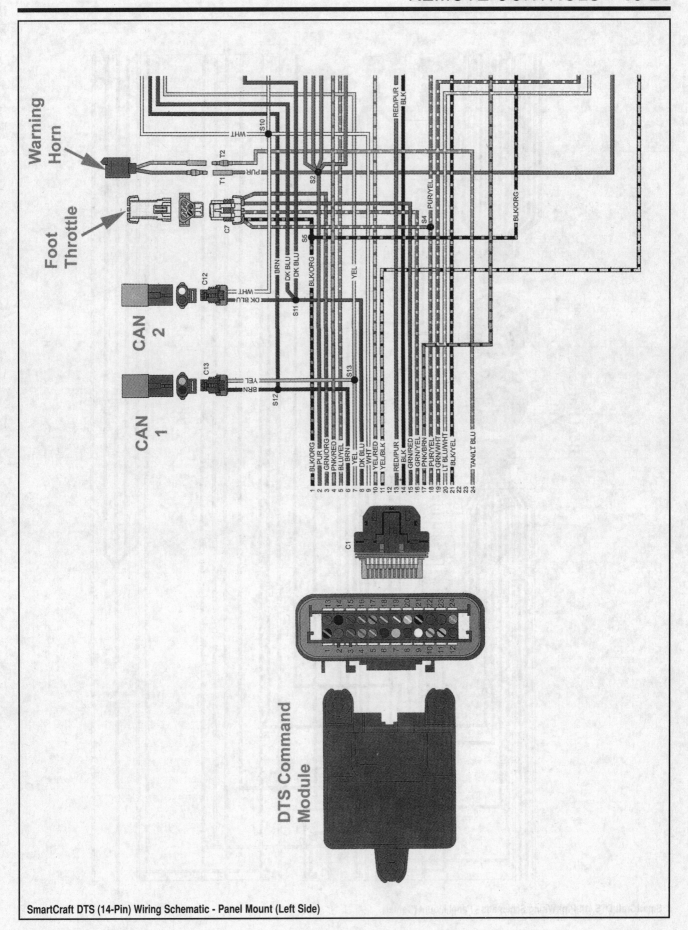

SmartCraft DTS (14-Pin) Wiring Schematic - Panel Mount (Left Side)

15-22 REMOTE CONTROLS

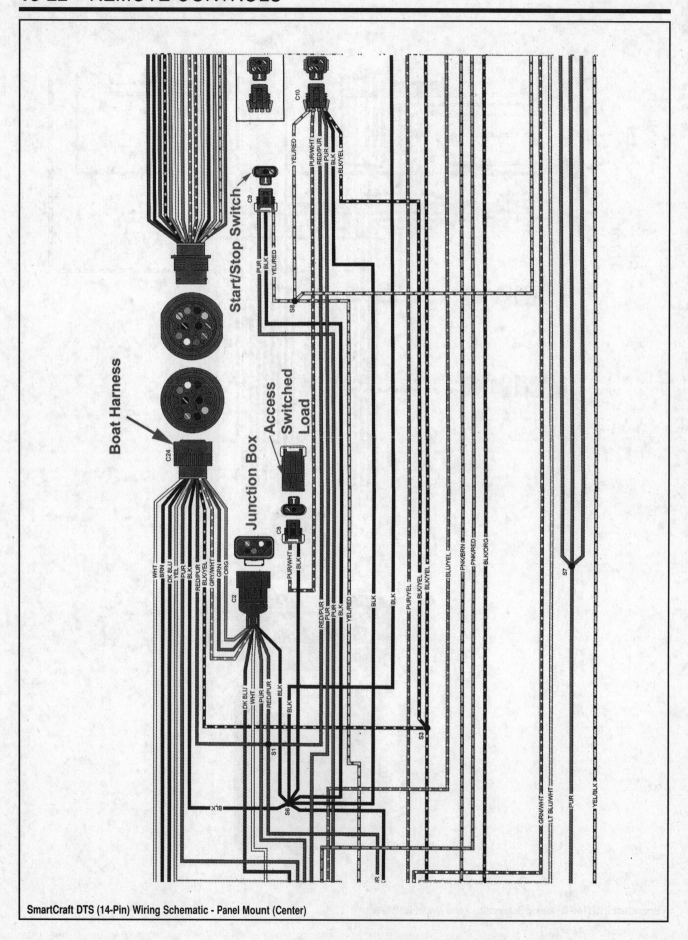

SmartCraft DTS (14-Pin) Wiring Schematic - Panel Mount (Center)

REMOTE CONTROLS 15-23

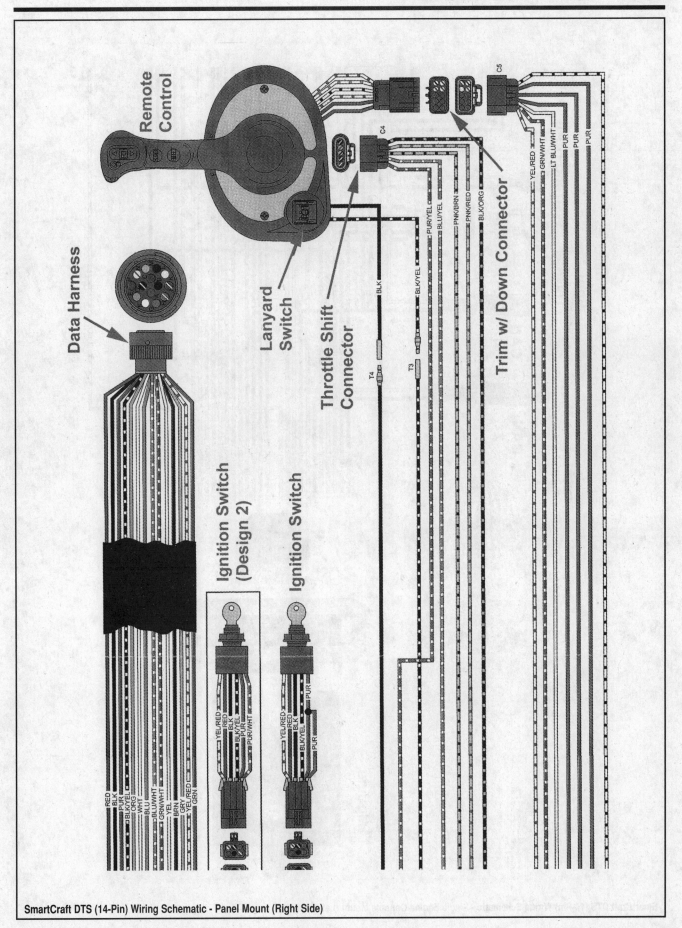

SmartCraft DTS (14-Pin) Wiring Schematic - Panel Mount (Right Side)

15-24 REMOTE CONTROLS

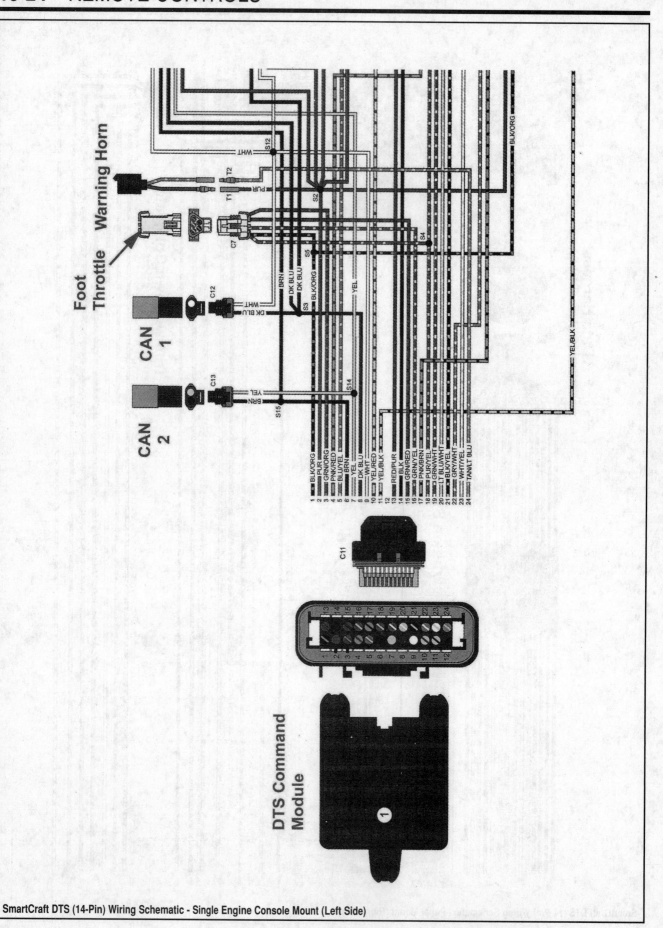

SmartCraft DTS (14-Pin) Wiring Schematic - Single Engine Console Mount (Left Side)

REMOTE CONTROLS 15-25

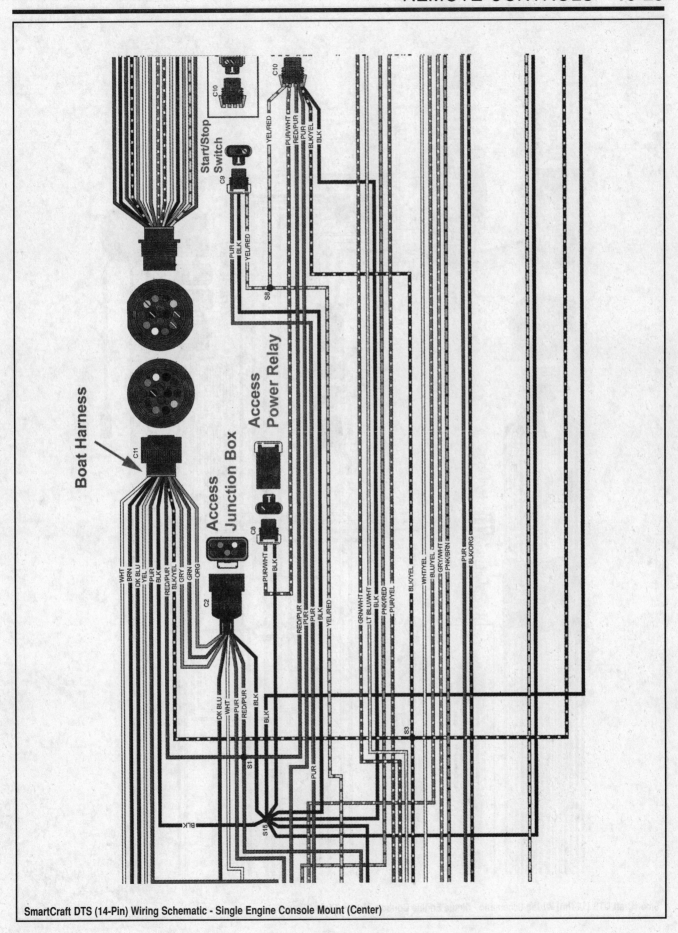

SmartCraft DTS (14-Pin) Wiring Schematic - Single Engine Console Mount (Center)

15-26 REMOTE CONTROLS

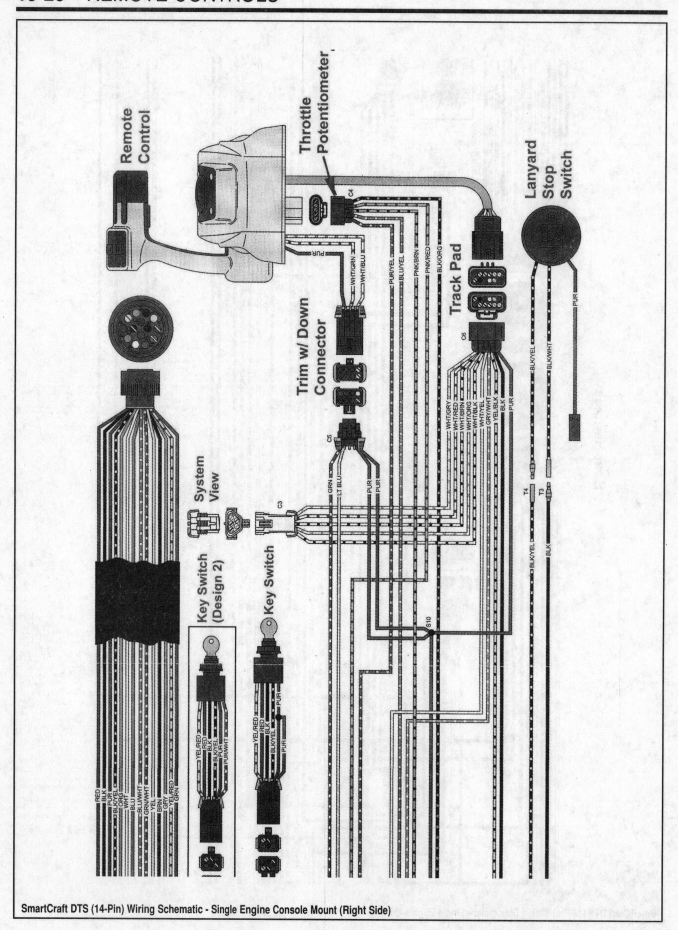

SmartCraft DTS (14-Pin) Wiring Schematic - Single Engine Console Mount (Right Side)

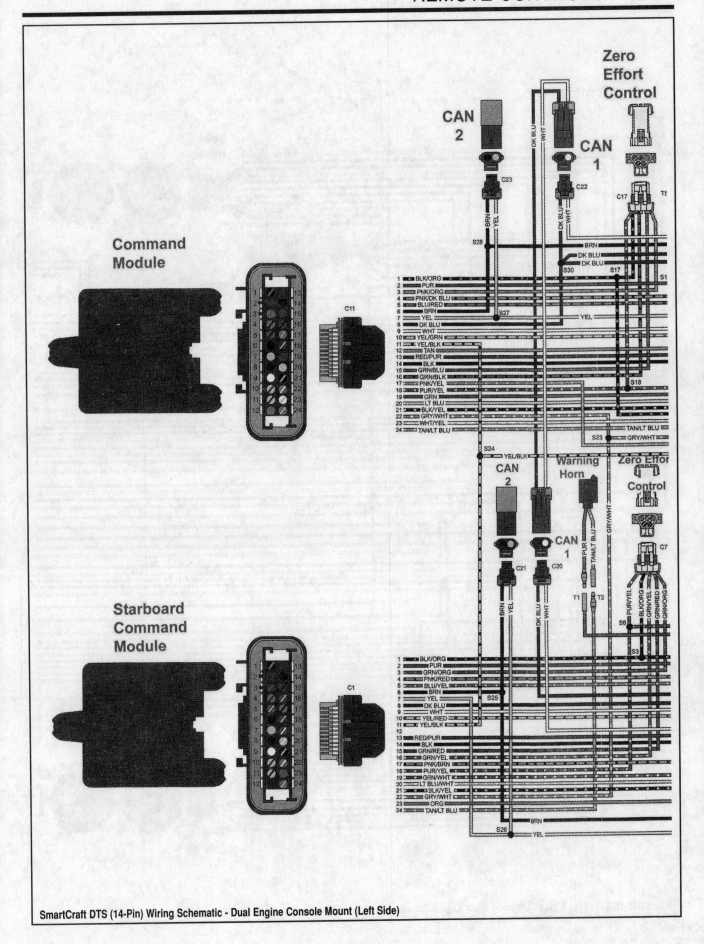

SmartCraft DTS (14-Pin) Wiring Schematic - Dual Engine Console Mount (Left Side)

15-28 REMOTE CONTROLS

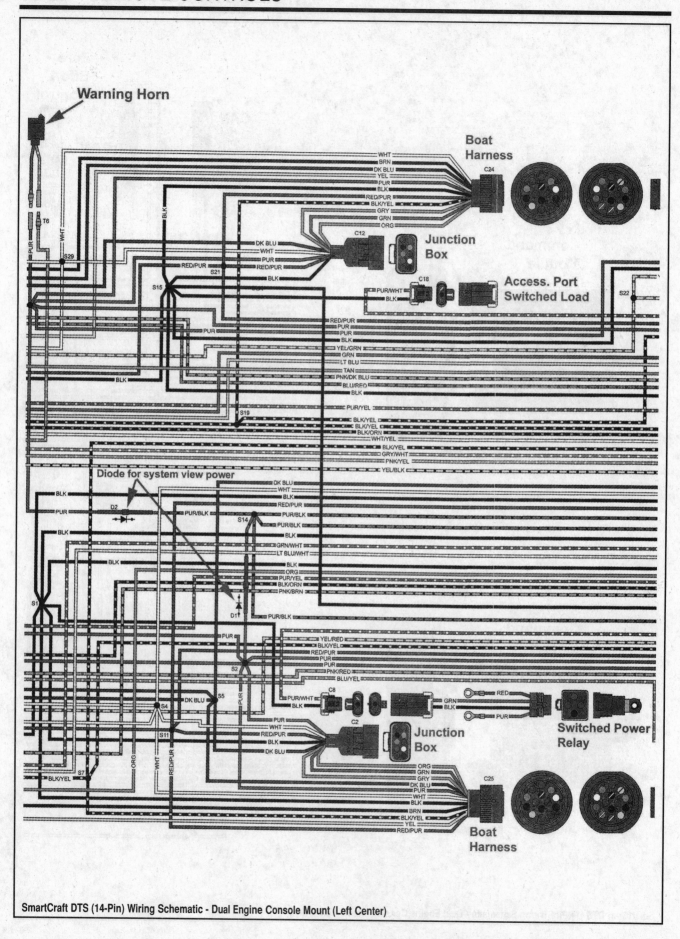

SmartCraft DTS (14-Pin) Wiring Schematic - Dual Engine Console Mount (Left Center)

REMOTE CONTROLS   15-29

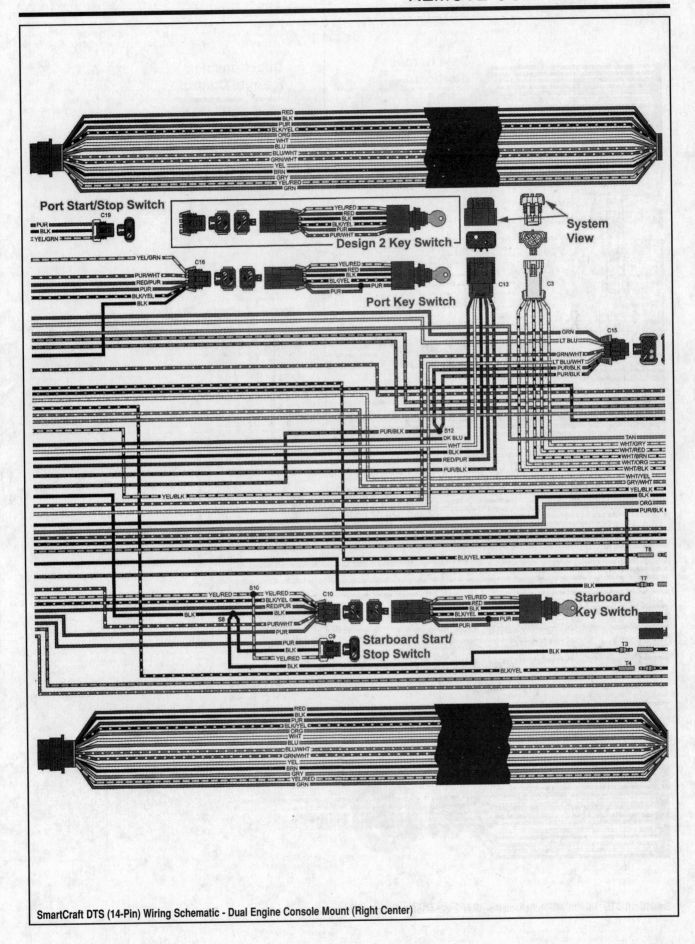

SmartCraft DTS (14-Pin) Wiring Schematic - Dual Engine Console Mount (Right Center)

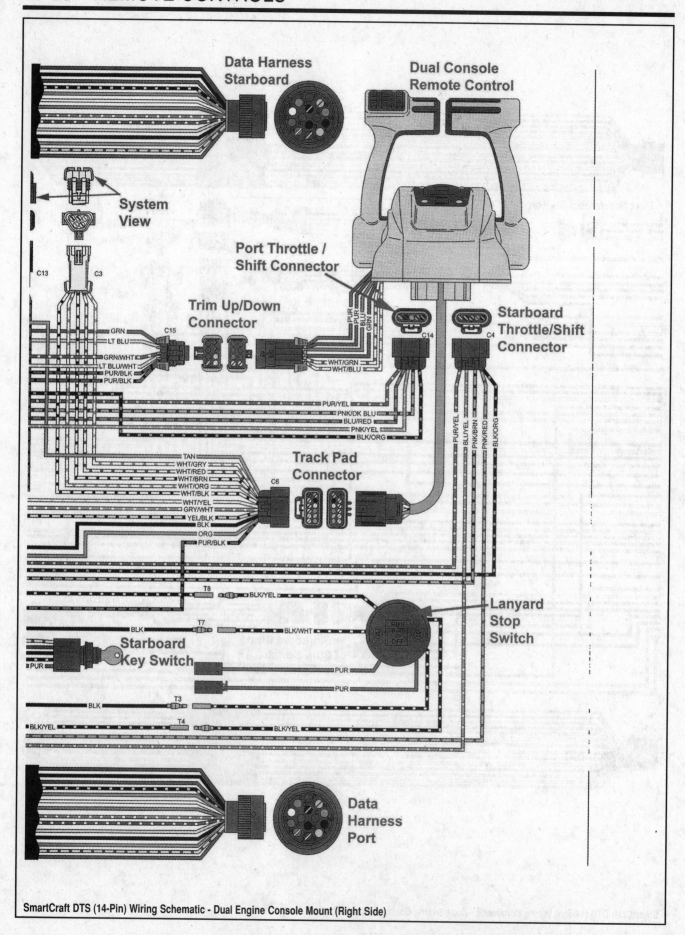

SmartCraft DTS (14-Pin) Wiring Schematic - Dual Engine Console Mount (Right Side)

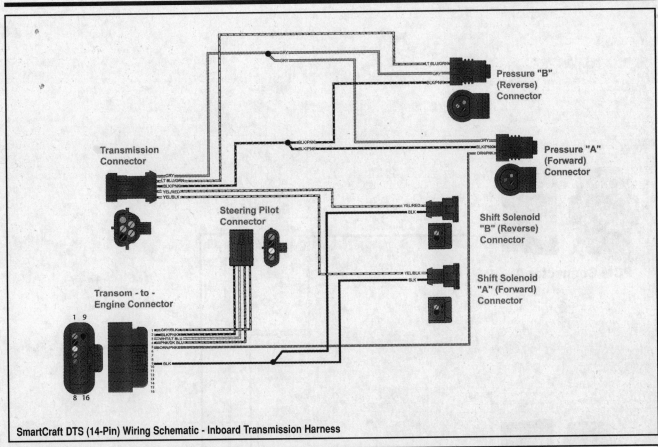

SmartCraft DTS (14-Pin) Wiring Schematic - Inboard Transmission Harness

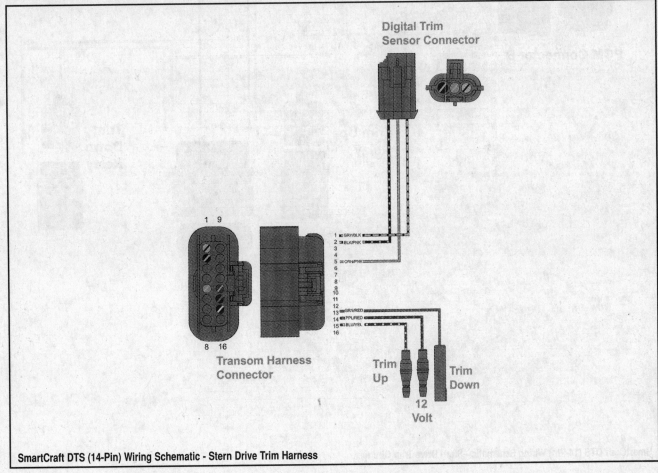

SmartCraft DTS (14-Pin) Wiring Schematic - Stern Drive Trim Harness

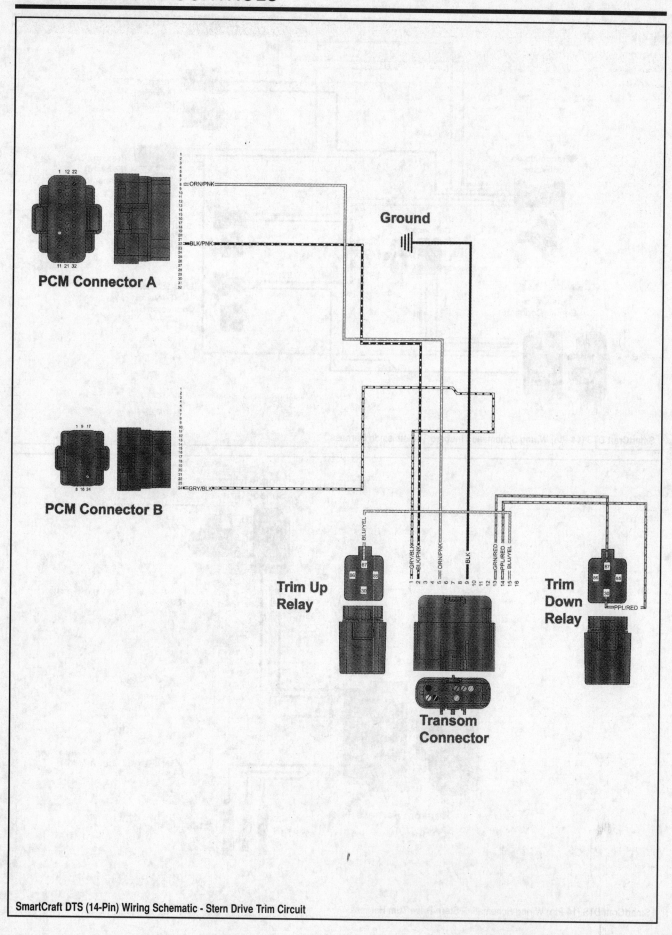

SmartCraft DTS (14-Pin) Wiring Schematic - Stern Drive Trim Circuit

REMOTE CONTROLS

PRECISION PILOT

Troubleshooting

Symptom	Cause	Action
1.0 Move joystick and nothing happens	1.0 Ignition switch off	1.0 Turn ignition switch on.
	1.1 Battery switch off	1.1 Turn battery switch on.
	1.2 36 volt battery pack discharged	1.2 Charge or replace batteries.
	1.3 12 volt battery discharged	1.3 Charge or replace battery.
	1.4 20 Amp fuse near 50 Amp breaker is blown	1.4 Replace 20 Amp fuse.
	1.5 Joystick harness not properly connected	1.5 Check harness connections.
	1.6 Purple wire from joystick not connected	1.6 Connect purple wire from joystick.
	1.7 50 Amp breaker tripped	1.7 Reset breaker.
	1.8 Faulty 50 Amp breaker	1.8 Close breaker, disconnect one end and check for continuity. If no continuity exists, replace breaker.
	1.9 Terminals on batteries or circuit breakers not making connection	1.9 Clean and tighten all terminal connections.
	1.10 Harness chafed or disconnected	1.10 Check all harnesses for proper connection and worn or broken spots.
	1.11 No ground to joystick	1.11 Check for 12 volts between engine ground and pin 10. If no voltage, check all purple wire connections.
	1.12 Faulty joystick	1.12 Disconnect joystick from joystick harness. Twist joystick clockwise. Measure continuity between pin 9 and pins 2, 4, 6 and 8. Twist joystick counterclockwise. Measure continuity between pin 9 and pins 1, 3, 5 and 7. If no continuity, replace joystick.
	1.13 Faulty control box	1.13 Disconnect motor at harness connection. Move the joystick CW and CCW and check voltage on motor harness coming from the control. If voltage is not 36 volts or −36 volts, replace control box.

Fig. 33 Troubleshooting the Precision Pilot system

Symptom	Cause	Action
2.0 Move joystick and motors jump, then stop and horn sounds	2.0 150 Amp breaker connection faulty in front of fuse	2.0 Check all connections to 150 Amp breaker.
3.0 Move joystick and intermittent horn sounds	3.0 150 Amp breaker tripped	3.0 Reset breaker.
	3.1 150 Amp breaker disconnected before fuse	3.1 Check connections.
	3.2 150 Amp breaker faulty	3.2 Close breaker, disconnect one end and check for continuity. If no continuity exists, replace breaker.
	3.3 Low voltage from 36 volt battery pack	3.3 Check voltage and charge batteries.
	3.4 Low voltage from 12 volt battery	3.4 Check voltage and charge battery.
	3.5 36 volt battery pack disconnected	3.5 Check 36 volt battery pack connections and terminals.
	3.6 Harness chafed or disconnected	3.6 Check all harnesses for proper connection and worn or broken spots.
4.0 Move joystick and solid horn sounds	4.0 Overcurrent circuit tripped. System will shut down for 15 to 30 seconds before resetting.	4.0 Remove boat from water and check for debris in impellers. The current draw of these motors should be approximately 1-2 amps no load and approximately 35-40 amps under load when stabilized. Check for system overheating. If no overheat condition exists, replace control box.
	4.1 Motors overheating. System will repeatedly shut down for 4 seconds and then operate for 6 seconds.	4.1 Move joystick and check for water movement on sides of boat indicating motors are operating. If no water movement is seen, check motor harness connections. Allow motors to cool down completely before operating system again. Remove boat from water and clean debris from impellers.
	4.2 Defective motor	4.2 Disconnect motor harness connections one at a time until system can be operated without horn sounding. Plug in a known good motor. If horn does not sound, replace defective motor. If horn still sounds after previous test, replace control box.

Fig. 34 Troubleshooting the Precision Pilot system (cont'd)

15-34 REMOTE CONTROLS

Symptom	Cause	Action
5.0 Boat does not move forward when joystick is activated	5.0 One or both rear motors are not operating	5.0 Move joystick forward and check for water movement on sides of boat indicating motor is operating. If no water movement is seen, check motor harness connections. Disconnect joystick from joystick harness. Twist joystick clockwise. Measure continuity between pin 9 and pins 2, 4, 6 and 8. Twist joystick counterclockwise. Measure continuity between pin 9 and pins 1, 3, 5 and 7. If no continuity, replace joystick. Disconnect non-functioning motor at harness connection. Move the joystick CW and CCW and check voltage on motor harness coming from the control. If voltage is not 36 volts or –36 volts, replace control box. Replace motor.
	5.1 One or both rear motor impellers are clogged	5.1 Remove boat from water and clean debris from impellers.
	5.2 One or both rear motor impellers are broken or missing	5.2 Remove boat from water and replace impellers.
6.0 Boat reverses sluggishly when joystick is activated	6.0 One or both rear motors are not operating	6.0 Move joystick forward and check for water movement on sides of boat indicating motor is operating. If no water movement is seen, check motor harness connections. Disconnect joystick from joystick harness. Twist joystick clockwise. Measure continuity between pin 9 and pins 2, 4, 6 and 8. Twist joystick counterclockwise. Measure continuity between pin 9 and pins 1, 3, 5 and 7. If no continuity, replace joystick. Disconnect non-functioning motor at harness connection. Move the joystick CW and CCW and check voltage on motor harness coming from the control. If voltage is not 36 volts or –36 volts, replace control box. Replace motor.
	6.1 One or both rear motor impellers are clogged	6.1 Remove boat from water and clean debris from impellers.
	6.2 One or both rear motor impellers are broken or missing	6.2 Remove boat from water and replace impellers.

Fig. 35 Troubleshooting the Precision Pilot system (cont'd)

Symptom	Cause	Action
7.0 Boat moves sluggishly when joystick is activated for sideways motion	7.0 Front motor is not operating	7.0 Move joystick side to side and check for water movement on sides of boat indicating motor is operating. If no water movement is seen, check motor harness connections. Disconnect non-functioning motor at harness connection. Move the joystick CW and CCW and check voltage on motor harness coming from the control. If voltage is not 36 volts or –36 volts, replace control box. Disconnect joystick from joystick harness. Twist joystick clockwise. Measure continuity between pin 9 and pins 2, 4, 6 and 8. Twist joystick counterclockwise. Measure continuity between pin 9 and pins 1, 3, 5 and 7. If no continuity, replace joystick. Replace motor.
	7.1 Front motor impeller clogged	7.1 Remove boat from water and clean debris from impeller.
	7.2 Front motor impeller broken or missing	7.2 Remove boat from water and replace impeller.
8.0 No voltage at equalizer	8.0 50 Amp breaker disconnected after the fuse	8.0 Check 50 Amp breaker connections.
9.0 36 volt battery pack discharges rapidly	9.0 Equalizer not providing 36 volt power	9.0 Check equalizer connections. Check voltage at battery pack. Voltage should be 36 volts. Check voltage on 36 volt side. If 36 volt side is lower than at battery pack, replace cable and recheck. Check voltage on 12 volt side. If voltage is less than 12 volts, replace equalizer.
	9.1 150 Amp breaker disconnected after fuse	9.1 Check connections.

Fig. 36 Troubleshooting the Precision Pilot system (cont'd)

REMOTE CONTROLS 15-35

Symptom	Cause	Action
10. 36 volt batteries will not charge properly	10.0 Terminals on batteries or circuit breakers not making connection	10.0 Clean and tighten all terminal connections.
	10.1 Harness chafed or disconnected	10.1 Check all harnesses for proper connection and worn or broken spots.
	10.2 20 Amp fuse near 150 Amp breaker is blown	10.2 Replace 20 Amp fuse.
	10.3 50 Amp breaker tripped	10.3 Reset breaker.
	10.4 Faulty 50 Amp breaker	10.4 Close breaker, disconnect one end and check for continuity. If no continuity exists, replace breaker.
	10.5 150 Amp breaker tripped	10.5 Reset breaker.
	10.6 150 Amp breaker faulty	10.6 Close breaker, disconnect one end and check for continuity. If no continuity exists, replace breaker.
	10.7 Faulty equalizer	10.7 Ensure 12 volts exist between the 12 volt terminal of the equalizer and ground. Ensure 36 volts exist between the 36 volt terminal of the equalizer and ground. Replace equalizer if voltage does not test correctly.
11. Loud vibrations coming from front of boat	11.0 Front motor impeller broken or cracked	11.0 Remove boat from water and replace impeller
12. Loud vibrations coming from rear of boat	12.0 One or both rear motor impellers broken or cracked	12.0 Remove boat from water and replace impeller

Fig. 37 Troubleshooting the Precision Pilot system (cont'd)

Exploded Views

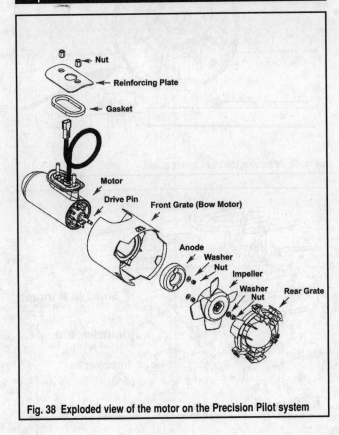

Fig. 38 Exploded view of the motor on the Precision Pilot system

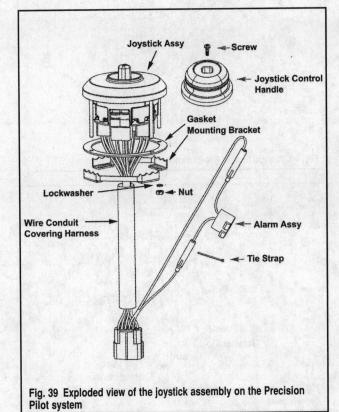

Fig. 39 Exploded view of the joystick assembly on the Precision Pilot system

15-36 REMOTE CONTROLS

Controller
REMOVAL & INSTALLATION

◆ See Figure 40

1. Disconnect the battery cables.
2. Tag and disconnect the electrical connections at the box.
3. Remove the mounting nuts and lift out the controller.

■ Do not remove the controller cover or you will void the warranty.

4. Install the unit and tighten the nuts securely.
5. Reconnect the electrical connections and the battery cables.

Equalizer
REMOVAL & INSTALLATION

◆ See Figure 41

1. Disconnect the battery cables.
2. Carefully pull back the heat shrink tubing on the electrical connections, tag them and then remove them.
3. Remove the mounting nuts and lift out the equalizer.
4. Position the unit and install the mounting nuts, tightening each securely.
5. Reconnect the 12V, Grd and 36V leads. Pull the heat shrink back over the connections and then heat it lightly.

Front Safety Guard
REMOVAL & INSTALLATION

◆ See Figure 42

1. Disconnect the battery cables.
2. Squeeze together the 3 sets of locking tabs on the perimeter of the guard and snap it out of the tunnel.
3. Line up the tabs on the guard with the small indents in the tunnel and snap the guard back into position.
4. Connect the battery cables.

Impeller
REMOVAL & INSTALLATION

◆ See Figure 43

1. Disconnect the battery cables.
2. Make sure that the ignition switch is in the **OFF** position or on keyless systems that the switch is off.
3. Remove any safety guards that may be in position and then insert a small piece of wood into the tunnel to hold the impeller in place.
4. Remove the impeller nut and slide the unit off of the shaft,
5. Position the impeller on the shaft and then tighten the nut securely. You'll need that same piece of wood to hold it in position.
6. Install any guards that were removed and connect the battery cables.

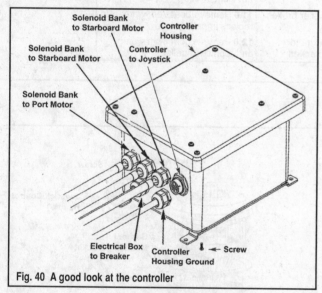

Fig. 40 A good look at the controller

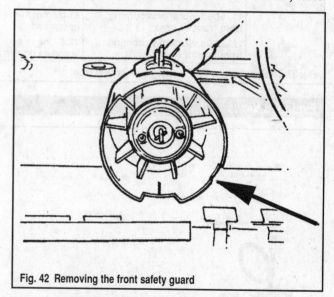

Fig. 42 Removing the front safety guard

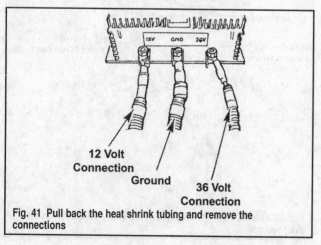

Fig. 41 Pull back the heat shrink tubing and remove the connections

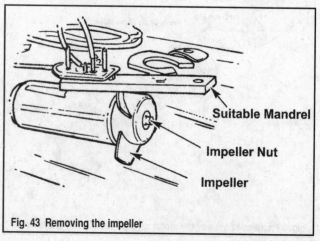

Fig. 43 Removing the impeller

REMOTE CONTROLS 15-37

Joystick

REMOVAL & INSTALLATION

◆ See Figure 39

1. Disconnect the battery cables.
2. Reach in and disconnect the harness at its connector below the joystick.
3. Remove the cover.
4. Remove the mounting nuts and brackets and pull the assembly out of the housing.
5. Drop the assembly back down into position and install the mounting brackets and nuts; tightening securely.
6. Reach in and reconnect the harness.
7. Install the cover and connect the battery cables.

Motors

REMOVAL & INSTALLATION

◆ See Figure 38

1. Disconnect the battery cables.
2. Tag and disconnect the motor harness connector.
3. Remove the safety guard if equipped. Note that some vessels may have one or two additional cover guards, particularly on the rear tunnels.
4. Remove the mounting nuts and ease the motor out of the tunnel.
5. Remove the gasket.
6. Remove the impeller and anode if desired.

To install:

7. Install the anode and impeller if removed; tighten the impeller nut securely.
8. Peel back the adhesive cover and position a new gasket onto the top of the motor.
9. Feed the harness leads through the thru-hull hole and slide the motor back into the tunnel. Position it under the thru-hull hole and press it up into position so the studs go through their respective holes. Tighten the nuts to 35 ft. lbs. (47 Nm).
10. Connect the motor harness and the battery cables.

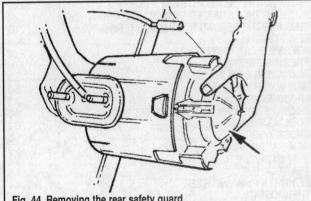

Fig. 44 Removing the rear safety guard

Rear Safety Guard

REMOVAL & INSTALLATION

◆ See Figure 44

1. Disconnect the battery cables.
2. Remove the 2 bolts securing the guard to the side entrance of the rear tunnel and pull out the guard(s).
3. Position the guard in the tunnel entrance and tighten the bolts securely.
4. Connect the battery cables.

Wiring Schematics

◆ See Figure 45

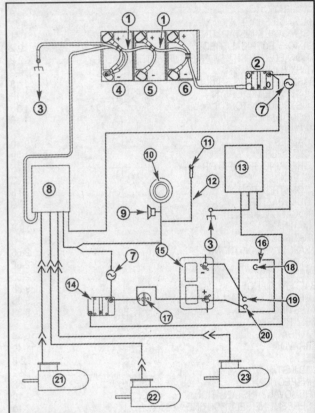

1 - Jumper Harness
2 - 150 Amp Circuit Breaker
3 - Engine Ground
4 - Battery 3 (36V Battery Bank)
5 - Battery 2 (36V Battery Bank)
6 - Battery 1 (36V Battery Bank)
7 - 20 Amp Fuse
8 - Controller
9 - Warning Horn
10 - Joystick
11 - Switched 12V Ignition
12 - Purple
13 - Equalizer
14 - 50 Amp Circuit Breaker
15 - Starting Battery
16 - Battery Charger
17 - Battery Switch
18 - AC 120V
19 - Ground
20 - 12V
21 - Starboard Motor
22 - Front Motor
23 - Port Motor

Fig. 45 Precision Pilot System Schematic

MASTER INDEX

ALTERNATOR ... 9-12
  CIRCUIT TESTING ... 9-13
  COMPONENT TESTING
    DELCO ... 9-16
    MANDO ... 9-14
  DISASSEMBLY & ASSEMBLY ... 9-16
  INSPECTION ... 9-12
  PRECAUTIONS ... 9-12
  REMOVAL & INSTALLATION ... 9-12
ANODES (ZINCS) ... 2-61
  INSPECTION ... 2-62
  LOCATIONS ... 2-62
  SERVICING ... 2-61
AUDIO WARNING SYSTEM ... 9-43
  TESTING ... 9-43
AUTO TRIM II SYSTEM ... 13-24
  CONTROL MODULE ... 13-24
  DESCRIPTION ... 13-24
  OPERATION ... 13-24
BALANCE SHAFT ... 4-41
  REMOVAL & INSTALLATION ... 4-41
BASIC ELECTRICAL THEORY ... 9-2
  HOW ELECTRICITY WORKS: THE WATER ANALOGY ... 9-2
  OHM'S LAW ... 9-2
BASIC OPERATING PRINCIPLES ... 1-13
  2-STROKE MOTORS ... 1-13
  4-STROKE MOTORS ... 1-15
  COMBUSTION ... 1-16
BATTERY
  CABLES ... 2-66, 9-10
  CHARGERS ... 2-66, 9-9
  CHECKING SPECIFIC GRAVITY ... 2-65
  CLEANING ... 2-63
  CONSTRUCTION ... 9-8
  DUAL BATTERY ... 9-9
  LOCATION ... 9-9
  MARINE ... 9-8
  RATINGS ... 9-8
  SAFETY PRECAUTIONS ... 2-66
  STORAGE ... 2-66
  TERMINALS ... 2-65
BATTERY GAUGE ... 9-38
  REMOVAL & INSTALLATION ... 9-38
  TESTING ... 9-38
BELL HOUSING
  ALPHA ... 10-44
  BRAVO ... 11-40
BELTS ... 2-8
  ADJUSTMENT ... 2-9
  INSPECTION ... 2-8
  REMOVAL & INSTALLATION ... 2-10
  SERPENTINE BELT ROUTING ... 2-10
BOAT MAINTENANCE ... 2-60
  ANODES (ZINCS) ... 2-61
  BATTERIES ... 2-63
  CONTINUITY CIRCUIT ... 2-63
  INSIDE THE BOAT ... 2-60
  THE BOAT'S EXTERIOR ... 2-60
BOATING SAFETY ... 1-4
  COURTESY MARINE EXAMINATIONS ... 1-10
  REGULATIONS FOR YOUR BOAT ... 1-4
  REQUIRED SAFETY EQUIPMENT ... 1-5
BOOSTER CYLINDER - POWER ... 14-12
  REMOVAL & INSTALLATION ... 14-12
  TESTING ... 14-13
BUY OR REBUILD? ... 5-3
CAMSHAFT, BEARINGS & GEAR - 3.0L ... 3-13
  CHECKING LIFT ... 3-13
  REMOVAL & INSTALLATION ... 3-14
CAMSHAFT & BEARINGS - V6/V8 ... 4-41
  CHECKING LIFT ... 4-41
  REMOVAL & INSTALLATION ... 4-42
CARBURETOR
  MERCARB 2BBL ... 6-8
  WEBER WFB 4BBL ... 6-17

CARBURETED FUEL SYSTEM ... 6-3
  CARBURETOR APPLICATIONS ... 6-3
  ELECTRIC FUEL PUMP ... 6-26
  MECHANICAL FUEL PUMP ... 6-24
  MERCARB 2BBL CARBURETOR ... 6-8
  OPERATION ... 6-5
  TROUBLESHOOTING ... 6-3
  WEBER WFB 4BBL CARBURETOR ... 6-17
CHARGING SYSTEM ... 9-10
  ALTERNATOR ... 9-12
  GENERAL INFORMATION ... 9-10
  TROUBLESHOOTING ... 9-11
CHEMICALS ... 1-17
  CLEANERS ... 1-18
  LUBRICANTS & PENETRANTS ... 1-17
  SEALANTS ... 1-17
COMBINATION MANIFOLD ... 3-7
  REMOVAL & INSTALLATION ... 3-7
COMBUSTION ... 6-2
COMPACT HYDRAULIC SYSTEM ... 14-19
  DESCRIPTION ... 14-19
  HYDRAULIC FLUID ... 14-19
  STEERING UNIT ... 14-19
COMPASS ... 1-10
  COMPASS PRECAUTIONS ... 1-11
  INSTALLATION ... 1-11
  SELECTION ... 1-10
CONTINUITY CIRCUIT ... 2-63
CONTROL MODULE ... 13-24
  ADJUSTMENT ... 13-24
CONTROL VALVE - POWER ... 14-10
  BALANCING ... 14-12
  DISASSEMBLY & ASSEMBLY ... 14-10
  REMOVAL & INSTALLATION ... 14-10
CONVENTIONAL REMOTE CONTROLS ... 15-2
  CABLE ADJUSTMENT ... 15-2
  EXPLODED VIEWS ... 15-2
COOL FUEL MODULE (GEN III) ... 7-12
  DISASSEMBLY & ASSEMBLY ... 7-12
  REMOVAL & INSTALLATION ... 7-12
COOL FUEL SYSTEM (GEN II) ... 7-10
  REMOVAL, DISASSEMBLY & INSTALLATION ... 7-10
COOLANT TEMPERATURE SENSOR (ECT) ... 7-22
  REMOVAL & INSTALLATION ... 7-22
COOLING SYSTEM
  DESCRIPTION & OPERATION ... 8-2
  DRAINING & FILLING ... 2-42
  DRAINING, FLUSHING & FILLING ... 8-2
  FLUSHING THE CLOSED SYSTEM ... 2-42
  FLUSHING THE SEAWATER SYSTEM - INBOARDS ... 2-40
  FLUSHING THE SEAWATER SYSTEM - STERN DRIVES ... 2-39
  LEVEL CHECK ... 2-38
  PRESSURE CAP TESTING ... 2-38
  PRESSURE TEST ... 2-39
  RAW WATER COOLING TEST ... 8-3
  RAW WATER FLOW TEST ... 8-3
  TROUBLESHOOTING ... 8-2
COOLING SYSTEM MAINTENANCE & TESTING ... 8-2
  COOLING SYSTEM ... 8-2
  CROSSOVER ... 8-3
  GENERAL INFORMATION ... 8-2
  HEAT EXCHANGER ... 8-4
  SEA STRAINER ... 8-5
  SEAWATER PUMP/IMPELLER ... 8-6
  THERMOSTAT ... 8-8
  WATER DISTRIBUTION HOUSING ... 8-8
CRANKSHAFT PULLEY - V6/V8 ... 4-39
  REMOVAL & INSTALLATION ... 4-39
CROSSOVER ... 8-3
  REMOVAL & INSTALLATION ... 8-3
CRUISELOG METER ... 9-39
  REMOVAL & INSTALLATION ... 9-39
  TESTING ... 9-39
CYLINDER COMPRESSION ... 2-20
  CHECKING ... 2-20

MASTER INDEX 15-39

CYLINDER HEAD
  ASSEMBLY 5-9
  DISASSEMBLY 5-5
  GENERAL INFORMATION 5-5
  INSPECTION 5-7
  REFINISHING & REPAIRING 5-8
  REMOVAL & INSTALLATION
    3.0L 3-15
    4.3L V6, 5.0L/5.7L/6.3L V8 4-43
    8.1L V8 4-44
CYLINDER HEAD COVER
  3.0L 3-5
  V6/V8 4-19
DESCRIPTION & OPERATION - EFI 7-4
  CLOSED COOLING SYSTEM 8-2
  MODES OF OPERATION 7-4
  MULTI-POINT INJECTION (MPI) 7-4
  SEAWATER COOLING SYSTEM 8-2
  THROTTLE BODY INJECTION (TBI) 7-4
DETERMINING ENGINE CONDITION
  COMPRESSION TEST 5-2
  OIL PRESSURE TEST 5-2
DIAGNOSTIC TROUBLE CODE (DTC) CHART - MPI 7-31
  CLEARING CODES 7-31
  READING CODES 7-30
DIAGNOSTIC TROUBLE CODE TEST SCHEMATICS 7-38
DIAGNOSTIC TROUBLE CODES (DTC) - MPI 7-30
DISTRIBUTOR
  DEI 9-35
  EST 9-27
  THUNDERBOLT 9-33
DISTRIBUTOR CAP
  DEI 9-34
  EST 9-27
  THUNDERBOLT 9-32
DISTRIBUTOR ELECTRONIC IGNITION (DEI) SYSTEM - ECM 555 9-34
  DISTRIBUTOR 9-35
  DISTRIBUTOR CAP 9-34
  GENERAL INFORMATION 9-34
  IGNITION COIL 9-36
DISTRIBUTORLESS ELECTRONIC IGNITION (DIS) SYSTEM - PCM 555 9-36
  GENERAL INFORMATION 9-36
  IGNITION COIL 9-36
DRIVESHAFT HOUSING - ALPHA 10-28
  DISASSEMBLY & ASSEMBLY 10-28
    DRIVEN GEAR/DRIVESHAFT ASSEMBLY 10-32
    EXPLODED VIEWS 10-28
    U-JOINT/DRIVE GEAR ASSEMBLY 10-28
  PRE-LOAD AND SHIM ADJUSTMENTS ... 10-34
    DRIVE (PINION) GEAR SHIMMING 10-35
    DRIVEN GEAR SHIMMING 10-34
    GENERAL INFORMATION 10-34
    UPPER DRIVESHAFT BEARING PRELOAD 10-34
  REMOVAL & INSTALLATION 10-28
DRIVESHAFT HOUSING (UPPER UNIT) - ALPHA 10-27
  DESCRIPTION 10-27
  DRIVESHAFT HOUSING (UPPER UNIT) ... 10-28
  GEAR RATIO IDENTIFICATION 10-27
DRIVESHAFT HOUSING (UPPER UNIT) - BRAVO 11-18
  DESCRIPTION 11-18
  DRIVESHAFT HOUSING 11-19
  GEAR RATIO IDENTIFICATION 11-18
  DRIVESHAFT OIL SEAL CARRIER 10-26
  REMOVAL & INSTALLATION 10-26
DUAL HANDLE CONSOLE MOUNT & SHADOW MOUNT CONSOLE REMOTE CONTROLS 15-10
  REMOVAL, DISASSEMBLY & INSTALLATION 15-10
DUAL POWER TRIM SYSTEM 13-22
  DIODE MODULE REPLACEMENT 13-23
  RELAY REPLACEMENT 13-23
  TESTING 13-22
    DIODE MODULE 13-23
    GENERAL 13-22
    PORT TRIM SWITCH 13-23

    RELAY NO. 1 13-22
    RELAY NO. 2 13-23
    STARBOARD TRIM SWITCH 13-23
    TRAILER SWITCH 13-23
  TRIM CONTROL SWITCH REPLACEMENT 13-24
ECM PIN LOCATIONS
  MPI 7-59
  TBI 7-43
ELECTRIC FUEL PUMP
  DESCRIPTION 6-26
  PRESSURE TEST 6-27
  REMOVAL & INSTALLATION
    CARB 6-26
    EFI 7-13
ELECTRICAL COMPONENTS 9-2
  CONNECTORS 9-4
  GROUND 9-2
  LOAD 9-4
  POWER SOURCE 9-2
  PROTECTIVE DEVICES 9-2
  SWITCHES & RELAYS 9-2
  WIRING & HARNESSES 9-4
ELECTRONIC CONTROL MODULE (ECM) ... 7-22
  REMOVAL & INSTALLATION 7-22
ELECTRONIC FUEL INJECTION 7-4
  COOL FUEL MODULE (GEN III) 7-12
  COOL FUEL SYSTEM (GEN II) 7-10
  DESCRIPTION & OPERATION 7-4
  ELECTRONIC CONTROL MODULE (ECM) ... 7-22
  ENGINE COOLANT TEMPERATURE SENSOR (ECT) 7-22
  FUEL BOOST PUMP (ELECTRIC) 7-13
  FUEL INJECTORS 7-14
  FUEL METER COVER 7-14
  FUEL PRESSURE REGULATOR 7-14
  FUEL PUMP - ELECTRIC 7-13
  FUEL PUMP RELAY 7-13
  FUEL RAIL & INJECTORS 7-15
  IDLE AIR CONTROL VALVE (IAC) 7-23
  KNOCK SENSOR (KS) 7-24
  KNOCK SENSOR MODULE 7-24
  MANIFOLD ABSOLUTE PRESSURE SENSOR (MAP/MAPT) 7-25
  OIL PRESSURE SWITCH/SENSOR 7-25
  PROPULSION CONTROL MODULE (PCM) 7-22
  RELIEVING FUEL PRESSURE 7-9
  THROTTLE BODY 7-17
  THROTTLE BODY ADAPTER PLATE 7-21
  THROTTLE POSITION SENSOR (TP) ... 7-25
ELECTRONIC SHIFT CONTROL (ESC) 15-8
  ACTUATOR ADJUSTMENT 15-9
  SHIFT ACTUATOR REMOVAL & INSTALLATION 15-8
  SHIFT OVERRIDE PROCEDURE 15-9
ELECTRONIC SPARK TIMING SYSTEM (EST) 9-24
  DESCRIPTION 9-24
  DISTRIBUTOR 9-27
  DISTRIBUTOR CAP 9-27
  IGNITION COIL 9-30
  TROUBLESHOOTING 9-26
ELECTRONIC THROTTLE CONTROL (ETC) 15-9
  REMOVAL & INSTALLATION 15-9
ELECTRONIC TOOLS 1-23
  BATTERY TESTERS 1-23
  BATTERY CHARGERS 1-23
  GAUGES 1-24
  MULTI-METERS (DVOMS) 1-23
EMERGENCY (LANYARD) STOP SWITCH 9-43
  TESTING 9-43
ENGINE 3-2
  PRIMING THE ENGINE 3-4
  REMOVAL & INSTALLATION
    3.0L 3-2
    V6/V8 - CARB/TBI 4-3
    V6/V8 - MPI 4-9
ENGINE & DRIVE MAINTENANCE 2-2
  BELTS 2-8
  CYLINDER COMPRESSION 2-20

15-40 MASTER INDEX

FLAME ARRESTOR 2-3
FUEL FILTER 2-6
IDLE SPEED & MIXTURE 2-27
IGNITION TIMING 2-25
IMPELLER/WATER PUMP 2-16
PCV VALVE 2-29
PROPELLER 2-30
SEAWATER STRAINER 2-16
SERIAL NUMBER IDENTIFICATION 2-2
SPARK PLUG WIRES 2-25
SPARK PLUGS 2-21
THERMOSTAT 2-13
VALVE ADJUSTMENT 2-29
WATER PUMP/IMPELLER 2-16
ENGINE BLOCK 5-10
   ASSEMBLY 5-13
   DISASSEMBLY 5-10
   GENERAL INFORMATION 5-10
   INSPECTION 5-11
   REFINISHING 5-13
ENGINE COOLANT TEMPERATURE SENSOR (ECT) 7-22
   REMOVAL & INSTALLATION 7-22
ENGINE COUPLER/U-JOINT SPLINES 2-59
ENGINE COUPLER & FLYWHEEL - 3.0L
   REMOVAL & INSTALLATION 3-11
ENGINE COUPLER/DRIVE PLATE - V6/V8 4-38
   REMOVAL & INSTALLATION 4-38
ENGINE IDENTIFICATION
   3.0L 3-2
   4.3L V6 & 5.0L/5.7L/6.2L V8 4-2
   8.1L V8 4-3
ENGINE MECHANICAL - 3.0L **3-2**
   CAMSHAFT, BEARINGS AND GEAR 3-13
   COMBINATION MANIFOLD 3-7
   CYLINDER HEAD 3-15
   CYLINDER HEAD COVER 3-5
   ENGINE 3-2
   ENGINE COUPLER AND FLYWHEEL 3-11
   ENGINE IDENTIFICATION 3-2
   EXHAUST ELBOW 3-9
   EXHAUST HOSE (BELLOWS) 3-9
   EXHAUST PIPE(S) 3-9
   EXHAUST VALVE (FLAPPER/SHUTTER) 3-9
   FRONT COVER AND OIL SEAL 3-12
   FRONT ENGINE MOUNTS 3-4
   GENERAL INFORMATION 3-2
   HYDRAULIC VALVE LIFTER 3-7
   OIL FILTER BYPASS VALVE 3-11
   OIL PAN 3-10
   OIL PUMP 3-10
   PUSH ROD COVER 3-5
   REAR ENGINE MOUNTS 3-5
   REAR MAIN SEAL 3-11
   ROCKER ARMS AND PUSH RODS 3-6
   TORSIONAL DAMPER 3-12
ENGINE MECHANICAL - V6/V8 .... **4-2**
   BALANCE SHAFT 4-41
   CAMSHAFT & BEARINGS 4-41
   CRANKSHAFT PULLEY 4-39
   CROSSOVER 4-47
   CYLINDER HEAD 4-43
   CYLINDER HEAD COVER 4-19
   ENGINE 4-3
   ENGINE COUPLER/DRIVE PLATE 4-38
   ENGINE IDENTIFICATION 4-2
   EXHAUST ELBOW/RISER 4-32
   EXHAUST HOSES (BELLOWS) & INTERMEDIATE EXHAUST PIPE ... 4-34
   EXHAUST MANIFOLD 4-28
   EXHAUST VALVE (FLAPPER/SHUTTER) 4-34
   FLYWHEEL 4-38
   FRONT COVER & OIL SEAL 4-39
   FRONT ENGINE MOUNTS 4-18
   GENERAL INFORMATION 4-2
   HARMONIC BALANCER 4-39
   HEAT EXCHANGER 4-46

   HYDRAULIC VALVE LIFTER 4-22
   INTAKE MANIFOLD 4-24
   LOWER EXHAUST PIPE (Y-PIPE) 4-34
   OIL FILTER BYPASS VALVE 4-37
   OIL PAN 4-35
   OIL PUMP 4-36
   REAR MAIN SEAL 4-44
   REAR MAIN SEAL RETAINER 4-46
   ROCKER ARM COVER 4-19
   ROCKER ARMS & PUSH RODS 4-19
   TIMING CHAIN & SPROCKETS/GEARS 4-40
   TORSIONAL DAMPER 4-39
   VALVE COVER 4-19
   WATER (ENGINE) CIRCULATING PUMP ... 4-46
ENGINE OVERHAUL TIPS 5-3
   CLEANING 5-3
   OVERHAUL TIPS 5-3
   REPAIRING DAMAGED THREADS 5-4
   TOOLS 5-3
ENGINE PREPARATION 5-4
ENGINE RECONDITIONING **5-2**
   BUY OR REBUILD? 5-3
   CYLINDER HEAD 5-5
   DETERMINING ENGINE CONDITION 5-2
   ENGINE BLOCK 5-10
   ENGINE OVERHAUL TIPS 5-3
   ENGINE PREPARATION 5-4
   ENGINE START-UP AND BREAK-IN 5-15
ENGINE START-UP AND BREAK-IN 5-15
   BREAKING IT IN 5-16
   KEEP IT MAINTAINED 5-16
   STARTING THE ENGINE 5-15
EQUALIZER 15-36
   REMOVAL & INSTALLATION 15-36
EQUIPMENT (NOT REQUIRED BUT RECOMMENDED) ... **1-10**
   ANCHORS 1-10
   BAILING DEVICES 1-10
   COMPASS 1-10
   FIRST AID KIT 1-10
   OAR/PADDLE 1-10
   TOOLS AND SPARE PARTS 1-12
   VHF-FM RADIO 1-10
EXHAUST BELLOWS
   ALPHA 10-42
   BRAVO 11-36
EXHAUST ELBOW/RISER
   3.0L 3-9
   8.1L 4-33
   ALL V6/V8 EXC. DRY JOINT & 8.1L 4-32
   DRY JOINT SYSTEMS 4-33
EXHAUST HOSE (BELLOWS)
   3.0L 3-9
   V6/V8 4-34
EXHAUST MANIFOLD 4-28
   ALL EXC. 8.1L V8 4-28
   8.1L V8 ENGINES 4-31
EXHAUST PIPE(S)
   3.0L 3-9
EXHAUST TUBE 11-36
   REMOVAL & INSTALLATION 11-36
EXHAUST VALVE (FLAPPER/SHUTTER)
   3.0L 3-9
   V6/V8 4-34
EXPLODED VIEWS
   3.0L 3-16
   4.3L V6 4-48
   5.0L, 5.7L & 6.2L V8 4-50
   8.1L V8 4-53
   BRAVO 11-38
   PRECISION PILOT 15-35
   REMOTE CONTROLS 15-2
FASTENERS, MEASUREMENTS AND CONVERSIONS ... **1-26**
   BOLTS, NUTS AND OTHER THREADED RETAINERS ... 1-26
   TORQUE 1-27
   STANDARD AND METRIC MEASUREMENTS ... 1-27

MASTER INDEX 15-41

FIRING ORDERS..2-69
FLAME ARRESTOR..2-3
  DESCRIPTION & OPERATION...2-3
  REMOVAL & INSTALLATION..2-4
FLOW DIAGRAMS..8-8
  INDEX..8-8
FLUID DISPOSAL...2-33
FLUIDS & LUBRICANTS...2-33
  COOLING SYSTEM..2-38
  FLUID DISPOSAL..2-33
  FUEL & OIL RECOMMENDATIONS......................................2-33
  POWER STEERING..2-37
  POWER TRIM PUMP...2-54
  STERN DRIVE UNIT..2-52
  TRANSMISSION FLUID..2-50
FLYWHEEL - V6/V8...4-38
  REMOVAL & INSTALLATION..4-38
FRONT COVER & OIL SEAL
  3.0L..3-12
  V6/V8...4-39
FRONT ENGINE MOUNTS
  3.0L...3-4
  V6/V8...4-18
FRONT SAFETY GUARD..15-36
  REMOVAL & INSTALLATION...15-36
FUEL & OIL RECOMMENDATIONS...2-33
  ENGINE OIL..2-33
  FUEL..2-33
  OIL & FILTER CHANGE...2-35
  OIL LEVEL CHECK...2-34
  PRIMING...2-33
FUEL & COMBUSTION
  APPLICATIONS - EFI...7-3
  COMBUSTION..6-2, 7-2
  FUEL..6-2, 7-2
FUEL
  ALCOHOL-BLENDED FUELS.......................................6-2, 7-2
  OCTANE RATING...6-2, 7-2
  RECOMMENDATIONS...6-2, 7-2
  VAPOR PRESSURE..6-2, 7-2
FUEL BOOST PUMP (ELECTRIC)...7-13
  REMOVAL & INSTALLATION..7-13
FUEL FILTER..2-6
  REMOVAL & INSTALLATION..2-6
FUEL GAUGE...9-37
  REMOVAL & INSTALLATION..9-38
  TESTING...9-38
  TROUBLESHOOTING...9-37
FUEL INJECTORS...7-14
  REMOVAL & INSTALLATION..7-14
    MPI ENGINES...7-15
    TBI ENGINES...7-14
FUEL METER COVER...7-14
  REMOVAL & INSTALLATION..7-14
FUEL PRESSURE REGULATOR..7-14
  REMOVAL & INSTALLATION..7-14
FUEL PUMP
  ELECTRIC
    CARB..6-26
    EFI...7-13
  MANUAL..6-24
FUEL PUMP RELAY..7-13
  REMOVAL & INSTALLATION..7-13
FUEL RAIL & INJECTORS..7-15
  BALANCE TEST..7-17
  REMOVAL & INSTALLATION..7-15
FUEL SYSTEM APPLICATIONS - EFI.....................................7-3
FUEL TANK SENDING UNIT...9-41
  TESTING...9-41
GEAR HOUSING - ALPHA..10-5
  ASSEMBLY...10-15
    COUNTER ROTATION...10-19
    STANDARD ROTATION..10-15
  CLEANING & INSPECTION..10-14
  DISASSEMBLY - STANDARD ROTATION.................................10-7

  BEARING CARRIER & REVERSE GEAR.................................10-10
  MAIN COMPONENTS..10-7
  PROPELLER SHAFT...10-11
  DISASSEMBLY - COUNTER ROTATION.................................10-12
  REMOVAL & INSTALLATION...10-5
GEAR HOUSING (LOWER UNIT) - ALPHA..............................10-5
  DESCRIPTION..10-5
  DRIVESHAFT OIL SEAL CARRIER...................................10-26
  GEAR HOUSING (LOWER UNIT)......................................10-5
  PROPELLER...10-23
  SEA WATER PUMP & IMPELLER.....................................10-26
GEAR HOUSING (LOWER UNIT) - BRAVO..............................11-4
  DESCRIPTION..11-4
  GEAR HOUSING - BRAVO I/II/X/XR/XZ..............................11-4
  GEAR HOUSING - BRAVO III......................................11-11
  PROPELLER(S)..11-18
GEAR RATIO IDENTIFICATION 10-27
GENERAL DIAGNOSTIC TEST SCHEMATICS - TBI...........................7-28
GENERAL DIAGNOSTIC TESTS
  MPI EXC 8.1...7-52
  8.1L MPI..7-56
  TBI...7-27
GIMBAL BEARING
  ALPHA..10-36
  BRAVO..11-32
GIMBAL HOUSING/TRANSOM PLATE......................................10-49
  ALPHA..10-49
  BRAVO..11-43
GIMBAL RING...10-45
  ALPHA..10-45
  BRAVO..11-43
HAND TOOLS...1-19
  BREAKER BARS..1-20
  HAMMERS...1-22
  PLIERS..1-21
  SCREWDRIVERS..1-21
  SOCKET SETS...1-19
  TORQUE WRENCHES...1-20
  WRENCHES..1-21
HEAT EXCHANGER...8-4
  CLEANING..8-5
  REMOVAL, DISASSEMBLY & INSTALLATION.............................8-4
  TESTING...8-5
HOW TO USE THIS MANUAL..1-2
  AVOIDING COMMON MISTAKES..1-3
  AVOIDING TROUBLE..1-2
  CAN YOU DO IT?..1-2
  DIRECTIONS AND LOCATIONS..1-2
  MAINTENANCE OR REPAIR?..1-2
  PROFESSIONAL HELP...1-3
  PURCHASING PARTS..1-3
  WHERE TO BEGIN..1-2
HYDRAULIC VALVE LIFTER
  NOISY LIFTERS...4-22
  REMOVAL & INSTALLATION
    3.0L...3-7
    V6/V8...4-22
IDLE AIR CONTROL VALVE (IAC).......................................7-23
  REMOVAL & INSTALLATION..7-23
IDLE SPEED & MIXTURE...2-27
  ADJUSTMENT..2-27
IGNITION COIL
  DEI...9-36
  DIS...9-36
  EST...9-30
IGNITION OR KNOCK CONTROL MODULES..................................9-33
  REMOVAL & INSTALLATION..9-33
IGNITION SWITCH..9-42
  TESTING...9-42
IGNITION SYSTEMS..9-24
  APPLICATIONS..9-24
  DEI...9-36
  DIS...9-24
  DISTRIBUTOR...9-35
  EST...9-24

MASTER INDEX

THUNDERBOLT	9-30
IGNITION TIMING	2-25
ADJUSTMENT	2-25
IMPELLER - PRECISION PILOT	15-36
REMOVAL & INSTALLATION	15-36
IMPELLER/WATER PUMP	
DISASSEMBLY & ASSEMBLY	8-6
ALPHA DRIVES	8-6
BRAVO/INBOARD - CARB/TBI	8-6
BRAVO/INBOARD - MPI	8-7
OUTPUT TEST	8-7
REMOVAL & INSTALLATION	2-16, 8-6, 10-26
SEAWATER PUMP PRESSURE SENSOR	8-8
INSIDE THE BOAT	2-60
INSTRUMENTS & GAUGES	**9-37**
BATTERY GAUGE	9-38
CRUISELOG METER	9-39
FUEL GAUGE	9-37
OIL & TEMPERATURE GAUGES	9-37
SPEEDOMETER	9-39
TACHOMETER	9-39
VACUUM GAUGE	9-40
INTAKE MANIFOLD	4-24
REMOVAL & INSTALLATION	4-24
ALL CARB & TBI	4-24
V6/V8 MPI (EXC. 8.1L)	4-25
8.1L V8	4-28
INTERMEDIATE EXHAUST PIPE	4-34
REMOVAL & INSTALLATION	4-34
INTERMITTENT FAULTS	
MPI	7-48
TBI	7-26
JOYSTICK	15-37
REMOVAL & INSTALLATION	15-37
KNOCK SENSOR (KS)	7-24
REMOVAL & INSTALLATION	7-24
KNOCK SENSOR MODULE	7-24
REMOVAL & INSTALLATION	7-24
LOWER EXHAUST PIPE (Y-PIPE)	4-34
REMOVAL & INSTALLATION	4-34
LOWER UNIT - ALPHA	**10-5**
ASSEMBLY	10-15
COUNTER ROTATION	10-19
STANDARD ROTATION	10-15
CLEANING & INSPECTION	10-14
DESCRIPTION	10-5
DISASSEMBLY - STANDARD ROTATION	10-7
BEARING CARRIER & REVERSE GEAR	10-10
MAIN COMPONENTS	10-7
PROPELLER SHAFT	10-11
DISASSEMBLY - COUNTER ROTATION	10-12
DRIVESHAFT OIL SEAL CARRIER	10-26
GEAR HOUSING (LOWER UNIT)	10-5
PROPELLER	10-23
REMOVAL & INSTALLATION	10-5
SEA WATER PUMP & IMPELLER	10-26
LOWER UNIT - BRAVO	**11-4**
DESCRIPTION	11-4
GEAR HOUSING - BRAVO I/II/X/XR/XZ	11-4
GEAR HOUSING - BRAVO III	11-11
PROPELLER(S)	11-18
LOWER UNIT - BRAVO I/II/X/XR/XZ	11-4
ASSEMBLY	11-10
CLEANING AND INSPECTION	11-9
DISASSEMBLY	11-5
REMOVAL & INSTALLATION	11-4
LOWER UNIT - BRAVO III	11-11
ASSEMBLY	11-16
CLEANING AND INSPECTION	11-16
DISASSEMBLY	11-12
REMOVAL & INSTALLATION	11-11
LUBRICATION POINTS	**2-55**
PROPELLER SHAFT	2-60
SHIFT CABLE & TRANSMISSION LINKAGE PIVOT POINTS	2-55
STARTER MOTOR	2-59
STEERING SYSTEM	2-57
THROTTLE CABLE	2-55
TRANSOM, GIMBAL ASSEMBLY, HINGE PINS & GIMBAL BEARING	2-58
MANIFOLD ABSOLUTE PRESSURE SENSOR (MAP/MAPT)	7-25
REMOVAL & INSTALLATION	7-25
MANUAL STEERING SYSTEM	**14-2**
GENERAL INFORMATION	14-2
STEERING CABLE	14-2
SWIVEL RING	14-2
MEASURING TOOLS	1-24
DEPTH GAUGES	1-26
DIAL INDICATORS	1-25
MICROMETERS & CALIPERS	1-25
TELESCOPING GAUGES	1-26
MECHANICAL FUEL PUMP	6-24
DESCRIPTION	6-24
FUEL LINE TEST	6-25
PRESSURE TEST	6-25
REMOVAL & INSTALLATION	6-25
MERCARB 2BBL CARB	**6-8**
ADJUSTMENT	6-9
ACCELERATOR PUMP LEVER	6-9
CHOKE SETTING	6-11
CHOKE UNLOADER	6-11
FLOAT LEVEL AND DROP	6-11
IDLE SPEED & MIXTURE	6-10
PUMP ROD	6-10
THROTTLE CABLE	6-10
ASSEMBLY	6-16
MODELS W/O TKS	6-16
MODELS W/TKS	6-17
CLEANING & INSPECTION	6-16
DESCRIPTION	6-8
DISASSEMBLY	6-12
MODELS W/O TKS	6-12
MODELS W/TKS	6-15
REMOVAL & INSTALLATION	6-9
OIL & TEMPERATURE GAUGES	9-37
REMOVAL & INSTALLATION	9-37
TESTING	9-37
TROUBLESHOOTING	9-37
OIL FILTER BYPASS VALVE	
3.0L	3-11
V6/V8	4-37
OIL PAN	
3.0L	3-10
V6/V8	4-35
OIL PRESSURE SENDING UNIT	9-40
GENERAL INFORMATION	9-40
TESTING	9-40
OIL PRESSURE SWITCH	9-43
TESTING	9-43
OIL PRESSURE SWITCH/SENSOR	7-25
REMOVAL & INSTALLATION	7-25
OIL PUMP	
3.0L	3-10
V6/V8	4-36
PANEL MOUNT REMOTE CONTROL	15-9
REMOVAL, DISASSEMBLY & INSTALLATION	15-9
PCV VALVE	2-29
INSPECTION	2-29
REMOVAL & INSTALLATION	2-29
POWER STEERING COOLER	14-16
REMOVAL & INSTALLATION	14-16
POWER STEERING PUMP	
BLEEDING	2-37, 14-13
DISASSEMBLY & ASSEMBLY	14-15
FLOW CONTROL VALVE SERVICE	14-14
FLUID LEVEL	2-37, 14-13
OIL SEAL REPLACEMENT	14-15
PUMP PULLEY	
REMOVAL & INSTALLATION	14-13
4 CYLINDER ENGINES	14-13
V6/V8 ENGINES (EXC. 8.1L)	14-13

MASTER INDEX

- 8.1L V8 ENGINES ... 14-14
- **POWER STEERING SYSTEM** ... **14-4**
  - BOOSTER CYLINDER ... 14-12
  - CONTROL VALVE ... 14-10
  - DESCRIPTION ... 14-4
  - POWER STEERING COOLER ... 14-16
  - POWER STEERING PUMP ... 14-13
  - POWER STEERING UNIT ... 14-9
  - STEERING CABLE ... 14-8
  - STEERING CABLE GUIDE TUBE ... 14-8
  - TESTING ... 14-4
  - TIE BAR - MULTIPLE DRIVES ... 14-17
- POWER STEERING UNIT ... 14-9
  - GENERAL INFORMATION ... 14-9
  - REMOVAL & INSTALLATION ... 14-9
- POWER TRIM PUMP ASSEMBLY ... 13-9
  - FLUID LEVEL ... 2-54, 13-9
  - DISASSEMBLY & ASSEMBLY ... 13-10
  - GENERAL INFORMATION ... 13-9
  - REMOVAL & INSTALLATION ... 13-9
- **POWER TRIM & TILT - STANDARD** ... **13-2**
  - DESCRIPTION & OPERATION ... 13-2
  - DUAL POWER TRIM SYSTEM ... 13-22
  - PUMP ASSEMBLY ... 13-9
  - SERVICE PRECAUTIONS ... 13-6
  - SYSTEM BLEEDING ... 13-7
  - SYSTEM TESTING ... 13-7
  - TRIM CYLINDER ... 13-17
  - TRIM LIMIT SWITCH/TRIM POSITION SENDER ... 13-8
  - TROUBLESHOOTING ... 13-3
- **PRECISION PILOT** ... **15-33**
  - CONTROLLER ... 15-36
  - EQUALIZER ... 15-36
  - EXPLODED VIEWS ... 15-35
  - FRONT SAFETY GUARD ... 15-36
  - IMPELLER ... 15-36
  - JOYSTICK ... 15-37
  - MOTORS ... 15-37
  - REAR SAFETY GUARD ... 15-37
  - TROUBLESHOOTING ... 15-33
  - WIRING SCHEMATICS ... 15-37
- PROPELLER
  - GENERAL INFORMATION ... 10-23, 11-18
    - CAVITATION ... 10-24
    - CUPPING ... 10-24
    - DIAMETER AND PITCH ... 10-23
    - PITCH ... 10-24
    - RAKE ... 10-24
    - ROTATION ... 10-25
    - SELECTION ... 10-23
    - SHOCK ABSORBERS ... 10-24
    - VIBRATION ... 10-24
  - INSPECTION ... 10-25
  - REMOVAL & INSTALLATION ... 2-30, 10-26, 11-18
- PROPELLER SHAFT ... 2-60
- PROPULSION CONTROL MODULE (PCM) ... 7-22
  - REMOVAL & INSTALLATION ... 7-22
- PUSH ROD COVER ... 3-5
  - REMOVAL & INSTALLATION ... 3-5
- REAR ENGINE MOUNTS ... 3-5
  - REMOVAL & INSTALLATION ... 3-5
- REAR MAIN SEAL
  - 3.0L ... 3-11
  - V6/V8 ... 4-44
- REAR MAIN SEAL RETAINER - V6/V8 ... 4-46
  - REMOVAL & INSTALLATION ... 4-46
- REAR SAFETY GUARD ... 15-37
  - REMOVAL & INSTALLATION ... 15-37
- REGULATIONS FOR YOUR BOAT ... 1-4
  - CAPACITY INFORMATION ... 1-4
  - CERTIFICATE OF COMPLIANCE ... 1-4
  - DOCUMENTING OF VESSELS ... 1-4
  - HULL IDENTIFICATION NUMBER ... 1-4
  - LENGTH OF BOATS ... 1-4
  - NUMBERING OF VESSELS ... 1-4
  - REGISTRATION OF BOATS ... 1-4
  - SALES AND TRANSFERS ... 1-4
  - VENTILATION ... 1-5
  - VENTILATION SYSTEMS ... 1-5
- RELIEVING FUEL PRESSURE ... 7-9
  - MPI ENGINES ... 7-9
  - TBI ENGINES ... 7-9
- **REMOTE CONTROLS**
  - CONVENTIONAL ... 15-2
  - SMARTCRAFT ... 15-7
- REQUIRED SAFETY EQUIPMENT ... 1-5
  - FIRE EXTINGUISHERS ... 1-6
  - PERSONAL FLOTATION DEVICES ... 1-7
  - SOUND PRODUCING DEVICES ... 1-8
  - TYPES OF FIRES ... 1-5
  - VISUAL DISTRESS SIGNALS ... 1-8
  - WARNING SYSTEM ... 1-7
- ROCKER ARM COVER
  - 3.0L ... 3-5
  - V6/V8 ... 4-19
- ROCKER ARMS & PUSH RODS
  - REMOVAL & INSTALLATION
    - 3.0L ... 3-6
    - 4.3L V6 ... 4-19
    - 5.0L, 5.7L & 6.2L V8 ... 4-20
    - 8.1L V8 ... 4-20
  - VALVE ADJUSTMENT
    - 3.0L ... 3-6
    - 5.0L, 5.7L & 6.2L V8 ... 4-21
    - 8.1L V8 ... 4-22
- ROTOR/SENSOR WHEEL ... 9-32
  - REMOVAL & INSTALLATION ... 9-32
- SAFETY GUARD
  - FRONT ... 15-36
  - REAR ... 15-37
- **SAFETY IN SERVICE** ... **1-12**
  - DO'S ... 1-12
  - DON'TS ... 1-12
- SAFETY TOOLS ... 1-16
  - EYE AND EAR PROTECTION ... 1-16
  - WORK CLOTHES ... 1-17
  - WORK GLOVES ... 1-16
- SEA STRAINER
  - CLEANING & INSPECTION ... 2-16
  - REMOVAL & INSTALLATION ... 8-5
- SEAWATER PUMP/IMPELLER
  - DISASSEMBLY & ASSEMBLY ... 8-6
    - ALPHA DRIVES ... 8-6
    - BRAVO/INBOARD - CARB/TBI ... 8-6
    - BRAVO/INBOARD - MPI ... 8-7
  - OUTPUT TEST ... 8-7
  - REMOVAL & INSTALLATION ... 2-16, 8-6, 10-26
  - SEAWATER PUMP PRESSURE SENSOR ... 8-8
- **SENDING UNITS & SWITCHES** ... **9-40**
  - AUDIO WARNING SYSTEM ... 9-43
  - EMERGENCY (LANYARD) STOP SWITCH ... 9-43
  - FUEL TANK SENDING UNIT ... 9-41
  - IGNITION SWITCH ... 9-42
  - OIL PRESSURE SENDING UNIT ... 9-40
  - OIL PRESSURE SWITCH ... 9-43
  - START/STOP SWITCH ... 9-43
  - STERN DRIVE GEAR LUBE MONITOR SWITCH ... 9-44
  - TRANSMISSION FLUID TEMPERATURE SWITCH ... 9-44
  - WATER TEMPERATURE SENDER ... 9-40
  - WATER TEMPERATURE SWITCH ... 9-44
- SENSOR ... 9-33
  - REMOVAL & INSTALLATION ... 9-33
  - TESTING ... 9-33
- SERIAL NUMBER IDENTIFICATION ... 2-2
  - ENGINE ... 2-2
  - STERN DRIVE ... 2-2
  - TRANSMISSION ... 2-3
  - TRANSOM ASSEMBLY ... 2-3
- SHIFT CABLE - ALPHA ... 10-37

MASTER INDEX

ADJUSTMENT	10-38
REMOTE CONTROL CABLE	10-38
REMOTE CONTROL CABLE - CHECKING OUTPUT	10-40
TRANSOM CABLE - CHECKING PLAY	10-40
TRANSOM CABLE - ISOLATING PLAY	10-41
REMOVAL & INSTALLATION	10-37
REMOTE CONTROL CABLE	10-37
TRANSOM CABLE	10-37
SHIFT CABLE - BRAVO	11-33
ADJUSTMENT	11-35
REMOVAL & INSTALLATION	11-33
REMOTE CONTROL CABLE	11-33
TRANSOM CABLE	11-33
SHIFT CABLE - VELVET	
ADJUSTMENT	12-2
EXCEPT 5000 SERIES	12-2
5000 SERIES	12-3
SHIFT CABLE - ZF/HURTH	
ADJUSTMENT	12-7
GENERAL INFORMATION	12-6
SHIFT CABLE & TRANSMISSION LINKAGE PIVOT POINTS	2-55
3.0L	2-55
4.3L V6 & 5.0L/5.7L/6.2L V8	2-55
8.1L V8	2-55
SHIFT CUT-OUT SWITCH	10-41
ADJUSTMENT	10-41
MODELS W/PLUNGER SWITCH	10-42
MODELS W/ROLLER SWITCH	10-41
SHOP EQUIPMENT	1-16
CHEMICALS	1-17
SAFETY TOOLS	1-16
SINGLE HANDLE CONSOLE MOUNT REMOTE CONTROL	15-10
REMOVAL, DISASSEMBLY & INSTALLATION	15-10
SLIM BINNACLE SINGLE HANDLE CONSOLE MOUNT REMOTE CONTROL	15-12
ADJUSTMENT	15-22
REMOVAL, DISASSEMBLY & INSTALLATION	15-12
SMARTCRAFT DIGITAL THROTTLE & SHIFT (DTS)	**15-7**
DESCRIPTION & OPERATION	15-7
DUAL HANDLE CONSOLE MOUNT REMOTE CONTROLS	15-10
ELECTRONIC SHIFT CONTROL (ESC)	15-8
ELECTRONIC THROTTLE CONTROL (ETC)	15-9
PANEL MOUNT REMOTE CONTROL	15-9
SHADOW MOUNT CONSOLE REMOTE CONTROLS	15-10
SINGLE HANDLE CONSOLE MOUNT REMOTE CONTROL	15-10
SLIM BINNACLE SINGLE HANDLE CONSOLE MOUNT REMOTE CONTROL	15-12
WIRING SCHEMATICS	15-13
ZERO EFFORT CONSOLE MOUNT REMOTE CONTROL	15-11
SPARK PLUG WIRES	2-25
REMOVAL & INSTALLATION	2-25
TESTING	2-25
SPARK PLUGS	2-21
HEAT RANGE	2-21
INSPECTION & GAPPING	2-25
READING PLUGS	2-23
REMOVAL & INSTALLATION	2-22
SERVICE	2-22
SPECIFICATIONS	
ALTERNATOR	9-74
CAPACITIES	2-77
CARB APPLICATIONS	6-3
CARB - 2 BBL MERCARB	6-30
CARB - 4 BBL WEBER	6-30
CONVERSION FACTORS	1-27
DRIVE UNIT APPLICATIONS	10-52,11-47
ENGINE	
3.0L	3-19
4.3L V6	4-59
5.0L & 5.7L V8	4-61
6.2L V8	4-63
8.1L V8	4-65
ENGINE MODEL APPLICATIONS	2-71
MAINTENANCE INTERVALS	2-79
PRESSURE - VELVET	12-12
STARTER	9-74
TORQUE	
3.0L	3-18
4.3L V6	4-56
5.0L, 5.7L & 6.2L V8	4-57
8.1L V8	4-58
ALPHA	10-52
BRAVO	11-46
EFI ENGINES	7-70
INBOARDS	12-11
TRANSMISSION IDENTIFICATION	12-10
TUNE-UP	2-74
TYPICAL TORQUE VALUES - METRIC BOLTS	1-28
TYPICAL TORQUE VALUES - U.S. STANDARD BOLTS	1-28
SPEEDOMETER	9-39
REMOVAL & INSTALLATION	9-39
TESTING	9-39
SPRING COMMISSIONING	2-69
SPRING COMMISSIONING CHECKLIST	2-69
SPRING COMMISSIONING CHECKLIST	2-69
START/STOP SWITCH	9-43
TESTING	9-43
STARTER CIRCUIT	**9-18**
DESCRIPTION & OPERATION	9-18
STARTER MOTOR	9-18
TROUBLESHOOTING	9-18
STARTER MOTOR	9-18
DESCRIPTION & OPERATION	9-18
DISASSEMBLY & ASSEMBLY	9-21
PRECAUTIONS	9-19
REMOVAL & INSTALLATION	9-21
TESTING	9-20
TROUBLESHOOTING	9-19
STEERING CABLE - MANUAL	14-2
REMOVAL & INSTALLATION	14-2
STEERING CABLE - POWER	14-8
REMOVAL & INSTALLATION	14-8
STEERING CABLE GUIDE TUBE - POWER	14-8
REMOVAL & INSTALLATION	14-8
STEERING SYSTEM	2-57
ALPHA	2-57
BRAVO	2-57
COMPACT HYDRAULIC	14-19
MANUAL	14-2
TIE BAR PIVOT POINTS	2-57
STEERING UNIT - COMPACT	14-19
REMOVAL & INSTALLATION	14-19
STERN DRIVE GEAR LUBE MONITOR SWITCH	9-44
STERN DRIVE UNIT	
ALPHA	10-3
BRAVO	11-2
STERN DRIVE UNIT - ALPHA	
CHECKING FOR WATER	2-53
DESCRIPTION	10-2
DRAIN AND REFILL	2-53
FLUID LEVEL	2-52
STERN DRIVE UNIT	10-3
TROUBLESHOOTING	10-2
STERN DRIVE UNIT - BRAVO	
CHECKING FOR WATER	2-53
DESCRIPTION	11-2
DRAIN AND REFILL	2-53
FLUID LEVEL	2-52
STERN DRIVE UNIT	11-2
TROUBLESHOOTING	11-2
SWIVEL RING - MANUAL	14-2
DISASSEMBLY & ASSEMBLY	14-3
REMOVAL & INSTALLATION	14-2
SYSTEM DIAGNOSIS - TBI ENGINES & EARLY 2001 MPI ENGINES	**7-26**
DIAGNOSTIC TROUBLE CODE (DTC) CHART	7-31
DIAGNOSTIC TROUBLE CODE TEST SCHEMATICS	7-38
DIAGNOSTIC TROUBLE CODES (DTC)	7-30
ECM PIN LOCATIONS & SYMPTOMS CHARTS	7-43
GENERAL DIAGNOSTIC TEST SCHEMATICS	7-28
GENERAL DIAGNOSTIC TESTS	7-27

MASTER INDEX 15-45

INTERMITTENT FAULTS.	7-26
PRECAUTIONS.	7-26
VACUUM DIAGRAMS	7-41
WIRING SCHEMATICS	7-41
SYSTEM DIAGNOSIS - 2001-08 MPI ENGINES	**7-47**
ECM PIN LOCATIONS.	7-59
GENERAL DIAGNOSTIC TESTS	
2001-08 4.3L, 5.0L, 5.7L & 6.2L MPI ENGINES	7-52
8.1L MPI ENGINES	7-56
INTERMITTENT FAULTS.	7-48
PRECAUTIONS	7-48
TROUBLESHOOTING.	7-49
WIRING SCHEMATICS	7-61
TACHOMETER.	9-39
REMOVAL & INSTALLATION	9-39
TESTING	9-39
TEST EQUIPMENT	9-4
JUMPER WIRES	9-4
MULTI-METERS.	9-5
TEST LIGHTS	9-5
THERMOSTAT	2-13
REMOVAL & INSTALLATION	2-13
TESTING	2-16
THROTTLE BODY.	7-17
REMOVAL & INSTALLATION	7-17
THROTTLE CABLE ADJUSTMENT	7-20
THROTTLE BODY ADAPTER PLATE	7-21
REMOVAL & INSTALLATION	7-21
THROTTLE CABLE	2-55
3.0L	2-55
4.3L V6 & 5.0L/5.7L/6.2L V8	2-55
8.1L V8	2-55
THROTTLE POSITION SENSOR (TP)	7-25
REMOVAL & INSTALLATION	7-25
THUNDERBOLT V IGNITION SYSTEM	**9-30**
DESCRIPTION.	9-30
DISTRIBUTOR.	9-33
DISTRIBUTOR CAP	9-32
IGNITION OR KNOCK CONTROL MODULES	9-33
ROTOR/SENSOR WHEEL	9-32
SENSOR	9-33
TROUBLESHOOTING	9-31
TIE BAR - MULTIPLE DRIVES	14-17
INSTALLATION	14-18
PORT CABLE.	14-18
STARBOARD CABLE.	14-18
TIE BAR LENGTH	14-17
TIMING CHAIN & SPROCKETS/GEARS - V6/V8	4-40
REMOVAL & INSTALLATION	4-40
TOOLS	1-18
ELECTRONIC TOOLS.	1-23
HAND TOOLS	1-19
MEASURING TOOLS	1-24
OTHER COMMON TOOLS	1-22
SPECIAL TOOLS.	1-23
TORSIONAL DAMPER	
3.0L	3-12
V6/V8	4-39
TRANSMISSION - VELVET	
PRESSURE TEST	12-6
REMOVAL & INSTALLATION	12-5
EXCEPT 5000 SERIES	12-5
5000 SERIES	12-5
TRANSMISSION - ZF/HURTH	
PRESSURE TEST	12-9
REMOVAL & INSTALLATION	12-9
TRANSMISSION FLUID	
DRAINING FLUID	2-50
FLUID LEVEL.	2-50
VELVET	12-2
ZF/HURTH	12-6
TRANSMISSION FLUID TEMPERATURE SWITCH	9-44
TESTING	9-44
TRANSOM ASSEMBLY - ALPHA	**10-36**
BELL HOUSING.	10-44
DESCRIPTION.	10-36
EXHAUST BELLOWS	10-42
GIMBAL BEARING	10-36
GIMBAL HOUSING/TRANSOM PLATE.	10-49
GIMBAL RING	10-45
SHIFT CABLE	10-37
SHIFT CUT-OUT SWITCH.	10-41
U-JOINT BELLOWS.	10-43
TRANSOM ASSEMBLY - BRAVO	**11-32**
DESCRIPTION.	11-32
EXHAUST BELLOWS	11-36
EXHAUST TUBE	11-36
EXPLODED VIEWS	11-38
GIMBAL BEARING	11-32
GIMBAL HOUSING/TRANSOM PLATE.	11-43
GIMBAL RING	11-41
SHIFT CABLE	11-33
U-JOINT BELLOWS.	11-37
WATER HOSE & FITTING	11-38
TRANSOM, GIMBAL ASSEMBLY, HINGE PINS & GIMBAL BEARING	2-58
TRIM CYLINDER	13-17
DISASSEMBLY & ASSEMBLY.	13-19
GENERAL INFORMATION	13-17
REMOVAL & INSTALLATION	13-18
TRIM LIMIT SWITCH/TRIM POSITION SENDER	13-8
ADJUSTMENT	13-8
TRIM LIMIT SWITCH	13-8
TRIM POSITION SENDER.	13-8
GENERAL INFORMATION	13-8
REMOVAL & INSTALLATION	13-8
TROUBLESHOOTING.	**1-13**
BASIC OPERATING PRINCIPLES	1-13
CARBURETOR	6-3
CHARGING SYSTEM	9-11
DRIVE UNIT WILL NOT SHIFT - SHIFT HANDLE DOES NOT MOVE.	10-3
DRIVE UNIT WILL NOT SHIFT - SHIFT HANDLE MOVES	10-2
DRIVE UNIT WILL NOT SLIDE INTO BELL HOUSING	10-2
DRIVESHAFT HOUSING NOISE	10-2
ENGINE OVERHEATS - MECHANICAL	8-2
ENGINE OVERHEATS - SEAWATER	8-2
EST IGNITION	9-26
GEAR HOUSING NOISE.	10-2
HARD SHIFTING.	10-3
JUMPS OUT OF GEAR.	10-3
MPI.	7-49
OVERHEATS - CLOSED, IN ADDITION TO ABOVE	8-2
PRECISION PILOT	15-33
STARTING SYSTEM	9-18
THUNDERBOLT IGNITION	9-31
TRIM & TILT.	13-3
WATER IN CYLINDERS	8-2
WATER IN OIL.	8-2
U-JOINT BELLOWS	
ALPHA	10-43
BRAVO	11-37
UNDERSTANDING & TROUBLESHOOTING ELECTRICAL SYSTEMS	**9-2**
BASIC ELECTRICAL THEORY	9-2
ELECTRICAL COMPONENTS	9-2
TEST EQUIPMENT	9-4
TROUBLESHOOTING THE ELECTRICAL SYSTEM	9-6
WIRE AND CONNECTOR REPAIR.	9-8
UPPER UNIT - ALPHA	**10-27**
DESCRIPTION.	10-27
DRIVESHAFT HOUSING (UPPER UNIT)	10-28
GEAR RATIO IDENTIFICATION	10-27
UPPER UNIT - ALPHA.	10-28
DISASSEMBLY & ASSEMBLY.	10-28
DRIVEN GEAR/DRIVESHAFT ASSEMBLY	10-32
EXPLODED VIEWS	10-28
U-JOINT/DRIVE GEAR ASSEMBLY	10-28
PRE-LOAD AND SHIM ADJUSTMENTS	10-34
DRIVE (PINION) GEAR SHIMMING	10-35
DRIVEN GEAR SHIMMING	10-34
GENERAL INFORMATION.	10-34
UPPER DRIVESHAFT BEARING PRELOAD	10-34

MASTER INDEX

REMOVAL & INSTALLATION 10-28
UPPER UNIT - BRAVO **11-18**
  DESCRIPTION. 11-18
  DISASSEMBLY & ASSEMBLY 11-23
  DRIVESHAFT HOUSING. 11-19
  EXPLODED VIEWS 11-19
  GEAR RATIO IDENTIFICATION 11-18
  REMOVAL & INSTALLATION. 11-19
    BRAVO I/II/Z/XR/XZ 11-19
    BRAVO III/X 11-22
VACUUM DIAGRAMS - TBI. 7-41
VACUUM GAUGE 9-40
  REMOVAL & INSTALLATION 9-40
  TESTING 9-40
VALVE ADJUSTMENT 2-29
VALVE COVER
  3.0L 3-5
  V6/V8 4-19
VELVET TRANSMISSIONS **12-2**
  GENERAL INFORMATION 12-2
  IDENTIFICATION. 12-2
  SHIFT CABLE 12-2
  TRANSMISSION 12-5
  TRANSMISSION FLUID 12-2
WATER (ENGINE) CIRCULATING PUMP. 4-46
  REMOVAL & INSTALLATION 4-46
WATER DISTRIBUTION HOUSING. 8-8
  REMOVAL & INSTALLATION 8-8
WATER HOSE & FITTING 11-38
  REMOVAL & INSTALLATION. 11-38
WATER PUMP/IMPELLER. 2-16
  REMOVAL & INSTALLATION 2-16
WATER TEMPERATURE SENDER. 9-40
  GENERAL INFORMATION 9-40
  TESTING 9-41
WATER TEMPERATURE SWITCH 9-44
  TESTING 9-44
WEBER WFB 4BBL CARB. **6-17**
  ADJUSTMENT 6-18
    ACCELERATOR PUMP LEVER 6-18
    ACCELERATOR PUMP 6-18
    ELECTRIC CHOKE 6-18
    FLOAT LEVEL AND DROP 6-18
    IDLE SPEED & MIXTURE 6-19
    PORTED VACUUM SWITCH (PVS). 6-19
    THROTTLE CABLE 6-20
  DESCRIPTION. 6-17
  DISASSEMBLY & ASSEMBLY 6-20
  REMOVAL & INSTALLATION 6-17
WINTER STORAGE. **2-66**
  GENERAL INFORMATION 2-66
  WINTER STORAGE CHECKLIST 2-66
WINTER STORAGE CHECKLIST 2-66
  CARBURETED & TBI ENGINES 2-66
  4.3L V6, 5.0L/5.7L/6.2L & 2006-08 8.1L MPI ENGINES .. 2-67
  2001-05 8.1L MPI ENGINES. 2-68
WIRE & CONNECTOR REPAIR 9-8
WIRING DIAGRAMS **6-27**
  AUTO TRIM II SYSTEM 13-27
  DUAL POWER TRIM 13-26
  GENERAL ENGINE 9-45
  HANDLE AND SEPARATE TRAILER SWITCH. 13-25
  MPI. 7-61
  PRECISION PILOT 15-37
  SMARTCRAFT DTS (10-PIN) 15-13
  SMARTCRAFT DTS (14-PIN) 15-16
  SMARTCRAFT DTS (14-PIN) - TRIM CIRCUIT 13-29
  SMARTCRAFT DTS (14-PIN) - TRIM HARNESS 13-28
  TBI 7-41
  TKS SYSTEM, 3.0L 6-27
  TKS SYSTEM, 4.3L/5.0L W/ALPHA 6-28
  TKS SYSTEM, 5.0L/5.7L W/BRAVO 6-29
  TRIM UP RELAY CIRCUIT 13-30
  3 BUTTON TRIM/TRAILER PANEL. 13-25
ZERO EFFORT CONSOLE MOUNT REMOTE CONTROL 15-11
  REMOVAL & INSTALLATION. 15-11
ZF/HURTH TRANSMISSIONS **12-6**
  GENERAL INFORMATION 12-6
  IDENTIFICATION. 12-6
  SHIFT CABLE 12-6
  TRANSMISSION 12-9
  TRANSMISSION FLUID 12-6
ZINCS (ANODES) **2-61**
  INSPECTION 2-62
  LOCATIONS. 2-62
  SERVICING 2-61

SELOC PUBLISHING'S FULL-LINE MASTER LIST

ISBN	PART NO.	TITLE/DESCRIPTION	YEARS
OUTBOARDS			
089330018-7	1000	Chrysler Outboards, All Engines	1962-84
089330055-1	1100	Force Outboards, All Engines	1984-99
089330048-9	1200	Honda Outboards, All Engines	1978-01
089330078-0	1202	Honda Outboards, All Engines	2002-08
089330007-1	1300	Johnson/Evinrude Outboards, 1-2 Cyl	1956-70
089330008-X	1302	Johnson/Evinrude Outboards, 1-2 Cyl	1971-89
089330009-8	1306	Johnson/Evinrude Outboards, 3-4 Cyl	1958-72
089330010-1	1308	Johnson/Evinrude Outboards, 3, 4 & 6 Cyl	1973-91
089330063-2	1311	Johnson/Evinrude Outboards - All V Engines	1992-01
089330052-7	1312	Johnson/Evinrude Outboards, All In-line Engines/2 & 4 Stroke	1990-01
089330071-3	1313	Evinrude Outboards, All 2-Stroke Engines	2002-06
089330072-1	1314	Johnson Outboards, All Engines/2 & 4-Stroke	2002-07
089330015-2	1400	Mariner Outboards, 1-2 Cyl	1977-89
089330016-0	1402	Mariner Outboards, 3, 4 & 6 Cyl	1977-89
089330012-8	1404	Mercury Outboards, 1-2 Cyl	1965-91
089330013-6	1406	Mercury Outboards, 3-4 Cyl	1965-89
089330014-4	1408	Mercury Outboards, 6 Cyl	1965-89
089330051-9	1416	Mercury/Mariner Outboards, All 2-Stroke Engines	1990-00
089330067-5	1418	Mercury Outboards, All 2-Stroke Models	2001-05
089330050-0	1600	Suzuki Outboards, All 2-Stroke Engines	1988-03
089330073-X	1602	Suzuki Outboards, All 4-Stroke Engines	1996-07
089330064-0	1701	Yamaha Outboards, All Engines	1984-96
089330065-9	1703	Yamaha Outboards, All Engines, All 2-Stroke Engines	1997-03
089330066-7	1705	Yamaha/Mercury & Mariner - All 4-Stroke Engines	1995-04
STERN DRIVES			
089330005-5	3200	Mercruiser Stern Drives	1964-91
089330053-5	3206	Mercruiser Stern Drives	1992-00
089330068-3	3208	Mercruiser Stern Drives	2001-08
089330004-7	3400	OMC Stern Drives	1964-86
089330056-X	3404	OMC Stern Drives	1986-98
089330011-X	3600	Volvo/Penta Stern Drives, All Gas Engines	1968-91
089330038-1	3602	Volvo/Penta Stern Drives, Volvo Engines	1992-93
089330057-8	3606	Volvo/Penta Stern Drives, All Gas Engines	1992-02
089330074-8	3608	Volvo/Penta Stern Drives, All Gas Engines	2003-07
INBOARDS			
089330049-7	7400	Yanmar Inboards	1975-98
PERSONAL WATERCRAFT			
089330032-2	9200	Kawasaki	1973-91
089330042-X	9202	Kawasaki	1992-97
089330045-4	9400	Polaris	1992-97
089330033-0	9000	Sea-Doo/Bombardier	1988-91
089330043-8	9002	Sea-Doo/Bombardier	1992-97
089330034-9	9600	Yamaha	1987-91
089330044-6	9602	Yamaha	1992-97
Seloc-On-Line (Internet Access)			
089330075-6	5000	One Mfg/Model - Subscription per user 3 years	1965-06

SelocOnLine

SelocOnLine is a maintenance and repair database accessed via the Internet. Always up-to-date, skill level and special tool icons, quick access buttons to wiring diagrams, specification charts, maintenance charts, and a parts database.
Contact your local marine dealer or see our demo at www.seloconline.com

SelocOnLine

Step 1
Select your manufacturer
Year / Model
Engine

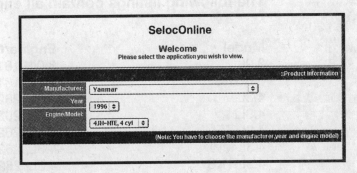

Step 2
Select an Engine System

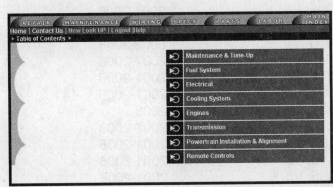

Step 3
Select the repair

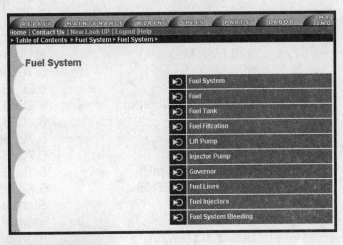

Step 4
Print the repair procedure

ENGINE FINDER

The following listings contain all engines covered in this manual

Model	Engine/Cylinder	Coverage Years
3.0L	3.0L, 181, 4 cyl	2001-2006
3.0 TKS	3.0L, 181, 4 cyl	2007-2008
4.3L	4.3L, 262, V6	2001-2006
4.3LH	4.3L, 262, V6	2001
4.3L EFI	4.3L, 262, V6	2001-2002
4.3L MPI	4.3L, 262, V6	2002-2008
4.3L TKS	4.3L, 262, V6	2007-2008
5.0L	5.0L, 305, V8	2001-2006
5.0L EFI	5.0L, 305, V8	2001-2002
5.0L MPI	5.0L, 305, V8	2002-2008
5.0L TKS	5.0L, 305, V8	2007-2008
5.7L	5.7L, 350, V8	2001-2006
5.7L EFI	5.7L, 350, V8	2001-2005
5.7L MPI	5.7L, 350, V8	2008
5.7L TKS	5.7L, 350, V8	2007-2008
6.2	377, V8	2001-2006
8.1S	8.1L, 496, V8	2001-2008
8.1S HO	8.1L, 496, V8	2001-2008
350 Mag MPI	5.7L, 350, V8	2001-2008
496 Mag	8.1L, 496, V8	2001-2006
496 Mag HO	8.1L, 496, V8	2001-2006
496 Mag MPI	8.1L, 496, V8	2007-2008
496 Mag HO MPI	8.1L, 496, V8	2007-2008
Black Scorpion	6.2L, 377, V8	2001-2006
MX6.2 Black Scorpion	6.2L, 377, V8	2002-2006
MX6.2 MPI	6.2L, 377, V8	2001-2008
Scorpion 350	5.7L, 350, V8	2007-2008
Scorpion 377	6.2L, 377, V8	2007-2008
Tow Sport 5.7 MPI	5.7L, 350, V8	2007-2008
Tow Sport 5.7 TKS	5.7L, 350, V8	2007-2008
Alpha	Sterndrive	2001-2008
Bravo I/II/III	Sterndrive	2001-2008
Bravo X/XR/XZ	Sterndrive	2001-2008
Hurth	Inboard	2001-2008
Velvet	Inboard	2001-2008
ZF	Inboard	2001-2008